INTERNATIONAL ASTRONOMICAL UNION

HIGHLIGHTS OF ASTRONOMY
VOLUME 16

AS PRESENTED AT THE
TWENTY EIGHTH GENERAL ASSEMBLY
BEIJING 20–31 AUGUST 2012

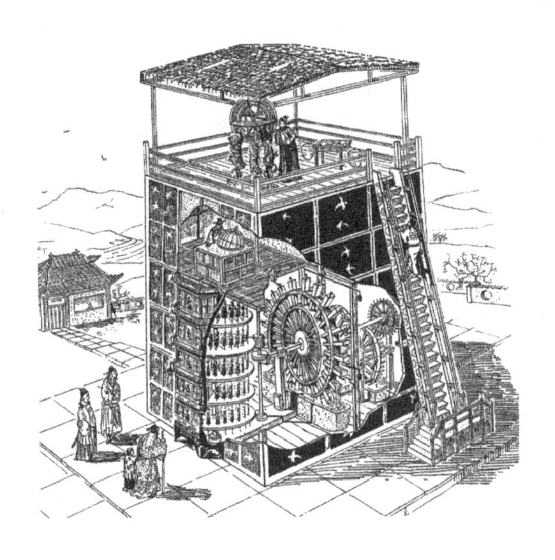

COVER ILLUSTRATION:

A pictorial reconstruction of the astronomical clock-tower built by Su Sung and his collaborators at Kaifeng in 1090. Original drawing by John Cristiansen in Needham *et al.* (1960) *Heavenly Clockwork. The Great Astronomical Clocks of Medieval China.* Cambridge University Press, Cambridge.

INTERNATIONAL ASTRONOMICAL UNION
UNION ASTRONOMIQUE INTERNATIONALE

HIGHLIGHTS OF ASTRONOMY VOLUME 16

Editors of Joint Discussions and Special Sessions held at the XXVIII General Assembly 2012

JD1: Diego F. Torres, Catherine Cesarsky, Helene Sol & Stefan Wagner
JD2: Jorick S. Vink
JD3: Lidia van Driel-Gesztelyi
JD4: Sugata Kaviraj, Sukyoung Yi & Martin Bureau
JD5: Junich Watanabe
JD6: Gabriele Giovannini, Teddy Cheung, Marcello Giroletti & Laura Maraschi
JD7: Nicole Capitaine, George Kaplan & Sergei Klioner

SpS1: Enrico Vesperini & Giampaolo Piotto
SpS2: Jan Vrtilek & Laurence David
SpS3: Ronald J. Buta & Daniel Pfenniger
SpS4: Marijke Haverkorn & JinLin Han
SpS5: Yaël Nazé
SpS6: Gianna Cauzzi & Alexandra Tritschler
SpS7: G. Valsecchi, A. Milani, W. Huebner & H. Rickman
SpS8: Andreas Zezas, Ann Hornschemeier & Daniela Calzetti
SpS9: Roger Davies
SpS10: Merav Opher, Abraham C.-L. Chian, Jean-Louis Bougeret & Xueshang Feng
SpS11: Kevin Govender & George Miley
SpS12: You-Hua Chu & Dieter Breitschwerdt
SpS13: Lucianne Walkopwicz & David Soderblom
SpS14: Dennis Crabtree, Lars Linberg Christensen & Pedro Russo
SpS15: Masatoshi Ohishi
SpS16: Sun Kwok
SpS17: W. Scott Kardel, Elizabeth Alvarez del Castillo, Rosa Ros & Magda Stavinschi

INTERNATIONAL ASTRONOMICAL UNION
UNION ASTRONOMIQUE INTERNATIONALE

HIGHLIGHTS OF ASTRONOMY VOLUME 16

AS PRESENTED AT THE XXVIII IAU GENERAL ASSEMBLY BEIJING, CHINA, 2012

Edited by

THIERRY MONTMERLE
Editor-in-Chief

CAMBRIDGE
UNIVERSITY PRESS

University Printing House, Cambridge CB2 8BS, United Kingdom

One Liberty Plaza, 20th Floor, New York, NY 10006, USA

477 Williamstown Road, Port Melbourne, VIC 3207, Australia

314-321, 3rd Floor, Plot 3, Splendor Forum, Jasola District Centre, New Delhi - 110025, India

79 Anson Road, #06-04/06, Singapore 079906

Cambridge University Press is part of the University of Cambridge.

It furthers the University's mission by disseminating knowledge in the pursuit of education, learning and research at the highest international levels of excellence.

www.cambridge.org
Information on this title: www.cambridge.org/9781107078840

First published 2014

A catalogue record for this publication is available from the British Library

This book has been printed on FSC-certified paper and cover board. FSC is an independent, non-governmental, not-for-profit organization established to promote the responsible management of the world's forests. Please see www.fsc.org for information.

ISBN 978-1-107-07884-0 Hardback
ISSN 1743-9213

Table of Contents

CHAPTER III. SPECIAL SESSIONS

SpS1: ORIGIN AND COMPLEXITY OF MASSIVE STAR CLUSTERS

SpS17: LIGHT POLLUTION: PROTECTING ASTRONOMICAL SITES AND INCREASING GLOBAL AWARENESS THROUGH EDUCATION

Preface

Highlights of Astronomy, vol. 16, contains the papers presented during the XXVIIIth IAU Genral Assembly, which took place in the new China National Convention Centre, situated in the Olympic Park, north of Beijing, from 20 to 31 August, 2012.

This Genetal Assembly was attended by 2710 astronomers, coming from 75 IAU member countries. The science program was established after a severe selection of proposals -for instance, no less than 23 Symposia were proposed, along with 18 Joint Discussions and 14 Special Sessions -in all 55 meetings! The Symposia, in particular, were of exceptional quality, which led the Selection Committee (composed of the Vice-Presidents and Division Presidents, and chaired by the General Secretary) to recommend to the Executive Committee holding 8 Symposia, in lieu of the usual six. In parallel, seven Joint Discussions and 17 Special Sessions were organized. While this large number of meetings did cause some logistical and scheduling problems, and daily scientific headaches for the participants to decide which sessions to attend, thanks to the good will, organization, and dedication of our hosts, the resulting scientific program was extremely broad, offering everyone, confirmed researchers and students alike, a fantastic forum to exchange news, discoveries, and ideas.

As usual, the Symposia have been published separately by Cambridge Univeristy Press (*IAU Symposium Proceedings Series*, vols. 288 to 295). The present volume contains the proceedings of the Joint Discussions and Special Sessions, along with four Invited Discourses that were delivered in plenary sessions. With only a few exceptions, all the texts are published here. As the attentive reader will notice, the format of each of these contributions varies, from one single, large synthetic paper, to shorter "highlights", down to more focused, short 2-page papers. Whatever the format chosen, it was a hard work to produce and I thank all the editors and meeting organizers for their dedication in assembling their respective Proceedings.

A companion volume, the Transactions of the International Union XXVIII A/B (also published by Cambridge University Press), contains the Proceedings of the General Assembly, including discourses, business meetings and reports by Commissions and Working Groups.

Financial support of a limited number of participants was provided by the IAU, and additional support from sponsors is gratefully acknowledged.

My deepest thanks go to Ian Corbett, my predecessor (and mentor in many ways), and to our Executive Secretary Vivien Reuter, who spent countless hours preparing and checking all the details of the organization of the XXVIIIth General Assembly, and other countless hours looking after the participants and handling grants.

Last but not least, our hearty thanks go also to our Chinese colleagues of the National Organizing Committee, and of the Local Organizing Committee who, under the leadership of Gang Zhao, offered us a very modern illustration of their traditional hospitality in all respects, either in the scientific organization, or in daily life. Thanks to the GA, we made many new friends.

Thierry Montmerle
IAU General Secretary
Paris, October 2014

Highlights of Astronomy, Volume 16
XXVIIIth IAU General Assembly, August 2012
T. Montmerle, ed.

doi:10.1017/S1743921314004608

A Zoo of Galaxies

Karen L. Masters[1,2]

[1]Institute for Cosmology and Gravitation, University of Portsmouth, Dennis Sciama Building, Burnaby Road, Portsmouth, PO1 3FX, UK
email: karen.masters@port.ac.uk

[2]South East Physics Network, www.sepnet.ac.uk

Abstract. We live in a universe filled with galaxies with an amazing variety of sizes and shapes. One of the biggest challenges for astronomers working in this field is to understand how all these types relate to each other in the background of an expanding universe. Modern astronomical surveys (like the Sloan Digital Sky Survey) have revolutionised this field of astronomy, by providing vast numbers of galaxies to study. The sheer size of the these databases made traditional visual classification of the types galaxies impossible and in 2007 inspired the Galaxy Zoo project (www.galaxyzoo.org); starting the largest ever scientific collaboration by asking members of the public to help classify galaxies by type and shape. Galaxy Zoo has since shown itself, in a series of now more than 30 scientific papers, to be a fantastic database for the study of galaxy evolution. In this Invited Discourse I spoke a little about the historical background of our understanding of what galaxies are, of galaxy classification, about our modern view of galaxies in the era of large surveys. I finish with showcasing some of the contributions galaxy classifications from the Galaxy Zoo project are making to our understanding of galaxy evolution.

1. What are Galaxies?

To say our understanding of the "zoo" of galaxies that are found in our Universe has changed a lot over the last century or two is a bit of an understatement. In 1845 the state of the art picture of an external galaxy, was an image of M51, or the Whirlpool galaxy drawn by William Parsons, Third Earl of Rosse (1800-1867), looking through what at the time was the largest telescope in the world - the "Leviathon of Parsontown" at his castle in Ireland (Left panel Figure 1; for a discussion of the history of picture see Steinicke 2012), At the time no-one fully understood that this was an external galaxy made of billions of stars. Although one of the motivations Parson is said to have had to build the telescope, was to resolve structure in distant nebula to see if they could be "island universe", rather similar in structure to our own galactic home. Today we have images of millions of galaxies from large surveys like the Sloan Digital Sky Survey (e.g. the last imaging release in Data Release Eight, Aihara *et al.* 2011), and extraordinarily high resolution images of hundreds of galaxies, including the Whirlpool, from the Hubble Space Telescope (Fig 1, right), as well as information on galaxies extending back more than half of the age of our Universe.

It took a long time for astronomers to understand that stars in the universe are organized in collections we now call galaxies. I can't help thinking that it was a major leap in understanding for astronomers to connect the uneven distribution of stars in the night sky with the three-dimensional structure of the galaxy that we live in. The first published example of this idea is the map shown in Figure 2, published by William Herschel in 1785 (Herschel 1785), and based on star counts made by himself and his sister Caroline. This diagram demonstrates an understanding of the Galaxy as a collection of stars, and while there is a a lot wrong with it (for example the Sun is at the centre, and the whole thing

Figure 1. Two views of M51 (the Whirlpool Galaxy). On the left is the original 1845 drawing by William Parsons (Figure 8 from Steinick 2012). On the right is an image from the Hubble Space Telescope (Credit: NASA, ESA, S. Beckwitch (STSci) and the Hubble Heritage Team (STSci/AURA)).

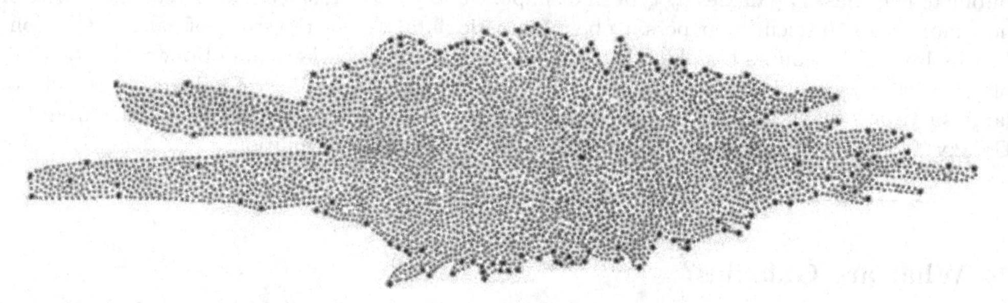

Figure 2. The first map of the Milky Way showing it as a collection of stars, and based on observational data. (Herschel 1785)

was much too small at ~ 7000 light years across) its an extraordinary piece of work as the first example of such a map based on actual astronomical data.

By 1900 astronomers understood quite a lot more about the basic structure of our galaxy. The map made by Cornelius Easton in 1900 (Fig 3) was the first to show our galaxy as having spiral structures (Trimble 1995). Easton used pictures of other spiral galaxies he saw in the sky to suggest the Milky Way might have this structure, although he still incorrectly placed the Sun in the centre of the Galaxy, and it's still too small.

The variety of different galaxies observed in the sky naturally caused people to wonder what they were. The scientific arguments surrounding this question at the start of the 20th century, are best represented perhaps, by the public debate held in 1920 between Heber Curtis and Harlow Shapley. Many other authors have described this debate, in much more detail that I intend to (for example Trimble 1995), and as is well known, both astronomers had it partly right. While Curtis held the opinion that the spiral nebula represented external galaxies, his measure of the size of our Milky Way was much too small (only 10 kpc across) and he still placed the Sun at its centre. Shapley believed the Milky Way was much larger, and the Sun offset (and in this was right), but because of the vast size of our Galaxy seemed unable to conceive that the Universe could be large enough to contain millions of other similar galaxies.

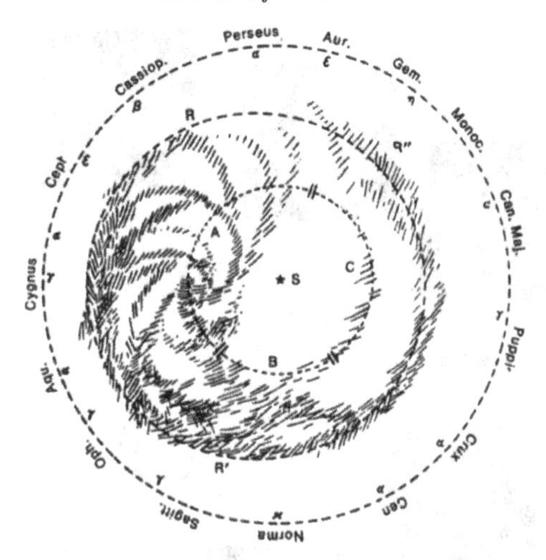

Figure 3. Cornelius Easton's model of the Galaxy in 1900 (reproduced from Fig 2 of Trimble 1995)

Figure 4. Modern view of the structure of our Galaxy. Image Credit: NASA/JPL-Caltech

The modern view of the structure of our Galaxy is presented in Figure 4, as an artists impression based on counting stars in data from the Spitzer Space Telescope observations (Benjamin 2008). Our Galaxy is a classic example of what we call a spiral galaxy, with most of its stars in a large, flat and very thin disc like structure which shows spiral arms, and a rounder region in the centre we call the "bulge". Our Galaxy also shows evidence for an elongated "bar" of stars across its central parts, a structure which is seen in many, but not all spiral galaxies. The Milky Way demonstrates just one of the two main kinds of big galaxies that are found in our Universe. Particularly in high density regions of the Universe, for example in the core of the nearby Virgo cluster of galaxies, many galaxies are large, smooth and spheroidal (or elliptical on the sky) in shape, and we call these types "elliptical" galaxies.

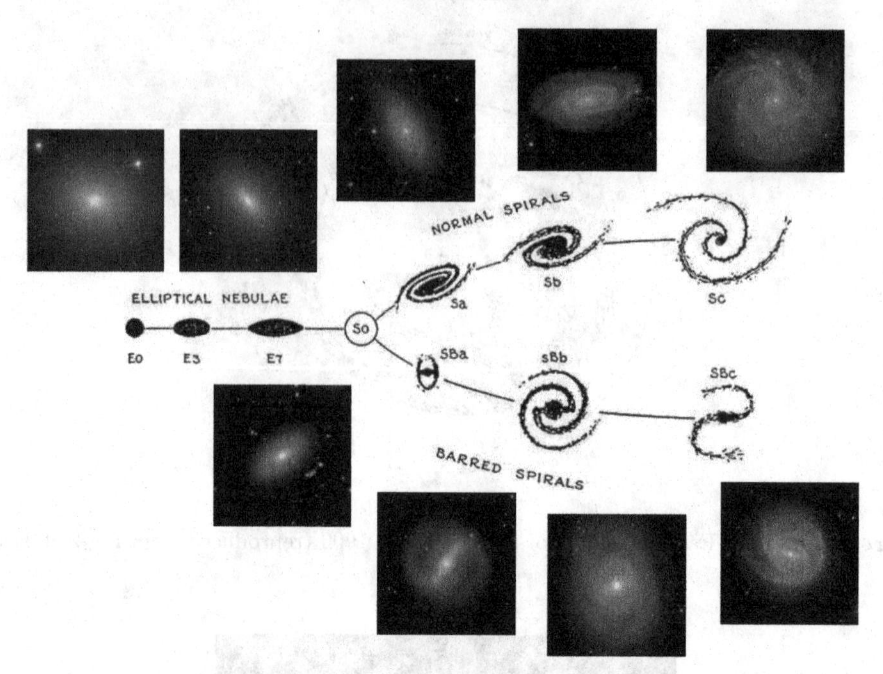

Figure 5. Hubble's original tuning fork as he presented in in Hubble (1936), with *gri* images of galaxies from the SDSS (similar to those used in the Galaxy Zoo project) added to further illustrate each galaxy type.

2. Classification of Galaxies

A common scientific approach to understanding a large collection of objects better is to classify it, then make categories. Much of the early effort in the field of extragalactic astronomy went into classifying the types of galaxies seen in the sky.

Probably the most famous of the early galaxy classifiers was Edwin Hubble (1889-1953). He was not the only person at the time to develop a classification scheme, but his scheme has been the most long lasting. It has been suggested that this was because he made such broad categories, that most galaxies can fit into one of them (Buta 2011). In a defence of the scheme against criticisms made by J. H. Reynolds (Reynolds 1927), Hubble claimed to have looked at more than a thousand images of galaxies, with only a small fraction not fitting the scheme, and uncertainty in placement in "less than ten per cent" (Hubble 1927).

In this same paper, Hubble described the scheme (which he first published in diagram form in *The Realm of the Nebula* in 1936, as shown in the centre of Figure 5). Hubble writes: "The classification under discussion arranges the extra-galactic nebulae in a sequence of expanding forms. There are two sections in the sequence, comprising the elliptical nebulae and the spirals respectively, which merge into one another." Ellipticals were ordered in how elliptical they appeared (from E0 being roundest, to E7 most elliptical), and spirals were ordered with reference to "conspicuous structural features", notably (1) the size of the central bulge, (2) the extent to which the arms wound, and (3) how distinct the spiral arms were. Hubble additionally notes that the "barred spirals" made up a distinct type of spiral ordered in the same fashion, but comprising "a small fraction of the numbers of normal spirals" (at least in the photographic plates he used). In between the spirals and ellipticals was the (then hypothetical) S0, or lenticular type, which shows a disc, but no spiral arms. The small number of galaxies Hubble found not fitting this scheme were called "irregular".

With only minor modifications and extensions (see e.g. Buta 2011 for an extensive discussion) the Hubble classification scheme as presented is still widely used in modern extragalactic astronomy. We don't just keep using it because its simple and it works well, but also because we have discovered that the actual physical properties of galaxies tend to vary in predictable ways along the sequence (e.g. as shown in Roberts & Haynes 1994). It's a common misconception that Hubble labelled galaxies at the left of the diagram "early" and those on the right "late" in an attempt to imply an evolutionary sequence. In fact Hubble himself writes that no such conclusion should be drawn (saying that "temporal connotations are made at one's peril" and that he set up the classification "without prejudice to theories of [galaxy] evolution", Hubble 1927), but rather he used the labels by analogy to what was then a commonly used classification scheme for stars. These terminology are still commonly used by astronomers today to discuss relative positions of galaxies on the diagram, with "early-type" galaxies used to mean elliptical and lenticular types, and "late-types" for spiral galaxies. Even among spirals, early and late will be used interchangeably with the Sa, Sb, Sc notation to describe relative location along the spiral sequence.

3. Physical Properties of Galaxies

A galaxy is a massive collection of stars, gas and interstellar dust. To first order, the optical colour of a galaxy tells about the types of stars which live in the galaxy (ignoring the complications of dust which can redden some types of galaxies quite significantly e.g. Masters *et al.* 2003, 2010a).When stars form, they form in a variety of masses. The most massive stars are extremely bright, and hot and so they look blue/white (in the same way very hot metal glows blue/white). They also live for very short times (astronomically), by which we mean just a few 100 million years, and when they are visible in a galaxy the galaxy will look blue because they totally outshine everything else.

If star formation ceases, after a short time time all the massive hot blue stars will die, and all that is left in the galaxy will be the less massive, cooler, red stars which can have lifetimes of tens of billions of years. So a galaxy which hasn't made new stars for while (more than a few 100 million years) will look red, as only red stars still shine in that galaxy.

Back to the Hubble diagram – the ellipticals on average are more massive (i.e. have more stars), redder, and tend to be found in clusters, while the spirals tend to be less massive, bluer and not in clusters (and increasingly so as you move along the Hubble Sequence away from the ellipticals). This tells us that the spiral galaxies are still forming stars, while the ellipticals have stopped, and more generally the Hubble sequence reveals a sequence of increasing star formation (and potential for future star formation) from left to right (Roberts & Haynes 1994), as well as decreasing (average) total galaxy stellar mass.

Hubble claimed almost all galaxies he looked at fitted his sequence, however we have observations of a lot more galaxies now. Thanks to improvements in detector technology and computers able to automatically process the data, very large surveys of the sky revealing millions of galaxies have become possible, including many much fainter than those available to Hubble, and revealing extra categories of dwarf, irregular ad low surface brightness galaxies.

4. How Many Galaxies are There?

In his 1927 defence of the classification scheme, Hubble claimed to have examined "upward of a thousand galaxies" in its construction. However for true physical understanding of a galaxy, more than just an image is required – we also need an estimate of its distance in order to reveal its size and mass. Hubble is most famous for a discovery which revolutionised extragalactic astronomy in its ability to provide a relatively quick and easy way to make these distance estimates. In 1929, Edwin Hubble used a sample of just 24 nearby galaxies to make this discovery. He used the Doppler shift of spectral lines to measure recessional velocities and used distances to these galaxies estimated based on the brightness of certain types of stars in them. By plotting these results against each other he demonstrated that the further away a galaxy is the faster it appears to be moving away from us (a finding now called "Hubble's Law", Hubble 1929). He got the proportionality constant drastically wrong (finding what we now call Hubble's constant to be 465±50 km/s/Mpc, while around 70 km/s/Mpc is the current accepted value, e.g. as we measured in Masters *et al.* 2006), but the main finding persisted, and in proceeding decades has generated an industry of using recessional velocities of galaxies (or "redshifts" since they are revealed by a reddening of known spectral lines via the Doppler shift) to map the universe. Even to this day it remains challenging to estimate redshift independent distances to galaxies, but the spectral measurements to indicate redshifts have become relatively routine. In the first 57 years after Hubble's original findings the number of galaxies mapped had increased by a factor of 50. In 1986 the state of the art in this field was the first slice of the CfA Redshift survey (de Lapparent, Geller & Huchra 1986). Using a sample of 1100 galaxies with redshifts this survey demonstrated that galaxies in our universe are not uniformly distributed, but rather clump in "large scale structure". Only 22 years after this survey, the state-of-the art was a sample 1000 times larger than this – the final release of the SDSS Main Galaxy Sample (Strauss *et al.* 2002; Abazajian *et al.* 2009) consisting of almost 1 million galaxy redshifts. Ongoing surveys (e.g. the Baryon Oscillation Spectroscopic Survey of SDSS-III, Dawson *et al.* 2012) are working on increasing this even more.

5. The "Zoo of Galaxies" in the Era of Large Surveys

Modern large surveys of hundreds of thousands of galaxies with detailed images and spectra (like the SDSS Main Galaxy Sample) have inspired a new field in extragalactic astronomy – the use of statistical analyses to reveal demographics of the population of galaxies. Perhaps the most famous example of this in the field of extragalactic astronomy is the use of the colour-magnitude diagram as a basic observation of galaxies (e.g. Strateva *et al.* 2001, Bell *et al.* 2003, Kauffmann *et al.* 2003, Baldry *et al.* 2004, Balogh *et al.* 2004). The galaxy population show a striking bimodality in this diagram, with galaxies mostly found in two regions, which have become known as the "red sequence" and the "blue cloud", with a sparsely populated "green valley" in between (e.g. a fraction of the galaxies of the SDSS Main Galaxy Sample are shown on this diagram in Fig 6). This diagram demonstrates a general trend among galaxies that bigger (or intrinsically brighter) galaxies tend to be optically redder (ie. having an older stellar population). This trend is most striking in the red sequence, but also apparent in the blue cloud. Hubble Space Telescope surveys have demonstrated that this colour magnitude diagram is in place since at least $z \sim 1$ (Bell *et al.* 2004). They also show that at earlier times in Universe more galaxies were found in the blue cloud – demonstrating that on average galaxies move from blue to red as cosmic time progresses. Much of extragalactic

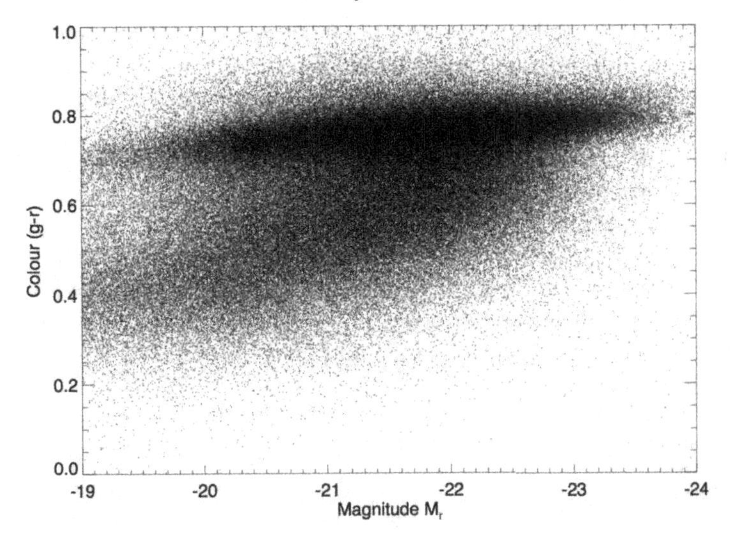

Figure 6. A colour magnitude diagram of a random selection of the SDSS Main Galaxy Sample (Strauss *et al.* 2002). This plots the $(g - r)$ colour versus the r-band absolute magnitude of almost 300,000 randomly selected galaxies in the Data Release 6 version (Adelman-McCarthy *et al.* 2008).

astronomy in recent decades has focused on developing an understanding of the mechanisms which shape the locations galaxies are found in diagram and the physical processes which move them around on it.

The bimodality in colour mirrors the two galaxy types discussed above, with the blue cloud hosting mostly spiral galaxies, and the red sequence mostly elliptical and lenticular galaxies. What had been missing in making this statement though was morphology (or type) for more than a few thousand objects (e.g. Fukugita *et al.* 2007 classified ~2000 of the SDSS galaxies). It seemed impossible to do that for the sample sizes being generated by astronomers in the early part of the 21st century (although the MOSES project tried – visually inspecting 50,000 galaxies to search for blue ellipticals, Schawinski *et al.* 2007). Without reliable information on galaxy morphology however, it's not possible to have the full picture of galaxy evolution. The colour of a galaxy is driven by the stellar (and gas and dust) content of the galaxy, while the shape or morphology of a galaxy reflects its dynamical history which could be very different (and have a different timescale). Therefore, one of the central motivations for the original Galaxy Zoo project was to construct a large sample of early and late type galaxy classifications from SDSS that were independent of colour.

6. Galaxy Zoo

Galaxy Zoo in its original form was launched on July 11, 2007 and introduced in a BBC online article that same day†. The original site simply asked volunteers to classify galaxies as either spiral or elliptical – the most basic morphological split among galaxies. Something in Galaxy Zoo resonated extraordinarily with the general public. The original projection estimated that if a few thousand members of the public got involved the 1 million galaxies might be classified in a couple of years. However, within twelve hours of the launch, the Galaxy Zoo site was receiving 20,000 classification per hour. After forty

† *Scientists seek galaxy hunt help*, by Christine McGourty (`http://news.bbc.co.uk/1/hi/sci/tech/6289474.stm`)

hours, the classification rate had increased to 60,000 per hour. After ten days, the public had submitted ~ 8 million classifications. By April 2008, when the Galaxy Zoo team submitted their first paper (Lintott *et al.* 2008), over 100,000 volunteers had classified each of the $\sim 900,000$ SDSS galaxy images an average of 38 times.

The popularity of Galaxy Zoo and the number of classifications received enabled science which simply would not have been possible without its contribution. Not only does Galaxy Zoo give a classification for each galaxy, but also by collecting ~ 40 independent classifications of each galaxy Galaxy Zoo produces an estimate of how likely that classification is to be be true (a classification "likelihood"). All of the data from this first phase of Galaxy Zoo were published in Lintott *et al.* (2011) and are available to download via the SDSS servers†. For an in-depth history and analysis of the Galaxy Zoo project see Fortson *et al.* (2012).

6.1. *Red Spirals*

One of the biggest contributions of Galaxy Zoo has been in finding large samples of rare classes of extragalactic object in the SDSS sample. An example of this are the "red spirals" – relatively rare spiral galaxies which are found on the red sequence (first discussed in Bamford *et al.* 2009, Skibba *et al.* 2009). Due to the reddening effects of dust, a significant number of edge-on spirals are found in the red sequence, and spiral galaxies with large central bulges can also be intrinsically very red due to old stars in the bulge (Masters *et al.* 2010a), however Galaxy Zoo revealed a signifiant fraction of even late-type spirals found on the red sequence (Masters *et al.* 2010b). These are galaxies which have small bulges, but intrinsically red discs yet still show clear spiral arms. The provide a direct probe of evolution affecting star formation but not morphology, revealing that processes exist which can turn spiral galaxies red without disturbing their morphology. Even though these are relatively rare objects in the galaxy population, studies have suggested they form a signifiant part of the route for most of the evolution from the blue cloud to the red sequence both with increasing density in the local universe (Bamford *et al.* 2009) and with redshift (Bundy *et al.* 2010). In Masters *et al.* (2010b) we revealed that while these objects are most common in intermediate density regions (as also shown in Bamford *et al.* 2009 and Skibba *et al.* 2009) they are found even at very low densities. We demonstrated that compared to blue cloud spirals, red spirals become more common as spiral galaxies become more massive, that they are not significantly more dusty (as revealed by Balmer decrements) but are significantly more likely to host LINER (Low Ionization Nuclear Emission Region) like emission and obvious bars. The Galaxy Zoo red spirals are not completely passive, but at fixed stellar mass show significantly less star formation and an older stellar population than their blue cousins.

The red spirals have provided part of the evidence prompting a move from the view of major mergers as the main route of galaxy evolution (e.g. as presented in Steinmetz & Navarro 2002), to more gentle and slower (secular) processes playing a significant role for most galaxies. In order to make a spiral galaxy red (by stopping star formation), something has happened to these galaxies to exhaust, or remove their supply of atomic hydrogen gas (the raw material for star formation). It's been discussed for many years that in our galaxy the amount of gas in the disc will be used up by the current rate of star formation in much less than a Hubble time (Larson, Tinsley & Caldwell 1980), and evidence exists that this gas is being replenished by the infall and cooling of hot gas from the halo (for a review see Putman *et al.* 2012). If this gas supply is shut off (either by being removed, or heated by processes like strangulation or harassment, Larson, Tinsley

† `http://skyserver.sdss3.org/CasJobs/`

& Caldwell 1980, Balogh, Navarro & Morris 2000, Bekki *et al.* 2002) star formation will cease and the disc redden on a timescale of ~ 1 Gyr. In fact recent work modelling the star formation histories of the Galaxy Zoo red spirals compared to blue spirals and red ellipticals supports this picture, showing that the average star formation history of red spirals only differs from blue spirals in the last 1 Gyr (Tojeiro *et al.* 2012).

One of the most striking observations about the red spirals was that such a large fraction of them were very strongly barred (75% Masters *et al.* 2010b). In much of the study of galaxy evolution, the division between barred and unbarred spirals has been ignored. It has often been argued that these types of galaxies are intrinsically the same, just caught with or without a bar (e.g. as recently discussed by van den Bergh 2011), even though most theoretical considerations now suggest bars must be very long lived. Around 30-60% of massive spirals host bars, with the exact value depending on how strict you are with the definition of what a bar is, which can vary from a linear feature stretching across most of the galaxy (as in the classic example NGC 1300), to mild oval distortions in the central regions.

6.2. *Galaxy Zoo 2 and Barred Spirals*

At the time we noticed the large bar fraction in red spirals, data from the second phase of Galaxy Zoo was starting to become available. Galaxies can show all sorts of interesting structures beyond the simple split between spiral and elliptical, and these features reveal more clues to the formation histories of the galaxies. Galaxy Zoo was so popular, and the results so reliable (agreeing with experts just as well as experts are able to agree with each other, as discussed in Lintott *et al.* 2008) that a new version was developed, asking for significantly more detail for the brightest quarter of the original sample. If a galaxy was identified as being a disk or showing features, this version asked questions about the number of spiral arms, size of the bulge, and most importantly for us, the presence of a bar. The full classification scheme for Galaxy Zoo 2 (GZ2) is shown in Figure 7, reproduced from Masters *et al.* (2011). This version of the site launched in February 2009 and ran for fourteen months, collecting in that time 60 million individual classifications of the images.

The first science result from GZ2 were a study of the bar fraction of disc galaxies as a function of colour. This work (Masters *et al.* 2011) demonstrated that the types of bars revealed by GZ2† were significantly more common in redder, more massive disc galaxies – with the extreme of this being the red spirals, among with 75% showed bars. This work also revealed a split in bar fraction between disc galaxies with large and small bulges (as revealed by fits to the light profile from SDSS) such that bulge dominated spirals showed more bars. These types of trends of bar fraction with galaxy colour have also been seen in the local universe by Nair & Abraham (2010) and Giordarno *et al.* (2010) and at higher redshift by Sheth *et al.* (2008). The GZ2 results also hinted at an upturn of the bar fraction in the bluest and lowest mass disc galaxies as observed more clearly by Barazza *et al.* (2009) and Aguerri *et al.* (2009) in disc galaxy samples dominated by these kinds of galaxies.

The striking thing about our theoretical understanding of bars in disc galaxies, is that forming a bar in a disc galaxy is extremely easy. The question is not why some galaxies make them, but why some do not (or have not yet).

While a bar is forming it enables the exchange of angular momentum in a disc galaxy – basically moving material around. This has the effect of growing central

† which have been shown to be very similar to the classical "strong" bars identified in early visual catalogues - see Masters *et al.* (2012) for a discussion.

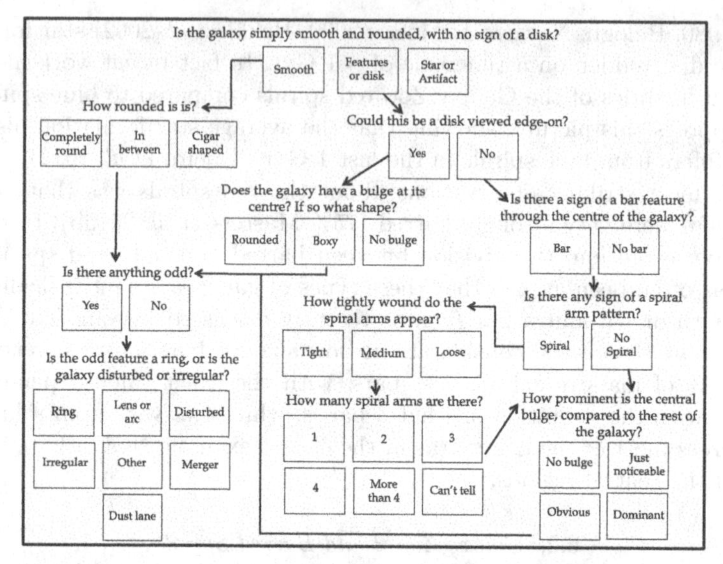

Figure 7. Full classification tree from Galaxy Zoo 2. First published in Masters *et al.* (2011).

concentrations (Kormendy & Kennicutt 2004), sparking central star formation (e.g. as observed by Sheth *et al.* 2005, Ellison *et al.* 2011), and possibly helping to feed central active galactic nuclei (although this is more controversial e.g. Ho *et al.* 1997, or Oh *et al.* 2012, Cardamone *et al.* in prep.). Theoretical considerations indicate that the gas content of a galaxy is important in this process (e.g. Athanassoula 2003; Combes 2008). By virtue of being able to dissipate energy, gas is an important sink of angular momentum, and any galaxy with a significant gas fraction in simulations struggles to develop a large bar (e.g. Villa-Vargas *et al.* 2010), while in a simulation of a gas poor disc, bars can be very stable and long lived.

The effective forces produced by the bar instability act to drive gas inwards from co-rotation (the point at which stars in the disk rotate with the same speed as the pattern speed of the bar) to the central regions. This gas looses its angular momentum which is transfered to the stars in the bar. Interestingly, the forces outside the ends of the bar may also act to inhibit inflow of gas from the outer regions of the disc, so that gas inflow of external gas onto a disc galaxy is inhibited in the presence of a strong bar (Combes 2008). It is possible that this process could play a role in the global reddening of disc galaxies, and while the concentration of gas in the centres of galaxies caused by the bar will spark central star formation, it evacuates gas from the rest of the disc (preventing star formation there) and potentially helps to use up the total gas content of the galaxy quicker.

6.3. *Bars and Atomic Gas Content*

The discovery that the type of strong, or obvious bars easily identified in GZ2 classifications are more common in more massive, redder disc galaxies, already implies that bars are more often found in gas poor disc galaxies, since optical colour is a good proxy for current star formation rate, which correlates well with gas content. However we wanted to investigate more directly the role of gas content on the likelihood of a galaxy hosting a bar.

The Arecibo Legacy Fast ALFA (Arecibo L-band Feed Array) survey (ALFALFA; Giovanelli *et al.* 2005) provide the ideal data to do this, matched well to the available GZ2 bar sample. ALFALFA has mapped all of the high Galactic latitude sky visible from the Arecibo Radio Telescope (which is located in Puerto Rico) in the 21cm line emitted by neutral hydrogen atoms (HI). This is giving an amazing census of the atomic hydrogen gas content of all galaxies in this part of the sky, which overlaps much of the SDSS Legacy Imaging area in which Galaxy Zoo galaxies are selected. In Masters *et al.* (2012) we constructed a sample of 2090 disc galaxies with bar identifications from GZ2 and neutral hydrogen detections from the 40% of ALFALFA available at the time (Haynes *et al.* 2011). Our main result was that we confirmed observationally that bars are more likely to be found in galaxies with less gas. In fact we showed that this is also true in galaxies which have less gas than is typical for their stellar mass and optical colour, and perhaps more interestingly that disc galaxies with both gas and bars are optically redder than similar disc galaxies with gas and no bar – perhaps the first direct evidence that the bar could be helping to globally redden its host galaxy.

This is a work in progress, with many more interesting results coming out on the GZ2 bars. For example in Hoyle *et al.* (2011) we demonstrated that redder disc galaxies not only host more bars, but that these bars are also longer (and therefore stronger). In Skibba *et al.* (2012) we provided the first clear evidence for an environmental effect on bar formation, finding more bars in galaxies clustered on certain scales even when corrected for galaxy mass and colour. Casteels *et al.* (2012) shows that in close pairs bar formation is suppressed, and Cardamone *et al.* (in prep.) will show that once corrected for galaxy mass, we observe no correlation between active galactic nuclei and bars identified in GZ2. All of this is providing vital clues to reveal the role of bars on the global evolution of disc galaxies.

7. What's Next for the Zoo?

The popularity of Galaxy Zoo was both immediate and long lasting, and raised questions very early on with the science team about why so many people would choose to spend so much time classifying galaxies online. A survey was launched to study the motivations of the citizen scientists participating in Galaxy Zoo, which was published by Raddick *et al.* (2010). For a scientist involved in research using the Galaxy Zoo classifications, the most striking things about this survey was that so many people identified a desire to help with scientific research as their main motivation. This means that the scientific results coming out of Galaxy Zoo are essential to the ongoing success of the project in attracting and retaining volunteer classifiers. Fortunately Galaxy Zoo was designed with specific and immediate science questions to answer. The first publications from Galaxy Zoo came out within a year of launch (Lintott *et al.* 2008, Land *et al.* 2008), and the total number of peer reviewed papers based on Galaxy Zoo data is now more than 30, with a growing number of results from scientists not involved in the Galaxy Zoo project, and using the publicly released GZ1 data. The classifications from GZ2 are also in the process of being prepared for public release. A selection of papers from the Galaxy Zoo team are listed in Table 1.

Galaxy Zoo was a pioneer in what has becoming a new methodology of involving "citizen scientists" in research. The Zooniverse† was launched in December 2009 on the back of the success of Galaxy Zoo, and provides a framework for collecting a variety of similar projects. This umbrella now includes several other astronomically themed projects (e.g. Milky Way Project, Moon Zoo, Supernova Zoo, Solar Stormwatch) but also projects in other areas of science, such as Old Weather (extracting climate data from handwritten ship logbooks), Whale.FM (comparing whale song), Cell Slider (identifying cancer cells

† `www.zooniverse.org`

Table 1. A selection of peer reviewed papers based on classifications collected from Galaxy Zoo (in order of publication).

Author & Year	Title - Galaxy Zoo:
Land *et al.* 2008	The large-scale spin statistics of spiral galaxies in the Sloan Digital Sky Survey
Lintott *et al.* 2008	Morphologies derived from visual inspection of galaxies from the SDSS
Slosar *et al.* 2009	Chiral correlation function of galaxy spins
Bamford *et al.* 2009	The dependence of morphology and colour on environment
Schawinski *et al.* 2009a	A sample of blue early-type galaxies at low redshift
Lintott *et al.* 2009	'Hanny's Voorwerp', a quasar light echo?
Skibba *et al.* 2009	Disentangling the environmental dependence of morphology and colour
Cardamone *et al.* 2009	Green Peas: discovery of a class of compact extremely star-forming galaxies
Darg *et al.* 2010a	The fraction of merging galaxies in the SDSS and their morphologies
Darg *et al.* 2010b	The properties of merging galaxies in the nearby Universe - local environments, colours, masses, star formation rates and AGN activity
Schawinski *et al.* 2010a	The Sudden Death of the Nearest Quasar
Schawinski *et al.* 2010b	The Fundamentally Different Co-Evolution of Supermassive Black Holes and Their Early- and Late-Type Host Galaxies
Masters *et al.* 2010a	Dust in spiral galaxies
Jimenez *et al.* 2010	A correlation between the coherence of galaxy spin chirality and star formation efficiency
Masters *et al.* 2010b	Passive red spirals
Banerji *et al.* 2010	Reproducing galaxy morphologies via machine learning
Lintott *et al.*2011	Data Release of Morphological Classifications for nearly 900,000 galaxies
Masters *et al.* 2011	Bars in disc galaxies
Smith *et al.* 2011	Galaxy Zoo Supernovae
Hoyle *et al.* 2011	Bar lengths in local disc galaxies
Darg *et al.* 2011	Multi-Mergers and the Millennium Simulation
Keel *et al.* 2012	The Galaxy Zoo survey for giant AGN-ionized clouds: past and present black hole accretion events
Wong *et al.* 2012	Building the low-mass end of the red sequence with local post-starburst galaxies
Kaviraj *et al.* 2012	Dust and molecular gas in early-type galaxies with prominent dust lanes
Shabala *et al.* 2012	Dust lane early-type galaxies are tracers of recent, gas-rich minor mergers
Skibba *et al.* 2012	The environmental dependence of bars and bulges in disc galaxies
Masters *et al.* 2012	Atomic gas and the regulation of star formation in barred disc galaxies
Hoyle *et al.* 2012	The fraction of early-type galaxies in low-redshift groups and clusters of galaxies
Teng *et al.* 2012	Chandra Observations of Galaxy Zoo Mergers: Frequency of Binary Active Nuclei in Massive Mergers
Simmons *et al.* 2012	Bulgeless Galaxies With Growing Black Holes
Casteels *et al.* 2012	Quantifying Morphological Indicators of Galaxy Interaction

in images) and even non science projects (e.g. Ancient Lives, which attempts to piece together scraps of papyrus). All of these and more can be accessed via the main Zooniverse website.

The Zooniverse team are also working on tools to improve the educational use of these sites. They provide a way to involve school children in real hands on science, but working out how to include this in lesson plans has proved difficult. Recently launched was Galaxy Zoo: Navigator† which allows a group to collect their classifications together, and use online tools to explore how these depend on other galaxy properties (from SDSS data). This increases the learning potential from Galaxy Zoo. The Zooniverse also hosts a website called "Zooteach"‡ for teachers and educators to share lesson plans and ideas for the use of Zooniverse projects in the classroom.

Galaxy Zoo itself remains one of the most successful and popular of the Zooniverse sites. Following GZ2, Galaxy Zoo: Hubble ran (April 2010-September 2012) collecting classifications on galaxies observed by the Hubble Space Telescope (e.g. as part of COSMOS, GOODS, EDS and other large surveys). Shortly after this Invited Discourse, Galaxy Zoo relaunched in its fourth version, now including images of galaxies from the HST CANDELS survey as well as new images of galaxies from the SDSS-III imaging area.

I'll finish by just saying thank you to all of the more than 200,000 volunteers who have helped make galaxy classifications via the Galaxy Zoo website by using the special

† `www.galaxyzoo.org/#/navigator`
‡ `www.zooteach.org/`

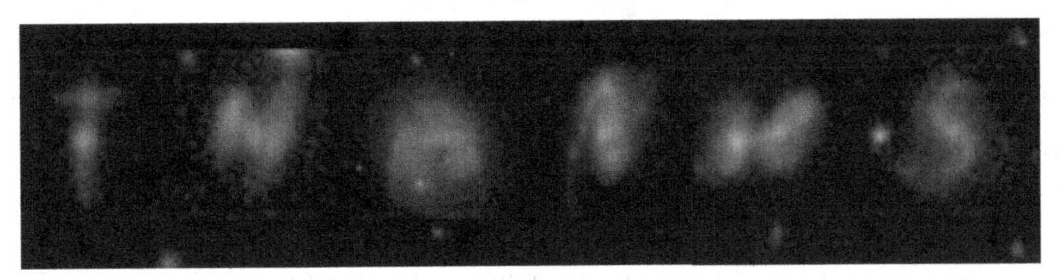

Figure 8. "Thanks" written in the galaxy alphabet found by Galaxy Zoo volunteers. Write your own words at `writing.galaxyzoo.org`

"galaxy font" they helped make (Fig 8)†. To anyone who has not yet tried out Galaxy Zoo, I'd like to encourage you to come and play in our "Xīng Xì Zǒng Dòng Yuán" ("Galaxy Zoo" in Chinese).

Acknowledgements

I'd like to thank the organisers of the 28th GA of the IAU for inviting me to give this talk.

This publication has been made possible by the participation of more than 200,000 volunteers in the Galaxy Zoo project. Their contributions are individually acknowledged at http://www.galaxyzoo.org/volunteers.

I acknowledge funding from the Peter and Patricia Gruber Foundation as the 2008 Peter and Patricia Gruber Foundation IAU Fellow, and from a 2010 Leverhulme Trust Early Career Fellowship, as well as support from the Royal Astronomical Society to attend the 28th GA of the IAU.

Thanks to Dongai for helping with my Chinese introductory remarks to the Invited Discourse.

References

Abazajian, K. N., Adelman-McCarthy, J. K., Agüeros, M. A., *et al.* 2009, *ApJS*, 182, 543
Adelman-McCarthy, J. K., Agüeros, M. A., Allam, S. S., *et al.* 2008, *ApJS*, 175, 297
Aguerri, J. A. L., Méndez-Abreu, J., & Corsini, E. M. 2009, *A&A*, 495, 491
Aihara, H., Allende Prieto, C., An, D., *et al.* 2011, *ApJS*, 193, 29
Athanassoula, E. 2003, *MNRAS*, 341, 1179
Baldry, I. K., Glazebrook, K., Brinkmann, J., Ivezić, Ž., Lupton, R. H., Nichol, R. C., & Szalay, A. S. 2004, *ApJj*, 600, 681
Bell, E. F., McIntosh, D. H., Katz, N., & Weinberg, M. D. 2003, *ApJS*, 149, 289
Bell, E. F., Wolf, C., Meisenheimer, K., *et al.* 2004, *ApJ*, 608, 752
Balogh, M. L., Navarro, J. F., & Morris, S. L. 2000, *ApJ*, 540, 113
Balogh, M. L., Baldry, I. K., Nichol, R., *et al.* 2004, *ApJL*, 615, L101
Bamford, S. P., *et al.* 2009, *MNRAS*, 393, 1324
Banerji, M., Lahav, O., Lintott, C. J., *et al.* 2010, *MNRAS*, 406, 342
Barazza, F. D., Jogee, S., & Marinova, I. 2008, *ApJ*, 675, 1194
Bekki, K., Couch, W. J., & Shioya, Y. 2002, *ApJ*, 577, 651
Benjamin, R. A. 2008, *Massive Star Formation: Observations Confront Theory*, 387, 375
Bundy, K., *et al.* 2010, *ApJ*, 719, 1969
Buta, R. J. 2011, in *Planets, Stars, and Stellar Systems*, Vol. 6, Series Editor T. D. Oswalt, Volume editor W. C. Keel, Springer (arXiv:1102.0550)
Cardamone, C., Schawinski, K., Sarzi, M., *et al.* 2009, *MNRAS*, 399, 1191

† Available at `writing.galaxyzoo.org`

Casteels, K. R. V., Bamford, S. P., Skibba, R. A., *et al.* 2012, *MNRAS* (submitted: arXiv:1206.5020)
Combes, F. 2008, in *Formation and Evolution of Galaxy Bulges*, Proceedings of IAU Symposium, 245, Eds M. Bureau, E. Athanassoula and B. Barbuy
Darg, D. W., Kaviraj, S., Lintott, C. J., *et al.* 2010, *MNRAS*, 401, 1043
Darg, D. W., Kaviraj, S., Lintott, C. J., *et al.* 2010, *MNRAS*, 401, 1552
Darg, D. W., Kaviraj, S., Lintott, C. J., *et al.* 2011, *MNRAS*, 416, 1745
Dawson, K. S., Schlegel, D. J., Ahn, C. P., *et al.* 2012, AJ (submitted; arXiv:1208.0022)
de Lapparent, V., Geller, M. J., & Huchra, J. P. 1986, *ApJL*, 302, L1
Ellison, S. L., Nair, P., Patton, D. R., *et al.* 2011, *MNRAS*, 416, 2182
Fortson, L., Masters, K., Nichol, R., *et al.* 2012, Advances in Machine Learning and Data Mining for Astronomy, CRC Press, Taylor & Francis Group, Eds.: Michael J. Way, Jeffrey D. Scargle, Kamal M. Ali, Ashok N. Srivastava, p. 213-236, 213 (arXiv:1104.5513)
Giordano, L., Tran, K. V. H., Moore, B., & Saintonge, A. 2012, *ApJ* (submitted; arXiv:1002.3167)
Giovanelli, R., *et al.* 2005, *AJ*, 130, 2598
Haynes, M. P., Giovanelli, R., Martin, A. M., *et al.* 2011, *AJ*, 142, 170 (α40)
Herschel, W. 1785, *Philosophical Transactions of the Royal Society of London*, Vol. 75., 213-266
Ho, L. C., Filippenko, A. V., & Sargent, W. L. W. 1997, *ApJ*, 487, 591
Hoyle, B., Masters, K. L., Nichol, R. C., *et al.* 2011, *MNRAS*, 415, 3627
Hoyle, B., Masters, K. L., Nichol, R. C., Jimenez, R., & Bamford, S. P. 2012, *MNRAS*, 423, 3478
Hubble, E. P. 1927, *The Observatory*, 50, 276
Hubble, E. P. 1936, *Realm of the Nebulae*, by E. P. Hubble. New Haven: Yale University Press, 1936. ISBN 9780300025002,
Jimenez, R., Slosar, A., Verde, L., *et al.* 2010, *MNRAS*, 404, 975
Kauffmann, G., Heckman, T. M., White, S. D. M., *et al.* 2003, *MNRAS*, 341, 54
Kaviraj, S., Ting, Y.-S., Bureau, M., *et al.* 2012, *MNRAS*, 423, 49
Keel, W. C., Chojnowski, S. D., Bennert, V. N., *et al.* 2012, *MNRAS*, 420, 878
Kormendy, J. & Kennicutt, R. C., Jr. 2004, *ARA&A*, 42, 603
Land, K., Slosar, A., Lintott, C. J., *et al.* 2008, *MNRAS*, 388, 1686
Larson, R. B., Tinsley, B. M., & Caldwell, C. N. 1980, *ApJ*, 237, 692
Lintott, C. J., *et al.* 2008, *MNRAS*, 389, 1179
Lintott, C. J., *et al.* 2011, *MNRAS*, 410, 166
Masters, K. L., Giovanelli, R., & Haynes, M. P. 2003, *AJ*, 126, 158.
Masters, K. L., Haynes, M. P., Giovanelli, R., & Springob, C. M. 2006, *ApJ*.
Masters, K. L., *et al.* 2010a, *MNRAS* 404, 792.
Masters, K. L., *et al.* 2010b, *MNRAS* 405, 783.
Masters, K. L., *et al.* 2011, *MNRAS*, 411, 2026
Masters, K. L., Nichol, R. C., Haynes, M. P., *et al.* 2012, *MNRAS*, 424, 2180
Messier, C. 1781, *Connoissance des Temps for 1784*, p. 227-267, 227
Nair, P. B. & Abraham, R. G. 2010b, *ApJL*, 714, L260
Oh, S., Oh, K., & Yi, S. K. 2012, *ApJS*, 198, 4
Putman, M. E., Peek, J. E. G., & Joung, M. R. 2012, *ARA&A*, 50, 491
Raddick, M. J., Bracey, G., Gay, P. L., *et al.* 2010, *Astronomy Education Review*, 9, 010103
Roberts, M. S. & Haynes, M. P. 1994, *ARA&A*, 32, 115
Schawinski K., Thomas D., Sarzi M., Maraston C., Kaviraj S., Joo S.-J., Yi S. K., Silk J. 2007, *MNRAS*, 382, 1415
Schawinski, K., Lintott, C., Thomas, D., *et al.* 2009, *MNRAS*, 396, 818
Schawinski, K., Urry, C. M., Virani, S., *et al.* 2010, *ApJ*, 711, 284
Simmons, B. D., Lintott, C., Schawinski, K., *et al.* 2012, *MNRAS* (submitted; arXiv:1207.4190)
Shabala, S. S., Ting, Y.-S., Kaviraj, S., *et al.* 2012, *MNRAS*, 423, 59
Sheth, K., Vogel, S. N., Regan, M. W., Thornley, M. D., & Teuben, P. J. 2005, *ApJ*, 632, 217
Sheth, K., *et al.* 2008, *ApJ*, 675, 1141
Slosar, A., Land, K., Bamford, S., *et al.* 2009, *MNRAS* 392, 1225

Skibba, R. A., *et al.* 2009, *MNRAS*, 399, 966
Skibba, R. A., Masters, K. L., Nichol, R. C., *et al.* 2012, *MNRAS*, 423, 1485
Smith, A. M., Lynn, S., Sullivan, M., *et al.* 2011, *MNRAS*, 412, 1309
Steinecke, W. 2012, *Journal and Astronomical History and Heritage*, 15(1), 19-29.
Steinmetz, M. & Navarro, J. F. 2002, *New Astronomy*, 7, 155
Strateva, I., *et al.* 2001, *AJ*, 122, 1861
Strauss, M. A., *et al.* 2002, *AJ*, 124, 1810
Teng, S. H., Schawinski, K., Urry, C. M., *et al.* 2012, *ApJ*, 753, 165
Tojeiro, R., Masters, K. L., *et al.* 2012, *MNRAS* (submitted)
Trimble, V. 1995, *PASP*, 107, 1133
van den Bergh, S. 2011, *AJ*, 141, 188
Villa-Vargas, J., Shlosman, I., & Heller, C. 2010, *ApJ*, 719, 1470
Wong, O. I., Schawinski, K., Kaviraj, S., *et al.* 2012, *MNRAS*, 420, 1684

Highlights of Astronomy, Volume 16
XXVIIIth IAU General Assembly, August 2012
T. Montmerle, ed.

doi:10.1017/S174392131400461X

Supernovae, the Accelerating Cosmos, and Dark Energy

Brian Schmidt

Reserach School of Astronomy & Astrophysics, Australia National University,
Mount Stromlo Observatory, Canberra ACT 2611, Australia
email: brian@mso.anu.edu.au

Abstract. Type Ia supernovae remain one of Astronomy's most precise tools for measuring distances in the Universe. I describe the cosmological application of these stellar explosions, and chronicle how they were used to discover an accelerating Universe in 1998 - an observation which is most simply explained if more than 70 % of the Universe is made up of some previously undetected form of 'Dark Energy'. Over the intervening 13 years, a variety of experiments have been completed, and even more proposed to better constrain the source of the acceleration. I review the range of experiments, describing the current state of our understanding of the observed acceleration, and speculate about future progress in understanding Dark Energy.

Text of presentation not available.

Highlights of Astronomy, Volume 16
XXVIIIth IAU General Assembly, August 2012
T. Montmerle, ed.

doi:10.1017/S1743921314004621

Past, Present and Future of Chinese Astronomy

Cheng Fang

School of Astronomy and Space Science, Nanjing University, 22 Han Kou road, Nanjing, China
email: fangc@nju.edu.cn

Abstract. Through out the ancient history, Chinese astronomers had made tremendous achievements. Since the main purpose of the ancient Chinese astronomy was to study the correlation between man and the universe, all the Emperors made ancient Chinese astronomy the highly regarded science throughout the history. After a brief introduction of the achievement of ancient Chinese astronomy, I describe the beginnings of modern astronomy research in China in the 20th century. Benefiting from the fast development of Chinese economy, the research in astronomy in China has made remarkable progress in recent years. The number of astronomers has doubled in the past ten years, and the number of graduate students has grown over 1300. The current budget for astronomy research is ten times larger than that ten years ago. The research covers all fields in astronomy, from galaxies to the Sun. The recent progress in both the instruments, such as the Guo Shoujing's telescope, a Large Sky Area Multi-Object Fiber Spectroscopic Telescope (LAMOST), and the theoretical research will be briefly presented. The ongoing and future projects on the space- and ground-based facilities will be described, including the Five Hundred Meter Aperture Spherical Radio Telescope (FAST), "Chang E" (Lunar mission) project, Hard X-ray Modulate Telescope (HXMT), DArk Matter Particle Explorer (DAMPE), Deep Space Solar Observatory (DSO), Chinese Antarctic Observatory (CAO), 65m steerable radio telescope, Chinese Spectral Radioheliogaph (CSRH) etc.

Keywords. history, ancient astronomy, present status, future astronomy, China

---

Through out the ancient history, Chinese astronomers had made tremendous achievements. They maintained the longest continuous records of all kinds of astronomical phenomena, which are still useful today for astronomical research. They invented various important astronomical instruments for astronomical observations and measurements. However, during the 18-19th century Chinese astronomy fell behind much due to the feudal and introverted Emperor's control.

1. Ancient Astronomy in China

In ancient times, the Emperors of various Dynasties nominated some special officers to observe stars. For instance, a legend said that astronomers Xi and He received commissions from Emperor Yao to observe stars and to make calendars. Figure 1 shows an ancient nomination ceremony before the Emperor.

There were mainly two reasons for observing stars: (1) The astrological and political reasons. Eclipse observations are good examples. People used to believe that during the eclipses, the Sun was devored by a celestial dog, an omen portending all kinds of disasters. So the Emperors and people would like to know when and how the eclipse would appear. (2) To make good astronomical systems. For instance, every dynasty needed astronomical chronology and made the calendars, etc.

For a period of over 3000 years, Chinese made tremendous progress in astronomy. They maintained the longest continuous records of all kinds of astronomical phenomena,

Figure 1. Ancient Chinese astronomers were nominated by Emperors.

which are still useful today for astronomical research. Ancient Chinese astronomers made more than 100 astronomical calendars, each being a sophisticated system of astronomical computation, more than 50 of them were officially used; They also constructed a large number of astronomical instruments. The construction of the Astronomical Clock Tower in the 11th century is a great example.

Ancient Chinese kept records of a wide range of celestial events, including solar and lunar eclipses, sunspots, comets and meteors, guest stars (nova or supernova), planetary events, aurora borealis etc.

1.1. *Historical eclipse records in China*

Solar eclipses have already been mentioned in oracle bone inscriptions since Shang Dynasty (1600 B.C-1046 B.C), as seen in the left panel in Figure 2. For instance, some records on oracle bones indicate that solar eclipses occurred on October 21, 1198 B.C, June 7, 1172 B.C, and October 31, 1161 B.C. etc. Generally the recorded data include the position of the Sun at the time of eclipse, the beginning and the end of the eclipse, the magnitude of the eclipse, the point of the first contact on the solar disc etc.

The ancient records of solar eclipses have been used to study the secular change of the Earth's rotation. A decreasing rotation rate has been found to be -22 - -26 seconds/century. The records can also be used for astronomical chronology. For example, the Bamboo Chronicles says: "During the first year of King Yi of the Zhou Dynasty, two daybreaks occurred.". According to the eclipse that occurred in 899 B.C, people can tell exactly the day of the year. By using these records, people have discovered the periodicity of eclipses. For instance, ancient Chinese already knew the 135 synodic months in 100 B.C. In 762 AD, a period of 458 months was discovered, which was confirmed by Newcomb 1100 years later. In 1199 AD, the Saros Cycle (223 months) was discovered.

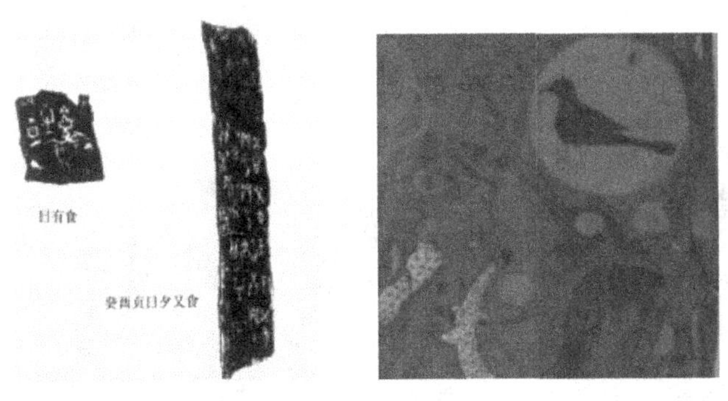

Figure 2. Solar eclipses recorded in oracle bones (left) and sunspot recorded as a gold bird (right).

Figure 3. The silk book from the Former Han (206 B.C-8 AD) had pictures of all types of comets.

1.2. *Ancient sunspot records*

In fact, sunspots had been recorded in the book of I Ching about around 800 B.C. The first well-recognized sunspot record was made back in 28 AD, Han Dynasty, describing the sunspot as "dark air like coin on the Sun". From Han to Ming Dynasty there were more than 100 sunspot records (Xu & Jiang 1986). Sunspots were depicted "like bird", "like coin" or "like chestnut", while the records of "disappeared in several months" and "disappeared in three days" indicate the evolution of sunspots. During Han Dynasty, there was a record of sunspot as a "Gold bird" (right penal in Fig. 2), which was discovered in a famous tomb in Hunan province.

1.3. *Pictures of all types of comets*

The silk book from the Former Han (206 B.C-8 AD) had pictures of all types of comets (Fig. 3). The shapes and states of the comets were meticulously represented.

1.4. *Ancient astronomical instruments*

The reason that ancient Chinese could made such great advancements in calendric astronomy was that they invented various important astronomical instruments for astronomical

Figure 4. A large gnomon in the beginning of the Yuan Dynasty (left) and the Armillary Sphere, a replica of Guo Shoujings instrument in 1442 (right).

observations and measurements. Among them, the Gnomon is perhaps the oldest astronomical instrument of ancient China. By measuring the gnomon shadow length at noon, one can tell the season of the year. To increase the accuracy of gnomon measurements, Guo Shoujing (1231-1316) constructed a large gnomon in the beginning of the Yuan Dynasty (left penal in Fig. 4).

The right penal in Figure 4 shows the Armillary Sphere, a replica of Guo Shoujings instrument in 1442. The positioning of the rings can be adjusted to reflect such celestial movements as precession and the retrograde of the nodes of the lunar path and the ecliptic. Another instrument made also by Guo Shoujing is the simplified Instrument, Jian yi.

The making of Chinese astronomical instrument reached a peak in the 11th century, during the North Song Dynasty, when Su Song (1020-1101) and his team constructed the water powered astronomical clock tower. It was an automated mechanical system powered by water. An escapement mechanism was used to control the movement of the driving wheel. Joseph Needham hailed it as the first astronomical clock in the world.

2. Beginnings of Modern Astronomy Research and Present Status in China

Several events marked the beginnings of modern astronomy in China: On 1922 October 30, more than 40 scholars held a meeting at the Beijing Ancient Astronomical Observatory to found the Chinese Astronomical Society; The Purple Mountain Observatory (PMO), the biggest observatory in East Asia at that time, was built in 1934; One year later, the fifth General Assembly of the international Astronomical Union (IAU) was held in Paris, and China became a member of the IAU .

Since the establishment of the People's Republic of China, and in particular, after the so-called great cultural revolution, Chinese astronomy has advanced quickly. People realized that the international collaboration was important for developing astronomy in China. Thus the first Chinese astronomer delegation visited Kitt Peak Observatory of the USA in 1976. In succession, the first USA astronomer delegation visited PMO in 1977. Six years later, the first international workshop on solar physics was successfully organized in Kunming in 1983.

Benefiting from the fast development of Chinese economy, the research of astronomy in China has achieved remarkable progress in recent years. The Chinese Astronomical Society now has 2481 members. There are about 400 researchers and professors, twice more than that ten years ago, and 1300 graduate students. The main astronomical organizations include Purple Mountain Observatory (PMO), National Astronomical Observatories of China (NAOC, Beijing), Yunnan Astronomical Observatory (YAO, NAOC), Xinjiang Astronomical Observatory (XAO, NAOC), Nanjing Astronomical Instrument Research Center (NAIRC, NAOC), Shanghai Astronomical Observatory (SAO), and National Time Service Center (NTSC).

In recent years, more than 20 universities have established astronomy education and research group. A complete series of educational programs of undergraduate, master, doctoral and post-doctoral levels have been set up. There are astronomy departments in five key universities, including Nanjing University (NJU, established in 1952), Beijing University (BJU, 1960), Beijing Normal University (BNU, 1960), University of Science and Technology of China (USTC, 1978), and Xiameng University at Fuzhou (2012).

Recently, Chinese astronomers have gotten many important results. Here only are some examples.

In the field of cosmology, Yipeng Jing and his group in SAO carried out a series of the state-of-the art N-body simulations of structure formation in the Universe, and obtained a number of important conclusions. Jing & Suto (2002) studied non-spherical modeling of dark matter halos in detail, and obtained a series of practically useful fitting formulae in applying the triaxial model. Zhao *et al.* (2009) found universal laws governing mass growth and structure evolution of dark matter halos. Their model is much more accurate than all previous works. By analyzing the shower development in the Advanced Thin Ionization Calorimeter (ATIC) instrument, which can separate electrons from large cosmic ray background, after four flights from 2000, the cosmic ray electron spectrum has shown an excess above 100 GeV compared to theoretical models. The excess of cosmic ray electrons is mysterious. It could be from near astrophysical objects or other exotic sources such as dark matter particles. Jin Chang and his group played an important role for these observations (Chang *et al.* 2006). Chinese astronomers have also developed evolutionary population synthesis and spectral fitting codes, which have been used to analyze the spectral parameters of galaxies.

In the field of high energy astrophysics, some authors proposed an universal light curves of GRB afterglows through analyzing the observational data of Swift satellite. A new direction of GRB cosmology was proposed and constrained the cosmological parameters and the properties of dark energy jointly with other cosmological probes (Dai *et al.* 2004). In the aspect of AGNs, a theory of advection-dominated accretion flow disk of black holes was developed and applied to the center black hole of the Galaxy. It can explain the observations successfully. Chinese scientists derived the magnetic field of some neutron stars through the "propeller" effect, suggested a method of direct measurement of the mass of black holes and neutron stars by observing the Doppler-shifted absorption lines from accretion disk winds.

In the field of the Galaxy research, using all HII regions and giant molecular clouds, Jinlin Han and his group derived the spiral arm structure, which is probably the best known structure of our Milky Way (Hou *et al.* 2009). They used pulsar rotation measurement to obtain the magnetic fields in the Galactic disk (Han *et al.* 2006) and used the rotation measure sky to deduce the magnetic fields in the Galactic halo (Han *et al.* 1999). Using VLBA and some maser sources in the Milky Way, Xingwu Zheng and his group made first parallax measurement and deduced the distances and the proper motions of the sources with the accuracy as high as 0.05 mas, so it provides the possibility to map the spiral structure of the Milky Way (Xu *et al.* 2006). By measuring the parallax and proper motions of high mass star forming regions, they found that the Milky Way was rotating about 15% faster than previously estimated (Xu *et al.* 2009). This implies that the mass of the Milky Way's dark matter halo would be increased by 50%.

In the field of stellar physics, a group led by Gang Zhao in the NAOC has gotten a batch of valuable results on stellar chemical abundance, especially in metal-poor stars (e.g. Zhao & Gehren 2000). Zhanwen Han *et al.* (e.g. Han & Podsiadlowski 2007) and Yan Li *et al.* (e.g. Li 2000; Li & Yang 2007) in Yunnan Observatory have made some achievements on the evolutions of single and binary stars, formation of extraordinary stars, and stellar oscillation theory.

In the field of solar physics, since the late 1980s extensive observations of solar vector magnetic fields have been made in NAOC. Lots of studies of magnetic energy accumulation and magnetic complexity in solar active regions have been carried out. Jingxiu Wang and his co-workers (Wang *et al.* 2007, Zhang *et al.* 2007) first recognized that there were large-scale source regions for the Earth-directed halo coronal mass ejections. Some global magnetic coupling has been identified by them for a few major solar activity events. In several major events trans-equatorial loops and filaments were identified to grow and erupt, which resulted in large-scale flaring.

Several important progresses have been made in the astrometric and planetary dynamics field. By the use of the 1.0/1.2m Near Earth Objects Space Telescope (NEOST), two large field of view Survey programs have been carried out and the staged achievements have been presented. Scientists in PMO have been engaged in the construction of a space debris detection system, improving notably the ability of detection and cataloging of space debrises. Astronomers in NJU studied the orbit design and control for lunar satellites, and the transfer orbit design and control for martian satellites. In the dynamics of extra-solar planetary system, astronomers in NJU provided an effective formation way of Earth-like planets, i.e. the combination between planetsimals by collisions after the formation and migration of the giant planet (Zhou *et al.* 2005). They also analyzed the stability of a planetary system containing N planets with equal mass. It was found that the essential reason of the orbital diffusion viscosity effect is the hyperbolic structure in conservative systems (Sun & Zhou 2009).

As for the astronomical facilities, a series of telescopes have been put into operation during the past three decades, including the 2.4m (Fig. 5), 2.16m, 1.56m, 1.2m optical telescopes, the 13.7m radio telescope, and the 21cm array radio telescope (21CMA). Furthermore, a Chinese VLBI network (Fig. 6) has been established since 1990. There are also a series of solar telescopes, including the 35cm magnetogragh at Huairou station, the 60cm solar tower of Nanjing university, the radiospectrometer with high-temporal resolution, the multi-wavelength spectrograph at PMO etc.

The Large Sky Area Multi-Object Fiber Spectroscopic Telescope (LAMOST, now named Guoshoujing telescope) is one of the National Major Scientific Projects undertaken by the Chinese Academy of Sciences. The telescope was installed at the Xinglong Observing Station of NAOC in 2010. The effective aperture of LAMOST is 4m, with a focal plane being 1.75m in diameter, corresponding to a 5 degree field of view. This may accommodate as many as 4000 optical fibers. LAMOST adopts the active optics technique both for thin mirror and segmented mirror on the Schmidt corrector, as well

Figure 5. The 2.4m telescope located at Guomeigu of Yunnan province.

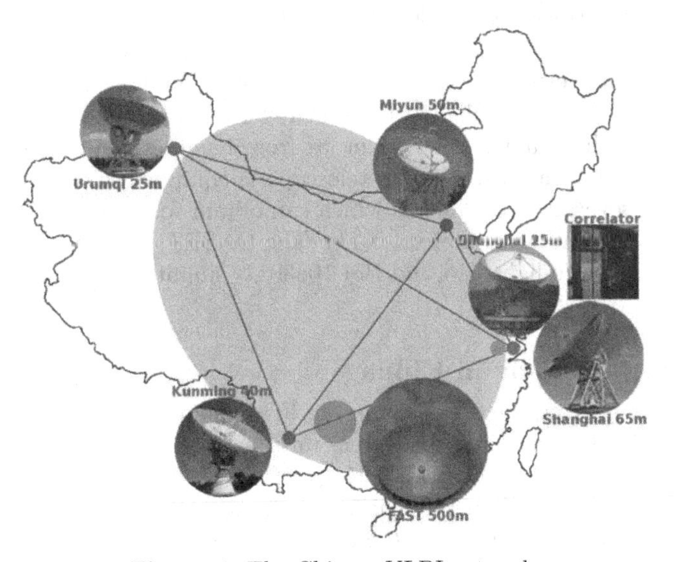

Figure 6. The Chinese VLBI network.

as the parallel controllable fiber positioning system. Thus the telescope will be the one that possesses the highest spectrum acquiring rate in the world (Cui *et al.* 2012). The Guoshoujing telescope will take Chinese astronomy to an advanced position in the large scale observations and the research field of optical spectra. Figure 7 gives an overview of the telescope.

The first satellite in the so-called Chang E project was launched on 2007 October 24. It obtained the first picture of the Moon on 2007 November 26. The second mission was launched in 2009 and got a full Moon high-resolution image with a resolution of 7m. The third mission will be carried out next year. A lunar roving vehicle will be softlanded on the Moon.

A 21 centimeter array (21CMA) has been installed at Tianshan mountains, west of China. It has 10287 antennas with 4km x 6km arms, working in the frequency range of 50-200 MHz. It aims to probe the epoch of reionization.

Figure 7. The Guoshoujing telescope (LAMOST) located at Xinlong, Beijing.

A 1m solar vacuum telescope (NVST) and an Optical and Near-infrared Solar Eruption Tracer (ONSET) were installed at the south station of Yunnan Observatory in 2011. The station is near the Wuxian lake, 60km far from Kunming, and is now the best site in China for solar observations. The 1m telescope is equipped with a multi-wavelength spectrograph and several CCD cameras which can obtain solar images at different wavelengths with high resolution (Liu *et al.* 2011). The ONSET has three tubes and can get the solar images at white-light, Hα and HeI 10830 Å simultaneously (Fang *et al.* 2012).

3. Future of Astronomy in China

The facilities under construction include the Five Hundred Meter Aperture Spherical Radio Telescope (FAST), the Chang E (Lunar mission) and Mars mission, the Hard X-ray Modulate Telescope (HXMT), the DArk Matter Particle Explorer (DAMPE), the Deep Space Solar Observatory (DSO), the Chinese Antarctic Observatory, the 65m steerable radio telescope, the Chinese Spectral Radioheliogaph (CSRH), etc.

The FAST, the biggest single-dish radio telescope in the world, has an active main reflector and cable-parallel robot feed support. Its working frequencies in GHz are 0.3-0.46, 0.46-0.92, 0.92-1.72, 2.15-2.35, 2.8-3.3, 4.5-5.1, 5.7-6.7, 8.0-8.8. The FAST project was approved by the Chinese government in 2010, and it is under construction in the southwest of China, GuiZhou provence, where an unique Karst depression can be used to put the large dish. FAST will be finished in 2016. Figure 8 depicts a sketch map of the FAST.

The HXMT aims to make hard X-ray sky (1-250 keV) survey with highest sensitivity and high precision pointed observations of high energy objects, such as black holes, AGN, SNR, neutron stars etc. Its orbit altitude is 550km with an inclination of 43°. The HXMT will be launched in 2014-2015.

The DAMPE will search for the exact cause of cosmic ray excess, which was first observed by ATIC, and then confirmed by PAMELA FERMI and PPB-BETS experiments. Compared with other relevant space projects in the world, DAMPE has the highest energy resolution, the widest energy measurement range and the lowest background in electron/gamma detection. It will be launched in 2015.

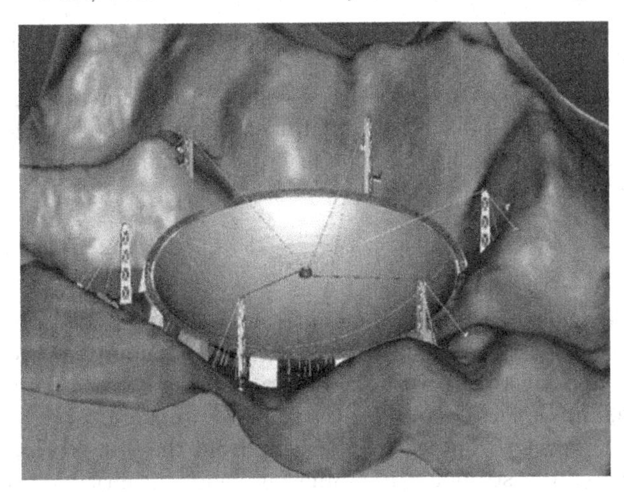

Figure 8. A sketch map of the Five Hundred Meter Aperture Spherical Radio Telescope (FAST), which is located at Guizhou province and is under construction.

The payloads of DSO include a 1m main optical telescope, an extreme ultraviolet telescope, a hard X- and γ-ray spectrometer, a Lyα coronagraph, a coronal imager etc. The DSO aims to observe solar magnetic field and other solar activities with high precision and high resolution. It will be launched to L1 point in about six years.

It should be mentioned that China now plans to construct a space station, on which there will be several astronomical facilities, including optical telescopes, X- and γ-ray spectrometer etc. It will certainly further strengthen much the ability of astronomical space observations in China.

The Dome A in Antarctica is possibly the best site for astronomical observations on the Earth, particularly in THz and infrared wavelength. The place has a very clear sky and low wind speed, with low thickness of ground layer and very good transmission from 150 to 800 μm. After a Chinese expedition team reached Dome A in 2005 January, Chinese astronomers followed on with international collaborations. Recently the Dome A observatory project is approved by the Chinese government. A 2.5m optical/infrared telescope and a 5m THz telescope will be installed at the observatory, aiming at studying dark energy and dark matter, exo-planets, time-domain subjects, formation and evolution of stars and galaxies etc.

A 65m fully steerable radio telescope is also under construction at Shanghai observatory. It will be working at the frequencies of L(1.6 GHz), S/X(2.3/8.4 GHz), C(5 GHz), Ku(15 GHz), K(22 GHz), Ka(32 GHz), and Q (43 GHz). It will be used in precise astrometry and astrophysics, such as observations of high-resolution VLBI, maser sources, molecular spectral line survey, radio stars etc. The telescope will be put into operation in 2015.

Another big project is the Chinese Spectral Radioheliogaph (CSRH), which has been under construction since 2009 at Zhengxiang Baiqi, Inner Mongolia. It will be working in the frequency range of 0.4-15 GHz, with 100 parabolic antennas, and achieve 1.3-50 arcsecond spatial resolution and better than 100ms temporal resolution. The CSRH will be put into operation in 2014.

Besides, some projects are under discuss and review. These include a 110m radio telescope, 20-30m optical/infrared Telescope, south LAMOST, large ground-based solar observatory (Chinese Giant Solar Telescope (CGST) and coronagraphs), new generation

Chinese VLBI network, X-ray Timing and Polarization Mission (XTP), advanced space-born solar observatory etc.

It should be mentioned that an extensive site survey both for the night- and day-time astronomy in the western China has been started since the beginning of this century. If the best site can be found, then China will build new large optical telescopes for the solar and stellar observations respectively.

4. Conclusions

Through out the ancient history, Chinese astronomers had made tremendous achievements. However, in the past several centuries, Chinese astronomy research has been much delayed behind comparing to the western countries. Since the beginning of the People's Republic of China, and in particular, after the so-called great cultural revolution, Chinese astronomy has been developed quickly. Benefiting from the fast development of Chinese economy, the research of astronomy in China has made remarkable progress in the recent years. The numbers of astronomers and the graduate students have been greatly increased. The budget for astronomy research is ten times larger than that ten years ago. Recently, Chinese astronomers have obtained many important results. A series of astronomical facilities have been put into operation, including the Guoshoujing telescope, which possesses the highest spectrum acquiring rate in the world. Particularly, China has many exciting projects to build advanced both ground-based and space-born telescopes. If these plans can be successfully put in practice, Chinese astronomy will enter into a new era! The future will be highly luciferous.

We understand that at present there is still a big gap between Chinese astronomy and that in advanced countries. We have a lot of catch-up to do and we believe that the international cooperation is essential during this journey. All suggestions and help from astronomers and friends over the world are highly welcome!

Acknowledgements

Thanks to the organizers for providing the opportunity to show Chinese astronomy from the past to the future. Thanks to my friends and colleagues for the use of their data and publications that help me prepare this paper. This work received support from the grant of the 973 project 2011CB811402 of China, and the National Natural Science Foundation of China (NSFC) under the grant numbers 10878002, 10610099, 10933003 and 10673004.

References

Chang, J., Adams, J. H., Ahn, H. S., *et al.* 2008, *Nature*, 456, 362
Cui, X. Q., Zhao, Y. H., & Chu, Y. Q., *et al.* 2012, *RAA*, 12, 1197
Dai, Z. G., Liang, E. W., & Xu, D. 2004, *ApJ*, 612, 101
Fang, C., Chen, P. F., Ding, M. D., *et al.* 2012, in proceeding of the 4th France-China meeting on solar physics, eds. M. Faurobert, C. Fang, T. Corbard, EAS Publication Series, Vol.55, 349
Han, J. L., Manchester, R. N., & Qiao, G. J. 1999, *MNRAS*, 306, 371
Han, J. L., Manchester, R. N., Lyne, A. G., Qiao, G. J., & van Straten, W. 2006, *ApJ*, 642, 868
Han, Z. & Podsiadlowski, Ph 2008, *IASS*, 252, 349
Hou, L. G., Han, J. L., & Shi, W. B. 2009, *A&A*, 499, 473
Jing, Y. P. & Suto, Y. 2002, *ApJ*, 574, 538
LI, Y. 2000, *ApJ*, 538, 346
Li, Y. & Yang, J. Y., 2007, *MNRAS*. 375, 388

Liu, Z. & Xu, J. 2011, in *First Asia-Pacific Solar Physics Meeting*, ASI Conference Series, eds. A. R. Choudhuri & D. Banerjee, Vol. 2, 9

Sun X. 2011, *IAU Symp.* 260, 98

Wang, J. X., Zhang, Y. Z., Zhou, G. P., Harra, L. K., Williams, D. R., & Jiang, Y. C. 2007, *Solar Phys.*, 244, 75

Xu, Y., Reid, M. J., Zheng, X. W., & Menten, K. M. 2006, *Science*, 311, 54

Xu, Y., Reid, M. J., Menten, K. M., Brunthaler, A., Zheng, X. W., & Moscadelli, L. 2009, *ApJ*, 693, 413

Xu Z. T. & Jiang Y. T. 1986, *Sunspots and Human*, Tianjin Science and Technic Publ.

Zhang, Y. Z., Wang, J. X., Attrill, G. D. R., Harra, L. K., Yang, Z. L., & He, X. T. 2007, *Solar Phys.*, 241, 329

Zhao, D., Jing, Y. P., Mo, H. J., & Borner, G. 2009, *ApJ*, 707, 354

Zhao, G. & Gehren, T. 2000, *A&A*, 362, 1077

Zhou, J. L., Aarseth, S. J., Lin, D. N. C., & Nagasawa, M. 2005, *ApJ*, 631, L85

Sun, Y. S. & Zhou, L. Y., 2009, *Celest Mech Dyn Astr*, 103, 119

Highlights of Astronomy, Volume 16
XXVIIIth IAU General Assembly, August 2012
T. Montmerle, ed.

doi:10.1017/S1743921314004633

The *Herschel* View of Star Formation

Philippe André
Laboratoire d'Astrophysique (AIM) Paris-Saclay, CEA Saclay, 91191 Gif-sur-Yvette, France
email: pandre@cea.fr

Abstract. Recent studies of the nearest star-forming clouds of the Galaxy at submillimeter wavelengths with the *Herschel* Space Observatory have provided us with unprecedented images of the initial conditions and early phases of the star formation process. The *Herschel* images reveal an intricate network of filamentary structure in every interstellar cloud. These filaments all exhibit remarkably similar widths - about a tenth of a parsec - but only the densest ones contain prestellar cores, the seeds of future stars. The *Herschel* results favor a scenario in which interstellar filaments and prestellar cores represent two key steps in the star formation process: first turbulence stirs up the gas, giving rise to a universal web-like structure in the interstellar medium, then gravity takes over and controls the further fragmentation of filaments into prestellar cores and ultimately protostars. This scenario provides new insight into the inefficiency of star formation, the origin of stellar masses, and the global rate of star formation in galaxies. Despite an apparent complexity, global star formation may be governed by relatively simple universal laws from filament to galactic scales.

Keywords. stars: formation – ISM: clouds – ISM: Filaments – ISM: structure – submillimeter

1. Introduction

Star formation is one of the most complex processes in astrophysics, involving a subtle interplay between gravity, turbulence, magnetic fields, feedback mechanisms, heating and cooling effects etc... Yet, despite this apparent complexity, the net products of the star formation process on global scales are relatively simple and robust. In particular, the distribution of stellar masses at birth or stellar initial mass function (IMF) is known to be quasi-universal (e.g. Kroupa 2001, Chabrier 2005, Bastian *et al.* 2010). Likewise, the star formation rate on both GMC and galaxy-wide scales is related to the mass of (dense molecular) gas available by rather well defined "star formation laws" (e.g. Kennicutt 1998, Gao & Solomon 2004, Lada *et al.* 2010). On the basis of recent submillimeter imaging observations obtained with the *Herschel* Space Observatory (Pilbratt *et al.* 2010) on Galactic interstellar clouds as part of the Gould Belt (André *et al.* 2010), HOBYS (Motte *et al.* 2010), and Hi-GAL (Molinari *et al.* 2010) surveys, the thesis advocated in this paper is that it may be possible to explain, at least partly, the IMF and the global rate of star formation in terms of the quasi-universal filamentary structure of the cold interstellar medium (ISM) out of which stars form.

In particular, the bulk of nearby ($d \lesssim 500$ pc) molecular clouds, mostly located in Gould's Belt (e.g. Guillout 2001, Perrot & Grenier 2003), have been imaged at 6 wavelengths between 70 μm and 500 μm as part of the *Herschel* Gould Belt survey (HGBS – André *et al.* 2010). Observationally, the molecular clouds of the Gould Belt are the best laboratories at our disposal to investigate the star formation process in detail, at least as far as solar-type stars are concerned. The $\sim$15 nearby clouds covered by the HGBS span a wide range of environmental conditions, from active, cluster-forming complexes such as the Orion A & B GMCs or the Aquila Rift cloud complex (e.g. Gutermuth *et al.* 2008) to quiescent regions with no star formation activity whatsoever such as the Polaris flare

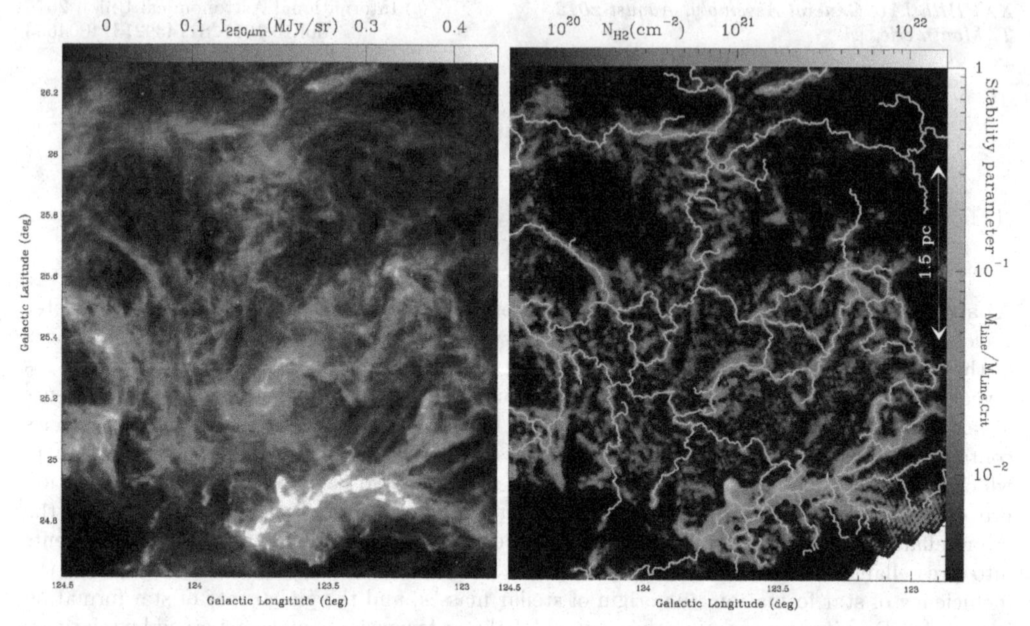

Figure 1. Left: *Herschel*/SPIRE 250 μm dust continuum map of a portion of the Polaris flare translucent cloud ($d \sim 150$ pc) taken as part of the HGBS survey (e.g. Miville-Deschênes *et al.* 2010, Ward-Thompson *et al.* 2010). **Right:** Corresponding column density map derived from *Herschel* data (André *et al.* 2010). The contrast of the filaments has been enhanced using a curvelet transform (cf. Starck *et al.* 2003). The skeleton of the filament network identified with the DisPerSE algorithm (Sousbie 2011) is shown in light blue. Given the typical filament width $\sim$0.1 pc (Arzoumanian *et al.* 2011 – see Fig. 4 below), this column density map is equivalent to a *map of the mass per unit length along the filaments* (see color scale on the right).

translucent cloud (e.g. Heithausen *et al.* 2002). The main scientific goals of the HGBS are to clarify the nature of the relationship between the prestellar core mass function (CMF) and the stellar IMF (cf. § 3 below) and to elucidate the physical mechanisms responsible for the formation of prestellar cores out of the diffuse ISM (cf. § 5 and § 7).

This paper presents an overview of the first results obtained with *Herschel* on nearby star-forming clouds. Section 2 emphasizes the universality of the filamentary structure revealed by *Herschel* in the cold ISM. Section 3 presents preliminary results obtained on the global properties of prestellar dense cores. Section 4 summarizes a few theoretical considerations on the gravitational instability of filamentary clouds. Section 5 presents the observational evidence of a column density threshold for the formation of prestellar cores and shows how this can be interpreted in terms of the gravitational instability threshold of interstellar filaments. Section 6 discusses implications of the *Herschel* results on filaments and cores for our understanding of the origin of the IMF and the global rate of star formation in galaxies. Finally, Sect. 7 concludes by summarizing the scenario of star formation emerging from the *Herschel* results.

2. Universality of the filamentary structure in the cold ISM

The high quality and dynamic range of the *Herschel* images are such that they provide key information on the structure of molecular clouds on a wide range of spatial scales from the size of entire cloud complexes ($\gtrsim$10 pc) down to the scale of individual dense cores ($<$ 0.1 pc). In particular, one of the most spectacular early findings made with *Herschel* is the omnipresence of long ($>$pc scale) filamentary structures in the cold ISM

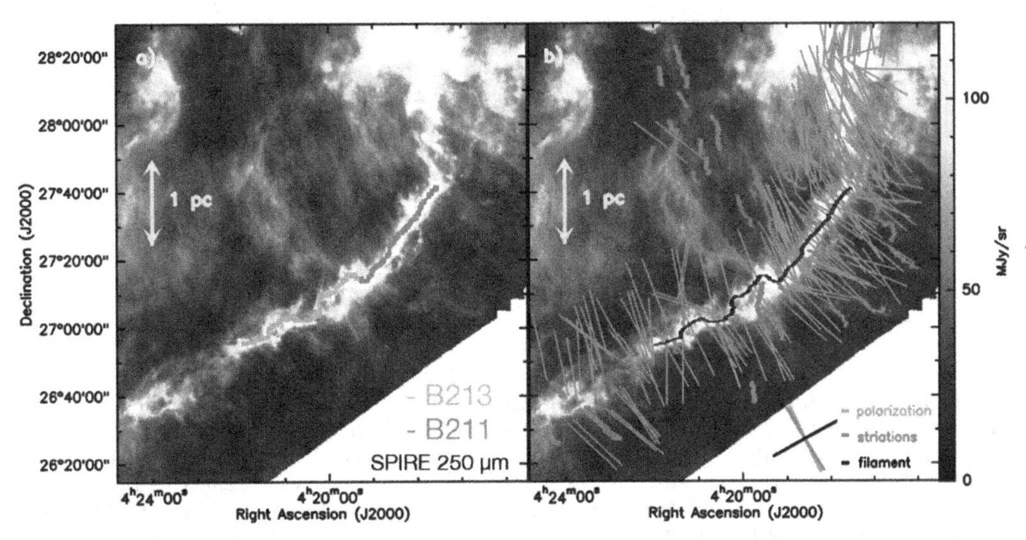

Figure 2. **(a)** *Herschel*/SPIRE 250 μm dust continuum image of the B211/B213/L1495 region in Taurus ($d \sim 140$ pc). The light blue and purple curves show the crest of the B213/B211 filament. **(b)** Display of optical and infrared polarization vectors from (e.g. Heyer *et al.* 2008; Chapman *et al.* 2011) tracing the magnetic field orientation in the same region, overlaid on the *Herschel*/SPIRE 250 μm image. The plane-of-the-sky projection of the magnetic field appears to be oriented perpendicular to the B211/B213 filament and roughly aligned with the general direction of the striations overlaid in blue. (From Palmeirim *et al.* 2013.)

and the apparently tight connection between the filaments and the formation process of dense cloud cores (e.g. André *et al.* 2010; Men'shchikov *et al.* 2010; Molinari *et al.* 2010). While interstellar clouds were already known to exhibit large-scale filamentary structures long before *Herschel* (e.g. Schneider & Elmegreen 1979; Abergel *et al.* 1994; Hartmann 2002; Hatchell *et al.* 2005; Myers 2009), *Herschel* now demonstrates that these filaments are truly ubiquitous in the giant molecular clouds (GMCs) of our Galaxy (Molinari *et al.* 2010) and provides an unprecedented large-scale view of the role of filaments in the formation of prestellar cores (see § 5 below). Filaments are omnipresent even in diffuse, non-star-forming complexes such as the Polaris translucent cloud (cf. Fig. 1 – Miville-Deschênes *et al.* 2010; Ward-Thompson *et al.* 2010), suggesting that the formation of filaments *precedes* star formation in the cold ISM. Importantly, the few high-resolution spectral line observations available to date suggest that the filaments seen in the *Herschel* dust continuum images are velocity-coherent structures (e.g. Hacar & Tafalla 2011; Li & Goldsmith 2012; Arzoumanian *et al.* 2013).

A very common filamentary pattern observed with *Herschel* is that of a main filament surrounded by a population of fainter "sub-filaments" or striations approaching the main filament from the side and apparently connected to it (see Fig. 2 and Palmeirim *et al.* 2013, Peretto *et al.* 2012, Hennemann *et al.* 2012, Cox *et al.* 2013). The morphology of these "sub-filaments" and striations is suggestive of accretion flows feeding the main filaments with surrounding cloud material.

More generally, in any given cloud complex, *Herschel* imaging reveals a whole network of filaments (see Fig. 1), making it possible to characterize their properties in a statistical manner. Furthermore, the *Herschel* maps have resolved the structure of nearby filaments with unprecedented detail. Detailed analysis of the radial column density profiles derived from *Herschel* data suggests that the shape of the filament radial profiles is quasi universal and well described by a Plummer-like function of the form (cf. Palmeirim *et al.*

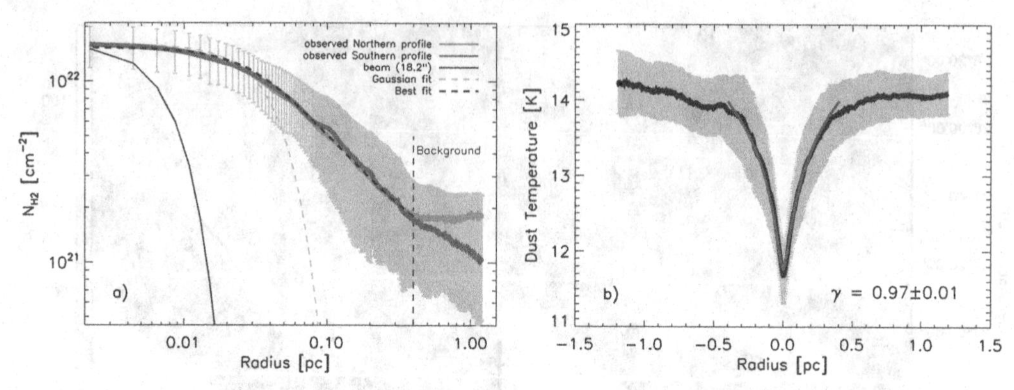

Figure 3. **(a)** Mean radial column density profile observed with *Herschel* perpendicular to the B213/B211 filament in Taurus (cf. Fig. 2), for both the Northern (blue curve) and the Southern part (red curve) of the filament. The yellow area shows the ($\pm 1\sigma$) dispersion of the distribution of radial profiles along the filament. The inner solid purple curve shows the effective 18″ HPBW resolution (0.012 pc at 140 pc) of the column density map used to construct the profile. The dashed black curve shows the best-fit Plummer model (convolved with the 18″ beam) described by Eq. (1) with p=2.0±0.4 and a diameter $2\times R_{\rm flat} = 0.07\pm0.01$ pc, which matches the data very well for $r \leqslant 0.4$ pc, **(b)** Mean dust temperature profile measured perpendicular to the B213/B211 filament. The solid red curve shows the best model temperature profile obtained by assuming that the filament has a density profile given by the Plummer model shown in **(a)** (with $p = 2$) and obeys a polytropic equation of state, $P \propto \rho^{\gamma}$, and thus $T(r) \propto \rho(r)^{(\gamma-1)}$. This best fit corresponds to a polytropic index γ=0.97±0.01. (From Palmeirim *et al.* 2013.)

2013 and Fig. 3):

$$\rho_p(r) = \frac{\rho_c}{\left[1+(r/R_{\rm flat})^2\right]^{p/2}} \longrightarrow \Sigma_p(r) = A_p \frac{\rho_c R_{\rm flat}}{\left[1+(r/R_{\rm flat})^2\right]^{\frac{p-1}{2}}}, \quad (1)$$

where ρ_c is the central density of the filament, $R_{\rm flat}$ is the radius of the flat inner region, $p \approx 2$ is the power-law exponent at large radii ($r \gg R_{\rm flat}$), A_p is a finite constant factor which includes the effect of the filament's inclination angle to the plane of the sky. Note that the density structure of an isothermal gas cylinder in hydrostatic equilibrium follows Eq. (1) with $p = 4$ (Ostriker 1964), instead of the observed $p \approx 2$ value.

Remarkably, the diameter $2 \times R_{\rm flat}$ of the flat inner plateau in the filament radial profiles appears to be roughly constant ~0.1 pc for all filaments, at least in the nearby clouds of Gould's Belt (cf. Arzoumanian *et al.* 2011). This is illustrated in Fig. 4 which shows that nearby interstellar filaments are characterized by a very narrow distribution of inner FWHM widths centered at about 0.1 pc.

The origin of this quasi-universal inner width of interstellar filaments is not yet well understood. A possible interpretation is that it corresponds to the sonic scale below which interstellar turbulence becomes subsonic in diffuse, non-star-forming molecular gas (cf. Padoan *et al.* 2001 and § 7 below). In this view, the observed filaments would correspond to dense, post-shock stagnation gas associated with shocked-compressed regions resulting from converging flows in supersonic interstellar turbulence. Interestingly, the filament width ~0.1 pc is also comparable to the cutoff wavelength $\lambda_A \sim 0.1\,{\rm pc} \times (\frac{\rm B}{10\mu{\rm G}}) \times (\frac{n_{\rm H_2}}{10^3\,{\rm cm}^{-3}})^{-1}$ for MHD waves in (low-density, primarily neutral) molecular clouds (cf. Mouschovias 1991), if the typical magnetic field strength is $B \sim 10\mu$G (e.g. Crutcher 2012). Alternatively, the characteristic width may also be understood if interstellar filaments are formed as quasi-equilibrium structures in pressure balance with a typical am-

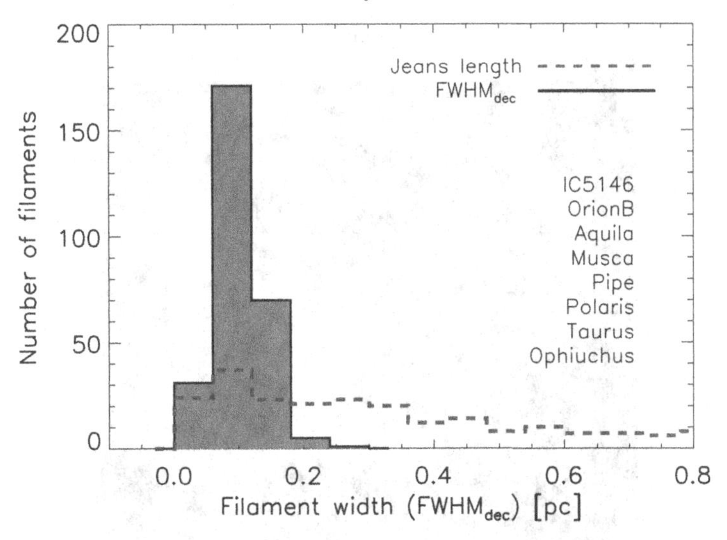

Figure 4. Histogram of deconvolved FWHM widths for a sample of 278 filaments in 8 nearby regions of the Gould Belt, all observed with *Herschel* (at effective spatial resolutions ranging from ~0.01 pc to ~0.04 pc) and analyzed in the same way. The distribution of filament widths is narrow with a median value of 0.09 pc and a standard deviation of 0.04 pc. In contrast, the distribution of Jeans lengths corresponding to the central column densities of the filaments (blue dashed histogram) is much broader. (Adapted from Arzoumanian *et al.* 2011.)

bient ISM pressure $P_{\rm ext} \sim 2\text{–}5 \times 10^4\,{\rm K\,cm^{-3}}$ (Fischera & Martin 2012; Inutsuka *et al.*, in prep.).

3. Dense cores and their ensemble properties as derived from *Herschel* observations

As prestellar cores and deeply embedded protostars emit the bulk of their luminosity at far-infrared and submillimeter wavelengths, *Herschel* observations are also ideally suited for taking a sensitive census of such cold objects in nearby molecular cloud complexes. This is indeed one of the main observational objectives of the *Herschel* Gould Belt survey (HGBS).

Conceptually, a dense core is an individual fragment or local overdensity which corresponds to a local minimum in the gravitational potential of a molecular cloud. A starless core is a dense core with no central protostellar object. A prestellar core may be defined as a dense core which is both starless and self-gravitating. In other words, a prestellar core is a self-gravitating condensation of gas and dust within a molecular cloud which may potentially form an individual star (or system) by gravitational collapse (e.g. Motte *et al.* 1998; André *et al.* 2000; Di Francesco *et al.* 2007; Ward-Thompson *et al.* 2007). Known prestellar cores are observed at the bottom of the hierarchy of interstellar cloud structures and depart from Larson (1981)'s self-similar scaling relations. They are the smallest units of star formation (e.g. Bergin & Tafalla 2007) and correspond to "coherent" regions of nearly constant and thermal velocity dispersion which do not obey Larson (1981)'s power-law linewidth vs. size relation (Myers 1983; Goodman *et al.* 1998; André *et al.* 2007). To first order, known prestellar cores have simple, convex (and not very elongated) shapes, and their density structure approaches that of Bonnor-Ebert isothermal spheroids bounded by the external pressure exerted by the background parent cloud (e.g. Johnstone *et al.* 2000, Alves *et al.* 2001). Apart from an unprecedented mapping speed,

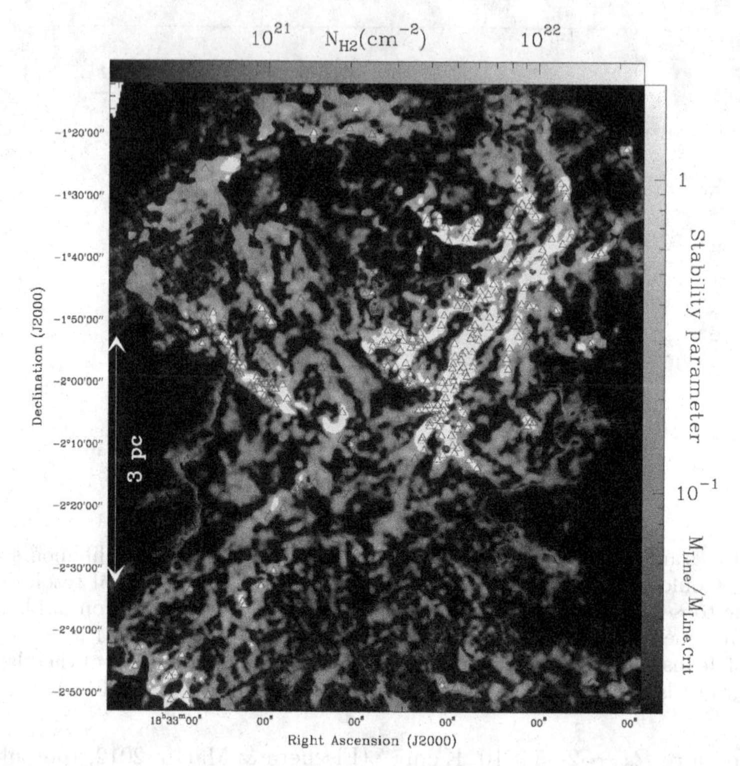

Figure 5. Column density map of a subfield of the Aquila star-forming region ($d \sim 260$ pc) derived from *Herschel* data (André *et al.* 2010). The contrast of the filaments has been enhanced using a curvelet transform (cf. Starck *et al.* 2003). Given the typical width ~0.1 pc of the filaments (Arzoumanian *et al.* 2011 – see Fig. 4), this map is equivalent to a *map of the mass per unit length along the filaments.* The areas where the filaments have a mass per unit length larger than half the critical value $2\,c_s^2/G$ (cf. Inutsuka & Miyama 1997 and § 4) and are thus likely gravitationally unstable have been highlighted in white. The bound prestellar cores identified by Könyves *et al.* (2010) in Aquila are shown as small blue triangles.

two key advantages of *Herschel* broad-band imaging for prestellar core surveys are 1) that dust continuum emission is largely optically thin at far-infrared/submillimeter wavelengths and thus directly tracing column density, and 2) that the ~18″ HPBW angular resolution of *Herschel* at $\lambda = 250\,\mu$m, corresponding to ~0.03 pc at a distance $d = 350$ pc, is sufficient to resolve the typical Jeans length in nearby clouds, which is also the characteristic diameter expected for Bonnor-Ebert-like cores.

While in general the gravitational potential cannot be inferred from observations, it turns out to be directly related to the observable column density distribution for the post-shock, filamentary cloud layers produced by supersonic turbulence in numerical simulations of cloud evolution (Gong & Ostriker 2011). In practical terms, this means that one can define a dense core as the immediate vicinity of a local maximum in observed column density maps such as the maps derived from *Herschel* imaging (see Fig. 5 and Fig. 6a for examples). In more mathematical terms, the projection of a dense core onto the plane of the sky corresponds to the "descending 2-manifold" (cf. Sousbie 2011) associated to a local maximum in column density, i.e., the set of points connected to the maximum by integral lines following the gradient of the column density distribution.

In practice, systematic source/core extraction in wide-field dust continuum images of highly structured molecular clouds is a complex problem which can be conveniently decomposed into two sub-tasks: 1) source/core detection, and 2) source/core measurement. In the presence of noise and background cloud fluctuations, the sub-task of detecting source/cores reduces to identifying statistically significant intensity/column density peaks based on the information provided by the finite-resolution far-infrared/submillimeter continuum image(s) observed with, e.g., *Herschel*. The main problem to be solved in the other sub-task of measuring detected sources/cores is to find the spatial extent or "footprint" of each source/core, corresponding to the "descending 2-manifold" of the "mathematical" definition given above. The *getsources* method devised by Men'shchikov *et al.* (2010, 2012) and used to identify cores in the HGBS data is a new approach to these two sub-tasks which makes full use of the multi-scale, multi-wavelength nature of the source extraction problem in the case of *Herschel* data. Once cores have been extracted from the maps, the *Herschel* observations provide a very sensitive way of distinguishing between protostellar and starless cores based on the presence or absence of 70 μm emission. The 70 μm flux is indeed known to be a very good tracer of the internal luminosity of a protostar (e.g. Dunham *et al.* 2008), and *Herschel* observations of nearby clouds have the sensitivity to detect "first protostellar cores" (cf. Pezzuto *et al.* 2012), the very first and lowest-luminosity ($\sim$0.01–0.1 $L_\odot$) stage of protostars (e.g. Larson 1969; Saigo & Tomisaka 2011).

Using *getsources*, more than 200 starless cores but no protostars were detected in the *Herschel* images of the Polaris flare region ($\sim$8 deg^2 field). The locations of the Polaris starless cores in a mass versus size diagram show that they are $\sim$2 orders of magnitude less dense than self-gravitating isothermal Bonnor-Ebert spheres and therefore cannot be gravitationally bound. The mass function of these unbound starless cores peaks at an order of magnitude smaller mass than the stellar IMF (André *et al.* 2010). In contrast, more than 200 (Class 0 & Class I) protostars could be identified in the *Herschel* images of the whole ($\sim$11 deg^2) Aquila region (Bontemps *et al.* 2010, Maury *et al.* 2011), along with more than 500 starless cores $\sim$ 0.01–0.1 pc in size (see Fig. 5 and Fig. 6a for some examples). Most (>60%) of the Aquila starless cores lie close to the loci of critical Bonnor-Ebert spheres in a mass versus size diagram, suggesting that they are self-gravitating and prestellar in nature (Könyves *et al.* 2010). The CMF derived for the entire sample of >500 starless cores in Aquila is well fit by a log-normal distribution and closely resembles the IMF (Fig. 6b – Könyves *et al.* 2010; André *et al.* 2010). The similarity between the Aquila CMF and the Chabrier (2005) system IMF is consistent with an essentially one-to-one correspondence between core mass and stellar system mass ($M_{\star\rm sys} = \epsilon_{\rm core}\, M_{\rm core}$). Comparing the peak of the CMF to the peak of the system IMF suggests that the efficiency $\epsilon_{\rm core}$ of the conversion from core mass to stellar system mass is between $\sim$0.2 and $\sim$0.4 in Aquila.

The first results of the HGBS survey on this topic therefore confirm the existence of a close relationship between the prestellar CMF and the stellar IMF, using data with already a factor of $\sim$ 2 to 9 better counting statistics than earlier ground-based studies (cf. Motte, André, Neri 1998; Johnstone *et al.* 2000; Stanke *et al.* 2006; Alves *et al.* 2007; Enoch *et al.* 2008). The efficiency factor $\epsilon_{\rm core} \sim 30\%$ may be attributed to mass loss due to the effect of outflows during the protostellar phase (Matzner & McKee 2000). More work is needed to derive a reliable prestellar CMF at the low-mass end and fully assess the potential importance of subtle observational biases (e.g. background-dependent incompleteness and blending of unresolved groups of cores). The results from the entire Gould Belt survey will also be necessary to fully characterize the nature of the CMF–IMF relationship as a function of environment. Our early findings with *Herschel* nevertheless seem to support models of the IMF based on pre-collapse cloud fragmentation such as the gravo-turbulent fragmentation picture (e.g. Larson 1985; Klessen & Burkert 2000;

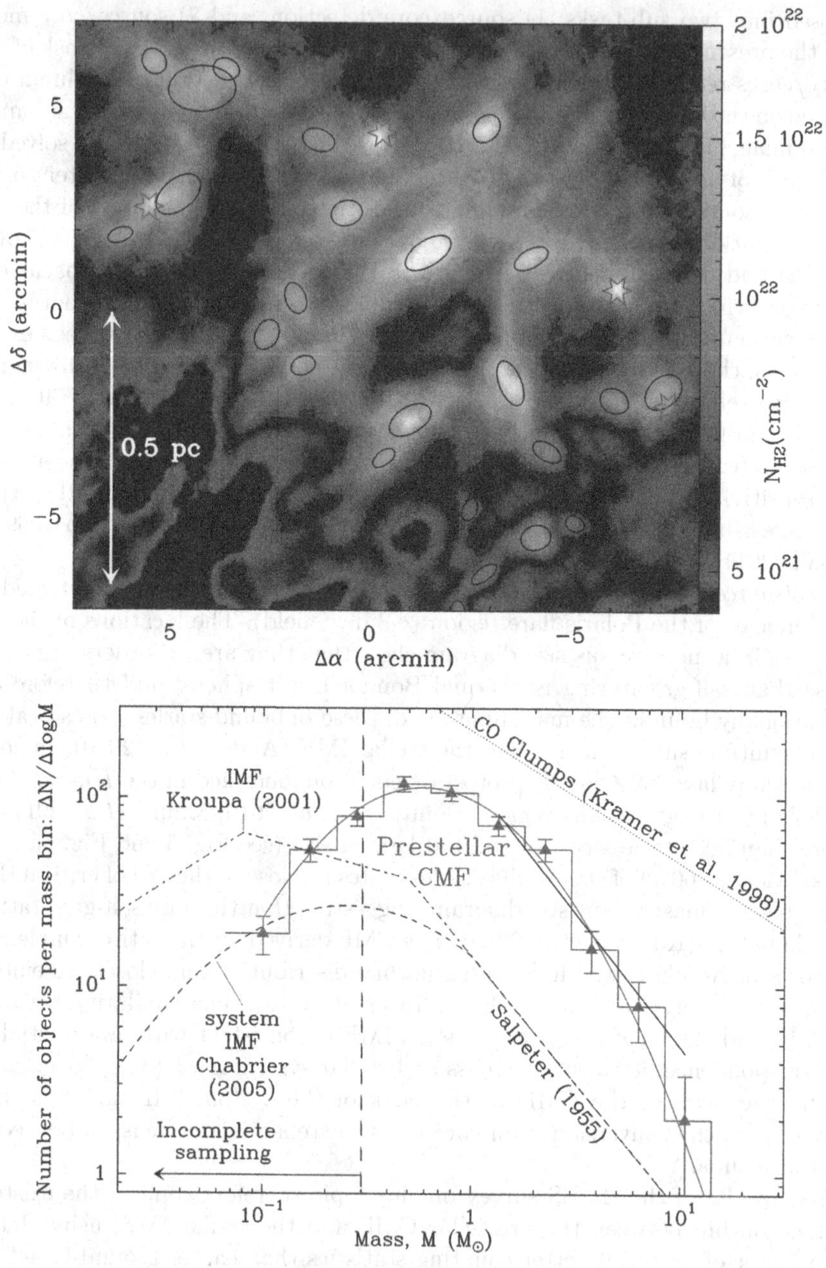

Figure 6. Top: Close-up column density image of a small subfield in the Aquila Rift complex showing several candidate prestellar cores identified with *Herschel* (adapted from Könyves *et al.* 2010). The black ellipses mark the major and minor FWHM sizes determined for these cores by the source extraction algorithm *getsources* (Menshchikov *et al.* 2010, 2012). Four protostellar cores are also shown by red stars. The effective resolution of the image is ~18" or ~0.02 pc at $d \sim 260$ pc. **Bottom:** Core mass function (blue histogram with error bars) of the ~500 candidate prestellar cores identified with *Herschel* in Aquila (André *et al.* 2010 and Könyves *et al.* 2010). The IMF of single stars (corrected for binaries – e.g. Kroupa 2001), the IMF of multiple systems (e.g. Chabrier 2005), and the typical mass spectrum of CO clumps (e.g. Kramer *et al.* 1998) are shown for comparison. A log-normal fit to the observed CMF is superimposed (red curve); it peaks at $\sim 0.6\,M_\odot$, close to the Jeans mass within marginally critical filaments at $T \sim 10$ K (cf. § 6).

Padoan & Nordlund 2002; Hennebelle & Chabrier 2008). Independently of any model, the *Herschel* observations suggest that one of the keys to the problem of the origin of the IMF lies in a good understanding of the formation mechanism of prestellar cores. This is true even if additional processes, such as rotational subfragmentation of prestellar cores into multiple systems during collapse (Bate *et al.* 2003; Goodwin *et al.* 2008) and "competitive" accretion from a larger-scale mass reservoir at the protostellar stage (e.g. Bate & Bonnell 2005), probably also play some role and help to populate the low- and high-mass ends of the IMF, respectively. In Sect. 6 below, we argue that the prestellar cores responsible for the peak of the CMF/IMF result primarily from filament fragmentation.

4. Theoretical considerations on filament collapse and fragmentation

The collapse and fragmentation properties of filaments under the assumption of cylindrical symmetry are well known theoretically (e.g. Nagasawa 1987) but have received renewed attention with the *Herschel* results. The gravitational instability of nearly isothermal filaments is primarily controlled by the value of their mass per unit length $M_{\rm line} \equiv M/L$. Above the critical value $M_{\rm line,crit} = 2\,c_s^2/G$ (where c_s is the isothermal sound speed) cylindrical filaments are expected to be globally unstable to both radial collapse and fragmentation along their lengths (e.g. Inutsuka & Miyama 1992, 1997), while below $M_{\rm line,crit}$ filaments are gravitationally unbound and thus expected to expand into the surrounding medium unless they are confined by some external pressure (e.g. Fischera & Martin 2012). Note that the critical mass per unit length $M_{\rm line,crit} \approx 16\,M_\odot/{\rm pc} \times (T_{\rm gas}/10\,{\rm K})$ depends only on gas temperature $T_{\rm gas}$ (Ostriker 1964). In the presence of non-thermal gas motions, the critical mass per unit length becomes $M_{\rm line,vir} = 2\,\sigma_{\rm tot}^2/G$, also called the virial mass per unit length, where $\sigma_{\rm tot} = \sqrt{c_s^2 + \sigma_{\rm NT}^2}$ is the total one-dimensional velocity dispersion including both thermal and non-thermal components (Fiege & Pudritz 2000). Furthermore, Fiege & Pudritz (2000) have shown that $M_{\rm line,vir}$ is only slightly modified in the case of magnetized filaments: $M_{\rm line,vir}^{\rm mag} = M_{\rm line,vir}^{\rm unmag} \times (1 - \mathcal{M}/|\mathcal{W}|)^{-1}$, where $\mathcal{M}$ is the magnetic energy (positive for poloidal magnetic fields and negative for toroidal fields) and $\mathcal{W}$ is the gravitational energy. In practice, since molecular clouds typically have $|\mathcal{M}|/|\mathcal{W}| \lesssim 1/2$ (Crutcher 1999, 2012), $M_{\rm line,vir}^{\rm mag}$ differs from $M_{\rm line,vir}^{\rm unmag} \equiv 2\,\sigma_{\rm tot}^2/G$ by less than a factor of 2.

Importantly, filaments differ from both sheets and spheroids in their global gravitational instability properties. For a sheet-like cloud, there is always an equilibrium configuration since the internal pressure gradient can always become strong enough to halt the gravitational collapse of the sheet independently of the initial state (e.g. Miyama *et al.* 1987; Inutsuka & Miyama 1997). In contrast, the radial collapse of an isothermal cylindrical cloud cannot be halted and no equilibrium is possible when the line mass exceeds the critical mass per unit length $M_{\rm line,crit}$. Conversely, if the line mass of the filamentary cloud is less than $M_{\rm line,crit}$, gravity can never be made to dominate by increasing the external pressure, so that the collapse is always halted at some finite cylindrical radius. Filaments also differ markedly from isothermal spherical clouds which can always be induced to collapse by a sufficient increase in external pressure (e.g. Bonnor 1956; Shu 1977). The peculiar behavior of the filamentary geometry in isothermal collapse is due to the fact that the isothermal equation of state ($\gamma = 1$) is a critical case for the collapse of a filament (e.g. Larson 2005): For a polytropic equation of state ($P \propto \rho^\gamma$) with $\gamma < 1$, an unstable cylinder can collapse indefinitely toward its axis, while if $\gamma > 1$ the pressure gradient increases faster than gravity during contraction and the collapse is always halted at a finite radius. For comparison, the critical value is $\gamma = 0$ for sheets and $\gamma = 4/3$ for spheres. Indefinite, global gravitational collapse of a structure can occur when γ is smaller than the critical value and is suppressed when γ is larger than the critical value. Gravitational fragmentation thus tends to be favored over global collapse when γ is close to or larger than the critical value. Since actual molecular clouds are well described by an effective equation of state with $\gamma \lesssim 1$ (see, e.g., Fig. 3b), this led Larson (2005) to suggest that the filamentary geometry may play a key role in cloud fragmentation leading to star formation (see also Nakamura 1998).

The fragmentation properties of filaments and sheets differ from those of spheroidal clouds in that there is a preferred scale for gravitational fragmentation which directly scales with the scale height of the filamentary or sheet-like medium (e.g. Larson 1985). In the spherical case, the largest possible scale or mode (i.e., overall collapse of the medium) has the fastest growth rate so that global collapse tends to overwhelm the local collapse of finite-sized density perturbations, and fragmentation is generally suppressed in the absence of sufficiently large initial density enhancements (e.g. Tohline 1982). It also well known that spherical collapse quickly becomes strongly centrally concentrated (Larson 1969; Shu 1977), which tends to produce a single central density peak as opposed to several condensations (e.g. Whitworth *et al.* 1996). In contrast, sheets have a natural tendency to fragment into filaments (e.g. Miyama *et al.* 1987) and filaments with line masses close to $M_{\rm line,crit}$ have a natural tendency to fragment into spheroidal cores (e.g. Inutsuka & Miyama 1997). The filamentary geometry is thus the most favorable configuration for small-scale perturbations to collapse locally and grow significantly before global collapse overwhelms them (Pon *et al.* 2011; Toalá *et al.* 2012). To summarize, theoretical considerations alone emphasize the potential importance of filaments for core and star formation.

5. The key role of filaments in the star formation process

The quasi "universal" filamentary structure of the cold ISM (cf. § 2) must be closely related to the star formation process since more than 70% of the prestellar cores identified with *Herschel* in nearby clouds appear to lie within filaments. The remarkable correspondence between the spatial distribution of compact cores and the most prominent filaments (see Fig. 5 and Men'shchikov *et al.* 2010) suggests that *prestellar dense cores form primarily along filaments.* More precisely, the prestellar cores identified with *Herschel* are preferentially found within the *densest filaments* with column densities exceeding $\sim 7 \times 10^{21}$ cm^{-2} (André *et al.* 2010 and Fig. 5). In the Aquila region, for instance, the distribution of background cloud column densities for the population of prestellar cores shows a steep rise above $N_{\rm H_2}^{\rm back} \sim 5 \times 10^{21}$ cm^{-2} (cf. Fig. 7) and is such that $\sim$90% of the candidate bound cores are found above a background column density $N_{\rm H_2}^{\rm back} \sim 7 \times 10^{21}$ cm^{-2}, corresponding to a background visual extinction $A_V^{\rm back} \sim 8$. The *Herschel* observations of the Aquila Rift complex therefore strongly support the existence of a column density or visual extinction threshold for the formation of prestellar cores at $A_V^{\rm back} \sim 5$–10, which had been suggested based on earlier ground-based studies of, e.g., the Taurus and Ophiuchus clouds (cf. Onishi *et al.* 1998; Johnstone *et al.* 2004). In the Polaris flare cirrus, the observations are also consistent with such an extinction threshold since the observed background column densities are all below $A_V^{\rm back} \sim 8$ and there are no examples of bound prestellar cores in this cloud. More generally, the results obtained with *Herschel* in nearby clouds suggest that a fraction $f_{\rm pre} \sim 15$–20% of the total gas mass above the column density threshold is in the form of prestellar cores.

These *Herschel* findings connect very well with the theoretical expectations for the gravitational instability of filaments (cf. § 4) and point to an *explanation* of the star

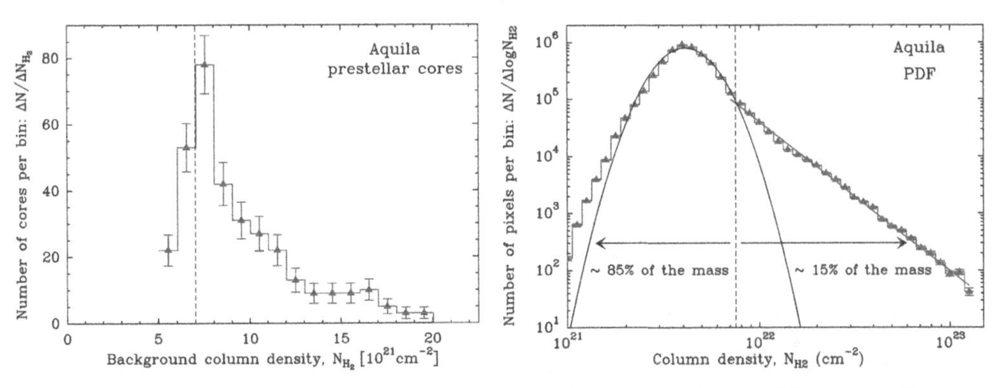

Figure 7. Left: Distribution of background column densities for the candidate prestellar cores identified with *Herschel* in the Aquila Rift complex (cf. Könyves *et al.* 2010). The vertical dashed line marks the column density or extinction threshold at $N_{\rm H_2}^{\rm back} \sim 7 \times 10^{21}$ cm^{-2} or $A_V^{\rm back} \sim 8$ (also corresponding to $\Sigma_{\rm gas}^{\rm back} \sim 130\ M_\odot\,{\rm pc}^{-2}$). **Right:** Probability density function of column density in the Aquila cloud complex, based on the column density image derived from *Herschel* data (André *et al.* 2011; Schneider *et al.* 2013). A log-normal fit at low column densities and a power-law fit at high column densities are superimposed. The vertical dashed line marks the same column density threshold as in the left panel.

formation threshold in terms of the filamentary structure of molecular clouds. Given the typical width $W_{\rm fil} \sim 0.1$ pc measured for interstellar filaments (Arzoumanian *et al.* 2011; see Fig. 4) and the relation $M_{\rm line} \approx \Sigma_0 \times W_{\rm fil}$ between the central gas surface density Σ_0 and the line mass $M_{\rm line}$ of a filament, the threshold at $A_V^{\rm back} \sim 8$ or $\Sigma_{\rm gas}^{\rm back} \sim 130\ M_\odot\,{\rm pc}^{-2}$ corresponds to within a factor of < 2 to the critical mass per unit length $M_{\rm line,crit} = 2\,c_s^2/G \sim 16\,M_\odot/{\rm pc}$ of nearly isothermal, long cylinders (see § 4 and Inutsuka & Miyama 1997) for a typical gas temperature $T \sim 10$ K. Thus, the core formation threshold approximately corresponds to the *threshold above which interstellar filaments are gravitationally unstable* (André *et al.* 2010). Prestellar cores tend to be observed only above this threshold (cf. Figs. 5 & 7) because they form out of a filamentary background and only the supercritical filaments with $M_{\rm line} > M_{\rm line,crit}$ are able to fragment into self-gravitating cores.

For several reasons, the column density threshold for core and star formation within filaments is not a sharp boundary but a smooth transition.
First, observations only provide information on the *projected* column density $\Sigma_{\rm obs} = \frac{1}{\cos i}\,\Sigma_{\rm int}$ of any given filament, where i is the inclination angle to the plane of the sky and $\Sigma_{\rm int}$ is the intrinsic column density of the filament (measured perpendicular to its long axis). For a population of randomly oriented filaments with respect to the plane of the sky, the net effect is that $\Sigma_{\rm obs}$ *overestimates* $\Sigma_{\rm int}$ by a factor $< \frac{1}{\cos i} > = \frac{\pi}{2} \sim 1.57$ on average. Likewise, the apparent masses per unit length tend to slightly overestimate the intrinsic line masses of the filaments. However, the probability density function of the correction factor is such that the median correction is only ~15% and that the correction is less than a factor of 1.5 for 75% of the filaments. Although systematic, this projection effect thus remains small and has little impact on the global classification of observed filamentary structures as supercritical or subcritical filaments.

Second there is also a spread in the distribution of filament inner widths of about a factor of 2 on either side of 0.1 pc (Arzoumanian *et al.* 2011 – cf. Fig. 4), implying a similar spread in the intrinsic column densities corresponding to the critical filament mass per unit length $M_{\rm line,crit}$.

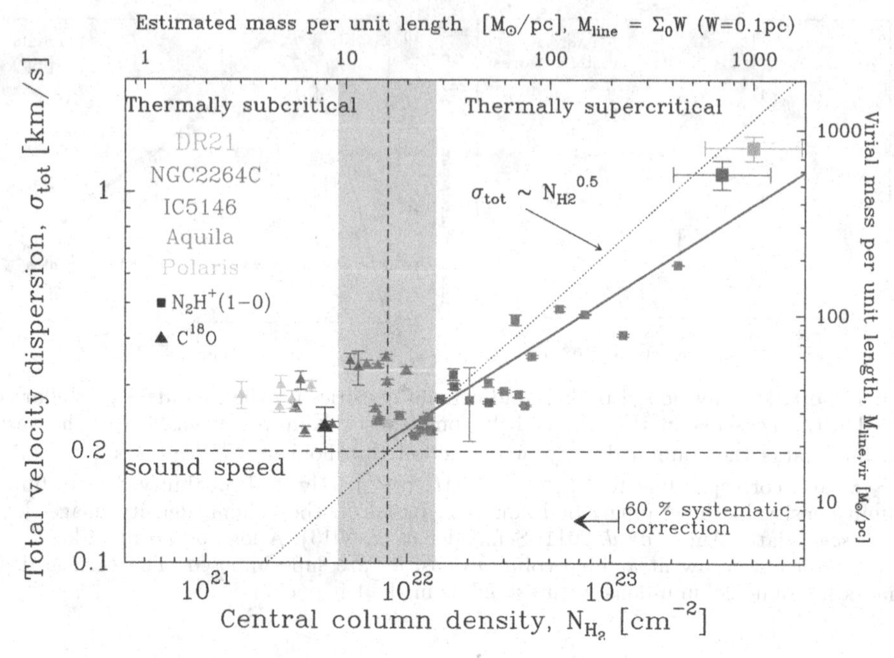

Figure 8. Total (thermal + nonthermal) velocity dispersion versus central column density for a sample of 46 filaments in nearby interstellar clouds (from Arzoumanian *et al.* 2013). The horizontal dashed line shows the value of the thermal sound speed ~0.2 km/s for $T = 10$ K. The vertical grey band marks the border zone between thermally subcritical and thermally supercritical filaments where the observed mass per unit length $M_{\rm line}$ is close to the critical value $M_{\rm line,crit} \sim 16\ M_{\odot}$/pc for $T = 10$ K. The blue solid line shows the best power-law fit $\sigma_{\rm tot} \propto N_{\rm H_2}{}^{0.36\ \pm\ 0.14}$ to the data points corresponding to supercritical filaments.

Third, interstellar filaments are not all exactly at $T = 10$ K and their internal velocity dispersion sometimes includes a small nonthermal component, $\sigma_{\rm NT}$, which must be accounted for in the evaluation of the critical or virial mass per unit length (Fiege & Pudritz 2000): $M_{\rm line,vir} = 2\,\sigma_{\rm tot}^2/G$, where $\sigma_{\rm tot}$ is the total internal velocity dispersion (see § 4). Recent molecular line measurements with the IRAM 30m telescope for a sample of *Herschel* filaments in the Aquila, IC5146, and Polaris clouds (Arzoumanian *et al.* 2013 – cf. Fig. 8) show that both thermally subcritical filaments and nearly critical filaments (with $M_{\rm line}$ within a factor of two of the thermal value of the critical mass per unit length $M_{\rm line,crit}$) have "transonic" internal velocity dispersions $\sigma_{\rm tot}$ such that $c_s \lesssim \sigma_{tot} < 2\,c_s$. Only the densest filaments (with $M_{\rm line} \gg M_{\rm line,crit}$) have internal velocity dispersions significantly in excess of the thermal sound speed $c_s \sim 0.2$ km/s (cf. Fig. 8). These velocity dispersion measurements confirm that there is a critical threshold in mass per unit length above which interstellar filaments are self-gravitating and below which they are unbound, and that the position of this threshold lies between $\sim 16\,M_{\odot}$/pc and $\sim 32\,M_{\odot}$/pc, i.e., is consistent to within a factor of 2 with the thermal value of the critical mass per unit length $M_{\rm line,crit}$ for $T = 10$ K (Arzoumanian *et al.* 2013). The IRAM results shown in Fig. 8 thus confirm that the thermal critical mass per unit length $M_{\rm line,crit}$ plays a fundamental role in the evolution of interstellar filaments. Combined with the *Herschel* findings summarized above and illustrated in Fig. 5 and Fig. 7, this supports the view that the gravitational fragmentation of filaments may control the bulk of core and star formation, at least in nearby Galactic clouds.

6. Implications for the IMF and the global star formation rate

Since most stars appear to form in filaments, the fragmentation of filaments at the threshold of gravitational instability is a plausible mechanism for the origin of (part of) the stellar IMF. We may expect local collapse into spheroidal protostellar cores to be controlled by the Jeans/Bonnor-Ebert criterion $M \geqslant M_{\rm BE}$ where the Jeans or critical Bonnor-Ebert mass $M_{\rm BE}$ (e.g. Bonnor 1956) is given by:

$$M_{\rm BE} \sim 1.3\, c_s^4/G^2\Sigma_{\rm cl} \sim 0.53\, M_\odot \times (T_{\rm eff}/10\,{\rm K})^2 \times \left(\Sigma_{\rm cl}/160\, M_\odot\, {\rm pc}^{-2}\right)^{-1}. \qquad (2)$$

If we consider a quasi-equilibrium isothermal cylindrical filament on the verge of *global* radial collapse, it has a mass per unit length equal to the critical value $M_{\rm line,crit} = 2\, c_s^2/G$ ($\sim$16 $M_\odot$/pc for $T_{\rm eff} \sim 10$ K) and an effective diameter $D_{\rm flat,crit} = 2\, c_s^2/G\Sigma_0$ ($\sim$0.1 pc for $T_{\rm eff} \sim 10$ K and $\Sigma_0 \sim 160\, M_\odot\, {\rm pc}^{-2}$). A segment of such a cylinder of length equal to $D_{\rm flat,crit}$ contains a mass $M_{\rm line,crit} \times D_{\rm flat,crit} = 4\, c_s^4/G^2\Sigma_0 \sim 3 \times M_{\rm BE}$ ($\sim$1.6 $M_\odot$ for $T_{\rm eff} \sim 10$ K and $\Sigma_0 \sim 160\, M_\odot\, {\rm pc}^{-2}$) and is thus potentially prone to *local* Jeans instability. Since local collapse tends to be favored over global collapse in the case of nearly isothermal filaments (see Sect. 4 and Pon *et al.* 2011), gravitational fragmentation into spheroidal cores is expected to occur along supercritical filamentary structures, as indeed found in both numerical simulations (e.g. Bastien *et al.* 1991; Inutsuka & Miyama 1997) and *Herschel* observations (see Fig. 5 and Sect. 5). Remarkably, the peak of the prestellar CMF at $\sim 0.6\, M_\odot$ as observed in the Aquila cloud complex (cf. Fig. 6b) corresponds very well to the Bonnor-Ebert mass $M_{\rm BE} \sim 0.5\, M_\odot$ within marginally critical filaments with $M_{\rm line} \approx M_{\rm line,crit} \sim 16\, M_\odot$/pc and surface densities $\Sigma \approx \Sigma_{\rm gas}^{\rm crit} \sim 160\, M_\odot\, {\rm pc}^{-2}$. Likewise, the median projected spacing $\sim$0.08 pc observed between the prestellar cores of Aquila roughly matches the thermal Jeans length within marginally critical filaments. All of this is consistent with the idea that gravitational fragmentation is the dominant physical mechanism generating prestellar cores within interstellar filaments. Furthermore, a typical prestellar core mass of $\sim$0.6 $M_\odot$ translates into a characteristic star or stellar system mass of $\sim$0.2 $M_\odot$, assuming a typical efficiency $\epsilon_{\rm core} \sim 30\%$ (cf. § 3).

Therefore, the *Herschel* results strongly support Larson (1985)'s interpretation of the peak of the IMF in terms of the typical Jeans mass in star-forming clouds. Overall, our *Herschel* findings suggest that the gravitational fragmentation of supercritical filaments produces the peak of the prestellar CMF which, in turn, may account for the log-normal "base" (cf. Bastian *et al.* 2010) of the IMF. It remains to be seen, however, whether the bottom end of the IMF and the Salpeter power-law slope at the high-mass end can also be explained by filament fragmentation. Naively, one would indeed expect gravitational fragmentation to result in a narrow prestellar CMF, sharply peaked at the median thermal Jeans mass. It should also be noted that a small ($< 30\%$) fraction of prestellar cores do not appear to form along filaments. A good example is the pre-brown dwarf core Oph B-11 recently identified in the Ophiuchus main cloud (Greaves *et al.* 2003, André *et al.* 2012). On the other hand, a Salpeter power-law tail at high masses may be produced by filament fragmentation if turbulence has generated an appropriate field of initial density fluctuations within the filaments in the first place (cf. Inutsuka 2001). More precisely, Inutsuka (2001) has shown that if the power spectrum of initial density fluctuations along the filaments approaches $P(k) \equiv |\delta_k|^2 \propto k^{-1.5}$ then the CMF produced by gravitational fragmentation evolves toward a power-law $dN/dM \propto M^{-2.5}$, similar to the Salpeter IMF ($dN/dM_\star \propto M_\star^{-2.35}$). Interestingly, the power spectrum of column density fluctuations along the filaments observed with *Herschel* in nearby clouds is typically $P(k) \propto k^{-1.8}$, which is close to the required spectrum. Alternatively, a CMF with a Salpeter power-law tail may result from the gravitational fragmentation of a population of filaments with a distribution of supercritical masses per unit length. Observationally, the supercritical filaments observed as part of the *Herschel* Gould Belt survey do seem to have a power-law distribution of masses per unit length $dN/dM_{\rm line} \propto M_{\rm line}^{-2.2}$ above $\sim$20 $M_\odot$/pc. Since the width of the filaments is roughly constant ($W_{\rm fil} \sim 0.1$ pc), the mass per unit length is directly proportional to the central surface density, $M_{\rm line} \sim \Sigma \times W_{\rm fil}$. Furthermore, the total velocity dispersion of these filaments increases roughly as $\sigma_{\rm tot} \propto \Sigma^{0.5}$ (Arzoumanian *et al.* 2013 – see Fig. 8), which means that their effective temperature scales roughly as $T_{\rm eff} \propto \Sigma$. Hence $M_{\rm BE} \propto \Sigma \propto M_{\rm line}$, and the observed distribution of masses per unit length directly translates into a power-law distribution of Bonnor-Ebert masses $dN/dM_{\rm BE} \propto M_{\rm BE}^{-2.2}$ along supercritical filaments, which is also reminiscent of the Salpeter IMF. Although these two alternative possibilities seem promising, more work is needed to assess whether the direct gravitational fragmentation of filaments can account for the high-mass end of the CMF/IMF.

The realization that prestellar core formation occurs primarily along gravitationally unstable filaments of roughly constant width $W_{\rm fil} \sim 0.1$ pc also has potential implications for our understanding of star formation on global Galactic and extragalactic scales. Remarkably, the critical line mass of a filament, $M_{\rm line,crit} = 2\,c_s^2/G$, depends only on gas temperature (i.e., $T \sim 10$ K for the bulk of molecular clouds, away from the immediate vicinity of massive stars) and is modified by only a factor of order unity for filaments with realistic levels of magnetization (Fiege & Pudritz 2000 – see Sect. 4). This may set a quasi-universal threshold for star formation in the cold ISM of galaxies at $M_{\rm line,crit} \sim 16\,M_\odot$/pc in terms of filament mass per unit length, or $M_{\rm line,crit}/W_{\rm fil} \sim 160\,M_\odot\,{\rm pc}^{-2}$ in terms of gas surface density, or $M_{\rm line,crit}/W_{\rm fil}^2 \sim 1600\,M_\odot\,{\rm pc}^{-3}$ in terms of gas density (the latter corresponding to a volume number density $n_{H_2} \sim 2 \times 10^4\,{\rm cm}^{-3}$). Indeed, recent near-/mid-infrared studies of the star formation rate as a function of gas surface density in both Galactic and extragalactic cloud complexes (e.g. Heiderman *et al.* 2010; Lada *et al.* 2010) show that the star formation rate tends to be linearly proportional to the mass of dense gas above a surface density threshold $\Sigma_{\rm gas}^{\rm th} \sim 120$–$130\,M_\odot\,{\rm pc}^{-2}$ and drops to negligible values below $\Sigma_{\rm gas}^{\rm th}$ (see Gao & Solomon 2004 for external galaxies). Note that this is the essentially *same* threshold as found with *Herschel* for the formation of prestellar cores in nearby clouds (cf. § 5 and Figs. 7 & 8). Moreover, the relation between the star formation rate SFR and the mass of dense gas $M_{\rm dense}$ above the threshold is estimated to be $SFR = 4.6 \times 10^{-8}\,M_\odot\,{\rm yr}^{-1} \times (M_{\rm dense}/M_\odot)$ in nearby clouds (Lada *et al.* 2010), which is close to the relation $SFR = 2 \times 10^{-8}\,M_\odot\,{\rm yr}^{-1} \times (M_{\rm dense}/M_\odot)$ found by Gao & Solomon (2004) for galaxies. Both of these values are very similar to the star formation rate per unit mass of dense gas $SFR/M_{\rm dense} = f_{\rm pre} \times \epsilon_{\rm core}/t_{\rm pre} \sim 0.15 \times 0.3/10^6 \sim 4.5 \times 10^{-8}\,M_\odot\,{\rm yr}^{-1}$ that we may derive based on the *Herschel* results in the Aquila complex, by considering that only a fraction $f_{\rm pre} \sim 15\%$ of the gas mass above the column density threshold is in the form of prestellar cores (cf. § 5), that the local star formation efficiency at the level of an individual core is $\epsilon_{\rm core} \sim 30\%$ (cf. § 3), and that the typical lifetime of the Aquila cores is $t_{\rm pre} \sim 10^6$ yr (Könyves *et al.*, in prep.). Despite relatively large uncertainties, the agreement with the extragalactic value of Gao & Solomon (2004) is surprisingly good, implying that there may well be a quasi-universal "star formation law" converting dense gas into stars above the threshold (see also Lada *et al.* 2012).

7. Conclusions: Toward a universal scenario for star formation?

The results obtained with *Herschel* on nearby clouds and summarized in the previous sections provide key insight into the first phases of the star formation process. They emphasize the role of filaments and support a scenario according to which the formation of prestellar cores occurs in two main steps (see, e.g., André *et al.* 2010). First, large-scale magneto-hydrodynamic (MHD) turbulence generates a whole network of filaments in the ISM (cf. Padoan *et al.* 2001); second, the densest filaments fragment into prestellar cores by gravitational instability (cf. Inutsuka & Miyama 1997) above a critical (column) density threshold corresponding to $\Sigma_{\rm gas}^{\rm crit} \sim 150\,M_\odot\,{\rm pc}^{-2}$ ($A_V^{\rm crit} \sim 8$) or $n_{\rm H_2}^{\rm crit} \sim 2 \times 10^4\,{\rm cm}^{-3}$.

That the formation of filaments in the diffuse ISM represents the first step toward core/star formation is suggested by the filaments *already* being omnipresent in a gravitationally unbound, non-star-forming cloud such as Polaris (cf. Fig. 1, Men'shchikov *et al.* 2010, and Miville-Deschênes *et al.* 2010). This indicates that interstellar filaments are not produced by large-scale gravity and that their formation must precede star formation. It is also consistent with the view that the filamentary structure results primarily from the dissipation of large-scale MHD turbulence (cf. Padoan *et al.* 2001; Hily-Blant & Falgarone 2007, 2009). In the picture proposed by Padoan *et al.* (2001), the dissipation of turbulence occurs in shocks and interstellar filaments correspond to dense, post-shock stagnation gas associated with compressed regions between interacting supersonic flows. One merit of this picture is that it can qualitatively account for the characteristic ~0.1 pc width of the filaments as measured with *Herschel* (cf. Fig. 4): the typical thickness of shock-compressed structures resulting from supersonic turbulence in the ISM is expected to be roughly the sonic scale of the turbulence which is ~0.1 pc in diffuse interstellar gas (cf. Larson 1981, Falgarone *et al.* 2009, and discussion in Arzoumanian *et al.* 2011). Direct evidence of the role of large-scale compressive flows has been found with *Herschel* in the Pipe Nebula in the form of filamentary structures with asymmetric column density profiles which most likely result from compression by the winds of the Sco OB2 association (Peretto *et al.* 2012). However, turbulent compression alone would tend to form layers and is unlikely to directly produce filaments. A more complete picture, recently proposed by Hennebelle (2013), is that interstellar filaments result from a combination of turbulent shear and compression. Hennebelle (2013) further suggests that the filament width ~0.1 pc may correspond to the dissipation length of MHD waves due to ion-neutral friction.

The second step appears to be the gravitational fragmentation of the densest filaments with supercritical masses per unit length ($M_{\rm line} \geqslant M_{\rm line,crit}$) into self-gravitating prestellar cores (cf. § 5). In active star-forming regions such as the Aquila complex, most of the prestellar cores identified with *Herschel* are indeed concentrated within supercritical filaments (cf. Fig. 5). In contrast, in non-star-forming clouds such as Polaris, all of the filaments have subcritical masses per unit length and only unbound starless cores are observed but no prestellar cores nor protostars (cf. Fig. 1). *Herschel* observations indicate that even star-forming, supercritical filaments maintain roughly constant inner widths ~0.1 pc while evolving (Arzoumanian *et al.* 2011 – see Figs. 3 & 4). At first sight, this seems surprising since supercritical filaments are unstable to radial collapse and are thus expected to undergo rapid radial contraction with time (e.g. Kawachi & Hanawa 1998 – see Sect. 4). The most likely solution to this paradox is that supercritical filaments are *accreting* additional background material while contracting. The increase in velocity dispersion with central column density observed for supercritical filaments (Arzoumanian *et al.* 2013 – see Fig. 8) is indeed suggestive of an increase in (virial) mass per unit length with time. More direct evidence of this accretion process for supercritical filaments exists in several cases in the form of low-density striations or sub-filaments observed perpendicular to the main filaments and apparently feeding them from the side. Examples include the B211/B213 filament in Taurus (Palmeirim *et al.* 2013 – see Fig. 2), the Musca filament (Cox *et al.* 2013, in preparation), and the DR21 ridge in Cygnus X (Schneider *et al.*

2010; Hennemann *et al.* 2012). In the case of the Taurus filament (Fig. 2), the estimated accretion rate is such that it would take $\sim$1–2 Myr for the central filament to double its mass (Palmeirim *et al.* 2013). This accretion process supplies gravitational energy to supercritical filaments which is then converted into turbulent kinetic energy (cf. Heitsch *et al.* 2009 and Klessen & Hennebelle 2010) and may explain the observed increase in velocity dispersion ($\sigma_{\rm tot} \propto \Sigma_0{}^{0.5}$ – cf. Fig. 8). The central diameter of such accreting filaments is expected to be of order the effective Jeans length $D_{\rm J,eff} \sim 2\,\sigma_{\rm tot}^2/G\Sigma_0$, which Arzoumanian *et al.* (2013) have shown to remain close to $\sim$0.1 pc. Hence, through accretion of parent cloud material, supercritical filaments may keep roughly constant inner widths and remain in rough virial balance while contracting. This process may effectively prevent the global (radial) collapse of supercritical filaments and thus favor their fragmentation into cores (e.g. Larson 2005), in agreement with the *Herschel* results (see Fig. 5).

The above scenario can explain the peak for the prestellar CMF and may account for the base of the stellar IMF (see Sect. 6 and Fig. 6b). It partly accounts for the general inefficiency of the star formation process since, even in active star-forming complexes such as Aquila (Fig. 5), only a small fraction of the total gas mass ($\sim$15% in the case of Aquila – see Fig. 7b) is above of the column density threshold, and only a small fraction $f_{\rm pre} \sim 15\%$ of the dense gas above the threshold is in the form of prestellar cores (see Sect. 5). Therefore, the vast majority of the gas in a GMC ($\sim$98% in the case of Aquila) does not participate in star formation at any given time (see also Heiderman *et al.* 2010 and Evans 2011). Furthermore, the fact that essentially the same "star formation law" is observed above the column density threshold in both Galactic clouds and external galaxies (see Sect. 6 and Lada *et al.* 2012) suggests that the star formation scenario sketched above is quasi-universal and may well apply to the ISM of all galaxies.

To conclude, the *Herschel* results discussed in this paper are extremely encouraging as they point to a unified picture of star formation on GMC scales in both Galactic clouds and external galaxies. Confirming and refining the scenario proposed here will require follow-up observations to constrain the dynamics of the filamentary structures imaged with *Herschel* as well as detailed comparisons with numerical simulations of molecular cloud formation and evolution. ALMA will be instrumental in testing whether this scenario, based on *Herschel* observations of nearby Galactic clouds forming mostly low-mass stars, is truly universal and also applies to high-mass star forming clouds and the GMCs of other galaxies.

Acknowledgements

It is a pleasure for me to acknowledge the important contributions of my colleagues of the *Herschel*/SPIRE SAG 3 working group, e.g., Doris Arzoumanian, James Di Francesco, Jason Kirk, Vera Könyves, Sasha Men'shchikov, Fred Motte, Pedro Palmeirim, Nicolas Peretto, Nicola Schneider, Derek Ward-Thompson, to the results presented in this paper. I am also grateful to Shu-ichiro Inutsuka, Fumitaka Nakamura, Patrick Hennebelle, Ralph Pudritz, Zhi-Yun Li, and Shantanu Basu for insightful discussions on filaments. This work has received support from the European Research Council under the European Union's Seventh Framework Programme (Grant Agreement no. 291294) and from the French National Research Agency (Grant no. ANR–11–BS56–0010).

References

Abergel, A., Boulanger, F., Mizuno, A., & Fukui, Y. 1994, *ApJ*, 423, L59
Alves, J. F., Lada, C. J., & Lada, E. A. 2001, *Nature*, 409, 159

Alves, J. F., Lombardi, M., & Lada, C. J. 2007, *A&A*, 462, L17
André, P., Belloche, A., Motte, F., & Peretto, N. 2007, *A&A*, 472, 519
André, Ph., Men'shchikov, A., Bontemps, S., *et al.* 2010, *A&A*, 518, L102
André, Ph., Men'shchikov, A., Könyves, V., & Arzoumanian, D. 2011, in *Computational Star Formation*, IAU Symp. 270, Eds. J. Alves *et al.*, p. 255
André, P., Ward-Thompson,D. & Barsony, M. 2000, in *Protostars and Planets IV*, Eds V. Mannings *et al.*, p.59
André, P., Ward-Thompson,D. & Greaves, J. S. 2012, *Science*, 337, 69
Arzoumanian, D., André, Ph., Didelon, P., *et al.* 2011, *A&A*, 529, L6
Arzoumanian, D., André, Ph., Peretto, N., & Könyves, V. 2013, *A&A*, 553, A119
Bastian, N., Covey, K. R., & Meyer, M. R. 2010, *ARA&A*, 48, 339
Bastien, P., Arcoragi, J.-P., Benz, W., Bonnell, I., & Martel, H. 1991, *ApJ*, 378, 255
Bate, M. R. & Bonnell, I. A. 2005, *MNRAS*, 356, 1201
Bate, M. R., Bonnell, I. A., & Bromm, V. 2003, *MNRAS*, 339, 577
Bergin, E. A. & Tafalla, M. 2007, *ARA&A*, 45, 339
Bonnor, W. B. 1956, *MNRAS*, 116, 351
Bontemps, S., André, Ph., Könyves, V., *et al.* 2010, *A&A*, 518, L85
Chabrier, G. 2005, in *The Initial Mass Function 50 years later*, Eds. E. Corbelli *et al.*, p.41
Chapman, N. L., Goldsmith, P. F., Pineda, J. L., *et al.* 2011, *ApJ*, 741, 21
Crutcher, R. M. 1999, *ApJ*, 520, 706
Crutcher, R. M. 2012, *ARA&A*, 50, 29
Di Francesco, J., Evans II, N. J., Caselli, P., *et al.* 2007, in *Protostars and Planets V*, p. 17
Dunham, M. M., Crapsi, A., Evans, N. J., *et al.* 2008, *ApJS*, 179, 249
Enoch, M. L., Young, K. E., Glenn, J., Evans, N. J., *et al.* 2008, *ApJ*, 684, 1240
Evans, N. J. 2011, in *Computational Star Formation*, IAU Symp. 270, Eds. J. Alves *et al.*, p. 25
Falgarone, E., Pety, J., & Hily-Blant, P. 2009, *A&A*, 507, 355
Federrath, C., Roman-Duval, J., Klessen, R. S., *et al.* 2010, *A&A*, 512, A81
Fiege, J. D. & Pudritz, R. E. 2000, *MNRAS*, 311, 85
Fischera, J. & Martin, P. G. 2012, *A&A*, 542, A77
Gao, Y. & Solomon, P. 2004, *ApJ*, 606, 271
Gong, H. & Ostriker, E. C. 2011, *ApJ*, 729, 120
Goodman, A. A., Barranco, J. A., Wilner, D. J., & Heyer, M. H. 1998, *ApJ*, 504, 223
Goodwin, S. P., Nutter, D., Kroupa, P., Ward-Thompson, D., & Whitworth, A. P. 2008, *A&A*, 477, 823
Greaves, J. S., Holland, W. S., & Pound, M. W. 2003, *MNRAS*, 346, 441
Guillout, P. 2001, in From Darkness to Light, Eds. T. Montmerle & P. André, ASP Conf. Ser., 243, p. 677
Gutermuth, R. A., Bourke, T. L., Allen, L. E., *et al.* 2008, *ApJ*, 673, L151
Hacar, A. & Tafalla, M. 2011, *A&A*, 533, A34
Hartmann, L. 2002, *ApJ*, 578, 914
Hatchell, J., Richer, J. S., Fuller, G. A., *et al.* 2005, *A&A*, 440, 151
Heiderman, A., Evans, N. J., Allen, L. E., *et al.* 2010, *ApJ*, 723, 1019
Heithausen, A., *et al.* 2002, *A&A*, 383, 591
Heitsch, F., Ballesteros-Paredes, J., & Hartmann, L. 2009, *ApJ*, 704, 1735
Hennebelle, P. 2013, *A&A*, submitted
Hennebelle, P. & Chabrier, G. 2008, *ApJ*, 684, 395
Hennemann, M., Motte, F., Schneider, N., *et al.* 2012, *A&A*, 543, L3
Heyer, M., Gong, H., Ostriker, E., & Brunt, C. 2008, *ApJ*, 680, 420
Hily-Blant, P. & Falgarone, E. 2007, *A&A*, 469, 173
Hily-Blant, P. & Falgarone, E. 2009, *A&A*, 500, L29
Inutsuka, S. 2001, *ApJ*, 559, L149
Inutsuka, S. & Miyama, S. M. 1992, *ApJ*, 388, 392
Inutsuka, S. & Miyama, S. M. 1997, *ApJ*, 480, 681
Johnstone, D., Wilson, C. D., Moriarty-Schieven, G., *et al.* 2000, *ApJ*, 545, 327

Johnstone, D., Di Francesco, J., & Kirk, H. 2004, *ApJ*, 611, L45
Kawachi, T. & Hanawa, T. 1998, *PASJ*, 50, 577
Kennicutt, R. 1998, *ApJ*, 498, 541
Klessen, R. S. & Burkert, A. 2000, *ApJS*, 128, 287
Klessen, R. S. & Hennebelle, P. 2010, *A&A*, 520, A17
Könyves, V., André, Ph., Men'shchikov, A., *et al.* 2010, *A&A*, 518, L106
Kramer, C., Stutzki, J., Rohrig, R., & Corneliussen, U. 1998, *A&A*, 329, 249
Kroupa, P. 2001, *MNRAS*, 322, 231
Lada, C. J., Lombardi, M., & Alves, J. 2010, *ApJ*, 724, 687
Lada, C. J., Forbrich, J., Lombardi, M., & Alves, J. F. 2012, *ApJ*, 745, 190
Larson, R. B. 1969, *MNRAS*, 145, 271
Larson, R. B., 1981, *MNRAS*, 194, 809
Larson, R. B. 1985, *MNRAS*, 214, 379
Larson, R. B. 2005, *MNRAS*, 359, 211
Li, D. & Goldsmith, P. F. 2012, *ApJ*, 756, 12
Matzner, C. D. & McKee, C. F. 2000, *ApJ*, 545, 364
Maury, A., André, Ph., Men'shchikov, A., Könyves, V., & Bontemps, S. 2011, *A&A*, 535, A77
Men'shchikov, A., André, Ph., Didelon, P., *et al.* 2010, *A&A*, 518, L103
Men'shchikov, A. André, Ph., Didelon, P., Motte, F., *et al.* 2012, *A&A*, 542, A81
Miville-Deschênes, M.-A., Martin, P. G., Abergel, A., *et al.* 2010, *A&A*, 518, L104
Miyama, S. M., Narita, S., & Hayashi, C. 1987, *Prog. Theor. Phys.*, 78, 1273
Molinari, S., Swinyard, B., Bally, J., *et al.* 2010, *A&A*, 518, L100
Motte, F., André, P., & Neri, R. 1998, *A&A*, 336, 150
Motte, F., Zavagno, A., Bontemps, S., *et al.* 2010, *A&A*, 518, L77
Myers, P. C. 1983, *ApJ*, 270, 105
Myers, P. C. 2009, *ApJ*, 700, 1609
Nagasawa, M. 1987, *Prog. Theor. Phys.*, 77, 635
Nakamura, F. 1998, *ApJ*, 507, L165
Onishi, T., Mizuno, A., Kawamura, A., *et al.* 1998, *ApJ*, 502, 296
Ostriker, J. 1964, *ApJ*, 140, 1056
Padoan, P. & Nordlund, A. 2002, *ApJ*, 576, 870
Padoan, P., Juvela, M., Goodman, A. A., & Nordlund, A. 2001, *ApJ*, 553, 227
Palmeirim, P., André, Ph., Kirk, J., *et al.* 2013, *A&A*, 550, A38
Peretto, N., André, P., & Belloche, A. 2006, *A&A*, 445, 979
Peretto, N., André, Ph., Könyves, V., *et al.* 2012, *A&A*, 541, A63
Perrot, C. A. & Grenier, I. A. 2003, *A&A*, 404, 519
Pezzuto, S., Elia, D., Schisano, E., *et al.* 2012, *A&A*, 547, A54
Pilbratt, G. L., Riedinger, J. R., Passvogel, T., *et al.* 2010, *A&A*, 518, L1
Pon, A., Johnstone, D., & Heitsch, F. 2011, *ApJ*, 740, 88
Saigo, K. & Tomisaka, K. 2011, *ApJ*, 728, 78
Schneider, N., André, Ph., Könyves, V., *et al.* 2013, *ApJL*, 766, L17
Schneider, N., Csengeri, T., Bontemps, S., *et al.* 2010, *A&A*, 520, A49
Schneider, S. & Elmegreen, B. G. 1979, *ApJS*, 41, 87
Shu, F. 1977, *ApJ*, 214, 488
Sousbie, T., 2011, *MNRAS*, 414, 350
Stanke, T., Smith, M. D., Gredel, R., & Khanzadyan, T. 2006, *A&A*, 447, 609
Starck, J. L., Donoho, D. L., & Candès, E. J. 2003, *A&A*, 398, 785
Toalá, J. A., Vázquez-Semadeni, E., & Gómez, G. C. 2012, *ApJ*, 744, 190
Tohline, J. E. 1982, *Fund. of Cos. Phys.*, 8, 1
Ward-Thompson, D., André, P., Crutcher, R., Johnstone, D., Onishi, T., & Wilson, C. 2007, *Protostars and Planets V*, Eds. B. Reipurth, D. Jewitt, K. Keil (Tucson: University of Arizona Press), p. 33
Ward-Thompson, D., Kirk, J. M., André, P., *et al.* 2010, *A&A*, 518, L92
Whitworth, A. P., Bhattal, A. S., Francis, N., & Watkins, S. J. 1996, *MNRAS*, 283, 1061

Highlights of Astronomy, Volume 16
XXVIIIth IAU General Assembly, August 2012
T. Montmerle, ed.

doi:10.1017/S1743921314004645

JD1: The highest-energy gamma-ray universe observed with Cherenkov telescope arrays

No contribution was received from this Joint Discussion.

Highlights of Astronomy, Volume 16
XXVIIIth IAU General Assembly, August 2012
T. Montmerle, ed.

doi:10.1017/S1743921314004657

Very Massive Stars in the local Universe

Jorick S. Vink[1], Alexander Heger[2], Mark R. Krumholz[3], Joachim Puls[4], S. Banerjee[5], N. Castro[6], K.-J. Chen[7], A.-N. Chenè[8,9], P. A. Crowther[10], A. Daminelli[11], G. Gräfener[1], J. H. Groh[12], W.-R. Hamann[13], S. Heap[14], A. Herrero[15], L. Kaper[16], F. Najarro[17], L. M. Oskinova[13], A. Roman-Lopes[18], A. Rosen[3], A. Sander[13], M. Shirazi[19], Y. Sugawara[20], F. Tramper[16], D. Vanbeveren[21], R. Voss[22], A. Wofford[23], Y. Zhang[24] and the participants of JD2

[1]Armagh Observatory, College Hill, BT61 9DG, Armaghm Northern Ireland, UK
email: jsv@arm.ac.uk
[2]Monash Centre for Astrophysics School of Mathematical Sciences, Building 28, M401 Monash University, Vic 3800, Australia
[3]Department of Astronomy & Astrophysics, University of California, Santa Cruz, CA 95064, USA
[4]Universitäts-Sternwarte, Scheinerstrasse 1, 81679, Munchen, Germany
[5]Argelander-Institut für Astronomie, Auf dem Hügel 71, D-53121, Bonn, Germany
[6]Institute of Astronomy & Astrophysics, National Observatory of Athens, I. Metaxa & Vas. Pavlou St. P. Penteli, 15236 Athens, Greece
[7]Minnesota Institute for Astrophysics, University of Minnesota, Minneapolis, MN 55455, USA
[8]Departamento de Física y Astronomía, Universidad de Valparaíso, Av. Gran Bretaña 1111, Playa Ancha, Casilla 5030, Chile
[9]Departamento de Astronomía, Universidad de Concepción, Casilla 160-C, Chile
[10]Dept. of Physics & Astronomy, Hounsfield Road, University of Sheffield, S3 7RH, UK
[11]Departamento de Astronomia do IAG-USP, R. do Matao 1226, 05508-090 Sao Paulo, Brazil
[12]Geneva Observatory, Geneva University, Chemin des Maillettes 51, CH-1290 Sauverny, Switzerland
[13]Institute for Physics and Astronomy, University Potsdam, 14476 Potsdam, Germany
[14]NASA Goddard Space Flight Center, Greenbelt, MD 20771, USA
[15]Instituto de Astrofisica de Canarias and Universidad de La Laguna, E-38200 La Laguna, Spain
[16]Astronomical Institute 'Anton Pannekoek', University of Amsterdam, Science Park 904, PO Box 94249, 1090 GE Amsterdam, The Netherlands
[17]Departamento de Astrofsica, Centro de Astrobiologia, (CSIC-INTA), Ctra. Torrejn a Ajalvir, km 4, 28850 Torrejon de Ardoz, Madrid, Spain
[18]Physics Department - Universidad de La Serena - Av. Cisternas, 1200 - La Serena - Chile
[19]Leiden Observatory, Leiden University, P.O. Box 9513, 2300 RA Leiden, The Netherlands
[20]Department of Physics, Faculty of Science & Engineering, Chuo University, 1-13-27 Kasuga, Bunkyo, Tokyo 112-8551
[21]Astrophysical Institute, Vrije Universiteit Brussel, Pleinlaan 2, 1050, Brussels, Belgium
[22]Department of Astrophysics/IMAPP, Radboud University Nijmegen, PO Box 9010, NL-6500 GL Nijmegen, the Netherlands
[23]Space Telescope Science Institute, 3700 San Martin Drive, Baltimore, MD, 21218, USA
[24]Department of Astronomy, University of Florida, Gainesville, FL 32611, USA

Abstract. Recent studies have claimed the existence of very massive stars (VMS) up to $300\,M_{\odot}$ in the local Universe. As this finding may represent a paradigm shift for the canonical stellar upper-mass limit of $150\,M_{\odot}$, it is timely to discuss the status of the data, as well as the far-reaching implications of such objects. We held a Joint Discussion at the General Assembly in Beijing to discuss (i) the determination of the current masses of the most massive stars, (ii) the formation of VMS, (iii) their mass loss, and (iv) their evolution and final fate. The prime aim was to reach broad consensus between observers and theorists on how to identify and quantify the dominant physical processes.

Keywords. Stars: massive stars, Stars: mass-loss, Stars: stellar evolution

---

1. Introduction

The last decade has seen a growing interest in the study of the most massive stars, as their formation seems to be favourable in the early Universe at low metalliciy (Z), and is thought to involve a population of objects in the range 100-300 $M_{\odot}$ (Bromm *et al.* 1999; Abel *et al.* 2002). The first couple of stellar generations may be good candidates for the reionization of the Universe. Notwithstanding the role of the first stars, the interest in the current generation of massive stars has grown as well. Massive stars are important drivers for the evolution of galaxies, as the prime contributors to the chemical and energy input into the interstellar medium (ISM) through stellar winds and supernovae (SNe). A number of exciting developments have recently taken place, such as the detection of a long-duration gamma-ray burst (GRB) at a redshift of 9.4, just a few hundred millions years after the Big Bang (Cucchiara *et al.* 2011). This provides convincing evidence that massive stars are able to form and die massive when the Universe was not yet enriched.

The specific reason for holding this JD was the recent evidence for the existence and subsequent deaths of very massive stars (VMS) up to 300 $M_{\odot}$. Gal-Yam *et al.* (2009) claimed the detection of a pair-instability SN (PSN) from a VMS. These explosions are thought to disrupt stars without leaving any remnants. Crowther *et al.* (2010) re-analyzed the most massive hydrogen-and nitrogen-rich Wolf-Rayet (WNh) stars in the center of R136, the ionizing cluster of the Tarantula nebula in the Large Magellanic Cloud (LMC). The conclusion from their analysis was that stars usually assumed to be below the canonical stellar upper-mass limit of 150 $M_{\odot}$ (of e.g. Figer 2005), were actually found to be much more luminous, and with initial masses up to $\sim$320 $M_{\odot}$. Prior to discussing the formation, evolution, and fate of VMS, and before we should explore the full implications of these findings, it is imperative to discuss the various lines of evidence for and against VMS.

VMS are usually found in and around young massive clusters, such as the Arches cluster in the Galactic centre and the local starburst region R136 in the LMC. Such clusters may harbor intermediate-mass black holes (IMBHs) with masses in the range of several 100 to several 1000 $M_{\odot}$ and may provide insight into the formation of supermassive black holes of order 10^5 $M_{\odot}$. Young clusters are also relevant for the unsolved problem of massive star formation.

For decades it was a struggle to form stars over 10 $M_{\odot}$, as radiation pressure on dust grains might halt and reverse the accretion flow onto the central object (e.g., Yorke & Kruegel 1977). Because of this problem, astrophysicists have been creative in forming massive stars via competitive accretion and merging in dense cluster environments (e.g., Bonnell *et al.* 1998). In more recent times several multi-D simulations have shown that massive stars might form via disk accretion after all (e.g., Krumholz *et al.* 2009; Kuiper *et al.* 2010). In the light of recent claims for the existence of VMS in dense clusters, however, we should redress the issue of forming VMS in extreme environments.

Massive clusters may be so dense that their early evolution is largely affected by stellar dynamics, and possibly form very massive objects via runaway collisions (e.g., Portegies Zwart *et al.* 1999; Gürkan *et al.* 2004), leading to the formation of VMS up to 1000 $M_{\odot}$ at the cluster center, which may produce IMBHs at the end of their lives, but only if

VMS mass loss is not too severe (see Belkus *et al.* 2007, Yungelson *et al.* 2008, Glebbeek *et al.* 2009, Pauldrach *et al.* 2012).

VMS are thought to evolve almost chemically homogeneously (Gräfener *et al.* 2011), implying that knowing the exact details of the mixing processes (e.g., rotation, magnetic fields) could be less relevant in comparison to their canonical ~10-60 $M_\odot$ counterparts. Instead, the evolution and death of VMS is likely dominated by mass loss. A crucial issue regards the relevance of episodes of super-Eddington, continuum-driven mass loss (such as occurs in Eta Carinae and other Luminous Blue Variable (LBV) star eruptions), which might be able to remove large amounts of mass – even in the absence of substantial line-driven winds (see Sect. 6).

A final issue concerns the fate of VMS, and more specifically whether VMS end their lives as canonical Wolf-Rayet (WR) stars giving rise to Type Ibc SNe, or do they explode prematurely during the LBV phase? Might some of the most massive stars even produce PSNe? And how do PSNe compare to the general population of super-luminous SNe (SLSNe) that have recently been unveiled by Quimby *et al.* (2011), and are now seen out to high redshifts (Cooke *et al.* 2012). Such spectacular events can only be understood once we have obtained a basic knowledge of the physics of VMS.

In Section 2 a definition for VMS is adopted, before we discuss the evidence for and against super-canonical stars, i.e. objects above the traditional 150 $M_\odot$ stellar upper mass limit (Sect. 3). After it is concluded that VMS probably exist, the next question to address is how to form such objects (Sect. 4). We then discuss the properties of VMS in Sect. 5, before discussing their mass loss (Sect. 6), evolution and fate (Sect. 7). We end with the implications (Sect. 8) and final words.

2. Definition of VMS

Before we can discuss the evidences for and against VMS, one of the very first issues we discussed during the JD was what actually constitutes a "very" massive star. One may approach this in several different ways. Theoretically, "normal" massive stars with masses above ~8 $M_\odot$ are those that produce core-collapse SNe (Smartt *et al.* 2009), but what happens at the upper-mass end (Nomoto 2012)?

Above a certain critical mass, one would expect the occurrence of PSNe, and ideally this could be the lower-mass limit for the definition of our VMS. However, in practice this number is not known a priori due to mass loss, and as a result the initial and final masses are not the same. In other words, the initial main-sequence mass for PSN formation is model-dependent, and thus somewhat arbitrary. Moreover, there is the complicating issue of pulsational pair-instability supernova (Puls-PSN) at masses below those of full-fedged PSNe (e.g. Woosley *et al.* 2007). One could of course resort to the mass of the helium (He) core for which objects reach the conditions of electron/positron pair-formation instability. Heger showed this minimum mass to be ~40 $M_\odot$ to encounter Puls-PSN and ~65 $M_\odot$ to encounter PSNe (see also Chatzopoulus & Wheeler 2012).

An alternative definition could involve the spectroscopic transition between normal main-sequence O-type stars and hydrogen-rich Wolf-Rayet stars (of WNh type), which have also been shown to be core H burning main-sequence objects. However, such a definition would also be model dependent.

For these reasons, we took the decision to follow a more pragmatic approach: we consider stars to be *very* massive when the initial masses are ~100 $M_\odot$, or higher.

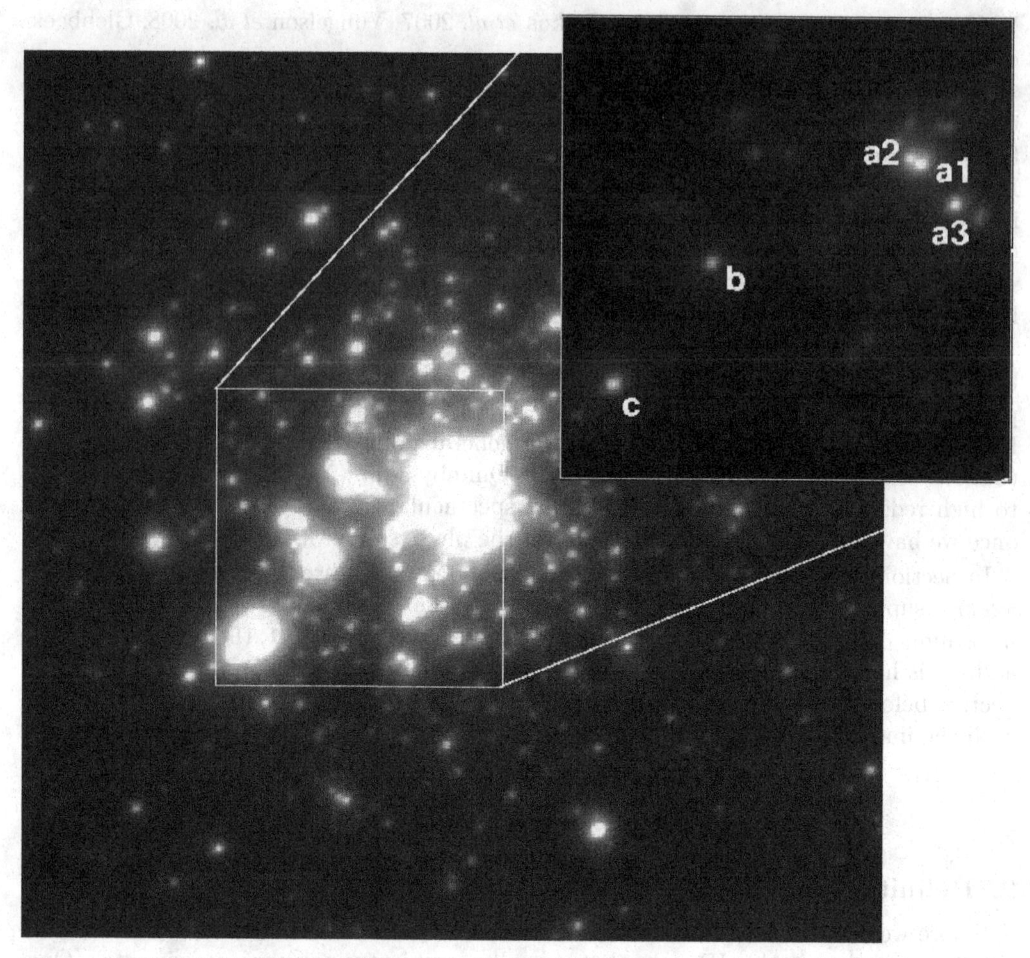

Figure 1. VLT MAD K_s-band 12 × 12 arcsec (3 × 3 parsec for the LMC distance of 49 kpc) image of R136 (Campbell *et al.* (2010) in conjunction with a view of the central 4 × 4 arcsec (1 × 1 parsec) in which the very massive WN5h stars discussed are labelled (component b is a lower mass WN9h star). Relative photometry agrees closely with integral field SINFONI observations of Schnurr *et al.* 2009). See Crowther *et al.* (2010) for details.

3. On the existence of VMS

With this definition, the question of whether *very* massive stars exist can convincingly be addressed with the one answer: *yes* they do! However, the pertinent issue for our JD was whether the widely held "canonical" upper-mass limit of $150\,M_\odot$ has recently been superseded.

Paul Crowther gave the first invited review presenting evidence for initial masses as high as $320\,M_\odot$. Crowther provided historical context to some of the astronomical community's sceptism regarding such high masses in R136, in particular the spectacular claim for the existence of a $2500\,M_\odot$ star R136 in the 30 Doradus region of the LMC some 3 decades ago (e.g. Cassinelli *et al.* 1981). Higher spatial resolution showed that R136 was actually not a single supermassive star, but the object broke up into a young cluster containing significantly lower mass objects, including the current record holder R136a1 (as well as R136a2, R136a3, and R136c). Therefore, over the last few decades, there was a consensus of a $150\,M_\odot$ stellar upper mass limit (Weidner & Kroupa 2004; Figer 2005;

Oey & Clarke 2005, Koen 2006), although the accuracy of this magic number of 150 was low (Massey 2011).

Crowther *et al.* (2010) re-analyzed the photometric and spectroscopic data of the VMS in R136. In comparison to the older WFPC2 data, they used ground-based adaptive optics photometry (see Fig. 1). In combination with their spectral analysis using the CMFGEN non-LTE atmosphere code of Hillier & Miller (1998), this lead to higher estimates for effective temperatures and bolometric corrections. In conclusion they claimed that the R136 cluster hosts several stars with masses as high as 200-300 $M_{\odot}$.

Crowther *et al.* also performed a "sanity check" on similar WNh objects in the Galactic starburst cluster NGC 3601. Although these objects were fainter, and less massive, than those in R136, the advantage was the available dynamical mass estimate by Schnurr *et al.* (2008) of the binary object NGC 3601-A1 of $116 \pm 31 + 89 \pm 16\ M_{\odot}$. This is important as the least model-dependent way to obtain stellar masses is through the analysis of the light-curves and radial velocities induced by binary motions. We also note that Rauw *et al.* (2004) and Bonanos *et al.* (2004) found both components of the eclipsing Wolf-Rayet binary WR20a to be particularly massive, with 83 ± 5 and $82 \pm 5\ M_{\odot}$, with small error bars.

During the lively discussion that followed Paul's review, some attendants argued that the luminosities derived by Crowther *et al.* are uncertain and these "single" objects might actually contain multiple sources. Whilst short-period binaries were not detected by Schnurr *et al.* (2009), longer period binaries are harder to exclude. One of the additional arguments by Crowther *et al.* was that X-rays have not been detected, whilst they may have been expected on the basis of colliding wind binary (CWB) simulations by Pittard & Stevens (2002). Oskinova however countered this argument on the basis that empirically a low X-ray luminosity cannot serve as a robust argument against a binary nature (see Sect. 6).

Najarro noted that even the best image so far of the Arches cluster with Keck (Fig. 2). has a limited spatial resolution of 50 milli-arcsec (mas), which corresponds to roughly 1/10 of the diameter of the circles marking the PSF reference stars in the figure. Given that the LMC is almost 7 times further away, and that the VLT is smaller than Keck, the circles from Fig. 2 would roughly correspond to the spatial resolution achieved with the VLT if the Arches clusters was in the LMC. In other words, the Arches stars would effectively "merge" with surrounding objects. This analogy suggests that we cannot exclude the possibility that the bright WNh stars in the R136 cluster core could still break up into several lower-mass WNh stars. However, for example a $\sim 300 M_{\odot}$ star could at best break up into a pair of $\sim 150 M_{\odot}$ stars (given the shallow slope of the stellar luminosity to mass ratio at the high-mass end).

Moreover, Bestenlehner noted that there is a near-identical twin of R136a3 WNh star in 30 Doradus: VFTS 682. Its key relevance is that it is found in isolation from the R136 cluster, some 30 pc away. For this object line-of-sight contamination is far less likely than for the R136 core stars. VFTS 682 thus offers another sanity check on the reliability of the luminosities for the R136 core stars. Bestenlehner *et al.* (2011) derived a high luminosity of $\log(L/L_{\odot}) = 6.5$ and a present-day mass of $150\ M_{\odot}$ for VFTS 682. This implied an initial mass on the zero-age main sequence (ZAMS) higher than the canonical upper-mass limit – likely $\sim 200\ M_{\odot}$.

In his talk Hamann showed results from another extremely luminous WN star in the Galactic center region: the Peony star (Barniske *et al.* 2008). The luminostiy of this star is determined from spectral analysis as $\log(L/L_{\odot}) = 6.5 \pm 0.2$ and initial mass between $150 \leqslant M_{\rm u}/M_{\odot} \leqslant 200$. The star is located above the Humphreys-Davidson limit, in the region populated by the LBV stars. However, the hydrogen content is lower in WR 102ka

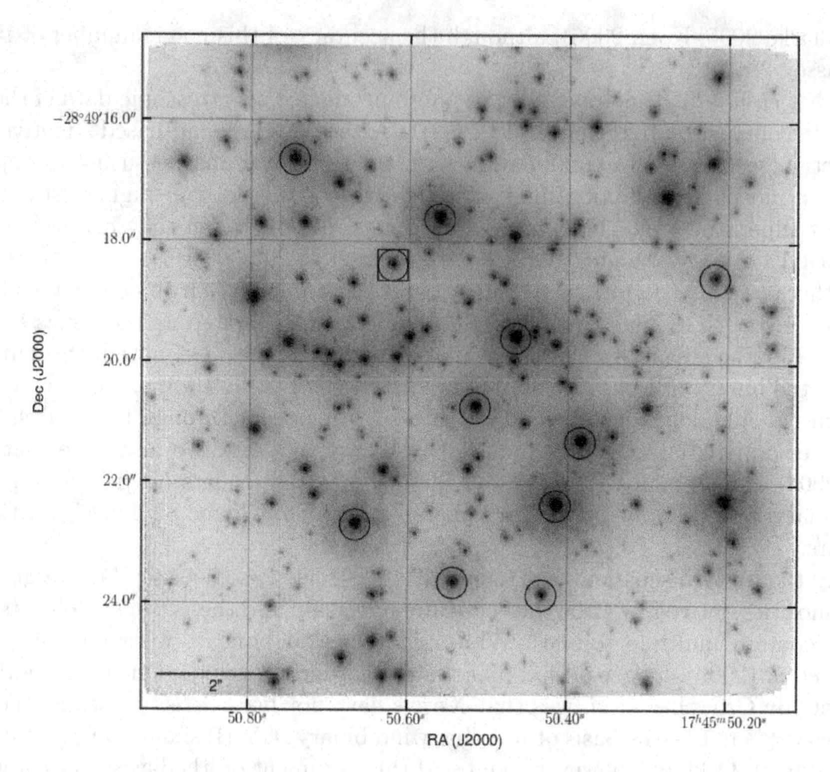

Figure 2. K-band mosaic of the core field of the Galactic Arches cluster at a resolution of 50 mas and with a sensitivity of $K_{\rm lim} = 20.6$mag. The figure is taken from Clarkson *et al.* (2012).

compared to the Pistol star, while helium is higher. This indicates a more advanced evolutionary stage of the former, compared to "normal" LBV stars.

In summary, whilst one cannot exclude that the object R136a1 claimed to be $\sim$300 $M_\odot$ in the R136 cluster might still turn out to "dissolve" when higher spatial resolution observations were to become available, a number of sanity checks involving binary dynamics and "isolated" objects make it rather convincing that stars with ZAMS masses of 200 $M_\odot$ exist. In any case, no fundamental reason was identified why "150" would be a magic number.

4. Formation of VMS

The key question to address was what is so special about the formation of 250 $M_\odot$ stars in comparison to 'normal' 50 $M_\odot$ O-type stars? Can these VMS only form inside dense cluster environments? Other relevant questions involve the time-scale and evolutionary stages when they finish formation/accretion. How fast do they rotate? What VMS fraction is in binaries, and what are the binary properties? Are the most massive stars all the result of stellar mergers?

4.1. *Theoretical considerations*

According to our second invited review speaker, Mark Krumholz, VMS formation is not a fundamentally different problem from the formation of massive stars in general. From an observational standpoint VMS in clusters appear as part of a continuous initial mass function (IMF), with no special features that mark them off as different from the remainder of the stellar population. From a theoretical standpoint, both the radiative and wind luminosities of stars are increasing functions of mass, but from the standpoint of star formation there is no natural dividing line that would put a star of, for example, 50 $M_\odot$ into one category and a star of 250 $M_\odot$ into another. For this reason, it makes sense *not* to treat the formation of very massive stars as a separate problem, and instead to embed it in the broader context of forming the upper end of the IMF.

There are three main challenges to forming massive stars: fragmentation, binarity, and radiation pressure. The fragmentation problem is simple to state: we certainly see gas over-densities ("cores") with masses of $\sim$100 $M_\odot$ or larger and radii of $\sim$0.1 pc or less that seem promising sites for massive star formation (e.g. Beuther *et al.* 2005; Bontemps *et al.* 2010). However, given these masses and radii, and typical molecular cloud temperatures of $\sim$10 K, the Jeans mass is below 1 $M_\odot$, so how can the objects avoid fragmenting and collapse to form an object containing 100 Jeans masses or more? The answer seems to be that the classical Jeans analysis of an isothermal gas does a rather poor job at predicting the behavior of a radiatively heated, magnetized fluid. Once small stars form and begin to accrete, their radiation heats the gas around them and suppresses fragmentation in it (Krumholz 2006; Krumholz *et al.* 2007; Krumholz & McKee 2008). Magnetic fields also make it much more difficult for the gas to fragment (Hennebelle *et al.* 2011), and the combined effects of radiation and magnetic fields seems to be particularly effective, as seen in Fig. 3 (Commerçon *et al.* 2011; Myers *et al.* 2012). These effects together seem to resolve the fragmentation problem, indicating that massive stars can form from the direct collapse of massive protostellar cores with properties similar to those observed.

The second challenge is explaining why so many massive stars appear to be binaries (e.g. Sana *et al.* 2012). The observed multiplicity fraction rises very sharply with stellar mass, reaching near unity for O and B stars (excluding runaways – e.g. Brown 2001). Radiation-hydrodynamic simulations of star cluster formation appear able to replicate the observed dependence of multiplicity on primary mass (Bate 2012; Krumholz *et al.* 2012), and digging into the physical origin for this result indicates that it combines two effects. The first is simple N-body processing: close encounters between stars in young clusters tend to put the most massive members into binaries even if they are not born that way, while stripping companions from less massive stars. The second is disk fragmentation, as explored by (Kratter & Matzner 2006) and (Kratter *et al.* 2008, 2010). Massive stars form with high accretion rates, and these accretion rates tend to produce disks with masses that can approach that of the primary. When this happens, disks are likely to fragment, and a common outcome of this process is that the disk produces a massive companion to the primary.

The third challenge is radiation pressure. Dusty interstellar gas has a high opacity, and the radiation force is proportional to this opacity. As a result, the radiative force exerted by a star's light can exceed its gravitational force for all stars larger than $\sim$20 $M_\odot$ (Wolfire & Cassinelli 1987; Krumholz *et al.* 2009). However, recent numerical and analytic work has shown that this problem is mostly an illusion; the radiation pressure barrier can be circumvented in numerous ways. First, in the presence of an accretion disk, radiation can be beamed away from the bulk of the accreting matter (Nakano *et al.* 1995; Jijina & Adams 1996; Yorke & Sonnhalter 2002; Kuiper *et al.* 2010, 2011). Second, radiation-driven Rayleigh-Taylor instabilities can break up bubbles of radiation and allow matter to accrete through optically-thick fingers (Krumholz *et al.* 2009; Jacquet & Krumholz 2011). Figure 4 shows an example. Third, protostellar jets can punch holes in accreting cores that allow radiation to leak out, reducing the net radiation force in most directions and allowing accretion to continue even though, averaged over 4π sr, the radiation force is stronger than gravity (Krumholz *et al.* 2005; Cunningham *et al.* 2011). The takeaway

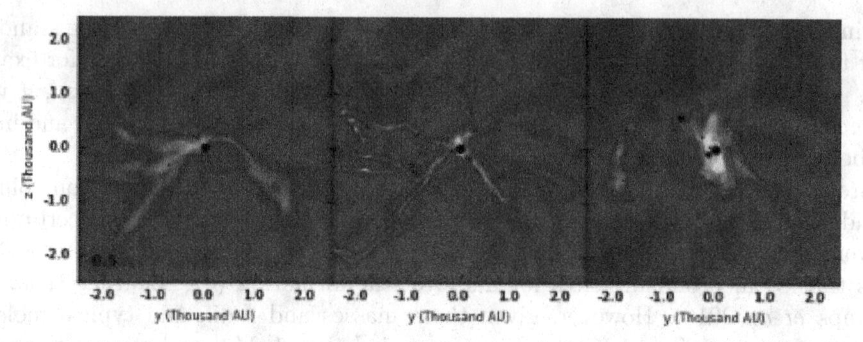

Figure 3. Column density projections from three simulations of a collapsing high mass core, from (Myers *et al.* 2012). All three simulations use identical initial conditions, but the leftmost one includes radiation and magnetic fields, the center one includes magnetic fields but not radiative transfer, and the rightmost one includes radiation but not magnetic fields. Note the dramatic reduction in number of stars (black circles) in the radiation plus magnetic fields case. The color scale runs from $10^{-0.75} - 10^{3.25}$ g cm^{-2}.

message from all of these simulations is that radiation pressure poses no barrier to the formation of stars up to arbitrary masses.

Given that modern theoretical models have removed all of the serious objections to forming massive stars in the same way that low mass stars form, i.e. by accretion from a collapsing gas core, there is no need to resort to exotic processes like stellar collisions (e.g. Bonnell *et al.* 1998). This does not mean, however, that stellar collisions cannot happen and also contribute to the massive star population. The question is: under what circumstances do we expect collisions to be important? The most recent and comprehensive papers to address this question are (Moeckel & Clarke 2011) and (Baumgardt & Klessen 2011) who both conducted N-body simulations of stars confined by a gaseous potential, and who came to similar conclusions. They find that the collisional formation of very massive stars is only significant if the stellar density is extremely high, in excess of 10^7 pc^{-3}, with surface densities reaching 10^5 pc^{-2}. These numbers are so high that even the Arches cluster would not be expected to have significant contributions to its massive star population by collisions. Moreover, if collisions are significant, they do not produce an IMF that looks like what we observe. Instead, because collisions tend to occur among the most massive stars, they produce an IMF that has a Salpeter-like slope at intermediate masses, then a deficit of stars at higher masses, and finally one or a few very massive stars. IMFs with dips of this sort have not been observed, again suggesting that collisions are not likely to be significant contributors to the massive star populations.

4.2. *The potential role of dynamically induced mergers*

An alternative VMS formation scenario was presented by Sambaran Banerjee (with Pavel Kroupa and Seungkyung Oh; Banerjee *et al.* 2012a) who argued that super-canonical stars can be formed out of a dense stellar population – with a canonical IMF *and* with a 150 $M_{\odot}$ upper limit – through dynamically induced mergers of the most massive binaries. Banerjee *et al.* performed direct N-body computations (NBODY6; Aarseth 2003) of a fully mass-segregated star cluster mimicking R136 in which all the massive stars are in primordial binaries. Banerjee *et al.* account for the mass evolution of the super-canonical stars and the resulting shortened ($\approx$ 1.5 Myr) lifetimes in their super-canonical phases using stellar evolutionary models by Köhler & Langer (2012) that incorporate Vink *et al.* (2000) mass-loss rates for the main sequence and Hamann *et al.* (1995) for the He burning WR phase.

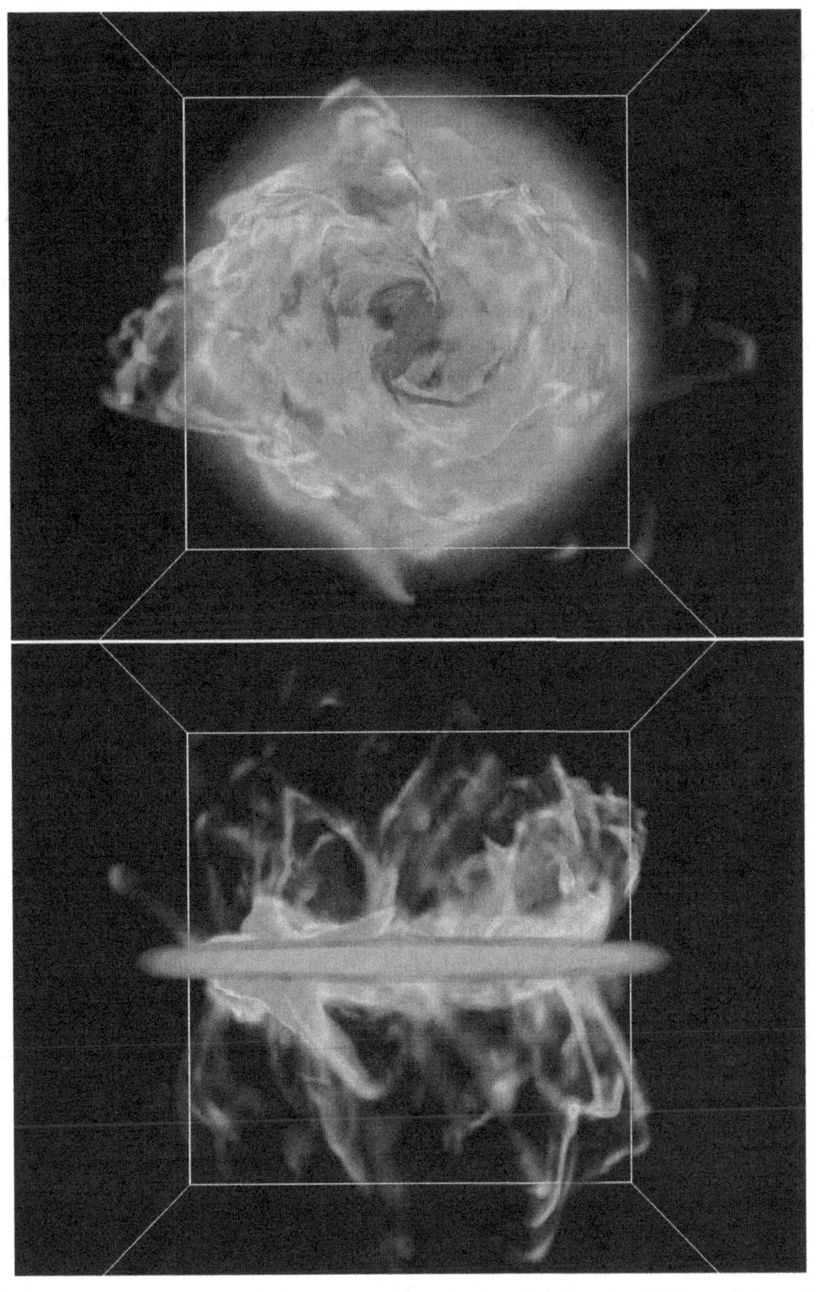

Figure 4. Volume renderings of a simulation from (Krumholz *et al.* 2009) involving the density field in a $(4000\text{AU})^3$ region at 55.0 kyr of evolution. The color scale is logarithmic and runs from $10^{-16.5} - 10^{-14}$ g cm^{-3}. The left panel shows the polar view, and the right one denotes the edge-on view. The figure highlights how Rayleigh-Taylor instability fingers channel matter onto a massive binary star system.

Banerjee *et al.* find that super-canonical stars begin to form via dynamical mergers of massive binaries from ≈1 Myr cluster age, obtaining stars with initial masses up to ≈250 $M_\odot$. Multiple super-canonical stars are found to remain bound to the cluster simultaneously within a super-canonical lifetime. Banerjee also noted that some of these objects can be formed at runaway velocities which escape the cluster at birth. For instance, the most massive apparently isolated WNh star VFTS 682 might be an expected slow runaway (Banerjee *et al.* 2012b; see also Fujii & Portegies-Zwart 2012).

The Banarjee *et al.* models indicate that had super-canonical stars formed primordially alongside the rest of the R136 cluster, i.e. violating the canonical upper limit, they would have evolved below the canonical 150 $M_{\odot}$ limit by $\approx$ 3 Myr, the likely age of the bulk of R136 according to Andersen *et al.* (2009). In other words, Banerjee *et al.* argue that primordially-formed super-canonical stars should not be observed at the present time in R136, whilst it is quite plausible that a collection of dynamically formed super-canonical VMS would be observed in the centres of young massive starburst clusters.

A fully self-consistent N-body computation incorporating detailed accurate evolutionary and mass-loss recipes would be needed to confirm these scenarios.

4.3. *Rotation rates as a constraint on massive star formation*

Returning to the more conventional ways of forming massive stars, Anna Rosen addressed the question of what sets the initial rotation rates of massive stars. The physical mechanisms that set the initial massive star rotation rates are a crucial unknown in current star-formation theory. Observations of young, massive stars provide evidence that they form in a similar fashion to their lower mass counterparts. The magnetic coupling between a star and its accretion disk may be sufficient to spin down low-mass pre-main-sequence (PMS) stars to well below breakup at the end stage of their formation when the accretion rate is low. However, Anna showed that these magnetic torques are insufficient to spin down massive PMS stars due to their short formation times and high accretion rates. Anna developed a model for the angular momentum evolution of stars over a wide range in mass, considering both magnetic and gravitational torques. She finds that magnetic torques are unable to spin down either low-mass or high-mass stars during the main accretion phase, and that massive stars cannot be spun down significantly by magnetic torques during the end stage of their formation either. Spin-down occurs only if massive stars' disk lifetimes are substantially longer or their magnetic fields are much stronger than current observations suggest (Rosen *et al.* 2012).

4.4. *Observations of massive star formation*

Heavy extinction hides the birthplaces of massive stars from view, and the short formation timescales set a strong limitation to the sample of objects that can be studied. So far, our physical knowledge of massive young stellar objects (YSOs) has been derived from near-IR imaging and spectroscopy, revealing populations of young OB-type stars, some still surrounded by a disk, others apparently 'normal' main sequence stars powering H II regions. The most important spectral features of OB-type stars are however located in the ultraviolet (UV) and optical range.

Kaper (with Ellerbroek, Ochsendorf and Bik) showed that with the new optical/near-infrared spectrograph X-shooter on ESO's Very Large Telescope (VLT), it is possible to extend the spectral coverage of these massive YSOs into the optical range. First results are very promising, although they seem to probe the intermediate-mass ($\sim$2 – 8 $M_{\odot}$) range rather than the massive star range. Ellerbroek *et al.* (2011) discovered a jet (HH 1042) produced by the massive YSO nr292 in the massive star forming region RCW 36, demonstrating that the object is still actively accreting. The mass of this star is likely less than 6 $M_{\odot}$ and remains uncertain as photospheric features are not detected. The first firm spectral classification of B275, a massive YSO in M17, results in its precise location on a PMS track for a $\sim 7\,M_{\odot}$ star (see Fig. 5; Ochsendorf *et al.* 2011), has the size of a

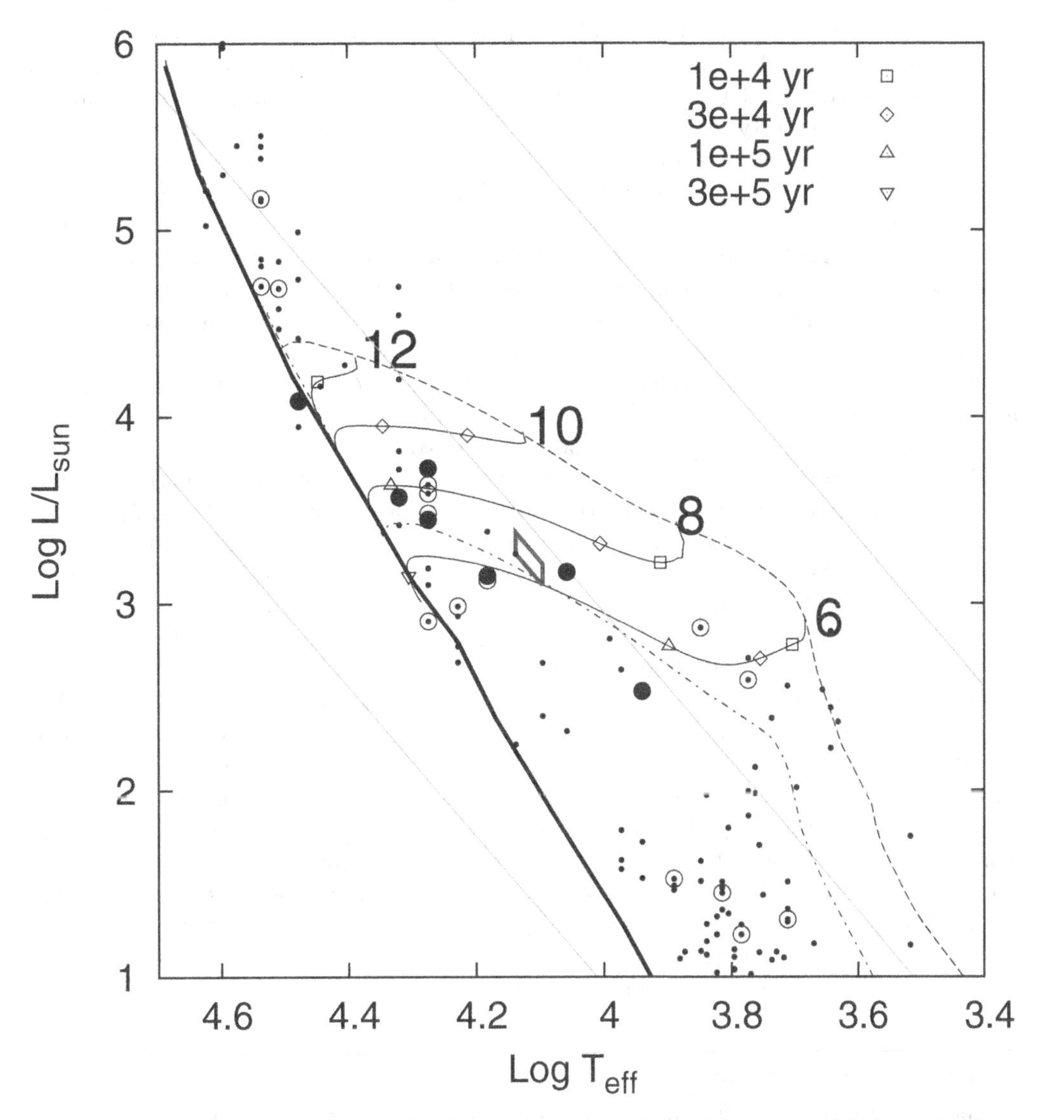

Figure 5. The location of B275 (red parallelogram) in the HRD next to PMS tracks from Hosokawa *et al.* (2010) with the ZAMS mass labeled and open symbols indicating lifetimes. The thin dashed and thin dot-dashed lines are the birth lines for accretion rates of 10^{-4} $M_\odot$ yr^{-1} and 10^{-5} $M_\odot$ yr^{-1}, respectively; the thick solid line is the ZAMS (Schaller *et al.* 1992). The filled and open circles represent stars in M 17 for which a spectral type has been determined (Hoffmeister *et al.* 2008); dots are other stars in M 17. B275 is on its way to becoming a 6 8 $M_\odot$ ZAMS star, so far one of the most massive pre-main-sequence star known. The figure has been adapted from Ochsendorf *et al.* (2011).

bloated giant, as predicted by models of Hosokawa *et al.* (2010), and is still surrounded by a disk.

It remains unclear whether the progenitors of the most massive VMS will be detectable this way.

4.4.1. *Radiation Transfer Modeling: From massive to very massive star formation*

In order to properly interpret the observations of massive protostars, Zhang *et al.* (with Tan, McKee, and de Buizer) presented a radiation transfer (RT) model for a massive core in a high pressure environment that forms a massive star through core accretion (Zhang & Tan 2011; Zhang, Tan & McKee, 2012, ApJ submitted). This RT model is based on the Turbulent Core model of McKee & Tan (2003): massive stars form in dense clumps with high surface density. Assuming the rotational-to-gravitational energy ratio is 2% in the core, whilst the disk has a diameter of ~1000 AU. The high accretion rate leads to a disk mass comparable to the stellar mass. In Zhang *et al.* (2012), the treatment of the disk was improved by allowing radially varying accretion rates due to a supply of mass and angular momentum from the infall envelope and their loss to the disk wind. The transfer of accretion power to mechanical power of the wind was also accounted for. An approximate disk wind solution was developed partly based on the Blandford & Payne (1982) model. The simulation was performed with the latest version of the Monte Carlo RT code by Whitney *et al.* (2003). Corrections made by adiabatic cooling/heating and advection were included, and so were the gas opacities.

The model was compared to the massive protostar G35.2-0.74N. At a distance of 2.2 kpc, radio continuum emission indicates there is a bipolar outflow from this source. In recent SOFIA-FORCAST observations of the massive protostar G35.2-0.74N at 31 and 37m (Zhang *et al.* 2012, in prep.) both the near- and far-facing sides of the outflow can be seen. The latter is missing in mid-IR continuum. By fitting both the observed SED and the outflow-axis intensity profiles with the RT model, the bolometric luminosity was inferred to be $\sim 10^5\,L_\odot$ after correcting for foreground extinction and the dependence of luminosity on inclination (flashlight effect).

The fitting model suggests a massive $\sim 30\,M_\odot$ protostar is forming from a $\sim 240\,M_\odot$ core with a high surface density of $\sim 1\mathrm{g/cm}^2$, via relatively ordered collapse and accretion, and driving powerful bipolar outflows. These results seem to support the core accretion theory which predicts massive stars may form in similar ways to their low-mass counterparts. Simply extending the model to protostars with higher masses up to $100\,M_\odot$, Zhang *et al.* noticed a shift of the SED peaks to shorter wavelengths and a change of mid-IR slope. The flashlight effect turns out to be huge for such sources. In the mid-IR, the luminosity from face-on view can be higher than that from edge-on view by as much as 3 dex – suggesting that a large correction factor needs to be applied when using IR observations for deducing source luminosities.

5. Properties of VMS

Once a star has formed it starts to burn hydrogen on the ZAMS. For massive stars, the main sequence band probably consists of two spectroscopically distinct groups of objects: the O stars, and the WNh stars. The first group are thought to form the normal massive star sequence with masses of up to at least $\sim 60 M_\odot$. For the most massive "O stars on steroids" WNh (H-rich WR stars), their masses may be up to $300\,M_\odot$. or higher.

The transition mass between O and WNh is not exactly known but is probably lies somewhere in the range 60-120$M_\odot$. Of course these numbers are model (mass loss and Z) dependent. The WNh stars are presumably in close proximity to the Eddington limit, and so are their descendant LBV and classical WN, WC, WO Wolf-Rayet stars. Some of these objects are thought to be in an evolutionary phase just prior to their final demise.

At the end of this section, we also consider extra-galactic VMS properties, both at very low (e.g. IC 1613) and high metal content (e.g. M33).

5.1. *WNh H-rich Wolf-Rayet stars – "O stars on steroids"*

WNh stars were discussed by Crowther and Hamann. In their analysis they make use of non-LTE model atmospheres, such as CMFGEN (Hillier & Miller 1998) and PoWR

(Hamann *et al.* 2006), whilst FASTWIND (Puls *et al.* 2005) is often used for normal O stars with weaker winds. These are 1D spherical symmetric codes that include line-blanketing and stellar winds allowing for micro-clumping (optically thick macro-clumping allowing for porosity is non-standard). Another important aspect of the analysis is that of accurate infrared photometry (see the infrared SpS on atmosphere modelling). This is especially true for those objects in the cores of dense clusters such as R136. The upshot of recent analyses is that WNh stars are more luminous, and more massive, than previously thought (see Sect. 3). Hamann presented some early results from their spectral analysis of WN stars in the Magellanic Clouds (Hainich *et al.* in prep.) – complementing the published Galactic study of Hamann *et al.* (2006). Whilst the LMC objects are found to be bright with $\log L/L_{\odot}$ values in the range 5.3-5.9, Hamann warns that many of the brighter allegedly single WNh stars may actually be binaries, and not accounting for this fact may overestimate the luminosities of WNh stars.

5.2. *LBVs: unstable massive stars close to the Eddington limit*

Very massive stars are thought to evolve through the unstable Luminous Blue Variable phase, when enormous amounts of mass are lost. While LBVs have been classically thought to be rapidly evolving massive stars in the transitory phase from O-type to Wolf-Rayet stars, recent studies have suggested that LBVs might surprisingly explode prematurely as a core-collapse supernova (Kotak & Vink 2006, Gal-Yam *et al.* 2007, Mauerhan *et al.* 2012). Such a striking result highlights that the evolution of VMS through the LBV phase is far from being understood.

Groh discussed the recent advances in understanding LBVs, in particular how to distinguish them from the normal B supergiants and hypergiants (as discussed by Negueruela). LBVs can be recognized either from Giant Eruptions like Eta Car and P Cygni, or through their S-Dor variability, sometimes imprinted on peculiar looking double-peaked absorption profiles, in just a single epoch spectrum (Groh & Vink 2011). LBVs do not always present both S Dor and giant outburst phenomena, leaving room for quite a heterogeneous class of objects.

Groh emphasized that LBVs are characterized on a phenomenological basis and, therefore, LBVs are neither a spectroscopic nor an evolutionary classification. Particular emphasis was given to describe the main properties of the S-Dor type variability (also performed by Stringfellow). These are changes in the hydrostatic radius and bolometric luminosity. Finally, the role of rapid rotation on LBVs was discussed (Groh *et al.* 2009).

5.3. *The Galactic WC stars*

WC star spectra are well known for their broad emission lines from helium, carbon and oxygen. Due to the absence of hydrogen these stars have to be core-helium burning and have as such been identified as the late evolutionary stage of massive and very massive stars. Sander *et al.* (2012) analyzed the optical and UV spectra of over 50 Galactic WC single stars and derived their parameters including their mass-loss rates. Sander *et al.* showed that the positions of the Galactic WC stars in the HR-diagram do not fit with the assumption that the most massive stars will pass the WC stage. Instead Sander *et al.* argue that their results indicate that WC stars come from an initial mass range between 20 and somewhere around $50\,M_{\odot}$. The stellar evolution models of Vanbeveren *et al.* (1998) that include enhanced mass loss during the red sugergiant (RSG) phase appear to properly account for the location of the WC stars in the observed HR-diagram. It also seems that stars with higher initial masses do not reach the WC stage but instead explode after passing a WNL and probably an LBV stage.

5.4. *WO stars*

Sander *et al.* also showed that the Galactic WO2 stars WR 142 and WR 102 have significantly different parameters from the WC stars. The WO positions in the HR diagram suggest that these stars could be close to or already in the stage of carbon burning which makes them interesting SN Ic candidates (Georgy *et al.* 2012; Yoon *et al.* 2012).

Tramper *et al.* presented VLT/X-Shooter spectroscopy of DR1, a WO3 star in the low-metallicity galaxy IC 1613. A preliminary spectroscopic analyis using CMFGEN indicates a high temperature (of ~150 kK) which is also supported by the very strong nebular He II emission. The oxygen abundance does not seem to be enhanced compared to values found for early-type WC stars, suggesting that the strong oxygen emission is likely a temperature effect, rather than being caused by an increased oxygen abundance.

5.5. *Rapidly rotating WR stars*

The WR rotation issue is especially relevant in view of the suggested link between rotating WR stars and long-duration GRBs. WR 2 (WN2) is a well known oddball. It is the most compact and the hottest (~140 kK) pop I WN star known in the Galaxy. It is also one of the best candidate for strange-mode pulsations (Glatzel *et al.* 1999), but they are not observed in photometric observations. The spectrum displays bowler-hat shaped emission lines, in contrast with the more normal Gaussian or even flat-top and triangular profiles of most other WR winds. Remarkably, Hamann *et al.* (2006) have analyzed WR 2 using their latest model-atmosphere code, and the model spectrum failed to reproduce its line-profiles, unless it was folded with a rotation curve near the break-up limit, i.e. 1900 km/s.

However, Chenè presented the polarized spectrum of WR 2, which shows no sign of wind asymmetry expected for such rapid rotation. Interestingly, WR 2 appears to display clumps that are moving in a similar fashion as in the optically thinner wind of WR 3 (Chené *et al.* 2008). Hence, the shape of WR 2 spectra line cannot be the result of extreme opacity either.

5.6. *Massive stars in the low metallicity galaxy IC1613*

Because low metallicity environments may favour the formation of massive and very massive stars, we switch our attention to extra-galactic properties. Herrero discussed the selection of massive star candidate stars in the low-Z galaxy IC 1613 from the catalog of Garcia *et al.*(2009). IC 1613 has a metallicity Z= 0.13 $Z_\odot$ according to the analysis of B-supergiants by Bresolin *et al.* (2007), or Z= 0.08–0.15 $Z_\odot$ according to several analyses of HII regions (see Herrero *et al.* 2012). Herrero *et al.* observed the selected stars with OSIRIS/GTC at a resolution $R = 1000$ to determine their spectral types, as a previous step for a more detailed analysis. They were able to classify 12 new OB stars and confirmed one more known O-type star.

The spectra of the O stars were good enough for quantitative analysis (albeit with errors slightly larger than in typical analyses, see Repolust *et al.* 2004). Combining these results to those from the literature (Tramper *et al.* 2011; Herrero *et al.* 2012), Herrero presented the first effective temperature scale for sub-SMC metallicities. This temperature scale is slightly hotter than that derived for SMC stars from the data of Massey *et al.* (2009), Mokiem *et al.* (2007), and Trundle *et al.* (2007).

5.7. *Stellar abundances from massive stars in M33*

With its distance less than a Mpc (Bonanos *et al.* 2006) and its favourable inclination angle, this makes M33 an ideal galaxy to study the chemical evolution in spiral galaxies.

Not long ago, the only way to carry out detailed chemical analyses of nearby galaxies was through the quantitative studies of H II regions. However, the abundances derived from massive OB stars are the tracers of the present-day chemical composition, providing information that cannot be obtained H II regions (e.g. the silicon abundance). Moreover, a simultaneous characterization of the stellar parameters may address important aspects of their evolution, and in particular the role of environmental factors.

Castro presented the results of a spectroscopic survey in M 33 involving 59 supergiants with spectral types between B9 and O9, and a quantitative analysis according to the steps described by Castro *et al.* (2012). A thorough comparison between optical spectra and new FASTWIND (Puls *et al.* 2005) grids resulted in both stellar parameters and chemical composition. The parameters derived in conjunction with the evolutionary tracks of Brott *et al.* (2011) hints at the presence of evolved stars with masses in the range 15 and 50 M⊙. New routines for deriving the chemical abundances automatically through a process optimization showed an oxygen distribution along M 33 that is compatible with previous H II region studies (e.g. Rosolowsky & Simon 2008).

6. Mass loss mechanisms for VMS

The evolution of Very Massive Stars is presumably dominated by mass loss, which thus needs to be understood both qualitatively and quantitatively. Joachim Puls reviewed different mass-loss mechanisms relevant in this context. Because of the high luminosities, only radiation-driven mass loss was considered, and time-dependence, rotation and magnetic fields were not accounted for.

6.1. *Theoretical considerations*

Basic considerations. The equation of motion for the transonic/supersonic regime of an expanding wind can be approximated by

$$v\Big(1-\frac{a^2}{v^2}\Big)\frac{\mathrm{d}v}{\mathrm{d}r} \approx g_{\rm grav}(r)+g_{\rm rad}^{\rm tot}(r) = -\frac{GM}{r^2}(1-\Gamma(r)), \quad \Gamma(r)=\frac{\bar\kappa(r)L_*}{4\pi GMc}=\frac{\bar\kappa(r)}{\sigma_{\rm e}}\Gamma_{\rm e}$$

where all quantities have their usual meaning, a is the isothermal sound speed, $g_{\rm rad}^{\rm tot}$ is the *total* radiative acceleration (lines + continuum), $\bar\kappa$ is the flux-mean opacity per unit mass, and Γ the corresponding Eddington parameter ($\Gamma_{\rm e}$ w.r.t. electron scattering only). At the sonic point, r_s, $v=a$, and thus $g_{\rm rad}^{\rm tot}=-g_{\rm grav}$ implying $\Gamma(r_s)=1$. To allow for an accelerating wind, $\frac{\mathrm{d}v}{\mathrm{d}r}|_s>0$, which then requires $\frac{\mathrm{d}\bar\kappa}{\mathrm{d}r}|_s>0$ and $\Gamma(r)<1$ below and $\Gamma(r)>1$ above the sonic point, respectively.

Photon tiring limit. The mechanical luminosity of the wind at 'infinity' is given by

$$L_{\rm wind}=\dot M\Big(\frac{{v_\infty}^2}{2}+\frac{GM}{R}\Big)=\dot M\Big(\frac{{v_\infty}^2}{2}+\frac{{v_{\rm esc}}^2}{2}\Big) \quad \text{with} \quad v_{\rm esc}=\sqrt{\frac{2GM}{R}},$$

and the maximum mass-loss rate follows from the condition that $L_{\rm wind}=L_*$ (when the star would become invisible): $\dot M_{\rm max}=2L_*/({v_\infty}^2+{v_{\rm esc}}^2)$. Following (Owocki & Gayley 1997), $\dot M_{\rm tir}$ then is the maximum mass-loss rate when the wind just escapes the gravitational potential, with $v_\infty\to 0$, and is much larger than typical mass-loss rates from line-driven winds,

$$\dot M_{\rm tir}=\frac{2L_*}{{v_{\rm esc}}^2}=0.032\frac{M_\odot}{\rm yr}\frac{L_*}{10^6L_\odot}\frac{R}{R_\odot}\frac{M_\odot}{M}=0.0012\frac{M_\odot}{\rm yr}\Gamma_{\rm e}\frac{R}{R_\odot}.$$

Continuum driven winds. To drive a wind by pure continuum acceleration ($\Gamma = \Gamma^{\rm cont}$) when the photosphere is sub-Eddington ($\Gamma(r) < 1$ for $r < r_s$) requires substantial fine-tuning to reach and maintain $\Gamma^{\rm cont}(r) \geqslant 1$ for $r \geqslant r_s$, and is rather unlikely. Anyhow, in such a situation $g_{\rm rad}^{\rm cont}$ is almost density-independent, and large mass-loss rates could be accelerated, only limited by photon tiring, which needs to be considered in the equation of motion (for details, see Owocki & Gayley 1997).

Super-Eddington winds. If, on the other hand, the complete atmosphere is super-Eddington, $\Gamma(r) > 1$, continuum driving (mostly due to electron-scattering) might become possible. When atmospheres approach or exceed the Eddington limit, non-radial instabilities arise making them inhomogeneous (clumpy). Photons on their way out avoid regions of enhanced density, and the medium becomes *porous*. In this case, the photospheric radiative acceleration decreases compared to an unclumped medium, leading to an *effective* Eddington parameter *below unity*. In the outer regions, where the clumps become optically thin due to expansion, porosity decreases and $\Gamma_{\rm eff}^{\rm cont} \to \Gamma(r) > 1$. Thus, an accelerating wind can be initiated (Shaviv 2000, 2001a,b). (Owocki *et al.* 2004) expressed the effective opacity in terms of a 'porosity length', and showed that associated mass-loss rates can become substantial when this length is on the order of the pressure scale height, amounting to a few percent of the tiring limit. Invoking a power-law distributed porosity length, they showed that the 'observed' mass loss from the giant outburst of η Car might be explained by this (metallicity-independent!) mechanism, and that $\dot{M}$ scales with $\dot{M} \propto \Gamma^{1/\alpha_p - 1}$, as long as the exponent of the power-law, $\alpha_p < 1$ and $\Gamma > 3\ldots4$.

Line-driven winds. The standard theory of line-driven winds (Castor *et al.* 1975 and later refinements) assumes that the continuum is still optically thin at the sonic point (valid for OB-stars, A-supergiants and LBVs in their quiet phase), and that $\Gamma^{\rm cont} < 1$ everywhere, with $\Gamma^{\rm cont} \to \Gamma_{\rm e}$ for $r \geqslant r_s$. The radiative acceleration exerted on a shell of mass $\Delta m = 4\pi r^2 \rho$ by *one* optically thick line can be expressed as

$$g_{\rm rad}^{\rm one\ line} = \frac{\Delta P}{\Delta t \Delta m} \propto \frac{L_\nu \nu}{4\pi r^2} \frac{{\rm d}v}{\rho {\rm d}r},$$

where ΔP is the transferred momentum, and the term dv arises because of the Doppler-shift within dr. Summing up over all lines using a line-strength distribution function and accounting for optical depth effects, the total line acceleration results in

$$g_{\rm rad}^{\rm all\ lines} \propto N_0 \frac{L_*}{4\pi r^2} \left(\frac{{\rm d}v}{\rho {\rm d}r}\right)^\alpha \to N_0 \frac{L_*}{4\pi r^2} \left(\frac{4\pi}{\dot{M}}\right)^\alpha \left(r^2 v \frac{{\rm d}v}{{\rm d}r}\right)^\alpha,$$

with N_0 the effective number of driving lines (depending on spectral type and metallicity), and $0 < \alpha < 1$ related to the slope of the line distribution function. Inserting this expression into the equation of motion and neglecting photon tiring, a unique (maximum) mass-loss rate can be calculated,

$$\dot{M} \propto N_0^{1/\alpha} L_*^{1/\alpha} \left(M(1-\Gamma_{\rm e})\right)^{1-1/\alpha} = N_0^{1/\alpha} L_* \left(\frac{\Gamma_{\rm e}}{1-\Gamma_{\rm e}}\right)^{1/\alpha - 1} = \left(\frac{N_0 \Gamma_{\rm e}}{1-\Gamma_{\rm e}}\right)^{1/\alpha} M(1-\Gamma_{\rm e}),$$

which dramatically increases for $\Gamma_{\rm e} \to 1$ (in this case, photon-tiring needs to be accounted

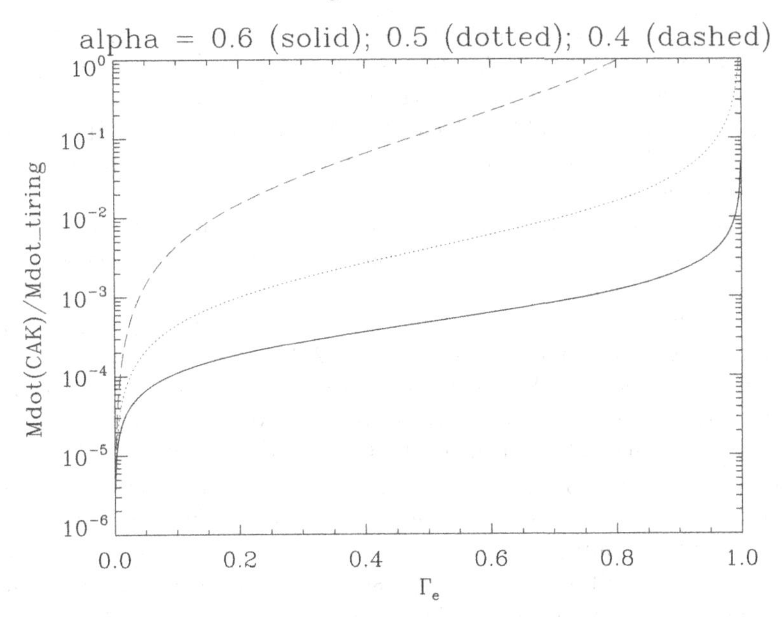

Figure 6. $\dot{M}/\dot{M}_{\rm tir}$ for a typical O-star wind with $v_{\rm esc}$= 600 km s^{-1}, as a function of $\Gamma_{\rm e}$ and α.

for). On the other hand, the terminal velocity scales with

$$v_\infty \propto v_{\rm esc}{}^{\rm eff} = \sqrt{\frac{2GM(1-\Gamma_{\rm e})}{R}} \rightarrow 0 \text{ for } \Gamma_{\rm e} \rightarrow 1.$$

Figure 6 displays the ratio of $\dot{M}/\dot{M}_{\rm tir}$ as a function of $\Gamma_{\rm e}$, and shows that this ratio is strongly sensitive to α. For $\alpha = 0.4$ (somewhat lower than the typical value for OB-stars), the tiring limit would already be reached at $\Gamma_{\rm e} = 0.8$.

Optically thick winds. The large mass-loss rates from WR-stars, being typically a factor of 10 higher compared to OB-star winds at the same luminosity, cannot be explained by the standard theory from above. The observed terminal velocities (similar to OB-stars) can be reached only when line-overlap effects become efficient. In such dense winds, the ionization equilibrium decreases outwards, and photons on their way out can interact with lines from different ions, whilst any 'gaps' between lines (as present in OB-stars because of an almost frozen-in ionization) are closed (see Lucy & Abbott 1993; Springmann 1997). The initiation of the mass loss, on the other hand, is supposed to rely on the condition that the winds are already optically thick at the sonic point, and that the (quasi-static) photospheric line acceleration due to the iron peak opacities around 150 kK (for WNEs) or 40 kK (for WNLs) is sufficient to overcome gravity (Nugis & Lamers 2002). (Gräfener & Hamann 2005, 2008) calculated self-consistent models for WNEs and WNLs, and showed that this mechanism actually allows for large $\dot{M}$, where the major prerequisite is a high Γ. Such optically thick winds might be present also in VMS (Vink *et al.* 2011 and these proceedings), although (Pauldrach *et al.* 2012) argue that VMS winds might remain optically thin.

6.2. *Monte Carlo mass-loss rates*

Vink discussed three relatively new aspects concerning mass-loss calculations from the Monte Carlo method (Abbott & Lucy 1985) - as previously used to predict $\dot{M}$ for canonical OB-type stars (e.g. Vink *et al.* 2000). The first one concerned the wind dynamics. Until

2008, the methodology was semi-empirical, as a velocity law was assumed that reached a certain empirical v_∞. Müller & Vink (2008) suggested a line-force parametrization that explicitly depends on radius (rather than the velocity gradient, as in CAK theory), and predicted v_∞ values in reasonable agreement with observations. Muijres *et al.* (2012) tested the Müller & Vink approach, and as both methods gave similar results, it was used in the following.

Secondly, a new parameter space was probed, i.e. that of the VMS. Vink *et al.* (2011) $\dot{M}$ predictions show a *kink* in the $\dot{M}$ - Γ relation. For "low" Γ optically-thin O-star winds, the $\dot{M} \propto \Gamma^x$ relation is shallow, with $x \simeq 2$, whilst there is a steepening at high Γ, with $x \simeq 5$. At high Γ the objects show optically thick WR-like winds, with optical depths and wind efficiencies above unity. Gräfener *et al.* (2011) provided empirical evidence for such a steep exponent ($x \simeq 5$), but there are still issues with the predictions of absolute v_∞ values in this high Γ range. Critical comparisons between observations and theory are underway by Bestenlehner *et al.* in the context of the VFTS survey (Evans *et al.* 2011)

In another study Vink & Gräfener (2012) calibrated wind mass-loss rates using an analytic method to find that the wind efficiency number equals unity right at the transition point between optically thin and thick winds: $\eta = \tau = 1$. Application of this relation to the most massive stars in the Arches cluster suggests there is little room for additional mass loss during e.g. LBV eruptions, and current wisdom would suggest that PSN explosions are unlikely, unless one were to move to lower Z galaxies (e.g. Langer *et al.* 2007, Yoshida & Umeda 2011, Yusof *et al.* in prep.).

6.3. *Alternative mass loss: eruptions and mass transfer*

Solar metallicity VMS likely evaporate as the result of stellar wind mass loss. However, alternative mass loss may also be important, especially for the lower initial mass and sub-solar metallicity ranges. Furthermore, we know Eta Car analogs and supernova impostors exist in external galaxies (e.g. Van Dyck *et al.* 2005, Pastorello *et al.* 2010, Kochanek *et al.* 2012), but quantitative estimates on the integrated amount of such eruptive mass loss are hard to obtain as both the eruption frequency, and the amounts of mass lost per eruption span a wide range with LBV nebular mass estimates varying from $\sim$0.1 $M_\odot$ in P Cygni to $\sim$10 $M_\odot$ in Eta Car (Smith & Owocki 2006).

The energies required to produce such giant mass eruptions are very high ($\sim 10^{50}$ erg), and their energy source is unknown. Soker (2004) discussed that the energy and angular momentum required for Eta Car's great eruption cannot be explained with a single-star scenario.

There is a growing amount of evidence that the most massive stars are oftentimes found in binary systems, and binary evolution with mass loss is pursued by many groups around the globe (Vanbeveren 1998; Eldridge *et al.* 2008). What has become particularly clear from recent spectroscopic radial velocity surveys (e.g. Sana *et al.* 2012) is that there is a particularly large number of short-period binaries, which might merge still during core H burning, and subsequently evolve as seemingly single stars. For these reasons one of the most famous recent phrases in the massive star community has been "binary stars might actually be the best single stars" (de Mink *et al.* 2011). However, also after the main sequence, there are still many physical processes involving mass loss through Roche lobe overflow (see e.g. Langer 2012) and common envelope evolution (Ivanova *et al.* 2012), which remain as yet ill-understood.

6.4. *Mass-loss diagnostics*

The traditional ways of determining mass-loss rates of (very) massive stars involve recombination lines (such as Hα and He II 4686), as well as radio and sub-mm continuum measurements that measure the amount of free-free emission (Wright & Barlow 1975, Lamers & Cassinelli 1999). Especially the free-free method may be used in the near future with new facilities such as ALMA coming online. The drawback of the above diagnostics is that they depend on an uncertain amount of wind clumping (Puls *et al.* 2008). The unsaturated resonance lines of trace elements, such as P V, located in the far UV, has been considered a more accurate tool for mass-loss diagnostics (e.g. Fullerton *et al.* 2006), because their formation depends linearly on density such that inhomogeneities average out. However, it was shown by Oskinova *et al.* (2007) that the formation of resonance lines is also affected by wind clumping, if the line opacity makes the individual clumps optically thick. Neglecting this effect may lead to *under*estimations of the true $\dot{M}$ (see also Sundqvist *et al.* 2010).

For these reasons it is important to (i) gain a greater understanding of both the physics and the diagnostics of wind clumping throughout the stellar atmosphere and wind, ideally as a function of radial distance, and (ii) to develop diagnostics that are *not* dependent on wind clumping.

6.4.1. *X-ray diagnostics for VMS*

Massive stars of most (but not all) spectral types are sources of X-ray emission. In single stars, the X-rays most likely originate in the gas heated by the strong shocks resulting from the line-driven instability of stellar winds (e.g. Lucy 1980, Owocki *et al.* 1988). Therefore, the properties of X-ray emission are sensitive to the wind driving mechanism. Because of the proximity of VMSs to the Eddington limit, the details of wind driving and line-driving instability growth may be different from lower mass massive stars.

The X-ray luminosity of VMS stars is challenging to predict. The X-ray luminosity of Galactic OB stars follows the trend $L_{\rm X} \propto 10^{-7} L_{\rm bol}$. While some binary O stars with colliding winds have X-ray luminosity significantly higher than $10^{-7} L_{\rm bol}$, the majority of O star binaries follow this correlation as well (Oskinova 2005, Naze 2009). In some cases, the binary O stars have X-ray luminosity significantly lower than the expected for a single star of similar spectral type. Oskinova produced a diagram showing the dependence of the X-ray luminosity in binary O stars on the period. No correlation was seen, and the short period binaries can have low X-ray luminosity.

Oskinova concluded that a low X-ray luminosity cannot serve as a robust argument against a binary nature of an O star. In other words, a binary luminous VMS could potentially have a low $L_{\rm X}$. However, the story for the WR-like VMS binaries (with strong winds) might be different from the weaker-winded O-star binaries.

6.4.2. *A new wind measurement approach using X-rays from colliding wind binaries*

Sugagawa presented *Suzaku* observations of the WR binary WR 140, taken at four different times around periastron passage in 2009 January. The X-ray spectra changed in shape and flux at each phase. As periastron approached, the column density of the low-energy absorption increased, indicating that the emission from the wind-wind collision plasma was absorbed by the dense WR wind. The luminosity of the dominant hot component from the wind-wind collision is not inversely proportional to the (variable) distance between the two stars. In the case of the mass-loss ratio $\dot{M}_{\rm O}/\dot{M}_{\rm WR}$ =0.04, Sugagawa could explain this discrepancy if the O-star wind collides with the WR wind before it has reached its terminal velocity, leading to a reduction in its wind momentum flux.

Sugagawa presented these mass-loss rates, which were calculated using the absorptions and variations of the spectra (Sugawara *et al.* 2012, submitted).

Daminelli showed that the He II 4686 line in Eta Carinae displays two peaks before periastron in good correlation with the X-ray intensity. There even is a third peak, in close coincidence with periastron, which is anti-correlated with X-rays intensity. This may be interpreted as a collapse in the wind-wind collision structure, when most of the energy escapes in the extreme UV, which would be possible if the eccentricity is larger than $e > 0.9$.

What is clear is that the approach presented by Sugagawa using X-ray observations is applicable to other massive CWBs with elliptical orbits. In addition, unexpected X-ray brightening of very massive CWBs (such as WR 21a) may be helpful for understanding VMS mass loss.

6.4.3. *Mass loss at very low metallicity*

Whilst there are still several uncertainties in our empirical knowledge of Solar metallicity mass loss, rates at low Z may provide additional constraints on the driving mechanisms. In this context, Tramper presented the results of a quantitative spectroscopic analysis of VLT/X-Shooter observations of six O-type stars in the low-metallicity galaxies IC 1613, WLM and NGC 3109 (Tramper *et al.*, 2011; Herrero *et al.* 2011, but see also Herrero *et al.* 2012). The obtained stellar and wind parameters can be used to probe the mass loss versus metallicity dependence at metallicities below that of the SMC. Tramper *et al.* compared their derived mass-loss rates with the empirical results from Mokiem *et al.* (2007) for the Galaxy, LMC and SMC, and with the theoretical prediction from Vink *et al.* (2001), and argued that the mass-loss rates appear to be higher than expected.

It is clear that the analysis of a larger sample of stars at sub-SMC metallicities is needed to confirm or disprove these results.

7. VMS Evolution and Fate

Alexander Heger (with Woosley & Chen) presented the fourth review talk on (very) massive star evolution. An introduction to massive star evolution can be found in (Woosley *et al.* 2002; Heger 2012). One of the most exciting questions concerning massive stars is how they will die, which of them will explode, and how. A range of outcomes is possible in terms of observational signatures, but the key question is how massive the star is at the time of death. Therefore, according to Heger understanding the mass-loss rates of massive stars is one of the key uncertainties that needs to be resolved.

Other stumbling blocks along the way involve rotation and binary evolution. For the sake of simplicity, Heger only discussed single stars. However, it is clear that some close binaries will merge early during their evolution, already during central hydrogen burning. These would likely evolve in a fashion similar to that of single stars - though likely rapidly rotating. Rotation however is a second parameter that makes the picture more complex. So for the sake of simplicity Heger dealt "just" with single stars and deferred the reader to the rotation works of (Yoon *et al.* 2006, 2012). In some aspects, rapidly rotating single stars may evolve in a fashion similar to non-rotating stars of a different (usually higher) mass, maybe if mass-loss rates are varied (artificially) similar results can be obtained.

For the final death, the explosion mechanism and magnitude, knowing the size of the (helium) core is the key ingredient. Heger discussed outcomes in terms of initial masses for Pop III stars where it was also assumed there is no mass loss by stellar winds, taking the key uncertainty out of the equation, and hence allows one to draw a clearer picture – though obviously not realistic. However, it allows one to understand and formulate the possible outcomes.

Supernovae from lower mass massive stars generally produce neutron stars within the 'classical' core collapse SN scenario. For higher masses, again really meaning for higher mass cores at the end of evolution, one eventually has to deal with core structures that no longer allow for an efficient explosion and expect the star to collapse to a black hole. Depending on the rotation of the star, a GRB may result. Again, this may require binary stars or rapidly rotating single stars, but here the focus is on the dependence on star/core mass for *possible* outcomes. A recent review on such possible outcomes can also be found in (Woosley & Heger 2012).

When the mass of the star exceeds some $90 - 100\,M_{\odot}$ ($\sim 42\,M_{\odot}$ He core), pulsational pair instability (Puls-PSN) can occur. These are violent nuclear-powered pulsations during the final burning stages, usually powered by oxygen or silicon burning. They may produce as much energy as a SN, or more. However, it could also be significantly less. Heger discussed this in more detail (see below). After the pulsing is over and the star forms an iron core, the fate should be similar to the more massive stars discussed above. For higher initial masses, above some $140\,M_{\odot}$ ($65\,M_{\odot}$ He core) the first pulse is already powerful enough to disrupt the star and a pair instability (PSN) results; for even higher initial masses, above some $260\,M_{\odot}$ ($133\,M_{\odot}$ He core) it is expected that the star collapses to a black hole. A brief discussion on what could happen in that case may be found in (Fryer *et al.* 2001).

One of the striking problems with the PSN theory is that they produce a unique chemical signature - at least for the non-rotating primordial stars. Simulations of the formation of the first stars suggest that they should have been rather massive (e.g., Abel *et al.* 2002), hence making PSN, and their nucleosynthesis products should have been incorporated into the next generation of stars, but to date, no stars with such nucleosynthesis pattern have been found.

7.1. *Pulsational pair instability (Puls-PSN)*

Figure 7 depicts the dynamics of a Puls-PSN simulation in 1D: after the pulse there is a ring-down phase in which radiative dampening by neutrino losses brings the star back into hydrostatic equilibrium. It is noted that the core of the star is much more extended after the pulse than it was before (the reader may follow convective boundaries as a reference). In other words, the energy deposited by the burning during the pulse leads to an expansion and cooling of the stars - as we know from textbooks, stars have a negative (gravothermal) heat capacity. But in order to get ready for the next pulse, it needs to get hot enough, and the energy needs to be lost again. The more energetic, the cooler the core of the star will be. But now the end stages of massive stars are dominated by neutrino losses, and these are known to depend steeply with temperature. Therefore, the cooler the star, the longer it takes to cool.

In extreme cases of rather powerful explosions, the temperature could drop enough such that the core has to cool on the classical Kelvin-Helmholtz time for radiative losses from the surface, as it has become so cool that neutrinos are no longer efficient. This can increase the recurrence time by several orders of magnitude. In summary, we generally expect that more powerful pulses result in longer delay times, whereas weak pulses result in short delay times for the next pulse. The recurrence times can range from hours to more than 10,000 yr, and the energies from 0.0001 B to several B ($1\text{ B} = 1 \times 10^{51}$ erg).

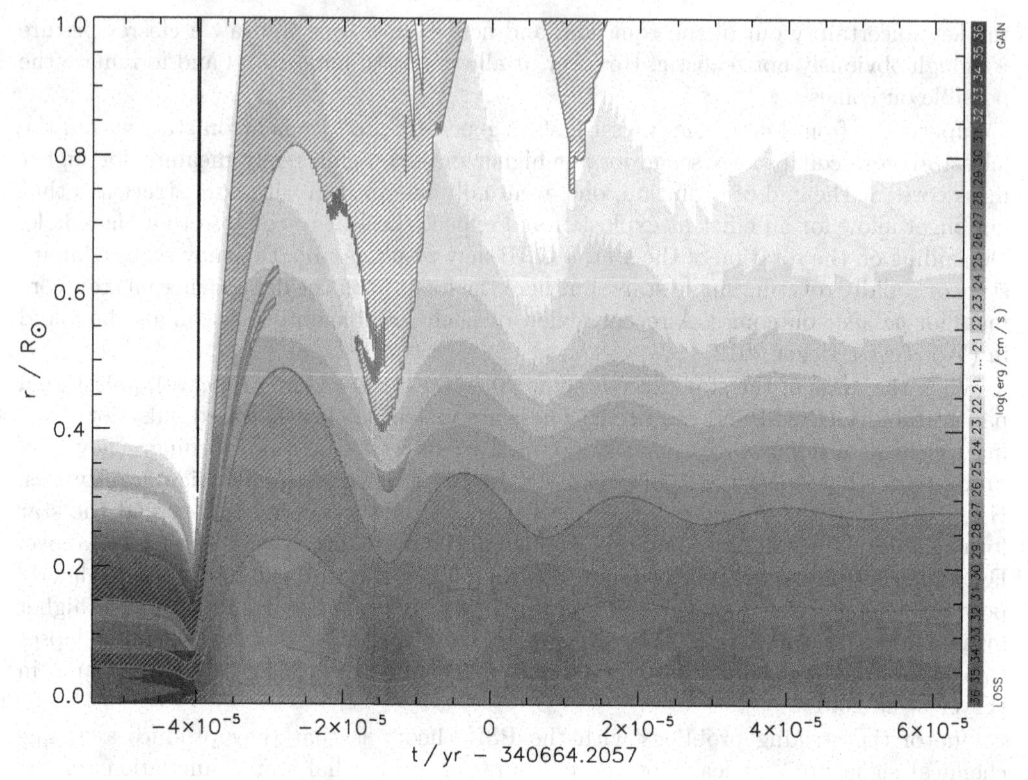

Figure 7. Energy loss (purple), energy generation (blue), and convection (green hatching) during a pulsational pair instability pulse of a 100 $M_{\odot}$ star. The x-axis indicates the time, *t*, in years since the beginning of helium burning, the *y*-axis shows the radius coordinate from the center of the star in solar radii.

Why is all this relevant? Because we wish to understand whether some giant eruption LBVs could be the result of Puls-PSN. In the case of Eta Car we dealt with a fairly weak explosion, as some of the H remained bounded to the star, so we would have expected the next pulse to happen soon after Eta Car's giant eruption. However, this was not observed, or so it appears. Therefore, at this point it seems unlikely that Eta Car's giant eruption was a Puls-PSN event.

7.2. *Stellar Envelope Inflation*

Alternative explanations for LBV outbursts and eruptions have been proposed over the years (see Humphreys & Davidson 1994; Vink 2009). Gräfener *et al.* (2012) proposed the possibility that envelope inflation near the Eddington limit may play a key role in explaining the radius increases during S Dor cycles. The peculiar structure of inflated envelopes, with an almost void region beneath a dense shell could mean that many in reality compact stars are hidden below inflated envelopes, displaying much lower effective temperatures (see also Ishii *et al.* 1999; Petrovic *et al.* 2006).

During the JD, Gräfener discussed the inflation effect for WR stars, whose observed radii are up to an order of magnitude larger than predicted by theory. Based on a new analytical formalism, he described the radial inflation as a function of a dimensionless parameter W, which largely depends on the topology of the Fe-opacity peak, i.e., on material properties. For $W > 1$, an instability limit is found for which the stellar envelope becomes gravitationally unbound, i.e. there no longer exists a static solution. Within this framework one may also be able to explain the S Doradus-type instabilities for LBVs like AG Car (discussed by Groh during the meeting). Moreover, due to the additional effect of sub-photospheric clumping, it may be possible to bring the observed WR radii in agreement with theory (see Sander *et al.* results discussed earlier).

It should be noted that stellar effective temperatures in the upper HR diagram may be strongly affected by the inflation effect. This may have particularly strong effects on the evolved massive LBV and WR stars just prior to their final collapse, as the progenitors of supernovae (SNe) Ibc, SNe II, and long-duration GRBs.

7.3. *3D Simulations of Thermonuclear Supernovae from VMS*

Ke-Jung Chen (with Heger & Woosley) presented results from numerical simulations of the demise of VMS with initial masses between 140 $M_{\odot}$ and 250 $M_{\odot}$ that can die as powerful PSN explosions. Chen *et al.* used CASTRO, a new multidimensional radiation-hydrodynamics code, to study the evolution of PSNe. The 3D simulations start with the collapse phase and follow the explosion until the shock breaks out from the stellar surface. Unlike the iron-core collapse SNe, PSNe are powered by thermonuclear runaway without leaving compact remnants. Much Ni is forged, up to 30 $M_{\odot}$, and its decay energy powers the PSN luminosity for several months. During the explosion, the emergent fluid instabilities cause the mixing of PSN ejecta, and the amount of mixing is related to PSN progenitors. The red supergiant progenitors demonstrate strong mixing, altering the spectrum and light curves.

8. Implications

8.1. *Population synthesis models*

The implications for the existence of VMS may be far-reaching, as VMS may dominate both the kinetic wind energy input and the ionizing radiation in the Universe. Because of higher temperatures at lower metallicities, VMS may be increasingly UV bright. However, their higher luminosities might imply higher mass-loss rates and terminal wind velocities, which would account for an increased kinetic wind energy.

Voss presented his recent population models (Voss *et al.* 2009) which follow both the energy – in the form of kinetic wind energy as well as radiation – and the ejection of radio-active isotopes (such as ^{26}Al) simultaneously. Below 120 $M_{\odot}$, stellar evolution calculations predict a strong increase in the ejected mass of ^{26}Al with stellar mass. The ejection of ^{26}Al from stars above this limit has not been studied in detail, but as the mass-loss increases dramatically towards higher masses, it is reasonable to expect the ^{26}Al to do the same. The short evolutionary timescale of VMSs mean that all the ^{26}Al is ejected 2-4 Myr after the star-forming episode, and if present, the VMSs will dominate the signal for 2-5 Myr populations, but due to the decay their signal will be negligible for older populations. Comparing the ^{26}Al signal from 2-5 Myr massive open clusters to their 5-10 Myr counterparts is therefore a promising way to probe the evolution of VMSs (Voss *et al.* 2010; 2012).

8.2. *Wolf-Rayet Stars in the Extraordinary Star Cluster NGC 3125-A1*

The powerful radiative and mechanical feedback from very massive stars ($\geqslant 100\,\mathrm{M}_{\odot}$) shape the evolution of star-forming galaxies and their environments. Nearby galaxies

($\lesssim$ 10 Mpc) provide excellent laboratories for studying populations of such stars in sufficient detail in a variety of astrophysical environments.

Aida Wofford (with Leitherer and Chandar) studied the massive star populations of clusters A1, A2, B1, and B2 in blue compact dwarf galaxy NGC 3125, which is located 11.5 Mpc away and has an LMC-like metallicity. It is unclear from past studies if cluster A1 hosts an extreme population of WR stars. In addition, the WR star populations of the other clusters are not well characterized. Wofford *et al.* obtained HST/STIS 1200-9000 Å spectra of these four clusters, and higher resolution HST/COS 1200-1450 Å spectra of cluster A1, on which Wofford focused. The STIS spectrum of this cluster shows that the equivalent width of He II λ1640 is three times the mean of local starburst galaxies (Chandar *et al.* 2004) and three times the value of the strongest Lyman Break Galaxy (Erb *et al.* 2010). This suggests that A1 must have a large fraction of WR stars relative to the number of O stars. Either A1 has a top heavy IMF or it contains a few massive stars with very strong winds. The COS spectrum of A1 shows the strongest O V + Fe V absorption feature at 1371 Å from a starburst in the local universe. The O V line originates in the most massive stars and is sensitive to clumping in the stellar wind. The analysis of the O V + Fe V using CMFGEN stellar atmosphere models is underway.

8.3. *Nebular He II 4686 emission: an indirect tracer of massive stars at low metallicities*

Shirazi (with Brinchmann) presented a carefully selected sample of 189 star-forming galaxies with strong nebular He II 4686 emissions in Sloan Digital Sky Survey Data (SDSS) Release 7. They used this sample to investigate the origin of this high ionization line in star-forming galaxies where the ionizing continuum almost certainly arises from massive stars. The current stellar population models can predict He II 4686 emission only for instantaneous bursts of 20% solar metallicity or higher, and only for ages of 4-5 Myr, the period when the extreme-ultraviolet continuum is dominated by emission from WR stars.

Shirazi & Brinchmann find however that 83 of the star-forming galaxies (70% at oxygen-abundance lower than 8.2) of their sample do not have WR features in their spectra despite showing strong nebular He II 4686 emission. Nevertheless, at higher metallicities He II is always seen with WR features. Shirazi went on to show that the stacked spectra of the non-WR He II emitters do not show WR features either, which suggests that the non-detection of WR features in these galaxies is *not* due to low signal-to-noise data, i.e. it is probably real.

Shirazi proposed that a possible explanation for the discrepancy between the model predictions and the observed data at very low redshifts could be the result of a spatial offset between the location of the WR stars and the region where the He II emission arises from. Alternatively, as the non-WR He II emitters appear to be preferentially present in younger starbursts, (quasi)-chemically homogeneous stellar evolution could provide a possible explanation, as this may lead to higher stellar temperatures, and perhaps result in an elevated He II emission even for main-sequence O stars.

Shirazi & Brinchmann are currently attempting to disentangle these explanations by analyzing higher signal-to-noise spectra of a sub-sample of these galaxies that were followed up with the WHT.

8.4. *Very Massive Stars in I Zw 18*

I Zw 18 is a blue compact dwarf (BCD) galaxy with the lowest metallicity known (at 1/30 - 1/50 the solar value), and is therefore thought to be the best local galaxy template to galaxies at high redshift. Although I Zw 18 is 15-19 Mpc away, Hubble/STIS imagery resolves stars in the galaxy.

Heap showed that UV color-magnitudes diagrams indicate that the most massive stars in the northwest cluster of I Zw 18 are as massive as $150\,M_{\odot}$. Heap also showed that Hubble/COS far-UV spectra reveal that the mass-loss rates from stars in the NW cluster must be very low, as only the N V 1240 doublet has a P Cygni profile, and it is very weak. The C IV 1549 doublet is resolved with an edge velocity of only ~250 km/s. The emission component of the C IV doublet is quite possibly of nebular origin. Most of the 12 most luminous stars are bluer than ZAMS models, suggesting that the evolution of the most massive stars is affected by rapid rotation. A comparison of the observed CMD and FUV spectra with new evolutionary models including rotation will yield valuable information about chemical enrichment of the ISM, injected energy via ionizing radiation, and types of SN explosions in the NW cluster of I Zw 18.

9. Final words

One of the key science goals for the James Webb Space Telescope (JWST) is going to be the identification of the first galaxies with Pop III stars. These objects may have been very massive (up to $1000\,M_{\odot}$). At this epoch black hole formation may have been more common than at the current time involving solar metallicities. The first couple of stellar generations may also have been responsible for the reionization of the Universe: an important cosmological epoch that is soon to be probed via the 21 cm line with instruments such as LOFAR.

With the E-ELT, *individual* (very) massive stars may be observed out to the Virgo cluster of galaxies in the Hubble flow, at ever larger distances, and for an increasing range of metallicities. Basic insights into mass loss at very low Z involve issues such as the self-enrichment of metals through mixing (rotation, overshooting, magnetic fields, binarity) and mass loss in close proximity to the Eddington limit, which does not necessarily require metals, if the winds are continuum driven. We also need to consider the potential mass-loss-Z-dependence for the angular momentum evolution and the quest for pinpointing the progenitor stars of long-duration GRBs.

The fate of massive stars is important for our understanding of the chemical enrichment of the Universe. Whilst the deaths of stars up to 15 $M_{\odot}$ now seem to be pretty well known; those of VMS up to 300 $M_{\odot}$ are as yet a complete mystery. Their fates may involve PSN, Puls-PSN, or just normal hydrogen-poor SNe Ibc, either with our without an accompanying GRB. Key unknown aspects in this quest involve the strength and geometry of the progenitor stellar wind. Equatorial winds would remove angular momentum during evolution, whilst spherical winds would not. Linear spectropolarimetry should become a particularly powerful tool to study the geometry of winds and disks during the evolution of (very) massive stars towards explosion.

Similar arguments can be provided for the evolution of rotation rates during star formation. T Tauri and Herbig Ae/Be stars are the optically visible PMS up to ~15 $M_{\odot}$, and the formation of such objects is thought to proceed via disks. The latest observations and theoretical developments seem to suggest that ever more massive stars may form in similar fashions, i.e. via disk accretion, but before such far-reaching conclusions can be drawn, large stellar samples over the full IMF are needed. It is particularly important to explore the near IR part, in order to diminish the complications by dust extinction.

GAIA is widely expected to provide key information about the massive star formation history of our own Milky Way. This will involve issues of stellar dynamics, cluster formation, the role of massive star binarity, and the physical links between massive stars and their lower mass (T Tauri) siblings.

Once we understand the upper IMF and the rotational/multiplicity properties of massive stars, we can concentrate on the evolution of the mass loss and rotation properties of VMS. Which objects could make PSNe and which ones might produce GRBs? May these transient phenomena become star-formation tracers at high redshift? If their luminosity functions could be mapped with respect to their redshifts, individual VMS may even allow astronomers to constrain galaxy formation models.

The future of VMS formation and evolution looks bright!

Acknowledgements

JSV would like to the the Royal Astronomical Society (RAS), the IAU, the UK Science and Technologies Facility Council (STFC) as well as the Northern Ireland Department of Culture Arts and Leisure (DCAL) for financial support.

References

Aarseth, S. J. 2003, *Gravitational N-Body Simulations*, by Sverre J. Aarseth, pp. 430, ISBN 0521432723, Cambridge University Press
Abbott, D. C. & Lucy, L. B. 1985, *ApJ*, 288, 679
Abel, T., Bryan, G. L., & Norman, M. L. 2002, *Science*, 295, 93
Andersen, M., Zinnecker, H., Moneti, A., *et al.* 2009, *ApJ*, 707, 1347
Barniske, A., Oskinova, L. M., & Hamann, W.-R. 2008, *A&A*, 486, 971
Banerjee, S., Kroupa, P., & Oh, S., 2012a, *MNRAS*, 426, 1416
Banerjee, S., Kroupa, P., & Oh, S., 2012b, *ApJ*, 746, 15
Bate, M. R. 2012, *MNRAS*, 419, 3115
Baumgardt, H. & Klessen, R. S. 2011, *MNRAS*, 413, 1810
Belkus, H., Van Bever, J., & Vanbeveren, D. 2007, *ApJ*, 659, 1576
Bestenlehner, J. M., Vink, J. S., Gräfener, G., *et al.* 2011, *A&A*, 530, L14
Beuther, H., Sridharan, T. K., & Saito, M. 2005, *ApJ*, 634, L185
Blandford, R. D. & Payne, D. G. 1982, *MNRAS*, 199, 883
Bonanos, A. Z., Stanek, K. Z., Udalski, A., *et al.* 2004, *ApJ*, 611, L33
Bonanos, A. Z., Stanek, K. Z., Kudritzki, R. P., *et al.* 2006, *ApJ* , 652, 313
Bonnell, I. A., Bate, M. R., & Zinnecker, H. 1998, *MNRAS*, 298, 93
Bontemps, S., Motte, F., Csengeri, T., & Schneider, N. 2010, *A&A*, 524, A18+
Bresolin, F., Urbaneja, M. A., Gieren, W., Pietrzynski, G., & Kudritzki, R.-P. 2007, *ApJ* 671, 2028
Bromm, V., Coppi, P. S., & Larson, R. B. 1999, *ApJ*, 527, L5
Brott, I., de Mink, S. E., Cantiello, M., *et al.* 2011, *A&A*, 530, A115+
Brown, A. 2001, Astronomische Nachrichten, 322, 43
Campbell, M. A., Evans, C. J., Mackey, A. D., *et al.* 2010, *MNRAS*, 405, 421
Cassinelli, J. P., Mathis, J. S., & Savage, B. D. 1981, *Science*, 212, 1497
Castor, J. I., Abbott, D. C., & Klein, R. I. 1975, *ApJ*, 195, 157
Castro, N., Urbaneja, M. A., Herrero, A., *et al.* 2012, *A&A*, 542, A79
Chandar, R., Leitherer, C., & Tremonti, C. A. 2004, *ApJ*, 604, 153
Chatzopoulos, E. & Wheeler, J. C. 2012, *ApJ*, 760, 154
Chené, A.-N., Moffat, A. F. J., & Crowther, P. A., 2008, cihw.conf, 163
Clarkson, W. I., Ghez, A. M., Morris, M. R., *et al.* 2012, *ApJ*, 751, 132
Cooke, J., Sullivan, M., Gal-Yam, A., *et al.* 2012, *Nature*, 491, 228
Cucchiara, A., Cenko, S. B., Bloom, J. S., *et al.* 2011, *ApJ*, 743, 154
Commerçon, B., Teyssier, R., Audit, E., Hennebelle, P., & Chabrier, G. 2011, *A&A*, 529, A35+
Crowther, P. A., Schnurr, O., Hirschi, R., *et al.* 2010, *MNRAS*, 408, 731
Cunningham, A. J., Klein, R. I., Krumholz, M. R., & McKee, C. F. 2011, *ApJ*, 740, 107

Eldridge, J. J., Izzard, R. G., & Tout, C. A. 2008, *MNRAS*, 384, 1109

Ellerbroek, L. E., Kaper, L., Bik, A., *et al.* 2011, *ApJ*, 732, L9

Erb, D. K., Pettini, M., Shapley, A. E., *et al.* 2010, *ApJ*, 719, 1168

Evans, C. J., Taylor, W. D., Hénault-Brunet, V., *et al.* 2011, *A&A*, 530, A108

Figer, D. F. 2005, *Nature*, 434, 192

Fryer, C. L., Woosley, S. E., & Heger, A. 2001, *ApJ*, 550, 372

Fujii, M. S. & Portegies Zwart, S. 2012, arXiv:1210.3732

Fullerton, A. W., Massa, D. L., & Prinja, R. K. 2006, *ApJ*, 637, 1025

Gal-Yam, A., Leonard, D. C., Fox, D. B., *et al.* 2007, *ApJ*, 656, 372

Gal-Yam, A., Mazzali, P., Ofek, E. O., *et al.* 2009, *Nature*, 462, 624

Garcia, M., Herrero, A., Vicente, B., Castro, N., Corral, L. J., Rosenberg, A., & Monelli, M. 2009, *A&A* 502, 1015

Georgy, C., Ekström, S., Meynet, G., *et al.* 2012, *A&A*, 542, A29

Glatzel, W., Kiriakidis, M., Chernigovskij, S., & Fricke, K. J. 1999, *MNRAS*, 303, 116

Glebbeek, E., Gaburov, E., de Mink, S. E., Pols, O. R., & Portegies Zwart, S. F. 2009, *A&A*, 497, 255

Gräfener, G. & Hamann, W.-R. 2005, *A&A*, 432, 633

Gräfener, G. & Hamann, W.-R. 2008, *A&A*, 482, 945

Gräfener, G., Vink, J. S., de Koter, A., & Langer, N. 2011, *A&A*, 535, A56

Gräfener, G., Owocki, S. P., & Vink, J. S. 2012, *A&A*, 538, A40

Groh, J. H. & Vink, J. S. 2011, *A&A*, 531, L10

Groh, J. H., Hillier, D. J., Damineli, A., *et al.* 2009, *ApJ*, 698, 1698

Gürkan, M. A., Freitag, M., & Rasio, F. A. 2004, *ApJ*, 604, 632

Hamann, W.-R., Gräfener, G., & Liermann, A. 2006, *A&A*, 457, 1015

Heger, A. 2012, in Astrophysics and Space Science Library, Vol. 384, Astrophysics and Space Science Library, ed. K. Davidson & R. M. Humphreys, 299

Hennebelle, P., Commerçon, B., Joos, M., *et al.* 2011, *A&A*, 528, A72+

Herrero, A., Garcia, M., Uytterhoeven, K., *et al.* 2011, *IAU Symposium*, 272, 292

Herrero, A., Garcia, M., Puls, J., Uytterhoeven, K., Najarro, F., Lennon, D. J., & Rivero-Gonzlez, J. G. 2012, *A&A* 543, A85

Hillier, D. J. & Miller, D. L. 1998, *ApJ*, 496, 407

Hoffmeister, V. H., Chini, R., Scheyda, C. M., *et al.* 2008, *ApJ*, 686, 310

Hosokawa, T., Yorke, H. W., & Omukai, K. 2010, *ApJ*, 721, 478

Humphreys, R. M. & Davidson, K. 1994, *PASP* 106, 1025

Ishii, M., Ueno, M., & Kato, M. 1999, *PASJ* 51, 417

Ivanova, N., Justham, S., Chen, X., *et al.* 2012, arXiv:1209.4302

Jacquet, E. & Krumholz, M. R. 2011, *ApJ*, 730, 116

Jijina, J. & Adams, F. C. 1996, *ApJ*, 462, 874

Koen, C. 2006, *MNRAS*, 365, 590

Kotak, R. & Vink, J. S. 2006, *A&A*, 460, L5

Kratter, K. M. & Matzner, C. D. 2006, *MNRAS*, 373, 1563

Kratter, K. M., Matzner, C. D., & Krumholz, M. R. 2008, *ApJ*, 681, 375

Kratter, K. M., Matzner, C. D., Krumholz, M. R., & Klein, R. I. 2010, *ApJ*, 708, 1585

Krumholz, M. R. 2006, *ApJ*, 641, L45

Krumholz, M. R., Klein, R. I., & McKee, C. F. 2007, *ApJ*, 656, 959

—. 2012, *ApJ*, 754, 71

Krumholz, M. R., Klein, R. I., McKee, C. F., Offner, S. S. R., & Cunningham, A. J. 2009, *Science*, 323, 754

Krumholz, M. R. & McKee, C. F. 2008, *Nature*, 451, 1082

Krumholz, M. R., McKee, C. F., & Klein, R. I. 2005, *ApJ*, 618, L33

Kuiper, R., Klahr, H., Beuther, H., & Henning, T. 2010, *ApJ*, 722, 1556

—. 2011, *ApJ*, 732, 20

Lamers, H. J. G. L. M. & Cassinelli, J. P. 1999, Introduction to Stellar Winds, by Henny J. G. L. M. Lamers and Joseph P. Cassinelli, pp. 452. ISBN 0521593980. Cambridge, UK: Cambridge University Press, June 1999.,

Langer, N., Norman, C. A., de Koter, A., *et al.* 2007, *A&A*, 475, L19

Lucy, L. B. & White, R. L. 1980, *ApJ*, 241, 300

Lucy, L. B. & Abbott, D. C. 1993, *ApJ*, 405, 738

Mauerhan, J. C., Smith, N., Filippenko, A., *et al.* 2012, arXiv:1209.6320

Massey, P., 2011, ASPC 440, 29

Massey, P., Zangari, A. M., Morrell, N. I., Puls, J., DeGioia-Eastwood, K., Bresolin, F., & Kudritzki, R.-P. 2009, *ApJ* 692, 618

McKee, C. F. & Tan, J. C. 2003, *ApJ*, 585, 850

de Mink, S. E., Langer, N., & Izzard, R. G. 2011, *Bulletin de la Societe Royale des Sciences de Liege*, 80, 543

Moeckel, N. & Clarke, C. J. 2011, *MNRAS*, 410, 2799

Mokiem, M. R., de Koter, A., Vink, J. S., Puls, J., Evans, C. J., Smartt, S. J., Crowther, P. A., Herrero, A., Langer, N., Lennon, D. J., Najarro, F., & Villamariz, M. R. 2007, *A&A* 473, 603

Mokiem, M. R., de Koter, A., Evans, C. J., *et al.* 2007, *A&A*, 465, 1003

Muijres, L. E., Vink, J. S., de Koter, A., Müller, P. E., & Langer, N. 2012, *A&A*, 537, A37

Müller, P. E. & Vink, J. S. 2008, *A&A*, 492, 493

Myers, A. T., Cunningham, A. J., Klein, R. I., Krumholz, M. R., & McKee, C. F. 2012, *ApJ*, in preparation

Nakano, T., Hasegawa, T., & Norman, C. 1995, *ApJ*, 450, 183

Nazé, Y. 2009, *A&A*, 506, 1055

Nomoto, K. 2012, *IAU Symposium*, 279, 1

Nugis, T. & Lamers, H. J. G. L. M. 2002, *A&A*, 389, 162

Ochsendorf, B. B., Ellerbroek, L. E., Chini, R., *et al.* 2011, *A&A*, 536, L1

Oey, M. S. & Clarke, C. J. 2005, *ApJL*, 620, L43

Oskinova, L. M. 2005, *MNRAS*, 361, 679

Oskinova, L. M., Hamann, W.-R., & Feldmeier, A. 2007, *A&A*, 476, 1331

Owocki, S. P. & Gayley, K. G. 1997, in *Astronomical Society of the Pacific Conference Series*, Vol. 120, Luminous Blue Variables: Massive Stars in Transition, ed. A. Nota & H. Lamers, 121

Owocki, S. P., Castor, J. I., & Rybicki, G. B. 1988, *ApJ*, 335, 914

Owocki, S. P., Gayley, K. G., & Shaviv, N. J. 2004, *ApJ*, 616, 525

Pauldrach, A. W. A., Vanbeveren, D., & Hoffmann, T. L. 2012, *A&A*, 538, A75

Petrovic, J., Pols, O., & Langer, N. 2006, *A&A*, 450, 219

Pittard, J. M. & Stevens, I. R. 2002, *A&A*, 388, L20

Portegies Zwart, S. F., Makino, J., McMillan, S. L. W., & Hut, P. 1999, *A&A*, 348, 117

Puls, J., Urbaneja, M. A., Venero, R., *et al.* 2005, *A&A*, 435, 669

Puls, J., Vink, J. S., & Najarro, F. 2008, *A&ARv* 16, 209

Quimby, R. M., Kulkarni, S. R., Kasliwal, M. M., *et al.* 2011, *Nature*, 474, 487

Rauw, G., De Becker, M., Nazé, Y., *et al.* 2004, *A&A*, 420, L9

Rosen, A. L., Krumholz, M. R., & Ramirez-Ruiz, E. 2012, *ApJ*, 748, 97

Rosolowsky, E. & Simon, J. D. 2008, *ApJ*, 675, 1213

Sana, H., de Koter, A., de Mink, S. E., *et al.* 2012, arXiv:1209.4638

Sander, A., Hamann, W.-R., & Todt, H. 2012, *A&A*, 540, A144

Schaller, G., Schaerer, D., Meynet, G., & Maeder, A. 1992, *A&AS*, 96, 269

Schnurr, O., Moffat, A. F. J., St-Louis, N., Morrell, N. I., & Guerrero, M. A. 2008, *MNRAS*, 389, 806

Schnurr, O., Moffat, A. F. J., Villar-Sbaffi, A., St-Louis, N., & Morrell, N. I. 2009, *MNRAS*, 395, 823

Shaviv, N. J. 2000, *ApJ*, 532, L137

Shaviv, N. J. 2001a, *ApJ*, 549, 1093

Shaviv, N. J. 2001b, *MNRAS*, 326, 126

Smartt, S. J., Eldridge, J. J., Crockett, R. M., & Maund, J. R. 2009, *MNRAS*, 395, 1409
Springmann, U. 1997, PhD thesis, Thesis, Ludwig-Maximilians-Universität München, (1997)
Sundqvist, J. O., Puls, J., & Feldmeier, A. 2010, *A&A*, 510, A11
Tramper, F., Sana, H., de Koter, A., & Kaper, L. 2011, *ApJ* Letters 741, L8
Trundle, C., Dufton, P. L., Hunter, I., Evans, C. J., Lennon, D. J., Smartt, S. J., & Ryans, R. S. I. 2007, *A&A* 471, 625
Vanbeveren, D., De Donder, E., van Bever, J., van Rensbergen, W., & De Loore, C. 1998, New Astronomy 3, 443
Vink, J. S. 2009, arXiv:0905.3338
Vink, J. S. & Gräfener, G. 2012, *ApJ*, 751, L34
Vink, J. S., de Koter, A., & Lamers, H. J. G. L. M. 2000, *A&A*, 362, 295
Vink, J. S., de Koter, A., & Lamers, H. J. G. L. M. 2001, *A&A*, 369, 574
Vink, J. S., Muijres, L. E., Anthonisse, B., *et al.* 2011, *A&A*, 531, A132
Voss, R., Diehl, R., Hartmann, D. H., *et al.* 2009, *A&A*, 504, 531
Voss, R., Diehl, R., Vink, J. S., & Hartmann, D. H. 2010, *A&A*, 520, A51
Voss, R., Martin, P., Diehl, R., *et al.* 2012, *A&A*, 539, A66
Weidner, C. & Kroupa, P. 2004, *MNRAS*, 348, 187
Whitney, B. A., Wood, K., Bjorkman, J. E., & Wolff, M. J. 2003, *ApJ*, 591, 1049
Wolfire, M. G. & Cassinelli, J. P. 1987, *ApJ*, 319, 850
Woosley, S. E. & Heger, A. 2012, *ApJ*, 752, 32
Woosley, S. E., Heger, A., & Weaver, T. A. 2002, *Reviews of Modern Physics*, 74, 1015
Woosley, S. E., Blinnikov, S., & Heger, A. 2007, *Nature*, 450, 390
Wright, A. E. & Barlow, M. J. 1975, *MNRAS*, 170, 41
Yoon, S.-C., Dierks, A., & Langer, N. 2012, *A&A*, 542, A113
Yoon, S.-C., Langer, N., & Norman, C. 2006, *A&A*, 460, 199
Yoon, S.-C., Gräfener, G., Vink, J. S., Kozyreva, A., & Izzard, R. G. 2012, *A&A*, 544, L11
Yorke, H. W. & Kruegel, E. 1977, *A&A*, 54, 183
Yorke, H. W. & Sonnhalter, C. 2002, *ApJ*, 569, 846
Yoshida, T. & Umeda, H. 2011, *MNRAS*, 412, L78
Yungelson, L. R., van den Heuvel, E. P. J., Vink, J. S., Portegies Zwart, S. F., & de Koter, A. 2008, *A&A*, 477, 223
Zhang, Y. & Tan, J. C. 2011, *ApJ*, 733, 55

Highlights of Astronomy, Volume 16
XXVIIIth IAU General Assembly, August 2012
T. Montmerle, ed.

doi:10.1017/S1743921314004669

JD3 – 3D Views of the Cycling Sun in Stellar Context: Overview

Lidia van Driel-Gesztelyi[1,2,3] and Carolus J. Schrijver[4]

[1]University College London, Mullard Space Science Laboratory, Dorking, UK
[2]Observatoire de Paris, LESIA, CNRS, UPMC Univ. Paris 06, Univ. Paris-Diderot, Meudon, France
email: Lidia.vanDriel@obspm.fr
[3]Konkoly Observatory, Hungarian Academy of Sciences, Budapest, Hungary

[4]Lockheed Martin Advanced Technology Center, Palo Alto, California, USA
email: schryver@lmsal.com

Abstract. We summarise the motivations and main results of the joint discussion "3D Views of the Cycling Sun in Stellar Context", and give credit to contributed talks and poster presentations, as due to the limited number of pages, this proceedings could only include contributions from the keynote speakers.

Keywords. Sun: rotation, Sun: interior, Sun: activity, stars: rotation, stars: interiors, stars: activity

---

1. Motivation

This joint discussion meeting was marking the importance of a historical achievement in astronomy: the first instantaneous 3D view of a star, our Sun. Hence the words "3D views" in the title. This achievement has importance for both the solar and stellar activity communities. Another motivation for bringing together the solar and stellar activity communities was that presently we appear to experience the start of a lower-activity cycle of solar activity following several strong cycles as well as the longest and deepest solar cycle minimum of the space age, which accounts for the motivation to look at the "cycling Sun". The latter phenomenon is one reason that brings solar physicists to look at other active stars and their cycles for guidance. Stellar astronomers, in turn, make use of the detailed observations of the Sun when interpreting stellar data. Moreover, both communities appreciate the breathtaking details brought by solar and inner-heliospheric observations provided by, e.g., Hinode, SDO, SoHO, and the STEREO spacecraft. Space-borne stellar observations by the Kepler mission provide unprecedented details of temporal variability in stars, fascinating both communities.

At the time of the start of the XXVIIIth IAU General Assembly, and in fact the very days JD3 took place in Beijing between 20-22 August 2012, we were in the unique position to have a full 3D view of the Sun provided by the twin STEREO spacecraft, which have been drifting away from the Earth since their launch in 2006, one of the spacecraft being 115° behind, while its twin was 123° ahead of the Earth. Their observations combined with those by other spacecraft around the Earth (SOHO, Hinode, SDO) as well as ground-based observatories, gave us instantaneous 3D view of the Sun. This is a historical achievement in stellar physics. As the STEREO spacecraft move about 22° around the Sun in opposite directions relative to the Earth, this unique 3D view will be maintained for as long as nearly six years, except for some periods in 2015 when one or the other will be out of contact because its apparent angular distance from the Sun will be too small for efficient communication, or will be physically occulted by the Sun. The temporal overlap of these out-of-contact periods is small, only 11 days in March 2015.

Solar activity is on the increase following the unexpectedly deep and long solar minimum in 2009. This long minimum period was monitored in great detail both on the Sun and the heliosphere, revealing a wide range of activity phenomena, which were observed with no disturbance from or overlap with other events. SDO has been showing us breathtaking details of the entire active Sun as viewed from Earth with temporal cadence of 12 seconds, monitoring solar activity, which has nearly reached maximum of cycle 24, with unprecedented observing details of various activity phenomena ranging from microflares and penumbral jets through quiet-sun mini-eruptions to the largest filament eruptions and coronal mass ejections. Although longer-term forecast of solar activity does not seem to indicate that a Maunder-type Grand Minimum would start in the next (few) hundred years, looking at the number of solar-type stars in Maunder-type deep activity minima is thought-provoking and crucial until we have a validated predictive model of the dynamo action of the Sun and its peers.

New insights brought by the combinations of these novel observational views are continuously posing new questions, inspiring and advance theoretical analysis and modelling, improving our understanding of the physics underlying magnetic activity phenomena. Therefore state-of-the-art observational and modelling results were presented side-by-side at the meeting.

2. Outline of the meeting

The programme was divided into seven sessions, each led by two keynote talks on the topic, presenting solar, then stellar state-of-the-art overviews side by side. The keynote talks combined insights gained from observations and modelling. Only the last session had a single keynote presentation, addressing stellar results by focusing on Kepler observations. All 13 keynote presentations are included in this proceedings, following this Overview.

2.1. *Evolution of solar and stellar magnetic fields*

Chair: van Driel-Gesztelyi (France, Hungary, UK). After a brief introduction by the Chair, the keynote presentations by Todd Hoeksema (USA) and Manuel Güdel (Austria) gave overview of the evolution of solar and then stellar magnetic fields (see Hoeksema, 2013 and Güdel, 2013). Three contributed talks followed. First, Jingxiu Wang (China) presented a work on "Vector magnetic field characteristics of the super-active regions with major flare activity", co-authored by Anqin Chen. This observational talk was followed by a theoretical one on the dynamo, presented by Arnab Choudhuri (India) on "Theoretical modelling of grand minima of solar activity using the flux transport dynamo model" (co-author: Bidya Karak), which invited many questions. The session ended with a lively presentation by Peng-Fei Chen (China) on "3D view of 'EIT waves' in the solar corona".

2.2. *Driving magnetic activity: differential rotation from seismology and patterns in surface activity*

Chair: John Leibacher (USA). The two keynote talks were given by Mark Miesch (USA) and Klaus Strassmeier (Germany) (see Miesch, 2013, and Strassmeier, 2013), who presented the role of convection and large-scale mean flows in the Sun and active stars and their implications for cyclic solar and stellar activity. Two solicited contributed talks were given by Krisztián Vida (Hungary) on "Activity in low-mass stars: spatial correlation and evolution of the observed features" and by Shaolan Bi (China) on " Revised solar models with rotation and magnetic fields". The latter talk was brilliantly presented by the PhD student of the author, Tanda Li.

2.3. *Magnetic activity from microflares to megaflares*

Chair: Mark Miesch (USA) The keynotes by Lyndsay Fletcher (UK) and Adam Kowalski (USA) introduced this exciting subject (see Fletcher, 2013, Kowalski & Hawley, 2013). Then Hugh Hudson (UK, USA) presented a thought-provoking solicited talk on "The X-ray limb of the Sun", calling into question our understanding of the solar limb structure. This was followed by a theoretical work presented by Jun Lin (China) on "Solar Flare, CME, and the Reconnecting Current Sheet in Between" (co-authors: Zhixing Mei, Chengcai Shen). The session concluded with two talks on flare spectra by Janusz Sylvester (Poland) "Solar Flare Sulphur Abundance" (co-authors: Barbara Sylwester, Kenneth J.H. Phillips) and Barbara Sylvester "Flare Differential Emission Measure from RESIK and RHESSI Spectra" (co-authors: Janusz Sylwester, Anna Kepa, Tomasz Mrozek, Kenneth J.H. Phillips).

2.4. *3D views of the Sun and active stars surfaces and interiors*

Chair: Takashi Sakurai (Japan). From the keynotes in this session by Allan Sacha Brun (France) and Heidi Korhonen (Denmark) the audience learned about spectacular results produced by recent solar dynamo modelling (Brun & Strugarek, 2013) as did 3D reconstructions of active stars (Korhonen, 2013). Two contributed talks were presented by Debi Prasad Choudhary (USA) on "Chromospheric Properties of Sun as a Star" and Junwei Zhao (USA) on "Helioseismic Studies of Solar Far-Side Active Regions and Emerging Active Regions" (co-authors: Stathis Ilonidis, Alexander Kosovichev). The session ended with oral presentations of posters (see full list below) by Hugh Hudson (P1), Ting Li (P4), Nariman Ismailov (P9), Valery Krivodubskij (P8), Jie Jiang (P7), and van Driel-Gesztelyi (P2, P3).

2.5. *3D views of the Sun and active stars – atmospheres and astrospheres*

Chair: Peng-Fei Chen (China). The keynotes by Alexis Rouillard (France) and Moira Jardine (UK) presented STEREO results on coronal mass ejections and their propagation in the heliosphere (Rouillard, 2013) and magnetic surface and even coronal structures from active star, e.g. filaments (Jardine, 2013). These were followed by a solicited talk by Etienne Pariat (France) on "Magnetic topology of solar activity events" (co-authors: Guillaume Aulanier, Antonia Savcheva, Sophie Masson, Pascal Démoulin) and another solicited talk by Lucie M. Green (UK) on "Multi-wavelength observations of solar eruptions". Then Dibyendu Nandy (India) spoke about a new initiative on "Solar-Stellar Cycles and their Implication on the Astrospheres" (co-authors: Allan Sacha Brun, Ed Cliver, Sarah E. Gibson, Margit Haberreiter, Andres Munoz-Jaramillo, Steven H. Saar, Adriana Silva-Valio, Ilya Usoskin). The session's final talk was presented by Hui Li (China) on "Magnetic energy evolution and its relation to solar flares".

2.6. *Solar and stellar cycles*

Chair: Hugh Hudson (UK, USA). This important session was expertly introduced by keynote talks by Robert Cameron (Germany) who emphasised that understanding nonlinearities in the solar dynamo is the way forward in cycle forecast, and Emre Işık (Turkey) who gave a taste how multiple stellar cycles can be modelled and understood (see Cameron, 2013, and Işık, 2013). Contributed talks were presented by David Webb (USA) on "Solar Cycle variations of Coronal Mass Ejections", and Aline Vidotto (UK) on “The stellar wind cycles and planetary radio emission of the Tau Boo system” (co-authors: Rim Fares, Moira Jardine, Jean-Francois Donati, Merav Opher, Claire Moutou, Claude Catala, Tamas Gombosi).

2.7. *New results from SDO and Kepler*

Chair: Karel Schrijver (USA). In the final session of JD3 in Jon M. Jenkins's (USA) keynote exciting new results on stellar variability from the Kepler mission were presented (Jenkins *et al.* 2013). SDO results were shown in contributed talks by Marcelo Emilio (Brasil) on “P and R modes in solar limb shape HMI-SDO observations” (co-authors: Jeff Kuhn, Rock Bush, Isabelle Scholl) and Jie Chen (China) on “An analysis of a Transequatorial Loop”, Yingna Su (USA) Observations and magnetic field modelling of a solar polar crown prominence (co-author: Aad van Ballegooien). Then Kiyoto Shibasaki (Japan) gave a talk on “Relation between solar activities at low and high latitudes inferred from microwave observations”, and Yuzong Zhang (China) presented “Revision of solar spicule classification”. Finally, the Chair summarised the main results of the meeting.

3. Conclusions

Bringing the solar and stellar activity communities together in this JD provided us with an opportunity to share our excitement and new insights with colleagues outside of the usual circles. At the JD solar and stellar talks were presented side by side, which kept both the solar and stellar communities together and led to lively discussions. The talks were outstanding, inspirational not only to one, but both communities, as stellar talks received many questions from the solar community and vice versa. For the three days of JD3 the lecture room was filled and buzzing. We conclude that the attempt has worked and it is a worthy path to follow in the future within the broad international context of the IAU.

Scientific Organising Committee

Lidia van Driel-Gesztelyi (France, Hungary, UK; Chair), Carolus J. Schrijver (USA, co-chair), Gianna Cauzzi (Italy), Peng-Fei Chen (China), John Leibacher (USA), Katalin Oláh (Hungary), Rachel Osten (USA).

Poster presentations

P1 Hugh Hudson: First Flare Results from EVE.

P2 L. van Driel-Gesztelyi, D. Baker, T. Trk, E. Pariat, L.M. Green, D.R. Williams, G. Valori, P. Démoulin, S.A. Matthews, E. Pedram, B. Kliem, P.F. Chen: Observations and modeling of a massive filament eruption, CME, and coronal wave.

P3 L. van Driel-Gesztelyi, J. L. Culhane, D. Baker, P. Démoulin, C.H. Mandrini, M.L. DeRosa, A.P. Rouillard, A. Opitz, G. Stenborg, A. Vourlidas, D.H. Brooks: Plasma outflows from active regions: are they sources of the slow solar wind?

P4 Ting Li: Interaction of solar global EUV wave with coronal structures.

P5 Zhi'e Liu: Solar-like oscillation in KIC 10516096,KIC 10644253 and KIC 11771760.

P6t Li Feng, Bernd Inhester, Yong Wei, Marilena Mierla, Weiqun Gan, Tielong Zhang, Mingyuan Wang: Morphological evolution of a 3D CME cloud reconstructed from three viewpoints and its comparison with other four reconstruction methods.

P7 Jie Jiang, Emre Işık, Robert Cameron, Dieter Schmitt, Manfred Schüssler: The Effect of Activity-related Meridional Flow Modulation on the Strength of the Solar Polar Magnetic Field.

P8 Valery Krivodubskij, Alexej Loginov, Oleg Cheremnykh, Nikolaj Salnikov: Hydrodynamic model of spatial and temporal variations of poloidal and toroidal components of 3D Sun flows.
P9 Nariman Ismailov, Peter Shustarev, Fekhrende Alimardanova, Gunel Bahaddinova: Planet formation processes in young solar-type stars.

Acknowledgements

We thank the members of the scientific organising committee, the chairs of the sessions, all the speakers, poster presenters, and participants for making this meeting a success. This Joint Discussion was coordinated by IAU Division II (Sun and Heliosphere) and supported by IAU Division V (Variable Stars), as well as Commissions 10 (Solar Activity), 12 (Solar Radiation and Structure) and 27 (Variable Stars). We thank the IAU for their help and support of this meeting.

References

Brun, A. S. & Strugarek, A. 2013, *Highlights of Astron.* 16, in this issue
Cameron, R. H. 2013, *Highlights of Astron.* 16, in this issue
Fletcher, L. 2013, *Highlights of Astron.* 16, in this issue
Güdel, M. 2013, *Highlights of Astron.* 16, in this issue
Hoeksema, J. T. 2013, *Highlights of Astron.* 16, in this issue
Işık, E. 2013, *Highlights of Astron.* 16, in this issue
Jardine, M. 2013, *Highlights of Astron.* 16, in this issue
Jenkins, J. M., Gilliland, R. L., Meibom, S., Walkowicz, L., Borucki, W. J., & Caldwell, D. A., the Kepler Science Team 2013, *Highlights of Astron.* 16, in this issue
Korhonen, H. 2013, *Highlights of Astron.* 16, in this issue
Kowalski, A. F. & Hawley, S. L. 2013, *Highlights of Astron.* 16, in this issue
Miesch, M. S. 2013, *Highlights of Astron.* 16, in this issue
Rouillard, A. P. 2013, *Highlights of Astron.* 16, in this issue
Strassmeier, K. G. 2013, *Highlights of Astron.* 16, in this issue

Highlights of Astronomy, Volume 16
XXVIIIth IAU General Assembly, August 2012
T. Montmerle, ed.

doi:10.1017/S1743921314004670

The Evolution of the Solar Magnetic Field

J. Todd Hoeksema

W.W. Hansen Experimental Physics Laboratory, Stanford University, 466 Via Ortega, Cypress C13, Stanford, CA 94305, USA
email: JTHoeksema@spd.aas.org

Abstract. The almost stately evolution of the global heliospheric magnetic field pattern during most of the solar cycle belies the intense dynamic interplay of photospheric and coronal flux concentrations on scales both large and small. The statistical characteristics of emerging bipoles and active regions lead to development of systematic magnetic patterns. Diffusion and flows impel features to interact constructively and destructively, and on longer time scales they may help drive the creation of new flux. Peculiar properties of the components in each solar cycle determine the specific details and provide additional clues about their sources. The interactions of complex developing features with the existing global magnetic environment drive impulsive events on all scales. Predominantly new-polarity surges originating in active regions at low latitudes can reach the poles in a year or two. Coronal holes and polar caps composed of short-lived, small-scale magnetic elements can persist for months and years. Advanced models coupled with comprehensive measurements of the visible solar surface, as well as the interior, corona, and heliosphere promise to revolutionize our understanding of the hierarchy we call the solar magnetic field.

Keywords. Sun: magnetic field, Sun: activity cycles

Of course comprehensively reviewing the evolution of the solar magnetic field in such a short report is impossible, so this discussion focuses on a few of the more salient features of large-scale and long-term evolution. The 22-year magnetic cycle drives most variations of the Sun, from active regions to the corona and heliosphere, and from the polar field reversal to the total solar irradiance. See for example discussions of the most recent deep solar minimum framed by Gibson *et al.* (2011) and references therein. The Sun's activity is fairly typical for stars of its age and rotation rate, thus what we learn about the drivers of solar dynamo activity will inform and be informed by what we see on other suns.

Figure 1 shows how the axial and equatorial dipole components of the Sun changed during the last 3.5 solar cycles, from 1976-2012, as measured by the Wilcox Solar Observatory. The axial dipole peaks a couple years before the solar minima in 1976, 1986, 1996, and 2009. The equatorial dipole reflects the large-scale average of active regions and peaks more broadly during solar maximum. There is a strong secular trend in the overall amplitude of the dipole, with the most recent extended minimum having less than half the strength of the previous three. The current Cycle 24 and the extended prior minimum is unprecedented in the space age, but is not unlike Cycle 14.

Flux transport models suggest that the photospheric field pattern emerges from the creation and decay of active regions. Active regions drive the most dramatic expressions of the cycle, from flares and coronal mass ejections to particle acceleration and high

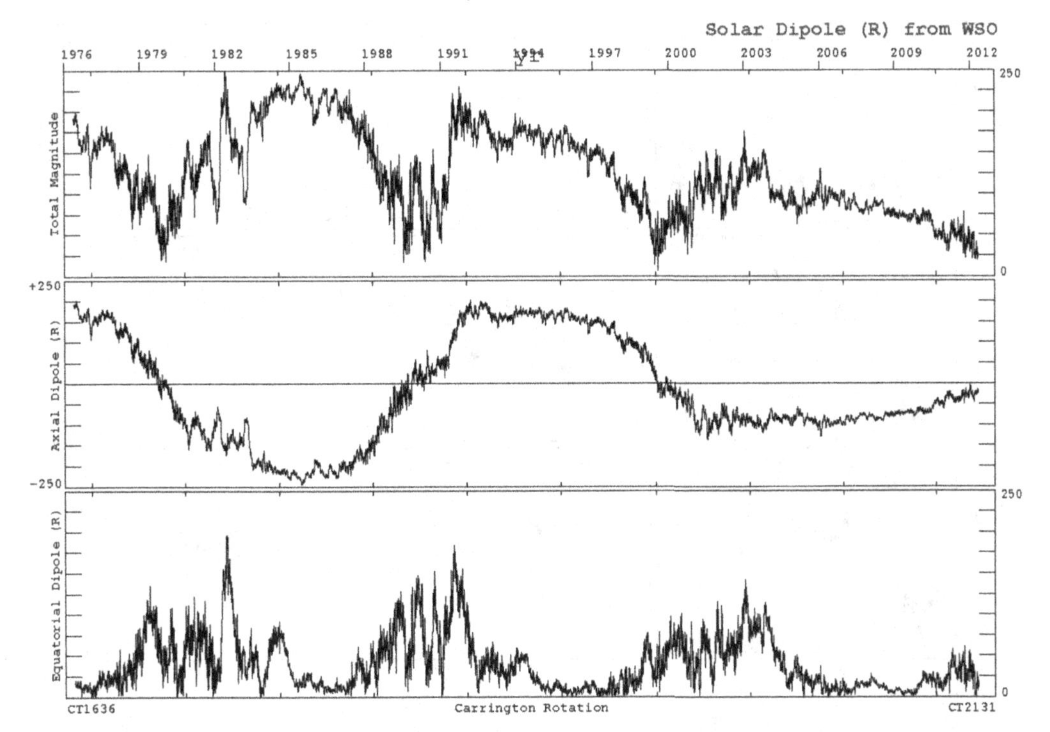

Figure 1. The Sun's dipole field from 1976 - 2012 as observed by the Wilcox Solar Observatory. The top panel shows the total dipole. The center and bottom show the axial and equatorial components. Note the phase of the components relative to the solar cycle and the secular decrease particularly in the axial dipole.

energy radiation. AR 11158 produced the first X-class flare of the cycle at 01:56 UT on February 15, 2011. Figure 2 shows how the region emerged and evolved over 5 days. The complex region demonstrated strong shearing motions and the currents computed from the vector field were strongly peaked at the flare site. The three columns in the figure show the photospheric radial field strength from HMI, the AIA 304A intensity in the overlying atmosphere, and the computed current density. The transverse component of the field increased sharply and permanently at the time of the flare (Sun *et al.* 2012). The current cycle has so far been weak and this may be related to changes not only in the number and complexity of sunspot groups, but also to an apparent decline in the maximum field strength observed in active regions that cause changes in the long-term relationship between sunspot number and F10.7 cm flux (e.g. Livingston *et al.* 2012).

As active regions decay, the trailing flux moves poleward, ultimately reversing the dominant magnetic polarity a little after maximum in each hemisphere. Figure 3 shows the decaying magnetic field pattern of a large southern hemisphere spot on July 12, 2012. The red-yellow negative pattern will eventually merge with flux from other regions to reverse the south polar field (e.g. Wang & Sheeley, 1991; Schrijver & DeRosa, 2003). The northern hemisphere has been active earlier and will likely have reversed at the end of 2012. Zonally averaging over all longitudes shows how the net flux emerges and moves poleward over the cycle. Figure 4 shows the dominant polarity at high latitude reversing in 2001 as a consequence of surges of flux from decaying low-latitude

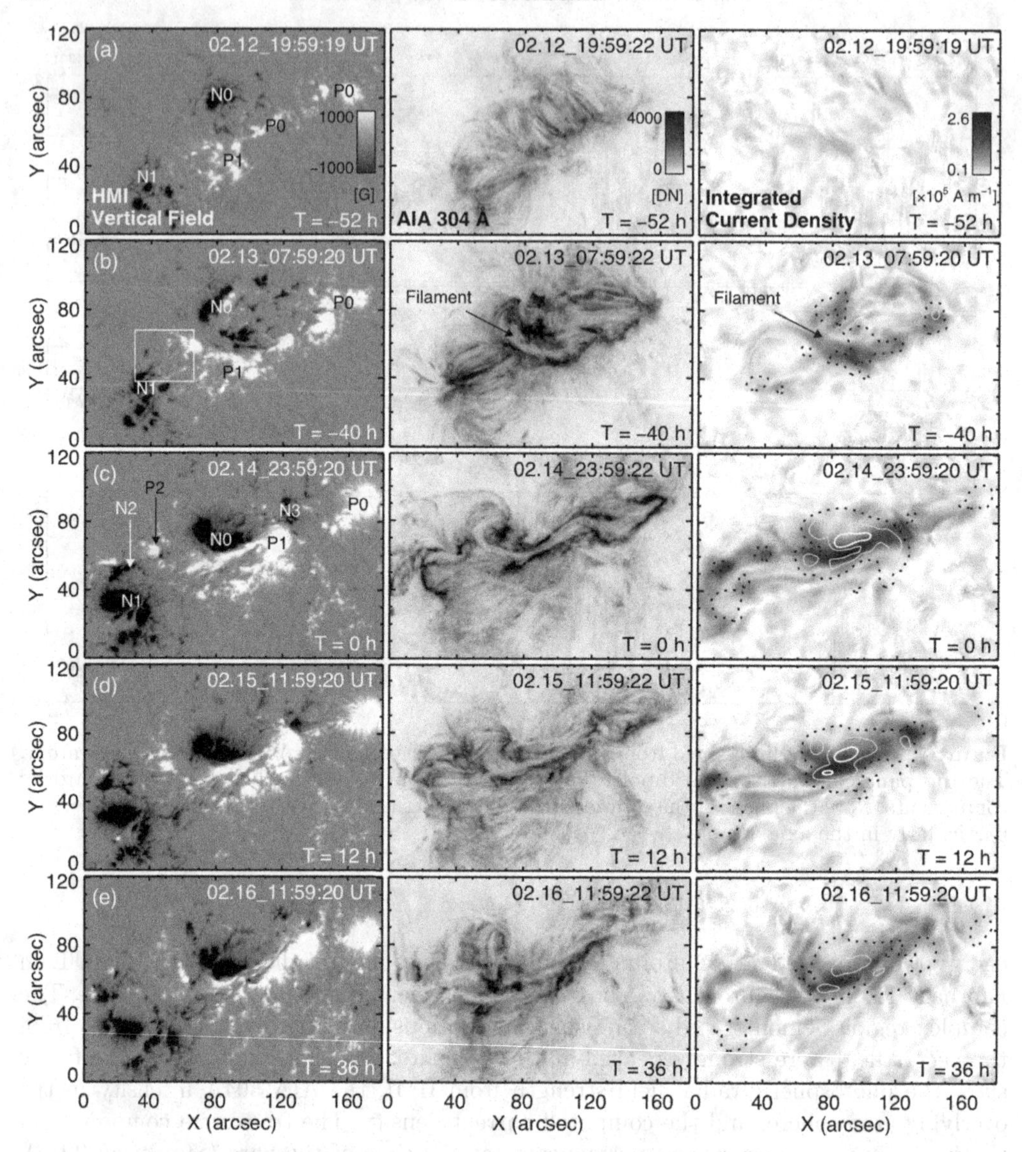

Figure 2. The evolution of the radial component of the magnetic field for AR 11158 from 12-16 February 2011 is shown in the left column. The current density, calculated from the HMI vector field, peaks at about the time of the x-class flare at 01:56 UT on 15 Feb. (right column). The center column shows the simultaneous 304A emission in the overlying atmosphere observed by AIA. (Sun *et al.*, 2012).

active regions carried poleward by meridional flows and diffusion. Cycle 24 has begun slowly and the surges are small. Since the polar field was weak, the reversal has proceeded relatively quickly in any case. The peculiarities of the cycle depend on the characteristics of the emerging active regions produced by the solar dynamo and their interactions with the changing flows. How the magnetic dross in the photosphere influences subsequent cycles is unclear. Improving continuous, long-term synoptic observations of the Sun are critical for understand this example of a stellar dynamo.

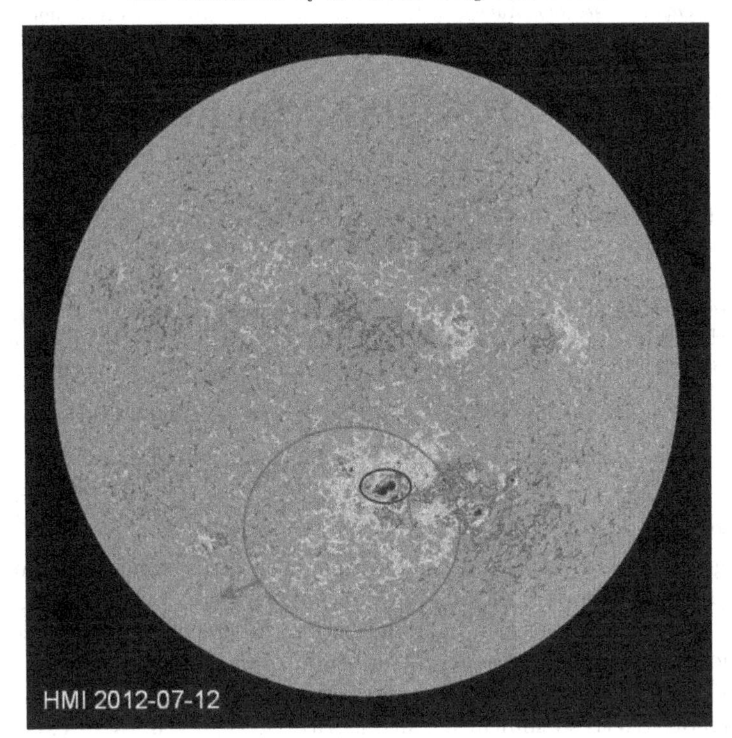

Figure 3. The extensive blue/yellow positive magnetic field region shown in this HMI magnetogram from 12 July 2012 will expand and merge with left-over flux from other decaying active regions and be carried poleward by meridional flows to reverse the southern cap.

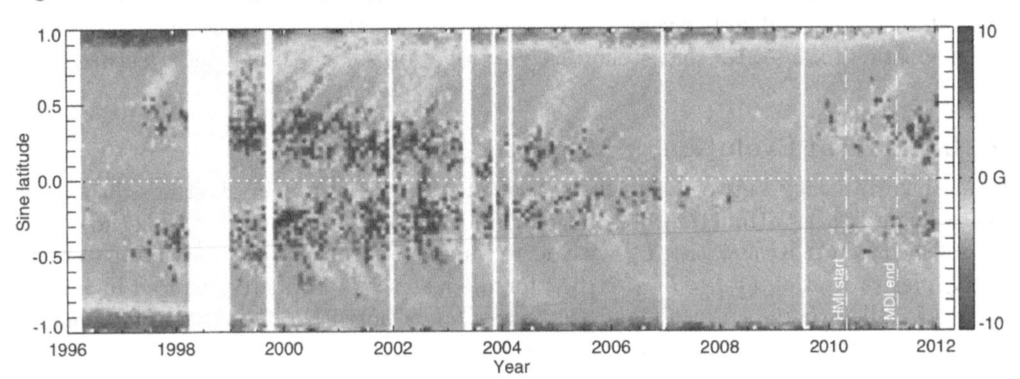

Figure 4. The zonal magnetic field from 1996 - 2012 as observed by MDI and HMI. The net flux in each sine latitude bin is computed for each Carrington Rotation and plotted as a function of time. The positive (blue) north pole reverses in 2001. Surges of following polarity from decaying active regions migrate poleward in 12-18 months to cause the reversal. In the current cycle weaker surges are sufficient to erode the weaker polar field. (Figure courtesy X. Sun)

References

Gibson, S., *et al.* 2011, *Solar Phys.* 274, 5
Livingston, W., Penn, M., & Svalgaard, L. 2012, *ApJL* 757, 8
Schrijver, C. J. & DeRosa, M. 2003, *Solar Phys.* 212, 165
Sun, X., *et al.* 2012, *ApJ* 748, 77
Wang, Y.-M., Sheeley, N., *et al.* 1991, *ApJ* 375, 761

Highlights of Astronomy, Volume 16
XXVIIIth IAU General Assembly, August 2012
T. Montmerle, ed.

doi:10.1017/S1743921314004682

Evolution of Stellar Magnetic Fields

Manuel Güdel
Department of Astrophysics, University of Vienna, Türkenschanzst. 17, 1180 Vienna, Austria
email: manuel.guedel@univie.ac.at

Abstract. Stellar magnetic fields can reliably be characterized by several magnetic activity indicators, such as X-ray or radio luminosity. Physical processes leading to such emission provide important information on dynamic processes in stellar atmospheres and magnetic structuring.

Keywords. Stars: activity, stars: coronae, stars: flare, stars: magnetic fields

1. Introduction

Stellar magnetic activity relies on "surface" magnetic fields expanding into the chromosphere and corona. Although methods have been developed to measure such fields (e.g., Zeeman line splitting resp. broadening, polarization, frequency analysis of coherent radio maser emission, models for gyrosynchrotron radio radiation), they remain challenging, with intricate problems making accurate measurements even of the average magnetic flux density ambiguous. Activity indicators such as Hα, Ca II, UV lines, or extreme ultraviolet, X-ray, and non-thermal radio fluxes are easier to measure for large samples of stars. How representative are they? Pevtsov *et al.* (2003) found a nearly linear correlation between total (unsigned) magnetic flux and X-ray luminosity of various solar features and the full-disk solar and stellar values, suggesting not only that L_X serves as a good proxy for magnetic flux but also that stellar activity relies on solar-like magnetic features.

2. Long-Term Evolution of Magnetic Activity

A similar analogy between activity and magnetic flux is seen in diagrams displaying these quantities as a function of rotation period P or Rossby number. The now well established trends of increasing L_X with decreasing P (approx. $L_X \propto P^{-2}$, Pizzolato *et al.* 2003) plus saturation at $L_X/L_{\rm bol} \approx 10^{-3}$ for the fastest rotators are mirrored in similar trends for the average surface magnetic flux $< Bf >$ notwithstanding complications in Zeeman broadening measurements involved here (Reiners 2012).

This then suggests L_X (or L_{UV}) measurements as convenient proxies for the long-term evolution of magnetic flux. For solar analogs, cluster and field star studies show a systematic decrease of L_X with age, approximately like $\propto t^{-1.5}$, accompanied by a decrease of average coronal temperature from about 10 MK for very young solar analogs to 2 MK for the present Sun (Güdel *et al.* 1997). Similar trends are found for non-thermal radio emission although the decaying trend seems to be steeper, while trends in the ultraviolet are more moderate (Ribas *et al.* 2005, Güdel 2004).

3. Magnetic Activity: Minimum States

How low, then, can magnetic activity go? Schröder *et al.* (2012) found the extremely low solar Ca II S index in 2009 comparable to "Maunder minimum" stars, and further suggested that magnetic heating is insufficient for chromospheres during such times while

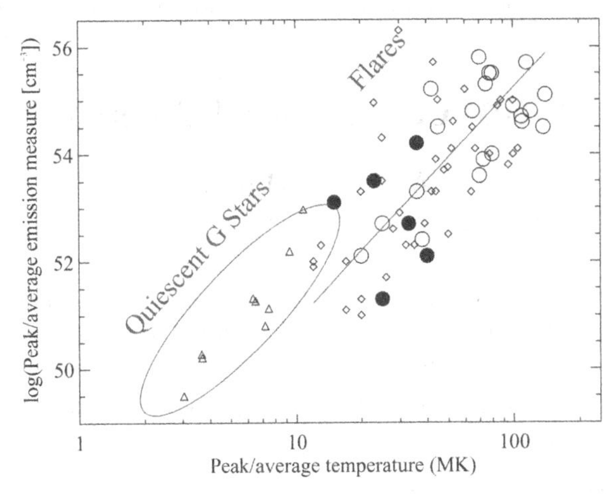

Figure 1. Correlation between peak/average emission measure and peak/average temperature for coronal plasma during flares (circles)/quiescence (triangles in red area; Güdel 2004.)

unproblematic for the weak X-ray emission. The low L_X in that episode (Sylwester *et al.* 2012) is comparable to or lower than the X-ray weakest main-sequence stars (Schmitt 1997). Magnetic activity does not disappear during a cool star's main-sequence lifetime.

4. Magnetic Activity: High Activity and Flaring

The other extreme of magnetic activity found in magnetically active stars is reminiscent of solar/stellar flares: high-temperature plasma, variability, non-thermal radio emission, and in particular a correlation between total luminosity (resp. flare luminosity) and average electron temperature (Fig. 1). This latter coincidence (Güdel 2004) may suggest that a large number of continuously occurring flares are at the origin of the continuous X-ray emission of active stars. This idea gets support from flare statistics (occurrence rate vs. flare energy) indicating that the sum of numerous, weak flares dominates the X-ray radiative output from active coronae (Audard *et al.* 2000). A "micro-flaring" corona would make all concepts of static coronal magnetic structures obsolete and introduce a dynamic magnetic atmosphere, with continuous mass flows, reconnection events, particle acceleration and frequent episodic heating keeping the corona at high temperatures.

Acknowledgements

I thank Ansgar Reiners for helpful discussions on stellar magnetic fields.

References

Audard, M., Güdel, M., Drake, J. J., & Kashyap, V. 2000, *ApJ*, 541, 3967
Güdel, M., Guinan, E. F., & Skinner, S. L. 1997, *ApJ*, 483, 947
Güdel, M. 2004, *A&AR*, 12, 71
Pevtsov, A. A., Fisher, G. H., Acton, L. W., *et al.* 2003, *ApJ*, 598, 1387
Pizzolato, N., Maggio, A., Micela, G., *et al.* 2012, *A&A*, 397, 147
Reiners, A. 2012, *Liv. Rev. Solar Phys.*, 9, 1
Ribas, I., Guinan, E. F., Güdel, M., & Audard, M. 2005, *ApJ*, 622, 680
Schmitt, J. H. M. M. 1997, *A&A*, 318, 215
Schröder, K.-P., Mittag, M., Pérez Martínez, M. I., *et al.* 2012, *A&A*, 540, 130
Sylwester, J., Kowalinski, M., Gburek, S., *et al.* 2012, *A&A*, 751, 111

Highlights of Astronomy, Volume 16
XXVIIIth IAU General Assembly, August 2012
T. Montmerle, ed.

doi:10.1017/S1743921314004694

Solar Convection and Mean Flows

Mark S. Miesch

HAO/NCAR†, Boulder, CO, 80301 email: miesch@ucar.edu

Abstract. We briefly review our current understanding of how the solar differential rotation and meridional circulation are maintained, which has important implications for understanding cyclic magnetic activity in the Sun and stars.

Keywords. Sun: rotation, Sun: interior, Sun: activity, stars: rotation, stars: interiors, stars: activity, convection, MHD

Differential rotation and meridional circulation play an essential role in all current dynamo models of solar and stellar activity cycles so a thorough understanding of how mean flows are established is an essential prerequisite to a comprehensive dynamo model.

Our current understanding of how the solar differential rotation and meridional circulation are established rests heavily on two key concepts: thermal wind balance and gyroscopic pumping. The first is derived from the zonal component of the vorticity equation under the assumption that the inertia of the differential rotation dominates over the Reynolds, Lorentz, and viscous stresses (e.g. Miesch 2005)

$$\frac{\partial \Omega^2}{\partial z} = \frac{g}{r \lambda C_P} \frac{\partial \langle S \rangle}{\partial \theta} \tag{0.1}$$

where Ω is the mean angular velocity, g is the gravitational acceleration, S is the specific entropy, C_P is the specific heat at constant pressure, and brackets denote averages over longitude and time. Both spherical polar (r, θ, ϕ) and cylindrical (λ, ϕ, z) coordinates are used, with $\lambda = r \sin\theta$ and $z = r\cos\theta$.

The second key equation, expressing gyroscopic pumping, is derived from the zonal momentum equation and can be written as

$$\langle \rho \mathbf{v}_m \rangle \cdot \nabla \mathcal{L} = \mathcal{F} \tag{0.2}$$

where ρ is the density, $\mathbf{v}_m$ denotes the meridional components of the velocity field, $\mathcal{L} = \lambda^2 \Omega$ is the specific angular momentum, and $\mathcal{F}$ represents the net torque due principally to the convective Reynolds stress (Miesch & Hindman 2011).

Much insight can be obtained merely from these two simple equations and the nature of the mean flows inferred from photospheric observations and helioseismic inversions. In particular, the conical nature of the solar rotation profile is attributed to (0.1), though the origin of the latitudinal entropy gradient is still unclear (Kitchatinov & Rüdiger 1995; Rempel 2005; Miesch *et al.* 2006; Balbus *et al.* 2009; Balbus & Schaan 2012). However, the differential rotation cannot be established solely by baroclinic forcing; the Reynolds stress must account for the observed super-rotation at the equator (Miesch *et al.* 2012, hereafter MFRT). Moreover, the meridional circulation cannot be strictly baroclinic either; rather, it must be maintained by the inertia of the differential rotation. This follows from equations (0.1) and (0.2) together with the observed sense of the rotational shear

† The National Center for Atmospheric Research is Sponsored by the National Science Foundation

($\partial\Omega/\partial z < 0$ in the northern hemisphere) and the meridional flow (poleward in the upper convection zone; MFRT).

Equation (0.2) can also be used to probe the nature of the elusive deep convection that sustains mean flows. Since the left-hand-side is known in the upper convection zone (CZ) from helioseismic inversions, this equation can be used to estimate the amplitude of the turbulent stresses on the right-hand-side, and thus the convective amplitude (MFRT). This yields values of at least 30 m s^{-1} in the upper CZ ($r \sim 0.95R$), in apparent contradiction with the results of Hanasoge *et al.* (2012) who place an upper limit of less than 1 m s^{-1} on persistent (lifetime $\tau > 96$ hrs), large-scale (spherical harmonic degree $\ell < 60$) convective motions at $r \sim 0.95R$ based on local helioseismic inversions. The resolution of this apparent inconsistency poses a significant challenge to both observations and models of solar convection and mean flows (MFRT).

Further clues on the nature of solar convection are provided by the existence of the near-surface shear layer (NSSL), which is a "smoking gun" signifying a transition from high to low Rossby number R_o (Miesch & Hindman 2011). The Rossby number is a nondimensional measure of the influence of rotation on convection, $R_o = U/(2\Omega L)$, where U and L are characteristic convective velocity and length scales. Taking the estimate of $U \sim 30$ m^{-1} in the previous paragraph and assuming L is of order the density scale height suggests that this transition occurs at $R_o \gtrsim 0.3$. This value is consistent with that estimated from global convection simulations (Featherstone & Miesch 2013).

The mean flows in global convection simulations generally exhibit the same dynamical balances that are thought to prevail in stars, namely equations (0.1) and (0.2), although the detailed dynamics may differ. At high latitudes, inward angular momentum transport by convective plumes induces a single-celled meridional circulation profile in radius, with poleward flow in the upper CZ and equatorward flow in the lower CZ. Meanwhile, the low-latitude dynamics is dominated by "banana cells", convective columns aligned with the rotation axis that transport angular momentum cylindrically outward and thereby induce multiple-celled meridional circulation profiles in radius. The balance between these two regimes changes with Rossby number; Fast rotators ($R_o < 0.3$) exhibit solar-like differential rotation profiles (fast equator, slow poles) with multi-celled circulation profiles while slow rotators ($R_o > 0.9$) exhibit anti-solar differential rotation profiles (slow equator, fast poles) and single-celled circulation profiles (Featherstone & Miesch 2013). The Sun lies near the transition, which suggests that its cyclic dynamo may operate somewhat differently relative to other stars.

Acknowledgements

This work is supported by NASA grants NNH09AK14I and NNX08AI57G.

References

Balbus, S. A., Bonart, J., Latter, H. N., & Weiss, N. O. 2009, *MNRAS* 400, 176
Balbus, S. A. & Schaan, E. 2012, *MNRAS* 426, 1546
Featherstone, N. A. & Miesch, M. S. 2013, *in preparation*
Hanasoge, S. M., Duvall, T. L. Jr. & Sreenivasan, K. R. 2012, *PNAS* 109 (30), 11928
Kitchatinov, L. L. & Rüdiger. 1995, *A&A* 299, 446
Miesch, M. S. 2005, *Liv. Rev. Solar Phys.* 2, 1, http://www.livingreviews.org/lrsp-2005-1
Miesch, M. S., Brun, A. S., & Toomre, J. 2006, *ApJ* 641, 618
Miesch, M. S., Featherstone, N. A., Rempel, M., & Trampedach, R. 2012, *ApJ*, 757, 128 (MFRT)
Miesch, M. S. & Hindman, B. W. 2011, *ApJ* 743, 79
Rempel, M. 2005, *ApJ* 622, 1320

Highlights of Astronomy, Volume 16
XXVIIIth IAU General Assembly, August 2012
T. Montmerle, ed.

doi:10.1017/S1743921314004700

Driving magnetic activity: differential rotation, flow structures, and surface patterns

Klaus G. Strassmeier

Leibniz-Institute for Astrophysics Potsdam (AIP), An der Sternwarte 16, D-14482 Potsdam, Germany
email: kstrassmeier@aip.de

Abstract. The interplay between stellar rotation and turbulent flows is a major ingredient for vertical angular momentum transport in stellar convection zone. Combined with the centrifugal force and the buoyancy force due to pole-equator temperature gradients one can expect a large-scale flow structure that is usually referred to as differential rotation and meridional flows. I review such observations for stars other than the Sun, mostly for stars significantly more active, and ask the question whether such observations can constrain the dynamo process.

Keywords. Stars: activity, Stars: rotation, Stars: spots, Sun: rotation, Magnetic fields

---

1. The Sun as a role model

The solar differential rotation is the result of the interplay between angular momentum transport by the Reynolds stress of rotating convection, a large-scale meridional flow, and a slight anisotropy in the convective heat transport. Nowadays, the Reynolds stress theory of the solar differential rotation reproduces the solar rotation law as known from helioseismology in great detail (Kitchatinov & Rüdiger 2005). In order to compute the differential rotation of a solar-type star the stratification of pressure, density and temperature are needed as input parameters. Küker & Stix (2001) computed first differential rotation models for the Sun for rotation periods between 56 days to 7 days. The stratifications were taken from a solar model but did not yet include the complex surface layers. These and subsequent computations (e.g. Küker *et al.* 2011) suggest an increased horizontal shear for a 56-day rotating Sun by ≈70% more than for today's Sun. The meridional flow shows one cell per hemisphere and points toward the equator on the surface and toward the pole at the bottom of the convection zone. A maximum meridional flow velocity of 9.2 m s^{-1} at the surface is predicted. For a 28-day rotating Sun the model of the meridional flow pattern consist of two cells per hemisphere, with the flow directed toward the equator at both the bottom and the top of the convection zone and poleward at intermediate depths. For a 14-day rotating Sun the normalized horizontal shear further decreases to ≈50% of today's value. There are still two flow cells per hemisphere, but the outer cell is restricted to a very shallow surface layer. The surface meridional flow is polewards with a flow speed of 3.8 m s^{-1}. Further decreasing the rotation period to seven days leads to a further reduction of the shear to less than a factor four with respect to today's Sun. Unfortunately, no stellar detections of meridional flows on stars other than the Sun exist to date.

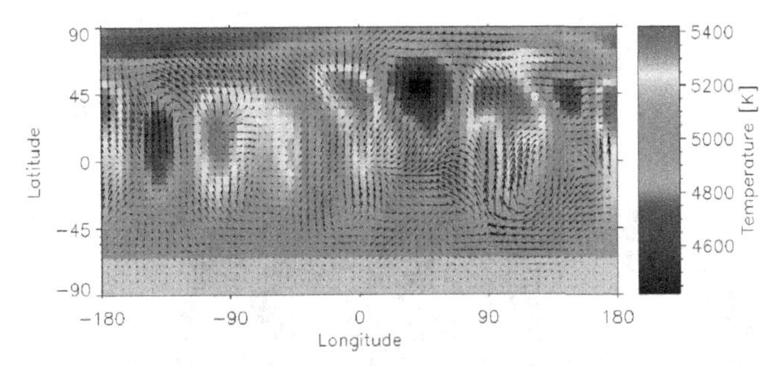

Figure 1. Surface velocity map (arrows) of the K2 dwarf LQ Hydrae from local-correlation tracking. The background color/shading is the average temperature map from the full epoch of Doppler imaging.

2. Differential rotation and meridional flows on the Sun and other stars

We typically simplify the "solar-type" differential rotation law by

$$\Omega(b) \;=\; \Omega_{\rm eq} \;-\; d\Omega\,\sin^2 b\,, \tag{2.1}$$

where b is the surface latitude, $\Omega(b)$ the respective rotation rate at a particular latitude, $\Omega_{\rm eq}$ the rotation rate at the equator, and $d\Omega$ the difference in rotation rate between the pole and the equator. The paper by Wöhl *et al.* (2010) gives an enlightening summary of the various solar differential rotation laws derived from a large range of tracers. Their closest representation of the observational data was obtained with

$$\Omega_{\odot}(b) \;=\; 14.499 \pm 0.006 - 2.54 \pm 0.06 \sin^2 b - 0.77 \pm 0.09 \sin^4 b\,. \tag{2.2}$$

$\Omega(b)$ represents the sidereal rotation velocity in degree per day.

I also present the current version of the ZDI code *iMap* in which the sheared image analysis is applied to temperature structures for the WTTS V410 Tau (Carroll *et al.* 2012). Only a very weak solar-like differential rotation is present, if at all. The formal value for the angular velocity at the equator is $\Omega_{eq} = 3.356 \pm 0.005$, well within the errors of the photometrically determined rotation period, and for the differential rotation rate between the equator and the pole $d\Omega = 0.007 \pm 0.009$. We conclude that differential rotation of this very active single WTTS must be very small, likely smaller than a factor 30 than on the Sun. For another very active star, the K0 giant XX Tri, we found a shear of 0.023 with a corresponding lap time of 1050 d, thus moderately strong differential rotation for a 10.7-$R_{\odot}$ star. We also present the results of an analysis of the very active K2 dwarf LQ Hya (Flores-Soriano *et al.* 2012) where we found a shear of 0.01 and a lap time of 150 d. The time series Doppler-imaging data even allowed the application of a solar-like, local-correlation tracking as shown in Fig. 1. Meridional flows were also computed for a variety of stars by Küker *et al.* (2011). Observations like in Fig. 1 would principally allow to search for such flows but no conclusive evidence is present and likely better data are required.

3. The link to the dynamo

A dynamo mechanism comparable to the Sun's converts mechanical energy into magnetic flux and transports it up to the surface where it is observable as spots, plages, flares, etc.. It has been proposed that the morphology of the emerging magnetic flux, i.e. flux tubes, allows a glimpse back into the star (e.g. Schüssler *et al.* 1996). Despite that

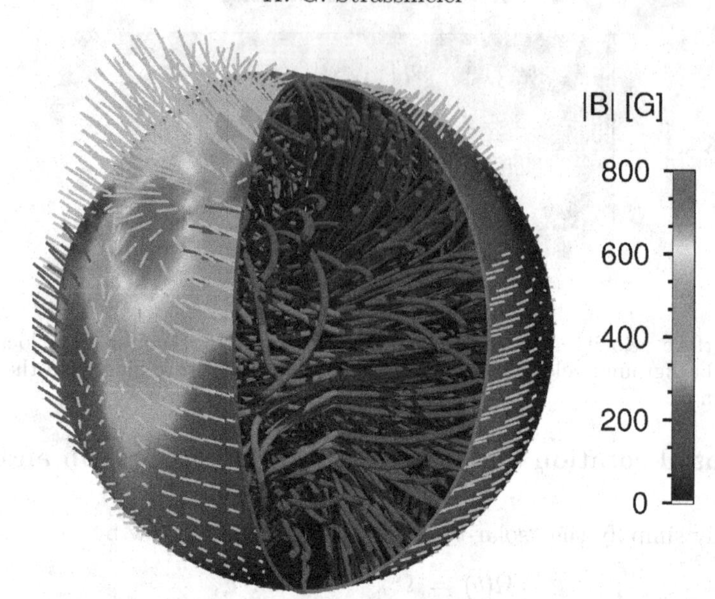

Figure 2. A combination of a mean-field dynamo simulation with a Zeeman-Doppler image of II Pegasi, a K2 subgiant. The simulation and the ZDI were obtained completely independent but the figure shows the potential of using ZDI images as the outer boundary condition for global numerical dynamo simulations.

many folks do not "believe" in flux tubes in the first place, they are an excellent modeling agent for (magnetic) flux and momentum transport to the surface. We demonstrate this visually in Fig. 2 for the active subgiant II Peg. There is mounting evidence that even the quiet solar photosphere is not field-free but rather permeated by small-scale, mixed polarity magnetic fields that show no significant variation with the solar activity cycle (see e.g. Trujillo Bueno *et al.* 2004). It has been suggested that a considerable part of the quiet Sun magnetic fields may be produced by a self-excited turbulent dynamo operating in the near-surface layers of the Sun, even though the flow in these layers is not noticeably affected by rotation and thus exhibits no net helicity (Cattaneo 1999, and subsequent papers). In principle, near-surface magneto-convection simulations could provide a self-consistent description of a local dynamo, including a prediction of its contribution to the mean magnetic flux of the quiet Sun and other stars.

Acknowledgements

I would like to thank the organizers for the invitation to speak.

References

Carroll, T. A., Strassmeier, K. G., Rice, J. B., & Künstler, A. 2012, *A&A*, in press
Cattaneo, F. 1999, *ApJ 515*, L39
Flores-Soriano, M., Strassmeier, K. G., & Küker, M. 2012, *Poster paper Cool Stars 17*, Barcelona
Kitchatinov, L. L. & Rüdiger, G. 2005, *AN* 326, 379
Küker, M. & Stix, M. 2001, *A&A* 366, 668
Küker, M., Rüdiger, G., & Kitchatinov, L. L. 2011, *A&A* 530, A48
Schüssler, M., Caligari, P., Ferriz-Mas, A., Solanki, S. K., & Stix, M. 1996, *A&A* 314, 503
Trujillo Bueno, J., Shchukina, N., & Asensio Ramos, A. 2004, *Nature* 430, 326
Wöhl, H., Brajsa, R., Hanslmeier, A., & Gissot, S. F. 2010, *A&A* 520, A29

Highlights of Astronomy, Volume 16
XXVIIIth IAU General Assembly, August 2012
T. Montmerle, ed.

doi:10.1017/S1743921314004712

Microflares to megaflares: Solar observations and modeling

Lyndsay Fletcher

School of Physics and Astronomy, SUPA, University of Glasgow,
Glasgow G12 8QQ, United Kingdom
email: lyndsay.fletcher@glasgow.ac.uk

Abstract. The observationally determined properties of solar flares such as overall energy budget and distribution in space, time and energy of flare radiation, have improved enormously over the last cycle. This has enabled precision diagnostics of flare plasmas and nonthermal particles in large and small events, informing and driving new theoretical models. The theoretical challenges in understanding flare are considerable, involving MHD and kinetic processes operating in an environment far from equilibrium. New observations have also provided some challenges to long-standing models of flare energy release and transport. This talk overviewed recent observational and theoretical developments, and highlighted some important questions for the future

Keywords. Sun: flares, activity

Overview Solar flares, defined as rapid (few minutes), localised, transient increases in the solar radiation, are the biggest energy releases in the solar system. The energy release results from the reconfiguration of the coronal magnetic field, most likely by magnetic reconnection. There is a reasonably concensus on the 'big picture' of a flare. Non-linear force free magnetic field reconstructions suggest pre-flare energy is stored in stressed fields less than 10,00 km above the photosphere, and strongly concentrated around the magnetic neutral line (Sun *et al.* 2012). The majority of the energy is released and converted, in the few minutes of the flare impulsive phase, into heat, mass motion, and the kinetic energy of non-thermal particles which go on to produce radiation across the electromagnetic spectrum (Emslie *et al.* 2012). Most - but not all - large flares have an associated CME, and the peak CME acceleration occurs on average about a minute before the peak of the impulsive phase hard X-ray (HXR) energy release (Berkebile-Stoiser *et al.* 2012). The bulk of the flare radiation is emitted in the UV and optical part of the spectrum, which is a common flare feature (Kretzschmar 2011) though very poorly observed. Fletcher *et al.* (2011) gives an up-to-date overview of solar flare observations.
Relationship to the evolving magnetic field Chromospheric flare emission is usually ordered into two ribbons of emission, separating in time. It is well known that these ribbons associate well with the calculated locations of the photospheric projections of magnetic quasi-separatrix layers, indicating the importance of the magnetic topology and its evolution. The product of ribbon velocity with the strength of the underlying magnetic field gives a measure of the magnetic reconnection rate. Peaks in this are found to be well-correlated with HXR peaks (Qiu 2009), emitted by a few compact HXR sources. The reasons for the location of the HXR sources within the ribbons is not clear; it may reflect the evolution of 'slip-running' reconnection within a QSL (Masson *et al.* 2009), or it may be linked to particular topological features called separators (Des Jardins *et al.* 2009).
Footpoint properties The flare footpoints - i.e. the HXR and optical emission regions, are the source of most of the impulsive-phase radiant energy. There have not been any UV measurements of these since the 1980s, and optical spectroscopy - except in strong

lines such as Hα - is almost entirely lacking. So we do not have a clear view of footpoint temperature and density structure, where the energy density is highest. With EUV spectroscopy the picture is better. The Extreme ultraviolet Imaging Spectrometer (EIS) on *Hinode* has been used for footpoint density and emission measure diagnostics (Graham *et al.* 2011), flow speed (also available from the Coronal Diagnostic Spectrometer on SOHO) and non-thermal line broadening measurements. The overall picture is that 'coronal' temperature (up to 10 MK) plasma can also be found in flare footpoints, with densities up to 10^{11}cm^{-3}. 'Sun-as-a-star' observations with SDO/EVE help constrain the continuum (Milligan *et al.* 2012) but resolved UV and optical observations are a priority

Electrons in flares An extensive review by Zharkova *et al.* (2011) describes the many theoretical models for flare electron acceleration. Flare HXR is usually interpreted in terms of a beam of electrons arriving from the corona, stopping collisionally in the chromosphere and emitting bremsstrahlung. With imaging spectroscopy and WL imaging, very firm lower estimates of the rate of electron arrival per unit area are made, and values up to 10^{20}electrons cm^{-2}s^{-1} determined Krucker *et al.* (2011). For $\sim$ 30 keV electrons this implies a beam density of 10^{10}electrons cm^{-3}. This presents a challenge, as such a beam should be too dense to propagate stably with its return-current. Significant effort is being expended on simulations of beam-return current propagation in the corona. It appears that a large fraction - perhaps 75% - of the beam energy is dissipated as plasma heating, but the beam continues at the thermal speed, and some energy may even go to electron re-acceleration(Lee & Büchner 2011; Karlický & Kontar 2012).

Flares through the cycle Flares are a way for the Sun to shed free energy, and are linked to helicity loss via CME eruptions. At least down to GOES A1 level, flares occur entirely within active regions, based on RHESSI imaging observations (Christe *et al.* 2008). The flare distribution is roughly a power-law, but the slope of the distribution depends on whether an assessment of thermal or non-thermal energy is made, and the proxy used for the thermal energy. Non-thermal signatures are probably a more robust measure of flare energy, though they do suffer from an unknown low-energy cutoff. The amplitude of the distribution of flare energies varies through the cycle, but its slope does not (Hannah *et al.* 2011). The flare rate per day is greater at maximum than at minimum. At the end cycle 23 the flare rate per active region was also anomalously high Hudson *et al.* (2012).

References

Berkebile-Stoiser, S., Veronig, A. M. , Bein, B. M., & Temmer, M. 2012, *ApJ* 753, 88
Christe, S., Hannah, I. G., Krucker, S., McTiernan, J., & Lin, R. P. 2008, *ApJ* 677, 1385
Des Jardins, A. & Canfield, R., Longcope *et al.* 2009, *ApJ* 693, 1628
Emslie, A. G. & Dennis, B. R., Shih A., *et al.* 2012, *ApJ* 759, 71
Fletcher, L., Dennis, B. R., Hudson, H. S., *et al.* 2011, *Space Sci. Revs.* 159, 19
Graham, D. R., Fletcher, L., & Hannah, I. G. 2011, *A&A* 532, A27
Hannah, I. G., Hudson, H. S., Battaglia, M., *et al.* 2011, *Space Sci. Revs.* 159, 263
Hudson, H. S., Fletcher, L., & McTiernan, J. 2012, *Solar Phys.* submitted
Karlický, M. & Kontar, E. P. 2012, *A&A* 544, A148
Kretzschmar, M. 2011, *A&A* 530, A84
Krucker, S., Hudson, H. S., Jeffrey, N. L. S., *et al.* 2011, *ApJ* 739, 96
Lee, K. W. & Büchner, J. 2011, *A&A* 535, A61
Masson, S., Pariat, E., Aulanier, G., & Schrijver, C. J. 2009, *ApJ* 700, 559
Milligan, R. O., Chamberlin, P. C., Hudson, H. S., *et al.* 2012, *ApJL* 748, L14
Qiu, J. 2009, *ApJ* 692, 1110
Sun, Z., Hoeksema, J. T. Liu, Y., *et al.* 2012, *ApJ* 748, 77
Zharkova, V. V., Arzner, K., Benz, A. O., *et al.* 2011, *Space Sci. Revs.* 159, 357

Highlights of Astronomy, Volume 16
XXVIIIth IAU General Assembly, August 2012
T. Montmerle, ed.

doi:10.1017/S1743921314004724

State-of-the-Art Observations and Modeling of Stellar Flares

Adam F. Kowalski† and Suzanne L. Hawley

University of Washington, Department of Astronomy, Box 351580, Seattle, WA 98195, USA
email: adam.f.kowalski@nasa.gov

Flares are observed on a wide variety of stellar types, ranging from closely orbiting binary systems consisting of an evolved member (RS CVn's) and young, nearby super-active M dwarfs (dMe's). The timescales and energies of flares span many orders of magnitude and typically far exceed the scales of even the largest solar flares observed. In particular, the active M dwarfs produce an energetic signature in the near-UV and optical continuum, which is often referred to as the white-light continuum. White-light emission has been studied in Johnson $UBVR$ filters during a few large-amplitude flares, and the best emission mechanism that fits the broadband color distribution is a $T \sim 10^4$ K blackbody (Hawley & Fisher 1992). Time-resolved blue spectra have revealed a consistent picture, with little or no Balmer jump and a smoothly rising continuum toward the near-UV (Hawley & Pettersen 1991). However, the most recent self-consistent radiative-hydrodynamic (RHD) models, which use a solar-type flare heating function from accelerated, nonthermal electrons, do not reproduce this emission spectrum. Instead, these models predict that the white-light is dominated by Balmer continuum emission from Hydrogen recombination in the chromosphere (Allred *et al.* 2006). Moreover, Allred *et al.* (2006) showed that the Johnson colors of the model prediction exhibit a broadband distribution similar to a blackbody with $T \sim 9000$ K.

To critically test these models, and to break the degeneracy using broadband colors for constraining white-light emission processes, we obtained time-resolved blue optical spectra ($\lambda = 3400-9200$Å) during a large sample of flares using the ARC 3.5-m telescope at the Apache Point Observatory (APO). We also obtained simultaneous broadband (typically, U band) photometry using the NMSU 1-m and ARCSAT 0.5-m at APO. We analyzed the line and continuum properties during eighteen flares, and we supplemented the sample with two flares that have been previously analyzed using data from other telescopes (Hawley & Pettersen 1991; Schmidt *et al.* 2012). The largest flare in our sample was a $\Delta U = -5.8$ mag (at peak) flare on the dM4.5e star, YZ CMi (Kowalski *et al.* 2010). The spectral monitoring started about an hour after the peak of the flare, when the U band was still elevated at ~ 25 times the quiescent level. The continuum analysis revealed *three* primary continuum components, which varied in relative strength during the flare decay. The energetically dominant component was found to be a moderately hot blackbody component with $T \sim$8500 K, like that observed in previous studies using broadband colors. However, we also found evidence for a Balmer continuum emission component that matched the shape of the RHD model prediction from Allred *et al.* (2006) for a heating flux of 10^{11} erg s^{-1} cm^{-2}. The third continuum component accounts for the gradual phase, excess emission above the extrapolation of the $T = 8500$ K blackbody component to redder wavelengths (e.g., at $\lambda = 6000$Å).

We applied our continuum analysis to the entire flare sample, which consisted of flares covering 2.5 orders of magnitude in peak flux. In particular, we found that the Balmer

† Present address: NPP Fellow, NASA-GSFC, Mail Code 671, Greenbelt, MD 20771, USA

continuum emission is ubiquitous at peak times but is much more dominant in the flares with longer impulsive phase timescales in the U-band light curves. The hot, blackbody emission component exhibits a range of peak temperatures, from $T \sim 9000 - 15\,000$ K; moreover, we discovered absorption features in the Balmer lines and continuum. These absorption features provide additional clues for the origin of the hot, blackbody emission component.

To attempt to reproduce the observed properties of the white-light emission, we began a new suite of RHD flare models with the RADYN code (Carlsson & Stein 1995; in collaboration with M. Carlsson, University of Oslo). Our new simulations use a larger flare heating flux, 10^{12} erg s^{-1} cm^{-2}, and include a gradual phase with the flare heating turned off. We find that the predicted white-light continuum experiences the same shortcomings as the simulations with 10^{11} erg s^{-1} cm^{-2} from Allred *et al.* (2006).

Broadband colors indicate that the peak of the white-light is located in the near-UV; we are seeking to precisely locate the peak and determine the detailed near-UV continuum properties using spectra that extend down to the atmospheric limit. Future work includes analysis of spectra at $\lambda \sim 3200$Å obtained with the South African Large Telescope during a "megaflare" on YZ CMi (in collaboration with B. Brown, University of Wisconsin-Madison, and M. Mathioudakis, Queen's University, Belfast). These observations will provide the next-generation of constraints for more successful RHD models. We are working on advanced RHD models, which include an even larger heating flux and a "stellar-type" flare heating mechanism deduced from X-ray observations reported in Osten *et al.* 2010.

Acknowledgements

A.F.K. would like to gratefully acknowledge the AAS International Travel Grant, the National Science Foundation and Solar Physics Division Studentship Award, and the IAU Travel Grant for travel and lodging funds that made travel to the IAU in Beijing possible.

References

Allred, J. C., Hawley, S. L., Abbett, W. P., & Carlsson, M. 2006, *ApJ*, 644, 484
Carlsson, M. & Stein, R. F., 1995, *ApJ*, 440L, 29
Hawley, S. L. & Pettersen, B. R. 1991, *ApJ*, 378, 725
Hawley, S. L. & Fisher, G. H. 1992, *ApJS*, 78, 565
Kowalski, A. F., Hawley, S. L., Holtzman, J. A., Wisniewski, J. P., & Hilton, E. J. 2010, *ApJL*, 714, L98
Osten, R. A., Godet, O., Drake, S., Tueller, J., Cummings, J., Krimm, H., Pye, J., Pal'shin, V., Golenetskii, S., Reale, F., Oates, S. R., Page, M. J., & Melandri, A., *ApJ*, 721, 785
Schmidt, S. J., Kowalski, A. F., Hawley, S. L., Hilton, E. J., Wisniewski, J. P., & Tofflemire, B. M. 2012, *ApJ*, 745, 14

Highlights of Astronomy, Volume 16
XXVIIIth IAU General Assembly, August 2012
T. Montmerle, ed.

doi:10.1017/S1743921314004736

Simulating Solar Global Magnetism in 3-D

A. S. Brun and A. Strugarek

Laboratoire AIM Paris-Saclay, CEA/Irfu Université Paris-Diderot CNRS/INSU,
91191 Gif-sur-Yvette, France
email: sacha.brun@cea.fr

Abstract. We briefly present recent progress using the ASH code to model in 3-D the solar convection, dynamo and its coupling to the deep radiative interior. We show how the presence of a self-consistent tachocline influences greatly the organization of the magnetic field and modifies the thermal structure of the convection zone leading to realistic profiles of the mean flows as deduced by helioseismology.

Keywords. Sun, dynamo, tachocline

---

1. The Global Sun in 3-D

Solar magnetism manifests itself on a broad range of scales up to the largest global ones. In this brief report we will summarize the progress we recently made in trying to understand how the Sun generates and maintain its large scale flows and its magnetic field and how these two aspects are linked.

On Figure 1 (left panel) we display the differential rotation realized in our most recent 3-D solar model using the ASH code (Brun *et al.* 2011). This model couples dynamically for the first time a convective envelope to a deep radiative interior. What is striking in that model is the presence of a tachocline of shear at the base of the convective envelope. Indeed the action of the Coriolis force on the convective motions results in the establishment of a large scale differential rotation. If the Rossby number ($Ro = v/2\Omega_0 L$, where v, L are characteristics velocity and length and Ω_0 the rotation rate of the star) is less than 1, which is the case for the Sun, Reynolds stresses transport angular momentum equatorward yielding a prograde differential rotation (fast equator, slow poles). The absence of convective motions in the deep interior on the contrary leaves the initial solid body rotation profile unchanged (on the short time scales considered here). A transition between both rotation profile must thus exist, i.e the tachocline. The thermal coupling of both zone further influences the profile of the differential rotation in the convective envelope by tilting the iso-contour of Ω from cylindrical to conical as inverted by helioseismology. Hence we see that the direct coupling of both zones yield a better description of the inner rotation profile of the Sun. In such coupled models convective overshooting also leads to the excitation of internal waves that can further transport non locally angular momentum thus possibly enforcing solid body rotation but on much longer time scales (Brun *et al.* 2011).

Such a tachocline will diffuse inward if no processes oppose its slow propagation as was shown by Spiegel & Zahn in 1992. It has been proposed by Gough & McIntyre (1998) that a large scale fossil dipolar magnetic field could oppose such spread. In Strugarek *et al.* 2011a,b, we have studied in 3-D MHD with the ASH code if such a scenario could work. We display on Figure 1 (right panel) the 3-D rendering of the magnetic field from r=0.07 up to 2.5 R. What is directly clear is that the inner fossil field pervades through the convective envelope. This leads to a direct coupling between the differentially rotating

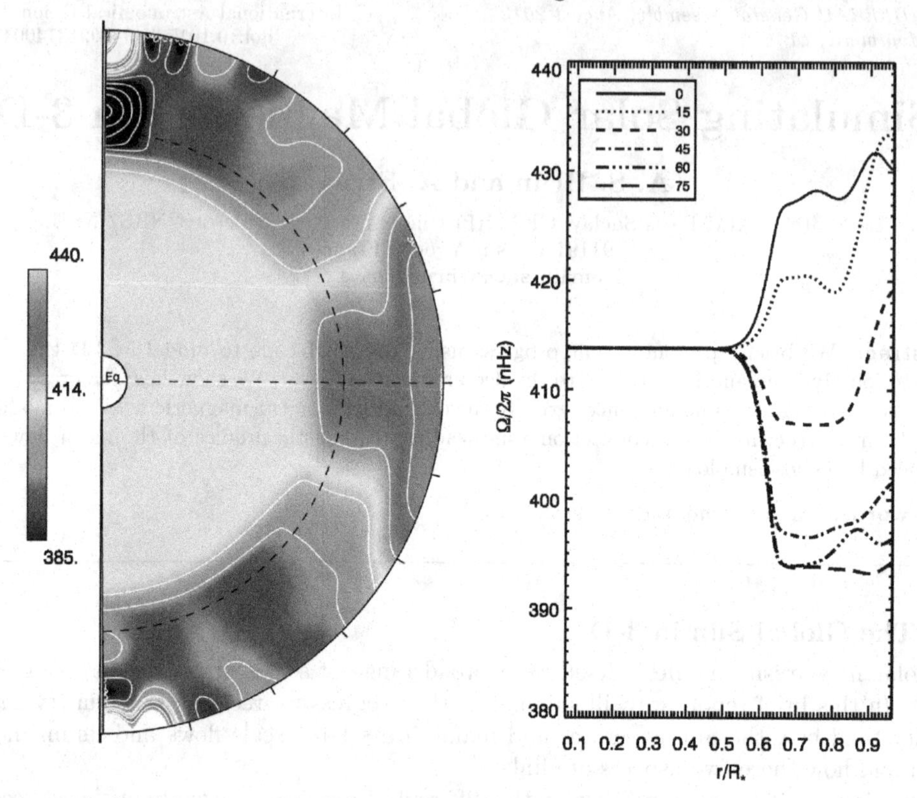

Figure 1. Left: Color contours of the azimuthally and temporally averaged angular velocity realized in a 3-D global solar model coupling the convective envelope to the radiative interior (Brun *et al.* 2001). Right: 3-D rendering of the magnetic field lines in a global MHD solar model from $r = 0.07R$ up to $r = 2.5R$, using potential field beyond $r = 0.97R$ (Strugarek *et al.* 2011a).

convective envelope and the deep interior resulting to an efficient angular momentum transport. The end state is an omega profile constant along the poloidal magnetic field (i.e the so called Ferraro's law of iso-rotation) which is not observed. In fact a purely axisymmetric dipolar field is unlikely in the deep interior of the Sun as MHD instabilities exist there and are likely to yield more complex magnetic topologies. In an attempt of taking that into account we have considered non axisymmetric field configurations (oblique and perpendicular dipole), but found that it does not make much differences (Strugarek *et al.* 2011b). We thus need to consider even more realistic setting or search for different solutions (turbulence, waves) to explain the thinness of the tachocline.

Finally such coupled models are ideal to study the solar global dynamo in particular to understand how the large scale field is being maintained by turbulent flows (Strugarek *et al.* 2012). Preliminary results show that large scale magnetic field are indeed organized at the base of the convective envelope and in the tachocline confirming the results of Browning *et al.* (2006). Fast dynamo action is found to operate in the convective envelope with efficient magnetic pumping of field at the base of the convective envelope and in the tachocline, where it is subsequently stretched into large wreaths of horizontal magnetic field. In these new models, a tapering of the diffusivities in the stable zone is present that will hopefully help getting these structures to become buoyant and rise to the surface, much as in the recent work of Nelson *et al.* (2011) in fastly rotating Suns.

References

Browning *et al.* 2006, *ApJ* 648, L157
Brun, A. S., Miesch, M. S., & Toomre 2011, *ApJ* 742, 79
Gough, D. & McIntyre, M. 1998, *Nature* 394, 755
Nelson *et al.* 2011, *ApJ*, 739, L38
Strugarek, A., Brun, A. S., & Zahn, J.-P. 2011a, *A&A* 532, 34
Strugarek, A., Brun, A. S., & Zahn, J.-P. 2011b, *AN* 332, 891
Strugarek, A., Brun, A. S. , Mathis, S., & Sarazin 2012, *ApJ* in press
Spiegel, E. A. & Zahn, J. P. 1992, *A&A* 265, 106

Highlights of Astronomy, Volume 16
XXVIIIth IAU General Assembly, August 2012
T. Montmerle, ed.

doi:10.1017/S1743921314004748

3-D reconstructions of active stars

Heidi Korhonen[1,2,3]

[1]Niels Bohr Institute, University of Copenhagen, Juliane Maries Vej 30, DK-2100 Copenhagen, Denmark

[2]Centre for Star and Planet Formation, Natural History Museum of Denmark, University of Copenhagen, Øster Voldgade 5-7, DK-1350, Copenhagen, Denmark

[3]Finnish Centre for Astronomy with ESO (FINCA), University of Turku, Väisäläntie 20, FI-21500 Piikkiö, Finland

Abstract. Stars are usually faint point sources and investigating their surfaces and interiors observationally is very demanding. Here I give a review on the state-of-the-art observing techniques and recent results on studying interiors and surface features of active stars.

Keywords. stars: activity, atmospheres, interiors, rotation, spots

1. Introduction

The only star we can study easily with any spatial resolution is the Sun, and now with the twin STEREO spacecraft we can do it even in 3-D. For other stars we are usually restricted to using indirect methods for getting information of their surface structures. Here a short review on methods for studying stellar interiors and surfaces and the main recent results are presented.

2. Interiors

Kepler has not only revolutionised the exoplanet research, but also stellar astrophysics. The short, one minute, cadence data is uniquely suited for studying stellar oscillations and thus the interiors of stars (see, e.g., Gilliland *et al.* 2010). Recent Kepler results, based on a sample of 500 solar-like stars, show that even though distribution of observed radii is similar to the one predicted by models of synthetic stellar populations in the Galaxy, the distribution of observed masses seems to deviate significantly from the predicted distribution (Chaplin *et al.* 2011a). This result raises interesting questions on the star formation rate and initial mass function. Studying the same sample of solar-like stars further reveals that the number of stars showing oscillations decreases significantly with increasing levels of activity (Chaplin *et al.* 2011b). This on the other hand implies that the magnetic fields can inhibit the amplitudes of solar-like oscillations.

3. Surfaces

The strength of the dynamo-created stellar magnetic fields and the frequency of the associated phenomena (e.g., starspots and flares) is strongly linked to the rotation of the star, with rapid rotation enhancing the field generation (e.g., Pallavicini *et al.* 1981). Many active stars, rapid rotation usually induced either by youth or binarity, show large starspots which are detectable even using photometric observations from small ground-based telescopes (e.g., Kron 1947). In some cases the monitoring has become almost nightly after the implementation of automatic photometric telescopes in the late 1980's (e.g., Rodonò & Cutispoto 1992; Strassmeier *et al.* 1997). These observations give crucial information on the stellar activity cycles (e.g., Oláh *et al.* 2009) and long-term spot longitudes (e.g., Lehtinen *et al.* 2012), i.e., 2-D information where one dimension is longitude and the other time. With observations using multiple filters crude spot latitudes can be obtained (e.g., Roettenbacher *et al.* 2011), making the reconstruction 3-D (longitude, latitude and time). Similarly, the continuous, high precision, space-based light-curves allow for recovering latitudinal and longitudinal information in addition to the temporal variations.

For obtaining more detailed stellar surface structures Doppler imaging techniques have to be used (e.g., Vogt, Penrod & Hatzes 1987), giving a 2-D snapshot of the surface temperature with a reasonable resolution (e.g., Korhonen *et al.* 2010). Combining many observations from different epochs adds the third dimension, namely time (e.g., Skelly *et al.* 2010; Hackman *et al.* 2012). Doppler images can also be used for studying stellar surface rotation and its dependence on latitude (for a recent review see Korhonen 2012). If spectral lines formed at different heights in the stellar atmosphere are used in the reconstruction, also vertical structures in starspots can be mapped (Berdyugina 2011), coming closer to what is possible in the solar case.

4. Conclusions and outlook

Obtaining multi-dimensional infromation on the active stars, which are just point sources, is challenging but not impossible. Observations give us detailed information, both temporal and spatial, on occurrence of starspots. In the future long baseline infrared/optical interferometry can offer a direct check on some of the results. Furthermore, while the interferometric imaging is restricted to the brightest and nearest stars, the method does not require large projected rotational velocities, like Doppler imaging does. This will open up new targets that have not been approachable earlier, significantly widening the parameter space for activity studies.

Acknowledgments H.K. acknowledges the support from European Commission's Marie Curie IEF Programme, and an IAU travel grant to participate the General Assembly.

References

Berdyugina S. V. 2011, in: J. R. Kuhn, *et al.* (eds), *ASP Conf. Ser.* 437, 219
Chaplin, W. J., Kjeldsen, H., Christensen-Dalsgaard, J., *et al.* 2011a, *Science* 332, 213
Chaplin, W. J., Bedding, T. R., Bonanno, A., *et al.* 2011b, *ApJ (Letters)* 732, L5
Gilliland, R. L., Brown, T. M., Christensen-Dalsgaard, J., *et al.* 2010, *PASP* 122, 131
Hackman, T., Mantere, M. J., Lindborg, J., *et al.* 2012, *A&A* 538, A126
Kron, G. E. 1947, *PASP* 59, 261
Lehtinen, J., Jetsu, L., Hackman, T., Kajatkari, P., & Henry, G. W. 2012, *A&A* 542, A38
Korhonen, H. 2012, in C. H. Mandrini, & D. F. Webb (eds), *Proceedings of IAU Symposium 286* (Cambridge University Press), p. 268
Korhonen, H., Wittkowski, M., Kővári, Zs, *et al.* 2010, *A&A* 515, A14
Oláh, K., Kolláth, Z., Granzer, T., *et al.* 2009, *A&A* 501, 703
Rodonò, M. & Cutispoto, G. 1992, *A&AS* 95, 55
Roettenbacher, R. M., Harmon, R. O., Vutisalchavakul, N., & Henry, G. W. 2011, *AJ* 141, 138
Skelly, M. B., Donati, J.-F., Bouvier, J., *et al.* 2010, *MNRAS* 403, 159
Strassmeier, K. G., Bartus, J., Cutispoto, G., & Rodono, M. 1997, *A&AS* 125, 11
Vogt, S. S., Penrod, G., & Donald, Hatzes, A. P. 1987, *ApJ* 321, 496

Highlights of Astronomy, Volume 16
XXVIIIth IAU General Assembly, August 2012
T. Montmerle, ed.

doi:10.1017/S174392131400475X

3-D views of the expanding CME: from the Sun to 1AU

Alexis P. Rouillard[1,2]

[1]Université de Toulouse; UPS-OMP; IRAP; Toulouse, France
[2]CNRS, IRAP, 9 Av. colonel Roche, BP 44346, F-31028 Toulouse cedex 4, France
email: arouillard@irap.omp.eu

Abstract. Three-dimensional information on Coronal Mass Ejections (CMEs) can be obtained from a wide range of in-situ measurements and remote-sensing techniques. Extreme ultraviolet (EUV) and white-light imaging sensed from several vantage points can be used to infer the 3-D geometry of the different parts that constitute a CME. High-resolution and high-cadence coronal imaging provides detailed information on the formation and release phase of a magnetic flux rope, the lateral expansion of the CME and the reconfiguration of the corona associated with the effects of pressure variations and reconnection. The evolution of the CME in the interplanetary medium and the connection of its various substructures with in-situ measurements can be obtained from multi-point heliospheric imaging.

Keywords. Solar Wind, Solar Activity, Coronal Mass Ejections, Corotating Interaction Regions

1. Introduction

We discuss recent observations, made by the Solar-Terrestrial Relations Observatory (STEREO) and the Solar Dynamics Observatory (SDO), of the formation, interaction and propagation of CMEs from the Sun to 1AU. We divide the CME structure into several parts: the prominence, the cavity and the bright fronts as envisioned by Illing and Hundhausen (1985). Only a small fraction of CMEs have these clearly distinct parts, at most 30% of all events (see e.g. Munro *et al.* 1979), however it provides a convenient framework to discuss in general terms the release and propagation of a CME. Recent results have provided new clues on the nature of cavities and bright fronts, leading to the concept of the five-part structure CME (Vourlidas *et al.* 2012).

2. Dark cavity

The formation and release of a CME can take from several minutes to several days. ProtoCMEs can be observed as cavities in the million-degree corona. When these cavities remain in suspension for several hours, high-cadence EUV imaging reveals that material spins inside them (e.g. Gibson *et al.* 2010). This spinning material likely streams along helical magnetic fields embedded inside these cavities (e.g. Wang and Stenborg 2010). Lower temperature plasma at temperatures near 50kK, called prominence material, is commonly observed at the bottom of this cavity (Berger *et al.* 2012). This prominence material is not always static but the observed motions remain spatially confined to the bottom (sunward edge) of the cavity (Gibson *et al.* 2006). The presence of prominence material is expected since cavities are always situated below helmet streamers, right above polarity inversion lines determined by photospheric magnetograms. During fast and impulsive CME events cavities appear and disappear from the EUV corona within minutes. SDO images permit unprecedented measurements of the lateral and radial expansions of cavities in the lower corona during these few minutes (Patsourakos *et al.* 2010a, Cheng *et al.* 2012). They show that cavities can undergo an intial phase of strong over-expansion (Patsourakos *et al.* 2010b) with measured expansion speeds in excess of 700 km s^{-1} (Cheng *et al.* 2012; Rouillard *et al.* 2013). These sudden cavity expansions tend to originate in strong magnetic field regions, likely some distance from the center of the peak magnetic distribution of active regions (Wang and Zhang 2007). Indeed, the presence of too strong confinement magnetic fields is thought to prevent cavity expansions/eruptions and the subsequent manifestation of a CME (e.g. Török and Kliem 2005). For events that do erupt, the question of whether the flux rope is formed during the eruption, or whether the flux rope existed prior to the eruption, remains controversial.

The extension of the cavity in white-light images (e.g. coronagraphs) is a dark circular or elliptical (depending on the viewing angle) region located within the larger CME (Thernisien *et al.* 2009). 3-D reconstructions of the white light scattered by electrons located on the surface of hollow and therefore dark cavities show that the latter are in fact the cross section of a three-dimensional croissant-shaped structure (Chen *et al.* 1997, Thernisien *et al.* 2009, 2011). It had long been suggested that dark cavities in white-light images are the location of the twisted magnetic field that forms the flux rope (e.g. Chen *et al.* 1997). Continuous tracking of the dark cavities of gradual CMEs to 1AU, using STEREO images, has confirmed that the in-situ signature of a cavity passing over a measuring spacecraft is a rotation of the magnetic vector that can be interpreted as a magnetic flux rope (Rouillard *et al.* 2009, Möstl *et al.* 2009). For well-defined and simple CME events, the 3-D orientation of the croissant reconstructed from white-light images is in good agreement with that of the magnetic flux rope inferred from in-situ measurements (Rouillard *et al.* 2009, Wood *et al.* 2010). These slow cavities retained their circular cross-section as they propagated to 1AU, suggesting a self-similar 3-D expansion of the magnetic flux rope during its propagation to 1AU (Rouillard *et al.* 2009, Wood *et al.* 2010). For some very gradual and slow CMEs, cavities are either absent in EUV or poorly defined in white-light images. The emergence over several days of distinct loops is instead observed in white-light images until the sudden catastrophic anti-sunward release of the back-end of the CME (Sheeley and Wang 2007, Wood *et al.* 2012). This catastrophic release is often associated with loops collapsing back towards the Sun in the phenomenon known as in-out pairs (Sheeley and Wang 2007).

3. The bright fronts

STEREO imaging, during quadrature orbital configuration in 2009, has demonstrated that the so-called Extreme-Ultraviolet (EUV) waves are clearest during the lateral expansion of the cavity. They originate near the strong pressure gradients generated on the flanks of these cavities (Patsourakos and Vourlidas 2009). The strong lateral expansion phase has generally ceased by the time a CME reaches 2 Rs, the EUV wave is then no longer confined to the edges of the cavity and becomes more freely propagating. Further deflections of ambient coronal material are observed simultaneously in white-light images, together with the release over a wide longitude band of solar energetic particle events (Rouillard *et al.* 2012). Two layers, of often different brightness intensities, are observed in outer coronagraphs and in numerical simulations. These layers appear to surround the dark cavity (Ontiveros and Vourlidas 2009; Rouillard *et al.* 2012). The inner layer corresponds to the plasma accumulated directly on the surface of the cavity (likely with the lifted overlying fields), termed N-shell or plasma pile-up (e.g. Hundhausen 1972). While the dark cavity is the likely location of the core field of the CME, originating from the vicinity of a neutral line, the regions surrounding the cavity, in particular its antisunward edge, must at first contain large-scale overlying (or else background) magnetic fields (Riley *et al.* 2008). In contrast to the inner layer that remains confined to the edges of the CME cavity, the fainter outer layer can engulf the entire Sun (Rouillard *et al.* 2013). Numerical simulations suggest that the anti-sunward boundary of this outer layer can mark the location of the coronal shock when its speed exceeds the local ambient characteristic speed (Vourlidas *et al.* 2012). For the very fast 2012 July 23 event multi-point observations clearly showed that the fastest section of the outer layer is directly ahead of the CME (>3000 kms^{-1}) where a shock is continuously driven (Rouillard *et al.* 2013). As the CME progresses in the outer corona, multi-point (heliospheric) images show that the different layers become a single bright leading front that can take the shape of a bow wave (Wood *et al.* 2011, Rouillard *et al.* 2011). The in-situ signature of this bow wave is, as expected, the location of an interplanetary shock (Rouillard *et al.* 2011).

4. Conclusion

Significant progress has been made concerning the nature and evolution of the dark cavity and its surrounding brightness variations. Additional progress remains to be done with regards to the nature and in-situ properties of post-cavity material by tracking prominences or post-CME current sheets continuously to 1AU and studying the composition of this material (e.g. Lepri and Zurbuchen 2010).

References

Berger T. E., Liu,W., & Low, B. C. 2012, *ApJ* 758, L37
Chen, J., Howard, R. A., Brueckner, G. E., *et al.* 1997, *ApJ* 490, L191
Cheng X., Zhang, J., Olmedo, O., Vourlidas, A., Ding, M. D., & Liu, Y. 2012, *ApJ* 745, L5
Gibson S., Foster, D., Burkepile, J., de Toma, G., & Stanger, A. 2006, *ApJ* 641, 590
Gibson, S. E., Kucera, T. A., Rastawicki, D., Dove, J., de Toma, G., *et al.* 2010, *ApJ* 724, 1133
Hundhausen, A. 1972, *Springer, 1st Edition*
Illing, R. M. E. & Hundhausen, A. J. 1985, *Solar Phys.* 90, 275
Lepri, S. T. & Zurbuchen, T. H. 2010, *ApJ* 723, L22
Möstl, C., Farrugia, C. J., Temmer, M., *et al.* 2009, *ApJ* 705, L180
Munro, R. H., Gosling, J. T., Hildner, *et al.* 1979, *Solar Phys.* 61, 201
Patsourakos, S. & Vourlidas, A. 2009, *ApJ* 700, L182
Patsourakos, S., Vourlidas, A., & Kliem, B. 2010a, *A&A* 552, id.A100
Patsourakos, S., Vourlidas, A., & Stenborg, G. 2010b, *ApJ* 724, L188
Riley, P., Lionello, R., & Mikić, Linker, J. 2008, *ApJ* 672, 1221
Rouillard, A. P., Davies, J. A., Forsyth, R. J., *et al.* 2009, *J. Geophys. Res.* 114, A07106
Rouillard, A. P., Odstrcil, D., Sheeley, N. R., *et al.* 2012, *ApJ* 735, id7
Rouillard, A. P., Sheeley, N. R., Tylka, A., *et al.*. 2012, *J. Geophys. Res.* 114, A07106
Rouillard, A. P., Sheeley, N. R., Tylka, A., Vourlidas, A., & Ng, C. K. 2013, *ApJ*, In Preparation
Sheeley, N. R. & Wang, Y.-M. 2007, *ApJ* 655, 1142
Thernisien, A., Vourlidas, A., & Howard, R. A. 2009, *ApJ* 256, 111
Thernisien, A., Vourlidas, A., & Howard, R. A. 2011, *JASTP* 73, 1156
Török, T. & Kliem, B. 2005, *ApJ* 630, L97
Vourlidas, A., Lynch, B. J., Howard, R. A., & Li, Y. 2012, *Solar Phys.*, Online First
Wang, Y.-M. & Stenborg, G. 2010, *ApJ* 719, L181
Wood, B. E., Howard, R. A., & Socker, D. G. 2010, *ApJ* 715, 1524
Wood, B. E., Rouillard, A. P., Mstl, C. *et al.* 2012, *Solar Phys.* Online First.
Zhang, Y. & Zhang, J. 2007, *ApJ* 665, 1438

Highlights of Astronomy, Volume 16
XXVIIIth IAU General Assembly, August 2012
T. Montmerle, ed.

doi:10.1017/S1743921314004761

3D Perspectives of Stellar Activity: Observation and Modelling

Moira Jardine

SUPA, University of St Andrews, School of Physics and Astronomy, North Haugh, St Andrews, Fife, KY16 9SS, UK
email: mmj@st-andrews.ac.uk

Abstract. As the Sun emerges from a period of unprecedented low activity, the nature of the Sun's magnetic field compared to that of other stars is a particularly timely question. Just as observations of the full 3D structure of the solar magnetic field are becoming available through STEREO and SDO, advances in spectropolarimetric techniques now allow us to map the surface magnetic fields of other stars, revealing the great diversity of magnetic geometries that stars of different masses and rotation rates can display. This has now been possible for over 60 main sequence stars, with a smaller number of younger, pre-main sequence stars also mapped. Modelling of coronal structures based on these observations is revealing the full nature of stellar magnetic activity and its possible impact on orbiting planets.

Keywords. stars: imaging, stars: low mass, brown dwarf, stars: winds, outflows

1. Introduction

Our understanding of the magnetic activity of the Sun has now reached the stage where we can relate magnetic field evolution at the solar surface to changes in the morphology of the large-scale corona and wind. Studies of stellar activity are of course hampered by the lack of resolved observations and in-situ measurements, but they have the advantage that a greater range of stellar parameters can be sampled. The more traditional studies of the activity-rotation relation have been augmented by studies of stellar winds and prominences whose ejection may contribute significantly to angular momentum loss in young stars (Jardine & van Ballegooijen 2005; Skelly *et al.* 2008; Skelly *et al.* 2009, Aarnio *et al.* 2012). Spectropolarimetric techniques such as Zeeman-Doppler imaging also now allow us to map all three vector components of the magnetic field at the surface of a star (Donati & Landstreet 2009). Coupled with field-extrapolation methods which provide the 3D structure of the coronal magnetic field and plasma, this makes it possible to study the morphology of stellar coronae in ways that were not possible in the past (Jardine *et al.* 2002a,b; Hussain *et al.* 2002).

2. Overview

A comprehensive study of the magnetic fields of low mass stars (Donati *et al.* 2008, Morin *et al.* 2008; Morin *et al.* 2010) shows the trends with stellar mass and rotation rate of the magnetic field strength, axisymmetry and the departure from the lowest-energy (potential) state. Since stars spin down with age, this also sheds light on the changes that may have occurred in the solar magnetic field over the time period when the Earth and other planets were forming. In particular, the decay with time in the ram pressure of the solar wind led to changes in the Earth's magnetosphere and the bow shock around it. Recent observations of an early ingress of the UV transit relative to the optical transit of the exoplanet WASP-12b suggest that such bow shocks may be detectable (Lai *et al.* 2010). Subsequent modelling suggests such detections may allow us to probe the strengths of exoplanetary magnetic fields and the conditions around exoplanets (Vidotto *et al.* 2010; Vidotto *et al.* 2011a,b).

Low mass stars are indeed popular choices as hosts for habitable exoplanets, but studies of their winds based on this survey of their magnetic field topologies suggests that the combination of rapid stellar rotation and high field strengths may lead to very non-solar winds, characterised by a low plasma β and centrifugal driving (Vidotto *et al.* 2011c). The high ram pressure of these winds may crush planetary magnetospheres unless planets have a significant magnetic field. This factor, combined with the very long spin-down times for these stars, may suggest that they are unlikely to be habitable.

Of course, surveys such as this only sample the large-scale field of these stars, as small-scale polarity changes are undetected. Zeeman broadening studies, such as that of Reiners *et al.* (2009), suggest that as much as 85% of the surface magnetic flux may be undetected. A recent study of the impact of this small-scale field on the large-scale coronal topology and the wind demonstrates however that the effect is modest, with a small reduction in the open magnetic flux that carries the wind and influences the spin-down times (Lang *et al.* 2012).

In summary then, studies of stellar activity have progressed enormously in the last 15 years, but with the results from Kepler's survey of around 150,00 stars, there is a very bright future ahead. This promises to allow us to progress from studies of individual stars to statistical studies of a large sample.

References

Aarnio, A., Llama, J., Jardine, M., & Gregory, S. G. 2012, *MN*, 421, 1797

Donati, J.-F., Morin, J., Petit, P., Delfosse, X., Forveille, T., Aurière, M., Cabanac, R., Dintrans, B., Fares, R., Gastine, T., Jardine, M. M., Lignières, F., Paletou, F., Ramirez Velez, J. C., & Théado, S. 2008, *MN*, 390, 545

Donati, J.-F. & Landstreet, J. D. 2009, *Ann. Rev. Astr. Astroph.*, 47, 333

Hussain, G. A. J., van Ballegooijen, A. A., Jardine, M., & Collier Cameron, A. 2002, *ApJ*, 575, 1078

Jardine, M., Collier Cameron, A., & Donati, J.-F. 2002, *MN*, 333, 339

Jardine, M., Wood, K., Collier Cameron, A., Donati, J.-F., & Mackay, D. H. 2002, *MN*, 336, 1364

Jardine, M. & van Ballegooijen, A. A. 2005, *MN*, 361, 1173

Lai, D., Helling, C., & van den Heuvel, E. P. J. 2010, *MN*, 721, 923

Lang, P., Jardine, M., Donati, J.-F., Morin, J., & Vidotto, A. 2012, *MN*, 424, 1077

Morin, J., Donati, J.-F., Petit, P., Delfosse, X., Forveille, T., Albert, L., Aurière, M., Cabanac, R., Dintrans, B., Fares, R., Gastine, T., Jardine, M. M., Lignières, F., Paletou, F., Ramirez Velez, J. C., & Théado, S. 2008, *MN*, 390, 567

Morin, J., Donati, J.-F., Petit, P., Delfosse, X., Forveille, T., & Jardine, M. M. 2010, *MN*, 407, 2269

Reiners, A. & Basri, G. 2009, *A&A*, 496, 787

Skelly, M. B., Unruh, Y. C., Barnes, J. R., Lawson, W. A., Donati, J.-F., & Collier Cameron, A. 2008, *MN*, 399, 1829

Skelly, M. B., Unruh, Y. C., Collier Cameron, A., Barnes, J. R., Donati, J.-F., Lawson, W. A., & Carter, B. D. 2009, *MN*, 385, 708

Vidotto, A. A., Jardine, M., & Helling, C. 2010, *ApJL*, 722, L168

Vidotto, A. A., Jardine, M., & Helling, C. 2011, *MN*, 411, L46

Vidotto, A. A., Jardine, M., & Helling, C. 2011, *MN*, 414, 1573

Vidotto, A. A., Jardine, M., Opher, M., Donati, J. F., & Gombosi, T. I. 2011, *MN*, 412, 351

Highlights of Astronomy, Volume 16
XXVIIIth IAU General Assembly, August 2012
T. Montmerle, ed.

doi:10.1017/S1743921314004773

The Solar cycle: looking forward

Robert H. Cameron

Max-Planck-Institut für Sonnensystemforschung, 37191 Katlenburg-Lindau, Germany

Abstract. We discuss predictions for cycle 24 and the way forward if progress is to be made for cycle 25 and beyond.

Keywords. Sun: activity

---

1. Introduction

The level of solar activity waxes and wanes with an approximate timescale of 11 years. The amplitudes of each maximum is observed to vary strongly from cycle to cycle. Numerous attempts to predict each of the recent maxima have been made. For cycle 24 a number of predictions were collected by Pesnell(2008) and the resulting distribution is shown in Figure 1. Also shown is the distribution of cycle maxima for cycles 1-23. From this figure it appears that, as of 2008, what could be said was that the amplitude of cycle 24 wil (probably) be drawn from the same distribution as cycles 1 to 23.

This result should not be surprising: with only 23 cycles, any number of patterns can be found in the data, and used as a basis for extrapolating the data to cycle 24. Furthermore since the predictions cover a large range it is obvious that some of them will be close to the actual level of maximum activity for cycle 24 - so the apparent success of some of these models is gauranteed (although which models will be successful was obviously not agreed on before the cycle began). Apparent success in this regard need not indicate the models will have predictive success in future cycles.

2. The way forward

To make progress it is essential to understand the non-linearity which is driving the solar cycle. This understanding needs to be based on observations: dynamo theory in the abstract is non-specific enough to allow predictions covering the whole range of predictions shown in figure 1 – compare for example the predictions of Dikpati & Gilman(2006) with those of Jiang *et al.*(2007).

A non-linearity in this context involves the magnetic field driving a flow. For a long time it was thought that the changes were likely to be on small spatial scales where the creation of the poloidal flux from toroidal flux would be inhibited (this was called 'alpha quenching'). This form of saturation does not however work for the Babcock-Leighton type of dynamo which is the favoured explanation for the global solar dynamo, see Kitchatinov & Olemskoy(2011). Consequently the theoretical discussion has recently focussed on the dynamo saturation being due to changes in the large-scale velocity field. While a significant focus of the theoretical work has been on changes in the global meridional circulation, e.g. Karak & Choudhuri(2012), the observations actually support only near-surfce changes to the flow field, see Cameron & Schüssler(2010). Specifically the observations, e.g. by Gizon(2004), support a shallow, time-dependent, inflow into the active region latitudes. Cameron & Schüssler(2012) have modelled the effect of this inflow

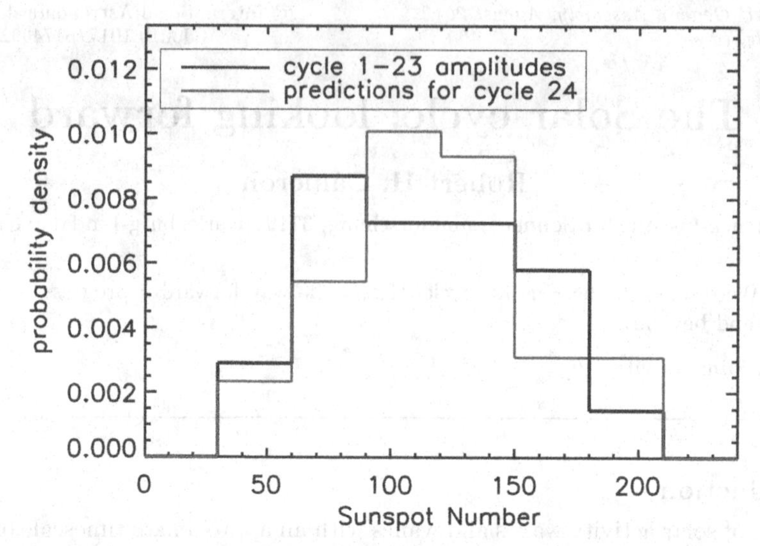

Figure 1. Distribution of predictions for the amplitude of cycle 24 given in Pesnell (2008) and of the observed amplitudes of cycles 1 to 23.

using the surface flux transport model and find that it is able to account for most of the observed cycle-to-cycle changes in activity levels.

The inflows into the active region belt are an observationally clear nonlinearity for the dynamo which is of the correct sign and magnitude to explain most of the observed cycle-to-cycle variations. The driving of the inflows has been studied theoretically by Spruit(2003) and modelled by Gizon & Rempel(2008): they are though to be a response to the cooling of the plasma by the excess radiance associated with regions of plage.

3. Conclusion

An understanding of the non-linearity which saturates the solar dynamo is a necessary precondition for believable predictions which extend beyond about 1/2 a cycle – for shorter times precursors might be effective because of the Waldmeier effect, see Cameron & Schüssler(2007). It seems that on the Sun the non-linearity occurs at the surface where it can be studied in great detail.

References

Cameron, R. & Schüssler, M. 2007, *ApJ* 659, 801
Cameron, R. H. & Schüssler, M. 2010, *ApJ* 720, 1030
Cameron, R. H. & Schüssler, M. 2012, *A&A*, accepted
Dikpati, M. & Gilman, P. A. 2006, *ApJ* 649, 498
Gizon, L. 2004, *Solar Phys.* 224, 217
Gizon, L. & Rempel, M. 2008, *Solar Phys.* 251, 241
Jiang, J., Chatterjee, P., & Choudhuri, A. R. 2007, *MNRAS* 381, 1527
Karak, B. B. & Choudhuri, A. R. 2012, *Solar Phys.* 278, 137
Kitchatinov, L. L. & Olemskoy, S. V. 2011, *Astron. Lett.* 37, 656
Pesnell, D. 2008, *Solar Phys.* 252, 209
Spruit, H. C. 2003, *Solar Phys.* 213, 1

Highlights of Astronomy, Volume 16
XXVIIIth IAU General Assembly, August 2012
T. Montmerle, ed.

doi:10.1017/S1743921314004785

Theoretical Models of Stellar Activity Cycles

Emre Işık

Department of Physics, Istanbul Kültür University, Bakırköy, 34156, Istanbul, Turkey
e-mail: e.isik@iku.edu.tr

Abstract. We discuss possible mechanisms underlying the observed features of stellar activity cycles, such as multiple periodicities in very active stars, non-cyclic activity observed in moderately active stars, and spatial distribution of stellar magnetic regions. We review selected attempts to model the dependence of stellar activity cycles on stellar properties, and their comparison with observations. We suggest that combined effects of dynamo action, flux emergence and surface flux transport have substantial effects on the long-term manifestations of stellar magnetism.

Keywords. stars: activity, stars: late-type, stars: magnetic fields

1. Modelling stellar magnetic cycles

Cool stars with convective envelopes exhibit long-term photometric and spectroscopic variations, which indicate long-term changes of surface magnetic flux. Many stars show single or multiple cycles with some irregularities, while others are characterised by relatively unchanging total radiative flux, hence they are called *flat activity stars.* Whether these stars represent grand minimum states is not yet certain.

The current paradigm concerning the physics of activity cycles in cool stars is based on solar dynamo models, which incorporate the generation and transport of magnetic flux within the convection zone. There are several unknowns regarding rapidly rotating cool stars, in particular the internal rotation, flows, magnetic field geometry, and their complex interplay. Therefore it is probably more appropriate to extrapolate the solar paradigm *gradually* towards faster rotating Sun-like stars or to cooler dwarfs. We expect the dynamo strength to increase with decreasing Rossby number, in parallel with the empirical rotation-activity relation. Simple mean-field dynamo models suggest that the cycle period decreases as the dynamo number increases. However, recent flux transport dynamo models by Jouve., *et al.* (2010) have shown that, to match the observed dependence, meridional flow should either increase its strength with the rotation rate (hereafter Ω), which is incompatible with 3D full-sphere simulations, or have preferred patterns which are difficult to justify. As an alternative explanation for the observed relation, Do Cao & Brun (2011) have shown that turbulent pumping of magnetic flux throughout the convection zone can decrease the cycle period with increasing Ω and decreasing meridional flow speed. However, it should be noted that recent mean-field models of differential rotation indicate an increase of meridional flow speed with Ω (Kitchatinov & Olemskoy 2012). Magnetoconvection simulations in 3D of the entire stellar convection zone have been presented by Brown *et al.* (2011), who demonstrated self-organisation of large-scale toroidal magnetic structures in the midst of the convection zone of a Sun-like star. These structures turn out to be anti-symmetric around the equator, stationary for $3\Omega_\odot$, and exhibit occasional polarity reversals for $5\Omega_\odot$.

We have recently presented numerical simulations combining generation, buoyant rise, and surface transport of magnetic flux in various cool star convection zone configurations (Işık *et al.* 2011). Our composite models are based on a mean-field overshoot dynamo, the period of which decreases with Ω. With increasing Ω the latitudinal distribution of the emerging flux at the surface deviates significantly from that of the dynamo waves at the base of the convection zone, owing to a poleward deflection of rising flux tubes.

When the rotation period is 9 days for a Sun-like star, our models indicate that the periodic dynamo in the deep convection zone can lead to a non-cyclic surface activity. The existence of such moderately active but non-cycling stars were reported by Hall & Lockwood (2004). A similar situation can occur for coronal X-ray cycles in some cool stars: McIvor *et al.* (2006) have shown that sufficiently overlapping magnetic cycles in the surface activity can lead to the absence of X-ray cycles.

Considering mean-field dynamo models, Moss., *et al.* (2011) have suggested that the dynamo must be operating in two distinct layers within the convection zone, to explain the observed poleward moving starspots in the K1-subgiant component of HR 1099. Our simulations of a K1-type subgiant star with Ω and surface shear adopted from HR 1099 have demonstrated that a deep-seated stellar dynamo with a well-defined period and amplitude can co-exist with fluctuating cycles of surface magnetic flux (Işık *et al.* 2011). Further analysis indicates that the short-period cycle at the surface results from the underlying dynamo, while the long-period cycle signal is caused by stochastic emergence of bipolar regions at very high latitudes, where the effects of surface transport are relatively weak (Işık 2012). This can be an explanation for the long-term 'cycle' in HR 1099 reported by Oláh., *et al.* (2009).

2. Open problems and outlook

It is clear that we still lack a rigorous physical understanding of the relations between the emerging flux, rotation rate, convection zone structure, and cycle properties of cool stars (Rempel 2008). How does the link between a tachocline dynamo and surface transport change with rotation rate and convection zone structure, as we depart from solar parameters? How do turbulent transport properties change with Ω? How does the dynamical disconnection of rising magnetic flux tubes vary with stellar parameters?

Such problems will remain as challenges in the future, but observations will reduce the theoretical degrees of freedom. The recent successes of the Babcock-Leighton type solar dynamo models in explaining various aspects of solar magnetism is encouraging for stellar modelling. Based on our recent models, we conclude that the combined effects of flux generation, emergence and transport should be considered in models of stellar magnetic fields, because their interrelations are sensitive to changes in stellar parameters.

References

Brown, B. P., Miesch, M. S., Browning, M. K., Brun, A. S., & Toomre, J. 2011, *ApJ*, 731, 69
Do Cao, O. & Brun, A. S. 2011, *AN*, 332, 907
Hall, J. C. & Lockwood, G. W. 2004, *ApJ*, 614, 942
Işık, E., Schmitt, D., & Schüssler, M. 2011, *A&A*, 528, A135
Işık, E. 2012, in: C. H. Mandrini & D. F. Webb (eds.), *Comparative Magnetic Minima: Characterizing Quiet Times in the Sun and Stars*, Proc. IAU Symposium No. 286 (Cambridge University Press), p. 291
Jouve, L., Brown, B. P., & Brun, A. S. 2010, *A&A*, 509, A32
Kitchatinov, L. L. & Olemskoy, S. V. 2012, *MNRAS*, 423, 3344
McIvor, T., Jardine, M., Mackay, D., & Holzwarth, V. 2006, *MNRAS*, 367, 592
Moss, D., Sokoloff, D., & Lanza, A. F. 2011, *A&A*, 531, A43
Oláh, K., Kolláth, Z., Granzer, T., *et al.* 2009, *A&A*, 501, 703
Rempel, M. 2008, *Journal of Physics Conf. Ser.*, 118, 012032

Highlights of Astronomy, Volume 16
XXVIIIth IAU General Assembly, August 2012
T. Montmerle, ed.

doi:10.1017/S1743921314004797

Stellar Variability Observed with *Kepler*

Jon M. Jenkins[1], Ronald L. Gilliland[2], Soeren Meibom[3], Lucianne Walkowicz[4], William J. Borucki[5], Douglas A. Caldwell[1] and the *Kepler* Science Team

[1]SETI Institute/NASA Ames Research Center, M/S 244-30, Moffett Field, CA 94035, USA
email: jon.jenkins@nasa.gov

[2]Center for Exoplanets and Habitable Worlds, Penn State University, PA 16802, USA

[3]Harvard-Smithsonian Center for Astrophysics, 60 Garden Street, Cambridge, MA 02138, USA

[4]Department of Astrophysical Sciences, Princeton University, Princeton, NJ 08544, USA

[5] NASA Ames Research Center, M/S 244-30, Moffett Field, CA 94035, USA

Abstract. The *Kepler* photometer was launched in March 2009 initiating NASA's search for Earth-size planets orbiting in the habitable zone of their star. After three years of science operations, *Kepler* has proven to be a veritable cornucopia of science results, both for exoplanets and for astrophysics. The phenomenal photometric precision and continuous observations required in order to identify small, rocky transiting planets enables the study of a large range of phenomena contributing to stellar variability for many thousands of solar-like stars in *Kepler*'s field of view in exquisite detail. These effects range from <1 ppm acoustic oscillations on timescales from a few minutes and longward, to flares on timescales of hours, to spot-induced modulation on timescales of days to weeks to activity cycles on timescales of months to years. Recent improvements to the science pipeline have greatly enhanced *Kepler*'s ability to reject instrumental signatures while better preserving intrinsic stellar variability, opening up the timescales for study well beyond 10 days. We give an overview of the stellar variability we see across the full range of spectral types observed by *Kepler*, from the cool, small red M stars to the hot, large late A stars, both in terms of amplitude as well as timescale. We also present a picture of what the extended mission will likely bring to the field of stellar variability as we progress from a 3.5 year mission to a 7.5+ year mission.

Keywords. techniques: photometric, stars: activity, stars: flare, stars: spots, stars: rotation

---

The launch of the *Kepler* Mission in March 2009 opened up a new window on the detailed photometric behavior of ∼190,000 stars on timescales from a few minutes to a few years and longer with ultra high precision of ∼30 ppm for a G2V star at a *Kepler* magnitude $K_p = 12$ in 6.5 hours. *Kepler* observes ∼150,000 stars at half hour intervals and up to 512 stars at 1 minute intervals. With this level of precision and data completeness above 90%, *Kepler* is a near perfect stellar variability observatory. *Kepler* was awarded an extended mission with operations continuing through September 2016, significantly enhancing the timescales available for study. The number and type of astrophysical phenomena contributing to stellar variability observed with *Kepler* include:

• Acoustic p-mode oscillations in hundreds of solar-like stars and red giants on timescales of minutes to hours and up (see Chaplin, Kjeldsen, Christensen-Dalsgaard, *et al.* 2011)

• Flares on timescales of hours (see, e.g., Balona 2012, Walkowicz, Basri, Batalha, *et al.* 2011)

• Gravity waves in red giants that have already begun to fuse Helium in their core (Mosser, Goupil, Belkacem, *et al.* 2012)

• Spot signatures (see, e.g., Désert, Charbonneau, Demory, *et al.* 2011, Sanchis-Ojeda, Fabrycky, Winn, *et al.* 2012)

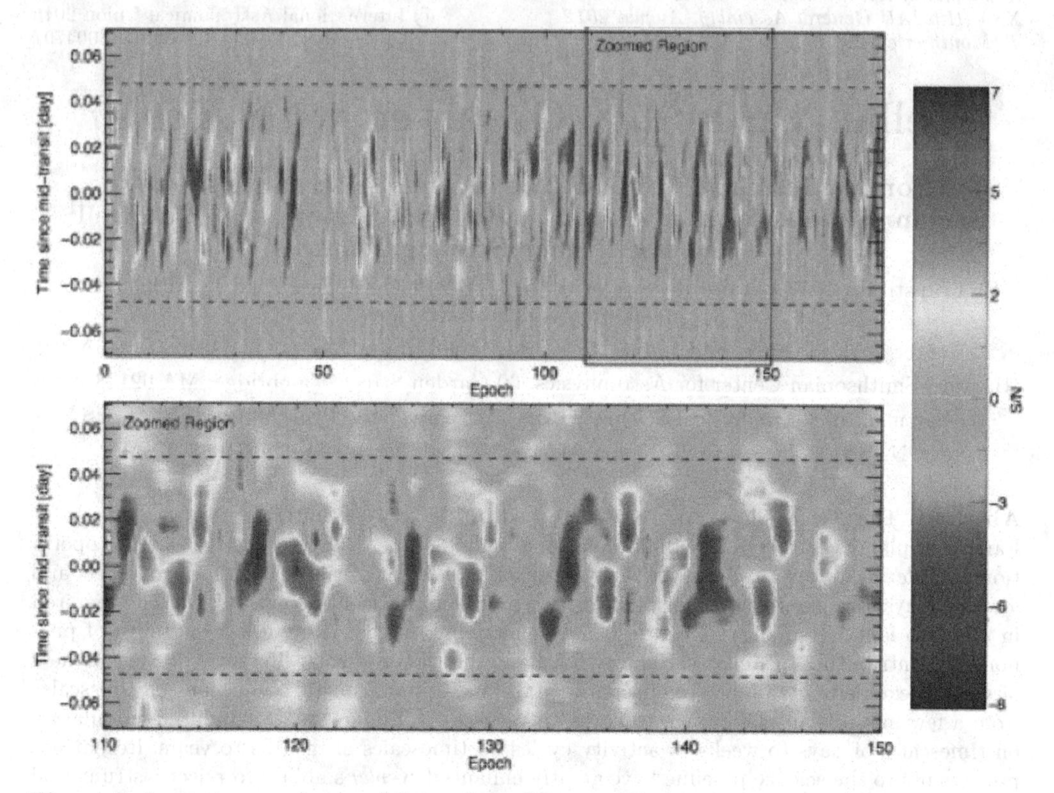

Figure 1. Spatio-temporal map of the surface of the star Kepler-17 scanned stroboscopically by its transiting planet, which has an orbital period exactly $1/8^{\text{th}}$ that of the star (Sanchis-Ojeda, Fabrycky, Winn, *et al.* 2012). This time-map image corresponds to the scatter (S/N) measured across the residuals in each transit epoch, relative to mid-transit as a function of the epoch.

- Stellar rotation and gyrochronology for star clusters from 0.5 to 9 GY (Meibom, S., Barnes, S. A., Latham, D. W., *et al.* 2011)
- Stellar variability of solar-like stars (Gilliland, Chaplin, Dunham, *et al.* 2011)
- Stellar cycles (in the 7.5-year extended mission)

Acknowledgements

Kepler was selected as the 10th mission of the Discovery Program. Funding for this mission is provided by NASA's Science Mission Directorate.

References

Balona, L. A. 2012, *MNRAS* 423, 3420
Chaplin, W. J., H. Kjeldsen, J. Christensen-Dalsgaard, *et al.* 2011, *Science* 332, 213
Desert, J-M., D. Charbonneau, B-O. Demory, *et al.* 2011, *ApJS* 197, 14
Fröhlich, C., Crommelynck, D. A., Wehrli, C., *et al.* 1997, *Solar Phys.* 175, 267
Gilliland, R. L., Chaplin, W. J., Dunham, E. W., *et al.* 2011, *ApJS* 197, 6
Meibom, S., Barnes, S. A., Latham, D. W., *et al.* 2011, *ApJ* 733, L9
Mosser, B., Goupil, M. J., Belkacem, K., *et al.* 2012, *A&A* 540, A143
Sanchis-Ojeda, R., Fabrycky, D. C., Winn, J. N.,*et al.* 2012, *Nature* 487, 449
Walkowicz, L. M., Basri, G., Batalha, N., *et al.* 2011, *AJ* 141, 50

Highlights of Astronomy, Volume 16
XXVIIIth IAU General Assembly, August 2012
T. Montmerle, ed.

doi:10.1017/S1743921315001556

Preface: Joint Discussion JD4 Ultraviolet emission in early-type galaxies

The presence of enhanced ultraviolet (UV) emission in early-type galaxies, which are dominated by old stellar populations, has been a puzzle for more than 40 years. The observed UV flux plausibly contains contributions from both evolved stars (traditionally referred to as the 'UV upturn' phenomenon) and young stars. While the UV upturn is well-characterised observationally, a firm understanding of its origin still eludes us. Similarly, while recent results from GALEX and deep optical surveys strongly indicate the presence of small mass fractions of young stars across low and intermediate redshift, the exact mechanism (e.g. stellar mass loss, minor mergers) that drives this star formation is not fully understood. The role of active galactic nuclei in regulating the low-level star formation at late epochs also remains relatively unexplored in galaxy formation models.

This joint discussion meeting aimed to deliver a review of each of the topics outlined above. It discussed the observational evidence for the UV-upturn, for example its correlation with other bulk galaxy properties such as metallicity, and to explore the various theories of its origin. Primarily, these are single-star channels (e.g. stars on the extreme horizontal branch and the post-asymptotic giant branch) and binary stars, both with extended distributions.

New UV results that demonstrate the presence of small mass fractions of young stars in early-type galaxies were explored, especially in the context of the principal driver of this low-level star formation. Traditional recipes for AGN feedback, which typically truncate star formation in massive galaxies at early epochs, must be revisited to accomodate the low levels of star formation observed in nearby early-type galaxies. The meeting reviewed the role of AGN feedback in the evolution of early-type galaxies and examined how the most recent UV results can further constrain the feedback recipes employed in the models.

With the recently renewed capability of HST (via the Wide Field Camera 3) to image the UV with a high spatial resolution and a wide field-of-view, many exciting results are expected over the next few years. For example, recent star formation should result in UV 'fine structure' on small spatial scales, as opposed to the smooth light profile expected from the UV upturn (since it is produced by the old, underlying stellar population). With the advent of the JWST and the extremely large telescopes, it will be possible to trace rest-frame UV emission in early-type galaxies both to unprecedentedly high redshift and at higher spatial resolution than is possible with current instruments. This meeting aimed to revisit the old issue of UV emission in early-type galaxies in a new light, with a view to preparing the ground for more definitive studies with forthcoming instrumentation.

Sugata Kaviraj
SOC Chair JD4

Highlights of Astronomy, Volume 16
XXVIIIth IAU General Assembly, August 2012
T. Montmerle, ed.

doi:10.1017/S1743921314004803

Molecular gas properties in early-type galaxies

Estelle Bayet[1], Martin Bureau[1], Tim Davis [2], Lisa Young[3], Alison Crocker[4] and the ATLAS 3D team

[1]Sub-Dept. of Astrophysics, Dept. of Physics at University of Oxford, Denys Wilkinson Building, Keble Road, Oxford OX1 3RH, U.K.
email: bayet@physics.ox.ac.uk

[2]European Southern Observatory, Karl-Schwarzschild-Str. 2, 85748 Garching bei Muenchen, Germany
email: tdavis@eso.org

[3]New Mexico Tech, 801 Leroy Place, Socorro, NM 8780, USA
email: lyoung@physics.nmt.edu

[4]Department of Astronomy, University of Massachusetts, Lederle Graduate Research Tower B 619E, 710 North Pleasant Street, Amherst, MA 01003-9305, USA
email: crocker@astro.umass.edu

1. Summary

We present detailed study of the physical properties of the molecular gas in a sample of 18 gas-rich early-type galaxies -ETGs- from the ATLAS^{3D} sample. Our goal is to better understand the star formation processes occurring in those galaxies, starting in this work by determining the properties of their molecular star-forming gas. Such study has never been performed before for ETGs and open a new window for exploring the star formation processes in the Universe. We use here the ^{12}CO(1-0, 2-1), ^{13}CO(1-0, 2-1), HCN(1-0) and HCO^+(1-0) transitions already obtained (Krips *et al.* 2010 and Crocker et al. 2012 and new detections of the ^{12}CO(3-2) line that we present too (see Bayet *et al.* 2012, resubmitted). From this dataset, we derive the average H_2 kinetic temperature, H_2 volume density and column density of the emitting gas using a non-Local Thermodynamical Equilibrium theoretical approach. For most of the gas-rich ETGs included in our sample, the CO transitions can be reproduced by gas kinetic temperatures between 10-20 K, densities of 10^{3-4} cm^{-3} and CO column densities of 10^{18-20} cm^{-2}. Since the CO lines trace different physical conditions than those required to emit the HCN and HCO^+ lines, they are treated separately. The physical parameters corresponding to the HCN and HCO^+ gas component suffer from large uncertainties and have to be considered as indicative only. In this study, for the first time, we also compare the predicted CO spectral line energy distributions of our gas-rich ETGs and their gas properties with those of a sample of nearby well-studied disc galaxies. The gas excitation conditions in 12/18 early-type galaxies appear analogous to those in the center fo Milky Way, hence the star formation activity driving these conditions may be of a similar strength and nature. The conclusions drawn have to be nevertheless considered carefully since they are based only on a limited number of observations. We show from our models that the ^{12}CO(6-5) line emission is particularly useful for improving these results.

References

Crocker, A., Krips, M., Bureau, M., *et al.* 2012, *MNRAS*, 421, 1298
Krips, M., Crocker, A. F., Bureau, M., Combes, F., & Young, L. M. 2010, *MNRAS*, 407, 2261

Highlights of Astronomy, Volume 16
XXVIIIth IAU General Assembly, August 2012
T. Montmerle, ed.

doi:10.1017/S1743921314004815

Tracing the evolution within nearby galaxy groups: a multi-wavelength approach

Daniela Bettoni[1], Antonina Marino[2], Roberto Rampazzo[1], Henri Plana[3], Margarita Rosado[4], Giuseppe Galletta[2], Paola Mazzei[1], Luciana Bianchi[5], Lucio M. Buson[1], Patricia Ambrocio-Cruz[6] and Ruslan Gabbasov[4]

[1]Inaf - Osservatorio Astronomico di Padova, Italy
email: daniela.bettoni@oapd.inaf.it

[2]Dept. of Physics & Astronomy University of Padova, Italy
email: antonina.marino@unipd.it

[3]Laboratorio de Astrofisica Teorica e Observacional, Universidade de Santa Cruz, Brasil

[4]Instituto de Astronomia Universidad National Autonoma de Mexico, Mexico

[5]Dept of physics and Astronomy, Johns Hopkins University, Baltimore, USA

[6]Universidad Autonoma del Estado de Hidalgo, area academica de Ciencias de la Tierra y Materiales, Mexico

Abstract. Evolutionary scenarios suggest that several mechanisms (from inner secular evolution to accretion/merging) may transform galaxy members, driving groups from an active star forming phase to a more passive, typical of dense environments. We are investigating this transition in a nearby group sample, designed to cover a wide range of properties (see also Marino *et al.* (2010), Bettoni *et al.* (2011) and Marino *et al.* (2012)). We study two groups, USGC U268 and USGC U376 located in different regions of the Leo cloud, through a photometric and kinematic characterization of their member galaxies. We revisit the group membership, using results from recent red-shift surveys, and we investigate their substructures. U268, composed of 10 catalogued members and 11 new added members, has a small fraction (~24%) of early-type galaxies (ETGs). U376 has 16 plus 8 new added members, with ~38% of ETGs. We find the significant substructuring in both groups suggesting that they are likely accreting galaxies. U268 is located in a more loose environment than U376. For each member galaxy, broad band integrated and surface photometry have been obtained in far-UV (FUV) and near-UV (NUV) with GALEX, and in u, g, r, i, z (SDSS) bands. H_α imaging and 2D high resolution kinematical data have been obtained using PUMA Scanning Fabry-Perot interferometer at the 2.12 m telescope in San Pedro Mártir (Baja California, México). We improved the galaxy classification and we detected morphological and kinematical distortions that may be connected to either on-going and/or past interaction/accretion events or environmental induced secular evolution. U268 appears more active than U376, with a large fraction of galaxies showing interaction signatures (60% vs. 13%). The presence of bars among late-type galaxies is ~10% in U268 and 29% in U376. The cumulative distribution of (FUV - NUV) colors of galaxies in U268 is significantly different (bluer) than that of U376's galaxies. Most (80%) of the early-type members in U376 inhabits the red sequence, a large fraction of galaxies, of different morphological types, are located in the green valley, while the blue sequence is under-populated with respect to U268.

Keywords. galaxies: evolution, galaxies: groups, galaxies: photometry, ultraviolet: galaxies

References

Bettoni, D., Galletta, G., Rampazzo, R., Marino, A., Mazzei, P., & Buson L. 2011, *A&A*, 534, A24

Marino A., Bianchi L., Rampazzo R., Buson L. M., & Bettoni D. 2010, *A&A*, 511, A29

Marino A., Plana, H., Rampazzo R., *et al.* 2012, *MNRAS*, in press. arXiv:1209.4278

Highlights of Astronomy, Volume 16
XXVIIIth IAU General Assembly, August 2012
T. Montmerle, ed.

doi:10.1017/S1743921314004827

Extreme Horizontal Branch Stars in Passively Evolving Early Type Galaxies

Fabiola Hernández-Pérez[1] and Gustavo Bruzual[2]

[1]Centro de Investigaciones de Astronomía CIDA, A.P. 264, Mérida, Venezuela
email: fhernandez@cida.ve

[2]Centro de Radioastronomía y Astrofísica, UNAM, Campus Morelia, México
email: g.bruzual@crya.unam.mx

Abstract. We study the effects of including binary star evolution in population synthesis models. We use the Hurley *et al.* (2002) code to compute binary star evolutionary tracks, and follow the procedure by Han *et al.* (2002), in particular, the two 2HeWD merger channel, to form EHB stars from a binary pair. We apply the resulting models to study UV excess ETGs.

Keywords. UV upturn: galaxies – Elliptical: galaxies – EHB: stars – Binaries: stars.

Using binary star evolutionary tracks computed with a modified version of the Hurley *et al.* (2002) code, we add the effects of binaries to standard population synthesis models (e.g. Bruzual & Charlot 2003). Following the isochrone synthesis scheme, we compute a series of isochrones at different ages which include blue stragglers and EHB stars. The resulting isochrones reproduce the principal characteristics of observed CMD of globular clusters. However, EHB in our models tend to be hotter and less luminous than observed in real CMDs. We derive the integrated spectral energy distributions and the colors corresponding to these populations.

The initial mass function (IMF) controls the number of progenitors, affecting the total number of EHB produced in a model that includes binary stars. Thus, Salpeter IMF models result in half the number of EHB as compared to Chabrier IMF models. Metallicity seems to have little effect on the EHB formation rate in these models. In low Z models, the FUV-NUV color is redder and the NUV-r color is bluer, because blue HB stars contribute to the NUV flux.

We compare our model with a crossmatching sample of $\sim$ 3400 early type galaxies from SDSS/DR8-GALEX/GR6. We study the color evolution of model populations and evaluate possible scenarios to explain the UV upturn. Population synthesis models including binary evolution cannot reproduce the observed SED's of all early-type galaxies (ETGs). Models remain in the blue zone of the diagram, and the red sequence cannot be reproduced by a model in which binaries are always present. We conclude that it could happen that binary stars are not the only progenitors of EHB stars, and that standard stellar population synthesis models, with no binaries, still remain a valid tool to study ETGs (Yi 2008). However, more realistic treatments of HB and binary star evolution are needed for a better understanding of the UVX phenomenon in ETGs.

References

Bruzual G. & Charlot S. 2003, *MNRAS*, 344, 1000
Han Z., Podsiadlowski P., Maxted P., Marsh T. & Ivanova N. 2002, *MNRAS*, 336, 449
Hurley J., Tout C. & Pols O. 2002, *MNRAS*, 329, 897
Yi S. K. 2008, *ASP Conference Series*, 392, 3

Highlights of Astronomy, Volume 16
XXVIIIth IAU General Assembly, August 2012
T. Montmerle, ed.

doi:10.1017/S1743921314004839

The effect of Helium-enhanced stellar populations on the ultraviolet upturn phenomenon of early-type galaxies

Chul Chung, Suk-Jin Yoon and Young-Wook Lee

Center for Galaxy Evolution Reasearch and Department of Astronomy,
Yonsei University, Seoul 120-749, Republic of Korea
email: chung@galaxy.yonsei.ac.kr

Abstract. We present new population synthesis models (Chung *et al.* 2011) for quiescent early-type galaxies (ETGs) with UV-upturn phenomenon using relatively metal-poor and helium-enhanced subpopulations in the model. We find that the presence of helium-enhanced subpopulations in ETGs can naturally reproduce the strong UV-upturns observed in giant elliptical galaxies (Figure 1. left panel), without invoking unrealistically old ages (Park & Lee 1997). Our models with helium-enhanced subpopulations also predict that the well-known Burstein relation can be explained by the fraction of helium-enhanced subpopulation, the mean age, and the mean metallicity of the underlying stellar populations (Figure 1. right panel).

Keywords. galaxies: elliptical and lenticular, CD — galaxies: evolution — galaxies: stellar content — ultraviolet: galaxies

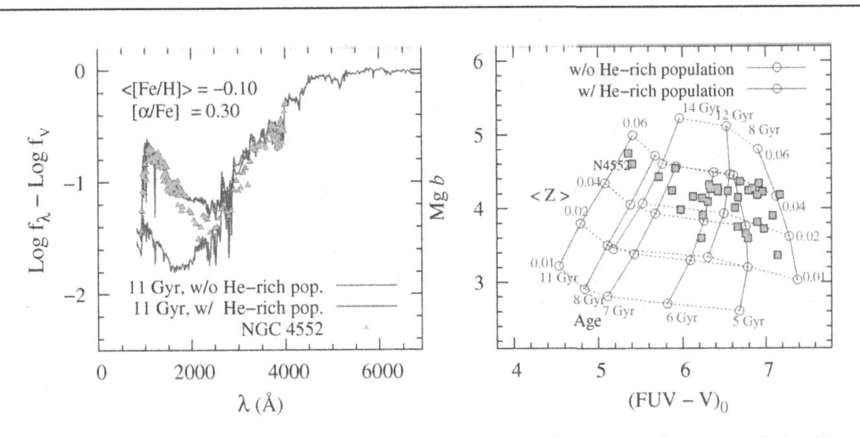

Figure 1. (*Left*) Comparison of observed SEDs of NGC 4552 with our models. (*Right*) FUV–V color vs. Mg b correlation for the sample of quiescent ETGs from Bureau *et al.* (2011).

References

Bureau, M., *et al.* 2011, *MNRAS*, 414, 1887
Chung, C., Yoon, S.-J., & Lee, Y.-W. 2011, *ApJ* (Letters) 740, L45
Park, J.-H. & Lee, Y.-W. 1997, *ApJ*, 476, 28

Highlights of Astronomy, Volume 16
XXVIIIth IAU General Assembly, August 2012
T. Montmerle, ed.

doi:10.1017/S1743921314004840

Spatially resolved molecular gas in early-type galaxies

T. A. Davis,[1] K. Alatalo,[2] M. Bureau,[3] L. Young,[4] L. Blitz,[2] A. Crocker,[5] E. Bayet,[3] M. Bois,[6] F. Bournaud,[7] M. Cappellari,[3] R. L. Davies,[3] P-A. Duc,[7] P. T. de Zeeuw,[1,8] E. Emsellem,[1,9] J. Falcon-Barroso,[10] S. Khochfar,[11] D. Krajnovic,[1] H. Kuntschner,[1] P.-Y. Lablanche,[1] R. M. McDermid,[12] R. Morganti,[13] T. Naab,[14] M. Sarzi,[15] N. Scott,[16] P. Serra,[13] and A. Weijmans[17]

[1]ESO, Germany; [2]UC Berkeley, USA; [3]Univ. Oxford, UK; [4]New Mexico Tech, USA; [5]Univ. Toledo, USA; [6]Observatoire de Paris, France; [7]Univ. Paris Diderot, France; [8]Leiden Univ., The Netherlands; [9]Univ. de Lyon, France; [10]IAC, Spain; [11]MPE, Germany; [12]Gemini Observatory, USA; [13]ASTRON, The Netherlands; [14]MPA, Germany; [15]Univ. Hertfordshire, UK; [16]Swinburne Univ., Australia; [17]Univ. Toronto, Canada;

Abstract. In around ≈25% of early-type galaxies (ETGs) UV emission from young stellar populations is present. Molecular gas reservoirs have been detected in these systems (e.g. Young *et al.* 2011), providing the fuel for this residual star-formation. The environment in which this molecular gas is found is quite different than that in spiral galaxies however, with harsher radiation fields, deeper potentials and high metallicity and alpha-element abundances. Here we report on one element of our multi-faceted programme to understand the similarities and differences between the gas reservoirs in spirals and ETGs. We use spatially resolved observations from the CARMA mm-wave interferometer to investigate the size of the molecular reservoirs in the the CO-rich ATLAS^{3D} ETGs (survey described in Alatalo *et al.* 2012, submitted). We find that the molecular gas extent is smaller in absolute terms in ETGs than in late-type galaxies, but that the size distributions are similar once scaled by the galaxies optical/stellar characteristic scale-lengths (Fig 1, left). Amongst ETGs, we find that the extent of the molecular gas is independent of the kinematic misalignment, despite the many reasons why misaligned gas might have a smaller extent. The extent of the molecular gas does depend on environment, with Virgo cluster ETGs having less extended molecular gas reservoirs (Fig 1, right). Whatever the cause, this further emphases that cluster ETGs follow different evolutionary pathways from those in the field. Full details of this work will be presented in Davis *et al.* (2012), submitted.

Reference

Young, L. M., Bureau, M., Davis, T. A., *et al.* 2011, *MNRAS*, 414, 940

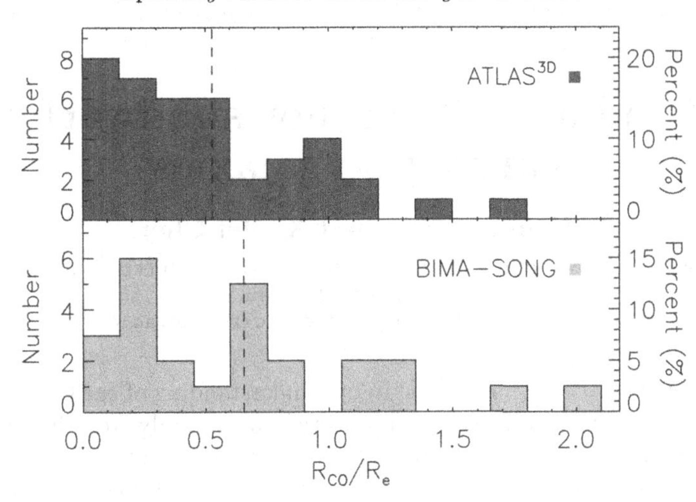

Figure 1. Molecular sizes for the ATLAS^{3D} ETGs normalised by the stellar effective radius. Compared to BIMA-SONG spirals (left) and as a function of environment (right). The mean radius is marked with a dashed line.

Highlights of Astronomy, Volume 16
XXVIIIth IAU General Assembly, August 2012
T. Montmerle, ed.

doi:10.1017/S1743921314004852

Far-UV radiation from hot subdwarf stars in early-type galaxies

Zhanwen Han and Xuefei Chen

Key Laboratory for the Structure and Evolution of Celestial Objects, Yunnan Observatory, Kunming, 650011, China
email: zhanwenhan@ynao.ac.cn, cxf@ynao.ac.cn

Far-ultraviolet (FUV) excess is crucial to our understanding of early-type galaxies and it is widely believed that the FUV radiation originates mainly from hot subwarf stars. Hot subdwarf stars may form from binary interactions or from single star evolution. In the binary channel, a star near the tip of the first giant branch (FGB) may get its envelope removed by its companion via stable Roche lobe overflow or common envelope ejection, and then evolves to a hot subdwarf star (Han *et al.* 2002, 2003, 2007). Such a process does not depend much on metallicity. In the single star channel, a low mass star may lose its envelope at the tip of FGB to become a hot subdwarf if the star is metal rich (over solar) and its envelope binding energy becomes positive (Han *et al.* 1994). We conclude from the combination of the two channels that

- FUV excess is universal (i.e. from dwarf to giant ellipticals),
- the FUV excess does NOT depend on metallicity or redshift if the metallicity is less than solar and DOES depend on metallicity or redshift if the metallicity is over solar.

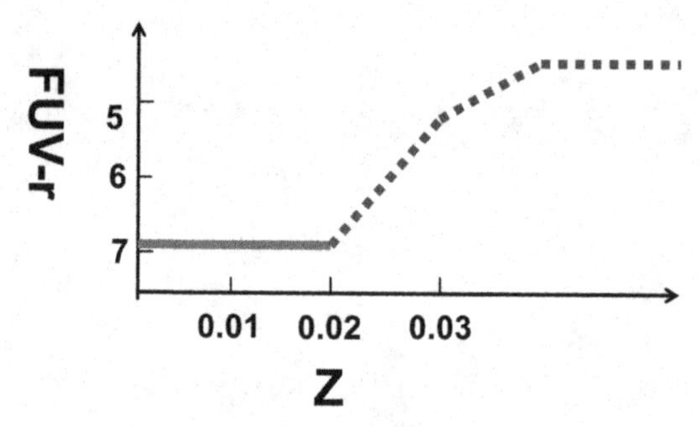

Figure 1. A schematic view of the FUV excess - metallicity relation as predicted by our theory. Binary channel dominates FUV for low metallicity (solid line) while single star channel dominates for high metallicity (dotted line)

References

Han, Z., Podsiadlowski, Ph., & Eggleton, P. P. 1994, *MNRAS*, 270, 121
Han, Z., Podsiadlowski, Ph., Maxted, P. F. L., Marsh, T. R., & Ivanova, N. 2002, *MNRAS*, 336, 449
Han, Z., Podsiadlowski, Ph., Maxted, P. F. L., & Marsh, T. R., 2003, *MNRAS*, 341, 669
Han, Z., Podsiadlowski, Ph., & Lynas-Gray, A. E. 2007, *MNRAS*, 380, 1098

Highlights of Astronomy, Volume 16
XXVIIIth IAU General Assembly, August 2012
T. Montmerle, ed.

doi:10.1017/S1743921314004864

Young stars in nearby early-type galaxies: The GALEX-SAURON perspective

Hyunjin Jeong[1], Sukyoung K. Yi[2], Martin Bureau[3] and Roger L. Davies[3]

[1]Korea Astronomy and Space Science Institute, Daejeon 305-348, Korea
email: hyunjin@kasi.re.kr

[2]Department of Astronomy, Yonsei University, Seoul 120-749, Korea
[3]Sub-Department of Astrophysics, University of Oxford, Oxford OX1 3RH, UK

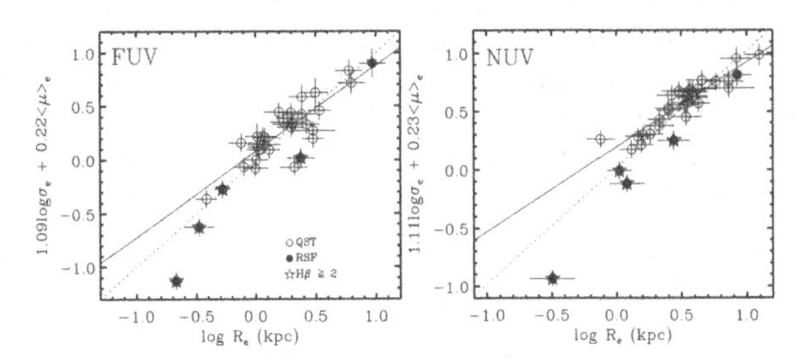

Figure 1. Fundamental Planes in the FUV and NUV bands. Using the empirical criterion (NUV−V=5.0), we divided the sample into quiescent (QST; open symbols) early-type galaxies and recent star formation (RSF, filled symbols) galaxies. Linear fits to the two subsamples (whole sample, quiescent galaxies only) are also shown as dotted and solid lines, respectively.

Recent studies from the Galaxy Evolution Explore (GALEX) ultraviolet (UV) data reveal that the recent star formation is more common in early-type galaxies (ETGs) that we used to believe (Jeong *et al.* 2007). Here we used the unique GALEX UV data on existing SAURON IFU-studied galaxies and combined these two datasets (UV and IFU) to find where photometric anomalies occur. One of the highlights of our study is the work on the Fundamental Plane (FP). The tilt and scatter found in optical FPs have been an issue. From our sample of 34 ETGs, we found that most of the tilt and scatter are caused by the minority ETGs which have been forming stars recently at very low level (see figure 1). Using our UV FPs, we found a strong evidence for star formation history being the main source of the mystery (Jeong *et al.* 2009).

References

Jeong, H., Bureau, M., Yi, S. K., Krajnovic D., & Davies, R. L. 2007, *MNRAS* 376, 1021
Jeong, H., *et al.* 2009, *MNRAS* 398, 2028

Highlights of Astronomy, Volume 16
XXVIIIth IAU General Assembly, August 2012
T. Montmerle, ed.

doi:10.1017/S1743921314004876

Ultraviolet Emission from Star-formation in Selected Gas-rich Early-type Galaxies

Lerothodi L. Leeuw[1,2]

[1]College of Graduate Studies, UNISA, P. O. Box 392, UNISA Pretoria, 0003, South Africa
[2]SETI Institute, 189 Bernardo Avenue, Suite 100, Mountain View, CA 94043, USA
email: lerothodi@alum.mit.edu

Abstract. We present *GALEX* ultraviolet (UV) emission results of star-formation in a small sample of nearby, gas-rich early-type galaxies. The first observational evidence of star-formation in this sample was presented by Leeuw *et al.* 2008, using 350 μm continuum data. The measured far-infrared (far-IR) excess of these galaxies showed that the most likely and dominant heating source of the observed 350 μm continuum emission from dust is star-formation, that could have been triggered by an accretion or merger event. Consistent with starbursts that are less than 1 Gyr (e.g., Kaviraj 2010), the *GALEX* near-UV (NUV) minus SDSS r-band emission of the galaxies is < 5.5. The UV results corroborate those of mid-IR to radio data for the sample.

Keywords. galaxies: elliptical and lenticular, cD – galaxies: evolution – galaxies: ISM – infrared: general – submillimeter – ultraviolet: general

---

Synopsis

We summarize ultraviolet (UV) emission results of on-going star-formation determined from archival *GALEX* and SDSS data for a small sample of nearby early-type galaxies, that were known to have CO in centrally located gas disks. Like classic ellipticals, the luminosity profiles of these galaxies follow the de Vaucoleur, $r^{1/4}$ law. However, they represent a spread of merger tracers or ages, from galaxies that have been morphologicaly classified as on-going or early-age major mergers to very-late accretion or quiescent systems (Leeuw *et al.* 2008; 2011). Using 350 μm continuum data, Leeuw *et al.* 2008 presented the first observational evidence of star-formation in this sample. The measured far-infrared (far-IR) excess of these galaxies showed that the most likely and dominant heating source of the observed 350 μm continuum emission from dust is star-formation, that could have been triggered by an accretion or merger event and is stronger in the "most recent mergers". Consistent with starbursts that are less than 1 Gyr (e.g., Kaviraj 2010), the *GALEX* near-UV (NUV) minus SDSS r-band emission of the galaxies is < 5.5. Further, the sample dwarf elliptical NGC 4476 and intermediate-age elliptical NGC 5666, that respectively have only a hint and a dense ring of star-formation, based on their far-IR data, have NUV-r = 4.58 and 2.64. This UV analysis corroborates the star-formation results from those of mid-IR to radio data for this very small sample.

References

Kaviraj, S. 2010, *MNRAS*, 408, 170

Leeuw, L. L., Davidson, J., Dowell, C. D., & Matthews, H. E. 2008, *ApJ*, 677, 249

Leeuw, L. L., Bregman, J., Davidson, J., Temi, P., & Im, S. S. 2011, in: C. Carignan, F. Combes & K. C. Freeman (eds.), *Tracing the Ancestry of Galaxies (on the land of our ancestors)*, Proceedings of the International Astronomical Union, *IAU Symposium*, Volume 277, p. 30

Highlights of Astronomy, Volume 16
XXVIIIth IAU General Assembly, August 2012
T. Montmerle, ed.

doi:10.1017/S1743921314004888

The UV-upturn in brightest cluster galaxies

S. I. Loubser[1] and P. Sánchez-Blázquez[2]

[1]Centre for Space Research, North-West University, Potchefstroom 2520, South Africa
email: Ilani.Loubser@nwu.ac.za

[2]Departamento de Física Teórica, Universidad Autónoma de Madrid, E28049, Spain
email: psanchezblazquez@googlemail.com

Abstract. This study is part of a series devoted to the investigation of a large sample of brightest cluster galaxies (BCGs), their properties and the relationships between these and the properties of the host clusters. In this paper, we compare the stellar population properties derived from high signal-to-noise, optical long-slit spectra with the *GALEX* ultraviolet (UV) colour measurements for 36 nearby BCGs to understand the diversity in the most rapidly evolving feature in old stellar systems, the UV-upturn. We investigate: (1) the possible differences between the UV-upturn of BCGs and those of a control sample of ordinary ellipticals in the same mass range, as well as possible correlations between the UV-upturn and other general properties of the galaxies; (2) possible correlations between the UV-upturn and the properties of the host clusters; (3) recently proposed scenarios where helium-sedimentation in the cluster centre can produce an enhanced UV-upturn. We find systematic differences between the UV-colours of BCGs and ordinary ellipticals, but we do not find correlations between these colours and the properties of the host clusters. Furthermore, the observations do not support the predictions made by the helium-sedimentation model as an enhancer of the UV-upturn.

Keywords. galaxies: evolution – galaxies: elliptical and lenticular, cD – galaxies: stellar content – ultraviolet: galaxies

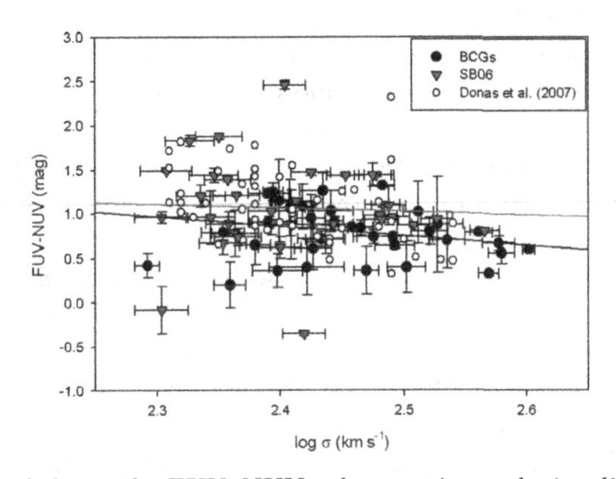

Figure 1. The top panel shows the FUV–NUV colour against velocity dispersion for BCGs and non-BCGs (see Loubser & Sánchez-Blázquez 2011). The BCGs are indicated with the filled circles, and the non-BCGs are indicated with the filled triangles and empty circles. The correlation for the BCGs is the lower line of the two lines, the upper line is the correlation for non-BCGs. We find that the UV colour distributions of the two elliptical samples agree very well, whereas the distribution of the BCG sample is significantly different from both elliptical samples.

Reference

Loubser, S. I. & Sánchez-Blázquez, P 2011, *MNRAS*, 410, 2679

Highlights of Astronomy, Volume 16
XXVIIIth IAU General Assembly, August 2012
T. Montmerle, ed.

doi:10.1017/S174392131400489X

Evolution of Massive Galaxy Structural Properties and Sizes via Star Formation

Jamie R. Ownsworth[1], Christopher J. Conselice[1], Alice Mortlock[1], William G. Hartley[1] and Fernando Buitrago[1,2]

[1]University of Nottingham, School of Physics and Astronomy, Nottingham, NG7 2RD, U.K.
[2]SUPA, Institute for Astronomy, University of Edinburgh, Royal Observatory, Edinburgh, EH9 3HJ, U.K.

We investigate the resolved star formation properties of a sample of 45 massive galaxies ($M_* > 10^{11} M_\odot$) within a redshift range of $1.5 \leqslant z \leqslant 3$ detected in the GOODS NICMOS Survey (Conselice *et al.* 2011), a HST H_{160}-band imaging program. We derive the star formation rate as a function of radius using rest frame UV data from deep z_{850} ACS imaging. The star formation present at high redshift is then extrapolated to $z = 0$, and we examine the stellar mass produced in individual regions within each galaxy. We also construct new stellar mass profiles of the in situ stellar mass at high redshift from Sérsic fits to rest-frame optical, H_{160}-band, data. We combine the two stellar mass profiles to produce an evolved stellar mass profile. We then fit a new Sérsic profile to the evolved profile, from which we examine what effect the resulting stellar mass distribution added via star formation has on the structure and size of each individual galaxy.
In summary: We find three different profiles of star formation within the massive galaxies in this sample, Non-significant Star Formation Growth (NG), Outer Star Formation Growth (OG) and Inner Star Formation Growth (IG) (see Ownsworth *et al.* 2012). With most of this sample of massive galaxies falling in to NG class using the derived tau model of evolution.We find that the star formation we observe at high redshift, and its effects on galaxy sizes, is not large enough to fully explain the observed size evolution of ~300 – 500%. Star formation alone alone can only produce an increase in effective radius on the order of ~16% over the whole sample. This value can vary as much a a factor of 4.5 by using different evolution mechanisms but is always insufficient to fully explain the observations. We find that over the whole sample of massive galaxies the stellar mass added via star formation has a slight effect on the surface brightness of the evolved galaxy profile such that their Sérsic indices decrease. This indicates that the star formation within these galaxies follows the same radial distribution as the original stellar mass profile. This also implies that star formation evolution has a minimal effect on structural evolution between $z \sim 3$ and the present day. The increase in effective radius can be enhanced by adding in the effects of stellar migration to the stellar mass created via star formation. This increases the total effective radius growth to ~55%, which is still however much smaller than the total observed size increase.

We conclude that due to the lack of sufficient size growth and Sérsic evolution by star formation other mechanisms such as merging must contribute a large proportion to account for the observed structural evolution from $z > 1$ to the present day.

References

Conselice, C. J., Bluck, A. F. L., Buitrago, F., *et al.* 2011, *MNRAS*, 413, 80
Ownsworth, J. R., Conselice, C. J., Mortlock, A., *et al.* 2012, *MNRAS*, 426, 764

Highlights of Astronomy, Volume 16
XXVIIIth IAU General Assembly, August 2012
T. Montmerle, ed.

doi:10.1017/S1743921314004906

Star Formation History of Early-Type Galaxies with Tidal Debris in the S^4G

Beatriz H. F. Ramos[1,2], Karín Menéndez-Delmestre[2], Taehyun Kim[3,4], Kartik Sheth[5] and S^4G team

[1]Instituto de Biociências, Departamento de Ciências Naturais, Universidade Federal do Estado do Rio de Janeiro, Rio de Janeiro, Brazil
email: `ramos@astro.ufrj.br`

[2]Observatório do Valongo, Universidade Federal de Rio de Janeiro, Rio de Janeiro, Brazil
[3]European Southern Observatory, Santiago, Chile
[4]Dept. Physics & Astronomy, Seoul National University, Seoul, Republic of Korea
[5]National Radio Astronomy Observatory, Charlottesville, VA, USA

Abstract. Local early-type galaxies (ETGs), despite typically being associated to old stellar populations and passive evolution, have been in some cases observed to present peculiarities in their stellar structure, like disks and shells (e.g., Kormendy *et al.* 1997, Rix, Carollo & Freeman 1999). Moreover, it has been observed that ETGs with such tidal features may present UV emission (Rampazzo *et al.* 2007, Salim & Rich 2010). These properties make them relevant constraints to galaxy formation models. We are analysing the structure of nearby ETGs observed in the Spitzer Survey of Stellar Structure in Galaxies (S^4G; Sheth *et al.* 2010), which comprises the largest mid-IR survey of galaxies in the local Universe. We perform a 2D GALFIT decomposition of the 3.6μm images of 146 ETGs and examine their residual images. We identify tidal features in 17% of these, suggesting that a non-negligible ETGs fraction may have experienced (after the formation of the bulk of their stellar budget) merger events that have left signatures (Canalizo *et al.* 2007). For 6 of these peculiar ETGs, we also applied GALFIT decomposition to public GALEX/UV and SDSS/optical images. With measurements in multiple bands, we applied SED fitting techniques to estimate star formation rates (SFR) and stellar masses for the galaxies and their tidal features. We find that these 6 peculiar ETGs present masses in agreement with the population of non-peculiar ETGs. However, SFRs are higher than what has been measured for the average ETG population (Shapiro *et al.* 2010, SDSS MPA-JHU catalog). Based on the Kaviraj (2010) relation, we find that for these peculiar ETGs the estimated age of the most recent star formation event is less than 3Gyrs. Despite this indication of recent star formation, we have not found evidence of prominent UV emission in the tidal features (Marino *et al.* 2011). We are currently extending our work to the full sample of peculiar ETGs identified in our sample.

Keywords. galaxies: elliptical and lenticular, galaxies: structure, galaxies: evolution

References

Canalizo, G., *et al.* 2007, *ApJ* 669, 801
Kaviraj, S. 2010, *MNRAS* 408, 170
Kormendy, J., *et al.* 1997, *ApJ* 482, L139
Marino, A., *et al.* 2011, *MNRAS* 411, 311
Rampazzo, R., *et al.* 2007, *MNRAS* 381, 245
Rix, H.-W., Carollo, C. M., & Freeman, K. 1999, *ApJ* 513, L25
Salim, S. & Rich, R. M. 2010, *ApJ* 714, L290
Shapiro, K. L., *et al.* 2010, *MNRAS* 402, 2140
Sheth, K., *et al.* 2010, *PASP* 122, 1397

Highlights of Astronomy, Volume 16
XXVIIIth IAU General Assembly, August 2012
T. Montmerle, ed.

doi:10.1017/S1743921314004918

UV color-color relation of early-type galaxies

Chang H. Ree[1], Hyunjin Jeong[1], Kyuseok Oh[2], Chul Chung[2], Joon Hyeop Lee[1], Sang Chul Kim[1] and Jaemann Kyeong[1]

[1]Korea Astronomy and Space Science Institute, Daejeon 305-348, Republic of Korea
email: chr@kasi.re.kr

[2]Dept. of Astronomy, Yonsei University, Seoul 120-749, Republic of Korea

Abstract. The ultraviolet (UV) color-color relation of early-type galaxies (ETGs) in the nearby universe ($0.05 < z < 0.12$) is re-examined with the latest GALEX GR6 and SDSS DR7 data. By drawing the FUV – NUV (as a measure of UV temperature) versus FUV – r (as a measure of UV amplitude) color-color diagram for the *morphologically-cleaned, spectroscopically-cleaned* sample of ~3700 quiescent ETGs, we find that the "old and dead" ETGs consist of a well-defined sequence in UV colors, the "UV red sequence", so that the stronger UV excess galaxies should have a harder UV spectral shape systematically. However, the observed UV spectral slope is too steep to be reproduced by the canonical models in which the UV flux is mainly controlled by age or metallicity parameters. The observed data support the helium enhancement scenario in which the UV spectral shape of UV upturn (FUV – NUV < 0.9; FUV – r ~ 6) galaxies may be governed by the minority population of helium-enhanced horizontal-branch (HB) stars.

Keywords. galaxies: elliptical and lenticular, cD, galaxies: evolution, galaxies: stellar content

The (FUV – NUV) vs. (FUV – r) color-color diagram is a powerful, photometric tool to trace the UV spectral shape and UV amplitude of ETGs, discriminating effectively the contaminations from young stars or AGNs. From the latest GALEX–SDSS matched catalog, we find that the *quiescent* ETGs in the nearby universe consist of the "UV red sequence" (Ree *et al.* 2012, *ApJ*, 744, L10) which was not well identified in previous studies. The observed steep slope of UV color-color relation, (FUV – NUV) = 1.36 (FUV – r) - 8.35, indicates that the hot HB stars in the UV upturn galaxies should have hotter temperature systematically than those in the UV weak galaxies, at a given UV amplitude. By using two-component model SED syntheses with different assumptions on the UV source populations (young stars, normal or enhanced helium abundances; Fig. 1), we show that the observed UV color distribution is better explained with the additional contribution from minority populations of helium-enhanced HB stars, besides the underlying old populations, as suggested by Chung *et al.* (2011, *ApJ*, 740, L45).

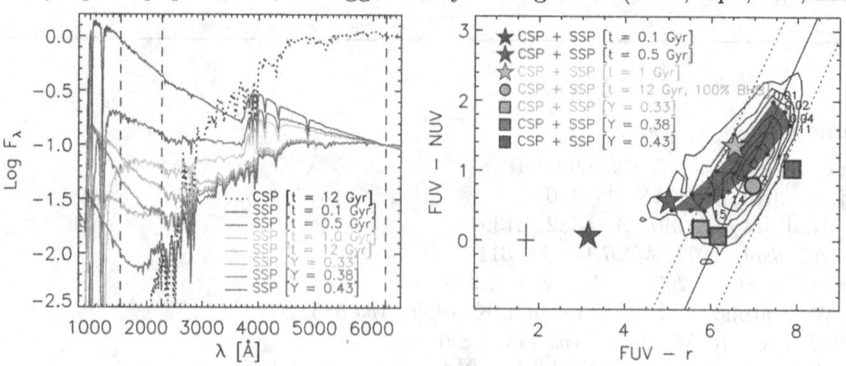

Figure 1. Two-component model SEDs (left) and comparison with the observation (right).

Highlights of Astronomy, Volume 16
XXVIIIth IAU General Assembly, August 2012
T. Montmerle, ed.

doi:10.1017/S174392131400492X

Herschel-ATLAS: Dusty early-type galaxies

K. Rowlands[1]†, L. Dunne[1,2], S. Maddox[1,2] and the Herschel-ATLAS and GAMA collaborations

[1]School of Physics & Astronomy, University of Nottingham, University Park, Nottingham, NG7 2RD, UK

[2]Department of Physics and Astronomy, University of Canterbury, Private Bag 4800, Christchurch, New Zealand

Early-type galaxies (ETGs) are thought to be devoid of dust and star-formation, having formed most of their stars at early epochs. We present the detection of the dustiest ETGs in a large-area blind submillimetre survey with *Herschel* (H-ATLAS, Eales *et al.* 2010), where the lack of pre-selection in other bands makes it the first unbiased survey for cold dust in ETGs. The parent sample of 1087 H-ATLAS galaxies in this study have a $\geqslant 5\sigma$ detection at 250μm, a reliable optical counterpart to the submillimetre source (Smith *et al.* 2011) and a spectroscopic redshift from the GAMA survey (Driver *et al.* 2011). Additionally, we construct a control sample of 1052 optically selected galaxies undetected at 250μm and matched in stellar mass to the H-ATLAS parent sample to eliminate selection effects. ETGs were selected from both samples via visual classifications using SDSS images. Further details can be found in Rowlands *et al.* (2012). Physical parameters are derived for each galaxy using the multiwavelength spectral energy distribution (SED) fitting code of da Cunha, Charlot and Elbaz (2008), Smith *et al.* 2012, using an energy balance argument. We investigate the differences between the dusty ETGs and the general ETG population, and find that the H-ATLAS ETGs are more than an order of magnitude dustier than the control ETGs. The mean dust mass of the 42 H-ATLAS ETGs is $5.5 \times 10^7 M_\odot$ (comparable to the dust mass of spirals in our sample), whereas the dust mass of the 233 control ETGs inferred from stacking at optical positions on the 250μm map is $(0.8 - 4.0) \times 10^6 M_\odot$ for 25-15 K dust. The average star-formation rate of the H-ATLAS ETGs is 1.0 dex higher than that of control ETGs, and the mean r-band light-weighted age of the H-ATLAS ETGs is 1.8 Gyr younger than the control ETGs. The rest-frame $NUV - r$ colours of the H-ATLAS ETGs are 1.0 magnitudes bluer than the control ETGs, and some ETGs may be transitioning from the blue cloud to the red sequence. Some H-ATLAS ETGs show signs of morphological disturbance and may have undergone recent rejuvenation of their ISM via gas and dust delivered by mergers. It is found that late-type stars cannot produce enough dust to account for that observed in the H-ATLAS ETGs. This indicates that either an external source of dust from mergers is required, a substantial amount of dust grain growth must occur in the ISM, or dust destruction by hot X-ray gas is less efficient than predicted.

Keywords. galaxies: evolution, galaxies: elliptical and lenticular, cD, (ISM:) dust, extinction, infrared: galaxies, submillimeter

References

da Cunha, E., Charlot S., & Elbaz, D. 2008, *MNRAS*, 388, 1595
Driver, S., *et al.* 2011, *MNRAS*, 413, 971
Eales, S., *et al.* 2010, *PASP*, 122, 499
Rowlands, K., *et al.* 2012, *MNRAS*, 419, 2545
Smith, D. J. B., *et al.* 2011, *MNRAS*, 416, 857
Smith, D. J. B., *et al.* 2012, *MNRAS*, accepted, arXiv:1208.3079

† E-mail:ppxkr@nottingham.ac.uk

Highlights of Astronomy, Volume 16
XXVIIIth IAU General Assembly, August 2012
T. Montmerle, ed.

doi:10.1017/S1743921314004931

Recent star formation in intermediate redshift (0.35< z <1.5) early–type galaxies

M. J. Rutkowski[1], H. Jeong[2], S. Yi[3], S. Kaviraj[4], S. H. Cohen[1] and R. A. Windhorst[1]

[1]Arizona State University, [2]Korean Astronomy and Space Science Institute, [3]Yonsei University, [4]Imperial College London

Abstract. We measured the UV–optical–near-IR spectral energy distributions (SEDs) of redshift $z \sim 0.3-1.5$ early-type galaxies (ETGs) with the *Hubble Space Telescope* (HST) Wide Field Camera 3 (Rutkowski *et al.* 2012). We searched for young stellar populations and morphological signatures of the mechanisms driving recent star formation (RSF) in these ETGs in order to provide observational constraints on models of galaxy evolution.

Keywords. elliptical & lenticular galaxies; star formation; galaxy evolution

1. Analysis

We measured the best-fit two-component ("young": t<1 Gyr; 'old": $t \gtrsim$4Gyr) stellar population model for the SED of each ETG and derive the mass fraction and age of the young stellar population (Figure a; Jeong *et al.* 2007). We also measured the Sérsic profile and characterized the visual morphology of each ETG (Figure b) and its local (d<100 kpc) environment.

2. Conclusions

We determined the following:

• ~40% of low redshift ($z < 1$) ETGs experienced RSF ($f \simeq$1–5%; 0.1< $t[Gyr]$ <0.5);

• ~30% of low redshift ETGs that were identified with disturbed morphologies or galaxies were also found to host RSF.

The morphologies of the ETGs and their companions suggests that RSF in ETGs is not driven by a single, *dominant* process. Instead, RSF is likely motivated by a variety of secular and merger processes, in general agreement with observations at the local ($z < 0.1$) and high ($z \gtrsim$1.5) redshift universe.

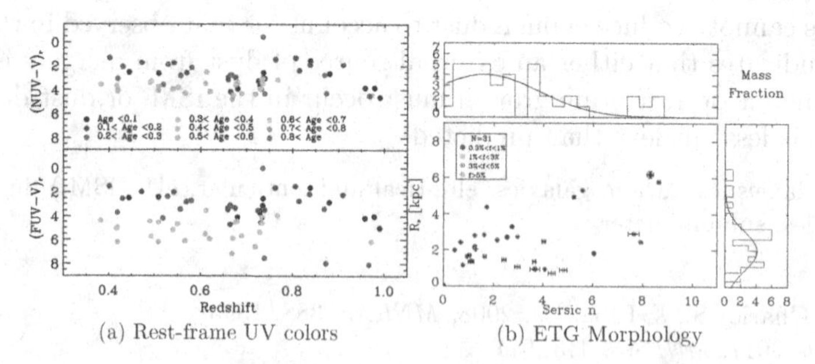

(a) Rest-frame UV colors (b) ETG Morphology

References

H. Jeong, M. Bureau, S. K. Yi, D. Krajnovíc, & R. L. Davies, 2007, *MNRAS*, 376, 1021

M. J. Rutkowski, S. H. Cohen, S. Kaviraj, R. W. O'Connell, N. P. Hathi, & WFC3 SOC, 2012, *ApJS*, 199, 1, 4

Highlights of Astronomy, Volume 16
XXVIIIth IAU General Assembly, August 2012
T. Montmerle, ed.

doi:10.1017/S1743921314004943

Positive AGN feedback in Centaurus A

Stanislav Shabala[1], R. Mark Crockett[2] and Sugata Kaviraj[2,3]

[1]School of Mathematics & Physics, Univ. of Tasmania, Private Bag 37, Hobart 7001, Australia
email: Stanislav.Shabala@utas.edu.au
[2]Department of Physics, University of Oxford, Keble Road, Oxford, OX1 3RH, UK
[3]Blackett Laboratory, Imperial College London, London SW7 2AZ, UK

Abstract. We observed the inner filament of NGC 5128 (Centaurus A) with the *Hubble Space Telescope Wide Field Camera 3* (WFC3), using the $F225W$, $F657N$ and $F814W$ filters. We find a young stellar population near the south-west tip of the filament. We constrain the ages of these stars to 1-3 Myrs. No further recent star formation is found along the filament.

We propose an updated explanation for the origin of the inner filament. It has been suggested (Sutherland *et al.* 1993) that radio jets can shock the surrounding gas, giving rise to the observed optical line emission. We argue that such shocks can naturally arise due to a weak cocoon-driven bow shock (rather than from the radio jet directly) propagating through the diffuse interstellar medium. We suggest such a shock has overrun a molecular cloud, triggering star formation in the dense molecular core. The outer, more diffuse parts of the cloud are then ablated and shock heated, giving rise to the observed optical line and X-ray emission.
Keywords: ultraviolet: galaxies; galaxies: individual: NGC 5128; galaxies: active

---

Figure 1 shows the colour composite image of the inner filament. Young stars, traced by NUV light, are only present at the south-west tip. Using stellar population synthesis models and single star isochrones, we find these stellar populations to have ages $\lesssim 10$ Myrs, and best-fit ages of $1-3$ Myrs.

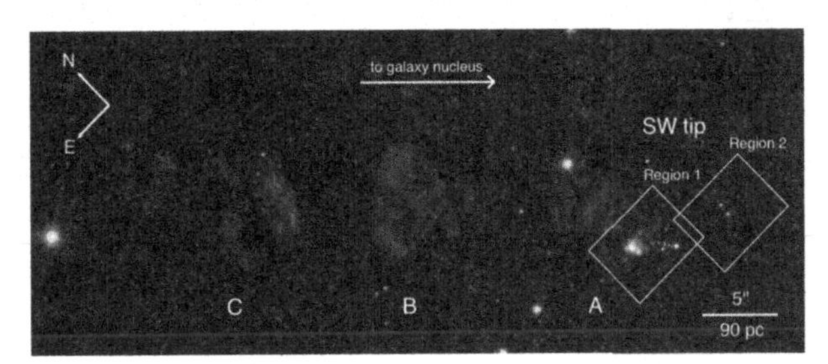

Figure 1. The inner filament in $F225W$ (blue), $F606W$ (green) and $F814W$ (red) filters.

The lack of recent star formation away from the tip of the filament, together with diffuse Hα and X-ray emission, suggests the inner filament may have been formed by shocking and ablation of a gas cloud by a passing weak shock. The dense inner parts of the cloud are radiative, and collapse to form stars. The diffuse outer parts are shock heated and ablated. Our simulations show that such a shock can be driven by the expansion of an AGN-inflated cocoon of radio plasma. The age of the young stellar population is consistent with both the radio AGN age and the travel time along the filament. We estimate the total mass of the disrupted cloud to be $\sim 6 \times 10^5 M_\odot$.

References

Crockett, R. M., Shabala, S. S., Kaviraj, S., *et al.* 2012, *MNRAS*, 421, 1603
Sutherland, R. S., Bicknell, G. V., & Dopita, M. A. 1993, *ApJ*, 414, 510

Highlights of Astronomy, Volume 16
XXVIIIth IAU General Assembly, August 2012
T. Montmerle, ed.

doi:10.1017/S1743921314004955

UV Emission in Type Ia Supernova Elliptical Host Galaxies

Brad E. Tucker
Mt. Stromlo Observatory, via Cotter Road,
Weston Creek, ACT 2611 Australia
email: brad@mso.anu.edu.au

The current use of Type Ia supernova (SN Ia) as standard candles is to measure the dark energy equation-of-state to better than 10%. However, we still lack a clear understanding of their progenitor systems. We analyze the host galaxies of type Ia Supernova (SN Ia) discovered by the ESSENCE survey using UV and optical data, as studying the environments of SN Ia is a great way to understand the progenitors. We developed a new method for determining the SED and rest-frame magnitudes of the host galaxies and we use empirical relations to derive stellar mass and star-formation rate (SFR) measurements of the host galaxies. We find a high rate of UV emission in our passive galaxies, suggesting current star-formation in these galaxies. Specifically, we have found that at $z > 0.3$, $\approx 60\%$ of the elliptical host galaxies have star-formation while with nearby hosts, it is $\approx 40\%$, with both samples having an additional $\approx 10 - 15\%$ of hosts being AGN. We also find a connection between the time required for the progenitor to evolve and explode as an SN and UV emission in the elliptical hosts, suggesting that UV emission in elliptical galaxies is indeed star-formation and linked to the production of SN Ia's.

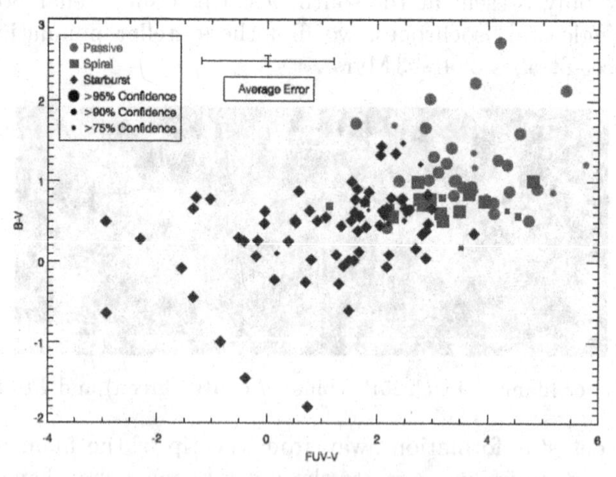

Figure 1. The rest-frame $FUV-V$ and $B-V$ colors of the host galaxies in the ESSENCE survey from Tucker *et al.* (2012a). The red circles are passive galaxies, purple squares are spiral galaxies, and the blue diamonds are active galaxies. The size of the point represents the confidence in the fit, with the largest points having a fit $> 95\%$, the middle-size points $90\% <$ fit $< 95\%$, and the smallest points $75\% <$ fit $< 90\%$. We also show the corresponding average error-bar in the top..

References

Tucker, B. E., *et al.* 2012a, *ApJ*, submitted
Tucker, B. E., *et al.* 2012b, *ApJ*, submitted

Highlights of Astronomy, Volume 16
XXVIIIth IAU General Assembly, August 2012
T. Montmerle, ed.

doi:10.1017/S1743921314004967

The Ionization of the Warm Gas in Early-type Galaxies and Its UV Upturn

Renbin Yan[1,2] and Michael R. Blanton[2]

[1]Department of Physics and Astronomy, University of Kentucky, Lexington, KY, 40506, USA
email: renbin@pa.uky.edu
[2]Center for Cosmology and Particle Physics, Department of Physics, New York University, New York, NY, 10003, USA
email: michael.blanton@nyu.edu

It has long been known that the majority of early-type galaxies contain warm ionized gas producing optical emission lines. These warm ionized gas are spatially extended to kpc scales. Their line ratios satisify the criteria of Low-ionization Nuclear Emission-line Regions (LINERs) on all major line-ratio diagnostic diagrams. However, their ionization mechanism has been hotly debated. Multiple ionization mechanisms can produce the same kind of line ratios, including photoionization by a central AGN, photoionization by hot evolved stars (e.g. post-AGB stars), collisional ionization by fast shocks, photoionization by hot X-ray emitting gas, and conductive heating or turbulent mixing. Therefore, determining the ionization mechanism requires other information.

Measuring the spatial gradient in a line ratio that is sensitive to ionization parameter will allow us to distinguish a central ionizing source from distributed ionizing sources. We utilized the SDSS survey data to measure this. SDSS used a fixed angular aperture of 3", which corresponds to different physical scales at different distances. By selecting the same population of galaxies at all distances, we can statistically study the spatial distribution of line emission and line ratio gradient. We found that the median line ratio gradient in [O III]/[S II] among non-star-forming red galaxies changes very slowly with increasing physical scale covered, which rules out AGN as the ionizing source and strongly favors a distributed ionizing sources that follow the stellar density profile. Additionally, the stellar photoionizing model predicts that the line ratio gradient is dependent on luminosity. Since the stellar density profile is different between bright and faint early-type galaxies, they should produce different ionizing flux profiles and different line ratio gradients. By separting the sample accoring to broadband luminosity, we verified this prediction using the SDSS data. This strongly suggests that the ionizing source is spatially distributed like the stars. For details, see Yan & Blanton (2012).

To evaluate the likelihood of the shock ionization and turbulent mixing models, we measured the temperature of the gas in coadded spectra. We selected non-star-forming red galaxies in SDSS and separated them into three gas-metallicity bins according to [N II]/[O II] ratio. After carefully removing the stellar continuum, we were able to measure the weak [N II] λ5755 line. The [N II] 6584/5755 ratio yields a temperature measurement of $15,000$K for the low-metallicity sample and $8,000$K for the high-metallicity sample. These temperatures are consistent with the photoionization but are inconsistent with shock models or turbulent mixing models.

Therefore, we conclude that the warm ionized gas in the majority of early-type galaxies are most likely ionized by the UV-bright old stellar population.

Reference

Yan, Renbin, Blanton, & Michael R., 2012 *Astrophys. J.*, 747, 61

Highlights of Astronomy, Volume 16
XXVIIIth IAU General Assembly, August 2012
T. Montmerle, ed.

doi:10.1017/S1743921314004979

Correlation of morphological fraction with redshift in galaxy clusters

Qi-Rong Yuan[1], Qiang-Qiang Dang[1], Peng-Fei Yan[1], Wei Chen[1], Zhong-Lue Wen[2], Jin-Lin Han[2] and Xu Zhou[2]

[1]Physics Department, Nanjing Normal University,
1 WenYuan Road, Nanjing 210046, China
email: yuanqirong@njnu.edu.cn

[2]National Astronomical Observatories, Chinese Academy of Science,
20A Datun Road, Beijing 100012, China

Abstract. Based on 187 galaxy clusters identified from the photometric redshifts of galaxies in the Cosmic Evolution Survey (COSMOS) field(Wen & Han 2011), cluster galaxies brighter than $M_V = -20.5$ are classified into four categories according to their best-fitting templates of the spectral energy distributions (SEDs) provided by Ilbert *et al.* (2009): early-type (including elliptical and lenticular) galaxies (E+S0), spiral galaxies (S), irregular galaxies (Irr), and starbursts (SB). The fractions of these four SED types are presented as the functions of redshift in Figure 1. Fraction of each category varies remarkably with cluster redshift: fractions of normal galaxies (E+S0+S+Irr) tend to decrease with redshift, whilst the starburst proportion tends to increase with redshift. For the normal galaxies, there exists a sequence for the decreasing slopes of morphological fractions. Majority of the galaxies in high-redshift clusters ($z > 1.0$) are experiencing strong star-formation activities, which leads to a very high proportion of starburst.

Keywords. galaxies: classification, galaxies: evolution, galaxies: clusters: general

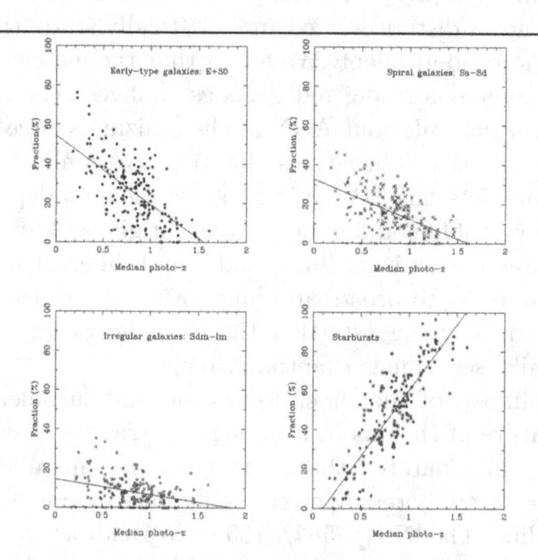

Figure 1. Fractions of four SED categories as the functions of redshift.

This work was supported by National Natural Science Foundation of China (No. 11173016).

References

Ilbert, O., *et al.* 2009, *ApJ*, 690, 1236
Wen, Z. & Han, J. 2011, *ApJ*, 724, 68

Highlights of Astronomy, Volume 16
XXVIIIth IAU General Assembly, August 2012
T. Montmerle, ed.

doi:10.1017/S1743921315001568

Preface: Joint Discussion JD5 From Meteors and Meteorites to their Parent Bodies: Current Status and Future Developments

The Joint Discussion 5 entitled "From Meteors and Meteorites to their Parent Bodies: Current Status and Future Developments" within the IAU GA 2012 was organized with the coordination of the IAU Division III Planetary Systems Sciences and the IAU Commission N. 22 Meteors, Meteorites & Interplanetary Dust, together with the supports by Divisions I Fundamental Astronomy, Division XII Union-Wide Activities, Commission 4 Ephemerides, Commission 6 Astronomical Telegrams, Commission 8 Astrometry, Commission 15 Physical Study of Comets & Minor Planets, and Commission 20 Positions & Motions of Minor Planets, Comets & Satellites.

This Joint Discussion was aimed to share the latest knowledge on the small solar system bodies and also the possible parent bodies of meteors, meteorites & interplanetary dusts from as wide a perspective as possible. The latest results were expected to come from international campaigns of ground-based observations (2011 Draconids meteor shower, fireball network etc.), space missions to comets & asteroids (HAYABUSA, DAWN, EPOXI, Post-Stardust, Rosetta etc.) and meteorite falls & recoveries. The scope extended to the meteor showers, meteorite falls, and comet appearance recorded in this oriental region over centuries. One of the purpose of this JD was to be dedicated to late Brian Marsden, who served as a leader of this field long time, and special invited talk on his work was set. We had 11 invited talks, 15 oral contributions along with 16 poster presentations in a wide range, from August 22 through 24. The most active researchers in this area attended the Joint Discussion and discussed the latest developments especially from Asian region. We wish to thank all participants to the meeting for their enthusiasm and for sharing their most recent results. I wish to thank the Scientific Organizing Committee for its advice and help in organizing the scientific program. The SOC (* Co-chairs) was composed of

Michael AfHearn (United States), Peter Brown (Canada), Tadeusz Jopek (Poland), Karen Meech (United States), Sho Sasaki (Japan), Caroline Smith (United Kingdom), Mitsuru Soma (Japan), Pavel Spurny (Czech Republic), Jeremie Vaubaillon (France), Hitoshi Yamaoka (Japan), Makoto Yoshikawa (Japan), Hajime Yano (Japan), Masateru Ishiguro (Republic of Korea), Daisuke Kinoshita (China Taipei), *Peter Jenniskens (United States), *Jun-ichi Watanabe (Japan), *Iwan Williams (United Kingdom), *Jin Zhu (China Nanjing).

I would like to thank Dr. Reiko Furusho for her kind effort to maintain the WEB site of our JD5 in the National Astronomical Observatory of Japan.

Junichi Watanabe
National Astronomical Observatory of Japan
email: jun.watanabe@nao.ac.jp

Highlights of Astronomy, Volume 16
XXVIIIth IAU General Assembly, August 2012
T. Montmerle, ed.

doi:10.1017/S1743921314004980

Phaethon-Gemind complex by Pan-STARRS

Shinsuke Abe

Institute of Astronomy, National Central University
email: shinsuke.avell@gmail.com

Abstract. The physical nature such as orbital distribution of asteroids is fundamental to understanding how our solar system has been evolved. The connection between Near-Earth Objects (NEOs) and Earth impactors such as meteorites and fireballs are still under debate, since there is no meteorite orbit whose parent NEO was identified. The orbital distribution of NEOs has been investigated by comprehensive sky surveys including Pan-STARRS (The Panoramic Survey Telescope And Rapid Response System). Here we focus on the Phaethon-Gemind complex detected by Pan-STARRS PS1 Prototype Telescope and our follow-up lightcurve observations.

Highlights of Astronomy, Volume 16
XXVIIIth IAU General Assembly, August 2012
T. Montmerle, ed.

doi:10.1017/S1743921314004992

Puzzling Snowballs: Main Belt Comets

Bin Yang and Karen Meech

Institute for Astronomy, University of Hawaii, USA
email: yangbin@ifa.hawaii.edu

Abstract. Main belt comets (MBCs) are a class of newly discovered objects that exhibit comet-like appearances and yet are dynamically indistinguishable from ordinary main belt asteroids. The measured size and albedo of MBCs are similar to those of classical comets. At present, six MBCs have been discovered, namely 133P/Elst-Pizarro, 176P/LINEAR, 238P/Read, P/2008 R1, P/La Sagra and P/2006 VW139. The total number of active MBCs is estimated to be at the level of a few hundreds (Hsieh & Jewitt, 2006). Several explanations for the activity of MBCs have been suggested. These include impact ejection, sublimation and rotational instability. However, since renewed activity has been observed in 133P and 238P at successive perihelion passages, the most likely explanation may be a thermally-driven process - e.g sublimation of exposed surface ice. Although the proximity of MBCs to the Sun (r $\sim$ 3 AU) makes the survival of surface ice improbable, thermal models have shown that water ice is thermally stable under a regolith layer a few meters thick. The study of MBCs has recently been complicated by the discoveries of two asteroid collisional events (P/2010 A2 (LINEAR) and (596) Scheila) in 2010, where comet-like dust coma/tail have been attributed to recent impacts. If MBCs are indeed icy, they represent the closest and the third established reservoir of comets (after the Oort cloud and the Kuiper belt). As such, they may have been an important source of water for the Earth's oceans. I will review the current state of MBC studies, present the latest observational results and discuss possible mechanisms that could produce the observed activity. I will also talk about current and future space missions that are dedicated or closely related to MBC studies.

Highlights of Astronomy, Volume 16
XXVIIIth IAU General Assembly, August 2012
T. Montmerle, ed.

doi:10.1017/S1743921314005006

The influx rate of long-period comets in the Earth's neighborhood and their debris contribution to the interplanetary medium

Julio Angel Fernández

Departamento de Astronomia, Facultad de Ciencias, Igua 4225, 11400 Montevideo, Uruguay
email: julio@fisica.edu.uy

Abstract. We analyze the flux of new and evolved long-period comets (LPCs) reaching the Earth's neighborhood (perihelion distances q < 1.3 AU), their physical lifetimes, and their implications as regards to the amount of meteoritic matter that is being deposited in the near-Earth region. The flux of LPCs with q < 1.3 au is found to be of about 340 ± 40, brighter than absolute total magnitude 8.6 (radius R $\sim$ 0.6 km) (Fernández and Sosa 2012). Bearing in mind that most of these comets disintegrate into meteoritic matter, this represents a large contribution to the interplanetary dust complex which requires an amount of matter of about 10 tons s^{-1} to keep it in steady state. These aspects, as well as the impact rate with Earth of meteoroids of LPC origin, will be discussed in this presentation.

Highlights of Astronomy, Volume 16
XXVIIIth IAU General Assembly, August 2012
T. Montmerle, ed.

doi:10.1017/S1743921314005018

The Legacy of Brian G. Marsden (1937-2010)

Daniel Green
Harvard University
email: dwe_green@eps.harvard.edu

Abstract. An outline of Brian Marsden's work in the field of small-body solar-system astronomy was presented.

Highlights of Astronomy, Volume 16
XXVIIIth IAU General Assembly, August 2012
T. Montmerle, ed.

doi:10.1017/S174392131400502X

Meteor Showers: which ones are real and where do they come from?

Peter Jenniskens
SETI Institute
email: petrus.m.jenniskens@nasa.gov

Abstract. The IAU Meteor Shower Working List contains 369 showers, of which only 64 are considered established (per February 28, 2012). In this invited review, we will give an overview of international efforts to validate the remaining showers. We report on the showers that were validated in this triennium and proposed to receive the predicate "established" at the present General Assembly. The meteoroid orbit surveys characterize the meteoroid streams in terms of orbital elements and their dispersions, which is ground truth for efforts to identify their parent comets, study the fragmentation history of the (mostly dormant) comet population in the inner solar system, and understand the origin of the zodiacal cloud.

Highlights of Astronomy, Volume 16
XXVIIIth IAU General Assembly, August 2012
T. Montmerle, ed.

doi:10.1017/S1743921314005031

Stream and sporadic meteoroids associated with Near Earth Objects

Tadeusz J. Jopek[1] **and Iwan P. Williams**[2]

[1]Astronomical Observatory Institute, Faculty of Physics, A.M. University, Poznan, Poland
[2]School of Physics and Astronomy, Queen Mary University of London, E1 4NS, UK

Abstract. NEOs come close to the Earth's orbit so that any dust ejected from them, might be seen as a meteor shower. Orbits evolve rapidly, so that a similarity of orbits at one given time is not sufficient to prove a relationship, orbital evolution over a long time interval also has to be similar. Sporadic meteoroids can not be associated with a single parent body, they can only be classified as cometary or asteroidal. However, by considering one parameter criteria, many sporadics are not classified properly therefore two parameter approach was proposed.
Keywords. meteors, meteoroids, meteoroids, methods: numerical

1. Introduction

The association between meteoroid streams and comets is well-established, though the idea that some originate from asteroids is quite old (see Williams, 2011). The formation mechanism of a meteor stream from a comet is well-understood. Ices sublimate generating a gas outflow that carries away small dust grains. Inter-asteroid collisions, as was seen in 2010A2 (Linear) (Jewitt ., *et al.*, 2010) is the most obvious formation process, though there are other possibilities. In all cases, the ejection velocity is small compared to the orbital speed so that a stream is formed on an orbit that is similar to that of the parent (the physics and mathematics involved are given in Williams, 2004).

Individual meteoroid orbits can evolve in a significantly different way from the parent. The obvious mechanism causing this is a close planetary encounter (Hughes Williams & Fox, 1981) but collisions between stream meteoroids, the parent or interplanetary dust particles can also play a part. These processes feed the sporadic population, making it hard to associate such meteoroids with a given parent.

The major tool for determining a pairing of parent and meteoroid stream has been orbit similarity. To quantify this, Southworth & Hawkins (1963) and Drummond (1981) proposed orbital similarity D-criteria, both being widely used. Using such criteria, many pairings between comets and meteor streams have been established and also between minor showers and near-Earth asteroids (NEAs). Because of the vast increase in the number of known asteroids, the probability of a chance similarity of orbits is high. For this reason, Porubčan, Kornoš & Williams (2004) proposed that orbits need to remain similar for 5000 years before any association can be claimed.

Three major showers, the Geminids, the Quadrantids and the Taurids all have asteroids associated with them. Whipple (1983) pointed out the orbital similarity between the Geminid stream and (3200) Phaethon, while Fox, Williams & Hughes (1983) showed that Phaethon could be the parent of the Geminid meteoroid stream assuming the ejection mechanism was Cometary. Battams & Watson (2009) reported that Phaethon brightened by at least 2 magnitudes in 2009. The asteroids 2005 UD and 1999 YC have orbits that are similar to that of Phaethon, all suggesting that a fragmenting large comet may be the actual parent.

Jenniskens (2004) noted the similarity between the orbits of the Quadrantids and asteroid 2003 EH1(now 196256). In the activity curve of the Quadrantids there is a broad background and a sharp narrow central peak suggesting that an old comet (probably 96P/Machholz), produced the background while a recent outburst produced the sharp peak. This could the fragmentation of C/1490 Y1 a few hundred years ago (Williams *et al.* 2004), producing asteroid 2003EH1.

The Taurid stream is a complex of several smaller streams and filaments. There are many associated asteroids and lists are given in Porubčan, Kornoš & Williams (2006). The Taurid complex has an active comet associated with it, 2P/Encke. Asher, Clube & Steel (1993) suggested that the family of meteoroid streams, comet 2P/Encke and associated Apollo asteroids could all have formed by the fragmentation of a giant comet 20-30 Ky ago.

2. Sporadic meteoroids

The sporadic meteors have lost the identity of their parent's orbit. In order to discriminate between the orbits of comets and asteroids, several one parameter criteria have been proposed, they are listed in Jopek & Williams (2013). They have been used by several authors: Whipple, 1954, found that cometary types dominated within photographic meteors, Jones & Sarma, 1985; Steel, 1996, found that the proportion of cometary and asteroidal were about equal within TV meteors; according to Steel, 1996 asteroidal meteoroids dominated among Kharkov radio meteors, while Voloshchuk *et al.*, 1997, claimed the opposite.

Applying several criteria to 780 comets and 7830 NEAs, Jopek & Williams (2013) found that the Q- and E-criterion were the most reliable. They also classified $\sim$ 78000 sporadic meteoroids collected from several sources. Using one parameter Q- (or E-) criterion the authors found that amongst all sporadic meteoroids 22-23% can not be classified correctly. The orbits of such objects have asteroidal aphelia $Q < 4.6$ (or semi-major axes $a < 2.8$) and cometary inclinations $i > 75$ deg. The ecliptic radiants of the meteors corresponding to these objects concentrate around the position of the apex of the Earth motion. Such meteoroids as was shown by Jones *et al.* 2001, Nesvorny *et al.* 2011 can be of cometary origin. Therefore to classify the sporadic meteoroids Jopek & Williams proposed that two parameter criteria such as *Q-i* are used. For cometary orbits we have:

$$Q = a(1+e) > 4.6[\mathrm{AU}], \quad \text{or} \quad \mathrm{i} > 75\,\mathrm{deg} \tag{2.1}$$

where: a — semi-major axis; e — eccentricity; i— inclination of the meteoroid orbit.

References

Asher, D. J., Clube, S. V. M., & Steel, D. I., 1993, *MNRAS*, 264, 93

Battams, K. W. & atson, A., 2009, *IAU Circ.*, 9054

Drummond, J. D. 1981, *Icarus*, 45, 453

Fox K., Williams I. P., & Hughes D. W., 1983, *MNRAS*, 205, 1155

Hughes D. W., Williams I. P., & Fox K., 1981, *MNRAS*, 195, 625

Jenniskens P., 2004, *AJ*, 127, 3018

Jewitt, D., Weaver, H., Agarwal, J., Mutchler, M., & Drahus, M., 2010, *Nature*, 467, 817

Jones, J. & Sarma, T., 1985, Bull. Astron. Inst. Czechosl., 36, 103

Jones, J., Campbell-Brown, M. & Nikolova, S., 2001, *in Meteors 2001*, Ed Warmbeim B., ESA-SP495 575

Jopek, T. J. & Williams, I. P., 2013, *MNRAS*, 430, 2377

Nesvorny, D., Vokrouhlicky, D., Pokorny, P., & Janches, D., 2011, *ApJ*, 747, 1

Porubčan, V., Kornoš, L., & Williams, I. P., 2004, *EM&P*. 95, 697

Porubčan, V., Kornoš, L., & Williams, I. P., 2006, *Contr. Astron. Obs. Skalnate Pleso*, 36, 103

Southworth, R. B. & Hawkins, G. S., 1963, *Smiths. Contr. Astroph.*,7, 261
Steel, D. I., 1996, *Space Sci. Rev.*, 78, 507
Voloshchuk, Yu. I., Vorgul', A. V., & Kashcheev, B. L., 1997. *Astronomicheskii Vestnik*, 31, 345
Whipple, F. L., 1954, *AJ*, 59, 201
Whipple, F. L. 1983, *IAU Circ* 3881.
Williams, I. P. 2004, *WGN J. IMO*, 32, 11
Williams, I. P. 2011, *A & G*, 52, 2.20
Williams, I. P., Hughes, D. W., McBride, N., & Wu, Z., 1993, *MNRAS*, 260, 43
Williams, I. P., Ryabova, G. O., Baturin, A. P., & Chernitsov, A. M., 2004, *MNRAS*, 355, 1171

Highlights of Astronomy, Volume 16
XXVIIIth IAU General Assembly, August 2012
T. Montmerle, ed.

doi:10.1017/S1743921314005043

Results from the EPOXI and StardustNExT Missions – A Changing View of Comet Volatiles and Activity

Karen Meech[1], Michael F. A'Hearn[2] and Joseph Veverka[3]

[1]Institute for Astronomy, Univ. of Hawaii, NASA Astrobiology Institute
email: meech@ifa.hawaii.edu

[2]University of Maryland, [3]Cornell University, USA

Abstract. Within a period of ~3 months there were two extended mission flybys of comets. Both encounters have provided an exciting new view of comet activity and volatile composition that is changing our paradigm of these small early solar system remnants. The EPOXI mission flew past the nucleus of comet 103P/Hartley 2 on 4 Nov. 2010. This small nucleus was known to be exceptionally active prior to the encounter, by virtue of a very large water production rate relative to its surface area. Both the encounter and ground-based data showed that comet Hartley 2fs perihelion activity was dominated by sub-surface CO2 outgassing rather than by water, suggesting our classic comet formation picture is not correct. The gas flow carried large grains (up to >10 cm in diameter) from the nucleus, and the icy grains contributed to the large observed water production. The CO2 abundance relative to water varies with rotation between 10-20% between the two lobes of the nucleus. The bi-lobed nucleus is rotating in an excited state, with a period that varied rapidly from ~16.5 hrs to longer than 18.5 hrs over 3 months. The nucleus morphology was different from that of other nuclei visited by space craft, with some regions of rough topography in which surface ice was visible. On 2011 Feb. 14 the Stardust-NExT spacecraft flew past the nucleus of comet 9P/Tempel 1, the target of the Deep Impact (DI) experiment in July 2005. The mission goal was to look at the nucleus after and intervening perihelion passage, extending the surface area imaged during the DI encounter and also image the 2005 impact site. The layering seen during the DI flyby was exhibited over the areas newly imaged in the NExT flyby, and it was found that 30% of the nucleus was covered by smooth deposits that were likely caused by eruption of subsurface materials. Although it has long been known that comets lose on average ~ a meter of their surface per perihelion passage, it was surprising to see that in the regions imaged by both DI and NExT there was little change in the surface photometric properties and morphology with the exception of the prominent smooth flow edges. As seen from both the spacecraft and ground-based campaign, the comet continued its trend of decreasing activity from previous perihelion passages. We will present highlights from both missions and discuss implications for formation scenarios.

Highlights of Astronomy, Volume 16
XXVIIIth IAU General Assembly, August 2012
T. Montmerle, ed.

doi:10.1017/S1743921314005055

Meteorites – The Significance of Collection and Curation and Future Developments

Caroline Smith
The Natural History Museum
email: c.l.smith@nhm.ac.uk

Abstract. Meteorites are some of the most important and valuable rocks available for scientific study. Approximately 43,000 meteorites are known on Earth and are egeologicalf samples of extraterrestrial bodies - meteorites are known to originate from asteroids, the Moon, Mars and possibly comets. With expanding exploration of our Solar System, meteorites provide the eground truthf to compare data collected by robotic missions with results gained from a variety of more accurate and precise techniques using laboratories on Earth. This talk will give an introduction to the history of meteorite science and the importance of meteorite collections to the field of meteoritics, planetary and solar system science. Curation of extraterrestrial samples is a particularly pertinent issue, especially with regards to particularly rare samples such as those from Mars like the recent Tissint meteorite. Future sample return missions to asteroids and Mars also pose siginificant challenges around the curation of these precious materials. Issues surrounding the curation of samples and how curation and curatorial actions can influence scientific studies will also be discussed.

Highlights of Astronomy, Volume 16
XXVIIIth IAU General Assembly, August 2012
T. Montmerle, ed.

doi:10.1017/S1743921314005067

A list of historical comets observed at plural sites: the beginning of astronomy in Japan and Korea

Kiyotaka Tanikawa and Mitsuru Sôma

National Astronomical Observatory of Japan
2-21-1 Osawa, Mitaka, Tokyo, 181-8588, Japan
email: tanikawa.ky@nao.ac.jp, Mitsuru.Soma@nao.ac.jp

Abstract. Comets generally stay long in the sky and can be seen from many places on the surface of the Earth. We are interested in historical comets which were observed at plural sites. We have shown in a previous work (Tanikawa & Sôma, 2008) that in the seventh century, five comets were observed independently in China and Japan. From this fact and other data, we deduced that Japanese observational astronomy started in the seventh century. We know that, other than China and Japan, Korea and Vietnam had observational astronomy before the 9th century. We look for historical comets observed at plural sites by surveying the existing literature of respective countries. Examining the independence of the records, we provide a list of comets observed independently at plural sites. This strengthens the reliability of the records of comets. The list can be used for other purposes.

Keywords. Japanese and Korean astronomy, seventh century, plural observations

---

1. Introduction

The astronomy until the end of the 17th century was the kinematics of bodies in the Solar System. More concretely, main contents of astronomy were (1) motion of the Sun; (2) motion of the Moon; (3) motion of five planets among constellations; (4) maintenance of constellations as a coordinate system on the celestial sphere. Dynamics has been incorporated since 1687 when I. Newton published Principia. The Solar System was the unique world until 1838 when the parallax of stars was measured.

Careful astronomers observed the sky day and night. They noticed unusual phenomena in the sky. Among them were gueststars (novae) and comets. These objects stayed long in the sky, so some of them were observed in plural civilized countries. We present a list of these comets until the end of the eighth century. Chinese and Korean Data have been taken from the list compiled by Hasegawa (1979). Vietnamese data are from Ho Peng-Yoke (1964). We divide Korean data into those of three kingdoms.

The beginning of observational astronomy in the east Asian countries, in particular, in Japan and Korea are interesting. The main relevant data are solar eclipses, lunar eclipses, and occultations of planets by the moon. These are useful since the existence of these phenomena can be checked by modern calculations. In this sense, comet records are auxiliary. However, records of comets observed in different places are important because the reliability becomes high. We have shown that Japanese observational astronomy began in the 7th century using solar and lunar eclipses, occultations, and five common comets with China and/or Korea. We tried to show the beginning year of the astronomy in the three kingdoms of Korea. We have not been successful. We guess that the data were lost.

2. A list of comets observed at Plural sites until the eighth century

Ho Peng-Yoke admits that earlier Vietnamese data are taken from Chinese sources. Some of Korean data of three Kingdoms may have been also taken from Chinese sources.

No.	Date of discovery		Country	No.	Date of discovery		Country
12	-1495		Syria, Chaldea, India(?)	245	302	May-June	China, Baekje
59	-203	Aug.-Sept	China, Vietnam, Europe	249	305	Nov. 21	China, Rome
74	-156	Oct.	China, Vietnam	253	336	Feb. 16	China, Baekje, Rome
75	-154	Winter	China, Vietnam	259	363	Aug.-Sept.	China, Rome
77	-153	Feb.	China, Vietnam	266	383	Oct.-Nov.	Goguryeo, Europe
78	-147	May	China, Vietnam	270	390	Aug. 7	China, Baekje, Rome
80	-146	Aug.6	China, Syria	275	400	Mar. 19	China, Baekje, Rome
81	-146	Nov.	China, Vietnam	284	415	June 24	China, Baekje
84	-137	Aug.	China, Vietnam	287	418	June 24	China, Europe
90	-134	Sept.	China, Vietnam	288	418	Sept. 15	China, Europe
95	-118	May	China, Vietnam, Asia Minor	289	419	Feb. 17	China, Baekje
				291	421	Jan.-Feb.	China, Europe
107	-86	Aug.	China, Rome	292	422	Mar. 26	China, Europe
119	-48	Apr.	China, Silla	301	442	Nov. 10	China, Europe
122	-43	May-June	China, Silla	305	451	June-July	China, Rome
127	-31	Feb.	China, Greece	307	453	Feb.-Mar.	China, Rome?
132	-11	Aug. 26	China, Rome	315	467	Feb. 6	China, Europe
135	-3	Apr.	China, Silla	319	498	Dec.	China, Europe
144	54	June 9	China, Silla, Europe	326	530	Aug. 29	China, Rome
146	60	Aug. 9	China, Silla?, Rome	333	539	Nov. 17	China, Europe
147	61	Sept. 27	China, Rome	334	541	Feb.-Mar.	China, France
148	64	May 3	China, Rome	354	582	Jan. 15	China, France
157	76	Oct. 7	China, Rome	361	595	Jan. 9	China, Baekje, Europe
159	79	Oct. 7	Silla, Rome	376	626	Mar. 26	China, Europe
161	85	June 1	China, Baekje, Silla	379	634	Aug.-Sept.	China, Japan
171	128	Sept.-Oct.	Silla, Asia Minor	381	639	Mar. 5	Baekje, Japan
178	149	Oct. 19	China, Silla	392	668	May-June	China, Goguryeo
181	158	Mar.-Apr.	Goguryeo, Europe	396	676	Sept. 4	China, Silla, Japan, Europe
182	161	June 14	China, Rome				
185	182	Aug.-Sept.	China, Goguryeo	398	681	Oct. 17	China, Japan
191	191	Oct.	China, Baekje, Silla, Rome	399	683	Apr. 20	China, Silla
				400	684	Sept. 6	China, Japan
196	200	Nov. 6	China, Silla (Error?)	402	684	Dec.-685 Jan.	Japan, Europe
197	204	Dec.-205 Jan.	China, Baekje, Rome	412	712	July-Aug.	China, Europe
203	217	Nov.-Dec.	China, Goguryeo	424	744	Winter	Korea, Syria
204	218	Apr.-May	China, Rome	428	760	May 16	China, Europe
228	269	Oct.-Nov.	China, Baekje	435	770	May 26	China, Silla, Japan
243	300	Dec.-301 Jan.	China, Goguryeo	436	773	Jan. 15	China, Japan

Dates in the table are given by the Julian Calendar.

References

Ho, Peng-Yoke, 1964, *J. Amer. Orient. Soc.* **84**, 127–149.
Ichiro, Hasegawa, 1979, *Publ. Astron. Soc. Japan* **31**, 257–270.
Tanikawa & Sôma, 2008, *Rep. Nat. Astron. Obs. Japan* **11**, 31–55.

Highlights of Astronomy, Volume 16
XXVIIIth IAU General Assembly, August 2012
T. Montmerle, ed.

doi:10.1017/S1743921314005079

Analysis of Historical Meteor and Meteor shower Records: Korea, China and Japan

Hong-Jin Yang[1], **Changbom Park**[2] **and Myeong-Gu Park**[3]
[1]Korea Astronomy and Space Science Institute, Korea
email: hjyang@kasi.re.kr
[2]Korea Institute for Advanced Study, Korea
3 Kyungpook National University, Korea

Abstract. We have compiled and analyzed historical meter and meteor shower records in Korean, Chinese, and Japanese chronicles. We have confirmed the peaks of Perseids and an excess due to the mixture of Orionids, north-Taurids, or Leonids through the Monte-Carlo test from the Korean records. The peaks persist for almost one thousand years. We have also analyzed seasonal variation of sporadic meteors from Korean records. Major features in Chinese meteor shower records are quite consistent with those of Korean records, particularly for the last millennium. Japanese records also show Perseids feature and Orionids/north-Taurids/Leonids feature, although they are less prominent compared to those of Korean or Chinese records.

Keywords. meteors, meteor showers, historical records

We have compiled and analyzed the meteor and meteor shower records in official Korean history books (Kim *et al.* 1145; Kim *et al.* 1451; Chunchugwan 1392-1863) dating from 57 B.C. to A.D. 1910, covering the Three Kingdoms period (from 57 B.C. to A.D. 918), Goryeo dynasty (from A.D. 918 to A.D. 1392), and Joseon dynasty (from A.D. 1392 to A.D. 1910). The books contain only a small number of meteor shower records in contrast to abundant meteor records.

The earliest records during the Three Kingdoms period are too few and have mostly year and month information only. However, the meteor records during Goryeo and Joseon dynasties, spanning roughly one thousand years, show several statistically significant features when rearranged in sidereal years. Most prominent is the peak of the number of records at the day 220, which coincides with the modern major periodic meteor shower, Perseids. The peak persists from Goryeo dynasty to Joseon dynasty, almost one thousand years. This implies that the comet 109P/Swift-Tuttle, the parent comet of Perseids, has been active for more than one thousand years without significant perturbation on its orbit. The change in the number of records for Perseids during Goryeo and Joseon dynasties suggests that the activity of Perseids has decreased over the same period. Another collection of peaks appear around the day 300, which can be associated with Orionids, north-Taurids, or Leonids. These peaks also suggest the activities of some of these major periodic meteor showers have increased over the last millennium.

We also find the evidence of seasonal variation of sporadic meteors in Korean records. According to Yrjola & Jenniskens (1998), the activity of sporadic meteors is expected to peak at the Autumnal equinox and to be lowest at the Vernal equinox in the northern hemisphere. We confirm the seasonal variation of sporadic meteors from the records of Joseon dynasty, with the maximum number roughly 1.7 times the minimum. The least-square sinusoidal fit shows a phase shift of 15 days, toward winter, with respect to the simple geometrical expectation. We can think of two possible causes for the marginally detected phase difference. One factor could be the residual contamination by meteor showers. Another is the real innate seasonal variation of sporadic meteors. It is possible for the debris of each comet not to completely dissipate in a few years, resulting in excess of sporadic meteors near periodic meteor shower. In fact, there are many periodic meteor shower events in the winter, and the scattered meteoroids related with the meteor showers may have shifted the peak of the seasonal variation of sporadic meteors towards the winter.

We also analyzed Chinese and Japanese meter shower and meter records in their compilation books (Beijing Obs. 1988; Kanda 1935; Kanda1936; Ohsaki 1994)) for the same periods like Korea. The major features in the distribution of Chinese meteor shower records are quite consistent with those in Korean meteor records for the last one thousand years. Chinese records also suggest decrease in Perseids activity and increase in Orionids/north-Taurids/Leonids activity as the Korean records do. Japanese records also show Perseids feature and Orionids/north-Taurids-Leonids feature, although they are less prominent compared to Korean or Chinese records. One of the modern major periodic meteor shower, eta-Aquarids, which is one of the pair remnant of the comet Halley, does not have any corresponding feature in Korean, Chinese, and Japanese records while another pair remnant of Halley, Orionids, appears somewhat strongly at the day 295. This may suggest eta-Aquarids has been active only recently.

Babadzhanov (1994) has suggested five stages in the evolution of meteoroid stream. At its final stage, meteor stream becomes almost indistinguishable from the sporadic background. Evolution of the stream left by a comet depends on the condition of the comet. However, bright meteors decrease year by year after the comet passes. According to Wu (2002), Leonid showers would never be seen for more than about four consecutive years. Yeomans (1981) on the other hand has suggested that significant Leonid meteor showers were maintained for roughly 2500 days. We have looked for this kind of change in meteor activity in Korean records, wishing to estimate the evolution time of identified periodic meteor showers. For example, 231 records associated with Perseids are folded by using its period of 135 years, and distribution of the number of records as a function of phase is constructed. We have found no significant change of activity over cycles.

References

Babadzhanov, P. B. 1994, Yoshihide, K., Richard P. B., & Tomohiro, H. (Eds.), *SFYHAF ASP Conference Series.* 63, 168

Beijing observatory 1988, *General Compilation of Chinese Ancient Astronomical Records*, Jiangsu, China

Chunchugwan (The Office for Annals Compilation) 1392-1863, *Joseonwangjosillok* (The Annals of the Joseon Dynasty)

Kanda, S. 1935, *Japanese Historical Astronomical Records.* Tokyo, Japan

Kanda, S. 1936, *General Inventory of Historical Astronomical Records.* Tokyo, Japan

Kim, B. S., *et al.* 1145, *Samguksagi* (The History of the Three Kingdoms)

Kim, J. S., *et al.* 1451, *Goryeosa* (The History of the Goryeo Dynasty)

Ohsaki, S. 1994, *Japanese Historical Astronomical Records after 1600.* Tokyo, Japan

Wu, G. J. 2002, *Chinese Astronomy and Astrophysics*, 26, 40

Yeomans, D. K. 1981, *Icarus*, 47, 492

Yrjola, I.& Jenniskens, P. 1998, *A& A*, 330, 739

Highlights of Astronomy, Volume 16
XXVIIIth IAU General Assembly, August 2012
T. Montmerle, ed.

doi:10.1017/S1743921314005080

Future Small Body Exploration after the Investigation of Asteroid Itokawa by Remote Sensing and Returned Sample Analyses

Hajime Yano

Japan Aerospace Exploration Agency/Institute of Space and Aeronautical Science
email: yano.hajime@jaxa.jp

Abstract. This paper outlines current achievements of the Hayabusa mission and future small body missions with an emphasis on scientific prospects by both remote sensing in the vicinity of target objects and retuned sample analyses of them. First, the Hayabusa spacecraft aimed as technology demonstration for the worldfs first deep space round trip and sample return from an asteroid and it was launched via the M-V rocket in May of 2003. Soon after the touchdown on Asteroid Itokawa, a sub-km, S-type NEO in November 2005, the spacecraft lost its attitude control due to the leak of RCS propellant; the communication link was lost for 46 days. While the ion engine thrusters reached their lifetime by November of 2009 owing to either of an ion source or neutralizers at each engine, a challenging combination of the neutralizer-A with the ion source-B was devised to resume the spacecraftfs propulsion. This enabled the spacecraft to have returned to the Australian desert on the Earth in June 2010. The sample return capsule (SRC) was successfully recovered and returned to Japan for initial inspection of the Itokawa samples. After the announcement of initial sample analysis results, international announcement of sample distributions has started in the spring of 2012. Following up the original Hayabusa mission, JAXA has approved the Hayabusa-2 project in 2011, an asteroid sample return mission to 1999 JU3, a sub-km, C-type NEO aiming for 2014-5 launch, 2018-9 remote sensing including artificial impactor excavation and 2020 Earth return of both surface and sub-surface samples of the asteroid. C-type asteroid is thought to be abundant in organic matters and hydrated compound, so it has important clues to solve the origin and evolution of the life. NASAfs OSIRIS-Rex and ESAfs Marco Polo-R missions are also carbonaceous asteroid sample return missions in 2010fs-2020fs. Cometary nucleus or/and D-type asteroid sample returns like Hayabusa-Mk-II concept are natural progression of this type of the endeavor. JAXAfs solar power sail mission aims for eventual rendezvous with Jovian Trojan asteroids, reservoir of D/P-type asteroids as either leftovers of Jupiter system formation or the second generation intruders from the Kuiper belt regions.

Highlights of Astronomy, Volume 16
XXVIIIth IAU General Assembly, August 2012
T. Montmerle, ed.

doi:10.1017/S1743921314005092

Near-Earth objects from the cometary flux

Vacheslav Emel'yanenko

Institute of Astronomy RAS
e-mail: vvemel@inasan.ru

Abstract. We analyze the orbital distribution of objects captured to near-Earth space from the flux of comets coming from the outer Solar system. For this purpose, we use the model of the cometary cloud developed earlier (Emelfyanenko, Asher, Bailey, 2007). This model is consistent with the broad dynamical characteristics of observed near-parabolic comets, short-period comets, Centaurs and high-eccentricity trans-Neptunian objects. We show that the observed distributions of both large and small near-Earth objects are different from the modeled distribution formed dynamically by the action of planetary perturbations. In particular, while the distributions of arguments of perihelion for observed Jupiter-family comets and modeled cometary asteroids follow a sinusoidal law with pronounced maxima around 0 and 180 degrees, it is not the case for observed cometary asteroids of any size. We conclude that there exist many unobserved extinct short-period comets among near-Earth objects of various sizes.

Highlights of Astronomy, Volume 16
XXVIIIth IAU General Assembly, August 2012
T. Montmerle, ed.

doi:10.1017/S1743921314005109

The Possible Interrelation of TNO and Long-Period Comets by MOID Distribution

A. S. Guliyev[1], Sh. A. Nabiyev[1,2], R. A. Guliyev[1] and A. S. Dadashov[1]

[1]Shamakhy Astrophysical Observatory NAS, Azerbaijan, email: ayyub54@yahoo.com,
[2]Qafqaz University, Azerbaijan, email: snebiyev@qu.edu.az

The study objects of our work were 91 TNO with diameters greater than 200 km. On the other hand, the paper used the data for 1048 and comets with aphelion and perihelion distances $Q > 30$ AU and $q > 0.1$AU, respectively, were observed until 2012. Short-perihelion comets (sporadic and concentrated in separate groups) were excluded from the analysis. If some comet split into several parties, we have taken data for only one fragment, which is marked with the letter A. Data for comets are taken from the catalog [4] and the individual Circulars International Astronomical Union, issued in period after 2008. The data for TNO, mostly borrowed from the website [5], as well as the issues of the same circulars.

Research methodology is the same as in [1]. By number of comet passes from traffic zone selected TNO is compared to the modeled 67 zones. That differs from it only on the parameters Ω (longitude of the ascending node) and I (inclination). The Ω ranges from 0^0 to 330^0 in increments of 30^0 and I - from 0^0 to 90^0 with such steps to pole the planes were equidistant from each other ([2] and[3]). Ultimately, defines the following values: N - number of comet passes in traffic zone selected TNO; n, σ, t and α - average comet passes for 67 zones, normalized difference ($t = (N - n)/\sigma$), the variance a confidence interval of t, respectively. As a validation of the values t and α appropriate use the one-sample t-test, which leads to values of 1.67 and 0.95, respectively.

To calculate the value of r (i.e. MOID) used the following formula:

$$r^2 = \left(R^2(dn) + \left[\frac{q(1+e)}{1+e\cos\nu}\right]^2 - \frac{2R(dn)(q(1+e))}{1+e\cos\nu}\sqrt{1-\sin^2 i \times \sin^2(\omega-\nu)} \right)$$

Here $R(dn)$ is the distance TNO in a direction of the ascending nodecomet's orbit, q and e-non-variant elements comet's orbits, i and ω angular elements of the comets orbitsrelatively to moving plane of selected TNO, ν- true anomaly of the comet.

Naturally, the formula contains some inaccuracies, because the distance of the planet towards remote host comets orbits might differ slightly from the true direction, where the distance between TNO and comet is minimal. However, for this issue, and its ultimate objective, such inaccuracy does not play a significant role. Anyway, it is extremely improbable, that redundancy of comet passes to zones of interesting us can be results of such discrepancy, most likely to the contrary, the result could be more significant when taking into account the exact distance of the planet.

Thus, at the first stage of our calculations for 91 TNO and 1048 comets are calculated values r (only 95368 values). It is difficult to assess what law should obey their distribution, but it whenever a selected should cover the intervals from TNO 0 to Q (aphelion distance of a planetary body). In a preliminary study of the set of values of r especially noteworthy existence of a large number of small values. It is possible that a more rigorous calculations, and in the case if comets and planetary data contained greater accuracy, the result would be more impressive.

Anyhow, for definition of a degree of redundancy of comets intersections in zones of movement TNO we use the above described way of comparison. Data analysis for the case $r < 1$ AU shows that 13 values of t are likely significant, greater than 0.95. Also pays attention to fact that there is a significant gap between positive and negative t values. Their ratio is 56:36, which is not by accident. In the case of $r < 0.5$ AU almost the same patter is observed (ratio 54:38), and the number of reliable t is 9. And in the case $r < 0.1$ AU large variations of t in particular, the ratio of 67:25 is likely a reflection of their uncertainty. Apparently this case will be of interest only when the number of planets and comets, as well as their accuracy, will be much more than present.

During the further analysis the correlation has been found between values t and I trans-Neptune bodies in case of $r < 1$ AU, and in case of $r < 0.5$ AU. In either case, the correlation coefficient is less than - 0.4, and its significance is greater than 0.95. Perhaps these relationship indicate that a selected planetary bodies or some of them "in contact" with the comet near the ecliptic than at lager ecliptic latitudes. Moreover, for case of $r < 1$ AU in considering the various options for comparing the values of t and H, we found that there is definite relationship only for $H < 3^m.7$.

RESULTS

1. There is significant number of comets having distant nodes of orbits in zones of orbital moving TNO;
2. In 12 cases correspondence of distant nodes in zones of orbital moving TNO does not randomness according of methods of mathematical statistics.
3. 13 from 91 values of t in the case MOID < 1AU have statistical reliability more than 0.95.
4. 10 from 91 values of t in the case MOID < 0.5AU have statistical reliability more than 0.95.
5. In collection of t obtained by calculations on MOID positive values dominate over negative ones obviously;
6. There is significant correlation dependence t and I;
7. There is significant correlation dependence t and H from some number bright TNO.

References

Guliyev A. S. & Guliyev R. A., 2011, On the question of relation of trans-Neptunian objects and comets. Azerbaijan Astronomical Journal.V.6, p.5

Guliyev A. S. & Nabiyev Sh. A., 2002, Pluto and comets: 1.Does exist a group of comets associated with Pluto? Kinematics and Dynamics of Celestial Bodies, V.18, p. 525

Guliyev A. S. & Nabiyev Sh. A., 2004, Pluto and comets: 2.Some peculiarities of the group of comets having a possible association with Pluto. Kinematics and Dynamics of Celestial Bodies, V.20, p.283

Marsden B. G. & Williams G. V. 2008, Catalogue of Cometary Orbits, 17th edition, IAU, Cambridge, p.197

http://www.minorplanetcenter.net/iau/lists/TNOs.html

Highlights of Astronomy, Volume 16
XXVIIIth IAU General Assembly, August 2012
T. Montmerle, ed.

doi:10.1017/S1743921314005110

Comets: extremal states and their observational manifestations

Ibadov Subhon[1] and Firuz S. Ibodov[2]

[1]Institute of Astrophysics, Dushanbe, Tajikistan
email: ibadovsu@yandex.ru

[2]Lomonosov Moscow State University, Moscow, Russia
email: mshtf@sai.msu.ru

Abstract. Current status and prospects of further investigations are considered for: 1.Anomaly distribution of Na-atoms emission in the cometary heads; 2. Mechanisms of X-ray generation in comets; 3. Evolution of sungrazing comets near the Sun.

Keywords. Comets:general; Na-emission, anomaly distribution; X-rays, high-temperature plasma, solar wind, multicharge ions; cometary impacts, the Sun.

Modern models of cometary nuclei and their atmospheres as well as parameters of orbital motion of comets indicate the presence of extremal states in such objects of the Solar System. 1. The temperature distribution of the cometary gas has the minimum, in the range of 5-100 K, in the inner coma. This may cause the observed anomaly distribution of emission of Na-atoms in the heads of such bright comets like comet Mrkos 1957 d/1957 V and, possibly, in jets of comet Halley 1986 III, due to effect of depression of the temperature of cometary dust by extremely cold and dense gas in the near-nuclear region. 2. High-velocity collisions between cometary and interplanetary dust particles can lead to generation of short-living high-temperature (>100 000 K) plasma clots in the dust coma of comets, that will be accompanied by generation of soft, 0.1-1 keV, X-rays from comets, as well as multicharge ions, in the inner heliosphere. 3. Impacts of sungrazing comet nuclei with the Sun with velocities more 600 km/s will be accompanied, due to preliminary aerodynamic crushing of comet nuclei and transversal expansion of crushed matter, by impulsive generation of hot sub-photospheric plasma and "blast" shock wave induced ejection of hot plume into space above the solar chromosphere. Observations of mentioned phenomena by modern large telescopes, having high spatio-temporal resolutions, from ground-based and space observatories, together with in situ measurements during space missions in the future, are of interest not only in the study of comets but also in related fields.

References

Dobrovolsky, O. V. 1966, in Comets, Nauka, Moscow, 288

Greenstein, F. L. & Arpigny, C. 1962, *Astrophys. J.* 135, 892

Grigorian, S. S., Ibadov, S., & Ibodov, F. S. 2000, Dokl. Akad. Nauk 374, 40 [Engl. Transl.: *Phys.-Dokl.* 45, 463]

Ibadov, S. 1990, *Icarus* 86, 283

Ibadov, S. 1996, in Physical Processes in Comets and Related Objects, Cosmosinform Publ. Comp., Moscow, 181

Ibadov, S. 2000, *Astron. Astrophys. Transac.* 18:6, 799

Ibadov, S. 2011, in: Advances in Plasma Astrophysics, Cambridge Univ. Press, 76

Ibodov, F. S. & Ibadov, S. 2011, in: Advances in Plasma Astrophysics, CUP, 119

Highlights of Astronomy, Volume 16
XXVIIIth IAU General Assembly, August 2012
T. Montmerle, ed.

doi:10.1017/S1743921314005122

Location of the upper border of the cavity excavated after the *Deep Impact* collision

Sergei I. Ipatov[1,2] †,

[1]Space Research Institute, Moscow, Russia
email: siipatov@hotmail.com

[2]Catholic University of America, Washington DC, USA

Abstract. The distance between the pre-impact surface of Comet 9P/Tempel 1 and the upper border of the largest cavity with dust and gas under pressure excavated after the collision of the impact module of the *Deep Impact* spacecraft with the comet is estimated to be about 6 m if the diameter of the transient crater was about 200 m. This result suggests that cavities containing dust and gas under pressure located a few meters below surfaces of comets can be frequent.

Keywords. comets: individual (Comet 9P/Tempel 1), comets: general

---

In 2005 the impact module of the *Deep Impact* (DI) spacecraft collided with Comet 9P/Tempel 1. Based on analysis of the images made during the first 13 minutes after this impact, Ipatov & A'Hearn (2011) concluded that the triggered outburst of small particles and excavation of a large cavity with dust and gas under pressure began at $t_e = 8$ s, where t_e is the time after the DI collision. The largest cavity excavated after the collision could be relatively deep because a considerable excess ejection lasted during about 50 s. Schultz *et al.* (2013) analyzed the images of Comet 9P/Tempel 1 made by the *Stardust* spacecraft and concluded that the diameter d_{tc} of the DI transient crater was about 200 m. Some authors support smaller values of d_{tc} (up to 50 m). Supposing the depth of a growing crater to be proportional to t_e^{γ} (where γ is about 0.25-0.4) during the intermediate stage of crater excavation and adding 1 m for the initial stage, Ipatov (2012b) estimated the depth of the DI crater at $t_e = 8$ s (at the time of the beginning of excavation of the main cavity) to be about 6 m for $d_{tc} = 200$ m and 4 m for $d_{tc} = 100$ m. This depth is in accordance with the depth (4-20 m) of the initial sublimation front of the CO ice in the models of the explosion of Comet 17P/Holmes considered by Kossacki & Szutowicz (2011). The porous structure of comets provides enough space for sublimation and testifies in favor of existence of cavities. Natural outbursts were observed for several comets (see references in Ipatov 2012a). Similarity of velocities of particles ejected at triggered and natural outbursts shows that these outbursts could be caused by similar internal processes in comets. Our studies testify in favor of that cavities with dust and gas under pressure located a few meters below surfaces of comets can be frequent.

References

Ipatov, S. I. 2012a, in: P. G. Melark (ed.), *Comets: Characteristics, Composition and Orbits* (Nova Science Publishers), p. 101 (http://arxiv.org/abs/1103.0330)
Ipatov, S. I. 2012b, *MNRAS*. 423, 3474
Ipatov, S. I. & A'Hearn, M. F. 2011, *MNRAS*. 414, 76
Kossacki, K. J. & Szutowicz, S. 2011, *Icarus*. 212, 847
Schultz, P. H., Hermalyn, B., & Veverka, J. 2013, *Icarus*. 222, 502

† Present address: Alsubai Est. for Scientific Studies, Doha, Qatar

Highlights of Astronomy, Volume 16
XXVIIIth IAU General Assembly, August 2012
T. Montmerle, ed.

doi:10.1017/S1743921314005134

Disk-Resolved Spectra of (25143) Itokawa with *Hayabusa*/AMICA observations

Masateru Ishiguro

Department of Physics and Astronomy, Seoul National University , 1 Gwanak-ro, Gwanak-gu, Seoul 151-742, Republic of Korea
email: ishiguro@astro.snu.ac.kr

Abstract. We introduce a new data reduction method for *Hayabusa*/AMICA data to subtract a scattered light inside the optics. As the result, we obtained a map of space weathering on Itokawa for all available channels.

Keywords. minor planets, asteroids

---

1. Introduction of *Hayabusa*/AMICA

Hayabusa is a spacecraft developed by JAXA to return a sample of material from a near-Earth asteroid named 25143 Itokawa to Earth. It was launched on 9 May 2003 and rendezvoused with Itokawa in September 2005. After the arrival, *Hayabusa* studied the physical properties of Itokawa, such as shape, spin, color, and composition before the sampling in November. AMICA is one of the optical navigation cameras (ONC) designed not only for the optical navigation but also the scientific observations. It thus equiped broadband filters originally contrived for ground-based telescopic observations.

2. Outline of Data Reduction Method

AMICA acquired more than 1400 multispectral and high-resolution images during the mission phase. Our group established the calibration method for AMICA images. The major steps in calibration include corrections for linearity, modeling and subtraction of bias, dark current, read-out smear, pixel-to-pixel responsivity variations, and flux calibration. It was reported that AMICA images are contaminated by light scattered inside the optics in the longer wavelength (Ishiguro *et al.* 2010). Unlike telescopic observations carried out in dark places, off-axis light from the Sun or target bodies may be an inherent problem with space exploration data. It is reported that approximately ten percent of the data taken by spacecrafts onboard cameras was contaminated by scattered light (NEAR/MSI and Galileo/SSI). Since the brightness of off-axis light depends on wavelengths and positions on the detector, it could mislead reflectance spectra. We examined the point-spread functions (PSFs) of AMICA using lunar images and Itokawa's images taken during the approaching phase, and developed a new technique to subtract the scattered light by PSFs. As the result, the background sky level is suppressed down to 10 DN, which corresponds to <1 % of Itokawa signal (1000-3000DN). The color map using us-band, x-band, p-band, zs-band is available for all channels. We plan to publish the results with this method in near future.

Reference

Ishiguro, *et al.* 2010, *Icarus*, 207, 2, 714–731.

Highlights of Astronomy, Volume 16
XXVIIIth IAU General Assembly, August 2012
T. Montmerle, ed.

doi:10.1017/S1743921314005146

C/2002 VQ94 (LINEAR) and 29P/Schwassmann- Wachmann 1 - CO^+ and N_2^+ rich comets

Aleksandra Ivanova[1], Pavlo Korsun[1] and Viktor Afanasiev[2]
[1]MAO of NASU,
email: sandra@mao.kiev.ua
[2]SAO of RAS

Abstract. We investigated comets active at large heliocentric distances using observations obtained at the 6-m BTA telescope (SAO RAS, Russia). Long-slit and photometric modes of the focal reducer SCORPIO were used. Two of the comets, 29P/Schwassmann-Wachmann 1 (SW1) and C/2002 VQ94 (LINEAR) were observed to be emission rich. Detection of CO^+ and N_2 + emissions in the comae of these comets is evidence that they were formed in the outer regions of the Solar System or in a pre-solar interstellar cloud in a low temperature environment with T $\sim$ 25K. The ratio of N_2^+/CO^+ is equal to 0.011 and 0.027 for SW1 and LINEAR, respectively. Comet LINEAR is the most distant object in the Solar System (7.332 AU) for which CO^+ and N_2^+ are measured. The photometric maximum of the isolated CO^+ coma in comet LINEAR is shifted by 1.4 arcsec (7.44 $\times$ 103 km) relative to the photometric maximum of the dust coma. This shift deviates from the sunward direction by 63 degrees.

Highlights of Astronomy, Volume 16
XXVIIIth IAU General Assembly, August 2012
T. Montmerle, ed.

doi:10.1017/S1743921314005158

Development of fully depleted CCD imager NCUcam-1 and follow-up observations for PS1 sky surveys

Daisuke Kinoshita
National Central University, Taiwan
email: kinoshita@astro.ncu.edu.tw

Abstract. Visible 4-color simultaneous imager is being developed for a new 2-m telescope at Lulin observatory in Taiwan. The main task of this instrument is quick and efficient follow-up observations for large scale surveys, such as PS1 sky surveys. NCUcam-1 is one of unit cameras for this instrument. It is equipped with a new type of CCD chip, fully depleted CCD, and have superb sensitivity at the wavelength of 1 micron. Igneous asteroids exhibit prominent absorption feature around 1 micron due to minerals, and NCUcam-1 is able to deliver reliable taxonomic classifications. NCUcam-1 has achieved the first-light using 1-m telescope at Lulin in July 2011. Results of the characterization work for NCUcam-1 and tests observations of asteroids are reported. Upcoming plan for the study of pair asteroids search and identification is also presented.

Highlights of Astronomy, Volume 16
XXVIIIth IAU General Assembly, August 2012
T. Montmerle, ed.

doi:10.1017/S174392131400516X

Chemical Enrichment of the Solar System by Stellar Ejecta

Sun Kwok

Faculty of Science, The University of Hong Kong, Hong Kong, China
sunkwok@hku.hk

Abstract. Spectroscopic observations of evolved stars have shown signatures of aromatic and aliphatic compounds. This suggests that complex organics with chemical structures similar to those of insoluble organic matter (IOM) found in carbonaceous meteorites are made in stars. This raises the possibility that in addition to known pre-solar grains such as silicon carbide, organic star dust may also have traveled across the Galaxy to the Solar System.

Through remote and in-situ observations, astronomers and space scientists have discovered that organic molecules and solids are widely present in comets, meteorites, asteroids, interplanetary dust particles, and in planets and their satellites. Almost all biologically relevant organic compounds have been identified in the soluble component of carbonaceous meteorites (Schmitt-Kopplin *et al.* 2010). The insoluble organic matter (IOM) component in meteorites have structures similar to that of kerogen (Cody *et al.* 2011). The excesses in D, ^{13}C, and ^{15}N suggest that the IOM in meteorites could be of interstellar origin.

Recent infrared and submm-wave spectroscopic observations have found over 70 different kinds of gas-phase molecules, including cyanopolyyenes and acetylene, in the stellar winds of stars in the late stages of evolution. In the subsequent proto-planetary nebulae and planetary nebulae phases, complex organics with aromatic and aliphatic structures are formed (Kwok 2004, Kwok & Zhang 2011). Since the expanding circumstellar envelopes have dynamical life times of $\sim 10^4$ yr, these discoveries suggest that organics are made naturally by stars over very short periods of time. Through stellar winds, these organics are ejected into the diffuse interstellar medium and spread all over the Galaxy. Isotopic analysis of meteorites has identified several different kinds of inorganic (e.g., SiC) star dust, showing that stellar solid materials have reached the Solar System.

Observations of distant galaxies also show spectroscopic signatures of aromatic and aliphatic compounds. It is now evident the complex organics were made as early as 10 billion years ago during the early days of the Universe (Kwok 2011). If complex organics are indeed prevalent in the Universe, to what extent the Solar System, and by implication, the early Earth, have been enriched by organic star dust? This is an interesting question that deserves serious further studies.

References

Cody, G. D., Heying, E., Alexander, C. M. O., Nittler, L. R., Kilcoyne, A. L. D., Sandford, S. A., & Stroud, R. M. 2011, *PNAS*, 108, 19171
Kwok, S. 2004, *Nature*, 430, 985
Kwok, S. 2011, *Organic Matter in the Universe*, Wiley
Kwok, S. 2013, *Stardust: the cosmic seeds of life*, Springer
Kwok, S. & Zhang, Y. 2011, *Nature*, 479, 80
Schmitt-Kopplin, P., *et al.* 2010, *PNAS*, 107, 2763

Highlights of Astronomy, Volume 16
XXVIIIth IAU General Assembly, August 2012
T. Montmerle, ed.

doi:10.1017/S1743921314005171

Temperature shocks at the origin of regolith on asteroids

Patrick Michel[1], Marco Delbo[1], Guy Libourel[2], Clément Ganino[3], Chrystèle Verati[3] and Benjamin Rémy[4]

[1]Univ. Nice/CNRS/Côte d'Azur Observatory, F
email: michelp@oca.eu

[2]CRPG/CNRS, [3]Univ. Nice, F, [4]ENSEM, F

Abstract. Space-based and remote sensing observations reveal that regolith - a layer of loose unconsolidated material - is present on all asteroids, including very small, subkm-sized near-Earth asteroids (NEAs) such as (25143) Itokawa. Classically, regolith is believed to be produced by the ejecta of impact craters produced by small particles hitting asteroid surfaces. Such an explanation works for bodies whose gravity field is strong enough for substantial reaccretion of impact debris, but it fails to account for the ubiquitous presence of regolith also on small asteroids with weaker gravity. Several works have proposed that the thermal fatigue due to a huge number of day/night temperature cycles is a process that contributes to the formation of regolith on the Moon, Mercury, and on the NEA (433) Eros by fracturing boulders and rocks on their surfaces. However, this process lacks a demonstration: in order to study under which conditions rock cracking on NEAs occurs, we calculated typical temperature cycles for NEAs and we performed laboratory experiments of similar thermal cycling on meteorites taken as analogue of asteroid surface material. We will present results of these experiments and discuss their implications regarding regolith formation on asteroids.

Highlights of Astronomy, Volume 16
XXVIIIth IAU General Assembly, August 2012
T. Montmerle, ed.

doi:10.1017/S1743921314005183

MarcoPolo-R: Near Earth Asteroid Sample Return Mission candidate as ESA-M3 class mission

Patrick Michel[1], Luisa-M. Lara[2], Bernard Marty[3], Detlef Koschny[4], Maria Antonietta Barucci[5], Andy Cheng,[6] Hermann Bohnhardt[7], John R. Brucato[8], Elisabetta Dotto[9], Pascale Ehrenfreund[10], Ian A. Franchi[11] and Simon F. Green[11]

[1]Univ. Nice/CNRS/Côte d'Azur Observatory, F
email: michelp@oca.eu

[2]IAA/CSIC, S, [3]CRPG, F, [4]ESA/ESTEC, NL, [5]LESIA/Obs. Paris, F,[6]APL/JHU, USA, [7]MPS, D, [8]INAF/Obs. of Arcetri, I, [9]INAF/Obs. of Roma, I,[10]Univ. of Leiden, NL, [11]The Open University, UK

Abstract. *MarcoPolo-R* is a sample return mission to a primitive Near-Earth Asteroid (NEA) selected in February 2011 for the Assessment Study Phase at ESA in the framework of ESAfs Cosmic Vision 2 program. *MarcoPolo-R* is a European-led mission with a proposed NASA contribution. *MarcoPolo-R* takes advantage of three industrial studies completed as part of the previous Marco Polo mission (see ESA/SRE (2009)3). The aim of the new Assessment Study is to reduce the cost of the mission while maintaining its high science level, on the basis of advanced studies and technologies, as well as optimization of the mission. *MarcoPolo-R* will rendezvous with a unique kind of target, a primitive binary NEA, scientifically characterize it at multiple scales, and return a unique pristine sample to Earth unaltered by the atmospheric entry process or terrestrial weathering. The baseline target of *MarcoPolo-R* is the primitive binary NEA (175706) 1996 FG3, which offers a very efficient operational and technical mission profile. A binary target also provides enhanced science return: the choice of this target will allow new investigations to be performed more easily compared to a single object, and also enables investigations of the fascinating geology and geophysics of asteroids that are impossible to obtain from a single object. Precise measurements of the mutual orbit and rotation state of both components can be used to probe higher-level harmonics of the gravitational potential, and therefore the internal structure. A unique opportunity is offered to study the dynamical evolution driven by the YORP/ Yarkovsky thermal effects. Possible migration of regolith on the primary from poles to equator allows the increasing maturity of asteroidal regolith with time to be expressed as a latitude-dependent trend, with the most-weathered material at the equator matching what is seen in the secondary. *MarcoPolo-R* will allow us to study the most primitive materials available to investigate early solar system formation processes. Moreover, *MarcoPolo-R* will provide a sample from a known target with known geological context. Direct investigation of both the regolith and fresh interior fragments is also impossible by any means other than sample return. The main goal of the *MarcoPolo-R* mission is to return unaltered NEA material for detailed analysis in ground-based laboratories. The limited sampling provided by meteorites does not offer the most primitive material available in near-Earth space. More primitive material, having experienced less alteration on the asteroid, will be more friable and would not survive atmospheric entry in any discernible amount. Only in the laboratory can instruments with the necessary precision and sensitivity be applied to individual components of the complex mixture of materials that forms an asteroid regolith, to determine their precise chemical and isotopic composition. Such measurements are vital for revealing the evidence of stellar, interstellar medium, pre-solar nebula and parent body processes that are retained in primitive asteroidal material, unaltered by atmospheric entry or terrestrial contamination. It is no surprise therefore that sample return missions are considered a priority by a number of the leading space agencies.

Highlights of Astronomy, Volume 16
XXVIIIth IAU General Assembly, August 2012
T. Montmerle, ed.

doi:10.1017/S1743921314005195

Supplemental ancient Chinese meteor, meteorite fall and comet records with Zhongguo gudai tianxiang jilu zongji (1)

Nogami Nagatoshi
Sumitomo Chemical
e-mail:nogamin@sc.sumitomo-chem.co.jp

Abstract. Zhongguo guidai tianxiang jilu zongji pressed in 1988 containes ancient Chinese astronomical records including that of meteor, meteorite fall and comet until 1911 from the Standard Histories and local gazetteers existed in China. On the other hand, many local gazetteers lost in China at present have been collected in university and public libraries in Japan. Especially the library of Chinese section in the Research Institute for Humanistic Studies in Kyoto University and the Oriental Library in Tokyo have big collection. This presentation will give a few dozens supplemential ancient records with that big book from local gazetteers in above mentioned libraries.

Highlights of Astronomy, Volume 16
XXVIIIth IAU General Assembly, August 2012
T. Montmerle, ed.

doi:10.1017/S1743921314005201

Brief Introduction of Promoting the Chinese Program For Exploring the Martian System

Jinsong Ping[1], Xian Shi[2], Nianchuan Jian[2], Sujun Zhang[2], Mingyuan Wang[1], Kun Shang[1] and *Yinghuo-1* VLBI team[1]

[1]National Astronomical Observatory of China, CAS
email: pjs@shao.ac.cn

[2]Shanghai Astronomical Observatory, CAS

Abstract. Following the progress of Chinese deep space exploration step, since 2006 we started a Mars mission, Yinghuo-1, by join in the Phobos-Grunt mission of Russia. A satellite bus platform and onboard payloads as well as an innovative open-loop radio tracking system have been developed by mission team. Also, together with Russian and German colleagues, we developed a kind of in-beam tracking method for measuring the rotation and nutation of Phobos, and developed the 1st Phobos global gravity field for the mission. We are promoting the Chinese new mission for Mars exploration. Although the joint YH-1 & Phobos-Grunt mission failed, the new techniques and knowledge developed by mission teams may benifit the future missions. In fact, the open-loop technique have been applied into lunar and other planetary missions, and the method in developing Phobos global gravity field will be used in the study of Rosetta mission and future Chinese mission for small body.

Highlights of Astronomy, Volume 16
XXVIIIth IAU General Assembly, August 2012
T. Montmerle, ed.

doi:10.1017/S1743921314005213

Jovian impact flashes and their implication to small bodies

Junichi Watanabe

National Astronomical Observatory of Japan
e-mail: jun.watanabe@nao.ac.jp

Abstract. Optical flashes on the surface of Jupiter were observed by amateur astronomers in June and August 2010. It is thought that these phenomena were bright meteors caused by the collision of small celestial bodies of a few to 10-m, and that they seemed to be more frequent than expected. If the frequency and the scale of these phenomena are investigated, the size distribution down to size of a few m can be decided at around the giant planet region. If the systematic observation is achieved, it will be a unique attempt to use the giant planets as a natural detector of small bodies.

Optical flashes on the surface of Jupiter were observed by amateur astronomers in June and August 2010. Four amateur astronomers in Japan recorded a flash at 3h 22m 12s on August 21 (UT). The coordinate is 140.4 and +21.1 degrees in system II. The duration was about two seconds, and the brightness was 6.2 magnitudes. The preliminary light curve is analyzed. This is presumed to be an equivalent or slightly smaller scale than that in June(Hueso *et al.* 2010). It is thought that these phenomena were bright meteors caused by the collision of small celestial bodies of a few to 10 m, and that they seemed to be more frequent than expected. If the frequency and the scale of these phenomena are investigated, the size distribution down to size of a few m can be decided at around the giant planet region. In case of Earth, the brightness of meteors depends not only on sizes but also on the entry velocity. However, in the case of Jupiter, the entry velocity becomes almost similar value (60-64km per second) which is almost independent on the direction of the orbits of bodies because of the strong gravity of Jupiter. We do not have any uncertainty for estimating size of impacting bodies from the brightness of the flashes. On the other hand, we have large uncertainty in the size distribution of small bodies in the giant planet region, because we cannot see directly any bodies of less than 1km. Therefore, if the systematic observation is achieved, it will be a unique attempt to use the giant planets as a natural detector of small bodies.

Reference

Hueso, R., *et al.*, 2010, *ApJ* 721, L129–L133

Highlights of Astronomy, Volume 16
XXVIIIth IAU General Assembly, August 2012
T. Montmerle, ed.

doi:10.1017/S1743921314005225

The quinquennial grand shrine festival with the Nogata meteorite

Hitoshi Yamaoka
Faculty of Science/ICSWSE, Kyushu Univ., Japan
email: yamaoka@phys.kyushu-u.ac.jp

Though there are many forklore concerning meteorite falls, a few genuine meteorites have been identified. One of them, the Nogata meteorite is related to the oldest recorded meteorite fall (A.D. 861) (Shima *et al.*, 1983).

Nogata meteorite is kept as a "shrine treasure" at Suga Jinja shrine, where the meteorite has fallen. It is publicly demonstrated only when the grand shrine festival is held once in five years.

We introduce the quinquennial festival held on 2011 October. The meteorite is carried by the cart (Fig. 1) at the head of a parade, and many decorated carts follow. The parade continued over an hour.

Such a religious treatment would be remarkable feature of the Japanese animism. I do not know other example than Japanese shrine of such keeping of meteorite in the religious place around the world. Japanese people feel the "spilit of nature" from the meteorite fall, which may lead such keeping.

Reference

Shima, *et al.*, 1983, *Meteoritics*, 18, 87

Figure 1. Nogata meteorite is put on the front of the cart.

Highlights of Astronomy, Volume 16
XXVIIIth IAU General Assembly, August 2012
T. Montmerle, ed.

doi:10.1017/S1743921314005237

Micrometeoroid Detection in the Inner Planetary Region by the IKAROS-ALADDIN

Hajime YANO[1], Takayuki HIRAI[2], Chisato OKAMOTO[3], Masayuki FUJII[4] and Makoto TANAKA[5]

[1]Japan Aerospace Exploration Agency/ISAS, Japan
email: yano.hajime@jaxa.jp

[2]Graduate University for Advanced Studies, Japan, [3]Japan Aerospace Exploration Agency/JSPEC, Japan, [4] FAM Science,Japan, [5]Tokai University, Japan

Abstract. The ALADDIN (Arrayed Large-Area Dust Detectors in INterplanetary space) made of 0.54 m^2 PVDF sensors was deployed on the anti-Sun face of the thin polyimide sail membrane of the deep space solar sail spacecraft gIKAROS (Interplanetary Kitecraft Accelerated by the Radiation Of the Sun)h. It has measured micrometeoroid flux between the Earthfs orbit and Venusf orbit (i.e., 1.0 $\sim$ 0.7 AU of heliocentric distance) for 1.5 revolutions from June 2010 until October 2011. The ALADDIN dust detector is arrayed by 8 channels of 9-20 micron-thick PVDF sensors, which are capable of detecting hypervelocity impacts of micrometeoroids at $>\sim 10^{-12}$g, according to ground calibration impact experiments. The sensors filter electronic, thermal and vibration noises and can record time, peak hold value above its threshold, and relaxation duration of each impact signal. In total, its cruising measurements counted more than 3000 dust impacts after screening noise signals. The ALADDIN flux in the 2010-2011 epoch was compared with fluxes at similar mass range of micrometeoroids and in similar heliocentric distances measured by Helios in 1970fs and Galileo in 1990fs, both of which were composed of much less number of impact data. Then, it suggested enhancement of dust flux in the trailing edge of circumsolar orbits of the Earth and Venus, which are consistent with previous reports of larger dust grain enhancements observed by infrared telescopes. This also implies that the temporal flux enhancement of large micrometeoroids in the blob may have caused a cascading effect to produce smaller dust by collisions with sporadic meteoroids. Also it is apparent that the micrometeoroid flux increases by approximately one order of magnitude from 1 AU to 0.7 AU during the 2010-2011 epoch. The temporal variance of the Helios flux data in 1976-80 in the same region of 1 AU may be associated with difference of averaged solar activities during both epochs. Since the solar activity in the years 2010-2011 was around the minimum of the solar cycle, smaller micrometeoroids, which are more affected by solar radiation pressure than larger ones, may have survived longer than those in the Helios epoch, which covered from the minimum to the maximum of the solar cycle in late 1970fs.

Highlights of Astronomy, Volume 16
XXVIIIth IAU General Assembly, August 2012
T. Montmerle, ed.

doi:10.1017/S1743921314005249

Meteor Showers of the Earth-crossing Asteroids

Babadzhanov Pulat and Kokhirova Gulchekhra

Institute of Astrophysics, Ac. of Sci. of Tajikistan
email: kokhirova2004@mail.ru

Abstract. The results of search for meteor showers associated with the asteroids crossing the Earthfs orbit and moving on comet-like orbits are given. It was shown that among 2872 asteroids discovered till 1.01.2005 and belonging to the Apollo and Amor groups, 130 asteroids have associated meteor showers and, therefore, are the extinct cometary nuclei.

Highlights of Astronomy, Volume 16
XXVIIIth IAU General Assembly, August 2012
T. Montmerle, ed.

doi:10.1017/S1743921314005250

New Outburst of Centaur Comet (60558) 174P/Echeclus

Young-Jun Choi[1], Masateru Ishiguro[2] and Hong-Kyu Moon[1]

[1]Korea Astronomy and Space Science Institute,
email: yjchoi@kasi.re.kr

[2]Seoul National University

Abstract. We report observations of new outburst of Centaur (60558) 174P/Echeclus, using Suprime-Cam of Subaru tele-scope. The outburst was detected by Jager on May 30, 2011 (Jaeger *et al.*, 2011). We made several follow-up observations for this outburst with 1m telescope at Mt. Lemon Optical Astronomy Observatory located in US and 60cm telescope at Sobaeksan Optical Astronomy Observatory in Korea. The very first presence of coma around (60558) 2000 EC98 has been detected by Choi and Weissman (2006) on 2005 December 30.50 UT with the Palomar 5m telescope. Soon after, the object was given the periodic comet designa-tion 174P/Echeclus. We will present the characterstics and discuss the reason of this recursive outburst.

Highlights of Astronomy, Volume 16
XXVIIIth IAU General Assembly, August 2012
T. Montmerle, ed.

doi:10.1017/S1743921314005262

The physical-chemical properties of substance of the bright fireball EN171101 Turyi Remety

Klim Churyumov[1], Rudolf Belevtsev[2], Emlen Sobotovich[2], Svitlana Spivak[2] and Tetyana Churyumova[3]

[1]Astronomical Observatory of Kyiv Shevchenko National University
email: klimchur@ukr.net

[2]Institute of of environmental geochemistry of NAS and MES of Ukraine [3]Kyiv Shevchenko National University

Abstract. In 2007-2011 searches were conducted for mineralogical and geochemical studies of the soil in the region of fall down of a bright fireball EN171101 "Turyi Remety" matter in Perechyn district of Transcarpathian. In the assumed location of the fall of a meteorite material for analysis was taken from the bottom of streams of Transcarpathian Mountains. In this matter we have been found numerous small magnetic spheres (microspherul) and fused segments, which have enough large sizes - up to 5 mm in diameter, which probably are fragments of the Turyi Remety meteoroid. One of the known signs of fireballs are sand-sized magnetic balls (by diameter 0.1-1.0 mm), which are often found in the magnetic concentrate fraction. This small balls, together with fragments of fused iotsit (FeO) are formed during the ablation of the meteoroid, and their sizes decreases during the motion of the meteoroid in the Earths atmosphere. From the east to the west, the radius of the balls in the study area decreased from an average of 0.7-0.5 mm to 0.1-0.3 mm. The sizes of such balls, as glowing molten particles of the meteoroid, are in good agreement with calculations based on the energy loss of the Turyi Remety meteoroid. This confirms the cosmic origin of these found small balls. Pre-calculated physical parameters of the Turyi Remety meteoroid are the velocity, mass, kinetic energy, the resistance force during ablation, the average fireball particle radius along trajectory path of a meteoroid fragments depending from the mass and size. Rapid mass loss of the meteoroid in more than 10 times, stronger, shorter ablation and damping fireball at the high altitude say about instability and the participation of the meteoroid gas in ablation. Perhaps the presence of ice, and other fireball gases in the meteoroid composition shows that its composition was close to comet one or to a chondrite with ice (gas hydrates). Especially likely gaseous hydrates of heavy gases such as CO2, H2S, hydrocarbons (propane, butane, etc.).

Highlights of Astronomy, Volume 16
XXVIIIth IAU General Assembly, August 2012
T. Montmerle, ed.

doi:10.1017/S1743921314005274

Fireball on 6 July 2002 over the Mediterranean Sea is a fragment of the comet's nucleus

Klim Churyumov, Vitaly G. Kruchinenko, Tetyana Churyumova and Alyona Mozgova

Astronomical Observatory of Kyiv Shevchenko National University
email: klimchur@ukr.net

Abstract. Today has been known for a considerable number of cases, the explosion of large meteoroids in Earth's atmosphere. This is confirmed by the data of registrations of fireballs by devices and the results of measurements in the atmosphere of bright light flashes by photodiodes Corporation "Sandia Laboratories", which were installed on geostationary satellites of the United States, and also by data of measurements of acoustic-gravitational waves from the thermal explosions of meteoroids [ReVelle D.O. Historical detection of atmospheric impacts by large bolides using acoustic-gravity waves, Near-Earth Objects, Ed. Remo J. Annals of the New York Academy of Sciences 882, 284-302, 1997]. The work [Brown P., Spalding R.E., ReVelle D.O. *et al.* The flux of small near-Earth objects colliding with the Earth, *Nature* 420, 314-316, 2002.] shows the results of processing the observations of flashes of large meteoroids in Earth's atmosphere, obtained with the help of geostationary satellites of the United States. Over 8.5 years (from February 1994 to September 2002) 300 such events were registered. On July 6, 2002 r over the Mediterranean Sea a bright fireball was registered. The energy of the meteoroid explosion that caused the phenomenon of the car, was 26 kilotons of TNT [Brown *et al.*, 2002]. We believe that this energy refers to the height of the full bracking of the meteoroid. At a speed of 20.3 km /s adopted by the authors, body mass at this height is 5×10^8 g, and when entering the Earth's atmosphere, it was about 7×10^8 g. Based on the obtained values of the mass, we conclude that the exploded meteoroid, causing a phenomenon of the fireball was a fragment of the comet nucleus. In processing the density of the body were taken 1 g/cm^3 and the initial velocity (~30 km/s).

Highlights of Astronomy, Volume 16
XXVIIIth IAU General Assembly, August 2012
T. Montmerle, ed.

doi:10.1017/S1743921314005286

Influence of thermal models on the YORP effect

Oleksiy Golubov[1,2] and Yurij N. Krugly[2]

[1]Astronomisches Rechen-Institut, Zentrum für Astronomie der Universität Heidelberg, Mönchhofstraße 12-14, 69120 Heidelberg, Germany,
[2]Institute of Astronomy of Kharkiv National University, Sumska Str. 35, 61022 Kharkiv, Ukraine,
email: golubov@ari.uni-heidelberg.de

Abstract. In 1-D heat conductivity model, the YORP acceleration is proved to be independent of the asteroid's heat properties. Considering small structures on the surface of an asteroid breaks 1-D model. A new force appears, pulling the asteroid's surface parallel to itself. Its effect on the asteroid's rotation is called tangential YORP, or TYORP.

Keywords. asteroids: rotation

Normal YORP effect (or NYORP) is produced by recoil forces normal to the surface of an asteroid, if the asteroid is asymmetric enough, so that these forces do not compensate each other. This effect was first considered by Rubincam (2000) in the model of zero heat conductivity. Later it was proved that the NYORP acceleration is the same for any 1-dimensional heat conductivity model (Golubov & Krugly 2010; Bottke *et al.* 2006)). It means that Rubincam's approximation gives precise results for any heat conductivity, and allows to derive simple analytic expression for the NYORP acceleration (Golubov & Krugly 2010; Steinberg & Sari 2011).

In contrast, if small structures on the surface of an asteroid are considered (e. g. decimetre-sized stones), 1-D heat conductivity model is no longer valid, and new behaviour arizes. Even though eastern and western sides of a stone absorb the same amount of energy, they can emit different energies (due to nonlinearity of boundary conditions of the heat conductivity equation). Thus the two sides experiences different recoil forces, adding up to push the stone in the direction tangential to the surface, and to produce the so-called tangential YORP (or TYORP), acting on the surface as a whole.

The tangential force experiencing by each surface element of an asteroid is much smaller than the normal force, but influences of the two forces on the rotation rate are estimated to be comparable (Golubov & Krugly 2012). This has two reasons: first, NYORP forces have bigger lever arms; second, NYORP torques of different surface elements add up rather than subtract, so that even a perfectly symmetric asteroid can experience TYORP. Therefore, TYORP and NYORP may be equally importand for evolution of asteroids.

References

Breiter, S., Bartczak, P., & Czekaj, M. 2010,*MNRAS* 408, 1576
Golubov O. & Krugly Yu. 2012, *APJL*, 752, L11
Golubov A. & Krugly Yu., 2010. In: A. M. Finkelstein, W. F. Huebner, V. A. Shor (Eds.), Protecting the Earth against collisions with asteroids and comet nuclei. Proceedings of Asteroid-Comet Hazard, St. Petersburg, 21-25 Sept., 2009. Nauka, St. Petersburg, pp. 90-94
Rubincam, D. P. 2000, *Icarus*, 148, 2
Steinberg, E. & Sari, R. 2011, AJ 141:55

Highlights of Astronomy, Volume 16
XXVIIIth IAU General Assembly, August 2012
T. Montmerle, ed.

doi:10.1017/S1743921314005298

The comet disintegration and meteor streams.

A. S. Guliyev and U. J. Poladova

Shamakha Astrophysical Observatory (www.shao.az), Azerbaijan,
email: ayyub54@yahoo.com

Possibility of disintegration of proto-comet nucleus of sungraser comets in three zones of Solar System predicted by one of authors is considered. Testing of parameters of 118 split comets confirms the basic idea. Results of the statistical analysis of comet outbursts gave us additional argument in favor of this assumption. Almost twenty years have passed since, as a result of the search for host phases of isotopically unusual noble gases, the first discovery in 1987 of surviving pre-solar minerals (diamond and silicon carbide) in primitive meteorites. These were followed by others (graphite, refractory oxides, silicon nitride, and finally silicates) in the years since. Pre-solar grains occur in even higher abundance than in meteorites in interplanetary dust particles (IDPs). The result is a kind of 'new astronomy' based on the study of pre-solar condensates with all the methods available in modern analytical laboratories. According to Guliev's hypothesis (Guliyev (2010)) comet families of Kreutz, Meyer and Kracht have been appeared regarding to falling of three proto-comet nuclear into hypothetical meteor streams in zones ($I_p = 76^0.34$; $\Omega_p = 267^0.15$; 1.5 a.u.$\leqslant r_1 \leqslant$ 2.5 a.u.); ($I_p = 84^0.68$;$\Omega_p = 106^0.03$; $1.5a.u. \leqslant r_2 \leqslant 2.5a.u.$) and ($I_p = 14^0.93$; $\Omega_p = 54^0.26$; $0.4a.u. \leqslant r_3 \leqslant 0.6a.u.$). These zones numbered as 1,2 and 3. We will try to find out is there are overpopulation of nearest and distant nodes of other splitted comets and having outburst ones in zones (1-3). Data of 118 split comets discovered up to 2012 have been used in our work. We have calculated the numbers of distant nodes for all 118 comets relative to the selected planes (1-3) and found number of nodes (N) corresponding to the interval (r_1 and r_2) from zones 1-3. We have also used the method of testing (Guliyev 2010) to demonstrate of excessiveness of N. On the base of data regarding to 67 comparison planes we have determined parameters: n - midrange value of nodes; σ - rms deviation; t - normalized difference (t = (N-n)/σ) and α - confidential probability of t. Calculation and analyze of distant nodes regarding to the zone 1,2 and 3 gave us following numerical results: Zone 1 - distant nodes: N=20; n=14.4478; σ =3.87; t = 1.43; α = 0.85. Zone 1 - nearest nodes: N=29; n=22.63; σ =3.54; t = 2.36; α = 0.99. Zone 2 - distant nodes: N=15; n=11.36; σ =3.02; t=1.21; α = 0.75. Zone 2 - nearest nodes N=40; n=34.72; σ = 2.71; t=1.95; α = 0.97. Zone 3 - nearest nodes N=9; n=5.64; σ =1.49; t=2.24; α = 0.99. Zone 3 - distant nodes N=3; n=0.49; σ = 0.75; t=3.36; α = 0.99. Thus in all three cases we have obtained confirmation of excess of comet nodes in zones (1-3). In zones 1,2 and 3 number of comet outbursts have to be more than in other ones. For checking of this consequence we demonstrate below (Fig.1. and Fig.2.)distribution of q of long-period comets and r (distant where comet outbursts have been observed), which confirms basis idea of (Guliyev (2010)).

Reference

Guliyev, A. S. 2010, *Origin of short-perihelion comets.Publishing company,"Elm", Baku*, 151

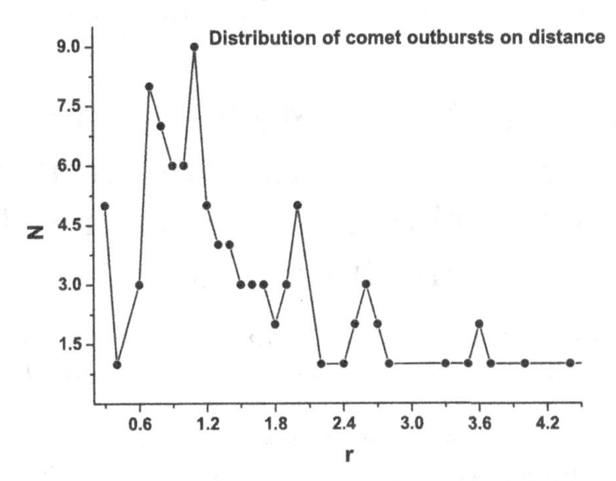

Figure 1. Distribution of comet outbursts as a function of distance from the Sun.

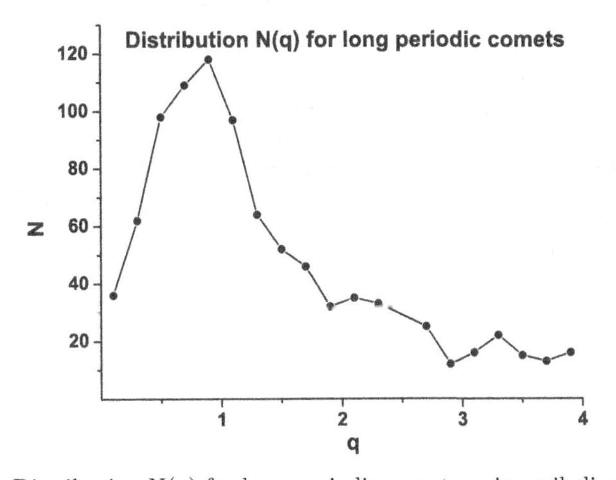

Figure 2. Distribution N(q) for long periodic comets. q is perihelion distance.

Highlights of Astronomy, Volume 16
XXVIIIth IAU General Assembly, August 2012
T. Montmerle, ed.

doi:10.1017/S1743921314005304

Determination of the rotational period of the comet 29P/Schwassmann-Wachmann-1 using dynamics of the dust structures (jets) in the coma

Aleksandra Ivanova[1], Viktor Afanasiev[2], Pavlo Korsun[1], Aleksandr Baransky[3], Maksim Andreev[4] and Vasyliy Ponomarenko[5]

[1]MAO of NASU,
email: sandra@mao.kiev.ua

[2]SAO of RAS, [3]Astronomical observatory Kyiv Shevchenko National University, [4]International Center for Astronomical, Medical and Ecological Research NASU, [5]Astronomical observatory Kyiv Shevchenko National University

Abstract. We present analysis of the photometric data of the distant comet 29P/Schwassmann-Wachmann-1, obtained at the 6-m BTA telescope (SAO RAS, Russia) and at the 2-meter telescope Zeiss-2000 (ICAMER, KB). The comet shows significant jets activity at large heliocentric distances, beyond the zone of water ice sublimation. Various digital filters were applied to increase the contrast of the jets and separate them. The rotation period of the nucleus was derived using cross-correlation method. The value of the rotation period is 12.1 ± 1.2 days for observations made in 2008 and 11.7 ± 1.5 days for observations made in 2009.

Highlights of Astronomy, Volume 16
XXVIIIth IAU General Assembly, August 2012
T. Montmerle, ed.

doi:10.1017/S1743921314005316

The activity of autumn meteor showers in 2006-2008

Anna Kartashova

Institute of Astronomy of the Russian Academy of Sciences
e-mail: akartashova@inasan.ru

Abstract. The purpose of meteor observations in INASAN is the study of meteor showers, as the elements of the migrant substance of the Solar System, and estimation of risk of hazardous collisions of spacecrafts with the particles of streams. Therefore we need to analyze the meteor events with brightness of up to 8 m, which stay in meteoroid streams for a long time and can be a hazardous for the spacecraft. The results of our single station TV observations of autumn meteor showers for the period from 2006 to 2008 are presented. The high-sensitive hybrid camera (the system with coupled of the Image Intensifier) FAVOR with limiting magnitude for meteors about 9m...10m in the field of view 20 × 18 was used for observations. In 2006-2008 from October to November more than 3 thousand of meteors were detected, 65% from them have the brightness from 6m to 9m. The identification with autumn meteor showers (Orionids, Taurids, Draconids, Leonids) was carried out. In order to estimate the density of the influx of meteor matter to the Earth for these meteor showers the Index of meteor activity (IMA) was calculated. The IMA distribution for the period 2006 - 2008 is given. The distributions of autumn meteor showers (the meteors with brightness of up to 8 m) by stellar magnitude from 2006 to 2008 are also presented.

Highlights of Astronomy, Volume 16
XXVIIIth IAU General Assembly, August 2012
T. Montmerle, ed.

doi:10.1017/S1743921314005328

Identification of radiants of low-light-level meteors from double station TV observations during autumnal equinox of 2001 and 2003

Pavlo M. Kozak, Olexander O. Rozhilo and Yuriy G. Taranukha

Astronomical Observatory, Kyiv National Taras Shevchenko University, 3, Observatorna Str., Kyiv, 04053, Ukraine
email: `kozak@observ.univ.kiev.ua`

Abstract. Results of double-station TV meteor observations which were carried out during autumnal equinox in 2001 and 2003 are used for the confirmation of existing meteor showers and for a search for probable new micro-showers. Seven existing September showers were confirmed, two of them by meteors from both years. Fourteen new groups were proposed to be considered as new meteor micro-showers. Taking into account similar kinematical parameters two of them can be supposed to be a part of already existing showers.

Keywords. TV observations of meteors, meteor radiant, new meteor stream.

Basing on double station TV observations of meteors on September 21-22 of 2001 and September 19-24 of 2003 we try to confirm known meteor showers in accordance with Meteor Data Center of International Astronomical Union (MDC IAU) Jopek & Jenniskens (2010), and to search for probable new micro-showers. The observational data were obtained with the help of TV systems of super-isocon type, briefly described in Kozak (2008). The amount of double station meteors which were used for the identification was 18 meteors of 2001 and 80 meteors of 2003. Confirmation of known mini-showers of meteors was carried out separately for each year. Elements of cluster analysis were used for the identification of meteor showers. The coordinates of geocentric radiant and velocity, and additionally angular elements of heliocentric orbit were used as parameters for the selection. In spite of small statistics seven known micro-showers were confirmed by a few meteors: September iota Cassiopeiids, nu Draconids, South delta Piscids, omega Piscids, kappa Aquariids, October Capricornids and sigma Orionids. The showers of nu Draconids and South delta Piscids were confirmed by meteors of both years. Fourteen compact meteor groups were selected as possible new micro-showers. Later we identified two of them with established showers (September epsilon Perseids and North delta Piscids) but proposed to specify the kinematical parameters of these showers. Other twelve groups are proposed to be considered as possible new meteor micro-showers.

References

Jopek, T. J. & Jenniskens P. M. 2011, in W. J. Cooke, D. E. Moser, B. F. Hardin, and D. Janches (eds.), *Proc. Meteoroids Conf. Breckenridge, Colorado, USA, NASA/CP-2011-216469.* p.7
Kozak, P. 2008, *EM* & *P* 102, 1-4, 277

Highlights of Astronomy, Volume 16
XXVIIIth IAU General Assembly, August 2012
T. Montmerle, ed.

doi:10.1017/S174392131400533X

Photometric Properties of Vesta

Jian-Yang Li[1], L. Jorda[2], H. U. Keller[3], N. Mastrodemos[4], S. Mottola[5], A. Nathues[6], C. Pieters[7], V. Reddy[6], C. A. Raymond[4], T. Roatsch[5], C. T. Russell[8], B. J. Buratti[4], S. E. Schroder[6], M. V. Sykes[1], T. Titus[9], F. Capaccioni[10], M. T. Capria,[10], L. Le Corre[6], B. W. Denevi[11], M. De Sanctis[12], M. Hoffmann[6] and M. D. Hicks[4]

[1]Planetary Science Institute
email: jyli@psi.edu

[2]Laboratoire d'Astrophysique de Marseille, [3]University Braunschweig, IGEP, [4]Jet Propulsion Laboratory, California Institute of Technology, [5]DLR, Inst. of Planetary Research, [6]Institute for Solar System Research, Max-Planck, [7]Brown University, Planetary Geosciences Group, [8]UCLA, Institute of Geophysics, [9]US Geological Survey, Astrogeology Science Center, [10]INAF, Istituto di Astrofisica Spaziale e Fisica Cosmica, [11]Johns Hopkins University Applied Physics Laboratory, [12]INAF, Istituto di Astrofisica Spaziale e Fisica Cosmica

Abstract. The Dawn spacecraft orbited Asteroid (4) Vesta for a year, and returned disk-resolved images and spectra covering visible and near-infrared wavelengths at scales as high as 20 m/pix. The visible geometric albedo of Vesta is $\sim$ 0.36. The disk-integrated phase function of Vesta in the visible wavelengths derived from Dawn approach data, previous ground-based observations, and Rosetta OSIRIS observations is consistent with an IAU H-G phase law with H=3.2 mag and G=0.28. Hapke's modeling yields a disk-averaged single-scattering albedo of 0.50, an asymmetry factor of -0.25, and a roughness parameter of $\sim$20 deg at 700 nm wavelength. Vesta's surface displays the largest albedo variations observed so far on asteroids, ranging from $\sim$0.10 to $\sim$0.76 in geometric albedo in the visible wavelengths. The phase function of Vesta displays obvious systematic variations with respect to wavelength, with steeper slopes within the 1- and 2-micron pyroxene bands, consistent with previous ground-based observations and laboratory measurement of HED meteorites showing deeper bands at higher phase angles. The relatively high albedo of Vesta suggests significant contribution of multiple scattering. The non-linear effect of multiple scattering and the possible systematic variations of phase function with albedo across the surface of Vesta may invalidate the traditional algorithm of applying photometric correction on airless planetary surfaces.

Highlights of Astronomy, Volume 16
XXVIIIth IAU General Assembly, August 2012
T. Montmerle, ed.

doi:10.1017/S1743921314005341

Disk-Resolved Photometry of Cometary Nuclei: Results from DIXI and Stardust-NExT

Jian-Yang Li[1], Peter C. Thomas[2], Joe Veverka[2], Michael F. A'Hearn[3], Sebastien Besse[3], Michael J. S. Belton[4], Tony L. Farnham[3], Kenneth P. Klaasen[5],Carey M. Lisse[6], Lucy A. McFadden[7] and Jessica M. Sunshine[3]

[1]Planetary Science Institute
email: jyli@psi.edu

[2]Cornell University, [3]University of Maryland at College Park, [4]Belton Space Exploration Initiatives, LLC, [5]Jet Propulsion Laboratory, California Institute of Technology,[6]JHU-APL,[7]NASA Goddard Space Flight Center

Abstract. Previous comet flyby missions enabled detailed studies of the photometric properties of several cometary nuclei from disk-resolved images, including 9P/Tempel 1, 19P/Borrelly, and 81P/Wild 2. Two recent missions, DIXI and Stardust-NExT, encountered Comets 103P/Hartley 2 and Tempel 1 respectively, expanding the pool of sampled cometary nuclei in their unique ways: Hartley 2 is a hyperactive comet; Tempel 1 was visited and impacted by the Deep Impact dual-spacecraft during its previous perihelion passage. Photometric modeling shows that the global photometric properties of the nuclei of Hartley 2 and Tempel 1 are similar to those of other cometary nuclei. The photometric variation of the hyperactive nucleus of Hartley 2 is about 15%, similar to that of weakly active comets Tempel 1 and Wild 2. The photometric properties of Tempel 1 measured by NExT suggest little change from those measured by DI. These results, together with the photometric properties of Wild 2 and Borrelly, indicate that the photometric properties of cometary nuclei are independent of the activity level and gross geomorphology of cometary nuclei. Instead, cometary nucleus photometric properties might be determined by its outgassing, which leaves low-albedo deposit on the surface and forms similar photometric texture. The time scale for the photometric alteration on cometary nuclei due to outgassing should be much shorter than the dynamic time scale.

Highlights of Astronomy, Volume 16
XXVIIIth IAU General Assembly, August 2012
T. Montmerle, ed.

doi:10.1017/S1743921314005353

Meteor studies applying incoherent scatter radar instruments

Ingrid Mann[1], Asta Pellinen-Wannberg[2] and Anders Tjulin[3]

[1]EISCAT Scientific Association, Kiruna and Umea University, Sweden
email: ingrid.mann@eiscat.se

[2]IRF Kiruna and Umea University, Umea, Sweden, [3]EISCAT Scientific Association, Kiruna, Sweden

Abstract. One of the interesting aspects of optical meteor studies is that the spectral composition of the brightness reveals information about the element composition of the solid particles that enter Earthfs atmosphere from interplanetary space. Deriving composition from optical spectra requires understanding the entry process during which the entering solid interacts with atmospheric species. This is especially so, because most meteors are observed at altitudes where the mean free path changes from tens of meters to millimeters, that is in the 120 km to 80 km altitude range within the atmosphere. The ionization that causes optical emission also reflects radio waves, so that meteors are observed with different kinds of radar instruments. Incoherent scatter radar facilities are in particular designed to study the upper atmosphere by using the backscattering from free electrons and are basically High Power Large Aperture radars. During the past 15 years they have been increasingly used for meteor studies. The phased-array incoherent scatter radars that are currently under development, such as the planned EISCAT-3Dsystem in northern Scandinavia, will further improve the spatial and time resolution of radar observations and will allow simultaneous measurements of the meteors and of the parameters of the surrounding ionosphere. Radar backscattering is also sensitive to objects that are smaller than those detected optically, so that the observations also permit studying the extension of the meteoroid size distribution to smaller sizes. In this presentation we consider the possibilities for measurements with the future EISCAT-3D as a new path of studying the physics of meteor phenomena with high accuracy.

Highlights of Astronomy, Volume 16
XXVIIIth IAU General Assembly, August 2012
T. Montmerle, ed.

doi:10.1017/S1743921314005365

Spectrophotometric properties of Moon's and Mars's surfaces exploration by shadow mechanism

Alexandr Morozhenko, Anatolij Vidmachenko and Nadiia Kostogryz
Main Astronomical Observatory of NAS of Ukraine,
email: avshulga@mail.ru

Abstract. Typically, to analyze the data of the phase dependence of brightness atmosphereless celestial bodies one use some modification of the shadow mechanism involving the coherent mechanism. There are several modification of B.Hapke [2] model divided into two groups by the number of unknown parameters: the first one with 4 parameters [3,4] and the second one with up to 10 unknown parameters [1] providing a good agreement of observations and calculations in several wavelengths. However, they are complicated by analysing of the colorindex $C(\alpha)$ dependence and photometric contrast of details with phase $K(\alpha)$ and on the disk (μ o = cos i). We have got good agreement between observed and calculated values of $C(\alpha) = U(\alpha)$-$I(\alpha)$, $K(\alpha)$, K(muo) for Moon and Mars with a minimum number of unknown parameters [4]. We used an empirical dependence of single scattering albedo (ω) and particle semi-transparency(æ): æ = (1-ω)n. Assuming that $[\chi(0^{\circ})/\chi(5^{\circ})] = \chi(5^{\circ})/\chi(0^{\circ})]$,where $\chi(\alpha)$ is scattering function, using the phase dependence of brightness and opposition effect in a single wavelength, we have defined ω,$\chi(\alpha)$,g (particle packing factor), and the first term expansion of $\chi(\alpha)$ in a series of Legendre polynomials x1. Good agreement between calculated and observed data of $C(\alpha) = U(\alpha)$-$I(\alpha)$ for the light and dark parts of the lunar surface and the integral disk reached at n $\approx$ 0,25, g = 0,4 (porosity 0,91), x1 = -0,93, ω = 0,137 at λ = 359nm and 0,394 at λ = 1064nm;, for Mars with n $\approx$ 0,25,g = 0,6 (porosity 0,84), x1 $\approx$ 0, ω = 0,210 at λ = 359nm and ω = 0,784 at λ = 730nm.

1. Bowell E., Hapke B., Domingue D., Lumme K., *et al.* Applications of photometric models to asteroids, in Asteroids II. Tucson: Univ. Arizona Press. p.524-556. (1989)
2. Hapke B. A theoretical function for the lunar surface, *J.Geophys.Res.* 68, No.15., 4571-4586(1963).
3. Irwine W. M., The shadowing effect in diffuse reflection, *J.Geophys.Res.* 71,No.12, 2931-2937(1966).
4. Morozhenko A. V., Yanovitskij E.G. ,An optical model of the Martian surface in the visible region of spectrum, *Astronomy Reports* 48, No.4, 795-809(1971).

Highlights of Astronomy, Volume 16
XXVIIIth IAU General Assembly, August 2012
T. Montmerle, ed.

doi:10.1017/S1743921314005377

The current state and prospects for meteors observations in RI NAO

A. Shulga, N. Kulichenko, V. Vovk, Y. Kozyryev and Y. Sybiryakova

RI "Nikolaev Astronomical Observatory" Mykolaiv, 54030, Ukraine
email: avshulga@mail.ru

The observations of meteors using optical telescopes equipped with TV-cameras and the observations in radio signals using FM radio and TV translators are started in RI NAO since 2010.

The main purpose of observations is

– in optics — statistics of meteors at night time, the calculation of flux maximum, the calculation of the coordinates of the radiant flux,

– in the radio range — statistics of meteors within 24 hours, the calculation of flux maximum, and its width, the definition of the velocity of meteor.

Optical observations of meteors at the RI NAO are conducted using Meteor patrol, which includes two optical telescopes (D = 47mm, F = 85mm) equipped with a TV CCD cameras WAT-902H2 (768×576 , $8.6 \times 8.3\mu$). The field of view of each telescope is $4.2° \times 3.2°$.

The process of observation and coordinate-photometric processing is automated using original software developed by the RI NAO. Software implements the method for automatical detecting of meteors using the search of cell of its image according signal/noise ratio which exceeds the liminal value on previous frames. In addition to the signal/noise ratio, the velocity of objects that allows not to register objects, such as satellites, airplanes, birds, and noises of CCD camera is also analyzed.

In 2011-2012, regular observations were conducted at the meteor patrol in RI NAO, more than 1200 meteors with the longitude of $(0.5 - 4.5)°$ and brightness (1-5)m were recorded. The error of reference system is (1-6)". Error in determining of meteor trajectories arc in right ascension and declination was (10-12)". Error estimate of the pole of a large circle of the meteor trajectory was (3-13)'.

In the radio band the observations of meteors are carried out by detecting the horizon FM radio signal, which is reflected from the meteor trail. FM station in Kielce (Poland) is used as signal source.

Hardware and software package is designed. It includes a highly directional antenna, PC with TV/FM-tuner and software for automatic registration of meteors. Automatic extraction of fragments of audio signal of FM radio stations in the time-frequency space using the FFT algorithm is used. The original software is developed by the NAO.

The data processing of observations all of the data is carried out and more than 100,000 meteoric phenomena are given. The characteristic bursts of meteors for all major meteor showers are detected from the time series of observations.

A comparison of data of automatic processing of radio observations in the NAO with observations of meteors by video observations of meteors IMO network were conducted, the correlation coefficient was > 0.5.

Highlights of Astronomy, Volume 16
XXVIIIth IAU General Assembly, August 2012
T. Montmerle, ed.

doi:10.1017/S1743921314005389

Present State and Prospects for the Meteor Research in Ukraine

O. Shulga[1], Y. Voloshchuk[2], S. Kolomiyets[2], Y. Cherkas[2], I. Kimakovskay[3], S. Kimakovsky[3], E. Knyazkova[3], Y. Kozyryev[4], Y. Sybiryakova[4], Y. Gorbanev[3], I. Stogneeva[3], V. Shestopalov[3], P. Kozak[5], O. Rozhilo[5] and Y. Taranukha[5]

[1]Nikolaev Astronomical Observatory, email: avshulga@mail.ru

[2]Kharkiv National University of Radioelectronics, [3]Astronomical observatory of I.I.Mechnikov Odessa National University, [4]Nikolaev Astronomical Observatory, [5]Astronomical Observatory of Kyiv National Taras Shevchenko University

Abstract. ODESSA. Systematical study of the meteor events are being carried out since 1953. In 2003 complete modernization of the observing technique was performed, and TV gmeteor patrolh on the base of WATEC LCL902 cameras was created. @ wide variety of mounts and objectives are used: from Schmidt telescope F = 540 mm, F/D = 2.25 (field of view FOV = (0.68x0.51) deg, star limiting magnitude SLM = 13.5 mag, star astrometric accuracy 1-2 arcsec) up to Fisheye lenses F = 8 mm, F/D = 3.5 (FOV = (36x49) deg, SLM = 7 mag). The database of observations that was collected between 2003 and 2012 consists of 6176 registered meteor events. Observational programs on basis and non-basis observations in Odessa (Kryzhanovka station) and Zmeiny island are presented. Software suite of 12 programs was created for processing of meteor TV observations. It enables one to carry out the whole cycle of data processing: from image preprocessing up to orbital elements determination. Major meteor particles research directions: statistic, areas of streams, precise stream radiant, orbit elements, phenomena physics, flare appearance, wakes, afterglow, chemistry and density. KYIV. The group of meteor investigations has been functioning more than twenty years. The observations are carried out simultaneously from two points placed at the distance of 54 km. Super-isocon low light camera tubes are used with photo lens: F = 50mm, F/D = 1.5 (FOV = (23.5 x 19.0) deg, SLM = 9.5 mag), or F = 85, F/D = 1.5 (FOV = (13x11) deg, SLM = 11.5 mag). Astrometry, photometry, calculation of meteor trajectory in Earth atmosphere and computation of heliocentric orbit are realized in developed gFalling Starh software. KHARKOV. Meteor radio-observations have begun in 1957. In 1972, the radiolocation system MARS designed for automatic meteor registration was recognized as being the most sensitive system in the world. With the help of this system 250 000 faint meteors (up to 12 mag) were registered between 1972 and 1978 (frequency 31.1 MHz, particle masses $10^{-3} \sim 10^{-6}$ g). Simultaneously, millions of reflections were registered for even fainter meteors (up to 14 mag). Information about 250 000 meteors and 5160 meteor streams is included in database. This is an unique material that can be used for hypotheses testing, as well as for creation new theories about meteor phenomena. Models of the meteor matter distribution in the Earthfs atmosphere, near-Earth space and in the Solar system, influence on surface of spacecrafts were developed. NIKOLAEV. The optical and radio observations of meteors have begun in 2011. Two WATEC LCL902 cameras are used with photo lens F = 85 mm, F/D = 1.8 (FOV = (3.2x4.3), SLM = 12 mag, star astrometric accuracy 1-6 arcsec). Original software was developed for automatic on-line detection of meteor in video stream. During 2011 year 105 meteor events were registered (with angular length (0.5-4.5) deg and brightness (1-5) mag). Error of determination of the meteor trajectory arc $\sim$ (10-12) arcsec. Error of determination of the large circle pole of the meteor trajectory is $\sim$ (3-13) arcmin. In the radio band observations of meteors are performed by registration of signal reflected from the meteor wake. As a signal source the over-the-horizon FM station in Kielce (Poland) is used. Narrow-beam antenna, computer with TV/FM tuner and audio recording software are used to perform radio observations. Original software was developed for automatic detection of meteor in audio stream.

Highlights of Astronomy, Volume 16
XXVIIIth IAU General Assembly, August 2012
T. Montmerle, ed.

doi:10.1017/S1743921314005390

Strategy for NEO follow-up observations

Milos Tichy, Michaela Honkova, Jana Ticha and Michal Kocer

Klet Observatory, Zatkovo nabrezi 4, CZ-370 01 Ceske Budejovice, Czech Republic

Abstract. The Near-Earth Objects (NEOs) belong to the most important small bodies in the solar system, having the capability of close approaches to the Earth and even possibility to collide with the Earth. In fact, it is impossible to calculate reliable orbit of an object from a single night observations. Therefore it is necessary to extend astrometry dataset by early follow-up astrometry. Follow-up observations of the newly discovered NEO candidate should be done over an arc of several hours after the discovery and should be repeated over several following nights. The basic service used for planning of the follow-up observations is the NEO Confirmation Page (NEOCP) maintained by the Minor Planet Center of the IAU. This service provides on-line tool for calculating geocentric and topocentic ephemerides and sky-plane uncertainty maps of these objects at the specific date and time. Uncertainty map is one of the most important information used for planning of follow-up observation strategy for given time, indicating also the estimated distance of the newly discovered object and including possibility of the impact. Moreover, observatories dealing with NEO follow-up regularly have prepared their special tools and systems for follow-up work. The system and strategy for the NEO follow-up observation used at the Klet Observatory are described here. Methods and techniques used at the Klet NEO follow-up CCD astrometric programme, using 1.06-m and 0.57-m telescopes, are also discussed.

Keywords. NEO, follow-up, astrometry, observations

1. FUTURE PLANS FOR NEO ACTIVITIES

The International Astronomical Union is just been preparing Resolution B3 of the GA IAU in Beijing (2012) on the establishment of an International NEO early warning system. It could say that there is now ample evidence that the probability of catastrophic impacts of Near-Earth Objects (NEOs) onto the Earth, potentially highly destructive to life, and for humankind in particular, is not negligible and that appropriate actions are being developed to avoid such catastrophes. Because NEOs are a threat to all nations on Earth, all nations should contribute to avert this threat. The IAU has been working in close cooperation with the United Nations Committee on the Peaceful Uses of Outer Space (UNCOPUOS) and the International Council for Science (ICSU) to coordinate and collaborate on the establishment of an International NEO early warning system.

Also, since January 2009, the European Space Agency (ESA) has been preparing its so-called Space Situational Awareness(SSA) programme.

The main idea of the KLENOT NEXT GENERATION PROJECT is to take active part in these initiatives and to work as a dedicated NEO follow-up station in close world-wide cooperation with the highest priority given to PHAs and VIs.

References

Ticha, J., Tichy, M., Kocer, M., & Honkova, M., 2009, *Met.Pl.Sci.* 44, 1889
IAA Planetary Defence Conference, Bucharest at http://www.pdc2011.org/ 2011

Highlights of Astronomy, Volume 16
XXVIIIth IAU General Assembly, August 2012
T. Montmerle, ed.

doi:10.1017/S1743921314005407

JD6 - The Connection between Radio Properties and High Energy Emission in AGNs

Gabriele Giovannini[1,2] and Teddy Cheung[3]

[1]Department of Physics and Astronomy, Bologna University, Italy
email: ggiovann@ira.inaf.it

[2]Istituto di Radioastronomia-INAF, Italy

[3]Naval Research Laboratory, Washington, DC 20375, USA
email: Teddy.Cheung@nrl.navy.mil

Abstract. While observations in the radio band are providing essential information on the innermost structures of relativistic jets in active galactic nuclei (AGN), the recent detection by *Fermi* of gamma-ray emission from many hundreds of blazars shows that the maximum jet power is emitted at high energies. Multi-wavelength monitoring observations further allow variability studies of the AGN spectral energy distributions over 13 orders of magnitude in frequency. The Joint Discussion offered the possibility for a comprehensive discussion of advances in the observational domain and stimulated theoretical discussion about our current understanding of jet physics.

Keywords. galaxies: active, BL Lacertae objects: general, galaxies: jets, quasars: general, gamma rays: observations, radio continuum: general

---

1. Introduction

A Joint Discussion (JD) dedicated to the connection between radio properties and high energy emission in active galactic nuclei (AGN) took place on the 23rd and 24th of August 2012 during the XXVIII IAU General Assembly in Beijing, China. It was a great opportunity to confront observational multiwavelength data with theoretical models for AGN. The JD included four main sessions covering the AGN population in the radio and gamma-ray bands (§ 2), high resolution core and jet properties (§ 3), multiwavelength correlations and variability (§ 4), and a final discussion on jet physics and the role of black hole (BH) spin and accretion (§ 5). Eight invited plus 11 contributed oral talks were presented together with 34 contributed posters to an interesting scientific discussion. In the following we present a brief summary for the talks in the four sessions.

All presentations are available at: `http://www.ira.inaf.it/meetings/iau2012jd6/Program.html` and poster abstracts in: `http://www.ira.inaf.it/meetings/iau2012jd6/Posters.html`.

2. Session 1: The AGN population as seen in the Radio and Gamma-ray bands

2.1. *T. Cheung – The AGN population in Radio and Gamma-rays: Origins and Present Perspective (invited)*

Radio observations present a rich phenomenology in studies of AGN – extensive multi-frequency lightcurves (e.g., Fig 1), observations and statistics of superluminal motions, radio spectral polarimetric imaging on all scales (jets, hotspots, lobes), and an abundant variety of source types including young radio sources and radio-quiet objects. In the EGRET era, there were 66 high-confidence blazars identified with gamma-ray sources (27 lower-confidence) and only a few radio galaxies (e.g., Cen A, 3C 111). Now we are in the era of the *Fermi* Gamma-ray Space Telescope with the increased capabilities provided by the Large Area Telescope (LAT). The 2nd LAC AGN catalog (2LAC) "clean" sample included: 310 flat spectrum radio quasars (FSRQs), 395 BL Lacs, 156 blazars of unknown type, and 24 AGNs (include radio galaxies). FSRQs have been detected with $L_\gamma \sim 10^{48}$ erg/s up to $z \sim 3$. Significant correlations are present between radio and gamma-ray properties. FSRQs are on average brighter and apparently more luminous in the radio band than BL Lacs (but a redshift incompleteness is present).

Thanks to the 1st *Fermi*-LAT Hard source Catalog (1FHL) which includes data from August 2008 through July 2011, possibly softer AGN gamma-ray spectra appear with increasing redshift, and many new targets are now available for current and future TeV telescopes. The high energy emission site for AGN can be probed with gamma-ray imaging in exceptional cases, like in the giant radio lobes of Centaurus A (and possibly in NGC 6251, and Fornax A). Moreover *Fermi* detected giant gamma-ray bubbles in our Galaxy (§ 5.4) and a possible young radio source (4C+55.17). In the coming years, it will be possible to extend the radio/gamma-ray correlations to low fluxes/luminosities, to continue to identify possible sites of gamma-ray emission, to test if radio-quiet AGN are also gamma-ray quiet, and to look for bubble sources in nearby galaxies.

References: Abdo *et al.* (2010a), Ackermann *et al.* (2011b), McConville *et al.* (2011), Paneque *et al.* (2012), Takeuchi *et al.* (2012).

2.2. *Ł. Stawarz – The AGN population in Radio and Gamma rays: theoretical perspectives (invited)*

Radio-loud systems appear to be the only AGN loud in gamma-rays. In addition to blazars and radio galaxies, the only newly established class of gamma-loud AGN are the radio-loud narrow line Seyfert 1s (NLS1s). Radio-quiet Seyferts seem gamma-quiet as well. Radio properties of nuclear jets may be directly related to gamma-ray properties. However, nuclear relativistic jets are not the only relevant gamma-ray production site as lobes and bubbles are observed in gamma-rays. Correlations between radio and gamma-ray properties seem to be present, but more data and better statistics are necessary to understand if they are real or not. In flux-flux limited samples, artificial correlations are expected. Evidence of co-spatiality is not firmly established.

References: Abdo *et al.* (2009b), Ackermann *et al.* (2012), Ghirlanda *et al.* (2011), HESS Collaboration (2012)

2.3. *H. Sol – Very high energy gamma-ray radiogalaxies and blazars (contribution)*

At present, 53 firmly known very high energy (VHE; > 0.1 TeV) AGN have been observed. Among them we have 47 blazars, 4 radio galaxies and 1 AGN of unknown type and possibly Sgr A*. The blazar sample includes: 34 high-frequency peaked BL Lacs (HBL), 4 intermediate-frequency peaked BL Lacs (IBL), 4 low-frequency peaked BL Lacs (LBL), 3 FSRQ, and 2 BL Lacs. The four radio galaxies are: M87, Cen A, NGC 1275, IC 130. Most of them are beamed sources with a strong Doppler boosting as expected since it helps to accommodate fast variability and to avoid strong intrinsic absorption. Variability time scales are from a few minutes to months and years. Multi-zone synchrotron self-Compton (SSC) models can reproduce most of HBL stationary state spectra.

References: Acciari *et al.* (2011), Aleksić *et al.* (2012), Abramowski *et al.* (2012).

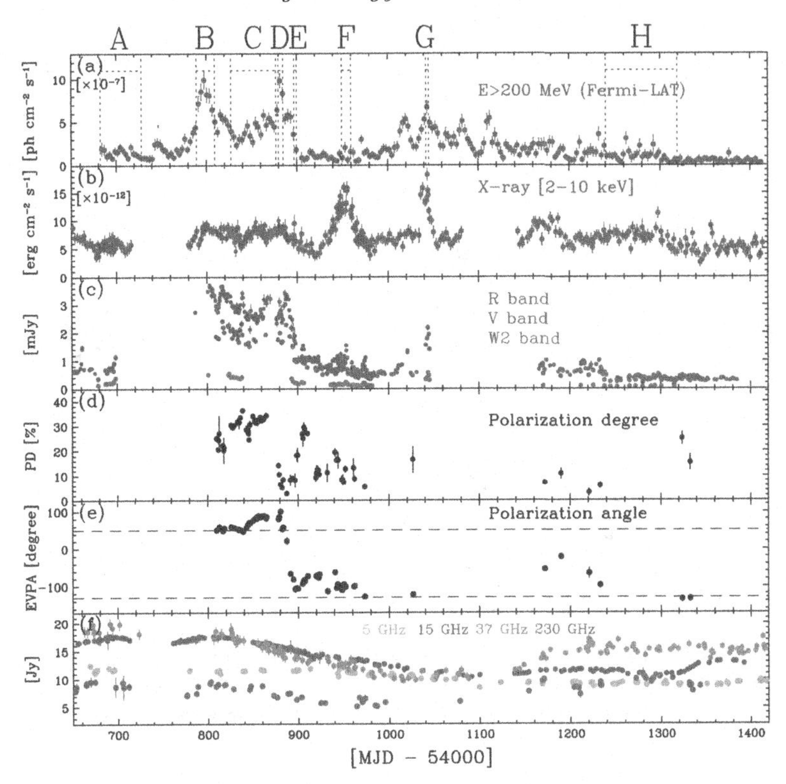

Figure 1. Multi-band (gamma-ray, X-ray, optical, including polarization degree and EVPA, and radio lightcurves of 3C 279 from 2008 August to 2010 August. Taken from Hayashida *et al.* (2012).

2.4. *D. McConnell – Counterparts to Fermi-LAT sources from the ATPMN 5 and 8GHz catalogue of southern radio sources (contribution)*

The Australia Telescope Parkes-MIT-NRAO (ATPMN) catalogue gives accurate positions and flux density measurements at 4.8 and 8.6 GHz for 8385 sources, with a typical image rms noise of 2 mJy/beam. Significant long term variability for 85% of the sources was detected using the 10-14 year baseline between the ATPMN and AT20G observations. Short term variations were gauged by inspecting ATPMN scans separated by up to 4-8 hrs, with about 30% showing intra-day variability. Polarization is available for 9040 sources. Among them ~126 sources have been identified with 2FGL sources and 112 have an optical identification within 3″ from the UK Schmidt survey.

Reference: McConnell *et al.* (2012).

2.5. *M. Karouzos – Gamma-rays in flat-spectrum AGN: revisiting the fast jet hypothesis (contribution)*

Based on studies using EGRET and the early *Fermi*-LAT AGN samples, gamma-ray detected AGN were found to show on average faster apparent speeds with respect to other AGN not detected in gamma-rays. To confirm and investigate this, the very long baseline interferometry (VLBI) proper motion results for 198 QSOs and 33 BL Lacs from the Caltech Jodrell Bank Flat-spectrum (CJF) 5 GHz VLBI survey were analyzed (Fig. 2). Among these 61 sources have been detected by *Fermi*-LAT, consisting of 32 FSRQs, 24 BL Lacs, and 5 radio galaxies. Conclusions are: i) no strong link is present between fast jets and gamma-ray detection; ii) AGN class and gamma-ray variability are

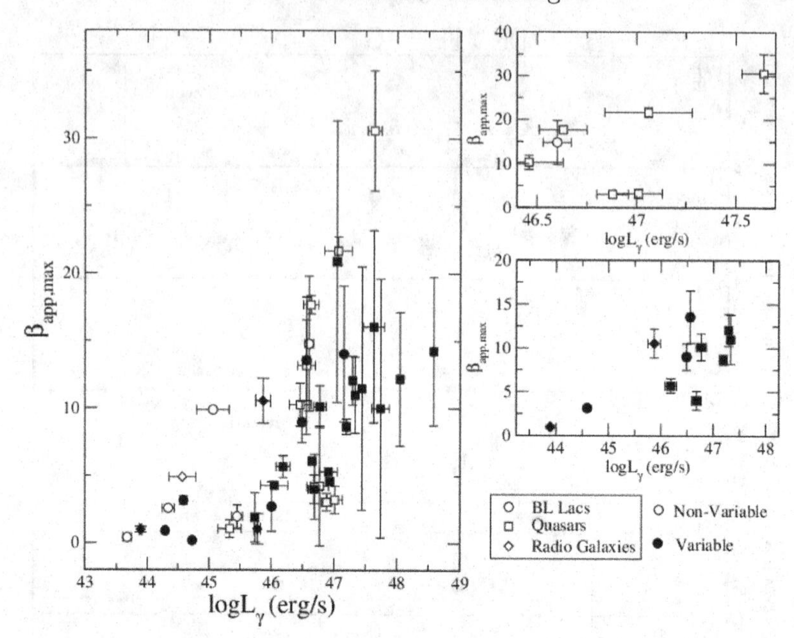

Figure 2. Maximum VLBI measured apparent velocities for gamma-ray detected sources from the CJF sample versus gamma-ray luminosity (filled symbols are gamma-ray variable while open symbols are non-variable in gamma-rays). Taken from Karouzos *et al.* (2011).

connected to jet speeds; iii) a correlation between gamma-ray luminosity and apparent velocity is found (higher velocity for stronger gamma-variable sources); iv) gamma-ray detected sources appear wider and with larger jet distortions. Different findings with respect to the previous studies may be due to different observing frequency (probing either different jet regions or structures) or the difference in sampling of the proper motion data.

Reference: Karouzos *et al.* (2011).

2.6. *F. D'Ammando – To be or not to be a blazar. The case of gamma-ray narrow line Seyfert 1 SBS 0846+513 (contribution)*

In 2008, the first NLS1 PMN J0948+0022 was detected by the *Fermi*-LAT. After that another four NLS1s were detected in gamma-rays. These results confirm the presence of relativistic jets also in NLS1s even though these sources are typically thought to be hosted in spiral galaxies. Their average spectral energy distributions (SEDs) are similar to FSRQs, but at lower luminosity. SBS 0846+513 is a new gamma-ray NLS1 clearly detected during the third year of *Fermi* operation, in particular during a flaring state in 2011 June-July. The gamma-ray peak on daily timescale corresponds to an isotropic luminosity of about 10^{48} erg s^{-1}, comparable to that of luminous FSRQs. While the kpc-scale structure is unresolved in VLA images, there is a core-jet structure seen in VLBA images. The mechanism at work for producing a relativistic jet in NLS1s is not clear. Fundamental parameters should be the BH mass and the BH spin. This source could be a blazar with a BH mass at the low end of the blazar's BH mass distribution. Gamma-ray NLS1s have larger masses with respect to the entire sample of NLS1s. Moreover prolonged accretion episodes could spin-up the SMBH leading to a relativistic jet formation.

References: Abdo *et al.* (2009a), D'Ammando *et al.* (2012).

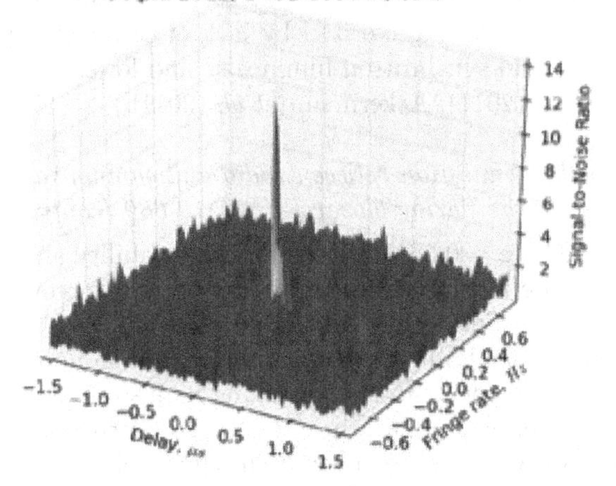

Figure 3. Early fringe obtained for OJ 287 at 6.2cm between the RadioAstron space radio telescope and the ground-based Effelsberg 100m with a baseline of 7.2 Earth diameters. Taken from Kardashev *et al.* (2013).

3. Session 2: High resolution core and jet properties

3.1. *N. Kardashev – RadioAstron Space VLBI mission: early results (invited)*

RadioAstron was launched on 18th July 2011. The orbital period is 8.5 days, with a perigee radius = 67,000 km and an apogee = 282,000 km. Observations are carried out at P, L, C, and K bands in dual-polarization, with finest angular resolutions (300,000 km baseline) of 580, 113, 39, and 7.5-10 μ-arcsec, respectively. Among the first sources detected was a single giant pulse in PSR B0950+08 at 92 cm, with significant variations in 1 hr due to ISM plasma scintillations. The first fringes in K-band were observed together with baselines to the Effelsberg 100m from the quasar 2013+370; see Fig. 3 for another example. Imaging of the quasar 0716+714 at 4.8 GHz was obtained using the space radio telescope and eight ground telescopes showing the core-jet structure as seen in previous VLBI images. A status and more recent summary of the RadioAstron results appeared recently.

Reference: Kardashev *et al.* (2013).

3.2. *M. Lister – Blazars at high resolution: what large multi-epoch VLBI studies can tell us (invited)*

Blazar studies suffer heavy sample selection effects: e.g., obscuration in optical and X-ray; spectral contamination from accretion disk emission and from lobe (unbeamed) emission; non simultaneous observations; and more. To address blazar sample biases it is important to concentrate on 'uncontaminated' bands. The *Fermi* 2LAC AGN catalog has no contamination from the host galaxy, even if may still be incomplete due to issues with source associations. The MOJAVE VLBA program provides regular observations of radio-bright AGNs at 15 GHz. With 24 hrs observing sessions every 3 weeks, it assures continuous time baseline data on many sources back to 1994. Among the main results we quote: the brightest gamma-ray and radio-selected quasars have similar redshift distributions; gamma-ray selected blazars have an additional sub-population of low-redshift HSP BL Lacs that are intrinsically very bright in gamma-rays; lowest luminosity BL Lacs (HSPs) all have high gamma-ray loudness (due to SED peak location). In BL Lacs (HSP and LSP) the photon index is well correlated with the Compton peak location. This trend could not exist if the gamma-ray and pc scale radio jet emission were fully independent. Analyzing kinematics of 889 discrete features in 201 jets from 1994 to 2011, we derive that: jets of HSP BL Lacs are characterized by lack of compact superluminal features; BL Lac jets have lower radio synchrotron luminosity and lower speeds.

References: Lister *et al.* (2011), Ackermann *et al.* (2011b).

3.3. *M. Orienti – On the connection between radio and gamma rays. The extraordinary case of the flaring blazar PKS 1510-089 (contribution)*

The FSRQ, PKS 1510-089 ($z = 0.361$), shows strong variability and highly superluminal jet components found close in time with gamma-ray flares. Moreover, it was detected at VHE gamma-rays, shows a high level of polarized emission, and a large rotation of the electric vector position angle (EVPA) close in time with a gamma-ray flare. PKS 1510-089 underwent a very active period in 2011 reaching its historical maximum flux density in October 2011. The gamma-ray flare in July 2011 occurred after a rotation of 380 degree of the optical EVPA suggesting a common region for the optical and gamma-ray emission. The new jet component is likely evidence of a shock propagating downstream along the jet. If the gamma-ray flare in October 2011 is related to the radio outburst, it would strongly support the idea that some gamma-ray flares are produced parsecs aways from the nucleus. Note that not all flares have the same characteristics, suggesting shocks with different properties. Follow-up in the mm regime with a high sensitivity VLBI array including ALMA will be crucial in determining the high-energy emitting region.

Reference: Orienti *et al.* (2013).

3.4. *Z. Abraham – The radio counterparts of the 2009 exceptional gamma-ray flares in 3C 273 (contribution)*

A very strong and complex gamma-ray flare was observed in 3C 273 in September 2009. The flare was related to the formation of superluminal components studied at 43 GHz and to a radio flare observed with the Itapetinga radio telescope at 43 GHz. The gamma-ray flux increased by a factor of 20, while only by a factor two in radio. We explain this fact as a change in the Doppler factor, due to a change in the angle between the jet and the line of sight. This fact was predicted by the precessing jet model of Abraham & Romero (1999). The precessing model based on the jet curvature and proper motions is compatible with a period of 16 years.

Reference: Abraham & Romero (1999), Jorstad *et al.* (2011).

3.5. *Z. Shen – High precision position measurements of the cores in 3C 66A and 3C 66B (invited)*

3C 66A is a low frequency peaked BL Lac object at $z = 0.444$. It is characterized by prominent variability at radio, IR, and optical frequencies. It shows a one-sided core-jet structure with detected superluminal motion. The core shift with frequency has been estimated in 2001 and 2006. A large difference has been found between the two measurements possibly due to a strong flux density increase at 15 GHz in 2006. This radio flare is possibly due to the core activity as shown by a new component that emerged from the central core region. Because of their small angular distance in the plane of the sky, 3C 66A and the radio galaxy 3C 66B, are an ideal pair to obtain a combined core-shift measurement of the two sources. Comparison with data at different epochs is confusing: the difference in core shift result cannot be simply explained by the core flare activity. Other parameters apart from core flux variability may influence the core shift. More data are required to explore this point.

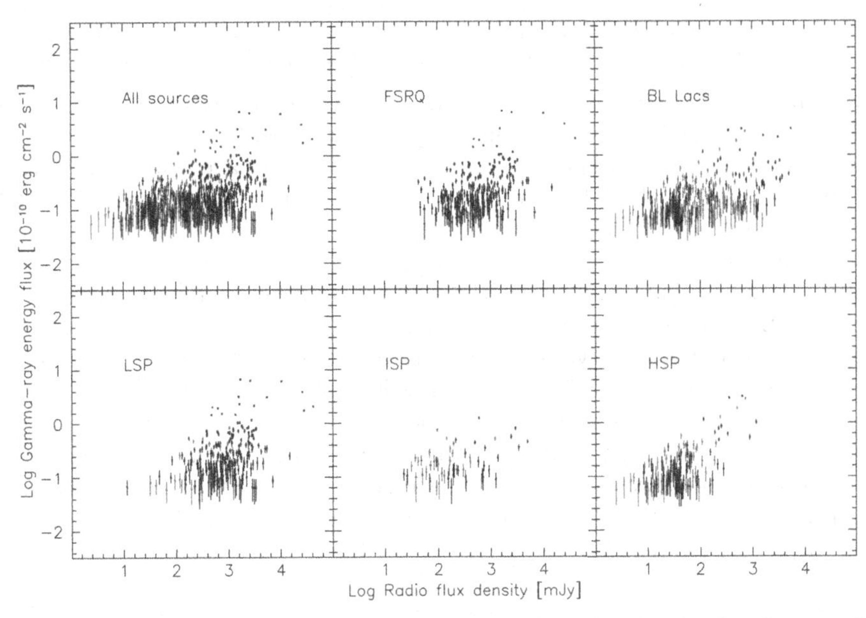

Figure 4. Broad band gamma-ray energy flux vs. 8 GHz archival radio flux densities for the 1LAC sample separated by source type. Taken from Ackermann *et al.* (2011a).

References: Cai *et al.* (2007), Sudou & Iguchi (2011), Zhao *et al.* (2011), Zhao *et al.* (2013).

4. Session 3: Multi-wavelength correlations and variability

4.1. *F. Tavecchio – Variability of blazars: probing emission regions and acceleration processes (invited)*

Blazars show variability on timescale of days suggesting a parsec scale emission region. Comparing the radio-gamma variability, there is evidence that this active region should be inside the broad line region (BLR). Moreover, in some sources the GeV emission shows spectral breaks that could be due to absorption effects inside the BLR. In this context, a relevant case to consider is that of PKS 1222+216 (4C+21.35) which doubled its TeV flux on a timescale on the order of 10 minutes (Aleksić *et al.* 2011). In this source the location of the VHE emission inside the BLR is problematic because the too strong absorption expected (e.g., Liu *et al.* 2008, Reimer 2007, Tavecchio & Mazin 2009, Poutanen & Stern 2010) due to the huge optical depth of the BLR. Possibilities to reconcile rapid variability in regions at large distances from the core (outside the BLR) require the presence of jet substructure (e.g., Ghisellini & Tavecchio 2008). Many models have been proposed but not all problems are solved – e.g., the presence of mini-jets from fast reconnection in a highly magnetized jet (Giannios *et al.* 2009, 2010), or narrow electron beams from magnetocentrifugal acceleration (e.g., Ghisellini *et al.* 2009), beams from relativistic reconnection (e.g., Cerutti *et al.* 2012, Nalewajko *et al.* 2012), or ultra-high energy (UHE) neutral beams (Dermer *et al.* 2012). In conclusion, rapid variability is perhaps currently the most compelling issue in high-energy astrophysics. The idea of a unique, large, "relaxed" emission region is, at least sometimes, inadequate.

4.2. *T. Hovatta – Assessing AGN Variability and Cross-waveband Correlations in the Era of High-Quality Monitoring Data in Low and High Energies (invited)*

A relevant problem in AGN is where the high-energy emission is located: close to the BH within the BLR or further down in the parsec-scale jet? Correlations can be used to locate the unresolved gamma-ray emission site. We can have a flux-flux correlation (amplitude domain; e.g., Fig. 4) or a light curve cross-correlation (time domain). Using simultaneous data, an intrinsic radio/gamma-ray flux density correlation is confirmed. FSRQs and BL Lacs show a different behavior (a possible selection effect?). Using archival non-contemporaneous data to increase the statistics, the correlation persists (it is even stronger for BL Lacs because of more sources).

More difficult is to tell if individual events are correlated and what are the time delays. Light curve correlations are difficult to establish in single sources. Opacity effects are important in the radio bands, moreover, we could still have too short time series. Stacked correlations show statistically significant time delays with increasing delays for longer wavelengths. Good multiwavelength coverage is really necessary to address this issue.

References: Kovalev *et al.* (2009), Ackermann *et al.* (2011a), Mahony *et al.* (2010), Nieppola *et al.* (2011), Max-Moerbeck (2012), Pavlidou *et al.* (2012).

4.3. *X. Liu – VLBI core flux density and position angle analysis of the MOJAVE AGN (contribution)*

The two-fluid jet model assumes that: 1) the outflow consists of an electronproton plasma (the jet), moving at mildly relativistic speed, 2) an electron-positron plasma (the beam) is moving at highly relativistic speed, and 3) the magnetic field lines are parallel to the flow in the beam and the mixing layer, and are toroidal in the jet.

To confirm and investigate this jet structure we model-fit the MOJAVE blazar core sample which includes blazars with more than 10 years of VLBA monitoring, and more than 15 observed epochs with a good time distribution. This sample consists of 104 sources, 77 of which are quasars, 27 BL Lacs, in which 82 are *Fermi*-LAT detected, and 22 non-detected sources. Of these, nine are also TeV sources. The model-fit result of the cores of 104 blazars from the MOJAVE monitoring data, suggests that *Fermi* LAT-detected blazars have wider position angle changes of the inner-jet than LAT non-detected blazars, and are preferentially associated with higher variable blazars.

A two-zone jet model can explain the correlations in the model-fitted parameters. The *Fermi* GeV gamma-ray detection rate show equally similar fraction for sources dominated by the innermost jet (zone-1) and sources dominated by the outer jet (zone-2). But importantly, TeV gamma-ray sources are associated mostly with blazars dominated by the outer part of inner-jet (zone-2).

Reference: Lister *et al.* (2009).

4.4. *E. Valtaoja – Gamma-ray emission along the radio jet: studies with Planck, Metsähovi and Fermi data (contribution)*

If the region emitting at VHE is close to the black hole - accretion disk region (i.e., it is inside the BLR), we expect that gamma-ray flares precede radio variations (assuming as VLBI zero epoch the beginning of a millimeter flare), and little or no correlation with radio variations. If the VHE region is distant, at or downstream of the radio core (i.e. outside the BLR), we expect gamma-ray flares simultaneous, or after, the beginning of radio variations and a correlation between VHE and radio variations.

Comparing data from *Fermi* and the Metsähovi radio sample, the case for "distant" gamma-ray origin appears much stronger. Direct observational evidence for "close" origin

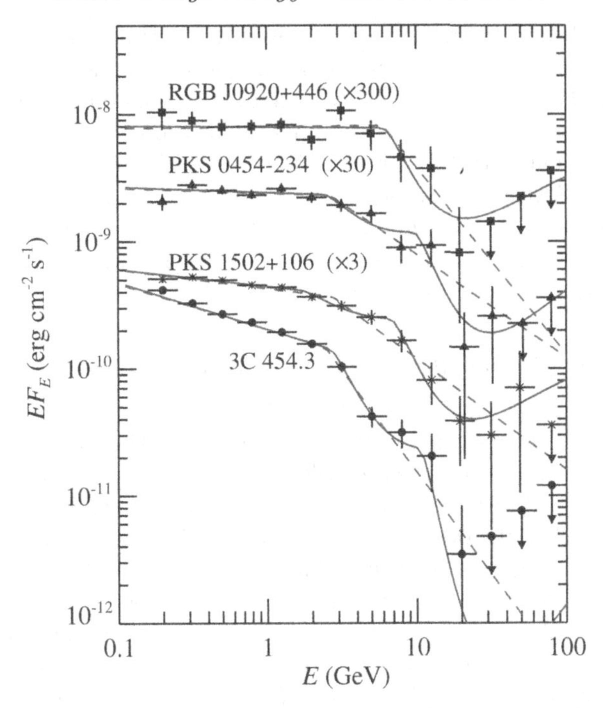

Figure 5. *Fermi*-LAT SEDs of four blazars with best-fit broken power law (blue dash lines) and power law with absorber model (red lines). Figure taken from Poutanen & Stern (2010).

was not found; observations point towards distant origins: in the radio-gamma correlations, there are evident delays from radio to gamma. A confirmation of this result will be possible from the final *Planck* data to model the radio to gamma-ray SEDs with unprecedented accuracy.

Reference: León-Tavares *et al.* (2011).

5. Session 4: Jet physics and the role of BH spin and BH accretion

5.1. *A. Tchekhovskoy – What Sets the Power of Jets from Accreting Black Holes? (invited)*

It is crucial to understand what sets the maximum power of jets, if jets are powered by black holes (BHs) or inner regions of the accretion disks. Jet power depends on magnetic field topology: dipolar geometry gives powerful jets, quadripolar or toroidal gives weak or no jets. Jet power increases with increasing BH magnetic flux. BH and a large magnetic flux give a magnetically-arrested accretion (MAD): the BH is saturated with flux, and the B-field is as strong as gravity. Radio-loud AGN have MADs with BH spins (a) near to 1 and radio-quiet AGN shows MADs with $a < 0.1$. Retrograde BHs appear to have less powerful jets while thicker disks show more powerful jets.

References: Tchekhovskoy *et al.* (2010), Tchekhovskoy *et al.* (2011), McKinney *et al.* (2012), Tchekhovskoy & McKinney (2012).

5.2. *J. Poutanen – Fermi observations of blazars: implications for gamma-ray production (contribution)*

The presence in the blazar spectra of GeV breaks (see Fig. 5 for examples) produced by photon-photon absorption on BLR photons proves that gamma-rays are produced within the BLR. These spectral breaks are relatively stable during flares. Most GeV breaks cannot be produced by Ly-α photons, but can be produced by Lyman continuum of ionized helium. The gamma-ray emitting region is located at the boundary of the He^{++} zone and moves away from the BLR high-ionization zone (moving region model). This model is consistent with the arrival of photons with energy larger than 10 GeV in the end of the flare. Alternative interpretations that gamma-rays are produced together or even after radio flares are ambiguous.

References: Abdo *et al.* (2010b), Abdo *et al.* (2011), Poutanen & Stern (2010), Stern & Poutanen (2011).

5.3. *S. Trippe – Probing the mm/radio polarization of active galactic nuclei (contribution)*

The Plateau de Bure Interferometer (PdBI) is located in the south-east of France, consists of six 15-m antennas, and observes with dual-polarization single-band receivers in the frequency range 80 371 GHz, with a maximum baseline of 760 m ($\sim 0.2'' - 7''$ angular resolution). Each PdBI observation collects polarization information on calibration quasars (since 2007), thus resulting in a large mm polarization survey at 1.3 / 2 / 3 mm "for free." Four years of polarization monitoring data of AGN were used for this study. Polarization is detected in almost all sources (73 out of 86 quasars), consisting of 316 detections out of 441 source measurements. Fast variability in flux and polarization are observed with almost identical fluctuation rates, implying similar spatial scales are probed. Polarization variations indicate strong shocks and allow for estimates of the shock parameters. Very large rotation measures (up to 400,000 radians m^{-2}) are measured and allow an estimate of the outflow geometries (spherical/conical outflows).

References: Trippe *et al.* (2010), Trippe *et al.* (2012).

5.4. *M. Su – Fermi Gamma-Ray bubbles, jets, and lines in the Milky Way (contribution)*

The Fermi bubbles are giant gamma-ray structures with sharp edges discovered using data from the *Fermi*-LAT. They rise up & down from the Galactic center with extents of $\sim$ 50 degrees ($\sim$8.5 kpc), are well centered on longitude zero and close to latitude zero, and imply the acceleration of TeV electron energy particles. They could be related to a jet or outflow activity from the Galactic center. The gamma-ray bubbles have counterparts at microwave frequencies (the WMAP haze, confirmed by *Planck*). Sharp edges are observed in X-ray also utilizing ROSAT data, and several small *XMM-Newton* pointings have recently been obtained in order to study this in more detail. The basic question surrounding the origin of the bubbles is whether they are jet or wind/outflow driven. A gamma-ray jet / cocoon feature was recently detected in the *Fermi*-LAT data offering additional clues. Overall, the Fermi bubbles are analogous to large-scale structures observed in more distant AGN, e.g., Cen A, and connections are being sought.

References: Su *et al.* (2010), Su & Finkbeiner (2012).

Acknowledgements

We would like to acknowledge the support from the IAU, the JD6 participants for their contributions, and members of the organizing committee: Ed Fomalont, Luigi Foschini, Marcello Giroletti, Xiaoyu Hong (co-chair), Seiji Kameno, Matthias Kadler, Yuri Kovalev, Laura Maraschi (co-chair), David Paneque, Maria Rioja, Eduardo Ros, Łukasz Stawarz, Meg Urry, and Anton Zensus. T.C. was supported at NRL by NASA DPR S-15633-Y.

References

Abdo, A. A., Ackermann, M., Ajello, M., *et al.* 2009a, *ApJ*, 699, 976 (PMN J0948+0022)
Abdo, A. A., Ackermann, M., Ajello, M., *et al.* 2009b, *ApJL*, 707, L142 (LAT radio-loud NLS1s)
Abdo, A. A., Ackermann, M., Ajello, M., *et al.* 2010a, *Science*, 328, 725 (Cen A lobes)
Abdo, A. A., Ackermann, M., Ajello, M., *et al.* 2010b, *ApJ*, 710, 1271 (LAT blazar spectra)
Abdo, A. A., Ackermann, M., Ajello, M., *et al.* 2011, *ApJL*, 733, L26 (3C 454.3 flare)
Abraham, Z. & Romero, G. E. 1999, *A&A*, 344, 61
Abramowski, A., Acero, F., Aharonian, F., *et al.* 2012, *ApJ*, 746, 151 (M87)
Acciari, V. A., Aliu, E., Arlen, T., *et al.* 2011, *ApJ*, 738, 25 (Mrk 421)
Ackermann, M., Ajello, M., Allafort, A., *et al.* 2011a, *ApJ*, 741, 30 (radio-gamma connection)
Ackermann, M., Ajello, M., Allafort, A., *et al.* 2011b, *ApJ*, 743, 171 (2LAC)
Ackermann, M., Ajello, M., Allafort, A., *et al.* 2012, *ApJ*, 747, 104 (LAT Seyferts)
Aleksić, J., Alvarez, E. A., Antonelli, L. A., *et al.* 2012, *A&A*, 542, A100 (Mrk 421)
Cai, H.-B., Shen, Z.-Q., Sudou, H., *et al.* 2007, *A&A*, 468, 963
Cerutti, B., Werner, G. R., Uzdensky, D. A., & Begelman, M. C. 2012, *ApJL*, 754, L33
D'Ammando, F., Orienti, M., Finke, J., *et al.* 2012, *MNRAS*, 426, 317
Dermer, C. D., Murase, K., & Takami, H. 2012, *ApJ*, 755, 147
Ghirlanda, G., Ghisellini, G., Tavecchio, F., Foschini, L., & Bonnoli, G. 2011, *MNRAS*, 413, 852
Ghisellini, G. & Tavecchio, F. 2008, *MNRAS*, 386, L28
Ghisellini, G., Tavecchio, F., Bodo, G., & Celotti, A. 2009, *MNRAS*, 393, L16
Giannios, D., Uzdensky, D. A., & Begelman, M. C. 2009, *MNRAS*, 395, L29
Giannios, D., Uzdensky, D. A., & Begelman, M. C. 2010, *MNRAS*, 402, 1649
Hayashida, M., Madejski, G. M., Nalewajko, K., *et al.* 2012, *ApJ*, 754, 114
HESS Collaboration: Abramowski, A., Acero, F., *et al.* 2012, *A&A*, 545, A103
Jorstad, S., Marscher, A., Agudo, I., & Harrison, B. 2011, 2011 Fermi Symposium, eConf C110509, arXiv:1111.0110
Kardashev, N. S., Khartov, V. V., Abramov, V. V., *et al.* 2013, *Astronomy Reports*, 57, 153
Karouzos, M., Britzen, S., Witzel, A., Zensus, J. A., & Eckart, A. 2011, *A&A*, 529, A16
Kovalev, Y. Y., Aller, H. D., Aller, M. F., *et al.* 2009, *ApJL*, 696, L17
León-Tavares, J., Valtaoja, E., Tornikoski, M., Lähteenmäki, A., & Nieppola, E. 2011, *A&A*, 532, A146
Lister, M. L., Cohen, M. H., Homan, D. C., *et al.* 2009, *AJ*, 138, 1874
Lister, M. L., Aller, M., Aller, H., *et al.* 2011, *ApJ*, 742, 27
Liu, H. T., Bai, J. M., & Ma, L. 2008, *ApJ*, 688, 148
Mahony, E. K., Sadler, E. M., Murphy, T., *et al.* 2010, *ApJ*, 718, 587
Max-Moerbeck, W. 2012, *BAAS*, 219, #321.05
McConnell, D., Sadler, E. M., Murphy, T., & Ekers, R. D. 2012, *MNRAS*, 422, 1527
McConville, W., Ostorero, L., Moderski, R., *et al.* 2011, *ApJ*, 738, 148
McKinney, J. C., Tchekhovskoy, A., & Blandford, R. D. 2012, *MNRAS*, 423, 3083
Nalewajko, K., Begelman, M. C., Cerutti, B., Uzdensky, D. A., & Sikora, M. 2012, *MNRAS*, 425, 2519
Nieppola, E., Tornikoski, M., Valtaoja, E., *et al.* 2011, *A&A*, 535, A69
Orienti, M., Koyama, S., D'Ammando, F., *et al.* 2013, *MNRAS*, 428, 2418
Paneque, D., Ballet, J., Burnett, T., *et al.* 2012, 4th Fermi Symposium, eConf C121028, arXiv:1304.4153
Pavlidou, V., Richards, J. L., Max-Moerbeck, W., *et al.* 2012, *ApJ*, 751, 149
Poutanen, J. & Stern, B. 2010, *ApJL*, 717, L118
Reimer, A. 2007, *ApJ*, 665, 1023
Sudou, H. & Iguchi, S. 2011, *AJ*, 142, 49
Stern, B. E. & Poutanen, J. 2011, *MNRAS*, 417, L11
Su, M., Slatyer, T. R., & Finkbeiner, D. P. 2010, *ApJ*, 724, 1044
Su, M. & Finkbeiner, D. P. 2012, *ApJ*, 753, 61
Takeuchi, Y., Kataoka, J., Stawarz, Ł., *et al.* 2012, *ApJ*, 749, 66
Tavecchio, F. & Mazin, D. 2009, *MNRAS*, 392, L40

Tchekhovskoy, A., Narayan, R., & McKinney, J. C. 2010, *ApJ*, 711, 50
Tchekhovskoy, A., Narayan, R., & McKinney, J. C. 2011, *MNRAS*, 418, L79
Tchekhovskoy, A. & McKinney, J. C. 2012, *MNRAS*, 423, L55
Trippe, S., Neri, R., Krips, M., *et al.* 2010, *A&A*, 515, A40
Trippe, S., Neri, R., Krips, M., *et al.* 2012, *A&A*, 540, A74
Zhao, G.-Y., Chen, Y.-J., Shen, Z.-Q., *et al.* 2011, *Journal of Astrophysics & Astronomy*, 32, 61
Zhao, G.-Y., Chen, Y.-J., Shen, Z.-Q., *et al.* 2013, *IAU Symposium*, 290, 367

Highlights of Astronomy, Volume 16
XXVIIIth IAU General Assembly, August 2012
T. Montmerle, ed.

doi:10.1017/S174392131500157X

Preface: Joint Discussion JD7 Space-time reference systems for future research

The Joint Discussion on Space-time reference systems for future research (JD7) was held at the XXVIIIth General Assembly of the IAU in Beijing, on 27–29 August 2012. It was organized by IAU Division I (Fundamental Astronomy), with the support of Division III (Planetary Systems Sciences), Division IX (Optical & Infrared Techniques), Division XI (Space & High Energy Astrophysics), and Division XII (Union-Wide Activities). The scientific organizing committee was composed of Nicole Capitaine (France; co-Chair), George H. Kaplan (USA), Sergei Klioner (Germany; co-Chair), Zoran Knezevic (Republic of Serbia), Dafydd Wyn Evans (UK), Dennis McCarthy (USA; co-Chair), Harald Schuh (Austria), Richard N. Manchester (Australia) and Gérard Petit (France).

The goal of this Joint Discussion was to coordinate ongoing efforts in the field of the reference frames, models and algorithms used by the international astronomical and space sciences community and to develop effective strategies for future cooperative work to ensure that the reference systems will be sufficiently accurate to meet the future needs. JD7 therefore reviewed currently planned projects devoted to improving reference systems and the links between the various reference frames.

Future astronomical, astrophysical and geophysical research will demand reference systems of the highest accuracy. Past efforts in the development of reference systems have already enabled unprecedented advances in these areas. These advances have, in turn, spurred new investigations that demand even higher accuracy reference systems. Benefits of recent improvements include the determination of the masses of planets from pulsar timings, determination of PPN parameters, accurate stellar distances, and dynamics of the Earth's motion. The validity of the most promising investigations can be limited by the reference frames, models and algorithms used to make the required observations and to analyze the results. Investigations of exoplanets and near-Earth objects demand highly precise reference frames. Future space missions and pulsar timing projects will require the development of relativistic time transfer algorithms and highly accurate planetary ephemerides. Investigations of terrestrial phenomena such as the rise of ocean levels require the most precise reference frames to enable the required astro-geodetic measurements.

The scientific programme of the meeting included 15 invited papers, 33 contributed and 30 poster presentations. It was composed of the seven following sessions:

1. Theoretical aspects of reference systems (Monday 27 August)
2. Reference timescale requirements (Monday 27 August)
3. Topics in celestial mechanics (Monday 27 August)
4. Space mission requirements (Tuesday 28 August)
5. Future requirements for planetary ephemerides (Tuesday 28 August)
6. Relative reference systems (Tuesday 28 August)
7. Concluding discussion and recommendations (Wednesday 29 August)

The Joint Discussion gathered over 100 participants. Presentations covered both completed works and prospects for the future. Session 7 concluded the Joint Discussion with a general discussion and recommendations regarding such topics as a redefinition of the astronomical unit, definition and realization of the future ICRS, leap seconds and pulsar time scales. This resulted in the proposal of establishing new Division I working groups; this also resulted in improvements of the wording of the IAU resolution proposal on the re-definition of the astronomical unit, which was submitted to be voted on during the second session of the 28th IAU GA. IAU 2012 Resolution B2 recommends that the astronomical unit be re-defined as a fixed number of Système International d'Unités (SI) metres through a defining constant. For continuity that constant should be the value of the current best estimate in metres as adopted by IAU 2009 Resolution B2 (i.e. 149 597 870 700 m).

The JD7 proceedings are composed of the preface, the list of invited, contributed and poster presentations and the summaries of 13 of the invited papers (denoted in the list by paper titles in bold face). Additional Information on the Joint Discussion as well as part of the presentations given during the JD can be found on the JD7 web page at `http://www.referencesystems.info/iau-joint-discussion-7.html`.

Acknowledgements

We express our thanks to the SOC for its valuable help in preparing the scientific programme and chairing the sessions, and to all the authors of presentations for their very interesting contributions. We thank the local organizers for the very efficient help during the meeting.

Nicole Capitaine, George Kaplan, Sergei Klioner
JD7 Proceedings co-editors
23 November 2012

List of JD7 invited, contributed and poster presentations

(Papers titles in bold face denote papers published in these Proceedings.)

Session 1. Theoretical aspects of reference systems

- **Relativity in fundamental astronomy: status and prospects** *by Michael Soffel* (invited)
- **Celestial dynamics and astrometry in an expanding universe** *by Sergei Kopeikin* (invited)
- Extension of the DSX-formalism to 2PN order for the problem of light propagation *by Chongming Xu, Sergei Klioner, Michael Soffel, Xuejun Wu*
- A dynamical reference frame for geophysics and experimental gravitation *by Pacôme Delva, Christophe Le Poncin-Lafitte*
- Observational evidences for the propagation speed of gravity from Earth tides *by Keyun Tang, Wu Wen, Changcai Hua, Shunliang Chi, Qingyu You, Dan-Yu Yu*
- Explicit expressions for the global metric and coordinate transformation with local multipole moments *by Yi Xie*
- New approach to relativistic celestial reference frames *by Olivier Minazzoli*
- 2PN light propagation in the scalar-tensor theory: an N-point mass case *by Xue-Mei Deng, Yi Xie*
- Relativistic spherical multipole moments in astronomy *by Jin-He Tao, Wen-Biao Han*
- Frequency shift at the post-post-Minkowskian approximation *by Christophe Le Poncin-Lafitte, Aurélien Hees, Stefano Bertone*
- Relativistic modeling for high precision space astrometry at post-post Minkowskian approximation *by Stefano Bertone, Christophe Le Poncin-Lafitte, Marie-Christine Angonin*

- Transformative relation of kinematical descriptive quantities defined by different spatial referential frame, its property and application *by Luo Ji*

Session 2. Reference timescale requirements

- **A pulsar-based timescale** *by George Hobbs* (invited)
- **Long-term stability of atomic time scales** *by Gérard Petit and Felicitas Arias* (invited)
- Perspectives for time and frequency transfer *by Philip Tuckey, Pierre Uhrich*
- Developments of optical clocks and their comparisons for future time reference *by Yasuhiro Koyama, Nobuyasu Shiga, Atsushi Yamaguchi, Tadahiro Goto, Yuko Hanado, Mizuhiko Hosokawa, Miho Fujieda, Hidekazu Hachisu, Tetsuya Ido, Reiko Kojima, Motohiro Kumagai, Ying Li, Kensuke Matsubara, Shigeo Nagano*
- Connecting kinematic and dynamic reference frames by D-VLBI *by Harald Schuh, Lucia Plank, Matthias Madzak, Johannes Boehm*
- Link of reference frames by pulsar observations *by Rodin Aleksandr, Sekido Mamoru*
- A convention for Coordinated Universal Time *by Dennis D. McCarthy*

Session 3. Topics in celestial mechanics

- Requirements on space-time reference systems for the BepiColombo and Juno missions *by Andrea Milani, Giacomo Tommei, Linda Dimare* (invited)
- The trajectory monitoring of spacecraft via VLBI in China's lunar exploration project *by Jinling Li*
- Advanced dynamical models for very well observed asteroids: relativity, non-gravitational effects, perturbations from small bodies *by Fabrizio Bernardi, Davide Famocchia, Andrea Milani* (invited)
- Phoebe's orbit from ground-based and space-based observations *by Josselin Desmars, Shan-Na Li, Radwan Tajeddine, Zheng-Hong Tang*
- The geoid computed from a new generalized theory of the figure of the Earth *by Cheng-Li Huang, Cheng-Jun Liu, Yu Liu*
- May small digital PZT and radio beacons improve the LPhl for future lunar missions? *by Jinsong Ping, Xiaoli Su, Hideo Hanada, Alexander Gusev, Jinling Li, Xian Shi, Qinghui Liu*
- Progress of astrometric research in Nikolaev Observatory *by Anatoliy Ivantsov, Nadia Maigurova, Maxim Martynov, Gennadiy Pinigin*

Session 4. Space mission requirements

- **Time and frequency transfer with the ESA/CNES ACES-PHARAO mission** *by Pacôme Delva, Peter Wolf, Frédéric Meynadier, Christophe Le Poncin-Lafitte, Philippe Laurent* (invited)
- **Celestial reference frame realizations at multiple radio frequency bands** *by Chris Jacobs* (invited)
- **Status of Gaia and early operation plans** *by François Mignard* (invited)
- Current status of the Celestial Reference Frame and future prospects *by Ralph Gaume*
- Next-generation VLBI model: higher accuracy and larger baselines *by Sergei Klioner*
- New Pulkovo combined catalogues of the radio source positions *by Yulia Sokolova, Zinovy Malkin*
- Morphology of QSOs; the grid points of the Gaia celestial reference frame *by Alexandre Humberto Andrei, Jean Souchay, Roberto Vieira Martins, Sonia Anton, Francois Taris, Sebastien Bouquillon, Marcelo Assafin, Christophe Barache, Julio Ignacio Bueno Camargo, Bruno Coelho, Dario Nepomuceno da Silva Neto*
- Systematic effect of the galactic aberration on the ICRS realization and the Earth orientation parameters *by Jia-Cheng Liu, Nicole Capitaine, Sébastien Lambert, Zinovy Malkin, Zi Zhu*

- Dipole systematic effect in proper motion of the reference radio sources *by Oleg Titov*
- Towards ICRF3: preparing the VLBI frame for future synergy with the Gaia frame *by Patrick Charlot, Géraldine Bourda*
- The Large Quasar Astrometric Catalog V.2, LQAC-2 *by Jean Souchay, Alexandre Andrei*
- The Gaia reference frame and the acceleration of the solar system barycenter in the presence of quasar variability *by Rajesh Kumar Bachchan, David Hobbs, Lennart Lindegren*
- Gaia's astrometric global iterative solution: method overview and results *by Uwe Lammers, Lennart Lindegren, David Hobbs, William O'Mullane, Ulrich Bastian, Jose Hernandez*
- An intensive consideration for the Galactic coordinate system *by Zi Zhu, Jia-Cheng Liu*
- Reconsidering the International Celestial Reference System based on the effect of the secular aberration *by Minghui Xu, Guangli Wang, Ming Zhao*
- VLBI application for time and frequency transfer, and comparison with other techniques *by Mamoru Sekido, Hiroshi Takiguchi, Thomas Hobiger, Ryuuichi Ichikawa, Miho Fujieda, Tadahiro Goto, Jun Amagai, Kazuhiro Takefuji*
- Observations of ERS which are visible in optical domain using 2 m telescope *by Goran Damljanovic, Ivana Milic*
- Radio-optical reference frame offsets from CTIO and UCAC4 data *by Marion Zacharias, Norbert Zacharias, Charlie T. Finch*
- Assessment of stochastic errors of radio source position catalogues *by Zinovy Malkin, Julia Sokolova*
- Source structure and VLBI position instabilities *by Romuald Bouffet, Patrick Charlot, Sébastien Lambert*

Session 5. Future requirements for planetary ephemerides

- **INPOP: evolution, applications and perspectives** *by Jacques Laskar, Agnès Fienga, Hervé Manche, Ashok Verma, Mickael Gastineau* (invited)
- **Linking the planetary ephemerides to the ICRF** *by William Folkner* (invited)
- **EPM– the high-precision planetary ephemerides of IAA RAS for scientific research, astronavigation on the Earth and in space** *by Elena Pitjeva* (invited)
- A new approach to asteroid modeling in a planetary ephemeris *by Petr Kuchynka*
- New developments in spacecraft raw data direct analysis for the INPOP planetary ephemerides *by Ashok Verma, Agnès Fienga*
- The re-definition of the astronomical unit of length: reasons and consequences *by Nicole Capitaine, Sergei Klioner, Dennis McCarthy*
- Sensitivity studies of HEO Mars orbiters using perturbation theory *by Zhi-Zhou He, Cheng-Li Huang, Mian Zhang*
- Global solution of the pulsar clock model and the Earth ecliptic position based on millisecond pulsar timing *by Feng Tian, Zheng-Hong Tang*
- Asteroids Dynamic Site - AstDyS *by Zoran Knezevic, Andrea Milani*

Session 6. Relative reference systems

- **Connecting terrestrial to celestial reference frames** *by Zinovy Malkin* (invited)
- **SOFA, an IAU Service for the future** *by Catherine Hohenkerk* (invited)
- **The IERS Conventions (2010): reference systems and new models** *by Brian Luzum and Gérard Petit* (invited)
- Link the EOC4 catalog to the ICRS *by Cyril Ron, Jan Vondrák*
- A new numerical theory of Earth rotation *by Enrico Gerlach, Sergei Klioner, Michael Soffel*
- Asymmetric effects in polar motion excitation *by Christian Bizouard*
- Evaluation of the accuracy of the IAU 2006/2000 precession-nutation *by Nicole Capitaine, Sébastien Lambert, Kunliang Yao, Jia-Cheng Liu*

- Impact of IERS conventions (2010) on VLBI-derived reference frames *by Robert Heinkelmann, Harald Schuh*
- Influence of the inner core on the rotation of the Earth revisited *by Alberto Escapa, Jose M. Ferrandiz, Juan Getino*
- Researches on predictions of Earth orientation parameters *by Xueqing Xu*
- A linear operator method to compute the rotational modes of asymmetric 3D Earth by vector spherical harmonics *by Mian Zhang, Cheng-Li Huang*
- A preliminary study of the relationship between gravity change and plumb line variation on ground in case of anomalous mass underground *by Yongzhang Yang, Zhengxin Li, Jinsong Ping*
- A study of discovered non-tidal vertical variations in China *by Zhengxin Li, Yongzhang Yang*
- Excitations of length-of-day variations determined from GPS, SLR and GRACE *by Shuanggen Jin, Xinggang Zhang*
- IAU SOFA software *by Catherine Hohenkerk*
- Naval Observatory Vector Astrometry Software (NOVAS) Version 3.1: Fortran, C, and Python Editions *by George H. Kaplan, John A. Bangert, Eric G. Barron, Jennifer L. Bartlett, Wendy Puatua, William Harris, Paul Barrett*
- Nutation determination using the Global Positioning System *by Kunliang Yao, Nicole Capitaine, Jean-Yves Richard*
- Progress in SLR - GPS co-location at San Juan (Argentina) station *by Hernan Alvis Rojas, Ana Pacheco, Ricardo Podestá, Eloy Actis, Jinzeng Li, Zhiqiang Yin, Rui Wang, Dongping Huang, Raùl Márquez, Sonia Adarvez, Johana Quinteros, Pablo Cobos, Andrés Aracena*
- Russian Astronomical Software *by Marina V. Lukashova, Nina I. Glebova, Ilja N. Netsvetaev, Galina A. Netsvetaeva, Ekaterina Ju. Parijskaja, Elena V. Pitjeva, Michael L. Sveshnikov, Vladimir I. Skripnichenko*
- SLR and GPS spatial techniques in ITRF. Argentine results *by Eloy Vicente Actis, Ricardo Podestá, Ana M. Pacheco, Hernán Alvis Rojas, Zhiqiang Yin, Jinzeng Li, Yanben Han, Weidong Liu, Rui Wang, Dongping Huang, Raùl Márquez, Sonia Adarvez, Matias Flores, Diego Brizuela, Jesica Nievas*
- The geopotential computed from global crustal models *by Cheng-Jun Liu*
- The interior structure of the Earth constrained from gravity field data and the generalized theory of the figure of the Earth *by Cheng-Jun Liu, Cheng-Li Huang*
- The new expansion of annual aberration into trigonometric series *by Cyril Ron, Jan Vondrák*
- New role of astrometry technique in sciences and technologies *by Zhengxin Li*

Highlights of Astronomy, Volume 16
XXVIIIth IAU General Assembly, August 2012
T. Montmerle, ed.

doi:10.1017/S1743921314005419

Relativity in fundamental astronomy

M. Soffel†
Lohrmann Observatory, Dresden Technical University, 01062 Dresden, Germany

Abstract. An overview is given over the broad field of Relativity in Fundamental Astronomy. The present status is recalled and deficiencies are pointed out that might lead to future work within IAU Commission 52.

Keywords. relativity, gravitation, reference systems, astrometry, ephemerides

---

The field of Relativity in Fundamental Astronomy comprises, at least, relativity in i) astronomical space-time reference systems, ii) in celestial mechanics, iii) in astrometry, and iv) in metrology. In addition, it concerns certain astronomical concepts; the ecliptic is an example.

In the field of astronomical reference systems, the Barycentric and Geocentric Celestial Reference Systems, BCRS and GCRS, were adopted by IAU 2000 resolutions (e.g., Soffel *et al.*, 2003). They were constructed in the 1st post-Newtonian approximation to Einstein's theory of gravity and all matter and cosmic energy outside the solar system was neglected, as was the influence of cosmic expansion. In Klioner & Soffel (2004) it was shown that the cosmic expansion has practically no influence on local physics, e.g. solar system ephemerides. Associated time scales have been discussed in great detail. First attempts to construct similar reference systems at the 2nd post-Newtonian approximation for astrometry have been published (e.g., Minazzoli & Chauvineau, 2009). Topocentric Celestial Reference systems, either directly related with the BCRS or coming from the GCRS with direct relation to the ITRS, have been constructed.

Every modern solar system ephemeris uses the post-Newtonian (PN) dynamical equations for mass monopoles (EIH equations). In the literature one finds discussions of spin-orbit and spin-spin couplings, the translational equations of motion with all mass and spin multipole moments as well as rotational equations of motion to post-Newtonian (PN) order (e.g., Damour *et al.*, 1991–1994). For practical applications such as a post-Newtonian treatment of Earth's rotation a PN model of rigidly rotating multipoles has been worked out as a starting point for further perturbative treatments (Klioner *et al.*, 2001). The dynamics of two mass monopoles (point masses) has been worked out to order $1/c^7$ (3.5 PN) (e.g., Königsdörffer *et al.*, 2003). For satellite motion relativistic potential coefficients have been adopted by IAU2000 resolutions. Relativistic inertial forces (Lense-Thirring and geodesic precession/nutation) have been discussed extensively, relativistic tidal forces have been discussed in Damour *et al.* (1993).

In the field of astrometry the relativistic model of Klioner (2003) has been refined over the years. It is basically a coordinate picture but it has been demonstrated in Klioner (2004) that with a choice of adequate local coordinates moving with the observer, results agree with those of a tetrad formalism, i.e., they are coordinate-independent. The intrinsic accuracy of the Klioner model is of order 0.1 μas.

In the field of relativity in metrology we face the following issues: clock synchronization and time dissemination, GPS - GLONASS - GALILEO, VLBI, SLR - LLR, Laser Gyros, Doppler measurements and Pulsar Timing. The problem of clock synchronization and

† E-mail: michael.soffel@tu-dresden.de

time dissemination by means of electromagnetic signals has been discussed extensively in the literature (e.g., Klioner 1991; Wolf & Petit, 1995; Petit & Wolf, 2005). The intrinsic accuracy here is below 1 ps. For the problem of relativity in GPS (GLONASS - GALILEO) the standard reference is Ashby (2003). The present relativistic VLBI model is a consensus model (Eubanks 1991) with an accuracy of about 1 ps. A post-Newtonian treatment of SLR and LLR can be found in many places (e.g., Soffel 1989). A ring-laser gyroscope consists basically of a closed tube in which laser activity is excited so that two laser beams, one traveling in clockwise the other one in counter-clockwise direction, interfere behind a beam splitter where the interference fringes can be analyzed. Such a gyroscope is an inertial device. In contrast to the geodetic space techniques it is sensitive to the instantaneous rotation vector of the Earth. A post-Newtonian theory of ring-laser gyroscopes can be found in Scully *et al.* (1981), Soffel (1989) and Bosi *et al.* (2011). Relativity terms are of order 7×10^{-10}, far below the present level of achievable accuracies. Martin *et al.* (1985) is the classical paper where relativity in Doppler measurements to spacecraft is treated. A large number of papers is devoted to a relativistic description of pulsar timing (e.g., Blandford *et al.* 1976; Haugan 1985, Damour & Taylor 1992; Kopeikin 1999).

Comparing the intrinsic theoretical accuracies with presently achievable ones in observations/measurements one finds that present relativistic models seem to be sufficient for the forthcoming decades. Nevertheless, there is the wish to work out a relativistic VLBI model with an accuracy of about 0.1 ps (presently the accuracy of VLBI observations is of order of a few ps); it should be consistent and 'complete' at this level and well documented. The concept of a post-Newtonian rigidly rotating multipole model as a useful starting point for a relativistic description of global geodynamics should be critically examined.

References

Ashby, N. 2003, *Living Rev. Relativity* 6, 1

Blandford, R. & Teukolsky, S. 1978, *Astrophys. J.* 295, 580

Bosi, F., *et al.* 2011, *Phys.Rev.* D84, 122002

Damour, T., Soffel, M., Xu, C. 1991-1994, *Phys.Rev.* D43, 3273 (1991); D45, 1017 (1992); D47, 3124 (1993); D49, 618 (1994)

Damour, T. & Taylor, J. 1992, *Phys.Rev.* D45, 1840

Eubanks, T. M. 1991, Proc. of the U.S. Naval Observatory Workshop on Relativistic Models for Use in Space Geodesy, USNO

Haugan, M. 1985, *Astrophys. J.* 296, 1

Klioner, S. 1991, *Celest.Mech.* 53, 81

Klioner, S. 2003, *Astron. J.* 125(3), 1580

Klioner, S. 2004, *Phys. Rev.* D69, 124001

Klioner, S., Soffel, M., Xu, C., & Wu, X. 2001, In: Proceedings of Journées'2001, N. Capitaine (ed.), Paris Observatory, Paris, 232; arXiv:astro-ph/0303376

Klioner, S. & Soffel, M. 2004, ESA SP-576, 305; arXiv:astro-ph/0411363v2

Königsdörffer, C., Faye, G., & Schäfer, G. 2003, arXiv:gr-qc/0305048v2

Kopeikin, S. 1999, *Mon. Not. R. Astron. Soc.* 305, 563

Martin, C., Torrence, M., & Misner, C. 1985, *J.Geophys.Res.* 90, 9403

Minazzoli, O. & Chauvineau, B. 2009, *Phys.Rev.* D79, 084027

Petit, G. & Wolf, P. 2005, *Metrologia* 42, S138

Scully, M., Zubairy, M., & Haugan, M. 1981, *Phys.Rev.* A24, 2009

Soffel, M. 1989, Relativity in Astrometry, Celestial Mechanics and Geodesy, Spinger, Berlin

Soffel *et al.* 2003, *Astron. J.*, 126, 2687

Wolf, P. & Petit, G. 1995, *Astron. Astrophys.* 304, 653

Highlights of Astronomy, Volume 16
XXVIIIth IAU General Assembly, August 2012
T. Montmerle, ed.

doi:10.1017/S1743921314005420

Celestial dynamics and astrometry in an expanding universe

S. Kopeikin

Department of Physics & Astronomy, University of Missouri, 322 Physics Building, Columbia, MO 65211, USA

Abstract. The mathematical concept of the Newtonian limit of Einstein's equations in an expanding universe is formulated. The equations of motion of planets and light are compared.

Keywords. gravitation, relativity, ephemerides, cosmology: theory

---

The current paradigm of the IAU 2000 resolutions assumes that the background spacetime is flat with the Minkowski metric tensor $\eta_{\alpha\beta}$. A more adequate model is the Friedmann spacetime with the metric $\bar{g}_{\alpha\beta} = a^2(\eta)\eta_{\alpha\beta}$, where $a(\eta)$ is the scale factor describing the Hubble expansion. The coordinate chart is $x^\alpha = (c\eta, x^i)$, where the conformal (non-physical) time η relates to the cosmic (physical) time t measured by the Hubble observers, by the equation $a(\eta)d\eta = dt$. We assume that $\eta = 0$ corresponds to the present epoch. The solar system is a localized astronomical system. It makes sense to introduce the *local* coordinates, $r^\alpha = (c\lambda, r^i)$, related to the *global* coordinates, x^α, by a *special conformal* transformation, $r^\alpha = (x^\alpha - b^\alpha x^2)/\sigma(x)$, where b^α is a constant vector, $\sigma(x) = 1 - 2b_\mu x^\mu + x^2$, and $x^2 \equiv \eta_{\mu\nu}x_\mu x^\nu$, $b^2 \equiv \eta_{\mu\nu}b_\mu b^\nu$ (Fulton *et al.* 1962). The vector b^α is chosen so that $a^2(\eta)\eta_{\alpha\beta}dx^\alpha dx^\beta = \eta_{\alpha\beta}dr^\alpha dr^\beta$ is a valid equation up to the second order in the Hubble constant, $H = \dot{a}/a$. Transformation to the local metric is achieved with $b^\alpha = (H/2c)\bar{u}^\alpha$, where $\bar{u}^\alpha$ is four-velocity of the Hubble flow. Approximating $a(\eta) = 1 + H\eta$ yields, $\lambda = a(\eta)[\eta - (1/2)H(c^2\eta^2 - \boldsymbol{x}^2)]$, and $\boldsymbol{r} = a(\eta)\boldsymbol{x}$. We have to relate the coordinate time λ to the cosmic time t. In the linearized approximation $\eta = t - (1/2)Ht^2$. For particles at rest, the conformal transformation of time yields $\lambda = t$. However, light moves along null geodesics, $c^2\eta^2 - \boldsymbol{x}^2 = 0$. The conformal time transformation yields the relation of parameter λ of the light geodesics with the cosmic time, $\lambda = t + (1/2)Ht^2$. It means that in the Newtonian limit the metric for slowly-moving particles is $\eta_{\alpha\beta}$ but the optical metric for light is $\hat{g}_{\alpha\beta} = \eta_{\alpha\beta} + (1 - a^2)\bar{u}_\alpha\bar{u}_\beta$, and depends on the scale factor, $a = a(t)$. This is interpreted as an index of refraction, $n = 1/a$, that makes the coordinate speed of light, $c_l = c/n = ac$, growing as time goes from past to future, and decreases otherwise. The equations of motion of particles and light are reduced to their Newtonian form for both metrics. However, equations for particles and for light differ by terms of the first order in the Hubble constant (Kopeikin 2012). This leads to the important conclusion that the correct light-propagation equation for ephemeris calculations is $\lambda_2 - \lambda_1 = (1/c)|\boldsymbol{r}_2 - \boldsymbol{r}_1|$. It eliminates the so-called Pioneer anomaly (Anderson *et al.* 2002) which may really have a cosmological origin.

References

Fulton, T., Rohrlich, F., & Witten, L. 1962, *Rev. Mod. Phys.*, 34, 442

Kopeikin, S. 2012, *Phys.Rev.*, D86, 064004

Anderson, J D., Laing, P. A., Lau, E. L., Liu, A. S., Nieto, M. M., & Turyshev, S. G. 2002, *Phys.Rev*, D65, 082004

Highlights of Astronomy, Volume 16
XXVIIIth IAU General Assembly, August 2012
T. Montmerle, ed.

doi:10.1017/S1743921314005432

Developing a pulsar-based time standard

G. Hobbs†

CSIRO Astronomy and Space Science, Australia Telescope National Facility,
PO Box 76, Epping NSW 1710, Australia

Abstract. We describe how observations of pulsars from the Parkes Pulsar Timing Array (PPTA) project have been used to develop a pulsar-based timescale. This is the first such timescale that has a precision comparable to uncertainties in international atomic timescales.

Keywords. time, pulsars: general

High precision timing observations of pulsars are being carried out at many observatories around the world. The main goal of these "pulsar timing array" (PTA) projects is to detect ultra-low frequency gravitational waves. As reviewed in this paper, these data sets can also be used to develop a pulsar-based time standard.

We have made use of observations from the Parkes radio telescope that were obtained as part of the Parkes Pulsar Timing Array (PPTA) project (Manchester *et al.* 2012). Initially the pulse times-of-arrival (ToAs) for each observation of each pulsar are determined. These ToAs are converted from the observatory time standard to a realisation of Terrestrial Time (TT). Barycentric arrival times are subsequently calculated and are compared with predictions of the arrival times using a model for the pulsar rotational and orbital parameters. The differences between the actual measurements and the predictions are known as the "pulsar timing residuals". This technique, known as "pulsar timing", is widely used in pulsar astronomy and is described in detail by Edwards, Hobbs & Manchester (2006).

Various phenomena such as gravitational waves, unexplained timing irregularities or errors in terrestrial time standards will induce timing residuals. They can be distinguished by searching for correlations between the timing residuals of multiple pulsars. For instance, pulsar timing irregularities, glitch events or interstellar medium variations will lead to timing residuals that are uncorrelated between different pulsars. In contrast an error in the terrestrial time standard will lead to timing residuals that are identical for different pulsars (assuming that all pulsars have been observed over the same time span).

The incredible stability of millisecond pulsar rotation leads to the possibility of developing a time scale based on the pulsar rotation which is analogous to the free atomic scale, Échelle Atomique Libre (EAL). The Ensemble Pulsar Scale (EPS) can be used to detect fluctuations in atomic timescales and therefore can lead to a new realisation of TT.

Earlier attempts to develop a pulsar timescale have been made by Guinot & Petit (1991), Petit & Tavella (1996), Rodin (2008) and Rodin & Chen (2011). Our new algorithm has been published in Hobbs *et al.* (2012) and is implemented as part of the TEMPO2 software package (Hobbs, Edwards & Manchester 2006). This algorithm accounts for various features of the observations such as: 1) irregular sampling, 2) different data spans for different pulsars and 3) different fitting parameters for different pulsars. Our result is reproduced in Fig. 1. We successfully follow features known to affect the

† E-mail: george.hobbs@csiro.au

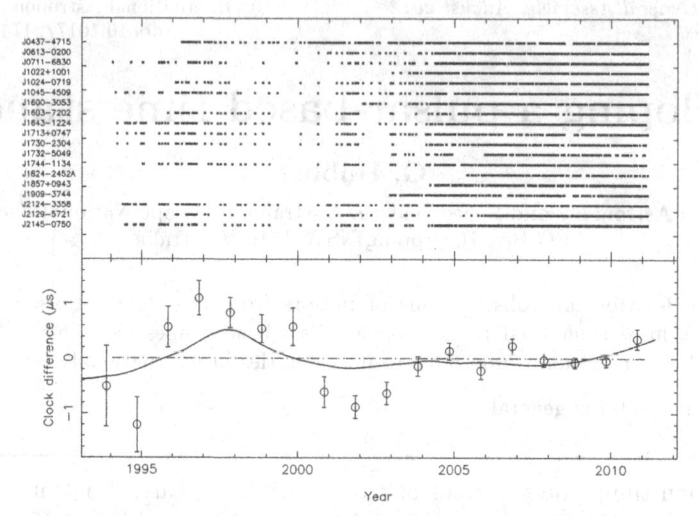

Figure 1. This figure is reproduced from Hobbs *et al.* (2012). The top panel shows the sampling for each of the pulsars in our sample. The lower panel shows the difference between the pulsar timescale and TT(TAI) as points with error bars. The solid line indicates the difference between TT(TAI) and TT(BIPM11) after a quadratic polynomial has been fitted and removed. Full details are available in Hobbs *et al.* (2012).

frequency of the International Atomic Timescale (TAI) and we find marginally significant differences between our pulsar time scale, TT(PPTA11), and TT(BIPM11).

This work is being continued by combining the Parkes observations with data from other observatories (see Hobbs *et al.* (2010) for a description of the International Pulsar Timing Array project). The new analysis will confirm or deny the tentative discrepancies between TT(PPTA11) and TT(BIPM11) whilst significantly improving the stability and precision of the pulsar scale. In the longer term it is expected that a future pulsar time scale will be combined with the best atomic timescale to give the world's most stable time scale that will be valid effectively forever.

Acknowledgements

The Parkes radio telescope is part of the Australia Telescope which is funded by the Commonwealth of Australia for operation as a National Facility managed by CSIRO.

References

Edwards, R., Hobbs, G., & Manchester, R. 2006, *MNRAS*, 372, 1549
Guinot, B. & Petit, G. 1991, *A&A*, 248, 292
Hobbs, G., Edwards, R., & Manchester, R. 2006, *MNRAS*, 369, 655
Hobbs, G., *et al.* 2010, *CQGra*, 27, 4013
Hobbs, G., *et al.* 2012, accepted by MNRAS (arXiv.1208.3560)
Manchester, R. N., *et al.* 2012, *submitted to MNRAS*
Petit, G. & Tavella, P. 1996, *A&A*, 308, 290
Rodin, A. 2008, *MNRAS*, 387, 1583
Rodin, A. & Chen, D. 2011, *Astronomy Reports*, 55, 622

Highlights of Astronomy, Volume 16
XXVIIIth IAU General Assembly, August 2012
T. Montmerle, ed.

doi:10.1017/S1743921314005444

Long term stability of atomic time scales

G. Petit† and F. Arias

Bureau International des Poids et Mesures
92312 Sèvres Cedex, France

Abstract. We review the stability and accuracy achieved by the reference atomic time scales TAI and TT(BIPM). We show that they presently are in the low 10^{-16} in relative value, based on the performance of primary standards, of the ensemble time scale and of the time transfer techniques. We consider how the 1×10^{-16} value could be reached or superseded and which are the present limitations to attain this goal.

Keywords. time, reference systems

1. Introduction

International Atomic Time TAI gets its stability from some 400 atomic clocks worldwide that generate the free atomic scale EAL, and its accuracy from about ten primary frequency standards (PFS) that are used to steer the EAL frequency. Terrestrial Time is a coordinate time in the geocentric reference system defined by the International Astronomical Union. TAI provides one realization of TT but it is not optimal because of operational constraints. The BIPM therefore computes in deferred time another realization, TT(BIPM), which is based on a weighted average of the evaluations of the TAI frequency by the PFS. A new version is computed each January, the latest available being TT(BIPM11), available at ftp://tai.bipm.org/TFG/TT(BIPM).

We review the stability and accuracy achieved by the reference atomic time scales TAI and TT(BIPM). We show that they presently are at the level of a few 10^{-16} in relative value, based on the performance of primary standards, of the ensemble time scale and of the time transfer techniques. We consider how the 1×10^{-16} value could be reached or superseded and which are the present limitations to attain this goal.

2. Achieving low-10^{-16} accuracy

The stability of atomic time scales, the PFS accuracy and the capabilities of frequency transfer all achieve a performance in the low 10^{-16} in relative frequency stability.

The stability and accuracy of the BIPM time scales has been studied in (Petit, G., 2007, *Proc. 21st EFTF conference*, 391–394). The 1-month stability of all scales is that of EAL, at $3 - 4 \times 10^{-16}$ and slightly improves with the number of participating clocks. Using TT(BIPM11) as a reference shows that the drift affecting EAL in past years has disappeared since a change of algorithm effective in August 2011. It is expected that this will improve the long-term (6 months and above) stability of TAI to well below 1×10^{-15} but this needs to be ascertained in future studies. The estimated accuracy of TT(BIPM) over recent years is at 3×10^{-16} or below (Petit, G., Panfilo, G., 2012, *IEEE Tans. IM*, accepted). This is due to the ever increasing number of Cs fountain evaluations, about 50 every year since 2008, and to the improved accuracy of each evaluation.

The time transfer techniques presently most used in TAI are dual frequency Global

† E-mail: gpetit@bipm.org

Positioning System (GPS) phase and code measurements in a mode called precise point positioning (PPP), and two-way time transfer (TW) using telecommunication satellites. A combination of these two techniques can provide best performance at all averaging times (Jiang, Z., Petit, G., 2009, *Metrologia*, 46-3, 305–314), reaching the low 10^{-16} at 30 day averaging. They are used for many TAI time links so that the contribution of frequency transfer to the instability of TAI is estimated to be in the low 10^{-16}.

3. Reaching 1×10^{-16} and beyond

From numerous recent publications, see e.g. in parts II, IV and V of (Maleki L. (ed.), 2009, *Proc 7th Symp. Freq. Standards and Metrology*, World Scientific, 308–313), it is clear that some frequency standards have reached a level where all systematic effects may be estimated with an uncertainty at 1×10^{-16} or below. Their accuracy budget is thus better than that of the Cs transition providing the definition of the SI second. Some of these transitions have been recognized as secondary representations of the second (SFS) by the Consultative Committee for Time and Frequency (CCTF). As of 2009, this list includes the hyperfine transition of ^{87}Rb and optical transitions of Sr, Hg^+, Sr^+ and Yb^+, and three of these transitions state total uncertainties in the low 10^{-15}, a value dominated by the uncertainty in the realization of the SI second itself. This list is expected to expand with time and the reported uncertainties of the transition frequencies should correspondingly decrease. The first update of this list will happen at the September 2012 meeting of the CCTF.

Submitting formal evaluations to the BIPM for the best SFS would allow providing information on these SFS to the community in an homogeneous and complete manner. It also ensures that all available evaluations for a given secondary transition can be compared to the PFS in a consistent manner by the BIPM, so as to allow an optimal determination of the reference value of the transition frequency. Eventually, SFS evaluations could also contribute, through the reference value of the transition frequency, to estimate the accuracy of TAI and to generate TT(BIPM). In January 2012, the first evaluations of an SFS have been transmitted to the BIPM. It concerns the ^{87}Rb transition realized in the dual atom fountain SYRTE-FO2 at the LNE-SYRTE (Guéna J. *et al.*, 2010, *IEEE Trans. Ultrason. Ferroelectr. Freq. Control*, 57 (3), pp. 647–653).

The techniques presently used for time transfer may progress, but they will still limit the performance of time scales, specially for short and moderate averaging times (typically up to one month). Several new techniques have emerged that should be able to provide frequency transfer uncertainty in the low 10^{-17} and possibly below. Some are based on microwave or optical links to a low Earth orbit payload and should provide global coverage provided a stable clock is available in the space payload. Another technique transfers a stable laser frequency over a standard fiber link and provides even better performance in frequency transfer, although presently limited in spatial extension.

4. Conclusions

We have shown that the present performance of the reference atomic time scales TAI and TT(BIPM) is in the low 10^{-16} in stability and accuracy. New frequency standards already promise that the performance of 1×10^{-16} and below is possible but a better reliability and wider availability of these are needed for use in time scale formation. Improving the techniques for time and frequency transfer is also an important issue. Ultimately,a new approach to the problem of elaborating an ensemble time scale may be needed.

Highlights of Astronomy, Volume 16
XXVIIIth IAU General Assembly, August 2012
T. Montmerle, ed.

doi:10.1017/S1743921314005456

Time and frequency transfer with the ESA/CNES ACES-PHARAO mission

P. Delva†, C. Le Poncin-Lafitte, P. Laurent, F. Meynadier and P. Wolf

LNE/Syrte - Observatoire de Paris,CNRS,UPMC Univ Paris 06,UMR8630, 75005 Paris, France

Abstract. We have written a theoretical description of one-way and two-way satellite time and frequency transfer and developed a model of the micro-wave link in the frame of the ACES/PHARAO mission. This is used to write a data analysis software and a simulation to test it. A very short description of the mission and of the micro-wave link is given here. A detailed description can be found in Delva *et al.*, 2012, Proceedings of the EFTF, Gothenburg, Sweden, arXiv:1206.6239.

Keywords. gravitation, time, methods: data analysis

1. The ACES-PHARAO mission

The Atomic Clocks Ensemble in Space – ACES-PHARAO mission (C. Salomon *et al.*, 2007, IJMPD, 16, 2511) – will be installed on board the International Space Station (ISS) in 2015 and will realize in space a time scale of very high stability and accuracy. This time scale will be compared to a ground clock network thanks to a dedicated two-way microwave link. The ACES mission will demonstrate the capability to perform phase/frequency comparison between space and ground clocks with a resolution at the level of 0.4 ps over one ISS pass (300 s), 8 ps after 1 day, and 25 ps after 10 days of integration time (see Fig. 1). For that purpose our team is developing advanced time and frequency transfer algorithms.

The altitude difference between the ACES-PHARAO clock and ground clocks will allow to measure the gravitational redshift with an unprecedented level of accuracy, as well as looking for a violation of Lorentz local invariance. Several ground clocks based on different atomic transitions will be compared to look for a drift of the fine structure constant. Moreover, the mission will pave the way to a new type of geodetic measurement: the gravitational redshift will be used to measure gravitational potential differences between distant clocks, with an accuracy of about 10 cm.

2. The micro-wave link

The micro-wave link will be used for space-ground time and frequency transfer. It is composed of three signals of different frequencies: one uplink at frequency $f_1 \simeq 13.5$ GHz, and two downlinks at $f_2 \simeq 14.7$ GHz and $f_3 = 2.2$ GHz. Measurements are done on the carrier itself and on a code which modulates the carrier. The link is asynchronous: a configuration can be chosen by interpolating observables. The so-called Λ-configuration minimizes the impact of the space clock orbit error on the determination of the desynchronisation (Duchayne *et al.*, 2009, *A&A*, 504, 653). We sketch on Fig. 2 the space-time diagram of the two signal f_1 and f_2 in a Λ-configuration. We define the SYRTE Team

† E-mail: Pacome.Delva@obspm.fr

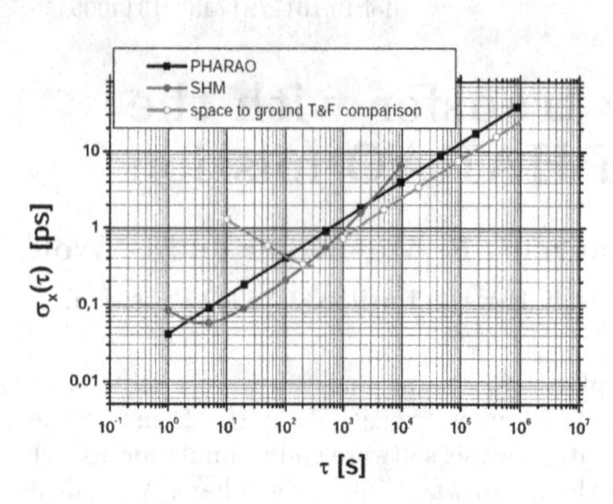

Figure 1. Performance objective of the ACES clocks and the ACES space-ground time and frequency transfer expressed in time deviation

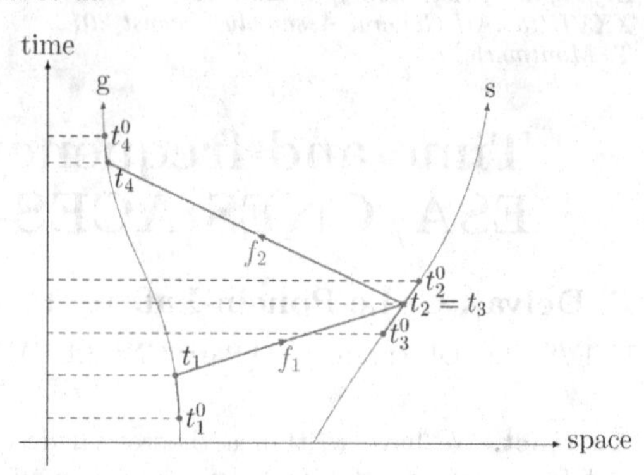

Figure 2. Signal f_1 is generated at coordinate time t_1^0, is emitted by the ground antenna at time t_1, received by the space antenna at t_2 and arrives at the receiver modem at t_2^0. For signal f_2 the sequence is $(t_3^0, t_3, t_4, t_4^0, t_8^0)$. In the Λ-configuration interpolation is such that $t_2 = t_3$.

(ST) observables by $\Delta\tau(\tau_e) = \tau_e - \tau_r$, where τ_e is the proper time of emission of the signal and τ_r the proper time of reception. It can be linked to desynchronisation:

$$\tau^s(t_2) - \tau^g(t_2) = \frac{1}{2}\left(\Delta\tau^g_{\text{mo}}(\tau^g(t_4^0)) - \Delta\tau^s_{\text{mo}}(\tau^s(t_2^0)) + [T_{34} - T_{12}]^g\right) \qquad (2.1)$$

where t is coordinate time, $\Delta\tau_{\text{mo}}$ are the ST observables corrected for the delays in the cable between the clock and the antenna at transmission and at reception, $T_{ij} = t_j - t_i$ and $[.]^g$ is the coordinate time to the ground clock proper time transformation.

The time-of-flights T_{34} and T_{12} can be calculated from the known orbits of the clocks, accounting for the tropospheric, ionospheric and Shapiro delays. The observables from the two downlinks can be used to determine the Total Electronic Content (TEC) of the atmosphere along the line-of-sight, in order to correct for the ionospheric delay. The two-way configuration cancels the tropospheric delay, which does not depend on the signal frequency at this level of accuracy.

The basic observables of the modem developed by TimeTech (TT observables) are different from the ST observables. A PPS signal (one Pulse Per Second), a 12.5 PPS (one pulse every 80 ms, the period of measurements), and a periodic signal (either code at 100 MHz or carrier) are generated in the emitter and the receiver. When received, the periodic signal is mixed with a local oscillator signal not far from the received frequency, and filtered to obtain the low frequency part of the beatnote. The beatnote frequency is around 195 kHz for code and 729 kHz for carrier. The receiver modem records the time of the first ascending zero-phase of the beatnote signal after the 12.5 PPS signal, and it counts the number of ascending zero-phase during one 80 ms sequence. The link between the TT and ST observables is detailed in Delva *et al.*, 2012, Proceedings of the EFTF, Gothenburg, Sweden, arXiv:1206.6239.

3. Conclusion

The ACES simulation of the micro-wave link is written in its first version, and used to assess our pre-processing software. The design of the data analysis software has been done in a modular way, and most of the building blocks are ready. Several questions remain, e.g. how to solve the phase ambiguity for the carrier observable.

Highlights of Astronomy, Volume 16
XXVIIIth IAU General Assembly, August 2012
T. Montmerle, ed.

doi:10.1017/S1743921314005468

Celestial reference frames at multiple radio wavelengths

C. S. Jacobs†

Jet Propulsion Laboratory, California Institute of Technology, Pasadena, CA 91109, USA

Abstract. In 1997 the IAU adopted the International Celestial Reference Frame (ICRF) built from S/X VLBI data. In response to IAU resolutions encouraging the extension of the ICRF to additional frequency bands, VLBI frames have been made at 24, 32, and 43 GHz. Meanwhile, the 8.4 GHz work has been greatly improved with the 2009 release of the ICRF-2. This paper discusses the motivations for extending the ICRF to these higher frequency radio bands. Results to date will be summarized including evidence that the high frequency frames are rapidly approaching the accuracy of the 8.4 GHz ICRF-2. We discuss current limiting errors and prospects for the future accuracy of radio reference frames. We note that comparison of multiple radio frames is characterizing the frequency dependent systematic noise floor from extended source morphology and core shift. Finally, given Gaia's potential for high accuracy optical astrometry, we have simulated the precision of a radio-optical frame tie to be ~10–15 μas (1-σ, per component).

Keywords. astrometry, reference systems, (galaxies:) quasars: general, instrumentation: interferometers, techniques: high angular resolution, radio continuum: galaxies

1. Introduction

The past decade has seen the expansion of global celestial reference frame work from S/X-band (e.g., Ma *et al.*, *AJ*, 1998) to three new frequencies: K-band and Q-band (Lanyi *et al.* AJ, 2010), and X/Ka-band (García-Miró *et al.* IVS, 2012). What changes when moving to these higher radio frequencies? Sensitivity worsens because (1) system temperatures and atmospheric absorption increase due to the H_2O 22 GHz and O_2 60 GHz lines, (2) antenna surface shape control becomes more difficult (3) antenna pointing becomes more difficult as the beam tightens (4) sources become more resolved. However, there is much to be gained. Sources become more compact. There is less extended structure as steep spectrum jet plumes fade (Charlot *et al.* AJ, 2010). Core shift decreases as one goes to higher frequencies (Sokolovsky *et al.* 2011, Kovalev *et al.* 2008, Porcas, *A&A*, 2009). The combined effect is that the astrophysical character of the sources becomes better for astrometry. Given that the sensitivity concerns raised can be solved by recording more bits using ever cheaper digital hard drive technology, moving to higher frequencies is on the whole a very attractive proposition for VLBI radio astrometry.

2. Overview of existing frames

At present, radio frames have been constructed at four frequencies. Their properties are summarized in Table 1. The S/X frame is the most well established. It has the longest history, the most data, and has undergone the most analyses. The current IAU standard celestial frame is the ICRF-2 based on S/X data. The (now dormant) work on K and Q-band frames from the last decade proved that high accuracy frames could be built above 8 GHz. The main limitation of K/Q frames arises from having data only from the

† E-mail: `Christopher.S.Jacobs@jpl.nasa.gov`

all-northern VLBA thereby limiting observations to $\delta > -40°$ and allowing $\Delta\delta$ vs. δ zonal errors of 100s of μas. The X/Ka frame has 482 sources well distributed for $\delta > -45°$. While it has no significant $\Delta\delta$ vs. δ zonal error, the precision becomes increasingly worse as one moves southward. X/Ka agreement with the ICRF-2 for 450 common sources is about 200 and 270 μas in $\alpha \cos\delta$ and δ, respectively.

Frame	Frequency (GHz)	wavelength (cm)	Accuracy (μas)	$N_{sources}$	Reference
S/X	2.3/8.4	13/3.6	~40–100	3414	Ma *et al.* IERS Tech Note 35, 2009
K	24	1.2	~100	268	Lanyi *et al.* AJ, 139, 5, 2010
Q	43	0.7	~300	131	Lanyi *et al.* AJ, 139, 5, 2010
X/Ka	8.4/32	3.6/0.9	~225	482	García-Miró *et al.* IVS, 2012

Table 1. Existing Celestial Reference Frames at Radio Frequencies

3. Future prospects for improvement

Four key aspects of VLBI can potentially be improved by the end of the decade.

(1) SNR: Modern digital recording (Whitney *et al.* IVS 2012) makes it affordable to increase data rates from current operations at 128–448 Mbps up to 2048–16,000 Mbps.

(2) Instrumentation: Digital Back Ends (Ruszczyk *et al.* Tuccari; Garcia-Miro *et al.* all IVS 2012) are replacing obsolete analog systems thereby improving channel uniformity, phase linearity and stability. Better frequency standards ('clocks') are possible (e.g. Petit & Arias, IAU 2012), but appear to be years away from operational use in VLBI.

(3) Troposphere: Many sites are moving to smaller, faster slewing antennas allowing faster sampling of the fluctuating troposphere. Improved Water Vapor Radiometer (WVR) calibrations have been demonstrated (Tanner *et al.* RSci, 2003; Bar-Sever *et al.* IEEE, 2007).

(4) Southern geometry: At S/X, the addition of 3 AuScope antennas plus Warwick, New Zealand creates more southern observing opportunities. At K-band, HART, S. Africa has an un-cooled receiver with yet-to-be-used potential. At X/Ka, the Malargue, Argentina 35-m (online Nov. 2012) gives immediate, significant improvements to network geometry.

4. Frame tie: VLBI radio frames to Gaia optical frame

The Gaia mission (Prusti; Mignard, IAU, 2012) plans (2013 launch) to acquire high accuracy optical astrometry for 10^9 stars including 500,000 quasars ($V < 20$ mag). Of these, a few thousand are expected to be both optically bright ($V < 18$ mag) and radio loud (30–300+ mJy) and thus suitable for optical-radio comparisons. Such comparisons will first require that the conventional 3-D orientation of the frames be aligned. Charlot & Bourda (IAU-JD7, 2012) have a strategy for maximizing the number of optically bright ($V<18$) quasars which are also detectable with S/X-band VLBI. García-Miró *et al.* (IVS, 2012), using existing X/Ka data and projected Gaia errors, estimate a tie precision of ~10 μas (1-σ) per 3-D component which could improve as more X/Ka data arrives.

Acknowledgements

I would like to acknowledge my colleagues who acquired and analyzed the data—especially the IVS, the ICRF2 working group, the KQ VLBI collaboration, and the X/Ka VLBI collaboration. This research was done in part under contract with NASA. Government sponsorship acknowledged.

Highlights of Astronomy, Volume 16
XXVIIIth IAU General Assembly, August 2012
T. Montmerle, ed.

doi:10.1017/S174392131400547X

Gaia promises for the reference frame

F. Mignard

University Nice Sophia Antipolis, CNRS, OCA, Le Mont Gros, 06304 Nice, France

Abstract. Expectations of the *Gaia* astrometry mission regarding the realisation of an optical kinematical reference frame based on extragalactic sources are summarized.

Keywords. astrometry, reference systems

The ESA space astrometry mission *Gaia* is due for launch before the end of 2013. It will survey the sky over a period of 5 years with a sensitivity limited instrument down to the 20th magnitude. The astrometric accuracy measured by the quality of the parallaxes is expected to be of 25 muas at 15 mag. *Gaia* will simultaneously perform multi-epoch photometry for all the sources and spectroscopy to 17th mag. The mission is optimised to observe stellar sources to produce a stereoscopic and kinematic census of about one billion stars in our Galaxy enabling to probe the formation and evolution of the Milky Way. The expected astrometric accuracy in position, annual proper motion and parallaxes is shown in Fig. 1 as a function of the G magnitude (very similar to R band).

The scientific core of the mission is dedicated to stellar and galactic physics, and will bring considerable high quality data for these fields from the combination of the kinematical information allowed by astrometry and the low resolution spectroscopy more or less equivalent to photometry over the visible range with at least 15 bands. Each star will be observed about 70 times during the mission, with a slight scatter about this mean as a function of the ecliptic latitude. An average sequence of observations will give two successive observations separated by just above 1.7h, followed by similar pairs of observations every 40 to 60 days, providing a good time sampling over the mission length of 5 years. This is required to properly sample the parallactic ellipse for disentangling the proper motion and parallax, but also for investigating more complex astrometric displacement in multiple systems or for solar system objects. A full reward of this sampling will show up also with the photometric variability analysis of at least 100 millions sources.

The internal detection system will allow also to observe measure the position of several 100 000s quasars, everywhere in sky except a small area at low galactic latitude. These consistent and repeated observations will lead to the realisation of a kinematically defined inertial frame in the optical wavelengths. Given the fact that the vast majority of these QSOs will be new we have developed a procedure to recognize the QSOs from the stars in an automatic and efficient way using the photometric and astrometric information. The survey mode will aim at identifying most of the QSOs, even if it means including a certain fraction of stellar contaminant in the set. In parallel a more restrictive setting of the selection parameters will enable to generate a clean sample of sources, incomplete but free of contaminant, to serve as defining sources for the establishment of the inertial frame.

From simulations based on realistic space density of optical QSOs as a function of magnitude and using the *Gaia* astrometric performance, *Gaia* should realise a quasi-inertial celestial reference frame with a residual rotation better than 0.3 μas per year. This is illustrated in Fig. 2 giving on the left scale the expected accuracy as a function

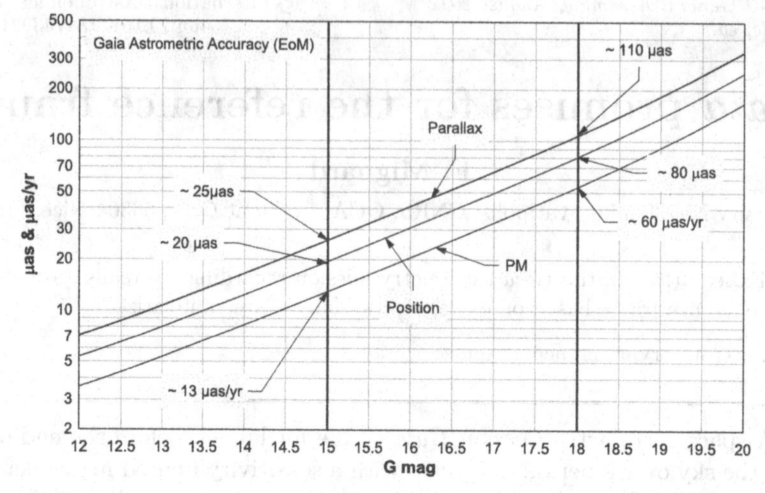

Figure 1. End-of-Mission astrometric accuracy expected with *Gaia* for the position, proper-motion and parallaxes for stars and extragalactic point-like sources.

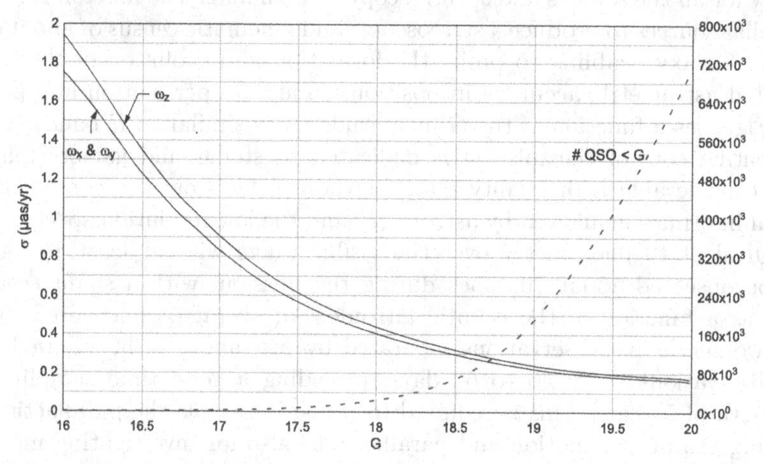

Figure 2. Inertiality of the final *Gaia* Celestial Reference Frame based on the QSO observations, expressed by the accuracy of the residual rotation in galactic coordinates. The precision for a G magnitude is computed with only sources brighter than G. A random instability has been taken equal to 20 μas yr^{-1} and added (quadratic sum) to the single star noise. The right scale gives the number of sources found brighter than G.

of the magnitude of the faintest sources kept in the sample. The number of sources is shown on the dashed line read on the right scale. Eventually the final *Gaia* product directly related to the Celestial Frame will include also all the stellar sources given in the same system with accurate proper motions. The whole set should be referred to as the *Gaia*-CRF (*Gaia* Celestial Reference Frame) which will comprise:

- A set of defining sources from the clean subset of QSOs used to fix the frame spin
- A larger set of secondary QSOs not used to remove the residual rotation
- A very small set of QSOs common to *Gaia* and ICRF2 used to tie the orientation
- The *Gaia*-SRF (Stellar Reference Frame):
  a. about one billion stars with positions, proper motions and parallaxes;
  b. an average density of $\sim$ 25 000 stars per square degree, highly variable with galactic latitude.

Highlights of Astronomy, Volume 16
XXVIIIth IAU General Assembly, August 2012
T. Montmerle, ed.

doi:10.1017/S1743921314005481

INPOP: evolution, applications, and perspectives

A. Fienga[1,2]†, J. Laskar[1], H. Manche[1], M. Gastineau[1] and A. Verma[2]

[1]Astronomie et Systèmes Dynamiques, IMCCE-CNRS UMR8028, Paris, France

[2]Institut UTINAM-CNRS 6213, Université de Franche-Comté, Besançon, France

Abstract. The INPOP ephemerides have undergone several improvements since the last INPOP10a release (Fienga *et al.* 2011). Improvements in the asteroid mass determinations have been made and the effect of the solar corona has also been investigated (Verma *et al.* 2012). Since 2010 and INPOP10a, in anticipation to IAU resolution B2, the au is fixed in the INPOP construction while the mass of the Sun is fitted. Descriptions of tests of general relativity made with INPOP10a are recalled here. Perspectives about Messenger data analysis and new gravity tests are finally introduced.

Keywords. solar system: general, ephemerides, reference systems, methods: n-body simulations, relativity, minor planets, asteroids

---

1. Improvements since INPOP10a

INPOP10e (Fienga *et al.* 2012) is the latest INPOP version. It is the release developed for the Gaia mission and is available for all users. Compared to INPOP10a, new sophisticated procedures related to the asteroid mass determinations have been implemented: bounded-values least-squares have been associated with *a priori* sigma estimators (Kuchynka 2010). Corrections to the solar corona perturbations on the radiometric measurements have been applied (Verma *et al.* 2012) allowing estimations of new electronic densities for slow and fast solar winds and 8% of supplementary data in the INPOP fit. Have also been added very recent Uranus observations (Viera Martins and Camargo 2012) and Pluto positions deduced from HST (Tholen *et al.* 2008).

In anticipation to the resolution B2 of IAU 2012 and since INPOP10a, the adjustment of the mass of the sun was made while the au was maintained at a fixed value. The solar oblateness (J_2) and the Earth-Moon mass ratio (EMRAT) were also fitted. Masses of the planets have also been updated to the values of the IAU best estimates (Luzum *et al.* 2012). Thanks to the added solar corrections and to the improvements in the fit procedure, 152 asteroid masses have been estimated showing good agreement with values found in the literature, especially for big perturbers (inducing perturbations bigger than 5 m over the Earth-Mars distances and the observational period) (see Figure 1(a)). Improvements in the INPOP extrapolation capabilities have also been achieved with this new version: over 32 months of extrapolation, a 30 m degradation arises with INPOP10e, while the degradation reaches up to 100 m with INPOP10a. Comparisons to other planetary ephemerides and postfit residuals can be found in Fienga *et al.* (2012).

2. Tests of general relativity

Estimates of intervals of PPN β and γ acceptable values (values inducing modifications of the postfit residuals smaller than 5%, see (Fienga *et al.* 2011) for more details) were

† E-mail: agnes.fienga@obs-besancon.fr

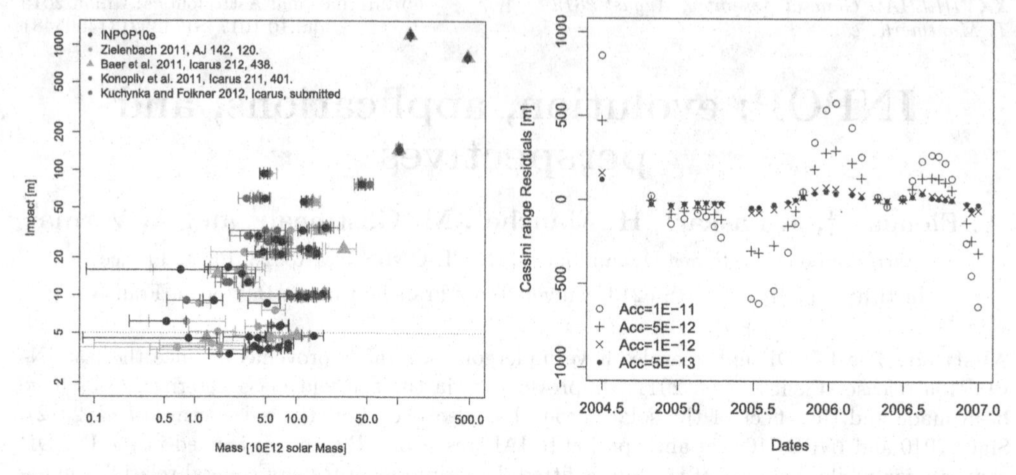

Figure 1. (a) INPOP10e asteroid masses compared to other published values, ranked according to their impact on the Earth-Mars distances over a 1970-2012 period. (b) Differences between Cassini range data and Earth-Saturn distances obtained with supplementary Pioneer-like acceleration on Saturn orbit.

done with INPOP08 (Fienga *et al.* 2009) and INPOP10a. We also estimated the intervals of the acceptable supplementary advances in perihelia and nodes of the planetary orbits, thus allowing some constraints on gravitational models (Blanchet and Novak, 2011). Tests of acceptable supplementary Pioneer-like accelerations were also done and compared to the most accurate observations of outer planets. As one can see in Fig. 1(b), only an acceleration with an amplitude below $5{\times}10^{-13}$ m s^{-2} is compatible with the observed geocentric Saturn distances deduced from the Cassini tracking data.

3. Perspectives

Complete analysis of the Messenger tracking data during its orbital phases are expected by the end of 2012. These data will be used for improving the Mercury orbit that, since INPOP10a, has been tied to only three positions deduced from the Messenger flybys of Mercury. As described in (Fienga *et al.* 2011), the Mercury positions are crucial for the tests of general relativity. New estimates of the β and γ intervals will be made as well as the first INPOP estimates of $d(GM_{\odot})/dt$.

References

Blanchet, L. & Novak, J., 2011, Testing MOND in the Solar System, *arXiv:1105.5815.*
Fienga, A., *et al.*, 2011, *Celestial Mechanics and Dynamical Astronomy*, 111, 363.
Fienga, A., *et al.*, 2011b, *EPSC-DPS Join Meeting 2011*
Fienga, A., *et al.*, 2012, *ArXiv e-prints.*
Fienga, A., *et al.*, 2009, *A & A*, 507,1675-1686
Kuchynka, P., 2010, *Observatoire de Paris, PhD in Astronomy*
Luzum, B., *et al.*, 2012, *Transactions of the International Astronomical Union, Series A*
Tholen, D. J., *et al.*, 2008, *AJ*, 135, 777-784
Verma, A. K., *et al. ArXiv e-prints*, 1206.5667
Viera Martins, R. & Camargo, J. I. B., 2012, *private communication*

Highlights of Astronomy, Volume 16
XXVIIIth IAU General Assembly, August 2012
T. Montmerle, ed.

doi:10.1017/S1743921314005493

Linking the planetary ephemeris to the International Celestial Reference Frame

W. M. Folkner† and J. S. Border

Jet Propulsion Laboratory, California Institute of Technology, Pasadena, CA 91109, USA

Abstract. The largest uncertainty in the ephemerides for the inner planets is in the orientation of the dynamical system to the celestial reference frame. A program of VLBI measurements of spacecraft in orbit about Venus and Mars has been performed to reduce the orientation uncertainty to 0.2 milliarcseconds.

Keywords. celestial mechanics, ephemerides, solar system: general

---

The dynamical orbits of the inner planets (Mercury, Venus, Earth, and Mars) have been determined through planetary radar ranging starting in the 1960's, and have been determined with increasing accuracy by radio range measurements to spacecraft in orbit about them and, in the case of Mars, landers on the surface (e.g. Reasenberg *et al.* 1979; Fienga *et al.* 2009; Konopliv *et al.* 2011; Smith *et al.*, 2012). The largest uncertainty in planetary positions comes from the determination of the overall orientation of the solar system with respect to the celestial reference frame, now defined by the positions of extragalactic radio sources (Ma *et al.* 2009). The earliest link between the dynamical solar system frame and the radio reference frame, with accuracy of $0.020''$, was obtained by use of narrow-band very-long-baseline interferometry (VLBI) measurements of the Mariner 9 spacecraft in orbit about Mars (Newhall *et al.* 1986). The accuracy was improved to $0.003''$ by a comparison of Earth orientation estimates from lunar laser ranging (LLR) and VLBI (Folkner *et al.* 1994).

In order to improve the accuracy of the orientation of the inner solar system to the ICRF, a series of measurements of the angular positions of planets using spacecraft in orbit about Venus and Mars has been performed, starting with observations of the Magellan spacecraft in orbit about Venus in 1990, continuing through 2012 with observations of Mars Odyssey and Mars Reconnaissance Orbiter. The measurements are based on VLBI instrumentation using a technique called delta differential one-way range (ΔDOR) (Thornton & Border 2000) to measure one component of the direction to a planet relative to radio sources. Measurements are made using pairs of tracking stations (baselines) from the NASA Deep Space Network, with one station near Goldstone, California, and a second station from near Madrid, Spain, or near Canberra, Australia.

The accuracy of the ΔDOR measurements has continued to improve. Early measurements were limited by several factors. Limited recording bandwidth required use of bright radio sources, which are more likely be be resolved. With resolved sources the position of the center of brightness varies depending on the observation geometry. The limited bandwidth also limited the measurement signal-to-noise ratio (SNR). Non-linearity of the phase response of the analog radio down-conversion electronics introduced significant errors in the determination of group delay. The delay due to the Earth's ionosphere was calibrated using Faraday rotation measurements from a small set of satellites.

Improved data transmission technology now allows the use of much wider signal band-

† e-mail: William.M.Folkner@jpl.nasa.gov

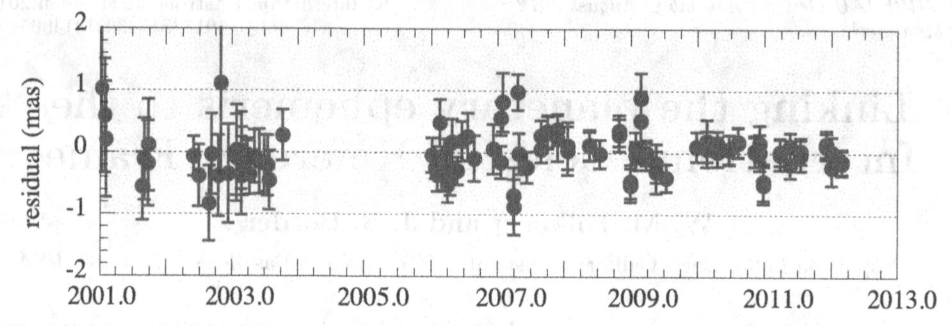

Figure 1. Residuals (milliarcseconds) of VLBI measurements of Mars-orbiting spacecraft on Goldstone-Madrid baseline

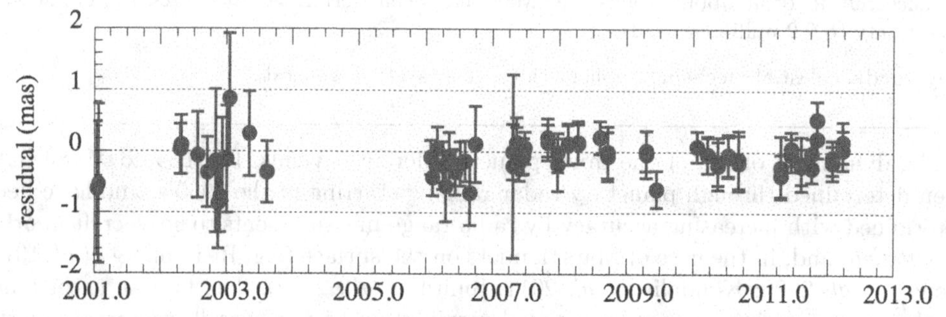

Figure 2. Residuals (milliarcseconds) of VLBI measurements of Mars-orbiting spacecraft on Goldstone-Canberra baseline

widths giving better SNR on more point-like radio sources. Digitization of radio signals at higher intermediate frequencies has reduced instrumental phase response errors. Measurements of dual-frequency radio signals from the Global Position System has improved the ionosphere calibration. As a result, measurements of the direction from Earth to Mars taken since early 2010 show residuals, after adjustment of the ephemerides of the inner planets, of less than 0.2 milliarcsecond, as shown in Figures 1 and 2. This improvement in accuracy enables increasingly precise targeting of planetary science missions.

Acknowledgements

The research described in this paper was carried out at the Jet Propulsion Laboratory, California Institute of Technology, under a contract with the National Aeronautics and Space Administration.

References

Fienga, A., *et al.* 2009, *A&A* 507, 1686

Folkner, W. M., *et al.* 1994, *A&A* 287, 279

Konopliv, A. S., *et al.* 2011, *Icarus* 211, 401

Ma, C., *et al.* 2009, *The Second Realization of the International Celestial Reference Frame by Very Long Baseline Interferometry* (IERS Technical Note No. 35, International Earth Rotation and Reference System Service

Newhall, X. X., *et al.* 1986, in: H. K. Eichhorn & R. J. Leacock (eds.), *Proc. 109th IAU Symposium*, Astrometric Techniques (Dordrecht: D. Reidel), p. 789

Reasenberg, R. D., *et al.* 1979, *ApJ* 234, L219

Smith, D. E., *et al.* 2012, *Science* 336, 214

Thornton, C. L. & Border, J. S. 2000, *Radiometric Tracking Techniques for Deep-Space Navigation* (Deep-Space Communications And Navigation Series), p. 47

Highlights of Astronomy, Volume 16
XXVIIIth IAU General Assembly, August 2012
T. Montmerle, ed.

doi:10.1017/S174392131400550X

EPM — High-Precision Planetary Ephemerides of IAA RAS for Scientific Research and Astronavigation on the Earth and in Space

E. V. Pitjeva†

Institute of Applied Astronomy RAS, 10, Kutuzova emb., St.-Petersburg, 191187, Russia

Abstract. The last version of the planet part of EPM's ephemerides of IAA RAS (EPM2011) is described briefly. At present EPM ephemerides are the basis for the Russian Astronomical and Nautical Astronomical Yearbooks and are used for scientific research.

Keywords. astrometry, ephemerides, reference systems, relativity, dark matter

---

The EPM ephemerides (**E**phemerides of **P**lanets and the **M**oon) of the IAA RAS originated in the 1970's and have been improved since that time. These ephemerides are based upon relativistic equations of motion for celestial bodies and light rays, as well as relativistic time scales. The numerical integration of the equations of motion of the celestial bodies has been performed in the Parameterized Post-Newtonian N-body metric for General Relativity in the TDB time scale. EPM ephemerides are computed in the barycentric coordinate frame of J2000.0 over a 400-year interval (1800–2200) using the program package ERA-7 (ERA: **E**phemeris **R**esearch in **A**stronomy) (Krasinsky, G. A., & Vasilyev, M. V., 1997, Proc. IAU Colloq. 165, 239–244).

For constructing planetary ephemerides using the best modern observations, it is necessary to take into account all influencing factors. The dynamical model of the planetary part of the EPM ephemerides includes mutual perturbations from the major planets, the Sun, the Moon; perturbations from 301 large asteroids and the 21 largest trans-neptunean objects (TNO); perturbations from a modeled massive asteroid ring with a uniform mass distribution; perturbations from a similar massive ring of TNO's in the ecliptic plane with a radius of 43 au; and perturbations due to the solar oblateness.

The EPM2011 ephemerides have been fitted to about 680 000 observations of different types, spanning 1913–2010, from classical meridian observations to modern planetary and spacecraft ranging. The ephemerides of the inner planets are based fully on radiotechnical observations (mostly, measurements of time delays). The ephemerides of the outer planets are mainly based on optical measurements taken since 1913. In addition to optical observations of these planets, positional observations of the satellites of the outer planets are used, as these observations are more precise and practically free from the phase effect, which is difficult to take into account. The reduction of the radar measurements includes all the relevant corrections. These observations have been reduced using relativistic corrections — the time delay of the propagation of radio signals in the gravitational fields of the Sun, Jupiter, Saturn (the Shapiro effect), and the reduction of observations from the coordinate time of the ephemerides to the proper time of the observer. In addition, the radar observations of Mercury, Venus, and Mars are corrected for their topography and for the extra delay of electromagnetic signals in the Earth's

† E-mail: evp@ipa.nw.ru

troposphere and in the solar corona. The main reductions of optical observations of planets involve the correction to the additional phase effect, the corrections for referring the observations to the ICRF reference frame, and the relativistic correction for light bending. For the transition from the time of observations (UTC) to Barycentric Dynamical Time (TDB), the time used in the construction of modern planet ephemerides, it is necessity to convert between Terrestrial Time (TT) and TDB. For the TT–TDB conversion, the differential equation from the paper by Klioner (Klioner, S. A., *et al.*, 2010, Proc. IAU Symp. 261, 112–123) has been used, and TT–TDB was obtained by numerically integrating the EPM.

For improvement of the planetary part of EPM2011, about 270 parameters are determined: orbital elements of the planets and the 18 satellites of the outer planets; the length of the astronomical unit or the value of the solar mass parameter; three angles of orientation with respect to the ICRF frame; parameters of the rotation of Mars and topography of the inner planets; masses of asteroids, the asteroid belt, and TNO's; as well as, some post-model parameters (β, γ, $\dot{G}/G$, $G\dot{M}_\odot/GM_\odot$, $\dot{\pi}_i$, $\dot{a}_i/a_i$, etc.).

EPM2011 has been oriented to the ICRF with an accuracy better than 1 mas by including into the total solution the 213 ICRF-based VLBI measurements of spacecraft taken from 1989–2010 near Venus, Mars, and Saturn (in mas): $\varepsilon_X = -0.000 \pm 0.042$, $\varepsilon_Y = -0.025 \pm 0.048$, $\varepsilon_Z = 0.004 \pm 0.028$. The present maximum errors of the coordinates of the Earth orbit determined from a comparison of the EPM2011 heliocentric X, Y, Z coordinates, velocities, and distances with those of DE424 in the 1950–2050 time interval are less than 250 m (coordinates), 0.05 mm s^{-1} (velocities), 6 m (distances).

At present EPM ephemerides are used for astronavigation on the Earth and space: they are the basis for the Russian Astronomical and Nautical Astronomical Yearbooks since 2006, and are used in the GLONASS and LUNA-RESOURCE programs and for various other investigations.

The masses of 21 largest asteroids have been estimated directly from spacecraft ranging, and masses of others in the group of 301 asteroids have been obtained from their known diameters and the estimated densities for the three taxonomic typess (C, S, M). From the estimates of the masses of all asteroids and the asteroid ring we obtain the value of the total mass of the asteroid main belt: $M_{\mathrm{belt}} = (12.3 \pm 2.1) \cdot 10^{-10} M_\odot$ ($\approx 3\, M_{\mathrm{Ceres}}$). From the mass estimate of the TNO ring and the known masses of the 21 largest TNO and Pluto we obtain the total mass of all TNO: $M_{\mathrm{TNO}} = 790 \cdot 10^{-10} M_\odot$ ($\approx 164\, M_{\mathrm{Ceres}}$ or $2\, M_{\mathrm{Moon}}$). These estimates illuminate the dynamics of the Solar System now and at its formation (Pitjeva, E. V., Proc. Inter. Conf. "Asteroid-Comet Hazard – 2009", 2010, 237-241).

New estimations of PPN parameters have also been obtained: β–1 $= -0.00002 \pm 0.00003$, γ–1 $= +0.00004 \pm 0.00006$. The good correspondence of the planetary motions and the propagation of light to the predictions of General Relativity narrows significantly the range of possibilities for alternative theories of gravitation (Pitjeva, E. V., Proc. IAU Symp. 261, 2010, 170-178).

It has been found that the solar mass parameter $GM_\odot$ decreases at a rate of $G\dot{M}_\odot/GM_\odot = (-5.04 \pm 4.14) \cdot 10^{-14}$ per year (3σ). The annual change of the gravitation constant G must fall within the interval $-4.2 \cdot 10^{-14} < \dot{G}/G < +7.5 \cdot 10^{-14}$ with a 95% probability (Pitjeva, E. V. & Pitjev, N. P., 2012, Solar System Research, 46, 78-87).

Using estimates of the additional perihelion advances obtained from observation for different planets, it has been found that the density of dark matter ρ_{dm} must be less than $1.1 \cdot 10^{-20}$ g cm^{-3} at the distance of Saturn's orbit, and the mass of dark matter inside Saturn's orbit must be less than $1.7 \cdot 10^{-10} M_\odot$, even if it concentrated toward the center (Pitjev, N. P. & Pitjeva, E. V., 2013, Astronomy Letters, 2, in press).

Highlights of Astronomy, Volume 16
XXVIIIth IAU General Assembly, August 2012
T. Montmerle, ed.

doi:10.1017/S1743921314005511

Connecting terrestrial to celestial reference frames

Z. Malkin†

Pulkovo Observatory and St. Petersburg State University, St. Petersburg, Russia

Abstract. In this paper we outline several problems related to the realization of the international celestial and terrestrial reference frames — the ICRF and ITRF — at the millimeter level of accuracy, with emphasis on ICRF issues. We consider here the current status of the ICRF, the interrelationship between the ICRF and ITRF, and considerations for future ICRF realizations.

Keywords. reference systems, astrometry

1. Introduction

There are several issues currently preventing the realization of the terrestrial and celestial reference systems (TRF and CRF, respectively) at the mm/μas level of accuracy:

- Insufficient number and non-optimal distribution of active and stable (systematically and physically) stations (VLBI and SLR) and suitable radio sources.
- Technological (precision) limitations of existing techniques;
- Incompleteness of the theory/models (e.g., reference systems definition, geophysics);
- Not fully understood and agreed-upon details of the processing strategy.

The latest ITRF realizations are derived from four space geodesy techniques: VLBI, GPS, SLR, and DORIS, whereas the ICRF is the result only of the global VLBI solution. The latter is tied to the ITRF datum using an arbitrary set of reference stations. VLBI also contributes, along with SLR, to the ITRF scale. Furthermore, all the techniques contribute to the positions and velocities of ITRF stations. As a consequence, we face systematic errors involving the connection between the ICRF and ITRF realizations, which cannot be fixed by datum correction during the current solution.

2. The connection between the ICRF and ITRF

A CRF realization obtained from a global VLBI solution depends on the tie to the ITRF. There are several problems that affect the VLBI-derived celestial reference frame, such as dependence on the ITRF datum, the set of reference stations used by the IVS Analysis Centers, and modeling of non-linear station motion.

The actual station movement cannot be represented to millimeter accuracy in the framework of the ITRF using a linear drift model with occasional jumps. Many stations show significant non-linear terms in their position time series. The most common are the exponential movement of stations due to post-seismic relaxation and seasonal signals; both cause a deviation from the ITRF model at the centimeter level. This has two consequences. First, using these stations may adversely affect the ICRF orientation. Secondly, their use causes errors in the daily/session Earth Orientation Parameter solution.

The impact of the ICRF on the ITRF seems not to be fully understood yet. Possible mechanisms include the nature of the global VLBI solution.

† E-mail: malkin@gao.spb.ru

3. The ICRF status and prospects

The current ICRF realization, ICRF2, created in 2009, provides much improvement over the first ICRF in terms of the total number of sources, the source position precision and accuracy, and the stability of the axes. However, there are severe problems preventing further ICRF improvement, especially with respect to systematic errors:

- Uncertainty in ICRS definition at the μas level;
- Uneven distribution of sources and source position errors over the sky (only about 2.5% of the observations are in the declination band -90° to -30°);
- Proper (physical) and apparent (instrumental and analysis) source motions;
- Source structure and its variability; and
- Dependence of source positions on wavelength (analogous to the color equation in optical astrometry), astronomical and geophysical models, observing network, and analysis strategy.

The ICRF history shows that successive versions were issued at intervals of about five years. It should therefore be reasonable to set a goal of completing the next ICRF versions in 2014 and 2019. It should be mentioned that the ICRF, ICRF-Ext. 1, ICRF-Ext. 2 represent, in fact, the same system based on the unchanged coordinates of the 212 defining sources. Preserving the positions of the defining (and, as a rule, most observed) sources, originally computed in 1995, may be the main reason for the ICRF systematic errors at a level of 0.2–0.25 mas. On the contrary, all the ICRF2 source positions were adjusted in a single global solution independently, and are tied to the ICRF only by the overall orientation (NNR constraint). One of several options would be to keep this strategy for the next realizations, and name them ICRF3 and ICRF4. In this regard, perhaps it would be better to use the term "core sources" instead of "defining sources", analogous to the ITRF core sources.

Two important motivations for the creating the next ICRF version in a few years are:

- It is expected that there will be about 9–10 million VLBI observations by 2014–2015, 1.5 times the number used for ICRF2. This will allow a substantial improvement in the position of many (now) poorly observed ICRF sources.
- Comparison of the ICRF2 with the latest CRF catalogues indicates that the ICRF2 may be affected by systematic errors at a level of 20–30 μas.

Of course, preparation for the next ICRF catalogue should involve an intensification of the observations of new and poorly observed southern sources.

4. Conclusion

In addition to the efforts of the IERS and technical services to improve the CRF and TRF, the following steps seem to be urgent.

1. For a timely ICRF update, it is important to establish continuous monitoring of the ICRF systematic errors by comparison to the latest CRF realizations.
2. We must identify new core ICRF sources in the southern hemisphere and start observations of them, along with sources that were poorly observed in the ICRF2. A practical way to do it would be to include more ICRF sources in the regular IVS sessions, such as R1 and R4. A trade-off between the practically insignificant degradation of the EOP precision and the long-term improvement of the ICRF can be easily found.
3. A method of describing and predicting the non-linear motion of TRF (and CRF?) objects at the μas/mm level of accuracy is needed.
4. An agreement on a standard set of VTRF (IVS TRF realization) core stations is needed.

Highlights of Astronomy, Volume 16
XXVIIIth IAU General Assembly, August 2012
T. Montmerle, ed.

doi:10.1017/S1743921314005523

SOFA—an IAU service fit for the future

C. Hohenkerk†, Chair, IAU SOFA Board
HM Nautical Almanac Office, UK Hydrographic Office, Taunton TA1 2DN, United Kingdom

Abstract. Standards of Fundamental Astronomy (SOFA) is an International Astronomical Union (IAU) service that provides accessible and authoritative algorithms and procedures that implement standard models used in fundamental astronomy. This paper summaries the current status, noting the changes during 2009-2012, and discusses issues that may arise in the future.

Keywords. time, reference systems, astrometry

1. Introduction

Standards of Fundamental Astronomy (SOFA) is an activity of IAU Division I. It was instigated in 1994 by the Working Group on Astronomical Standards, chaired by Toshio Fukushima. To accomplish its task SOFA is made up of three parts: a board of experts, a collection of software, and a web site—the SOFA Centre—that makes the software accessible to everybody. SOFA is described in *Scholarpedia* (6(1):11404).

2. The SOFA Board

In order to be authoritative and maintain IAU standards, SOFA has a Board of experts that produces and validates the material. Upon his retirement in March 2010, Patrick Wallace stepped down as chair (with acknowledgements and thanks) and Catherine Hohenkerk was elected as his replacement. The Board were saddened by the death of Anne-Marie Gontier, who provided a strong link with the Observatoire de Paris. Board members belong to most of the Commissions of IAU Division I as well as C5, Division XII and the Working Groups on FITS and Astronomical Data. Brian Luzum represents the International Earth Rotation and Reference Systems Service. SOFA welcomed new board members Nicole Capitaine, William Folkner, George Hobbs and webmaster Steve Bell.

3. The SOFA Center

The SOFA Centre—the public face of SOFA—is at **www.iausofa.org**. In January 2010, the web site was updated and streamlined and a lot of work was required on compliance, particularly to minimize accessibility issues for all types of users. New features include user registration, "News" items and a search facility. The terms and conditions have been relaxed so that it is not mandatory for installers to acknowledge SOFA, thus facilitating inclusion of the SOFA libraries in the Debian operating system release. However, we remind our users that their acknowledgement matters.

Over the last three years we have issued three major releases; introducing (1) transformations between geodetic and geocentric coordinates, (2) transformations between time scales, and (3) an update for the leap second at the end of June 2012, and minor release 9a in July. At the time of the 9th release on 2012 March 1 (release numbers unified) there were 619 registered users. On average, the web site receives over 1200 unique visits each

† E-mail: Catherine.Hohenkerk@UKHO.gov.uk

month. This average has risen from 789 in 2009 to 1244 now. There were 2836 downloads of a whole library for the 8th release (2010 December 31) and 1205 for the 9th release, split ∼40:60 for Fortran and ANSI C. The total size of the web site, including the archive, is ∼250 Mbytes.

4. The SOFA Software

The SOFA collection provides the building blocks that enable users to write their own applications using authoritative methods. The library of routines (Fortran and ANSI C) currently has 187 routines, of which 59 are *canonical* and support IAU resolutions. During the last triennium, 16 canonical routines were added to time scales, while five routines were added for geocentric/geodetic transformations. Routine `DTDB` has been degraded to "support" status. In total there were 23 additional routines.

Geodetic/Geocentric: In this category there are the routines `GD2GC`, `GC2GD`, `GD2GCE` and `GC2GDE` and the canonical routine `EFORM` that gives the parameters for the WGS 84, GRS 80 and WGS 72 ellipsoids.

Time Scales: SOFA recognizes seven time scales, namely TAI, UTC, UT1, TT, TCG, TDB and TCB. The strategy is to provide routines that link adjacent pairs of time scales (e.g. `TAIUTC` and `UTCTAI`). This is the simplest scheme, and gives the user the most flexibility. It also allows users to provide the supplementary quantities such as ΔT and UT1−UTC, which either cannot be predicted or for which there are model choices to make. The routines use SOFA's two-argument Julian date convention, which enables rounding errors to be minimised. The routines `DTF2D` and `D2DTF` handle the conversion between civil date and time and Julian date (or, in the case of UTC, quasi-JD) and vice versa. In the case of UTC this deals with leap seconds, when it is correct to report 60.... seconds. Importantly, the routines preserve precision by ensuring that the tiny differences are added to (or subtracted from) the smaller of the two date arguments.

Documentation & Tools: The manual contains the introductory comments of all the routines giving a complete specification. The "cookbooks" contain more descriptive material including examples. The latest is *SOFA Time Scale & Calendar Tools*, while the previous cookbook is *SOFA Tools for Earth Attitude*. SOFA provides the Unix `make` file to build the library as well as an improved validation program. This calls all the routines at least once and checks, to some specified precision, the calculated results against stored values, which have been produced independently using quadruple precision.

5. The Future

We are currently working on code for the effects of parallax, light-time, light-deflection and aberration, balancing the need for simple interfaces with accuracy requirements. Other coordinate transformations are being considered (e.g. equatorial/ecliptic/galactic).

Making SOFA available in other languages has been discussed. However, new code must be written to an agreed standard and the resulting output verified by the Board.

The position of SOFA within Division I will be reviewed during the 2012-2015 restructuring of the IAU. SOFA provides a service to all users requiring standard fundamental astronomical software. This activity is provided by a small and dedicated team.

Acknowledgements

SOFA thanks the UK Hydrographic Office for hosting the SOFA web site, acknowledges the work of Board members, and thanks their institutions, all of which are listed at www.iausofa.org/board.html.

Highlights of Astronomy, Volume 16
XXVIIIth IAU General Assembly, August 2012
T. Montmerle, ed.

doi:10.1017/S1743921314005535

The IERS Conventions (2010): reference systems and new models

B. Luzum[1] and G. Petit[2]†

[1]Earth Orientation Department, U.S. Naval Observatory, Washington, DC 20392, USA
[2]Bureau International des Poids et Mesures, Sèvres, France

Abstract. The IERS Conventions (2010) provides the international standard for models for use in the generation of celestial reference systems (CRS), terrestrial reference systems (TRS), and the Earth orientation parameters (EOPs) that relate the associated frames. Significant improvements over the previous IERS Conventions (2003) are outlined, and an overview of the latest adopted models and standards is shown. Finally, future plans for the Conventions are provided.

Keywords. standards, reference systems, time, Earth orientation, astrometry

1. Introduction

The IERS Conventions (2010) (Petit & Luzum 2010) is a set of constants, models, and algorithms, assembled and verified by subject matter experts, to be used in the analysis of Earth orientation and reference systems data. The Conventions strives to be internally consistent as well as consistent with IERS products including the ITRF and ICRF as well as the EOPs that link the two reference frames. Whenever possible, the Conventions are consistent with international standards. Because of the efforts to maintain internal and external consistency, implementation of the IERS Conventions in reduction and analysis software helps to minimize systematic errors in the resulting EOP and reference frame data.

2. Content

The IERS Conventions (2010) contains the latest adopted models and standards. Topics included are numerical standards, reference systems, transformations between systems, geopotential, displacement of reference points, tidal variations, atmospheric propagation delays, and relativistic models. The recommended models are reviewed continually, and suggested improvements are evaluated by an Advisory Board. Based on their recommendation, modifications are made to the content of the IERS Conventions Updates and ultimately result in a new registered edition.

3. Improvements in IERS Conventions (2010)

Several improvements were made to the IERS Conventions (2010). The most significant changes to the Conventions came with the adoption of the ITRF2008 (Altamimi *et al.* 2011), ICRF2 (Fey *et al.* 2009), and the IAU2000/2006 precession-nutation model (Mathews *et al.* 2002, Capitaine *et al.* 2003). Additional improvements in modeling were

† E-mail: brian.luzum@usno.navy.mil or gpetit@bipm.org

achieved through the adoption of a new conventional geopotential model and new tropospheric propagation model/zenith mapping function.

To reduce confusion, classification of models were introduced and the criteria for choosing models was stated explcitly. Tables with the magnitude of the modeled effects were provided and better consistency throughout the document was enforced. Significant efforts were made to improve the standardization of software including the creation of a new software template, improved documentation, improved robustness of the code, and the inclusion of test cases. In assembling the software, efforts were made to cooperate closely with the IAU Standards of Fundamental Astronomy (SOFA) software.

4. Future changes

To keep the IERS Conventions current, additional changes are being considered. Several models can be improved, and efforts are underway to incorporate, for example, an improved global pressure and temperature model, diurnal and semidiurnal EOP variation model, conventional mean pole, ionospheric correction to ray bending, and a geocenter model. The sections on displacement of reference points and on ranging techniques will be expanded. Also, non-tidal loading and a SINEX format for modeling will be investigated.

Acknowledgements

We would like to acknowledge the extensive work of the Advisory Board, led by Jim Ray, and all of the contributors who have helped make the IERS Conventions (2010) the high-quality document that it is.

References

Altamimi, Z., Collilieux, X., & Métivier, L. 2011, *J. Geod.* 85, 457

Capitaine, N., Wallace, P. T., & Chapront, J. 2003, *Astron. Astrophys.* 412, 567

Fey, A. L., Gordon, D., & Jacobs, C. S. 2009, *IERS Technical Note* 35, 204 pp

Mathews, P. M., Herring, T. A., & Buffett, B. A. 2002, *J. Geophys. Res.* 107, doi: 10.1029/2001JB000390

Petit, G. & Luzum, B. 2010, *IERS Technical Note* 36, 179 pp

Highlights of Astronomy, Volume 16
XXVIIIth IAU General Assembly, August 2012
T. Montmerle, ed.

doi:10.1017/S1743921315001581

Preface: Special Session 1 Origin and Complexity of Massive Star Clusters

In the last few years, numerous observational studies have provided us with strong evidence that star clusters host multiple stellar generations and subverted the standard view according to which these systems were the prototype of 'simple stellar populations'.

These findings have raised a number of fundamental questions concerning the origin and evolution of star clusters, the nature of the first generation of stars which must have provided the gas for the formation of subsequent stellar generations, the link between star clusters and other massive stellar systems (such as nuclear star clusters and dwarf galaxies), and the relationship between star clusters and field stellar populations.

The expertise of observational and theoretical researchers working on stellar evolution, nuclear astrophysics, hydrodynamics, stellar dynamics, high precision astrometric, photometric, and spectroscopic observations of Galactic and extragalactic globular clusters, massive young star clusters in external galaxies, nucleated galaxies and dwarf galaxies are absolutely crucial in any attempt to make progress towards an answer to these complex questions.

This Special Session addressed these issues and, by bringing together scientists spanning a broad range of expertise, allowed the exchange of ideas among scientists working in different fields which do not often overlap and who do not usually communicate with each other. The excellent oral and poster presentations allowed the numerous participants to have a global view of all these topics and to better appreciate the network of constraints and implications coming from observations and theoretical studies of these issues.

There was a lot of interest in this Special Session and all the sessions were well attended. We would like to thank the IAU for supporting this Special Session and all the members of the Scientific Organizing Committee (A. Aparicio, B. Barbuy, K. Bekki, T. Boeker, C. Charbonnel, C. Clarke, F. D'Antona, L. Deng, B. Elmegreen, R. Gratton, Y.W. Lee, S. Majewski, G. Piotto (co-Chair), E. Tolstoy, E. Vesperini (co-Chair), H. Zinnecker).

Giampaolo Piotto and Enrico Vesperini, co-chairs SOC

Highlights of Astronomy, Volume 16
XXVIIIth IAU General Assembly, August 2012
T. Montmerle, ed.

doi:10.1017/S1743921314005547

Spectroscopic evidence of multiple populations in globular clusters

R. Gratton[1], S. Lucatello[1], E. Carretta[2] and A. Bragaglia[2]
[1]INAF-Osservatorio Astronomico di Padova
[2]INAF-Osservatorio Astronomico di Bologna

Abstract. We review spectroscopic evidence of multiple stellar populations in globular clusters. First, we lay down the basic data: the C-N, Na-O, Mg-Al anti-correlations among red giants and main sequence stars, and discuss how they appear to be general properties of globular clusters, in spite of cluster-to-cluster differences. We will then describe what is currently known about He from spectroscopy. We will then present the implications and current observations for the interpretation of the horizontal branches, showing that the multiple population phenomenon is strongly related to the distribution of stars along them. We will briefly mention the spectroscopic evidence related to some less understood cases, like the clusters with multiple subgiant branches. Finally, we summarize the relation between multiple populations and general properties for globular clusters, and their implications for the formation scenario.

Keywords. (Galaxy:) globular clusters: general, stars: abundances, Galaxy: halo

---

The last ten years have seen a revolution in our concept of globular clusters (GCs) thanks to data coming from new efficient multi-object spectrographs on 8-m telescopes (e.g. Carretta *et al.* 2009a, b) and the exquisite photometric precision of HST (e.g. Bedin *et al.* 2004). These studies clarified the meaning of large star-to-star abundance variations for several light elements known to be present in GCs for almost forty years (see Gratton *et al.* 2004 and 2012 for recent reviews) and revealed multiple sequences on the red giant branch (RGB), subgiant branch (SGB), and main sequence (MS) in several GCs. Regarded for a long time as intriguing abundance anomalies restricted to some cluster stars, the peculiar chemical patterns only recently were explicitly understood as a universal phenomenon in GCs related to their very same nature/origin (Carretta 2006; Carretta *et al.* 2006, 2009a, b, 2010). The observational pattern is quite well assessed (see the reviews by Gratton *et al.* 2004, 2012, and Piotto 2009), thanks to several important milestones:

(i) Abundance variations of the heavier species (O-Na and Mg-Al anti-correlations) are restricted to the denser cluster environment. Evolutionary changes along the RGB additionally affect other light elements (Li, C, and N: Charbonnel *et al.* 1998; Gratton *et al.* 2000).

(ii) This pattern is established in $p-$capture reactions of the CNO, NeNa and MgAl chains in high temperature H-burning (Denisenkov & Denisenkova 1989).

(iii) The variations are found also among unevolved stars currently on the MS of GCs (Briley *et al.* 1996; Gratton *et al.*2001). This calls for a class of now-extinct stars, more massive than the low-mass ones presently evolving in GCs, as the site for the nucleosynthesis.

(iv) The observed variations in the abundances of $p-$capture elements are connected to that of He, i.e., stars that are Na-rich and O-poor are also He-rich, although the relation between these abundance variations may be complicated. Multiple MSs, attributed to populations with different He fraction Y, have been found in some GCs (Bedin *et al.*

2004, Piotto *et al.* 2007). There are indications that Y differences can be traced also on the RGB (Carretta *et al.* 2007, Bragaglia *et al.* 2010, Pasquini *et al.* 2011), and contribute to explaining the horizontal branch (HB) second parameter problem (Gratton *et al.* 2010).

In summary, in contrast with the common assumption, GCs are not simple stellar populations: they harbour various stellar generations, distinct by their chemistry. These populations may be separated using the patterns of anti-correlated Na-O, Mg-Al or photometric methods. They provide insights into the early phases of GCs, since the time scale for the release of matter processed by H-burning at high temperatures is less than 10^8 year. To decipher the relevant information, large and homogeneous datasets are needed, like the one recently gathered by Carretta *et al.* (2009a, 2009b), from which Carretta *et al.* (2010) were able to place the properties of different stellar generations in GCs and global cluster parameters into a general framework for the formation of GCs. The Na-O anti-correlation may be used to separate GCs (hosting multiple stellar populations) from open clusters (OCs, where a single stellar population is present), and first and second generation (FG and SG) stars within each GC. The emerging scenario for the formation of GCs includes various phases: a large episode of star formation (the FG) and then the formation of the current GC (mainly made of SG stars) within a cooling flow formed by the slow winds of a fraction of this FG (D'Ercole et al. 2008). Some stars of the FG remain in the cluster. Since the ejecta of only a small fraction of the FG stars are used to form the SG, a large fraction of the FG must have been lost by the proto-GCs and may represent the main component of halo field stars (see also Martell & Grebel 2010; Vesperini *et al.* 2010; Schaerer & Charbonnel 2011). This links directly the formation of GC to that of galaxies.

References

Bedin, L. R., *et al.*, 2004, *ApJ*, 605, L125
Bragaglia, A., *et al.*, 2010, *A&A*, 519, 60
Briley, M., *et al.*, 1996, *Nature*, 383, 604
Charbonnel, C., *et al.*, 1998, *A&A*, 332, 204
Carretta E., 2006, *AJ*, 131, 1766
Carretta, E., *et al.*, 2006, *A&A*, 450, 523
Carretta, E., *et al.*, 2007, *A&A*, 464, 927
Carretta, E., *et al.*, 2009a, *A&A*, 505, 117
Carretta, E., *et al.*, 2009b, *A&A*, 505, 139
Carretta, E., *et al.*2010, *A&A*, 516, 55
Denisenkov, P. A. & Denisenkova, S. N. 1989, A.Tsir., 1538, 11
D'Ercole, A., *et al.* 2008, *MNRAS*, 391, 825
Gratton, R. G., *et al.*, 2000, *A&A*, 358, 671
Gratton, R. G., *et al.*, 2001, *A&A*, 369, 87
Gratton, R. G., *et al.*, 2004, *ARA&A*, 42, 38
Gratton, R. G., *et al.*, 2010, *A&A*, 517, 8
Gratton, R. G., *et al.*, 2012, *A&ARv*, 20, 50
Martell, S. L. & Grebel, E. K. 2010, *A&A*, 519, 14
Pasquini, L., *et al.*, 2011, *A&A*, 531, 35
Piotto, G. 2009, in *IAUS* 258, 233
Piotto, G., *et al.*, 2007, *ApJ*, 661, L53
Schaerer, D. & Charbonnel, C., 2011, *MNRAS* 413, 2297
Vesperini, E., *et al.*, 2010, *ApJ*, 718, L112

Highlights of Astronomy, Volume 16
XXVIIIth IAU General Assembly, August 2012
T. Montmerle, ed.

doi:10.1017/S1743921314005559

Multiple Populations in Globular Clusters – The Spectroscopic View

Judith G. Cohen†

Abstract. I review the evidence supporting and characterizing multiple populations within globular clusters (GCs) based on spectroscopy, i.e. on abundance variations within the stellar population of an individual GC, which dates back to almost 40 years ago. I discuss some of my recent work in this area.

Keywords. globular clusters: general, globular clusters: individual (M13, M15, M92, NGC 2419), stars: abundances

---

I review some of the evidence accumulated over the past 20 years regarding abundance variations within Galactic globular clusters (GCs) as inferred from spectroscopy, beginning with early work on ω Cen dating back almost 40 years (see, e.g. Freeman & Rodgers 1975 and Norris, Freeman & Mighell 1996), and continuing with the extensive surveys of C and N abundances in GCs using molecular band spectral syntheses as illustrated in Briley, Cohen & Stetson (2004) and Cohen, Briley & Stetson (2005). Averaging out the significant star-to-star scatter for stars of similar luminosity, the depletion of C from below the main sequence turnoff to the RGB tip for two GCs, M13 and M15, is filled out in detail and modeled in Briley, Cohen & Harbeck (2008), where the depletion rates as a star evolves up the RGB towards the tip are estimated, and increasing rate of depletion which occurs after crossing the RGB bump luminosity is documented (see Fig. 1). The recent major survey of the Na/O anti-correlation in GCs (see Carretta *et al.* 2009 and related papers) demonstrates yet again with very large samples and exquisite data that this correlation is ubiquitous and reminds us yet again that most, but not all, GCs show no internal variation in [Fe/H].

I also present an update on the variation of r-process elements within GCs, first found in M15 by Sneden *et al.* (1997) and not seen in any other GC. My unpublished data on r-process abundances demonstrates that the variation of abundances of the heavy neutron capture elements among giants in M15 has a range of about a factor of 5 and within each star closely follows the r-process distribution based on a detection of typically 6 elements from Ba to Dy. Although Roederer & Sneden (2011) suggested that the same phenomena also occurs in M92, a GC of comparable low metallicity, I (Cohen 2011) demonstrated there is no such variation of the r-process elements within M92.

The bulk of the talk focuses on the extreme outer halo globular cluster NGC 2419. After our initial efforts, described in Cohen *et al.* (2010) and Cohen, Huang & Kirby (2011), we demonstrated in Cohen & Kirby (2012) that it harbors a population of stars, comprising about one third of its mass, that is depleted in Mg by a factor of 8 and enhanced in K by a factor of 6 with respect to the Mg-normal population, with the majority of the cluster stars appearing normal. The Mg-poor giants show abundances of K and Sc that are strongly anti-correlated with Mg, and some other elements (Si and Ca among others) are weakly anti-correlated with Mg. But the abundances of Fe-peak elements except Sc show no star-to-star variation. Although Mucciarelli *et al.* (2012) suggest that all the

† Present address: Mail Code 249-17, California Insititute of Technology, Pasadena, Ca. USA

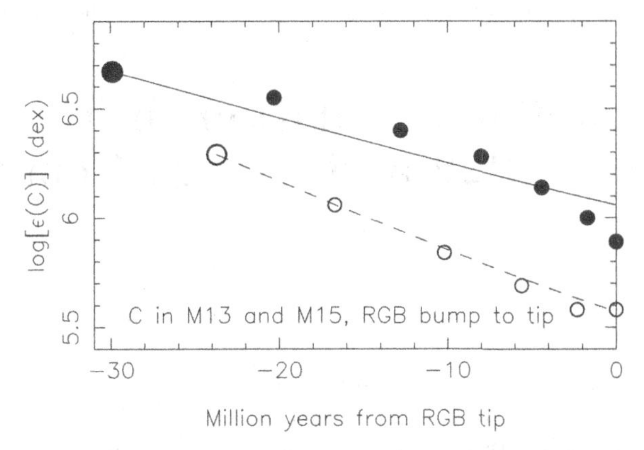

Figure 1. Carbon depletion as a function of time is shown for M13 (filled circles) and for M15 (open circles). The RGB tip defines $t = 0$, and the large point on the left end of each track is the RGB bump. (From Briley, Cohen & Harbeck 2008).

abundance anomalies except that of Mg are not real but rather the result of a reduction in P_e in the atmosphere due to the strong depletion of Mg, we believe they are all real, although we know of no nucleosynthetic source that satisfactorily explains them. Even the most extreme AGB hot bottom burning nucleosynthesis calculations published to date such as those of Ventura & D'Antona (2011), who study 8 $M_{\odot}$ AGB stars, fail to reproduce the chemical inventory of the peculiar population in NGC 2419.

Acknowledgements

I would like to acknowledge my collaborators along the way, Michael Briley and Peter Stetson for the C studies and Evan Kirby for NGC 2419, and NSF award AST-0908139 for partial financial support.

References

Briley, M. M., Cohen, J. G., & Stetson, P. G. 2004, *AJ* 127, 1579
Briley, M. M., Cohen, J. G., & Harbeck, D. 2008, never published
Carretta, E., Bragaglia, A., Gratton, R. & Lucatello, S. 2009 *A&A* 505, 117
Cohen, J. G., Briley, M. M., & Stetson, P. B. 2005, *AJ* 130, 1177
Cohen, J. G., Kirby, E. N., Simon, J., & Geha, M. 2010, *ApJ* 725, 288
Cohen, J. G. 2011 *ApJL* 740, L38
Cohen, J. G., Huang, W., & Kirby, E. N. 2011, *ApJ* 740, 60
Cohen, J. G. & Kirby, E. N. 2012, *ApJ* in press
Freeman, K. C. & Rodgers, A. W. 1975 *ApJ* 201, L71
Mucciarelli, A., Bellazzini, M., Ibata, R., Merle, T. & Chapman, S. C. *MNRAS* in press
Norris, J. E., Freeman, K. C. & Mighell, K. J. 1996 *ApJ* 462, 241
Roederer, I. U. & Sneden, C. 2011 *AJ* 142, A22
Sneden, C., *et al.* 1997 *AJ* 114, 1964
Ventura, P. & D'Antona, F. 2011, *MNRAS* 410, 2760

Highlights of Astronomy, Volume 16
XXVIIIth IAU General Assembly, August 2012
T. Montmerle, ed.

doi:10.1017/S1743921314005560

Multiple stellar populations in the massive clusters M22 and Omega Centauri

A. F. Marino

Max-Planck-Institut für Astrophysik, Karl-Schwarzschild-Str. 1, 85748 Garching, Germany
email: amarino@MPA-Garching.MPG.DE

Abstract. An intriguing discovery in the field of multiple stellar populations in globular clusters is that some of them show internal variations in the bulk of the heavy-element content. I summarize the chemical properties of one of these clusters, M22, in comparison with the most extreme ω Centauri, underlying the analogies and differences between the two objects.

Keywords. stars: abundances, globular clusters: individual (M22, ω Centauri).

---

1. Introduction

Since the '80s we know that light elements show peculiar patterns in globular clusters (GCs). On the other hand, variations in heavier elements were considered to be a trait of more massive systems capable to retain SN fast ejecta. In this respect the most massive Galactic GC ω Cen was always considered a peculiarity. In fact, to account for its well known huge metallicity variations, it has even been suggested that ω Cen is the remnant of a tidally disrupted dwarf galaxy rather than a *real* GC.

Surprisingly, recent discoveries have revealed that some GCs, besides ω Cen, have variations also in the bulk heavy element content, in analogy with more massive systems. Differently from the simple *normal GCs* that do not show evidence for SN-based self enrichment, in these objects successive generation(s) may need to be invoked, with SNe also playing a role in the pollution of intra-cluster medium.

Among these *anomalous* GCs, M22 is the first discovered (Marino *et al.* 2009, hereafter M09; Da Costa *et al.* 2009) and surely the one whose spectroscopic features more closely resemble ω Cen (Da Costa & Marino 2011).

2. M22 versus ω Centauri

The most striking similarity between M22 and ω Cen is the internal variation in the overall metallicity. In M22, M09 and Marino *et al.* (2011a, hereafter M11a) obtain a total metallicity spread of more than a factor of two: $-1.97 \leqslant$ [Fe/H] $\leqslant -1.57$, a range that cannot be explained by observational uncertainties. Note that, although metallicity variations are present in both clusters, in ω Cen the range in [Fe/H] is more than a factor of 20 larger, with stars from [Fe/H] ≈ -1.90 to [Fe/H] ≈ -0.60.

In addition to this, the abundance distribution for elements mainly produced in the slow (s) neutron-capture processes is clearly bimodal in M22 (M09, M11a). The two stellar groups with different s element content (s-rich and s-poor groups) are also characterized by: *(ii)* a mean different metallicity (M09); *(ii)* a mean different C+N+O content (M11a, Alves-Brito *et al.* 2012); *(iii)* in both groups internal variations in elements involved in the high temperature H-burning are present (M11a), so that each group individually traces the (anti)correlations in light elements found in *normal* GCs; *(iv)* on

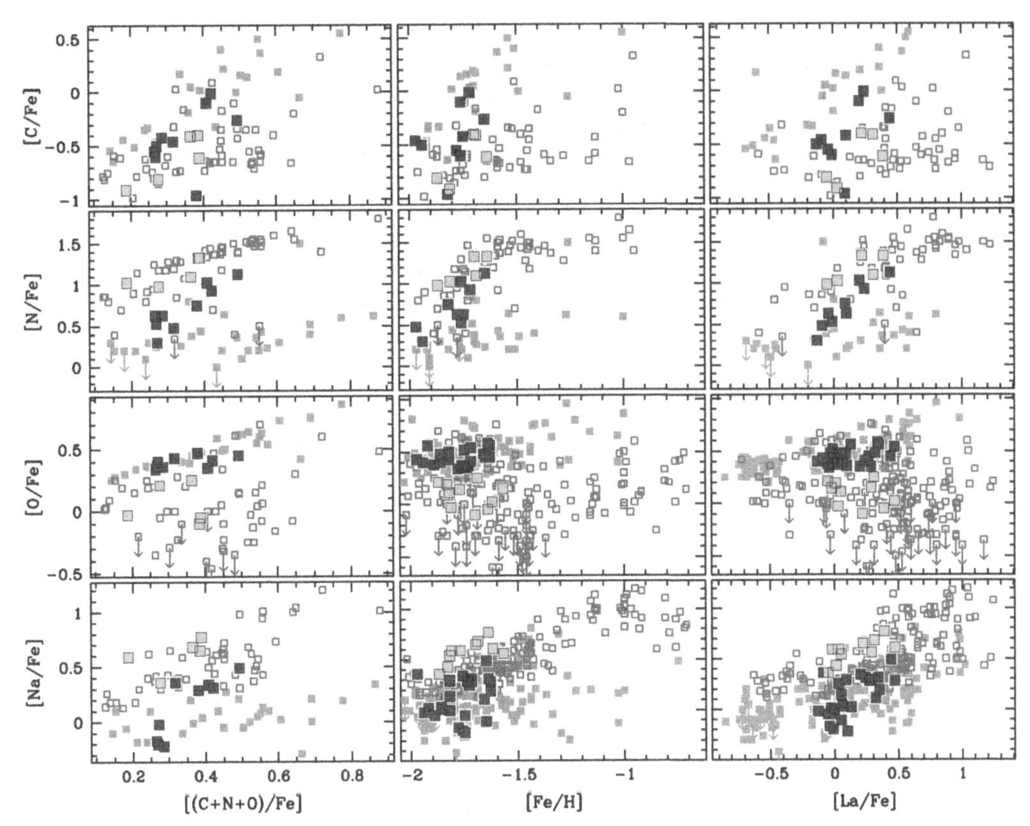

Figure 1. C, N, O Na relative to Fe as a function of [CNO/Fe], [Fe/H], and [La/Fe] for M22 and ω Cen stars. Dark-green and yellow points are the Na-poor and Na-rich stars in M22, and light-green and magenta points are the Na-poor and Na-rich stars in ω Cen, respectively.

the photometric side, M22 shows a split sub-giant branch (Piotto *et al.* 2012), whose sequences correspond to the two *s* groups (Marino *et al.* 2012a).

The properties observed in M22 are also present in ω Cen, but again, in this latter they are much more extreme. In Figure 1 a collection of chemical abundances for M22 (from M09 and M11a) and ω Cen (Marino *et al.* 2011b, 2012b, hereafter M11b, M12b) is shown. In both clusters, stars have been divided into two groups on the basis of their position along the O-Na anti-correlation (see M12b for details). Note that the separation of different stellar groups on the basis of their light-element abundances is just a possibility. Alternatively, the separation in stellar groups could be explored on the basis of metallicity or *s* content (as in M11a, and M11b). An inspection of Figure 1 immediately gives an idea of how more extreme are the chemical variations in ω Cen, and, at the same time, how similar are these objects in terms of chemical patterns: *(i)* all the *p*-capture elements (on the y-axis) have similar trends as a function of the CNO, Fe and *s* element La; *(ii)* Na-poor and Na-rich stars separately show similar patterns in the two clusters.

The most natural development of these findings is to understand where these objects formed. The analogies with ω Cen, considered the possible relict of a dwarf galaxy, suggest the fascinating idea that also M22 could be the surviving nuclei of more massive system.

References

Alves-Brito, A., Yong, D., Meléndez, J., Vásquez, S., & Karakas, A. I., 2012, *A&A*, 540, A3
Da Costa, G. S., Held, E. V., Saviane, I., & Gullieuszik, M., 2009, *ApJ*, 705, 1481
Da Costa, G. S., & Marino, A. F., 2011, *PASA*, 28, 28
Marino, A. F., Milone, A. P., Piotto, G., *et al.*, 2009, *A&A*, 505, 1099
Marino, A. F., Sneden, C., Kraft, R. P., *et al.*, 2011a, *A&A*, 532, A8
Marino, A. F., Milone, A. P., Piotto, G., *et al.*, 2011b, *ApJ*, 731, 64
Marino, A. F., Milone, A. P., Sneden, C., *et al.* 2012a, *A&A*, 541, A15
Marino, A. F., Milone, A. P., Piotto, G., *et al.* 2012b, *ApJ*, 746, 14
Piotto, G., Milone, A. P., Anderson, J., *et al.*, 2012, arXiv:1208.1873

Highlights of Astronomy, Volume 16
XXVIIIth IAU General Assembly, August 2012
T. Montmerle, ed.

doi:10.1017/S1743921314005572

Precision Chemical Abundance Measurements

David Yong[1], Frank Grundahl[2], Jorge Meléndez[3] and John E. Norris[1]

[1]Research School of Astronomy and Astrophysics, The Australian National University, Weston, ACT 2611, Australia
email: yong@mso.anu.edu.au

[2]Department of Physics and Astronomy, Aarhus University, Ny Munkegade, 8000, Aarhus C, Denmark,

[3]Departamento de Astronomia do IAG/USP, Universidade de Sao Paulo, Rua do Matao 1226, Sao Paulo, 05508-900, SP, Brasil

Abstract. This talk covers preliminary work in which we apply a strictly differential line-by-line chemical abundance analysis to high quality UVES spectra of the globular cluster NGC 6752. We achieve extremely high precision in the measurement of relative abundance ratios. Our results indicate that the observed abundance dispersion exceeds the measurement uncertainties and that many pairs of elements show significant correlations when plotting [X1/H] vs. [X2/H]. Our tentative conclusions are that either NGC 6752 is not chemically homogeneous at the $\simeq$0.03 dex level or the abundance variations and correlations signify star-to-star He abundance variations.

Keywords. Stars: abundances, globular clusters: individual: NGC 6752.

1. Introduction

The Milky Way globular clusters show enormous complexity in colour-magnitude diagrams (Piotto 2009) as well as a large diversity in the chemical compositions of stars within a given cluster (Gratton *et al.* 2012). The key question we seek to answer in the present work is "how low are the abundance dispersions for 'well-behaved' elements like Ti and Ni in the 'least complex' globular clusters"? The answer to this question may contain important new clues for understanding the formation and evolution of globular clusters.

A number of recent papers have conducted a so-called "line-by-line strictly differential chemical abundance analysis" (e.g., Meléndez *et al.* 2009, Nissen & Schuster 2010). These works have achieved very high precision in relative abundance measurements, in some cases as low as σ[X/H] = 0.01 dex. In this work, we apply the same analysis techniques to the globular cluster NGC 6752.

2. Analysis and Preliminary Results

We adopt the procedure outlined in Meléndez *et al.* (2012) in the analysis of high quality UVES spectra of NGC 6752 stars (i) near the luminosity bump (these data are described in Grundahl *et al.* 2002) and (ii) near the tip of the red giant branch (these data are described in Yong *et al.* 2003). That is, we conducted a line-by-line strictly differential chemical abundance study of the luminosity bump sample and a similar independent analysis of the red giant branch tip sample. The elements we studied included the light element Na, various alpha elements, Fe-peak elements and neutron-capture elements. Our analysis also includes neutral and ionised species of a given element (e.g., TiI and TiII as well as CrI and CrII). The main results can be summarised as follows:

1. Taking into account all error terms as described in Johnson (2002), we find that the abundance errors for a given element in a given star can be as low as ∼0.01 dex.
2. For most elements, the observed abundance dispersion in either the luminosity bump sample or the red giant branch tip sample exceeds the typical measurement uncertainty for a given sample. In some cases, the difference exceeds a factor of three.
3. When plotting pairs of elements [X1/H] vs. [X2/H] (e.g., [Ca/H] vs. [Si/H]), in the majority of cases there are positive correlations between the abundance ratios. Furthermore, in many cases the correlations are highly significant, 5-σ or higher.

If these results are real, then NGC 6752 is not chemically homogeneous as the $\simeq$0.03 dex level, for all elements studied in the present work. Star-to-star differences in He abundance may be another possible explanation for these results. We are continuing this work in order to explore these, and other, possibilities.

Acknowledgements

We would like to thank the Scientific Organizing Committee for their efforts in arranging this successful meeting.

References

Gratton, R. G., Carretta, E., & Bragaglia, A. 2012, *A&ARv*, 20, 50
Grundahl, F., Briley, M., Nissen, P. E., & Feltzing, S. 2002, *A&A*, 385, L14
Johnson, J. A. 2002, *ApJS*, 139, 219
Meléndez, J., Asplund, M., Gustafsson, B., & Yong, D. 2009, *ApJ*, 704, L66
Meléndez, J., Bergemann, M., Cohen, J. G., *et al.* 2012, *A&A*, 543, A29
Nissen, P. E. & Schuster, W. J. 2010, *A&A*, 511, L10
Piotto, G. 2009, *IAU Symposium*, 258, 233
Yong, D., Grundahl, F., Lambert, D. L., Nissen, P. E., & Shetrone, M. D. 2003, *A&A*, 402, 985

Highlights of Astronomy, Volume 16
XXVIIIth IAU General Assembly, August 2012
T. Montmerle, ed.

doi:10.1017/S1743921314005584

Photometric Techniques for Exploring Multiple Populations in Clusters

Jay Anderson

Space Telescope Science Institute, 3700 San Martin Drive, Baltimore, MD 21218 USA

Abstract. The advent of the Hubble Space Telescope (HST) and the development of new photometric algorithms that take advantage of its stable observing platform above the atmosphere have allowed us to study the populations in globular clusters with very high precision.

Keywords. (Galaxy:) globular clusters: general, ultraviolet: stars, methods: data analysis

---

1. Introduction

Globular clusters have been studied intensely for decades both photometrically and spectroscopically, yet only in the past few years have we realized their true multi-population nature. Spectroscopy has long hinted at non-singular abundances, but it was not until the high-precision photometry of unevolved main-sequence stars made possible by the Hubble Space Telescope (HST) that it was clear that the abundance variations indicated multiple (or gradual) formation episodes. This brief article will highlight the properties of HST that make these studies possible and the approaches that thus far have worked best to highlight the non-singular populations.

2. The promise and challenges of HST

Ground-based photometry has the advantages of well-sampled near-Gaussian PSFs and a large field of view (FOV), but must deal with variable seeing and resolution of $\sim 1''$ or worse. This makes crowding a significant issue for studying the faint stars near the cluster centers. HST has the benefits of being extremely stable, and also provides high-resolution access to both the UV and IR and the large color baselines these filter systems make possible. The downside of HST is that we must deal with (1) an undersampled PSF, (2) large field distortion, and (3) a small FOV. Thankfully these challenging issues have been addressed with focused efforts at characterizing the detector and optimizing the approaches to photometry.

The HST PSF varies by $\pm 20\%$ across the FOV of most detectors. Thankfully, HST's exquisite stability allows us to characterize this variation empirically once for all and then use it for all future analyses. Similarly, even though the PSF is undersampled, once we construct a model for the PSF from a well-observed starfield, we can use the PSF model to measure future observations. HST's detectors have a lot of distortion, but once the mapping from detector to sky coordinates has been done once, it can later be used to allow very precise differential astrometry.

Although HST's FOV is very small, it is compensated somewhat by its increased resolution. HST can study stars into the very centers of almost all clusters, so even within $3' \times 3'$ we can measure over a hundred thousand stars well. This has been a critical aspect of the multi-population study, since the second generation tends to be more concentrated than the first. HST's small FOV does, though, underline how important it is to combine space-based surveys of the central regions with ground-based surveys of the outer regions.

3. New approaches

In addition to photometric techniques that have been optimized to the undersampled, yet stable, HST detectors, several additional approaches have been necessary to tease out the fine-scale structure present in globular-cluster CMDs. Recent papers by Antonino Milone exhibit all of the following.

Selective finding. Not all stars can be measured with the same accuracy. Some have nearby neighbors that prevent a good measurement. If we focus our efforts on the stars that are isolated, then we can improve our overall photometry. It is certainly true that bright stars are more likely to be considered "isolated" than faint stars, but the fine structure we are interested in concerns variations of color at the 1% to 5% level; these small variations introduce no bias in our finding stars on (say) the blue or red parts of the various branches. This selective-finding approach works for studying *horizontal* structure, but not *vertical* structure (luminosity-function-type analyisis).

Corrections for Differential Reddening. The focus above is on measuring accurate colors for stars in hopes of measuring fine structure of 0.01 magnitude in color across the main sequence (or other population sequences). Although globular clusters do not have much gas or dust, the lines of sight to these clusters often do, and differential reddening can introduce field-dependent shifts of 0.05 magnitude or more in color. The scale over which the reddening is coherent tends to be an arcminute or so. Typically we have hundreds of HST stars within each coherent patch, so it is possible to determine a fiducial sequence line within each region, then determine how to shift the sequence for each region along the reddening line to account for the variations in reddening.

Multiple Independent Observations. Even with the best possible static empirical models of the PSF, short-term and long-term variations in focus can introduce some small perturbations in the PSF that affect photometry at the 1% level. It can be hard to confidently average down multiple exposures to better than 1% if we do not have large dithers to allow us to assess such systematic spatial issues. Furthermore, we often must get photometry from the archive and must live with the limitations of the original observing plan. In this case, it can help to have more than one pair of filters, or multiple pointings with the same filter, so that we can assess whether a star on the blue side of the MS in one observation is also on the blue side in another.

Filter optimization. It was first thought that HST's WFC3 would help the multiple population programs because of its larger color baseline, providing coverage from the UV to the IR. But it turns out that there is a particular combination of three colors with HST that allow multiple populations to be identified trivially in color-color diagrams. Recent papers by Milone show what can be done with F275W, F336W, and a visible filter (say, F814W). This separation works well on *all* branches: MS, SGB, RGB, and HB. Once populations are separated in one color system, we can use other color systems to explore extremely fine structure in the CMD, structure that is well below the typical 1% level needed to make clean separations. Ground-based studies have investigated their own filter optimization by exploring the synergy of spectroscopy and photometry to determine which filters best allow known populations to be separated.

4. What's next?

The obvious next thing to do is to repeat the Sarajedini GC Treasury clusters with the optimal HST filter system; this will allow all clusters to have their multiple populations assessed. This will be combined with spectroscopy to give us details about each population.

Highlights of Astronomy, Volume 16
XXVIIIth IAU General Assembly, August 2012
T. Montmerle, ed.

doi:10.1017/S1743921314005596

Multiple Sequences of M-dwarfs in NGC 2808 and ω Centauri.

A. P. Milone

Instituto de Astrofisica de Canarias and Department of Astrophysics, University of La Laguna, E-38200 La Laguna, Tenerife, Canary Islands, Spain;
email: milone@iac.es

Abstract. The infrared channel of the Wide-Field Camera 3 on the *Hubble Space Telescope* revealed multiple main sequences of very low-mass stars in the globular clusters NGC 2808 and ω Cen. In this paper I summarize the observational facts and provide a possible interpretation.

Keywords. stars: population II, globular clusters: individual (NGC 2808, ω Centauri).

1. Introduction

In the context of multiple stellar populations, NGC 2808 and ω Cen are certainly two of the most intriguing objects. The CMD of NGC 2808 shows three main sequences (MSs), with middle and blue MS being highly helium enhanced up to $Y \sim 0.39$ with respect to the red MS which has primordial helium (D'Antona *et al.* 2005, Piotto *et al.* 2007). Furthermore, spectroscopic studies have revealed significant star-to-star variations in the light-element abundances with the presence of an extreme Na-O anticorrelation (e.g. Norris 1981, Carretta *et al.* 2006).

The observational scenario for ω Cen is even more complex. Photometry shows a multimodal MS (Bedin *et al.* 2004, Bellini *et al.* 2010), which imply extreme helium enhancement (D'Antona *et al.* 2004, Norris *et al.* 2004). At odds with most globulars, which have homogeneous iron abundance, its stars span a wide interval of metallicity and define a multimodal distribution in [Fe/H], and *s*-elements. Large star-to-star light elements variations, Na-O and C-N anticorrelations are present within each metallicity interval (e.g. Marino *et al.* 2011).

Photometry of globular-cluster sequences extended to date over a limited spectral region, from the ultra-violet (UV, $\lambda \sim 2000$Å) to the near-infrared (NIR, $\lambda \sim 8000$Å). As such, multiple sequences are rarely detected along the lower part of the MS, because observational limits make it hard to get high-accuracy photometry of very faint and red stars in optical and UV colors. In the following we use *HST* to extend the study to the NIR passbands and investigate multiple sequences in NGC 2808 and ω Cen over a wide interval of stellar masses, from the turn off down to very low-mass MS stars ($\mathcal{M} \sim 0.2\ \mathcal{M}_{\odot}$).

2. Multiple populations of very low-mass stars

Figure 1a shows the NIR CMD for NGC 2808 from Milone *et al.* (2012a, left panel) and ω Cen (right panel). The upper MS of NGC 2808 is consistent with three stellar populations with different helium and light-element abundance, in agreement with previous observations based on visual photometry. The three MSs merge together at the luminosity of the MS bend while at fainter magnitudes, at least two MSs can be identified. A bluer, more populated $MS_{\rm I}$, which includes $\sim 65\%$ of MS stars, and a $MS_{\rm II}$ with $\sim 35\%$ of stars. The fractions of stars along $MS_{\rm I}$, and $MS_{\rm II}$ are very similar to the fraction of red-MS stars ($\sim$62%) and the total fraction of middle-MS and blu-MS stars ($\sim$24+14=38%, Milone *et al.* 2012b).

The observed CMD of NGC 2808 has been compared with appropriate evolutionary models for very low-mass stars and synthetic spectra that account for the chemical composition of the three stellar populations of this clusters (see Milone *et al.* 2012a for details). It comes out that $MS_{\rm I}$ is associated with the first stellar generation, which has primordial He, and O-C-rich/N-poor

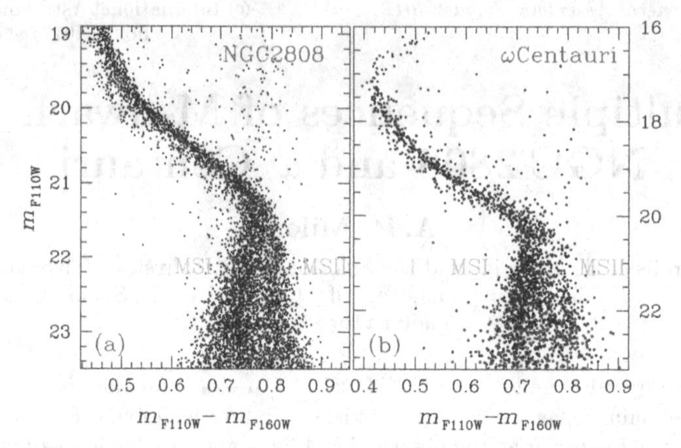

Figure 1. NIR CMD corrected for differential reddening for NGC 2808 and ω Cen.

stars, and that $MS_{\rm II}$, corresponds to a second-generation stellar population that is enriched in He and N and depleted in C and O. The $MS_{\rm I}$ of Fig. 1a is the faint counterpart of the red MS identified by Piotto *et al.* (2007), whereas the $MS_{\rm II}$ corresponds to the lower-mass counterpart of the middle MS and blue MS of Piotto *et al.* paper.

The brightest part of the CMD of ω Cen in Fig. 1b shows that the blue and the red MS, already detected in previous papers, merge together at the magnitude of the MS bend. At fainter luminosities, it appears a broad MS with a blue, more populated component ($MS_{\rm I}$), and a red tail ($MS_{\rm II}$). The similarities with NGC 2808, make it very tempting to associate the $MS_{\rm I}$ and the $MS_{\rm II}$ to the red and the blue MS, respectively. A comparison of the CMD with stellar models and synthetic spectra that account for the complex chemical composition of ω Cen stellar populations is mandatory to clarify this issue.

These results provide the first detection of multiple populations with different helium and light-element abundances among very low-mass stars. The fact that the signatures of abundance anticorrelation are also observed among fully-convective M-dwarfs definitely demonstrates that they have primordial origin and hence correspond to different stellar generations.

Acknowledgements

I warmly thank A. F. Marino, S. Cassisi, G. Piotto, L. R. Bedin, J. Anderson, F. Allard, A. Aparicio, A. Bellini, R. Buonanno, M. Monelli, A. Pietrinferni for their collaboration to this work. Support for this work has been provided by the IAC (grant 310394), and the Education and Science Ministry of Spain (grants AYA2007-3E3506, and AYA2010-16717).

References

Bedin, L. R., Piotto, G., Anderson, J., *et al.* 2004, *ApJ*, 605, L125
Bellini, A., Bedin, L. R., Piotto, G., *et al.* 2010, *AJ* 140, 631
Carretta, E., Bragaglia, A., Gratton, R. G., *et al.* 2006, *A&A*, 450, 523
D'Antona, F. & Caloi, V. 2004, *APJ*, 611, 871
D'Antona, F., Bellazzini, M., Caloi, V., *et al.* 2005, *ApJ*, 631, 868
Marino, A. F., Milone, A. P., Piotto, G., *et al.*, 2011, *ApJ*, 731, 64
Milone, A. P., Marino, A. F., Cassisi, S., *et al.* 2012, *ApJ*, 754, L34
Milone, A. P., Piotto, G., Bedin, L. R., *et al.* 2012, *A&A*, 540, A16
Piotto, G., Bedin, L. R., Anderson, J., *et al.* 2007, *ApJ*, 661, L53
Norris, J. 1981, *ApJ*, 248, 177
Norris, J. E. 2004, *ApJ*, 612, L25

Highlights of Astronomy, Volume 16
XXVIIIth IAU General Assembly, August 2012
T. Montmerle, ed.

doi:10.1017/S1743921314005602

Terzan 5: a pristine fragment of the Bulge

Francesco R. Ferraro

Dipartimento di Fisica e Astronomia, Università di Bologna (Italy)
email: francesco.ferraro3@unibo.it

Abstract. We have discovered that Terzan 5, a stellar system in the Galactic Bulge, harbors two stellar populations with different iron content (Δ[Fe/H] $\sim$ 0.5 dex) and possibly different ages. Moreover, the observed chemical patterns significantly differ from those observed in any known genuine globular cluster. These evidences demonstrate that, similarly to ω Centauri in the Halo, Terzan 5 is not a globular cluster, but a stellar system that was able to retain the gas ejected by violent supernova explosions. Moreover the striking chemical similarity with the Bulge stars suggests that Terzan 5 could be the relic of one of the massive clumps that contributed (through strong dynamical interactions with other pre-formed and internally-evolved sub-structures) to the formation of the Galactic Bulge.

Keywords. Stars: abundances – Globular clusters: individual (Terzan 5)

There is now a great deal of evidence for a significant spread in the abundance patterns of a few light-elements (as Na, O, Al, etc.) in Galactic globular clusters (GCs; see Carretta *et al.* 2010). This suggests that GC formation may have been more complex than previously thought. The multiple sequences observed in a few GCs are interpreted as the photometric manifestation of this phenomenon and of the connected variations in the He abundance (Piotto 2009 and references therein). However, GCs still are the best approximations of simple stellar populations we have in Nature, with their striking homogeneity in terms of the iron content still holding and indicating that their stellar populations formed in a parent potential well which was too shallow to retain the high-velocity gas ejected by violent supernova (SN) explosions. To date only two major exceptions to this rule are known: ω Centauri (Norris & Da Costa 1995) in the Galactic Halo, and Terzan 5 (Ferraro *et al.* 2009) in the Bulge.

The case of Terzan 5 - Thanks to high-resolution NIR imaging obtained with the multi-conjugate adaptive-optics demonstrator MAD@VLT, we revealed the presence of two distinct red clumps in Terzan 5 (Fig.1a), and from NIR spectroscopy with NIRSPEC@Keck we demonstrated that the two stellar populations are characterized by very different iron contents: [Fe/H] $= -0.2$ and $+0.3$ (Ferraro *et al.* 2009). In addition, follow-up spectroscopy (Origlia *et al.* 2011) revealed no evidence of the Al-O anti-correlation commonly found in GCs and showed that the metal-poor component has $[\alpha/\text{Fe}] = +0.34$, suggesting that it formed early and quickly from a gas mainly polluted by a huge amount of SNeII, while the metal-rich population has $[\alpha/\text{Fe}] = +0.03$, indicating that it experienced additional enrichment from both SNII and SNIa ejecta on longer timescales.

These observational facts clearly demonstrate that Terzan 5 is far from being a genuine GC. Instead, they can be naturally explained within a self-enrichment scenario, where the proto-Terzan 5 was much more massive in the past than today (its current mass being $10^6 M_\odot$; Lanzoni *et al.* 2010) and experienced two main and relatively short episodes of star formation with a separation of a few Gyr (a timescale which is much longer than assumed to account for multiple populations in genuine GCs): Terzan 5 was therefore able to retain the SN ejecta and gave rise to two distinct populations with different iron content. The high number of SNeII would have also produced a large population of

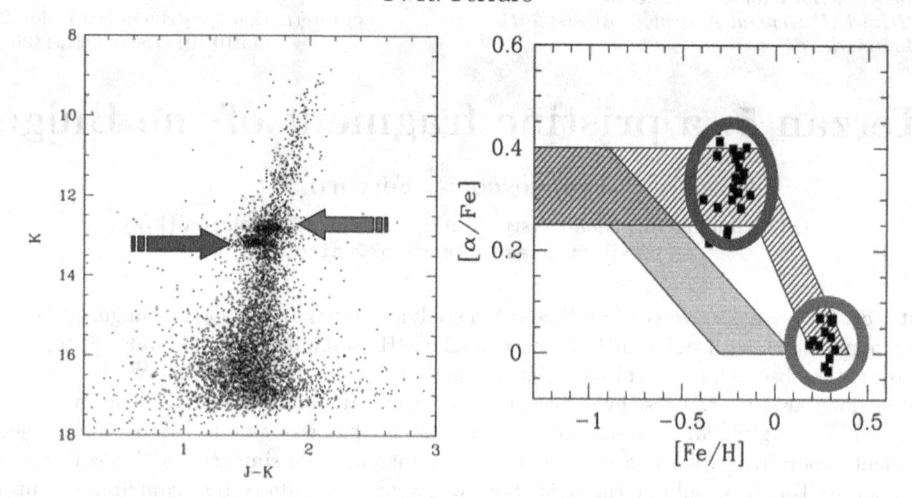

Figure 1. *left panel:* The two red clumps discovered in Terzan 5 thanks to K and J MAD observations. *right panel:* [α/Fe] and [Fe/H] ratios for a sample of giants in Terzan 5 (with the two populations highlighted by the blue and red circles, according to the left panel), compared to those of the Galactic Bulge (shaded region) and Halo+Disk giants (grey region).

neutron stars, most of which would have been retained within the deep potential well of the system. Then the high collision rate of Terzan 5 (the highest among all GCs, Lanzoni *et al.* 2010) could have favored the formation of binary systems containing neutron stars and promoted the recycling process that finally generated the huge population of millisecond pulsars now observed in Terzan 5 (which is the largest population ever detected in any stellar system; Ransom *et al.* 2005).

Indeed, Terzan 5 has all the characteristics expected for a system which had an original mass larger than a few $10^7 - 10^8 M_\odot$ and experienced a complex evolutionary history. Moreover its metal content and α-enhancement very closely resemble those of the Bulge stars (see Fig.1b; e.g. Rich *et al.* 2007; Zoccali *et al.* 2008). These striking similarities with the Bulge population strongly suggest that, at odds with ω Centauri that was probably accreted from outside the Galaxy, Terzan 5 formed and evolved within its current environment, the Bulge. Indeed these two stellar systems could have shared a common origin and evolutionary history, and Terzan 5 might well be the fossil fragment of a larger substructure that contributed to form the Galactic Bulge via strong dynamical interactions with other similar systems (e.g. Immeli *et al.* 2004).

This research is part of the project Cosmic-Lab funded by the European Research Council (under contract ERC-2010-AdG-267675).

References

Carretta, E., *et al.* 2009, *A&A*, 505, 117
Ferraro, F. R., *et al.* 2009, *Nature*, 462, 483
Immeli, A., *et al.* 2004, *A&A*, 413, 547
Lanzoni, B., *et al.* 2010, *ApJ*, 717, 653
Norris, J. E. & Da Costa, G. S. 1995, *ApJ*, 441, L81
Origlia, L., *et al.* 2011, *ApJ*, 726, 20
Piotto, G. 2009, *IAUS*, 258, 233
Ransom, S., *et al.* 2005, *Science*, 307, 892
Rich, M., *et al.* 2007, *ApJ*, 665, 119
Zoccali, M., *et al.* 2008, *A&A*, 486, 177

Highlights of Astronomy, Volume 16
XXVIIIth IAU General Assembly, August 2012
T. Montmerle, ed.

doi:10.1017/S1743921314005614

Multiple Stellar Populations: the evolutionary framework

Santi Cassisi

INAF - Astronomical Observatory of Teramo
email: cassisi@oa-teramo.inaf.it

Abstract. In these last years a huge amount of both spectroscopical and photometric data has provided a plain evidence of the fact that Galactic globular clusters (GCs) host various stellar sub-populations characterized by peculiar chemical patterns. The need of properly interpreting the various observational features observed in the Color-Magnitude Diagrams (CMDs) of these stellar systems requires a new generation of stellar models properly accounting for these chemical peculiarities both in the stellar model computations and in the color - T_{eff} transformations. In this review we discuss the evolutionary framework that is mandatory in order to trace the various sub-populations in any given GC.

Keywords. stars: Population II, globular clusters: general

1. The evolutionary framework

An extensive spectroscopical survey of several Galactic GCs has shown the existence of well-defined and peculiar chemical patterns among the stars belonging to the same cluster such as the existence of light-elements (anti-) correlations - the most famous being the Na-O anti-correlation - as discussed in the comprehensive reviews on this topic by Gratton *et al.* (2004) and Gratton *et al.* (2012) (see also the contribution of S. Lucatello in this volume). In the same time, very accurate photometric investigations performed by means of the HST have revealed the existence in the CMD of the various GCs of multiple sequences such as distinct Main Sequence (MS) and Sub-Giant Branch (SGB) loci, and multiple Red Giant Branch (RGB) loci. Actually the features observed in the CMD change significantly from a cluster to another one, and their properties strongly depend on the adopted photometric systems (see the review by Piotto (2010) for a detailed discussion on this issue and the contribution of G. Piotto in this volume).

The investigations performed so far in order to understand the physical reasons for the occurrence of these multiple evolutionary sequences in the CMD of GCs hosting various sub-populations have provided sound evidence of the fact that these distinct sequences can be interpreted as due to 'quantized' He abundances such as in the case of ω Cen and NGC 2808, distinct CNO abundance pattens as in the case of M 22 and (the still-debated case) of NGC1851, and the presence of light-elements anti-correlations.

It is evident that in order to properly trace the various sub-populations in a given GCs, the availability of an evolutionary theoretical framework properly - and self-consistently - accounting for the observed peculiar chemical patterns is mandatory.

In this last decade many theoretical investigations have been performed in order to investigate the effects of the chemical patterns characteristic of the multiple population phenomenon on both the evolutionary and structural properties of stars (Salaris *et al.* (2006), Cassisi *et al.* (2008), Pietrinferni *et al.* (2009), Vandenberg *et al.* (2012)) and model atmospheres and, hence on the corresponding color - T_{eff} transformations (Sbordone *et al.* (2011)).

2. A (very) short summary of theoretical results

Although a detailed review of the main outcomes of these analysis can not be done within the page limit of this contribution, we wish to briefly summarize the results obtained so far (for a more detailed discussion we refer to Sbordone *et al.* (2011)). Since, as mentioned, the photometric appearance of the multiple populations largely depends on the adopted photometric system, we discuss separately the case of optical, ultraviolet and Strömgren CMDs.

(*a*) *BVI CMDs*: a splitting (or a spread) of sequences along the MS up to the Turn-off (TO), and to a lesser degree of the RGB can only be achieved by varying the helium content. The CNONa anti-correlations influence neither the stellar models nor the spectrum sufficiently when the C+N+O abundance is unchanged. On the other hand, a variation of the C+N+O-abundance with respect the 'canonical' value leads to a split of the SGB; this is entirely an effect of the stellar models.

(*b*) *UBV- and uy CMDs*: anticorrelations in CNONa abundances as well as He differences may lead to multiple sequences from the MS to the RGB, where the effect tends to be larger, and may reach 0.2–0.3 mag. This multiplicity is independent of the sum of C+N+O. The individual element variations are decisive. Helium enhancement, however, works in the opposite direction than CNONa anticorrelations.

(*c*) *vy CMDs*: as in the case of the *BVI*-colours, a splitting of the MS up to the TO can be achieved only by a variation in helium. Similarly, after the TO, a split of the SGB is the result of a change in the C+N+O abundance. Additionally, a split along the RGB may result both from helium and from C+N+O variations; this is different from the *BVI*-case.

(*d*) $m_1 uy$ *CMDs*: CNONa anticorrelations lead to splits along the MS; along the SG and RGB the same anticorrelations, but also helium variations, lead to colour differences. However, the sign of the colour change is different for the lower and upper part of the RGB.

(*e*) $c_y V$ *CMDs*: here, all the evolutionary sequences in the CMD show the influence of both element anticorrelations and of helium variations, and a strong separation between the various sequences can be easily achieved.

References

Cassisi, S. Salaris, M., Pietrinferni, A., Piotto, G., Milone, A. P., Bedin, L. R., & Anderson, J. 2008, *ApJL*, 672, L115

Gratton, R. G., Carretta, E., & Bragaglia, A. 2012, *The Astronomy and Astrophysics Review, in press*, arXiv:1201.6526

Gratton, R., Sneden, C., & Carretta, E. 2004, *Annual Review Astronomy & Astrophysics*, 42, 385

Pietrinferni, A., Cassisi, S., Salaris, M., Percival, S., & Ferguson, J. W. 2009, *Astrophysical Journal*, 697, 275

Piotto, G. 2010, *Publications of The Korean Astronomical Society*, 25, 91

Salaris, M., Weiss, A., Ferguson, J. W., & Fusilier, D. J. 2006, *ApJ*, 645, 1131

Sbordone, L., Salaris, M., Weiss, A., & Cassisi, S. 2011, *Astronomy & Astrophysics*, 534, 9

VandenBerg, D. A., Bergbusch, P. A., Dotter, A., Ferguson, J. W., Michaud, G., Richer, J., & Proffitt, C. R. 2012, *ApJ*, 755, 15

Highlights of Astronomy, Volume 16
XXVIIIth IAU General Assembly, August 2012
T. Montmerle, ed.

doi:10.1017/S1743921314005626

Population Models for Massive Globular Clusters

Young-Wook Lee, Seok-Joo Joo, Sang-Il Han, Chongsam Na, Dongwook Lim and Dong-Goo Roh

Center for Galaxy Evolution Research and Department of Astronomy, Yonsei University, Seoul 120-749, Korea
email: ywlee2@yonsei.ac.kr

1. Stellar Models with Enhanced CNO Abundance

Increasing number of massive globular clusters (GCs) in the Milky Way are now turned out to host multiple stellar populations having different heavy element abundances enriched by supernovae. Recent observations have further shown that [CNO/Fe] is also enhanced in metal-rich subpopulations in most of these GCs, including ω Cen and M22 (Marino *et al.* 2011, 2012). In order to reflect this in our population modeling, we have expanded the parameter space of Y^2 isochrones and horizontal-branch (HB) evolutionary tracks to include the cases of normal and enhanced nitrogen abundances ([N/Fe] = 0.0, 0.8, and 1.6). The observed variations in the total CNO content were reproduced by interpolating these nitrogen enhanced stellar models. Our test simulations with varying N and O abundances show that, once the total CNO sum ([CNO/Fe]) is held constant, both N and O have almost identical effects on the HR diagram (see Fig. 1).

2. New Calcium Narrow-band Photometry and Low Resolution Spectroscopy

On the observational side, we are performing new Calcium narrow-band photometry using the new Calcium filter we designed to avoid CN contamination. The Calcium filter available at CTIO, the one used in Lee, J.-W. *et al.* (2009) and Roh *et al.* (2011), was suspected to be affected by some CN contamination, and this was recently confirmed, upon our request, by C. Johnson, D. Hölck, and A. Kunder (2012, private communication). Fortunately, we are still detecting two distinct red giant branches (RGBs) from the new filter for M22 and NGC 1851, suggesting that they are indeed different in Ca abundance. To confirm this, we have performed low-resolution spectroscopy for the stars on the two RGBs in each of these GCs, and have detected more than 4 sigma differences in spectroscopic *hk* index for both M22 and NGC 1851. The difference is about 0.2 dex in [Ca/H]. We have also detected more than 8 sigma differences in CN abundance, and therefore the second generation stars in these clusters are enhanced in both Ca and CN. For the NGC 288, however, the presence of two distinct RGBs (Roh *et al.* 2011) is now turned out to be due to the large difference in CN abundance, and the Ca abundance difference is not confirmed from both new Calcium photometry and the spectroscopy.

3. Population Models

In order to investigate the star formation histories of these peculiar GCs, we have constructed synthetic CMDs for ω Cen, M22, and NGC 1851 (see Joo & Lee 2013 for details). As described above, our models are based on the updated versions of Y^2 isochrones and

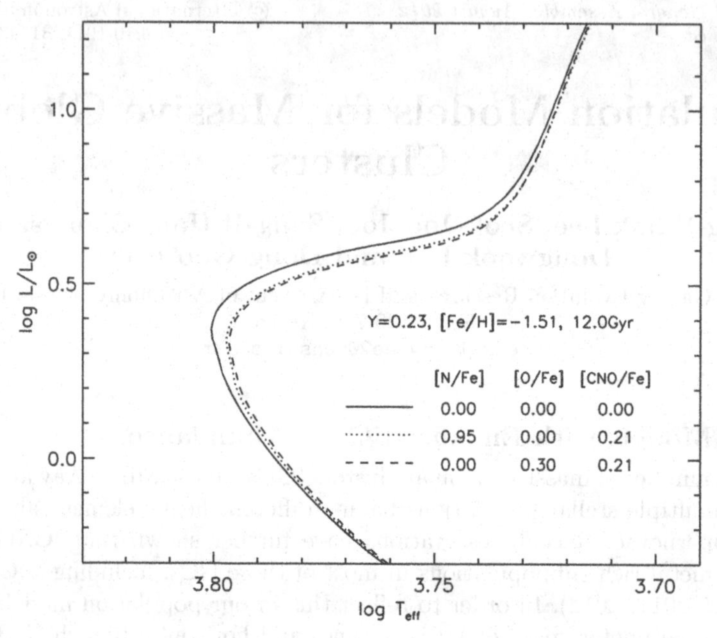

Figure 1. Two isochrones, one with enhanced N, and the other with enhanced O, compared to the reference case. In these two isochrones, N and O abundances are adjusted so that they all have the same value of [CNO/Fe]. Note that once the CNO sum is held constant, both N & O have almost identical effects on the HR diagram.

HB evolutionary tracks, with which the enhancements in both helium abundance and [CNO/Fe] can be tested self-consistently. To estimate ages and helium abundances of subpopulations in each GC, we have compared our models with the observations (including our new Calcium photometry) on the Hess diagram by employing a Chi-square minimization technique. We find that metal-rich subpopulations in each of these GCs are also enhanced in helium abundance, and the age differences between the metal-rich and metal-poor subpopulations are fairly small (0.3 - 1.7 Gyr), even in the models with the observed variations in the total CNO content. These are required to simultaneously reproduce the observed extended HB and the splits on the main sequence, subgiant branch, and RGB. Our results are consistent with the hypothesis that these GCs are the relics of more massive primeval dwarf galaxies that merged and disrupted to form the proto-Galaxy.

References

Joo, S.-J & Lee, Y.-W. 2013, *ApJ*, submitted
Lee, J.-W., Kang, Y.-W., Lee, J., & Lee, Y.-W. 2009, *Nature* 462, 480
Marino, A. F., Sneden, C., Kraft, R. P., *et al.* 2011, *A&A* 532, A8
Marino, A. F., Milone, A. P., Piotto, G., *et al.* 2012, *ApJ* 746, 14
Roh, D.-G., Lee, Y.-W., Joo, S.-J., Han, S.-I., Sohn, Y.-J., & Lee, J.-W. 2011, *ApJ* 733, L45

Highlights of Astronomy, Volume 16
XXVIIIth IAU General Assembly, August 2012
T. Montmerle, ed.

doi:10.1017/S1743921314005638

The pollution of the interstellar medium from AGB stars in Globular Clusters

Paolo Ventura[1] **and Roberta Carini**[1,2]

[1]INAF - Osservatorio Astronomico di Roma
Via Frascati 33, 00040, Monte Porzio Catone (RM), Italy
email: paolo.ventura@oa-roma.inaf.it

[2]Universitá di Roma "La Sapienza"
Piazzale Aldo Moro 5, 00135 Rome, Italy
email: roberta.carini@oa-roma.inaf.it

Abstract. We discuss the yields from Asymptotic Giant Branch stars, depending on their mass and metallicity. In agreement with previous investigations, we find that the extent of Hot Bottom Burning increases with mass. The yields of models with chemistry typical of high–metallicity Globular Clusters, i.e. Z = 0.008, show only a modest depletion of magnesium, and an oxgen depletion of $\sim$ 0.4dex. Low–metallicity yields show a much stronger magnesium depletion, and a dramatic drop in the oxygen content, $\sim$ 1.2dex smaller than the initial value. We suggest that the Globular Cluster NGC 2419 is a possible target to the hypothesis of the self–enrichment scenario of Globular Clusters by the winds of Asymptotic Giant Branch stars.

Keywords. stars: AGB and post-AGB, stars: abundances, globular clusters: general

1. Introduction

Intermediate mass stars, with mass $M \geqslant 4M_{\odot}$, have been proposed as the sources favoring the pollution of the interstellar medium in Globular Clusters (GC) (Ventura *et al.* 2001), giving origin to the formation of further stellar generations. The new stars would form in the environment polluted by the gas ejected by stars in the Asymptotic Giant Branch (AGB) phase, when they loose their external mantle. This gas will present traces of Hot Bottom Burning (HBB), showing up the patterns produced by p–capture nucleosynthesis.

In this contribution we discuss how the yields of massive AGBs change with mass and metallicity, and try to deduce the chemistry of stars of Globular Clusters belonging to the second generation (SG), depending on their metallicity.

2. Yields from AGBs and SAGBs

The yields from AGB models are shown in Fig. 1. The metallicities examined are $Z = 3 \times 10^{-4}$, $Z = 10^{-3}$, and $Z = 8 \times 10^{-3}$, which represent the chemistry of GCs with low, intermediate and high metallicity.

The oxygen and sodium yields are correlated: at large temperatures ($T > 70$MK) the destruction channel for sodium prevails, so that, while oxygen is consumed, part of the sodium previously accumulated via the second dredge–up and ^{22}Ne burning, is destroyed.

The low–Z models show a more extreme chemistry, i.e. more oxygen–poor; this is a clear indication that at low Z's HBB takes place at larger temperatures, favoring a more advanced nucleosynthesis.

The right panel of Fig. 1 shows the yields of AGB stars in the Mg–Al plane. The $Z = 3\times10^{-4}$ models show–up the greatest depletion of magnesium, because of the higher HBB

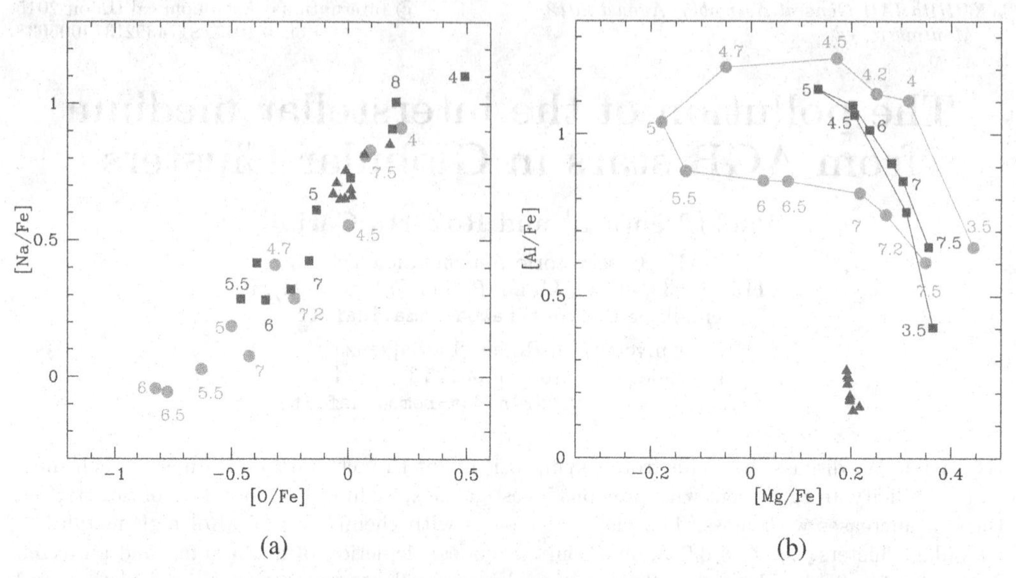

Figure 1. Yields of AGB stars in the O–Na and Mg–Al plane. The meaning of symbols is as follows. green points: $Z = 3 \times 10^{-4}$; blue squares: $Z = 3 \times 10^{-3}$; red triangles: $Z = 8 \times 10^{-3}$

temperatures. The maximum Al–enhancement reaches $\sim 1 - 1.2$dex, eventually becoming independent of the Mg–depletion. This finds a motivation in the saturation occurring at very large T's, exceeding $\sim$ 100MK, when production and destruction channels of Aluminium compensate.

In both panels of Fig. 1 we note that the stars producing the ejecta with the most extreme chemistry are those with masses around $\sim 5.5 - 6 M_{\odot}$, at the edge between the AGB and the SAGB regimes. The reason is that despite massive AGBs experience a stronger HBB, they loose mass with a higher rate: they loose their envelope before a great modification of the surface chemistry may occur (Ventura & D'Antona 2011).

3. The implications for the self–enrichment of Globular Clusters

Based on the results discussed above, we predict that low–metallicity clusters with a "pure" second generation (i.e. formed directly from the wind of AGB stars) should show the signature of a very advanced p–capture nucleosynthesis, with extremely O–poor chemistry. Sodium increase of these stars is also predicted to be modest.

NGC 2419 could be the ideal target to test these predictions: the Horizontal Branch of this cluster is known to harbor a group of blue stars separated from the main component (Ripepi *et al.* 2007), that, in analogy with the case of NGC 2808, is likely to be composed by a pure SG, made up of helium–rich stars.

Spectroscopic analysis of the the Main sequence counterparts of these stars could help accepting or disregarding the self–enrichment scenario hypothesis.

References

Ripepi, V., *et al.* 2007, *ApJ*, 667, L61

Ventura, P., D'Antona, F., Mazzitelli, I., & Decressin, T. 2001, *ApJ*, 550, L65

Ventura, P. & D'Antona, F. 2011, *MNRAS*, 410, 2760

Highlights of Astronomy, Volume 16
XXVIIIth IAU General Assembly, August 2012
T. Montmerle, ed.

doi:10.1017/S174392131400564X

Dynamics of Multiple Stellar Populations in Globular Clusters

Enrico Vesperini[1], Steve McMillan[2], Franca D'Antona[3] and Annibale D'Ercole[4]

[1]Department of Astronomy, Indiana University, Bloomington, IN 47401, USA

[2]Department of Physics, Drexel University, Philadelphia, PA, 19104, USA

[3]INAF- Osservatorio Astronomico di Roma, via di Frascati 33, I-00040 Monteporzio, Italy

[4]INAF- Osservatorio Astronomico di Bologna, via Ranzani 1, I-40127 Bologna, Italy

Abstract. The results of numerous spectroscopic and photometric studies have revealed the presence of multiple stellar populations in many globular clusters. In this paper we summarize the results of our recent studies on the dynamical evolution of multiple-population clusters, the implications of the structural properties of multiple-population clusters for the evolution of binary stars, and the possible contribution of globular clusters to the assembly of the Galactic halo.

Keywords. (Galaxy:) globular clusters: general, stellar dynamics, methods: n-body simulations

---

The discovery of multiple stellar populations in globular clusters (see e.g. Gratton *et al.* 2012 and references therein for a recent review) is extremely important for all the aspects of globular cluster formation and evolution.

In D'Ercole *et al.* (2008), we have studied the formation and dynamical evolution of multiple populations in globular cluster by means of hydrodynamical and N-body simulations. In our model second generation (SG) stars form from the ejecta of first generation (FG) AGB stars. Our simulations show that the AGB ejecta form a cooling flow and rapidly collect in the innermost regions of the cluster, forming a SG stellar subsystem segregated in the FG cluster inner regions. In order to form the numbers of SG stars observed today, the FG cluster must have been initially much more massive than it is now, and it must have been initially dominated by the FG population. By means of N-body simulations, in D'Ercole *et al.* (2008) we have shown that the early expansion triggered by the loss of SN II ejecta leads to the strong preferential loss of FG stars while, the inner SG stars, are unscathed by this early evolution: during this early evolutionary phase the cluster evolves from a configuration in which FG stars dominate to one with a similar number of of FG and SG stars (or even one in which the SG stars are now the dominant population), as observed in several Galactic globular clusters (see e.g. Carretta *et al.* 2009a,b). In our recent studies, we have explored the dynamical evolution of multiple-population clusters, the implications of the structural properties of multiple-population clusters for the evolution of a cluster binary star population and the connection between SG stars in clusters and in the Galactic halo.

• The structural properties of multiple population clusters differ from those of simple Plummer or King models usually adopted in the study of the dynamical evolution of globular clusters. In Vesperini *et al.* (2011), we have explored the implications of the presence of a concentrated SG subcluster on the evolution of FG and SG binary stars. Binary stars are extremely important for different aspects of the evolution of globular clusters and their stellar content (see e.g. Heggie & Hut 2003 for a review) and it is important to understand their evolution in multiple-population clusters. Our study shows that in multiple-population clusters, SG binaries are preferentially disrupted and that, more generally, binary disruption is enhanced compared to a standard cluster with similar mass and size but without an inner SG subsystem (see D'Orazi *et al.* 2010 for the first observational results indicating the preferential disruption of SG binaries).

• As a cluster evolves and enters the phase dominated by two-body relaxation some SG stars may escape and populate the Galactic halo. In Vesperini *et al.* (2010) we have carried out a study to estimate the number of SG stars that may have formed in Galactic globular clusters as a function of the cluster mass, structural properties and stellar initial mass function and the fraction of the Galactic halo composed of SG stars escaped from clusters. Our models show that the fraction of SG stars in clusters and in the Galactic halo can constrain both models for the formation and dynamical evolution of multiple stellar generations in globular clusters and the globular cluster contribution to the formation history of the Galactic halo.

• In Vesperini *et al.* (2012) we have carried out a survey of N-body simulations to explore the long-term evolution of multiple-population clusters. In particular we have focussed our attention on the study of the time scales and the dynamics of the spatial mixing of FG and SG stars. We have shown that the time scale of FG-SG complete mixing depends on the SG initial concentration but in all cases complete mixing is expected only for clusters in the advanced phases of their dynamical evolution. Unless a cluster is completely mixed, the SG-to-FG number ratio at a given distance from the cluster center is in general different from the global value. Our simulations indicate that, so long as mixing is not complete, the local value of N_{SG}/N_{FG} measured at $R \approx (1-2)R_h$ (where R_h is the cluster half-mass radius) is approximately equal to the global SG-to-FG number ratio.
Our simulations suggest that in many Galactic globular clusters, SG stars should still be more spatially concentrated than FG stars.

References

Carretta, E., *et al.* 2009a, *A&A*, 505, 117
Carretta, E., Bragaglia, A., Gratton, R., & Lucatello, S. 2009b, *A&A*, 505, 139
D'Ercole, A., Vesperini, E., D'Antona, F., McMillan, S. L. W., & Recchi, S., 2008, *MNRAS*, 391, 825
D'Orazi, V., Gratton, R., Lucatello, S., Carretta, E., Bragaglia, A., & Marino, A. F., 2010, *ApJ*, 719, L213
Gratton, R., Carretta, E., & Bragaglia, A., 2012, *A&ARv*, 20, 50
Heggie, D., & Hut, P. 2003, *The Gravitational Million-Body Problem*, Cambridge University Press
Vesperini, E., McMillan, S. L. W., D'Antona, F., & D'Ercole, A., 2010, *ApJ*, 718, L112
Vesperini, E., McMillan, S. L. W., D'Antona, F., & D'Ercole, A., 2011, *MNRAS*, 416, 355
Vesperini, E., McMillan, S. L. W., D'Antona, F., & D'Ercole, A., 2012, *MNRAS*, *in press*

Highlights of Astronomy, Volume 16
XXVIIIth IAU General Assembly, August 2012
T. Montmerle, ed.

doi:10.1017/S1743921314005651

Physical processes for the origin of globular clusters with multiple stellar populations

Kenji Bekki

ICRAR, M468, The University of Western Australia 35 Stirling Highway, Crawley Western Australia, 6009, Australia
email: bekki@cyllene.uwa.edu.au

Abstract. We numerically investigate the formation processes of globular clusters (GCs) in gas-rich dwarf galaxies at high redshifts. Our particular focus is on how the first and second generations of stars can be formed from high-density gas clouds in dwarf galaxies. We find that massive stellar clumps first form from massive gas clumps that are developed from local gravitational instability in gas-rich dwarfs. These stellar clumps with masses larger than $\sim 2 \times 10^6 M_\odot$ can finally become the first generation of stars in GCs. After supernova explosion expels the remaining gas in the clumps, stars can form from eject of AGB stars that is accreted onto the central regions of the clumps (i.e., first generation of stars). The compact clusters of these stars have much higher densities and a significant amount of internal rotation (~ 5 km s^{-1}) in comparison with the first generation and thus correspond to the second generation.

Keywords. stars: formation, globular clusters: general

---

1. Formation processes of originally massive GCs

A growing number of recent observational and theoretical studies on GCs with multiple stellar populations have shown that (i) most of the Galactic GCs investigated so far have characteristic anti-correlations between light elements, (ii) original GCs should be at least ~ 10 times more massive than the present ones, and (iii) physical properties of the first and second generations of stars in GCs can be significantly different (Bekki & Norris 2006, D'Ercole *et al.* 2008, Bekki 2010, Carretta *et al.* 2010, Bekki 2011, Bellazzini *et al.* 2012). Numerical simulations have shown that the second generation of stars can be formed from gas ejected from AGB stars in the first generation, if the first generation of stellar systems are as massive as $10^6 - 10^7 M_\odot$. Furthermore, radial distributions and rotational properties are demonstrated to be significantly different between the first and second generations of stars in simulated GCs. However, it remains unclear how the first generation stellar systems of GCs can be formed owing to the lack of extensive numerical studies of GC formation.

We thus construct a new model by which we can investigate both the formation process of the first generation of stars and that of the second in a fully self-consistent manner. We consider that GCs can be formed in the very early evolution of gas-rich dwarf galaxies, which could be fundamental building blocks of luminous galaxies like the Galaxy. We use our original hydrodynamical/Nbody numerical simulations with models for star formation and chemical evolution in order to investigate the formation processes and physical properties of massive stellar objects with masses larger than $2 \times 10^6 M_\odot$ that can correspond to masses of the first generation stellar systems. The simulations include the gas ejection processes of AGB stars and thus enable us to investigate how the second generation of stars can be formed from the AGB ejecta of the massive first generation. The total masses of baryonic and dark matter components in the initial dwarf galaxies

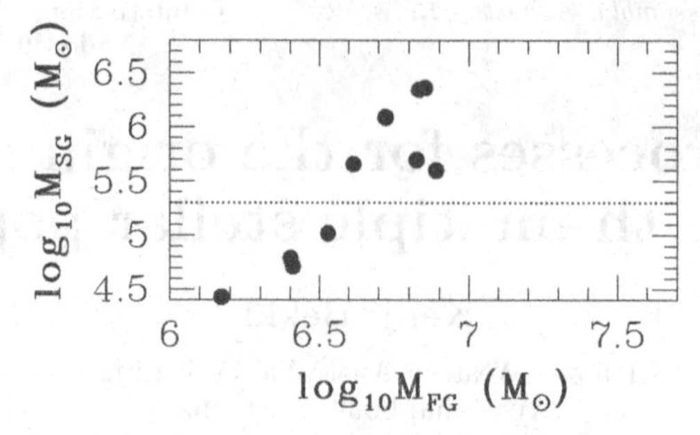

Figure 1. The dependence of the total masses of the second generations (M_{SG}) on those of the first (M_{FG}) in the simulated GCs of a gas-rich dwarf galaxy.The dotted line indicates the typical present GC mass ($2 \times 10^5 M_\odot$) in the Galaxy.

are set to be $2.5 \times 10^9 M_\odot$ and $2.5 \times 10^{10} M_\odot$, respectively. The details of the models will be described in our forthcoming papers.

2. Two-fold GC formation process

Fig. 1 shows the total masses of the first (M_{FG}) and second (M_{SG}) generations of stars in the selected GC candidate. We find the following principle results on physical properties of these GCs. Firstly, the formation of massive FG with $M_{FG} \geqslant 2 \times 10^6 M_\odot$ is due to the local gravitational instability in their host dwarfs. Secondly, the AGB ejecta from FG can not be escaped from the original FG stellar systems so that it can be accumulated in the central regions and consequently converted into new stars there, which correspond to SG. Thirdly, not all of massive FG systems have $M_{SG} \geqslant 2 \times 10^5 M_\odot$, which means that only some of them can finally become the present GCs as massive as $\sim 2 \times 10^5 M_\odot$ because most of FG stars can be lost during the long-term evolution of FG. Fourthly, SG stars clearly show rotational kinematics with the rotational amplitudes of ~ 5 km s^{-1}, which reflects the fact that SG is formed in the central regions of FG stellar systems through dissipative processes of gaseous ejecta from AGB stars. Fifthly, radial gradients of He can be clearly seen in the simulated GCs, in particular, in massive GCs ($\sim 10^7 M_\odot$). Sixthly, formation of GCs with both FG and SG is possible only for dwarfs with higher gas mass fractions (> 0.5) and masses larger than $\sim 5 \times 10^9 M_\odot$ (including dark halo masses). Seventhly, GC formation processes can proceed more efficiently in high-surface brightness dwarfs in comparison with low-surface brightness ones. Observational implications of these results will be discussed in our forthcoming papers.

References

Bekki, K. 2010, *ApJ* (Letters), 724, 99

Bekki, K. 2011, *MNRAS*, 412, 2241

Bekki, K. & Noris, J. E. 2006, *ApJ* (Letters), 637, 109

Bellazzini, M., Bragaglia, A., Carretta, E., Gratton, R. G., Lucatello, S., Catanzaro, G., & Leone, F. 2012, *A&A*, 538, 18

Carretta, E., Bragaglia, A., Gratton, R. G., Recio-Blanco, A., Lucatello, S., D'Orazi, V., & Cassisi, S. 2010, *A&A*, 516, 55

D'Ercole, A., Vesperini, E., D'Antona, F., McMillan, S. L. W., & Recchi, S. 2008, *MNRAS*, 391, 825

Highlights of Astronomy, Volume 16
XXVIIIth IAU General Assembly, August 2012
T. Montmerle, ed.

doi:10.1017/S1743921314005663

How did globular clusters lose their gas?

C. Charbonnel[1,2], M. Krause[1,3,4], T. Decressin[1], G. Meynet[1], N. Prantzos[5] and R. Diehl[4,3]

[1]Geneva Observatory, University of Geneva, Versoix, Swizerland
[2]IRAP, CNRS UMR 5277, Toulouse, France
[3] Excellence Cluster Universe, Technische Universität München, Garching, Germany
[4] Max-Planck-Institut für extraterrestrische Physik, Garching, Germany
[5]Institut d'Astrophysique de Paris, CNRS UMR7095, Paris, France

Abstract. We summarize the results presented in Krause *et al.* (2012a, K12) on the impact of supernova-driven shells and dark-remnant accretion on gas expulsion in globular cluster infancy.

Keywords. Globular clusters - ISM: bubbles

1. Introduction

Galactic globular clusters (GCs) today appear as large aggregates of long-lived low-mass stars, with little or even no gas. Yet, they must have formed as gas-rich objects hosting also numerous massive and intermediate-mass stars. Clues on this early epoch were revealed recently with the discovery of multiple stellar generations thanks to detailed spectroscopic and deep photometric investigations (see e.g. reviews by S.Lucatello and J.Anderson, this volume). In particular, abundance data for light elements from C to Al call for early self-enrichment of GCs by a first generation of rapidly evolving stars. Current scenarii involving either fast rotating massive stars or massive AGBs as potential polluters imply that GCs were initially much more massive and have lost most of their first generation low-mass stars (e.g. Decressin *et al.* 2007, 2010; D'Ercole *et al.* 2008; Vesperini *et al.* 2010; Schaerer & Charbonnel 2011).

2. Driving mechanisms for gas expulsion

Based on crude energetic arguments, gas expulsion by SNe was long thought to be the timely mechanism that could have unbound first generation stars and removed the bulk of gas together with metal-enriched SNe ejecta that are not found in second generation stars. In K12 we actually show that this mechanism does not generally work for typical GCs. Indeed, while the energy released by SNe usually exceeds the binding energy, SNe-driven shells turn out to be destroyed by Rayleigh-Taylor instability before they reach escape speed as shown on Figure 1 (left panels). Consequently the shell fragments that contain the gas remain bound to the cluster. This result, which is presented here for a typical protocluster of initial mass equal to 9×10^6 $M_\odot$ and initial half-mass radius of 3pc, holds for all but perhaps the initially least massive and most extended GCs (see K12 for details).

Instead, K12 propose that gas expulsion is launched thanks to the energy released by coherent and extremely fast accretion of interstellar gas onto dark-remnants. Due to sudden power increase in that case, the shell reaches escape speed before being affected by the Rayleigh-Taylor instability as depicted in Figure 1 (right panels). Consequently, the gas can be expelled from the cluster, and the sudden change of gravitational potential

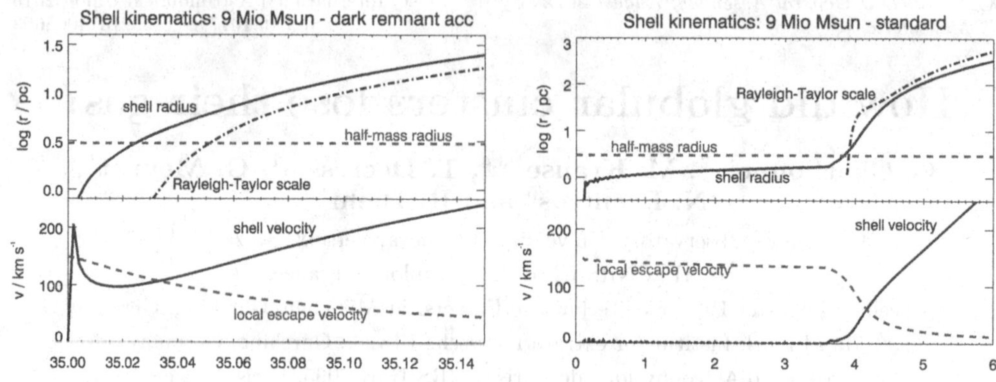

Figure 1. Superbubble kinematics within spherically symmetric thin-shell approximation. Two cases are depicted depending on the energy contributors: *(Left)* We consider SNe explosions for all first generation stars in the mass range between 9 and 120 $M_\odot$, assuming that each provides 10^{51} erg with an efficiency parameter of 0.2. *(Right)* We consider sudden accretion onto all first generation dark remnants (1.5 $M_\odot$ neutron stars for stars between 10 and 25 $M_\odot$, and 3 $M_\odot$ black holes for more massive stars) assuming they contribute to gas energy at a rate of 20% of the Eddington luminosity. For both cases the upper diagrams show the bubble radius, the Rayleigh-Taylor scale, and the GC half-mass radius (solid, dash-dotted, and dashed lines respectively), while the lower diagrams compare the shell velocity with the escape velocity at current bubble radius (solid and dashed lines respectively); note that in the left diagrams, after the Rayleigh-Taylor scale becomes larger than the shell radius, the continuous lines indicate the evolution of the shell radius and velocity that these two quantities would have in case Rayleigh-Taylor instability would not be present. The timescale is initialized at the birth of first generation stars. The abscissae indicate the respective times (in Myrs) when the considered energy sources become available; in the dark remnant accretion case this is assumed here to correspond to the moment when all massive stars have died.

is expected to unbind a large fraction of first generation low-mass stars sitting initially in the GC outskirts (see e.g. Decressin *et al.* 2010). Consequences of these results for the self-enrichment scenario will be presented in a forthcoming paper (Krause *et al.* 2012b, in preparation). The impact of SNe and stellar winds on their surroundings are among the current issues in understanding chemical evolution of galaxies, e.g. how interstellar gas is energized near those sources, and how ejecta are mixed into remaining gas. GCs serve as a laboratory to study this in a special, possibly extreme, environment of a smaller system, thus less complex than a galaxy as a whole.

Acknowledgements

We acknowledge support from the Swiss National Science Foundation, the French Society of Astronomy and Astrophysics, the cluster of excellence "Origin and Structure of the Universe", and the ESF EUROCORES Programme "Origin of the Elements and Nuclear History of the Universe".

References

Decressin, T., Charbonnel, C., & Meynet, G., 2007, *A&A* 475, 859
Decressin, T., Baumgardt, H., Charbonnel, C., & Kroupa, P., 2010, *A&A* 516, A73
D'Ercole, A., Vesperini, E., D'Antona, F., *et al.*, 2008, *MNRAS* 391, 825
Krause, M., Charbonnel, C., Decressin, T., Meynet, G., Prantzos, N., & Diehl, R., 2012a, *A&A* 546, L5 (K12)
Schaerer, D., & Charbonnel, C., 2011, *MNRAS* 413, 2297
Vesperini, E., McMillan, S. L. W., D'Antona, F., & D'Ercole, A., 2010, *ApJ* 718, L112

Highlights of Astronomy, Volume 16
XXVIIIth IAU General Assembly, August 2012
T. Montmerle, ed.

doi:10.1017/S1743921314005675

LAE Galaxies at High Redshift: Formation Sites for Low-Metal Globular Clusters

Bruce G. Elmegreen[1], **Sangeeta Malhotra**[2] **and James Rhoads**[2]

[1]IBM Research Division, T. J. Watson Research Center, Yorktown Heights, NY, USA
email: bge@us.ibm.com

[2]School of Earth and Space Exploration, Arizona State University, Tempe, AZ, USA

Abstract. Lyman-α emitting (LAE) galaxies observed at intermediate to high redshift have the correct size, mass, star formation rate, metallicity, and space density to have been the formation sites of metal-poor globular clusters. LAEs are typically small galaxies with transient starbursts. They should accrete onto spiral and elliptical galaxies over time, delivering metal-poor clusters into the larger galaxies' halos as they themselves get dispersed by tidal forces. The galaxy WLM is a good example of a dwarf remnant from a very early starburst that contains a metal-poor globular cluster but failed to get incorporated into the Milky Way or M31 because of its remote location in the local group.

Keywords. globular clusters: general, galaxies: dwarf, galaxies: high-redshift, galaxies: star clusters

---

1. Introduction

Elmegreen, Malhotra & Rhoads (2012) suggested that Lyman-α emitting (LAE) galaxies at redshifts of $z \sim 2$ to 7 (i.e., 10 – 13 Gyr ago) are the formation sites of metal-poor globular clusters (GCs). This suggestion followed from four main concepts: (1) metal poor GCs probably formed in dwarf galaxies (Searle & Zinn 1978, Zinnecker *et al.* 1988) and entered the halos of larger galaxies during hierarchical build-up, sometimes producing halo streams such as those observed by Mackey *et al.* (2010) and Romanowsky *et al.* (2012); (2) the low metallicity of these GCs is the result of the mass-metallicity relation (Mannucci *et al.* 2009, Chies-Santos *et al.* 2011) in galaxies at the GC formation redshift, rather than an earlier formation time for metal-poor GCs compared to metal-rich GCs, (3) the most heterogeneous GCs could have been dwarf galaxy nuclei (e.g., Bekki & Norris 2006, Carretta *et al.* 2010), and (4) LAEs are low-mass starburst galaxies with low metallicities (Finkelstein *et al.* 2011) that are observed at the redshifts where metal-poor GCs should have formed.

Clusters with the mass of a young GC probably formed among other clusters and stars having a total mass in excess of $3 \times 10^7\ M_\odot$ with $\sim 10^8 - 10^9\ M_\odot$ in molecules and a star formation rate of $\sim 3\ M_\odot\ \mathrm{yr}^{-1}$ or more. The associated emission would be $10^{10}\ L_\odot$, and in the Lyman-α line, 10^{42} erg s^{-1} if there is ~ 1 mag of extinction to absorb the local Lyman continuum radiation. This is in the range of emission rates for an LAE galaxy. With a more likely 25% escape fraction (Blanc *et al.* 2011, Zheng *et al.* 2012), the Lyα emission is 2.5×10^{41} erg s^{-1}. Integrating the LAE Luminosity function down to this luminosity gives a space density of 0.007 cMpc^{-3} in the GC-forming phase. If this phase lasts 10 Myr (Malhotra & Rhoads 2004), and each such burst makes 1 GC, then the space density of GCs formed is the space density of LAEs multiplied by the ratio of the range of times during which the LAEs are observed (2.5 Gyr from $z = 2$ to 7) to 10 Myr, giving ~ 2 cMpc^{-3}. This is comparable to present day space density of metal-poor GCs with $M > 2 \times 10^5\ M_\odot$. Longer lifetimes for the GC-forming phase that produce the same total young star mass correspond to weaker Lyα galaxies and more of them from the luminosity function, but the duty cycle of this phase increases in proportion, making the total number of GCs produced about the same. Forming more than 1 GC per event also leaves the total GC number unchanged, because each LAE galaxy would be either brighter or it would have been active for a longer time, reducing the number of such galaxies in proportion.

The WLM galaxy is a local dwarf with a metal-poor GC and a distance of ~ 1 Mpc from both the Milky Way and M31. The GC is ~ 14 Gyr old and has Fe/H $= -1.63 \pm 0.14$ and $M_V \sim -8.8$ mag (Hodge *et al.* 1999, Dolphin 2000), which suggests the GC mass is $\sim 10^6\ M_\odot$. The WLM stellar mass is now $1.6 \times 10^7\ M_\odot$ (Zhang *et al.* 2012) but it was $\sim 10^6\ M_\odot$ when the GC formed (Leaman *et al.* 2012). WLM is an example what is likely to be a remnant of an LAE that formed a metal-poor GC at the equivalent of $z \sim 2$ and never got incorporated into a spiral galaxy halo.

References

Bekki K. & Norris J. E. 2006, *ApJL*, 637, 109

Blanc, G. A., *et al.* 2011, *ApJ*, 736, 31

Carretta, E. *et al.* 2010, *ApJ*, 714, L7

Chies-Santos, A. L., *et al.* 2011, *A&A*, 525, A20

Dolphin, A. E. 2000, *ApJ*, 531, 804

Elmegreen, B. G., Malhotra, S., & Rhoads, J. 2012, *ApJ*, 757, 9

Finkelstein, S. L., *et al.* 2011, *ApJ*, 729, 140

Hodge, P. W., Dolphin, A. E., Smith, T. R., & Mateo, M. 1999, *ApJ*, 521, 577

Leaman, R., *et al.* 2012, *ApJ*, 750, 33

Mackey, A. D., *et al.* 2010, *ApJ*, 717, L11

Malhotra, S. & Rhoads, J. E. 2004, *ApJL*, 617, 5

Mannucci F. *et al.* 2009, *MNRAS*, 398, 1915

Romanowsky, A. J. *et al.* 2012, *ApJ*, 748, 29

Searle, L. & Zinn R. 1978, *ApJ*, 225, 357

Zhang, H.-X., Hunter, D. A., Elmegreen, B. G., Gao, Y., & Schruba, A. 2012, *AJ*, 143, 47

Zheng Z.-Y. *et al.* 2012, *ApJ*, 746, 28

Zinnecker, H., Keable, C. J., Dunlop, J. S., Cannon, R. D., & Griffiths, W. K. 1988, *IAUS* 126, 603

Highlights of Astronomy, Volume 16
XXVIIIth IAU General Assembly, August 2012
T. Montmerle, ed.

doi:10.1017/S1743921314005687

Rapid Mass Segregation in Massive Star Clusters

Stephen McMillan[1], Enrico Vesperini[2] and Nicholas Kruczek[1]

[1]Department of Physics, Drexel University, Philadelphia, PA 19104, USA
Contact email: steve@physics.drexel.edu

[2]Department of Astronomy, Indiana University, Swain Hall West, Bloomington, IN 47405, USA

Abstract. Several dynamical scenarios have been proposed that can lead to prompt mass segregation on the crossing time scale of a young cluster. They generally rely on cool and/or clumpy initial conditions, and are most relevant to small systems. As a counterpoint, we present a novel dynamical mechanism that can operate in relatively large, homogeneous, cool or cold systems. This mechanism may be important in understanding the assembly of large mass-segregated clusters from smaller clumps.

Keywords. globular clusters: general, galaxies: star clusters, stellar dynamics

---

Early mass segregation may be critical to the long-term survival of a stellar system (Vesperini *et al.* 2009a). It also defines the early cluster environment within which stars move and interact. In recent years, several dynamical studies have explored routes to early mass segregation that do not simply require that a cluster formed in that state. McMillan *et al.* (2007) found that mergers of mass-segregated "clumps" tend to preserve that segregation in the final merger product, so that, if the clumps are formed segregated, or have time to segregate before they merge, the result is a strongly mass-segregated cluster. Allison *et al.* (2009) found similar behavior, starting from fractal clumpy initial conditions in small, cool model clusters.

Ultimately, these scenarios rely on normal relaxation processes in small stellar systems. However, as illustrated in Figure 1, rapid segregation is also possible in significantly larger systems. The initial conditions of the simulation shown here consist of a cold (virial ratio $q = 0.001$), homogeneous sphere with a Kroupa (2002) mass distribution. No segregation is seen before the "bounce" at $t \sim 1.5$ initial dynamical times, while immediately afterward the highest mass groups are clearly ordered by radius. This behavior was also noted by Vesperini *et al.* (2006) and Vesperini *et al.* (2009b).

The phenomenon of rapid segregation cannot be due to enhanced relaxation around the high density bounce. This would only be possible if the system were still cold at that time, and our simulations clearly indicate that this is not the case. Instead, as shown in Figure 2, we find that the system fragments as it collapses, as discussed in detail by Aarseth *et al.* (1988), and the fragments mass segregate quite early on during the collapse process. Significant segregation within the clumps is already established by $t \sim 1$, well before the bounce, and is preserved when the clumps subsequently merge at $t = 1.5$, essentially as described by McMillan *et al.* (2007).

The phenomenon persists as we vary the initial system parameters, and is still measurable even for fairly "warm" initial conditions ($q \sim 0.1$), with and without initial clumping (fractal dimension $d \sim 2-3$), and for large systems, up to $N \sim 10^5$. Thus it may provide the basis of a viable mechanism for extending earlier dynamical segregation scenarios to substantially larger systems.

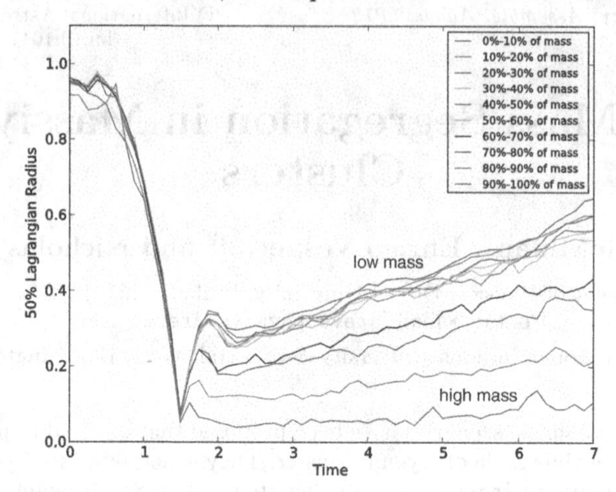

Figure 1. Collapse of an initially cold, homogeneous spherical system containing 10^4 particles with a Kroupa (2002) mass function. The half-mass radii of the particles making up the bottom 10 percent, 10–20 percent, 20–30 percent, etc., of the cumulative mass distribution are shown. The bottom four lines after the collapse represent the top four mass groups, their half-mass radii decreasing with mass, indicating strong mass segregation.

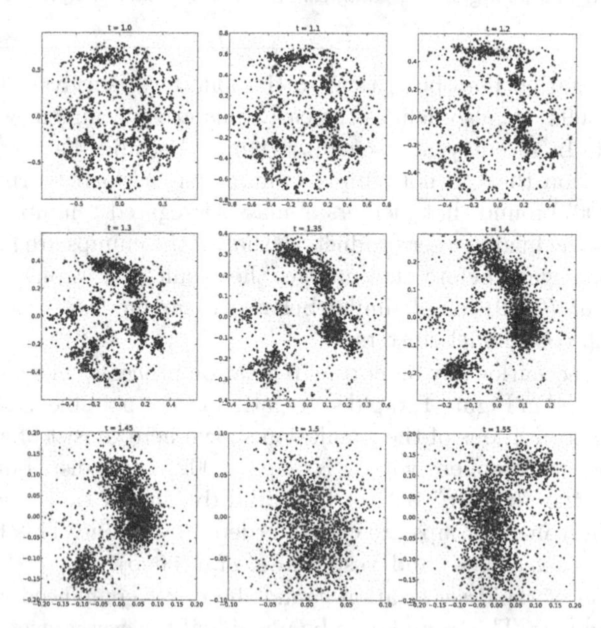

Figure 2. Fragmentation and mass segregation during the collapse can be seen in this sequence of frames running from $t = 1$ to $t = 1.55$, just after the moment of collapse. The spatial scale of the frames shrinks to follow the collapsing system, from ± 0.9 at top left ($t = 1$) to ± 0.4 at center ($t = 1.35$), to ± 0.1 at bottom center ($t = 1.5$) and ± 0.2 at bottom right ($t = 1.55$).

Acknowledgements

This work was supported in part by NSF grants AST-0708299 and AST-0959884, and NASA grant NNX08AH15G.

References

Aarseth, S. J., Lin, D. N. C., & Papaloizou, J. C. B. 1988, *ApJ*, 324, 288

Allison, R. J., Goodwin, S. P., Parker, R. J., de Grijs, R., Portegies Zwart, S. F., & Kouwenhoven, M. B. N. 2009, *ApJL*, 700, L99

Kroupa, P. 2002, *Science*, 295, 82

McMillan, S. L. W., Vesperini, E., & Portegies Zwart, S. F. 2007, *ApJL*, 655, L45

Vesperini, E., McMillan, S. L. W., & Portegies Zwart, S. F. 2006, *Joint Discussion 14*, 26th General Assembly of the IAU

Vesperini, E., McMillan, S. L. W., & Portegies Zwart, S. F. 2009a, *ApJ*, 698, 615

Vesperini, E., McMillan, S. L. W., & Portegies Zwart, S. F. 2009b, in Globular Clusters – Guides to Galaxies, ESO Astrophysics Symposia, T. Richtler and S. Larsen eds. (Springer: Berlin), p. 429

Highlights of Astronomy, Volume 16
XXVIIIth IAU General Assembly, August 2012
T. Montmerle, ed.

doi:10.1017/S1743921314005699

Nuclear Star Clusters Structure and Stellar Populations

Nadine Neumayer
ESO, Karl-Schwarzschild Str. 2, 85748 Garching bei München, Germany
email: nneumaye@eso.org

Abstract. This is an overview of nuclear star cluster observations, covering their structure, stellar populations, kinematics and possible connection to black holes at the centers of galaxies.

Keywords. galaxies: nuclei, star clusters, black holes

---

1. Nucleation Fraction, Stellar Populations, and Kinematics

Nuclear Star Clusters (NCs) are a very common structural component at the centers of galaxies. They are found in 77% of late type galaxies (Böker *et al.* 2002), 55% of spirals (Carollo *et al.* 1998), and at least 66% of (dwarf) ellipticals and S0s (Côté P *et al.* 2006). These studies find that the half-light radii of NCs are typically 5*pc*, and due to these small sizes, their detection requires very high spatial resolution observations, making the HST crucial for systematic searches. NCs are on average 4mag brighter than GCs (Böker *et al.* 2004), i.e. they are more massive but have similar half-light radii. This makes NCs the densest stellar systems in the universe (Walcher *et al.* 2005, Misgeld & Hilker 2011). They lie at the high-mass end of the star cluster mass function, and are structurally very different from bulges.

Nuclear star clusters truly occupy the centers of galaxies, both photometrically but also kinematically (Böker *et al.* 2002, Neumayer *et al.* 2011, respectively), and it may be this special location at the bottom of the potential well of the galaxies, that causes the star formation history of NCs to be rather complex. Several studies have shown that NCs have multiple stellar populations both in late type (Walcher *et al.* 2006, Seth *et al.* 2006, Rossa *et al.* 2006) and also early type galaxies (Seth *et al.* 2010). The NC seems to be typically more metal-rich and younger than the surrounding galaxy (Koleva *et al.* 2011), and the abundance ratios [α/Fe] show that NCs are more metal enriched than globular clusters (GCs) (Evstigneeva *et al.* 2007). This finding suggests that NCs cannot solely be the merger product of GCs, but need some gas for recurrent star formation. This finding is also supported by recent kinematical studies (Hartmann *et al.* 2011, De Lorenzi *et al.* 2012), where cluster infall alone cannot explain the dynamical state of the NC.

Recent studies of the kinematics of NCs with integral-field spectroscopy show that the cluster as a whole rotates (Seth *et al.* 2008, 2010). Combined with the superb spatial resolution of adaptive-optics, the 2D velocity maps resolve stellar and gas kinematics down to a few parsecs on physical scales. In addition, due to the extremely high central stellar density in NCs, it becomes possible to pick-up kinematic signatures for intermediate-mass black hole inside NCs (Seth *et al.* 2010, Neumayer *et al.* in prep).

2. Connection to Black Holes

Unlike black holes, NCs provide a visible record of the accretion of stars and gas into the center of a galaxy, and studying their stellar populations, structure and kinematics

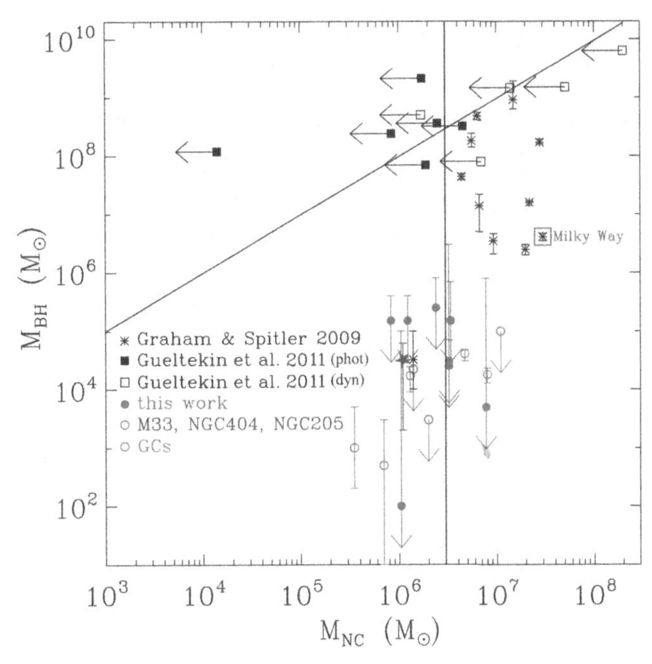

Figure 1. The mass of the BH mass vs. the NC mass. The two full lines indicate a NC mass of $3 \times 10^6 M_\odot$ and a MBH / MNC mass ratio of 100. These lines separate NC dominated galaxy nuclei (lower left of both lines) from BH dominated galaxy nuclei (upper left of both lines) and a transition region (to the right of both lines) (Figure 5 from Neumayer & Walcher 2012).

allow us to disentangle their formation history. NCs do co-exist with black holes. The best studied example is the NC in our own Galaxy (see R. Schödel this edition). But there are also NCs with very tight upper limits on the mass of a central black hole (see Neumayer & Walcher 2012 for an overview), and it is not yet clear under what conditions galaxies make NCs and/or BHs. Figure 1 shows a compilation of mass measurements of BHs and NCs. For the lowest mass NCs BHs are very hard to detect (if present), while for the highest mass BHs, the surrounding NCs seem to have been destroyed already. The underlying physical connection remains unclear for now, but it may well be that BHs grow inside NCs, that may thus be the precursors of massive BHs at the nuclei of galaxies.

Acknowledgements

I acknowledge the support by the DFG cluster of excellence 'Origin and Structure of the Universe', and the IAU for financial support.

References

Böker, T., Laine, S., van der Marel, R. P., *et al.* 2002, *AJ*, 123, 1389
Böker, T., Sarzi, M., McLaughlin, D. E., *et al.* 2004, *AJ*, 127, 105
Carollo, C. M., Stiavelli, M., & Mack, J. 1998, *AJ*, 116, 68
Côté, P., Piatek, S., Ferrarese, L., *et al.* 2006, *ApJS*, 165, 57
De Lorenzi, F., Hartmann, M., Debattista, V., *et al.* 2012, *arXiv:1208.2161*
Evstigneeva, E. A., Gregg, M. D., Drinkwater, M. J., & Hilker, M. 2007, *AJ*j, 133, 1722
Hartmann, M., Debattista, V. P., Seth, A., *et al.* 2011, *MNRAS*, 418, 2697
Koleva, M., Prugniel, P., de Rijcke, S., & Zeilinger, W. W. 2011, *MNRAS*, 417, 1643
Misgeld, I. & Hilker, M. 2011, *MNRAS*, 414, 3699

Neumayer, N., Walcher, C. J., Andersen, D., *et al.* 2011, *MNRAS*, 413, 1875
Neumayer, N. & Walcher, C. J. 2012, *Advances in Astronomy*, 2012,
Rossa, J., van der Marel, R. P., Böker, T., *et al.* 2006, *AJ*, 132, 1074
Seth, A. C., Dalcanton, J. J., Hodge, P. W., & Debattista, V. P. 2006, *AJ*, 132, 2539
Seth, A. C., Blum, R. D., Bastian, N., *et al.* 2008, *ApJ*, 687, 997
Seth, A. C., Cappellari, M., Neumayer, N., *et al.* 2010, *ApJ*, 714, 713
Walcher, C. J., Böker, T., Charlot, S., *et al.* 2006, *ApJ*, 649, 692
Walcher, C. J., van der Marel, R. P., McLaughlin, D., *et al.* 2005, *ApJ*, 618, 237

Highlights of Astronomy, Volume 16
XXVIIIth IAU General Assembly, August 2012
T. Montmerle, ed.

doi:10.1017/S1743921314005705

Formation, Growth, and Destruction of Nuclear Star Clusters

Torsten Böker
European Space Agency, Keplerlaan 1, 2200AG Noordwijk, Netherlands
email: tboeker@rssd.esa.int

Abstract. This talk is an attempt to combine recent insights into the nature of the nuclear star clusters in galaxies of various morphologies into a coherent (albeit simplistic) picture for their formation, growth, and eventual destruction.

Keywords. galaxies: nuclei; galaxies: star clusters

---

1. Formation and Growth of Nuclear Star Clusters

The structural properties, masses, and stellar populations of nuclear star clusters (NSCs) have been extensively discussed elsewhere, both for late-type spirals (Böker *et al.* 2004, Walcher *et al.* 2005, 2006) as well as for spheroidal galaxies (Côté *et al.* 2006). A key finding from these studies is that, in general, the star formation histories of NSCs are long and complex, but **only** NSCs in late-type spirals also contain a young (few hundred Myrs) stellar population. The NSCs in spheroidal galaxies, on the other hand, generally have luminosity-weighted ages of many Gyrs, but are on average 3.5 Gyrs younger than the bodies of their host galaxies (Paudel 2010).

This implies that in the present-day universe, any growth mechanism that causes the "rejuvenation" of NSCs occurs only in (gas-rich) disk galaxies. The NSCs in (gas-poor) spheroidals, on the other hand, must have evolved passively for at least a few Gyrs. If they have acquired stellar mass over this timescale, it can only have occurred via the accretion of evolved stars, e.g. through the merging of globular clusters onto the NSC.

The merging of star clusters is indeed one of the proposed mechanisms for the build-up of NSCs (Capuzzo-Dolcetta & Miocchi 2008). At least in disk galaxies, however, this is unlikely to be the whole story, because Hartmann *et al.* (2011) show that cluster infall alone cannot explain the dynamical properties of the NSC in the nearby edge-on disk NGC 4244 . They conclude that at least 50% of its mass must have been produced by in-situ star formation. The star formation is the result of gas infall into the central few pc, which can be caused by a number of mechanisms such as bar-induced torques (Schinnerer *et al.* 2006), compressive tidal forces (Emsellem & van der Ven 2008), or the magneto-rotational instability in galaxy disks (Milosavljevic 2004).

In any case, it is important to keep in mind that the *formation* of NSCs and their subsequent *growth* do not necessarily have to be governed by the same mechanism, and that more than one mechanism can contribute to the evolution of an NSC after its formation.

2. SMBHs and the Destruction of Nuclear Star Clusters

The co-evolution of NSCs and central super-massive black holes (SMBHs) is currently a very active field of research, triggered by the realization that both types of a central massive object (CMO) can co-exist at the low end of the SMBH mass range

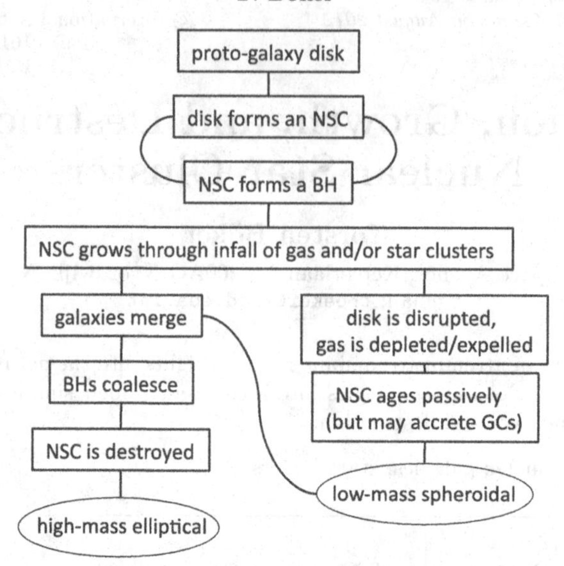

Figure 1. Possible evolutionary path of an NSC that forms early in the life of a gas-rich, disk-dominated "proto"-galaxy. As long as gas is available for infall into the center, the NSC (and any SMBH it may harbor) will grow. Once the gas is depleted or expelled (e.g. by the disruption of the galaxy disk through harassment or a close encounter), the CMO growth stops, and it will only evolve passively from then on. The host galaxy may progress along the merger tree, and its SMBH may well merge with the SMBHs of the merger partners, destroying their hosts' NSCs in the process.

($M_{\rm BH} < 10^9 \rm M_\odot$, Seth *et al.* 2008). Unfortunately, the exact mechanism for SMBH formation, and whether or not an NSC is *necessary* to form an SMBH remain unclear for now.

However, there appears to be a transition in what type of CMO dominates: at very low CMO masses, SMBHs are hard to identify (Satyapal *et al.* 2009), and their mass is usually less than that of the NSC. In galaxies hosting SMBHs with masses above $\approx 10^{10} \rm M_\odot$, on the other hand, NSCs are usually not observed (Graham & Spitler 2009), suggesting that the most massive SMBHs have destroyed their host NSCs. A similar transition is also evident in the surface brightness profiles of spheroidal galaxies, which smoothly transition from a pronounced light excess in the central few pc of low-mass systems to a clear light deficit in high mass system (Côté *et al.* 2007). This gradual change may be identified with a decreasing prominence of the NSC as the mass of the host spheroid grows.

The mass ratio between SMBH and NSC may well be imprinted early-on in the galaxy's life by their mutual feedback during the "competitive accretion" phase, as proposed by Nayakshin, Wilkinson & King (2009). However, it is also tempting to speculate that, as the evolution of the host galaxy progresses, the SMBH grows in mass at the expense of the NSC. For example, the merger of two BHs has been shown to efficiently destroy the surrounding NSC(s) via loss-cone depletion (e.g. Merritt 2006). While many details of NSC evolution remain to be addressed, the general picture outlined in Figure 1 appears to be consistent with what we currently know.

References

Böker, T., *et al.* 2004, *AJ*, 127, 105
Capuzzo-Dolcetta, R. & Miocchi, P. 2008, *MNRAS*, 388, 69
Côté, P., *et al.* 2006, *ApJS*, 165, 57
Côté, P., *et al.* 2007, *ApJ*, 671, 1456

Emsellem, E. & van der Ven, G. 2008, *ApJ*, 674, 653
Graham, A. W. & Spitler, L. R. 2009, *MNRAS*, 397, 2148
Hartmann, M., *et al.* 2011, *MNRAS*, 418, 2697
Merritt, D. 2006, *Reports on Progress in Physics*, 69, 2513
Milosavljevic, M. 2004, *ApJL*, 605, 13
Nayakshin S., Wilkinson M. I., King A. 2009, *MNRAS*, 398, 54
Paudel, S., Lisker, T., & Kuntschner, H. 2010, *MNRAS*, 413, 1764
Satyapal, S., *et al.* 2009, *ApJ*, 704, 439
Seth, A., *et al.* 2008, 678, 116
Schinnerer, E., *et al.* 2006, *ApJ*, 649, 181
Walcher, C. J., *et al.* 2005, *ApJ*, 618, 237
Walcher C. J., *et al.* 2006, *ApJ*, 649, 692

Highlights of Astronomy, Volume 16
XXVIIIth IAU General Assembly, August 2012
T. Montmerle, ed.

doi:10.1017/S1743921314005717

The Nuclear Star Cluster of the Milky Way

R. Schödel

IAA-CSIC, Glorieta de la Astronomía s/n, 18008 Granada, Spain

Abstract. This contribution briefly summarizes the current state of our knowledge about the Milky Way's nuclear star cluster.

---

1. The center of the Milky Way

The Galactic center (GC) is located at a mere 8 kpc from Earth, about a hundred times closer than the next comparable galaxy. It can thus provide us with unique data about the structure and evolution of a galactic stellar nucleus. Adaptive optics assisted observations with 8-10m-class telescopes in the near-infrared (NIR) allow us to resolve scales as small as 0.05", corresponding to about 8 mpc at the GC. On the downside, we must cope with the extreme interstellar extinction toward the GC that is caused by the line-of-sight through the Galactic Disk. At $A_V \approx 30$, observations in the visual regime are impossible. Fortunately, extinction decreases steeply throughout the NIR. With $A_K \approx 2.7$ toward the central parsec at $\lambda = 2.2\,\mu$m (Schödel *et al.*, 2010) we can study the distribution and kinematics of the stars in our Galaxy's nucleus.

Figure 1 presents an overview of the GC. The emission is dominated by stars in the the *nuclear bulge*, a disk-like region that stands out from the kpc-scale Galactic Bulge through intense far-infrared emission, ongoing star formation, a strong ionizing radiation field, large amounts of clumpy molecular gas, and an extremely high density of stars The nuclear bulge (sometimes also called *Nuclear Stellar Disk*) has a radius of ~230 pc, a scale height of ~45 pc and a total mass of $\sim 1.4 \times 10^9$ M$_\odot$ (Launhardt *et al.*, 2002).

2. Properties of the nuclear star cluster

The nuclear star cluster (NSC) is a compact entity at the center of the nuclear bulge. Its mass is estimated to be $3 \pm 1.5 \times 10^7$ M$_\odot$ and its half light radius to be ~3 pc (Launhardt *et al.*, 2002; Graham & Spitler, 2009). It must be stressed, however, that these values are highly uncertain because of the limitations of existing observations and the problem of differential extinction. We do not even know whether the NSC is spherical or not (its apparent elongation may in part be due to strong differential extinction, note the IR dark cloud south of the NSC in Fig. 1). The entire NSC rotates parallel to Galactic rotation (Trippe *et al.* 2008, Schödel *et al.* 2009).

The stellar density (e.g., Schödel *et al.* 2007) and velocity dispersion (e.g., Schödel *et al.* 2009) peak at the very center of the NSC, at the location of the 4×10^6 M$_\odot$ massive black hole Sagittarius A* (e.g., Gillessen *et al.*, 2009, Yelda *et al.* 2010). Hence, in the case of the Milky Way, there is no doubt that the NSC is truly central and that it co-exists with a massive black hole (MBH).

A question of special interest is the interaction between the stars and the MBH. Theoretical work predicts the formation of a *stellar cusp* within the sphere of influence of the MBH ($1 - 2$ pc) with a power-law number density of the form $\rho(r) \propto r^{-1.5...-1.75}$ if the system is older than a two-body relaxation time (e.g., Merritt 2006). The azimuthally

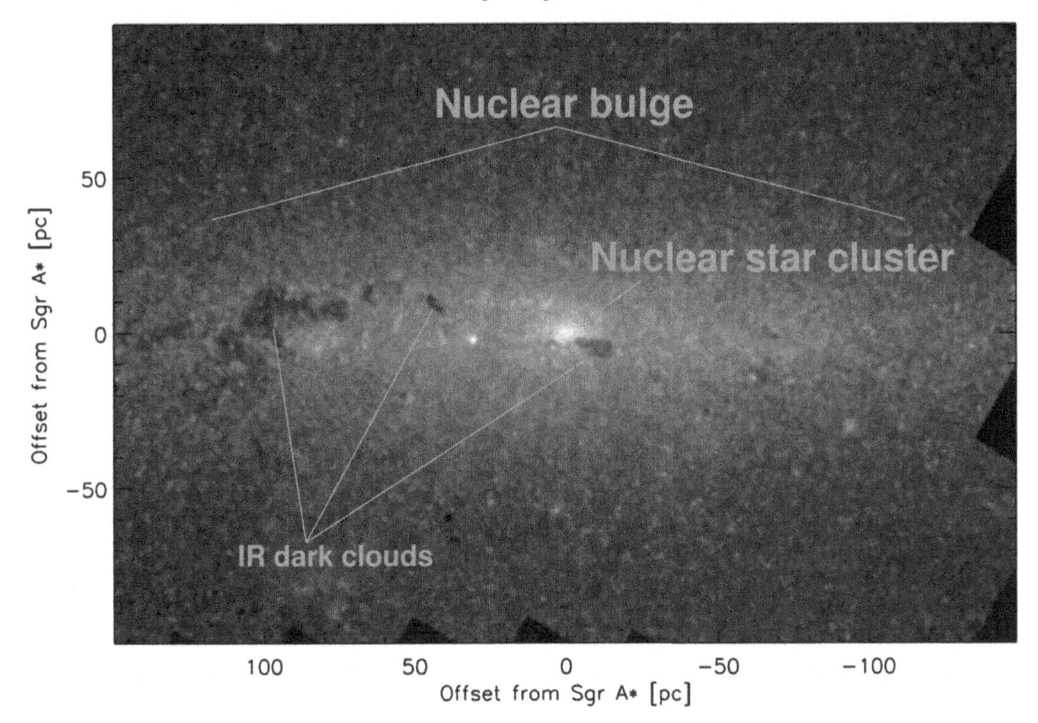

Figure 1. Extinction-corrected (using [3.6μm] − [4.5μm] colors) Spitzer/IRAC 4.5 μm image of the GC (see Stolovy *et al.*, 2006; re-binned to a 16" × 16" pixel size). Extinction reaches a minimum at this wavelength. The Galactic Plane runs horizontally across the middle of this image.

averaged number density of the NSC is $\rho(r) \propto r^{-1.75\pm0.1}$, i.e. close to the predicted value, but only at distances $r > 0.5$ pc from Sagittarius A* (Schödel *et al.* 2007). Further inside, the number density of the old – and therefore probably dynamically relaxed – stars is too flat to be consistent with the cusp scenario (Buchholz *et al.* 2009). This observation could be attributed to a number of reasons, like a very long relaxation time (Merritt 2010) or the effect of stellar collisions on giant stars (Dale *et al.* 2009). The presence (or not) of a stellar cusp at the GC must be considered ongoing research.

Similar to NSCs in other late-type galaxies, there are clear signs of recent star formation at the GC. Supergiants trace star formation about 100 Myr ago and young, massive stars within 0.5 pc of Sagittarius A* provide evidence of a star formation event a few Myr ago (Krabbe *et al.* 1995). There are indications that young, massive stars can be found throughout the entire NSC (Nishiyama & Schödel, 2012).

In summary, the Milky Way's NSC appears to be a close cousin to the NSCs found in other galaxies (Böker 2010).

References

Böker, T. 2010, *IAU Symposium* 266, 58

Buchholz, R. M., Schödel, R., & Eckart, A. 2009, *A&A* 499, 483

Dale, J. E., Davies, M. B., Church, R. P., & Freitag, M. 2009, *MNRAS* 393, 1016

Gillessen, S., Eisenhauer, F., Trippe, S., *et al.* 2009, *ApJ* 692, 1075

Graham, A. W. & Spitler, L. R. 2009, *MNRAS* 397, 2148

Krabbe, A., Genzel, R., Eckart, A., *et al.* 1995, *ApJL* 447, L95

Launhardt, R., Zylka, R., & Mezger, P. G. 2002, *A&A* 384, 112

Meyer, L., Ghez, A. M., Schödel, R., *et al.* 2012, *Science* 338, 84
Merritt, D. 2006, *Rep.Prog.Phys.* 69, 2513
Merritt, D. 2010, *ApJ* 718, 739
Nishiyama, S. & Schödel, R. 2012, *MNRAS* in press
Schödel, R., Eckart, A., Alexander, T., *et al.* 2007, *A&A* 469, 125
Schödel, R., Merritt, D., & Eckart, A. 2009, *A&A* 502, 91
Schödel, R., Najarro, A., Muzic, K., & Eckart, A. 2010, *A&A* 511, A18
Stolovy, S., *et al.* 2006, *J. Phys. Conf. S.*, 54, 176
Trippe, S., Gillessen, S., Gerhard, O. E., *et al.* 2008, *A&A*, 492, 419

Highlights of Astronomy, Volume 16
XXVIIIth IAU General Assembly, August 2012
T. Montmerle, ed.

doi:10.1017/S1743921314005729

Dwarf Galaxies and Globular Clusters

Michele Bellazzini

INAF - Osservatorio Astronomico di Bologna, via Ranzani 1, 40127, Bologna, Italy
email: michele.bellazzini@oabo.inaf.it

Abstract. I briefly explore some relevant connections and differences between the evolutionary paths of dwarf galaxies and globular clusters.

Keywords. galaxies: dwarf, globular clusters, stars: abundances

---

Two decades of deep photometric and spectroscopic studies of nearby dwarf galaxies has revealed that significant differences in the Star Formation (SF) and chemical evolution histories can be observed also among gas poor and predominantly old dwarf spheroidals (see Tolstoy *et al.* 2009). Still, a recent systematic study of 60 nearby dwarf galaxies showed that the average dwarf galaxy formed more than half of its stars by z=2, i.e. in the first 2-3 billion years from the big bang, irrespectively of its morphological type (Weisz et. al 2010). It is very interesting to note that *(a)* this was also the epoch when the large majority of the Galactic globular clusters (GC) were formed (Dotter *et al.* 2010), and *(b)* during which Milky Way (MW) sized galaxies are predicted to assemble most of their mass by hierarchical merging, according to the more recent Λ-CDM simulations (Abadi *et al.* 2003, Fakhouri *et al.* 2010). The fact that, in average, half of the stellar mass of dwarfs was produced in the earliest 2-3 Gyr and half in the following 10 Gyr, indicates that SF proceeded at a very different pace in two phases: it was very fast in the first phase, with chemical enrichment dominated by SNII and production of metal-poor stars with enhanced abundance of α elements, and slower (and more discontinuous) in the second phase, with self-enrichment dominated by SNI and production of more metal-rich stars and [α/Fe] declining with metallicity. It was in this second long-lasting phase that most of the observed dwarf-to-dwarf differences in the SF history took place.

Hence, dwarf galaxies - in their earliest times - appear as the ideal environment for the birth of GCs, also providing the gas pre-enriched in metals and α-enhanced that was required to produce the GCs that we observe today (there is no known GC with [Fe/H]< -2.5, hence none of them was formed from pristine gas). Now we have direct proofs that at least a significant fraction of GCs in MW-sized galaxies comes from disrupted/disrupting satellites (Bellazzini *et al.* 2003, Law & Majewski 2010, Mackey *et al.* 2010).

The recent discovery that also GCs were able to produce more than one generation of stars, with chemical self-enrichment, may be seen as an unexpected similarity between GCs and dwarf galaxies. In fact this marks a deep difference in the evolutionary path of the two classes of stellar systems:*(a)* the chemical evolution of dwarf galaxies was driven by supernovae, leading to correlated enrichment in iron, α elements, etc., while *(b)* the chemical evolution of the overwhelming majority of GCs was driven by low-energy polluters, leading to significant (and correlated) spread in the abundance of a few light elements (He, C, N, Na, O, Mg, Al) and *no enrichment in iron* (Gratton *et al.* 2012). The specific abundance pattern observed in GCs has been suggested as a *defining property* for these systems (Carretta *et al.* 2010b). This is a powerful approach, but the comparisons shown in Fig. 1 suggest that it may not be sufficient to embrace the whole diversity of observed cases. Stars from clusters associated with the disrupting Sgr dSph (as well

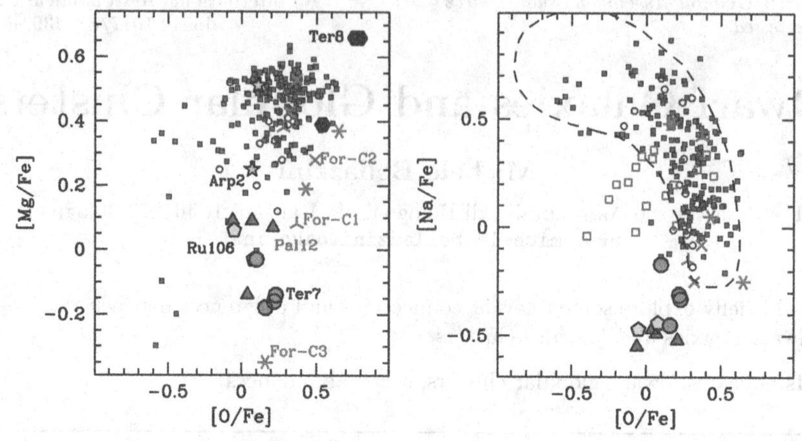

Figure 1. Distributions of Mg vs. O and Na vs. O abundances for stars in different star clusters. Small red dots are stars from 15 MW GCs (Carretta *et al.* 2009); large symbols (labelled) are stars from GCs in dSph satellites of the MW (Fornax and Sagittarius) plus Ru106 (Letarte *et al.* 2006, Brown *et al.* 1997, Cohen 2004, Mottini *et al.* 2008, Sbordone *et al.* 2005). Small open circles are stars from LMC GCs (Mucciarelli *et al.* 2010). Open squares are *mean* abundances of a set of open clusters from de Silva *et al.* (2009); the dashed contour enclose the distribution of stars in the GC M54, in the central nucleus of Sgr (Bellazzini *et al.* 2008, Carretta *et al.* 2010b).

as Ru 106, likely coming from a completely disrupted satellite) have Na, O, and Mg abundance very different from any other star from Galactic GCs and open clusters (see also K. Venn, these proceedings). The chemical initial conditions set by the evolutionary path of the host dwarf galaxy are likely another important factor in determining the subsequent chemical evolution of GCs.

Acknowledgements

I acknowledge the financial support of INAF through the PRIN Grant assigned to the project *Formation and evolution of massive star clusters*, P.I.: R. Gratton.

References

Abadi, M. G., Navarro, J. F., Steinmetz, M., & Eke, V. R. 2003, *ApJ* 591, 499
Bellazzini, M., Ferraro, F. R., & Ibata, R. A. 2003, *AJ*, 405, 577
Bellazzini, M., Ibata, R. A., & Chapman, S. C. 2008, *AJ*, 136, 1147
Brown, J. A., Wallerstein, G., & Gonzalez, G. 1997, *AJ*, 114, 180
Carretta, E., Bragaglia, A., Gratton, R. G., *et al.* 2009, *A&A* 505, 117
Carretta, E., Bragaglia, A., Gratton, R. G., *et al.* 2010a, *A&A* 516, 55
Carretta, E., Bragaglia, A., Gratton, R. G., *et al.* 2010b, *A&A* 52, 95
Cohen, J. G. 2004, *AJ*, 127, 1545
de Silva, G. M., Gibson, B. K., Lattanzio, J., & Asplund, M. 2009, *A&A*, 500, L25
Dotter, A., Sarajedini, A., Anderson, J. *et al.*(2010) *ApJ* 708, 698
Fakhouri, O., Ma, C.-P., & Boylan-Kolchin, M. 2010, *MNRAS*, 406, 2267
Gratton, R. G., Carretta, E., & Bragaglia, A. 2012, *A&ARv*, 20, 50
Law, D. R. & Majewski, S. R. 2010, *ApJ* , 714, 229
Letarte, B., Hill, V., Jablonka, P., Tolstoy, E., Francois, P., & Meylan, G. 2006, *A&A* 453, 547
Mackey, A. D., *et al.* 2010, *ApJ* 717, L11
Mottini, M., Wallerstein, G., & McWilliam, A. 2008, *AJ*, 136, 614
Mucciarelli, A., Origlia, L., & Ferraro, F. R. 2010, *ApJ* 717, 277
Sbordone, L., Bonifacio, P., Marconi, G., Buonanno, R., & Zaggia, S. 2005, *A&A* 437, 905
Tolstoy, E., Hill, V., & Tosi, M. 2009, *ARA&A*, 47, 371
Weisz, D. R., Dalcanton, J. J., Williams, B. F., *et al.* 2011, *ApJ* 739, 5

Highlights of Astronomy, Volume 16
XXVIIIth IAU General Assembly, August 2012
T. Montmerle, ed.

doi:10.1017/S1743921314005730

Seeking footprints of the primeval Universe in dwarf galaxies

Sebastian L. Hidalgo[1,2] and the LCID group.

[1]Instituto de Astrofísica de Canarias. Vía Láctea s/n. E38200 - La Laguna, Tenerife, Canary Islands, Spain
email:shidalgo@iac.es

[2]Department of Astrophysics, University of La Laguna. Vía Láctea s/n. E38200 - La Laguna, Tenerife, Canary Islands, Spain

Abstract. We present the star formation histories (SFHs) of four isolated dwarf galaxies, Cetus, Tucana, LGS-3, and Phoenix, as a function of galactocentric radius. Our results suggest that beyond some distance from the center, there are no significative differences in fundamental properties of these galaxies, such as the star formation rate (SFR) or age-metallicity relation (AMR). The stellar content of this region would be composed of old ($\gtrsim$ 10.5 Gyr) metal-poor stars only. In the innermost regions, dwarf galaxies appear to have formed stars during time intervals which duration varies from galaxy to galaxy. This extended star formation produces the dichotomy between dwarf spheroidal (dSph) and dwarf Transition (dTr) galaxy types.

Keywords. galaxies: dwarf, galaxies: evolution, Local Group.

We present the SFHs of four isolated dwarf galaxies, Cetus, Tucana, LGS-3, and Phoenix, as a function of galactocentric radius. The results are based on the observations of the LCID project (Local Cosmology form the Isolated Dwarfs). The aim of this project is to obtain the full SFHs of six isolated dwarf galaxies of the Local Group: Leo-A, Phoenix, Cetus, Tucana, LGS-3, and IC1613. The global SFHs of these galaxies has been already described in Cole, Skillman, Tolstoy, *et al.* (2007), Hidalgo, Aparicio, Martínez-Delgado, *et al.* (2009), Monelli, Hidalgo, Stetson, *et al.* (2010a), Monelli, Gallart, Hidalgo, *et al.* (2010b), Hidalgo, Aparicio, Skillman, *et al.* (2011), and Skillman *et al.* (2013), respectively. We have selected a subsample of the dwarf galaxies of the LCID project to obtain the SFH as a function of the galactocentric radius: two dSph galaxies, Cetus and Tucana, and two transition (dTr) galaxies, LGS-3 and Phoenix.

For a clearer and more consistent comparison of the results, four of the regions (R1 to R4) were selected in each galaxy. The three innermost ones are such that their equivalent radii approximately correspond to α_ψ (R_1), 1.5α_ψ (R_2), and 2α_ψ (R_3) where α_ψ is the scale length or the stellar mass distribution of each galaxy. R_4 corresponds to the area outer of 2α_ψ. This fourth area has not been used in Phoenix due to the low number of stars observed in it (less than 3% of the total).

We have used a CMD fitting technique (IAC-star/IAC-pop/MinnIAC) to obtain the SFHs as described in Aparicio & Hidalgo (2009) and Hidalgo, Aparicio, Skillman, *et al.* (2011). Figure 1 (left panel) shows the SFR as a function of time, $\psi(t)$, of Cetus, Tucana, LGS-3, and Phoenix obtained for the four regions defined above. The SFR has been normalized to its time integral for each region. There is star formation activity in all gallaxies for ages $\gtrsim$ 10 Gyr, regardless of distance to the center. Interestingly, in the outermost area (R_4) the three normalized SFRs are almost indistinguishable within error bars. However, $\psi(t)$ gradually decreases outwards for ages $\lesssim$ 9 Gyr. Few stars younger than this age exist in the R_4 regions, though some stars can still be observed at $\sim$ 3.5 Gyr

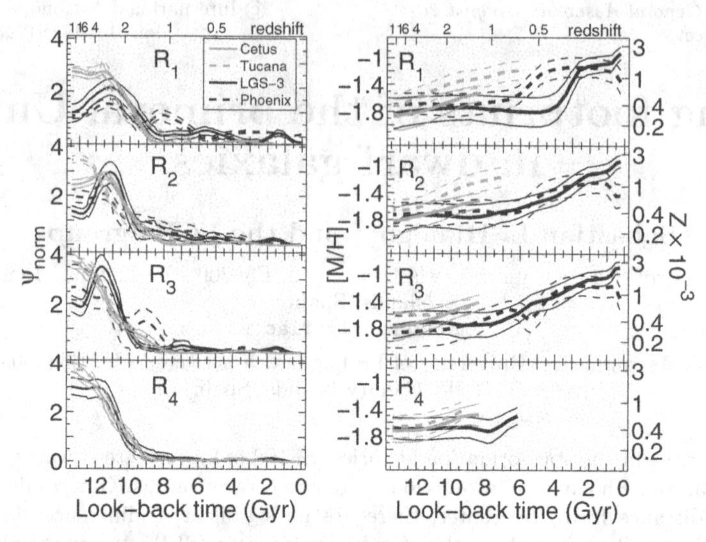

Figure 1. SFHs (left panel) and age-metallicity relations (rigth panel) as a function of galactocentric radius of four isolated dwarf galaxies of the Local Group.

for Cetus and Tucana and at ∼ 1.5 Gyr for LGS-3. These stars have been identified as a population of Blue-straggler stars (BSS) in Cetus and Tucana by Monelli, Cassis, Mapelli, *et al.* (2012). In the case of LGS-3, the expected BSSs may be mixed with some young and intermediate age stellar population, which produces the bump at 1.5 Gyr in $\psi(t)$.

Figure 1 (right panel) shows the AMR for the four regions in all the galaxies (three in the case of Phoenix). The overall trend is that metallicity increases with time. However, in the case of LGS-3 and Phoenix, the metallicity remains almost unchanged for most (LGS3) or half (Phoenix) of the lifetime of the galaxy in the central region. For these galaxies, the AMR relation can be separated into two distinct time periods: a first period involving the older stars where metallicity does not change significantly with time, and a final period where the metallicity increases steeply. The duration of the first period seems a function of radius, decreasing with increasing radius. In contrast, the metallicity of Cetus and Tucana increases steadily with time, showing no trend with radius. As a result, stars between 6 – 10 Gyr old are on average more metal poor in LGS-3 and Phoenix than in Cetus and Tucana.

References

Aparicio, A. & Gallart, C. 2004, *AJ*, 128, 1465
Aparicio, A. & Hidalgo, S. L. 2009, *AJ*, 138, 558
Cole, A. A., Skillman, E. D., Tolstoy, E., *et al.* 2007, *ApJL*, 659, L17
Hidalgo, S. L., Aparicio, A., Martínez-Delgado, D., & Gallart, C. 2009, *ApJ*, 705, 704
Hidalgo, S. L., Aparicio, A., Skillman, E., *et al.* 2011, *ApJ*, 730, 14
Monelli, M., Cassisi, S., Mapelli, M., *et al.* 2012, *ApJ*, 744, 157
Monelli, M., Hidalgo, S. L., Stetson, P. B., *et al.* 2010a, *ApJ*, 720, 1225
Monelli, M., Gallart, C., Hidalgo, S. L., *et al.* 2010b, *ApJ*, 722, 1864
Skillman *et al.* 2013, in preparation.

Highlights of Astronomy, Volume 16
XXVIIIth IAU General Assembly, August 2012
T. Montmerle, ed.

doi:10.1017/S1743921314005742

Connections between MWG Star Clusters and Dwarf Galaxies

Kim A. Venn
University of Victoria,
PO Box 3055 STN CSC, Victoria, BC, Canada V8W 3P6
email: kvenn@uvic.ca

1. Introduction

It seems that in the past decade, there have been two paradigm shifts regarding star clusters. Firstly, the observational evidence for multiple stellar populations requires more extended and often complex star formation histories in star clusters. Secondly, theoretical models that form globular clusters in dwarf galaxies that are accreted at very early epochs ($z > 5$) are able to reproduce the age-metallicity relations observed. For the accretion scenario to be viable, globular clusters should also resemble the chemistry of at least some dwarf galaxies.

2. A Chemical Comparison

We investigate the chemical connections between star clusters and dwarf galaxies here, but sadly, this is a short examination. At low metallicities, the chemical composition of globular clusters, Galactic halo stars, and dwarf galaxies are all similar (Venn *et al.* 2004, Tafelmeyer *et al.* 2010, Frebel & Norris 2011), with few clues as to potential differences in their origins or birth places. Only at intermediate metallicities ([Fe/H] > -1.0) do dwarf galaxies show unique chemical properties due to their slower star formation histories compared to most stars in the Galaxy, and nearly all of those are associated with the Sgr dwarf remnant. A few interesting exceptions include Rup 106 (see talk by Doug Geisler) and Pal 1 (Sakari *et al.* 2011), both showing unique chemical signatures but neither associated with the Sgr remnant.

When chemical abundances differ in dwarf galaxies, this can usually be traced through several elements, and not only the [α/Fe] ratios. For example, Pal 12 and Ter 7 both show lower abundances of [Na/Fe], [Al/Fe], [Cu/Fe], and [Zn/Fe] suggestive of metallicity dependent AGB and/or SNe Ia yields These lower abundances are also seen in dwarf galaxies, such as low Na in Fornax (Letarte *et al.* 2010), low Cu in the LMC (Pompeia *et al.* 2008), and all four are low in the Sgr field stars (Monaco *et al.* 2007)

3. Younger Clusters From Dwarf Galaxies?

A "vital diagram" of mass versus half light radius for Sgr clusters suggests that some open clusters could be associated with the Sgr dwarf remnant (Law & Majewski 2010). These include the peculiar young halo clusters, such as Whiting 1, and open clusters Berkeley 29 and Saurer 1. The ages ($\sim$5 Gyr) and metallicities ([Fe/H] ~ -0.6) of these clusters are similar to the Sgr clusters Pal 12 and Ter 7, however the chemistries of Be 29 and Sa 1 such as [α/Fe] and [Na/Fe] ratios are more similar to stars in the Galaxy (Yong *et al.* 2005, Carraro *et al.* 2004; see Figure 1). Upper limits on their Cu and Zn abundances do not add further constraints. Heavier elements such as La and Ba are inconclusive.

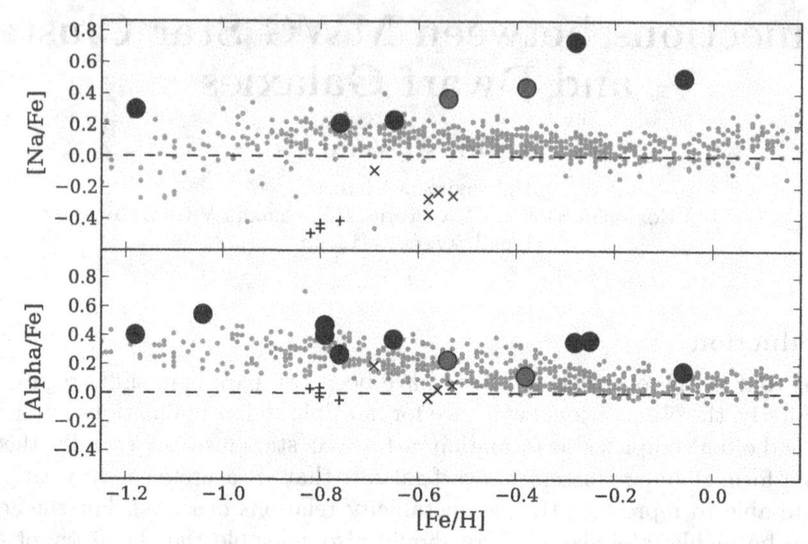

Figure 1. Elemental abundances for Berkeley 29 and Saurer 1 (red points; Yong *et al.* 2005, Carraro *et al.* 2004), and other Galactic globular clusters (black points, Pritzl *et al.* 2005) and Galactic field stars (grey points, Venn *et al.* 2004). The Sgr clusters Pal 12 and Ter 7 are shown by crosses and "x" marks (Cohen 2004, Sbordone *et al.* 2007), respectively, and are clearly lower than the Galactic stars including Be 29 and Sa 1. [Alpha/Fe] is an average of Mg and Ca.

4. Discussion

The similarities in age and metallicity of young stellar clusters in the Galactic halo with Sgr dwarf remnant clusters is striking, and seems unlikely to be only a coincidence. Perhaps these clusters are associated with the Sgr merger through gas deposition and compressed along the orbital plane (e.g., K. Bekki's talk). Or perhaps the young stellar clusters are associated with the Galactic disk, and their ages are similar to the Sgr clusters if they were triggered to collapse through tidal interactions during the accretion of Sgr. Thus, currently, there are no chemical abundances that link young stellar clusters near the Galactic disk with the Sgr stellar clusters. However, with higher order element abundances (Ba, Sr, Cu, Zn) and homogenized analyses (i.e., see comments by Friel *et al.* 2010), the origins of the MWG star clusters can be further clarified.

References

Carraro G., Bresolin F., Villanova S., *et al.* 2004, *AJ*, 128, 1676
Cohen J. 2004, *AJ*, 127, 1545
Friel E., Jacobson H. R., & Pilachowski C. A. 2010, *AJ*, 139, 1942
Frebel A. & Norris J. E. 2011, to appear in *Planets, Stars and Stellar Systems* Springer (arXiv:1102.1748)
Law D. & Majewski S. 2010, *ApJ*, 718, 1128
Letarte B., Hill V., Tolstoy E., *et al.* 2010, *A&A*, 523, 17
Monaco L., Bellazzini M., *et al.* 2007, *A&A*, 464, 201
Pompeia L., Hill V., Spite M., *et al.* 2008, *A&A*, 480, 379
Pritzl B. J., Venn K. A., & Irwin M. J. 2005, *AJ*, 130, 2140

Sakari C. M., Venn K. A., *et al.* 2011, *ApJ*, 740, 106
Sbordone L., Bonifacio P., *et al.* 2007, *A&A*, 465, 815
Tafelmeyer M., Jablonka P., Hill V., *et al.* 2010, *A&A*, 524, 58
Tolstoy, E., Hill V. & Tosi, M. 2009, *ARAA*, 47, 371
Venn K. A., Irwin M. J., Shetrone M. D. *et al.* 2004, *AJ*, 128, 1177
Yong D., Carney B. W., & Teixera de Almeida M. L. 2005, *AJ*, 130, 597

Highlights of Astronomy, Volume 16
XXVIIIth IAU General Assembly, August 2012
T. Montmerle, ed.

doi:10.1017/S1743921314005754

The Chemical Evolution of Milky Way Satellite Galaxies from Keck Spectroscopy

Evan N. Kirby†

Department of Physics, University of California, 4129 Reines Hall, Irvine, CA 92697, USA
email: ekirby@uci.edu

Abstract. The primary data product of a recent Keck/DEIMOS spectroscopic campaign of eight Milky Way dwarf spheroidal (dSph) satellite galaxies is a catalog of nearly 3000 red giants with spectral synthesis-based abundance measurements of Fe and the α elements Mg, Si, Ca, and Ti. The dSph metallicity distributions show that the histories of the less luminous dSphs were marked by massive amounts of metal loss. The [α/Fe] distributions reveal that the early star formation histories (at a lookback time of > 12 Gyr) of most dSphs were very similar and that Type Ia supernova ejecta contributed to the abundances of all but the most metal-poor ([Fe/H] < −2.5) stars. This large data set allows inferences of past outflows from the dSphs in order to determine their contribution to the metal content of the intergalactic medium.

Keywords. galaxies: dwarf, galaxies: abundances, galaxies: evolution, galaxies: stellar content, intergalactic medium, stars: abundances

---

The Milky Way dSphs are attractive targets for resolved stellar spectroscopy because their stellar populations are at uniform distances in small patches of the sky. Spectroscopically derived chemical abundances can reveal their early star formation histories, where color-magnitude diagrams have limited sensitivity to relative ages. Kirby, Guhathakurta, Simon, *et al.* (2010) observed about 3000 red giants in eight dSphs. Using spectral synthesis, they measured metallicities ([Fe/H]) for each of these stars. When possible, they also measured the abundances of four individual α elements: Mg, Si, Ca, and Ti.

The average metallicities of dSphs range from one tenth to one hundredth of the solar metallicity. The lack of metals indicates that the galaxies suffered a great deal of metal loss. Although the missing metals cannot be observed directly, the degree of loss can be inferred from the stellar mass and current metallicity. The stellar mass—combined with an assumed initial mass function and theoretical supernova yields—dictates the total amount of metals that the stellar population produced. The amount of metals lost, shown in Figure 1, is the difference between the amount produced and the present metal content. The dSphs lost the vast majority of their metals. However, they were not significant contributors to the intergalactic medium (IGM). Fornax alone injected more Fe into the IGM than all smaller dSphs combined. More massive galaxies presumably enriched the IGM even more. For more information, see Kirby, Martin, & Finlator (2011).

Despite the near inability of the dSphs to retain any metals, they did experience chemical evolution. Figure 2 shows that all of the [α/Fe] ratios decline with increasing [Fe/H]. This pattern indicates that Type Ia supernovae exploded during the star formation lifetimes of all of the dSphs. Therefore, the dSphs must have formed stars over at least the delay time of a Type Ia supernova, which is at least 100 Myr. Kirby, Cohen, Smith, *et al.* (2011) developed a chemical evolution model that suggests that even Ursa Minor, the dSph with the shortest star formation lifetime, actually lasted for at least 500 Myr.

† Center for Galaxy Evolution Fellow

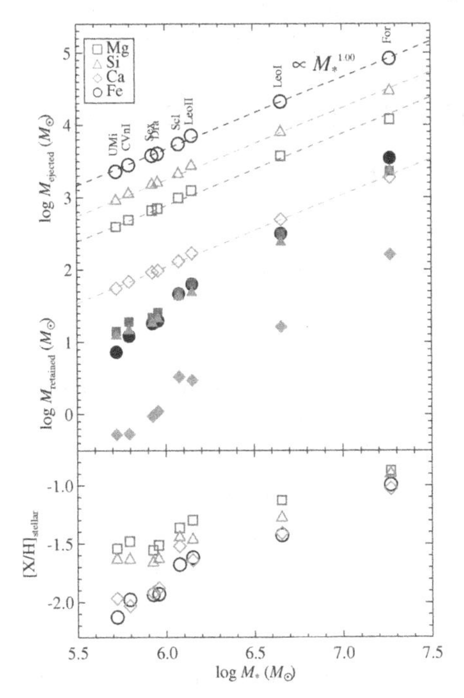

Figure 1. *Top:* Metals presently in dSphs (*filled symbols*) and inferred to have been lost (*open symbols*) versus stellar masses. *Bottom:* The dSph mass-metallicity relation that Kirby, Lanfranchi, Simon, *et al.* (2011) measured from Keck/DEIMOS spectroscopy.

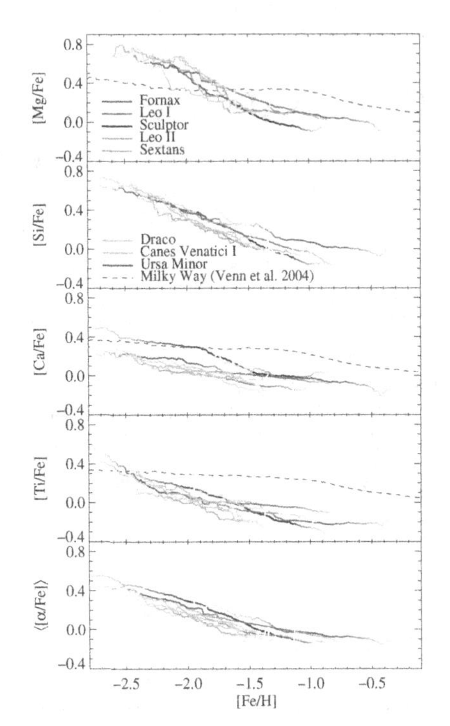

Figure 2. The moving averages of [α/Fe] ratios versus metallicity compared to the Milky Way halo. The bottom panel shows the averages from the top four panels. Different dSphs are indicated by color or shade.

The [α/Fe] paths in Figure 2 show no plateau in the metallicity range $-2.5 <$ [Fe/H] $<$ -1.0. The only exception is [Ca/Fe] in Sculptor. This pattern shows that the ejecta from Type Ia supernovae polluted the dSphs even when the metallicity was only [Fe/H] $= -2.5$. There are two possibilities: (1) Type Ia supernovae can be extremely prompt, or (2) the star formation rates were so low in dSphs that over 100 Myr (the minimum delay time for a Type Ia supernova) elapsed before the metallicity of the dSphs reached [Fe/H] $= -2.5$.

Acknowledgements

I thank the conference organizers for their kind invitation. Support for this work was provided by NASA Hubble Fellowship grant 51256.01 and by the Southern California Center for Galaxy Evolution sponsored by the University of California Office of Research.

References

Kirby, E. N., Cohen, J. G., Smith, G. H., Majewski, S. R., Sohn, S. T., & Guhathakurta, P. 2011, *ApJ* 727, 79

Kirby, E. N., Guhathakurta, P., Simon, J. D., Geha, M. C., Rockosi, C. M., Sneden, C., Cohen, J. G., Sohn, S. T., Majewski, S. R., & Siegel, M. 2010, *ApJS* 191, 352

Kirby, E. N., Lanfranchi, G. A., Simon, J. D., Cohen, J. G., & Guhathakurta, P. 2011, *ApJ* 727, 78

Kirby, E. N., Martin, C., & Finlator, K. 2011, *ApJ* (Letters) 742, L25

Venn, K. A., Irwin, M., Shetrone, M. D., Tout, C. A., Hill, V., & Tolstoy, E. 2004, *AJ* 128, 1177

Highlights of Astronomy, Volume 16
XXVIIIth IAU General Assembly, August 2012
T. Montmerle, ed.

doi:10.1017/S1743921314005766

The Galactic halo: stellar populations and their chemical properties

John E. Norris

Research School of Astronomy and Astrophysics, The Australian National University, Weston, ACT 2611, Australia
email: jen@mso.anu.edu.au

Abstract. Below [Fe/H] = –3.0, there is an enormous range in [C/Fe]. We discuss the properties of C-rich ([C/Fe] > +0.7) and C-normal ([C/Fe] $\leqslant$ +0.7) stars in this regime, and suggest that there existed two different gas cooling channels in the very early Universe.

Keywords. stars: abundances, Galaxy: abundances, Galaxy: halo, cosmology: early Universe

---

1. Introduction

The stellar populations of the Galactic halo provide insight into the manner in which the Milky Way formed, while its most metal-poor stars have the potential to constrain the nature of the first stars, believed to have formed some 100 Myr after the Big Bang. We refer the reader to Frebel & Norris (2011) and Carollo (this volume), and references therein, for the rich background to these topics. Here we shall concentrate on the origins of C-rich and C-normal stars having [Fe/H] $\lesssim$ –3.0.

2. The C-rich and C-normal populations below [Fe/H] $\sim$ –3.0

2.1. *Chemical abundances for stars with [Fe/H] < –3.1*

Many researchers have contributed to the search for the most metal-poor stars. Here we utilize the recent work of Yong *et al.* (2012), to whom we refer the reader for the chemical abundances (and related references)of some 85 stars with [Fe/H] < –3.1. Suffice it to say that Yong *et al.* report new abundances for some 16 elements in $\sim$ 20 stars in this abundance range, together with abundances re-determined for a further $\sim$ 65 objects from the literature. Of this sample, some 18 are C-rich, with [C/Fe] > +0.7. Of the remainder, 35 have [C/Fe] $\leqslant$ +0.7, which we shall refer as C-normal stars.

Norris *et al.* (2012) have used these data to investigate the abundance trends and other relationships between the two groups. We refer the reader to that work for details. Their main results include:

- All of the C-rich stars belong to, or appear related to, the CEMP-no subclass of Carbon-Enhanced Metal-Poor stars (Beers & Christlieb 2005). None are CEMP-s, -r, or r/s.
- The C-rich stars are oxygen rich; the light elements Na, Mg, and Al are enhanced relative to Fe in half the sample; and for Z > 20 (Ca) there is little evidence for enhancements relative to solar values.
- While more radial-velocity data are required, there is no support for the hypothesis that the C-rich stars are all members of binary systems. That is to say, the binary statistics for CEMP-no stars are decidedly different from those of CEMP-s stars.

2.2. *Possible explanations for the abundance patterns*

Here are suggestions that may be relevant for an explanation of the observations:

- Fine-structure line transitions of C II and O I as a major cooling agent in the early Universe (Bromm & Loeb 2003)
- Supermassive ($M > 100\ M_\odot$), rotating stars (Fryer *et al.* 2001)
- "Mixing and fallback" Type II SNe ($M \sim 10 - 40\ M_\odot$) (Umeda & Nomoto 2003)
- Rotating, massive ($\sim 60\ M_\odot$) stars (Meynet *et al.* 2006)

In particular, the chemical abundances of the C-rich stars are best explained in terms of the admixing and processing of material from H-burning and He-burning regions as achieved by nucleosynthesis in zero-heavy-element models of "mixing and fallback" supernovae (SNe), and of rotating massive stars. The existence of a large fraction of C-rich and O-rich stars at lowest Fe abundances is suggestive of a strong role by carbon and oxygen in the formation of stars at the earliest times.

2.3. *A scenario for the earliest times*

We suggest that the C-rich and C-normal populations result from two different gas cooling channels in the very early Universe, of material that formed the progenitors of the two populations. The first was cooling by fine-structure line transitions of C II and O I (to form the C-rich population); the second, while not well-defined (dust-induced cooling? e.g., Schneider *et al.* 2006), led to the C-normal group. Here is a possible sequence:

- The first stars form in "mini dark halos" from material comprising only H and He; the cooling is provided by molecular hydrogen; and the mass function of these objects is top-heavy ($M \gtrsim 20 - 300\ M_\odot$). None of these objects survives until the present time.
- Some fraction of the first population synthesizes large amounts of C and O, as described by the above stellar evolutionary models (the rotating $60 - 300\ M_\odot$ stars of Meynet *et al.* 2006 and Fryer *et al.* 2001) and/or the 'mixing-and-fallback" models (Umeda & Nomoto 2003). During subsequent star formation, material with large enhancements of C and O cools via the fine structure lines of C II and O I, and fragments to form low-mass, long-lived stars still observed today. This is the C-rich population.
- The remainder of the first generation stars does not produce large amounts of carbon, but rather more solar-like abundance patterns. This material has more difficulty in cooling and fragmenting, but several possibilities exist (e.g. dust-induced star formation). A second channel forms carbon normal low-mass, long-lived stars, on a longer timescale. This is the C-normal population.

Acknowledgements

The author would like to gratefully acknowledge his collaborations on this topic, over many years, with M. Asplund, T. C. Beers, M. S. Bessell, N. Christlieb, A. Frebel, G. Gilmore, S. G. Ryan, R. F. G. Wyse and D. Yong.

References

Beers, T. C. & Christlieb, N. 2005, *ARA&A*, 43, 531
Bromm, V. & Loeb, A. 2003, *Nature*, 425, 812
Frebel, A. & Norris, J. E. 2011, arXiv:1102.1748
Fryer, C. L., Woosley, S. E., & Heger, A. 2001, *ApJ*, 550, 372
Norris, J. E., Yong, D., Bessell, M. S., *et al.* 2012, *ApJ*, submitted
Schneider, R., Omukai, K., Inoue, A. K., & Ferrara, A. 2006, *MNRAS*, 369, 1437
Umeda, H. & Nomoto, K. 2003, *Nature*, 422, 871
Yong, D., Norris, J. E., Bessell, M. S., *et al.* 2012, *ApJ*, in press

Highlights of Astronomy, Volume 16
XXVIIIth IAU General Assembly, August 2012
T. Montmerle, ed.

doi:10.1017/S1743921314005778

CN Anomalies in the Halo System

Daniela Carollo[1,2]

[1]Dept. of Physics and Astronomy - Astronomy Astrophysics and Astrophotonic Research Center
Macquarie University - North Ryde, 2109, NSW, Australia
email: daniela.carollo@mq.edu.au

[2]INAF - Astronomical Observatory of Turin - Italy

Abstract. I present an evaluation of the kinematic properties of halo red giants thought to have formed in globular clusters based on the strength of their UV/blue CN and CH absorption features. The sample has been selected from the catalog of Martell *et al.* (2011). The orbital parameters of CN-strong halo stars are compared to those of the inner and outer halo populations, and to the orbital parameters of globular clusters with well-studied Galactic orbits. It has been found that both the clusters and the CN-strong field stars exhibit kinematic and orbital properties similar to the inner halo population, indicating that globular clusters could be a significant source of inner halo field stars, and suggesting that both globular clusters and CN-strong stars could belong primarily to the inner halo population of the Milky Way.

Keywords. Galaxy: halo, Galaxy: formation, Galaxy: structure, Galaxy: kinematics and dynamics, Galaxy: stellar content, Globular Clusters: general

1. Selection of the sample

The Martell & Grebel (2010) and Martell *et al.* (2011) studies of halo field giants drew their data from the SDSS-II/SEGUE (Abazajian *et al.* 2009; Yanny *et al.* 2009) and SDSS-III/SEGUE-2 (Aihara *et al.* 2011; Eisenstein *et al.* 2011) surveys, respectively. The sample have been further refined to include only those stars with available proper motions, which means that the star satisfies additional criteria designed to eliminate spurious reported motions. Also, stars belonging to the SDSS/SEGUE fields that fall in the direction of the Sagittarius stream were removed, in order to remove possible contaminants. A sample of Galactic globular clusters (hereafter, GCs) with available proper motions from the literature (http://www.astro.yale.edu/dana/gc.html), has also been selected in order to compare the properties of these GCs with those of the CN-strong stars, and discuss them in the context of the inner- and outer-halo populations described in Carollo *et al.* (2007, 2010). The sample comprises 59 GCs for which positions, absolute proper motions, distances, and radial velocities are listed.

2. Results

The analysis shows that the CN-strong stars are concentrated in the Inner Halo Region (IHR; Carollo *et al.*, in prep., Tissera *et al.*, in prep.), and their frequency drops rapidly beyond 20 kpc, as previously pointed out by Martell *et al.* (2011). Another remarkable feature is that the great majority of the orbits of the CN-strong stars are located within $Z_{max} < 10$ kpc, again corresponding to the IHR, where the Inner Halo Population (IHP; Carollo *et al.*, in prep., Tissera *et al.*, in prep.) dominates in the metallicity range $-2.0 <$ [Fe/H] < -1.5. It has been found that the mean rotational velocity and dispersion for the metal poor CN-normal stars is $\langle V_\phi \rangle = 25 \pm 6$ km s^{-1}, and $\sigma_{V_\phi} = 100 \pm 4$ km s^{-1}, consistent with membership in the inner-halo population ($\langle V_\phi \rangle = 7 \pm 4$ km s^{-1}, and $\sigma_{V_\phi} = 95 \pm 2$ km s^{-1}; Carollo *et al.* 2010). Application of a two-sample Kolmogorov-Smirnoff test to the distributions of rotational velocity for the low-metallicity CN-normal and CN-strong stars is unable to reject the hypothesis that they were drawn from the same parent population ($p = 0.62$). Note that the number of stars with highly retrograde velocities in the low-metallicity subsample is very small, $N_{retr} = 33$ at $V_\phi < -100$ km s^{-1}, and $N_{retr} = 7$, at $V_\phi < -200$ km s^{-1}, respectively. Among these groups of stars, none of them are CN-strong. The general properties of the GCs in our sample with available absolute proper motions are typical of the Milky Way's cluster population in terms of their spatial and metallicity distributions. The values of rotational velocity and dispersion for the sub-sample at low metallicity are consistent with membership in the IHP, perhaps with some contamination from a higher-dispersion population.

3. Implications

The fact that the CN-strong stars exhibit spatial distributions, rotational velocities, and orbital properties in agreement with the IHP provides important clues on the origin and fate of GCs in the Milky Way. The primordial sub-Galactic fragments of higher mass and gas content presented favorable conditions to form GCs in the inner cores of giant high-density clouds. By way of contrast, smaller-mass fragments may not have had sufficient masses of gas to form GCs. These lower-mass mini-halos would likely have had a truncated star-formation history, relative to the higher-mass mini-halos, since they would not have been able to retain gas once star-formation commenced. Although further investigation is required, this may account in a natural way for the apparent lack of Galactic GCs with metallicity below [Fe/H] ~ -2.3. Since it is expected that CN-strong stars *require* the dense environment of GCs in order to form in the first place, their observed properties strongly suggest that a significant fraction of GCs have been stripped or disrupted in the IHR. In this context, the similarity of the global properties of the metal-poor GCs, including their spatial, kinematical and orbital properties, to those of the CN-strong stars is especially intriguing. Our present data certainly suggest a strong relationship between these two samples – both appear to be associated with the IHP of the Milky Way.

References

Aihara, H., *et al.* 2011, *AJ* 193, 29
Abazajian., *et al.* 2009, *ApJS* 182, 543
Carollo, D., *et al.* 2007, *ApJ* 450, 1020
Carollo, D., *et al.* 2010, *ApJ* 712, 692
Martell, S. L. & Grebel, E. K. 2010, *A&A* 519, 14
Martell, S. L., Smolinski, J. P., Beers, T. C., & Grebel, E. K. 2011, *A&A* 534, 136
Yanny, B., *et al.* 2009, *AJ* 137, 4377
Eisenstein, D., *et al.* 2011, *AJ* 142, 72

Highlights of Astronomy, Volume 16
XXVIIIth IAU General Assembly, August 2012
T. Montmerle, ed.

doi:10.1017/S174392131400578X

Extremely Metal Poor Stars in the Galaxy

P. François
Paris-Meudon Observatory, 61 Avenue de l'Observatoire, F-75014 PARIS
email: patrick.francois@obspm.fr

Abstract. Globular Clusters and metal poor stars represent two aspects of the population II stars in our Galaxy. The presentation will focus on the metal poor end of population II of galactic field stars.

Keywords. stars : abundances, stars : population II, Galaxy : halo.

---

1. Introduction

It is well know that the metallicity distribution of galactic halo stars and globular clusters is very different and the link between the two is not trivial. Recent observations have shown that the single metallicity population paradigm for globular clusters is no more valid for some Globular Clusters for which very accurate photometry can be performed (Piotto *et al.*, 2007). What can we say about galactic metal poor stars ?

2. Metal Poor Stars and Extremely Metal Poor Stars

The history of the study of metal poor stars is closely linked with the progress in observing facilities. As example, the early 80's abundance diagrams [Ca/Fe] vs [Fe/H] showed an over-solar value of this ratio in metal poor stars with some hints of an increasing value as metallicity decreases. A 10 years observational effort has permitted to confirm this progressive rise of the [Ca/Fe] as [Fe/H] decreases (Edvardsson *et al.* 1993). A more careful selection of metal poor stars thanks to kinematics has revealed the existence of substructures in the abundance diagrams [α/[Fe] (Nissen & Schuster 2010). These two distinct populations stars in the solar neighborhood belong to the metal-rich end of the halo metallicity distribution function. The standard single zone models of chemical evolution of the Galaxy where no or a small abundance dispersion is found at a given metallicity have to be extended to inhomogeneous models able to include these substructures. The recent discovery of extensive sub-structures (Belokurov *et al.* (2006) reveal that the halo has a more complex history than previously thought (Eggen *et al.* 1962, Searle *et al.*, 1978). Carollo (2007) has recently shown that the halo can be divided of two broadly overlapping structural components-an inner and an outer halo with different spatial density profiles, stellar orbits and stellar metallicities. As extremely metal poor stars are rare, the extension of this work towards very low metallicities requires ambitious large scale spectroscopic surveys. Among them, we can mention the LAMOST project for which a dedicated telescope equipped with 16 spectrographs have been built with the aim of acquiring 10^6 spectra. Another ongoing project is the Gaia-ESO survey (GES) which will use the FLAMES and UVES spectrographs with a goal of more than 2. $\times 10^5$ spectra for mostly F and G stars. Similar and complementary projects are foreseen in the near future (HERMES, APOGEE, 4MOST, MOONS to cite few of them).

These large spectroscopic surveys are also of paramount interest for their use to find the most metal poor stars in our Galaxy. These peculiar objects are identified from low to medium resolution spectra from large surveys (mainly H&K BPS, Hamburg-ESO Survey & Sloan Digital Sky Survey with SEGUE) by analyzing the strength of the Ca H& K lines used a metallicity indicator and photometric or spectro-photometric indices used as a temperature estimator. In 2005, the most metal poor star known was HE 1327-2326 (Frebel *et al.* 2005). This star has a very high enhancement in Carbon. This enhancement has been also found in the top five most metal poor stars. Frebel *et al.* (2007) concluded that forming low-mass stars should have a minimum amount of Carbon and Oxygen in order to allow the cooling of the forming low mass star. The consequence was that all the most metal poor stars should be C and O enriched. Several programs to find these most metal poor stars are ongoing. They are based on a selection of candidates on the low/med resolution spectroscopic surveys which are observed with high resolution spectroscopic observations (CASH, OZ, ...). This star formation cooling theory of low mass stars has been challenged by the discovery of a star Caffau *et al.* (2011) which has a metallicity $[Fe/H] = -5$ and no C enhancement falling in the range where stars could not form in the frame of Frebel's model of low-mass star formation.

3. The ESO Large Programme: Turn Off PrimOrdial Stars (ToPoS)

An ESO large program (ToPoS) led by E. Caffau has been initiated in 2012. This project has three main goals: to determine the metal-weak tail of the halo metallicity distribution function, below [M/H]=-3.5, where the low resolution SDSS spectra are inadequate, to determine the relative abundance of the elements in Extremely/Ultra metal poor stars, signature of the massive First stars and to determine the trend of the lithium abundance in the matter at the beginning of the Galaxy. Among the 9 stars observed so far, 4 have a [Fe/H] below -3.5 dex in which a star with [Fe/H] $<$ -4.6 dex. The results which will be published in a forthcoming paper demonstrate the efficiency of the target selection process.

References

Belokurov, V., Zucker, D. B., Evans, N. W., Gilmore, G., Vidrih, S., Bramich, D. M., Newberg, H. J., Wyse, R. F. G., Irwin, M. J., Fellhauer, M., Hewett, P. C., Walton, N. A., Wilkinson, M. I., Cole, N., Yanny, B., Rockosi, C. M., Beers, T. C., Bell, E. F., Brinkmann, J., Ivezić, Ž., & Lupton, R. 2006 *ApJ* 642, L137

Caffau, E., Bonifacio, P., François, P., Sbordone, L., Monaco, L., Spite, M., Spite, F., Ludwig, H.-G., Cayrel, R., Zaggia, S., Hammer, F., Randich, S., Molaro, P., & Hill, V. 2006 *Nature* 477, 67

Carollo, D., Beers, T. C., Lee, Y. S., Chiba, M., Norris, J. E., Wilhelm, R., Sivarani, T., Marsteller, B., Munn, J. A., Bailer-Jones, C. A. L., Fiorentin, P. R., & York, D. G. 2007 *Nature* 450, 1020

Edvardsson, B., Andersen, J., Gustafsson, B., Lambert, D. L., Nissen, P. E., & Tomkin, J. 1993, *A&A* 275, 101

Eggen, O. Lynden-Bell, D., & Sandage, A. 1962, *ApJ* 137,748

Frebel, A., Aoki, W., Christlieb, N., Ando, H., Asplund, M., Barklem, P. S., Beers, T. C., Eriksson, K., Fechner, C., Fujimoto, M. Y., Honda, S., Kajino, T., Minezaki, T., Nomoto, K., Norris, J. E., Ryan, S. G., Takada-Hidai, M., Tsangarides, S., & Yoshii, Y. 2005, *Nature* 434, 871

Frebel, A., Johnson, J. L., & Bromm, V. *MNRAS* 380, 40

Nissen, P. E., & Schuster, W. J. (2010) 2010, *A&A* 511, L10

Piotto, G., Bedin, L. R., Anderson, J., King, I. R., Cassisi, S., Milone, A. P., Villanova, S., Pietrinferni, A., & Renzini, A. 2007, *ApJ* 661, L53

Searle, L. & Zinn, R., 1978 1978, *ApJ* 225,357

Highlights of Astronomy, Volume 16
XXVIIIth IAU General Assembly, August 2012
T. Montmerle, ed.

doi:10.1017/S1743921314005791

Globular cluster contributions to Galactic halo assembly

Sarah L. Martell

Australian Astronomical Observatory, North Ryde NSW 2122, Australia
email: sarahmartell@aao.gov.au

Abstract. I discuss a search for red giant stars in the Galactic halo with light-element abundances similar to second-generation globular cluster stars, and discuss the implications of such a population for globular cluster formation models and the balance between *in situ* star formation and accretion for the assembly of the Galactic halo.

Keywords. Galaxy:formation, Galaxy:halo, Galaxy:stellar content, globular clusters:general

1. Introduction

We presently interpret the C-N, O-Na and Mg-Al abundance anticorrelations in globular cluster (GC) stars as a result of two-generation star formation and stellar-mode (i.e., supernova-free) chemical feedback (e.g., Carretta *et al.* 2010; D'Ercole *et al.* 2008). The present-day ratio of second- to first-generation stars in GCs is roughly 1:1, which leads to what is known as the "mass budget problem": there is simply not enough mass in stellar winds from the first generation to produce an equally massive second generation. Proposed solutions have included a top-heavy IMF for the first generation (Decressin *et al.* 2007) and a truncated mass function for the second generation (D'Ercole *et al.* 2008), but currently favored models require that the first generation was originally more massive (by a factor of 10-20) than it currently is. These massive GC formation models then require that the excess first-generation stars be preferentially removed from the cluster to reduce the ratio of second- to first-generation stars to its present-day level.

There are GCs that are currently losing stars to the halo field through extended tidal tails (e.g., Palomar 5, Odenkirchen *et al.* 2003; NGC 5466, Belokurov *et al.* 2006), and there is a theoretical expectation that many more GCs should have dissolved at earlier times as a result of tidal interactions with the Galaxy, internal 2-body interactions, and stellar evolution (Gnedin & Ostriker 1997). If these globular clusters contained second-generation stars at the point of dissolution, then some fraction of halo field stars should carry the second-generation light-element abundance pattern.

2. Globular cluster migrants in the halo field

To look for halo field stars with second-generation abundances, Martell & Grebel (2010) and Martell *et al.* (2011) searched the Sloan Digital Sky Survey (SDSS) SEGUE and SEGUE-2 low-resolution spectroscopic databases, respectively. Selecting red giants with $-1.8 \leqslant [Fe/H] \leqslant -1.0$, reasonably well-determined stellar parameters and clean spectra, they identified a total of 2519 halo giants, 65 of which ($\sim 2.5\%$) appear to have second-generation carbon and nitrogen abudances. For these stars, the 3883Å CN band is strong and the 4320Å CH G-band is weak, relative to other field stars at similar metallicity and evolutionary phase.

Ongoing work (Carollo *et al.*, in prep.) is finding that these CN-strong field giants, presumably second-generation globular cluster stars that have been lost to the halo, have orbits and kinematics consistent with the inner halo population (Carollo *et al.* 2007; 2010). In that work, we are also finding that globular clusters with proper motion measurements available in the literature† have orbits and kinematics similar to the inner halo population, making them a reasonable potential source for *in situ* formation of the inner halo.

3. Conclusions

The fraction of stars in the Galactic halo with light-element abundances similar to second-generation globular cluster stars is small, roughly 2.5%, but high-mass models for globular cluster formation require that they should be accompanied by several times as many first-generation stars, chemically indistinguishable from halo stars that formed outside GCs. This implies that globular clusters, as a major site of star formation 12 Gyr ago, are a significant contributor to Galactic halo assembly.

Acknowledgements

SDSS-III is managed by the Astrophysical Research Consortium for the Participating Institutions of the SDSS-III Collaboration including the University of Arizona, the Brazilian Participation Group, Brookhaven National Laboratory, University of Cambridge, University of Florida, the French Participation Group, the German Participation Group, the Instituto de Astrofisica de Canarias, the Michigan State/Notre Dame/JINA Participation Group, Johns Hopkins University, Lawrence Berkeley National Laboratory, Max Planck Institute for Astrophysics, New Mexico State University, New York University, Ohio State University, Pennsylvania State University, University of Portsmouth, Princeton University, the Spanish Participation Group, University of Tokyo, University of Utah, Vanderbilt University, University of Virginia, University of Washington, and Yale University.

References

Belokurov, V., Evans, N. W., Irwin, M. J., Hewett, P. C., & Wilkinson, M. I. 2006, *ApJL* 637, L29

Carollo, D., Beers, T. C., Lee, Y.-S., Chiba, M., Norris, J. E., Wilhelm, R., Sivarani, T., Marsteller, B., Munn, J. A., Bailer-Jones, C. A. L., Re Fiorentin, R., & York, D. G. 2007, *Nature* 450, 1020

Carollo, D., Beers, T. C., Chiba, M., Norris, J. E., Freeman, K. C., Lee, Y.-S., Ivezić, Ž., Rockosi, C. M., & Yanny, B. 2010, *ApJ* 712, 692

Carretta, E., Bragaglia, A., Gratton, R. G., Recio-Blanco, A., Lucatello, S., D'Orazi, V., & Cassisi, S. 2010, *A&A* 516, 55

D'Ercole, A., Vesperini, E., D'Antona, F., McMillan, S. L. W., & Recchi, S. 2008, *MNRAS* 391, 825

Decressin, T., Meynet, G., Charbonnel, C., Prantzos, N., & Ekström, S. 2007, *A&A* 464, 1029

Gnedin, O. Y. & Ostriker, J. P. 1997, *ApJ* 474, 223

Martell, S. L. & Grebel, E. K. 2010, *A&A* 519, 14

Martell, S. L., Smolinski, J. P., Beers, T. C., & Grebel, E. K. 2011, *A&A* 534, 136

Nissen, P. E. & Schuster, W. J. 2010, *A&A* 511, L10

Odenkirchen, M., Grebel, E. K., Dehnen, W., Rix, H.-W., Yanny, B., Newberg, H. J., Rockosi, C. M., Martínez-Delgado, D., Brinkmann, J., & Pier, J. R. 2003, *AJ* 126, 2385

Ramírez, I., Meléndez, J., & Chanamé, J. 2012, *ApJ* 757, 164

† Available at http://www.astro.yale.edu/dana/gc.html

Highlights of Astronomy, Volume 16
XXVIIIth IAU General Assembly, August 2012
T. Montmerle, ed.

doi:10.1017/S1743921314005808

Special Session 2: Cosmic Evolution of Groups and Clusters

J. M. Vrtilek[1] **and L. P. David**[2]

[1]Harvard-Smithsonian Center for Astrophysics, Cambridge, MA 02138, USA
email: jvrtilek@cfa.harvard.edu

[2]Harvard-Smithsonian Center for Astrophysics, Cambridge, MA 02138, USA,
email: ldavid@cfa.harvard.edu

Session chairs and section authors:
Yipeng Jing (Shanghai Astronomical Observatory)
Christine Jones (Center for Astrophysics)
Elke Roediger (Hamburg Observatory)
Sebastian Heinz (University of Wisconsin)
Jeremy Lim (Hong Kong University)
Daisuke Nagai (Yale University)
Jan Vrtilek (Center for Astrophysics)
Larry David (Center for Astrophysics)
Diana Worrall (University of Bristol)
Paulo Lopes (Universidade Federal do Rio de Janeiro)
Matthew Colless (Australian Astronomical Observatory)
Simon Lilly (ETH Zürich)

Abstract. During the past decade observations across the electromagnetic spectrum have led to broad progress in the understanding of galaxy clusters and their far more abundant smaller siblings, groups. From the X-rays, where *Chandra* and *XMM* have illuminated old phenomena such as cooling cores and discovered new ones such as shocks, cold fronts, bubbles and cavities, through rich collections of optical data (including vast and growing arrays of redshifts), to the imaging of AGN outbursts of various ages through radio observations, our access to cluster and group measurements has leaped forward, while parallel advances in theory and modeling have kept pace.

This Special Session offered a survey of progress to this point, an assessment of outstanding problems, and a multiwavelength overview of the uses of the next generation of observatories. Holding the symposium in conjuction with the XXVIIIth General Assembly provided the significant advantage of involving not only a specialist audience, but also interacting with a broad cross-section of the world astronomical community.

Keywords. black hole physics, galaxies: clusters: general, galaxies: cooling flows, galaxies: evolution, galaxies: intergalactic medium, galaxies: jets

---

1. Preface

In the following pages represent an attempt to summarize the presentations and discussion at Special Session 2 held at the XXVIIth IAU General Assembly. The intent, where possible, has been to emphasize discussion of the next generation of problems on which each area of investigation might profitably focus in the near future.

The twelve sessions, spread across five days in the first week of the General Assembly, focused in turn on the four major topics of cosmology and cluster formation, cooling flows and AGN feedback, the non-thermal properties of clusters, and the environmental impact of galaxy evolution in clusters. The Special Session offered 74 invited and contributed talks and over 60 posters. We particularly wish to express appreciation to the many presenters of talks and posters from whose research, contributions, and discussion this meeting was built.

We thank the session chairs whose contributions included the preparation of the session remarks presented below, and the Local Organizing Committee and the IAU management, who made the entire meeting possible.

We especially acknowledge the SpS2 organizing committee, without whose work this session would have been impossible. They are:
Monique Arnaud (Commissariat a l'energie atomique/Saclay, France)
Paulo Lopes (Universidade Federal do Rio de Janeiro, Brazil)
D J Saikia (National Centre for Radio Astrophysics, India)
Omar Lopez-Cruz (INAOE, Mexico)
Eugene Churazov (Space Research Institute (IKI), Russia)
Sabine Schindler (University of Innsbruck, Austria)
Diana Worrall (University of Bristol, UK)
Matthew Colless (Australian Astronomical Observatory, Australia)
Manolis Plionis (National Observatory of Athens, Greece)
Yipeng Jing (Shanghai Astronomical Observatory, China)
Jeremy Lim (Hong Kong University, China)
Laurence David (Center for Astrophysics, USA)
Jan Vrtilek (Center for Astrophysics, USA)

2. Cosmology: SZ Studies

Session chair and section author: Yipeng Jing

We highlight the recent results presented in the "Cosmology-SZ studies" session on cosmological applications of the Sunyaev-Zel'dovich (SZ) observations of clusters of galaxies. The South Pole Telescope Sunyaev-Zel'dovich Cluster Survey has led to several major cosmological results on dark energy equation of state , neutrino mass, and number of neutrino species. Observations with MUSTANG on the Green Bank Telescope have resulted in the highest resolution SZ effect images to date, and are revealing complex pressure substructures in intermediate redshift clusters. Future observations with ALMA, with Planck, and with the next generation of instruments on SPT will reveal much better complicated cluster thermodynamics, provide important clues to gas processes in clusters, and put more stringent constraints on cosmological parameters.

The session began with an introductory talk by Mark Birkinshaw on the Sunyaev-Zel'dovich (SZ) effect and on the cluster and group surveys using this effect. The SZ observations can provide an almost mass-limited and distance-independent measure of the gas content of a cluster of galaxies. Recent surveys have found many clusters using the effect. The observations are powerful probes to cosmological models and to formation and physical processes of clusters of galaxies. His talk focused on the significant differences between typical clusters found through the Sunyaev-Zeldovich effect, X-ray emission, and galaxy content, and discussed how recent imaging in the X-ray and SZ effects illustrates the origins of some of the these differences and indicates the underlying physical processes. Other speakers have presented recent surveys and new results from these surveys.

Recent results

Bradford Benson presented recent results based on the South Pole Telescope Sunyaev-Zel'dovich Cluster Survey. The South Pole Telescope is a 10-meter telescope optimized for sensitive, high-resolution measurements of cosmic microwave background (CMB) anisotropy and of the millimeter-wavelength sky. In November 2011, the SPT completed the 2500 deg^2 SPT-SZ survey. The survey has led to several major cosmological results, derived from measurements of the fine angular scale primary and secondary CMB anisotropies, the discovery of galaxy clusters via the Sunyaev-Zel'dovich (SZ) effect and the resulting mass-limited cluster catalog, and the discovery of a population of distant, dusty star forming galaxies (DSFGs). Notably, the SPT-SZ survey has led to the best current measurement of the primordial fine-scale CMB anisotropy power spectrum, and recently released a catalog of 158 SZ-selected optically-confirmed clusters from the first 720 deg^2 of the survey, more than doubling the number of comparably massive clusters discovered at redshift $z > 0.5$. The SPT-SZ data have been used to demonstrate significant improvements in the constraints on the dark energy equation of state, w, and the sum of the neutrino masses, Σm_ν, measuring $w = -0.973 \pm 0.063$ and $\Sigma m_\nu < 0.28$ eV at 95% confidence, a factor of 1.25 and 1.4 improvement, respectively, over the constraints without SPT-SZ cluster data. Adding the number of relativistic particle species as a free parameter, the SPT-SZ data has been used to measure $N_{\rm eff} = 3.91 \pm 0.42$ and constrain $\Sigma m_\nu < 0.63$ eV at 95% confidence, intriguingly 2σ higher than the value of 3.046 expected from only three neutrinos and early energy injection from electron-positron annihilation at the end of neutrino freeze-out. The full SPT-SZ 2500 deg^2 catalog is expected to be released in 2013, which consists of $\sim$400 SZ-selected clusters, $\sim$80% of them new discoveries. In combination with multi-wavelength X-ray and weak lensing observations, he expected the combined CMB power spectrum and SPT-SZ cluster data sets to constrain $\delta(w)/w = 5\%$, a constraint on dark energy from the growth of structure comparable to current geometric-based constraints, a powerful systematic test of our standard dark energy paradigm.

Suzanne Staggs presented results based on the Atacama Cosmology Telescope (ACT) observations. The Atacama Cosmology Telescope (ACT) is a 6-m special purpose telescope designed to measure the CMB. Its first receiver had channels at 148 GHz, 218 GHz and 277 GHz. ACT observes from a site at 5300 m in the Atacama Desert in Chile. This midlatitude site allows ACT to map regions of the sky in which there exist substantial data from surveys at other wavelengths. The talk focued on ACT's measurements of clusters and groups via their Sunyaev-Zeldovich effect, through direct blind detection as well as statistical analyses involving stacking, measurement of the 3-point function in the maps, and the impact on the angular power spectrum.

Tony Mroczkowski presented recent high angular resolution (9") Sunyaev-Zel'dovich effect (SZE) observations with MUSTANG, a 90-GHz bolometric receiver on the 100-meter Green Bank Telescope (GBT). MUSTANG is currently imaging a sample of clusters with complementary Chandra X-ray observations, HST optical observations that probe the mass distribution through strong and weak lensing, radio observations that probe the non-thermal component of the intracluster gas, and lower resolution SZE observations that can recover larger scales (>1'). The MUSTANG observations, which are used to assess the impact of substructure on SZE scaling relations, are among the highest resolution SZE images to date, and are revealing complex pressure substructures in intermediate redshift clusters. Combined, these observations reveal complicated cluster thermodynamics, which must be understood in order to use clusters as cosmological probes.

Ongoing and future surveys

Ruediger Kneissl reported the status of the Sunyaev-Zeldovich observations of clusters with the Atacama Large Millimeter/submillimeter Array (ALMA). The radio array, with an array of 50 12m antennas, the Atacama Compact Array of 12 7m antennas and 4 single dish 12m antennas, will offer high sensitivity and high angular resolution for SZ observations of galaxy clusters at frequencies below and above the peak of the cosmic microwave background spectrum, i.e., as decrements and increments. The interferometric technique, the large range of angular scales and the wide frequency coverage should provide high quality, direct images of the small scale gas pressure distribution, and at the same time allow for the rejection of confusing signals, in particular the emission from compact sources. He presented the test data taken during commissioning activities, and outlined future opportunities for science observations with ALMA. He emphasized that the observations will be very powerful for observing substructures in the clusters of galaxies.

Bradford Benson has briefly outlined future plans of the SPT. In 2012, the SPT was equipped with a new polarization sensitive camera, SPTpol, that will detect the contribution to the CMB polarization power spectrum from lensing by large scale structure (the so-called "lensing B-modes") and, on larger angular scales, detect, or set an improved upper limit on, the primordial inflationary signal ("gravitational-wave B-modes"), thereby constraining the energy scale of inflation. The SPTpol survey will survey about one-quarter the area, three times as deep as SPT-SZ, and expects to discover an $\sim$equal number of clusters, typically finding higher-redshift, lower-mass clusters. Development is underway for SPT-3G, the third-generation camera for SPT. In 2016, they will begin the SPT-3G survey, which we expect to find a factor of $\sim$20 more clusters than SPT-SZ and be sensitive enough to calibrate cluster masses to $\sim$3% from CMB cluster lensing.

Jose Diego, on behalf of Planck Collaboration, present some of the latest results of Planck on galaxy clusters through the Sunyaev-Zel'dovich effect. This talk described some new results found combining Planck SZ data with X-ray data on galaxy clusters.

Melanie Johnston-Hollit described the growing importance of lower-frequency radio observations in the study of clusters, with emphasis on relics, halos, and the role of AGN, with overarching goals including the understanding of the generation of magnetic fields, the origin of cosmic rays, and the effects of merger bias on cluster cosmology. The diffuse nature of the sources makes observations particularly challenging. Systems of future interest include the Australia SKA pathfinder and the Murchison Wide Field Array.

3. Cosmology: X-rays and high-z clusters

Session chair: Christine Jones

This session focused on the role of X-ray observations and the study of high-z clusters in modern cosmology. It included the following talks:
Steven Allen on"Cosmology using clusters of galaxies";
Rene Fassbender on "The X-ray luminous galaxy cluster population at $0.9 < z \lesssim 1.6$";
Masayuki Tanaka on "Quiescent early-type galaxies in $z > 1.5$ groups";
Yuying Zhang on "Probing substructures in galaxy clusters and resulting systematic in cosmological applications with X-ray and optical surveys"; and
Joana Santos on "Deep Chandra observation of the galaxy cluster WARPJ1415.1+3612 at z = 1: an evolved cool- core cluster at high redshift".

4. Large scale properties of clusters and cluster merging

Session chair and section author: Elke Roediger

The past decade has made it clear that clusters of galaxies are not relaxed objects. Many clusters show substructure of various kinds and merger or AGN activity. Despite observational challenges, deep observations of cluster outskirts have made the regions of ongoing cluster growth accessible.

Large-scale structure

Studying the large-scale properties of clusters requires observing their outskirts. This is challenging, because both the galaxy density as well as the ICM density are low. Consequently, the ICM emission is faint, and decreases below 30% of the background level. Moreover, large field of views (FOVs) are required to cover a significant fraction of the outskirts of nearby clusters. Reliable spectroscopic measurements in cluster outskirts require careful background modelling. Miller *et al.* (2012) combine the virtues of different X-ray observatories, i.e. the low and stable background of Suzaku, the sensitivity of XMM-Newton and the high spatial resolution of Chandra. These and other work (Akamatsu *et al.* 2011, Walker *et al.* 2012) find declining temperature profiles beyond $\sim 0.2 r_{200}$, in broad agreement with theoretical predictions from cosmological simulations (e.g., Burns *et al.* 2010). The entropy profiles appear to flatten around $0.5 r_{200}$ and even decline beyond $\sim 0.7 r_{200}$ in some cases (Walker *et al.* 2012). Furthermore, the clusters show an increased degree of azimuthal asymmetry in their outskirts, because accretion takes place preferentially along the large-scale filaments (see also Eckert *et al.* 2012). In addition to the faint ICM emission, the unknown nature of the ICM at large radii complicates the interpretation of the data. For example, gas clumping can lead to an over-estimate of the ICM density and an under-estimate of the entropy. Thus, some observations suggest declining entropy profiles in the very outskirts and even *apparent* baryon fractions above the cosmic average (e.g., in Perseus, Simionescu *et al.* 2011). Other observations agree well with theoretical predictions (e.g. RXCJ 0605, Miller *et al.* 2012) and show increasing entropy profiles out to the last data point. Recent work also started taking into account the baryons trapped in stars inside galaxies (Lin *et al.* 2012). While the stellar mass content does not seem to have evolved since redshift 0.6, the ICM fraction has increased in agreement with the self-similar model.

Mergers

Understanding cluster mergers, their signatures and their effects on clusters is essential if we want to use clusters as cosmological probes. Multiple approaches are currently made, both observationally and theoretically. Mergers are known to leave traces in the ICM that are detected in X-ray observations. Textbook examples include the bullet cluster (Markevitch *et al.* 2002) and Abell 3667 (Vikhlinin & Markevitch 2002). Mergers must lead to bulk motions and turbulence in the ICM. The direct observation of radial motions in the ICM requires, however, high spectral resolution. The current X-ray observatories provide only upper limits on turbulence (Sanders *et al.* 2010). Bulk motions have been measured for the merging cluster Abell 2256, where the radial velocity of the main core and the subcore could be separated (Tamura *et al.* 2011, 2012). While the relative velocity of 1500km/s between both subclusters is large, measurements of the upcoming ASTRO-H observatory will be significantly more sensitive (Shang & Oh 2012). The radio band has provided the link between mergers and the acceleration of relativistic particles by the observation of radio halos and relics (e.g., van Weeren *et al.* 2010). These observations will give insights in the physics of particle acceleration mechanisms, turbulence and shocks in the dilute ICM plasma as well. Additionally, mergers can be traced in the galactic or stellar components. Tidal stripping of stars from galaxies leads to a faint, but ubiquitous intra-cluster light, including plumes, shells and streams around cluster galaxies (e.g., Mihos *et al.* 2005; Janowiecki *et al.* 2010 for Virgo galaxies). Arnaboldi *et al.* (2012) present a recent study on the diffuse light distribution in the center of the Hydra I cluster. An off-center envelope and two tidal streams are found in both the diffuse light and in the kinematic data, which enables us to witness the tidal disruption of two companion galaxies. Cluster mergers may even indirectly influence the evolution of their galaxies due to the related shocks. Owers *et al.* (2012) find "jellyfish" galaxies in Abell 2744. Chandra observations reveal the ICM in Abell 2744 to be disturbed by an ongoing merger. The member galaxies of this cluster separate into two kinematically and spatially distinct subgroups. The "jellyfish" galaxies are star forming galaxies with blue, star forming tails with knots and filaments. Three of these galaxies are found in the region where the ICM must have been shocked recently by the merger, and the increase in ICM pressure may have caused an increased star formation in these galaxies and their tails. However, such a shock-triggered star formation enhancement seems to be a rare phenomenon, since only 4 jellyfish galaxies known so far.

AGN

The impact of AGN on the cluster cores has been discussed extensively in the recent decade (e.g., review by McNamara & Nulsen 2007). However, AGN outbursts can reach cluster-wide scales, e.g. in Hydra A (Nulsen *et al.* 2005) or MS0735 (McNamara *et al.* 2009). Now that observations can measure the ICM properties well beyond the cluster core, we can truly start to use the ICM as a calorimeter. Chaudhuri *et al.* (2012) successfully compare a simple model of energy deposition into the ICM, which consists of pre-heating as well as central AGN heating, to observational data.

5. Cluster simulations and theory: Computational Modeling of Clusters

Session chair and section author: Sebastian Heinz

Cluster cosmology has entered an era where precision matches that of other cosmological probes, and where accurate theoretical understanding of observational signatures of clusters is critical for cluster cosmology to advance. Computational models are integral for this next step, allowing a detailed understanding of non-linear physics that will affect X-ray and SZ signatures of clusters. We will highlight some of the recent advances and key next steps in this field.

Cluster Dynamics: Cluster Outskirts

The natural bias of X-ray observations to observe the cluster core (field of view and, more critically, emission measure) has focused much of the attention in cluster studies on the inner 100 kpc or so. The session on feedback addressed this important region. However, the advent of SZ observations has focused attention outward toward scales approaching the Virial radius. Understanding cluster density and temperature profiles out to large radii is therefore critical, both observationally and theoretically.

Observationally, Suzaku has provided the necessary sensitivity to allow such studies Simionescu *et al.* (2011). A key result from the observations of Perseus are a clear increase in the Baryon fraction with cluster radius, exceeding the cosmic mean. Clearly, this effect has to be understood for proper mass models of clusters from X-ray data to be derived.

The natural interpretation of this effect has been that the medium in the outskirts of clusters is clumpy (leading to a bias in the gas density estimate). Additional evidence for a clumpy gas distribution was presented for the cluster A133 from Chandra data by Vikhlinin *et al.* (in prep), as well as from a stacking analysis of Rosat data, which also show an enhanced baryon fraction (Eckert *et al.* 2012).

Simulations have, in fact, predicted clumping Nagai & Lau (2011). Understanding the extent to which clusters are thus out of equilibrium at large scales and what the effect of clumping is on other cluster observables is one of the key next steps in cluster modeling.

Given that turbulence and clumping may well happen on small scales, it will be important to benchmark numerical results from different codes against each other. In a sense, this question is similar to the study of "cold flows", where mass resolution as well as the proper spatial resolution of dynamical instabilities are required, requirements for which the capabilities of SPH and grid codes classically diverge. It will thus be interesting to see how moving mesh codes like Arepo address this important problem Springel (2011).

Dynamical State: Cold Fronts and Mergers

Probing cluster emission at larger scales is also helping to illuminate the dynamical response of clusters to recent mergers. While Chandra detected a surprising number of "cold front" contact discontinuities in clusters, a larger field of view of the cluster allows a much more robust identification of the underlying dynamical origin of prominent cold fronts.

The current understanding is that sloshing in response to gravitational perturbations of the cluster center is the primary mechanism for the formation of cold fronts Markevitch *et al.* (2001), while ram pressure stripping in direct cluster mergers with un-even mass ratios can also lead to cold fronts in some cases Vikhlinin *et al.* (2001).

Both zoomed and taylored hydrodynamic cluster simulations show that this process works (ZuHone *et al.* 2013). Recent simulations show that sloshing often leads to the development of characteristic sprial density waves. Emission measure maps of the density distribution from simulations of this effect show a clear set of alternating surface brightness peaks.

Such off-set brightness peaks were recently confirmed by deep XMM imaging of the Perseus cluster Simionescu *et al.* (2011), but had, in fact, already been seen in Rosat HRI images Churazov *et al.* (2000). Abell 496 also shows the tell-tale signature of merger induced surface brightness peaks Roediger *et al.* (2011, 2012). Like tidal tails in galaxy merger studies, these features have the potential to illuminate the merger history of clusters, and provide, in my view, a very promising avenue for detailed cluster studies.

A second important set of recent studies on the observational signatures of dynamical interactions show that radio relics can be produced in merger shocks and trace the merger morphology well. For example, in the case of Abell 3667 Rottgering *et al.* (1997), the original shock-induced cold front, a set of radio relics is aligned with the merger axis and presumably indicates the current location of the shock. Other clusters show similarly aligned relics, and simulations match the observed relic morphologies well (Hoeft *et al.* 2011; Skillman *et al.* 2012).

Baryon Physics and Dark Matter Profiles

A third important thread discussed during the cluster theory session is the effect that baryon physics can have on the dark matter profiles in clusters.

While collisionless dark matter is typically assumed to evolve only under the influence of its own self gravity, and that dark-matter-only simulations represent the dark matter profiles resulting from the hierarchical assembly of a halo well, a number of recent studies suggest that gas physics might be more important in the final dark matter profile of a halo than previously assumed.

In the case of galaxy sized halos, it was recently argued that the inclusion of stellar feedback significantly flattens the centeral portion of not just the gas mass profile, but the dark matter profile as well into a cored profile Governato *et al.* (2012).

Martizzi *et al.* (2012) presented recent cluster simulations that include AGN feedback, in the form of thermal feedback similar to the supernova feedback used in galactic scale simulations, and found that cluster gas and dark matter mass profiles are affected, reducing the central dark matter densities by about an order of magnitude.

Given that the underlying dark matter physics is scale free, and that comparable feedback prescriptions were employed, it is, perhaps, not surprising that both approaches yield similar results. It does, however, highlight the importance of baryon physics for the dynamical evolution of not just the gas but even the dark matter. Larger samples of simulations that explore these effects will be required to fully understand the effects at play.

6. Cooling flows/AGN feedback — Observation

Session chair and section author: Jeremy Lim

In the classical picture where X-ray-emitting gas comprising the intragroup medium (IGM) or intracluster medium (ICM) cools and flows inwards ("X-ray cooling flow") without being subjected to any external energy injection ("reheating"), the predicted mass-deposition rates from such X-ray cooling flows can reach values in excess of 1000 $M_{\odot}$ yr^{-1} (Fig. 1, left panel). Today, we know that X-ray cooling flows must be strongly mitigated, if not entirely quenched, in all the galaxy groups and clusters so far observed. The most direct evidence comes from the lack of detectable X-ray gas in the cores of galaxy groups or clusters at temperatures below about one-third the bulk ambient temperature, implying that any X-ray cooling flows must have mass-deposition rates at least one-tenth smaller than the classically predicted value. (Alternatively, the X-ray gas penetrates into and mixes with the optical line-emitting gas, so that, effectively, the gas does not cool through X-ray temperatures; e.g., Fabian's talk.) Additional evidence comes from the smaller-than-predicted levels of gas below X-ray temperatures, and rate of star formation (~1% of the classically predicted mass-deposition rate; e.g., McNamara's talk, presented by Nulsen), in the central dominant elliptical galaxies of groups and clusters that, based on the strong X-ray radiative loss at their cores, ought to to harbor X-ray cooling flows.

Jets of relativistic particles (most easily detectable by their synchrotron emission at radio wavelengths) from the central supermassive black hole in the central dominant group or cluster elliptical galaxy have for some time now been invoked to reheat the cooling X-ray gas. The mechanical energy contained in such jets, as inferred from their synchrotron emission (Leith's talk), has long been known to be much more than sufficient

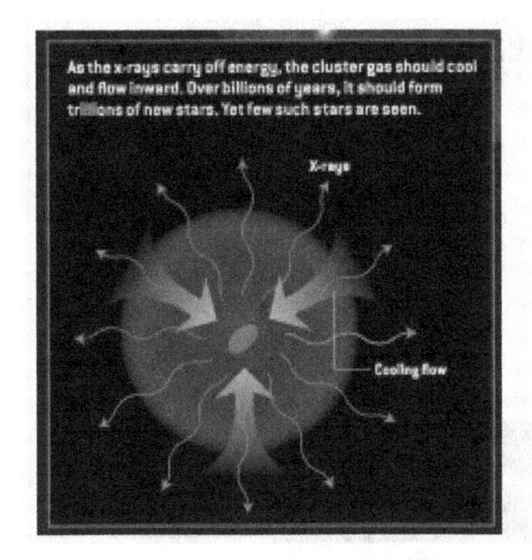

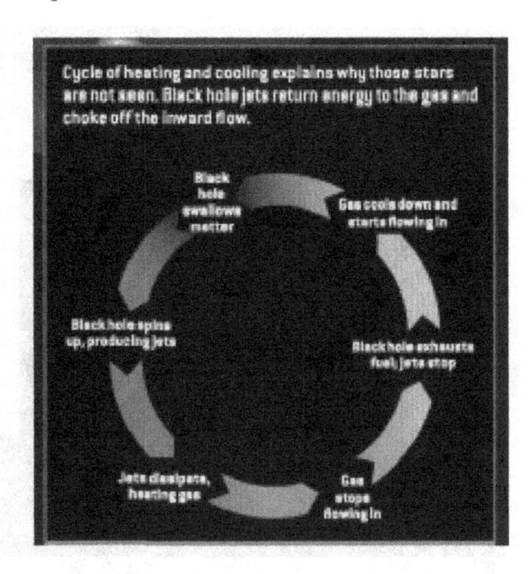

Figure 1. Left panel – X-ray cooling flow in the absence of reheating. Right panel – Feedback between X-ray cooling flow and AGN reheating.

to quench X-ray cooling flows. The primary issues that remain incompletely resolved — some would argue entirely unresolved — are:

- How do the jets actually inject energy into the X-ray gas?
- Given that jets are highly collimated, how do they inject energy more or less isotropically into the X-ray gas?
- Why is there an apparent fine tuning such that the inferred AGN heating-rate is approximately equal to the classically predicted X-ray cooling-flow rate?
- Do the jets completely quench the X-ray cooling flow, or is there a residual but still significant X-ray cooling flow?
- If there is a significant residual cooling flow, what is the nature of the cooled gas? Is this gas involved in star formation or fueling of the AGN?

Early ideas that jets inject energy into the IGM or ICM through shocks, and which invoke jet precession to deliver this energy more-or-less isotropically throughout group/cluster cores, have largely — but by no means entirely (e.g., Worrall's talk) — given way to a more nuanced view of reheating thanks to detailed X-ray and radio observations of galaxy groups and clusters (e.g., McNamara's talk and O'Sullivan's talk). X-ray observations reveal cavities in the IGM/ICM in the vicinity of the central dominant elliptical galaxies of groups/clusters expected to harbor X-ray cooling flows; radio observations, in particular those at relatively low frequencies (below 1 GHz) that probe lower-energy but longer-lived synchrotron-emitting electrons, reveal relativistic electrons filling the X-ray cavities, implying that radio jets from the AGN in the central dominant group/cluster elliptical galaxy are responsible for inflating these cavities (Figs. 2-3). The mechanical energy required to inflate the observed cavities (henceforth referred to as the AGN heating-rate) is comparable with the X-ray radiative loss in group/cluster cores, and can therefore strongly mitigate if not entirely quench their X-ray cooling flows. In this manner, jets are able to inject the required energy to counteract cooling over large volumes in the surrounding X-ray gas. The exact manner in which relativistic particles contained in the X-ray cavities transfer their energy to the surrounding X-ray gas, however, remains a subject of debate.

Remarkably, the inferred AGN heating-rate is correlated with the ICM radiative luminosity within the cooling radius (where the cooling time is shorter than the Hubble

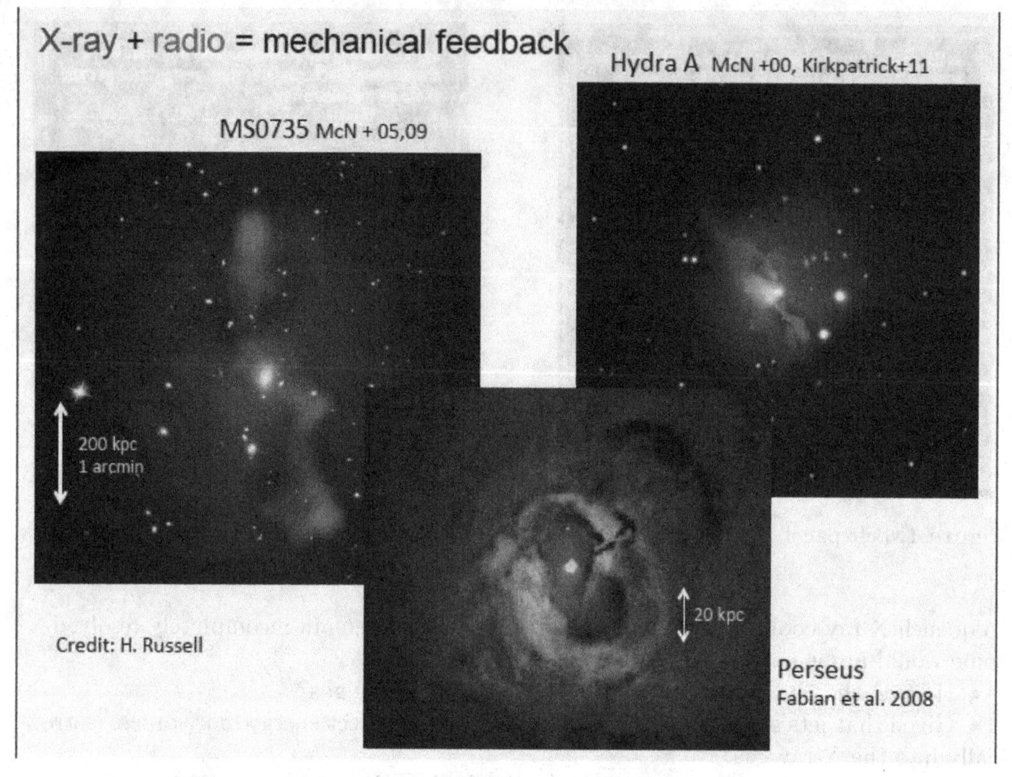

Figure 2. Radio-jet-filled X-ray cavities in galaxy clusters. From McNamara's talk.

time) over seven decades in heating rate/radiative luminosity (McNamara's talk) (Fig. 4, left panel). Such a correlation is also seen in galaxy groups, which of course have lower IGM radiative luminosities, despite the fact that jets often penetrate far beyond the cooling radius in galaxy groups because of their much lower IGM gas pressures; in galaxy groups, less energy contained within the cavity is required to be transferred to entirely reheat the surrounding X-ray gas (O'Sullivan's talk) (Fig. 4, right panel). How does the AGN know about cooling of the ICM given that fueling of the AGN occurs on a spatial scale that is imperceptibly small compared with the volume of the cooling ICM? There must presumably be a feedback process at work (e.g., as proposed in Fig. 1, right panel), but how does this feedback process operate in such a way as to be fined tuned over a tremendously large range in spatial scales and therefore dynamical timescales? Or are we simply witnessing the cumulative effects of repeated episodes of reheating that, on average, produces an apparent relationship between the AGN heating-rate in a given episode (that is comparable to the AGN heating-rate in any given episode) and the time-averaged response of the IGM/ICM? A basic understanding of this feedback process, of vital importance for informing theoretical simulations on the growth and evolution of massive galaxies, is one of the most difficult yet urgent challenges we now face.

Are X-ray cooling flows completely quenched? Galaxy clusters that ought to harbor strong X-ray cooling flows in the absence of reheating have long been known preferentially to exhibit luminous optical line-emitting gas closely associated with their central elliptical galaxies, and this optical line-emitting gas is usually spatially coincident with the coolest X-ray gas component. This situation is true also in galaxy groups (O'Sullivan's talk). The inferred mass of the optical line-emitting gas, however, is relatively small ($\lesssim 10^7$ M$_\odot$). On the other hand, over the past decade, surveys in CO at millimeter wavelengths have

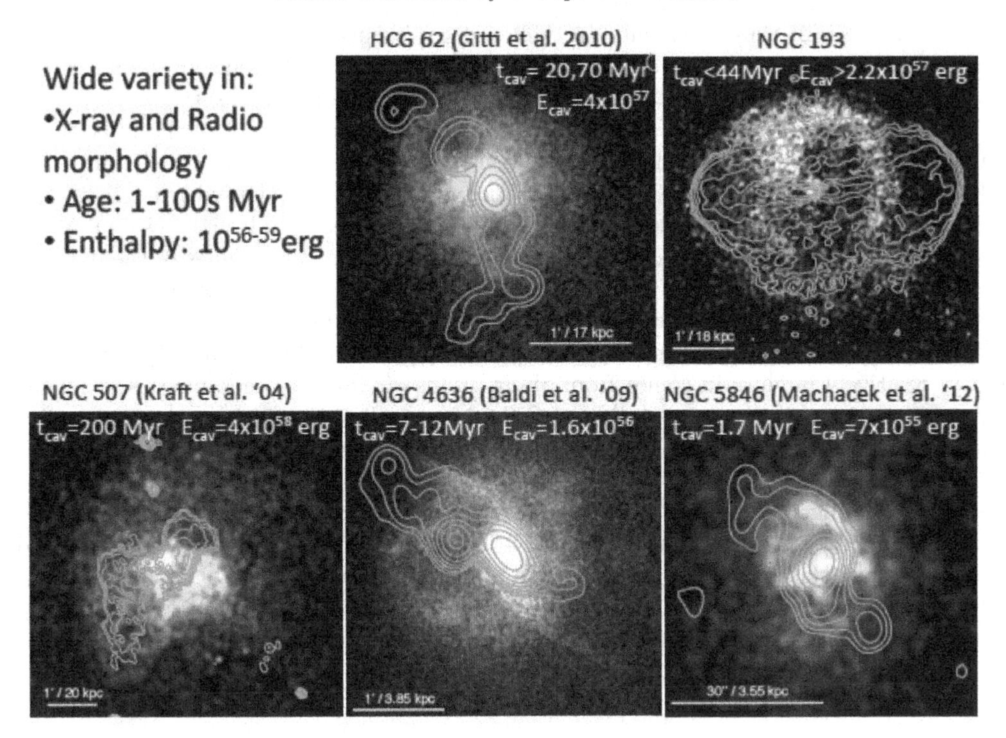

Figure 3. Radio-filled X-ray cavities in galaxy groups. From O'Sullivan's talk.

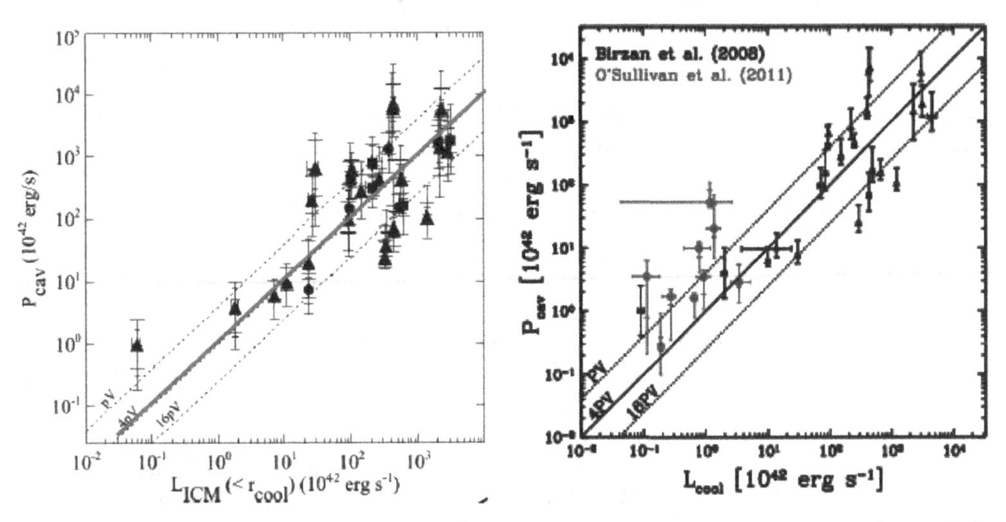

Figure 4. Left panel — Correlation between ICM luminosity within the cooling radius and the mechanical power required to inflate X-ray cavities in clusters. From McNamara's talk. Right panel — Same as left panel, but including galaxy groups. From O'Sullivan's talk.

revealed large masses of cool molecular gas (up to $\sim 10^{11}$ $M_\odot$) in the central elliptical galaxies of the same clusters. Is all or a portion of this gas deposited by X-ray cooling flows, and if so was this gas deposited in the distant past or still forming from a residual X-ray cooling flow at the present time? In the core of the Perseus cluster, both atomic and molecular gas are distributed in filaments that are mostly aligned approximately radially although there also are tangential as well as curved filaments. At least some, and perhaps all, of these filaments are believed to comprise gas either dragged out by rising X-ray bubbles (i.e., jet-blown X-ray cavities that have become buoyant) or gas draining back into the galaxy after being dragged out by rising bubbles (Chan's talk). In this picture, much if not all of the cool molecular gas, which comprises the bulk of the gas detectable below X-ray temperatures, may have been deposited in the distant past. Alternatively, a portion of the cool molecular gas may have condensed from X-ray gas compressed at the outer surface of rising bubbles, or more simply from a residual X-ray cooling flow, in the recent past. Progress on this front will require studies of the physical properties and relationship between the individual components of the multiphase gas, as well as their spatial relationship with X-ray cavities and bubbles.

Indirect evidence that buoyant X-ray bubbles drag gas from the central elliptical galaxy out to relatively large radii comes from observations of enhanced metallicities along the axis of radio jets (McNamara's talk and O'Sullivan's talk) (Fig. 5), and that the radius over which metal enrichment is observed in galaxy clusters is correlated with the jet mechanical power (McNamara's talk) (Fig. 6). In Abell 2597, relatively cool X-ray gas is seen extending far beyond the north-eastern jet associated with the central cluster elliptical galaxy, suggesting that even cool X-ray gas from the cluster core can be dragged out to relatively large radii by rising X-ray bubbles (Tremblay's talk). Such observations present an alternative picture, or at least an additional pathway, for how the IGM or ICM can be enriched in metals other than through galactic outflows driven by stellar winds or supernova explosions.

7. Cooling flows/AGN feedback — Theory

Session chair and section author: Daisuke Nagai

Introduction: Recent Chandra and XMM-Newton X-ray observations reveal direct evidence of gas heating in the central regions of galaxy clusters, through the interaction of jets and bubbles produced by active galactic nuclei (AGN) with the surrounding intracluster medium (ICM) (Fabian 1994; Peterson & Fabian 2006; McNamara & Nulsen 2007). While it is widely believed that AGN can produce enough energy to heat the core, the mechanism of how the energy of the AGN jets and bubbles are transported to the ICM remains unclear.

Numerical simulations of AGN jets and bubbles in galaxy clusters have been performed by a number of authors to investigate this issue. The simulations performed to date are broadly classified into two categories: (i) idealized simulation of 3D spherically symmetric clusters (Quilis *et al.* 2001; Brüggen & Kaiser 2002; Omma *et al.* 2004; Sijacki & Springel 2006), and (ii) cosmological simulations with sub-grid models of AGN accretion and energy injection (Sijacki *et al.* 2007; Booth & Schaye 2009; Teyssier *et al.* 2011; Yang *et al.* 2012). While many of the AGN feedback models have been studied using pure hydrodynamic simulations, some of the recent work has investigated the roles of non-ideal and/or non-thermal phenomena on the AGN-ICM interactions, including viscosity (Ruszkowski *et al.* 2004; Reynolds *et al.* 2005; Sijacki & Springel 2006), magnetic fields (Dubois *et al.* 2009) and cosmic-rays (Sijacki *et al.* 2008), finding that these effects play important roles in modeling the AGN-ICM interactions, interpreting radio and X-ray observations of non-thermal phenomena in clusters, as well as constraining cosmological parameters using X-ray and Sunyaev-Zeldovich (SZ) cluster surveys.

In SpS2 session 6, we discussed recent progress in theoretical modeling of AGN feedback in galaxy clusters. We briefly summarize some of the key findings reported in this section.

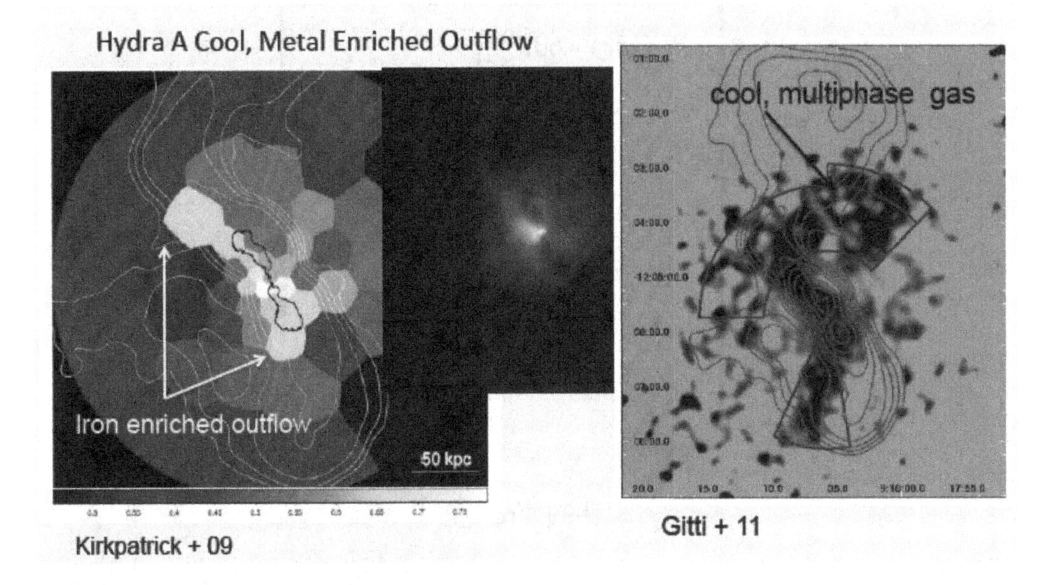

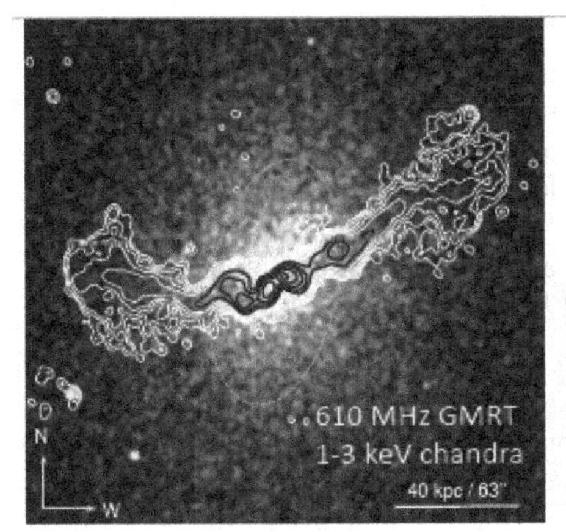

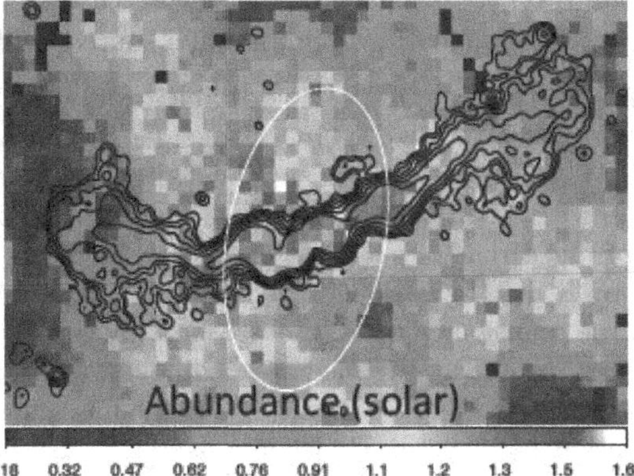

Figure 5. Upper panels — Metal enrichment along the axis of the radio jet from the central elliptical galaxy of a cluster. From McNamara's talk. Lower panel — Metal enrichment along the axis of the radio jet from the central elliptical galaxy of a group. From O'Sullivan's talk.

Simulating the Interaction of Jets with Dynamic Cluster Atmospheres: Heinz *et al.* presented the FLASH hydrodynamical simulations of the AGN-ICM interaction in realistic, cosmologically evolved cluster atmospheres, focusing on modes of energy transfer from non-thermal to thermal gas and the development of diagnostics one can use to connect simulations to observations (Heinz *et al.* 2006; Brüggen *et al.* 2007). Their simulations highlighted that the dynamic state of the cluster is critical for the development of this interaction; i.e., the internal cluster dynamics, excited by the action of the jets themselves, residual motions from previous dynamical encounters, and turbulence

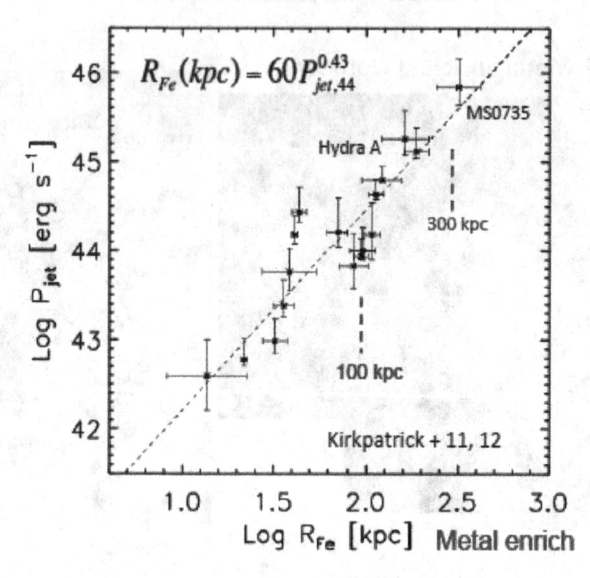

Figure 6. Correlation between metal enrichment radius and inferred jet mechanical power. From McNamara's talk.

induced by cluster substructure (e.g., moving galaxies), can redistribute the gas in the cluster center in such a way that it erases any channels that were previously carved by the jet. This means that subsequent powerful jet outbursts can efficiently couple with the dense gas of the inner cluster, making it possible for jets to heat cold gas near the centers of clusters. Simulated radio and X-ray maps bear a striking resemblance to observations of prototypical sources, indicating that we are approaching a level of accuracy where detailed re-simulations of individual sources will be possible.

Simulating the AGN Feedback of Cool-Core Clusters: Li & Bryan carried out high-resolution Enzo AMR simulations of an isolated cool core cluster with and without an AGN jet. The cluster is modeled as a 3D spherically symmetric NFW dark matter halo with gas, BCG, and SMBH, where the initial gas density and temperature is set to reproduce that of the Perseus Cluster, while resolving the flow from Mpc scales down to pc scales (Li & Bryan 2012). Following Omma *et al.* (2004), the AGN jet is modeled as the addition of mass, momentum, and energy to cells within the x-y plane of the jet launching region, (defined as a cylinder with $r = 100$ pc and $h = 300$ pc), with the total jet power ($dE/dt = 2 \times 10^{44}$ erg/s), mass injection rate ($dM/dt = 6M_{\odot}$/yr), kinetic fraction ($f = 0.1 - 1$), jet speed ($v_{\rm jet} = 10^4$km/s for f=1), and precessing jet period ($P = 5$ Myr).

In their pure "cooling-flow" simulations (without AGN), they find that the model cluster develops a global cooling catastrophe and forms a thick accretion disk at the center inside a transition radius of about 50 pc from the SMBH, while local thermal instabilities do not grow outside of this region. When the AGN jet is turned on, their simulations formed cold filaments extending out to ~ 10 kpc, mostly along jet directions. There is a distinct lack of gas below a few keV in simulations with and without AGN.

Magnetized outflows from AGN in galaxy clusters: Sutter *et al.* (2012) presented 3D adaptive mesh refinement (AMR) magnetodydrodynamic (MHD) simulations of an isolated galaxy cluster that included injection of kinetic, thermal, and magnetic energy via a central AGN in order to study and evaluate the role that AGN might play in producing the observed cluster-wide magnetic fields. Using the MHD solver in the FLASH

N-body+hydrodynamic code, they compared several sub-grid models of the evolution of AGN, focusing especially on large-scale AGN jets and bubbles, and examined the effects of magnetized outflows on the accretion history of the black hole and cluster thermodynamic properties. Their study showed that magnetized jet models cause significant reduction in BH accretion rate compared to hydrodynamic jet models, while the accretion rates in the bubble models remain largely unaffected by the introduction of magnetic fields. Both jet and sporadically-placed bubble models have difficulty reproducing the observed strength and topology of cluster magnetic fields. While the bubble models are generally very unstable, the jet models seem successful in producing weakly magnetized entire clusters and strongly magnetized cluster cores.

Stable Heating of Cluster Cool Cores by Cosmic-Ray Streaming: Fujita & Ohira (2011) discussed heating of cool cores in galaxy clusters by cosmic-ray (CR) streaming using numerical simulations. In this model, CRs are injected by the central AGN and move outward with Alfvén waves. In this mechanism, CR streaming in the ICM excites Alfvén waves. The CRs interact and move outwards with the waves. The PdV work done by the CRs effectively heats the ICM. Using their simulations, the authors showed that CR streaming can stably heat both high and low temperature clusters for a long time without the assistance of thermal conduction, and it can prevent the development of massive cooling flows. The reason for the stability is that CR pressure is insensitive to changes in the ICM and that the density dependence of the heating term is similar to that of radiative cooling. Moreover, CRs created by the central AGN can stream into larger regions, providing a stable heat source for the entire cluster core. Fujita & Ohira (2012) also showed that the observed radial profiles of radio mini-halos are remarkably consistent with the predictions of their model.

Impact of AGN Feedback on the Observable-Mass Relations of Galaxy Clusters: Paul Ricker presented Flash AMR simulations of galaxy clusters in both cosmological and isolated contexts to address the influence of feedback from AGN on scatter in cluster X-ray and SZ observable-mass relations (Yang *et al.* 2012). By performing a detailed parameter sensitivity study in a single cluster using several commonly-adopted AGN accretion and feedback models with FLASH, they quantified the model uncertainties in predictions of cluster integrated properties, including the normalization, slope, and scatter. The study showed that quantities that are more sensitive to gas density have larger uncertainties ($\sim 20\%$ for $M_{\rm gas}$ and a factor of 2 for L_X at R_{500}), while T_X, $Y_{\rm sz}$, and Y_X are more robust ($\sim 10-20\%$ at $r = R_{500}$). By studying the impact of AGN feedback on the scaling relations, they found that an anti-correlation exists between $M_{\rm gas}$ and T_X, which is another reason why Y_{SZ} and Y_X are excellent mass proxies. This anti-correlation also implies that AGN feedback is likely to be an important source of intrinsic scatter in the $M_{\rm gas} - T_X$ and $L_X - T_X$ relations. Using a sensitivity analysis of subgrid model parameters, they identify mechanical heating efficiency and the size of the feedback region as parameters which, if better constrained by observations, could have the largest impact on model uncertainties in the scatter in the mass-observable relations.

8. Cold gas and star formation in CFs

Session chair and section author: Jan Vrtilek

The results of strong central cooling in cluster gas — the establishment of a powerful cooling flow and the consequent accumulation of large amounts of cold gas at massive central ellipticals — have long been known (e.g., Fabian 1994), as has the absence of the expected large amount of star formation. The "problem" was further illuminated by the absence of predicted X-ray spectral cooling features that should have been seen by the sensitive XMM-Newton RGS. The resolution, as discussed in the previous session, is now widely thought to be found in AGN feedback, even if numerous details remain to be understood. Nonetheless, some gas does cool, and the corresponding evidence is to be found in reduced levels of star formation, Hα filaments, and detections of molecular gas. The focus of this session was on the observation, origin, physical state, and fate of that cold material in cluster cores.

It has since the late 1980s been known that it is principally in cool core clusters that cold material — whether atomic or molecular — is to be found. Jeremy Lim reviewed the history of the investigation of this subject, including the correlations between atomic material (as seen in Hα) and molecular material (detected in CO rotational lines). The detection of even the strongest molecular lines (i.e., low-lying rotational lines of CO) in nearby clusters has not been easy, and surveys have been confined to single-dish detections which, while valuable for multiple purposes (including as pathfinders for future interferometry), do not fully exploit the strength of imaging spectroscopy that radio spectral line work can bring to the subject. Such imaging spectroscopy offers for molecular gas what X-ray spectroscopy cannot do with the current generation of instruments for hot gas: a spatial map of velocities, and the subsequent insight into the movement of material at the centers of cooling flows. To this point interferometry, with PdB or the SMA, has required substantial integration time, even on the strongest and most nearby sources (e.g., Persus or Abell 1795). The advent of ALMA promises a revolution in our understanding by enabling, with practical integration times, the study of a vastly broader array of targets. The first results are now arriving, both in clusters (see, e.g., recent work by McNamara and collaborators on Abell 1664 and Abell 1835) and in groups (NGC 5044: David *et al.*, in preparation). A complication in the interpretation of these results is the X-factor — the conversion beween observed CO and the total molecular column density, which has been calibrated for the particular circumstances common in our Galaxy, but may be quite different elsewhere.

A number of well-studied nearby clusters — e.g., Perseus, Abell 1795, Centaurus — exhibit, most clearly in HST images, long (up to 6 kpc) and very narrow (only 70 pc) Hα filaments. In Perseus these have been shown to adjoin star forming regions. Andy Fabian reviewed these observations, and argued that the stars cannot be the excitation source for the filaments, as the stars are dynamically unconnected to the gas and will "fall out" of the gas in $\sim 10^6$ years. The origin of the filaments is likely from rising bubbles that drag cold gas upwards. Spectra of these filaments are unlike any seen in Galactic objects: there is, for example, no forbidden line emission, such as from [OIII]. If the source of the excitation were from a hot blackbody, its temperature would have to be $\sim$150,000 K. Excitation by energetic particles is instead proposed: these produce showers of secondary electrons on impact.

Based on HST UV images of the CLASH sample (25 massive, dynamically-relaxed clusters), the ACCEPT archival analysis of clusters, and in keeping with the above themes, Megan Donahue noted that BCGs are in general *not* "red and dead". For example, CLASH has revealed clusters characterized by familiar phenomenology, only at greater redshift: e.g., RXJ 1532+30, similar to Perseus but at $z = 0.363$. From ACCEPT, a central entropy $K_0 < 30\text{keV}\,\text{cm}^{-2}$ implies a cool core; lower resolution worsens the limit at higher redshift. When star formation is seen, where does the star-forming gas come from? — H_2 lines uncorrelated with IR emission from dust are found in such cases that are much stronger than those detected in star-forming galaxies. Is there a "warm" phase?

A new approach to the examination of cold matter in cluster cores is now available from Herschel data. Francoise Combes reviewed some of the recent findings, based in part on the observation of a dozen clusters in the course of a Herschel key project (PI: A. Edge). This work has revealed gas-to-dust ratios around 100 and temperatures of a few 10s K, remarkably similar to the Milky Way. This could be a surprise: why is the dust not destroyed by sputtering? Lines of [CII] and [OI] are the major coolants of atomic gas; they tend to be spatially associated with CO emission. The FIR to CO ratios are typical of those of star-forming galaxies. In Perseus, the [OI] and [CII] line velocities suggest inflow rather than rotation. The [CII] emission tends to be spatially coincident with Hα.

At redshifts greater than 0.5 to 0.75, cool cores as defined by a cuspy surface brightness distribution become rare, but cooling flows identified by cooling time and low central entropy are about as abundant as at low-z. Michael McDonald noted this, and raised the question whether optical line emission might be added as a method of CC/NCC classification, a question supported by the recent discovery that the Phoenix cluster supports an 800 M_{sun} yr^{-1} starburst and a large and complex emission-line nebula: SZ-selected clusters are by no means all highly -disturbed "train wrecks" — the Phoenix cluster is not the only one to show strong central cooling.

Based on slit spectroscopy observations along Hα filaments, Jeffrey Chan discussed the structure of the velocity field in Perseus/NGC 1275, revealing a complex structure occasionally confused by overlapping filaments along the line-of-sight. Some filaments may be show rising outer portions and infalling inner portions. Entrainment behind a rising bubble could be involved in some of the observed phenomena.

9. Ellipticals and Groups of Galaxies

Session chair and section author: Larry David

The motion of galaxies through cluster atmospheres can have dramatic affects on the interstellar medium within galaxies as shown by Kraft, who presented X-ray observations and simulations of infalling ellipticals in the Virgo cluster. Kraft presented preliminary results from a deep Chandra observation of the hot ISM of NGC 4552. NGC 4552 lies 350 kpc from M87 in projection, and is presently falling into the Virgo cluster. X-ray observations have shown that the interstellar gas in NGC 4552 has been been stripped during the infall and produced a 20 kpc long tail of stripped gas behind the galaxy. The temperature in the ram-pressure stripped gas increases from 0.5 keV near the galaxy to 1 keV at the edge of the Chandra FOV (see Fig. 7). Specifically tailored viscous hydrodynamical simulations of the infall and stripping process of NGC 4552 were created. The simulations show that if the flow is viscid, then a long, cool tail is formed, similar to what is observed in NGC 4552. However, if the flow is inviscid, small scale Kelvin-Helmholtz instabilities efficiently mix the stripped gas with the ambient ICM and no tail is formed. The presence of the tail behind NGC 4552 thus suggests that the viscosity of the Virgo cluster ICM is at least a few percent of the Spitzer value.

The Galaxy and Mass Assembly (GAMA) survey provides an opportunity to search for analogues of the Local Group. Robotham gave a presentation based on the analysis of all GAMA galaxies within a factor of two of the stellar mass of the Milky Way (MW) and found that there is a 11.9 close companion within a projected separation of 70 kpc, a radial separation of 400 km/s, and at least as massive as the Large Magellanic Cloud (LMC). There is only a 3.4 finding two close companions at least as massive as the Small Magellanic Cloud (SMC). Only two analogues of the MW-LMC-SMC system were found in GAMA. One example is shown in Fig. 8. These results suggest that such a

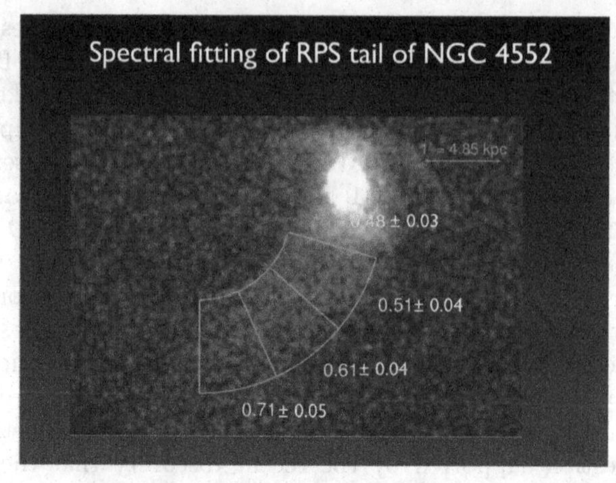

Figure 7. Chandra image of NGC 4552 showing the temperature variation along the tail of ram-pressure stripped gas. (R. Kraft)

Figure 8. An Analogue to the Local Group

combination of near-by, late-type, star-forming galaxies is rare. Only 0.4 where LMC and SMC mass galaxies could be detected) are embedded in such a system. In summary, the MW-LMC-SMC system is a 2.7σ event (when recast into Gaussian statistics).

A question which has been actively debated recently is whether the relationship between halo mass and stellar mass is more fundamental than the relationship between halo mass and velocity dispersion. Li gave a presentation summarizing his work on this question in Li *et al.* (2012a) and in Li *et al.* (2012b). Li *et al.* (2012a) recently determined the velocity dispersion profile (VDP) for galaxy groups in SDSS/DR7. They also estimated the redshift-space cross-correlation function (CCF) between central galaxies of a given mass (or luminosity) and a reference galaxy sample. The VDP is then measured by modeling the redshift distortion in the CCF. Within the virial radius, the VDP shows

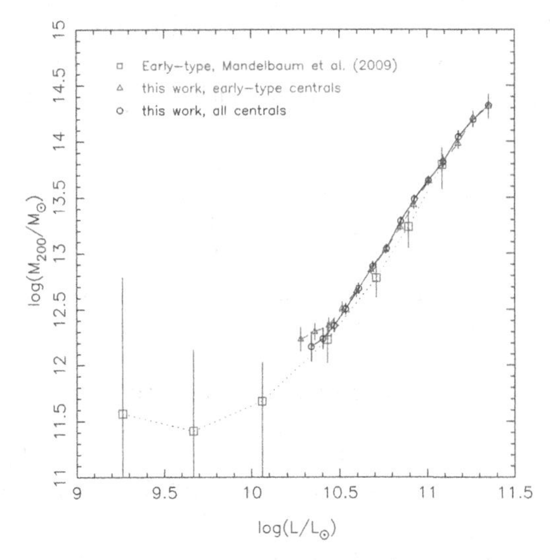

Figure 9. Dark matter halo mass as a function of central galaxy luminosity. Black circles and red triangles are estimated from the velocity dispersion measurements for all central galaxies being studied and for the subset of early-type central galaxies, respectively. Blue squares are the results obtained by Mandelbaum *et al.* (2006) by stacking the gravitational lensing signals of the early-type central galaxies in the SDSS (Li *et al.* 2012b)

a roughly flat profile, while the average velocity dispersion (VD) is a strongly increasing function of the central galaxy mass. By applying the same method to N-body cosmological simulations, Li *et al.* showed that the velocity dispersion vs. stellar mass relation can be used to constrain both the density fluctuation parameter (σ_8) and the dispersion in stellar mass at fixed halo mass for central galaxies. Since the velocity dispersion is caused by the local gravitational field, the VDP measurements also provides a direct measure of the dark matter halo mass for central galaxies of different masses and luminosities. As can be seen from Fig. 9, the halo mass vs. luminosity relation agrees well with the results from the weak lensing analysis of Mandelbaum *et al.* (2006). Li *et al.* (2012b) extended this work by estimating the halo mass as a function of both the stellar mass (M_*) and stellar velocity dispersion (σ_*), and found a much tighter correlation between the halo mass and M_* than that between the halo mass and σ_*. This demonstrates that, for central galaxies in halos, the stellar mass is still a good (if not the best) indicator of the host halo mass and is better than the stellar velocity dispersion, which was recently suggested by Wake *et al.* (2012).

10. Global radio properties, relativistic particles, and magnetic fields

Session chair and section author: Diana Worrall

Although the non-thermal content of clusters makes up only a small fraction of their overall energy, it is key in several areas, such as tracing magnetic field and sites of particle (cosmic ray) acceleration, reflecting merger history, identifying and measuring large-scale outflows from AGN, and indicating AGN life-cycles. This was the subject of Topic 3. It contained a single session with the title "Global Radio Properties, Relativistic Particles, and Magnetic Fields". Two of the invited speakers were unable to attend at the last minute, and while it had been possible to make arrangements for one of the presentations to be given by another speaker, the planned talk from the X-ray perspective was not given.

Discussion was therefore biased towards what was learned from radio observations, and with a recent growth in radio-astronomy facilities able to map clusters (e.g., GMRT and LOFAR), and upgrades to major facilities such as JVLA and ATCA, there was no shortage of interesting material.

The radio emission of clusters can be grouped into three main types of structures: radio halos, radio relics, and individual active galaxies. They were described in invited talks by Tiziana Venturi and Tracy Clarke (the latter presented by Tiziana Venturi), and Rowan Miller discussed the case study Abell 3266. There were several interesting posters associated with the session, including one by Chernyshov, Dogiel and Ko that discussed the parameter space within which stochastic acceleration of electrons might be responsible for the hard-X-ray excess reported in some clusters.

Radio halo sources cover a large fraction of their host clusters, with overall unpolarized structures that roughly correspond to the morphology mapped in X-ray emitting gas. More than 20 giant halos are known. They are detected in bigger, brighter, hotter, more massive clusters, and correlations have been reported between radio power and X-ray temperature and luminosity.

Radio relic sources have a tendency to lie at the edges of clusters, and are bright where other evidence suggests recent merger activity. About 25 such cases are known. The radio polarization that is seen suggests the presence of strong shocks and the restructuring of magnetic field. Power-law spectra of $\nu^{-\alpha}$ with $\alpha = 1.2$–1.5 are typical. The larger relics appear to have flatter spectra related to larger Mach numbers and reside at larger cluster radii. There are examples that are interpreted as adiabatically compressed fossil plasma, and simulations can reproduce the sort of filamentary structure that is seen.

While the existence of radio halos and relics has been known for as long as 50 years, they pose a problem in that the diffusion time across them is typically 100 times longer than the electron energy-loss time. This mandates in-situ electron acceleration. Given that such acceleration must be possible, why are halos and relics a relative rarity? One clear point to emerge is that halos and relics exist only in unrelaxed clusters.

Alternative models have been proposed for particle acceleration in relics and halos. It may arise as a result of strong shocks and MHD turbulence resulting from a cluster merger. Of note are relic structures that are impressively long with very uniform spectral indices. The alternative suggestion is that the electrons are secondaries from charged pion decay after cosmic-ray protons have interacted with thermal protons, although this may disagree with the steep spectrum that is observed, and the accompanying neutral pions would be a source of gamma-rays, in possible contradiction with Fermi results.

A more recent development has been the detection of mini halos (although these can still be of order 100 to 400 kpc in size) in clusters where a central radio galaxy can be the source of relativistic particles. There is an apparent correlation with the presence of cold fronts, often interpreted in the context of gas sloshing, which may provide the source of turbulence for particle acceleration.

Radio galaxies in clusters tend to be distorted in morphology, with more narrow-angle-tail (NAT) and wide-angle-tail (WAT) examples. Rotation measures of radio galaxies provide estimates of cluster magnetic field strength (typically a few μG). The enhanced gas density presumably helps to limit adiabatic energy losses, but while the brightest clusters are most likely to host radio galaxies, clusters have little effect on the overall radio-galaxy luminosity function. Abell 3266 is an interesting cluster where there is evidence from spectroscopic redshifts of a merging subcluster and from radio mapping of possible relics. It also hosts a radio galaxy of severely bent NAT morphology that appears to be in-falling from the cluster's edge. Polarization is strongly ordered along the tails, but appears mixed with turbulent field belonging to the cluster itself.

An obvious conclusion is that clusters are complex beasts, particularly when caught in phases of merging! Deeper radio spectral index and polarization mapping are needed, especially to higher redshift where perhaps NATs and WATs could be used as tracers for clusters. The future is bright for radio mapping, particularly at the low frequencies needed for the steep-spectrum extended structures (e.g., JVLA is replacing the 74 and 330 MHz receivers with improved broad-band capability over 56-470 MHz).

My thoughts from the session are that it might be nice to think there is the potential of discriminating between merging and relaxed clusters through their non-thermal radio content, and that the construction of useful mass scaling relations using the radio might be possible, but all that will depend on findings from deeper radio observations. A better understanding of the locations and mechanisms of particle acceleration is very likely to occur in the near future. A more profound problem is the origin of the magnetic fields.

11. Galaxy evolution — Part 1 (Clusters)

Session chair and section author: Paulo Lopes

Galaxy clusters have been important for galaxy evolution studies in a variety of forms, related to the star formation rate and morphological transformation. In particular, cluster studies based on the colour-magnitude (CM) relation at high-z ($z = 1.5$) show that massive red galaxies with no star formation were fully assembled in mass at $z > 2-3$ (Mei *et al.* 2009). At intermediate to high redshifts ($z < 1$) studies of the evolution in the fraction of blue galaxies indicate the presence of a large population of non-active galaxies. The investigation of the red sequence (RS) faint-to-luminous ratio also points to a progressive assembling of the red sequence for faint objects, indicating that star formation ends in a downsizing way (Stott *et al.* 2007). The evolution of the morphology-density relation indicates that a large number of spiral galaxies are transformed into S0s in the last billion years (from $z = 1$ to $z = 0$; Dressler *et al.* 1997). A morphological change is also seen in the field, as the early-type fraction grows from high to low-z (Capak *et al.* 2007).

In the local Universe it has been found that galaxy morphologies and star formation activity do not depend of the parent halo mass (indicated by the cluster velocity dispersion). That is true at least for objects with $\sigma_{cl} > 500$ km s^{-1} (Poggianti *et al.* 2009, based on the Wide-Field Nearby Galaxy-Cluster Survey, WINGS). Also based on the WINGS, Valentinuzzi *et al.* 2011 showed that a number of parameters from the red sequence (RS slope and scatter, red luminous-to-faint fraction, blue fraction, and fractions of ellipticals, S0s, and spirals) correlates with different tracers of cluster mass or substructure indicators (such as velocity dispersion, X-ray luminosity, number of cluster substructures, BCG prevalence and spatial concentration of ellipticals). On the contrary, these authors found that all the above parameters correlate with local density, in agreement with Peng *et al.* 2012.

These works suggests that not only galaxy mass, but also local density is a key parameter driving galaxy evolution. The dependence on local density of the galaxy stellar mass function (Vulcani *et al.* 2012) and of the morphology-mass relation (Calvi *et al.* 2012) also support that idea (see Figure 1 below). It is well know that morphological transformation happens preferentially on the galaxy group scale (Helsdon & Ponman 2003; Hoyle *et al.* 2012). Nonetheless, galaxy clusters still are great places to study galaxy evolution, as cluster specific processes may have a frosting effect on top of the pre-processing on a group level.

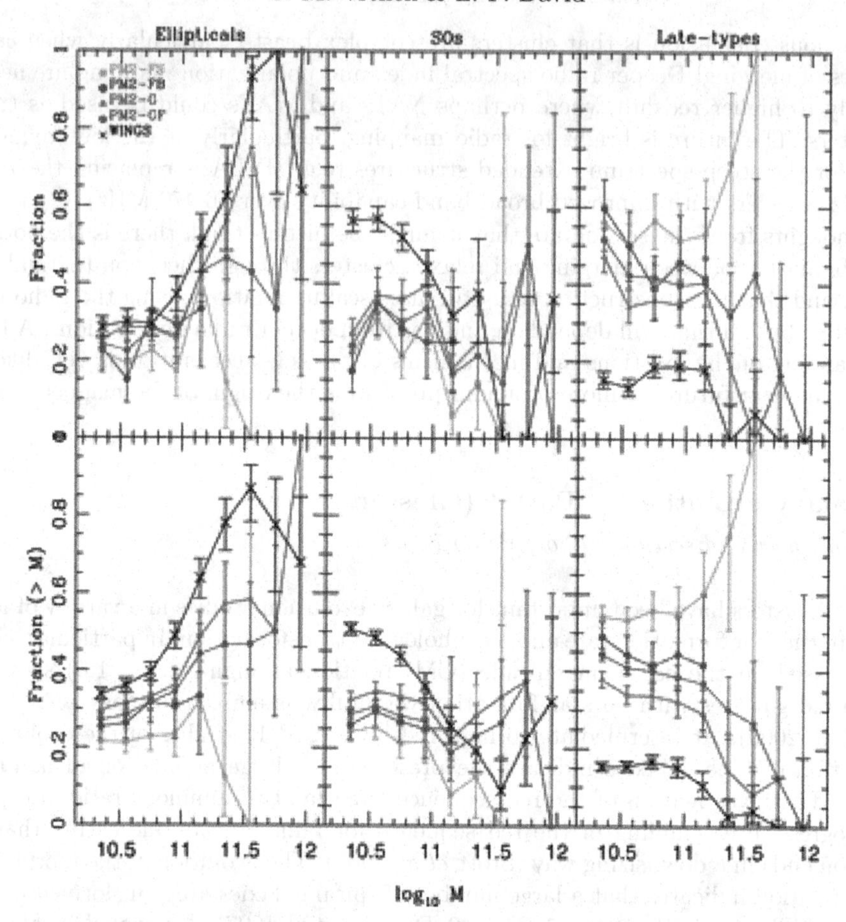

Figure 10. Top panels: fractions of elliptical (left), S0 (central), and late-type (right) galaxies in different environments as a function of stellar mass. Bottom panels: Cumulative distributions of ellipticals, S0s and late-type galaxies. Figure from Calvi *et al.* 2012.

On what regards high redshift studies it is known that a large fraction of the passive and massive galaxies at $z \sim 2$ end up in clusters at $z = 0$. That has implications, for instance, on the evoltuion of galaxy sizes. Investigation of the object Cl J1449+0856 ($z \sim 2$) show the existence of a concentration of candidate members, classified as early-type galaxies, that are already massive, red, quiescent (Figure 2). Those also are a factor of 2-3 smaller than $z = 0$ objects (Strazzullo *et al.* 2012). On the other hand, the authors find that interlopers at the same redshift ($z \sim 2$) are more compact than candidate members. The main difficulties on this type of work regards membership determination, as most of the work is based on photometric redshifts. The conclusions above are supported by the work of Strazzullo *et al.* 2010, based on the cluster XMMU J2235-2557 at $z = 1.39$. They find this object already has a tight central red sequence composed of massive early-type galaxies, with negligible star formation.

At $z \sim 1$ Burke *et al.* 2012, measured for the first time the intra-cluster light (ICL), showing that it represents 1-4% of the total cluster light. The comparison to nearby objects indicate a strong evolution (growth of 2-4 times) of the ICL component. That is in contrast to the small growth of the BCGs, what suggests that stripping may be dominant over merging in the cluser cores. The authors further sugest that this stripped material should be accounted in cluster evolution simulations. An ongoing study based on

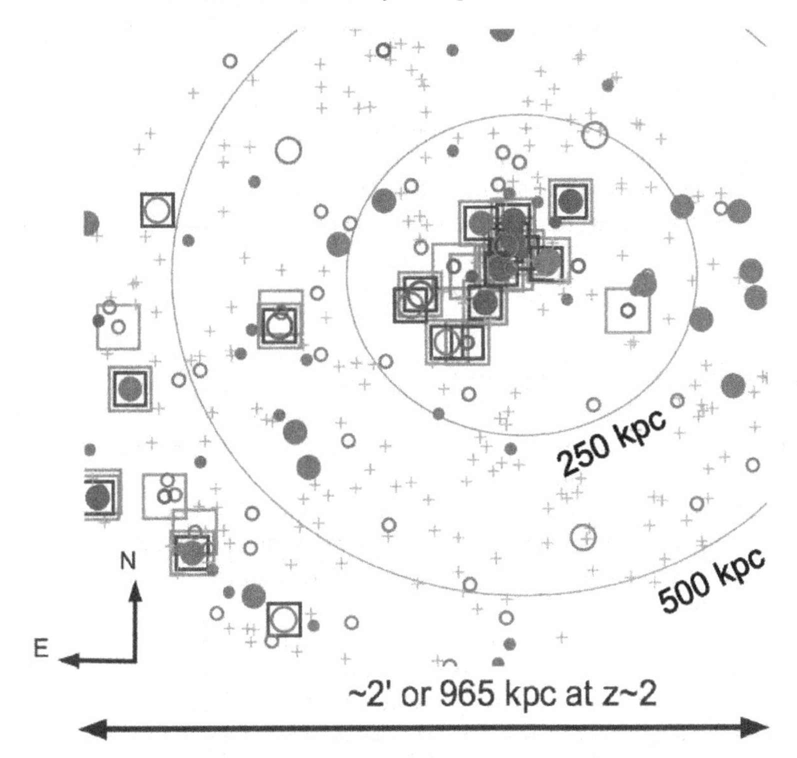

Figure 11. Galaxy distribution in the Cl J1449+0856 cluster field, at $z \sim 2$. In blue are shown star-forming candidate members and in red, quiescent candidate members. Figure from Strazzullo *et al.* 2012.

IFU observations of galaxies in different local densities shows that in dense environments ($\Sigma_5 > 1$ gals/Mpc^2; typical of groups) when the star formation has decreased the gas has been removed from the outskirts. That indicates stripping and/or slow starvation may be acting on these galaxies (Brough *et al.* 2012).

12. Galaxy evolution — Part 2 (Groups)

Session chair: Matthew Colless

This session focused on the general area of galaxy evolution in groups rather than clusters, and included the following talks:
Xiaohu Yang on "Galaxy Evolution in Groups";
Eric Wilcots on "The Assembly and Evolution of Groups of Galaxies";
Marcella Carollo on "Which environment affects galaxy evolution?";
Yingjie Peng on "The quenching of satellite galaxies as the origin of environmental effects";
Simon Lilly on "Galaxy evolution in zCOSMOS groups to z   1"; and
Areg Mickaelian on "Multiple galaxies and groups among the BIG objects".

13. Galaxy evolution — Part 3

Session chair: Simon Lilly

This session completed the presentations portion of the meeting with a variety of talks on galaxy groups:
Duncan Forbes on "Galaxy Groups: Results from the GEMS Survey";
Paulo Lopes on "Segregation Effects in Galaxy Groups";
Claudia Mendes de Oliveira on "The Tully-Fisher relation for interacting and field galaxies at z=0: lessons to be learned for high- redshift TF work";
Mike Hudson on "When, where and how star formation is quenched on cluster infall"; and
Thais Idiart on "Galaxy Evolution through cosmic time".

14. Concluding panel discussion

Andy Fabian (chair), Mark Birkinshaw, Marcella Carollo, Ralph Kraft, Daisuke Nagai, and Tiziana Venturi kindly agreed to lead a concluding summary and discussion of open questions to close the meeting.

Among the questions and opportunities for future investigation raised by the panelists and in subsequent discussion with the attendees were the following:

- What is the prevalence of cool cores as a function of redshift? Is this the same in clusters and groups?
- In just a a few years, new low-frequency radio telescopes will begin to have major impact, and will change our view of AGN outbursts. There is much to be done in the comparison of these data and deep targeted X-ray observations.
- The new radio facilities will also open the study of low-mass systems at high redshifts.
- The kind of study that has in the past been applied to massive clusters is moving toward lower masses, with a focus on the statistics of substantial samples and the refinement of physical understanding.
- Two new methods for finding clusters at high redshifts are promising: the SZ effect and the search for radio halos.
- The comparison of optical and X-ray data promises to illuminate group and cluster assembly history.
- The detection of the kinematic SZ effect opens new opportunities.
- N-body and hydrodynamic simulations are already highly valuable, but need to and will improve at fine scales, with the inclusion of better physics. The close connection between observation and simulation is vital.
- The imminent launches of eROSITA and ASTRO-H offer remarkable opportunities for cluster surveys and direct measurements of turbulence in cluster gas, respectively.
- Integral field spectroscopy of large numbers of galaxies is likely to have major impact.
- The combination of surveys using multiple techniques and wavelengths has growing importance.
- Selection effects still plague much of our work; a comprehensive understanding of populations needs increasing attention.
- There are still asymmetries in coverage of the northern and southern skies; these may soon be equalized or reversed, with numerous major southern hemisphere facilities starting up.
- Cluster studies continue to have major impact in cosmology: twenty years ago it was on the determination of Ω_m, ten years ago on σ_8: what will be next?

Your chapter editors would like to conclude with their thanks to the members of the organizing committee, to the session chairs, to the organizers from the Local Committee and from the IAU, and most of all to the many presenters whose research, commitment, and energetic participation has made this Special Session a memorable summary of current progress in the field and of the outlook for future discovery.

References

Akamatsu, H., Hoshino, A., Ishisaki, Y., *et al.* 2011, PASJ

Arnaboldi, M., Ventimiglia, G., Iodice, E., Gerhard, O., & Coccato, L. 2012, *A&A*, 545, A37

Booth, C. M. & Schaye, J. 2009, *MNRAS*, 398, 53, 0904.2572

Brough, S., *et al.* 2012, *in preparation*

Brüggen, M., Heinz, S., Roediger, E., Ruszkowski, M., & Simionescu, A. 2007, *MNRAS*, 380, L67, 0706.1869

Brüggen, M. & Kaiser, C. R. 2002, *Nat*, 418, 301, arXiv:astro-ph/0207354

Burke, C., *et al.* 2012, *MNRAS*, 425, 2058

Burns, J. O., Skillman, S. W., & O'Shea, B. W. 2010, *ApJ*, 721, 1105

Calvi, R., *et al.* 2012, MNRASL, 419, L14

Capak, P., *et al.* 2007, *ApJS*, 172, 284

Chaudhuri, A., Nath, B. B., & Majumdar, S. 2012, *ApJ*, 759, 87

Churazov, E., Forman, W., Jones, C., & Boehringer, H. 2000, *A&A*, 356, 788

Dressler, A., *et al.* 1997, *ApJ*, 490, 577

Dubois, Y., Devriendt, J., Slyz, A., & Silk, J. 2009, *MNRAS*, 399, L49, 0905.3345

Eckert, D., Vazza, F., Ettori, S., *et al.* 2012, *Astronomy & Astrophysics*, 541, A57

Fabian, A. C. 1994, *ARA&A*, 32, 277

Fujita, Y. & Ohira, Y. 2011, *ApJ*, 738, 182, 1106.5790

——. 2012, *ApJ*, 746, 53, 1111.4208

Governato, F., Zolotov, A., Pontzen, A., *et al.* 2012, *MNRAS*, 422, 123

Heinz, S., Brüggen, M., Young, A., & Levesque, E. 2006, *MNRAS*, 373, L65, arXiv:astro-ph/0000004

Helsdon, S. & Ponman, T., 2003, *MNRAS*, 339, 29

Hoeft, M., Nuza, S. E., Gottlöber, S., *et al.* 2011, Journal of Astrophysics and Astronomy, 32, 509

Hoyle, B., *et al.* 2012, *MNRAS*, 423, 3478

Janowiecki, S., Mihos, J. C., Harding, P., *et al.* 2010, *ApJ*, 715, 972

Li, Y. & Bryan, G. L. 2012, *ApJ*, 747, 26, 1112.2701

Li, C., Jing, Y. P., Mao, S., *et al.* 2012a, *ApJ*, 758, 50

Li, C., Wang, L. X., & Jing, Y. P. 2012, arXiv:1210.5700

Lin, Y.-T., Stanford, S. A., Eisenhardt, P. R. M., *et al.* 2012, *ApJ*, 745, L3

Mandelbaum, R., Seljak, U., Kauffmann, G., Hirata, C. M., & Brinkmann, J. 2006, *MNRAS*, 368, 715

Markevitch, M., Vikhlinin, A., & Mazzotta, P. 2001, *ApJ*, 562, L153

Markevitch, M., Gonzalez, A. H., David, L., *et al.* 2002, *ApJ*, 567, L27

Martizzi, D., Teyssier, R., Moore, B., & Wentz, T. 2012, *MNRAS*, 422, 3081

McNamara, B. & Nulsen, P. E. J. 2007, ARAA, 45, 117

McNamara, B. R., Kazemzadeh, F., Rafferty, D. A., *et al.* 2009, *ApJ*, 698, 594

McNamara, B. R. & Nulsen, P. E. J. 2007, *ARA&A*, 45, 117, 0709.2152

Mei, S., *et al.* 2009, *ApJ*, 690, 42

Mihos, J. C., Harding, P., Feldmeier, J., & Morrison, H. 2005, *ApJ*, 631, L41

Miller, E. D., Bautz, M., George, J., *et al.* 2012, in SUZAKU 2011: Exploring the X-ray Universe: Suzaku and Beyond. AIP Conference Proceedings, Vol. 1427, 13–20

Nagai, D. & Lau, E. T. 2011, *ApJ*, 731, L10

Nulsen, P. E. J., McNamara, B. R., Wise, M. W., & David, L. P. 2005, *ApJ*, 628, 629

Omma, H., Binney, J., Bryan, G., & Slyz, A. 2004, *MNRAS*, 348, 1105, arXiv:astro-ph/0307471

Owers, M. S., Couch, W. J., Nulsen, P. E. J., & Randall, S. W. 2012, *ApJ*, 750, L23

Peng, Y., *et al.* 2012, *ApJ*, 757, 4

Peterson, J. R. & Fabian, A. C. 2006, *Phys. Rep.*, 427, 1, arXiv:astro-ph/0512549

Poggianti, B., *et al.* 2009, *ApJ*, 697, 137
Quilis, V., Bower, R. G., & Balogh, M. L. 2001, *MNRAS*, 328 , 1091, arXiv:astro-ph/0109022
Reynolds, C. S., McKernan, B., Fabian, A. C., Stone, J. M., & Vernaleo, J. C. 2005, *MNRAS*, 357, 242, arXiv:astro-ph/0402632
Roediger, E., Brüggen, M., Simionescu, A., *et al.* 2011, *MNRAS*, 413, 2057
Roediger, E., Lovisari, L., Dupke, R., *et al.* 2012, *MNRAS*, 420, 3632
Rottgering, H. J. A., Wieringa, M. H., Hunstead, R. W., & Ekers, R. D. 1997, *MNRAS*, 290, 577
Ruszkowski, M., Brüggen, M., & Begelman, M. C. 2004, *ApJ*, 615, 675, arXiv:astro-ph/0403690
Sanders, J. S., Fabian, A. C., & Smith, R. K. 2010, *MNRAS*, 410, no
Shang, C. & Oh, S. P. 2012, *MNRAS*, 426, 3435
Sijacki, D., Pfrommer, C., Springel, V., & Enßlin, T. A. 2008, *MNRAS*, 387, 1403, 0801.3285
Sijacki, D. & Springel, V. 2006, *MNRAS*, 371 , 1025, arXiv:astro-ph/0605301
Sijacki, D., Springel, V., Di Matteo, T., & Hernquist, L. 2007, *MNRAS*, 380, 877, 0705.2238
Simionescu, A., Allen, S. W., Mantz, A., *et al.* 2011, *Science*, 331, 1576
Simionescu, A., Werner, N., Urban, O., *et al.* 2012, *ApJ*, 757, 182
Skillman, S. W., Xu, H., Hallman, E. J., *et al.* 2012, arXiv:1211.3122
Springel, V. 2011, arXiv:1109.2218
Sutter, P. M., Yang, H.-Y. K., Ricker, P. M., Foreman, G., & Pugmire, D. 2012, *MNRAS*, 419, 2293, 1108.3344
Tamura, T., Sekiya, N., Hayashida, K., Ueda, S., & Nagai, M. 2011, in SUZAKU 2011: Exploring the X-ray Universe: Suzaku and Beyond. AIP Conference Proceedings, Vol. 1427, 332–333
Tamura, T., Hayashida, K., Ueda, S., & Nagai, M. 2012, *PASJ*, 1427, 332
Simionescu, A., Allen, S. W., Mantz, A., *et al.* 2011, Science (New York, N.Y.), 331, 1576
Stott, J., *et al.* 2007, *ApJ*, 661, 95
Strazzullo, V., *et al.* 2010, *A&A*, 524, 17
Strazzullo, V., *et al.* 2012, *in preparation*
Teyssier, R., Moore, B., Martizzi, D., Dubois, Y., & Mayer, L. 2011, *MNRAS*, 414, 195, 1003.4744
Valentinuzzi, T., *et al.* 2011, *A&A*, 536, 34
van Weeren, R. J., Röttgering, H. J. A., Brüggen, M., & Hoeft, M. 2010, *Science*, 330, 347
Vikhlinin, A., Markevitch, M., & Murray, S. S. 2001, *ApJ*, 551, 160
Vikhlinin, A. A. & Markevitch, M. 2002, Astron. Let., 28, 495
Vulcani, B., *et al.* 2012, *MNRAS*, 420, 1481
Wake, D. A., Franx, M., & van Dokkum, P. G. 2012 (arXiv:1201.1913)
Walker, S. A., Fabian, A. C., Sanders, J. S., & George, M. R. 2012, *MNRAS: Letters*, 427, 45
Yang, H.-Y. K., Sutter, P. M., & Ricker, P. M. 2012, ArXiv e-prints, 1207.6106
ZuHone, J. A., Markevitch, M., Ruszkowski, M., & Lee, D. 2013, *ApJ*, 762, 69

Highlights of Astronomy, Volume 16
XXVIIIth IAU General Assembly, August 2012
T. Montmerle, ed.

doi:10.1017/S1743921315001593

Preface: Special Session SpS3 Galaxy Evolution through Secular Processes

The meeting "Galaxy Evolution through Secular Processes," hosted as Special Session 3 at the XVIIIth General Assembly of the IAU in Beijing, brought together a broad community of astronomers actively working on topics related to the evolution of galaxies. This evolution is currently thought to have two phases: (1) a formative chaotic phase, where baryonic matter collected into dark halo "seeds" with the occurrence of frequent mergers; and (2) a slower secular evolutionary phase where internal perturbations such as bars and spirals interacted (and likely still interact) with a galaxy's stars and gas clouds to redistribute material and slowly change the morphology, such as growing bulges or pseudobulges from disk material.

At the present epoch, secular evolution is thought to be the dominant process that changes galaxies. Major mergers since z=1 are far less important than they were in the past, and minor mergers can explain mainly some peculiar morphologies. For the majority of normal galaxies, secular evolution is likely to be mostly an internal process of change, although the meeting also brought attention to environmental secular evolution and the role it plays in changing magellanic spirals and irregulars into what John Kormendy calls "spheroidal galaxies." This and studies of the rotation of early-type galaxies and the structural components of S0 galaxies has led to a serious modification of the old Hubble "tuning fork" of galaxy morphologies. The goal of the meeting was to examine these issues and others from a wide variety of viewpoints, to better establish the role secular processes have played on galaxy evolution.

The program of the meeting included 20 invited reviews and 43 contributed talks presented in 12 sessions over a five day period, plus a few dozen related posters. The meeting covered dynamical mechanisms of secular evolution, including radial migration, external gas accretion, and the roles of collective effects and resonances; the origins and lifetimes of internal perturbations that drive secular evolution and the features they may generate, such as rings, lenses, and secondary bars; characteristics of classical bulges versus pseudobulges; the rise of parallel-sequencing as an alternative view of galaxy morphology; the role secular evolution plays on stellar populations, chemical abundances, and star formation; how secular evolution affects vertical disk structure; secular processes in the Milky Way, including new models of the Galactic bar; the structure of isolated galaxies whose properties may be driven only by secular evolution; and finally, secular evolution in a cosmological context.

We are grateful to the SOC of IAU SpS3 and to the speakers and poster presenters for making this meeting a great success. We are also grateful to the LOC for the smooth operation of the meeting venue.

Ronald J. Buta and Daniel Pfenniger, co-chairs SOC,
November 30, 2012

Highlights of Astronomy, Volume 16
XXVIIIth IAU General Assembly, August 2012
T. Montmerle, ed.

doi:10.1017/S174392131400581X

Internal and environmental secular evolution of disk galaxies

John Kormendy

Department of Astronomy, University of Texas at Austin, Austin, TX 78712, USA;
Max-Planck-Institut für Extraterrestrische Physik, Garching bei München, Germany;
Universitäts-Sternwarte, Ludwig-Maximilians-Universität, München, Germany;
email: kormendy@astro.as.utexas.edu

Abstract. This Special Session is devoted to the secular evolution of disk galaxies. Here 'secular' means 'slow'; i. e., evolution on time scales that are generally much longer than the galaxy crossing or rotation time. Internal and environmentally driven evolution both are covered.

I am indebted to Albert Bosma for reminding me at the 2011 Canary Islands Winter School on Secular Evolution that our subject first appeared in print in a comment made by Ivan King (1977) in his introductory talk at the Yale University meeting on *The Evolution of Galaxies and Stellar Populations*: 'John Kormendy would like us to consider the possibility that a galaxy can interact with itself.... I'm not at all convinced, but John can show you some interesting pictures.' Two of the earliest papers that followed were Kormendy (1979a, b); the first discusses the interaction of galaxy components with each other, and the second studies these phenomena in the context of a morphological survey of barred galaxies. The earliest modeling paper that we still use regularly is Combes & Sanders (1981), which introduces the now well known idea that box-shaped bulges in edge-on galaxies are side-on, vertically thickened bars.

It is gratifying to see how this subject has grown since that time. Hundreds of papers have been written, and the topic features prominently at many meetings (e. g., Block *et al.* 2004; Falcoń-Barroso & Knapen 2012, and this Special Session). My talk here introduces both internal and environmental secular evolution; a brief abstract follows. My Canary Islands Winter School review covers both subjects in more detail (Kormendy 2012). Kormendy & Kennicutt (2004) is a comprehensive review of internal secular evolution, and Kormendy & Bender (2012) covers environmental evolution. Both of these subject make significant progress at this meeting.

Secular evolution happens because self-gravitating systems evolve toward the most tightly bound configuration that is reachable by the evolution processes that are available to them. They do this by spreading – the inner parts shrink while the outer parts expand. Significant changes happen only if some process efficiently transports energy or angular momentum outward. The consequences are very general: evolution by spreading happens in stars, star clusters, protostellar and protoplanetary disks, black hole accretion disks and galaxy disks. This meeting is about disk galaxies, so the evolution most often involves the redistribution of angular momentum.

We now have a good heuristic understanding of how nonaxisymmetric structures rearrange disk gas into outer rings, inner rings and stuff dumped onto the center. Numerical simulations reproduce observed morphologies very well. Gas that is transported to small radii reaches high densities that are seen in CO observations. Star formation rates measured (e. g.) in the mid-infrared show that many barred and oval galaxies grow, on timescales of a few Gyr, dense central 'pseudobulges' that are frequently mistaken for classical (elliptical-galaxy-like) bulges but that were grown slowly out of the disk (not made rapidly by major mergers). Our resulting picture of secular evolution accounts for the richness observed in morphological classification schemes such as those of de Vaucouleurs (1959) and Sandage (1961). State-of-the art morphology discussions include the *de Vaucouleurs Atlas of Galaxies* (Buta *et al.* 2007) and Buta (2012, 2013).

Pseudobulges as disk-grown alternatives to merger-built classical bulges are important because they impact many aspects of our understanding of galaxy evolution. For example, they are observed to contain supermassive black holes (BHs), but they do not show the well known, tight correlations between BH mass and host properties (Kormendy *et al.* 2011). We can distinguish between classical and pseudo bulges because the latter retain a 'memory' of their disky origin. That is, they have one or more characteristics of disks: (1) flatter shapes than those of classical bulges, (2) correspondingly large ratios of ordered to random velocities, (3) small velocity dispersions σ with respect to the Faber-Jackson correlation between σ and bulge luminosity, (4) spiral structure or nuclear bars in the 'bulge' part of the light profile, (5) nearly exponential brightness profiles and (6) starbursts. *None of the above classification criteria are 100 % reliable. Published disagreements on (pseudo)bulge classifications usually result from the use of diffferent criteria. It is very important to use as many classification criteria as possible. When two or more criteria are used, the probability of misclassification becomes very small.*

I also review environmental secular evolution – the transformation of gas-rich, star-forming spiral and irregular galaxies into gas-poor, 'red and dead' S0 and spheroidal ('Sph') galaxies. I show that Sph galaxies such as NGC 205 and Draco are not the low-luminosity end of the structural sequence (the 'fundamental plane') of elliptical galaxies. Instead, Sph galaxies have structural parameters like those of low-luminosity S+Im galaxies. Spheroidals are continuous in their structural parameters with the disks of S0 galaxies. They are bulgeless S0s. S+Im → S0+Sph transformation involves a variety of internal (supernova-driven baryon ejection) and environmental processes (e. g., ram-pressure gas stripping, harassment, and starvation). Improved evidence for galaxy transformation is presented in several papers at this meeting.

Keywords. galaxies: bulges; galaxies: evolution; galaxies: structure

---

Acknowledgements

My work is supported by the Curtis T. Vaughan, Jr. Centennial Chair in Astronomy at the University of Texas at Austin and by the Max-Planck-Institut für Extraterrestrische Physik (MPE). This review was written during a visit to MPE and to the Universitäts-Sternwarte of the Ludwig-Maximilians-Universität. I thank MPE Managing Director Ralf Bender and to the staffs of both institutions for their hospitality and support.

References

Block, D. L., Puerari, I., Freeman, K. C., *et al.* (eds.) 2004, *Penetrating Bars Through Masks of Cosmic Dust: The Hubble Tuning Fork Strikes a New Note* (Dordrecht: Kluwer)

Buta, R. 2012, in *XXIII Canary Islands Winter School of Astrophysics, Secular Evolution of Galaxies*, ed. J. Falcón-Barroso & J. H. Knapen (Cambridge: Cambridge University Press), in press

Buta, R. 2013, in *Planets, Stars, and Stellar Systems, Vol. 6, Extragalactic Astronomy and Cosmology*, ed. W. C. Keel (New York: Springer-Verlag), in press (arXiv:1102.0550)

Buta, R. J., Corwin, H. G. & Odewahn, S. C. 2007, *The de Vaucouleurs Atlas of Galaxies* (Cambridge: Cambridge University Press)

Combes, F. & Sanders, R. H. 1981, *A&A*, 96, 164

de Vaucouleurs, G. 1959, *Handbuch der Physik*, 53, 275

Falcón-Barroso, J. & Knapen, J. H. (eds.) 2012, *XXIII Canary Islands Winter School of Astrophysics, Secular Evolution of Galaxies* (Cambridge: Cambridge University Press)

King, I. R. 1977, in *The Evolution of Galaxies and Stellar Populations*, ed. B. M. Tinsley & R. B. Larson (New Haven: Yale University Observatory), p. 1

Kormendy, J. 1979a, in *Photometry, Kinematics and Dynamics of Galaxies*, ed. D. S. Evans (Austin: Department of Astronomy, University of Texas at Austin), p. 341

Kormendy, J. 1979b, *ApJ*, 227, 714

Kormendy, J. 2012, in *XXIII Canary Islands Winter School of Astrophysics, Secular Evolution of Galaxies*, ed. J. Falcón-Barroso & J. H. Knapen (Cambridge: Cambridge Univ. Press), p. 1

Kormendy, J. & Bender, R. 2012, *ApJS*, 198, 2

Kormendy, J., Bender, R. & Cornell, M. E. 2011, *Nature*, 469, 374

Kormendy, J. & Kennicutt, R. C. 2004, *ARAA*, 42, 603

Sandage, A. 1961, *The Hubble Atlas of Galaxies* (Washington: Carnegie Inst. of Washington)

Highlights of Astronomy, Volume 16
XXVIIIth IAU General Assembly, August 2012
T. Montmerle, ed.

doi:10.1017/S1743921314005821

Overview of dynamical mechanisms of secular evolution

Daniel Pfenniger

Geneva Observatory, Geneva, Switzerland
email: daniel.pfenniger@unige.ch

Abstract. Gravity-bound isolated systems, from stars, planetary systems, star clusters to galaxies, share common properties where evolution is the rule. Typically if they start forming at a well defined epoch they tend to change significantly over a timescale comparable to their present age. So evolution is never truly stopped, it just proceeds slower and slower: after a rapid, violent phase a slower, secular phase follows. In galactic astronomy for many decades the paradigm was rather that after a short violent time galaxies would settle in a stable steady state just consuming gas into stars. Actually today it appears that the progressive appearance of galaxy systematic morphologies and the slowing pace of mergers indicate that common intrinsic dynamical factors continue to shape galaxies towards similar properties irrespective of their largely different formation histories and initial conditions. Newtonian physics supplemented by a weakly dissipative component provides an amazing amount of explanations for the galaxy properties, like exponential stellar disks, spirals, bars, and peanut-shaped bulges. The purpose of this talk is to review these mechanisms of dynamical secular evolution.

Keywords. galaxies: evolution; galaxies: structure

Highlights of Astronomy, Volume 16
XXVIIIth IAU General Assembly, August 2012
T. Montmerle, ed.

doi:10.1017/S1743921314005833

The role of collective effects and secular mass migration on galactic transformation

Xiaolei Zhang[1] **and Ronald J. Buta**[2]

[1]Department of Physics and Astronomy, George Mason University, Fairfax, VA 22030, USA
email: xzhang5@gmu.edu

[2]Department of Physics & Astronomy, University of Alabama,
Box 870324, Tuscaloosa, AL, USA

Abstract. During the lifetime of a galaxy, secular radial mass redistribution is expected to gradually build up a bulge and transform the Hubble type from late to early. The dominant dynamical process responsible for this transformation is a collective instability mediated by density-wave collisionless shocks (Zhang 1996, 1998, 1999). The ability of this new mechanism to secularly redistribute the STELLAR mass provides a general pathway for the formation and evolution of the majority of Hubble types, ranging from late type disk galaxiess to disky ellipticals. ATLAS^{3D} results (Cappellari *et al.* 2013) showed that spirals and S0s and disky ellipticals form a continuous trend of evolution which also coincides with the aging of the stellar population of galactic disks. The importance of stellar accretion is also revealed in the results of the COSMOS team which showed that the evolution of the black-hole-mass/bulge-mass correlation since $z = 1$ was mainly due to the mass redistribution on pre-existing STELLAR disks which were already in place by $z = 1$ (Cisternas *et al.* 2011). The weaker correlation between the masses of *late-type* bulges and AGNs observed at any given epoch in our view is a result of the quicker initial onset of accretion events in AGN disks compared to that in galactic disks, since the dynamical timescale is shorter for smaller AGN accretion disks.

The same secular dynamical process can produce and maintain the well-known scaling relations and universal rotation curves of observed galaxies during their Hubble-type transformation (Zhang 2008), as well as reproduce many other observed structural and kinematic properties of galaxies such as the size-line-width relation of the interstellar medium and the age-velocity dispersion relation of solar neighborhood stars in our own Galaxy. A by-product of this analysis is a powerful new method for locating the multiple corotation resonances in galaxies (Zhang & Buta 2007; Buta & Zhang 2009).

The current work also highlights the connection between collective effects in galactic dynamics and nonequilibrium phase transition processes in other branches of physics such as fluid turbulence and spontaneous breaking of gauge symmetry in high-energy physics. The continuous build-up of the Hubble sequence of galaxies through secular mass accretion also hints at the baryonic nature of galactic dark matter and poses challenges to the existing LCDM paradigm, since the well-known adiabatic compression process during baryonic mass inflow produced by secular evolution would lead to a concentration of the cold dark matter to the central region of early-type galaxies, which is not observed.

References

Buta, R. J. & Zhang, X. 2009, *ApJS* 182, 559
Cappellari. M., *et al.* 2013, *MNRAS* 432, 1862
Cisternas, M., *et al.* 2011, *ApJ* 726, 57; *ApJ* 741, L11
Zhang, X. 1996, *ApJ* 457, 125; 1998, *ApJ* 499, 93 ; 1999, *ApJ* 518, 613
Zhang, X. 2008, *PASP* 120, 121
Zhang, X. & Buta, R. J. 2007, *AJ* 133, 2584

Highlights of Astronomy, Volume 16
XXVIIIth IAU General Assembly, August 2012
T. Montmerle, ed.

doi:10.1017/S1743921314005845

The role of resonances in the evolution of galactic disks

Jacques Lepine[1], Sergio Scarano Jr.[1], Sergei Andrievsky[2], Douglas A. de Barros[1] and Thiago C. Junqueira[1]

[1]University of Sao Paulo Brazil, email: jacques@astro.iag.usp.br
[2]Odessa National University

Abstract. Resonances play an important role in the evolution of the disks of spiral galaxies, and in particular in the chemical abundance evolution. The dominant effect is that of corotation; this effect can be even used as a tool to estimate the age of the present spiral arm pattern, which are usually found to be long-lived, contrary to a recent common belief. We investigated a sample of galaxies for which the corotation radius is known and for which there are available in the literature measurements of abundance gradients for Oxygen. A very good correlation is found between corotation radii and the radii at which there is a break in the slope of the gradients. The gradients are usually decreasing in the inner regions and become flat or rising at larger radii. In several galaxies, including the Milky Way, one observes not only a change in the slope of the abundance gradient, but also an abrupt step in metallicity, at corotation. This step is due to the fact that corotation separates the disk of a galaxy in two regions (inside corotation and outside corotation) which are isolated one from the other, so that the two sides evolve in an independent way. The barrier between the two regions is produced by the flow of gas in opposite directions in the two sides and by the ring-shaped void of gas observed at corotation. Besides this, an independent effect of corotation is a minimum of star formation associated with the minimum velocity at which the spiral arms (seen as potential wells) are fed with interstellar gas. Still another effect is the scattering of stars by the resonance, which causes their migration to different galactic radii. Other resonances, like 4:1, have properties almost opposite to corotation; they stimulate star-formation, and tend to gather the stars in the resonant orbit, instead of scattering them out, as shown by numerical simulations. Due to this property, one can see arms which have the shape of resonant stellar orbits, which depart from logarithmic spirals.

Keywords. galaxies: evolution; galaxies: structure

Highlights of Astronomy, Volume 16
XXVIIIth IAU General Assembly, August 2012
T. Montmerle, ed.

doi:10.1017/S1743921314005857

The Lifetimes of Spirals and Bars

J. A. Sellwood

Department of Physics & Astronomy, Rutgers University,
136 Frelinghuysen Road, Piscataway, NJ 08854-8019, USA
email: `sellwood@physics.rutgers.edu`

Simulations of isolated galaxy disks that are stable against bar formation readily manifest multiple, transient spiral patterns. It therefore seems likely that some spirals in real galaxies are similarly self-excited, although others are clearly driven by tidal interactions or by bars. The rapidly changing appearance of simulated spirals does not, however imply that the patterns last only a fraction of an orbit. Power spectrum analysis reveals a few underlying, longer-lived spiral waves that turn at different rates, which when super-posed give the appearance of swing-amplified transients. These longer-lived waves are genuine unstable spiral modes; each grows vigorously, saturates and decays over a total of several orbit periods. As each mode decays, the wave action created as it grew drains away to the Lindblad resonances, where it scatters stars. The resulting changes to the disk create the conditions for a new instability, giving rise to a recurring cycle of unstable modes.

Transient spiral modes are one of the most important agents of secular evolution in disk galaxies. They redistribute angular momentum, cause the random motions of stars to increase over time, and changes at corotation cause extensive radial mixing of both the stars and the gas, as discussed elsewhere at this conference. I also show here that they are able to smooth small-scale irregularities in the radial mass profile of the disk.

We still do not understand why some galaxies have strong bars, while others do not. A mild deficiency of bars in galaxies with massive bulges (Barazza *et al.* 2009) seems consistent with Toomre's stabilizing mechanism, but some barred galaxies should have been stable while other supposedly unstable cases, such as M33, lack strong bars! Skibba *et al.* (2012) have resurrected the idea that mild interactions may trigger some bars.

We have long known that steady, long-lived bars are built from stars pursuing elongated orbits that all turn at the bar pattern speed, and the majority remain within the bar. A significant minority of stars, however, cross and re-cross corotation but are unable to escape to the outer disk unless the bar weakens, or changes in some other way. Some stars in thick disk of the Milky Way have migrated from the inner Galaxy, but the outward flux of stars from the very innermost regions must have been curtailed when the bar formed. Thus identifying the oldest stars in the thick disk having chemical abundances characteristic of the inner Milky Way may allow us to estimate when the bar formed.

Bars in purely stellar particle simulations are very robust. Gas dynamics has been invoked to suggest that they could form and decay several times in the life of a galaxy, but recent cosmological simulations (Kraljic *et al.* 2012) seem to show that this does not happen, except possibly in the tumultuous early stages. Kormendy makes a strong case that bars dissolve to make pseudo-bulges, but the mechanism by which this could happen is unclear as yet.

References

Barazza, F. D. *et al.* 2009, *A&A*, 497, 713
Kraljic, K., Bournaud, F., & Martig, M. 2012, *ApJ*, 757, 60
Skibba, R. A., *et al.* 2012, *MNRAS*, 423, 1485

Highlights of Astronomy, Volume 16
XXVIIIth IAU General Assembly, August 2012
T. Montmerle, ed.

doi:10.1017/S1743921314005869

Origin of structures in disc galaxies: internal or external processes?

E. Athanassoula
LAM/OAMP France, email: lia@oamp.fr

Abstract. Disc galaxies have a number of structures, such as bars, spirals, rings, discy bulges, m = 1 asymmetries, thick discs, warps etc. I will summarise what is known about their origin and in particular whether it is due to an external or an internal process. The former include interactions, major or minor mergers etc, while the latter include instabilities, or driving by another component of the same galaxy, as e.g. the bar or the halo. In cases where more than one process is eligible, I will analyse whether it is possible to distinguish between different origins, and what it would take to do so. This discussion will show that, at least in some cases, it is difficult to distinguish between an internal and an external origin.

Highlights of Astronomy, Volume 16
XXVIIIth IAU General Assembly, August 2012
T. Montmerle, ed.

doi:10.1017/S1743921314005870

Signatures of long-lived spiral patterns: The color gradient trend

Eric E. Martínez-García[1] and Rosa Amelia González-Lópezlira[2]

[1] Instituto de Astronomía, Universidad Nacional Autónoma de México, AP 70-264, Distrito Federal 04510, México; e-mail: martinez@astro.unam.mx. [2] Centro de Radioastronomía y Astrofísica, UNAM, Campus Morelia, Michoacán, México; e-mail: r.gonzalez@crya.unam.mx

Abstract. Based both on observations and simulations, recent works propose that the speed of the spiral pattern in disk galaxies may decrease with increasing radius; the implications are that patterns are actually short-lived, and that the azimuthal color/age gradients across spiral arms predicted by density wave theory could not be produced. We, however, have consistently found such gradients, and measured spiral pattern speeds by comparing the observations with stellar population synthesis models (González & Graham 1996; Martínez-García *et al.* 2009a,b; Martínez-García & González-Lópezlira 2011). Here, we summarize our previous results in non-barred and weakly barred spirals, together with six new, as yet unpublished, objects. On the other hand, we have indeed found a trend whereby pattern speeds at smaller radii are larger than expected from a model that assumes purely circular orbits (cf. Figure 1), likely due to the effect of spiral shocks on the orbits of newborn stars. The results suggest that spirals may behave as steady long-lived patterns.

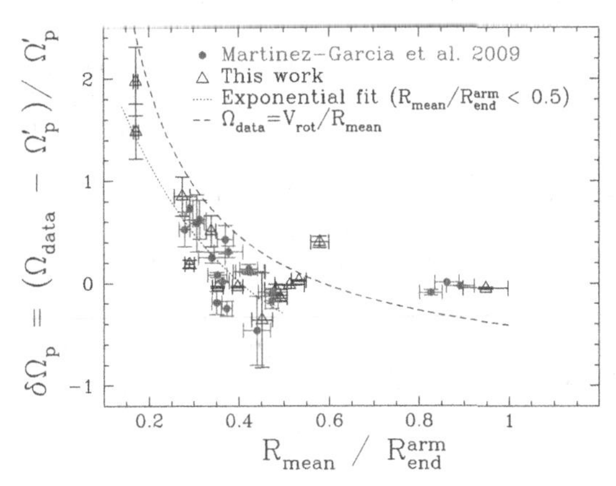

Figure 1. The quantity $\delta\Omega_p$ (cf. Martínez-García *et al.* 2009b) vs. $\frac{R_{mean}}{R^{arm}_{end}}$ for 20 regions analyzed in Martínez-García *et al.* 2009a (red filled circles), and 14 unpublished regions (black empty triangles). The dotted line indicates an exponential fit for $\frac{R_{mean}}{R^{arm}_{end}} < 0.5$. Long-dashed line: rotation curve.

References

González, R. A. & Graham, J. R. 1996, *ApJ*, 460, 651
Martínez-García, E. E., González-Lópezlira, R. A., & Bruzual-A, G. 2009a, *ApJ*, 694, 512
Martínez-García, E. E., González-Lópezlira, R. A., & Gómez, G. C. 2009b, *ApJ*, 707, 1650
Martínez-García, E. E., & González-Lópezlira, R. A. 2011, *ApJ*, 734, 122

Highlights of Astronomy, Volume 16
XXVIIIth IAU General Assembly, August 2012
T. Montmerle, ed.

doi:10.1017/S1743921314005882

Revealing galactic scale bars with the help of Galaxy Zoo

Karen L. Masters[1,2] and the Galaxy Zoo Team

[1]Institute for Cosmology and Gravitation, University of Portsmouth, Dennis Sciama Building, Burnaby Road, Portsmouth, PO1 3FX, UK
email: karen.masters@port.ac.uk

[2]South East Physics Network, www.sepnet.ac.uk

Abstract. We use visual classifications of the brightest 250,000 galaxies in the Sloan Digital Sky Survey Main Galaxy Sample provided by citizen scientists via the Galaxy Zoo project (www.galaxyzoo.org, Lintott *et al.* 2008) to identify a sample of local disc galaxies with reliable bar identifications.

These data, combined with information on the atomic gas content from the ALFALFA survey (Haynes *et al.* 2011) show that disc galaxies with higher gas content have lower bar fractions.

We use a gas deficiency parameter to show that disc galaxies with more/less gas than expected for their stellar mass are less/more likely to host bars. Furthermore, we see that at a fixed gas content there is no residual correlation between bar fraction and stellar mass. We argue that this suggests previously observed correlations between galaxy colour/stellar mass and (strong) bar fraction (e.g. from the sample in Masters *et al.* 2011, and also see Nair & Abraham 2010) could be driven by the interaction between bars and the gas content of the disc, since more massive, optically redder disc galaxies are observed to have lower gas contents.

Furthermore we see evidence that at a fixed gas content the global colours of barred galaxies are redder than those of unbarred galaxies. We suggest that this could be due to the exchange of angular momentum beyond co-rotation which might stop a replenishment of gas from external sources, and act as a source of feedback to temporarily halt or reduce the star formation in the outer parts of barred discs.

These results (published as Masters *et al.* 2012) combined with those of Skibba *et al.* (2012), who use the same sample to show a clear (but subtle and complicated) environmental dependence of the bar fraction in disc galaxies, suggest that bars are intimately linked to the evolution of disc galaxies.

Keywords. galaxies: evolution, galaxies: spiral, galaxies: fundamental parameters, galaxies: statistics, galaxies: structure, surveys, ISM: atoms, radio lines: galaxies

---

Acknowledgements

This publication has been made possible by the participation of more than 200,000 volunteers in the Galaxy Zoo project. Their contributions are individually acknowledged at http://www.galaxyzoo.org/volunteers. We also acknowledge the many members of the ALFALFA team who have contributed to the acquisition and processing of the ALFALFA data set over the last six years.

KLM acknowledges funding from the Peter and Patricia Gruber Foundation as the 2008 Peter and Patricia Gruber Foundation IAU Fellow, and from a 2010 Leverhulme Trust Early Career Fellowship, as well as support from the Royal Astronomical Society to attend the 28th GA of the IAU .

References

Haynes, M. P., *et al.* 2011, *AJ*, 142, 170
Lintott, C. J., *et al.* 2008, *MNRAS*, 389, 1179
Masters, K. L., *et al.* 2011, *MNRAS*, 411, 2026
Masters, K. L., *et al.* 2012, *MNRAS*, 424, 2180
Nair, P. B. & Abraham, R. G. 2010, *ApJL*, 714, L260
Skibba, R. A., Masters, K. L., *et al.* 2012, *MNRAS*, 423, 1485

Highlights of Astronomy, Volume 16
XXVIIIth IAU General Assembly, August 2012
T. Montmerle, ed.

doi:10.1017/S1743921314005894

Bar properties as seen in the *Spitzer* Survey of Stellar Structure in Galaxies

Kartik Sheth[1] and The *Spitzer* Survey for Stellar Structure in Galaxies (S^4G) Team[2]

[1]National Radio Astronomy Observatory United States, email: astrokartik@gmail.com
[2]Various Institutions world-wide

Abstract. Bars serve a crucial signpost in galaxy evolution because they form quickly once a disk is sufficiently massive and dynamically cold. Although the bar fraction in the local Universe is well-established since the mid-60s, a variety of studies have concluded varying bar fractions due to different definitions of bars, use of low quality data or different sample selection. The *Spitzer* Survey of Stellar Structure in Galaxies (S^4G) offers us the ideal data set for resolving this outstanding issue once and for all. S^4G consists of over 2000 nearby galaxies chosen based on optical brightness, distance, galactic latitude and size in a 40 Mpc volume. With a 4 minute integration time per pixel over $>1.5\times$D25 diameter for each galaxy, the data provide the deepest, homogenous, mid-infrared (3.6 and 4.5 microns) data on the nearby Universe. The data are so deep that we are tracing stellar surface densities $<<$ 1 solar mass per square parsec. With these data we can confidently constrain the bar fraction and thus shed important light on the evolutionary state of galaxies as a function of mass, environment and other galaxy host properties.

Highlights of Astronomy, Volume 16
XXVIIIth IAU General Assembly, August 2012
T. Montmerle, ed.

doi:10.1017/S1743921314005900

Galactic rings and secular evolution in barred galaxies

Johan H. Knapen[1,2]

[1]Instituto de Astrofísica de Canarias, E-38200 La Laguna, Spain
[2]Departamento de Astrofísica, Universidad de La Laguna, E-38205 La Laguna, Tenerife, Spain
email: jhk@iac.es

Abstract. Rings are common in galaxies. Several kinds of rings are known: collisional, polar, and resonance rings, of which the latter is by far most common. Resonance rings are prime tracers of the underlying dynamical structure of disk galaxies, in particular of orbital resonances and of manifolds. Rings are also indicators of angular momentum transport, and this is a key factor in secular evolution (see the various reviews in Falcón-Barroso & Knapen 2012).

Resonance rings come in three flavours, primarily defined by their size, namely nuclear, inner, and outer rings. From studies like those of Buta (1995), Knapen (2005) and Comerón *et al.* (2010, 2013) we know that the radii of nuclear rings range from a few tens of parsec to some 3.5 kpc, while inner rings and outer rings have typical radii of 1.2 and 2.5–3 times the length of the bar. Many host galaxies of rings are barred, but so are most galaxies in general. Some 20% of all rings occur in non-barred galaxies, which implies that rings do not, or hardly, occur preferentially in barred galaxies (Knapen 2005, Comerón *et al.* 2010, 2013). In most non-barred ringed galaxies an oval, a past interaction, or even a prominent spiral pattern lies at the dynamical origin of the ring, but this needs additional scrutiny.

From an inventory of all known *nuclear* rings, Comerón *et al.* (2010) reach the following conclusions. Star-forming nuclear rings occur in $20 \pm 2\%$ of disk galaxies with $-3 < T < 7$; 18/96 occur in disk galaxies without a bar (19%); they are found in S0 to Sd galaxies, peaking in types Sab Sb; when nuclear rings occur in barred galaxies, the ring radius is limited to one quarter of the bar radius; and stronger bars host smaller rings (cf. Knapen 2005).

We are now using the *Spitzer* Survey of Spiral Structure in Galaxies (S^4G; Sheth *et al.* 2010) to expand our survey to inner and outer rings (Comerón *et al.* 2013). We aim to study the relations between ring and host properties – as we did before for nuclear rings. We will use the S^4G sample size and image depth to reach further insight into the secular evolution of galaxies by measuring structural properties of rings, as well as those of components like bars and disks. We will then be able to tackle outstanding questions such as the origin of rings in non-barred galaxies, and how exactly ring properties are determined by the bar.

Keywords. galaxies: evolution, galaxies: formation, galaxies: structure, galaxies: spiral, galaxies: kinematics and dynamics

---

References

Buta, R. J. 1995, *ApJS*, 96, 39

Comerón, S., Knapen, J. H., Beckman, J. E., Laurikainen, E., Salo, H., Martínez-Valpuesta, I., & Buta, R. J. 2010, *MNRAS*, 402, 2462

Comerón, S., Knapen, J. H., *et al.* 2013, *A&A*, 562, 121

Falcón-Barroso, J. & Knapen, J. H. 2012, Eds., *Secular Evolution in Galaxies* (Cambridge: Cambridge University Press)

Knapen, J. H. 2005, *A&A*, 429, 141

Sheth, K., *et al.* 2010, *PASP*, 122, 1397

Highlights of Astronomy, Volume 16
XXVIIIth IAU General Assembly, August 2012
T. Montmerle, ed.

doi:10.1017/S1743921314005912

Multiple bars and secular evolution

Juntai Shen
Shanghai Astronomical Observatory, CAS China, email: jshen@shao.ac.cn

Abstract. Bars are the most important driver of secular evolution. A significant fraction of barred galaxies also harbor small secondary bars. Secondary bars are visible even in near-infrared images, so they are not just dusty and blue, but stellar features (Erwin & Sparke 2002). Since they are quite common, secondary bars are probably long-lived stellar features. The random relative orientation of the two bars indicates that they are dynamically decoupled with different pattern speeds (Buta & Crocker 1993). Corsini *et al.* (2003) presented conclusive direct kinematic evidence for a decoupled secondary bar in NGC 2950. Dynamically decoupled secondary bars have long been hypothesized to be a mechanism to drive gas past the ILR of primary bars to feed active galactic nuclei (Shlosman *et al.* 1989). However, the dynamics of secondary bars are still not well understood, and it is still unclear what role secondary bars play in the AGN fueling process.

Numerical simulations offer the best approach to understanding double-barred systems. Decoupled secondary bar in the earlier gaseous simulations only last a short time (< 1 Gyr, e.g. Friedli & Martinet 1993). Orbital studies of double-barred systems discovered a family of loop orbits that may be building blocks of long-lived nuclear stellar bars (Maciejewski & Sparke 1997, 2000). To complement orbital studies, which are not fully self-consistent, N-body simulations are preferred to further our understanding of double-barred systems. Debattista & Shen (2007) and Shen & Debattista (2009) managed to form long-lived double-barred systems with purely collisionless simulations, where a pre-existing rotating pseudo-bulge is introduced initially. The shape and size of secondary bars in the models are comparable to observed ones. They found that the rotation of the two bars is not rigid. The amplitude and pattern speed of the secondary bars oscillate as they rotate through their primary counterparts. Although the secondary bar rotates faster than the primary bar in this model, the stellar velocity field in the central region only shows a weakly twisted kinematic minor axis.

Recently more simulations of double-barred galaxies with simpler initial conditions are explored (Du, Shen & Debattista 2014). We expect that the new models can be used to cross-check with the kinematic properties of double-barred galaxies from IFU observations such as SAURON and Atlas3D.

References

Buta, R. & Crocker, D. A. 1993, *AJ*, 105, 134 4
Corsini, E. M., Debattista, V. P., & Aguerri , J. A. L. 2003, *ApJ*, 599, L29
Debattista, V. P. & Shen, J. 2007, *ApJ*, 654, L127
Du, M., Shen, J., & Debattista, V. P. 2014, in preparation
Erwin, P. & Sparke, L. S. 2002, *AJ*, 124, 6 5
Friedli, D. & Martinet, L. 1993, *A&A*, 277, 27
Shen, J. & Debattista, V. P. 2009, *ApJ*, 690, 758
Shlosman, I., Frank, J., & Begelman, M. C. 1989, *Nature*, 338, 45
Maciejewski, W. & Sparke, L. S. 1997, *ApJL*, 484, L117
Maciejewski, W., & Sparke, L. S. 2000, *MNRAS*, 313, 745

Highlights of Astronomy, Volume 16
XXVIIIth IAU General Assembly, August 2012
T. Montmerle, ed.

doi:10.1017/S1743921314005924

Kinematical evidence for secular evolution in *Spitzer Survey of Stellar Structure in Galaxies* (S^{4}G) spirals

Santiago Erroz-Ferrer,[1,2,3] Johan H. Knapen,[1,2] Joan Font,[1,2] John E. Beckman[1,2] and the S^{4}G team

[1]Instituto de Astrofísica de Canarias, Vía Láctea s/n 38205 La Laguna, Spain
[2]Departamento de Astrofísica, Universidad de La Laguna, 38206 La Laguna, Spain
[3]email: serroz@iac.es

Abstract. We present a study of the kinematics of a sample of isolated spiral galaxies in the *Spitzer* Survey of Stellar Structure in Galaxies (S^{4}G). We use Hα Fabry-Perot data from the GHαFaS instrument at the William Herschel Telescope (WHT) in La Palma, complemented with images at 3.6 microns, in the R band and in the Hα filter. The resulting data cubes and velocity field maps allow a complete study of the kinematics of a galaxy, including in-depth investigations of the rotation curve, velocity moment maps, velocity residual maps, gradient maps and position-velocity (PV) diagrams. We find clear evidence of the secular evolution processes going on in these galaxies, such as asymmetries in the velocity field in the bar zone, and non-circular motions, probably in response to the potential of the structural components of the galaxies, or to past or present interactions.

Keywords. galaxies: kinematics and dynamics - galaxies: spiral - galaxies : individuals : NGC 864

Results

In Erroz-Ferrer *et al.* (2012) (arXiv:1208.1409) we have presented the first results of this survey, a kinematical analysis of NGC 864. In the paper we have mainly analysed the kinematic data cubes. Also, we have used other ancillary data, like *R*-band and Hα images taken with the instrument ACAM in the WHT, IFU data with the SAURON instrument, also in the WHT, and we used the 3.6 micron S^{4}G image.

The data cubes and velocity maps allow the study of the kinematics of every galaxy, including in-depth investigations of the rotation curve, velocity moment maps, velocity residual maps and position-velocity diagrams.

In the residual maps, we have found that there are deviations from the circular rotation velocity, confirming the presence of non-circular motions along the bar. These are probably caused by the non-axisymmetrical potential created by the bar. We can also observe the non-circular motions by creating a position-velocity diagram along the kinematic minor axis. In an ideal case without non-circular motions, the velocity profile along the kinematic minor axis would have been completely flat. This confirms that the bar has a significant influence on the kinematics of the galaxy, causing the velocities to deviate from the circular, rotational, motion. The presence of the non-axisymmetric gravitational potential of the bar can be recognised in the deviations from circular motion across the galaxy.

Reference

Erroz-Ferrer, S., Knapen, J. H., *et al.* 2012, *MNRAS*, 427, 2938

Highlights of Astronomy, Volume 16
XXVIIIth IAU General Assembly, August 2012
T. Montmerle, ed.

doi:10.1017/S1743921314005936

Rotation of classical bulges during secular evolution of barred galaxies

Kanak Saha and Ortwin Gerhard

Max-Planck-Institut für Extraterrestrische Physik, Giessenbachstraße, D-85748 Garching, Germany,
email: saha@mpe.mpg.de

Abstract. Bar driven secular evolution plays a key role in changing the morphology and kinematics of disk galaxies, leading to the formation of rapidly rotating boxy/peanut bulges. If these disk galaxies also hosted a preexisting classical bulge, how would the secular evolution influence the classical bulge, and also the observational properties.

We first study the co-evolution of a bar and a preexisting non-rotating low-mass classical bulge such as might be present in galaxies like the Milky Way. It is shown with N-body simulations that during the secular evolution, such a bulge can gain significant angular momentum emitted by the bar through resonant and stochastic orbits. Thereby it transforms into a cylindrically rotating, anisotropic and triaxial object, embedded in the fast rotating boxy bulge that forms via disk instability (Saha *et al.* 2012). The composite boxy/peanut bulge also rotates cylindrically.

We then show that the growth of the bar depends only slightly on the rotation properties of the preexisting classical bulge. For the initially rotating small classical bulge, cylindrical rotation in the resulting composite boxy/peanut bulge extends to lower heights (Saha & Gerhard 2013). More massive classical bulges also gain angular momentum emitted by the bar, inducing surprisingly large rotational support within about 4 Gyrs (Saha *et al.* in prep).

Keywords. galaxies:bulges, galaxies: evolution, galaxies: structure

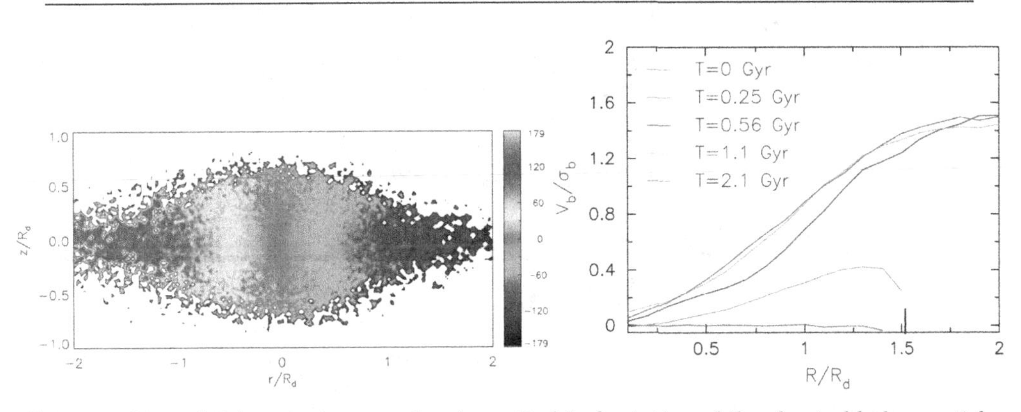

Figure 1. Line-of-sight velocity map showing cylindrical rotation of the classical bulge particles at time 2.1 Gyr in the Saha *et al.* (2012) model (left). Variation of rotation velocity with radius and time during secular evolution (right).

References

Saha, K. & Gerhard, O. 2013, *MNRAS*, 430, 2039
Saha, K, Martinez-Valpuesta, I., & Gerhard, O. 2012, *MNRAS* 421, 33

Highlights of Astronomy, Volume 16
XXVIIIth IAU General Assembly, August 2012
T. Montmerle, ed.

doi:10.1017/S1743921314005948

Parallel-sequencing of early-type and spiral galaxies

Michele Cappellari

Sub-department of Astrophysics, Department of Physics, University of Oxford
Denys Wilkinson Building, Keble Road, Oxford OX1 3RH
email: cappellari@astro.ox.ac.uk

Abstract. Since Edwin Hubble introduced his famous tuning fork diagram more than 70 years ago, spiral galaxies and early-type galaxies (ETGs) have been regarded as two distinct families. The spirals are characterized by the presence of disks of stars and gas in rapid rotation, while the early-types are gas poor and described as spheroidal systems, with less rotation and often non-axisymmetric shapes. The separation is physically relevant as it implies a distinct path of formation for the two classes of objects. I will give an overview of recent findings, from independent teams, that motivated a radical revision to Hubble's classic view of ETGs. These results imply a much closer link between spiral galaxies and ETGs than generally assumed.

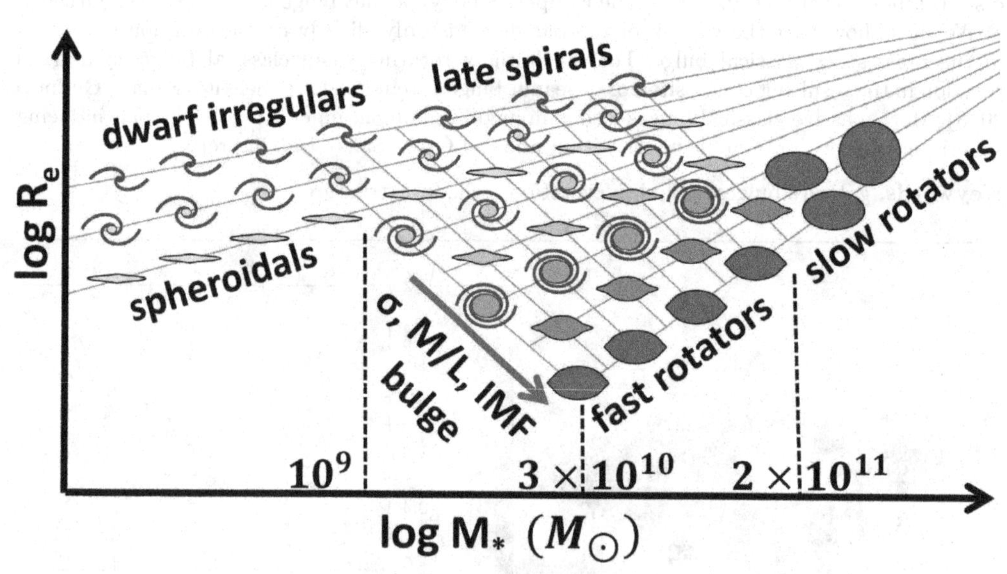

Figure 1. ETGs properties like shape, dynamics, population and IMF, merge smoothly with the properties of spiral galaxies on the mass-size diagram. All trends appear driven by an increase of the bulge fraction, which greatly enhance the likelihood for a galaxy to have his star formation quenched. This parallelism between the properties of spirals and ETGs motivated a proposed revision of Hubble's tuning-fork diagram. The same symbols are used in this figure (taken from Cappellari *et al.* 2012) as in the 'comb' morphological classification diagram proposed in Cappellari *et al.* (2011).

References

Cappellari, M., *et al.*, 2011, *MNRAS*, 416, 1680
Cappellari, M., *et al.*, 2013, *MNRAS*, 432, 1862

Highlights of Astronomy, Volume 16
XXVIIIth IAU General Assembly, August 2012
T. Montmerle, ed.

doi:10.1017/S174392131400595X

NIRS0S: Observations of early-type galaxy secular evolution spanning the Sa/S0/disky-E boundaries

Eija Laurikainen[1], Heikki Salo[2], Ronald Buta[3] and Johan Knapen[4]

[1]Finnish Centre for Astronomy with ESO (FINCA), Univ. of Oulu/Turku
email: eija.laurikainen@oulu.fi

[2]Dept. of Physics, Univ. of Oulu, [3]Dept. of Physics and Astronomy, Univ. of Alabama,
[4]Instituto de Astrofísica de Canarias, Dept. de Astrofísica, Univ. de La Laguna

Abstract. NIRS0S (Near-IR S0 galaxy Survey), is a K-band survey of $\sim$ 200 early-type disk galaxies, mainly S0s, 2-3 mag deeper than the 2Micron All Sky Survey. In depth morphological analysis was done, in which multi-component structural decompositions played an important role. Possible implications to internal dynamical galaxy evolution were discussed. S0s were suggested to be former spirals in which star formation has ceased, forming a parallel sequence with spirals (see Fig. 1). If that evolution is faster among the brighter galaxies, the observed magnitude difference between the barred and non-barred S0s could be understood. Bars are suggested to play a critical role in such evolution. For example, the inner lenses in the bright non-barred S0s can be explained as former barlenses (inner parts of bars), in which the elongated bar component has dissolved. We suggest that the last destructive merger event happened at a fairly large redshift.

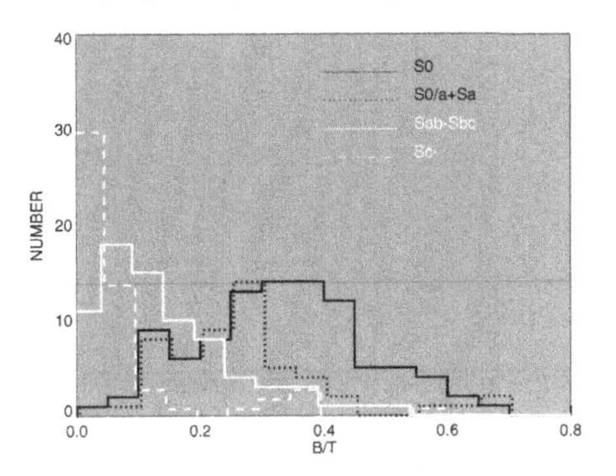

Figure 1. Many S0s have small B/T flux-ratios, overlapping even with the late-type spirals (Laurikainen *et al.* 2010). This evidence points to a 'parallel sequence', where S0s are spread throughout the Hubble sequence in a similar manner as spirals (S0a, S0b, S0c, see van den Bergh 1976; Cappellari *et al.* 2011; Kormendy & Bender 2012).

References

Cappellari, M. *et al.* 2011, *MNRAS* 416 1680
Kormendy, J. & Bender, R. 2012, *ApJS* 198, 2
Laurikainen, E. *et al.* 2010, *MNRAS* 405, 1089
van den Bergh, S. 1976, *ApJ* 206, 883

Highlights of Astronomy, Volume 16
XXVIIIth IAU General Assembly, August 2012
T. Montmerle, ed.

doi:10.1017/S1743921314005961

Comparison of NIRS0S K_s-band and S^{4}G 3.6 micron data: Fourier amplitudes, force profiles and color maps

Heikki Salo[1], Eija Laurikainen[2] and the S^{4}G Collaboration[3]

[1]Univ. Oulu, Finland Finland, email: `heikki.salo@oulu.fi`
[2]FINCA, Univ. Turku, Finland
[3]Various institutions

Abstract. Near-IR observations are considered to give an extinction-free view of the old stellar population in galaxies, thus ideal for the analysis of gravitational torques associated with bar and spiral structures. In the past, H or K_s band data have often been employed (Buta *et al.* 2010, Salo *et al.* 2010). S^{4}G (Spitzer Survey of Stellar Structure in Galaxies, Sheth *et al.* 2010) provides new deep homogenious 3.6 and 4.5 micron data for over 2000 nearby galaxies, allowing to probe the bar and spiral properties over a wide range of morphological types and environments. Here we compare the Fourier-amplitude profiles derived from S^{4}G data for about 50 early-type disk galaxies (SO and S0/a), with those from NIRSOS K_s data (Near-IR S0 Survey, Laurikainen *et al.* 2011). We also make detailed K_s-3.6 micron color maps. Interestingly, nuclear ring features stand up very clearly in these maps, indicating significantly different contributions of recent star formation in the K_s and 3.6 micron bands. However, the effect of these detailed differences on the overall force profiles is fairly small: this confirms that the S^{4}G data can be confidently used for estimation of bar torques.

Highlights of Astronomy, Volume 16
XXVIIIth IAU General Assembly, August 2012
T. Montmerle, ed.

doi:10.1017/S1743921314005973

Characterization of peculiar early-type galaxies in the local universe

Beatriz H. F. Ramos[1,2], Karín Menéndez-Delmestre[2], Taehyun Kim[3,4], Kartik Sheth[5] and S^4G team

[1]Instituto de Biociências, Departamento de Ciências Naturais, Universidade Federal do Estado do Rio de Janeiro, Rio de Janeiro, Brazil
email: ramos@astro.ufrj.br

[2]Observatório do Valongo, Universidade Federal de Rio de Janeiro, Rio de Janeiro, Brazil
[3]European Southern Observatory, Santiago, Chile
[4]Dept. Physics & Astronomy, Seoul National University, Seoul, Republic of Korea
[5]National Radio Astronomy Observatory, Charlottesville, VA, USA

Abstract. Early-type galaxies (ETGs) have been characterized as objects dominated by old stellar populations, containing little or no cold gas and dust, and thus, non-existent star formation. However, there are indications in the literature that some ETGs deviate from this: some have significant amounts of gas and dust, are forming stars, and/or display stellar substructures (tidal features, disks or shells, e.g., Kormendy *et al.* 1997, Rix, Carollo & Freeman 1999). A better understanding of the evolution of ETGs and the details of their "peculiarities" is critical to properly constrain models of galaxy formation. We present preliminary results on a photometric analysis of substructures in local ETGs, based on 3.6μm IRAC images from the Spitzer Survey of Stellar Structure in Galaxies (S^4G; Sheth *et al.* 2010), which comprises one of the largest mid-IR photometric surveys of the local Universe. Relatively unhindered by extinction and dominated by the low-mass stellar populations that dominate a galaxy's stellar mass budget, the IR is the ideal waveband to trace the details of stellar structures in galaxies. Based on 2D GALFIT (Peng *et al.* 2002) decomposition, we find tidal features in 17% of 146 ETGs from S^4G. For both the GALFIT model and the galaxy residual images, we calculate the total counts inside an annular region centered on the galaxy, where the inner radius is the effective radius of the galaxy. Assuming that a tidal feature and its host galaxy have the same mass-to-luminosity ratio (M/L), the ratio of the residual counts over model counts translates into the ratio of their stellar masses. We find that the tidal features in the majority of peculiar ETGs in our sample account for no more than 11% of the galaxy's total stellar mass. Considering that simulations (Canalizo *et al.* 2007) suggest an upper limit in relative stellar mass of 25% for shells resulting from a past major merger, the values we find support a merger origin. We are in the process of applying the decomposition method to GALEX UV images and optical SDSS images of these peculiar ETGs in order to characterize the underlying substructure and provide constraints on astrophysical properties such as star formation rates and stellar masses associated to these tidal features, based on broad-band SED template fitting techniques.

Keywords. galaxies: elliptical and lenticular, galaxies: structure, galaxies: evolution

---

References

Canalizo, G., *et al.* 2007, *ApJ* 669, 801
Kormendy, J., *et al.* 1997, *ApJ* 482, L139
Peng, C. Y., *et al.* 2002, *AJ* 124, 266
Rix, H.-W., Carollo, C. M., & Freeman, K. 1999, *ApJ* 513, L25
Sheth, K., *et al.* 2010, *PASP* 122, 1397

Highlights of Astronomy, Volume 16
XXVIIIth IAU General Assembly, August 2012
T. Montmerle, ed.

doi:10.1017/S1743921314005985

3D view on Virgo and field dwarf elliptical galaxies: late-type origin and environmental transformations

Agnieszka Ryś[1,2], Jesús Falcón-Barroso[1,2] and Glenn van de Ven[3]
[1]Instituto de Astrofísica de Canarias, 38200 La Laguna, Tenerife, Spain
[2]Departamento de Astrofísica, Universidad de La Laguna, 38205 La Laguna, Tenerife, Spain
[3]Max Planck Institute for Astronomy, Königstuhl 17, 69117 Heidelberg, Germany
email: arys@iac.es

Abstract. In our contribution we show the effects of environmental evolution on cluster and field dwarf elliptical galaxies (dEs), presenting the first large-scale integral-field spectroscopic data for this galaxy class. Our sample con sists of 12 galaxies and no two of them are alike. We find that the level of rotation is not tied to flattening; we observe kinematic twists; we discover large-scale kinematically-decoupled components; we see varying gradient s in line-strength maps: from nearly flat to strongly peaked in the center. The great variety of morphological, kinematic, and stellar population parameters seen in our data supports the claim that dEs are defunct dwarf spiral/irregular galaxies and points to a formation scenario that allows for a stochastic shaping of galaxy properties. The combined influence of ram-pressure stripping and harassment fulfills these requirements, still, the exact impact of the two is not yet understood. We further investigate the properties of our sample by performing a detailed comprehensive analysis of its kinematic, dynamical, and stellar population parameters. The combined knowledge of the dynamical properties and star-formation histories, together with model predictions for different formation mechanisms, will be used to quant itatively determine the actual transformation paths for these galaxies.

Keywords. galaxies: dwarf – galaxies: evolution – galaxies: kinematics a nd dynamics

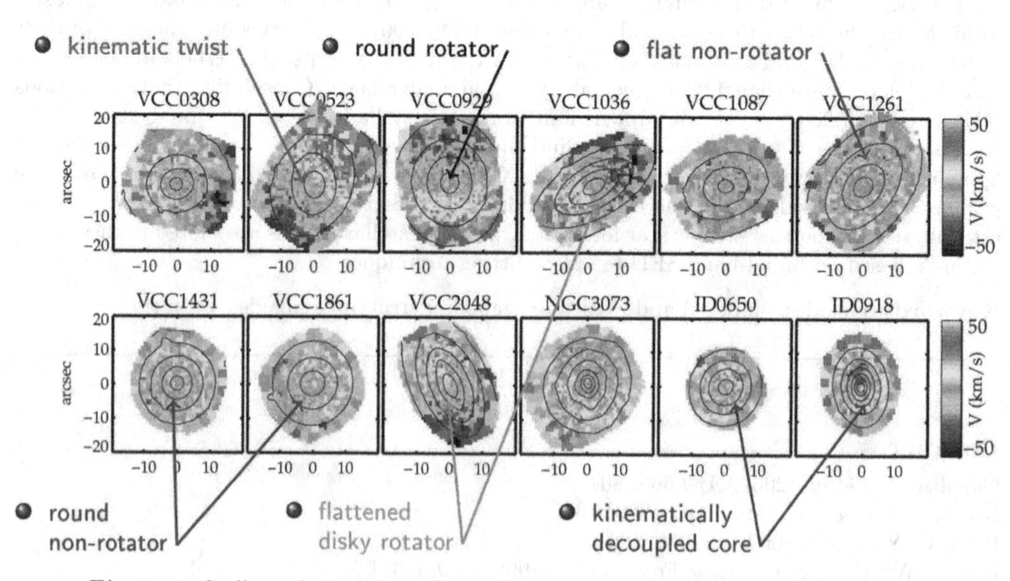

Figure 1. Stellar velocity maps showing the kinematic variety seen in our data.

Highlights of Astronomy, Volume 16
XXVIIIth IAU General Assembly, August 2012
T. Montmerle, ed.

doi:10.1017/S1743921314005997

Kinematic properties and dark matter fraction of Virgo dwarf early-type galaxies

E. Toloba[1,2] †, A. Boselli[3], R. Peletier[4] and J. Gorgas[5]

[1]UCO/Lick Observatory, University of California, Santa Cruz, CA 95064

[2]Observatories of the Carnegie Institution of Washington, Pasadena, CA 91101
[3]Laboratoire d'Astrophysique de Marseille-LAM, 13388 Marseille
[4]Kapteyn Astronomical Institute, University of Groningen, the Netherlands
[5]Departamento de Astrofísica y CC. de la Atmósfera, Universidad Complutense de Madrid

Abstract. What happens to dwarf galaxies as they enter the cluster potential well is one of the main unknowns in studies of galaxy evolution. Several evidence suggests that late-type galaxies enter the cluster and are transformed to dwarf early-type galaxies (dEs). We study the Virgo cluster to understand which mechanisms are involved in this transformation. We find that the dEs in the outer parts of Virgo have rotation curves with shapes and amplitudes similar to late-type galaxies of the same luminosity (Fig. 1). These dEs are rotationally supported, have disky isophotes, and younger ages than those dEs in the center of Virgo, which are pressure supported, often have boxy isophotes and are older (Fig. 1). Ram pressure stripping, thus, explains the properties of the dEs located in the outskirts of Virgo. However, the dEs in the central cluster regions, which have lost their angular momentum, must have suffered a more violent transformation. A combination of ram pressure stripping and harassment is not enough to remove the rotation and the spiral/disky structures of these galaxies. We find that on the the Faber-Jackson and the Fundamental Plane relations dEs deviate from the trends of massive elliptical galaxies towards the position of dark matter dominated systems such as the dwarf spheroidal satellites of the Milky Way and M31. Both, rotationally and pressure supported dEs, however, populate the same region in these diagrams. This indicates that dEs have a non-negligible dark matter fraction within their half light radius.

Keywords. Virgo cluster, dwarf galaxies, kinematics, dynamics, and dark matter.

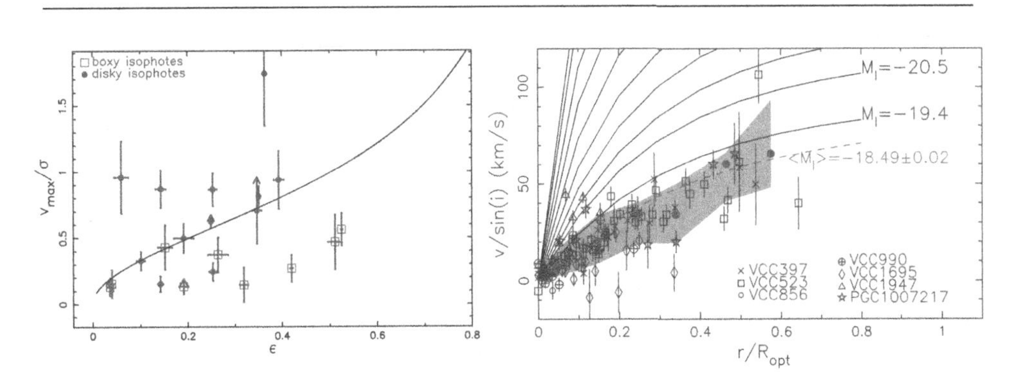

Figure 1. Right panel: Squares indicate dEs with boxy isophotes in the inner r $\leqslant 2°$ region of the Virgo cluster. Dots show dEs with disky isophotes located in the range $2° \leqslant r \leqslant 6°$. Rotationally/pressure supported dEs are those above/below the line (model for an isotropic oblate system flattened by rotation), respectively. **Left panel:** Rotationally supported dEs (open symbols, the filled dots represent their median observed rotation curve, 1σ deviation in the shaded area) are compared to rotation curves of late-type galaxies (solid and dashed lines).

† Fulbright Fellow, email: `toloba@ucolick.org`

Highlights of Astronomy, Volume 16
XXVIIIth IAU General Assembly, August 2012
T. Montmerle, ed.

doi:10.1017/S1743921314006000

Stellar populations in bulges and disks and the secular evolution connection

Reynier Peletier

Kapteyn Institute, Groningen Netherlands, email: peletier@astro.rug.nl

Abstract. In recent years a considerable amount of new data has become available about stellar populations in the central regions of galaxies. I will discuss what stellar populations can tell us about the formation origin of bulges, and how this relates to the formation of the central regions in giant and dwarf ellipticals. In particular I will concentrate on results from integral field spectroscopy and Spitzer imaging at 3.6, 4.5 and 8 microns.

Highlights of Astronomy, Volume 16
XXVIIIth IAU General Assembly, August 2012
T. Montmerle, ed.

doi:10.1017/S1743921314006012

The gas and star formation in bulges

David Fisher

Department of Astronomy, University of Maryland, email: dbfisher@astro.umd.edu

Abstract.

Highlights of Astronomy, Volume 16
XXVIIIth IAU General Assembly, August 2012
T. Montmerle, ed.

doi:10.1017/S1743921314006024

The growth of mass and metallicity in bulges and disks: CALIFA perspective

R. M. González Delgado[1], E. Pérez[1], R. Cid Fernandes[1,2], R. García-Benito[1], A. de Amorim[2], S. F. Sánchez[1], B. Husemann[3], R. López Fernández[1], C. Cortijo[1], E. Lacerda[2], D. Mast[1] and the CALIFA collaboration

[1]Instituto de Astrofísica de Andalucia, Granada, Spain; [2]Universidade Federal de Santa Catarina, Florianópolis, Brazil; [3] Leibniz-Institut fur Astrophysik, Postdam, Germany

Abstract. CALIFA (Calar Alto Legacy Integral Field Area) is a 3D spectroscopic survey of 600 nearby galaxies that we are obtaining with PPaK@3.5m at Calar Alto (Sánchez *et al.* 2012; Husemann *et al.* 2012). This pioneer survey is providing valuable clues on how the mass and metallicity grow in the different galactic spatial sub-components ("bulge" and "disk"). Processed through spectral synthesis techniques, CALIFA datacubes allow us to, for the first time, spatially resolve the star formation history of galaxies (Cid Fernandes *et al.* 2012). The richness of this approach is already evident from the results obtained for the first ~ 100 galaxies of the sample (Pérez *et al.* 2012). We have found that galaxies grow inside-out, and that the growth rate depends on a galaxy's mass. Here, we present the radial variations of physical properties sorting galaxies by their morphological type (Figure 1). We have found a good correlation between the stellar mass surface density, stellar ages and metallicities and the Hubble type, but being the the early type spirals (Sa-Sbc) the galaxies with strong negative age and metallicity gradient from the bulge to the disk.

Keywords. galaxies; stellar populations; structure; evolution

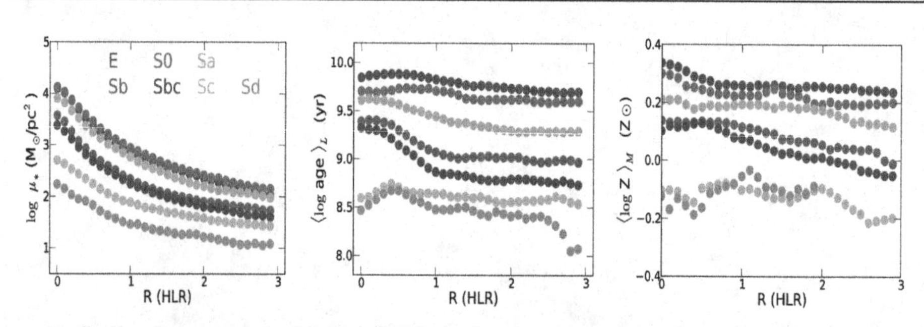

Figure 1. Stellar mass surface density *(left)*, the luminosity-weighted mean stellar age *(middle)*, the mass-weighted mean stellar metallicity *(right)*, radial profiles of 107 galaxies that are stacking by their morphological type.

References

Cid Fernandes, R., *et al.* 2012, *A&A* submitted
Husemann, B., *et al.* 2012, *A&A* in press, (arXiv:1210.8150)
Pérez, E., *et al.* 2012, *ApJ* submitted
Sánchez, S. F., *et al.* 2012, *A&A* 538, 8

Highlights of Astronomy, Volume 16
XXVIIIth IAU General Assembly, August 2012
T. Montmerle, ed.

doi:10.1017/S1743921314006036

Rejuvenation of bulges by bars: evidence from stellar population analysis

Dimitri A. Gadotti[1] and Paula Coelho[2]

[1]European Southern Observatory
email: dgadotti@eso.org

[2]Universidade Cruzeiro do Sul
email: paula.coelho@cruzeirodosul.edu.br

Abstract. We obtained stellar ages and metallicities via spectrum fitting for a sample of 575 bulges with spectra available from the SDSS. Using the whole sample, where galaxy stellar mass distributions for barred and unbarred galaxies are similar, we find that the distribution of bulge ages in barred galaxies shows an excess of populations younger than 4 Gyr, when compared to bulges in unbarred galaxies. KS statistics confirm that the age distributions are different with a significance of 99.94%. If we select sub-samples for which the bulge stellar mass distributions are similar for barred and unbarred galaxies, this excess vanishes for galaxies with low-mass bulges, while for more massive bulges we find a bimodal stellar age distribution for barred galaxies only, corresponding to two normal distributions with mean ages of 10.4 and 4.7 Gyr (see Fig. 1). These results lend strong support to models in which bars trigger star formation activity in the centers of galaxies. We also find twice as much AGN among barred galaxies, as compared to unbarred galaxies, for low-mass bulges. Full results are in Coelho & Gadotti (2011).

Keywords. galaxies: bulges, galaxies: evolution, galaxies: stellar content

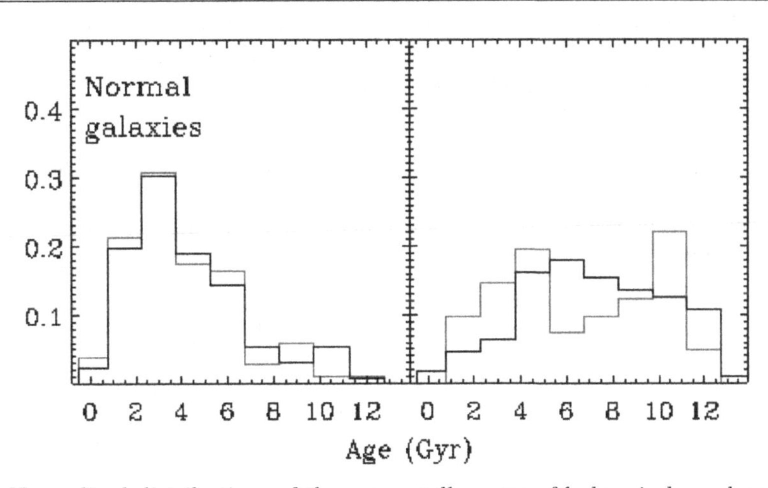

Figure 1. Normalized distributions of the mean stellar ages of bulges in barred and unbarred galaxies. For low mass bulges ($< 10^{10.1} M_\odot$ – left) the distributions are similar, whereas for high mass bulges ($> 10^{10.1} M_\odot$ – right) only barred galaxies have a bimodal distribution of bulge mean stellar ages, with a significant young component. *Adapted from Coelho & Gadotti (2011).*

Reference

Coelho, P. & Gadotti, D. A. 2011, *ApJ Letters* 743, 13

Highlights of Astronomy, Volume 16
XXVIIIth IAU General Assembly, August 2012
T. Montmerle, ed.

doi:10.1017/S1743921314006048

Stellar populations of bulges in galaxies with a low surface-brightness disc

L. Morelli[1,2]† **E. M. Corsini**[1,2], **A. Pizzella**[1,2], **E. Dalla Bontà**[1,2], **L. Coccato**[3], **J. Méndez-Abreu**[4,5] **and M. Cesetti**[1,2]

[1]Dipartimento di Fisica e Astronomia, Università di Padova, vicolo dell'Osservatorio 3, I-35122 Padova, Italy

[2]INAF–Osservatorio Astronomico di Padova, vicolo dell'Osservatorio 5, I-35122 Padova, Italy

[3]European Southern Observatory, Karl-Schwarzschild-Straβe 2, D-85748 Garching bei München, Germany

[4]Instituto Astrofísico de Canarias, C/ Vía Láctea s/n, E-38200 La Laguna, Spain

[5]Departamento de Astrofísica, Universidad de La Laguna, Universidad de La Laguna, C/ Astrofísico Francisco Sánchez, E-38205 La Laguna, Spain

Abstract. The radial profiles of the H_β, Mg, and Fe line-strength indices are presented for a sample of eight spiral galaxies with a low surface-brightness stellar disc and a bulge. The correlations between the central values of the line-strength indices and velocity dispersion are consistent to those known for early-type galaxies and bulges of high surface-brightness galaxies. The age, metallicity, and α/Fe enhancement of the stellar populations in the bulge-dominated region are obtained using stellar population models with variable element abundance ratios. Almost all the sample bulges are characterized by a young stellar population, on-going star formation, and a solar α/Fe enhancement. Their metallicity spans from high to sub-solar values. No significant gradient in age and α/Fe enhancement is measured, whereas only in a few cases a negative metallicity gradient is found. These properties suggest that a pure dissipative collapse is not able to explain formation of all the sample bulges and that other phenomena, like mergers or acquisition events, need to be invoked. Such a picture is also supported by the lack of a correlation between the central value and gradient of the metallicity in bulges with very low metallicity. The stellar populations of the bulges hosted by low surface-brightness discs share many properties with those of high surface-brightness galaxies. Therefore, they are likely to have common formation scenarios and evolution histories. A strong interplay between bulges and discs is ruled out by the fact that in spite of being hosted by discs with extremely different properties, the bulges of low and high surface-brightness discs are remarkably similar.

Keywords. galaxies : abundances – galaxies : bulges – galaxies : evolution – galaxies : stellar content – galaxies : formation – galaxies : Kinemaitics and Dynamics

---

† email: `lorenzo.morelli@unipd.it`

Highlights of Astronomy, Volume 16
XXVIIIth IAU General Assembly, August 2012
T. Montmerle, ed.

doi:10.1017/S174392131400605X

Evolution of the star formation efficiency in galaxies

Jonathan Braine

Laboratoire d'Astrophysique de Bordeaux France, email: braine@obs.u-bordeaux1.fr

Abstract. The physical and chemical evolution of galaxies is intimately linked to star formation, We present evidence that molecular gas (H_2) is transformed into stars more quickly in smaller and/or subsolar metallicity galaxies than in large spirals – which we consider to be equivalent to a star formation efficiency (SFE). In particular, we show that this is not due to uncertainties in the $N(H_2)/I_{co}$ conversion factor. Several possible reasons for the high SFE in galaxies like the nearby M33 or NGC 6822 are proposed which, separately or together, are the likely cause of the high SFE in this environment. We then try to estimate how much this could contribute to the increase in cosmic star formation rate density from $z = 0$ to $z = 1$.

Keywords. Stars: Formation, Interstellar Medium, Galaxies: M33, Galaxies: Local Group

References

Braine J., Gratier P., Kramer C., Schuster K. F., Tabatabaei F., & Gardan E., 2010, *A&A*, 520, 107
Braine J., *et al.*, 2010, *A&A*, 518, L69
Combes F., García-Burillo S., Braine J., Schinnerer E., Walter F., & Colina L., 2013, *A&A*, 550, 41
Combes F., García-Burillo S., Braine J., Schinnerer E., Walter F., & Colina L., 2011, *A&A*, 528, 124
Dib S., Piau L., Mohanty S., & Braine J., 2011, *MNRAS*, 415, 3439
Gardan E., Braine J., Schuster K. F., Brouillet N., & Sievers A., 2007, *A&A*, 473, 91
Gratier P., *et al.*, 2012, *A&A*, 542, 108
Gratier P., *et al.*, 2010, *A&A*, 522, 3
Gratier P., Braine J., Rodriguez-Fernandez N. J., Israel F. P., Schuster K. F., Brouillet N., & Gardan E., 2010, *A&A*, 512, 68

Highlights of Astronomy, Volume 16
XXVIIIth IAU General Assembly, August 2012
T. Montmerle, ed.

doi:10.1017/S1743921314006061

The origin of thick discs

Sébastien Comerón

Department of Physics/Astronomy Division, University of Oulu, FI-90014, Finland
email: seb.comeron@gmail.com

Abstract. Thick discs are defined to be disc-like components with a scale height larger than that of the classical discs. They are ubiquitous (Yoachim & Dalcanton 2006; Comerón *et al.* 2011a), they are made of mostly old and metal-poor stars and are most easily detected in close to edge-on galaxies. Their origin has been considered mysterious and several formation theories have been proposed:

- The thick disc being formed secularly by thin disc stars heated by disc overdensities such as giant molecular clouds or spiral arms (Villumsen 1985, *ApJ*, 290, 75) and by stars moved outwards from their original orbits by radial migration mechanisms (Schönrich & Binney 2009).
- The thick disc being formed by the heating of the thin disc by satellites (Quinn *et al.* 1993) and the tidal stripping of them (Abadi *et al.* 2003).
- The thick disc being formed fast and already thick at high redshift in an highly unstable disc. Inside that thick disc, a thin disc would form afterwards as suggested by Elemgreen & Elmegreen (2006).
- The thick disc being formed originally thick at high redshift by the merger of gas-rich protogalactic fragments and a thin disc forming afterwards within it (Brook *et al.* 2007).

The first mechanism is a secular evolution mechanism. The time-scale of the second one is dependent on the merger history of the main galaxy. In the two last mechanisms, the thick disc forms already thick in a short time-scale at high redshift.

Recent Milky Way studies, (see, e.g., Bovy *et al.* 2012), have shown indications that there is no discontinuity between the thin and the thick disc chemical and kinematic properties. Instead, those studies indicate the presence of a monotonic distribution of disc thicknesses. This would suggest a secular origin for the Milky Way thick disc.

Studies in external galaxies (Yoachim & Dalcanton 2006; Comerón *et al.* 2011b), have shown that low-mass disc galaxies have thick disc relative masses much larger than those found in large-mass galaxies. Because low-mass galaxies are dynamically younger than their larger counterparts, it seems difficult for their thick discs to have a secular evolution origin, but simulations show that their thick disc masses are compatible with those of a thick disc formed at high redshift.

Thus, recent studies seem to indicate that large-mass galaxies have their thick discs formed mainly due to secular evolution and that low-mass galaxies have them formed at high redshift.

Keywords. galaxies: evolution – galaxies: formation – galaxies: kinematics and dynamics

References

Abadi, M. G., Navarro, J. F., Steinmetz, M., & Eke, V. R. 2003, *ApJ*, 597, 21
Bovy, J., Rix, H.-W., & Hogg, D. W. 2012, *ApJ*, 751, 131
Brook, C., Richard, S., Kawata, D., Martel, H., & Gibson, B. K. 2007, *ApJ*, 658, 60
Comerón, S., Elmegreen, B. G., Knapen, J. H., *et al.* 2011b, *ApJ*, 741, 28
Comerón, S., Knapen, J. H., Sheth, K., *et al.* 2011a, *ApJ*, 729, 18
Elmegreen, B. G., & Elmegreen, D. M. 2006, *ApJ*, 650, 644
Quinn, P. J., Hernquist, L., & Fullagar, D. P., 1993, *ApJ*, 403, 74
Schönrich, R., & Binney, J. 2009, *MNRAS*, 396, 203
Villumsen, J. V. 1985, *ApJ*, 290, 75
Yoachim, P. & Dalcanton, J. J., 2006, *AJ*, 131, 226

Highlights of Astronomy, Volume 16
XXVIIIth IAU General Assembly, August 2012
T. Montmerle, ed.

doi:10.1017/S1743921314011156

Vertical structure of stellar populations in galaxy disks

David Streich, Roelof S. de Jong and the GHOSTS team

Leibniz-Institut für Astrophysik Potsdam (AIP), An der Sternwarte 16, 14486 Potsdam, Germany
email: dstreich@aip.de, rdejong@aip.de

Abstract. Stellar populations are most useful for disentangling formation and evolution histories of galaxies. We present here results obtained using data from the GHOSTS survey (Radburn-Smith *et al.*, 2011) which uses HST photometry to resolve stellar populations in nearby massive disk galaxies. Using color magnitude diagrams we can distingiush stellar populations of different ages and analyse the spatial structure of each population seperately.

We have examined the vertical disk structure in six edge-on galaxies. We find a general heating of disk, i.e. larger scaleheights for older populations. The scaleheight of each population is constant over most of radial extent of each galaxy.

In massive galaxies ($V_{rot} > 150\,km/s$) we clearly see a thick component (i.e. there are more stars at large distances from the plane than expected from a single disk model). These thick components consist of intermediate-aged and old stars (>1 Gyr), and the (thick) scaleheight of the old population (>4 Gyr) is significantly larger than the (thick) scaleheight of the intermediate aged (1-2 Gyr) population.

This finding argues against a rapid formation of the thick components and favors a more secular formation of these components.

Keywords. galaxies: spiral, galaxies: stellar content, galaxies: structure, galaxies: evolution

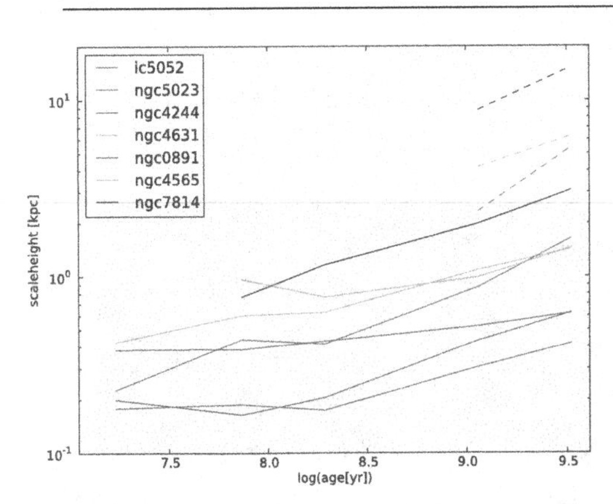

Figure 1. Scaleheights as a function of age. solid line: (thin) disk, dashed line: thick component. A clear increase with age is seen for both the thin and the thick components. *(A color version of this figure is available in the online journal.)*

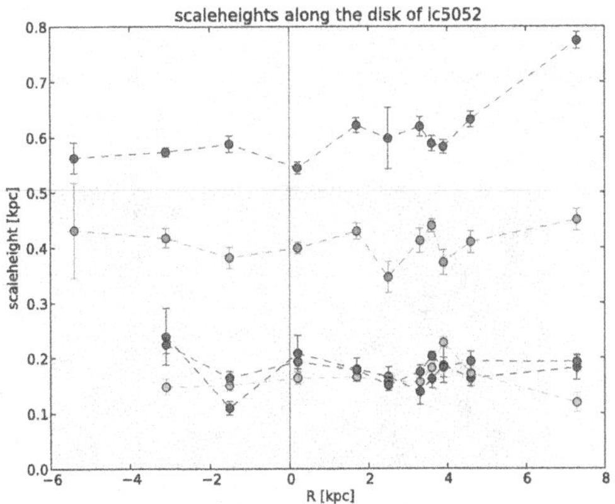

Figure 2. Scaleheights as a function of radius in IC 5052. Different lines show different population. With the exception of the outermost point if the old populations, the scaleheights are constant. *(A color version of this figure is available in the online journal.)*

Reference

Radburn-Smith, D. J., *et al.* 2011, *ApJS* 195, 18

Highlights of Astronomy, Volume 16
XXVIIIth IAU General Assembly, August 2012
T. Montmerle, ed.

doi:10.1017/S1743921314011168

The downplayed role of secular processes in the co-evolution of galaxies and black holes

Mauricio Cisternas[1] **and Knud Jahnke**[2]

[1]Instituto de Astrofísica de Canarias, E-38205 La Laguna, Tenerife, Spain
email: mauricio@iac.es

[2]Max-Planck-Institut für Astronomie, Königstuhl 17, D-69117 Heidelberg, Germany

Abstract. According to the current co-evolution picture, most present-day galaxies experienced at least one phase of vigorous black hole (BH) activity in the past, during which a tight link between galaxy and BH gets established. While during the last two decades we have witnessed tremendous progress in the field, additional robust observational constraints are required on how galaxy and BH related at earlier times, and which mechanisms are responsible for triggering these BH growth phases. In our recent studies, we analyzed a large sample of active galactic nuclei (AGN) out to $z \sim 1$ from the COSMOS survey (Scoville *et al.* 2007), allowing us to study in detail growing BHs together with their host galaxies. In Cisternas *et al.* (2011b) we found that, for a sample of 32 active galaxies at $z \sim 0.7$, BH mass scales with total galaxy stellar mass in the same way as it does locally, at $z = 0$, with galactic bulge mass. I will argue that for these galaxies to obey the local relation only a disk-to-bulge stellar mass redistribution is needed, likely driven by passive secular evolution. I will also present the results from Cisternas *et al.* (2011a), aiming to understand the relevance of major mergers as AGN activity triggering mechanisms. By looking for merging signatures on the morphologies of 140 AGN (some examples shown in Figure 1), and comparing them with a sample of over 1200 matched inactive galaxies, we found that the merger fraction between samples is statistically the same, at roughly 15%. Together with the fact that the majority of the AGN host galaxies are disk-dominated, unlikely relics of a recent major merger, these results are the strongest evidence to date that secular evolution rather than major merging has dominated BH fueling at least since $z \sim 1$.

Keywords. galaxies: evolution, galaxies: active, galaxies: bulges, galaxies: interactions

---

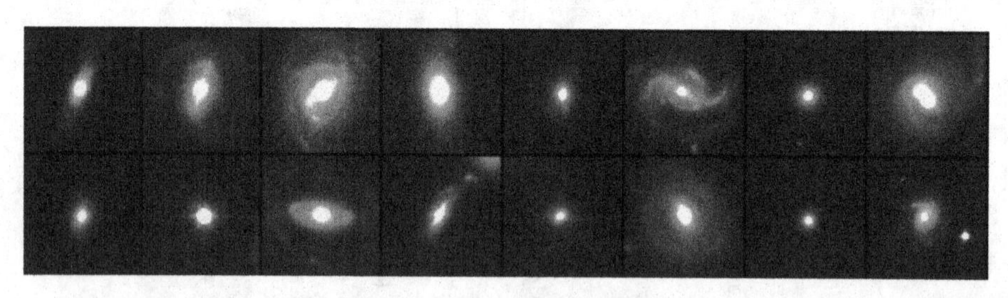

Figure 1. Representative examples of AGN host galaxies out to $z \sim 1$ imaged with *HST*/ACS illustrating that interacting galaxies are a minority and disks are highly common.

References

Cisternas, M., Jahnke, K., Inskip, K. J., *et al.* 2011a, *ApJ*, 726, 57
Cisternas, M., Jahnke, K., Bongiorno, A., *et al.* 2011b, *ApJ*, 741, L11
Scoville, N., Aussel, H., Brusa, M., *et al.* 2007, *ApJS*, 172, 1

Highlights of Astronomy, Volume 16
XXVIIIth IAU General Assembly, August 2012
T. Montmerle, ed.

doi:10.1017/S174392131401117X

A longslit spectroscopic survey of bulges in disc galaxies

Maximilian Fabricius[1], **Roberto Saglia**[1], **David Fisher**[2], **Niv Drory**[3], **Ralf Bender**[1] **and Ulrich Hopp**[1]

[1]Max Planck Institute for Extraterrestrial Physics & University Observatory, Munich, Germany, email: mxhf@mpe.mpg.de
[2]Laboratory of Millimeter Astronomy, University of Maryland
[3]Instituto de Astronomia, Universidad Nacional Autonoma de Mexico (UNAM)

Abstract. We use the Marcario Low Resolution Spectrograph (LRS) at the Hobby-Eberly-Telescope (HET) to study the kinematics of pseudobulges and classical bulges in 45 S0-Sc type galaxies in the nearby universe. Our high-resolution (instrumental $\sigma \approx 39$ km s^{-1}) spectra allo only to resolve the typical velocity dispersions of our targets but also to derive the h3 and h4 Gauss-Hermite moments. We demonstrate for the first time that purely kinematic diagnostics of the bulge dichotomy agree systematically with those based on Sérsic index. Low Sérsic index bulges have both increased rotational support (higher v/σ values) and on average lower central velocity dispersions. Pseudobulges have systematically shallower velocity dispersion profiles. The same correlation also holds when visual morphologies are used to diagnose bulge type. Finally, we present evidence for formerly undetected counter rotation in the two systems NGC 3945 and NGC 4736. With these, a total of 16% of the systems in or sample show signs for stellar counter rotation.

Highlights of Astronomy, Volume 16
XXVIIIth IAU General Assembly, August 2012
T. Montmerle, ed.

doi:10.1017/S1743921314011181

Tidal evolution of dwarf galaxies with shallow dark matter density profiles

Ewa L. Łokas

Nicolaus Copernicus Astronomical Center, Bartycka 18, 00-716 Warsaw, Poland,
email: lokas@camk.edu.pl

Abstract. One of the scenarios for the formation of dwarf spheroidal galaxies in the Local Group proposes that the objects formed from late type dwarfs via tidal interaction with bigger galaxies such as the Milky Way and Andromeda. The scenario naturally explains the morphology-density relation observed for dwarf galaxies in the Local Group. Using N-body simulations we study the long-term tidal evolution of dwarf galaxies in the vicinity of the Milky Way. The dwarf galaxies were initially composed of stellar disks embedded in dark matter haloes of different inner density slopes including shallow ones recently obtained in N-body+hydro simulations of dwarf galaxy formation in isolation. Such progenitors were placed on five different orbits around the Milky Way and their evolution was followed for 10 Gyr. The outcome of the evolution, in terms of the mass loss, morphological transformation and randomization of stellar orbits depends very sensitively on the inner density slope of dark matter. The effects of tides are stronger for dwarfs with shallower slopes; they are more heavily stripped, in some cases down to the scale of ultra-faint satellites of the Milky Way or even dissolved completely with obvious implications for the missing satellites problem. The morphological evolution of the stellar component, from rotationally supported disks to spheroids dominated by random motions, also proceeds faster. In addition, bars which usually form at the first pericenter passage are created more easily and live longer in dwarfs with shallow dark matter density profiles on extended orbits.

Highlights of Astronomy, Volume 16
XXVIIIth IAU General Assembly, August 2012
T. Montmerle, ed.

doi:10.1017/S1743921314011193

Secular Evolution in the Milky Way

Victor Debattista

Jeremiah Horrocks Institute, University of Central Lancashire, Preston, Lancashire, PR1 2HE, United Kingdom, email: vpdebattista@gmail.com

Abstract. I will review the secular evolution of the Milky Way disk and bulge with particular attention to the bulge and disk. Evidence for the importance of stellar migration in the Solar neighbourhood will be presented. The upcoming Gaia satellite will lead to a wealth of new data with which to explore these effects.

Highlights of Astronomy, Volume 16
XXVIIIth IAU General Assembly, August 2012
T. Montmerle, ed.

doi:10.1017/S174392131401120X

The Digital Sky Survey of the Galactic Anti-center (DSS-GAC)

X.-W. Liu[1,2], H.-B. Yuan[1,3], Z.-Y. Huo[4], M.-S. Xiang[2], H.-H. Zhang[2], Y. Huang[2], H.-W. Zhang[2], H.-B. Zhao[5], J. S. Yao[5], H. Lu[5] *et al.*

[1]Kavli Institute for Astronomy and Astrophysics, Peking University, Beijing 100871, China
email: x.liu@pku.edu.cn
[2]Department of Astronomy, Peking University, Beijing 100871, China [3]LAMOST Fellow
[4]National Astronomical Observatories, Chinese Academy of Sciences, Beijing 100012, China
[5]Purple Mountain Observatory, Chinese Academy of Sciences, Nanjing 210008, China.

Abstract. As an integral component of the LAMOST Experiment for Galactic Understanding and Evolution (LEGUE; Deng *et al.* 2012), the LAMOST Galactic anti-center spectroscopic survey (Liu *et al.* in preparation) will survey over three thousand square degree sky area centered on the Galactic anti-center (150d $\leqslant l \leqslant$ 210d, $-$30d $\leqslant b \leqslant$ +30d) and obtain low resolution ($R \sim 1800$) optical spectra for a statistically complete sample of more than three million stars down to a limiting magnitude of 18.5 in r band, distributed in a spatially contiguous area and probing a significant volume of the Galactic thin/thick disks, halo and their interface. Sample stars of the LAMOST survey of the Galactic anti-center are derived from a recently completed CCD imaging photometric survey utilizing the newly built 1.0/1.2m Schmidt Telescope at the Xuyi Station of the Purple Mountain Observatory. The Xuyi imaging survey (Yuan *et al.*, in preparation; Zhang *et al.* 2012) provides high quality photometry (~2 per cent) in the SDSS g, r and i bands and astrometry (~0.1 arcsec) for about a hundred million stars down to a limiting magnitude of about 19 (10 sigma) for over six thousand square degree sky area (3h $\leqslant$ RA $\leqslant$ 9h, $-$10d $\leqslant$ Dec $\leqslant$ +60d) that envelopes the LAMOST spectroscopic survey area of the Galactic anti-center , plus an extension to the M 31 and M 33 region.

This Digital Sky Survey of the Galactic Anti-center (DSS-GAC) with the Xuyi Schmidt and LAMOST telescopes will yield for the first time optical photometry and spectra for millions of stars in the Galactic disk(s), the defining component of the Milky Way as a typical spiral galaxy that contains most Galactic baryonic material and angular momentum. DSS-GAC will deliver classification, extinction, radial velocity and stellar parameters ($T_{\rm eff}$, log g, [Fe/H], probably also [α/Fe], and in some cases, [C/Fe]), for each sample star. Together with the accurate proper motions and distances to be obtained with the forthcoming GAIA mission, DSS-GAC offers a unique opportunity for major breakthroughs in studies of the Galactic structure, formation and evolution. In particular, DSS-GAC will generate a huge data set to 1) study the stellar populations, chemical composition and kinematics of the thin and thick disks and their interface with the halo; 2) Understand how resilient galaxy disks are to gravitational interactions/perturbations and study the temporal and secular evolution of the disks; 3) identify tidal streams and debris of disrupted dwarfs and clusters; 4) probe the gravitational potential and dark matter distribution; 5) map the three-dimensional distribution and extinction of the interstellar medium; 6) search for rare objects (e.g. stars of peculiar chemical composition, hyper-velocity stars); and 7) ultimately advance our understanding of the formation and evolution of stars and galaxies.

Following a two-year commissioning, the LAMOST pilot survey was initiated in October, 2011 and completed in June, 2012. In total, about 370,000 spectra of 270,000 stars have been obtained for DSS-GAC, with 70 per cent of the spectra reaching a spectral S/N ratio per resolution element at 7150 Å higher than 20. The formal LAMOST DSS-GAC survey will commence in September, 2012, and is expected to complete in five years.

Keywords. Galaxy: disk, structure, formation, evolution, Surveys, LAMOST

Highlights of Astronomy, Volume 16
XXVIIIth IAU General Assembly, August 2012
T. Montmerle, ed.

doi:10.1017/S1743921314011211

Frequency maps as a probe of secular evolution in the Milky Way

Monica Valluri

Department of Astronomy, University of Michigan, Ann Arbor, MI 48104, USA
email: mvalluri@umich.edu

Abstract. The frequency analysis of the orbits of halo stars and dark matter particles from a cosmological hydrodynamical simulation of a disk galaxy from the MUGS collaboration (Stinson *et al.* 2010) shows that even if the shape of the dark matter halo is nearly oblate, only about 50% of its orbits are on short-axis tubes, confirming a previous result: under baryonic condensation all orbit families can deform their shapes without changing orbital type (Valluri *et al.* 2010). Orbits of dark matter particles and halo stars are very similar reflecting their common accretion origin and the influence of baryons. Frequency maps provide a compact representation of the 6-D phase space distribution that also reveals the history of the halo (Valluri *et al.* 2012). The 6-D phase space coordinates for a large population of halo stars in the Milky Way that will be obtained from future surveys can be used to reconstruct the phase-space distribution function of the stellar halo. The similarity between the frequency maps of halo stars and dark matter particles (Fig. 1) implies that reconstruction of the stellar halo distribution function can reveal the phase space distribution of the unseen dark matter particles and provide evidence for secular evolution. MV is supported by NSF grant AST-0908346 and the Elizabeth Crosby grant.

Keywords. methods: n-body simulations, Galaxy: evolution, galaxies: halos, galaxies: dark matter

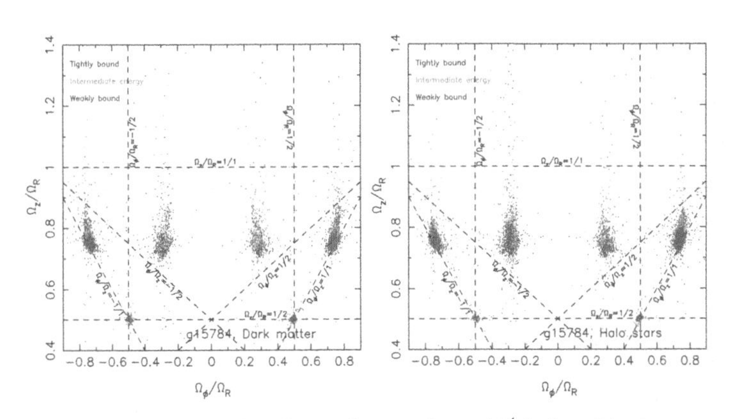

Figure 1. Frequency maps in cylindrical coordinates for $\sim 10^4$ halo orbits in a cosmological disk galaxy. Color represents total energy of each particle. The resonances at $\Omega_z/\Omega_R = |\Omega_\phi/\Omega_R| = 0.5$ correspond to a "thin shell" family: its prominence even in this hierarchical simulation is evidence of secular resonant trapping of halo particles by the growth of the stellar disk.

References

Stinson, G. S., *et al.* T. 2010, MNRAS, 408, 812
Valluri, M., Debattista, V. P., Quinn, T., & Moore, B. 2010, MNRAS, 403, 525
Valluri, M., Debattista, V. P., Quinn, T. R., Roškar, R., & Wadsley, J. 2012, MNRAS, 419, 1951

Highlights of Astronomy, Volume 16
XXVIIIth IAU General Assembly, August 2012
T. Montmerle, ed.

doi:10.1017/S1743921314011223

A new model for the Milky Way bar

Yougang Wang[1,2]†, Hongsheng Zhao[2,3,4], Shude Mao[2,5] and R. M. Rich[6]

[1]Key Laboratory of Optical Astronomy, National Astronomical Observatories, Chinese Academy of Sciences, Beijing 100012, China
[2]National Astronomical Observatories, Chinese Academy of Sciences, Beijing 100012, China
[3]SUPA, University of St Andrews, KY16 9SS, UK
[4]Vrije Universiteit, De Boelelaan 1081, 1081 HV Amsterdam, The Netherlands
[5]Jodrell Bank Centre for Astrophysics, University of Manchester, Manchester M13 9PL, UK
[6]Department of Physics and Astronomy, University of California, Los Angeles, CA 90095-1562, USA

Abstract. We use Schwarzschild's orbit-superposition technique to construct self-consistent models of the Galactic bar. Using χ^2 minimisation, we find that the best-fit Galactic bar model has a pattern speed $\Omega_\mathrm{p} = 60$ km s^{-1} kpc^{-1}, disk mass $\mathrm{M_d} = 1.0 \times 10^{11}\mathrm{M_\odot}$ and bar angle $\theta_\mathrm{bar} = 20°$ for an adopted bar mass $\mathrm{M_{bar}} = 2 \times 10^{10}\mathrm{M_\odot}$. The model can reproduce not only the three-dimensional and projected density distributions but also velocity and velocity dispersion data from the BRAVA survey. We also predict the proper motions in the range $l = [-12°, 12°]$, $b = [-10°, 10°]$, which appear to be higher than observations in the longitudinal direction. The model is stable within a timescale of 0.5 Gyr, but appears to deviate from steady-state on longer timescales. Our model can be further tested by future observations such as those from GAIA.

Keywords. Galaxy: bulge - Galaxy: centre - Galaxy: kinematics and dynamics

---

† email: wangyg@bao.ac.cn

Highlights of Astronomy, Volume 16
XXVIIIth IAU General Assembly, August 2012
T. Montmerle, ed.

doi:10.1017/S1743921314011235

A secularly evolved model for the Milky Way bar and bulge

Inma Martinez-Valpuesta and Ortwin Gerhard

Max Planck Institute for Extraterrestrial Physics, 85748, Garching, Germany
email: imv@mpe.mpg.de

Abstract. Bars are strong drivers of secular evolution in disk galaxies. Bars themselves can evolve secularly through angular momentum transport, producing different boxy/peanut and X-shaped bulges. Our Milky Way is an example of a barred galaxy with a boxy bulge. We present a self-consistent N-body simulation of a barred galaxy which matches remarkably well the structure of the inner Milky Way deduced from star counts. In particular, features taken as signatures of a second "long bar" can be explained by the interaction between the bar and the spiral arms of the galaxy (Martinez-Valpuesta & Gerhard 2011). Furthermore the structural change in the bulge inside $l = 4^{\circ}$ measured recently from VVV data can be explained by the high-density near-axisymmetric part of the inner boxy bulge (Gerhard & Martinez-Valpuesta 2012). We also compare this model with kinematic data from recent spectroscopic surveys. We use a modified version of the NMAGIC code (de Lorenzi *et al.* 2007) to study the properties of the Milky Way bar, obtaining an upper limit for the pattern speed of ~ 42 km/sec/kpc. See Fig. 1 for a comparison of one of our best models with BRAVA data (Kunder *et al.* 2012).

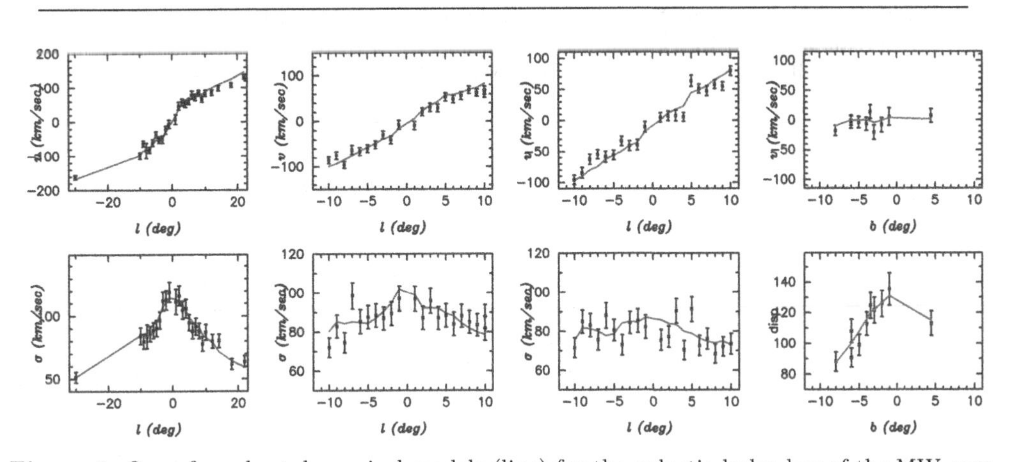

Figure 1. One of our best dynamical models (line) for the galactic bulge-bar of the MW compared to BRAVA data (points). From left to right: mean velocities (top) and velocity dispersions (bottom) in fields at latitudes $b = -4^{\circ}, -6^{\circ}, -8^{\circ}$ with longitude, and along the minor axis with latitude.

References

de Lorenzi, F., Debattista, V. P., Gerhard, O., & Sambhus, N. 2007, *MNRAS*, 376, 71
Kunder, A., Koch, A., Rich, R. M., *et al.* 2012, *AJ*, 143, 57
Gerhard, O. & Martine z-Valpuesta, I. 2012, *ApJ*, 744, L8
Martinez-Valpuesta, I. & Gerhard , O. 2011, *ApJ*, 734, L20

Highlights of Astronomy, Volume 16
XXVIIIth IAU General Assembly, August 2012
T. Montmerle, ed.

doi:10.1017/S1743921314011247

Chemical fingerprinting of stellar populations in the Milky Way halo

Mei-Yin Chou

Institute of Astronomy and Astrophysics, Academia Sinica, Taipei 10617, Taiwan,
email: cmy@asiaa.sinica.edu.tw

Abstract. The idea of "chemically fingerprinting" stars to their birth systems has been discussed over the last decade. Here we present an investigation of the chemical abundance patterns of halo substructures using high-resolution spectra. In particular, we study the abundances of the α-like element titanium (Ti) and the s-process elements yttrium (Y) and lanthanum (La) for M giant candidates of the Galactic Anticenter Stellar Structure (GASS, also known as the Monoceros Ring) and the Triangulum-Andromeda (TriAnd) Star Cloud. We apply "chemical fingerprinting" to the GASS/Monoceros Ring and TriAnd Star Cloud, to explore the origins of the two systems and the hypothesized connections between them. GASS has been debated either to originate from a part (e.g., warp) of the Galactic disk or tidal debris of a disrupted Milky Way (MW) satellite galaxy. Our exploration shows that GASS is indeed made of stars from a dwarf spheroidal (dSph) galaxy, although we still can not rule out the possibility that GASS was dynamically created out of a previously formed outer MW disk. And whereas the TriAnd Star Cloud has been assumed to come from the tidal disruption of the same accreted MW satellite as the GASS/Monoceros Ring, our comparison of the abundance patterns in GASS and TriAnd M giants suggests that the TriAnd Star Cloud is likely an independent halo substructure unrelated to the GASS/Monoceros Ring. Furthermore, our findings also suggest that the MW may have accreted other satellites in addition to the on-going, well-known Sagittarius (Sgr) dwarf galaxy.

Highlights of Astronomy, Volume 16
XXVIIIth IAU General Assembly, August 2012
T. Montmerle, ed.

doi:10.1017/S1743921314011259

Quantifying the mixing due to bars

Patricia Sanchez-Blazquez
Universidad Autonoma de Madrid, email: psanchezblazquez@gmail.com

Abstract. We will present star formation histories and the stellar and gaseous metallicity gradients in the disk of a sample of 50 face-on spiral galaxies with and without bars observed with the integral field unit spectrograph PMAS. The final aim is to quantify the redistribution of mass and angular momentum in the galactic disks due to bars by comparing both the gas-phase and star-phase metallicity gradients on the disk of barred and non-barred galaxies. Numerical simulations have shown that strong gravitational torque by non-axisymmetric components induce evolutionary processes such as redistribution of mass and angular momentum in the galactic disks (Sellwood & Binney 2002) and consequent change of chemical abundance profiles. If we hope to understand chemical evolution gradients and their evolution we must understand the secular processes and re-arrangement of material by non-axisymmetric components and vice-versa. Furthermore, the re-arrangement of stellar disk material influences the interpretation of various critical observed metrics of Galaxy evolution, including the age-metallicity relation in the solar neighborhood and the local G-dwarf metallicity distribution. Perhaps the most obvious of these aforementioned non-axisymmetric components are bars - at least 2/3 of spiral galaxies host a bar, and possibly all disk galaxies have hosted a bar at some point in their evolution. While observationally it has been found that barred galaxies have shallower gas-phase metallicity gradients than non-barred galaxies, a complementary analysis of the stellar abundance profiles has not yet been undertaken. This is unfortunate because the study of both gas and stars is important in providing a complete picture, as the two components undergo (and suffer from) very different evolutionary processes.

Reference

Sellwood, J. & Binney, J. 2002, *MNRAS*, 336, 785

Highlights of Astronomy, Volume 16
XXVIIIth IAU General Assembly, August 2012
T. Montmerle, ed.

doi:10.1017/S1743921314011260

The outskirts of spiral galaxies: probing stellar migration theory

Judit Bakos and Ignacio Trujillo

Instituto de Astrofísica de Canarias Spain, email: bakosjud@gmail.com

Abstract. One of the most outstanding predictions of the radial migration theory was a progressive increase in the age of the stellar populations toward the galaxy outskirts. This gradual change in the age is caused by a net flux of old stars formed in the inner disk moving toward the outskirts. Thanks to the fact that the age of stellar populations can be interpreted in terms of colors, this prediction was confirmed observationally a few years ago with deep surface brightness profiles of a large sample of nearby galaxies. Our group has now taken a step forward on this as we have explored the properties of the stellar populations beyond the star formation threshold using ultra deep data from SDSS Stripe82. These data allowed us to study the faint outskirts of disks in detail down to a surface brightness level of 30 mag/arcsec2 in the r'-band. At these surface brightness levels spiral galaxies reveal new and exciting structural components which could not have been seen by regular SDSS imaging: outer disks extend farther out, tidal streams and satellites become visible. We will present deep color profiles of the regions of the galaxies where the disks starts to be confused with the stellar haloes. We confront these colors with model predictions. For some galaxies, the very outer regions are so red that conventional IMFs can not explain their colors. We will discuss whether our new results can be explained within the radial migration scheme.

Highlights of Astronomy, Volume 16
XXVIIIth IAU General Assembly, August 2012
T. Montmerle, ed.

doi:10.1017/S1743921314011272

Radial migration in barred galaxies

P. Di Matteo[1], M. Haywood[1], F. Combes[2], B. Semelin[2], C. Babusiaux[1] and A. Gomez[1]

[1]GEPI, Observatoire de Paris, CNRS, Université Paris Diderot,
5 place Jules Janssen, 92190 Meudon, France email: `paola.dimatteo@obspm.fr`

[2]LERMA, CNRS, UPMC, Observatoire de Paris,
61 Avenue de l'Observatoire, 75014 Paris, France

Abstract. In this talk, I will present the result of high resolution numerical simulations of disk galaxies with various bulge/disk ratios evolving isolated, showing that:

- Most of migration takes place when the bar strength is high and decreases in the phases of low activity (in agreement with the results by Brunetti *et el.* 2011, Minchev *et al.* 2011).
- Most of the stars inside the corotation radius (CR) do not migrate in the outer regions, but stay confined in the inner disk, while stars outside CR can migrate either inwards or outwards, diffusing over the whole disk.
- Migration is accompanied by significative azimuthal variations in the metallicity distribution, of the order of 0.1 dex for an initial gradient of $\sim$-0.07 dex/kpc.
- Boxy bulges are an example of stellar structures whose properties (stellar content, vertical metallicity, [α/Fe] and age gradients, ..) are affected by radial migration (see also Fig. 1).

Keywords. Galaxies: abundances, Galaxies: evolution, Galaxies: dynamics

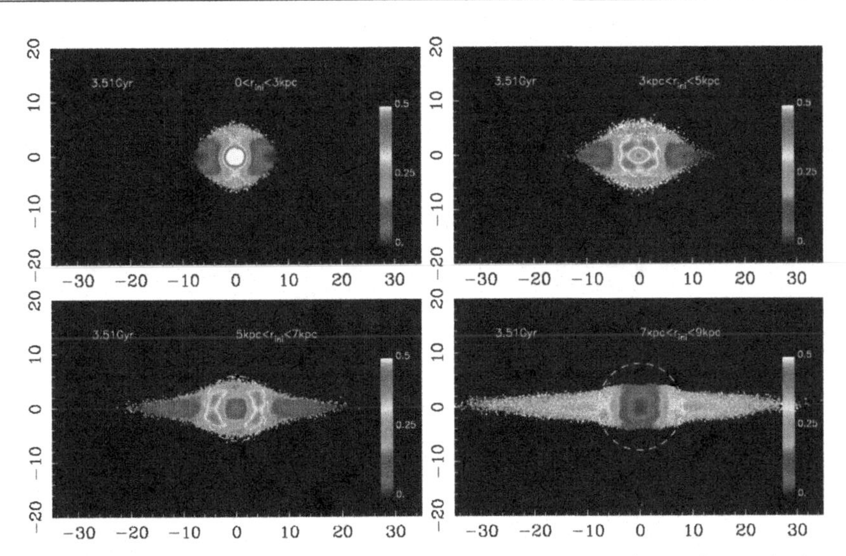

Figure 1. Fractional contribution of stars born at different radii (top-left panel: $r_{ini} < 3$kpc; top-right panel: 3 kpc $< r_{ini} <$ 5kpc; bottom-left: 5 kpc $< r_{ini} <$ 7kpc; bottom-right: 7 kpc $< r_{ini} <$ 9kpc) to the stellar content of a boxy bulge and disk. Axis units are in kpc.

References

Brunetti, M., Chiappini, C., & Pfenniger, D. 2011, *A&A* 534, 75
Minchev, I., *et al.* 2011, *A&A* 527, 147

Highlights of Astronomy, Volume 16
XXVIIIth IAU General Assembly, August 2012
T. Montmerle, ed.

doi:10.1017/S1743921314011284

Searching for observational evidence of radial mixing in the Milky Way disk

Misha Haywood

Observatoire de Paris, 61 Av de l'Observatoire, F-75014 Paris, France,
email: Misha.Haywood@obspm.fr

Abstract. Secular evolution in disks through angular momentum redistribution of stars induce radial mixing of their orbits. While theoretical studies and simulations now abound on the subject - with various predicted effects : disks growth, flattening of metallicity gradients, possible reversing of the mean age as a function of radius in disk, etc, observational evidences remain sparse. In the Galaxy, possible signatures are searched for in the local distributions of velocities, abundances and ages, or in the variation of large scale chemical gradients with time. I will present the current state of affairs and discuss what kind of evidences is available from data in the Milky Way.

Highlights of Astronomy, Volume 16
XXVIIIth IAU General Assembly, August 2012
T. Montmerle, ed.

doi:10.1017/S1743921314011296

A test for radial mixing using local star samples

Jincheng Yu[1], Jerry Sellwood[2], Carlton Pryor[2], Li Chen[1] and Jinliang Hou[1]

[1]Shanghai Astronomical Observatory, Chinese Academy of Sciences, China,
email: `yujc@shao.ac.cn`
[2]Department of Physics and Astronomy, Rutgers University

Abstract. We use samples of local main-sequence stars to show that the radial gradient of [Fe/H] in the thin disk of the Milky Way decreases with mean effective stellar temperature. We use the angular momentum of each star about the Galactic center to eliminate the effects of epicyclic motion, which would otherwise blur the estimated gradients. We use the effective temperatures as a proxy for mean age, and conclude that the decreasing gradient is consistent with the predictions of radial mixing due to transient spiral patterns. We find some evidence that the trend of decreasing gradient with increasing mean age breaks to a constant gradient for samples of stars whose main-sequence life-times exceed the likely age of the thin disk.

Highlights of Astronomy, Volume 16
XXVIIIth IAU General Assembly, August 2012
T. Montmerle, ed.

doi:10.1017/S1743921314011302

Disk structures in the CGS Survey

Zhao-Yu Li[1], Luis Ho[2], Aaron Barth[3] and Chien Peng[2]

[1]Shanghai Astronomical Observatory China, email: lizy@shao.ac.cn
[2]Observatories of the Carnegie Insititution for Science
[3]University of California, Irvine

Abstract. The Carnegie-Irvine Galaxy Survey (CGS) is a long term program to investigate the photometric and spectroscopic properties of a statistically complete sample of 605 bright (BT ¡ 12.9 mag), southern (delta ¡ 0 deg) galaxies using the facilities at Las Campanas Observatory. For each galaxy, we have broadband images (BVRI) with good seeing ($\approx 1''$) and deep surface brightness ($\approx$27.5 B-band). Using the IRAF task ELLIPSE and the fourier decomposition method, we measured the bar and the lopsidedness properties of disk galaxies in the CGS sample. Our results show that the bar fraction is lower in the early-type galaxies than that in the late-type ones. The (relative) bar length is longer in early-type ones, and strong bars are rare (the one with large ellipticity). We find that the lopsidedness is independent on the galaxy environment, and correlation studies suggest that the lopsided disk may have helped drive gas inward to form stars.

Highlights of Astronomy, Volume 16
XXVIIIth IAU General Assembly, August 2012
T. Montmerle, ed.

doi:10.1017/S1743921314011314

Dynamical evolution of star clusters in transient spiral arms

Michiko Fujii[1] **and Junichi Baba**[2]

[1]Leiden Observatory Netherlands, email: `fujii@strw.leidenuniv.nl`
[2]Tokyo Institute of Technology

Abstract. Star clusters are one of fundamental building blocks of galactic disks. They form in a potential well of spiral arms and travel in the disk. We performed N-body simulation of star clusters in stellar disk with "live" spiral arms. In this simulation, both star clusters and stellar disks are modeled as N-body systems. We found that star clusters migrated in the galactic disk in a timescale of their galactic rotation. The tidal tails spread over a few kpc, but they might be detectable if we are able to measure their velocity.

Highlights of Astronomy, Volume 16
XXVIIIth IAU General Assembly, August 2012
T. Montmerle, ed.

doi:10.1017/S1743921314011326

How well can we identify pseudobulges?

Alister Graham

Swinburne University of Technology Australia, email: agraham@astro.swin.edu.au

Abstract. Since the discovery of rotating galaxy bulges (e.g. Pease 1918; Babcock 1938, 1939), especially in the 1970s (e.g. Rubin, Ford & Kumar 1973; Pellet 1976; Bertola & Capaccioli 1977; Peterson 1978; Mebold *et al.* 1979; Kormendy & Illingworth 1979), coupled with early computer simulations of disks which formed rotating, exponential-like "pseudobulges" (e.g. Bardeen 1975; Hohl 1975, and references therein), a number of often over-looked problems pertaining to the identification of real "pseudobulges" have arisen. Drawing on my recent review article of disk galaxy structure and modern scaling laws (Graham 2012), some of these important issues are presented. Topics include: classical spheroids with exponential light distributions; curved but continuous scaling relations involving the 'effective' structural parameters; the old age of most bulge stars (e.g. Thomas & Davies 2006; MacArthur *et al.* 2009); that most disk galaxies have bulge-to-disk flux ratios $< 1/3$ (Graham & Worley 2008); rotation in simulated merger remnants (e.g. Bekki 2010; Keselman & Nusser 2012) plus many other frustrating yet interesting reasons why rotation may not be a definitive signature of bulges built via secular processes (e.g. Babusiaux *et al.* 2010; Williams *et al.* 2010, Qu *et al.* 2011; Saha *et al.* 2012)

References

Babcock, H. W. 1938, *PASP*, 50, 174
Babcock, H. W. 1939, *Lick Observatory Bulletin*, 19, 41
Babusiaux, C., *et al.* 2010, *A&A*, 519, A77
Bekki, K. 2010, *MNRAS*, 401, L58
Bardeen, J. M. 1975, IAU Symp., 69, 297
Bertola, F. & Capaccioli, M. 1977, *ApJ*, 211, 697
Graham A. W., Worley, C. C. 2008, *MNRAS*, 388, 1708
Graham, A. W. 2012, in *Planets, Stars, and Stellar Systems, Vol. 6, Extragalactic Astronomy and Cosmology*, ed. W. C. Keel (New York: Springer-Verlag), in press (arXiv:1108.0997)
Hohl, F. 1975, IAU Symp., 69, 349
Keselman, J. A. & Nusser, A. 2012, *MNRAS*, 424, 1232
Kormendy, J. & Illingworth, G. 1979, *Photometry, Kinematics and Dynamics of Galaxies*, 195
MacArthur, L. A., González, J. J., & Courteau, S. 2009, *MNRAS*, 395, 28
Mebold U., Goss W. M., Siegman B., van Woerden H., & Hawarden T. G. 1979, *A&A*, 74, 100
Pease, F. G. 1918, *Proc. Nat. Acad. Sci.*, 4, 21
Pellet, A. 1976, *A&A*, 50, 421
Peterson, C. J. 1978, *ApJ*, 221, 80
Qu, Y., Di Matteo, P., Lehnert, M. D., & van Driel, W. 2011, *A&A*, 530, A10
Rubin, V. C., Ford, W. K., Krishna, Kumar C. 1973, *ApJ*, 181, 61
Saha, K., Martinez-Valpuesta, I., & Gerhard, O. 2012, *MNRAS*, 421, 333
Thomas, D. & Davies R. L. 2006, *MNRAS*, 366, 510
Williams, M. J., *et al.* 2010, *MNRAS*, 414, 2163

Highlights of Astronomy, Volume 16
XXVIIIth IAU General Assembly, August 2012
T. Montmerle, ed.

doi:10.1017/S1743921314011338

What Disc Brightness Profiles Can Tell us about Galaxy Evolution

John Beckman[1,2,3], **Peter Erwin**[4] **and Leonel Gutiérrez**[5]

[1]Instituto de Astrofísica de Canarias, c/ Vía Láctea, s/n, E38205, La Laguna, Tenerife, Spain.
email: jeb@iac.es

[2]Departamento de Astrofísica. Universidad de La Laguna, Tenerife, Spain.

[3]Consejo Superior de Investigaciones Científicas, Spain.

[4]Max Planck Institut fur Extraterrestrische Physik, Germany

[5]UNAM, Ensenada, México.

Abstract. Azimuthally averaged surface brightness profiles of disc galaxies provide a most useful practical classification scheme which gives insights into their evolution. Freeman (1970) first classified disc profiles into Type I, with a single exponential decline in surface brightness, and Type II, having a split exponential profile, whose inner radial portion is shallower than its outer section. Van der Kruit & and Searle, (1981) drew attention to sharply truncated profiles of outer discs observed edge-on, but more recently Pohlen *et al.* (2004) showed that if these same galaxies were observed face-on their profiles would be of Type II. Finally in Erwin, Beckman and Pohlen (2005) we found a significant fraction of profiles with inner portion steeper than the outer portion, which we termed "antitruncations" or Type III profiles. In Erwin, Pohlen and Beckman (2008), we produced a refined classification, taking into account those Type II's produced by dynamical effects at the outer Lindblad resonance, and those Type III's caused by the presence of an outer stellar halo. In Gutiérrez *et al.* (2011) we showed the distribution of the three main profile types along the Hubble sequence. In early type discs Types I and III predominate, while in late types, Sc and later, Type II predominates.

The evolution of Type II's over cosmic time was studied by Azzollini *et al.* (2008a, 2008b) who obtained four key results: (a) between z = 1 and z = 0 the break radius between the inner (shallower) and outer (steeper) profile has increased systematically, by a factor 1.3; (b) the inner profile has steepened while the outer profile is shallower at lower z; (c) the extrapolated central surface brightness has fallen by over two magnitudes; (d) the discs in the full redshift interval are always bluest at the break radius. While this behaviour can be qualitatively explained via evolutionary models including stellar migration plus gas infall, such as that by Roskar *et al.* (2008), and while Type III profiles may have a qualitative explanation via mergers and/or accretion, the widespread existence of Type I's is still a major conceptual challenge.

References

Azzollini, R., Trujillo, I., & Beckman, J. E. 2008a, *ApJ. Letters* 679, L69
Azzollini R., Trujillo, I. & Beckman, J. E. 2008b, *ApJ* 684, 1026
Erwin, P. E., Beckman, J. E., & Pohlen, M. 2005, *ApJ. Letters* 626, L81
Erwin, J. E., Pohlen, M., & Beckman, J. E. 2008, *AJ* 135, 20
Freeman, K. C. 1970, *ApJ* 160, 767
Van der Kruit, P. & Searle, L. 1981 *A&A* 95, 105
Gutiérrez, L., Erwin, P., Aladro, R., & Beckman, J. E. 2011, *AJ* 142, 145
Pohlen, M., Beckman, J. E., *et al.* 2004,in: Block, D., *et al.* (eds.), *ASSL*, Lecture Notes in Physics (Kluwer, Dordrecht), vol. 317, p. 713
Roskar, R. & Debattista, P. 2008, *ApJ. Letters* 675, L65

Highlights of Astronomy, Volume 16
XXVIIIth IAU General Assembly, August 2012
T. Montmerle, ed.

doi:10.1017/S174392131401134X

Quantifying secular evolution through structural decomposition

Lee Kelvin[1,2,3]

[1]School of Physics & Astronomy, University of St Andrews, St Andrews, KY16 9SS, UK
[2]ICRAR, The University of Western Australia, 35 Stirling Hwy, WA 6009, Australia
[3]Inst für Astro- u Teilchenphysik, Universität Innsbruck, Techstr 25, 6020 Innsbruck, Austria
email: lee.kelvin@uibk.ac.at

Abstract. Structure within a galaxy is not random, instead emerging as a direct function of its evolutionary path. It is thought that secular evolutionary processes leave behind distinct structural tracers in the form of bars, pseudo-bulges and rings. We have developed a robust automated structural analysis pipeline (Kelvin *et al.*, 2012) able to accurately map structure across a range of ground and space-based datasets. Using reprocessed SDSS and UKIDSS data from the GAMA survey: an imaging and spectroscopic survey with over 300, 000 redshifts across 300 square degrees (Driver *et al.*, 2009); we measure the relative abundance and stellar mass locked up within these structures in the local ($z < 0.06$) Universe. Future robust calculations of the stellar mass budget within bulges, bars, disks and pseudo-bulges should allow us to measure the relative importance of secular evolution against other mechanisms across cosmic time.

Keywords. galaxies: evolution, galaxies: structure, techniques: image processing, astronomical data bases: miscellaneous, galaxies: bulges, galaxies: spiral, galaxies: fundamental parameters

---

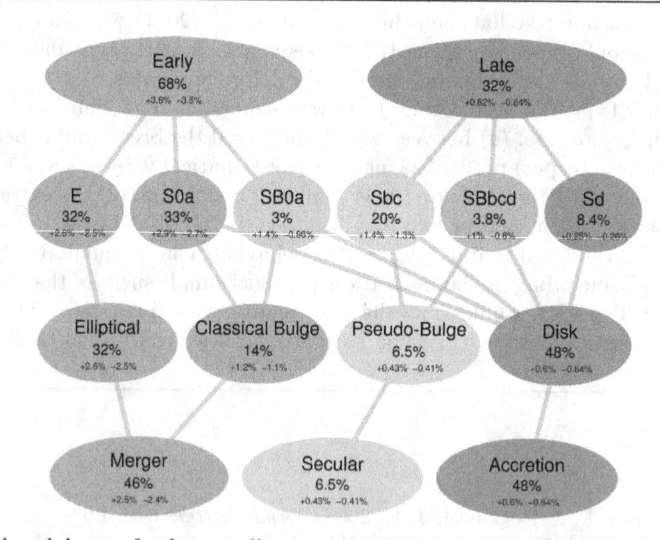

Figure 1. The breakdown of galaxy stellar mass in the local Universe by (top to bottom): morphological type; morphological class; galaxy structure, and; evolutionary processes. Percentages represent the fraction of mass within that division, with 2σ errors shown below for reference.

References

Driver. S. P. *et al.*, 2009, *Astron. Geophys.* 50, 050000
Kelvin, L. S. *et al.*, 2012, *MNRAS* 421, 1007

Highlights of Astronomy, Volume 16
XXVIIIth IAU General Assembly, August 2012
T. Montmerle, ed.

doi:10.1017/S1743921314011351

Bar-driven evolution of fast rotators: the role and fate of bars in early and late-type galaxies

Eric Emsellem[1] and Renaud Florent[2]

[1]ESO-CRAL Germany, email: eric.emsellem@eso.org
[2]CEA-Saclay France

Abstract. We have performed state-of-the-art high resolution simulations of early-type galaxies with bars, including (multi-phase) gas, star formation and feedback. The aim of this programme is to better understand the observed morphology, kinematical structures, (2D) metallicity distribution, observed in fast rotators with bars. Our simulations were designed via a newly developed code allowing us to build a library of initial conditions closely mimicking barred galaxies in the Atlas3D sample. We will present the role and importance of bars on the gas fueling, redistribution of angular momentum, and overall secular evolution of fast rotators. These results are compared with actual observations (IFU, CO maps, stellar population distributions) obtained in the course of the Atlas3D project. The results from these "early-type" simulations will also be compared in the context of recently conducted simulations of later-type barred galaxies, including one of a Milky-Way type object with a resolution down to 0.05 parsec.

Highlights of Astronomy, Volume 16
XXVIIIth IAU General Assembly, August 2012
T. Montmerle, ed.

doi:10.1017/S1743921314011363

Dissecting early-type dwarf galaxies into their multiple components

Joachim Janz[1,3,*], Eija Laurikainen[2], Thorsten Lisker[3] and Heikki Salo[1]

[1]Division of Astronomy, Department of Physics, University of Oulu, Finland

[2]Astronomisches Rechen-Institut, ZAH, University of Heidelberg, Germany

[3]Finnish Centre for Astronomy with ESO (FINCA), University of Turku, Finland

[*] Fellow of the Gottlieb Daimler and Karl Benz Foundation, email: `jjanz@ari.uni-heidelberg.de`

Abstract. Early-type dwarf galaxies are often thought to be either more diffuse versions of giant ellipticals or to be low-mass disk galaxies that were quenched and heated by the environment. In both cases, the picture that most astronomers have in mind probably is that of a dynamically hot, regular shaped galaxy, in which any previous substructure has either been smeared out, or has never been there. However, the early-type dwarfs are not that simple.

We analyzed ~100 such objects in the Virgo cluster using deep near-infrared images and found that the majority has a multi-component structure, sometimes even with bars or lenses. The study was done by applying GALFIT to images from the SMAKCED collaboration (*S*tellar content, *MA*ss and *K*inematics of *C*luster *E*arly-type *D*warfs, http://www.smakced.net). The sample comprises early-type galaxies in the Virgo cluster in a brightness range of $-19 \leqslant M_r \leqslant -16$ mag, and the data is complete down to $M_r = -16.73$ mag. The images typically reach a signal-to-noise ratio of 1 per pixel of ~0.25" at a surface brightness of ~22.5 mag/arcsec^2 in the H-band. The galaxies were fitted with two-dimensional models, either with a simple Sérsic model or inner and outer components, as well as bars and lenses. Only a fraction of 31% of the galaxies can be fitted with a single Sérsic function. This fraction of "simple" galaxies turns out to be a strong function of luminosity, with a smaller fraction for brighter objects. The bar fraction is 14% and also in 14% of the galaxies lenses were fitted.

When comparing the flattening distribution, the early-type dwarfs are more similar to spiral galaxies than to elliptical or lenticular galaxies. It is disputable whether or not the dwarfs follow a common relation with the bright elliptical galaxies, e.g. in the brightness versus size diagram. At the same time, they appear as smooth continuation of bright late-type galaxies in this diagram. The inner and outer components, as well as the simple galaxies have similar flattening distributions. The inner components are mostly fitted with Sérsic-n values close to 1, i.e. with nearly exponential profiles. We argue that the inner components in the early-type dwarfs are not be bulges but may form parts of the disks, in which the matter was re-distributed during the transformation process from a late-type progenitor.

Keywords. galaxies: dwarf — galaxies: photometry — galaxies: structure — galaxies: clusters: individual: (Virgo Cluster)

References

Janz, J., *et al.* 2012, *ApJ*, 745, L24
Janz, J., *et al.* 2012, *ApJ*, submitted

Highlights of Astronomy, Volume 16
XXVIIIth IAU General Assembly, August 2012
T. Montmerle, ed.

doi:10.1017/S1743921314011375

Galaxies driven only by secular evolution?

Lourdes Verdes-Montenegro on behalf of the AMIGA team
Instituto de Astrofisica de Andalucia (CSIC), email: lourdes@iaa.es

Abstract. The AMIGA project (Analysis of the interstellar Medium of Isolated GAlaxies, http://amiga.iaa.es) has identified a significant sample of very isolated (T_{cc}(nearest-neighbor) $\approx$2-3Gyr) galaxies in the local Universe and revealed that they have different properties than galaxies in richer environments. Our analysis of a multiwavelength database includes quantification of degree of isolation, morphologies, as well as FIR and radio line/continuum properties.

Properties usually regarded as susceptible to interaction enhancement show lower averages in AMIGA–lower than any galaxy sample yet identified. We find lower MIR/FIR measures (Lisenfeld *et al.* 2007), low levels of radio continuum emission (Leon *et al.* 2008), no radioexcess above the radioFIR correlation (0%, Sabater *et al.*2008), a small number of AGN (22%, Sabater *et al.* 2012), and lower molecular gas content (Lisenfeld *et al.* 2011). The late-type spiral majority in our sample show very small bulge/total ratios (largely <0.1) and Sersic indices consistent with an absence of classical bulges (Durbala *et al.* 2008). They show redder $g - r$ colors and lower color dispersion for AMIGA subtypes (Fernandez-Lorenzo *et al.* 2012) and show the narrowest (gaussian) distribution of HI profile asymmetries of any sample yet studied.

This work has been supported by Grant AYA2011-30491-C02-01 co-financed by MICINN and FEDER funds, and the Junta de Andalucia (Spain) grants P08-FQM-4205 and TIC-114.

References

Durbala, A., *et al.* 2008, *MNRAS* 390, 881
Fernandez Lorenzo, M., *et al.* 2012, *A&A* 540, 47
Leon, S., *et al.* 2008, *A&A* 485, 475
Lisenfeld, U., *et al.* 2007, *A&A* 462, 507
Lisenfeld, U., *et al.* 2011, *A&A* 534, 102
Sabater, J., *et al.* 2008, *A&A* 486, 73
Sabater, J., *et al.* 2012, *A&A* 545, 15

Highlights of Astronomy, Volume 16
XXVIIIth IAU General Assembly, August 2012
T. Montmerle, ed.

doi:10.1017/S1743921314011387

The role of external gas accretion on galaxy transformations, and evidence of such accretion

Françoise Combes
LERMA, CNRS, Observatoire de Paris, 61 Av de l'Observatoire, F-75014 Paris, France
email: francoise.combes@obspm.fr

Abstract. Continuously accreting matter from cosmic filaments is one of the main way to assemble mass for galaxies (Keres *et al.* 2005, Dekel *et al.* 2009). This external accretion accelerates secular processes, and maintain star formation, but also bar and spiral formation (Bournaud & Combes 2002), and consequent radial migration. Secular evolution may alleviate the problem of too massive bulge formation in the standard LCDM hierarchical scenario. Inside out formation of galaxies may account for the evolution of the size-mass relation and evolution with redshift. I will show how gas accretion from the inter galactic medium can mimick perturbations due to galaxy interactions (cf Figure 1), and I will describe evidence of such accretion, through warps, polar rings or damped Lyman-α systems.

Keywords. hydrodynamics, instabilities, stellar dynamics, turbulence, waves, galaxies: evolution, galaxies: formation

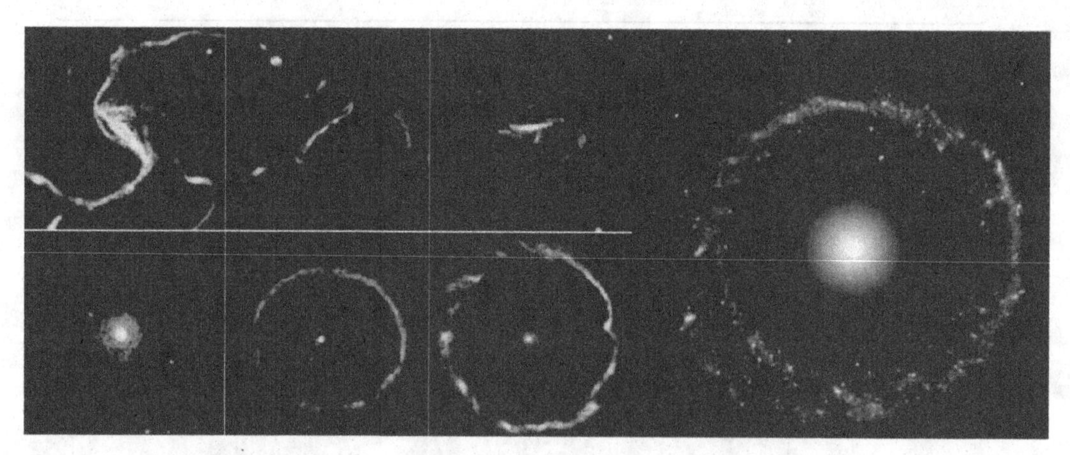

Figure 1. Formation through gas accretion of a ring of gas and new stars around an isolated galaxy (Mastropietro *et al.*in prep). This could be an explanation for the formation of Hoag-type rings, as shown at right.

References

Bournaud F. & Combes F. 2002, *A&A* 392, 83
Dekel, A., *et al.* 2009, *Nature* 457, 451
Keres, D., *et al.* 2005, *MNRAS* 363, 2

Highlights of Astronomy, Volume 16
XXVIIIth IAU General Assembly, August 2012
T. Montmerle, ed.

doi:10.1017/S1743921314011399

Effects of secular evolution on the star formation history of galaxies

M. Fernández Lorenzo[1], J. Sulentic[1], L. Verdes–Montenegro[1], M. Argudo-Fernández[1], J. E. Ruiz[1], J. Sabater[2] and S. Sánchez–Expósito[1]

[1]Instituto de Astrofísica de Andalucía, Granada, IAA-CSIC, Granada, Spain,
email: `mirian@iaa.es`
[2]Institute for Astronomy, University of Edinburgh, Edinburgh, UK

Abstract. We report the study performed as part of the AMIGA (Analysis of the interstellar Medium of Isolated GAlaxies; http://www.amiga.iaa.es) project, focused on the SDSS $(g-r)$ colors of the sample. Assuming that color is an indicator of star formation history, this work better records the signature of passive star formation via pure secular evolution. Median values for each morphological type in AMIGA were compared with equivalent measures for galaxies in denser environments. We found a tendency for AMIGA spiral galaxies to be redder than galaxies in close pairs, but no clear difference when we compare with galaxies in other (e.g. group) environments. The $(g-r)$ color of isolated galaxies presents a Gaussian distribution, as indicative of pure secular evolution, and a smaller median absolute deviation (almost half) compared to both wide and close pairs. This redder color and lower color dispersion of AMIGA spirals compared with close pairs is likely due to a more passive star formation in very isolated galaxies. In Fig. 1, we represent the size versus stellar mass for early and late–type galaxies of our sample, compared with the local relations of Shen *et al.* (2003). The late–type isolated galaxies are ~1.2 times larger or have less stellar mass than local spirals in other environments. The latter would be in agreement with the passive star formation found in the previous part.

We acknowledge Grant AYA2011-30491-C02-01, P08-FQM-4205 and TIC-114.

Keywords. galaxies: general, galaxies: fundamental parameters, galaxies: evolution

---

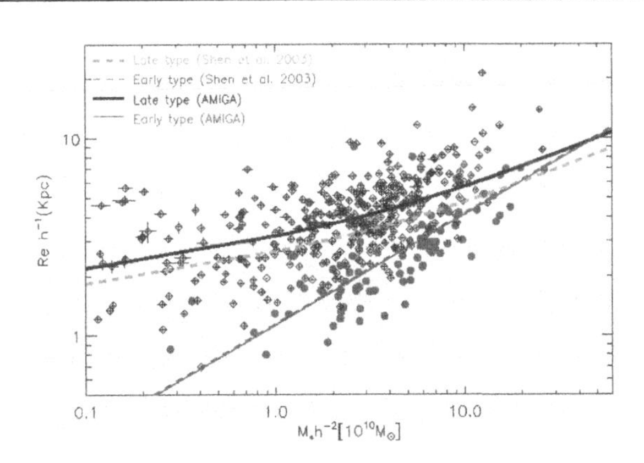

Figure 1. Stellar mass size relation for the AMIGA and Shen *et al.* (2003) samples. Open diamonds are the AMIGA late–types and solid points are the AMIGA early–types.

References

Shen, S., Mo, H. J., White, S. D. M., *et al.* 2003, *MNRAS*, 343, 978

Highlights of Astronomy, Volume 16
XXVIIIth IAU General Assembly, August 2012
T. Montmerle, ed.

doi:10.1017/S1743921314011405

Hoag's object: the quintessential ring galaxy

Noah Brosch[1], Ido Finkelman[1] and Alexei Moiseev[2]

[1]Tel Aviv University, Israel, email: noah@wise.tau.ac.il
[2]Special Astrophysical Observatory of the Russian Academy of Sciences

Abstract. We present new observations of Hoag's Object, known as "the most perfect ring galaxy," that show that a preferred explanation for this object is (a) the formation of a triaxial elliptical galaxy some 10 Gyr ago, (b) the accretion of a large disk of neutral hydrogen at about the same time, (c) low-level star formation in the HI disk for all the time since that event triggered by the triaxial potential of the core.

Highlights of Astronomy, Volume 16
XXVIIIth IAU General Assembly, August 2012
T. Montmerle, ed.

doi:10.1017/S1743921314011417

The role of close pair interactions in triggering stellar bars and rings

Preethi Nair[1], Sara Ellison[2] and David Patton[3]

[1]Space Telescope Science Institute, Baltimore, MD 21218 USA, email: nair@stsci.edu
[2]University of Victoria
[3]Trent University

Abstract. Recent works which have looked at bars in clusters versus the field have found no significant difference in bar fraction. However, other works (Nair & Abraham 2010, Lee *et al.* 2012) have found that bar fractions depend sensitively on the mass, morphology and color of the galaxy. In addition, simulations suggest that bar formation may depend on the merger ratio of close pair interactions as well as on the separation between the pairs. In this work, we analyze the bar fractions in a complete sample of ≈23,000 close pairs derived from the Sloan Digital Sky Survey Data Release 7. We will present results illustrating the dependence of bar and ring fractions as a function of merger mass ratio, pair separation, galaxy morphology, and stellar mass. I will further compare the role of bars and close pairs in triggering central star formation and AGN.

References

Nair, P. & Abraham, R. G. 2010, *ApJ* 714, L260
Lee, J. H., *et al.* 2012, ArXiv 1211.3973

Highlights of Astronomy, Volume 16
XXVIIIth IAU General Assembly, August 2012
T. Montmerle, ed.

doi:10.1017/S1743921314011429

Role of massive stars in the evolution of primitive galaxies

Sara Heap
NASA Goddard Space Flight Center, email: sara.r.heap@NASA.gov

Abstract. An important factor controlling galaxy evolution is feedback from massive stars. It is believed that the nature and intensity of stellar feedback changes as a function of galaxy mass and metallicity. At low mass and metallicity, feedback from massive stars is mainly in the form of photoionizing radiation. At higher mass and metallicity, it is in stellar winds. I Zw 18 is a local blue, compact dwarf galaxy that meets the requirements for a primitive galaxy: low halo mass $<10^9 \, M_\odot$, strong photoionizing radiation, no galactic outflow, and very low metallicity, log(O/H)+12=7.2. We will describe the properties of massive stars and their role in the evolution of I Zw 18, based on analysis of ultraviolet images and spectra obtained with HST.

Highlights of Astronomy, Volume 16
XXVIIIth IAU General Assembly, August 2012
T. Montmerle, ed.

doi:10.1017/S1743921314011430

The influence of halo evolution on galaxy structure

Simon White

Max Planck Institute for Astrophysics Germany, email: swhite@mpa-garching.mpg.de

Abstract. If Einstein-Newton gravity holds on galactic and larger scales, then current observations demonstrate that the stars and interstellar gas of a typical bright galaxy account for only a few percent of its total nonlinear mass. Dark matter makes up the rest and cannot be faint stars or any other baryonic form because it was already present and decoupled from the radiation plasma at z = 1000, long before any nonlinear object formed. The weak gravito-sonic waves so precisely measured by CMB observations are detected again at z = 4 as order unity fluctuations in intergalactic matter. These subsequently collapse to form today's galaxy/halo systems, whose mean mass profiles can be accurately determined through gravitational lensing. High-resolution simulations link the observed dark matter structures seen at all these epochs, demonstrating that they are consistent and providing detailed predictions for all aspects of halo structure and growth. Requiring consistency with the abundance and clustering of real galaxies strongly constrains the galaxy-halo relation, both today and at high redshift. This results in detailed predictions for galaxy assembly histories and for the gravitational arena in which galaxies live. Dark halos are not expected to be passive or symmetric but to have a rich and continually evolving structure which will drive evolution in the central galaxy over its full life, exciting warps, spiral patterns and tidal arms, thickening disks, producing rings, bars and bulges. Their growth is closely related to the provision of new gas for galaxy building.

Highlights of Astronomy, Volume 16
XXVIIIth IAU General Assembly, August 2012
T. Montmerle, ed.

doi:10.1017/S1743921314011442

Shaping Disk Galaxy Stellar Populations via Internal and External Processes

Rok Roškar

Institute for Theoretical Physics, University of Zürich, 8057 Zürich, Switzerland
email: roskar@physik.uzh.ch

Abstract. In recent years, effects such as the radial migration of stars in disks have been recognized as important drivers of the properties of stellar populations. Radial migration arises due to perturbative effects of disk structures such as bars and spiral arms, and can deposit stars formed in disks to regions far from their birthplaces. Migrant stars can significantly affect the demographics of their new locales, especially in low-density regions such as in the outer disks. However, in the cosmological environment, other effects such as mergers and filamentary gas accretion also influence the disk formation process. Understanding the relative importance of these processes on the detailed evolution of stellar population signatures is crucial for reconstructing the history of the Milky Way and other nearby galaxies. In the Milky Way disk in particular, the formation of the thickened component has recently attracted much attention due to its potential to serve as a diagnostic of the galaxy's early history. Some recent work suggests, however, that the vertical structure of Milky Way stellar populations is consistent with models that build up the thickened component through migration. I discuss these developments in the context of cosmological galaxy formation.

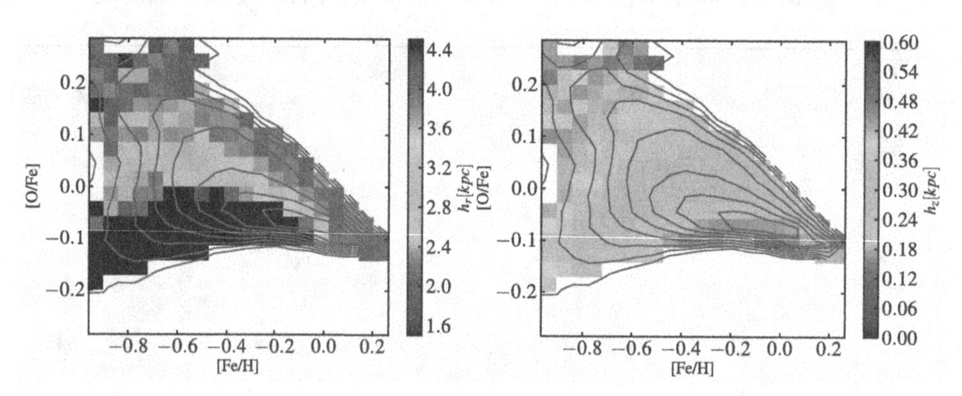

Figure 1. The simulated distribution of stars in the [O/Fe] vs. [Fe/H] plane, coloured by scale length h_r on the left and scale height h_z on the right (from Roškar 2012). The stars lie between $4 < R$ [kpc] < 9 and $|z| < 3$ [kpc]. The smooth transition of the structural parameters in the abundance-metallicity plane is in broad agreement with the results of Bovy *et al.* 2012.

References

Bovy J., *et al.* 2012, *ApJ*, 753, 148
Roškar R. *et al.* 2008, *ApJL*, 675, L65
Roškar R., *et al.* 2010, *MNRAS*, 408, 783
Roškar R., Debattista V. P., & Loebman S. R. 2012, arXiv 1211.1982

Highlights of Astronomy, Volume 16
XXVIIIth IAU General Assembly, August 2012
T. Montmerle, ed.

doi:10.1017/S1743921314011454

Bars in a cosmological context

Marie Martig[1], Katarina Kraljic[2] and Frédéric Bournaud[2]

[1] Centre for Astrophysics & Supercomputing, Swinburne University of Technology, P.O. Box 218, Hawthorn, VIC 3122, Australia
email: mmartig@astro.swin.edu.au

[2]Laboratoire AIM Paris-Saclay, CEA/IRFU/SAp CNRS Université Paris Diderot, 91191 Gif-sur-Yvette Cedex, France

Abstract. We study the properties of bars in a series of zoom cosmological simulations (Martig *et al.* 2012, Kraljic *et al.* 2012). We find that bars are almost absent from galaxies at $z > 1$, and if they form they tend to be quickly destroyed by mergers and instabilities. On the contrary, at $z < 1$ bars are long-lived, and the fraction of barred galaxies rises steadily. Bars are eventually found in $\sim 80\%$ of $z = 0$ spiral galaxies. This redshift evolution is quantitatively consistent with existing data from the COSMOS survey (Sheth *et al.* 2008), although the detectability of bars is presently limited to $z < 0.8$ because of band-shifting and resolution effects. We predict later bar formation in lower-mass galaxies, also in agreement with existing data (e.g., Sheth *et al.* 2012). We actually find that the characteristic epoch of bar formation is the epoch of massive thin disk formation, corresponding to the transition between an early violent phase at $z > 1$ and a later secular phase. Bar formation thus traces the emergence of the disk-dominated morphology of today's spirals.

Keywords. galaxies: bulges, galaxies: evolution, galaxies: formation, galaxies: spiral, galaxies: structure

References

Kraljic, K. Bournaud, F., & Martig M. 2012, *ApJ* 757, 60
Martig, M., Bournaud, F., Croton, D. J., Dekel, A., & Teyssier R. 2012, *ApJ* 756, 26
Sheth, K., *et al.* 2008, *ApJ* 675, 1141
Sheth, K., *et al.* 2012, arXiv:1208.6304

Highlights of Astronomy, Volume 16
XXVIIIth IAU General Assembly, August 2012
T. Montmerle, ed.

doi:10.1017/S1743921314011466

Star formation history: secular processes in "main sequence" galaxies versus merger-driven starbursts

Matthieu Bethermin

CEA Saclay France, email: matthieu.bethermin@cea.fr

Abstract. Some recent works indicate that most star-forming galaxies follow a main sequence in the SFR-stellar mass plane with a surprisingly low scatter of $\approx$0.2 dex, suggesting that the star formation in these objects is driven by secular processes. Nevertheless, Herschel identified a population of starbursting galaxies, probably triggered by mergers, which display a large excess of specific star formation rate (sSFR=SFR/Mstar) compared to the main sequence. We will present a new set of models for the contribution of these two populations to the IR/sub-mm luminosity function, but also to source counts selected at various wavelengths.

Our model is based on the stellar mass function of star-forming galaxies, the distribution of sSFR measured at $z = 2$ and its double-Gaussian decomposition, and the observed evolution of the main sequence in the sSFR-Mass plane as a function of redshift. We found that the non-Schechter bright-end of the LF is due to the starbursting galaxies, which represent only 4% in number density and 15% in luminosity density. This fraction of starbursts is remarkably constant with the redshift at $0 < z < 2$, contrary to naive expectation from hierarchical merging. It thus suggests that the majority of stars in the Universe were formed through secular processes. We will then discuss the contribution of starbursting and main sequence galaxies to the number counts and the selection effects towards starbursts sources for various flux-limited IR/sub-mm samples.

We will also present studies of the clustering properties of the main sequence and starburst galaxies at z 2. These measurements suggest strong links between star formation rate, stellar mass and halo mass in the main sequence galaxies. In addition, we will present some clues suggesting that main sequence and starbursting galaxies follows the same M*-Mhalo relation."

Highlights of Astronomy, Volume 16
XXVIIIth IAU General Assembly, August 2012
T. Montmerle, ed.

doi:10.1017/S1743921314011478

Secular evolution in young galaxies

Bruce G. Elmegreen

IBM T.J. Watson Research Center, 1101 Kitchawan Road, Yorktown Heights, NY 10598 USA
email: bge@us.ibm.com

Abstract. Young galaxies viewed at high redshift have high turbulent velocities, high star formation rates, high gas fractions, and chaotic structures, suggesting wild instabilities during which giant gas clumps form and make stars in their dense regions, stir other disk stars and gas, and transport angular momentum outward with a resulting net mass flow inward (e.g., Ceverino *et al.* 2010). At $z = 1.5$, 40% of star-forming galaxies have significant clumps (Elmegreen *et al.* 2007; Wuyts *et al.* 2012), and in these, 10%-20% of the stellar mass is in clumps that last $\sim$150 Myr (Elmegreen *et al.* 2009; Wuyts *et al.* 2012). The thick disk and bulge in modern galaxies could form in this phase. The similarity in the α/Fe ratio (Meléndez *et al.* 2008), K-giant abundances (Bensby *et al.* 2010) and ages for the Milky Way bulge and thick disk suggest they formed at the same time. High dispersion gas at $z \sim 1.5$ can do this because it makes the young disk thick and the SF clumps big enough to drive fast secular evolution (Elmegreen *et al.* 2006; Genzel *et al.* 2008; Bournaud *et al.* 2009). Local analogues might be present in dynamically young galaxies like BCDs (Elmegreen *et al.* 2012). The high fraction of $z \sim 1.5$ galaxies with massive clumps suggests clump formation is a long-lived phase and that clump torques should last $\sim$ 1 Gyr or more even if individual clumps come and go on shorter timescales. Clump formation may cease when stars finally dominate the disk mass (Cacciato *et al.* 2012).

Keywords. galaxies: bulges, galaxies: dwarf, galaxies: evolution, galaxies: high-redshift

References

Bensby, T., Alves-Brito, A., Oey, M. S., Yong, D., & Meléndez, J. 2010, *A&A* 516, L13
Bournaud, F., Elmegreen, B. G., & Martig, M. 2009, *ApJ* 707, L1
Cacciato, M., Dekel, A., & Genel, S. *MNRAS* 421, 818
Ceverino, D., Dekel, A., & Bournaud, F. 2010, *MNRAS* 404, 2151
Elmegreen, B. G. & Elmegreen, D. M. 2006, *ApJ* 650, 644
Elmegreen, D. M., Elmegreen, B. G., Ravindranath, S., & Coe, D. A. 2007, *ApJ* 658, 763
Elmegreen, B. G., Elmegreen, D. M., Fernandez, M. X., & Lemonias, J. J. 2009, *ApJ* 692, 12
Elmegreen, B. G., Zhang, H.-X., & Hunter, D. A. 2012, *ApJ* 747, 105
Genzel, R. *et al.* 2008, *ApJ* 687, 59
Meléndez, *et al.* 2008, *A&A* 484, L21
Wuyts, S., *et al.* 2012, *ApJ* 753, 114

Highlights of Astronomy, Volume 16
XXVIIIth IAU General Assembly, August 2012
T. Montmerle, ed.

doi:10.1017/S174392131401148X

Hydrodynamical simulations of the barred spiral galaxy NGC 1097

Lien-Hsuan Lin[1], **Hsiang-Hsu Wang, Pei-Ying Hsieh, Ronald E. Taam, Chao-Chin Yang and David C. C. Yen**

[1]Institute of Astronomy and Astrophysics, Academia Sinica, P.O. Box 23-141, Taipei 10617, Taiwan, R.O.C.
email: `lhlin@asiaa.sinica.edu.tw`

Abstract. NGC 1097 is a nearby barred spiral galaxy believed to be interacting with the elliptical galaxy NGC 1097A located to its northwest. It hosts a Seyfert 1 nucleus surrounded by a circumnuclear starburst ring. Two straight dust lanes connected to the ring extend almost continuously out to the bar. The other ends of the dust lanes attach to two main spiral arms. To provide a physical understanding of its structural and kinematical properties, two-dimensional hydrodynamical simulations have been carried out. Numerical calculations reveal that many features of the gas morphology and kinematics can be reproduced provided that the gas flow is governed by a gravitational potential associated with a slowly rotating strong bar. By including the self-gravity of the gas disk in our calculation, we have found the starburst ring to be gravitationally unstable which is consistent with the observation in Hsieh *et al.* (2011). Our simulations also show that gas can flow into the region within the starburst ring even after its formation, leading to the coexistence of both a nuclear ring and a circumnuclear disk.

Keywords. hydrodynamics, accretion, instabilities, galaxies: individual(NGC 1097) — galaxies: evolution — galaxies: spiral — galaxies: structure

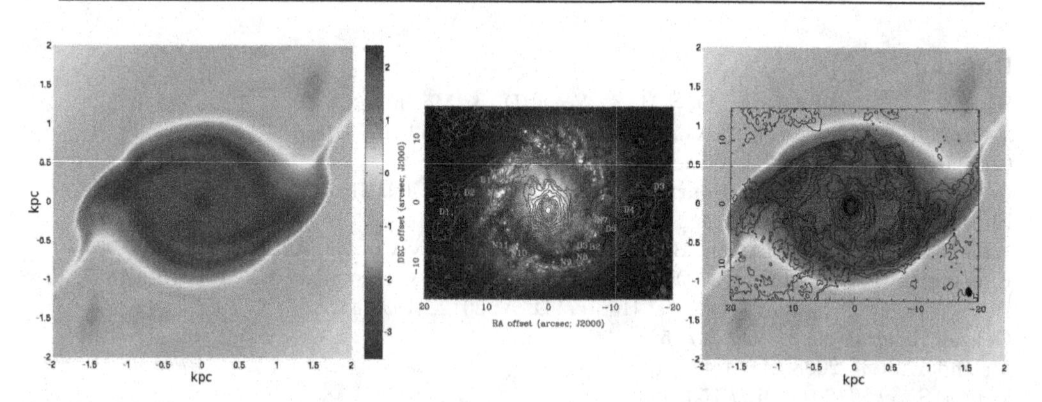

Figure 1. Left: The central part of the projected simulated density map. Middle: ^{12}CO(2-1)-integrated map (contours) overlaid on the *HST I*-band image (grayscale) (Fig. 4 in Hsieh *et al.* 2011). Right: The superposition of the left panel and the ^{12}CO(2-1) emission (contours). The starburst ring in the simulation is of the same size as that in the observation. The two groups of strongest emission knots in the northeast and the southwest parts of the starburst ring in the observation coincide with the densest regions in the simulation.

Reference

Hsieh, Pei-Ying, Matsushita, Satoki, Liu, Guilin, Ho, Paul T. P., Oi & Nagisa; Wu, Ya-Lin 2011, *ApJ* 736, 129

Highlights of Astronomy, Volume 16
XXVIIIth IAU General Assembly, August 2012
T. Montmerle, ed.

doi:10.1017/S1743921314011491

Galaxies in most dense environments at z ∼ 1.4

V. Strazzullo

Irfu/SAp, CEA - Saclay, Orme des Merisiers, 91191 Gif sur Yvette Cedex, France

Abstract. The X-ray luminous system XMMU J2235-2557 at z∼1.4 is among the most massive of the very distant galaxy clusters, and remains a unique laboratory to observe environment-biased galaxy evolution already 9 Gyr ago (Lidman *et al.* 2008, Rosati *et al.* 2009, Strazzullo *et al.* 2010). At a cosmic time when cluster cores start showing evidence of a still active galaxy population, star-forming ($M > 10^{10} M_\odot$) galaxies in XMMU J2235-2557 are typically located beyond ∼250kpc from the cluster center, with the cluster core already effectively quenched and dominated by massive galaxies on a tight red sequence, showing early-type spectral features and bulge-dominated morphologies. While masses and stellar populations of these red-sequence galaxies suggest that they have largely completed their formation, their size is found to be typically smaller that similarly massive early-type galaxies in the local Universe, in agreement with many high-redshift studies. This would leave room for later evolution, likely through non-secular processes, changing their structure to match their local counterparts. On the other hand, uncertainties and biases in the determination of masses and sizes, as well as in the local mass-size relation, and the possible effect of progenitor bias, still hamper a final conclusion on the actual relevance of size evolution for early-type galaxies in this dense high-redshift environment.

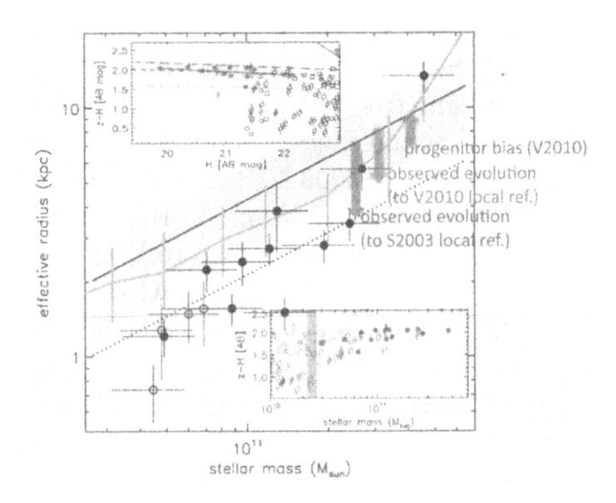

Figure 1. The main panel shows the effective radius vs stellar mass of quiescent early-type galaxies on the red sequence (small insets) of XMMU J2235-2557. Downward arrows show the measured size evolution with respect to the Shen *et al.* (2003) and Valentinuzzi *et al.* (2010) local relations, while the upward arrow shows the effect of progenitor bias as estimated by Valentinuzzi *et al.* (2010). Adapted from Strazzullo *et al.* (2010), see original paper for details.

References

Lidman, C., *et al.* 2008, *A&A* 489, 981
Rosati, P., *et al.* 2009, *A&A* 508, 583
Strazzullo, V., *et al.* 2010, *A&A* 524, 17
Shen, S., *et al.* 2003, *MNRAS* 343, 978
Valentinuzzi, T., *et al.* 2010, *ApJ* 721, L19

Highlights of Astronomy, Volume 16
XXVIIIth IAU General Assembly, August 2012
T. Montmerle, ed.

doi:10.1017/S1743921314011508

ALHAMBRA survey: morphological classification

M. Pović[1], M. Huertas-Company[2], I. Márquez[1], J. Masegosa[1], J. A. López Aguerri[3], C. Husillos[1], A. Molino[1], D. Cristóbal-Hornillos[4] and ALHAMBRA team

[1]Instituto de Astrofísica de Andalucía (IAA-CSIC), Granada, Spain
email: mpovic@iaa.es

[2]GEPI, Paris Observatory, Paris, France

[3]Instituto de Astrofísica de Canarias (IAC), La Laguna, Tenerife, Spain

[4]Centro de Estudios de Física del Cosmos de Aragón (CEFCA), Teruel, Spain

Abstract. The Advanced Large Homogeneous Area Medium Band Redshift Astronomical (ALHAMBRA) survey is a photometric survey designed to study systematically cosmic evolution and cosmic variance (Moles *et al.* 2008). It employs 20 continuous medium-band filters (3500 - 9700 Å), plus JHK near-infrared (NIR) bands, which enable measurements of photometric redshifts with good accuracy. ALHAMBRA covers $> 4\,\mathrm{deg}^2$ in eight discontinuous regions ($\sim 0.5\,\mathrm{deg}^2$ per region), of theseseven fields overlap with other extragalactic, multiwavelength surveys (DEEP2, SDSS, COSMOS, HDF-N, Groth, ELAIS-N1). We detect > 600.000 sources, reaching the depth of R(AB) ∼ 25.0, and photometric accuracy of 2 - 4% (Husillos *et al.*, in prep.). Photometric redshifts are measured using the Bayesian Photometric Redshift (BPZ) code (Benítez *et al.* 2000), reaching one of the best accuracies up to date of $\delta z/z \leqslant 1.2\%$ (Molino *et al.*, in prep.).

To deal with the morphological classification of galaxies in the ALHAMBRA survey (Pović *et al.*, in prep.), we used the galaxy Support Vector Machine code (galSVM; Huertas-Company 2008, 2009), one of the new non-parametric methods for morphological classification, specially useful when dealing with low resolution and high-redshift data. To test the accuracy of our morphological classification we used a sample of 3000 local, visually classified galaxies (Nair & Abraham 2010), moving them to conditions typical of our ALHAMBRA data (taking into account the background, redshift and magnitude distributions, etc.), and measuring their morphology using galSVM. Finally, we measured the morphology of ALHAMBRA galaxies, obtaining for each source seven morphological parameters (two concentration indexes, asymmetry, Gini, M_{20} moment of light, smoothness, and elongation), probability if the source belongs to early- or late-type, and its error. Comparing ALHAMBRA morph COSMOS/ACS morphology (obtained with the same method) we expect to have qualitative separation in two main morphological types for ∼ 20.000 sources in 8 ALHAMBRA fields. For early-type galaxies we expect to recover ∼ 70% and 30-40% up to magnitudes 20.0 and 21.5, respectively, having the contamination of late-types of < 7%. For late-type galaxies, we expect to recover ∼ 70%, 60 - 70%, and ∼ 30% of sources up to magnitudes 22.0, 22.5, and 23.0, respectively, having the contamination of early-types of ⩽ 10%. These data will be used to study the evolution of active and non-active galaxies respect to morphology and morphological properties of galaxies in groups and clusters.

Keywords. surveys, catalog, galaxies: fundamental parameters (morphological classification)

References

Benítez, N., *et al.* 2000, *ApJ*, 536, 571
Huertas-Company, M., *et al.* 2008, *A&A*, 478, 971
Moles, M., *et al.* 2008, *AJ*, 136, 1325
Nair, P. & Abraham, R., 2010, *ApJS*, 186, 427

Highlights of Astronomy, Volume 16
XXVIIIth IAU General Assembly, August 2012
T. Montmerle, ed.

doi:10.1017/S174392131401151X

Testing galaxy formation models with the GHOSTS survey: The stellar halo of M81

A. Monachesi[1], E. Bell[1], D. Radburn-Smith[2], M. Vlajić[3], R. de Jong[3], J. Bailin[1], J. Dalcanton[2], B. Holwerda[4] and D. Streich[3]

[1]Department of Astronomy, University of Michigan, 830 Dennison Bldg., 500 Church St., Ann Arbor, MI 48109, USA, [2]Department of Astronomy, University of Washington, Seattle, WA 98195, USA, [3]Leibniz-Institut für Astrophysik Potsdam, D-14482 Potsdam, Germany, [4]ESTEC, Keplerlaan 1, 2200 AG Noordwijk, The Netherlands. Email: antonela@umich.edu

Abstract. The GHOSTS survey is the largest study to date of the resolved stellar populations in the outskirts of disk galaxies (Radburn-Smith *et al.* 2011). The sample currently consists of 16 nearby disk galaxies, whose outer disks and halos are imaged with the Hubble Space Telescope (HST). I will present new results obtained from the study of 19 GHOSTS fields in M81's outermost part. The observed fields probe the stellar halo of M81 out to projected distances of ~50 kpc, an unprecedented distance for halo studies outside the Local Group. The 50% completeness levels of the color magnitude diagrams are typically at 2.5 mag below the tip of the red giant branch. When considering only fields located at galactocentric radius $R > 15$ kpc, we detect no color gradient in the stellar halo of M81. We compare these results with model predictions for the colors of stellar halos formed purely via accretion of satellite galaxies (Bullock & Johnston 2005). When we analyze the cosmologically motivated models in the same way as the HST data, we find that they predict no color gradient for the stellar halos, in good agreement with the observations (see Fig. 1).

Keywords. galaxies: halos, galaxies: individual (M81), galaxies: spiral, galaxies: formation

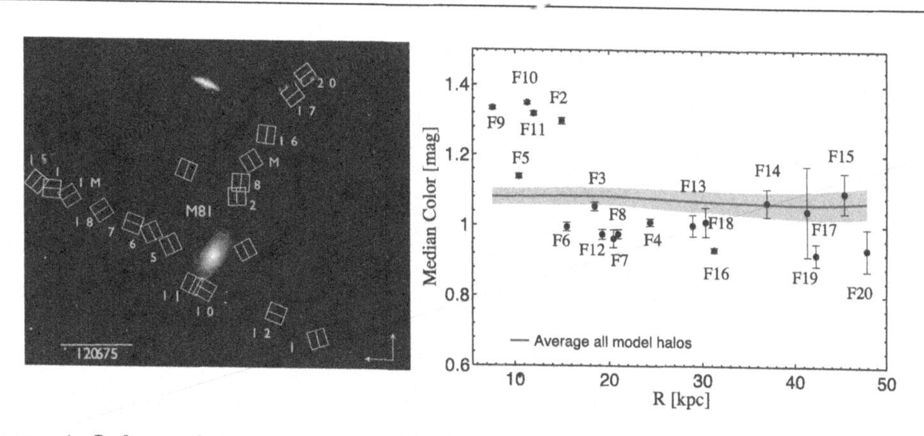

Figure 1. Left panel: Location of the 19 observed HST/ACS GHOSTS fields. Right panel: Median color profile of the M81 observed fields (dots). There is no color gradient for fields at $R > 15.5$ kpc, which we assume here to be M81's halo populations. The solid line shows the average color profile of the halo models analyzed and the shaded area indicates the 1σ model-to-model deviations.

References

Bullock, J. S. & Johnston, K. V., 2005, *ApJ* 635, 931
Radburn-Smith, D. J., *et al.* 2011, *ApJS* 195, 18

Highlights of Astronomy, Volume 16
XXVIIIth IAU General Assembly, August 2012
T. Montmerle, ed.

doi:10.1017/S174392131500160X

Preface: Special Session SpS4 New era for studying interstellar and intergalactic magnetic fields

Magnetic fields dominate the universal energy balance on a wide variety of spatial scales; preserving life on Earth from extinction by cosmic rays, regulating star formation in giant molecular clouds, regulating the enrichment of the intergalactic medium by galactic winds and possibly regulating the growth of individual galaxies and filaments of galaxies. The structure of magnetic fields is determined by ubiquitous astrophysical turbulence and critically affects transport processes, including propagation and acceleration of cosmic rays and transfer of heat. Turbulent magnetic fields play an important role in magnetic field generation via dynamo processes, and must be understood to separate galactic foregrounds from the Cosmic Microwave Background signal.

Despite their importance and ubiquity, magnetic fields remain one of the most poorly understood components of the cosmos due to the challenges involved in both their measurement and theoretical description.

Recent years have been marked by two significant developments. First of all, advances in instrumentation, both in the more traditional radio frequency portion of the spectrum, and in the sub-mm and even the near-IR and optical, are dramatically advancing our knowledge of the incidence, strength and topology of magnetic fields in astrophysics. For the first time, there is clear evidence for an all-pervasive intergalactic magnetic fields according to HESS/FERMI observations. Second, but equally important, significant advances in numerical techniques provided possibilities to simulate magnetized plasmas with realistic turbulent structure and to test theoretical models of how turbulent magnetic fields interact with fully and partially ionized gases and cosmic rays. Combined with progress in the development of techniques to compare numerical and observational data, this makes the field ripe for a breakthrough in understanding of astrophysical magnetic fields, their properties and effects on key astrophysical processes.

At the same time, advances in numerical simulations are providing a greatly improved context in the interpretation of the observations and testing theoretical predictions. The tremendous increase of computational power helps to describe the complex processes of magnetic field generation, evolution and its effects on astrophysical processes.

Recent insights emerging from studies addressing magnetism on a wide range of specific scales and employing a wide range of techniques call for closer synergetic interactions of observers and theorists in order to synthesize a deeper understanding of the astrophysical magnetic phenomenon. Observers from various different wavelength ranges, theorists and computational astrophysicists converge to discuss the results in hand and plan for a new epoch of observing interstellar and intergalactic magnetic fields.

Given the very wide range of new data and results to present and discuss, there have been a lot of interest presentattions, including 31 invited talks, 21 talks and 23 posters in 8 sessions. All these sessions were well attended by assmebly participants, because the meeting room of 150 seats is always very full.

We would like to thank the IAU for supporting this Special Session and all the members of the Scientific Organizing Committee:

Beck, Rainer (Max-Planck Institute for Radioastronomy, Germany);
Braun, Robert (CSIRO Astronomy and Space Science, Australia. Co-Chair);
Brown, Jo-Anne (University of Calgary, Canada);
de Gouveia Dal Pino, Elisabete (Universidade de Sao Paulo, Brazil);
Ensslin, Torsten (MPI fuer Astrophysik, Germany);
Feretti, Luigina (INAF Istituto di Radioastronomia, Italy);
Gaensler, Bryan M. (University of Sydney, Australia);
Han, JinLin (National Astronomical Observatories, China. Chair);
Haverkorn, Marijke (ASTRON, Netherlands, co-Chair);
Troland, Tom (University of Kentucky, USA) ;
Lazarian, Alex (University of Wisconsin-Madison, USA) ;
Magalhaes, Antonio Mario (Universidade de Sao Paulo, Brazil);
Novak, Giles (Northwestern University, USA);
Ostriker, Eve (University of Maryland College Park, USA);
Ryu, Dongsu (Chungnam National University, Korea);
Subramanian, Kandaswamy (Inter-University Centre for A&A, India).

JinLin Han, Robert Braun, and Marijke Haverkorn, co-chairs SOC

Highlights of Astronomy, Volume 16
XXVIIIth IAU General Assembly, August 2012
T. Montmerle, ed.

doi:10.1017/S1743921314011521

The history of polarisation measurements: their role in studies of magnetic fields

R. Wielebinski

Max-Planck-Institut für Radioastronomie, Auf dem Hügel 69, 53121 Bonn, Germany

Radio astronomy gave us new methods to study magnetic fields. Synchrotron radiation, the main cause of comic radio waves, is highly linearly polarised with the 'E' vector normal to the magnetic field. The Faraday Effect rotates the 'E' vector in thermal regions by the magnetic field in the line of sight. Also the radio Zeeman Effect has been observed.

At first radio sources were shown to have a non-thermal spectrum. Solar observers detected polarisation in earliest observations. The 'Rosetta stone' of the magnetic fields was the Crab Nebula. The prediction that the Crab Nebula was a synchrotron source was confirmed first in 1954 by optical polarisation observations and in 1957 at radio frequencies. The year 1962 was a great year for magnetic field research: the polarisation of the Galaxy was detected as well as the polarisation of radio galaxies. In 1968 pulsars were discovered, highly polarised sources. Both pulsars and compact radio sources can be used in conjunction with the Faraday Effect to probe interstellar medium. Also in 1968 the detection of the first radio Zeeman Effect was announced.

In the 1970s several new large radio telescopes became operational. All of these telescopes were capable of polarisation observations. The Westerbork radio telescope first detected polarisation in the galaxy M51. The Effelsberg 100-m dish was ideal for studying the diffuse polarised emission of the Galaxy, galaxies and radio galaxies. The Ryle telescope and the VLA made polarisation maps radio galaxies. For some time only the Parkes radio telescope was available for polarisation studies of southern sources. The Australia Telescope came later with good polarisation mapping facilities.

Polarised sources were used to delineate magnetic fields of the Galaxy. Pulsars were favoured for such studies. More recently new data became available from synthesis instruments for extragalactic sources along the Galactic plane to allow more detailed studies. Also the increase of the numbers of Rotation Measures at high Galactic latitudes allow studies of the halo magnetic field.

All-sky surveys in polarisation were made by the WMAP satellite at high radio frequencies (from 23 GHz to 94GHz) with medium angular resolution. Also a 1.4 GHz all-sky polarisation survey was made. High resolution low frequency polarisation observations showed rapid variation of the 'E' vector implying high rotation measures leading to complete depolarisation. Zeeman Effect observations have been extended to various molecular species, some at mm-wavelengths. The advent of new digital devices led to the development of the Rotation Measure synthesis. This development allows the search in depth of interstellar magnetic field features.

The future of magnetic field research is in good hands. All the important new 'mega-telescopes' have magnetism as key science. The first E-VLA polarization map of NGC 4631 has been published. The LOFAR and ALMA radio telescopes have polarisation capabilities. The advent of the two Square Kilometre Array pathfinders ASCAP and MeerKAT, followed by the final SKA array one day will enhance greatly the studies of cosmic magnetic fields.

Highlights of Astronomy, Volume 16
XXVIIIth IAU General Assembly, August 2012
T. Montmerle, ed.

doi:10.1017/S1743921314011533

Characterizing the correlation between column density structure and magnetic fields

J. D. Soler[1], P. Hennebelle[2], P. G. Martin[3], M. A. Miville-Deschenes[4], B. Netterfield[1] and The BLASTpol Collaboration[5]

[1]Dept. of Astronomy and Astrophysics, University of Toronto, Toronto, ON M5S 3H4, Canada
email: soler@astro.utoronto.ca, [2]Lab. de Radioastronomie, ENS et Observatoire de Paris, 75231 Paris, France [3]CITA, Toronto, ON M5S 3H8, Canada [4]Institut d'Astrophysique Spatiale, CNRS/Universit Paris-Sud 11, 91405 Orsay, France [5]http://blastexperiment.info

Abstract. We present a statistical tool to characterize correlation between the column density structure and the polarized emission from a molecular cloud. This tool uses the gradient as an estimator of the directionality of the structure in order to systematically relate the orientation of filaments and cores to the orientation of the magnetic field inferred from the polarization vectors.

Keywords. ISM: magnetic fields, ISM: clouds, polarization, submillimeter

---

The importance of magnetic fields in the process of star formation is not yet determined (Crutcher, 2012). Recent studies try to relate the morphology of the magnetic field in molecular clouds to the processes which lead to the formation of filaments and subsequently to the formation of stars (e.g. Sugitani, *et al.* 2011 and Girart, *et al.* 2006). In this context, we introduce the Histograms of Oriented Gradients (HOGs), a statistical tool to quantitatively compare the orientation of structures with respect to the magnetic field (**B**) inferred from dust polarization observations. We developed this tool by characterizing 3D data cubes and 2D maps from simulations of molecular clouds. The results of the analysis are histograms of the angles between **B** and the gradient of the density ($\nabla\rho$) or column density ($\nabla\Sigma$). We calculate the gradient using Gaussian derivative kernels to sample different scales in the map and make the measurement robust against noise. Additionally we segment the cubes or the maps by intervals of density and magnetic field intensity.

Conclusion

We applied HOGs to column density and polarization maps built from simulations, assuming perfect grain alignment. We find two distinct modes of orientation between **B** and filament direction correlated with the column density.

After producing simulated BLASTPol observations (Pascale, *et al.* 2008) of our polarization maps, we show this correlation should be measurable with BLASTPol. Potentially, this method can also be applied to the observations of other submillimeter instruments such as Planck, POL-2 and PILOT.

References

Crutcher, R. M., 2012, *ARA&A* 50, 29C
Girart, J. M., Rao, R., & Marrone, D. P., 2006, *Science* 313, 812G
Pascale, E., *et al.* 2008, *ApJ* 681, 400
Sugitani, E., *et al.* 2011, *ApJ* 734, 63S

Highlights of Astronomy, Volume 16
XXVIIIth IAU General Assembly, August 2012
T. Montmerle, ed.

doi:10.1017/S1743921314011545

Dust properties and magnetic field geometry towards LDN 1570

C. Eswaraiah, G. Maheswar and A. K. Pandey

Aryabhatta Research Institute of Observational Sciences, Manora Peak, Nainital 263 129, India

Abstract. We have performed both optical linear polarimetric and photometric observations of an isolated dark globule LDN 1570 aim to study the dust polarizing and extinction properties and to map the magnetic field geometry so as to understand not only the importance of magnetic fields in formation and evolution of clouds but also the correlation of the inferred magnetic field structure with the cloud structure and its dynamics. Dust size indicators (R_V and λ_{max}) reveal for the presence of slightly bigger dust grains towards the cloud region. The inferred magnetic field geometry, which closely follows the cloud structure revealed by *Herschel* images, suggest that the cloud could have been formed due to converging material flows along the magnetic field lines.

Keywords. (ISM:) dust, extinction, ISM: magnetic fields, techniques: polarimetric, ISM:clouds, ISM: individual-LDN 1570

---

1. Results

LDN 1570 (Lynds, 1962) is same as Barnard 227 (Lynds, 1962) and CB44 (Clemens & Barvainis, 1998) which is a isolated small dark globule located at a distance of ~400 pc (Stutz *et al.*, 2009). Small dark globules are the unique laboratories to study the early evolutionary stages that precede core collapse and subsequent star formation (Kane *et al.* 1995). The mean values of P_{max} and λ_{max} are found to be 3.27±1.02 per cent and 0.59±0.08 μm respectively, slightly higher than the value, 0.545 μm, corresponding to the general ISM. The value of R_V estimated using the λ_{max} is found to be 3.3±0.5. Using $(V-I)$, $(V-J)$, $(V-H)$, $(V-K)$ vs. $(B-V)$ two-colour diagrams, we evaluated the value of R_V as 3.64±0.01. The R_V values derived from both these methods show that the grain size in LDN 1570 is slightly bigger that those found in the diffuse ISM. The magnetic field geometry of LDN 1570 seems to follow the large scale structure seen in the 250 μm image produced by the *Herschel.* Towards the southern parts the field seems to be almost parallel to the Galactic parallel ($b = -0.62°$) whereas towards the northern parts the field lines are bend by ≈20° towards the Galactic plane ($b = 0°$). The filamentary structure seen towards the northern condensation are found to be aligned with the magnetic field lines. Based on the morphology of the magnetic field lines with respect to the cloud structure, we believe that LDN 1570 could have formed due to the converging flow of material along the field lines.

References

Clemens, D. P. & Barvainis, R. 1988, *ApJS*, 68, 257
Kane, B. D., Clemens, D. P., Leach, R. W., & Barvainis, R. 1995, *ApJ*, 445, 269
Lynds, B. T. 1962, *ApJS*, 7, 1
Stutz, A. M., Rieke, G. H., Bieging, J. H., *et al.* 2009, *ApJ*, 707, 137

Highlights of Astronomy, Volume 16
XXVIIIth IAU General Assembly, August 2012
T. Montmerle, ed.

doi:10.1017/S1743921314011557

Structure and Dynamics of Magnetized Dark Molecular Clouds

P. S. Li[1], C. F. McKee[1,2] and R. I. Klein[1,3]

1Astronomy Department, University of California, Berkeley, CA 94720, USA email: psli@berkeley.edu
2Physics Department, University of California, Berkeley, CA 94720, USA
3Lawrence Livermore National Laboratory, P.O. Box 808, L-23, Livermore, CA 94550, USA

Massive infrared dark clouds (IRDCs) are believed to be the precursors to star clusters and massive stars (e.g. Bergin & Tafalla 2007). The supersonic, turbulent nature of molecular clouds in the presence of magnetic fields poses a great challenge in understanding the structure and dynamics of magnetized molecular clouds and the star formation therein. Using the high-order radiation-magneto-hydrodynamic adaptive mesh refinement (AMR) code ORION2 (Li *et al.* 2012), we perform a large-scale driven-turbulence simulation to reveal the 3D filamentary structure and dynamical state of a highly supersonic (thermal Mach number = 10) and strongly magnetized (plasma $\beta = 0.02$) massive infrared dark molecular cloud. With the high resolution afforded by AMR, we follow the dynamical evolution of the cloud in order to understand the roles of strong magnetic fields, turbulence, and self-gravity in shaping the cloud and in the formation of dense cores.

The most massive dark cloud filament at half a free-fall time in the simulation has a length of about $4 - 5$ pc with a mass of about $600 - 650 M_\odot$ ($\sim 20\%$ of the total mass in the simulated region). The mean density n(H) and column density N(H) are $\sim 2 \times 10^4$ cm^{-3} and $\sim 4 \times 10^{22}$ cm^{-2}, respectively. The velocity dispersion is $1 \sim 1.5$ km s^{-1}. We observe a very complex magnetic field structure around the cloud filament, with most magnetic field lines piercing through the dark cloud filament in a direction roughly normal to the filament axis, consistent with polarization observations of filamentary IRDCs (e.g. Pereyra & Magalhães 2004, Alves *et al.* 2008). The column density profiles of the cloud filament are well fitted by the power law $N \propto r^{-k_\rho}$ with $k_\rho \sim 0.6 - 0.8$, as observations have found (e.g. Hill *et al.* 2011, Arzoumanian *et al.* 2011). We use RADMC-3D (Dullemond 2012) to generate molecular line emission images. Emission line profiles show both single and multiple line peaks and appear to be consistent with the braided filaments observed in dark clouds (e.g. Moriarty-Schieven *et al.* 1997, Hill *et. al.* 2011). We also observe several gravitationally collapsing cores that show a disklike structure with aspect ratios as high as $4 - 5$. The maximum density-weighted line-of-sight, the volume-weighted root-mean-square, and the highest magnetic field strength in the dense cores in our simulation are about 0.2, 0.4, and 5 mG, respectively.

References

Alves, F. O., Franco, G. A. P., & Girart, J. M. 2008, *A&A* 486, 13
Arzoumanian, D., Andre, P., Didelon, P., Konyves, V., Schneider, N., *et al.* 2011, *A&A* 529, L6
Bergin, E. A. & Tafalla, M. 2007, *ARAA* 45, 339
Dullemond, C. P. 2012 http://www.ita.uni-heidelberg.de/ dullemond/software/radmc-3d
Hill, T., Motte, F., Didelon, P., Bontemps, S., Minier, V., *et al.* 2011, *A&A* 533, A94
Li, P. S., Martin, D., Klein, R. I., & McKee, C. F. 2012, *ApJ* 745, 139
Moriarty-Schieven, G. H., Anderson, B. G., & Wannier, P. G. 1997, *ApJ* 475, 642
Pereyra, A. & Magalhães, A. M. 2004, *ApJ* 603, 584

Highlights of Astronomy, Volume 16
XXVIIIth IAU General Assembly, August 2012
T. Montmerle, ed.

doi:10.1017/S1743921314011569

Near-infrared Polarimetry and Interstellar Magnetic Fields in the Galactic Center

S. Nishiyama[1], H. Hatano[2], T. Nagata[3] and M. Tamura[1]

[1]National Astronomical Observatory of Japan, Mitaka, Tokyo 181-8588, Japan

[2]Department of Astrophysics, Nagoya University, Nagoya 464-8602, Japan

[3]Department of Astronomy, Kyoto University, Kyoto 606-8502, Japan
email: shogo.nishiyama@nao.ac.jp

Abstract. We present a large-scale view of the magnetic field (MF) in the central $3^\circ \times 2^\circ$ region of our Galaxy. There is a smooth transition of the large-scale MF configuration in this region.

Keywords. infrared: ISM, techniques: polarimetric, magnetic fields

---

We have carried out polarimetric observations using the near-infrared polarimetric camera SIRPOL on the 1.4 m telescope IRSF, and have obtained a large-scale view of the magnetic field in the central $3^\circ \times 2^\circ$ region of our Galaxy. We find that near the Galactic plane, the magnetic field is almost parallel to the Galactic plane (i.e., *toroidal* configuration) but at high Galactic latitudes ($|b| > 0.4^\circ$), the magnetic field is nearly perpendicular to the plane (i.e., *poloidal* configuration). For more detail, see Nishiyama *et al.* (2009), Nishiyama *et al.* (2010).

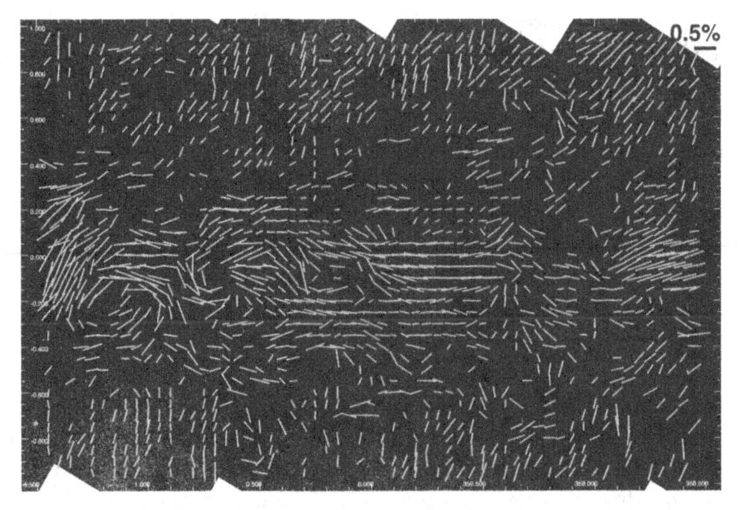

Figure 1. Near-infrared (J, H, K_S) mosaic image of the Galactic center region covering $3^\circ \times 2^\circ$ in the Galactic coordinate. Observed directions of the magnetic fields *at* the Galactic center are also plotted with cyan bars whose length indicates the degree of polarization in the K_S band.

References

Nishiyama, S., *et al.* 2009, *ApJ*, **690**, 1648
Nishiyama, S., *et al.* 2010, *ApJ*, **722**, L23

Highlights of Astronomy, Volume 16
XXVIIIth IAU General Assembly, August 2012
T. Montmerle, ed.

doi:10.1017/S1743921314011570

Intense velocity-shears, magnetic fields and filaments in diffuse gas

Edith Falgarone[1], Pierre Hily-Blant[2], François Levrier[1], Manuel Berthet[1], Pierre Bastien[3] and Dan Clemens[4]

[1]LERMA/LRA, Ecole Normale Supérieure & Observatoire de Paris, 24 rue Lhomond, 75005 Paris, France
email: edith.falgarone@ens.fr

[2]IPAG, Observatoire de Grenoble, St Martin d'Hères, France

[3]University of Montréal, Montréal, Canada

[4]Institute for Astrophysical Research, Boston University, Boston, USA

Abstract. The dissipation of turbulence is a key process in the evolution of diffuse gas towards denser structures. The vast range of coupled scales and the variety of dissipative processes in interstellar turbulence make it a complex system to analyze. Observations now provide powerful statistics of the gas velocity field, density and magnetic field orientations, opening a rich field of investigation. On-going comparisons of the orientation of intense velocity-shears, magnetic field and tenuous filaments of matter in a turbulent high-latitude cloud are promising.

Keywords. Turbulence, Magnetic Fields, ISM: molecules, ISM: kinematics and dynamics, ISM: general, ISM: evolution

---

Turbulent dissipation is intermittent in space and time and it is now identified in the ISM via the non-Gaussian statistics of the velocity field. Parsec-scale coherent structures of intense velocity–shears have been found in the CO line emission of a diffuse molecular cloud in the Polaris Flare (Hily-Blant *et al.* 2008). One of them is structured into narrow shear-layers down to the milliparsec–scale with straight projections on the plane of the sky (POS) and widely different orientations (Falgarone *et al.* 2009). Each of the main shear orientations in the mpc-field can be found in the pc-scale structures.

The POS projections of the magnetic field have been studied via the polarization of starlight: (i) at the 30 pc-scale, in the visible, with the Beauty and the Beast polarimeter. The distribution of the 50 position–angles (PA) in the 30 pc field is also broad, (ii) at the 0.1 pc–scale with the Mimir polarimeter (Clemens *et al.*, 2012) in the near IR. Interestingly the 7 measured polarization PAs cover the same broad range as those of the 30 pc field, and a few of them are parallel to the local structure of intense velocity-shear. Last, the orientations of the striations in dust emission maps of the large scale field have also been studied. The remarkable similarity of the three PA distributions (magnetic field, velocity-shears and dust filaments) supports a close connection not only between large and small scales but also between the topology of **B** with that of the most dissipative structures and tenuous filaments of matter. Recent *Herschel*/SPIRE observations further support this view.

References

Clemens, D. P., Pinnick, A. F. & Pavel, M. D., 2012 *ApJS*, 200, 20
Falgarone, E., Pety, J., & Hily-Blant, P., 2009, *A&A*, 507, 355
Hily-Blant, P., Falgarone, E. & Pety, J., 2008 *A&A* 481, 367

Highlights of Astronomy, Volume 16
XXVIIIth IAU General Assembly, August 2012
T. Montmerle, ed.

doi:10.1017/S1743921314011582

Effects of Magnetic Fields on Bar Substructures in Barred Galaxies

Woong-Tae Kim

Center for the Exploration of the Origin of the Universe (CEOU), Department of Physics & Astronomy, Seoul National University, Seoul 151-742, Republic of Korea
email: wkim@astro.snu.ac.kr

Abstract. To study the effects of magnetic fields on the properties of bar substructures, we run two-dimensional, ideal MHD simulations of barred galaxies under the influence of a non-axisymmetric bar potential. In the bar regions, magnetic fields reduce density compression in the dust-lane shocks, while removing angular momentum further from the gas at the shocks. This evidently results in a smaller and more distributed ring, and a larger mass inflows rate to the galaxy center in models with stronger magnetic fields. In the outer regions, an MHD dynamo due to the combined action of the bar potential and background shear operates, amplifying magnetic fields near the corotation resonance. In the absence of spiral arms, the amplified fields naturally shape into trailing magnetic arms with strong fields and low density. The reader is refereed to Kim & Stone (2012) for a detailed presentation of the simulation outcomes.

Keywords. magnetohydrodynamics, galaxies: ISM, galaxies: spiral, shock waves

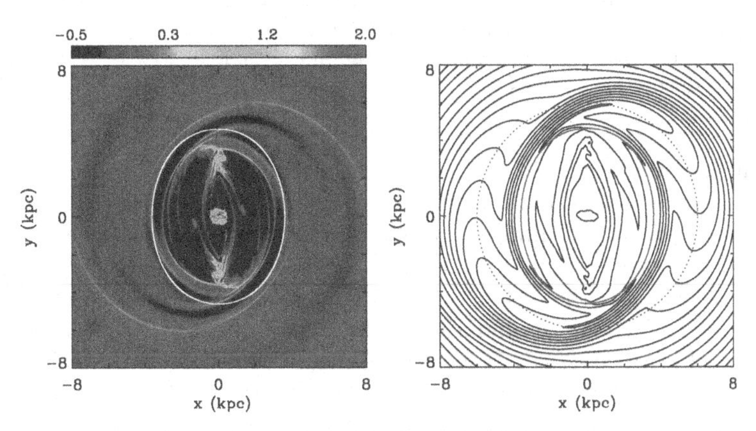

Figure 1. Distribution of density in logarithmic scale (left) and magnetic fields (right) at $t = 0.3$ Gyr in the $\beta = 3$ model, where β is the plasma parameter. The solid oval represents the outermost x_1-orbit, relative to which gas responses are different dramatically. The dotted line in the left panel draws the corotation radius, around which magnetic fields are amplified.

Acknowledgements

This work was supported by the National Research Foundation of Korea (NRF) grant funded by the Korean government (MEST), No. 2010-0000712.

Reference

Kim, W.-T. & Stone, J. M. 2012, *ApJ*, 751, 124

Highlights of Astronomy, Volume 16
XXVIIIth IAU General Assembly, August 2012
T. Montmerle, ed.

doi:10.1017/S1743921314011594

Magnetic Field Structure in Molecular Clouds by Polarization Measurements

W. P. Chen[1], B. H. Su[1], C. Eswaraiah[2], A. K. Pandey[2], C. W. Wang[3], S. P. Lai[3], M. Tamura[4] and S. Sato[5]

[1]Graduate Institute of Astronomy, National Central University, Taiwan
[2]Aryabhatta Research Institute of Observational Sciences, India
[3]Graduate Institute of Astronomy, National Tsing Hua University, Taiwan
[4]National Astronomical Observatory of Japan, Japan
[5]Department of Astrophysics, Nagoya University, Japan

Abstract. We report on a program to delineate magnetic field structure inside molecular clouds by optical and infrared polarization observations. An ordered magnetic field inside a dense cloud may efficiently align the spinning dust grains to cause a detectable level of optical and near-infrared polarization of otherwise unpolarized background starlight due to dichroic extinction. The near-infrared polarization data were taken by SIRPOL mounted on IRSF in SAAO. Here we present the SIRPOL results in RCW 57, for which the magnetic field is oriented along the cloud filaments, and in Carina Nebula, for which no intrinsic polarization is detected in the turbulent environment. We further describe TRIPOL, a compact and efficient polarimer to acquire polarized images simultaneously at g', r', and i' bands, which is recently developed at Nagoya University for adaption to small-aperture telescopes. We show how optical observations probe the translucent outer parts of a cloud, and when combining with infrared observations probing the dense parts, and with millimeter and submillimeter observations to sutdy the central embedded protostar, if there is one, would yield the magnetic field structure on different length scales in the star-formation process.

We present near-infrared JHKs imaging polarimetry of RCW 57A (NGC 3576) and the Carina Nebula (NGC 3372), both among the the brightest Galactic H II nebulae, with a wealth of massive stars and infrared excess stars, suggestive of recent and ongoing star formation. By measuring the polarization of background stars seen through the molecular clouds, the magnetic field structure in the clouds can be diagnosed. In the central part of the Carina Nebula, around Eta Carina where a cavity has been created, only a moderate level of polarization is measured, mainly by the general Galactic magnetic field. In contrast, in RCW 57A, which is associated with copious molecular clouds, we infer an hour-glass shaped field that governs the cloud morphology.

An optical imaging polarimeter has been recently acquired at Lulin Observatory. The Triple-Range Imaging Polarimeter (TRIPOL) uses dichroic mirrors to split the beam into three SBIG ST-9XE camera detectors and a wire-grid polarizer, providing an efficient, flexible, compact, and economic solution for polarization measurements with small telescopes. The ST-9XE camera uses has 512 20-micron pixels on a side, rendering a field-of-view of 4.4 arcmin when adapting to the 1-m telescope. We show a variety of TRIPOL measurements of T Tauri stars, classical Be stars, AGNs, and solar system bodies, along with polarized and unpolarized standard stars. A few Bok globules have been observed to delineate the magnetic field geometry in the periphery of a cloud, which will be corroborated with infrared results to probe the inner part of the cloud, and with submillimeter data to study the field structure close in to the embedded protostar.

Highlights of Astronomy, Volume 16
XXVIIIth IAU General Assembly, August 2012
T. Montmerle, ed.

doi:10.1017/S1743921314011600

Magnetic field components analysis of the SCUPOL 850 microns polarization data catalog

Frédérick Poidevin, Diego Falceta-Gonçalves, Grzegorz Kowal, Elisabete De Gouveia Dal Pino and Antonio-Mário Magalhães

University College London, Kathleen Lonsdale Building, Department of Physics & Astronomy, Gower Place, London WC1E 6BT, United Kingdom
email: Poidevin@star.ucl.ac.uk

Abstract. The SCUPOL catalog is a compilation of 83 regions that were observed at the JCMT between 1997 and 2005. For sufficiently sampled maps, we conduct an analysis for characterizing the polarization and magneto-turbulent properties of the observed regions. The same analysis is done on 2D polarization maps produced by 3D MHD 1024 pixel grid simulations that have been scaled on a sample of observed maps. Each scaled MHD cube of simulated data is used to calculate the mean turbulent regime of each observed regions.

Keywords. ISM: magnetic fields, ISM: molecular clouds, polarimetry, submm, MHD: simulations

---

We present an analysis of the SCUPOL catalog produced by Matthews *et al.* (2009). For sufficiently sampled maps of star-forming regions, inferred parameters (<p>, γ, b) are estimated to characterize the polarization properties, the depolarization properties and the turbulent-to-mean magnetic field ratio of each region as seen on the plane-of-sky. Statistical studies show no specific correlation of each parameters with each other suggesting that they provide information of different nature about the region considered, at least at the 14″ resolution of the observations. Similar set of parameters are calculated from 2D polarization maps produced with 3D MHD 1024 pixel grid simulations (see Falceta-Gonçalves *et al.* (2008)) performed for different MHD regimes. Such MHD regimes have been estimated for 4 regions with 3D MHD cubes properly scaled on the observed maps. Our results shown in Tab. 1 and those obtained by Crutcher (1999)'s analysis of Zeeman Measurements toward subregions of the clouds are consistent within a factor 2.

Table 1. Description of the simulations - MHD, 1024^3.

Model	$M_S^{(a)}$	$M_A^{(b)}$	Object Name
1.........	2.0	0.3	S106
2.........	3.0	0.5	OMC-2/3
3.........	5.0	0.7	W49
4.........	5.0	0.7	DR21

Notes: [a] Sonic Mach Number. [b] Alfvénic Mach Number.

References

Crutcher, R. 1999, *ApJ*, 520, 706
Falceta-Gonçalves, D., Lazarian, A., & Kowal, G. 2008, *ApJ*, 679, 537-551
Matthews, B. C., McPhee, C., Fissel, L., & Curran, R. L. 2009, *ApJSS*, 182, 143

Highlights of Astronomy, Volume 16
XXVIIIth IAU General Assembly, August 2012
T. Montmerle, ed.

doi:10.1017/S1743921314011612

Magnetic field morphologies at mpc scale

Ya-Wen Tang[1,2], Patrick M. Koch[3], Paul T. P. Ho[3,4], Stephane Guilloteau[1,2] and Anne Dutrey[1,2]

[1]LAB, France; [2]CNRS, France; [3] ASIAA, Taiwan; [4] CfA, USA

We report our new results of the magnetic field (B) morphologies toward W51 North,traced with the linear polarization of the dust continuum at wavelengths of 870 μm. The B morphologies are resolved with an angular resolution of typically 1" using the Submillimeter Array (SMA). Dense structures with a number density 10^5 to 10^7 cm^{-3} are traced. In comparison, the B morphologies of sources at different evolutionary stages, from the collapsing core in W51 e2 (Tang *et al.* 2009a) and part of Orion BN/KL (Tang *et al.* 2010) to the ultra-compact HII region G5.89-0.39 (Tang *et al.* 2009b) clearly exhibit different morphologies, likely suggesting different roles of the B fields at different stages.

In the W51 North region we analyze field structures at three different physical scales (Tang *et al.* 2012). In a sequence of increasingly higher resolution observations - from CSO/JCMT single dish at 2 pc to the SMA highest resolution at about 10 mpc - it becomes manifest how the field morphologies change from the envelope surface layer to the inner core and disk. Structures vary from uniform to cometary and hourglass-like. We quantify these changes, providing evidence that the interplay of the B field with other forces, such as gravity, evolves with scale. Additionally, new analysis methods to interpret these observational results and to derive B field strength maps are also discussed (Koch *et al.* 2012a,b,c).

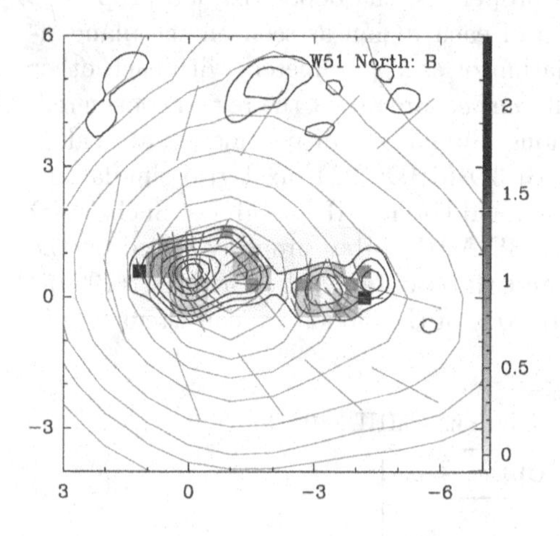

Figure 1. The figure in the left presents the B field morphologies observed in two different SMA configurations, which trace the B fields at two different scales. The dust continuum emission traced at these two scales is shown in contours. The magnetic-field-to-gravity force ratio map, derived from our newly developed method, is shown in grey scale. The ratio is smaller toward the denser regions, suggesting a varying role of the B field as a function of scale.

References

Koch, P. M., Tang, Y.-W., & Ho, P. T. P. 2012a, *ApJ*, 747, 79
Koch, P. M., Tang, Y.-W., & Ho, P. T. P. 2012b, *ApJ*, 747, 80
Koch, P. M., Tang, Y.-W., & Ho, P. T. P. 2012c, *ApJ*, submitted
Tang, Y.-W., Ho, P. T. P., Girart, J. M., *et al.* 2009a, *ApJ*, 695, 1399
Tang, Y.-W., Ho, P. T. P., Koch, P. M., *et al.* 2009b, *ApJ*, 700, 251
Tang, Y.-W., Ho, P. T. P., Koch, P. M., *et al.* 2010, *ApJ*, 717, 1262
Tang, Y.-W., Ho, P. T. P., Koch, P. M., *et al.* 2012, *ApJ*, submitted

Highlights of Astronomy, Volume 16
XXVIIIth IAU General Assembly, August 2012
T. Montmerle, ed.

doi:10.1017/S1743921314011624

CGPS studies of the Galactic Magnetic Field

Joern Geisbuesch, R. Kothes and T. L. Landecker

DRAO, NSI-NRC, Penticton, V2A 6J9, Canada; email: Joern.Geisbuesch@nrc-cnrc.gc.ca

Abstract. The Canadian Galactic Plane Survey (CGPS) is the largest effort of its kind to study and understand the Galactic Magnetic Field (GMF) and Interstellar Medium (ISM) in our Galaxy (see e.g. Taylor *et al.* 2003). The CGPS has mapped the Galactic plane visible from DRAO on all spatial scales down to arcminute resolution in total intensity and polarized emission at $\nu_{\rm obs} = 1.4$ GHz (see Landecker *et al.* 2010). The latest results invoking Faraday rotation and polarization gradient studies of the CGPS are discussed.

Keywords. Radio continuum: ISM, Polarization, ISM: magnetic fields, Methods: statistical

---

Faraday rotation measures (RMs) and polarization gradients (PGs), a method advocated by Gaensler *et al.* 2011, have been derived for the CGPS and its northern and southern latitude extensions (NLE & SLE). Stokes Q and U maps have been corrected for instrumental effects. The PG maps have been statistically analyzed by utilizing higher order moments of the probability density functions of selected regions within the CGPS, namely the skewness and excess kurtosis. In the NLE, the state of the ISM turbulence shows a latitude dependence. In particular, at the disk-halo transition around $\sim 10.5°$ the magneto-ionic ISM appears often trans- and at times even super-sonic, while it is rather sub- to trans-sonic at lower latitudes. Possible explanations are projection effects, disk warping and/or Galactic in- and outflows. The chosen areas for the PG analysis exclude structures caused by compact and discrete extended polarized emitters to avoid biased statistics. On the basis of the CGPS data, we demonstrate the potential of the method when applied to single-frequency interferometer data and perform a resolution study on the combined Effelsberg, DRAO 26m and ST data to show how PGs probe different structures in the field of view. Under the assumption of negligible polarization angle rotation on scales of half a degree and above, which the angular regularity of the polarization angle on these scales at $\nu_{\rm obs}$ supports, we obtain fairly low RM estimates ($|{\rm RM}| \lesssim 100$ rad/m^2) for most of the diffuse polarized emission around the Cygnus X region suggesting that the region resembles a polarization horizon, whose boundary is the nearside of the nearby very active star forming region. Hereby, implications on deviations from a simple Faraday structure along the sightline are obtained by using chi-squared statistics and the Bayesian Information Criterion. RM estimates on a grid towards compact and confined extended sources with implied simple Faraday structures along their sightlines are derived for the SLE. Its large-scale RM structure and the angular distribution of Hα emission in the region are correlated. The average RM value is -170 at $b \approx -2°$ and decreases to -70 rad/m^2 at $b \approx -12°$. The disk-halo transition as indicated by diffuse polarization and PG structures occurs in the SLE at a lower absolute latitude than in the NLE ($b \sim -7°$). PGs of HB9 show structures caused by shocked ionized gas and indicate a possible outflow. In regions with less complex polarized emission along sightlines, spatial correlations between polarized emission and HI features can be identified.

References

Gaensler, B., Haverkorn, M., Burkhart, B., Newton-McGee, K. J., *et al.* 2011, *Nature*, 478, 214
Landecker, T., Reich, W., Reid, R., Reich, P., Wolleben, M., *et al.* 2010, *A&A* 520, 80
Taylor, A. R., Gibson, S., Peracaula, M., Martin, P., Landecker, T. L., *et al.* 2003, *ApJ* 125, 3145

Highlights of Astronomy, Volume 16
XXVIIIth IAU General Assembly, August 2012
T. Montmerle, ed.

doi:10.1017/S1743921314011636

The Sino-German λ6cm polarization survey of the Galactic plane

J. L. Han[1], W. Reich[2], X. H. Sun[1,2], X. Y. Gao[1], L. Xiao[1], P. Reich[2], W. B. Shi[1] and R. Wielebinski[2]

[1]National Astronomical Observatories, Chinese Academy of Sciences,Jia-20 Datun Road, Chaoyang District, Beijing 100012, China. hjl@nao.cas.cn
[2]Max-Planck-Institut für Radioastronomie, Auf dem Hügel 69, 53121 Bonn, Germany

After Prof. R. Wielebinski visited China in 1999, we started to plan the Sino-German λ6 cm polarization survey of the Galactic plane, using the Urumqi 25-m radio telescope of Xinjiang (formerly Urumqi) Astronomical Observatory, Chinese Academy of Sciences. It is a high-frequency complement of previous Effelsberg 21-cm and 11-cm surveys, using the same observing and processing methods. The telescope is located at an altitude of 2029 m above sea level at geographic longitude of 87°E and latitude 43°N. The dual-channel λ6 cm receiver with a polarimeter and a bandwidth of 600 MHz was designed by O. Lochner and constructed at the MPIfR in Germany with involvements by the Urumqi engineers M.Z. Chen and J. Ma. In August 2004, the receiver was installed at the secondary focus of the Urumqi 25-m telescope.

We surveyed the polarized emission of the Galactic plane from $10° \leqslant l \leqslant 230°$ in Galactic longitude and $|b| \leqslant 5°$ in Galactic latitude, scanning section by section with a 9.5′ beam and observing the calibrator sources 3C286 and 3C295 before and after each scan section. The observations took more than 4000 hours in 5 years until April 2009, sometimes only a few tens of hours in a month. Four former PhD students (X.H. Sun – even a few years after his PhD, W.B. Shi, L. Xiao, X.Y. Gao) have been heavily involved in observations and data processing, and a few other PhD students (H. Shi, C. Wang and J.W. Xu *et al.*) were involved in observations and studies of some individual objects. All survey data have been properly processed (Sun *et al.* 2007, Gao *et al.* 2010, Sun *et al.* 2011a, Xiao *et al.* 2011) and released at http://zmtt.bao.ac.cn/6cm/ and http://www.mpifr.de/survey.html. The sensitivity is about 1 mK in total power and 0.5 mK in polarization. The instrumental polarization is below 1%.

It is a 10 years work with long-term support from the Xinjiang Observatory. The new 6-cm polarization data not only reveal new properties of the diffuse magnetized interstellar medium (Gao *et al.* 2010, Sun *et al.* 2011a, Xiao *et al.* 2011), but are also very useful for studying individual objects such as HII regions, which may act as Faraday screens with strong regular magnetic fields inside, and for determining the flux densities at λ6 cm and radio spectra of supernova remnants (Gao *et al.* 2011a, Sun *et al.* 2011b). The high sensitivity of the survey enabled us to discover two large ($\sim 1^o$) SNRs G178.2−4.2 and G25.3−2.1 (Gao *et al.* 2011b).

References

Gao, X. Y., Reich, W., Han, J. L., *et al.* 2010, *A&A* 515, A64
Gao, X. Y., Han, J. L., Reich, W., Reich, P., Sun, X. H., & Xiao, L. 2011a, *A&A* 529, A159
Gao, X. Y., Sun, X. H., Han, J. L., Reich, W., Reich, P., & Wielebinski, R. 2011b, *A&A* 532, A144
Sun, X. H., Han, J. L., Reich, W., *et al.* 2007, *A&A* 463, 993
Sun, X. H., Reich, W., Han, J. L., *et al.* 2011a, *A&A* 527, A74
Sun, X. H., Reich, P., Reich, W., Xiao, L., Gao, X. Y., & Han, J. L. 2011b, *A&A* 536, A83
Xiao, L., Han, J. L., Reich, W., *et al.* 2011, *A&A* 529, A15

Highlights of Astronomy, Volume 16
XXVIIIth IAU General Assembly, August 2012
T. Montmerle, ed.

doi:10.1017/S1743921314011648

Probing Magnetic Fields With SNRs

Roland Kothes

DRAO, NSI-NRCC, Penticton, Canada; email: roland.kothes@nrc-cnrc.gc.ca

Abstract. As supernova remnants (SNRs) expand, their shock waves freeze in and compress magnetic field lines they encounter; consequently we can use SNRs as magnifying glasses for interstellar magnetic fields. A simple model is used to derive polarization and rotation measure (RM) signatures of SNRs. This model is exploited to gain knowledge about the large-scale magnetic field in the Milky Way. Three examples are given which indicate a magnetic anomaly, an azimuthal large-scale magnetic field towards the anti-centre, and a chimney that releases magnetic energy from the plane into the halo.

Keywords. ISM: magnetic fields, supernova remnants

Introduction: Recently, there have been many studies of the Milky Way's large-scale magnetic field utilizing RM observations of compact polarized objects such as extragalactic point sources and pulsars. However, the magnetic field is averaged along the line of sight weighted by the electron density; field reversals are averaged out. This ambiguity can be solved, using polarization and RM studies of SNRs as anchor points, utilizing polarization surveys such as the Canadian Galactic Plane Survey (CGPS, Taylor *et al.* (2003), Landecker *et al.* (2010)).
The Model: I assume spherical, mature SNRs expanding into an environment of constant density and magnetic field. Emission and RM structure are computed as described in Kothes & Brown (2008). The simulations show that ambient magnetic field directions can easily be determined from measurements of polarization angle corrected for Faraday rotation and the RM gradient on the bright SNR shells.
CTB104a: A CGPS radio polarization study of the SNR CTB104a (G93.7−0.2) indicated that the ambient magnetic field direction is opposite to the direction expected for this part of the Galaxy. A fit with the new model confirms this. The ambient magnetic field points towards us at an angle between 5-10° with the plane of the sky and an ambient magnetic field strength of at least 10 μG.
G182.4+4.3: The SNR G182.4+4.3, discovered by Kothes, Fürst & Reich (1998), is located close to the anti-centre of our Galaxy. If we assume that the magnetic field here is constant along the line of sight at about 4μG, a fit to the SNRs RM structure reveals that we are looking perpendicular to the large-scale magnetic field at a Galactic longitude of 180°, which indicates an azimuthal magnetic field for the Outer Galaxy.
DA 530: The SNR DA 530 (G93.3+6.9) is located at high Galactic latitude above an area of the plane which is rich in SNRs, HII, and star formation regions. A model fit to this SNR's polarization and RM structure indicates that it is expanding into an area with twisted magnetic field lines. This is the kind of magnetic field structure that we would expect to find inside a Galactic chimney.

References

Kothes, R., Fürst, E., & Reich, W. 1998, *A&A* 331, 661
Kothes, R. & Brown, J.-A. 2008, *Proc. IAU Symposium* 259, 75
Landecker, T. L., Reich, W., Reid, R., *et al.* 2010, *A&A* 520, 80
Taylor, A. R., Gibson, S. J., Peracaula, M., *et al.* 2003, *AJ* 125, 3145
Uyanıker, B., Kothes, R., & Brunt, C. M. 2002, *ApJ* 565, 1022

Highlights of Astronomy, Volume 16
XXVIIIth IAU General Assembly, August 2012
T. Montmerle, ed.

doi:10.1017/S174392131401165X

Theoretical understanding of Galactic magnetic fields

Katia M. Ferrière

IRAP, Université de Toulouse, CNRS, 14 avenue Edouard Belin, F-31400 Toulouse, France

Abstract. I discuss our current theoretical understanding of the origin and evolution of interstellar magnetic fields in our Galaxy, with a focus on dynamo theory.

Keywords. ISM: magnetic fields

It is likely that the very first cosmic magnetic fields were generated before the formation of galaxies – for instance, through a battery effect. These pre-galactic magnetic fields would then have been amplified in galaxies by several mechanisms, starting with (1) the compression of field lines during the collapse of protogalaxies, and continuing with (2) the stretching of field lines by the large-scale differential rotation of galaxies and (3) the production of magnetic field perpendicular to the prevailing field by small-scale helical turbulence (a process known as the alpha effect). Together the last two mechanisms can amplify magnetic fields at a very fast (initially exponential) rate: this is the essence of dynamo action in galaxies.

Mean-field dynamo theory provides an evolution equation for the mean (or large-scale) magnetic field, $\langle \vec{B} \rangle$ – the dynamo equation – wherein the effects of turbulent motions are parameterized by means of two tensors: α, which mainly describes the alpha effect, and β, which mainly describes turbulent magnetic diffusion.

When the large-scale velocity and the α, β tensors are prescribed, the dynamo equation is linear in $\langle \vec{B} \rangle$, which, therefore, grows (or decays) exponentially with time. In addition, $\langle \vec{B} \rangle$ generally has the following properties: (1) In galactic disks, the azimuthal component of $\langle \vec{B} \rangle$ dominates, while its vertical component is very small. (2) In an axisymmetric galaxy, axisymmetric ($m = 0$) modes are easiest to amplify; in the presence of an external disturbance, bisymmetric ($m = 1$) modes can possibly be amplified, while in the presence of a spiral or bar, quadrisymmetric ($m = 2$) modes can possibly be amplified. (3) Under typical galactic conditions, both symmetric and antisymmetric (with respect to the midplane) modes are amplified; if the disk dominates, symmetric modes tend to grow faster, while if the halo dominates, antisymmetric modes tend to grow faster.

When the magnetic field has grown strong enough, it starts having a back-reaction on the turbulence, which "quenches" α and causes the field growth to saturate. The conventional approach is to consider that back-reaction starts when the mean field approaches equipartition with the turbulence. However, several authors have pointed out that the small-scale field grows much faster than the mean field, such that back-reaction actually starts when the small-scale field approaches equipartition, i.e., when the mean field is still a huge factor ($\sim \sqrt{\mathrm{Rm}}$, with Rm the magnetic Reynolds number) below equipartition (catastrophic α-quenching). In reality, the mean field can keep growing after the small-scale field has reached equipartition, with the temporal evolution of α governed by the conservation of magnetic helicity (dynamical α-quenching).

The galactic dynamo problem has also been tackled through direct numerical simulations, the most recent of which cover a full galaxy (disk + halo).

Highlights of Astronomy, Volume 16
XXVIIIth IAU General Assembly, August 2012
T. Montmerle, ed.

doi:10.1017/S1743921314011661

Detection of Linear Polarization from SNR Cassiopeia A at Low Radio Frequencies

Wasim Raja and Avinash A. Deshpande
Raman Research Institute, Bangalore, India

We report detection of the weak but significant linear polarization from the Supernova Remnant Cas A at low radio frequencies (327 MHz) using the GMRT. The spectro-polarimetric data (16 MHz bandwidth with 256 spectral channels) was analyzed using the technique of Faraday Tomography. Ascertaining association of this weak polarization to the source is non-trivial in the presence of the remnant instrumental polarization (<1% in our case) – the expected anti-correlation $\rho_{lp,x}$, between the linear polarized intensity and the soft X-ray counts gets masked by the correlation between the Stokes-I dependent instrumental leakage and the X-radiation that is spatially correlated with Stokes-I, if $\rho_{lp,x}$ is computed naively. Hence, we compute $\rho_{lp,x}$ using pixels within *ultra narrow* bins of Stokes-I within which the instrumental leakage is expected to remain constant, and establish the anti-correlation as well as the correspondence of this correlation with the mean X-ray profile (Figure 1). Given the angular and RM-resolution in our data, the observed depolarization relative to that at higher frequencies, implies that the mixing of thermal and non-thermal plasma within the source might be occurring on spatial scales $\sim 1000AU$, assuming random superposition of polarization states.

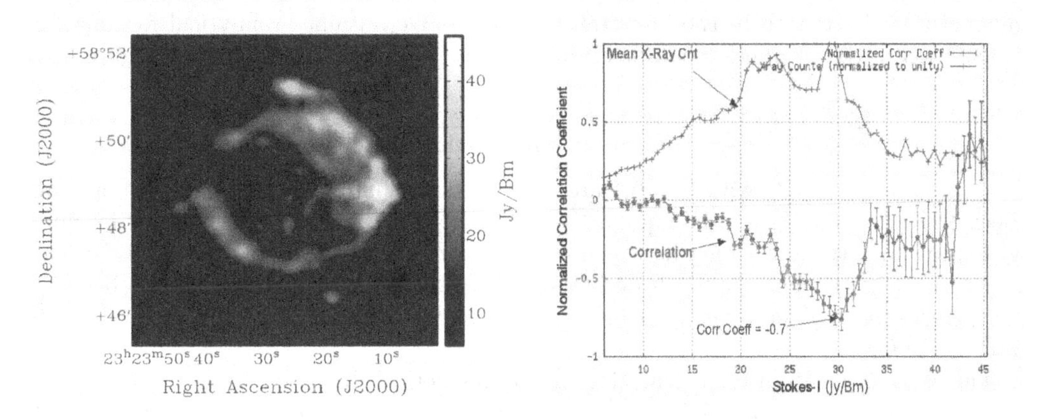

Figure 1. Left: 327 MHz Stk-I image of Cas A. Right: Profile of mean X-ray counts (max normalized to unity) (blue) & $\rho_{lp,x}$ (at RM=-110 rad/m^2) (red) as a function of binned Stk-I.

References

Anderson, *et al.*, 1995, *ApJ*, 441, 300
Brentjens, M. A. & de Bruyn, A. G. 2005, *A&A*, 441, 1217
Burns, B. J. 1966, *MNRAS*, 133, 67
Downs, G. S. & Thompson, A. R., 1972, *AJ*, 77, 120
Ramkumar, P. S. & Deshpande A. A., 1999, *J. Astrophys. Astr.*, 20, 37

Highlights of Astronomy, Volume 16
XXVIIIth IAU General Assembly, August 2012
T. Montmerle, ed.

doi:10.1017/S1743921314011673

The modified equipartition calculation for supernova remnants with the spectral index $\alpha = 0.5$

Dejan Urošević, Marko Z. Pavlović, Bojan Arbutina and Aleksandra Dobardžić

Department of Astronomy, Faculty of Mathematics, University of Belgrade
Studentski trg 16, 11000 Belgrade, Serbia
email: dejanu@matf.bg.ac.rs

Abstract. Recently, the modified equipartition calculation for supernova remnants (SNRs) has been derived by Arbutina *et al.* (2012). Their formulae can be used for SNRs with the spectral indices between $0.5 < \alpha < 1$. Here, by using approximately the same analytical method, we derive the equipartition formulae useful for SNRs with spectral index $\alpha = 0.5$. These formulae represent next step upgrade of Arbutina *et al.* (2012) derivation, because among 30 Galactic SNRs with available observational parameters for the equipartition calculation, 16 have spectral index $\alpha = 0.5$. For these 16 Galactic SNRs we calculated the magnetic field strengths which are approximately 40 per cent higher than those calculated by using Pacholczyk (1970) equipartition and similar to those calculated by using Beck & Krause (2005) calculation.

Keywords. ISM: magnetic fields, ISM: supernova remnants, radio continuum: general

---

This paper represents next step upgrade of Arbutina *et al.* (2012) determination of the magnetic field strength in the interstellar medium. We obtained analytical formula for calculation of the magnetic field for SNRs with the radio spectral index $\alpha = 0.5$:

$$B[\mathrm{G}] = 9.1 \cdot 10^{-5} \left[(1 + \kappa_{\gamma=2}) \frac{\mathrm{S}_\nu[\mathrm{Jy}]}{\mathrm{f}\theta[\mathrm{arcmin}]^3 \mathrm{d}[\mathrm{kpc}]}\right]^{2/7} \nu[\mathrm{GHz}]^{1/7} \tag{0.1}$$

Our values for SNRs with $\alpha = 0.5$ are slightly lower than those obtained by applying Beck & Krause (2005) revised equipartition formula for $\alpha = 0.5$. Applying our formula yields to magnetic field strengths which are approximately 40 % higher than those calculated by using classical approach derived by Pacholczyk (1970). This is mostly because we were using a wider interval of integration than standard Pacholczyk interval of 10 MHz – 10 GHz.

The Web application for calculation of the magnetic field strengths of SNRs for the spectral indices $0.5 < \alpha < 1$ is available at http://poincare.matf.bg.ac.rs/~arbo/eqp/ and value $\alpha = 0.578$ should be used instead of $\alpha = 0.5$.

During the work on this paper, the authors were financially supported by the Ministry of Education and Science of the Republic of Serbia through the project number 176005.

References

Arbutina, B., Urošević, D., Andjelić, M. M., Pavlović, M. Z., & Vukotić, B. 2012, *ApJ*, 746, 79
Beck, R. & Krause, M. 2005, *Astron. Nachr.*, 326, 414
Pacholczyk, A. G. 1970, *Radio Astrophysics* (San Francisco, CA:Freeman)

Highlights of Astronomy, Volume 16
XXVIIIth IAU General Assembly, August 2012
T. Montmerle, ed.

doi:10.1017/S1743921314011685

Magnetic fields in spiral galaxies

Marita Krause
Max-Planck-Institut für Radioastronomie, Auf dem Hügel 69, 53121 Bonn, Germany

The magnetic field structure in edge-on galaxies observed so far shows a plane-parallel magnetic field component in the disk of the galaxy and an X-shaped field in its halo. The plane-parallel field is thought to be the projected axisymmetric (ASS) disk field as observed in face-on galaxies. Some galaxies addionionally exhibit strong vertical magnetic fields in the halo right above and below the central region of the disk. The mean-field dynamo theory in the disk cannot explain these observed fields without the action of a wind, which also probably plays an important role to keep the vertical scale heights constant in galaxies of different Hubble types and star formation activities, as has been observed in the radio continuum: At $\lambda 6$ cm the vertical *scale heights* of the thin disk and the thick disk/halo in a sample of five edge-on galaxies are similar with a mean value of 300 ± 50 pc for the thin disk and 1.8 ± 0.2 kpc for the thick disk (a table and references are given in Krause 2011) with our sample including the brightest halo observed so far, NGC 253, with strong star formation, as well as one of the weakest halos, NGC 4565, with weak star formation. If synchrotron emission is the dominant loss process of the relativistic electrons the outer shape of the radio emission should be dumbbell-like as has been observed in several edge-on galaxies like e.g. NGC 253 (Heesen *et al.* 2009) and NGC 4565. As the synchrotron lifetime t_{syn} at a single frequency is proportional to the total magnetic field strength $B_t^{-1.5}$, a cosmic ray bulk speed (velocity of a galactic wind) can be defined as $v_{CR} = h_{CR}/t_{syn} = 2h_z/t_{syn}$, where h_{CR} and h_z are the scale heights of the cosmic rays and the observed radio emission at this freqnency. Similar observed radio scale heights imply a self regulation mechanism between the galactic wind velocity, the total magnetic field strength and the star formation rate SFR in the disk: $v_{CR} \propto B_t^{1.5} \propto SFR^{\approx 0.5}$ (Niklas & Beck 1997).

However, recent determination of the scaleheights in two other nearby spirals (M82 and NGC 4631) yielded different values: the scaleheights in M82 are significantly smaller than the mean values mentioned above with larger values north of the disk than south of the disk (Adebahr *et al.* 2013). The scaleheights for both, the thin and thick disk in NGC 4631 vary strongly within the galaxy, being significantly larger in some areas than the mean values (Mora & Krause, in prep.). Both galaxies show -different from the other five galaxies- strong signs of tidal interaction like HI tails and bridges (Yun *et al.* 1993, Rand 1994) where M82 might even have lost its outer HI disk by tidal disruption. Hence, from present observations of edge-on galaxies we can conclude that while star formation and even starbursts in the disk alone do not significantly change the scaleheights of the disk and halo, (strong) tidal interactions may well modify these parameters.

References

Adebahr, B., Krause, M., Klein, U., *et al.* 2013, *A&A* in press
Heesen, V., Beck, R., Krause, M., & Dettmar, R.-J. 2009, *A&A* 494, 563
Niklas, S. & Beck, R. 1997, *A&A* 320, 54
Krause, M. 2011, ArXiv:1111.708
Rand, R. J. 1994, *A&A* 285, 833
Yun, M. S., Ho, T. P., & Lo, K. Y. 1993, *ApJ* (Letters) 411, L17

Highlights of Astronomy, Volume 16
XXVIIIth IAU General Assembly, August 2012
T. Montmerle, ed.

doi:10.1017/S1743921314011697

Fluctuation dynamos and their Faraday rotation signatures

Pallavi Bhat and Kandaswamy Subramanian

IUCAA, Post Bag 4, Ganeshkhind, Pune 411007, India.
email: palvi@iucaa.ernet.in, kandu@iucaa.ernet.in

We study fluctuation dynamo (FD) action in turbulent systems like galaxy-clusters focusing on the Faraday rotation signature. This is defined as $\mathrm{RM} = K \int_L n_e \mathbf{B} \cdot \mathbf{dl}$ where n_e is the thermal electron density, $\mathbf{B}$ is the magnetic field, the integration is along the line of sight from the source to the observer, and $K = 0.81$ rad m^{-2} cm^{-3} μG^{-1} pc^{-1}. We directly compute, using the simulation data, $\int \mathbf{B} \cdot \mathbf{dl}$, and hence the Faraday rotation measure (RM) over $3N^2$ lines of sight, along each x, y and z-directions. We normalise the RM by the rms value expected in a simple model, where a field of strength B_{rms} fills each turbulent cell but is randomly oriented from one turbulent cell to another. This normalised RM is expected to have a nearly zero mean but a non-zero dispersion, $\bar{\sigma}_{RM}$. We show in Fig. 1a and 1b, that a suite of simulations, on saturation, obtain the value of $\bar{\sigma}_{RM} = 0.4 - 0.5$, and this is independent of P_M, R_M and the resolution of the run. This is a fairly large value for an intermittent random field; as it is of order 40%–50%, of that expected in a model where B_{rms} strength fields volume fill each turbulent cell, but are randomly oriented from one cell to another. We also find that the regions with a field strength larger than $2B_{rms}$ contribute only 15–20% to the total RM (see Fig. 1a). This shows that it is the general 'sea' of volume filling fluctuating fields that contribute dominantly to the RM produced, rather than the the high field regions.

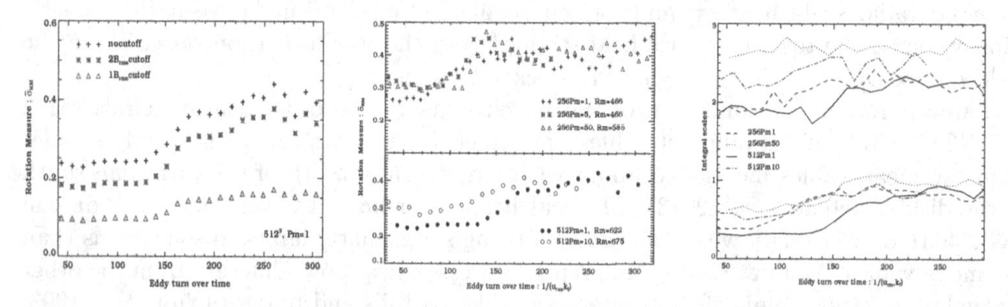

Figure 1. Figure (a) on the left shows the effective RM after removing the regions with $|\mathbf{B}| > 2B_{rms}$ and $1B_{rms}$ fields. Fig. (b) in the middle shows RM evolution for runs with different R_M and P_M. Fig (c) on the right shows magnetic integral scales in the lower half and velocity integral scales in the upper half.

Interestingly, the magnetic integral scale, L_{int} (see Fig. 1c) starts to increase in all the runs, as Lorentz forces become important to saturate the dynamo. It appears that on saturation, the magnetic integral scale, L_{int} tends to a modest fraction $1/2 - 1/3$ of the integral scale of the velocity field for all our runs. Finally, we find that $\bar{\sigma}_{RM} \sim 0.4 - 0.5$ obtained, implies a dimensional $\sigma_{RM} \sim 180$ rad m^{-1}, for parameters appropriate for galaxy clusters. This is sufficiently large to account for the observed Faraday rotation seen in these systems.

Highlights of Astronomy, Volume 16
XXVIIIth IAU General Assembly, August 2012
T. Montmerle, ed.

doi:10.1017/S1743921314011703

Multiwavelength Magnetic Field Modeling

T. R. Jaffe

Université de Toulouse & CNRS; UPS-OMP; IRAP; Toulouse, France
email: tess.jaffe@irap.omp.eu

Abstract. We model the large-scale Galactic magnetic fields, including a spiral arm compression to generate anisotropic turbulence, by comparing polarized synchrotron and thermal dust emission. Preliminary results show that in the outer Galaxy, the dust emission comes from regions where the fields are more ordered than average while the situation is reversed in the inner Galaxy. We will attempt in subsequent work to present a more complete picture of what the comparison of these observables tells us about the distribution of the components of the magnetized ISM and about the physics of spiral arm shocks and turbulence.

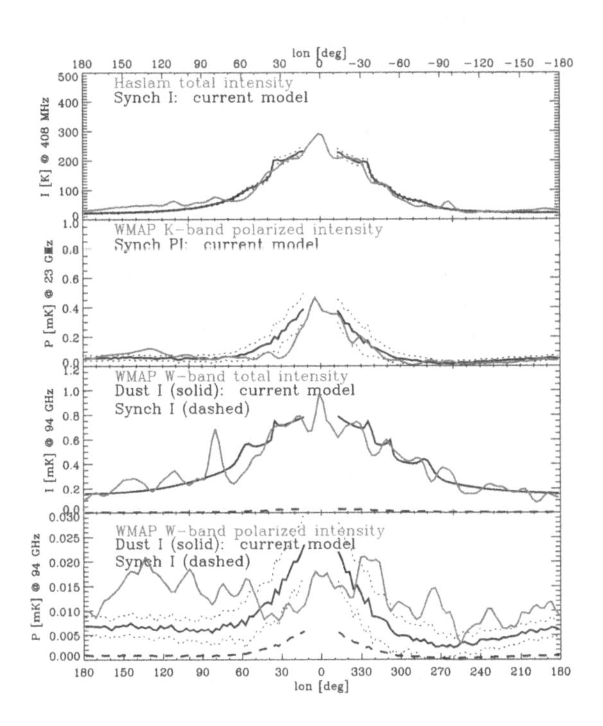

Figure 1. Observables *(lighter curves))* compared to model *(darker curves)* similar to Jaffe *et al.* (2010,2011), now adding polarized thermal dust emission (following Fauvet *et al.* 2011). From top to bottom: synchrotron total intensity at 408 MHz (Haslam *et al.* 1982); *WMAP* synchrotron polarized intensity at 23 GHz (Jarosik *et al.* 2011); thermal dust total intensity at 94 GHz (id.); thermal dust polarized intensity at 94 GHz (id.). From the underprediction of its polarization in the outer Galaxy, we infer that the dust emission is coming from regions where the fields are more ordered than average. By contrast, in the inner Galaxy near the Galactic center, the overprediction implies to the contrary that the dust emission is coming from regions that are less ordered than average.

References

Fauvet, L., Macías-Perez, J. F., Aumont, J., *et al.* 2011, *A&A*, 526, 145
Haslam C. G. T., Stoffel H., Salter C. J., & Wilson W. E., 1982, *A&AS*, 47, 1
Jaffe, T. R., Leahy, J. P., Banday, A. J., *et al.* 2010, *MNRAS*, 401, 1013
Jaffe, T. R., Banday, A. J., Leahy, J. P., Leach, S., & Strong, A. W. 2011, *MNRAS*, 416, 1152
Jarosik, N., *et al.*, 2011, *ApJS*, 192, 14

Highlights of Astronomy, Volume 16
XXVIIIth IAU General Assembly, August 2012
T. Montmerle, ed.

doi:10.1017/S1743921314011715

MAGMO: Mapping the Galactic Magnetic field through OH masers

James A. Green[1], Naomi M. McClure-Griffiths[1], James L. Caswell[1], Tim Robishaw[2], Lisa Harvey-Smith[1] and Sui Ann Mao[3,4]

[1]CSIRO Astronomy and Space Science, Australia Telescope National Facility,
PO Box 76, Epping, NSW 1710, Australia
email: james.green@csiro.au

[2]National Research Council, Herzberg Institute of Astrophysics,
Dominion Radio Astrophysical Observatory,
PO Box 248, Penticton, BC V2A 6J9, Canada

[3]Jansky Fellow, National Radio Astronomy Observatory, P.O. Box O, Socorro, NM 87801, USA

[4]Department of Astronomy, University of Wisconsin, Madison WI 53706, USA

Abstract. We are undertaking a project (MAGMO) to examine large-scale magnetic fields pervading regions of high-mass star formation. The project will test if the orientations of weak large-scale magnetic fields can be maintained in the contraction (and field amplification) to the high densities encountered in high-mass star forming regions. This will be achieved through correlating targeted observations of ground-state hydroxyl (OH) maser emission towards hundreds of sites of high-mass star formation spread throughout the spiral arms of the Milky Way. Through the Zeeman splitting of the OH maser emission these observations will determine the strength and orientation of the in-situ magnetic field. The completion of the southern hemisphere Methanol Multibeam survey has provided an abundance of targets for ground-state OH maser observations, approximately 1000 sites of high-mass star formation. With this sample, much larger and more homogeneous than previously available, we will have the statistics necessary to outweigh random fluctuations and observe an underlying Galactic magnetic field if it exists. We presented details of the overall progress of the project illustrated by the results of a pilot sample of sources towards the Carina-Sagittarius spiral arm tangent, where a coherent field is implied.

Keywords. masers, stars: formation, ISM: magnetic fields, Galaxy: structure

---

We described the completion of the southern hemisphere observations with the Australia Telescope Compact Array utilising the Compact Array Broadband Backend to observe all four ground-state transitions of hydroxyl maser emission towards sites of high-mass star formation, as signposted by 6.7-GHz methanol masers from the Methanol Multibeam (MMB) survey (Green *et al.* 2009, Caswell *et al.* 2010). We presented the results of the pilot observations towards the Carina-Sagittarius spiral arm tangent, where a coherent field is seen across six sites of high-mass star formation (Green *et al.* 2012).

References

Green, J. A., Caswell, J. L., & Fuller, G. A., 2009, *MNRAS*, 392, 783
Caswell, J. L., Fuller, G. A., & Green, J. A., 2010, *MNRAS*, 404, 1029
Green, J. A., McClure-Griffiths, N. M., Caswell, J. L., Robishaw, T., & Harvey-Smith, L., 2012, *MNRAS*, 425, 2530

Highlights of Astronomy, Volume 16
XXVIIIth IAU General Assembly, August 2012
T. Montmerle, ed.

doi:10.1017/S1743921314011727

Magnetic Fields in the Milky Way Halo

S. A. Mao[1,2], N. M. McClure-Griffiths[3], B. M. Gaensler[4], J. C. Brown[5], C. L. van Eck[5], M. Haverkorn[6], P. P. Kronberg[7,8], J. M. Stil[5], A. Shukurov[9] and A. R. Taylor[5]

[1] Jansky Fellow, National Radio Astronomy Observatory,
P.O. Box O, Socorro, NM 87801, USA
email: mao@astro.wisc.edu
[2] Department of Astronomy, University of Wisconsin, Madison, WI 53706, USA
[3] CSIRO Astronomy and Space Science, Australia Telescope National Facility,
PO Box 76, Epping, NSW 1710, Australia
[4] Sydney Institute for Astronomy, School of Physics,
The University of Sydney, NSW 2006, Australia
[5] Department of Physics and Astronomy, and Institute for Space Imaging Science,
University of Calgary, AB, Canada
[6] Department of Astrophysics, Radboud University,
P.O. Box 9010, 6500 GL Nijmegen, The Netherlands
[7] Los Alamos National Laboratory, P.O. Box 1663, Los Alamos, NM 87545, USA
[8] Department of Physics, University of Toronto, 60 St. George Street,
Toronto, M5S 1A7, Canada
[9] School of Mathematics and Statistics, University of Newcastle,
Newcastle upon Tyne, NE1 7RU, UK

Abstract. We present a study of the Milky Way halo magnetic field, determined from observations of Faraday rotation measure (RM) of extragalactic radio sources (EGS) in Galactic longitude range 100°–117° within 30° of the Galactic plane. We find negative median RMs in both the northern and southern Galactic hemispheres for $|b|>15°$, outside the latitude range where the disk field dominates. This suggest that the halo magnetic field towards the outer Galaxy does not reverse direction across the mid-plane. An azimuthal magnetic field at heights 0.8–2 kpc above/below the Galactic plane between the local and the Perseus spiral arm can reproduce the observed trend of RM against Galactic latitude. We propose that the Milky Way could have a halo magnetic field similar to that observed in M51.

Keywords. Galaxy: halo, ISM: magnetic fields, polarization

We describe a study of the Galactic halo magnetic field towards the Perseus arm in the second Galactic quadrant using Faraday rotation measure of extragalactic radio sources. This work is based on 302 new EGS RMs measured using the Very Large Array and 339 EGSs RMs from the Canadian Galactic Plane Survey latitude extension region (J. C. Brown *et al.* in preparation). For $|b|<15°$, we find a symmetric RM distribution about the mid-plane, which is consistent with an even parity disk magnetic field. For $|b|>15°$, we demonstrate that none of the existing halo magnetic field models can reproduce the observed RM distribution. We then propose a simple halo magnetic field model that can reproduce the behavior of RM as a function of Galactic latitude in the observed region. We suggest that the Milky Way halo magnetic field could have a geometry similar to that observed in M51 (Mao *et al.* 2012).

References

Mao, S. A., McClure-Griffiths, N. M., Gaensler, B. M., Brown, J. C., van Eck, C. L., Haverkorn, M., Kronberg, P. P., Stil, J. M., Shukurov, A., & Taylor, A. R. 2012, *ApJ* 755, 21

Highlights of Astronomy, Volume 16
XXVIIIth IAU General Assembly, August 2012
T. Montmerle, ed.

doi:10.1017/S1743921314011739

Observations of magnetic fields in intracluster medium

Federica Govoni
INAF - Osservatorio Astronomico di Cagliari, Strada 54, Loc. Poggio dei Pini, 09012 Capoterra (Ca), Italy
email: fgovoni@oa-cagliari.inaf.it

Abstract. The presence of μG-level magnetic fields associated with the intracluster medium of galaxy clusters is now widely acknowledged. Our knowledge of their properties has greatly improved in the recent years thanks to both new radio observations and the developments of new techniques to interpret data.

Keywords. magnetic fields, galaxies: clusters: general, polarization

1. Radio halos and Faraday rotation measures

Most of what we know about intracluster magnetic fields derives from the study of radio halos and Faraday rotation measures of polarized radio galaxies located inside or behind galaxy clusters (see e.g. the reviews by Carilli & Taylor 2002, Govoni & Feretti 2004, Ferrari *et al.* 2008, Feretti *et al.* 2012).

Sensitive radio observations have revealed diffuse emission from the central regions of some merging galaxy clusters. These radio sources, which extend over Mpc scales and are called radio halos, are diffuse, low-surface-brightness, and steep-spectrum synchrotron sources with no obvious optical counterparts. To date, approximately 50 radio halos are known. They demonstrate the existence of relativistic electrons and magnetic fields spread in the intracluster medium. The assumption that radio halos have the magnetic energy density comparable to the energy density in relativistic electrons, requires a volume-averaged magnetic field $\sim 0.1{-}1\ \mu$G. Sometimes the intracluster magnetic fields can be ordered on scales of hundreds of kpc, as revealed in A2255 (Govoni & Feretti 2005, Pizzo *et al.* 2011) and MACS J0717.5+3745 (Bonafede *et al.* 2009), where a polarized signal, possibly associated to the radio halo, has been detected. Actually, total intensity and polarization intensity radio halo surface brightness fluctuations are strictly related to the magnetic field power spectrum (Tribble 1991, Murgia *et al.* 2004). Thus, observations of radio halos have been used to study the structure of the cluster wide magnetic fields by comparing observations with mock halos from turbulent magnetic fields (Murgia *et al.* 2004, Govoni *et al.* 2006, Vacca *et al.* 2010, Xu *et al.* 2012).

High resolution, detailed RM images of extended cluster radio galaxies have been obtained (e.g. Eilek & Owen 2002, Taylor *et al.* 2007, Guidetti *et al.* 2008, Guidetti *et al.* 2010, Bonafede *et al.* 2010, Vacca *et al.* 2012). These data are usually consistent with central magnetic field strengths of a few μG, but stronger fields, with values exceeding $\simeq 10\,\mu$G, are derived in the inner regions of relaxed cooling core clusters. By analyzing the RM of radio galaxies located at different projected distance from the cluster center, it is possible to investigate the radial decrease of the magnetic field with the gas density (e.g. Dolag *et al.* 2001). The RM distributions seen across the radio galaxies present patchy structures. The observed RM fluctuations indicate that the intracluster magnetic field is not regularly ordered but turbulent on scales ranging from tens of kpc to $\lesssim$ 100 pc. Dedicated software tools and semi-analytical approach have been developed to constrain the magnetic field power spectrum parameters (Enßlin & Vogt, Murgia *et al.* 2004, Laing *et al.* 2008, Kuchar & Enßlin 2011). The magnetic field power spectrum can be approximated with a power law with the slope close to the Kolmogorov index in some clusters but shallower index are also observed.

In addition to detailed RM studies focused on single clusters, magnetic fields in galaxy clusters can be investigated statistically. Clarke *et al.* (2001) analyzed the average RM values as a function of source impact parameter for a sample of Abell clusters. They found a clear broadening of the RM distribution toward small projected distances from the cluster center (see also e.g. Johnston-Hollitt *et al.* 2004, Govoni *et al.* 2010), clearly indicating that most of the RM contribution comes from the intracluster medium and proving that magnetic fields are present in all galaxy clusters. Bonafede *et al.* (2011) selected a sample of massive galaxy clusters and used the NRAO VLA Sky Survey (Condon *et al.* 1998) to analyze the fractional polarization of hundreds radio sources lying at different projected distances from the cluster center. They detected a clear trend of the fractional polarization, being smaller for sources close to the cluster center and increasing with increasing distance form the cluster central regions. This trend is interpreted as the result of an higher depolarization, occurring because of the higher magnetic field and gas density at the cluster center, and can be reproduced by a magnetic field model with a central value of few μG.

References

Bonafede, A., Feretti, L., Giovannini, G., *et al.* 2009, *A&A* 503, 707
Bonafede, A., Feretti, L., Murgia, M., *et al.* 2010, *A&A* 513, A30
Bonafede, A., Govoni, F., Feretti, *et al.* 2011, *A&A* 530, A24
Carilli, C. L. & Taylor, G. B. 2002, *ARA&A* 40, 319
Clarke, T. E., Kronberg, P. P., & Böhringer, H. 2001, *ApJ* 547, L111
Condon, J. J., Cotton, W. D., Greisen, E. W., *et al.* 1998, *AJ* 115, 1693
Dolag, K., Schindler, S., Govoni, F., & Feretti, L. 2001, *A&A* 378, 777
Eilek, J. A. & Owen, F. N. 2002, *ApJ* 567, 202
Enßlin T. A. & Vogt C. 2003, *A&A* 401, 835
Feretti, L., Giovannini, G., Govoni, F., & Murgia, M. 2012, *A&A Rev.* 20, 54
Ferrari, C., Govoni, F., Schindler, S., *et al.* 2008, *Space Science Reviews* 134, 93
Govoni, F. & Feretti, L. 2004, *Int. J. Mod. Phys. D*, Vol. 13, N.8, p. 1549
Govoni, F., Murgia, M., Feretti, L., *et al.* 2005, *A&A* 430, L5
Govoni, F., Murgia, M., Feretti, *et al.* 2006, *A&A* 460, 425
Govoni, F., Dolag, K., Murgia, M., *et al.* 2010, *A&A* 522, A105
Guidetti, D., Murgia, M., Govoni, F., *et al.* 2008, *A&A* 483, 699
Guidetti, D., Laing, R. A., Murgia, M., *et al.* 2008, *A&A* 514, A50
Johnston-Hollitt, M., Hollitt, C. P., & Ekers, R. D. 2004, *The Magnetized Interstellar Medium, Eds: B. Uyaniker, W. Reich, and R. Wielebinski p.13*
Kuchar, P. & Enßlin, T. A. 2011, *A&A* 529, A13
Laing, R. A., Bridle, A. H., Parma, P., & Murgia, M. 2008, *MNRAS* 391, 521
Murgia M., Govoni F., Feretti L., *et al.* 2004, *A&A* 424, 429
Pizzo, R. F., de Bruyn, A. G., Bernardi, G., & Brentjens, M. A. 2011, *A&A* 525, A104
Taylor G. B., Fabian, A. C., Gentile, G., *et al.* 2007, *MNRAS* 382, 67
Tribble, P. C. 1991, *MNRAS* 253, 147
Vacca, V., Murgia, M., Govoni, F., *et al.* 2010, *A&A* 514, 71
Vacca, V., Murgia, M., Govoni, F., *et al.* 2012, *A&A* 540, 38
Xu, H., Govoni, F., Murgia, M., *et al.* 2012, *ApJ* 759, 40

Highlights of Astronomy, Volume 16
XXVIIIth IAU General Assembly, August 2012
T. Montmerle, ed.

doi:10.1017/S1743921314011740

MHD turbulence in the intracluster medium

Diego Falceta-Gonçalves, G. Kowal, E. de Gouveia Dal Pino, R. Santos-Lima, S. Nakwacki and A. Lazarian
Universidade de São Paulo, Rua Arlindo Bettio 1000, CEP 03828-000, São Paulo, Brazil
email: dfalceta@usp.br

Abstract. In this work we discuss the turbulent evolution of structures in the intracluster medium based on the two fluid approximations: MHD and collisionless plasma under Chew Goldberger Low (CGL) closure. Turbulence excited by galactic motions and gas inflow in intracluster medium will develop in very different ways considering the two fluid approaches. Statistics of density distributions, and velocity and magnetic fields are provided. Compared to the standard MHD case, the instabilities that arise from CGL-MHD models strongly modify the probability distribution functions of the plasma velocity and density, basically increasing their dispersion. Moreover, the spectra of both density and velocity show increased power at small scales, due to the instabilities growth rate that are larger as smaller scales. Finally, in high beta plasmas, i.e. $B^2 << P$, a fast increase of the magnetic energy density is observed in the CGL-MHD models, faster than the standard MHD turbulent dynamo that operates at timescales $\tau \sim L/v_L$. The signatures of the increased power at small scales and the increase of magnetic field intensity from CGL-MHD models could be observed at radio wavelengths. A comparison of the structure function of the synchrotron emission, as well as the statistics of Faraday rotation effects on the synchrotron polarization, for both the MHD and CGL-MHD models is provided.

Keywords. galaxies: clusters: general, galaxies: intergalactic medium, galaxies: magnetic fields

---

Several observational studies of Faraday rotation of synchrotron emission at radio wavelengths have reveal the presence of μG magnetic fields in the intracluster medium (ICM). Standard cosmological dynamo theories predict maximum amplitudes of 10^{-9}G at intergalactic scales, at $z \rightarrow 0$. The origin of the relatively strong magnetic field of the ICM is still a matter of debate. Turbulence at the ICM, triggered by AGNs and galactic dynamics within the cluster, could help solve the issue (Falceta-Gonçalves *et al.* 2010). The turbulent kinetic energy density of the ICM is estimated as 10^{11}erg cm^{-3}, large enough to explain the observed magnetic fields. However, the timescales for the turbulent dynamo to occur are still too large.

The ICM is collisionless, i.e. present anisotropic pressures. Pressure anisitropies may result in instabilities such as the mirror and firehose, studied numerically in Kowal *et al.* (2011). We show that in a turbulent ICM both instabilities are triggered naturally and that the firehose instability result in magnetic field amplification in timescales much shorter than the turbulent dynamo. Statistical analysis of the simulated density and magnetic field distributions may also be directly compared to observations (Falceta-Gonçalves *et al.* 2008, Burkhart *et al.* 2009).

References

Kowal, G., Falceta-Gonçalves, D. A., & Lazarian, A. 2011, *NJPh* 13, 3001
Falceta-Gonçalves, D., de Gouveia Dal Pino, E. M., Gallagher, J. S., & Lazarian, A. 2010, *ApJL* 708, 57
Falceta-Gonçalves, D., Lazarian, A., & Kowal, G. 2008, *ApJ* 679, 537
Burkhart, B., Falceta-Gonçalves, D., Kowal, G., & Lazarian, A. 2009, *ApJ* 693, 250

Highlights of Astronomy, Volume 16
XXVIIIth IAU General Assembly, August 2012
T. Montmerle, ed.

doi:10.1017/S1743921314011752

RM due to magnetic fields in the cosmic web and SKA observations

Takuya Akahori[1,2] and Dongsu Ryu[3]

[1]Korea Astronomy and Space Science Institute, Daejeon 305-348, Korea, akahori@kasi.re.kr
[2]School of Physics, The University of Sydney, NSW 2006, Australia
[3]Chungnam National University Daejeon 305-764, Korea, ryu@canopus.cnu.ac.kr

Abstract. We estimated that rotation measure (RM) due to the intergalactic magnetic field (IGMF) in the cosmic web is $\sim 1 - 10$ rad m^{-2}. The RMs could be tested with the Square Kilometer Array (SKA) and SKA pathfinders.

Keywords. large-scale structure of universe magnetic fields polarization

---

The nature and origin of the IGMF are not well understood. Akahori & Ryu (2010) studied RMs through filaments of galaxies using a model of the IGMF based on MHD turbulence simulations (Ryu *et al.* 2008). They found that the inducement of RMs through filaments is a random walk process, and the root mean square (rms) value of the RMs for the present-day local universe is ~ 1 rad m^{-2} (Fig. 1 left). Akahori & Ryu (2011) studied cosmological contribution of RMs through filaments, and found that the rms value of the RMs reaches $\sim$ several-10 rad m^{-2} (Fig. 1 middle). They also found that structure functions (SFs) of the RMs are $\sim$ 100-200 rad^2 m^{-4} down to 0.2° scale (Fig. 1 right). Akahori *et al.* (2012) recently found that SFs for Galactic RMs toward high Galactic latitudes have values substantially smaller than the observed ones in angular separations less than a few degrees, suggesting that the contribution of the IGMF to the observed RMs could be significant, particularly at small angular scales. SFs at small angular scales could be tested with the SKA and SKA pathfinders with dense extragalactic sources and low instrumental noise. We suggest that image processing such as high-pass filters could be a promising method to extract the RM due to the IGMF from observed one.

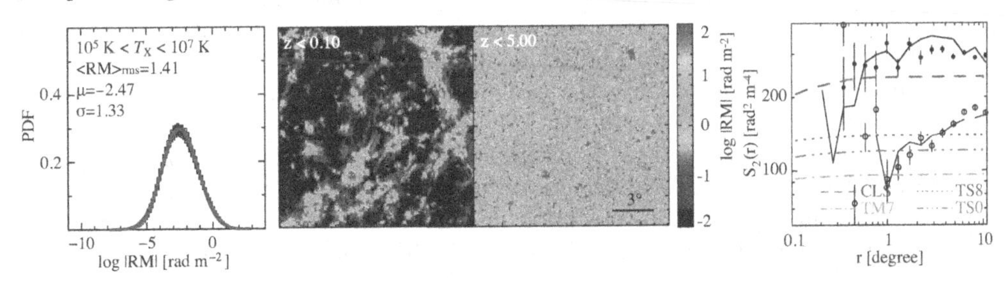

Figure 1. (Left) Probability distribution function of RMs through local filaments. (Middle) Two dimensional RM maps integrated up to redshifts of 0.1 and 5.0. (Right) SFs of RMs through filaments up to redshift of 5.0. Black lines and marks are observed SFs toward high Galactic latitudes, and color lines are expected SFs of the IGMF (see Akahori & Ryu 2011 for details).

References

Akahori, T. & Ryu, D. 2010, *ApJ*, 723, 476
Akahori, T. & Ryu, D. 2011, *ApJ*, 738, 134
Akahori, T., Ryu, D., Kim, J., & Gaensler, B. M. 2012, *ApJ*, submitted
Ryu, D., Kang, H., Cho, J., & Das, S. 2008, *Science*, 320, 909

Highlights of Astronomy, Volume 16
XXVIIIth IAU General Assembly, August 2012
T. Montmerle, ed.

doi:10.1017/S1743921314011764

The mystery of cosmic magnetogenesis

Christos G. Tsagas

Section of Astrophysics, Astronomy and Mechanics, Department of Physics
Aristotle University of Thessaloniki, Thessaloniki 54124, Greece
email: tsagas@astro.auth.gr

Abstract. The origin of the large-scale magnetic fields remains an open question, despite the efforts and the continual detection of magnetic fields in the universe. Primordial magnetism could answer the question of cosmic magnetogenesis, but there are still obstacles to overcome.

Keywords. magnetic fields, cosmology: theory, early universe

Magnetic fields appear to be everywhere in the universe. The Milky Way and other spiral galaxies have coherent magnetic fields of μG-order magnitude, while galaxy clusters and remote protogalactic clouds also carry fields of similar strength (Han & Wielebinski 2002, Vallée 2011). Moreover, recently, there have been independent reports of relatively strong magnetic fields (around 10^{-15} G) in intergalactic voids (Neronov 2010, Tavecchio *et al.* 2010, Ando & Kusenko 2010). Despite this widespread presence, however, the origin of cosmic magnetism remains a mystery and an open question.

It is generally believed that the galactic fields have been amplified and sustained by some kind of dynamo action (Brandenburg & Subramanian 2005). Dynamos, however, cannot work without an initial seed field. The latter can be the result of astrophysical processes, operating after recombination, or have primordial origin. Early-time magnetogenesis is attractive because it can in principle explain all the magnetic fields seen in the universe, especially those at high redshift and in empty intergalactic space. Nevertheless, generating primordial fields that could sustain the galactic dynamo is not a problem-free exercise (Kandus, Kunze & Tsagas 2011). Seed fields created between the end of inflation and recombination are generally too small in size. The inverse cascade of magnetic helicity could increase the coherence length of the post-inflationary fields, but requires highly helical magnetic seeds. Inflation can solve the size problem, but magnetic fields that survived a period of standard inflation are typically too weak. Over the years there have been many attempts to solve the weakness problem of the inflationary magnetic fields. Success, however, comes at a cost. Almost all of the proposed scenarios operate outside classical electromagnetism, or introduce some other kind of new physics.

Overall, although the continual detection of magnetic fields in the universe makes the questions regarding their origin, evolution and implications increasingly pressing, the issue of cosmic magnetism remains open and the subject of ongoing debate.

References

Han, J. & Wielebinski, R. 2002 *Chin. J. Astron. Astrophys.*, 2, 293
Vallée, J. P. 2011 *New Astron. Rev.*, 55, 91
Neronov, A. 2010 *Science*, 328, 73
Tavecchio, F., *et al.*2010 *Mon. Not. R. Astron. Soc.* 406, L70
Ando, S. & Kusenko, A. 2010 *Ap. J. Lett.* 722, L39
Brandenburg, A. & Subramanian, K. 2005 *Phys. Rep.* 417, 1
Kandus, A., Kunze, K. E. Tsagas, C. G. 2011 *Phys. Rep.*, 505, 1

Highlights of Astronomy, Volume 16
XXVIIIth IAU General Assembly, August 2012
T. Montmerle, ed.

doi:10.1017/S1743921315001611

Preface: Special Session SpS5 The IR view of massive stars: the main sequence and beyond

Though multiwavelength astronomy was born over fifty years ago, the wide-spread use of multiwavelength diagnostics is a more recent phenomenon. Even in the last decade, astronomers continued to rely on the optical domain for the bulk of their analysis. However, this is certain to change, as most of the current and future instruments are increasingly dedicated to observations in the infrared, from the near- to the far-infrared bands.

While the infrared domain is well established in research on low-mass stars, especially the very low-mass ones, the enormous potential for the study and analysis of infrared emission from high-mass stars has yet to be realized. Many advantages of the infrared must however be acknowledged, like its strong potential for circumstellar material and atmosphere diagnostics, and its insensitivity to obscuration. This is important when one considers the typical distances one works to locate massive stars, often in the plane of our Galaxy. The use of infrared diagnostics is particularly relevant with regards to the first generation of stars, thought to be very massive.

This Special Session provided an opportunity to discuss the results obtained for massive stars from existing infrared facilities (VLTs/VLTI, Spitzer, Herschel, CRIRES, GAIA,...) as well as tools for interpreting infrared data (e.g. atmosphere modeling) and observing capabilities of future facilities (ELTs, JWST,...). It was split into three topics.

The first topic of this Special Session dealt with obscured and distant clusters. To improve the knowledge of the (rare) massive objects, the infrared domain is crucial as it reveals obscured and/or distant clusters, like those close to the Galactic Center. Such studies, by providing many new objects to work on, enable us to better understand the massive stars as a population and to reveal the strong impact of massive stars on the environment and clusters themselves, thanks to the eroding effect of their energetic radiation and dynamical interactions.

The Session continued with presentations on the determination of stellar and wind parameters of (mostly evolved) massive stars, which remain poorly constrained. However, new wind diagnostics extended to include infrared data are being developed. They provide additional leverage to select between competing wind models (including different clumping scenarios). Metallicity studies also benefit from access to infrared wavelengths, particularly with regards to tracing the chemistry of circumstellar environments.

Finally, the third topic of this Session considered matter ejection by massive stars (winds, LBV eruptions, supernovae). With the recent reduction in 2̆01cobserved2̆01d mass-loss rates (by a factor of about 3), episodic matter ejection represents crucial, but poorly understood mechanisms needed for understanding the evolution of massive stars.

The SOC - Jura Borissova (Chile), Margaret Hanson (USA), Fabrice Martins (France), Paco Najarro (Spain), Yaël Nazé (Belgium, chair), Barbara Whitney (USA)

Highlights of Astronomy, Volume 16
XXVIIIth IAU General Assembly, August 2012
T. Montmerle, ed.

doi:10.1017/S1743921314011776

SpS5 - I. Obscured and distant clusters

M. M. Hanson[1], D. Froebrich[2], F. Martins[3], A.-N. Chené[4], C. Rosslowe[5], A. Herrero[6] and H.-J. Kim[7]

[1]University of Cincinnati, USA email: hansonmm@uc.edu
[2]University of Kent, U.K. [3]LUPM, CNRS & Université Montpellier II, France
[4]Universidad de Concepcion/Valparaiso, Chile [5]University of Sheffield, U.K.,
[6]Instituto de Astrofisica de Canarias and Universidad de La Laguna, Spain,
[7]Seoul National University, Korea

Abstract. This first part of Special Session 5 explored the current status of infrared-based observations of obscured and distant stellar clusters in the Milky Way galaxy. Recent infrared surveys, either serendipitously or using targeted searches, have uncovered a rich population of young and massive clusters. However, cluster characterization is more challenging as it must be obtained often entirely in the infrared due to high line-of-sight extinction. Despite this, much is to be gained through the identification and careful analysis of these clusters, as they allow for the early evolution of massive stars to be better constrained. Further, they act as beacons delineating the Milky Way's structure and as nearby, resolved analogues to the distant unresolved massive clusters studied in distant galaxies.

Keywords. infrared: stars, stars: early-type, galaxies: clusters: general, stars: luminosity function, mass function, stars: fundamental parameters

---

1. Introduction

Two decades ago, the study of large star forming regions was limited to distant objects, with R136 in the LMC as the closest example. However, now, several massive young clusters† have been identified in our Galaxy. First, deep near-infrared studies of the Galactic center region serendipitously identified at least three spectacular and very massive clusters with unique stellar and cluster properties. Then near-infrared surveys, such as the Two Micron All Sky Survey (2MASS) began uncovering massive open clusters deep in the Galaxy. Numerous new clusters have been found starting with several 2MASS-based targeted near-infrared surveys (Dutra & Bica 2000, 2001, Froebrich *et al.* 2007), GLIMPSE (Mercer *et al.* 2005) and DENIS (Reyle & Robin 2002). The search continues with still newer surveys, such as the VISTA Variables in the Vía Láctea (VVV, Borissova *et al.* 2011). Moreover, it appears young massive clusters were already known to us, as several long-known optical clusters. Westerlund 1 & 2, Stephenson 2, Cygnus OB2 have been found to be far more massive than previously thought (Hanson 2003).

The reasons to study large populations of young clusters are numerous. First, such clusters help us to better reveal galactic structure (Fig. 1) and they constrain the history of the chemical evolution throughout the Milky Way. They also provide critical benchmarks to test stellar evolution. For example, having many clusters imply sampling different environments, an essential tool for testing theoretical models. Massive clusters are especially important in this context, as they allow us to beat the small-number statistics that plague typical intermediate-mass cluster studies. With a massive stellar cluster

† Defined as being a few Myr and about 10^4 $M_\odot$ in mass - objects with smaller masses would not be detected if distant or obscured, objects with larger masses are few in number.

they may display several red supergiants, or several very massive core H-burning stars simultaneously. Finally, the most massive clusters in our Galaxy, and where there is access to study the individual stars, are close analogs to distant starbursts for which only integrated properties are known. A more in depth knowledge of nearby massive stellar clusters will enhance our understanding of extragalactic stellar clusters.

To improve further our knowledge of distant and obscured clusters, this field will benefit from the forthcoming advent of several new facilities: Gaia, JWST, ALMA, SKA. Besides the detection of new objects, they will provide a better characterization of the identified cases. Notably, an improvement of the distance determinations is awaited, not only for calibration purposes, but also to understand the Galactic structure. Discrepancies have been noted between trigonometric, kinematic and spectrophotometric estimates (Moisés *et al.* 2011). Kinematic values may be wrong if the cluster's velocity is not only fixed by the global kinematic field of the Milky Way (*e.g.* if the cluster formation results from the collision of two moving clouds).

2. The search and characterization of obscured and distant young stellar clusters

Young and massive clusters of stars are rare, distant and often obscured by surrounding gas and dust. However, their energy and momentum input into the ISM via e.g. stellar winds, ionizing radiation or SN explosions is extensive. A first step to understand the details of the evolution and feedback from massive stars is to find and characterize their birth clusters

Infrared-based searches for obscured and distant massive clusters in the Milky Way has progressed greatly since the first 2MASS cluster searches. Higher spatial resolution, deeper photometry, and numerous new algorithms have been obtained and applied to discover ever more candidate stellar clusters. With so many searches underway, will the numbers ever reach the 30,000 stellar clusters in the Milky Way as predicted by Portegies Zwart *et al.* (2010)? Currently, the cluster mass function in the Milky Way is underrepresented in young massive clusters older than a few tens of millions of years and less than 100 million years. This has driven a boom of stellar cluster searches in the Milky Way using recent, current and future surveys such as 2MASS, Spitzer, WISE, VISTA. They provide outstanding, homogenous large-scale data sets and are ideally suited for such searches.

Unbiased, targeted and serendipitous methods have already been mentioned as a means for locating new young clusters. Also, many "special" targets point towards massive cluster candidates: γ-ray or X-ray sources, and Luminous Blue Variables (LBV) or Wolf-Rayet (WR) Stars might be identified and are typically located in or near a massive young stellar cluster (Homeier *et al.* 2003). Locating any kind of stellar remnant or highly massive star provides a high likelihood of locating a massive young cluster nearby.

Despite the abundance search methods, targeted, simple visual (by eye) searches, sophisticated algorithms searching for increased stellar densities or photometric color ranges, no method can be sure to catch all clusters of a certain distance or luminosity (or mass) limit and each has its own biases. Variations, predominantly in extinction, but also the complexity of the background star field, the concentration level of the stellar cluster, and the age and mass of the cluster will strongly influence the detectability of real clusters (Hanson *et al.* 2010). This makes the derivation of such values as the cluster luminosity function and the cluster destruction timescales nearly impossible to obtain confidently for the Milky Way. A number of cluster databases now exist on the web (WEBDA, Dias, Kharchenko, SAI Catalog, FSR listing, etc.), yet they represent

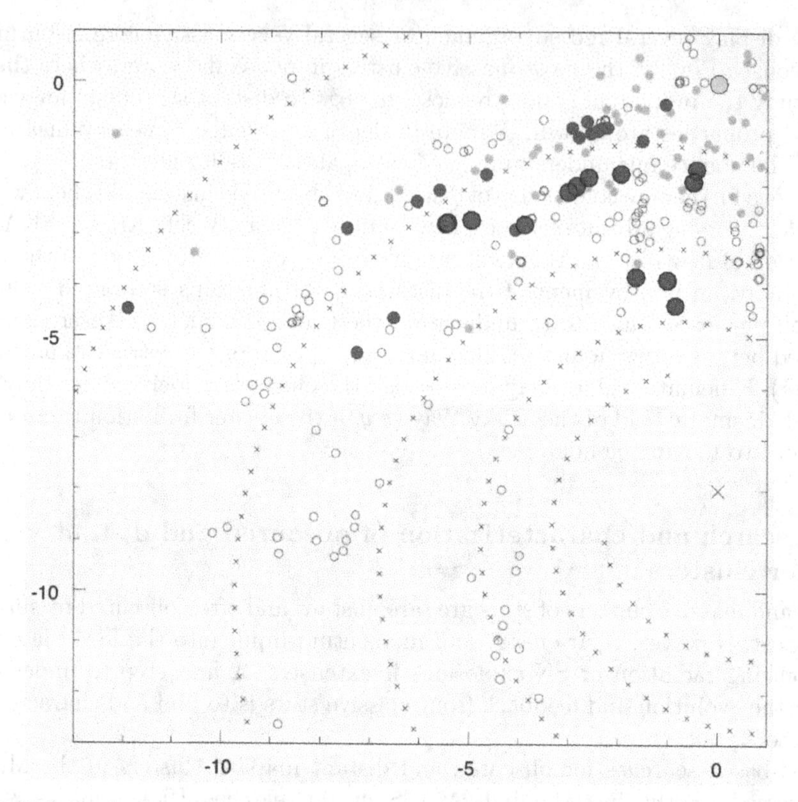

Figure 1. Infrared clusters as spiral arm tracers. Vallée (2008) model for the Milky Way spiral arms is given in x's, the Sun is at [0,0]. Small open red circles mark HII tracers, small green small filled circles mark Cepheid distances. The larger, filled blue and even larger filled red dots represent young clusters, optical and infrared, respectively, being studied by the VVV survey (figure from Chené *et al.* 2012).

just 10% of what is expected to exist in the Milky Way. Moreover, false positives are expected to be rather high with the deepest searches.

Many of the same methods used for obtaining age, distance and extinction of optical clusters are applied in the infrared for these clusters with some success: color-magnitude diagrams, spectral classification, etc (*e.g.* Borissova *et al.* 2011, Chené *et al.* 2012b). However, characterization of infrared clusters is more challenging than in the optical due to the weaker dependence of infrared colors with temperature variation and the strong contamination by red-giant field stars. For clusters in the age range of a few to tens of millions of years, their most massive stars may exist as exceedingly bright, red supergiants. Their blaring luminosity impedes study of the rest of the cluster (including identification of the cluster in the first place). Radial velocities combined with galactic rotation curves can provide some independent values for distance (though see Moisés *et al.* 2011). Maser parallaxes are most promising. In massive young clusters, there are sure to be red supergiants and these can provide strong radio masers, useful for parallax and proper motion studies to derive distance and cluster membership (Zhang *et al.* 2012).

Cluster mass is always a difficult characteristic to determine as it is strongly dependent on assumptions of the initial mass function, lower mass cutoff and binary fraction. Virial masses can sometimes provide better insight, but not many infrared clusters have been studied in this way, due to the lack of high resolution spectra of constituent stars. Virial mass has been used as a mass measure with integrated spectra of extragalactic clusters

(Mengel *et al.* 2002). Promising future surveys include exploring the time domain (*e.g.* VVV and LSST) for variability and proper motions, as well as narrow band surveys to locate variable (LBVs) or emission-line stars (WRs) expected to be associated with massive stellar clusters.

3. Massive stars in the Galactic Center

The center of the Galaxy is a unique environment to study massive stars. On a scale of about 100 parsecs, three young massive clusters are encountered: the Arches, the Quintuplet and the central cluster. In addition to the massive clusters, a number of "isolated" massive stars are also found. In total, about 90 Wolf–Rayet stars and more than 250 OB stars have been identified in the last 20 years. The Galactic Center is hidden behind 25 to 30 magnitudes of visual extinction, rendering optical observations unfeasible. Hence, it is perfectly suited for infrared studies. In addition, its distance is well constrained: 8 ± 0.5 kpc (Gillessen *et al.* 2009). This is important since it allows good luminosity determinations.

The Arches cluster: It was first discovered by Nagata *et al.* (1995) and Cotera *et al.* (1996). It is the youngest and also the most compact of the three clusters. The current census of the massive star population includes 13 Wolf–Rayet stars and more than 100 OB stars. All Wolf–Rayet stars are late–type WNh stars. There is no WC or red supergiant. The quantitative analysis of their fundamental properties, as well as of the brightest O supergiants, confirms that the population is young: 2–3 Myr (Figer *et al.* 2002, Martins *et al.* 2008). The analysis of the H, He, C and N abundances of the WNh stars indicates that they are still core–hydrogen burning objects. They are very luminous. Some of them might have masses in excess of 120 $M_{\odot}$ (Martins *et al.* 2008, Crowther *et al.* 2010). The nitrogen content of the WNh stars is a metallicity indicator (Najarro *et al.* 2004). A value of $Z = 1.2–1.4\ Z_{\odot}$ is preferred. The present-day mass function is still debated: early studies by Figer *et al.* 1999 and Stolte *et al.* (2005) indicated a top–heavy mass function ($\Gamma = -0.7..-0.9$), but the recent analysis by Espinoza *et al.* (2009) points to a value almost consistent with the Salpeter slope ($\Gamma = -1.1 \pm 0.2$). What is clear from all studies is that the cluster is mass segregated: the most massive stars are located in the central region. This segregation could be due to dynamical interactions (Kim *et al.* 2006, Harfst *et al.* 2010) or could be primordial (Dib *et al.* 2007).

The Quintuplet cluster: The stellar content comprises 21 Wolf–Rayet stars (13 WN + 8 WC) and about 85 OB stars (Liermann *et al.* 2012). Several WN stars are the same type of objects as the Arches WNh stars: very luminous with a high hydrogen content (Leirmann *et al.* 2010). The analysis of the OB population and of WR/O ratios indicate an age of 4 ± 1 Myr (Liermann *et al.* 2012), slightly older than the Arches. Hußman *et al.* (2012) derived the mass function and found a slope slightly steeper than the Salpeter mass function ($\Gamma = -0.68 \pm 0.13$). The Quintuplet clusters hosts a number of peculiar objects: Tuthill *et al.* (2006) showed that at least two of the five WC stars from which the cluster was named are binaries showing "pinwheel nebulae"; the Pistol star, a Luminous Blue Variable, is also located very close to the cluster. Once thought to be the most massive star in the Galaxy, it is now considered to be a binary (Martayan *et al.* 2011) with the most massive component of which being a ~100 $M_{\odot}$ star. Detailed analysis of its infrared spectrum shows that it has a solar Fe content, and α elements abundances larger by a factor of two compared to the Sun (Najarro *et al.* 2009).

The Central cluster: It hosts the supermassive black hole at the position of the radio source SgrA*. The current census of the massive star population includes 31 Wolf–Rayet stars (18 WN + 13 WC) and more than 140 OB stars. Three red supergiants are also located in the cluster. Based on the positions of OB stars in the HR diagram, Paumard *et al.* (2006) estimated an age of 6±2 Myr. The detailed analysis of the physical properties of most of the Wolf–Rayet population allowed Martins *et al.* (2007) to define the following evolutionary sequence for stars with masses $\sim$ 50 $M_\odot$:

$$(\mathrm{Ofpe/WN9} \leftrightarrow \mathrm{LBV}) \rightarrow \mathrm{WN8} \rightarrow \mathrm{WN8/WC9} \rightarrow \mathrm{WC9}$$

The mass function has been analyzed by Bartko *et al.* (2010): it is extremely top–heavy between 1" and 12" from the position of SgrA*, and becomes consistent with Salpeter beyond. This is consistent with the simulations of star formation in a gas cloud falling in the Galactic Center (Bonnell & Rice 2008). Binaries are present in the cluster (Martins *et al.* 2006) and can explain the spectral type of the bow–shock star IRS8, much earlier than any other O star in the cluster (Geballe *et al.* 2006)

Isolated massive stars: Two main techniques have been used to identify massive stars outside the three clusters: Pα narrow-band imaging (Cotera *et al.* 1999, Homeier *et al.* 2003, Mauerhann *et al.* 2010) and cross–correlation between Chandra and 2MASS catalogs (Muno *et al.* 2006, Mikles *et al.* 2006). Follow-up spectroscopy has revealed the presence of Wolf–Rayet and OB stars of all types. Currently, 26 WR stars (16 WN + 10 WC) are known outside the clusters. They might have been ejected from those clusters by dynamical interactions. But they might as well result from a constant, low efficiency star formation in the Galactic Center, on top of which extreme starbursts events leading to massive clusters happen from time to time. The Galactic Center is a unique environment for the infrared studies of massive stars properties, formation and evolution.

4. Young stellar clusters in the VVV Survey

VISTA Variables in the Vía Láctea (VVV) is one of the six ESO Public Surveys selected to operate with the new 4-meter Visible and Infrared Survey Telescope for Astronomy (VISTA). It is currently performing unprecedented deep infrared observations of the Galaxy's bulge and an adjacent section of the mid-plane. This unique database is being used to achieve a large (700-800, including some discovered by us in VVV data; Borissova *et al.* 2011, Bonatto *et al.* in prep), statistically significant sample of star clusters with homogeneously derived (i.e. all observed with the same instrument and set-up) physical parameters (including angular sizes, radial velocities, reddening, distances, masses, and ages). This sample will represent a real gold mine for testing and constraining the theories of star cluster formation and evolution.

The first phase of the VVV survey was completed last year and deep near infrared images in all the filters of the whole area covered by the survey have been obtained. These data and the analysis methods have been presented in a recent paper (Chené *et al.* 2012a). For all the clusters and cluster candidates included in the survey statistically decontaminated and analyzed the color-magnitude diagrams (CMDs) have been calculated. A spectroscopic follow-up has been conducted to confirm their cluster nature, and to derive the spectral types and distances of the brighter cluster members.

As a first step, efforts were focused on young open clusters in their first few Myrs. During this period, which corresponds to Phase I in the recent classification of Portegies Zwart *et al.* (2010), stars are still forming and the cluster contains a significant amount of gas. The cluster evolution during this phase is governed by a complex mixture of gas dynamics, stellar dynamics, stellar evolution, and radiative transfer, and is not completely understood (Elmegreen 2007, Price & Bate 2009). Thus many basic (and critical) cluster properties, such as the duration and efficiency of the star-formation process, the cluster survival probability and the stellar mass function at the beginning of the next phase are uncertain. A subsample of fairly massive young clusters have been recently studied and

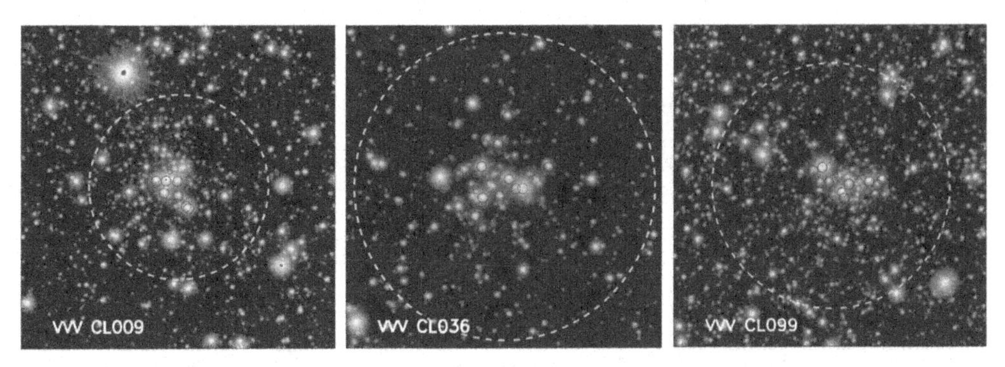

Figure 2. New infrared stellar clusters, identified by the VVV Survey (Chené *et al.* 2012b).

presented by Chené *et al.* (2012a, 2012b) and Baume *et al.* (in prep). On a hybrid map of of the Galaxy, plotting this subsample in combination with data from previous papers, it shows that clusters detected in the optical regime form a group offset and parallel to the group of clusters studied in the IR based on VVV data (Fig. 1). In fact, both traces may reveal the internal structure of only one arm (the Carina arm), as can be seen in other galaxies (e.g. M51). Clusters discovered in the VVV data would be embedded in dark clouds behind the young (optical) populations at the front end of the arm.

5. Distribution of WR stars in the Milky Way from near-IR surveys

The past two decades have seen a three-fold increase in the number of known Galactic WR stars. As the majority of these newly discovered objects are visually obscured, it is essential to develop our capability to analyze these rare and valuable stars at IR wavelengths. Furthermore, a comprehensive knowledge of the Galactic WR distribution will facilitate important tests of massive star evolution theory. A near-IR classification scheme has been developed for Nitrogen (WN) and Carbon (WC) sequence WR stars from IRTF SpeX 1-5μm spectra of 29 robustly (optically) classified WR stars. The resulting scheme is based primarily on emission line equivalent width ratios throughout this spectral range. For WN stars, an exact spectral type is achievable with a medium/low resolution 1-5μ m spectrum. An accuracy of ± 1 is achievable with a J- or K-band spectrum alone, and ± 2 is achievable with only a H- or L-band spectrum. For WC stars, it is possible to obtain an exact spectral type based on a medium/low resolution $J-$band spectrum only. Accuracies of ± 1 and 2 are achievable using K-band and H- or L-band spectra, respectively.

Using this new near-IR classification capability, revised spectral types of a large number of recently reported WR discoveries have been obtained. With accurate spectral types in place, an absolute magnitude-subtype calibration has been performed comprising of 83 and 22 Galactic WN and non-dusty WC (figure 3) stars with distance measurements taken from the literature. The extent of reddening to each object in this calibration sample is calculated by a color excess method invoking the intrinsic WR colours of Crowther *et al.* (2006). As can be seen in figure 3, a factor of ~ 5 increase in the WR sample size over Crowther *et al.* (2006) provides, for the first time, an absolute magnitude value for each individual WN and WC subtype.

The availability of reliable M_K values for each WR subtype has allowed us to map the spatial distribution of the known Galactic WR population. So far this has been accomplished for 332 Galactic Wolf-Rayet stars which appear to be without a massive companion. With knowledge of the Galactic metallicity gradient, this population of 332

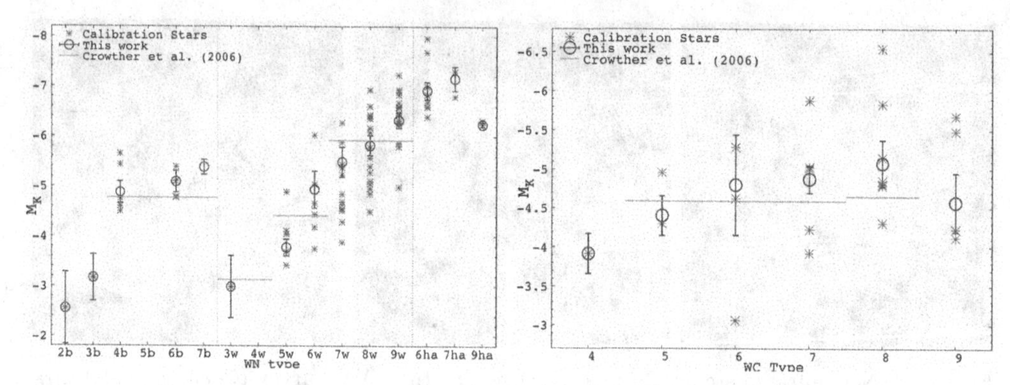

Figure 3. Absolute magnitude-subtype calibration of 83 WN and 22 WC Wolf-Rayet stars with distance measurements from the literature

WR stars has been divided into sub-solar ($\log(\frac{O}{H}) + 12 \simeq 8.5$), solar ($\log(\frac{O}{H}) + 12 \simeq 8.7$) and super-solar ($\log(\frac{O}{H}) + 12 \simeq 8.9$) metallicity regions. Comparing the WC:WN number ratios across this metallicity range to the predictions of non-rotating evolutionary models by Eldridge & Vink (2006) and rotating evolutionary models by Maeder & Meynet (2005), shows that at solar metallicity and above, the predicted WC:WN number ratio is too low in the rotating case and too high in the non-rotating case. At this time, the number of WR stars in the sub-solar Galactic population is too low to provide a meaningful comparison to evolutionary predictions. A possible interpretation of the discordance of our number ratios with evolutionary predictions will be attempted after the inclusion of further Galactic WR stars, specifically those in binary systems.

6. Massive clusters in the Milky Way

The number of massive young stellar clusters ($M_{cluster} \geqslant 10^4$ $M_\odot$) known in the Milky Way has dramatically increased in the last years thanks to recent advances in NIR observations and the exploitation of large NIR photometric surveys. Rough estimates, however, indicate that the actual number of massive young stellar clusters may exceed by at least a factor of five the number of known ones (Hanson & Popescu 2008). And the number of intermediate mass clusters is much larger.

The MASGOMAS (**MA**ssive **S**tars in **G**alactic **O**bscured **MA**ssive cluster**S**), a project aimed at studying the stars in these clusters (see Marín-Franch et al. 2009) was started. this effort was begin with a search of massive young obscured clusters in a limited region of the Milky Way using photometric cuts optimized for the search of OB-stars (Ramírez-Alegría, Marín-Franch & Herrero, 2012). Moreover, the search was limited to stars with Ks < 12.5 to allow a spectroscopic follow-up with LIRIS@GTC.

In this way, two new clusters have been found in the Milky Way. The first, MASGOMAS-1 (Ramírez-Alegría *et al.* 2012), was identified as candidate by the presence of a concentration of candidate OB-stars in the photometric diagrams and was confirmed by the presence of 17 stars classified as O9-B1V, 3 M supergiants and one A supergiant among the 28 stars observed spectroscopically in the follow-up programme (of course, this is a minimum for the cluster population). Based on this spectroscopically confirmed population, a distance to MASGOMAS-1 was found of 3.53+/−1.48 kpc, with a mass of $1.94 \times 10^4 M_\odot$ (using the IMF from Kroupa 2001) and an age between 6.5 and 10 Myr, based on the simultaneous presence of red supergiants and O9V stars.

The second cluster, MASGOMAS-4, has two cores surrounded by strong nebulosity at 5.8 μm. One of the cores is the HII region Sh2-76 E, known as an active massive star forming region (Wu *et al.* 2007). It is unclear whether the two cores are physically related. Spectroscopic studies of the stellar population of both cores has found that they are at the same distance from the Sun and have the same average extinction. This indicates that both cores are physically related, and that the whole cluster, at a distance of 1.9 kpc, has a mass of $2.19 \times 10^3 M_\odot$, and an age < 3 Myr, based on the presence of the HII region. However, there is a significant difference between both cores: core B displays in its color-color diagram a population of young, possible pre-main sequence objects, while this population is absent in core A. The reason for this difference is not yet known.

7. Massive stars with infrared excesses in G54.1+0.3

G54.1+0.3 is a young (~3,000 yr) Crab-like supernova remnant (SNR) that hosts numerous point-like, strong infrared-excess sources in its outer IR loop. Color-color diagrams reveal a cluster of possibly hot, massive stars, with $A_V = 7 - 9$. Spectral energy distributions (SED) of these sources show large mid-infrared excesses, with a strong dip in emission between $6 - 10$ μm. Infrared classification spectroscopy indicates the brightest are late-O and early-B, with a distance of about 6 kpc. Did the supernova trigger this star formation? Models of the SED indicate the dust is located at a distance from the stars, probably originating from the SN, indicating a causal connection.

8. The Scutum Complex

The region of the Galactic Plane between $l = 24$ and $l = 30$ contains a number of clusters of red supergiants, which are believed to be very massive. Neg have developed techniques to identify red supergiants from infrared photometric catalogues and used them to search for such objects over a wide field. Hundreds of red supergiants have been found with radial velocities close to the terminal value corresponding to the Scutum tangent. Their distribution has been analyzed to characterize this vast region of star formation, known as the Scutum Complex. Red supergiants in such numbers can be used to trace Galactic structure better than any other usual tracer. Finally, numerous methods are being explored to detect and observe red supergiants through high extinction.

9. Finding Red Supergiants in the Galaxy and Measuring their Distances

Red supergiants evolve from moderate to fairly high-mass stars. They are thus tracing recently formed, high-mass clusters. They are extremely bright in the infrared, and serve as beacons amongst the dimmer main-sequence objects in a heavily extinguished cluster. Most uniquely, they have the potential to be naturally strong radio masers, created by simple molecules in their outer circumstellar envelopes (Chapman & Cohen 1986). These masers make excellent kinematic probes, as their line-of-sight velocities follow directly from the centroid of the line profiles, revealing kinematic distances when used with a rotation curve. Moreover, very long baseline interferometric observations can provide direct determination of distances and proper motions (Messineo *et al.* 2002). More sophisticated methods for locating these very useful objects in the Milky Way is well underway. They use JHK colors and look for the telltale sign of an extended circumstellar shell where a radio maser might be detected (Messineo *et al.* 2012).

10. Poster contributions to this session

A. Bonanos presented new results from an ongoing survey to discover and characterize massive binaries in young massive clusters through out the Milky Way, with targeted efforts in Westerlund 1, the Danks and Arches clusters. Massive binaries are important to constrain theoretical models of star formation and evolution in single and binary stars.

A. Damineli presented a near-infrared study of the stellar content of 35 HII regions in the galactic plan, 24 of which are giant HII regions. Their goal is to confirm, using color-magnitude diagrams, if the kinematic distances seem consistent with the stellar luminosities. About half are in agreement, but the remaining half appear to be closer than their kinematic distance indicates (Moisés *et al.* 2011). Several are further confirmed to be closer than their kinematic distances based on additional spectrophotometric parallaxes.

T. Geballe presented new $J-$ and $H-$band spectra of several of the bright stars in the young, very massive galactic center Quintuplet cluster. This allowed the photospheres to be clearly seen for the first time, as longer wavelength spectra had been dominated by emission from their dusty, warm cocoons. GCS3-1, GCS3-4 and GCS4 all resemble late-type WC stars. $H-$band spectra of GCS3-2 reveal lines of He I and C II initially in 2009, but then missing in 2011, possibly due to increased dust production.

C.C. Lin presented a star-counting algorithm to search the 2MASS point source catalog for density enhancements indicative of new stellar cluster candidates. One such enhancement located the open cluster G144.9 + 0.4, which they went further to analyze using proper motions to constrain membership. Several classical T-Tauri stars were found to be members of the cluster. Along with fits to their spectral energy distribution, these indicate a very young age for the cluster.

N. Panwar presented a multi-wavelength study of the HII region IC 1805. They analyzed the spatial extend and structure of the young stellar objects, identified by their spectral energy distributions. These were found to be mainly intermediate-high mass stars.

11. Conclusions

The identification and characterization of obscured, distant massive clusters is still in its early phases, marked by an initial excitement over the awareness of the existence of young massive clusters being in and an important component of the Milky Way. How might these clusters help us constrain the structure and evolutionary history of the Milky Way? What critical questions do Milky Way massive clusters uniquely answer to increase our understanding of the formation and evolution of massive stellar clusters in *all galaxies*? Now is the time for the community to assess how the upcoming new facilities, combined with our own innovative ideas might be applied in the next decade to make this a very promising research direction.

References

Bartko, H., Martins, F., Trippe, S., *et al.*, 2010, *ApJ* 708, 834
Bonnell, I. A. & Rice, W. K. M. 2008, *Science*, 321, 1060
Borissova, J., Bonatto, C., Kurtev, R., *et al.* 2011, *A&A*, 532A, 131
Chapman, J. M. & Cohen, R. J. 1986, *MNRAS*, 220, 513
Chené, A.-N., Borissova, J., Clarke, J. R. A., *et al.* 2012a *A&A*, 545A, 54
Chené, A.-N., Borissova, J., Bonatto, C., *et al.* 2012b *A&A*, accepted
Cotera, A. S., Erickson, E. F., Colgan, S. W. J., *et al.* 1996, *ApJ* 461, 750
Cotera, A. S., Simpson, J. P., Erickson, E. F., *et al.* 1999, *ApJ* 510, 747

Crowther P. A., Hadfield, L. J., Clark, J. S., *et al.* 2006, *A&A*, 372, 1407
Crowther, P. A., Schnurr, O., Hirschi, R., *et al.*2010, *MNRAS* 408, 731
Dib, S., Kim, J. & Shadmehri, M. 2007, *MNRAS* 381, L40
Dutra, C. M. & Bica, E. 2000, *A&A*, 359L, 9
Dutra, C. M. & Bica, E. 2001, *A&A*, 376, 434
Eldridge, J. J. & Vink, J. S. 2006, *A&A*, 452, 295
Elmegreen, B. G. 2007, *ApJ*, 668, 1064
Espinoza, P., Selman, F. J., & Melnick, J. 2009, *A&A* 501, 563
Figer, D. F., Kim, S. S., Morris, M., *et al.* 1999, *ApJ* 525, 750
Figer, D. F., Najarro, F., Gilmore, D., *et al.* 2002, *ApJ* 581, 258
Froebrich, D., Scholz, A., & Raftery, C. L. 2007, *MNRAS*, 374, 399
Geballe, T. R., Najarro, F., Rigaut, F., & Roy, J.-R. 2006, *ApJ* 652, 370
Gillessen, S., Eisenhauer, F., Trippe, S., *et al.* 2009, *ApJ* 692, 1075
Hanson, M. M. 2003, *ApJ*, 597, 957
Hanson, M. M. & Popescu, B., 2008, Proc. of the IAU Symp. 250 on *"Massive stars as cosmic engines"*, ed. F. Bresolin P. A. Crowther & J. Puls, CUP, p. 307
Hanson, M. M., Popescu, B., Larsen, S. S., & Ivanov, V. D. 2010, *HiA*, 15, 794
Harfst, S., Portegies Zwart, S., & Stolte, A. 2010, *MNRAS* 409, 628
Homeier, N. L., Blum, R. D., Pasquali, A., Conti, P. S., & Damineli, A. 2003, *A&A* 408, 153
Hußmann, B., Stolte, A., Brandner, W., Gennaro, M., & Liermann, A. 2012, *A&A* 540, A57
Kim, S. S., Figer, D. F., Kudritzki, R. P., & Najarro, F.: 2006, *ApJ* 653, L113
Kroupa, P. 2001, *MNRAS*, 322, 231
Liermann, A., Hamann, W.-R., & Oskinova, L. M.: 2012, *A&A* 540, A14
Liermann, A., Hamann, W.-R., Oskinova, L. M., Todt, H., & Butler, K. 2010, *A&A* 524, A82
Maeder, A. & Meynet, G. 2005, *A&A*, 440, 1041
Marín-Franch, A., Herrero, A., Lenorzer, A., *et al.* 2009, *A&A*, 502, 559
Martayan, C., Blomme, R., Le Bouquin, J.-B., *et al.* : 2011, in C. Neiner, G. Wade, G. Meynet, G. Peters (eds.), *IAU Symposium*, Vol. 272, pp 616–617
Martins, F., Genzel, R., & Hillier, D. J., *et al.* 2007, *A&A* 468, 233
Martins, F., Hillier, D. J., Paumard, T., *et al.* 2008, *A&A* 478, 219
Martins, F., Trippe, S., Paumard, T., *et al.* 2006, *ApJ* 649, L103
Mauerhan, J. C., Morris, M. R., Cotera, A., *et al.* 2010b, *ApJ* 713, L33
Mercer, E., Clemens, D. P., Meade, M. R., *et al.* 2005, *ApJ*, 635, 560
Messineo, M., Habing, J. J., Sjouwerman, L. O., Omont, A., & Menten, K. M. 2002, *A&A* 393 115
Messineo, M., Menten, K. M., Churchwell, E., & Habing, H. 2012, *A&A* 537A, 10M
Mikles, V. J., Eikenberry, S. S., Muno, M. P., *et al.* 2006, *ApJ* 651, 408
Moisés, A. P. Damineli, A., Figuerêdo, E., *et al.* 2011, *MNRAS*, 411, 705
Muno, M. P., Bower, G. C., Burgasser, A. J., *et al.* 2006, *ApJ* 638, 183
Nagata, T., Woodward, C. E., Shure, M., & Kobayashi, N. 1995, *AJ* 109, 1676
Najarro, F., Figer, D. F., Hillier, D. J., Geballe, T. R., & Kudritzki, R. P. 2009, *ApJ* 691, 1816
Najarro, F., Figer, D. F., Hillier, D. J., & Kudritzki, R. P. 2004, *ApJ* 611, L105
Paumard, T., Genzel, R., Martins, F., *et al.* 2006, *ApJ* 643, 1011
Portegies Zwart, S. F, McMillan, S. L. W., & Gieles, M., *ARA&A*, 48, 431
Price & Bate 2009, *MNRAS*, 398, 33
Ramírez-Alegría, S., & Marín-Franch, A. Herrero, A. 2012, *A&A*, 541, A75
Reylé, C. & Robin, A. C. 2002, *A&A*, 384, 403
Stolte, A., Brandner, W., Grebel, E. K., Lenzen, R., & Lagrange, A.-M. 2005, *ApJ* 628, L113
Tuthill, P., Monnier, J., Tanner, A., *et al.* 2006, *Science* 313, 935
Vallée, J.-P. 2008, A, 135, 1301
Wu, Y., Henkel, C., Xue, R., Guan, X., & Miller, M. 2007, *ApJ*, 669, L37
Zhang, B., Reid, M. J., Menten, K. M., *et al.* 2012, *A&A* 544A, 42

Highlights of Astronomy, Volume 16
XXVIIIth IAU General Assembly, August 2012
T. Montmerle, ed.

doi:10.1017/S1743921314011788

SpS5 - II. Stellar and wind parameters

F. Martins[1], M. Bergemann[2], J. M. Bestenlehner[3], P. A. Crowther[4], W. R. Hamann[5], F. Najarro[6], M. F. Nieva[7], N. Przybilla[7], J. Freimanis[8], W. Hou[9] and L. Kaper[10]

[1]LUPM, CNRS & Université Montpellier II, Place Eugène Batailon, F-34095, Montpellier, email: fabrice.martins@univ-montp2.fr

[2]Max-Planck-Institute for Astrophysics, Karl-Schwarzschild-Str.1, D-85741 Garching, Germany

[3]Armagh Observatory, College Hill, Armagh BT61 9DG, United Kingdom

[4]Dept of Physics and Astronomy, Univ. of Sheffield, Hounsfield Road, Sheffield, UK, S3 7RH

[5]Lehrstuhl Astrophysik der Univ. Potsdam, Am Neuen Palais 10, 14469 Potsdam, Germany

[6]Centro de Astrobiología, Ctra. Torrejón Ajalvir km 4, 28850 Torrejón de Ardoz, Spain

[7]Dr. Karl Remeis-Observatory Bamberg & ECAP, University Erlangen-Nuremberg, Sternwartstr. 7, D-96049 Bamberg, Germany

[8]Ventspils International Radio Astronomy Centre, Ventspils University College, Inzenieru iela 101a, LV-3600 Ventspils, Latvia

[9]The National Astronomical observatories, Chinese Academy of Science

[10]Astronomical Institute Anton Pannekoek, University of Amsterdam, Science Park 904, 1098 XH Amsterdam, The Netherlands

Abstract. The development of infrared observational facilities has revealed a number of massive stars in obscured environments throughout the Milky Way and beyond. The determination of their stellar and wind properties from infrared diagnostics is thus required to take full advantage of the wealth of observations available in the near and mid infrared. However, the task is challenging. This session addressed some of the problems encountered and showed the limitations and successes of infrared studies of massive stars.

Keywords. infrared: stars, stars: early-type, stars: late-type, stars: fundamental parameters, stars: mass loss, stars: abundances

---

1. Introduction

The use of IR for deriving stellar and wind parameters has become a necessity, as many stars are obscured. Many diagnostics are available in this range, however, so that using IR does not necessarily imply restrictions of the output quality. Line EWs and the appearance of the spectra can be used to derive temperature or spectral types, though morphological spectral types so inferred are less precise. This range is also helpful for mass-loss diagnostics, especially for low-mass loss rates. Two caveats, however, must be noted: the NIR lines form in the wind acceleration zone and they are very sensitive to NLTE and 3D effects - they are thus extremely sensitive to modelling details. For most of the cases, observing K-band is sufficient, but diagnostics improve with J and H spectra, or - even better - if IR is complemented by optical and UV data (terminal velocities, are not constrained by IR, for example). Forbidden lines observed in the IR may arise from outer wind regions, so that also constitute sensitive problems of these poorly known zones.

In this context, an improved knowledge of the atomic and molecule parameters is needed. In addition, 3D wind modelling may become necessary, to take into account both small-scale structures (clumps) and large-scale features (magnetically-compressed winds, wind collisions, corotating regions...). Convection in red supergiants also requires the use of 3D atmosphere models.

2. Stellar analysis from the infrared wavelength range

Infrared analysis of massive stars is mandatory as long as there is a substantial amount of foreground extinction. In that case, the optical and UV flux are much lower than the infrared ones, despite the fact that hot massive stars emit most of their radiation at short wavelength. But infrared photometry and spectroscopy are also useful even if optical/UV data can be obtained, especially for luminosity and extinction determination. This is usually done by fitting the spectral energy distribution of synthetic models to flux-calibrated spectra and/or photometry (e.g., Barniske *et al.* (2008), Crowther *et al.* (2006)). When infrared spectrophotometry is available, a much better accuracy in the luminosity determination can be obtained. The drawback is that uncertainties in photometry lead to larger errors on the luminosity in the infrared than in the optical (Bestenlehner *et al.*, 2011 and this session). For red supergiants, infrared spectroscopy is extremely useful since these stars have their emission peak at those wavelengths.

Obtaining quantitative information on the stellar and wind parameters of obscured objects requires the use of quantitative analysis of infrared spectra. Historically, it is the near–infrared range (especially the H and K bands) which is favoured, since the contribution from thermal emission of the surrounding medium is weak at those wavelengths and increases at longer wavelengths. N. Przybilla presented a detailed view of the difficulties of modelling infrared spectra.

The first problem is related to non-LTE effects. Given the form of the source function, its variation with changes in the departure coefficients can be expressed as follows:

$$|\Delta(S_l)| = \left| \frac{S_l}{b_i/b_j - e^{-\frac{h\nu}{kT}}} \Delta(b_i/b_j) \right| \tag{2.1}$$

We see that at long wavelengths, the ratio $h\nu/kT$ becomes low. In that case, it can be shown that the source function is much more sensitive to changes in the departure coefficients than at optical or UV wavelengths. This implies that model atmosphere must account for non-LTE effects in great detail to correctly reproduce line profiles. Fig. 1 illustrates how different model atoms impact on the departure coefficients of a model with fixed stellar parameters. A simple change of the detailed input atomic data in the models has a significant effect on the departure coefficients, and thus on the line profiles. In Fig. 2, the profiles of Brackett lines are shown. First, the non-LTE effects are clearly seen: the LTE model produces much less absorption that the non-LTE models. But there is also a significant degree of variation among the non-LTE models depending on the type of collisional excitation formalism used. This stresses the need for accurate atomic data, even for hydrogen, to correctly model infrared spectra. For red supergiants, M. Bergemann showed that non-LTE effects are at least as important as for hot massive stars. A correct treatment of the radiative transfer leads to significant variations of the line profiles†.

Given the sensitivity of infrared lines to non-LTE effects and model atoms, it is not surprising to see that line–blanketing effects are also extremely important. N. Przybilla

† Related to this topic, J. Freimanis presented a poster dedicated to polarized continuum radiative transfer in various nontrivial astrophysical coordinate systems, especially relevant for circumstellar gas envelopes.

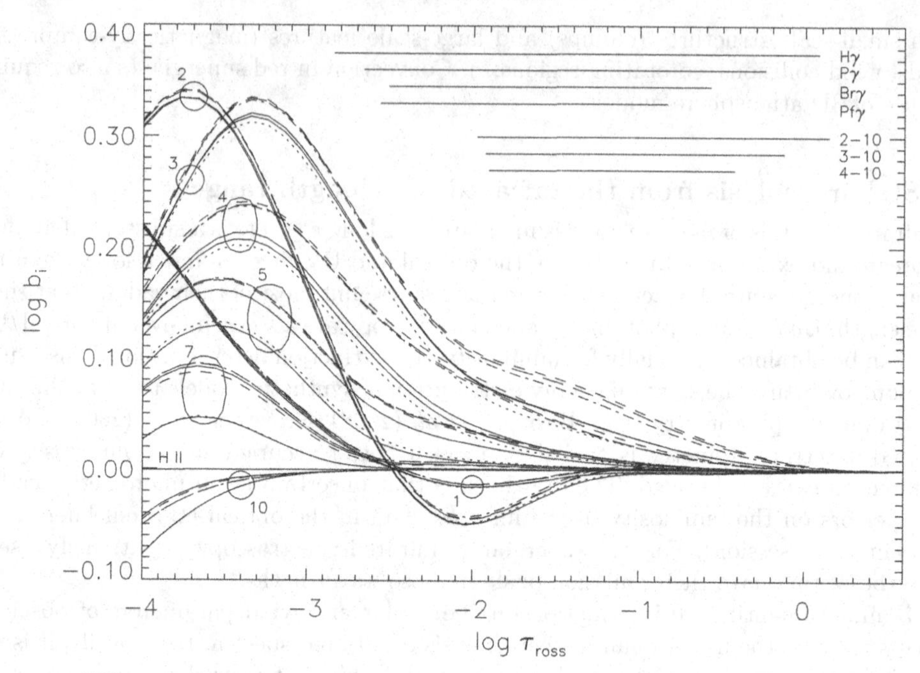

Figure 1. Effect of model atoms on departure coefficients of levels 1–5, 7 and 10 of hydrogen. The levels are indicated by their principal quantum number and are marked by circles. The line formation depth of several transition are indicated by the horizontal solid lines. From Przybilla & Butler (2004).

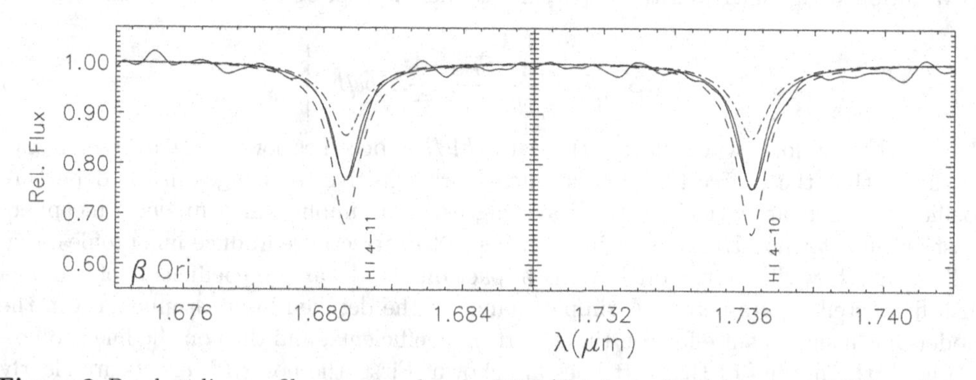

Figure 2. Brackett line profiles computed under LTE (dot-dashed lines) and non-LTE conditions with different collisional excitation prescriptions (dashed: approximate, according to Johnson 1972; dotted: data from new *ab-initio* computations). From Przybilla & Butler (2004).

illustrated the effect of line-blocking on the shape of the HeI 1.083 μm line. The emission is reduced by almost a factor of two when line-blocking is included, due to the different level populations and consequently different line source function. Najarro *et al.* (2006) showed that even the very detail of atomic physics of minor FeIV lines in the extreme UV could affect the shape of the HeI triplet lines, in particular HeI 2.058 μm. This is illustrated in Fig. 3. When reducing the oscillator strength of FeIV 584 Å, the absorption of HeI 2.058 μm decreases. This is caused by the higher and higher population of the

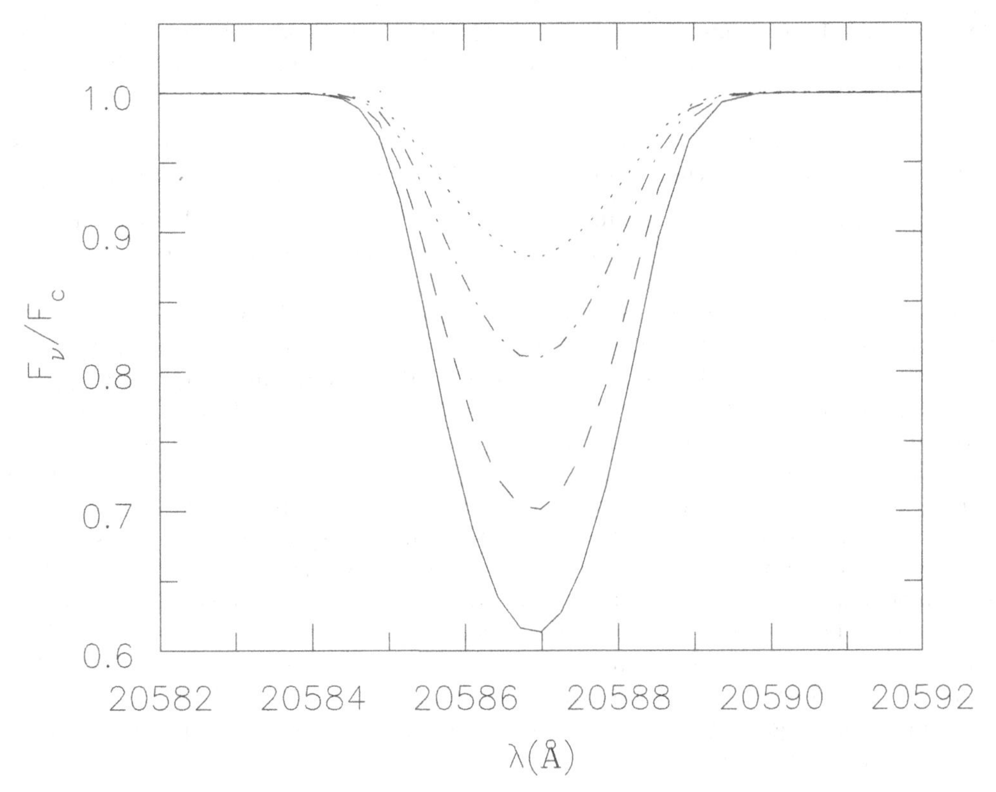

Figure 3. Effect of FeIV 584 Å on HeI 2.058 μm. The solid line is the initial model. The dashed (dot-dashed, dotted) line corresponds to the same model in which the oscillator strength of FeIV 584 Å has been reduced by a factor 2 (5, 10). From Najarro *et al.* (2006).

upper level of HeI 2.058 μm from which photons emitted in the HeI 584 Å line are not stolen by the FeIV 584 Å transition.

3. Key diagnostics

The use of infrared spectrophotometry is helpful to constrain the spectral energy distribution, and thus the luminosity and extinction. But it is the analysis of medium–high resolution spectroscopy that provides information on the other stellar and wind parameters.

Effective temperature: as at other wavelength ranges, the ionization balance is the most widely used method to constrain Teff. The HeI/HeII ratio is used when possible. W.R. Hamann pointed out in this session that the K-band is usually sufficient due to the presence of several HeI lines as well as HeII 2.189 μm. But below about 30000 K, the latter line disappears, and HeII lines present in the H and J bands need to be used. They also disappear rapidly when Teff decreases. N. Przybilla showed that for early B stars, the CII/CIII ionization ratio could be used, taking advantage of CII 0.9903 μm and CIII 1.1981–1.1987 μm (see also Fig. 5). Several examples of temperature determinations have been shown by J. Bestenlehner, N. Przybilla, W.R. Hamann, P. Crowther. N. Przybilla highlighted that comparison of parameters derived purely from infrared diagnostics are usually in agreement with those resulting from optical studies, a conclusion also reached by Repolust *et al.* (2005). N. Przybilla also indicated that for B stars, hydrogen lines can serve as secondary temperature indicators. M. Bergemann discussed different methods to determine T_{eff} for red supergiants, demonstrating that TiO bands severely underestimate temperatures when modeled in 1D LTE compared to more accurate full SED fits (Davies *et al.* in prep.).

Gravity: hydrogen lines of the Brackett and Pfund series are the preferred diagnostics. In general not only their wings (as Balmer lines) but also their line core vary with log g.

Surface abundances: H and He lines present in the JHK bands provide constraints on the He/H ratio. This is true for OB stars, but more importantly for Wolf-Rayet stars which shows significant He enrichment. W.R. Hamann and N. Przybilla illustrated how synthetic spectra could be used to derive the helium content. N. Przybilla also highlighted that for A supergiants, abundances for C, N, O, Mg, Si and Fe could be derived from high-resolution spectroscopy. For O stars, nitrogen and in some cases carbon and oxygen abundances can be obtained. In Wolf-Rayet stars, several CIII, CIV, NIII lines give access to carbon and nitrogen abundances (as shown by F. Martins in session 1). P. Crowther presented determination of oxygen and neon abundances in WC stars using mid-infrared data (Dessart *et al.* (2000), Crowther *et al.*, in prep). The Ne content seems consistent with prediction of stellar evolution, while there might be a problem with oxygen. Abundances of Mg, Ti, Si, and Fe can be obtained from the J-band spectra of red supergiants. The lines of neutral atoms of these elements are very strong, and can be reliably studied even in low-resolution spectra. However, the non-LTE effects in these lines, especially Si I and Ti I, are very large (from -0.4 to $+0.3$ dex), requiring that non-LTE is properly taken into account (Bergemann *et al.* (2012)). The relatively strong emission lines of Mg, Si, Na and Fe in cool LBV stars are also used to derive metallicities (N. Przybilla, F. Najarro).

Mass loss rate: H and He emission lines of Wolf–Rayet stars are good diagnostics of the wind density and thus of mass loss rate (J. Bestenlehner, W.R. Hamann, P. Crowther). For OB stars, F. Najarro introduced Brα as the best mass loss rate indicator (Najarro *et al.* (2011)). Other Brackett lines have been used (especially Brγ) but Brα reacts to density changes even for very low mass loss rates (in the so-called "weak wind" regime). The line core getting *stronger* in emission when mass loss rate decreases, while the line wings are weakening. This peculiar behaviour is illustrated in Fig. 4.

Clumping: F. Najarro showed that the Bracket and Pfund hydrogen lines could be used to constrain the clumping distribution. In particular, the combination of Pfγ formed in the inner wind and Brγ emitted at intermediate depth is a powerful tool to constrain the clumping law in dense wind O stars. P. Crowther presented similar conclusions for Wolf-Rayet stars.

Several diagnostics of the stellar and winds properties of all types of massive stars are thus available. They are used for pioneering and systematic studies of massive stars in obscured environments.

4. Analysis of various types of massive stars

During session 2, several examples of analysis of massive stars from infrared data have been presented. They are listed below.

A Supergiants: N. Przybilla showed pioneering studies of A supergiants using high resolution spectroscopy obtained with CRIRES on the ESO/VLT. The results are presented in Przybilla *et al.* (2009) and Przybilla *et al.* (in prep). He first introduced the results of an analysis based on optical spectra to get a set of stellar parameters for Galactic A

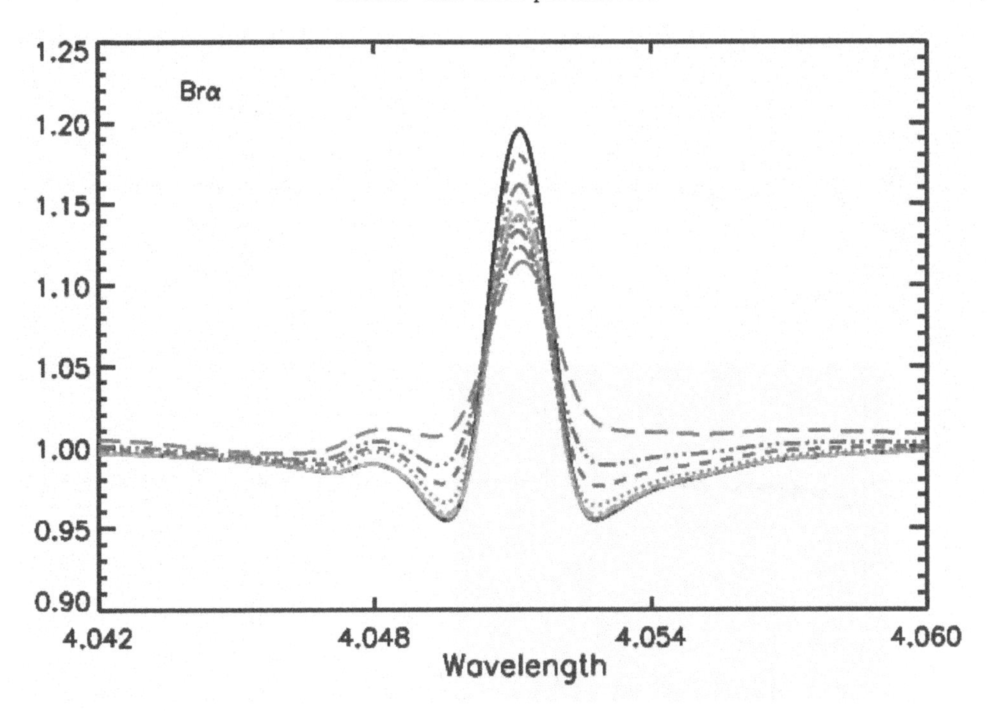

Figure 4. Sensitivity of Brα to mass loss rate. The dashed line is a model with $\dot{M} = 1.0 \times 10^{-7}$ $M_{\odot}$ yr^{-1} while the solid line is a model with $\dot{M} = 5.0 \times 10^{-10}$ $M_{\odot}$ yr^{-1}. Other lines corresponds to models with intermediate mass loss rates. From Najarro *et al.* (2011).

supergiants†. He then presented the results of the near-infrared analysis based on CRIRES spectra, highlighting that similar sets of parameters are found. This pilot study shows that future analysis of A supergiants in external galaxies, accessible with the new generation extremely large telescopes, will be feasible. This will provide information not only on the stars themselves, but also on galactochemical evolution and distance scales.

OB stars: M.F. Nieva also presented analysis of CRIRES spectra of a few B stars, again pointing that results consistent with the optical are found (Nieva *et al.* (2011)). Fig. 5 shows an example of the fit of carbon lines in the early B star τ Sco. Good agreement between model and observation is found if the LTE assumption is dropped. F. Najarro used Brackett and Pfund lines in several weak and dense wind O stars to constrain the clumping distribution and the mass loss rates. He confirmed previous results that late O dwarfs have mass loss rates as low as a few 10^{-10} $M_{\odot}$ yr^{-1}. For the dense wind star Cyg OB2 # 7, he concluded that the clumping law was probably constant throughout most of the wind. Ellerbroek & Kaper presented VLT/X-shooter spectroscopy of young massive stars.

Wolf-Rayet stars: J. Bestenlehner presented the analysis of Of/WN transition objects from a combined UV/optical/near-IR approach. Object VFTS 682 is very luminous but appears to be in isolation, questioning its formation process. Preliminary results on the analysis of O, Of/WN and WN stars indicate that Of/WN stars have wind properties in between those of O and WN stars. They also show a mass loss dependency on the Eddington factor, which indicate a larger L/M ratio for WN stars (Bestenlehner *et al.* in prep.) W.R. Hamann presented studies of several WN and WC stars from

† Hou *et al.* presented a poster on the optical analysis of A stars.

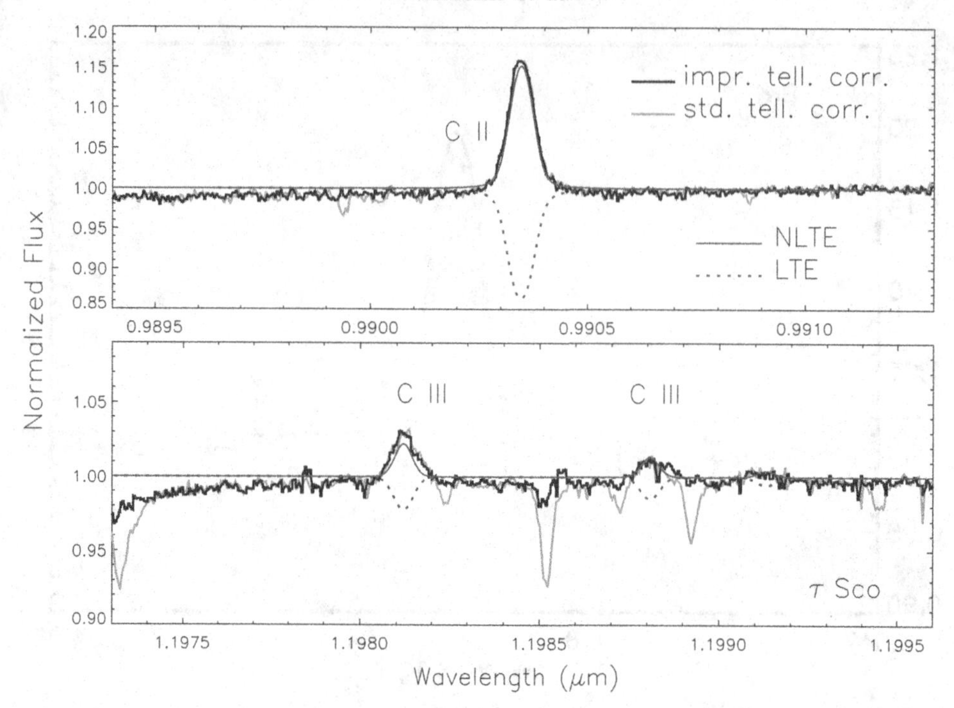

Figure 5. Near-infrared carbon lines of the early B star τ Sco (black and green) together with a LTE synthetic spectrum (blue dots) and a non-LTE synthetic model (red solid). The use of non-LTE models is mandatory. The ratio of CII to CIII can be used to constrain the effective temperature. From Nieva *et al.* (2011).

multiwavelength and pure IR analysis. Of special interest is the so-called "Peony" star in the Galactic Center since its high luminosity makes it possibly the most massive Galactic star (Barniske *et al.* (2008)). Fig. 6 shows the fit of the K-band spectrum of that star. A WN star in the Scutum-Crux arm was also presented, highlighting the need for the J and H bands to correctly constrain the effective temperature (Burgemeister *et al.*, in prep.). P. Crowther focused on mid-infrared data of WC and WO stars obtained with *Spitzer* and *Herschel.* He highlighted the interest of forbidden lines to constrain stellar evolution. Such lines being formed very far away from the photosphere (at heights from 10000 to 500000 stellar radius), they trace the outter wind. Ne abundances are derived in WC stars from *Spitzer* spectra. A mass fraction of about 1% is determined, in reasonable agreement with theoretical yields. In WO/WC stars, [OIII] 88.0 μm is used to constrain the oxygen content. From preliminary results on the binary γ Vel (WC8+O), the O/C ratio seems to be a factor of two lower than the predictions of Meynet & Maeder (2003).

Red supergiants: KM (red) supergiants: M. Bergemann presented analysis of red supergiants from pure near-infrared spectroscopy (Bergemann *et al.* (2012)). She raised the importance of non-LTE effects for the abundance determination of Fe, Ti, and Si and cautioned against using the TiO molecular fits for determination of effective temperature of RSG's, also because severe non-LTE effects in Ti ionization equilibrium will have an impact on the TiO formation. Corrections to LTE calculations have been quantified. Red supergiants are especially important for future telescopes/instruments since they are the brightest objects at infrared wavelengths.

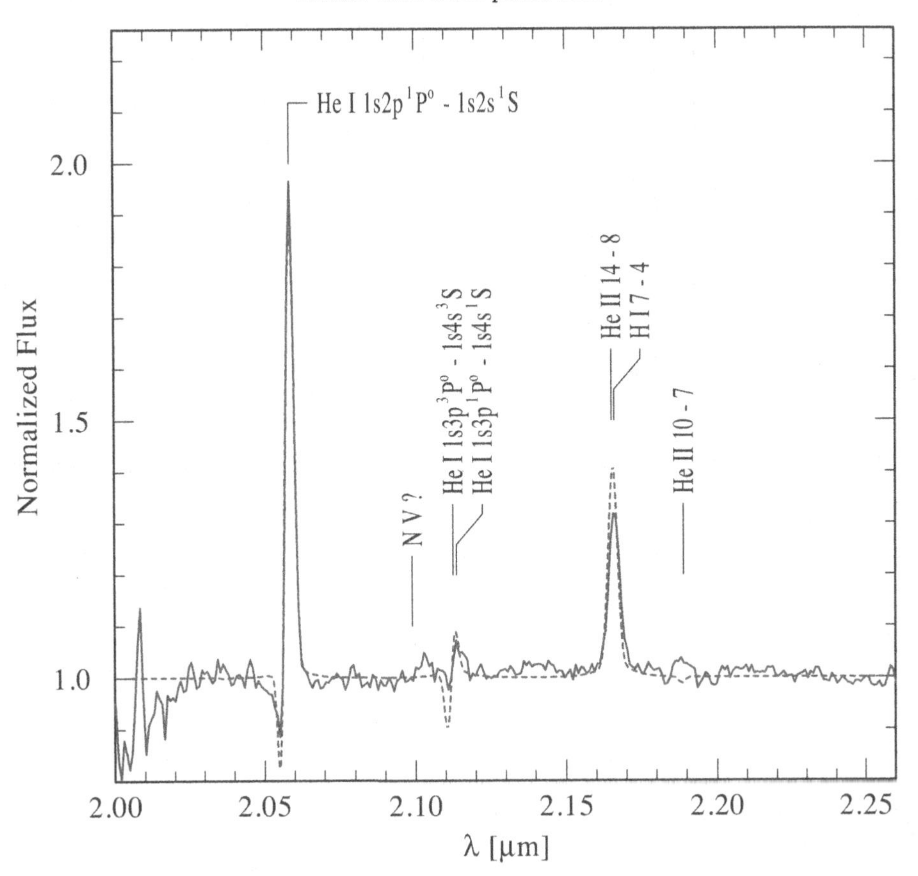

Figure 6. Observed K-band spectrum of the Galactic Center star WR102ka (blue) and best fit model (red). The main lines are indicated. From Barniske *et al.* (2008).

Luminous Blue Variables: N. Przybilla presented the results of Najarro *et al.* (2009) on the surface chemical abundances of LBVs in the Quintuplet cluster. Najarro *et al.* were able to derive a solar Fe content, and an abundance in Mg, Si and Na about twice the solar abundance. LBVs are well suited for such analysis in the infrared since they display a number of metallic emission lines. This is illustrated in Fig. 7.

5. Conclusion

Infrared phototemetry and spectroscopy can be used to determine stellar and wind properties of massive stars (hot and cool). The modelling of infrared spectra is more difficult than shorter wavelength spectra because the non-LTE effects are amplified. Accurate atomic data are necessary to correctly reproduce the observed line profile. Line-blanketing effects are also important. In spite of these difficulties, analysis based on pure infrared diagnostics usually give results consistent with optical studies. All types of massive stars can be studied. In addition to temperature, luminosity, gravity and mass loss rate, abundances of He, C, N, O, Mg, Si, Ne are feasible in cool stars and strong wind objects. Infrared studies are crucial to study stellar population in external galaxies with future extremely large telescopes.

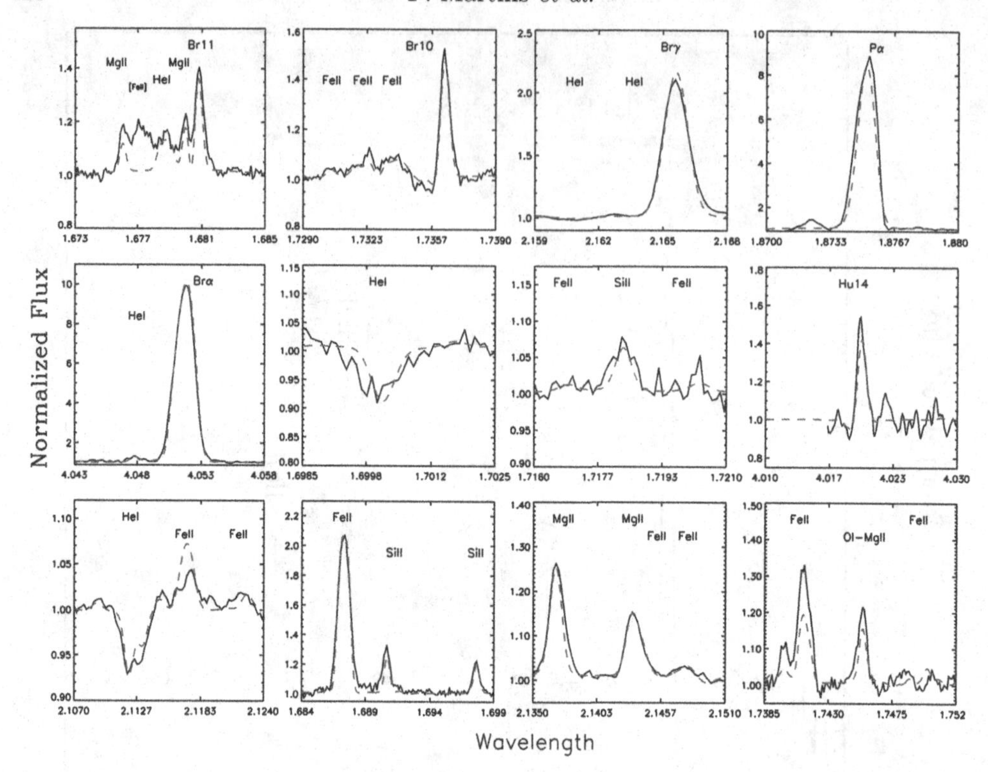

Figure 7. Determination of Fe, Mg, Si, Na abundances in the LBV star Pistol. The black solid line is the observed spectrum, the red dashed line is th ebest fit model. From Najarro *et al.* (2009).

Acknowledgements

FM thanks the french “Agence Nationale de la Recherche” for funding.

References

Barniske, *et al.* 2008, *A&A* 486, 971
Bergemann.,*et al.* 2012, *ApJ* 751, 156
Bestenlehner, *et al.* 2011, *A&A* 530, L14
Crowther, *et al.* 2006, *ApJ* 636, 1033
Dessart, *et al.* 2000, *MNRAS* 315, 407
Meynet & Maeder 2003, *A&A* 404, 975
Najarro, *et al.* 2011, *A&A* 535, A32
Najarro, *et al.* 2009, *ApJ* 691, 1816
Najarro, *et al.* 2006, *ApJ* 456, 659
Nieva, *et al.* 2011, *BSRSL* 80, 175
Przybilla, *et al.* 2009, in: *Science with the VLT in the ELT Era*, ed. A. Moorwood (Springer, Netherlands)
Przybilla & Butler 2004, *ApJ* 609, 1181
Repolust, *et al.* 2005, *A&A* 440, 261

Highlights of Astronomy, Volume 16
XXVIIIth IAU General Assembly, August 2012
T. Montmerle, ed.

doi:10.1017/S174392131401179X

SpS5 - III. Matter ejection and feedback

Yaël Nazé[1] †, and Xiao Che[2], Nick L. J. Cox[3], José H. Groh[4], Martin Guerrero[5], Pierre Kervella[6], Chien-De Lee[7], Mikako Matsuura[8], Sally Oey[2], Guy S. Stringfellow[9] and Stephanie Wachter[10]

[1]Department AGO, Allée du 6 Août 17, B5C, 4000-Liège, Belgium, email: naze@astro.ulg.ac.be

[2] Dpt of Astronomy, Univ. of Michigan, 830 Dennison Bldg, Ann Arbor, MI 48109-1042, USA
[3]Institute of Astronomy, KU Leuven, Celestijnenlaan 200D, Leuven, Belgium
[4] Geneva Observ., Univ. of Geneva, Chemin des Maillettes 51, CH-1290 Sauverny, Switzerland
[5] IAA-CSIC, Glorieta de la Astronomia s/n, 18008 Granada, Spain
[6] LESIA, Obs. de Paris, CNRS UMR 8109, UPMC, 5 place J. Janssen, 92195, Meudon, France
[7] Graduate institute of astronomy, National central university, Taiwan
[8] Department of Physics and Astronomy, UCL, Gower Street, London WC1E 6BT, UK
[9] Center for Astroph. and Space Astr., Univ. of Colorado, 389 UCB, Boulder, CO, USA
[10] IPAC, California Inst. of Technology, Pasadena, CA 91125, USA

Abstract. The last part of SpS5 dealt with the circumstellar environment. Structures are indeed found around several types of massive stars, such as blue and red supergiants, as well as WRs and LBVs. As shown in the last years, the potential of IR for their study is twofold: first, IR can help discover many previously unknown nebulae, leading to the identification of new massive stars as their progenitors; second, IR can help characterize the nebular features. Current and new IR facilities thus pave the way to a better understanding of the feedback from massive stars.

Keywords. infrared: stars, stars: early-type, stars: mass loss, circumstellar matter

1. Introduction

Circumstellar material holds clues about the mass-loss history of massive stars. Indeed, as the winds interact with the interstellar medium (wind-blown bubbles, bow shocks), they leave a characteristic signature that depends on the wind properties. Moreover, the material ejected during short eruptive phases is visible as nebulae around massive stars. The analysis of these features reveals which material was ejected and in which quantity. With the recent reduction in mass-loss rates, these episodes of enhanced mass-loss have gained more attention, as they seem more crucial than ever in the evolution of massive stars.

Another reason to study the close environment of massive stars is to better understand the evolution of supernova remnants (SNRs). Indeed, the famous rings of SN1987A may only be understood if one considers the previous mass-loss episodes of the progenitor. Morphology is not the only SNR parameter which is affected, as the SNR dynamics in an homogeneous medium or in winds and circumstellat ejecta is not identical.

For its study, the IR provides several key diagnostics. Continuum emission in this range is provided by heated dust, which may have a range of temperatures depending of the framework (very close hot features, large, old, and cool bubbles). In addition, IR lines probe the many phases of the material: molecules (e.g. PAHs) for the neutral material, ionized metals for HII regions,...

† FNRS Research Associate

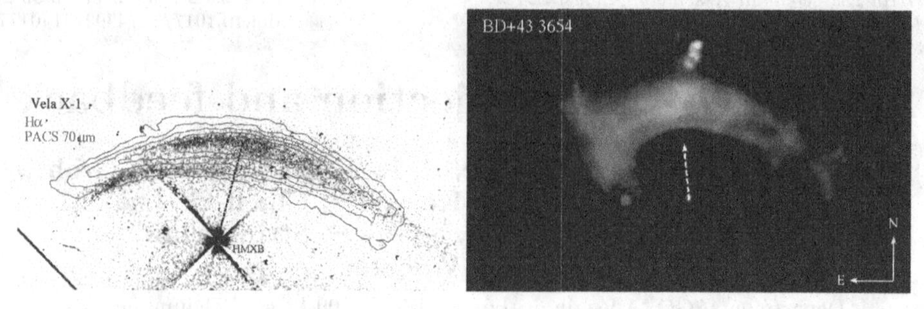

Figure 1. *Left:* Hα emission (greyscale) of Vela X-1 with PACS 70μm emission contours shown on top. *Right:* Colour composite image of bow shock of BD+43°3654 (WISE 12μm in blue, PACS 70μm in green, and PACS 160μm in red). The direction of proper motion is indicated by the arrow in both cases. From Cox *et al.* (in prep.).

This summary of SpS5 - part III examines each case of circumstellar environment in turn, and concludes with the potential offered by current and future facilities.

2. Blue supergiants

Circumstellar structures around BSGs have been predominantely identified as bow shocks around runaway stars. Originally discovered with IRAS (e.g. Van Buren & McCray, 1988, *ApJ*, 329, L93), such structures have also been seen with MSX and WISE (Peri *et al.* 2012). A more general survey of BSGs, i.e. not targeting runaway stars, with objects selected from Crowther *et al.* (2006) and Przybilla *et al.* (2010), reveals IR material around six of the 45 targets at 22μm with WISE, also mostly in the form of bow shocks (Wachter, in prep). Several examples of bipolar nebulae around BSGs are also known (e.g. Sher 25, Smartt *et al.* 2002; HD 168625, Smith 2007). However, this material could have also been ejected during an LBV phase, since LBVs can exhibit BSG spectra, and we will therefore concentrate on the bow shocks.

Runaway stars have large stellar velocities (above $30\,\mathrm{km\,s^{-1}}$) resulting from dynamical interactions in (dense) clusters or from a supernova explosion in a binary system. These stars can thus travel at supersonic speeds through the local medium giving rise to "bow shocks" as their stellar winds interact with the surrounding medium, which has been previously ionised by stellar photons from the hot star (Weaver 1977). The occurrence of such bow shocks has been shown to depend primarily on the ISM conditions (Huthoff & Kaper 2002). For example, even a runaway star may travel at subsonic speeds in the tenuous interior of a superbubble, where the sound speed can be as much as $100\,\mathrm{km\,s^{-1}}$, hence no (detectable) bow shock will be produced in that case. The filling factor of ISM with $v_{\mathrm{sound}} \leqslant 10\,\mathrm{km\,s^{-1}}$ is 20% and 75% of O-stars have velocities $\geqslant 10\,\mathrm{km\,s^{-1}}$, so the expected fraction of O-stars with bow shocks is ~15%. This is remarkably similar to the values derived from IRAS and WISE observations (Noriega-Crespo *et al.* 1997, Peri *et al.* 2012).

Once formed, the size, shape and morphology of a bow shock depends on both stellar (wind kinetic energy and stellar velocity) and interstellar parameters (density and temperature). In particular the ratio $v_\star/v_{\mathrm{wind}}$ indicates whether or not instabilities are likely to develop (Dgani *et al.* 1996), and the stand-off distance between the star and the apex of the shock is determined from the pressure balance between the stellar wind and the ISM (see analytical results by Wilkin 1996 and simulations by e.g. Comeron & Kaper 1998, Blondin & Koerwer 1998). Independent estimates of the wind parameters can thus be inferred from bow shocks, which serves as a useful check for atmosphere models, but the values are sensitive to the ISM properties, which are not always known with precision.

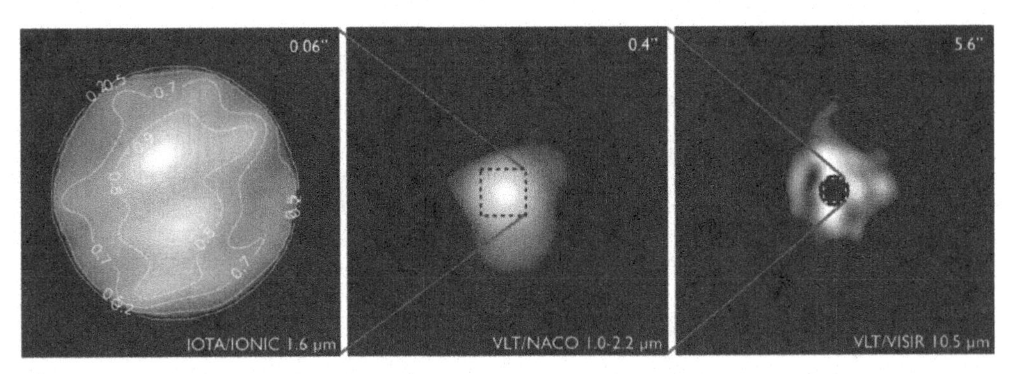

Figure 2. *Left:* Interferometric image of the photosphere of Betelgeuse obtained by Haubois *et al.* (2009), showing its inhomogeneous surface brightness. *Center:* VLT/NACO adaptive optics tricolor composite image (RGB = KHJ) obtained by Kervella *et al.* (2009), showing the emission from a compact molecular envelope. *Right:* VLT/VISIR image at 10.49 μm of the dust thermal emission obtained by Kervella *et al.* (2011). North is up, East to the left, and the field of view is given in the upper right corner of each image.

Currently, a small survey with Herschel-PACS of 5 runaways with known bow-shocks is ongoing: α Cam, ζ Oph, τ CMa, Vela X-1 and BD+43°3654 (Cox *et al.*, in preparation). For Vela X-1, the peak emission of the dust emission is co-spatial with the most prominent Hα arc seen in the supposed direction of space motion (Fig. 1): it is concluded that the outer shock is radiative, but the inner shock is adiabatic, though some Hα emission possibly related to (part of) the inner termination shock is also detected. From the analysis of its "puffed-up" bow shock (Fig. 1), the mass-loss rate of BD+43°3654 (O4If) was found to be 10^{-4} M$_\odot$ yr^{-1}: this is very high (by 2 orders of magnitude) in view of current mass-loss rate estimates of such stars, but the exact value strongly depends on ISM density, which need to be refined. The dust temperature, $\sim$ 45 K, is compatible with heating by stellar photons only, suggesting there is no additional shock-heating of grains. The thickness of a bow shock ($\sim$ 1 pc) suggests a Mach number close to unity, implying a ISM temperature of 10^3 - 10^4 K.

3. Red supergiants

Circumstellar structures on scales of a few arcseconds or less around RSGs have been revealed through interferometric techniques (e.g. Monnier *et al.* 2004). Stencel *et al.* (1988, 1989) reported the IRAS detection of resolved shells with typical radii of a few arcminutes around RSGs for a significant fraction (25%) of their sample. However, higher resolution Spitzer images fail to confirm several of these extended structures (Wachter, in prep), indicating that a systematic survey is needed to ascertain the occurrence of large scale circumstellar shells around RSGs.

A few (famous) cases have however been studied in depth. One of these is Betelgeuse, a cool (3600K), large (700 R$_\odot$), rather massive (10–15 M$_\odot$), luminous (> 10^5 L$_\odot$), and nearby (150 pc) star. Because of its distance, Betelgeuse can be probed on almost all scales, providing a unique panorama of stellar surroundings (Fig. 2). Space-based and interferometric instruments (e.g. HST, IOTA/Ionic and VLTI/Pionier) revealed the photosphere, notably the expected non-uniformities due to large convection cells. Adaptive optics imaging in the near-IR (e.g. NACO) and (radio or IR) interferometers unveiled the properties of the internal, compact molecular envelope (1–10 R$_*$). Precursors of dust have been found there, as well as an extended "plume" reaching 6R$_*$ and maybe linked to a hot spot on the photosphere. High-res imaging (e.g. VLT/VISIR) shows the envelope

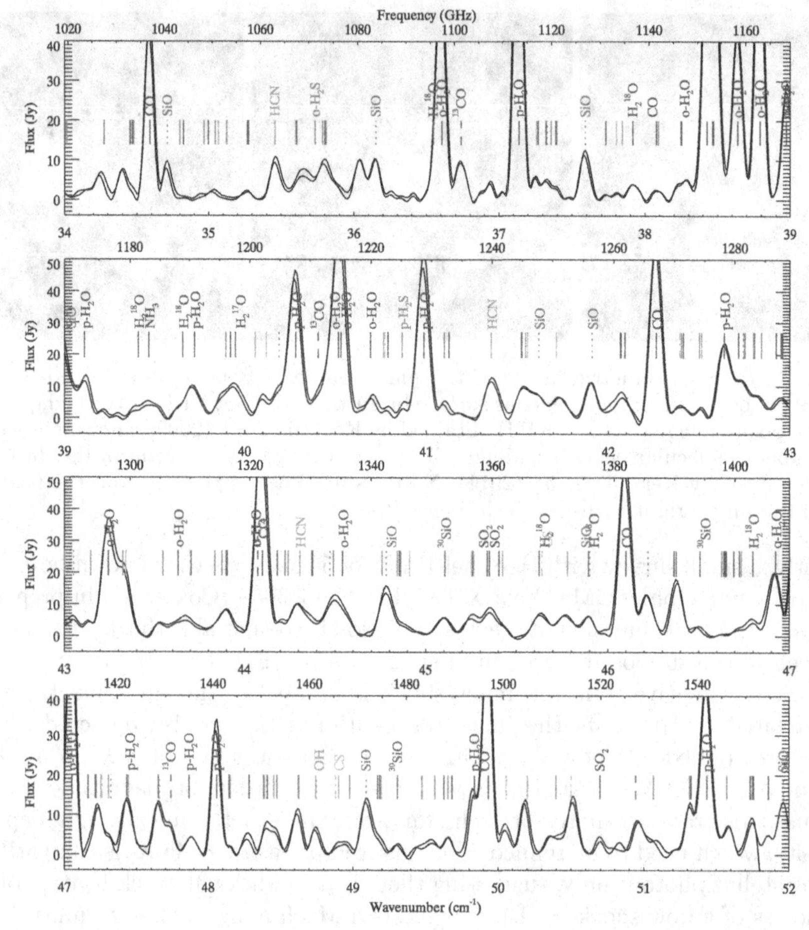

Figure 3. Continuum subtracted Herschel SPIRE spectrum of VY CMa from 294μm to 192μm. Multitude of molecular lines have been detected. From Matsuura *et al.* (in prep.).

at intermediate scales (10–100 R_*), where the dust forms (a possible signature of silicates has been found). At these small and intermediate scales, Betegeuse presents a complex circumstellar envelope (with knots and filaments) at all wavelengths, which implies an inhomogeneous spatial distribution of the material lost by the star. Finally, at the largest scale, IR imagers such as Herschel unveil the cool external envelope (100-10000 R_*), where a bow shock with the ISM is detected (Cox *et al.* 2012).

Herschel has also probed the envelope of other red supergiants (Groenewegen *et al.* 2011). Turning in particular to the case of VY CMa (Matsuura *et al.*, in prep.), the potential of IR spectroscopy is obvious. Herschel-SPIRE reveals a rich spectrum, with a dust continuum and hundreds of lines dues to molecules (one third linked to water, others to CO, CS, SiO,...), which constrain the envelope's properties. For example, the isotopic ratio $^{12}C/^{13}C$ is found to be 6.5, in agreement with observations of other RSGs but at odds with theoretical predictions which are four times higher at least. Very strong emission of submm molecular lines can be explained if a temperature gradient is present in the envelope, e.g. because of dust formation at a certain radius.

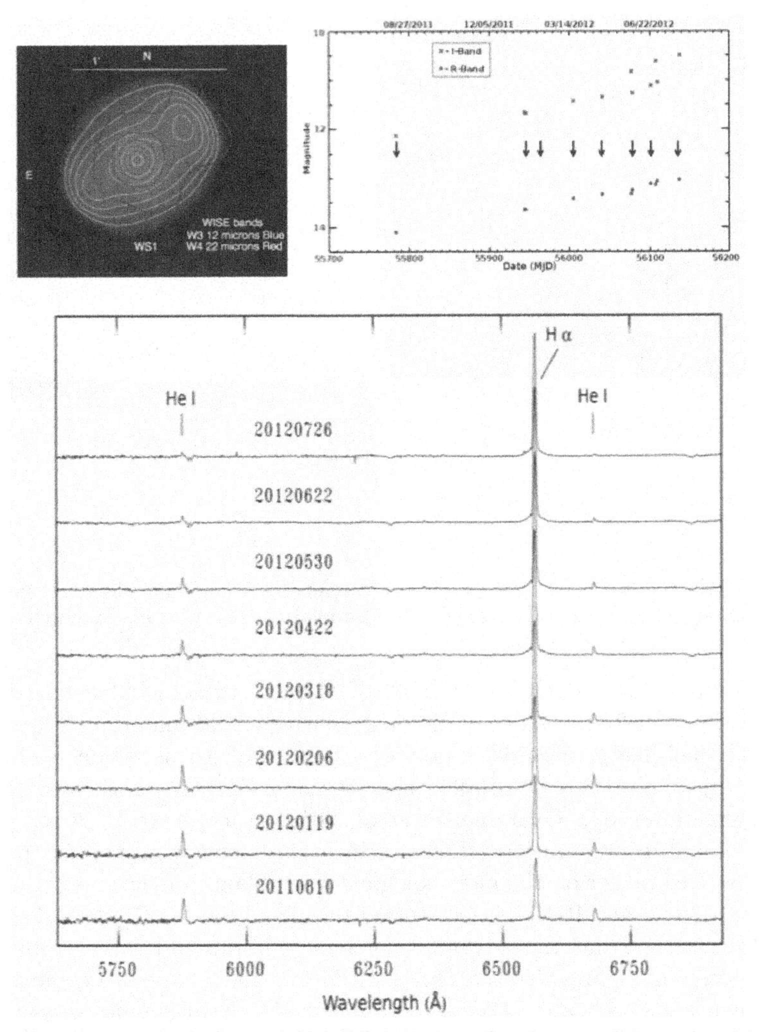

Figure 4. *Top left:* A WISE color composite of 12 μm (blue; blue contours) and 22 μm (red; yellow contours) of WS1, discovered and initially characterized by Gvaramadze *et al.* (2012). The contours for each band help illuminate the morphology of the nebular material, which has an overall SE-NW elongation, reminiscent of bipolar structure. *Bottom left:* Optical photometric monitoring since discovery show both the R and I light curves have brightened by about 1 magnitude over the last year. Arrows indicate times when same night spectroscopy were secured. *Right panel:* Optical spectroscopic monitoring indicates evolution to cooler temperature with near disappearance of the He I 5876 Å and 6678 Å and changes in the H_α line profile. Figures from Stringfellow *et al.* (in preparation).

4. Luminous Blue Variables

Because of their spectacular eruptions, LBVs are the most well-known cases of massive stars with ejecta. It is not yet certain, however, at what stage (BSG? after a RSG phase or not?) this material is ejected, and how (multiple events?). LBVs are rare: in the list of Clark *et al.* (2005), there are only 12 confirmed and 23 candidate LBVs. IR has played a key role in recent years. The search, through surveys like MIPSGAL, of round-shaped nebulae with luminous central stars resulted in the discovery of many new nebulae: 62 shells in Wachter *et al.* (2010), 115 shells and bipolar nebulae in Gvaramadze *et al.*

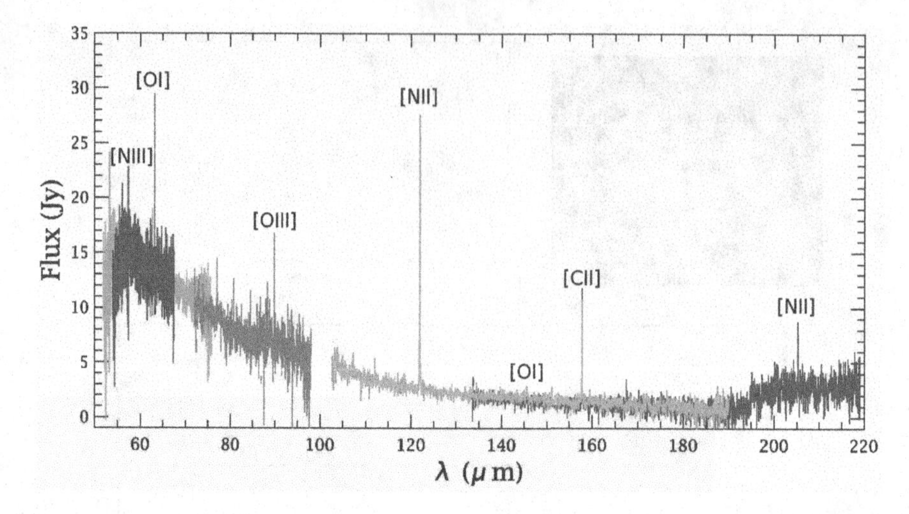

Figure 5. PACS spectrum (central spaxel) of WRAY 15-751 nebula, showing the lines from the ionized and neutral regions (from Vamvatira-Nakou *et al.*, submitted).

(2010), 416 structures in Mizuno *et al.* (2010). Many of these nebulae are preferentially detected with Spitzer 24 μm band, indicating relatively cold material.

Identifying shell-like structures is only the first step. To ascertain a cLBV status, the central object needs to be studied spectroscopically. This was done for many of these new detections (c.f., Gvaramadze *et al.* 2010; Wachter *et al.* 2010; Stringfellow *et al.* 2012a,b, and in preparation). The classification does not rely on the presence of a particular line, but rather on the morphological resemblance of the spectra to spectra of known LBVs - while not 100% perfect (some peculiar O and WR stars display similar features), this method has the advantage of being simple and rather robust. A more definitive answer can be provided through photometric and/or spectroscopic monitorings. Indeed, as their name indicate, LBVs should be *variable.* Near-simultaneous photometric and spectroscopic monitoring in the optical (and IR) of about a dozen newly identified candidate LBVs has revealed that WS1 (discovered by Gvaramadze *et al.* 2012) is indeed a bona fide LBV, presently displaying what appears to be S Dor type variability as shown in Fig. 4 (Stringfellow *et al.* 2012, in preparation).

IR is also useful in revealing details of particular objects. For example, a Herschel survey of LBVs undertaken at Liège yielded as first result a characterization of the surroundings of WRAY 15-751 (Vamvatira-Nakou *et al.*, submitted). IR photometry can only be explained if the star evolves at constant luminosity and dust grains are Fe-rich. Images also revealed the presence of a second shell, about 4 times larger than the previously known one, which most probably results from an older eruptive event. Considering both structures, there is about 0.075 $M_{\odot}$ of dust in the system. Ionized gas is responsible for several forbidden lines observed in the Herschel-PACS range (Fig. 5), which allow a N/O abundance of about 7 times solar and a mass of ionized gas of 1–2 $M_{\odot}$ (20 times that of dust), to be derived.

Dust can be well studied in the IR, so this range may provide clues on where dust come from in galaxies. Two examples of such feedback were presented in the session: η Car and SN1987A. The latter was observed with Herschel at 100-500μm wavelengths, and 0.4-0.7 $M_{\odot}$ of dust was detected - mostly silicates and amorphous carbon (Matsuura *et al.* 2011).

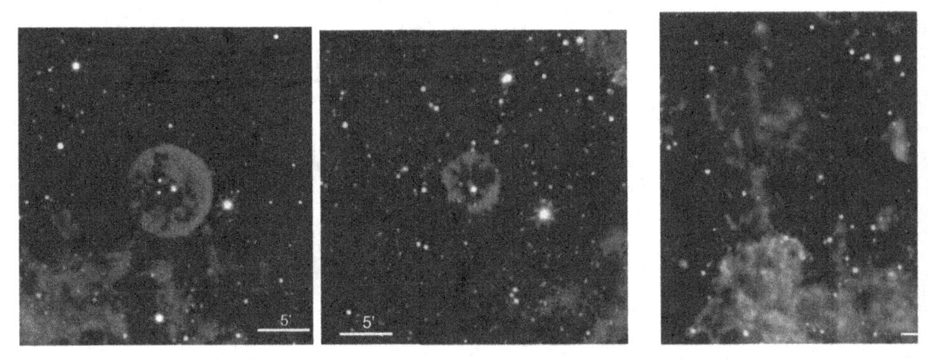

Figure 6. Examples of the three stages in WR nebula morphologies: from left to right - bubble (WR16), clumpy phase (WR8), and mixed phase (WR35b). From Toalá *et al.* (in prep.).

It is thought that this dust come from the explosion, but the role played by previous mass-loss episodes, in particular the LBV phase, is not yet clear. For example, about 0.12 $M_{\odot}$ of dust was detected, thanks to 30μm MiniTAO observations, in the famous LBV η Car. Up to 80% of that dust belonged to the torus, hence may not be related to the big 1843 event.

5. Wolf-Rayet stars

Only a few percentage (4-6%) of Wolf-Rayet stars displays surrounding nebulosities in the WISE survey, and most are found around WN stars (Wachter, in prep).

The morphological classification scheme of WR nebulae proposed by Chu (1981) has been revised in this meeting by Guerrero *et al.* (Toalá *et al.* in prep.) using *WISE* IR images and SDSS or Super Cosmos sky survey Hα images for 35 nebulae associated with WRs. Two *WISE* bands were particularly used: the one at 12μm, which encompasses PAH lines and lines of low excitation ions, and that at 22 μm, to which thermal emission from dust and lines of He I as well as high excitation ions contribute. Three phases are defined. In the first one, WR nebulae appear as complete shells or bubbles. It corresponds to the star just entering the WR stage, when its powerful wind sweeps up the previous slow and dense winds (from e.g. LBV or RSG stages). The second phase is the clumpy phase. At that point, the nebulae display knots of gas and dust connected by partial shells and arcs. It corresponds to an age of a few 10^4 yr, when instabilities break down the swept-up shell. The stellar motion through the ISM has an impact on the morphology, for example one-sided arc may be sometimes seen. Finally, the mixed nebular phase ends the cycle, with no definite morphology nor always a 1-to-1 correspondence between optical and IR images. It corresponds to the last stage, when the circumstellar nebula begins to dissolve into the ISM.

6. Studying the close environment of massive stars

The close environment of massive stars is the “missing link” between the star itself and the large circumstellar features. It plays a key role in understanding the mass-loss, but it is also difficult to probe directly.

Emission lines arising in the wind and circumstellar material are a classical way to study this region, as well as near-IR excess linked to disk-like features. In this context, Be and B[e] stars are targets of choice, and surprises are frequent. For example, Graus *et al.* (2012) found three new sgB[e] in the SMC: they display typical spectra, with forbidden

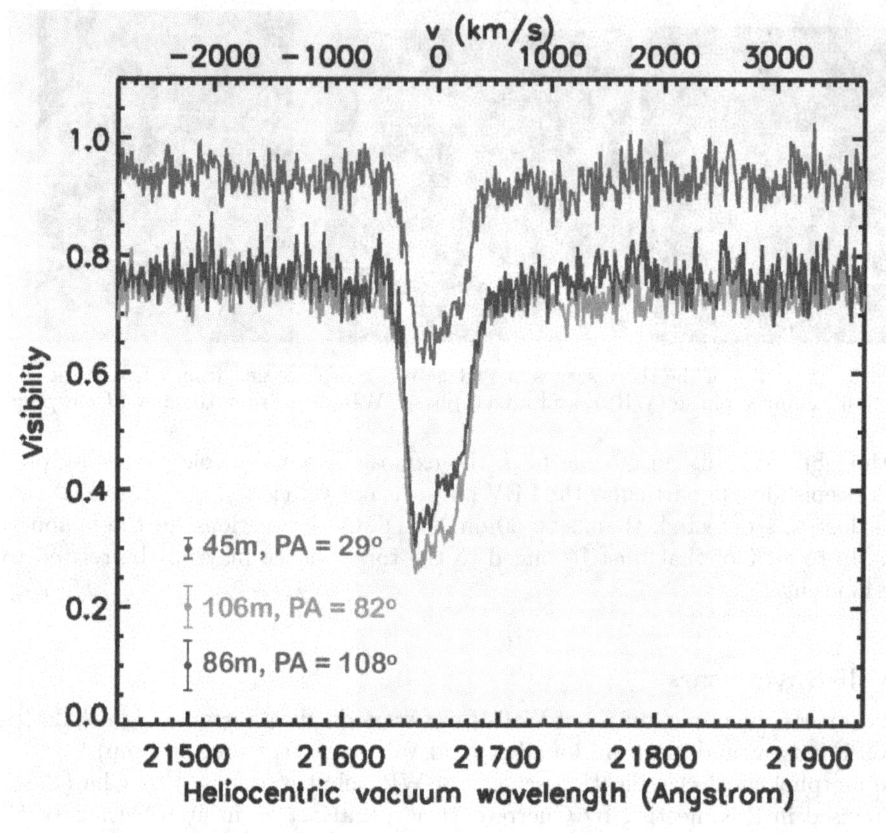

Figure 7. Visibility of HD 316285 as a function of wavelength for the three different telescope baselines measured with VLTI/AMBER. Notice the drop in visibility within the Bracket gamma line, indicating that the line is more extended than the neighboring K-band continuum. From Groh *et al.* (in prep.).

lines, but the line strengths as well as the IR excess appear reduced compared to usual objects of this class. It suggests that either the disks have less material or less dust than usual, or maybe that these stars are transitional objects. Another case intriguingly shows the opposite situation: CD$-49°3441$ displays forbidden lines and appreciable IR excess, but is a main-sequence Be star away from any star-forming region (Lee and Chen 2009). A possibility may be that this star is in fact a weak B[e], rather than a classical Be.

The environment close to the star can also be studied, directly, by means of interferometry, which is usually performed at long wavelengths. Most optical/IR interferometric measurements rely on the measurements of "visibilities", which are directly linked to the size of the object. Recently, several massive stars, including nine LBVs, were observed with the VLTI (Groh *et al.*, in prep.). Amongst these, HD316285 (Fig. 7): the recorded visibilities implied a size of 0.002" for the source of continuum radiation, and 0.004" for the source of the Brγ line. A CMFGEN fit to the spectrum yields a stellar model with which one can estimate the wind size in IR, and it agrees well with VLTI observations. The asymetric shape of the measured differential phases (red vs blue side) favors a prolate shape for the rotating star wind, but it could also be explained by clumps or binarity. Since the latter imply in time variability, a monitoring will be needed to ascertain the exact nature of the asymmetry.

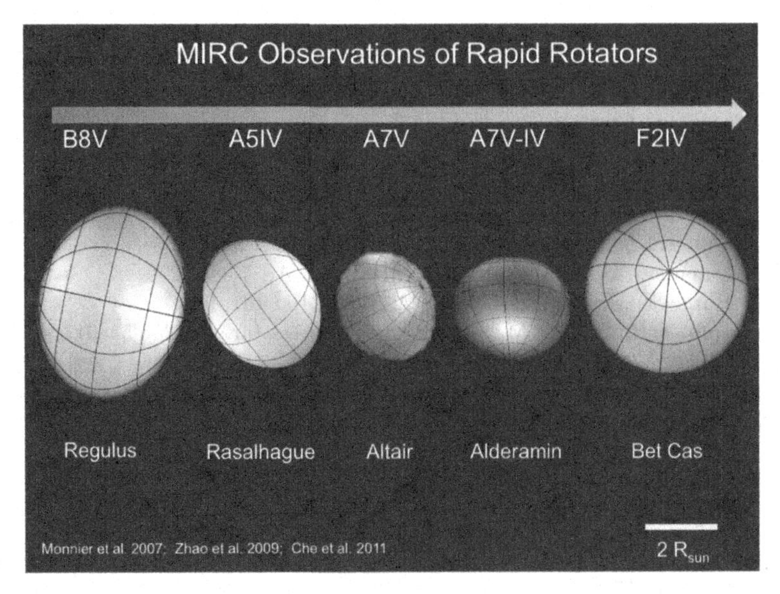

Figure 8. Images of rapid rotators derived from MIRC interferometric measurements. Each star shows a bright pole and dark equator, which is caused by gravity darkening effect. Modeling of Regulus gives gravity darkening coefficient $\beta = 0.19$, rather than 0.25 as von Zeipel predicted in 1924.

While visibilities provide valuable data, "real" images are always more impressive. Interferometric instruments such as Michigan Infrared Combiner (MIRC) and PIONIER are beginning to provide such data. MIRC was the first to image Altair (Monnier *et al.* 2007) and several other rapidly rotating stars as shown in Fig. 8. It has also imaged circumstellar disks and multi-object systems. For example, the disk contribution of δ Sco was shown to remain stable during the periastron in 2011 (Che *et al.* 2012), and the mass-exchange in the β Lyr system can be clearly imaged (Zhao *et al.* 2008), as well as the 3 components of Algol (Baron *et al.* 2012) or the disk of the eclipsing companion of ϵ Aur (Kloppenborg *et al.*, 2010).

7. Conclusion

This session has demonstrated the usefulness of studies in the IR in studying the environment of massive stars. Recent advances in this domain are notably provided by surveys, as they enable discovery of new objects to study, thereby improving the census of nebular features associated with hot stars. Furthermore, IR diagnostics unveil the properties of these neighbouring nebulosities: morphology, temperature, composition, density are the necessary keys paving the way of a better understanding of the mass-loss in massive stars.

Acknowledgements

YN acknowledges comments from Augusto Daminelli and support from FNRS and Prodex Herschel/XMM-Integral contracts. NLJC thanks FWO and Prodex-Herschel for financial support. JHG is supported by an Ambizione Fellowship of the Swiss National Science Foundation. CDL has been supported financially by grant NSC-101-2922-I-008-120 of the National Science Council of Taiwan.

References

Baron, F., *et al.* 2012, *ApJ* 752, 20

Blondin, J. M. & Koerwer, J. F. 1998, *New Astronomy* 3, 571

Che, X., *et al.* 2012, *ApJ* 757, 29

Chu, Y.-H. 1981, *ApJ* 249, 195

Clark, J. S., Larionov, V. M., & Arkharov, A. 2005, *A&A* 435, 239

Comeron, F. & Kaper, L. 1998, *A&A* 338, 273

Cox, N., Kerschbaum, F., van Marle, A.-J., *et al.* 2012, *A&A* 537, A35

Crowther, P. A., Lennon, D. J., & Walborn, N. R. 2006, *A&A* 446, 279

Dgani, R., van Buren, D., & Noriega-Crespo, A. 1996, *ApJ* 461, 927

Graus, A. S., Lamb, J. B., & Oey, M. S. 2012, *ApJ* in press (arXiv:1208.5486)

Groenewegen, M., *et al.* 2011, *A&A* 526, A162

Gvaramadze, V. V., Kniazev, A. Y., & Fabrika, S. 2010, *MNRAS* 405, 1047

Gvaramadze, V. V., *et al.* 2012, *MNRAS* 421, 3325

Haubois, X., *et al.* 2009, *A&A* 508, 923

Huthoff, F. & Kaper, L. 2002, *A&A* 383, 999

Kervella, P., Verhoelst, T., Ridgway, S. T., Perrin, G., Lacour, S., Cami, J., & Haubois, X. 2009, *A&A* 504, 115

Kervella, P., Perrin, G., Chiavassa, A., Ridgway, S. T., Cami, J., Haubois, X., & Verhoelst, T. 2011, *A&A* 531, A117

Lee, C. D. & Chen, W. P. 2009, *ASPC* 404, 302

Kloppenborg, B., *et al.* 2010, *Nature* 464, 870

Matsuura, M., *et al.* 2011, *Science* 333, 1258

Mizuno, D. R., *et al.* 2010, *AJ* 139, 1542

Monnier, J. D., *et al.* 2004, *ApJ* 605, 436

Monnier, J. D. 2007, *Science* 317, 342

Noriega-Crespo, A., van Buren, D., & Dgani, R. 1997, *AJ* 113, 780

Peri, C. S., Benaglia, P., Brookes, D. P., Stevens, I. R., & Isequilla, N. L. 2012, *A&A* 538, A108

Przybilla, N., Firnstein, M., Nieva, M. F., Meynet, G., & Maeder, A. 2010, *A&A* 517, A38

Smartt, S. J., Lennon, D. J., Kudritzki, R. P., Rosales, F., Ryans, R. S. I., Wright, N., *A&A* 391, 979

Smith, N. 2007, *AJ* 133, 1034

Stencel, R. E., Pesce, J. E., & Hagen Bauer, W. 1988, *AJ* 95, 141

Stencel, R. E., Pesce, J. E., & Hagen Bauer, W. 1989, *AJ* 97, 1120

Stringfellow, G. S., Gvaramadze, V. V., Beletsky, Y., & Kniazev, A. Y. 2012a, *ASP Con Series* in press, arXiv:1112.2686

Stringfellow, G. S., Gvaramadze, V. V., Beletsky, Y., & Kniazev, A. Y. 2012b, *IAU Symp.* 282, 267

Van Buren, D. & McCray, R. 1988, *ApJ* 329, L93

Wachter, S., Mauerhan, J. C., Van Dyk, S. D., Hoard, D. W., Kafka, S., & Morris, P. W. 2010, *AJ* 139, 2330

Weaver, R. P. 1977, *Ph.D. Thesis Colorado University, Boulder*

Wilkin, F. P. 1996, *ApJ Letters* 459, 31

Zhao, M., *et al.* 2008, *ApJ* 684, L95

Highlights of Astronomy, Volume 16
XXVIIIth IAU General Assembly, August 2012
T. Montmerle, ed.

doi:10.1017/S1743921314011806

Science with Large Solar Telescopes: Overview of SpS 6

Gianna Cauzzi[1], **Alexandra Tritschler**[2] **and Yuanyong Deng**[3]

[1]INAF-Osservatorio Astrofisico di Arcetri, Firenze, Italy,
email: gcauzzi@arcetri.astro.it

[2]National Solar Observatory, Sunspot, NM 88349, USA,
email: ali@nso.edu

[3]Key Laboratory of Solar Activity, National Astronomical Observatories,
Chinese Academy of Sciences, China,
email: dyy@nao.cas.cn

Abstract. With several large aperture optical and IR telescopes just coming on-line, or scheduled for the near future, solar physics is on the verge of a quantum leap in observational capabilities. An efficient use of such facilities will require new and innovative approaches to both observatory operations and data handling.

This two-days long Special Session discussed the science expected with large solar telescopes, and started addressing the strategies necessary to optimize their scientific return. Cutting edge solar science as derived from state-of-the-art observations and numerical simulations and modeling was presented, and discussions were held on the role of large facilities in satisfying the demanding requirements of spatial and temporal resolution, stray-light correction, and spectro-polarimetric accuracy. Building on the experience of recently commissioned telescopes, critical issues for the development of future facilities were discussed. These included operational issues peculiar to large telecopes as well as strategies for their best use.

Keywords. telescopes; techniques: image processing; techniques: polarimetric; methods: numerical; Sun: atmosphere, magnetic fields

---

1. Introduction

In the last decade, vast improvements have been obtained in our observational capabilities of the solar atmosphere. New spectral windows have been opened by space-based facilities such as the highly successful *Solar and Heliospheric Observatory* (SOHO, Domingo *et al.* (1995)), *Hinode* (Kosugi *et al.* (2007)), or the recent *Solar Dynamic Observatory* (SDO, Pesnell *et al.* (2012)), avoiding the influence of Earth's atmosphere even if with reduced telescope aperture and instrument flexibility. At ground-based observatories, efficient adaptive optics systems (AO, Rimmele & Marino (2011), Scharmer *et al.* (2010)) and novel instrumentation have allowed existing telescopes (e.g., VTT, DST, SST) to push the spatial resolution closer to diffraction limit from the visible to the near-infrared. Yet, when compared with realistic numerical 3D simulations of solar surface magneto-convection (e.g. Rempel (2011)), it is clear that we are still not resolving the fundamental scales at work in the solar atmosphere, which might be as small as few tens of km and of the order of seconds. At the same time, the critical role of the magnetic field as a main agent in shaping the solar atmosphere and its dynamics implies the need for accurate and precise high-cadence polarimetric measurements, as well as reliable

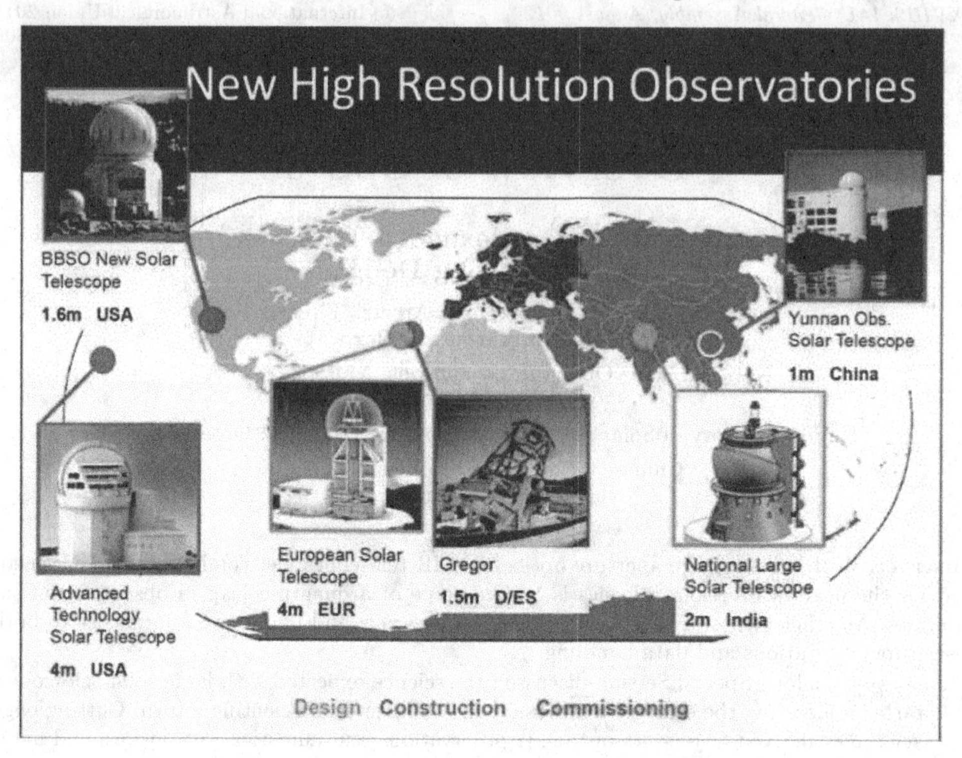

Figure 1. A world-map with the location of just commissioned, upcoming, or planned large, ground-based open solar telescopes. From Hasan (2012).

interpretational tools to obtain a quantitative knowledge of solar magnetism throughout the atmosphere.

To address these issues, in recent years a number of innovative solar telescopes, mostly working in the optical and infrared range, have been proposed by the international solar physics community to operate on the ground or from space. In particular, technological developments including the feasibility of air-cooled, open telescopes have allowed for the planning of ground-based facilities with apertures sensibly larger than the existing evacuated telescopes. These new facilities will provide for much increased spatio-temporal resolution and spectro-polarimetric sensistivity.

However, brand new challenges will accompany their operations. As an example, the foreseen use of multiple high speed, large format scientific cameras will increase enormously the data volume with respect to current standards, likely approaching hundreds of TB daily. As the pioneering efforts with the datastream of the *Atmospheric Imaging Assembly* (AIA, Lemen *et al.* (2012)) instrument on board SDO are currently showing, highly innovative solutions will be required in order to efficiently extract accurate scientific results from such large volumes (see e.g. the Heliophysics Events Knowledgebase site, at http://www.lmsal.com/hek/). Further, the scientific output of large telescopes will have to be optimized in order to justify the considerable resources invested. For ground-based telescopes, this might require developing robust data reduction pipelines to provide science-ready data to a larger user base, as well as adopting more efficient modes of operation, e.g. scheduling observations on a flexible basis in order to best match science programs to observing conditions.

Discussion on these and other issues is just starting. In the next few years however, we expect that the operation and scientific results of new facilities, as well as instrumental upgrades in existing telescopes (including developments of Multi-Conjugate AO, chromospheric polarimetry, and tests with different types of observing modes) will yield much novel insight into the peculiarities and possibilities of observations with large solar telescopes. This Special Session, held during the XXVIII IAU General Assembly, was meant to start addressing these topics with the community at large. About 100 scientists convened for its 2-day duration, providing for a thorough and very interactive discussion on the current status and results of new telescopes, observing strategies, and critical issues for the development of future facilities that will be at the forefront of solar astrophysics in the next decades.

This paper provides a summary of the presentations and discussions held during the Session. The meeting's program is reported in the Appendix. Most of the individual presentations are available from ADS under the first author's name, with the identifier "2012IAUSS...6E".

2. Key Scientific Questions

Next generation large solar telescopes (see Sect. 3) are expected to make breakthrough discoveries significantly advancing our knowledge in critical solar research areas. This vital science, however, translates into significant challenges for those future endeavours since it is intricately coupled to the ability to observe the relevant solar phenomena with high precision and accuracy on unprecedented spatial scales (of the order of 10 kilometers) and time scales (of the order of seconds).

It is therefore that the Special Session 6 started with the invited introductory talk of Steiner (2012), diving into some selected solar physics problems for which small-spatial scales and low signals prevail, and extrapolating from high-resolution observations and numerical simulations to derive sensible requirements for future instrumentation. Exemplified science throughout the session focused on the following topics: photospheric and chromospheric vector-polarimetry at small spatial scales in general and the horizontal magnetic field of the quiet Sun specifically; observational evidence for a turbulent surface dynamo; drivers for dynamic fibrils and spicules; small-scale vorticity and chromospheric swirls, and large-scale coronal vortex structures. During this first session several major conclusions (or better, challenges) directly flowing from most of the science cases stood out.

First, next generation large solar telescopes should provide vectorpolarimetric observations at a spatial resolution that is well below the pressure scale height and the photon mean-free-path *on the way to the dissipative scales*, with an accuracy of $I_{\rm pol}/I_{\rm cont} \leqslant 10^{-4}$. This is most important in order to to answer remaining questions regarding the horizontal magnetic field of the quiet Sun, to reveal the structure of the *hidden magnetic field*, and to determine the spatial spectrum of the magnetic energy (outside active regions) with the aim of learning more about its origin and the role of the turbulent dynamo (e.g. Ishikawa (2012), Wang *et al.* (2012), Steiner (2012)), as well as address magnetic dissipation in the chromosphere (Judge & Casini (2012)).

Second, it will be crucial to perfom multi-wavelength spectroscopic and spectropolarimetric observations from the photosphere up to the corona at the same time (or at least quasi-simultaneously) by creating synergies between and sensible coordination of ground-based and space-borne instruments. This capability will be increasingly important in the future since it should allow us to follow effects that small-scale events may have through each atmospheric layer, understand the coupling between those layers, and ultimately address the question why some stars (must) successfully sustain a net-chromosphere and coronae (e.g. Judge & Casini (2012)). For instance, it should be possible to identify the driving mechanism and cause of dynamic fibrils (e.g. Martinez-Sykora (2012)), spicules (e.g. Klimchuk (2012)), small-scale chromospheric swirls and large-scale coronal cyclones. Specifically with regards to the latter two phenomena, it should be possible to clarify if and how those are related to or caused by small-scale vortical flows (e.g. Kitiashvili *et al.* (2012)) in the solar atmosphere.

It was also emphasized and became clear from discussions that diagnostic and interpretational strategies that are based on simple physical models might not be able to address the complexities of current problems in solar physics, and that a better *interlocking of realistic numerical simulations with observations* was necessary although not without repeated critical analysises of the numerical models (Steiner (2012), Steiner & Rezaei (2012)). On the one hand, a solid understanding of the numerical deficiencies and the boundary and initial conditions are necessary when comparing results (e.g. in form of synthesized observables) with observations because even a so called "realistic" simulation is only a crude approximation to the reality on the Sun. Also it was pointed out that more simulations in general are needed *to gain intuition* with regards to the influence that changes of initial and boundary conditions have. As examples of such *interlocking* we refer to the invited talks of Georgobiani *et al.* (2012) and Martinez-Sykora (2012), and the contribution of Kitiashvili *et al.* (2012).

In brief, this first session, guided by the introductory presentation, took the audience from the quest for the three-dimensional structure and origin of the quiet-Sun magnetic field of the photosphere (plasma-$\beta > 1$) upwards through the enigmatic and complex chromosphere into the corona (plasma-$\beta < 1$) where recent observations of *solar magnetized tornadoes* and *EUV cyclones* have amazed solar observers.

2.1. The 3D Topography and Origin of the Quiet-Sun Magnetic Field

Since the original discovery of an omnipresent weak magnetic field outside of active regions by Livingston & Harvey (1971) (using the Zeeman effect) at the beginning of the seventies of the last century, the magnetic field of the quiet Sun has drawn increasing attention for essentially two reasons. First, it could be a major driver of the heating of the chromosphere and ultimately the corona (e.g. Schrijver *et al.* (1998), Trujillo Bueno *et al.* (2004)). Second, its origin could be intricately coupled to both, global and local dynamo processes. In order to clarify and address those conjectures the combination of high-spatial resolution with high-precision multi-wavelength spectropolarimetry is really needed. While spectropolarimetric observations with high sensitivity in the visible and near-infrared are possible from the ground, those observations are achieved only by long integration times and as a consequence the spatial resolution (even with adaptive optics systems) is degraded to a level that significantly impedes the detection of weak fields on sub-arcsecond spatial scales.

It is thus not surprising that observations with the Spectro-Polarimeter (SP, see e.g. Lites *et al.* (2001)) of the *Solar Optical Telescope* (SOT, e.g. Kosugi *et al.* (2007)) on board the *Hinode* satellite (seeing-free environment, steady spatial resolution of ~ 0.32 arcsec, good polarimetric precision) considerably contributed to the advances made. This is particularly true related to the magnetic field component that is *transverse* to the line-of-sight (LOS) and as such detectable in the linear polarization and the Stokes-Q and U signals, which in the weak-field limit are intrinsically diminshed when compared to the Stokes-V signal. The pre-*Hinode* results led to a picture where the internetwork (IN) field

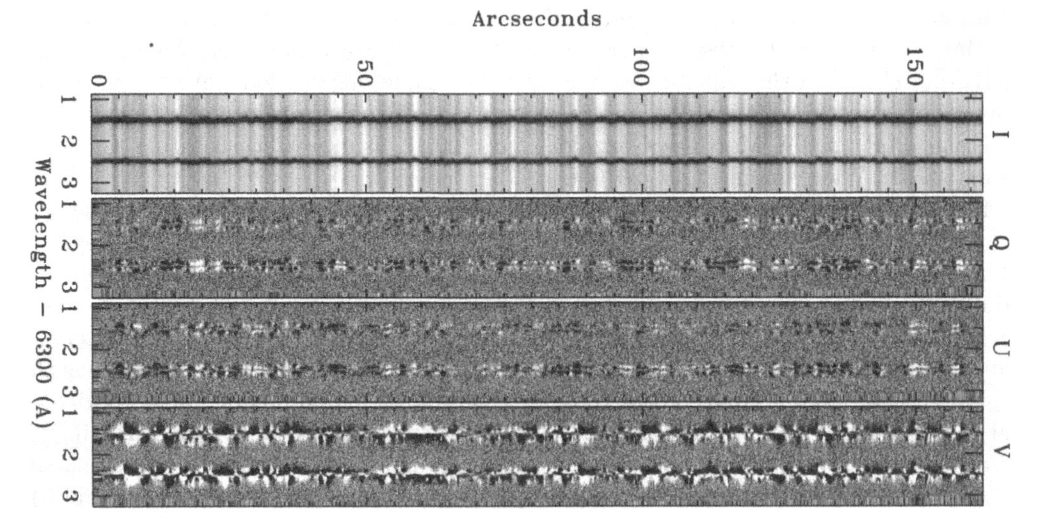

Figure 2. Typical deep mode, time-averaged Stokes I, Q, U, V spectrum taken at disk center. The effective integration time is 67.2 sec achieving a noise level in the continuum of $2.9 \times 10^{-4} I_c$. From Lites *et al.* (2008).

is characterized by dynamic small-scale mixed polarities rather randomly oriented with much disputed magnetic flux densities, intrinsic field strengths, and magnetic topologies, while the network appears as a relatively organized and stable magnetic structure harbouring the stronger kilo-Gauss magnetic fields outlining the borders of the super granulation.

The *deep mode* and temporally averaged Stokes spectra obtained with the SP/SOT, as visualized in Fig. 2, for the first time clearly revealed and demonstrated that the IN is heavily populated with inclined magnetic fields, evidencing a relatively large imbalance in the average apparent magnetic flux density between the vertical and horizontal component of the magnetic field (e.g. Lites *et al.* (2007), Orozco Suárez *et al.* (2007), Orozco Suárez *et al.* (2007), Lites *et al.* (2008)). It is important to remark that particularly this latter finding is much debated (see comparison of different results in Steiner (2012)). These IN *transverse* fields are "everywhere" (see e.g. Tsuneta *et al.* (2008) for a study of the Sun's polar region) and are likely the source of the *seething fields* discovered by Harvey *et al.* (2007). The total magnetic flux carried by those fields could be as high as 8×10^{23} Mx hr^{-1} (e.g. Ishikawa *et al.* (2010)) which is higher than what sunspot regions and ephemeral regions contribute combined.

There is general agreement that IN horizontal magnetic fields are very dynamic and rather transient in nature with lifetimes in the range ot $1 - 10$ min, smaller than the evolutionary timescales of the granulation, and they appear with sizes that are typically smaller than a granule with a lower size limit yet un-determined (see Ishikawa (2012) and references therein). Studies of the spatial distribution of the vertical and the horizontal flux density conclude that the vertical flux prefers intergranular lanes, while the horizontal flux emerges within or at the edges of granules close to but not co-spatial with the vertical flux (see e.g. Lites *et al.* (2008), Ishikawa (2012), Ishikawa & Tsuneta (2011b)). Most interestingly, vertical and horizontal flux components appear organized on mesogranular spatial scales coinciding with locations where the horizontal flow field shows a negative divergence (indicative of downflows) with "voids" in between (Lites *et al.* (2007), Lites *et al.* (2008), Ishikawa & Tsuneta (2010a), Yelles Chaouche *et al.* (2011)). However, it is not clear whether (all of) these "voids" really correspond to "field free" regions

(Ishikawa (2012)) or we just encounter the limitations with regards to sensitivity and spatial resolution of the observations (see Sect. 4.1 and contribution of Martinez Pillet (2012)). Related to the topology of the IN fields, some observations give evidence for tiny loopy structures of which some may even rise to higher atmospheric layers (e.g. Centeno *et al.* (2007), Martínez González *et al.* (2010), Ishikawa *et al.* (2010), Gömöry *et al.* (2010), Wiegelmann *et al.* (2010)).

The distribution of the inclination of IN magnetic fields is similarly debated as the apparent flux densities and intrinsic field strengths. No consensus has been reached so far. While numerical simulations of magneto-convection (independent of the orientation of the seed field) predict a predominance of the horizontal field increasing with height (Steiner *et al.* (2008), Schüssler & Vögler (2008), Danilovic *et al.* (2010)), results from observations using different methodology (inversions, weak-field approximation) and wavelength regimes (visible and infrared) do not converge. Some authors find predominantly horizontal IN fields (Lites *et al.* (2007), Lites *et al.* (2008), Orozco Suárez *et al.* (2007), Danilovic *et al.* (2010)), other studies point towards predominantly vertical fields (e.g. Khomenko *et al.* (2003), Beck & Rezaei (2009), Stenflo (2010)), and some find indication for (quasi) isotropic distributions (e.g. Martínez González *et al.* (2008), Asensio Ramos (2009), Bommier *et al.* (2009)).

It is argued that the different sensitivity in linear polarization and circular polarization to the magnetic field combined with the finite sensitivity of the photon noise inflicted measurements leading to selection effects (e.g. Kobel *et al.* (2011)), is the main reason for the many disparate results. However, Beck & Rezaei (2009) argue that the thermodynamic state of the atmosphere and different formation heights of the lines (visible and infrared) play an important role as well. The distribution of the magnetic flux and the angular distribution are of such special interest since they can provide clues about the origin and cause of the quiet-Sun internetwork magnetism. Mostly, observational results appear consistent with the notion that the action of a local dynamo driven by turbulent convection is a contributor to the generation and maintenance of the quiet-Sun's internetwork magnetic field (e.g. Cattaneo (1999), Vögler & Schüssler (2007), Schüssler & Vögler (2008), Pietarila Graham *et al.* (2009), Pietarila Graham *et al.* (2010)). In this context, it is particularly conspicuous that the amount of observed IN magnetic flux neither indicates a solar-cycle nor any latitudinal dependence (e.g. Hagenaar *et al.* (2003), Sánchez Almeida (2003), Trujillo Bueno *et al.* (2004), Harvey *et al.* (2007)). However, it remains to be answered to which degree a surface dynamo contributes to the IN magnetism and what fraction does result from the global magnetic field. Furthermore, the detection of "voids" in the IN magnetic field needs to be better understood in the context of a local dynamo working on granular spatial scales (see also Sect. 4.1). Specifically in this area more "realistic" simulations are needed that allow a comparison between the observations and the predictions of those simulations (hopefully in form of synthesized observables). Furthermore a proper understanding of the initial and boundary conditions of the individual simulations is needed. Georgobiani *et al.* (2012) presented "realistic" 3D simulations of solar magnetoconvection leading to flux emergence and the formation of pores and active regions without an initial coherent flux rope implying that a tachocline is not necessary. Performing a forward synthesis of Stokes spectra, Georgobiani *et al.* (2012) also point out that the mere limitations imposed by a finite aperture do already significantly alter the Stokes spectra.

In summary, as much progress has been made since the launch of *Hinode* satellite many questions remain un-answered. Most important, what is the structure of the magnetic field on spatial scales beyond the SP/SOT 0.32 arcsec benchmark? This addresses the *hidden magnetic flux* issue and the discrepancies between the results when using Zeeman-

or Hanle diagnostics. The intrinsic problem that the net-Stokes signal of co-existent opposite polarities within the same resolution element are hard to distinguish from a single polarity with small filling factor will only be solved when the spatial resolution is pushed further and further down to smaller scales. Hence, accurate and precise vectorpolarimetry of the photosphere (and chromosphere!) very close to the diffraction limit of future telescopes will play an incredible important role allowing for new discoveries and hopefully resolving some existing riddles.

2.2. **Photospheric Vorticity, Chromospheric Swirls and Coronal Tornadoes**

Another area of active research currently living through a rennaissance is concerned with small-scale vorticity in the photosphere and its consequences for the higher atmospheric layers. Vortex flows are nothing new, they have been theoretically predicted based on 3D numerical simulations of solar convection already back in the 80's by Nordlund (1985) introducing descriptive terms like *bathtub effect* and *inverted tornado.* Several observational attempts were undertaken to prove the existence of those flow structures but without success for only one exception. The first confirmation of a vortical flow although with a larger size (covering several granules, lifetime of $\sim$ 1.5 hours) than predicted must be attributed to Brandt *et al.* (1988) suggesting that vortical flows could introduce twisting motions of the footpoints of magnetic flux tubes and as such provide a mechanism for heating the chromosphere and corona (also Balmaceda *et al.* (2010)).

With increasing spatial resolution various studies were later able to infer vortex flows on small spatial scales ($\leqslant$ 0.5 Mm) based on local correlation tracking techniques calculating horizontal flow maps and/or tracing the proper motion of photospheric magnetic elements that get "dragged into" and "trapped" in the downdraft of a vortex (e.g. Bonet *et al.* (2008), Bonet *et al.* (2010), Balmaceda *et al.* (2010), Vargas Domínguez *et al.* (2011)). Those small-scale vortical flows or *whirlpools* show lifetimes up to 20 min with varying values for the space-time density in the range $1.8 - 3.1 \times 10^{-3}$ Mm^{-2} minute^{-1}. The observations are not conclusive about whether a preferred sense of rotation exist. Only Bonet *et al.* (2010) find a preferred rotation sense using high-resolution observations obtained with the IMaX instrument of the *SUNRISE* balloon. Vortical flows or "vortex tubes" are not restricted to the vertical but have been identified also oriented in the horizontal direction and being closely related to the sub-structure of granules (dark lanes and bright rims) and their "edgy" appearance (Steiner *et al.* (2010)). However, with the current spatial resolution it remains open whether those observable "granular lanes" are just the most prominent out of a wide spectrum of vortex tubes (Steiner (2012)).

The observational evidence of photospheric vortex flows is accompanied by recent numerical simulations examining small-scale vorticity and the consequences. For instance, Kitiashvili *et al.* (2011) find that simulated *whirlpools* are quite numerous, form in intergranular lanes with no preferred sense of rotation, are characterized often by very strong horizontal shear velocities (7-11 km sec^{-1}), strong downflows in the vortex cores (7 km sec^{-1}), and very low densities ("density holes"). However, the cores (where much of the motion takes place) of those simulated *whirlpools* are tiny in the range of 100 km which is challenging to spatially resolve (see Fig. 3). Most interesting, they find that their simulated whirlpool flows can attract other vortical flows with opposite vorticity which then interact with each other, partially annihilate and in the aftermath excite acoustic waves. Observational evidence for the latter has been found analyzing a time sequence of TiO images obtained with the New Solar Telescope (NST) of the Big Bear Solar Observatory (Kitiashvili *et al.* (2012)). The simulations of solar surface convection by Moll *et al.* (2011) provide a complex picture where regions of high "swirling strength" form an unsteady network of tangled filaments. Close to the optical surface, vertical and

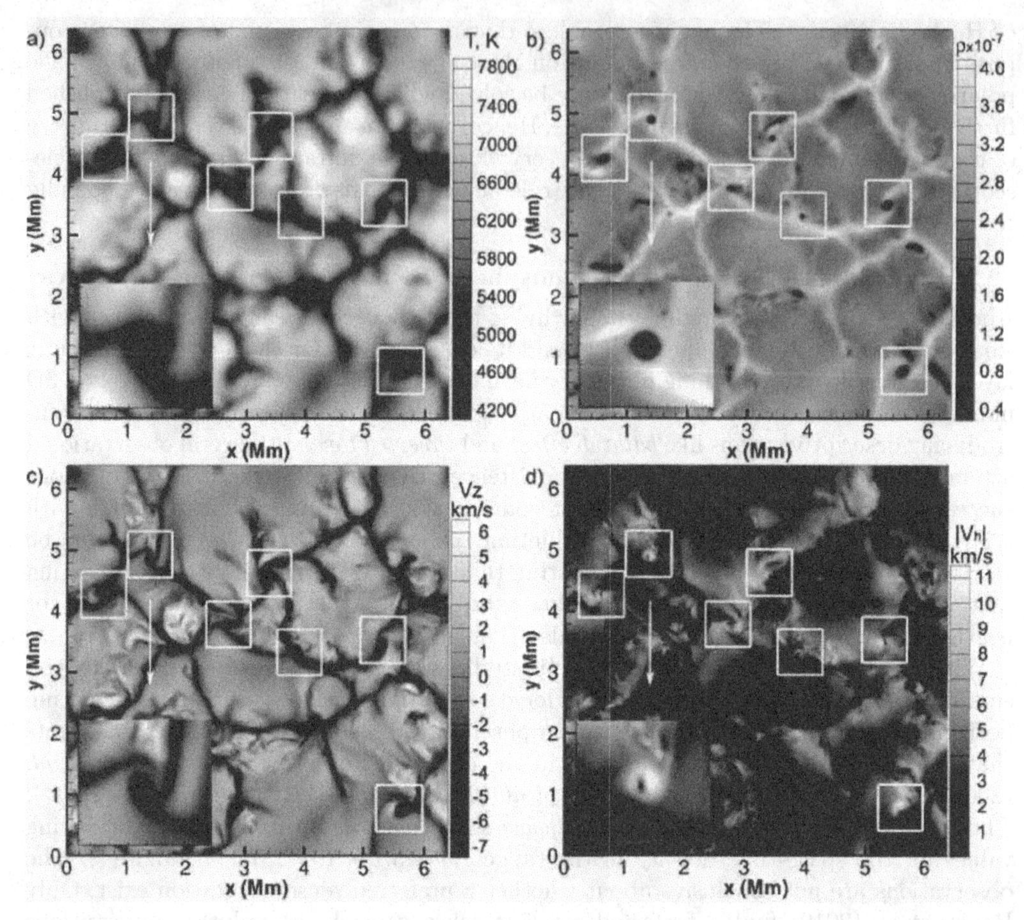

Figure 3. Snapshots of the simulation results at the solar surface for (a) temperature, (b) density, (c) vertical velocity, and (d) magnitude of the horizontal velocities. White squares indicate the largest whirlpools. From Kitiashvili *et al.* (2012).

horizontal swirls are preferentially found within the intergrnaular lanes and at the edges of granules, respectively. Above the surface, swirls appear in many shapes as small-scale bent and arc-like filaments with diameters of only 30 km which renders observational proof impossible with current solar telescopes.

It is interesting to remark that photospheric vortex flows do appear with sizes larger than just a couple of granules. For example, Attie *et al.* (2009) observed very long-lasting (1-2 hours) photospheric vortex flows extending up to 27 arcsec that are located at the supergranular junctions. Those observations specifically showed that opposite polarity magnetic elements co-exist in the same vortex where one magnetic element caught at the vortex core, is orbited by the opposite polarity magnetic element.

Recently, observations of the quiet chromosphere in a coronal hole demonstrated that whirling motions are not limited to the photosphere. From a time sequence of Ca II 854.21 nm line-core images, Wedemeyer-Böhm & Rouppe van der Voort (2009) were able to identify numerous (although subtle) chromospheric *swirls* made of dark and bright rotating patches in form of arcs, spiral arms, rings or ring fragments (also Wedemeyer *et al.* (2013)). Those *swirls* typically have a width of 2 arcsec (with the fragments being much smaller, $\sim$ 0.2 arcsec) and indicate fast upflows up to 7 km sec^{-1}. Most interestingly, below those chromospheric swirls, groups of magnetic elements move with respect to each other but there is no indication for the presence of a vortical flow. Wedemeyer-Böhm *et al.* (2012), however, identify the chromospheric *swirls* as observational signatures of rapidly rotating magnetic structures ("tornadoes") rooted in the photosphere in magnetic elements and extending through the tansition region (He II 30.4 nm) into the low corona (Fe IX 17.1 nm). The observations are supported by numerical simulations using the CO5BOLD code (Freytag *et al.* (2012), Beeck *et al.* (2012)) suggesting that chromospheric *swirls* are the (chromospheric) imprint of rotating magnetic elements braiding and twisting the magnetic field. A numerical experiment comparing three different simulation runs (field-free case, vertical or horizontal seed field of 50 G) show that chromospheric *swirls* develop in the vertical run but were absent in the field-free and horizontal run (see Steiner & Rezaei (2012)). Similarly, the simulations of Kitiashvili *et al.* (2012) show that the swirling motions of vertical vortex tubes can capture and twist magnetic fields forming magnetized vortex tubes penetrating into the chromosphere. The ubiquity of those structures and the fact that thy provide a direct link between the convection zone and the upper solar atmosphere suggests that they could considerably contribute to the channeling of energy into the chromosphere and corona.

In the corona and on much larger spatial scales, Zhang & Liu (2011) and Su *et al.* (2012) find *EUV cyclones* and *solar magnetized tornadoes*, respectively, based on observations with the AIA instrument of the SDO. The cyclones are quite abundant and are seen everywhere on the Sun with the sense of rotation showing a weak hemispheric preference, last for several hours and are found rooted in rotating network magnetic fields. The *solar magnetized tornadoes* typically appear in groups, are related to filaments/prominences, and are believed to be driven by underlying photospheric vortex flows.

In summary, small-scale vorticity generated in the photosphere by convective motions interacting with the magnetic field has implications throughout the solar atmosphere and seems to be the source of a many recently discovered phenomena. However, the small sub-arcsecond spatial scales over which the actual motions take place and disturbances propagate through the atmosphere make it extremely difficult to observe those phenomena with present day solar telescopes. Observationally, vortex motions are identified by local correlation tracking techniques and feature tracking algorithms, but not with spectroscopic methods. For instance, a direct measurement of the individual rotation of a magnetic element trapped in a vortex flow and the braiding and twisting of the magnetic field has not been achieved so far and will probably need the enhanced osberving capabilities of future large telescopes.

2.3. Chromospheric Structure and Dynamics

The chromosphere is the focus of much contemporary research. This boundary region hosts the critical transition from fluid to magnetic dominance on the atmospheric plasma, and its many complexities (partial ionization, non-LTE, time-dependent H ionization, large spatial and temporal gradients, role of mechanical heating) represent a formidable challenge to our understanding of a magnetized stellar atmosphere. Yet, as a key interface between the photospheric magneto-convection "driver" and the resulting outer solar atmosphere and heliosphere, as well as the main source of the ionizing UV radiation that drives the Earth's atmosphere, it is mandatory that we unravel its working.

In a sense, the presence of a chromosphere and a corona derives fairly straightforwardly from the existence of any non-radiative heating source at the solar surface (Judge & Casini (2012)). How this energy is actually transported at larger heights, and how it is dissipated in loco, remains however the problem in defining the solar chromospheric structure, even if observations clearly indicate that magnetism lies at the heart of the phenomenon. *Measuring* the vector magnetic field in the chromosphere and its variations in time and/or height appears as the proper way to obtain reliable estimates of the "free energy" available in the magnetic configuration for local dissipation (Judge (2009)), or a direct measure of the electric currents' density (Socas-Navarro (2005)) †.

This is arguably a very hard task with most chromospheric signatures (e.g. Uitenbroek (2011)); still, a very promising tool to this end is the use of the He I triplet at 1083 nm (see also Sect. 4.2), a spectral signature showing large sensitivity to chromospheric magnetic fields through both the Hanle and Zeeman effect (Asensio Ramos *et al.* (2008), Schad *et al.* (2013)). The required polarimetric precision must be high, of the order or better that 10^{-4}, to be able to properly measure the vector components of the field: this implies that new large aperture telescopes will have to function as large "light buckets", relaxing the requirement of diffraction-limited spatial resolution for the case of reliable chromospheric polarimetry (see Judge & Casini (2012) and Sect. 5).

A complementary, important approach in the study of the chromosphere is that provided by 3D-MHD numerical simulations. Notwithstanding the complications of the many non-linear equations that govern the physics of the solar magnetized atmosphere, and the proper treatment of radiation and scattering (see e.g. Skartlien (2000), Martinez-Sykora (2012)), recent advances in this area are bringing about "realistic" simulations which encompass the whole atmosphere, and the very different physical regimes present from the convection zone to the corona (see among others, Fang *et al.* (2010), Gudiksen *et al.*(2011)). Special care must be given in the chromosphere and transition region to physical processes inherent to the low density, magnetized, highly intermittent conditions of plasma, including thermal conduction, partial ionization effects and ion-neutral interaction, or non-equilibrium effects such as hydrogen ionization (Leenaarts *et al.* (2007), Martínez-Sykora *et al.* (2012), Khomenko & Collados (2012), Leake & Linton (2013)). Recent efforts in these directions show some very promising results about the reliability of the simulations; in particular the introduction of partial ionization (treated as multi-fluid or otherwise) in the simulations' domain has shown that the resulting Pedersen resistivity is of the same order of the numerical resistivity value normally used to stabilize the codes, an issue which had been long cause of concern (Martínez-Sykora *et al.* (2012), de la Cruz Rodríguez *et al.* (2013)). One of the most important limitations remains the limited spatial extension of the simulations, usually of the order of a super granular cell or less, which might hide important phenomena related to large scale magnetic connectivity (Martinez-Sykora (2012)).

The comparison of the time evolution of MHD variables in the simulations with observed quantities is an important tool to understand dynamic chromospheric phenomena or, conversely, to determine whether simulations are missing some important physics. Some intriguing small-scale phenomena like *dynamic fibrils* (De Pontieu *et al.* (2007)) have been successfully explained as due to magneto-acoustic shocks propagating along field lines in the upper atmosphere (Heggland *et al.* (2007), Martínez-Sykora *et al.* (2009)), even if some discussion is still ongoing about the origin of the driving piston. In particular, the simulations of Kato *et al.* (2011) identify the "turbulent pumping" mechanism as an alternate way to excite longitudinal waves along the magnetic flux tubes. This is an interesting example of how simulations carried out by different groups can

† While extrapolation of photospheric magnetic maps is often attempted, most notably to infer coronal conditions, the required assumption of force-free status for the magnetic field in the photosphere is usually not satisfied, leading to inconsistencies in the derived results, see e.g. De Rosa *et al.* (2009).

provide different insights (or results) into the physics of selected phenomena, and thus of the need to properly understand the role of, e.g., different numerical methods, boundary conditions etc., as remarked above. Extreme high resolutions observations such as provided by large-aperture telescopes, in this case of downflows around small magnetic network elements, could provide clear discriminant between conflicting hypotheses.

An intense debate is currently undergoing in the community about a similar small-scale, dynamical chromospheric phenomenon, namely "type II spicules". These were first defined through *Hinode* Ca II observations at the limb as very thin (at the limit of telescope resolution of 0.2 arcsec, rapidly evolving (50-150 sec) features, which are observed to propagate upward (velocities of 30-100 km sec^{-1}) and fade around their maximum length rather than fall down to the solar surface (de Pontieu *et al.* (2007)). Ever-present type-II spicules have been identified as main agents in the transfer of energy and mass from the photosphere to the corona in several works (e.g. De Pontieu *et al.* (2009), De Pontieu etal (2011)), attributing their fading from chromospheric signatures to heating to coronal temperatures. Arguments to the contrary argue that EUV spectroscopy indicates at most a very small fraction of coronal plasma originating from such features (Klimchuk (2012), Klimchuk (2012)), and that at least a fraction of type-II spicules could be due to warps in two-dimensional sheet-like structures, related to the magnetic tangential discontinuities naturally arising in low-β plasma (Judge *et al.* (2011), Judge *et al.* (2012)).

Numerical simulations are for now just hinting at the creation of type-II spicules, in rather particular conditions (Martinez-Sykora (2012)), a fact that seems in contrast with their prevalent appearance over the whole solar surface (Pereira *et al.* (2012)). Further comparisons of diagnostics synthesized from simulations adopting different magnetic topologies and spatial domains, with proper resolution observations of the chromosphere are expected to much advance our understanding of this fascinating phenomenon. Indeed, many of the technical requirements for the new, large solar telescopes derive from chromospheric science of this kind (see e.g. Sects. 3 and 4).

3. Existing and Planned Large Facilities

3.1. Ground-based optical telescopes

A large number of new ground based facilities have been proposed, constructed and commissioned in recent years; virtually all of them have been presented during the second session of SpS6, as listed in the Appendix. After many years of "stagnation" in the field, two main reasons are behind this striking development. The first one is a series of technological breakthroughs which have finally allowed the design of air-cooled, open large solar telescopes. By virtue of this design, one can plan for much larger apertures than for the case of "classical" evacuated telescopes, and aim at achieving spatial resolutions well below the current $\approx 0.2''$ (150 km) limit, thus approaching some of the important spatial scales predicted by simulations. The second one is the advent of reliable and efficient Adaptive Optics systems, coupled with powerful image-reconstruction techniques (e.g. Rimmele & Marino (2011), Wöger (2010)), both indispensable tools to achieve resolutions close to the actual diffraction limit of these large telescopes.

Facilities recently become operative, such as the German on-axis GREGOR Telescope on Tenerife (Schmidt *et al.* (2012), Denker *et al.* (2012), currently being commissioned) or the *New Solar Telescope* (NST, Goode & Cao (2012), Cao (2012) of the Big Bear Solar Observatory, online since 2010), have sizes around 1.5 m (Fig. 4). They represent a funda-

Figure 4. *Left*: the 1.6 m off-axis New Solar Telescope of the Big Bear Solar Observatory. The position on a pier within Big Bear Lake (CA) guarantees long periods of good to excellent seeing. NST has been operating since 2010. *Right*: the 1.5 m, on-axis German GREGOR telescope on Tenerife, inaugurated June 2012. The unique folding dome is visible, retracted on the floor of the building.

mental step in the quest for higher apertures, and such facilities share the honor and responsibility to demonstrate the feasibility and performances of large open solar telescopes. Future facilities plan on much larger sizes, up to the impressive 4 m aperture of the US *Advanced Technology Solar Telescope* (ATST, Keil *et al.* (2010), Rimmele *et al.* (2012), Rimmele *et al.* (2012)), currently undergoing construction on Haleakala, Hawaii and scheduled for first light in 2018 (http://http://www.nso.edu/press/ATSTConstruction). The *European Solar Telescope* (EST, Collados Vera & EST Team (2012), Socas-Navarro (2012)), currently in the design phase, will also have a 4 m diameter, while the ambitious *Chinese Giant Solar telescope* (CGST, Liu *et al.* (2012), whose design is currently being pursued by a consortium of Chinese Institutes with support from the Chinese Academy of Sciences foresees a giant ring telescope of a 1 m width with 8 m diameter. Such a telescope would offer many advantages in terms of low-instrumental polarization and thermal control, while still allowing a large collecting area (equivalent to a 5 m diameter solid mirror).

Extensive site-testing campaigns have been conducted ahead of installation of these facilities (or choice of their future site) to ensure that the best seeing characteristics are achieved over long periods of time. The ATST Team conducted a thorough survey of 6 possible sites (Socas-Navarro *et al.* (2005)), eventually settling on the Haleakala location due to its superior coronal characteristics, an important scientific goal of the project. Similar campaigns held in India and China respectively, have identified superb sites nearby high mountain lakes. At least two lake sites in the high altitude ($> 4,000$ m) dry Ladakh desert, at the foot of the Himalaya, have been marked as possible location of the planned 2 m, India's *National Large Solar Telescope* (NLST, Hasan (2012)), foreseen for first light in 2017. The Fuxian Lake Solar Observatory in the Yunnan province in China has instead been chosen for the recently commissioned 1 m *New Vacuum Solar Telescope* (NVST, Ji & Liu (2012)). The excellent seeing characteristics of the site can be appreciated in Fig. 5 which shows a speckle-reconstructed image of an active region acquired, in the TiO band at 706 nm (Ji & Liu (2012)).

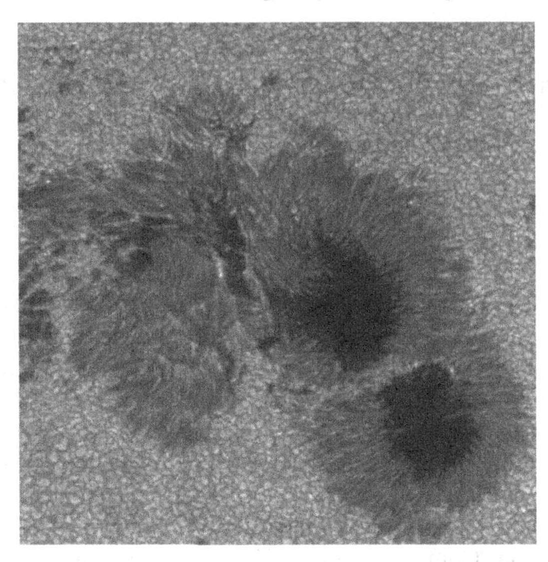

Figure 5. Active region observed in July 2011 at the NVST in the TiO band at 706 nm. The image has been reconstructed with speckle methods from bursts of 5 ms exposure. From Ji & Liu (2012).

As mentioned above, high-order AO systems are part of the original design/construction plan for all of the new ground-based telescopes, as they are necessary to achieve high Strehl ratio throughout the visible spectrum (see e.g. Richards *et al.* (2010), Berkefeld *et al.* (2012)). Still, the size of the isoplanatic patch, which determines the size of the field of view effectively diffraction limited, can be a severe limitation even in the best of conditions (e.g. Rimmele *et al.* (2010)). Much effort is thus currently being devoted to characterize the contributions to seeing from atmospheric layers up to 10-12 km (Scharmer & van Werkhoven (2010), Berkefeld *et al.* (2010), Kellerer *et al.* (2012)) in order to develop and optimize the so-called Multi-Conjugate Adaptive Optics systems, which couple (multiple) deformable mirrors to a number of relevant layers in the atmosphere and allows correction on larger fields of view. These systems are currently being tested at the DST (Rimmele *et al.* (2010)), the GREGOR telescope (Schmidt *et al.* (2012)), the NST (Goode & Cao (2012)), with the obvious scope to install them at the largest telescopes of the future (e.g. Rimmele *et al.* (2012), Socas-Navarro (2012)).

All the facilities described above are planned to run as multi-purpose, flexible observatories, with a large variety of scientific targets that can be addressed by versatile instrument suites (see Sect. 5 below). A somewhat different facility is the *Coronal Solar Magnetism Observatory* (COSMO, Tomczyk (2012), Gallagher *et al.* (2012)), a proposed US-China project currently completing the Preliminary Design Review. COSMO's principal goal is that of accurately, and continuously, measure the coronal field through Zeeman effect in the Fe XIII 1074.7 nm coronal line using the Large Coronograph, a 1.5 m refracting chronograph, designed to achieve a very low level of scattered light.

The balloon-borne 1 m telescope *SUNRISE* (Solanki *et al.* (2010)) is a unique facility, tailored for visible imaging spectra-polarimetry and near-UV imaging, taking advantage of the reduced atmospheric absorption at its operating $\approx$ 35 km of altitude. The first flight of *SUNRISE* in June 2009 provided a novel view of quiet-Sun magnetism at small spatial scales (Danilovic *et al.* (2010), see also Martinez Pillet (2012) and Sect. 4). *SUNRISE* is currently scheduled for a second flight in mid-2013.

3.2. Radio-telescopes

Technological development in the field of radio-astronomy has recently spawned many new projects worldwide, several of which are either solar dedicated, or of general astrophysical interest but capable of observing the Sun as well.

Much like the efforts in solar optical astrophysics to obtain full spectral information coupled with spatial and temporal information, the main emphasis in contemporary solar radio-astronomy is that of developing dynamic imaging spectroscopy of the Sun in a large range of frequencies (e.g. Pick & Vilmer (2008)). This implies the use of interferometric arrays with dynamic baseline arranging, as for example with the recent transformation of the Jansky VLA into the *Expanded Very Large Array* (EVLA, Perley *et al.* (2011)) which can currently perform, albeit sporadically, true solar imaging spectroscopy in the range 1-8 GHz (to be expanded to 1-18 GHz, see Bastian & Gary (2012)). The same frequency range will be monitored by the solar dedicated *Expanded Owens Valley Solar Array* (EOVSA, Gary *et al.* (2011)), an array of thirteen 2.1 m antennas which will allow imaging spectroscopy of flares and active regions with a few arcsec spatial realution and 1 sec temporal resolution. Completion of EOVSA is foreseen for October 2013.

The *Frequency Agile Solar Radiotelescope* (FASR, Bastian & Gary (2012)) represent the next development in the US, aiming at achieving dynamic imaging spectroscopy over an extremely large frequency bandwidth, thus allowing to track phenomena in the solar atmosphere simultaneously at many heights. FASR will comprise three separate arrays of antennas sweeping the frequency range 50 MHz – 21 GHz in less than 1 sec, with a max. spatial resolution of 1 arcsec. FASR, to be installed in Owens Valley, has been highly rated in the National Academies Decadal Surveys, but is currently awaiting development of a mid-scale funding line from the US National Science Foundation before it can proceed.

A similar project is that of the *Chinese Solar Radioheliograph* (CSRH, Yan *et al.* (2009)), currently under construction in a radio-quiet area of inner Mongolia, which will cover the 0.4-15 GHz range with about 100 antennas. The first portion of the project, CSHR-I, covers the range 0.4–2 GHz with 64 channels (40 antennas), and has been recently completed. (Li *et al.* (2012)). Solar observations have been obtained in early 2012 and are currently being processed for first results. The second portion of the project, CSRH-II, covering the 2-15 GHz range with 60 antennas, is foreseen for completion at the end of 2013 (Yan *et al.* (2012)).

The largest recent development in radio astronomy is the *Atacama Large Millimeter Array* (ALMA, e.g. Busatta *et al.* (2012)), which employes 66, 12 m antennas for observations of a wide range of astrophysical sources in the 86-720 GHz frequency range (3.4-0.3 mm). ALMA has been configured from the start to be able to observe the Sun, by using Sun-proof antennas and a set of specifically designed attenuators with potential for precise diagnostics of the quiet and active chromosphere, the low corona, as well as rapid transients such as flares (Karlický *et al.* (2011)). Test solar observations have been obtained with ALMA in 2011 and 2012 during the science verification phase, using an array of 13 antennas. The single-dish mosaicing procedure has been verified and calibrated, and has produced promising results in clearly identifying enhanced plage and active region emission, with spatial resolution of several arcsec (Fig. 6, Benz *et al.* (2012)). Interferometric observations, which will allow sub-arcsec spatial resolution at 0.3–9 mm, potentially resolving filamentary shock structures in the quiet chromosphere (e.g. Wedemeyer-Böhm *et al.* (2005)), are currently being tested.

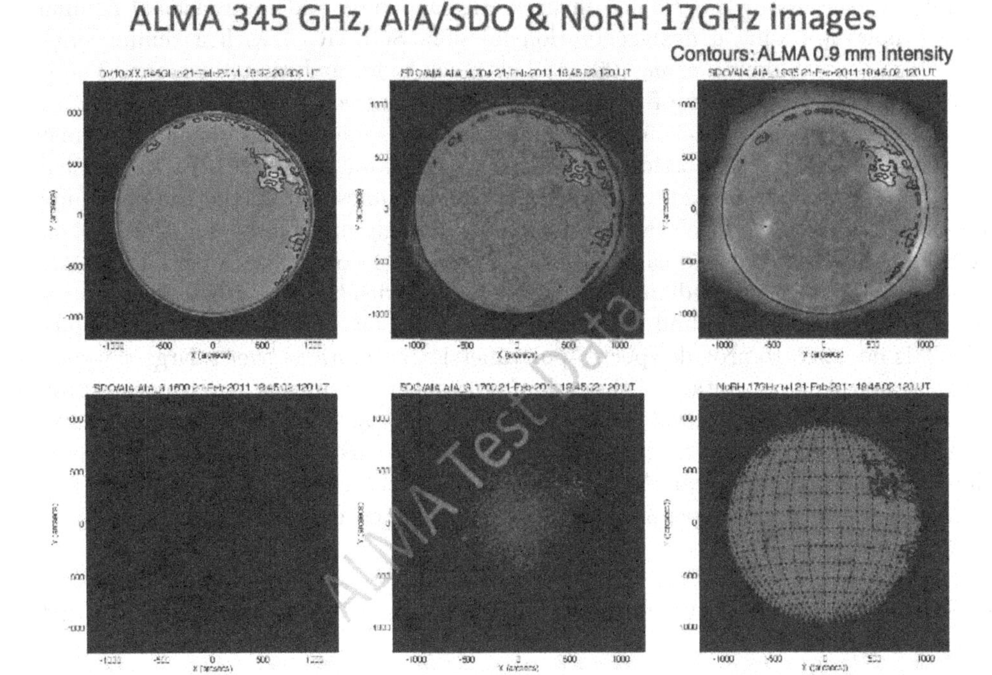

Figure 6. ALMA first solar data, from Science Verification program. From Benz *et al.* (2012).

3.3. **Space projects**

Several new space solar observatories are also being considered for construction in the next decade. The Chinese solar community is currently developing a project for a multi-purpose solar spacecraft, the *Deep-space Solar Observatory* (DSO, formerly Space Solar Telescope, see Deng (2012)) which has passed budget approval in mid-Aug 2011, and whose realization will be decided upon within 2013. If successful, this will be the first scientific spacecraft launched and operated by China. The DSO will be a multi-instrument project, with emphasis on a 1 m optical telescope devoted to magnetic measurements in the photosphere (Fe I 532.4 nm) and chromosphere (Hβ line, 486.1 nm) at 0.1-0.15 arcsec resolution with a temporal resolution of $\sim$ 30 sec. The core of the magnetograph is the Two-Dimensional Real-Time Spectrograph (2DS), a multi-element birifringent filter that allows fully simultaneous observations in 8 wavelengths within a spectral line, and which is currently being prototyped (Deng (2012)). The payload includes complementary instruments as well, such as an EUV imager, an Hα and WL imager, and a Lyα telescope.

The *Interhelioprobe solar observatory* is an out-of-the-ecliptic, inner heliosphere spacecraft (perihelion = 0.28 AU), currently being developed within the Federal Space Program of the Russian Federation (Bogachev *et al.* (2012)). It is aimed at studying the Sun from close distances in the inner heliosphere, and carries a suite of 5 remote instruments (plus 15 in situ ones. These include a multi-channel EUV imager, a white-light chronograph, and a heliospheric imager, with increasingly large FOVs, to continuously couple coronal structure to inner-heliosphere conditions. Interhelioprobe is scheduled for launch in the 2018-2020 time frame, and will perform stereoscopic observations of the Sun together with Earth-based instrumentation and *Solar Orbiter* (Müller *et al.* (2012)).

After the success of the *Hinode* mission, the Japanese and international communities are now proposing a next-generation mission, SOLAR-C, with a similar suite of instruments (see Shimizu *et al.* (2012)) but much more ambitious in scope. The goal of SOLAR-C is infact that of providing a comprehensive, seamless view of the solar atmosphere, from the photosphere to the corona, by simultaneously using instruments with complementary diagnostics capabilities (Katsukawa *et al.* (2012)). In particular, SOLAR-C will seek to provide diagnostics of the dynamics and magnetism of the upper chromosphere, an "interface" region which is still largely unexplored and that was missing from the *Hinode* payload capabilities. This will be accomplished with a large optical telescope, SUVIT, 1.5 m in diameter, which, coupled with the seeing-free environment of space, will rival modern ground-based instruments in terms of achieved spatial resolution. SUVIT is designed to provide spectro-polarimetric observations over a large wavelength range: again taking advantage of the lack of atmosphere, it aims at covering the chromospheric Mg II h&k lines around 280 nm, as well as the visible and near-infrared range up to the He I 1083 nm triplet (Suematsu *et al.* (2012)). Broad- and narrow-band filtergraphs will complement the spectral observations, although the cadence and multi-wavelength coverage achievable will depend on the actual telemetry, currently under discussion.

4. Steps Ahead: Future Science and what the New Facilities should not Forget

A number of state-of-the-art observational and theoretical advances are pushing current facilities to their limit, and can offer important insights and lessons for future developments. We report below some examples that were discussed during SpS6.

4.1. Quiet Sun magnetism

The need for high-precision polarimetry is well illustrated by the still puzzling structure of "quiet Sun" magnetic fields, as shown in Sect. 2. Given the different sensitivity of the Zeeman signal to longitudinal (LOS) or transverse fields, the typical polarimetric sensitivity of $10^{-3} I_c$ of modern instruments (such as the SP on *Hinode*, or the MTR at THEMIS, Bommier (2011)) translates into our capacity to much better discern LOS fields, with respect to the transverse ones – namely, few Gauss vs. many tens of Gauss (cf. Martinez Pillet (2012)). It is possible that such a "biased" view of solar magnetism could be the origin of some controversial results related to quiet-Sun magnetism, including the long-standing issue of why some isolated network flux patches appear to lose flux and disappear in 1-2 hours (Spruit *et al.* (1987)), or the existence of "flux voids" on mesogranular scales (e.g. Yelles Chaouche *et al.* (2011), Berrilli *et al.* (2013)), which would contrast with the likely exixtence of a small scale turbulent dynamo, operating at granular scales (Brandenburg (2011); see also Sect. ??).

Indeed, when pushing the data to higher sensitivity, a "sea" of field structures appears, as shown in Fig. 7. In particular, at the 10^{-4} sensitivity range all flux patches are constantly interacting with opposite polarity neighbors, which might explain the reported "in-place" disappearance of unipolar flux as a sensitivity issue (Martinez Pillet (2012)). Voids instead seem to be a persistent feature; a possible resolution of the small scale dynamo at granular scale might be waiting for systematic $10^{-4} I_c$ observations.

A problem intrinsic to the use of Zeeman polarimetry to study weak fields is the fact that the long integration times, necessary to reach the desired polarimetric accuracy, inevitably degrade spatial resolution, leading to polarity cancellation. The complemen-

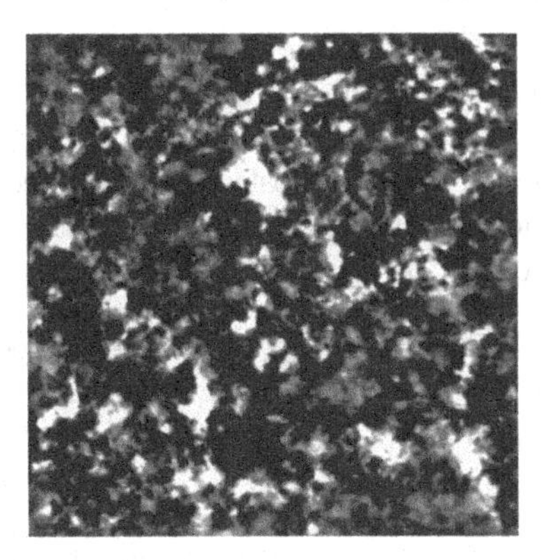
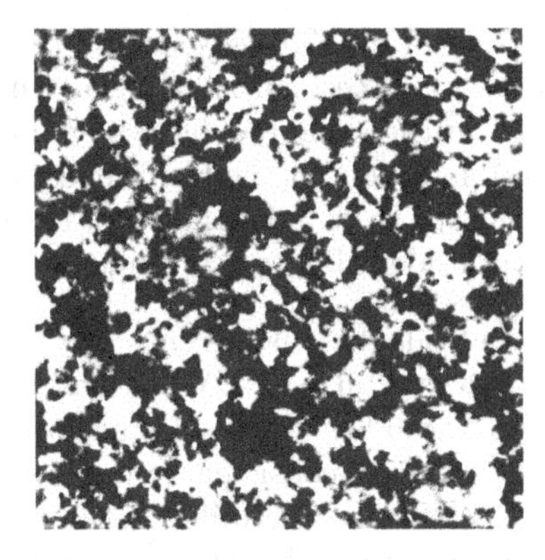

Figure 7. *Left*: IMaX magnetic field map of a quiet Sun region, saturated at a level of $10^{-3} I_c$. *Right*: the same map when data are pushed to $10^{-4} I_c$ sensitivity. From Martinez Pillet (2012).

tary use of the Hanle effect offers much promise in this respect. The Hanle effect is the modification, caused by the presence of a magnetic field, of the linear polarization produced by scattering processes in a spectral line (Belluzzi *et al.* (2012)). As the signal from the Hanle effect does not vanish in presence of mixed-polarities (it is proportional to B^2), and its sensitivity peaks between a few Gauss and $\approx$200 G, this diagnostic is well suited to investigate the magnetism of the quiet Sun (see also Faurobert (2012)). As Hanle polarimetry moved more into "mainstream" solar physics in the last years, many investigations have dealt with the presence of a weak turbulent magnetic field in the solar photosphere, whose strength has been estimated at a few Gauss up to several tens of G (e.g. Kleint *et al.* (2010), Faurobert *et al.* (2001), Bommier *et al.* (2005)). The use of differential Hanle diagnostics, i.e. the use of multiple lines with different sensitivity to magnetic fields (especially molecular lines such as those of CN, C_2, MgH), might be able to provide more robust results to settle such controversy. Investigations of this kind have recently suggested that the turbulent field has a strong dependence with atmospheric depth (Milić & Faurobert (2012)), as well as shown an apparent lack of relationship with solar cycle phase (e.g. Kleint *et al.* (2010)).

Observational difficulties related to the use of the Hanle effect lie in the weakness of the scattering linear polarization of solar lines (often a fraction of a percent); the quest for higher polarimetric sensitivity will thus much benefit the use of both Zeeman and Hanle polarimetry. It is also important to remark that many modeling assumption enter the derivation of magnetic fields once precise polarimetric measures are acquired. In particular, observed Stokes profiles are often asymmetric, indicating a variation of atmospheric parameters with height along the line of sight. This, however, has just begun to be considered in the inversion methodology using the Zeeman effect (e.g. Viticchié *et al.* (2011)). Analogously, the derivation of fields when using the Hanle effect relies on a number of assumptions, e.g. on the isotropy of the field, an hypothesis which has recently undergone scrutiny (e.g. Frisch *et al.* (2009), Faurobert (2012)).

4.2. **Polarimetry of the chromosphere and transition region**

The magnetism of the upper solar atmosphere - namely upper chromosphere and transition region, is of high scientific interest as a crucial force in shaping this complex interface between the photosphere and corona, where most of the non-thermal energy that creates the corona and solar wind is released (Judge (2009), see also Sect. 2.3). In a physical regime where most of the spectral diagnostics lie in the far-UV or EUV range, Zeeman polarimetry is of limited use because the (expected) small field values at these wavelengths produce a Zeeman splitting much smaller than the width of the lines. On the other hand, in the same regime a large degree of scattering polarization is expected in the spectral lines, so that the Hanle effect can be of great diagnostic potential because its magnetic sensitivity is independent of the wavelength and Doppler width of the spectral line under consideration (e.g. Trujillo Bueno (2012)).

Theoretical calculations are currently being performed to determine the degree of polarization expected in such UV spectral lines in the presence of magnetic fields, both in "classical" 1-D models of the solar atmosphere, and adopting physical parameters deriving from modern, self-consistent numeric simulations such as those presented in Leenaarts *et al.* (2012). Belluzzi & Trujillo Bueno (2012) analyzed for example the behavior of the Mg II lines at 280 nm, which form in the upper chromosphere, while Trujillo Bueno *et al.* (2012), Štěpán *et al.* (2012) investigated the polarization properties of the Lyα lines of H I and He II. These latter lines appear very good candidates for inferring fields between roughly 10 and 100 G within the transition region; in particular, their joint use might alleviate ambiguities in the retrieval of the fields, due to, e.g., to horizontal atmospheric inhomogeneities. Still, the expected linear polarization is of the order of 1-2% at most, again highlighting the need for very precise polarimetric measurements.

As already described in Sect. 2.3, a "special" spectral line for investigation of the upper chromosphere' magnetism is the He I triplet at 1083 nm, that lies in a completely different spectral range, namely the near-infrared. The line forms in chromospheric conditions, mostly through the photoionization-recombination mechanism (Andretta & Jones (1997)) which implies that it is mostly visible where abundant UV radiation is present, i.e. usually in magnetic, heated structures. Its sensitivity to magnetic fields through both Hanle and Zeeman effect has been well investigated in recent years (e.g. Asensio Ramos *et al.* (2008)), and its use for diagnostics of fields in prominences, active regions' chromosphere, and even spicules, is rapidly increasing (Casini *et al.* (2009), Centeno *et al.* (2010), Schad *et al.* (2011), Schad *et al.* (2012), Schad *et al.* (2013)).

4.3. **Scattered light issues**

While image restoration has improved enormously in the last decade, affording near-diffraction limited data from ground-based telescopes (e.g. van Noort *et al.* (2005)), the contrast of such images is still a controversial issue. Indeed, it appears that the measured granulation contrast at several wavelengths, as measured by the SOT/*Hinode* (taking into account its actual Point Spread Function, derived using transit and eclipse data), it is well matched to the values deriving from modern numerical simulations (Wedemeyer-Böhm & Rouppe van der Voort (2009)). On the contrary, the granulation contrast measured at ground-based observatories is still much lower than predicted (Uitenbroek *et al.* (2007), Scharmer *et al.* (2010)), the hypothesis being that several sources of straylight must be present, whether from the Earth's atmosphere or the combined telescope-instrumentation optical train (Löfdahl (2012)). Attempts to identify the possible sources of stray-light within a modern solar telescopes show that the fraction of scattered light in a signal can be surprisingly high, around 0.4, and that the dominant contribution appears to derive from wavefront aberrations, which possibly originate in the deformable mirror of the adaptive optics system (Löfdahl & Scharmer (2012)). These studies are still in their infancy, but they are paving the way for much needed investigations that will, hopefully, fully characterize the influence of scattered light and, eventually, drive our ability to correct for it, especially for what concerns chromospheric imaging and spectra.

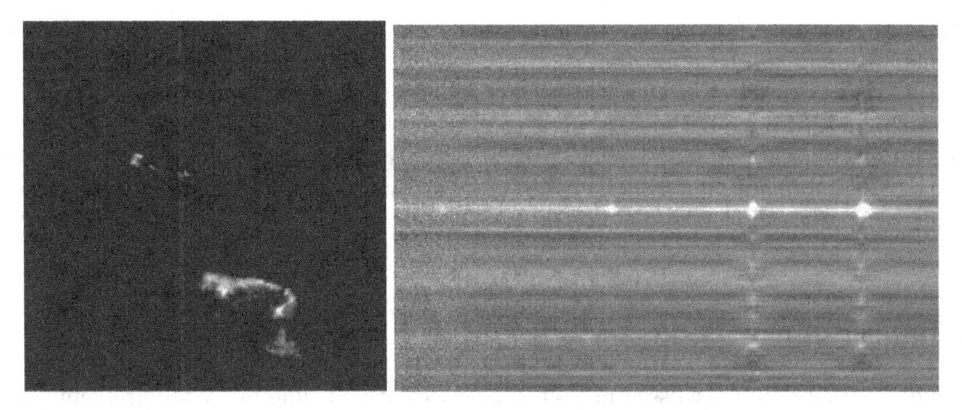

Figure 8. *Left*: IBIS image in Hα core during a C1.0 flare on August 18, 2011. The vertical line indicates the position of the spectrograph slit. *Right*: corresponding spectrum in Ca II H&K range (over a longer slit extension than the left image). A reference spectrum has been subtracted to enhance flare signature. Easily visible is the emission form Ca II H and nearby Hϵ, Ca II K, as well as H_8 and H_9 pushing into the blue limit of the spectrum. The bright strip indicates (putative) continuum emission (courtesy of A. Kowalski and G. Cauzzi).

4.4. Flare physics

Studies of the corona during solar flares have increased ten-fold since the advent of UV and X-ray imaging and spectroscopy from satellites such as Yohkoh, SOHO, Hinode, etc. Yet, comprehensive studies of the flaring chromosphere, the main source of the flares radiative output, are still lacking. Difficulties include the complex physics involved in the rapid development of a flare within a dense medium (non-equilibrium ionization, shock and plasma waves, return currents, etc.), as well as the observationally difficulty to correctly position the instruments' FOV on chromospheric flaring kernels that are small, rapidly evolving, and occurring in mostly unpredictable locations. While this difficulty exists for coronal instruments as well, the latter usually possess a larger FOV (albeit at lower spatial resolution) and often focus on the slower decay phase of a flare, while the main value of chromospheric diagnostics is certainly in the rapid impulsive phase.

Novel developments in this field hold promise of a renewed observational and theoretical interest in the physics of chromospheric flares. One the one hand, the availability of modern radiation-hydrodynamical calculations including energy dissipation from accelerated electron beams allows a more direct link from observables to physical conditions (e.g. Allred *et al.* (2005), Cheng *et al.* (2010)). On the other hand, the advent of imaging spectrographs with large FOVs, such as IBIS, CRISP or GFPI (Cavallini (2006), Scharmer *et al.* (2008), Puschmann *et al.* (2012)) will permit a more comprehensive coverage of regions with flaring potential, while maintaining full spectral information on relevant lines (Deng *et al.* (2013); see also Fig. 8).

Multi-line diagnostics are also a powerful tool to identify and discriminate physical processes in flares, for example spatio-temporal details of the chromospheric heating which are ultimately related to acceleration processes in the reconnection site (Cheng *et al.* (2006), Kašparová & Heinzel (2002)). Recent observations of enhancements in the infrared continuum during a particularly energetic flare (Xu *et al.* (2012)) have also prompted a revisit of the mechanisms that lead to the so-called white light flares (WLFs), and the recognition that broad-spectral observations at high spatial and temporal resolution are necessary for this task (e.g. Kowalski *et al.* (2012)). The (apparent) extreme fragmentation of the basic reconnection process, inferred in radio (Chen *et al.* (2013)), optical (Radziszewski *et al.* (2011)) and X-rays observations (Cheng *et al.* (2012)) highlights this need for high cadence, high spatial resolution observation while preserving spectral and spatial coverage (Deng (2012)). Large aperture telescopes appear thus necessary to reveal fine flaring structures at 0.1 arcsec and 0.01 sec resolution, to reveal the nature of WLFs and elementary bursts.

5. Impact of new Large Facilities on Key Scientific Questions

The scientific issues discussed in Sect. 2 and 4 clearly outline many requirements that new facilities will have to comply with, in order to properly address relevant science.

Routinely achieving a high polarimetric sensitivity, of the order of $10^{-4} I_c$ or better, as necessary for weak field studies and Hanle polarimetry, will require telescopes with apertures sensibly larger than what currently available. The 1.5 m class NST and GREGOR telescopes are just starting to tackle the issue of measuring polarization with dedicated instrumentation (Cao *et al.* (2010), Hofmann *et al.* (2012)), and will in the coming years reveal fascinating science as well as (probable) technical limitations. However, it is widely acknowledged that a quantum leap in this field will occur when the planned 4-m class telescopes will come online later in the decade, with their design for highest throughput (through all reflective optics, and minimal number of surfaces). Yet, one has to constantly remember that *any* telescope, when working at the diffraction limit, will collect the same number of photons no matter the aperture (see Martinez Pillet (2012))! Worse still, to maintain a higher spatial resolution, exposure times will need to be shorter – effectively reducing the signal-to-noise ratio of data acquired with larger facilities. This poses an interesting conundrum, whereas to provide the much needed improvement in polarimetric accuracy, one will have to purposely limit the spatial resolution to be achieved with large telescopes, using them more like "light bucket" rather than high resolution facilities. This is the approach taken for example for some of the polarimetric instruments planned for the ATST, like the near- to mid-infrared CryoNIRSP (optimized for coronal physics, see Kuhn *et al.* (2012)).

A second, important problem in the quest for high polarimetric accuracy will be the ability to characterize and control telescope and instrumental polarization (as already remarked more than 30 years ago! by Stenflo (1982)). Large telescopes are adopting different strategies to address this issue. Both GREGOR and the future EST will use on-axis designs to minimize polarization from the main optical elements. The preliminary design of EST is pushing further this concept by developing the optical path, all the way to the Coude' focus, with a 14-mirror "compensated" design so to minimize the instrumental polarization induced by the telescope and make it independent of the solar elevation and azimuth, at once (Collados Vera & EST Team (2012), Socas-Navarro (2012)). The NST and ATST have instead an off-axis design, chosen to optimize telescope PSF and minimize scattered light, with the latter property being especially relevant for use of ATST for coronal physics (Rimmele *et al.* (2012), see below). Recent investigations, using both theoretical models and actual solar data, indicate that techniques recently adopted to precisely infer polarization properties of the main optical elements of telescopes like the DST will be applicable to the ATST to achieve the required precision (Elmore (2012)).

Coronal polarimetry will push even further telescope requirements, due to the combination of intrinsically weak magnetic signals and brightness of the background with respect to the actual coronal intensity. Thus modern projects are striving to provide, together with larger apertures, very low levels of scattered light deriving from the main optics. To reach this goal, ATST, which has coronal physics among its main scientific objectives, adopted an off-axis design, strict figures for the main mirror, as well as procedures for contamination control of the primary and secondary mirrors (McMullin *et al.* (2012)). The much reduced scattered light will also benefit accurate measures of the physical properties of small structures in, for example, sunspots. The COSMO project is instead adopting a unique solution, i.e. a 1.5 m *refractive* telescope, as lenses provide better scattered light levels and control with respect to a mirror (Tomczyk (2012)).

The majority of the new and planned facilities are designed to function as full observatories, i.e. with multiple and flexible instrumentation that can address a variety of scientific issues. Indeed, the emphasis of many new facilities is the *simultaneous* use of varied instrumentation that can observe at different wavelengths and/or with different capabilities. In particular, the combined use of "classical" slit-spectrographs and imagers, either broadband, or tunable narrow-band ones, appears as the most promising approach (see e.g. Fig. 8, or Judge *et al.* (2010), using the DST imaging spectroscopy together with UV and IR spectroscopy), and is foreseen as the standard observing mode of large facilities (see e.g. Fig. 9).

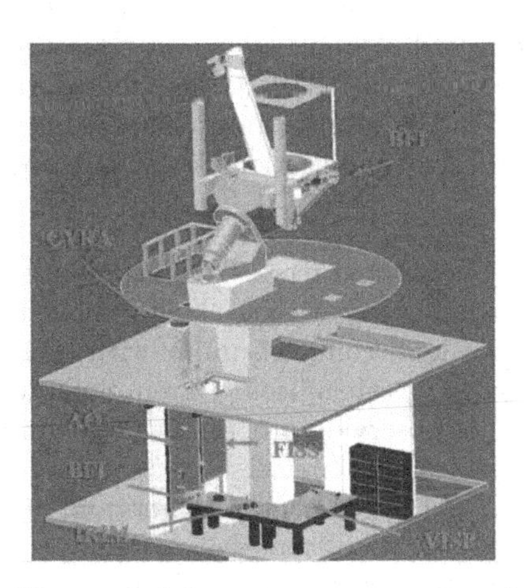

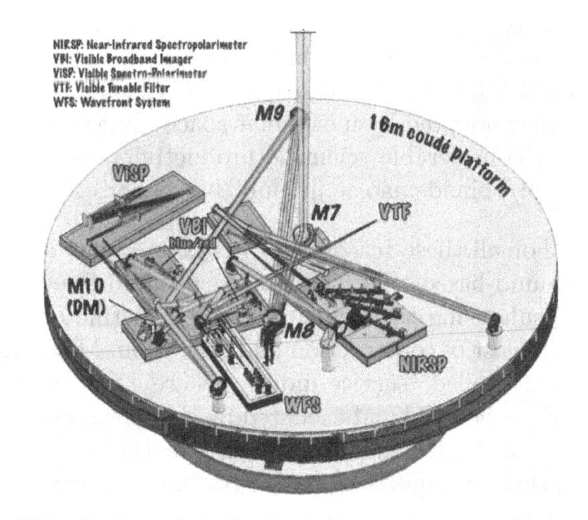

Figure 9. Schematic representation of the NST (left) and ATST (right) optical arrangement, showing capability to observe simultaneously with multiple instruments.

The large wavelength coverage is also a defining characteristics of most new telescopes. ATST will be able to operate from 350 nm up to thermal infrared (Rimmele *et al.* (2012)), exploiting, for example, the far-blue range for molecular spectroscopy (see previous Section), and the far-red range for coronal spectro-polarimetry (see Kuhn *et al.* (2012)). It is interesting to note that many of the new telescopes will enjoy diffraction-limited conditions at near-infrared wavelengths for very large fractions of the observing time (up to 80% for Haleakala! see Socas-Navarro *et al.* (2005))). It is thus expected that much emphasis will be give to (near) IR observations in the coming years.

6. Lesson Learned, and Future Directions for Telescope Operation

6.1. Observing strategies for future solar facilities

A very advantageous characteristics of ground-based solar telescopes devoted to high-resolution observations, has always been the large flexibility offered in instrumental configurations and observing programs. This permits the acquisition of unique observations easily adapted to ever-evolving targets and scientific questions (e.g. Uitenbroek (2006)). Traditionally, ground based telescopes have nurtured this flexibility by assigning fixed observing periods (of one to two weeks duration) to Principal Investigators traveling to the facility, during which the available instrumentation would be suited to the scientific goals of the peculiar observing program.

Such extreme flexibility can result however also in large inefficiencies of the system. First of all, the time necessary to modify the instrumental setup, sometimes substantially, accounts for a large fraction of the available observing time, of the order of 20% or more. Second, the fixed assigned dates might not optimally match with the desired observing target, whether an active region, an eruptive event, or specific values of sky transparency and turbulence levels ("seeing"), thus significantly reducing the observing duty cycle. Last, but not least, the "uniqueness" of each observing program makes it difficult to devise and utilize standard data reduction pipelines, forcing PIs to spend a large amount of time before obtaining science-ready data, and making it difficult to share data between scientists. Indeed, data acquired at modern (ground based) solar telescopes are usually proprietary, and very little of it gets ever used for more than one particular scientific objective, resulting in a generally modest record of publications/year for any given telescope (e.g. Reardon *et al.* (2009)). This is in stark contrast with the public-data policy adopted by most solar space observatories, which has resulted in wide-spread use and considerable scientific productivity (see e.g. the list of publications by Hinode, at http://hinode.nao.ac.jp/publ/hsc_paper_e_main.shtml).

For all these reasons, a discussion about different operational modes for future large ground-based solar facilities has been initiated in the community, with the long-term goal of enhancing their efficiency and scientific output to a level commensurable with their high cost of construction and operation. Much discussion has then been given in SpS6 to the so-called "service mode", where a prioritized list of observing programs is dynamically scheduled and executed by the telescope staff (resident astronomers and operators), optimizing the match between scientific requirements and instrumental, target, and atmospheric conditions (e.g. Silva (2001)). Service mode operations represent a significant fraction of the observing time at several night-time facilities, such as the VLT or Gemini; beside being very favorably judged by users, it has resulted in significantly improving the scientific output of these facilities (Comerón *et al.* (2006), Puxley & Jørgensen (2006)). The recently commissioned ALMA radio telescope has been built to operate exclusively in this mode (Andreani & Zwaan (2008)).

It must be noted however that a number of requirements must be satisfied for service mode operations to be highly effective, including a clear communications with the community about users' expectations vs. the real capabilities of the telescope and instrumentation, clear rules to allow the resident astronomers to undertake decisions. In particular, the VLT experience has shown the necessity of a software infrastructure for supporting the end-to-end operations of the facility, including tools for the preparation, scheduling and execution of observations, as well as archiving of data, pipeline data reduction and quality control (Peron (2012)).

ATST is expected to operate predominantly in service mode (Uitenbroek & Tritschler (2012), Berger (2012)). Many topics still need to be addressed and flushed-out, including the handling of "guaranteed time" for partner institutions, the communication and relationship in real time with a remote PI, and in general the design and implementation of a robust software infrastructure like that described above (Berger (2012)). However, the intervening time between now and actual ATST operation can be used to solve many of these issues and gain practical skills. Much can certainly be learned by analyzing the experience of night-time facilities, but the special circumstances of solar observations require also some "hands-on" experience at existing telescopes. After an early attempt at optimize coordinate DST observations with *Hinode* (Reardon *et al.* (2009)), the NSO has nowdevised a "service mode experiment" in preparation for ATST operations to explicitly start confronting the problems that might be encountered with this operational mode (Uitenbroek & Tritschler (2012)). This experiment has been planned to run for one month at the beginning of 2013, utilizing a selected subset of instruments at the DST, including IBIS, ROSA and FIRS. The availability of stable, well calibrated instrumentation that can be used in relatively standard configurations is an essential pre-requisite for service mode operations, as it allows for an easier planning of observations, the sharing of calibration between different observing programs, as well as the rapid switches required by the optimization of target vs. sky conditions.†

Another important aspect of successful service mode operations is the availability of standardized, robust data reduction pipelines which can minimize the effort needed by the PIs in order to obtain science-ready data. This has long been a problem in high-resolution solar data: on the one hand, both the ever-changing instrumental configurations and the vagaries of the seeing conspire against the development of stable and well distributed data pipelines; on the other, the extreme specialization required to analyze each and every high-resolution dataset has prevented the involvement of a larger user-base, with detriment to the scientific productivity of the field (Reardon *et al.* (2009)). Several groups are currently trying to overcome this limitation. In particular, the recently funded EU project SOLARNET (http://www.iac.es/proyecto/solarnet/) is planning a Europe-wide effort to increase impact of high-resolution data by offering science-ready data and facilitate their usage throughout a larger fraction of the solar community. A major part of this goal is the coordinate development of largely automatic data-reduction pipelines and archival data access for a wide variety of high resolution solar instrumentation (Uitenbroek & Tritschler (2012)).

6.2. **Access to large databases: the lessons from SDO**

At its best, service mode operation is foreseen to increase scientific output of future telescopes by optimizing the acquisition of data with respect to local and solar conditions, targets of opportunity, etc. A "collateral effect" of such an efficient data acquisition scheme, however, is that the future large telescopes will produce enormous data volumes, of the order of tens to hundreds of TB per day (e.g. Ermolli *et al.* (2012)). Such volumes are not yet commonly managed in current telescopes: although some instruments come close to achieving those lower limits, (see e.g. Jess *et al.* (2010)), they are not operated

† Note: The DST service mode experiment has been run from Jan 15 through Feb 15, 2013, and has attracted considerable attention from the community, with a total of 21 requests for observing programs. About half of them have been executed during the 30-day period (most of the remaining programs have not been executed due to weather conditions and lack of suitable targets). This can be contrasted with the typical number of programs run in a comparable period in "classical" mode, which is in the range of 3–4.

daily! A large effort in the development of viable large telescopes will then have to be spent in the intervening years before their actual operation in planning for their data storage, reduction, and analysis.

Probably the most appropriate comparison with existing facilities, in terms of data handling, is with the Solar Dynamics Observatory (Pesnell *et al.* (2012)). The SDO has been operating since spring of 2010, and archives daily about 1.5-2 TB (a factor of 3000 larger than what SOHO did!, Schrijver (2012)), for a projected total of 2 PB over the mission's lifetime. While virtually all of this database is freely accessible by the community, its sheer volume requires a different approach than to "normal" data. On the one hand, it is impossible for any scientist to just browse through all of the data to identify interesting events; on the other, it would be a waste of the enormous scientific potential of the SDO database if only a small fraction of it would ever be utilized because of limited manpower. Hence, barring a sensible increase of the budget devoted to data analysis (unlikely), the community needs an array of tools to optimize scientific efficiency to properly exploit this data bounty.

A large effort by the SDO Team has thus gone into optimizing the process of archive searches and data distribution, for example by distributing focused, smaller amounts of data to relevant institutes, or keeping tracks of data requests, maintaing the most popular data sets available for immediate export (Schrijver (2012)). Earlier missions like TRACE have paved the way for the use of standard software like SolarSoft (http://www.lmsal.com/solarsoft, originally used for Yohkoh data) and standard searches through the Virtual Solar Observatory (http://vso.nso.edu), based on parameters' search such as instrument, temporal intervals etc. While the Hinode mission pushed archival searches to the more complex case of high volume, spectroscopic imaging instruments and their flexible use (http://sdc.uio.no), the SDO Team has further expanded on these capabilities by introducing an "event search system" (iSolSearch, http://www.lmsal.com/isolsearch) that enables searching on data contents rather than instrument attributes and settings. The search is based on a list of "event classes", such as flares, filaments, active regions, etc, to which new events can constantly be added to be "recognized" by the system.

Similar efforts will have to be carried out for databases of future large facilities, with the ultimate goal of a fully-shared approach between archival of space missions and ground-based telescopes' data (Reardon (2012)). Many of the lessons learned from SDO can be incorporated in the design and implementation of such systems, including the need of involving actual users in the development of archives and data-retrieving tools, and adopting a design open to modifications and system changes to accommodate evolving technology and user needs (Schrijver (2012)).

7. Conclusions

Large solar telescopes hold promise of revolutionizing our knowledge of the solar atmosphere and its dynamics, but pose formidable challenges for what concern their full implementation, and future operation. The scale of these challenges is such that the solar community needs to gather all forces to be able to properly address technical and scientific hurdles, from the handling of telescope polarization to the development of proper interpretational tools to address observed new phenomena.

Special Session 6 exploited the interdisciplinary and international character of the IAU to start gather interested and involved scientists, and discuss some of the more pressing issues in this field. The enthusiastic response of the community made clear that such a shared effort is high in the list of priorities, and made for a very lively and fruitful discussion, which will continue in the years ahead. In particular, we expect that the next IAU General Assembly in Honolulu in 2015 will be a proper venue to host a sequel of this discussion, also considering the ongoing construction of ATST in the Hawaiian island of Maui.

Acknowledgements

Travel funds for several participants were provided by grants No. 11178005 and 10921303 of the National Natural Science Foundation of China, No. 2011CB811401 of the Ministry of Science and Technology of China, and No. KJCX2-EW-T07 of the Chinese Academy of Sciences, as well as from the Metcalf Travel Award of the US Solar Physics Division. We gratefully acknowledge their support.

The National Solar Observatory is operated under a cooperative agreement between the Association of Universities for Research in Astronomy, Inc. (AURA), and the National Science Foundation.

References

Allred, J. C., Hawley, S. L., Abbett, W. P., & Carlsson, M. 2005, *ApJ*, 630, 573

Andreani, P. & Zwaan, M. 2008, vol. 7016, *SPIE Conf. Ser.*

Andretta, V. & Jones, H. P. 1997, *ApJ*, 489, 375

Asensio Ramos, A. 2009, *ApJ*, 701, 1032

Asensio Ramos, A., Trujillo Bueno, J., & Landi Degl'Innocenti, E. 2008, *ApJ*, 683, 542

Attie, R., Innes, D. E., & Potts, H. E. 2009, *A&A* (Letters), 493, L13

Balmaceda, L., Vargas Domínguez, S., Palacios, J., Cabello, I., & Domingo, V. 2010, *A&A* (Letters), 513, L6

Bastian, T. & Gary, D. 2012, *IAU Special Session*, 6, E2.15

Beck, C. & Rezaei, R. 2009, *A&A*, 502, 969

Beeck, B., Collet, R., Steffen, M., Asplund, M., Cameron, R. H., Freytag, B., Hayek, W., Ludwig, H.-G., & Schüssler, M. 2012, *A&A*, 539, A121

Belluzzi, L. & Trujillo Bueno, J. 2012, *ApJ* (Letters), 750, L11

Belluzzi, L., Trujillo Bueno, J., & Štěpán, J. 2012, *ApJ* (Letters), 755, L2

Benz, A. O., Brasja, R., Shimojo, M., Karlicky, M., & Testi, L. 2012, *IAU Special Session*, 6, E2.05

Berger, T. 2012, *IAU Special Session*, 6, E2.01

Berkefeld, T., Schmidt, D., Soltau, D., von der Lühe, O., & Heidecke, F. 2012, *Astron. Nachr.*, 333, 863

Berkefeld, T., Bettonvil, F., Collados, M., López, R., Martín, Y., Peñate, J., Pérez, A., Scharmer, G. B., Sliepen, G., Soltau, D., Waldmann, T. A., & van Werkhoven, T. 2010, vol. 7733, *SPIE Conf. Ser.*

Berrilli, F., Scardigli, S., & Giordano, S. 2013, *Solar Phys.*, 282, 379

Bogachev, S., Kuzin, S., Shestov, S., & Perzov, A. 2012, *IAU Special Session*, 6, E2

Bommier, V. 2011, *A&A*, 530, A51

Bommier, V., Derouich, M., Landi Degl'Innocenti, E., Molodij, G., & Sahal-Bréchot, S. 2005, *A&A*, 432, 295

Bommier, V., Martínez González, M., Bianda, M., Frisch, H., Asensio Ramos, A., Gelly, B., & Landi Degl'Innocenti, E. 2009, *A&A*, 506, 1415

Bonet, J. A., Márquez, I., Sánchez Almeida, J., Cabello, I., & Domingo, V. 2008, *ApJ* (Letters), 687, L131

Bonet, J. A., Márquez, I., Sánchez Almeida, J., Palacios, J., Martínez Pillet, V., Solanki, S. K., del Toro Iniesta, J. C., Domingo, V., Berkefeld, T., Schmidt, W., Gandorfer, A., Barthol, P., & Knölker, M. 2010, *ApJ* (Letters), 723, L139

Brandenburg, A. 2011, *ApJ*, 741, 92

Brandt, P. N., Scharmer, G. B., Ferguson, S., Shine, R. A., & Tarbell, T. D. 1988, *Nature*, 335, 238
Busatta, A., Brunelli, A., Rampini, F., & Marchiori, G. 2012, vol. 8450, *SPIE Conf. Ser.*
Cao, W. 2012, *IAU Special Session*, 6, E2.02
Cao, W., Gorceix, N., Coulter, R., Ahn, K., Rimmele, T. R., & Goode, P. R. 2010, *Astron. Nachr.*, 331, 636
Casini, R., Manso Sainz, R., & Low, B. C. 2009, *ApJ* (Letters), 701, L43
Cattaneo, F. 1999, *ApJ* (Letters), 515, L39
Cavallini, F. 2006, *Solar Phys.*, 236, 415
Centeno, R., Socas-Navarro, H., Lites, B., Kubo, M., Frank, Z., Shine, R., Tarbell, T., Title, A., Ichimoto, K., Tsuneta, S., Katsukawa, Y., Suematsu, Y., Shimizu, T., & Nagata, S. 2007, *ApJ* (Letters), 666, L137
Centeno, R., Trujillo Bueno, J., & Asensio Ramos, A. 2010, *ApJ*, 708, 1579
Chen, B., Bastian, T. S., White, S. M., Gary, D. E., Perley, R., Rupen, M., & Carlson, B. 2013, *ApJ* (Letters), 763, L21
Cheng, J. X., Qiu, J., Ding, M. D., & Wang, H. 2012, *A&A*, 547, A73
Cheng, J. X., Ding, M. D., & Carlsson, M. 2010, *ApJ*, 711, 185
Cheng, J. X., Ding, M. D., & Li, J. P. 2006, *ApJ*, 653, 733
Collados Vera, M., EST Team 2012, *ASP Conf. Ser.*, 463, 413
Comerón, F., Mathys, G., Kaufer, A., Hainaut, O., Hanuschik, R., Romaniello, M., Silva, D., & Quinn, P. 2006, vol. 6270, *SPIE Conf. Ser.*
Danilovic, S., Beeck, B., Pietarila, A., Schüssler, M., Solanki, S. K., Martínez Pillet, V., Bonet, J. A., del Toro Iniesta, J. C., Domingo, V., Barthol, P., Berkefeld, T., Gandorfer, A., Knölker, M., Schmidt, W., & Title, A. M. 2010, *ApJ* (Letters), 723, L149
Danilovic, S., Schüssler, M., & Solanki, S. K. 2010, *A&A*, 513, A1
de la Cruz Rodríguez, J., De Pontieu, B., Carlsson, M., & Rouppe van der Voort, L. H. M. 2013, *ApJ* (Letters), 764, L11
Deng, N., Tritschler, A., Jing, J., Chen, X., Liu, C., Reardon, K., Denker, C., Xu, Y., & Wang, H. 2013, arXiv1304.4171
Deng, Y. 2012, *IAU Special Session*, 6, E2.09
Denker, C., Lagg, A., Puschmann, K. G., Schmidt, D., Schmidt, W., Sobotka, M., Soltau, D., Strassmeier, K. G., Volkmer, R., von der Lühe, O., Solanki, S. K., Balthasar, H., Bello Gonzalez, N., Berkefeld, T., Collados Vera, M., Hofmann, A., & Kneer, F. 2012, *IAU Special Session*, 6, E2.03
De Pontieu, B., McIntosh, S. W., Carlsson, M., *et al.* 2011, *Science*, 331, 55
De Pontieu, B., McIntosh, S. W., Hansteen, V. H., & Schrijver, C. J. 2009, *Apj* (Letters), 701, L1
De Pontieu, B., Hansteen, V. H., Rouppe van der Voort, L., van Noort, M., & Carlsson, M. 2007, *ApJ*, 655, 624
de Pontieu, B., McIntosh, S., Hansteen, V. H., *et al.* 2007, *PASJ*, 59, 655
De Rosa, M. L., Schrijver, C. J., Barnes, G., *et al.* 2009, *Apj*, 696, 1780
Domingo, V., Fleck, B., & Poland, A. I. 1995, *Solar Phys.*, 162, 1
Elmore, D. F. 2012, *ASP Conf. Ser.*, 463, 307
Ermolli, I., Cauzzi, G., Collados, M., Paletou, F., Reardon, K., Aboudarham, J., Cirami, R., Cosentino, R., Del Moro, D., Di Marcantonio, P., Giorgi, F., Lafon, M., Pietropaolo, E., & Romano, P. 2012, vol. 8448, *SPIE Conf. Ser.*
Fang, F., Manchester, W., Abbett, W. P., & van der Holst, B. 2010, *ApJ*, 714, 1649
Faurobert, M. 2012, *IAU Special Session*, 6, E3.01
Faurobert, M., Arnaud, J., Vigneau, J., & Frisch, H. 2001, *A&A*, 378, 627
Freytag, B., Steffen, M., Ludwig, H.-G., Wedemeyer-Böhm, S., Schaffenberger, W., & Steiner, O. 2012, *Journal of Computational Physics*, 231, 919
Frisch, H., Anusha, L. S., Sampoorna, M., & Nagendra, K. N. 2009, *A&A*, 501, 335
Gallagher, D., Tomczyk, S., Zhang, H., & Nelson, P. G. 2012, vol. 8444, *SPIE Conf. Ser.*
Gary, D. E., Hurford, G. J., Nita, G. M., White, S. M., Tun, S. D., Fleishman, G. D., & McTiernan, J. M. 2011, AAS/Solar Physics Division Abstracts #42, 102
Georgobiani, D., Stein, R., & Nordlund, A. 2012, *IAU Special Session*, 6, E1.02

Gömöry, P., Beck, C., Balthasar, H., *et al.* 2010, *A&A*, 511, A14

Goode, P. R. & Cao, W. 2012, vol. 8444, *SPIE Conf. Ser.*

Gudiksen, B. V., Carlsson, M., Hansteen, V. H., *et al.* 2011, *A&A*, 531, A154

Hagenaar, H. J., Schrijver, C. J., & Title, A. M. 2003, *ApJ*, 584, 1107

Harvey, J. W., Branston, D., Henney, C. J., & Keller, C. U., SOLIS and GONG Teams 2007, *ApJ* (Letters), 659, L177

Hasan, S. S. 2012, *IAU Special Session*, 6, E2.11

Heggland, L., De Pontieu, B., & Hansteen, V. H. 2007, *ApJ*, 666, 1277

Hofmann, A., Arlt, K., Balthasar, H., Bauer, S. M., Bittner, W., Paschke, J., Popow, E., Rendtel, J., Soltau, D., & Waldmann, T. 2012, *Astron. Nachr.*, 333, 854

Ishikawa, R. 2012, *IAU Special Session*, 6, E1.03

Ishikawa, R. & Tsuneta, S. 2010, *ApJ* (Letters), 718, L171

Ishikawa, R. & Tsuneta, S. 2011, *ApJ*, 735, 74

Ishikawa, R., Tsuneta, S., & Jurčák, J. 2010, *ApJ*, 713, 1310

Jess, D. B., Mathioudakis, M., Christian, D. J., Keenan, F. P., Ryans, R. S. I., & Crockett, P. J. 2010, *Solar Phys.*, 261, 363

Ji, H. & Liu, Z. 2012, *IAU Special Session*, 6, E2.04

Judge, P. & Casini, R. 2012, *IAU Special Session*, 6, E1.06

Judge, P. G., Reardon, K., & Cauzzi, G. 2012, *ApJ* (Letters), 755, L11

Judge, P. G., Tritschler, A., & Chye Low, B. 2011, *ApJ* (Letters), 730, L4

Judge, P. G. 2009, *ASP Conf. Ser.*, 415, 7

Judge, P. G., Centeno, R., Tritschler, A., Uitenbroek, H., Jaeggli, S., & Lin, H. 2010, *AGU Fall Meeting Abstracts*, A1783

Karlický, M., Bárta, M., Dąbrowski, B. P., & Heinzel, P. 2011, *Solar Phys.*, 268, 165

Kato, Y., Steiner, O., Steffen, M., & Suematsu, Y. 2011, *ApJ* (Letters), 730, L24

Katsukawa, Y., Watanabe, T., Hara, H., Ichimoto, K., Kubo, M., Kusano, K., Sakao, T., Shimizu, T., Suematsu, Y., & Tsuneta, S. 2012, *IAU Special Session*, 6, E2.07

Kašparová, J. & Heinzel, P. 2002, *A&A*, 382, 688

Keil, S. L., Rimmele, T. R., & Wagner, J., ATST Team 2010, *Astron. Nachr.*, 331, 609

Kellerer, A., Gorceix, N., Marino, J., Cao, W., & Goode, P. R. 2012, *A&A*, 542, A2

Khomenko, E. V., Collados, M., Solanki, S. K., Lagg, A., & Trujillo Bueno, J. 2003, *A&A*, 408, 1115

Kitiashvili, I., Abramenko, V., Goode, P. R., Kosovichev, A., Mansour, N., Wray, A., & Yurchyshyn, V. 2012, *IAU Special Session*, 6, E1.04

Kitiashvili, I. N., Kosovichev, A. G., Mansour, N. N., & Wray, A. A. 2011, *ApJ* (Letters), 727, L50

Kitiashvili, I. N., Kosovichev, A. G., Mansour, N. N., & Wray, A. A. 2012, *ApJ* (Letters), 751, L21

Kleint, L., Berdyugina, S. V., Shapiro, A. I., & Bianda, M. 2010, *A&A*, 524, A37

Klimchuk, J. 2012, *IAU Special Session*, 6, E1.07

Klimchuk, J. A. 2012, *Journal of Geophysical Research* (Space Physics), 117, 12102

Kobel, P., Solanki, S. K., & Borrero, J. M. 2011, *A&A*, 531, A112

Khomenko, E. & Collados, M. 2012, *ApJ*, 747, 87

Kosugi, T., Matsuzaki, K., Sakao, T., Shimizu, T., Sone, Y., Tachikawa, S., Hashimoto, T., Minesugi, K., Ohnishi, A., Yamada, T., Tsuneta, S., Hara, H., Ichimoto, K., Suematsu, Y., Shimojo, M., Watanabe, T., Shimada, S., Davis, J. M., Hill, L. D., Owens, J. K., Title, A. M., Culhane, J. L., Harra, L. K., Doschek, G. A., & Golub, L. 2007, *Solar Phys.*, 243, 3

Kowalski, A. F., Hawley, S. L., Holtzman, J. A., Wisniewski, J. P., & Hilton, E. J. 2012, *Solar Phys.*, 277, 21

Kuhn, J. R., Scholl, I. F., & Mickey, D. L. 2012, *ASP Conf. Ser.*, 463, 207

Leake, J. E. & Linton, M. G. 2013, *ApJ*, 764, 54

Leenaarts, J., Pereira, T., & Uitenbroek, H. 2012, *A&A*, 543, A109

Leenaarts, J., Carlsson, M., Hansteen, V., & Rutten, R. J. 2007, *A&A*, 473, 625

Lemen, J. R., Title, A. M., Akin, D. J., *et al.* 2012, *Solar Phys.*, 275, 17

Li, S., Yan, Y., Wang, W., & Liu, D. 2012, *IAU Special Session*, 6, E5

Lites, B., Socas-Navarro, H., Kubo, M., Berger, T., Frank, Z., Shine, R. A., Tarbell, T. D., Title, A. M., Ichimoto, K., Katsukawa, Y., Tsuneta, S., Suematsu, Y., & Shimizu, T. 2007, *PASJ*, 59, 571

Lites, B. W., Elmore, D. F., & Streander, K. V. 2001, *ASP Conf. Ser.*, 236, 33

Lites, B. W., Kubo, M., Socas-Navarro, H., Berger, T., Frank, Z., Shine, R., Tarbell, T., Title, A., Ichimoto, K., Katsukawa, Y., Tsuneta, S., Suematsu, Y., Shimizu, T., & Nagata, S. 2008, *ApJ*, 672, 1237

Liu, Z., Deng, Y., & Ji, H. 2012, *IAU Special Session*, 6, E2

Livingston, W. & Harvey, J. 1971, *IAU Symposium*, 43, 51

Löfdahl, M. 2012, *IAU Special Session*, 6, E3.05

Löfdahl, M. G. & Scharmer, G. B. 2012, *A&A*, 537, A80

Martínez González, M. J., Asensio Ramos, A., López Ariste, A., & Manso Sainz, R. 2008, *A&A*, 479, 229

Martínez González, M. J., Manso Sainz, R., Asensio Ramos, A., & Bellot Rubio, L. R. 2010, *ApJ* (Letters), 714, L94

Martinez Pillet, V. 2012, *IAU Special Session*, 6, E3.04

Martinez-Sykora, J. 2012, *IAU Special Session*, 6, E1.05

Martínez-Sykora, J., De Pontieu, B., & Hansteen, V. 2012, *ApJ*, 753, 161

Martínez-Sykora, J., Hansteen, V., De Pontieu, B., & Carlsson, M. 2009, *ApJ*, 701, 1569

McMullin, J. P., Rimmele, T. R., Keil, S. L., Warner, M., Barden, S., Bulau, S., Craig, S., Goodrich, B., Hansen, E., Hegwer, S., Hubbard, R., McBride, W., Shimko, S., Wöger, F., & Ditsler, J. 2012, vol. 8444, *SPIE Conf. Ser.*

Milić, I. & Faurobert, M. 2012, *A&A*, 547, A38

Moll, R., Cameron, R. H., & Schüssler, M. 2011, *A&A*, 533, A126

Müller, D., Marsden, R. G., St. Cyr, O. C., & Gilbert, H. R. 2012, *Solar Phys.*

Nordlund, Å. 1985, *Solar Phys.*, 100, 209

Orozco Suárez, D., Bellot Rubio, L. R., Del Toro Iniesta, J. C., Tsuneta, S., Lites, B., Ichimoto, K., Katsukawa, Y., Nagata, S., Shimizu, T., Shine, R. A., Suematsu, Y., Tarbell, T. D., & Title, A. M. 2007, *PASJ*, 59, 837

Orozco Suárez, D., Bellot Rubio, L. R., del Toro Iniesta, J. C., Tsuneta, S., Lites, B. W., Ichimoto, K., Katsukawa, Y., Nagata, S., Shimizu, T., Shine, R. A., Suematsu, Y., Tarbell, T. D., & Title, A. M. 2007, *ApJ* (Letters), 670, L61

Pereira, T. M. D., De Pontieu, B., & Carlsson, M. 2012, *ApJ*, 759, 18

Perley, R. A., Chandler, C. J., Butler, B. J., & Wrobel, J. M. 2011, *ApJ* (Letters), 739, L1

Peron, M. 2012, *ASP Conf. Ser.*, 461, 115

Pesnell, W. D., Thompson, B. J., & Chamberlin, P. C. 2012, *Solar Phys.*, 275, 3

Pick, M. & Vilmer, N. 2008, *A&ARv*, 16, 1

Pietarila Graham, J., Cameron, R., & Schüssler, M. 2010, *ApJ*, 714, 1606

Pietarila Graham, J., Danilovic, S., & Schüssler, M. 2009, *ApJ*, 693, 1728

Puschmann, K. G., Denker, C., Kneer, F., Al Erdogan, N., Balthasar, H., Bauer, S. M., Beck, C., Bello González, N., Collados, M., Hahn, T., Hirzberger, J., Hofmann, A., Louis, R. E., Nicklas, H., Okunev, O., Martínez Pillet, V., Popow, E., Seelemann, T., Volkmer, R., Wittmann, A. D., & Woche, M. 2012, *Astron. Nachr.*, 333, 880

Puxley, P. & Jørgensen, I. 2006, vol. 6270, *SPIE Conf. Ser.*

Radziszewski, K., Rudawy, P., & Phillips, K. J. H. 2011, *A&A*, 535, A123

Reardon, K. P. 2012, *American Astronomical Society Meeting Abstracts*, 220, #323.05

Reardon, K. P., Rimmele, T., Tritschler, A., Cauzzi, G., Wöger, F., Uitenbroek, H., Tsuneta, S., & Berger, T. 2009, *ASP Conf. Ser.*, 415, 332

Rempel, M. 2011, *ApJ*, 729, 5

Richards, K., Rimmele, T., Hegwer, S. L., Upton, R. S., Wöger, F., Marino, J., Gregory, S., & Goodrich, B. 2010, vol. 7736, *SPIE Conf. Ser.*

Rimmele, T. R., Keil, S., McMullin, J., Goode, P. R., Knölker, M., Kuhn, J. R., & Rosner, R., ATST Team 2012, *IAU Special Session*, 6, E2.06

Rimmele, T. R., Keil, S., McMullin, J., Knölker, M., Kuhn, J. R., Goode, P. R., Rosner, R., Casini, R., Lin, H., Tritschler, A., & Wöger, F., ATST Team 2012, *ASP Conf. Ser.*, 463, 377

Rimmele, T. R. & Marino, J. 2011, *Living Reviews in Solar Physics*, 8, 2
Rimmele, T. R., Wöger, F., Marino, J., Richards, K., Hegwer, S., Berkefeld, T., Soltau, D., Schmidt, D., & Waldmann, T. 2010, vol. 7736, *SPIE Conf. Ser.*
Sánchez Almeida, J. 2003, *A&A*, 411, 615
Schad , T. A., Jaeggli , S. A., Lin, H., & Penn, M. J. 2011, *ASP Conf. Ser.*, 437, 483
Schad, T. A., Penn, M. J., & Lin, H. 2013, *ApJ*, 768, 111
Schad, T. A., Penn, M. J., Lin, H., & Tritschler, A. 2012, *ASP Conf. Ser.*, 463, 25
Scharmer, G. B., Löfdahl, M. G., van Werkhoven, T. I. M., & de la Cruz Rodríguez, J. 2010, *A&A*, 521, A68
Scharmer, G. B., Narayan, G., Hillberg, T., de la Cruz Rodríguez, J., Löfdahl, M. G., Kiselman, D., Sütterlin, P., van Noort, M., & Lagg, A. 2008, *ApJ* (Letters), 689, L69
Scharmer, G. B. & van Werkhoven, T. I. M. 2010, *A&A*, 513, A25
Schmidt, W., von der Lühe, O., Volkmer, R., Denker, C., Solanki, S. K., Balthasar, H., Bello Gonzalez, N., Berkefeld, T., Collados, M., Fischer, A., Halbgewachs, C., Heidecke, F., Hofmann, A., Kneer, F., Lagg, A., Nicklas, H., Popow, E., Puschmann, K. G., Schmidt, D., Sigwarth, M., Sobotka, M., Soltau, D., Staude, J., Strassmeier, K. G., & Waldmann , T. A. 2012, *Astron. Nachr.*, 333, 796
Schrijver, C. 2012, *IAU Special Session*, 6, E4.02
Schrijver, C. J., Title, A. M., Harvey, K. L., Sheeley, N. R., Wang, Y.-M., van den Oord, G. H. J., Shine, R. A., Tarbell, T. D., & Hurlburt, N. E. 1998, *Nature*, 394, 152
Schüssler, M. & Vögler, A. 2008, *A&A* (Letters), 481, L5
Shimizu, T., Sakao, T., Katsukawa, Y., & Group, J. S. W. 2012, *ASP Conf. Ser.*, 454, 449
Silva, D. 2001, *The Messenger*, 105, 18
Skartlien, R. 2000, *ApJ*, 536, 465
Socas-Navarro, H. 2012, *IAU Special Session*, 6, E2.12
Socas-Navarro, H. 2005, *ApJ* (Letters), 633, L57
Socas-Navarro, H., Beckers, J., Brandt, P., Briggs, J., Brown, T., Brown, W., Collados, M., Denker, C., Fletcher, S., Hegwer, S., Hill, F., Horst, T., Komsa, M., Kuhn, J., Lecinski, A., Lin, H., Oncley, S., Penn, M., Rimmele, T., & Streander, K. 2005, *PASP*, 117, 1296
Solanki, S. K., Barthol, P., Danilovic, S., Feller, A., Gandorfer, A., Hirzberger, J., Riethmüller, T. L., Schüssler, M., Bonet, J. A., Martínez Pillet, V., del Toro Iniesta, J. C., Domingo, V., Palacios, J., Knölker, M., Bello González, N., Berkefeld, T., Franz, M., Schmidt, W., & Title, A. M. 2010, *ApJ* (Letters), 723, L127
Spruit, H. C., Title, A. M., & van Ballegooijen, A. A. 1987, *Solar Phys.*, 110, 115
Steiner, O. 2012, *IAU Special Session*, 6, E1.01
Steiner, O., Franz, M., Bello González, N., Nutto, C., Rezaei, R., Martínez Pillet, V., Bonet Navarro, J. A., del Toro Iniesta, J. C., Domingo, V., Solanki, S. K., Knölker, M., Schmidt, W., Barthol, P., & Gandorfer, A. 2010, *ApJ* (Letters), 723, L180
Steiner, O. & Rezaei, R. 2012, *ASP Conf. Ser.*, 456, 3
Steiner, O., Rezaei, R., Schaffenberger, W., & Wedemeyer-Böhm, S. 2008, *ApJ* (Letters), 680, L85
Stenflo, J. O. 1982, *Solar Phys.*, 80, 209
Stenflo, J. O. 2010, *A&A*, 517, A37
Štěpán, J., Trujillo Bueno, J., Carlsson, M., & Leenaarts, J. 2012, *ApJ* (Letters), 758, L43
Su, Y., Wang, T., Veronig, A., Temmer, M., & Gan, W. 2012, *ApJ* (Letters), 756, L41
Suematsu, Y., Katsukawa, Y., Ichimoto, K., & Shimizu, T. 2012, *IAU Special Session*, 6, E2.08
Tomczyk, S. 2012, *IAU Special Session*, 6, E2.14
Trujillo Bueno, J. 2012, *IAU Special Session*, 6, E3.02
Trujillo Bueno, J., Shchukina, N., & Asensio Ramos, A. 2004, *Nature*, 430, 326
Trujillo Bueno, J., Štěpán, J., & Belluzzi, L. 2012, *ApJ* (Letters), 746, L9
Tsuneta, S., Ichimoto, K., Katsukawa, Y., Lites, B. W., Matsuzaki, K., Nagata, S., Orozco Suárez, D., Shimizu, T., Shimojo, M., Shine, R. A., Suematsu, Y., Suzuki, T. K., Tarbell, T. D., & Title, A. M. 2008, *ApJ*, 688, 1374
Uitenbroek, H. 2006, *Organizations and Strategies in Astronomy*, Vol. 7, 335, 117
Uitenbroek, H. & Tritschler, A. 2012, *IAU Special Session*, 6, E4.01

Uitenbroek, H. 2011, *ASP Conf. Ser.*, 437, 439
Uitenbroek, H., Tritschler, A., & Rimmele, T. 2007, *ApJ*, 668, 586
van Noort, M., Rouppe van der Voort, L., & Löfdahl, M. G. 2005, *Solar Phys.*, 228, 191
Vargas Domínguez, S., Palacios, J., Balmaceda, L., Cabello, I., & Domingo, V. 2011, *MNRAS*, 416, 148
Viticchié, B., Sánchez Almeida, J., Del Moro, D., & Berrilli, F. 2011, *A&A*, 526, A60
Vögler, A. & Schüssler, M. 2007, *A&A* (Letters), 465, L43
Wang, J., Jin, C., & Zhou, G. 2012, *IAU Special Session*, 6, E1.08
Wedemeyer, S., Scullion, E., Steiner, O., de la Cruz Rodriguez, J., & Rouppe van der Voort, L. 2013, arXiv1303.0179
Wedemeyer-Böhm, S., Ludwig, H.-G., Steffen, M., Freytag, B., & Holweger, H. 2005, *ESA Special Publication*, 560, 1035
Wedemeyer-Böhm, S. & Rouppe van der Voort, L. 2009, *A&A*, 503, 225
Wedemeyer-Böhm, S. & Rouppe van der Voort, L. 2009, *A&A*, 507, L9
Wedemeyer-Böhm, S., Scullion, E., Steiner, O., Rouppe van der Voort, L., de La Cruz Rodriguez, J., Fedun, V., & Erdélyi, R. 2012, *Nature*, 486, 505
Wiegelmann, T., Solanki, S. K., Borrero, J. M., Martínez Pillet, V., del Toro Iniesta, J. C., Domingo, V., Bonet, J. A., Barthol, P., Gandorfer, A., Knölker, M., Schmidt, W., & Title, A. M. 2010, *ApJ* (Letters), 723, L185
Wöger, F. 2010, *Applied Optics*, 49, 1818
Xu, Y., Cao, W., Jing, J., & Wang, H. 2012, *ApJ* (Letters), 750, L7
Yan, Y., Zhang, J., Chen, Z., Wang, W., Liu, F., & Geng, L. 2012, *IAU Special Session*, 6, E2.16
Yan, Y., Zhang, J., Wang, W., Liu, F., Chen, Z., & Ji, G. 2009, *Earth Moon and Planets*, 104, 97
Yelles Chaouche, L., Moreno-Insertis, F., Martínez Pillet, V., Wiegelmann, T., Bonet, J. A., Knölker, M., Bellot Rubio, L. R., del Toro Iniesta, J. C., Barthol, P., Gandorfer, A., Schmidt, W., & Solanki, S. K. 2011, *ApJ* (Letters), 727, L30
Zhang, J. & Liu, Y. 2011, *ApJ* (Letters), 741, L7

Appendix: Program of Special Session 6 at IAU XXVIII General Assembly

SESSION 1 - KEY SCIENTIFIC QUESTIONS

OVERVIEW:

Oskar Steiner: Science challenges for large solar telescopes

STATE OF THE ART:
Dali Georgobiani, R. Stein, Å. Nordlund: Realistic numerical simulations of solar convection: emerging flux, pores, and Stokes spectra
Ryohko Ishikawa: Properties of transient horizontal magnetic fields, and their implication for the origin of quiet-Sun magnetism
Irina Kitiashvili. et al. Investigation of small scale turbulent MHD phenomena using numerical simulations and NST observations
Juan Martínez-Sykora (Metcalf Lecturer): Current status of self-consistent 3D radiative-MHD simulations of the solar atmosphere
Phil Judge and R. Casini: Using large telescopes to answer: why must the Sun have a chromosphere and corona?
James Klimchuck: The role of spicules in explaining the corona and the transition region

Jingxiu Wang et al.: Solar intranetwork magnetic elements - the weakest component of solar magnetism

SESSION 2 - EXISTING AND PLANNED LARGE FACILITIES: WHAT IS THEIR IMPACT ON KEY SCIENTIFIC QUESTIONS?

OVERVIEW:
Thomas Berger: Science with large solar telescopes: Addressing key science questions with new observing modes

NEWLY OPERATED FACILITIES:
Wenda Cao: The 1.6 m New Solar Telescope (NST) in Big Bear
Carsten Denker et al.: The GREGOR Solar Telescope
Haisheng Ji: The one meter aperture solar telescope in China
Arnold Benz: Observing the Sun with ALMA

UPCOMING FACILITIES:
Thomas Rimmele et al. (presented by T. Berger): Construction of the Advanced Technology Solar Telescope - a progress report
Yukio Katsukawa et al.: Next space solar observatory SOLAR-C: mission overview and science objectives
Yoshinori Suematsu et al.: Science and instrument design of 1.5 m aperture Solar Optical Telescope for the Solar-C mission
Yuanyong Deng (on behalf of the SST group): The Space Solar Telescope
Sergey Bogachev et al.: The set of imaging instruments for Interhelioprobe solar observatory
Shiraj Hasan: The National Large Solar Telescope (NLST) of India
Hector Socas Navarro: The European Solar Telescope
Zhong Liu et al.: An introduction to the Chinese Giant Solar Telescope
Steve Tomzcyk: The Coronal Solar Magnetism Observatory
Tim Bastian (delivered by A. Benz): Observing the Sun at radio-wavelengths: current status and future prospects
Yihua Yan et al.: On solar radio imaging-spectroscopy

SESSION 3 - STEP AHEAD: FUTURE SCIENCE AND WHAT THE NEW FACILITIES SHOULD NOT FORGET !
Marianne Faurobert: The quiet Sun magnetism: What can we learn from the Hanle effect?
Javier Trujillo Bueno: Polarized Radiation Diagnostics for Measuring the Magnetic Field of the Outer Solar Atmosphere
Valentin Martínez Pillet: Towards the next frontier in high precision solar polarimetry: 10^{-4}
Mats Löfdahl: Restoration of the contrast in solar images
MingDe Ding: Spectral diagnostics of the heating and dynamics of the solar chromosphere
Na Deng: Hα Imaging Spectroscopy of a C-class Flare with IBIS

SESSION 4 - LESSON LEARNED, AND FUTURE DIRECTIONS FOR TELESCOPE OPERATION
Han Uitenbroek, A. Tritschler: Observing strategies for future solar facilities: the ATST test case
Karel Schrjiver: Mining Solar Data: the experience with SDO, Hinode, and TRACE

GENERAL DISCUSSION

POSTERS:
A. Benz et al.: Spectrometer Telescope for Imaging X-rays (STIX)
L. Geng: The 0.4-2.0GHz Analogous Receiving System Performance of CSRH
S. Li et al.: Radiation pattern measurement for Chinese Spectral Radio Heliograph
A. Hady: Geomagnetic storms in the rising phase of solar cycle 24
A. Valio et al.: Solar Patrol Polarization Telescopes at 45 and 90 GHz
Y. Sun: The Measuring System of Near Infrared Fabry-Perot Etalon
J. Lin, Y. Deng: Automatic full disk vector magnetogram observing system in HSOS
J. Lin et al.: Research about the high precision temperature measurement
J. Guo: A statistical study on the soft X-ray flare in solar cycles 22 and 23
W. P. Zhou, Y. Wang: A new energy-efficient control approach for astronomical telescope drive system
W. P. Zhou, Y. Wang: The real-time motion control with high precision for the ground-based telescopes
F. Ibodov, S. Ibadov: High-resolution observations of solar explosive phenomena due to cometary impacts with the Sun
G. Cauzzi, K. Reardon: The IBIS Mosaic

Highlights of Astronomy, Volume 16
XXVIIIth IAU General Assembly, August 2012
T. Montmerle, ed.

doi:10.1017/S1743921315001623

Preface: Special Session SpS7 The impact hazard: current activities and future plans

In 2012, at the time of the Beijing GA, two decades had passed since the publication of the Spaceguard Report. Time has not passed in vain for the subject of NEO-related hazards, and we are currently in a totally different situation than in the early nineties. The amount of work done, and the level of awareness of the underlying problems, have both risen to such a level that a re-assessment by the astronomical community of its rôle and involvement was in order, and the GA Special Session 7 "The impact hazard: current activities and future plans" was aimed exactly at that.

Increasingly sophisticated sky surveys have succeeded in discovering thousands of NEOs, and much has been learned about the physical properties of two NEOs, Eros and Itokawa, from the NEAR-Shoemaker and Hayabusa rendezvous missions. However, while observational and modeling work aimed at investigating the physical nature of the NEO population as a whole is gathering pace, efforts in this area are still being outpaced by the discovery rate.

Groundbased NEO observational efforts suffer from the preference of telescope time-allocation committees for purely scientifically motivated proposals. The space-based observational programs Warm Spitzer and WISE are providing size, albedo, and potentially other mitigation relevant data for hundreds of NEOs but both are limited, short-lived programs. Progress in other (more expensive) areas, such as developing and testing mitigation techniques, has been more limited.

However, events like the Chelyabinsk superbolide of 15 February 2013, with its vast media coverage and worldwide echo, do remind us that the NEO hazard is a field that continues to require our attention.

G. B. Valsecchi, A. Milani and W. Huebner, co-chairs SOC

Highlights of Astronomy, Volume 16
XXVIIIth IAU General Assembly, August 2012
T. Montmerle, ed.

doi:10.1017/S1743921314011818

Near Earth Objects Research in Pulkovo Observatory

A. V. Devyatkin, E. A. Bashakova, D. L. Gorshanov, A. V. Ivanov, S. V. Karashevich, V. V. Kouprianov, V. N. L'vov, K. N. Naumov, E. S. Romas, V. Yu. Slesarenko, N. A. Shakht, E. N. Sokov, S. D. Tsekmeister, O. O. Vasilkova and I. A. Vereschagina

Pulkovo Astronomical Observatory of Russian Academy of Science
email: olyaov@mail.ru

More than 20000 observations of Near Earth asteroids and comets are collected and reduced in Pulkovo Observatory during last 10 years. For observations of these objects two robotic telescopes are used – ZA-320M (Cassegrain system, $D = 320$ mm, $F = 3200$ mm) at Pulkovo and MTM-500M (Maksutov – Cassegrain system, $D = 500$ mm, $F = 4100$ mm) at Kislovodsk mountain station. These telescopes perform CCD observations of objects up to 18.0 and 20.5 magnitude, correspondingly. The results of observations are regularly submitted to Minor Planet Center.

Two software packages developed in Pulkovo observatory are now in practice. The first one, APEX, is used for automatic CCD frame processing (Devyatkin *et al.* (2010)). The second, EPOS, provides independent ephemeris support, checkout of observations and estimation of their accuracy. It helps in objects identification, orbits determination and improvement, modeling the objects apparent motion on the sky and orbital motion in space (L'vov, Tsekmeister (2012)). Its recent facility - studying the motion of Troyan, Horseshoe and Quasi-satellite asteroids.

Several NEAs were observed in the frame of GAIA FUN-SSO program.

The Near Earth asteroid 2005 YU55 has approached to Venus, Earth and Mars. The EPOS software helped to model the sharp changes of its orbit during the close approaches. It was observed with ZA-320M and MTM-500M telescopes during its close approach to the Earth in November 2012. 926 positions with mean accuracies of 0".1 – 0".4 were obtained, results were sent to Minor Planet Center. Using the observations the asteroid orbit was improved. From its light-curves the period of rotation was estimated and the color-indices in $BVRI$ system were determined. To estimate the period (using Scargles method) all our observations which were made without filter were processed. The value of 16.3 hr was obtained, while the value determined earlier with radar observations is 18 hr (see http://ssd.jpl.nasa.gov/sbdb.cgi?sstr=308635).

During processing the data, one more period was noticed in observations made with both telescopes in several nights. Its value varies from 0.9 to 1.2 hr for different nights. The magnitude range is about 0.15^m. The long-time observations ($\approx$ 10 hr) include several period durations. If this period is real, it could meant that the asteroid is binary.

The weighted means of color-indices from our observations are the following: $B - V = 0.67^m \pm 0.07^m$, $V - R = 0.34^m \pm 0.09^m$, $R - I = 0.30^m \pm 0.07^m$. Our value of $V - R$ color-index is in agreement with the effective color ($V - R = 0.37^m$) measured from spectrum of the asteroid in works Hicks *et al.* (2011a), Hicks *et al.* (2011b).

We have tried to classify 2005 YU55 asteroid using our wide-band photometry. Dandy *et al.* (2003) classify asteroids on Tholens classification using wide-band photometry: $BVRI$ and additional Z band at $0.91 \mu m$. We have no Z filter. But using their method in this narrower ($0.44 - 0.83\mu m$) spectral diapason, the spectrum form of the asteroid is close to B, F, C, G (Tholen) classes – the classes with flat spectra in visible range. B class is the closest.

Hicks *et al.* (2011b) have built phase curve for 2005 YU55 asteroid in R band and determined its absolute magnitude and slope parameter. We put our R band observations into the sketch from this work. The observations lay on the same phase curve but have greater scattering. They are made with lower phase angles than in work Hicks *et al.* (2011b) and confirm the values of absolute magnitude and slope parameter of the asteroid.

Another asteroid observed in the frame of GAIA FUN-SSO program is TP3522. On January 20, 2013, we made 24 observations of this faint (about 20^m) asteroid with MTM-500M telescope. These observations amount about 1/3 of all observations for the object in MPC database (77 observations). Mean accuracy of our observations is 0".2 in each coordinate. These observations (along with observations of other observatories) confirmed discovery of the asteroid and allowed to improve its orbit.

The asteroid 2008 TC3 was observerd at Pulkovo with ZA-320M telescope. The collision of 2008 TC3 with Earth (in 19 hours after its discovery) was predicted, presumably, in Northern Sudan. About 800 observations were obtained in Pulkovo which amount about 1/3 of all observations available in MPC. Based on the analysis of these observations, the following physical parameters of asteroid were estimated: its absolute magnitude, $M = 30.6 \pm 0.4^m$; size, 4.8 ± 0.8 m; weight, 131 ± 5 t; and rotation period of 48.6 ± 0.6 sec. The trajectory of the asteroid was computed, as well as the place of its collision with the Earth (for the case when explosion does not occur). The point of explosion and place of debris fall correlate with computed trajectory.

For potentially hazardous asteroid 2009 WZ104, more than 1000 observations were made at Pulkovo. Using these, orbit of the asteroid was improved and color-indices were obtained: $B - V = 0.65^m \pm 0.7^m$, $V - R = 0.18^m \pm 0.05^m$, $R - I = -0.26^m \pm 0.07^m$. From its light curve, 2 periods were determined: $P_1 = 0.4$ and $P_2 = 2.76$ days.

With MTM-500M telescope, several faint ($> 20^m$) asteroids were discovered (2010 UP67, 2010 XL46, 2010 XM46, 2010 XA15) and rediscovered (2010 SY11 = 2004 TR356, 2010 VX29 = 2008 EG30, C20N117 = 2008 FM60, NA12044=2005 XP17).

References

Dandy, C. L. *et al.* 2003, *Icarus* 163, 363
Devyatkin, A. V. *et al.* 2010, *Solar System Research* 44, 66
Hicks, M. *et al.* 2011a, *The Astronomer's Telegram* 2571
Hicks, M. *et al.* 2011b, *The Astronomer's Telegram* 3763
L'vov, V. N. & Tsekmeister, S. D. 2012, *Solar System Research* 46, 177

Highlights of Astronomy, Volume 16
XXVIIIth IAU General Assembly, August 2012
T. Montmerle, ed.

doi:10.1017/S174392131401182X

The Near Earth Asteroid associations

Tadeusz J. Jopek
Astronomical Observatory Institute, Faculty of Physics, A.M. University, Poznan, Poland

Abstract. We have made an extensive search for grouping amongst the near Earth asteroids (NEAs). We used two D- functions and rigorous cluster analysis approach. We have found several new groups (associations) among the NEAs: the objects moving on similar orbits with small minimum orbital intersection distances (MOID) with the Earth trajectory. Reliability of some of these groups is quite high.
Keywords. minor planets, methods: data analysis

1. Introduction

The existence of the main belt asteroid families is beyond the doubts. However existence of groups among the NEAs, with members of common origin, suggested by similarity between their orbital parameters (Drummond (1991), Obrubov(1991), Fu *et al.*(2005), Jopek (2011), Schunová *et al.*(2012)) is not yet generally accepted. Many groups found by these authors might be attributed to chance alignments.

In this study we have searched for associations amongst 9004 NEA's osculating orbits (NEODyS, 2012) by two strict cluster analysis methods, similar to that used in Jopek (2011). Two D- functions were used: D_{SH} introduced in Southworth and Hawkins (1963) and D_H introduced in Jopek (1993). The orbital similarity thresholds corresponded to 1% probability that given group was found by chance alignment. The clusters were detected by a single neighbour linking technique.

2. Results and discussion

Using two searching methods we found 20 groups of ten or more members: 13 groups with the D_{SH} function and 9 groups using D_H function. Among 20 formally separated groups we have selected 10 associations. All but two has been found in both searches, however with different amount of members. Associations No 1 and 2 (SH -179 and SH-380) were detected only with D_{SH} function. Of course when the values of the similarity thresholds has been slightly increased, these groups were found also with D_H function. Associations No 3-7 were identified with both functions with considerable amount of members in common. In case of associations 8,9 and 10 the results shown to be more complicated: several overlapping subgroups were detected for which the common members were less numerous. In Table 1 we see that for all groups the mean orbital inclinations are smaller than 10 degrees. It is not unexpected, we should recall that all NEAs but two have $i < 75$ degrees, and for ~ 4500 orbits the inclinations are smaller than 10 degrees.

We have continued our study gradually decreasing thresholds. As was expected, amount of members of the group was decreasing; some groups were splitting and finally disappearing. Associations SH-306, SH-1385, H23 have proved to be the most robust one. On Figure 1, just for an example we have plotted the orbits of the members of association SH380.

In this study we have shown, that using similar rigorous method as for the meteoroid orbits, the NEA's associations can be found easily. However, to ensure about their common origin, the long term numerical integration is needed.

Table 1. The list of 10 NEAs associations of 10 or more members found in this study. The codes of the associations include the distance function tag and the ordinal number of the asteroid from the NEODyS list which was identified as a first group member. The name of this asteroid is also given. N is the amount of members in the group. In brackets the amount of the common members identified by both functions are given. Within each group the orbital elements were averaged by the method described in Jopek *et al.* (2006).

No	NEA	Code	N	a [AU]	q [AU]	e	i [deg]	ω [deg]	Ω [deg]
1	'13553' Masaakikiyama	SH-179	53	2.133	1.154	0.459	5.7	157.3	143.5
2	'89136' 2001US16	SH-380	24	1.375	1.025	0.254	1.0	16.0	227.3
3	'2368' Beltrovata	SH-35	77(74)	2.128	1.221	0.426	2.7	32.2	301.6
		H-35	101	2.137	1.222	0.428	3.0	31.2	305.6
4	'4660' Nereus	SH-77	63(40)	1.864	.936	0.498	1.2	37.2	82.1
		H-77	63	1.793	0.959	0.465	1.0	46.0	70.9
5	'8014' 1990MF	H-143	69	2.019	1.051	0.479	0.1	100.3	232.1
	'10860' 1995LE	SH-167	65(61)	2.135	1.111	0.480	1.3	160.9	167.4
6	'11054' 1991FA	H-168	68	1.513	0.984	0.350	0.9	84.7	349.1
	'190491' 2000FJ10	SH-866	25(23)	1.427	1.014	0.289	2.6	184.9	249.8
7	'36017' 1999ND43	SH-255	39(21)	1.425	1.012	0.290	2.6	148.5	227.1
	'256004' 2006UP	H-1050	35	1.421	1.013	0.287	2.8	157.4	217.0
8	'54509' YORP	H-295	78	1.045	1.031	0.013	1.0	321.6	224.2
	'65717' 1993BX3	SH-306	89	1.162	0.957	0.176	0.6	235.4	210.2
	'209215' 2003WP25	SH-910	23	1.082	0.926	0.144	0.8	95.9	188.8
9	'19356' 1997GH3	H-202	73	1.815	1.007	0.445	0.9	14.5	145.1
	'27002' 1998DV9	SH-236	27(21)	1.974	1.015	0.486	3.3	10.4	130.3
	'1994GV'	SH-1385	85	1.676	1.015	0.394	0.1	296.2	243.3
	'2003DW10'	H-2582	16	1.587	0.998	0.371	0.2	127.5	74.8
10	'136564' 1977VA	SH-423	17(1)	2.172	1.141	0.475	4.2	204.1	199.3
	'2061' Anza	H-23	181	2.202	1.046	0.525	0.2	201.2	182.1
	'4015' Wilson-Harrington	SH-60	196(134)	2.150	1.046	0.514	0.3	169.0	223.1

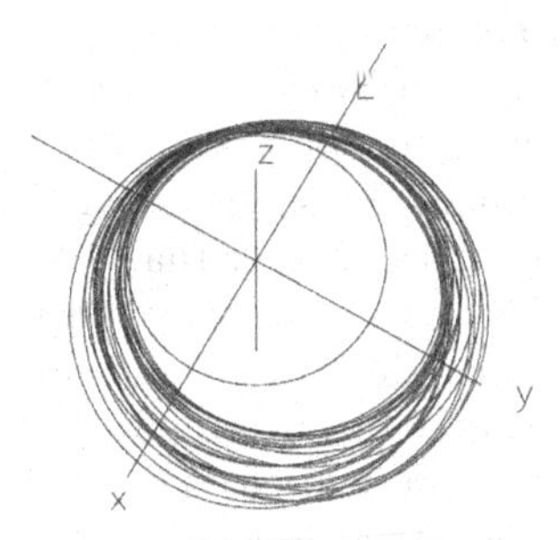

Figure 1. Association SH-380 plotted on the ecliptic plane. The orbits remarkably resemble a meteoroid stream. Association includes 24 NEAs: 89136, 26430 8,1994CJ1, 2003CC, 2003GA, 2004KG17, 2005HB4, 2005JT1, 2006HX30, 2006KL103, 2007HB15, 2008GR3, 2008LE, 2009DC12, 2009HG, 2009HH21, 2010CE55, 2011GR59, 2011GV9, 2011OK45, 2011OR5, 2011PU1, 2012GP1, 2012KT12. The Earth circular trajectory is seen inside the association.

References

Drummond, J. D. 1991, *Icarus* 89, 14

Drummond, J. D. 2000, *Icarus* 146, 453

Fu, H., Jedicke, R., Durda, D. D., Fevig, R., & Scotti, J. V., 2005, *Icarus* 178, 434

Jopek, T. J., 2011, *MemSAI* 82, 310

Jopek, T. J. 1993, *Icarus* 106, 603

Jopek, T. J., Rudawska, R., & Pretka-Ziomek, H., 2006, *MNRAS* 371, 1367

NEO Dynamic Site, 2012, August, http://newton.dm.unipi.it/neodys

Obrubov, Y. V., 1991, Complexes of Minor Solar System Bodies, *Soviet Astron.* 35, 531

Schunová, E., Granvik, M., Jedicke, R., Gronchi, G., Wainscoat, R., & Abe, S., 2012, *Icarus* 220, 1050

Southworth, R. B. & Hawkins, G. S., 1963, *Smithson. Contr. Astrophys.* 7, 261

Highlights of Astronomy, Volume 16
XXVIIIth IAU General Assembly, August 2012
T. Montmerle, ed.

doi:10.1017/S1743921314011831

The Role of Radar Astronomy in Assessing and Mitigating the Asteroid Impact Hazard

Jean-Luc Margot[1,2] and Jon D. Giorgini[3]

[1]Department of Physics and Astronomy, University of California, Los Angeles, CA 90095, USA

[2]Department of Earth, Planetary, and Space Sciences, University of California, Los Angeles, CA 90095, USA
email: jlm@astro.ucla.edu

[3]Jet Propulsion Laboratory, Pasadena, CA 91109, USA
email: Jon.Giorgini@jpl.nasa.gov

Abstract. Radar instruments play a critical role in studies of Near-Earth Asteroids (NEAs) on two distinct levels: trajectory prediction and physical characterization.

Keywords. minor planets, asteroids, astrometry, radar astronomy

---

1. Trajectory prediction

Because of superb ($\sim 10^{-8}$) fractional uncertainties in round-trip delay and Doppler measurements, which are orthogonal to plane-of-sky optical astrometry, radar measurements can decrease orbital parameter uncertainties by factors of 10^3-10^5 and increase the interval over which an object's Earth close approaches can be reliably known at the three-sigma level of confidence by factors of 5-10 (Ostro and Giorgini 2004). The addition of radar data is particularly valuable for single-apparition objects and for objects that have been observed with an optical arc shorter than 10-20 years. The radar benefits can be secured for about 10-25% of NEAs within five years of their detection (NASA 2007).

For NEAs having astrometry spanning two or more apparitions and future close planetary encounters, the primary source of uncertainty in trajectory prediction is usually the object-specific (and normally unmeasured) Yarkovsky effect. This has been demonstrated for the ~1.2 km diameter object (29075) 1950 DA (Giorgini et al 2002, Busch *et al.*2007) and the ~270 m diameter object (99942) Apophis (Giorgini et al 2008). Studies of Yarkovsky influences on near-Earth objects show that radar ranges on at least two apparitions improve the precision of the semi-major axis drift rate determinations on average by an order of magnitude compared to optical-only determinations (Nugent *et al.*2012). Therefore, radar data not only improve trajectory predictions but also improve our knowledge of uncertainties affecting the predictions, which allows us to better quantify the risk associated with NEAs.

2. Physical Characterization

Physical characterization of objects posing a significant risk will be paramount for any mitigation effort. For instance, effective mitigation efforts will require knowledge of multiplicity. Because 1 in 6 NEAs larger than 200 m is a binary (Margot et al 2002), radar observations to assess binarity will be essential. Optical observations of mutual events can reveal binarity in some geometries, but radar observations can detect satellites even if they do not cause detectable mutual events. Radar observations have uncovered the majority of known binary NEAs and all the known triple NEAs. Effective mitigation efforts

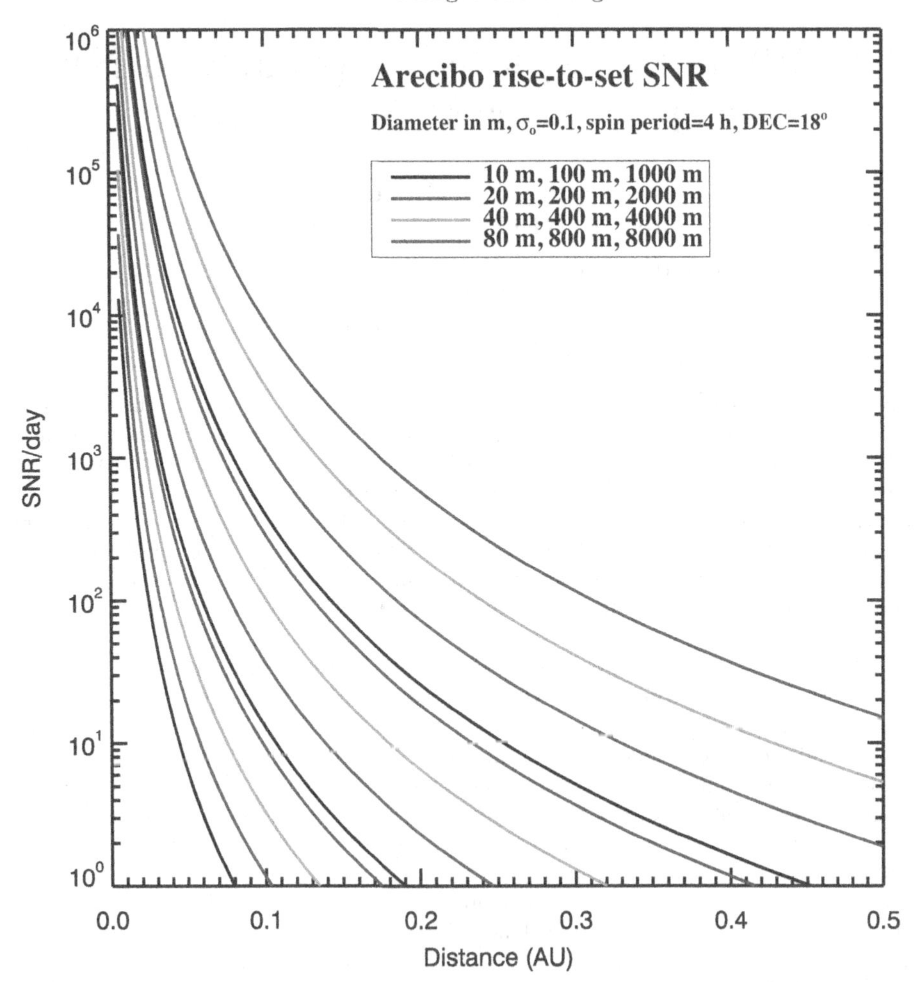

Figure 1. Arecibo signal-to-noise ratio (SNR) as a function of asteroid size and distance

would also likely be compromised without knowledge of the spin, shape, mass, density, and porosity. Radar observations provide the most realistic prospects, and sometimes the only realistic ground-based prospects, of securing estimates of all of these quantities. The combination of radar observations (Fig. 1) with observations at other wavelengths will provide the best possible ground-based characterization and will inform decisions about mitigation approaches.

References

Busch, M. W., *et al. Icarus*, 190:608, 2007.
Giorgini, J. D., *et al. Icarus*, 193:1, 2008.
Giorgini, J. D., *et al. Science*, 296:132, 2002.
Margot, J. L., *et al. Science*, 296:1445, 2002.
NASA, Near-Earth Object Survey and Deflection Analysis of Alternatives, 2007.
Nugent, C. R., Margot, J. L., Chesley, S. R., & Vokrouhlický, D., *AJ*, 144:60, 2012.
Ostro, S. J. & Giorgini, J. D. In *Mitigation of Hazardous Comets and Asteroids.* CUP, 2004.

Highlights of Astronomy, Volume 16
XXVIIIth IAU General Assembly, August 2012
T. Montmerle, ed.

doi:10.1017/S1743921314011843

NEOShield - A global approach to NEO Impact Threat Mitigation

Patrick Michel
and the NEOShield Consortium

Laboratoire Lagrange, University of Nice Sophia-Antipolis, Côte d'Azur Observatrory
BP 4229, 06304 Nice Cedex 4, France
email: michelp@oca.eu

Abstract. NEOShield is a European-Union funded project coordinated by the German Aerospace Center, DLR, to address near-Earth object (NEO) impact hazard mitigation issues. The NEOShield consortium consists of 13 research institutes, universities, and industrial partners from 6 countries and includes leading US and Russian space organizations. The project is funded for a period of 3.5 years from January 2012 with a total of 5.8 million euros. The primary aim of the project is to investigate in detail promising mitigation techniques, such as the kinetic impactor, blast deflection, and the gravity tractor, and devise feasible demonstration missions. Options for an international strategy for implementation when an actual impact threat arises will also be investigated.

The NEOShield work plan consists of scientific investigations into the nature of the impact hazard and the physical properties of NEOs, and technical and engineering studies of practical means of deflecting NEOs. There exist many ideas for asteroid deflection techniques, many of which would require considerable scientific and technological development. The emphasis of NEOShield is on techniques that are feasible with current technology, requiring a minimum of research and development work. NEOShield aims to provide detailed designs of feasible mitigation demonstration missions, targeting NEOs of the kind most likely to trigger the first space-based mitigation action.

Most of the asteroid deflection techniques proposed to date require physical contact with the threatening object, an example being the kinetic impactor. NEOShield includes research into the mitigation-relevant physical properties of NEOs on the basis of remotely-sensed astronomical data and the results of rendezvous missions, the observational techniques required to efficiently gather mitigation-relevant data on the dynamical state and physical properties of a threatening NEO, and laboratory investigations using gas guns to fire projectiles into asteroid regolith analog materials. The gas-gun investigations enable state-of-the-art numerical models to be verified at small scales. Computer simulations at realistic NEO scales are used to investigate how NEOs with a range of properties would respond to a pulse of energy applied in a deflection attempt.

The technical work includes the development of crucial technologies, such as the autonomous guidance of a kinetic impactor to a precise point on the surface of the target, and the detailed design of realistic missions for the purpose of demonstrating the applicability and feasibility of one or more of the techniques investigated. Theoretical work on the blast deflection method of mitigation is designed to probe the circumstances in which this last line of defense may be the only viable option and the issues relating to its deployment. A global response campaign roadmap will be developed based on realistic scenarios presented, for example, by the discovery of an object such as 99942 Apophis or 2011 AG5 on a threatening orbit. The work will include considerations of the timeline of orbit knowledge and impact probability development, reconnaissance observations and fly-by or rendezvous missions, the political decision to mount a mitigation attempt, and the design, development, and launch of the mitigation mission. Collaboration with colleagues outside the NEOShield Consortium involved in complementary activities (e.g. under the auspices of the UN, NASA, or ESA) is being sought in order to establish a broad international strategy.

We present a brief overview of the history and planned scope of the project, and progress made to date.

The NEOShield project (http://www.neoshield.net) has received funding from the European Union Seventh Framework Program (FP7/2007-2013) under Grant Agreement no. 282703.

Keywords. celestial mechanics, space vehicles: instruments, minor planets, asteroids.

Highlights of Astronomy, Volume 16
XXVIIIth IAU General Assembly, August 2012
T. Montmerle, ed.

doi:10.1017/S1743921314011855

AIDA: Asteroid Impact and Deflection Assessment

Patrick Michel[1], A. Cheng[2], A. Galvez[3], C. Reed[4], I. Carnelli[5], P. Abell[6], S. Ulamec[7], A. Rivkin[8], J. Biele[9] and N. Murdoch[10]

[1]Laboratoire Lagrange, University of Nice Sophia-Antipolis, Côte d'Azur Observatrory
BP 4229, 06304 Nice Cedex 4, France
email: michelp@oca.eu

[2]APL/JHU
11101 John Hopkins Road Laurel, MD 20723-6099, USA
email: andrew.cheng@jhuapl.edu

[3]European Space Agency
8-10 rue Mario Nikis, 75738 Paris Cedex 15, France
email: Andres.Galvez@esa.int

[4]APL/JHU
11101 John Hopkins Road Laurel, MD 20723-6099, USA
email: Cheryl.Reed@jhuapl.edu

[5]European Space Agency
8-10 rue Mario Nikis, 75738 Paris Cedex 15, France
email: Ian.Carnelli@esa.int

[6]NASA JSC
Mail Code KR, 2101 NASA Parkway Houston, TX 77058-3696, USA
email: paul.a.abell@nasa.gov

[7]DLR, RB-MC
Linder Hhe 1, 51147 Cologne, Germany
email: Stephan.Ulamec@dlr.de

[8]APL/JHU
11101 John Hopkins Road Laurel, MD 20723-6099, USA
email: Andy.Rivkin@jhuapl.edu

[9]DLR, RB-MUSC
Linder Hhe 29, 51147 Kln, Germany
email: jens.biele@dlr.de

[10]Institut Suprieur de l'Aronautique et de l'Espace (ISAE)
10 avenue Edouard Belin, BP 54032, 31055 Toulouse cedex 4, France
email: naomi.murdoch@isae.fr

Abstract. AIDA (Asteroid Impact and Deflection Assessment) is a project of a joint mission demonstration of asteroid deflection and characterisation of the kinetic impact effects. It involves the Johns Hopkins Applied Physics Laboratory (with support from members of NASA centers including Goddard Space Flight Center, Johnson Space Center, and the Jet Propulsion Laboratory), and the European Space Agency (with support from members of the french CNRS/Cte dAzur Observatory and the german DLR). This assessment will be done using a binary asteroid target. AIDA consists of two independent but mutually supporting mission concepts, one of which is the asteroid kinetic impactor and the other is the characterisation spacecraft. The objective and status of the project will be presented.

Keywords. celestial mechanics, space vehicles: instruments, minor planets, asteroids, etc.

Highlights of Astronomy, Volume 16
XXVIIIth IAU General Assembly, August 2012
T. Montmerle, ed.

doi:10.1017/S1743921314011867

Probing the interior of asteroid Apophis: a unique opportunity in 2029

Patrick Michel[1], J. Y. Prado[2], M. A. Barucci[3], O. Groussin[4], A. Hérique[5], E. Hinglais[6], D. Mimoun[7], W. Thuillot[8] and D. Hestroffer[9]

[1]Laboratoire Lagrange, University of Nice Sophia-Antipolis, Côte d'Azur Observatrory
BP 4229, 06304 Nice Cedex 4, France
email: michelp@oca.eu

[2]CNES
18 Avenue Edouard Belin, 31401 Toulouse Cedex 9, France
email: jean-yves.prado@cnes.fr

[3]LESIA, Paris Observatory
5 Place Jules Janssen, 92195 MEUDON Cedex, France
email: antonella.barucci@obspm.fr

[4]Laboratoire d'Astrophysique de Marseille
Technopole de Marseille-Etoile, 38 rue Frdric Joliot-Curie, 13388 Marseille Cedex 13, France
email: olivier.groussin@oamp.fr

[5] Institut de Plantologie et d'Astrophysique de Grenoble - IPAG
Bat D de Physique, BP. 53, 38041 Grenoble, France
email: alain.herique@obs.ujf-grenoble.fr

[6]CNES
18 Avenue Edouard Belin, 31401 Toulouse Cedex 9, France
email: emmanuel.hinglais@cnes.fr

[7]Institut Suprieur de l'Aronautique et de l'Espace (ISAE)
10 avenue Edouard Belin, BP 54032, 31055 Toulouse cedex 4, France
email: david.mimoun at isae.fr

[8]IMCCE
UMR 8028 CNRS, Observatoire de Paris, 77 Av. Denfert Rochereau, 75014 Paris, France
email: william.thuillot@imcce.fr

[9]IMCCE
UMR 8028 CNRS, Observatoire de Paris, 77 Av. Denfert Rochereau, 75014 Paris, France
email: Daniel.Hestroffer@imcce.fr

Abstract. The near Earth asteroid (99942) Apophis, discovered in 2004, (with a diameter of about 270 meters) will come back very close to the Earth on April 13, 2029.

The close approach of Apophis to the Earth in 2029 will present an unique opportunity for characterizing this object, serving both science and mitigation purposes. The object will be easily visible from the Earth and it can be expected that its shape and thermal properties will be well determined from ground based observations. However, the characterization of its interior will not be achievable from purely terrestrial observations. Such a characterization, beyond its high scientific value, is essential for planning any mitigation operation, should it be necessary in the future.

Near Earth objects are a precious source of information as they represent a mixture of different populations of small bodies containing fundamental information on the origin and early evolution of the solar system.

Monitoring the response of Apophis to the gravitational constraints induced by its close approach to the Earth may provide a way to access information on its internal structure.

A study to identify some affordable mission scenarii for such a mission is presently underway at CNES (the French Space Agency).

We will present the scientific and mitigation objectives of such a mission as well as the preliminary results of the mission analysis and the main system characteristics.

Keywords. celestial mechanics, space vehicles: instruments, minor planets, asteroids, etc.

Highlights of Astronomy, Volume 16
XXVIIIth IAU General Assembly, August 2012
T. Montmerle, ed.

doi:10.1017/S1743921314011879

MarcoPolo-R: Near Earth Asteroid Sample Return Mission candidate as ESA-M3 class mission

Patrick Michel
and the *MarcoPolo-R* Science Study Team
Laboratoire Lagrange, University of Nice Sophia-Antipolis, Côte d'Azur Observatrory
BP 4229, 06304 Nice Cedex 4, France
email: michelp@oca.eu

Abstract. *MarcoPolo-R* is a sample return mission to a primitive Near-Earth Asteroid (NEA) selected in 2011 for the Assessment Study Phase of M3-class missions in the framework of ESAs Cosmic Vision (CV) 2015-2025 programme. The phase A study started at the end of 2012 and will proceed throughout 2013. The final selection by ESA will occur in February 2014. *MarcoPolo-R* is a European-led mission with a possible contribution from other agencies. *MarcoPolo-R* will rendez-vous with the primitive NEA 2008 EV5. Before returning a unique sample to Earth, the asteroid will be scientifically characterized at multiple scales. *MarcoPolo-R* will provide detailed knowledge of the physical and compositional properties of a member of the population of Potentially Hazardous Asteroids (PHA), which is an important contribution to mitigation studies.

Keywords. celestial mechanics, space vehicles: instruments, minor planets, asteroids.

Highlights of Astronomy, Volume 16
XXVIIIth IAU General Assembly, August 2012
T. Montmerle, ed.

doi:10.1017/S1743921314011880

Whom should we call? Data policy for immediate impactors announcements

Andrea Milani[1] and Giovanni B. Valsecchi[2,3]

[1]Department of Mathematics, University of Pisa, Largo Pontecorvo 5, 56127 Pisa, Italy
email: milani@dm.unipi.it
[2]IAPS, INAF, Roma, Italy
[3]IFAC-CNR, Sesto Fiorentino, Italy

1. Impact Monitoring

After the PR disaster of 1997 XF_{11} (March 1998), we started a crash research program on impact predictions. The difficulty was the chaotic motion of Earth-crossing asteroids (orbit uncertainty increases exponentially with time); it can be solved by replacing a real asteroid with a swarm of Virtual Asteroids. In 1999 we introduced Geometric Sampling to replace Monte-Carlo methods (see Milani, Chesley & Valsecchi, A&A 346, 1999). In November 1999 the first Impact Monitoring system CLOMON was operational. From 2002 the second generation systems CLOMON2 at Pisa and SENTRY at JPL are operational: critical cases are scanned for possible impacts in the next 90–100 years.

1.1. *The current Data Policy*

In collaboration with the IAU executive we established a data policy, (technical verification): all the reliable impact information must be posted on the web; however, no critical case (Palermo Scale > -2) is posted before verification. The current data policy does not prescribe exactly what to do about press releases, but we are very careful. NEODyS leaves to the media to select newsworthy cases, JPL has a PR person in charge of announcements, but uses this seldom. This data policy is motivated by the need of astronomical follow up (astrometric, physical observations). As observational information increases most virtual impactors become contradicted by data. The list of need-to-know people is very long, and they are not under the control of any single organization: thus secret is both detrimental and dangerous, especially because of conspiracy theories.

2. Immediate Impactors

The current setting of Impact Monitoring (NEODyS + JPL) has been operating for >10 years. There has been progress in understanding after cases with new features, e.g., (99942) Apophis, which initially had a very high Impact Probability for 2029. An algorithm upgrade was needed, but the policy for impact announcements did not need any change. The map of possible impact area was not made public.

In October 2008 the very small asteroid/meteoroid 2008 TC3 was discovered 21 hours before impact. The system (observers, MPC, Impact Monitoring) was found to work very efficiently: impact was announced by many astronomers, also by NEODyS, the impact location was predicted by JPL, the atmospheric explosion was observed by Meteosat and seen by an airplane pilot; meteorites were recovered.

The proposed Wide Survey for NEOs (as defined by ESA SSA studies) would cover the entire dark sky ($> 30,000$ sq. deg., 3 images each night). The goal is discovery of a large fraction of impactors large enough for ground damage in time for mitigation action (evacuation of target area). A simulation of 10 years of Wide Survey shows the capability of warning at IP> 0.9 10 days before impact for most Tunguska-class impactors, 3 days for a majority of objects which could result in damage.

2.1. *Incomplete knowledge for the atmospheric segment*

As long as the orbit is in interplanetary space it is deterministic and can be predicted, with uncertainty depending only upon the amount, accuracy and time span of the available astrometry. Even over the much shorter time scale of an immediate impactor (few weeks to few days) the data policy has to be the same: given the need of collaboration by astronomers from all over the world, secrecy is impossible.

The situation changes when a small asteroid/meteoroid actually enters the Earth's atmosphere. Then the typically small uncertainty of the orbit(few km) grows significantly in the last minute of atmospheric flight. The most uncertain prediction is on where most of the kinetic energy of the impact is released, forming a fireball. This results, for low elevation entry, in a uncertainty of the ground zero location by tens of km.

The height of the energy main release can be such that strong thermal/pressure effect do not reach the ground. In the Tunguska event, with energy release in troposphere, a ~ 40 m diameter objects with $3 \sim 5$ MT energy generated a destruction zone of ~ 2000 sq. km. An impactor of the same composition but smaller diameter (like $20 \sim 30$ m) could result in no ground damage, or possibly a weak one. It is also possible for smaller impactors to cross the troposphere and impact in one piece. E.g., Barringer (Meteor) Crater was excavated by an iron meteorite about the same diameter of Tunguska.

2.2. *Evaluation of the risk*

When an immediate impactor is discovered, how much do we know about the object? If only photometry incidental to astrometry is available, the diameter may be uncertain by a factor 2.5. Estimates of mass include an additional uncertainty by a factor 1.5 in density, overall a range of about 20 for the uncertainty in mass. Thus a "damage prediction" could range from almost certainly not dangerous to 10 MT groundburst with very substantial damage: the evaluation of the risk is very poor.

The Wide Survey will need advanced follow up capabilities, also for physical observations, and these will have to be contributed by astronomers (as for 2008 TC3). This implies a significant list of need-to-know people and the impossibility of secrecy. The effect of these observations (spectra, polarimetry, infrared, radar) could decrease the uncertainty in the total energy to a much more limited range, such as $2 \div 3$. Thus the recommended data policy is to declare in public that we need more information in order to assess the damage, and thus ask for contributions.

2.3. *Data policy for the damage predictions*

The results of the modeling for the atmospheric reentry are far from certain, because they depend upon assumptions on composition and internal structure. This may result in uncertainties of tens of km on the location of ground zero, and several km in the altitude of the main energy release. Because of the poor information and of the nature of the problem, with no simple deterministic model, there is no guarantee that the information which would be needed by the emergency planning authorities would be available.

The only firm data policy rule applicable is that "thou shalt not lie". If it is not known what the ground effect of a given impactor would be, nobody should claim he/she knows. The danger is that someone could like to claim he knows what to do, when in fact this is not possible. This applies both to the communication with the public at large as well as to the briefing of a restricted group of authorities in charge of mitigation planning.

Highlights of Astronomy, Volume 16
XXVIIIth IAU General Assembly, August 2012
T. Montmerle, ed.

doi:10.1017/S1743921314011892

NAO and SHAO participation in the near-Earth space observations

A. Shulga[1], Y. Kozyryev[1], Y. Sybiryakova[1], Z. Tang[2], Y. Mao[2], Y. Li[2] and Y. Yu[2]

[1]Nikolaev Astronomical Observatory
Nikolaev, 54030, Ukraine
email: avshulga@mail.ru

[2]Shanghai Astronomical Observatory
Shanghai, 200030, China
email: zhtang@shao.as.cn

Abstract. The joint observations of space debris are conducted by NAO and SHAO. The main task of observations is precise estimation of the hazard collisions of SD with operating satellites. Observations of the near earth asteroids (NEA) on the distance of less than 0.05 AU, according to the NEODyS program, are conducted in NAO. The main task is to obtain the large number of precise observations of the NEAs during their closest approach to the Earth.

Keywords. space debris, near earth asteroids

1. Telescopes and method of observation

The observations are carried out by combined CCD observation method (developed by NAO in 2000). The combined CCD observation method is based on using of CCD time delay and integration mode (TDI) and obtaining images of object and reference stars in different frames. To obtain images of space debris or fast NEAs using TDI mode, the CCD matrix columns must be parallel to the object motion direction. To rotate the CCD camera around the optical axis of the lens a special mechanical device was developed in NAO (Shulga *et al.* 2008). Exposure time of the classical TDI imaging is equal to the time of object passage through the field of view, and is inefficient for space debris and NEAs observation. In most cases this fixed exposure is too long, and may cause the oversaturation of frame background and object image or distortion of the object image due to acceleration of apparent motion of the space debris or NEAs. Therefore, a special "short" TDI imaging technique was developed in NAO (Shulga *et al.* 2011).

Three telescopes in NAO and one telescope in SHAO were designed and used for observation of space debris and NEAs. The main parameters of telescopes are given in Table 1. All telescopes are equipped with U9000 Apogee CCD cameras, provided by SHAO.

Table 1. The main telescope parameters.

Name of telescope	Observatory	Diameter[mm]	Focal length [mm]
Fast robotic telescope (FRT)	NAO	300	1500
Mobitel-KT50	NAO	500	3000
Satellite tracking camera AFU-75	NAO	267	750
Large-field of view telescope RDS-30	SHAO	300	250

2. Results

Regular joint observations of low Earth space debris have been carried out in NAO and SHAO since 2008. The orbital elements and observation errors are calculated using software developed by NAO and SHAO. The following standard deviation for both coordinates: (1-3)" for the AFU-75, (2-5)" for the RDS-30, were obtained during calculations of the orbital elements.

During 2008-2012, 1702 positions of 109 NEAs were obtained in NAO (observatory code 089), and 30 of them were of potentially hazardous asteroids. The positions were sent to the IAU Minor Planet Center. The 178 positions of 5 NEAs with magnitude of (12.8-15.9), apparent motion of (4.6-68.3) "/min and at distance of (0.01-0.5) AU were obtained with the FRT. The standard deviations of position of 5 NEAs for both coordinates are (0.23-0.43)". The 1524 positions of 104 NEAs with magnitude of (9.5-18.5), apparent motion of (0.5-306.3) "/min and at distance of (0.009-0.98) AU were obtained with the Mobitel-KT50. The standard deviations of positions for both coordinates are (0.23-0.47)".

References

Shulga, O., Kozyryev, Y., & Sybiryakova, Y. 2008, *Proc. of IAU Symposium 248*, pp. 128-129.

Shulga, O., Kozyryev, Y., & Sybiryakova, Y. 2011, *Proc. Of Gaia follow-up network for solar system objects workshop* , pp. 97-100.

Highlights of Astronomy, Volume 16
XXVIIIth IAU General Assembly, August 2012
T. Montmerle, ed.

doi:10.1017/S1743921314011909

A space mission to detect imminent Earth impactors

G. B. Valsecchi[1,3], E. Perozzi[2,1] and A. Rossi[3]

[1]IAPS-INAF, Roma, Italy
[2]Deimos Space, Madrid, Spain
[3]IFAC-CNR, Sesto Fiorentino, Italy

One of the goals of NEO surveys is to discover Earth impactors before they hit. How much warning time is desirable depends on the size of the impactors: for the larger ones more time is needed to mount effective mitigation measures. Initially, NEO surveys were aimed at large impactors, that can have significant global effects; however, their typical time scale is orders of magnitude larger than human lifetime. At the other extreme, monthly and annual events, liberating energies of the order of 1 to 10 kilotons, are immaterial as a threat to mankind, not justifying substantial expenditure on them. Intermediate events are of more concern: in the megatons range, timescales are of the order of centuries, and the damage can be substantial. A classical example is the Tunguska event, in which a body with a diameter of about 30 to 50 m liberated about 5 megatons in the atmosphere, devastating 2 000 square kilometers of Siberian forest.

Tunguska-class impactors are faint, not easy to discover and track for ground-based surveys; if such an impactor were to approach the Earth from the direction of the Sun, it could not be discovered telescopically, as it would be in the daytime sky. However, if it would make pre-impact approaches to the Earth, it could be discovered during one of them, allowing for tracking and orbit determination. A NEO survey then would, after an initial period during which it would be "blind" with respect to immediate impactors coming from the direction of the Sun, have discovered and tracked a sufficiently large number of Tunguska-class NEOs that the next impactor would already be among the discovered ones at the time of fall. Farnocchia *et al.* (2012) simulated such a process, concluding that the survey envisaged for the NEO Segment of the SSA Program of ESA would almost reach a 90% probability of discovering the next Tunguska-class impactor before impact in about one century from start. Thus, there is room for an alternative survey strategy for discovering Tunguska-sized impactors during the period of time in which ground-based surveys complete the catalog of impactors coming from the direction of the Sun. This strategy has to spot impactors while they are just inside the Earth orbit.

Let us consider a NEA with a geocentric speed of 15 km/s; if it approaches the Earth from the direction of the Sun, it would spend about 11.5 days within the last 0.1 AU of its trajectory towards the Earth. If discovered from space, when still at 0.1 AU from our planet, this would give enough time for an effective evacuation of the impact area.

The crucial question is: in what orbit do we put the spacecraft? Earth-bound orbits allow to observe close to the Sun, but their advantage over ground telescopes in terms of warning time for an imminent impactor is negligible. Satellites at L_1 would be more efficient than those in Earth orbit, but the warning time would still be too short. If the spacecraft is put in a heliocentric orbit interior to that of the Earth, its sensor can point in the anti-solar direction, to spot objects when their brightness is near maximum due to the small phase angle. However, the drawback would be that the difference of heliocentric

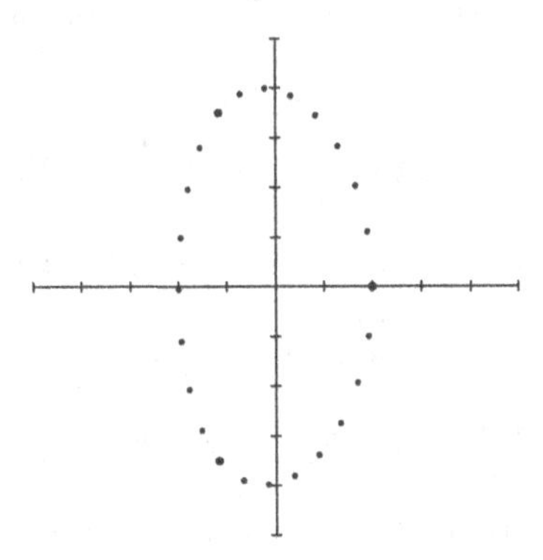

Figure 1. The motion of the 3 spacecraft in a geocentric rotating frame (Sun on the negative x-axis); the tic marks are spaced by 0.05 AU, and the 3 largest dots (at $x = 0.1, y = 0$, at $x = -0.06, y = -0.17$, at $x = -0.06, y = 0.17$) show the 3 spacecraft at a certain epoch. After that, their motion is clockwise; small dots show the positions every 5 days, and intermediate size dots show those every 15 days.

longitude between the spacecraft and the Earth would go from 0° to 180°, making the communications difficult and impairing to ability to spot imminent impactors.

We propose to put the spacecraft in an orbit like those studied by Hénon (1969, 1970), resembling a distant retrograde Earth satellite. These are heliocentric orbits in which the semimajor axis is almost exactly 1 AU and the inclination is small; the eccentricity can be chosen within a rather large range, and for our purposes we can choose a value of 0.1.

A spacecraft moving on such an orbit would:

(*a*) have the same orbital semimajor axis (1 AU), and hence the same orbital period (1 yr) as the Earth;

(*b*) undergo a radial excursion from the Sun equal to twice its eccentricity (for $e = 0.1$, it would have $q = 0.9$ AU, $Q = 1.1$ AU);

(*c*) undergo an excursion in longitude such that it would, at the extrema, precede or follow the Earth by about twice the value of its orbital eccentricity, expressed in AU.
Seen from Earth, the spacecraft would make a retrograde orbit about it, with minimum geocentric distance, at inferior and at superior conjunction, equal to the value of its orbital eccentricity expressed in AU, and maximum geocentric distance, at quadratures, equal to twice the value of its orbital eccentricity, again in AU. In such an orbit, even a relatively small telescope would be very efficient in monitoring the region immediately interior to the Earth orbit, when moving at heliocentric distance less than 1 AU, but would be inefficient otherwise. To address this problem, a possible solution is the insertion of three spacecraft, appropriately spaced, in the same orbit. In this way, there would be at any time at least 1 spacecraft optimally placed for impactor monitoring. In the following, we assume to have 3 spacecraft moving on an orbit with semimajor axis equal to 1 AU, $e = 0.1$, and very small i, of the order of 0.1°.

A small constellation of 3 satellites properly spaced along the same "distant retrograde satellite" orbit could be quite effective in monitoring imminent small impactors, ensuring a reasonable warning time. These satellites could of course be very effective also in helping to complete the catalogue of low-MOID small NEOs.

References

Farnocchia, D., Bernardi, F., & Valsecchi, G. B. 2012, *Icarus* 219, 41
Hénon, M. 1969, *A&A* 1, 223
Hénon, M. 1970, *A&A* 9, 24

Highlights of Astronomy, Volume 16
XXVIIIth IAU General Assembly, August 2012
T. Montmerle, ed.

doi:10.1017/S1743921314011910

Selection effects in the discovery of NEAs

G. B. Valsecchi[1,2], G. D'Abramo[3] and A. Boattini[4]

[1]IAPS-INAF, Roma, Italy
[2]IFAC-CNR, Sesto Fiorentino, Italy
[3]SpaceDyS, Navacchio, Italy
[4]LPL, University of Arizona, USA

To highlight discovery selection effects, we consider four NEA subpopulations:
(*a*) "Taurid asteroids", the Apollos with orbits similar to those of 2P/Encke and of the Taurid meteoroid complex;
(*b*) Atens, to which we add the Inner Earth Objects;
(*c*) non-Taurid Apollos;
(*d*) Amors.
The "Taurid asteroids" are identified by Asher *et al.* (1993) with a reduced version of the *D*-criterion (Southworth and Hawkins 1963), involving only a, e and i:

$$D = \sqrt{\left(\frac{a - 2.1}{3}\right)^2 + (e - 0.82)^2 + \left(2 \sin \frac{i - 4^\circ}{2}\right)^2} \leqslant 0.25.$$

It turns out that the distribution of the longitudes of perihelion ϖ of NEAs with $D < 0.25$ is significantly non-random, due to the existence of two groups whose apse lines are approximately aligned with those of 2P/Encke and of (2212) Hephaistos.

Asher *et al.* (1993) suggested that this finding supports a scenario in which a giant Jupiter family comet could have ended in an Encke-like orbit and suffered a hierarchical fragmentation in the last 20 000 yr, the ϖ concentrations being the signature of this process. Valsecchi *et al.* (1995) investigated the dynamics of 2P/Encke and of Taurid asteroids. If the hierarchical fragmentation of a single cometary progenitor had really taken place, traces of it should be recognizable in past orbital separations; this, however, turned out not to be the case.

Given that the dynamics in the Taurids region does not support the scenario of a common progenitor for the NEAs residing there, we are left with the problem of explaining the ϖ concentrations. Could they be due to observational selection effects?

To investigate this possibility, we reconstructed the geometry at the discovery of each NEA with Orbfit (http://adams.dm.unipi.it/ orbmaint/orbfit/), using the observations available at NEODyS (http://newton.dm.unipi.it/neodys2/) and the discovery dates of NEAs available at the Minor Planet Center (http://www.minorplanetcenter.net/iau/mpc.html).

Due to the small sizes of Near Earth Asteroids (NEAs), most of them are discovered when moving near the Earth. Two correlations can be established:

• between the discovery date and the longitude of node of the NEA's orbit;
• between the discovery date and a range of possible longitudes of perihelion of the NEA's orbit, depending on the semimajor axis and the eccentricity of the latter.

These correlations would have no effect on the distributions of the longitudes of nodes and perihelia on known NEAs if the rate of NEA discovery over the year were constant. However, as pointed out by Kresák and Klačka (1989), this is not the case: there are significant seasonal variations of observing conditions that result in a variable rate of asteroid discoveries over the year. As a consequence:

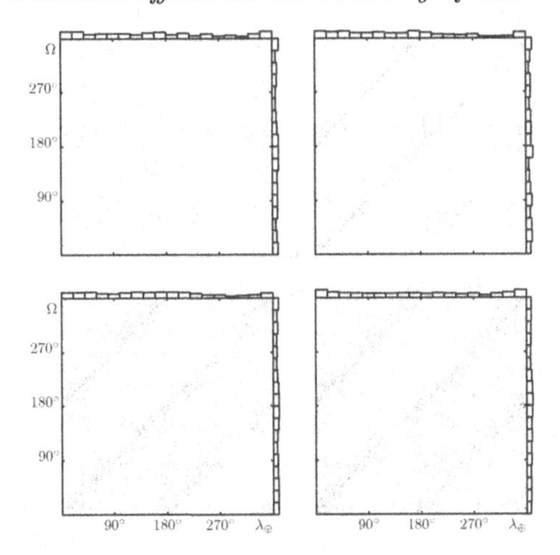

Figure 1. Plot of the longitude of node Ω vs longitude of Earth at discovery $\lambda_\oplus$ for (clockwise from top left): Taurid asteroids, Atens, Amors and non-Taurid Apollos.

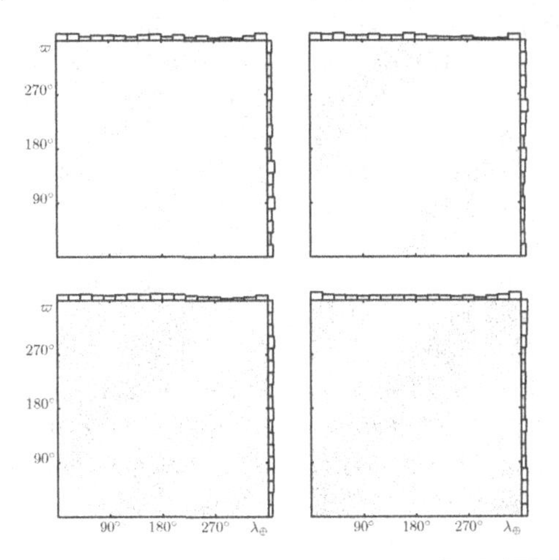

Figure 2. Same as Fig. 1 for the longitude of perihelion ϖ.

• small asteroids in inclined orbits tend to have the longitude of node close to Earth longitude at discovery;

• Amors tend to have the perihelion longitude close to Earth longitude at discovery;

• Atens tend to have the aphelion longitude close to Earth longitude at discovery;

• Taurids, due to their orbital size and shape, tend to be discovered when the Earth longitude is about 100° to 120° ahead of or behind their perihelion;

• non-Taurid Apollos show a similar selection effect, although to a much lesser degree. These selection effects must be taken into account when evaluating the completeness of a survey.

References

Asher, D. J., Clube, S. V. M., & Steel, D. I. 1993, *MNRAS* 264, 93

Kresák, Ľ. & Klačka, J. 1989, *Icarus* 78, 287

Southworth, R. B. & Hawkins, G. S. 1963, *Smithson. Contr. Astrophys.* 7, 261

Valsecchi, G. B., Morbidelli, A., Gonczi, R., Farinella, P., Froeschlé, Ch., & Froeschlé, Cl. 1995, *Icarus* 118, 169

Highlights of Astronomy, Volume 16
XXVIIIth IAU General Assembly, August 2012
T. Montmerle, ed.

doi:10.1017/S1743921314011922

The population of bright NEAs

G. B. Valsecchi[1,2] and G. F. Gronchi[3]

[1]IAPS-INAF, Roma, Italy
[2]IFAC-CNR, Sesto Fiorentino, Italy
[3]University of Pisa, Pisa, Italy

Our understanding of the orbital distribution of NEAs is influenced by discovery selection effects, so that it is likely that the orbital distribution of known NEAs differs from the true distribution. In fact, our ability to reconstruct the true distribution critically depends on the removal of discovery biases from the known population.

The reference model of the orbital distribution of NEAs is described in Bottke *et al.* (2002). It combines dynamics (a numerical reconstruction of how NEAs are transferred from source regions to their current orbits), with the observational record of Spacewatch, whose biases are well studied (Jedicke 1996). According to Bottke *et al.*, the proportion of orbital classes within the NEA population with $H \leqslant 18$ should be $32\% \pm 1\%$ Amors, $62\% \pm 1\%$ Apollos, $6\% \pm 1\%$ Atens, and $2\% \pm 0\%$ IEOs; moreover, the fraction of PHAs (i.e., objects with $MOID \leqslant 0.05$ AU) should be 21%, and Bottke *et al.* give also the proportions, within the various subclasses, of objects with $a < 2$ AU, $e < 0.4$, $e < 0.6$, $i < 10°$, $i < 20°$ and $i < 30°$.

About 1500 NEAs were known when the NEA population model just discussed was developed, and the estimated completeness for $H \leqslant 18$ was about 50%. More than 9000 NEAs are known now, the estimated completeness for $H \leqslant 18$ is much higher than 50%, and that for $H \leqslant 16$ is very close to 100% (Harris 2011, personal communication). Because of its near-completeness, the population of NEAs with $H \leqslant 16$ is nearly unbiased; it comprised, in April 2011, 192 objects (source: http://newton.dm.unipi.it/neodys2/), a number large enough to allow for an overall statistical analysis. The orbital elements of these NEAs are in general rather well determined ($\sigma(a) < 10^{-5}$ AU), with very few exceptions.

Our main working hypothesis is that, since we know practically all the NEAs with $H \leqslant 16$, their orbital distribution can be considered as representative of the true orbital distribution of NEAs; with this in mind, in Table 1 we compare them to the model.

Class	Bottke *et al.*	Bright NEAs
Amors	32%	50%
Apollos	62%	47%
Atens	6%	3%
IEOs	2%	0%
PHAs	21%	15%

Table 1. The proportions of the various orbital classes among bright NEAs differs from those of Bottke *et al.* (2002).

Bright NEAs exhibit a significantly larger proportion of Amors, and significantly smaller proportions of Apollos, Atens, IEOs and PHAs.

As evidenced in Table 2, also the orbital distribution of bright NEAs within the classes differs from Bottke *et al.* (2002). As the Table shows, bright NEAs exhibit significantly larger proportions of Amors and Apollos in high-inclination orbits.

Orbital region	Amors		Apollos	
	Bottke *et al.*	Bright NEAs	Bottke *et al.*	Bright NEAs
$a < 2$ AU	27%	26%	55%	49%
$e < 0.4$	25%	23%	9%	9%
$e < 0.6$	87%	91%	34%	37%
$i < 10°$	41%	25%	20%	16%
$i < 20°$	74%	49%	48%	38%
$i < 30°$	87%	72%	67%	59%

Table 2. The orbital distributions of bright NEAs differ from those of Bottke *et al.* (2002).

NEAs move in chaotic orbits, allowing them to encounter planets. However, this only happens for small values of the $MOID$ relative to a given planet. Unless encounters with Jupiter are possible, encounters with the terrestrial planets able to significantly alter the orbit are infrequent, so that in the time interval between such encounters the osculating orbital elements undergo significant variations due to secular perturbations. Thus, it is of interest to consider the evolution of a NEA sample subject only to secular perturbations. Under the action of the latter, NEAs with $a > 1.3$ AU may evolve to orbits with $q > 1.3$ AU, while some asteroids with $q > 1.3$ AU may evolve to orbits with $q < 1.3$ AU.

We therefore augmented our sample of bright NEAs by including the numbered and multi-opposition asteroids with $H \leqslant 16$ that can, due to secular perturbations, become NEAs (source: http://hamilton.dm.unipi.it/astdys2/). This sample, composed of bright NEAs and Potential NEAs (PNEAs), amounts to 277 objects. Its secular evolution has then been computed over a time span of 200 000 yr, starting from the present epoch, with the method of Gronchi and Milani (2001). It must be kept in mind that the secular propagator we used is not appropriate for NEAs in low-order resonances.

The result of the secular integration is that the proportions within the Apollo and Amor classes given for the bright NEAs in Table 2 are remain essentially constant in time, showing that the sample seems to be in a steady state, at least from the point of view of secular perturbations. Also the proportion of PHAs remains constant.

We can thus conclude that:

• the orbital distribution of bright NEAs ($H \leqslant 16$) is significantly different from the model by Bottke *et al.* (2002);

• the differences give interesting hints about the distribution of the real population: there should be more Amors, more high-inclination orbits, less PHAs compared to the model;

• for the sample of known bright NEAs, the action of secular perturbations does not significantly alter the proportions of orbital types and of PHAs.

References

Bottke, W. F., Morbidelli, A., Jedicke, R., Petit, J.-M., Levison, H. F., Michel, P., & Metcalfe, T. S. 2002, *Icarus* 156, 399

Gronchi, G. F. & Milani, A. 2001, *Icarus* 152, 58

Jedicke, R. 1996, *AJ* 111, 970

Highlights of Astronomy, Volume 16
XXVIIIth IAU General Assembly, August 2012
T. Montmerle, ed.

doi:10.1017/S1743921314011934

Calibration of Star-Formation Rate Measurements Across the Electromagnetic Spectrum

V. Buat[1], J. Braine[2], D. A. Dale[3], A. Hornschemeier[4], B. Lehmer[4,5], P. Kroupa[6] J. Pflamm-Altenburg[6], C. C. Popescu[7], H. Wu[8] and A. Zezas[9]

[1]Aix-Marseille Université, CNRS, LAM (Laboratoire d'Astrophysique de Marseille) UMR7326, 13388, France
email: veronique.buat@oamp.fr

[2]Observatoire de Bordeaux, Laboratoire d'Astrophysique de Bordeaux, 2 rue de l'Observatoire, BP 89, 33270 Floirac, France
email: Jonathan.Braine@obs.u-bordeaux1.fr

[3]Department of Physics & Astronomy, University of Wyoming, Laramie, WY 82070 USA
email: ddale@uwyo.edu

[4]NASA Goddard Space Flight Centre, Code 662, Greenbelt, MD 20771, USA
email: Ann.Hornschemeier@nasa.gov

[5]The Johns Hopkins University, Homewood Campus, Baltimore, MD 21218, USA
email: blehmer@pha.jhu.edu

[6]Argelander-Institute for Astronomy, Auf dem Hügel 71, 53121 Bonn, Germany
email: pavel, jpflamm@astro.uni-bonn.de

[7]Jeremiah Horrocks Institute, University of Central Lancashire, PR1 2HE, Preston, UK
email: cpopescu@uclan.ac.uk

[8]Key Laboratory of Optical Astronomy, National Astronomical Observatories, Chinese Academy of Sciences, Beijing 100012, China

[9]Department of Physics, University of Crete, P.O. Box 2208, 71003 Heraklion, Crete, Greece
Institute of Electronic stucture and LASER, FORTH, Heraklion, Crete, Greece
Harvard-Smithnonoan Center for Astrophysics, 60 Garden Street, Cambridge, MA 02138, USA
email: azezas@physics.uoc.gr

Abstract. Star-formation is one of the main processes that shape galaxies, and together with black-hole accretion activity the two agents of energy production in galaxies. It is important on a range of scales from star clusters/OB associations to galaxy-wide and even group/cluster scales. Recently, studies of star-formation in sub-galactic and galaxy-wide scales have met significant advances owing to: (a) developments in the theory of stellar evolution, stellar atmospheres, and radiative transfer in the interstellar medium; (b) the availability of more sensitive and higher resolution data; and (c) observations in previously poorly charted wavebands (e.g. Ultraviolet, Infrared, and X-rays). These data allow us to study more galaxies at ever-increasing distances and nearby galaxies in greater detail, and different modes of star formation activity such as massive star formation and low level continuous star formation in a variety of environments. In this contribution we summarize recent results in the fields of multi-wavelength calibrations of star-formation rate indicators, the Stellar Initial Mass function, and radiative transfer and modeling of the Spectrale Energy Disrtributions of galaxies.

Keywords. stars: luminosity function, mass function, binaries (including multiple): close, galaxies: star clusters, galaxies: fundamental parameters, infrared: galaxies, ultraviolet: galaxies, X-rays: binaries

1. Multi-wavelength Calibrations of SFR

1.1. *General Characteristics of Star Formation Calibrations*

Measuring accurate star formation rates (SFRs) is a crucial step in any study of galaxy evolution. As an example, the evolution of the SFR density with redshift is one of the major diagnostics to study galaxy evolution and any model of galaxy formation and evolution must reproduce this evolution. Before drawing any firm conclusions about the validity of theoretical models it must be stressed that large uncertainties come from the difficulty in measuring accurately the star formation rate (e.g., Davé 2008). Various calibrators are used, covering the full electromagnetic spectrum, from the X-ray through the ultraviolet (UV), the optical, the infrared (IR), all the way to the radio. Both continuum emission and emission lines have been calibrated in terms of SFR. The derivation of SFR relies on conversion factors depending on the indicator used. In this subsection we present the general way to derive SFR from direct stellar emission and then discuss the uncertainties of the calibrations due to variations of the star formation histories. The effect of dust attenuation will be discussed in Section 1.3. Complete reviews of different star formation indicators were published recently (Calzetti 2012; Kennicutt & Evans 2012; Boissier 2012).

1.1.1. *Basic Equation and Standard Calibrations*

The fundamental equation to link the intrinsic luminosity emitted by stars at wavelength λ and time t is

$$L(\lambda,t) = \int_0^t \int_{M_{low}}^{M_{up}} F_\lambda(m,\theta)\ SFR(t-\theta)\ \psi(m)\ \mathrm{d}m\ \mathrm{d}\theta \qquad (1.1)$$

where $F_\lambda(m,\theta)$ are the evolutionary stellar tracks, $\psi(m)$ the initial mass function, and $SFR(t)$ the star formation rate. From this fundamental equation there are two ways to

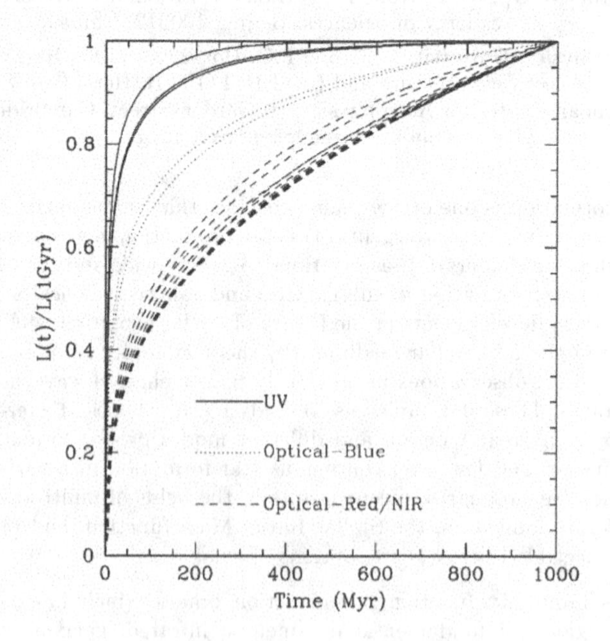

Figure 1. Evolution of luminosity for several filters from the UV to the NIR for a constant SFR of 1 $M_\odot$ yr $^{-1}$. The luminosity is normalised to its value obtained after 1 Gyr. See Boissier (2012) for more details.

derive SFRs. One can use stellar population synthesis models to fit a large set of data at different wavelengths; the current SFR is an output parameter of the fit (with the stellar mass and other parameters depending on the specific code used). Another popular way to proceed is to derive simple recipes. One assumes a SFR constant over a timescale τ and the SFR becomes simply proportional to the luminosity integrated over τ. The timescale τ is chosen in order that the luminosity at wavelength λ reaches a steady state:

$$SFR = \left(\int_0^t \int_{M_{low}}^{M_{up}} F_\lambda(m,\theta) SFR(t-\theta)\psi(m)\, \mathrm{d}m\, \mathrm{d}\theta \right)^{-1} \times L(\lambda) \tag{1.2}$$

The value of the conversion factor $C = SFR/L(\lambda)$ is calculated with a spectral synthesis code. The SFR calculated using a conversion factor can be different from the average of the star formation activity during τ if the actual SFR is strongly varying during this period. The assumption of a constant SFR is likely to be valid on short timescales only and SFR should be derived from the emission of stars with short lifetimes, preferentially in the ultraviolet or for recombination lines of ionizing photons. The evolution of the luminosity at different wavelength is illustrated in Figure 1. It is clearly seen that the timescale to reach a steady state increases with wavelength from $\sim$ 100 Myr in the ultraviolet to more than 1 Gyr in optical-near IR. The timescale found for ionizing photons (i.e., recombination lines) is of the order of few Myr (Boissier 2012; Kennicutt & Evans 2012). There is some evidence (see next subsection) for a SFR constant over a few hundred Myr, at least in the nearby universe; standard calibrations can be calculated, as was done in the popular review of Kennicutt (1998). The calibration depends also strongly on the initial mass function (e.g., Meurer *et al.* 2009, Pflamm-Altenburg *et al.* 2009) as discussed in Section 3 of this review.

1.1.2. *Impact of a Varying Star Formation Rate*

The assumption of a constant star formation rate is quite strong. It is important to check that the condition is fulfilled before using standard calibrations. In the nearby universe the tight correlation found between the Hα and UV luminosity of galaxies is a strong argument in favor of a rather constant star formation rate for whole galaxies over

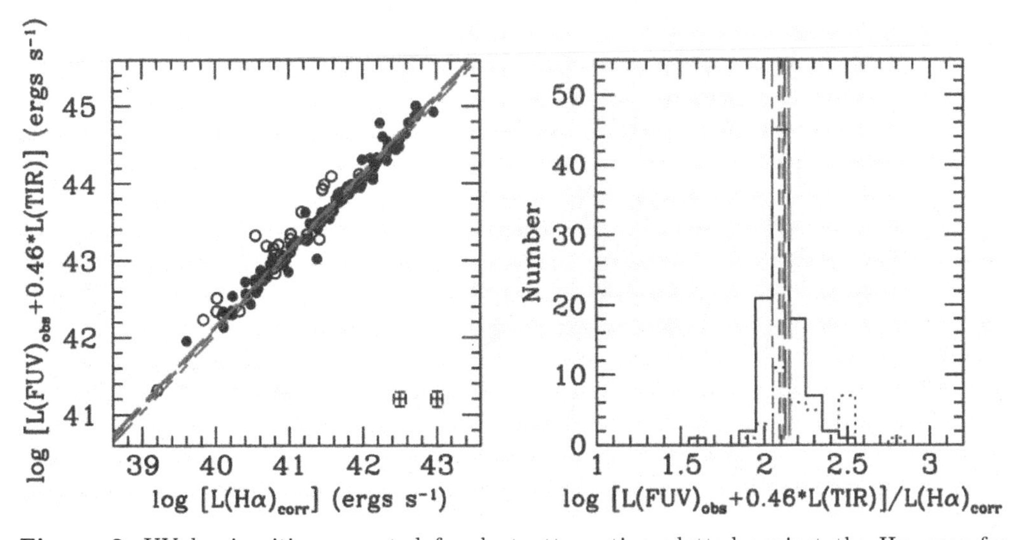

Figure 2. UV luminosities corrected for dust attenuation plotted against the Hα ones for galaxies in the nearby universe; both luminosities are corrected for dust attenuation. See Hao *et al.* (2011) for more details.

the typical timescale for the UV light to reach a steady state (i.e., ∼300 Myr) as illustrated in Figure 2. Nearby dwarf galaxies are likely to experience rapid changes in their star formation rates (Weisz *et al.* 2012), implying an apparent deficit of Hα emission with respect to the UV one. However, it was found difficult to distinguish between variations of the recent star formation history and that of the initial mass function by comparing the Hα and UV emission of these objects (Meurer *et al.* 2009; Boselli *et al.* 2009; Lee *et al.* 2009; Fumagalli *et al.* 2011; Weiz *et al.* 2012).

At high redshift, timescales become shorter and the assumption of a constant star formation rate over several hundreds of Myr may not be valid. It is nicely illustrated by Schaerer *et al.* (2013) who explored a large number of star formation histories described by declining or rising functions in addition to constant and delayed star formation. Their models and the corresponding ratio of the SFR to the bolometric stellar luminosity $L_{\rm bol}$ is shown in Figure 3. At $z \simeq 4$ Schaerer *et al.* show that the standard calibrations assuming constant star formation must be used with much caution, with strong implications on the SFR-stellar mass relation.

1.2. *Far-Infrared Calibrations*

In the *IRAS* and *ISO* eras (e.g., Hunter *et al.* 1986; Lehnert & Heckman 1996), infrared luminosities from wavelengths less than 200μm proved to be good star formation rate tracers. Since the successful launch of the *Spitzer Space Telescope*, infrared luminosities have been widely used as important tools to calculate the star formation rates of galaxies. Recent studies enabled by *Spitzer* observations indicate that mid-IR luminosities (such as 8μm and 24μm) of galaxies also have good correlations with 1.4GHz and Hα/Pα luminosities (e.g. Wu *et al.* 2005; Calzetti *et al.* 2007; Zhu *et al.* 2008; Kennicutt *et al.* 2009; Calzetti *et al.* 2010).

However, Alonso-Herrero *et al.* (2006) shows that the *Spitzer*-8μm-Pα relations are different for nuclear regions and HII regions of luminous infrared galaxies, possibly because of diffuse 8μm emission associated with nuclear regions. Also many dwarf galaxies do not follow the general correlation of star-forming galaxies (Wu *et al.* 2005). What

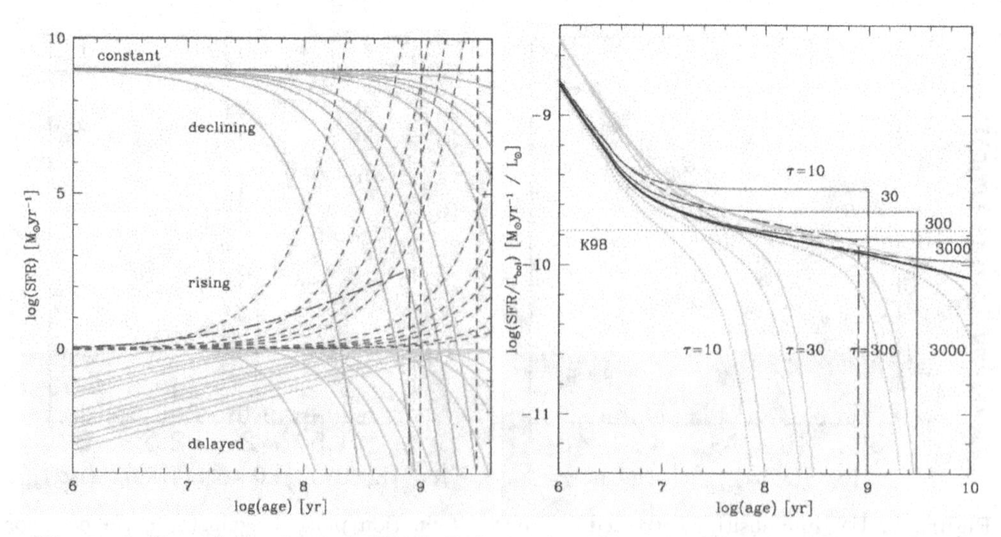

Figure 3. From Schaerer *et al.* (2012). Left panel: various star formation histories. Right panel: $SFR/L_{\rm bol}$ for the various star formation histories as a function of time.

would affect the mid-IR properties? Many efforts (Engelbracht *et al.* 2005; Wu *et al.* 2006; Wu *et al.* 2007; Siebenmorgen *et al.* 2004) have shown that both metallicity and AGN activity can affect the mid-IR properties of galaxies. This will bias the estimate of star formation rates. Some other factors, such as extinction and morphological type (Li *et al.* 2007) can also affect the calculation of star formation rates. In fact the most intrinsic factors involve the radiation field and dust. The strength and intensity of the radiation field, as well as the chemical content and the distribution of dust grain sizes, determine the deviation of the mid-IR-SFR relation for star-forming galaxies. Meanwhile, in many early-type galaxies, the contamination from the some diffuse regions, which are associated with old stellar populations rather than with current star formation, can also play important roles in the emission arising from mid-IR bands.

As the mid-IR luminosity can trace the dust-obscured star formation and the Hα luminosity can trace the unobscured star formation, a combined Hα/24μm luminosity is introduced by Calzetti *et al.* (2007) and can trace the star formation rates almost linearly (see also § 1.3). Zhu *et al.* (2008) and Kennicutt *et al.* (2009) confirmed Calzetti *et al.* (2007)'s results and obtain the similar relation for combined Hα/8μm luminosity. The combined Hα/mid-IR luminosities are more robust to flucgtuations in radiation field and dust content, and thus give an overall more reliable estimate of galaxy SFRs. For example, dwarf galaxies obey the same correlation as normal star-forming galaxies. However, as *Spitzer* was not a survey facility and its coverage was limited to a relatively small portion of the sky, the mid-IR bands cannot be widely used as star formation rate indicators. After the release of the *AKARI* full sky survey (Yuan *et al.* 2011), the mid-IR luminosity could be widely employed. Moreover, the *WISE* 22μm band is quite similar to the *Spitzer* 24μm band.

After cross-matching star-forming galaxy data between *WISE* and that of SDSS, a similar correlation was established between the combined Hα/22μm luminosity and extinction-corrected Hα luminosity (Figure 4) as Zhu *et al.* (2008) did for *Spitzer* 24μm. The coefficient $b = 0.020$ is similar to that of Zhu *et al.* (2008) and Kennicutt *et al.* (2009). For galaxies at higher redshift ($z > 0.35$) with only [OII]3727 emission line data available, the [OII]3727 luminosity was employed instead of the Hα luminosity; a good correlation between the combined [OII]3727/22μm luminosity and extinction-corrected Hα luminosity is shown in Figure 4.

At longer (sub-millimeter) wavelengths the properties of galaxies were still unclear before the *Herschel* era. The *Herschel* data opened a new window to studying the cool dust universe. For example, Dominguez *et al.* (2012) found the total IR luminosity is a good star formation tracer in nearby galaxies by using *Herschel* PACS data. However, whether or not a single SPIRE sub-millimeter band of *Herschel* can trace star formation of galaxies is still questionable. Using *Herschel* ATLAS science demonstration phase data cross-identified with SDSS DR7 spectra, a sample of 297 galaxies was constructed and classified into five morphological types; more than 40% of galaxies are peculiar/compact galaxies. The peculiar galaxies show higher far-infrared/sub-millimeter luminosity-to-mass ratios than other types. Wu and collaborators analyzed the correlations of far-infrared/sub-millimeter and Hα luminosities for different morphological types and different spectral types. The Spearman rank coefficient decreases and the scatter increases with the wavelength increasing from 100μm to 500μm. These trends indicate that a single *Herschel* SPIRE band is not adequate for tracing star formation activity in galaxies. Finally, AGNs contribute proportionally less to the far-infrared/sub-millimeter luminosity; AGN galaxies show little difference from star-forming galaxies in these diagnostics. However, the earlier type galaxies present significant deviations from the best-fitting of star-forming galaxies.

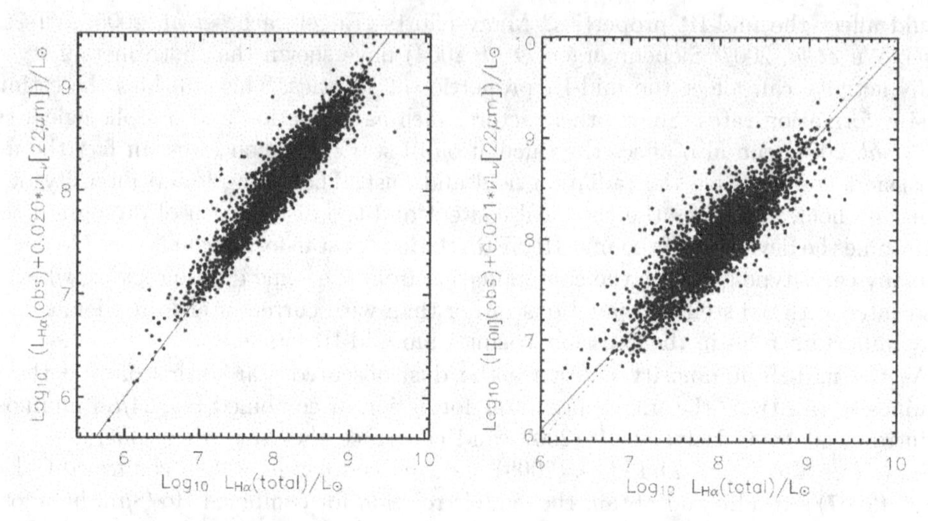

Figure 4. Correlations between the combined Hα/ *WISE*(22μm) luminosity, the combined [OII]3727/ *WISE*(22μm) luminosity, and the extinction-corrected Hα luminosity.

1.3. *Composite Indicators of Star Formation Rates*

Composite star formation rate indicators combine emission from unobscured star formation at wavelength λ_1 and emission from obscured star formation that reappears at wavelength λ_2,

$$L(\lambda_1)_{\text{corrected}} = L(\lambda_1)_{\text{observed}} + aL(\lambda_2), \quad (1.3)$$

where $L(\lambda_1)_{\text{corr}}$ has been previously calibrated as a star formation rate indicator (e.g., Kennicutt 1998; see also Figure 2). The ability to reliably combine two photometric measurements to yield an extinction-corrected SFR is a powerful tool for situations where spectroscopy-based extinction estimates are unavailable. Recent examples of such combinations include Hα+24 μm or IR+UV (Hirashita *et al.* 2003; Iglesias-Páramo *et al.* 2006; Buat *et al.* 2007; Calzetti *et al.* 2007; Elbaz *et al.* 2007; Daddi *et al.* 2007; Leroy *et al.* 2008; Zhu *et al.* 2008; Kennicutt *et al.* 2009; Liu *et al.* 2011; Wuyts *et al.* 2011). A direct corollary to this approach is that the ratio of the obscured and unobscured components can be used to estimate the amount of attenuation by dust (Prescott *et al.* 2007; Moore *et al.* 2010; Hao *et al.* 2011), i.e.,

$$A(\text{H}\alpha) = 2.5\log[1 + a(\text{H}\alpha)\nu L_\nu(24\mu\text{m})/L(\text{H}\alpha)] \quad (1.4)$$

$$A(\text{FUV}) = 2.5\log[1 + a(\text{FUV})L(\text{IR})/L(\text{FUV})]. \quad (1.5)$$

The latter relation involves the so-called "infrared excess" that is frequently used to quantify the amount of attenuation by dust in galaxies (e.g., Meurer *et al.* 1995; see Figure 5).

The dependence of the calibration constant (a(Hα) or a(FUV) in the above two instances) on region size is described in § 1.4. Table 1 summarizes these calibration coefficients from literature studies of either "global" (spatially-integrated) emission or localized emission, where available.

The scatter for all the different (global) composite indicators are similarly (and somewhat surprisingly) tight. In contrast, corrections for UV extinction via UV colors result in nonlinearities and scatters 2.5 times larger (Hao *et al.* 2011).

Table 1. Coefficients for Select Composite SFR Tracers

Composite SFR Relation	Coefficient a	Dispersion (dex)	Region	Reference
$L(\mathrm{H}\alpha)_{\mathrm{corr}} = L(\mathrm{H}\alpha)_{\mathrm{obs}} + aL(8\mu\mathrm{m})$	0.011 ± 0.003	0.11	Global	(1), (2)
$L(\mathrm{H}\alpha)_{\mathrm{corr}} = L(\mathrm{H}\alpha)_{\mathrm{obs}} + aL(24\mu\mathrm{m})$	0.020 ± 0.005	0.12	Global	(1), (2)
$L(\mathrm{H}\alpha)_{\mathrm{corr}} = L(\mathrm{H}\alpha)_{\mathrm{obs}} + aL(24\mu\mathrm{m})$	0.031 ± 0.006	0.3	Local	(3)
$L(\mathrm{H}\alpha)_{\mathrm{corr}} = L(\mathrm{H}\alpha)_{\mathrm{obs}} + aL(70\mu\mathrm{m})$	0.011 ± 0.001	0.24	Local	(4)
$L(\mathrm{H}\alpha)_{\mathrm{corr}} = L(\mathrm{H}\alpha)_{\mathrm{obs}} + aL(\mathrm{TIR})$	0.0024 ± 0.0006	0.09	Global	(1)
$L(\mathrm{FUV})_{\mathrm{corr}} = L(\mathrm{FUV})_{\mathrm{obs}} + aL(\mathrm{TIR})$	0.46 ± 0.12	0.09	Global	(5)
$L(\mathrm{FUV})_{\mathrm{corr}} = L(\mathrm{FUV})_{\mathrm{obs}} + aL(25\mu\mathrm{m})$	3.89 ± 0.15	0.13	Global	(5)

References: (1) Kennicutt *et al.* 2009; (2) Zhu *et al.* 2008; (3) Calzetti *et al.* 2007; (4) Li *et al.* 2013; (5) Hao *et al.* 2011

1.4. *SFR Calibrations and Their Dependence on Region Size*

Inspection of Table 1 shows a ~35% difference in the coefficient between the global and local Hα+24 SFR indicators. The reason for this discrepancy is tied to the sizescale over which the measurements are carried out. While the global coefficient stems from studies of spatially-integrated emission, the local coefficient derives from the Calzetti *et al.* (2007) study of "HII knots" of size ~200–600 pc. An important aspect of the Calzetti *et al.* study is the subtraction of the "background" emission from each local photometric measurement, which effectively removes the smooth galaxy emission that underlies each HII knot. Comparisons with models shows that perhaps up to 50% of a galaxy's infrared emission derives from interstellar dust grains heated by the diffuse radiation field that is unrelated to localized sites of star formation (Kennicutt *et al.* 2009).

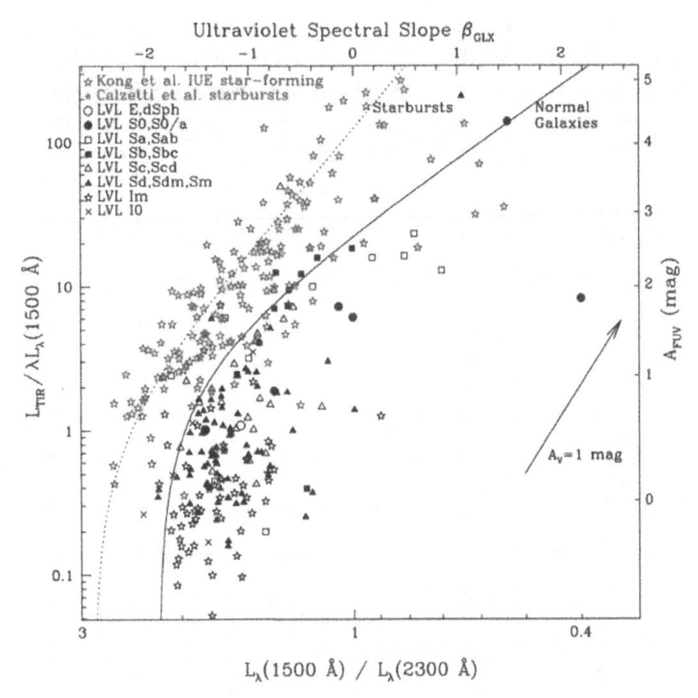

Figure 5. Example of the "IRX-β" relation for quantifying attenuation by dust as a function of observed UV color (from Dale *et al.* 2009). The relation is tighter for starbursting systems, and more loosely defined for normal star-forming galaxies.

Likewise, infrared monochromatic SFR indicators can also show a region size dependence, and for similar reasons as described above. For example, Li *et al.* (2013) find that the SFR calibration coefficient for 70 μm data changes by a factor ~2 for apertures spanning 0.1–10 kpc; their interpretation is that ~50% of the total 70 μm emission comes from dust that is heated by stellar populations not associated with current star formation (see Figure 6).

1.5. *Measuring star formation rates with Spectral Energy Distributions*

Star formation rates are basic output parameters of codes aimed at fitting spectral energy distributions (SED). By comparing data with models, one can attempt to derive some physical parameters related to the star formation history and dust attenuation in a homogeneous way and simultaneously for all galaxies of a given sample and at different redshifts. In practice, the current SFR, the stellar mass, and the amount of dust attenuation are commonly derived from broad-band photometry. Without spectral information and given the high degree of degeneracy in the SEDs, details about the star formation history are difficult to constrain implying an uncertainty on the determination of SFR. We refer to Walcher *et al.* (2010) for an exhaustive review on SED fitting techniques.

With the availability of mid- and far-IR data for large samples of galaxies, codes that combine stellar and dust emission to analyse SEDs are particularly useful. The standard approach consists of solving the radiation transfer in model galaxies to build self-consistent SEDs from the UV to the far-IR (see e.g. Bruzual & Charlot, 2003, for a detailed description of these models, or refer to Popescu *et al.* 2011, and Silva *et al.* 1998). These sophisticated models require complex calculations and are not directly applicable to large samples of galaxies. Another tool consists of models based on a simple energetic budget where the global dust emission corresponds to the difference between the emitted and observed stellar light (da Cunha *et al.* 2008; Noll *et al.* 2009). These codes, based on synthetic stellar population modeling and simple emission properties for the dust, are well suited to analysing large datasets at the expense of an oversimplification of the physical processes at work in galaxies. An example of spectral energy distributions generated by the CIGALE code (Noll *et al.* 2009; http://cigale.oamp.fr) is shown in Figure7.

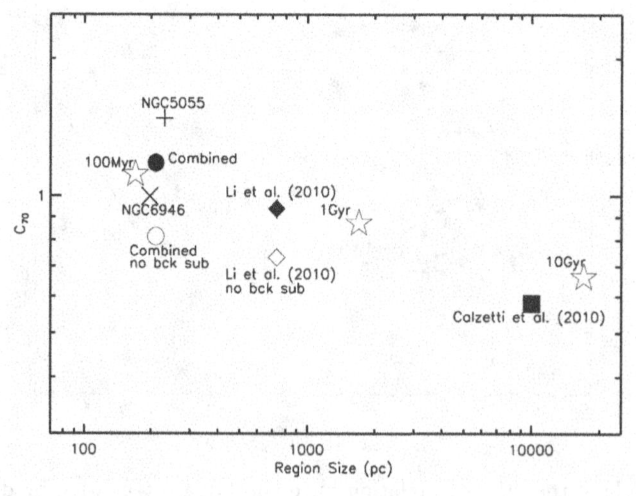

Figure 6. Region size dependence of the 70 μm SFR calibration coefficient (taken from Li *et al.* 2013).

1.6. *X-ray Emission as a Star-Formation Rate Tracer*

X-ray emission from normal galaxies (i.e., not dominated by luminous active galactic nuclei [AGN]) originates from a variety of components that vary dependent upon galaxy type and physical properties (see, e.g., Fabbiano 1989, 2006 for reviews). For late-type star-forming galaxies, X-ray emitting populations associated with both young and old stellar populations are expected to provide dominant contributions (i.e., high-mass and low-mass X-ray binaries, supernovae and their remnants, hot gas from starburst flows, and young stars). Passive early-type galaxies, without significant star-formation activity, have X-ray emission dominated by low-mass X-ray binaries (LMXBs) and hot X-ray emitting gas (e.g., O'Sullivan *et al.* 2001; Gilfanov 2004; Boroson *et al.* 2011). For star-forming galaxies, it has been observed that the X-ray power output is directly correlated with the emission from other wavebands that are associated with star-formation activity (e.g., UV, Hα, total IR, and radio), suggesting that X-ray emission may provide a useful measure of the current SFR in galaxies (hereafter the X-ray/SFR correlation). Such a prospect is attractive from a practical standpoint, since X-ray emission is penetrating in nature, and therefore has the potential to provide an independent view of star-formation activity in galaxies that does not suffer significantly from absorption. Understanding the calibration of the X-ray/SFR correlation, its scatter, and its cosmic evolution has been a rapidly advancing area of research. In this section we briefly highlight some recent results with emphasis on hard X-ray emission as a tracer of the SFR.

1.6.1. *The X-ray/SFR Correlation in the Local Universe*

The local X-ray/SFR correlation has been calibrated for star-forming galaxies over ~5 orders of magnitude in SFR (e.g., Ranalli *et al.* 2003; Colbert et al. 2004; Persic & Rephaeli 2007; Lehmer *et al.* 2010; Iwasawa *et al.* 2009; Pereira-Santaella *et al.* 2011; Mineo *et al.* 2012a,b). Given the multiple components contributing to the X-ray emis-

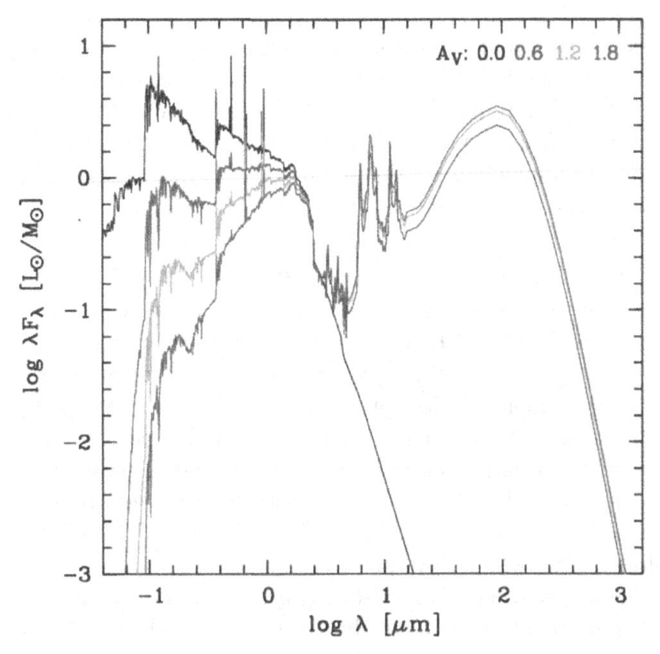

Figure 7. Example of SEDs generated by the CIGALE code assuming an age of 1 Gyr, an exponentially decreasing SFR with a decay time $\tau = 10$ Gyr and different dust attenuation in the V band (from Noll *et al.* 2009).

sion from normal galaxies, recent calibrations of the X-ray/SFR correlation have focused on separating these components either spectrally (e.g., Lehmer *et al.* 2010 make use of 2–10 keV emission dominated by X-ray binaries), or by spatially decomposing the emitting components into point sources (X-ray binaries) and diffuse emission (hot gas, unresolved binaries, and stars; see, e.g., Mineo et al. 2012a,b). Thanks to resolved imaging of nearby galaxies with *Chandra* and *XMM-Newton*, we know that the X-ray/SFR correlation at energies above ~1 keV is primarily driven by bright X-ray binaries. Of the bright X-ray binaries in star-forming galaxies, LMXBs and HMXBs are expected to provide substantial contributions, which are expected to trace the relatively old and young stellar populations, respectively.

To account for the LMXB and HMXB contributions to the galaxy-wide hard X-ray emission, Lehmer *et al.* (2010) gathered a *Chandra*-observed sample of 66 nearby normal galaxies with good measurements of SFR and $M_\star$ and fit their data to the following relation:

$$L_{2-10\ \mathrm{keV}} = \alpha M_\star + \beta \mathrm{SFR} \tag{1.6}$$

where $L_{2-10\ \mathrm{keV}}$ is the galaxy-wide integrated 2–10 keV power output, and α and β are constants, which to first order account for the emission of LMXBs and HMXBs, and their scaling with $M_\star$ and SFR, respectively. For a Kroupa (2001) IMF, the best-fit values of these constants are $\alpha = (9.05 \pm 0.37) \times 10^{28}$ ergs s^{-1} $M_\odot^{-1}$ and $\beta = (1.62 \pm 0.22) \times 10^{39}$ ergs s^{-1} $(M_\odot\ \mathrm{yr}^{-1})^{-1}$. The residual scatter in this relation is $\approx$0.34 dex, which is a $\approx$0.14 dex improvement over a direct scaling of a direct linear scaling of $L_{2-10\ \mathrm{keV}}$ vs. SFR. More recently, Mineo *et al.* (2012a) studied in detail nearby galaxies with high specific-SFRs (SFR/$M_\star > 10^{-10}$ yr^{-1}) and utilized spatial decomposition of X-ray point sources to directly compute the HMXB X-ray luminosity functions of each galaxy. Through this independent analysis, they provided a direct calibration for how the HMXB luminosity of galaxies scales linearly with SFR (i.e., β from Equation 1.6). This analysis produced a scaling similar to that found by Lehmer *et al.* (2010) once differences between X-ray bandpass and IMF were accounted for. Despite the general agreement between the X-ray/SFR correlation, there remains significant scatter on the order of $\approx$0.3–0.4 dex. The remaining scatter is likely to be both intrinsically statistical and physical in nature. Some unaccounted for physical dependencies to the correlation in Equation 1.6 include dependencies on stellar age and metallicity (see, e.g., the models from Fragos *et al.* 2013).

1.6.2. *Cosmic Evolution of the X-ray/SFR Correlation*

With the advent of ultradeep surveys with *Chandra* (e.g., the *Chandra* Deep Fields; see Brandt & Hasinger 2005 and Brandt & Alexander 2010 for reviews), it has become possible to study X-ray faint, cosmologically-distant normal galaxies, the majority of which are powered by star-formation processes. At the flux limit of the deepest X-ray survey to date, the 4 Ms *Chandra* Deep Field-South (CDF-S; Xue *et al.* 2011), the sky density of normal galaxies may already be larger than that of AGN (Lehmer et al. 2012). Initial investigations have placed first-order constraints on the evolution of the normal-galaxy X-ray luminosity function (e.g., Norman et al. 2004; Ptak *et al.* 2007; Tzanavaris & Georgantopolous 2008), which have provided evidence for a rapidly increasing normal-galaxy X-ray luminosity density with redshift (proportional to $[1+z]^{2-4}$), consistent with the rising star-formation rate density of the Universe (e.g., Hopkins et al. 2004).

Analysis of the X-ray detected normal galaxies have shown that the X-ray/SFR correlation appears to hold out to $z \approx 1$ (e.g., Cowie *et al.* 2012; Mineo *et al.* 2012c; Ranalli *et al.* 2012; Symeonidis *et al.* 2012; Vattakunnel et al. 2012). Powerful X-ray stacking

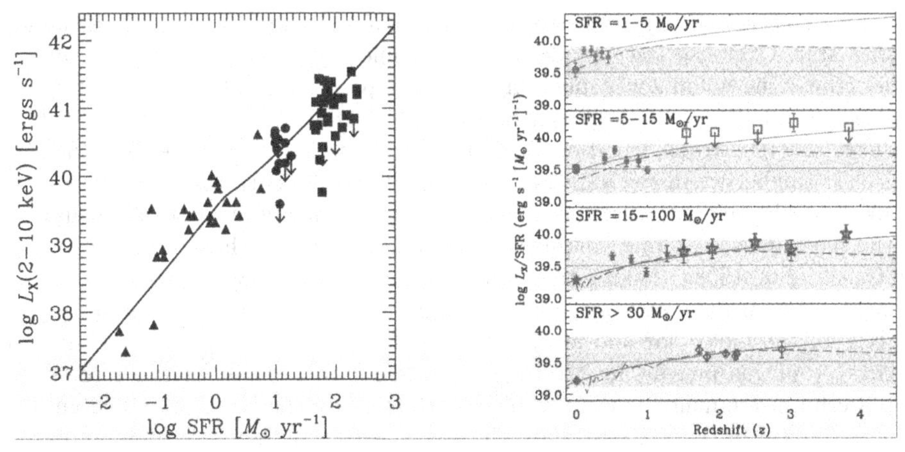

Figure 8. (*Left Panel*) The local X-ray/SFR correlation from Lehmer *et al.* (2010). The data shown in this plot are compiled from local normal galaxies (*triangles*; Colbert *et al.* 2004), LIRGs (*circles*; Lehmer *et al.* 2010), and LIRGs/ULIRGs (*black squares*; Iwasawa *et al.* 2009). The solid line represents the expected correlation given how HMXB and LMXB emission scale with SFR and $M_\star$, respectively, and how $M_\star$ scales with SFR. (*Right Panel*) Mean values of $L_{2-10 \text{ keV}}$/SFR from the stacking analyses presented in Basu-Zych *et al.* (2012). The best-fit redshift-dependent parameterization is shown as solid curves (Equation 1.6.2), and the population synthesis prediction from Fragos et al. (2013) is shown with dashed curves.

studies of large populations of normal star-forming galaxies selected by morphology and galaxy physical properties (e.g., SFR and stellar mass) have provided more sensitive insight into how the X-ray/SFR correlation has evolved over the majority of cosmic history. Using the 4 Ms CDF-S, Basu-Zych *et al.* (2012) utilized results from the literature (e.g., Reddy & Steidel 2004; Laird *et al.* 2006; Lehmer *et al.* 2008) and large samples of thousands of Lyman-break galaxies to show that the X-ray/SFR correlation undergoes very mild evolution out to $z \approx 4$, which can be parameterized as follows:

$$\log L_{2-10 \text{ keV}} = A \log(1+z) + B \log \text{SFR} + C, \tag{1.7}$$

where $A = 0.93 \pm 0.07$, $B = 0.65 \pm 0.03$, and $C = 39.80 \pm 0.03$ are fitting constants. Additional indirect support for mild evolution of the X-ray/SFR correlation with redshift has been provided by Dijkstra *et al.* (2012) who used the observed SFR density evolution of the Universe to show that the unresolved SB CXRB can be fully explained by a L_X/SFR ratio $\propto (1+z)^b$, where b is constrained to be less than 1.4; consistent with Equation 1.6.2. The mild increase in the X-ray/SFR correlation is predicted from the X-ray binary population synthesis modeling of Fragos *et al.* (2013) (see Fig. 1). The population synthesis models predict that since higher redshift galaxies have lower metallicities, this enables the production of a more massive and numerous compact object population and more luminous X-ray binaries for a given SFR (see, e.g., Linden *et al.* 2010). Such a dependence of the L_X/SFR ratio on metallicity may provide some physical explanation for some of the scatter observed in the local X-ray/SFR correlation. Future studies of both nearby and distant galaxies will be needed to more confidently parameterize the effects of metallicity (and other physical parameters) on X-ray binary emission.

2. Local Star Formation

In a Cosmological or Extragalactic context, we are interested in understanding why stars form where they do, whether the efficiency varies, and what factors influence the initial mass function (IMF) of the stars, rather than the details of the collapse of a cloud core to a star. Over the last decade, it has become very clear that the star formation rate per co-moving volume was much higher in the past, some 10 or 20 times the current rate at a redshift of $z \sim 0.5 - 1$ (e.g., Heavens *et al.* 2004; Wilkins *et al.* 2008). In turn, this shows that the transformation rate of gas into stars was considerably (factor of a few at least) higher when the universe was roughly half its current age. Galaxies at that time were smaller and of lower metallicity, such that naively at least one would expect that the molecular-to-atomic gas mass ratio would be lower than today (Young *et al.* 1989; Casoli *et al.* 1998), making the higher efficiency even more surprising. This may or may not be the case and one of the issues for the next generation of telescopes—ALMA and SKA for the molecular and atomic gas respectively—is the measure of the amount of each type of gas present in "average" spirals out to redshift unity or so. Even for such powerful instruments, it will not be easy to measure the HI, CO, and submillimeter continuum fluxes in order to estimate the gas masses. Furthermore, it is important to spatially resolve the galaxies, separating the central kpc from the disk due to the different gas temperatures and densities, in order to physically interpret the observational results. However, for galaxies in the Local Group, at less than 1 Mpc, molecular clouds can be resolved with large antennae and interferometers. Recent studies (e.g., Engargiola *et al.* 2003; Rosolowsky *et al.* 2008; Gardan *et al.* 2007; Gratier *et al.* 2010; Leroy *et al.* 2006; Nieten *et al.* 2006), following the precursor work by Neininger *et al.* (1998) and Wilson *et al.* (1997), have provided interesting results, enabling a comparison of the ISM of galaxies of a variety of sizes, morphologies, and metallicities. Fortunately, within 1 Mpc we have a number of small star-forming galaxies—M33, the Magellanic Clouds, NGC 6822, and IC10, in addition to the Milky Way and Andromeda (M31).

An equally interesting alternative is that the IMF has evolved substantially, such that the measured star formation rates simply show that the number of *massive* stars has increased. The parameter space for this alternative is being reduced due to the number of concordant studies made at different wavelengths and the construction of galaxy Star Formation Histories. Nonetheless, the assumption, often implicit, that the IMF remains constant with cosmological epoch, host galaxy, or local conditions, requires further testing whenever possible.

It is generally believed that the star formation cycle starts with clouds of atomic gas (but see Allen *et al.* 1997), some of which condenses to form molecular hydrogen (H_2), from which the stars form in the densest fragments. The most massive stars then create HII regions and photo-dissociation regions (PDRs) and finally disperse much of the cloud. In a way probably linked to the large-scale dynamics (e.g., spiral arms), the scattered material forms clouds anew.

The Atomic gas: The atomic hydrogen is quite extended in galaxies, extending to several times the optical extent in isolated objects. The HI has two stable phases—one warm ($\sim$ 5000K) and diffuse and the other cool ($\sim$100K) and much denser. Presumably it is this phase which is linked to molecular cloud formation and the subsequent stars. Deep high resolution HI measurements, notably by Braun *et al.* (1997), have shown that the cool phase dominates within the optically bright parts of spirals but that the outer parts (beyond R_{25}) contain essentially gas in the warm phase. However, the recent GALEX UV observations show that star formation proceeds further out in galactic disks than was believed previously and is spatially correlated with the HI, indicating that there must be a mechanism compressing the warm HI such that it passes into the cool dense phase and then into molecular gas and stars, even where the stellar content is very low. It is not currently clear whether the stellar IMF is necessarily truncated, because the UV is observed further out than Hα emission, or whether high mass stars are rare and short-lived enough that their probability of detection is very low (Boissier *et al.* 2007). Going to smaller, lower metallicity, and more irregular objects, the warm phase appears to dominate throughout as narrow HI line profiles are not seen. In general, molecular gas (see below) is not observed in these objects but star formation is clearly present.

In spirals, the vast majority of the star formation takes place within R_{25} but in isolated objects the rate of flow of gas from the outer disk to the inner disk is not sufficient to replenish the gas gone into stars. Recent very deep observations (Sancisi *et al.* 2008; Oosterloo *et al.* 2007) show that extra-planar gas is present in spirals. Does it come from the gradual inflow from past tidal encounters or is it gas that has cooled out from the warm ionized halo gas, becoming neutral and denser than its surroundings and falling towards the disk? The upcoming long-wavelength instruments, SKA and the pathfinders, will extend this work, particularly the southern instruments which have access to the Magellanic Clouds, enabling truly high-resolution observations of the atomic component and how its properties change in the vicinity of molecular clouds.

The Molecular gas: Molecular gas is the direct fuel for star formation. The primary tracer of H_2 is the CO molecule, the most abundant with an electric dipole moment, observed in its rotational transitions at 115 GHz and multiples thereof. A secondary tracer is the continuum emission of the cool dust near the Rayleigh-Jeans part of the spectrum. Because both are metal-dependent, it is unclear to what extent and how these tracers can be used to estimate the H_2 mass and thus the Star Formation Efficiency (SFE), the rate of star formation per unit molecular gas, especially in low-metallicity systems. One naturally expects that if the nature of star formation varies with e.g., galaxy morphology, then this will be apparent in the SFE, whether due to a change in the IMF (and thus the estimate of the SF rate) or to a real change in the efficiency of conversion of H_2 into stars. It is thus important to properly estimate the H_2 mass.

The correlation both in space and velocity between molecular clouds and the HI is excellent at large scales but breaks down when close to the cloud scale (~100 pc). A major outstanding question is what provokes the transformation of HI into H_2. Comparing the total gas surface density with the SFR as determined from the Hα brightness, Kennicutt *et al.* (1989) found that a global Schmidt law of $\sigma_{SF} \propto \Sigma_{gas}^{1.4}$ reproduced the data in his sample and that the Toomre (1964) stability criterion provided a good estimate of the gas surface density threshold for star formation. It now appears clear, with better CO data than was available at the time of those articles, that within individual galaxies the SFR follows the H_2 surface density at scales larger than that of GMCs (e.g., Bigiel *et al.* 2008). Many lines of reasoning point to a threshold HI column density for H_2 and star formation (e.g., Schaye *et al.* 2004; Martin & Kennicutt 2001) but so far this has not been unequivocally observed. And it is clear that while H_2 (CO in most cases) only forms where HI is present and is more likely to form where HI column densities are higher, for any given (high) HI column density there can be quite a bit of H_2 or often none at all. Gardan *et al.* (2007) showed that in M33 the likelihood of H_2 formation per unit gas surface density diminished with radius beyond about 4 kpc. In addition to the Toomre criterion, pressure has been proposed (Elmegreen & Parravano 1994; Blitz & Rosolowsky 2006) as the factor driving H_2 and star formation. While the formalisms proposed provide reasonable results in the bright regions of spirals, they do not predict discrete features such as cut-offs or rings of star formation.

An interesting result, mentioned in passing by Blitz & Rosolowsky (2006) and Leroy *et al.* (2006) and focussed on by Gardan *et al.* (2007) followed by Gratier *et al.* (2010) and Braine *et al.* (2010), is that the small, subsolar metallicity, galaxies M33 and IC10 seem to have a higher SFE than the large spirals that dominate the universe today (Murgia *et al.* 2002; Kennicutt 1998). Dib *et al.* 2011 have shown that in low metallicity environments the stellar winds are weaker, such that molecular clouds stay intact for longer and transform more of their H_2 into stars although this is probably not the only cause of the increased SFE. An increase in the SFE is at least partially responsible for the increase in the star formation rate per co-moving volume with redshift (Combes *et al.* 2012).

While the transformation of HI into H_2 appears to be a medium to large scale process, affected by spiral arms, pressure (related to the stellar surface density), and large-scale compression such as by SN shock waves, the collapse to form stars appears to be a rather local process; this is supported by the apparent constance of the stellar IMF. The basic scenario for low-mass (i.e., solar-like) stars—gravitational collapse of a cool dense core, formation of a disk and bipolar outflows perpendicular to the disk along the magnetic field lines accompanied by an increase of the central temperature, dispersal of disk from stellar winds and arrival of the star on the ZAMS—occurs at a tiny fraction of a parsec. Massive star formation is more problematic as for spherical accretion, the radiation pressure from the protostar should stop stellar growth. Several processes can overcome this problem but require a dense environment and high accretion rates, possibly driven by large scale (1–100 parsec) inflow along the filaments making up large-scale cloud structure. For these reasons, combined with the generally less massive clouds (Gratier *et al.* 2012; Rosolowsky *et al.* 2005) and clusters (Corbelli *et al.* 2011 in the outer parts of spirals, and the idea that the mass of the most massive star in a cluster varies with the cluster mass (Weidner *et al.* 2011), it appears that massive stars are more easily formed in the inner parts of spirals. Tracing the upper end of the stellar mass function is difficult because these stars have very short main-sequence lifetimes and remain embedded, making it difficult to separate a single very massive star from several less massive stars.

Star formation can be detected in many ways at many different wavelengths—(far) ultraviolet, Hα, mid- and far-infrared emission from dust grains, thermal and non-thermal radio continuum emission, as well as other recombination lines and broad-band colors. Some suffer from extinction or require the presence of metals or magnetic fields. It has been known for decades now that star formation generates bipolar outflows. Observing clear signatures of inflow or accretion, although necessary for star formation, has not proved so easy. As such, directly measuring accretion rates or identifying gas inflow is one of the limiting factors in studying star formation and particularly massive star formation.

Galactic and Extragalactic star formation studies are now meeting each other as galactic astronomers try to observe star formation throughout the Galaxy (Schuller *et al.* 2009; Molinari *et al.* 2010) and on large scales as extragalactic astronomers resolve star forming regions in nearby galaxies. The latter studies have lower spatial resolution but do not suffer from the distance uncertainties or extinction inherent to studies of our galactic plane. Instruments with high spatial and spectral resolution such as ALMA are expected to contribute greatly to the field of star formation, both by observing protostars in detail and by observing the nearest galaxies with the same linear resolution as earlier studies of the Milky Way.

3. Initial Mass Functions

The initial stellar mass function (IMF) determines the number of newly formed stars, dN, per (initial stellar) mass interval, dm. The IMF is not measurable (Kroupa *et al.* 2013) but has been inferred on a star-by-star basis only in star clusters and associations, i.e. on parsec scales. Access to the whole stellar mass range down to the regime of brown dwarfs is only possible in the local region of the Milky Way, whereas counting massive stars is possible to some extend in nearby Local Group galaxies. Because stars of different masses contribute differently to photometric pass-bands, the knowledge of the functional form as well as the mass limits of the IMF are of fundamental importance for the determination of star formation rates of galaxies. As these galaxies lie beyond the region where star counts are possible, it has to be assumed that the IMF determined in the Milky way can be applied on extragalactic systems.

3.1. *The IMF in correlated star-formation events*

Stars form in the densest regions of molecular clouds. These correlated star-formation events (CSFEs) comprise from a few to millions of stars formed within characteristically 1 Myr and within a region with a radius of less than a pc (Marks & Kroupa 2012). Some of these CFSEs, commonly also sinply known as embeeded star clusters, emerge to become long-lived, gravitationally bound star clusters.

The empirical IMF inferred from Milky Way CFSEs shows a remarkable universality in its functional form. The form of this universal or *canonical* IMF is most simply described by a two-part power-law in the stellar regime,

$$\xi(m) = \frac{dN}{dm} \propto m^{-\alpha_i} , \tag{3.1}$$

with slopes $\alpha_1 \approx 1.3$ between 0.07 $M_\odot$ and 0.5 $M_\odot$, and $\alpha_2 = 2.3$ above 0.5 $M_\odot$, where the value of α_2 is commonly referred to the "Massey-Salpeter" power-law index. Note that this definition of the IMF refers to single objects, i.e. all multiple systems such as binaries or triples are split up to single stars before counting (Kroupa *et al.* 2013). The upper mass limit is discussed in detail in Sec. 3.3.

Different functional forms are also used in the literature. The most widely used form beside the two-part power law is the log-normal form for stellar masses below 1 $M_\odot$ but with the same power-law part for $m > 1\,M_\odot$ (Kroupa *et al.* 2013). This leads to a mathematically more complex object without the gain of physical reality. The physical origin of the form of the IMF is still unclear and subject of theoretical and observational research. But Andre *et al.* (2010) point out the remarkable similarity between the pre-stellar core mass function and the stellar IMF, "suggesting a $\sim$ one-to-one correspondence between core mass and star/system mass with $M_{*,\mathrm{sys}} = \epsilon\, M_{\mathrm{core}}$ and $\epsilon \approx 0.4$ in Aquila."

3.2. *The sub-stellar mass regime*

It has been known for some time that brown dwarfs (BDs) are unlikely to form from direct gravitational collapse in a molecular cloud such that the observationally deduced mass function contains a significant surplus of brown dwarfs (Padoan & Nordlund 2002, Padoan *et al.* 2007, Andersen *et al.* 2011, Hennebelle 2012). Interestingly, several authors have claimed good agreement with the empirical (Chabrier 2003) IMF, but scrutiny of the published work shows consistently significant disagreement (Thies, Pflamm-Altenburg & Kroupa, in prep.). The reason is that the distribution of density maxima in a cold but turbulent molecular cloud has very few peaks which can collapse through eigengravity at the mass scale of a BD such that not much further material is accreted. Although Withworth *et al.* (2007) and Whitworth *et al.* (2010) argue that BDs form a continuous extension of the stellar distribution, the observational and theoretical evidence they provide strongly suggests that BDs and stars have different properties in terms of their pairing (Thies & Kroupa 2007, Thies & Kroupa 2008).

Recent work, however, came to the conclusion that the pairing properties of BDs and stars do not show that BDs are significantly different from stars (Reggiani & Meyer 2011). But this result originates from the usage of the low-mass IMF from Bochanski *et al.* (2010) who fitted a log-normal with $\sigma = 0.34$ to a sample of 14 million stars between 0.1 and 0.8 $M_{\odot}$. When extrapolating this very tight log-normal function into the BD-regime drawing BDs is highly suppressed and few star-BD binaries are created in Monte-Carlo simulations, in agreement with the observations. On the other hand, when using the IMF which describes the BD-mass regime properly, a significant fraction of BD-star binaries are expected (Kroupa *et al.* 2003, Thies & Kroupa 2007, Thies & Kroupa 2008). The usage of the Bochanski-IMF for BD Monte-Carlo experiments is not understandable as Bochanski *et al.* (2010) explicitly write in their abstract, "*We stress that our results should not be extrapolated to other mass regimes*".

Kroupa *et al.* (2003) have tested the hypothesis that BDs and stars follow the exact same distribution functions and exclude this hypothesis with very high confidence. The various flavours of BDs that can in principle arise (collisional, photo-evaporated, ejected embryos) have been discussed (Kroupa & Bouvier 2003b) with the result that in the present-day star-forming conditions mostly the ejected embryo flavour dominates. The original suggestion of this scenario has been updated by Stamatellos *et al.* (2007), Thies *et al.* (2010), and Basu & Vorobyov (2012) by the argument that the gravitationally pre-processed material in outer accretion disks is able to cool sufficiently rapidly upon compression to allow direct gravitational collapse at the BD mass scale. The resulting IMF of BDs compares remarkably well with the observationally deduced BD IMF ($\alpha_0 \approx 0.3$). The resulting binary properties of BDs are also accounted for naturally (Thies *et al.* 2010).

The BD IMF is thus a nearly flat power-law from the opacity limit for fragmentation ($m_l \approx 0.01\,M_{\odot}$) to an upper limit which transgresses the hydrogen burning limit. In principle, arbitrarily massive "BDs" can form in very massive disks around massive stars such that here the origin of stars vs BDs becomes blurred. Because massive stars are exceedingly rare the stellar population formed through this disk-fragmentation channel is negligible in comparison to the "normal" stellar population which results from direct molecular cloud fragmentation.

Thus in order to correctly account for a stellar population with BDs most of the BD population must be added in terms of a separate distribution function, as is also the case for planets which follow their own mass distribution. The BD IMF can be expressed as a nearly flat power-law with a continuous log-normal extension from the stellar regime being ruled out.

A single continuous log-normal form for the stellar/BD IMF such as the Chabrier form is therefore ruled out by the data.

3.3. *The high-mass end of the IMF — a size-of-sample effect?*

It is widely assumed that the IMF can be interpreted as an environmentally independent probability density distribution function. This IMF has a constant functional form (see Sec. 3.1) and a constant upper stellar mass limit of $\approx$150 $M_{\odot}$ (Weidner & Kroupa 2004, Oey & Clarke 2005, Figer 2005, Koen 2006). In a recent study Crowther *et al.* (2010) find that a few massive stars in the very young massive cluster R136 greatly exceed an upper limit of 150 $M_{\odot}$. However, these few outliers can be understood as the stellar merger product in the very dense central region of R136 although the initial stellar masses were smaller than 150 $M_{\odot}$ (Banerjee, Kroupa & Oh 2012).

In this context drawing an O star from the IMF is much less probable than drawing a low-mass star. As a consequence, in low-mass star clusters the IMF is not fully populated and they are void of massive stars due to the size-of-sample effect. However, serious doubts arose over the last decade that the IMF can be interpreted as a pure invariant probability density distribution function.

The high-mass slope, α_3 ($m > 1\,M_{\odot}$), fitted to the set of high-mass stars in individual CSFEs, shows a distribution centred around the Salpeter-Massey index of $\alpha_3 = 2.3$. At first sight this seems to be consistent with the expectation from random drawing. However, despite their youth the observed CSFEs are already dynamically evolved. Therefore, the initial α_3-spread due to random drawing gets even larger as a result of dynamical evolution. N-body simulations of star clusters show that the α_3-distribution resulting from random drawing and dynamical evolution is broader than the observed distribution of the high-mass slope (Kroupa 2001, Kroupa 2002). There are two possibilities for solving this discrepancy. The predominant finding of a Salpeter-Massey value of $\alpha_3 = 2.3$ is a sociological effect. Or the initial α_3-distribution must be narrower than expected from random drawing. But this means that the IMF in CSFEs can not be interpreted as a pure invariant probability density distribution function (Kroupa *et al.* 2013). Instead, if the IMF is a probability distribution function, it must be a constrained one. Or it may not be a probability density function at all.

If the most-massive stars in a CSFE are determined by the size-of-sample effect, as expected from random drawing, then low-mass clusters are typically void of massive stars. Very few CSFEs would exist in which the stellar populations are dominated by one massive O star. Furthermore, some massive stars would then have formed in complete isolation. De Wit *et al.* (2004, 2005) analyse the Galactic-O-star catalogue and found that 4 per cent of the Galactic O stars are candidates for massive star formation in isolation. Gvaramadze *et al.* (2012) analyse these candidates thoroughly. The majority are identified as being runaways based on bow-shock detections and using results from orbit back-tracing from Schilbach & Röser (2008). The remaining set of candidates is then reduced to one per cent. This little number of O-stars, for which a parent star cluster can not be found, is in agreement with the expectation of *two-step ejection*, where a massive binary is ejected dynamically from a young star cluster with a subsequent disintegration in the Galactic field due to supernova explosion of the primary. As the new moving direction of the secondary is arbitrary runaway can not be traced back to its parent star cluster (Pflamm-Altenburg & Kroupa 2010). There is therefore no significant evidence that massive stars can form in isolation.

If the most-massive-star in a CSFE is determined by the size-of-sample effect than a relation between the mass of the most-massive stars and the total stellar mass of the CSFE with a broad scatter emerges. In order to compare this prediction with the observations an unbiased set of CSFEs is required and are selected by the following two criteria: i) All star clusters must be younger than $\approx$4 Myr in order to exclude star clusters where the most-massive star has already exploded in a supernova (such clusters would have a less massive most-massive star). ii) All star clusters must be still embedded in their natal gas (otherwise the star clusters can have already expanded due to the effect of gas expulsion (Kroupa *et al.* 2001) and may have lost a significant fraction of their stars and are therefore reduced in mass). A thorough analysis by Weidner & Kroupa (2006) and Weidner *et al.* (2010) shows that the observed relation between the most-massive star and the stellar mass of the CSFE lies statistically significantly below the predicted relation from the random sampling ansatz, therewith ruling out an invariant probability density distribution interpretation of the IMF. Considering dynamical ejections of the most-massive stars, this conclusion is not affected (Oh & Kroupa 2012, Pflamm-Altenburg - in prep.).

Contrary to the results above, Maschberger & Clarke (2008) came to the conclusion that *"the data are not indicating any striking deviation from the expectations of random drawing"*. But they restricted their analysis to the low-mass star cluster data from Weidner & Kroupa (2006) and complemented them with additional low-mass star clusters including the data set from Testi *et al.* (1997, 1998). After finding a KS-probability of 10^{-14} for an agreement with random sampling they removed the Testi-data from their sample. After subsequently reducing the sample even further they finally end up obtaining a KS-probability of 20 per cent for an agreement with random sampling. But, the need of giving up the picture of a randomly sampled constant IMF described above is driven by the result from the whole data of most-massive stars in star clusters and not only by a tiny subset of low-mass star clusters as employed by Maschberger & Clarke (2008). Furthermore, they have tested no alternatives. And finding that model A explains the observations does not imply that model B is immediately shown to be wrong. It should be mentioned here that the work by Maschberger & Clarke (2008) has been wrongly cited by Lamb *et al.* (2010). They write: *"Maschberger & Clarke (2008) complement the WK06 data set with a sample of very small clusters from Testi et al. (1997) and find that the resultant ensemble of clusters does not significantly deviate from the expectations of a universal stellar IMF, when examining the correlation between the number of stars in a cluster (N) and* $m_{\max}$*"*. Such false citations to work which selects data in order to achieve a pre-conceived result, as seems to be the case, lead to appreciable damage to scientific progress.

The observational result that the formation of massive stars require a compact high-mass environment is supported by the recent work by Hsu *et al.* (2012). They examine the young stellar content of the low-density star-forming region L 1641 in the Orion A cloud and find that it is deficient in O and B stars to a 3–4 σ significance level. This is in-line with the result from hydrodynamical star-formation simulations, where the upper mass limit of stellar groups scales with the number of stars of these subgroups (Maschberger *et al.* 2010).

We can finally summarise that there is very strong evidence that the IMF can not be treated as a pure invariant probability density distribution function and that the formation of high-mass stars is not characterised by the size-of-sample effect but requires a massive and compact. ie. dense, environment.

3.4. *IMF variations under extreme conditions*

The above discussion considers the IMF in young star forming regions in the Milky Way. The functional slope of the IMF seems to be universal whereas the population of the high-mass regime can not be driven by pure statistics. However, recent analyses of observations show that the functional form of the IMF may have been different under extreme conditions, which are not present in the star forming regions in the vicinity of the Milky Way considered above.

Two indications exist that the IMF may have been bottom-heavy in cases of high metallicity: i) A hint at a possible variation of the IMF in the MW has emerged due to present-day star-formation events possibly producing more low-mass stars than previously. This has been quantified as a metallicity dependence, $\alpha \approx 1.3 + 0.5[\mathrm{Fe/H}]$ (Kroupa *et al.* 2013). ii) From the study of massive elliptical (E) galaxies, it has emerged that the IMF may have been significantly bottom heavy when the E galaxies formed. Based on spectral analyses Cenarro *et al.* (2003) inferred $\alpha = 3.41 + 2.78[\mathrm{Fe/H}] - 3.79[\mathrm{Fe/H}]^2$ (for $0.1 < m/M_\odot < 100$, although not explicitly stated in the paper). iii) A more recent analysis by van Dokkum & Conroy (2010, 2011) also suggests an increasingly bottom heavy IMF with increasingly massive E galaxies. This may be related to the postulated cooling-flow-accretion population of low-mass stars (Kroupa & Gilmore 1994).

An unusual IMF in massive ellipticals is also reported by Cappellari *et al.* (2012) based on their larger mass-to-light ratios. The authors conclude that this can be due to an bottom-heavy IMF (enriched by faint low-mass stars) as well as a top-heavy IMF (enriched by non-luminous remnants of high-mass stars).

Whitworth *et al.* (1994) had already suggested that massive stars may preferentially form in shocked gas. As reviewed in Kroupa *et al.* (2013) there has been much observational evidence for top-heavy IMFs in star-bursts. As these are observationally unresolved, this evidence was indirect and largely ignored. Observations of the assembly of the stellar population over cosmological epoch have also been pointing to top-heavy IMFs in the cosmological past, as otherwise there would be more low-mass stars locally than are observed. Three independent more-direct lines of evidence for the IMF becoming top-heavy with star-formation rate density have recently emerged:

Firstly: It is well known that ultra-compact dwarf galaxies (UCDs), which have a mass scale of $10^6 - 10^8\,M_\odot$, have larger dynamical mass-to-light (M/L) ratios than normal stellar populations. This is unlikely due to exotic dark matter as the phase-space available in UCDs would not accommodate significant amounts of dark matter. Instead, a top-heavy IMF would have lead to an overabundance of stellar remnants in UCDs which would enhance their dynamical M/L ratios. Thus, the variation of the required $\alpha_3, m > 1\,M_\odot$, can be sought to explain the dynamical M/L ratia (Dabringhausen *et al.* 2009).

Secondly: A larger-than-expected fraction of UCDs have low-mass X-ray bright sources (LMXBs). In globular clusters (GCs), LMXBs are known to be formed from the dynamical capture of stars by stellar remnants mostly in the core of the GCs. As the star evolves the remnant accretes part of the star's envelope thus becoming detectable with X-rays. The LMXB population is constantly depopulating and needs to be replenished by new capture events. Indeed, the theoretically expected scaling of the fraction of GCs with LMXB sources with GC mass is nicely consistent with the observed data assuming an invariant stellar MF. Applying the same theory to UCDs uncovers a break-down of this agreement as the UCDs have a surplus of LMXB sources. By adding stellar remnants through a top-heavy IMF when the UCDs were born, i.e. by allowing α_3 to vary with UCD birth mass, consistency with the data can be sought (Dabringhausen *et al.* 2012).

Thirdly: Low-concentration GCs have been found by de Marchi *et al.* (2007) to be depleted in low mass stars while high-concentration GCs have a normal MF. This is contrary to the energy-equipartition driven depopulation of low mass stars because more concentrated clusters ought to have lost more low mass stars. It is also not consistent with any known theory of star formation, because the low-concentration clusters are found to typically have a higher metallicity which would, if anything, imply a surplus of low-mass stars. The currently only physically plausible explanation is to suppose that the young GCs formed compact and mass segregated and that the expulsion of residual gas unbound a part of the low-mass stellar population. By constraining the necessary expansion of the proto-GCs, correlations between metallicity, α_3 and tidal field strength emerge which constrain the very early sequence of events that formed the Milky Way as well as the dependency of α_3 on density and metallicity of the CSFE which can lead to a bound long-lived globular cluster after emergence from the natal cloud (Marks *et al.* 2012).

Putting all this together (Marks *et al.* 2012, Kroupa *et al.* 2013), a consistent variation of α_3 with density and metallicity of the CSFEs emerges: for $m > 1\,M_\odot, x \geqslant -0.89$: $\alpha_3 = -0.41\,x + 1.94$ with $x = -0.14[\mathrm{Fe/H}] + 0.99\log_{10}(\rho_6)$, where $\rho_6 = \rho/(10^6\,M_\odot \mathrm{pc}^{-3})$ and ρ is the density in $M_\odot/\mathrm{pc}^3$. This translates into a star-formation rate density (SFRD) assuming the CSFEs form within 1 Myr.

Thus, CSFEs with $SFRD < 0.1\,M_\odot/(\mathrm{pc}^3\,\mathrm{Myr})$ can be assumed to have an invariant IMF with $\alpha_3 = \alpha_2$ (subject to the possible variation with metallicity discussed above), while CSFEs with larger SFRDs tend towards top-heavy IMFs whereby the trend is enhanced at lower-metallicities.

3.5. *The IGIMF*

The previous sections consider the stellar content in correlated star-forming events (CSFEs), i.e. essentially in embedded star clusters which need not be gravitationally bound on emergence from their natal clouds. But when converting galaxy-wide photometric and spectroscopic data into physical properties (e.g. SFRs) the galaxy-wide IMF has to be known.

As stars form in CSFEs, i.e. in embedded star clusters or clusterings (Lada & Lada 2003, Allen *et al.* 2007), the galaxy wide-IMF has to be calculated by adding all IMFs of all CSFEs in a galaxy, irrespectively of whether the distribution functions are pure probability density functions or whether they are, at the other extreme, optimally sampled distributions (Kroupa *et al.* 2013). *This is the basic tenet of the integrated galactic initial mass function (IGIMF) theory* (Kroupa & Weidner 2003, Weidner *et al.* 2004).

In deriving the analytical IGIMF theory, the following empirical star formation laws need to be considered: (i) The functional form of the IMF in each CSFE is invariant for $SFRD < 0.1\,M_{\odot}/$ (pc^3 Myr), becoming top-heavy for lager values of $SFRD$ (Sec. 3.4). (ii) The most-massive star in each CSFE scales with the stellar mass of the CSFE according to the observations (Sec. 3.3). (iii) Observations show that the mass of the most-massive CSFE (i.e. young star cluster) scales with the current SFR of a galaxy (Weidner *et al.* 2004). *This means that low-mass CSFEs are void of high-mass stars, and low-SFR galaxies do not host high-mass CSFEs.*

The combination of these empirical relations leads to an IGIMF which differs from the canonical IMF in a CSFE. The IGIMF becomes steeper in the stellar high-mass regime with decreasing SFR (Weidner & Kroupa 2005, Pflamm-Altenburg *et al.* 2007). Thus with decreasing SFR the fraction of high-mass stars among all newly formed stars decreases, which is called the IGIMF-effect. At high galaxy-wide SFRs, the IGIMF becomes top-heavy relative to the canonical IMF. At low SFRs the IGIMF becomes top-light relative to the canonical IMF.

Consequently, as low-SFR galaxies have a smaller fraction of ionising stars, these galaxies have a higher SFR–Hα-luminosity ratio than Milky Way type galaxies (Pflamm-Altenburg *et al.* 2007). Thus, if using a classical linear SFR-Hα-luminosity relation the SFRs of dwarf galaxies are underestimated by up to two orders of magnitude. Correcting the obtained SFRs for the IGIMF effect it emerges that dwarf galaxies are not inefficient in forming stars but have the same star-formation efficiency as Milky-Way type galaxies. That is, it emerges that the SFR scales linearly with the gas mass of the galaxy (Pflamm-Altenburg *et al.* 2009) and that galaxies have constant gas depletion time-scales of $\approx$3 Gyr.

In contrast to the Hα luminosity which has its origin in the presence of ionising high-mass stars, long-lived B-stars contribute most to the far ultraviolet (FUV) luminosity. Thus the IGIMF-effect is very much stronger for Hα-radiation than for the FUV-flux. This has been predicted in Pflamm-Altenburg *et al.* (2009) and is in remarkable agreement with the observations by Lee *et al.* (2009).

The IGIMF-model can be refined into a local surface-density framework, where the mass of the locally most-massive CSFE scales with the local gas density. This formulation of the local IGIMF (LIGIMF) explains simultaneously the observed Hα cut-off in the disks of star forming galaxies and their extended FUV-disks (Pflamm-Altenburg & Kroupa 2008).

Yields of chemical elements are effected by the IGIMF-effect as well. As the fraction of high-mass stars decreases with decreasing SFR the effective oxygen yield should decrease (Köppen *et al.* 2007). This automatically leads to a mass-metallicity relation of galaxies as observed (Tremonti *et al.* 2004). In order to explain the observed mass-metallicity relation of galaxies in terms of an invariant IMF, metal enriched outflows must be postulated to be active in dwarf galaxies to account for the required low effective oxygen yields. On first sight this might be plausible as lower-mass galaxies have shallower gravitational potentials than large disk galaxies, and expanding supernova shells containing freshly produced metals might escape easier from dwarf galaxies. In order to break the degeneracy between the IGIMF-model, on the one-hand side, and the constant IMF model plus metal enriched outflows, on the other hand side, one has to concentrate on galaxies with different SFRs but equal gravitational potentials. This can be done by comparing low-surface brightness galaxies (LSBs) with normal disk galaxies having the same rotational velocity which is a proxy for the deepness of the gravitational potential. For the same potential, the constant-IMF model combined with metal enriched outflows would predict that the effective yields are higher for LSBs than for normal disk galaxies, because normal disk galaxies have higher SFRs and larger feedback by supernovae and metals should escape easier reducing the effective, i.e. detectable, yields. The IGIMF-model, on the other hand, predicts that the effective yields are higher for normal disk galaxies than for LSBs because normal disk galaxies have higher SFRs and therefore flatter IGIMFs and a larger fraction of high-mass stars. An analysis of effective oxygen yields of high- and low-surface brightness galaxies show that the IGIMF model is in agreement with the observations (Pflamm-Altenburg *et al.* 2011).

Furthermore, as Fe and α-elements have their main contribution from supernovae for which the stellar progenitors have different masses, the IGIMF-effect predicts different yields for both types of elements. Observations show that the $[\alpha/\mathrm{Fe}]$ abundance ratio decreases with decreasing velocity dispersion of early type galaxies. As lower-mass elliptical (E) galaxies had smaller SFRs than high-mass E galaxies, the production rate of α-elements should decrease faster with decreasing velocity dispersion than the production rate of Fe. Recci *et al.* (2009) show that the observations are in agreement with the expectation from the IGIMF-theory.

3.6. *The IGIMF — ruled out?*

Despite the most remarkable success of the IGIMF theory to naturally explain a large variety of observational results, a number of authors claim to have falsified the theory. These works are often cited by others uncritically as evidence against the IGIMF theory. Here we briefly touch upon these works noting that a more thorough analysis will be presented elsewhere.

The successful prediction of the Hα/UV luminosity ratio in dwarf galaxies by the IGIMF theory has been challenged recently. Fumagalli *et al.* (2011) made Monte-Carlo simulations of star forming galaxies by randomly sampling both distribution functions: the star cluster mass function (CMF) of a star-forming galaxy and the IMF in each star cluster. Therewith they unknowingly apply the basic tenet of the IGIMF theory because they assume the IGIMF to be the result of the integration over all star clusters. They nevertheless conclude *"that a truncation in the IMF in clusters is inconsistent with the observations"*. But the description of their publicly available code (da Silva *et al.* 2012) reveals that the relation of the most-massive star and the mass of the star cluster has been implemented as a truncation limit of the IMF and not as the observed most-massive star of the particular star cluster. This means that the additional under-sampling effect in their calculations leads to wrong results. The IGIMF has been implemented wrongly by Fumagalli *et al.* (2011).

Calzetti *et al.* (2010) tested if low-mass star clusters are void of very massive-stars due to physical constraints, as stated by the IGIMF theory which in turn is based on the best observational material available from fully resolved star-forming regions, rather than being purely statistical, as would be the case if the IMF and IGIMF were identical and pure probability density distribution functions. In order to perform their test, they used extragalactic observations: if the upper mass regime of the IMF in star clusters is populated according to the size-of-sample effect then low-mass clusters must have the same Hα-luminosity/star-cluster-mass ratio as high-mass star clusters if the sample of low-mass star clusters is sufficiently large. They came to the conclusion that their *"results for NGC5194 show no obvious dependence of the upper mass end of the IMF on the mass of the star cluster down to* $\approx 10^3\ M_\odot$*"*. However, a thorough perusal of their Figure 1 shows that the 1σ error ellipse is slightly closer to the IGIMF-model than to the expectation of pure random sampling. Other issues with this work by Calzetti *et al.* (2010) will be published soon, and it is clear that their statements are based on an incorrect account of the biases at work in extragalactic observations of individual star clusters.

Weisz *et al.* (2012) compare the spread of Hα luminosities of galaxies with the same FUV-luminosity. They argue that *"these results demonstrate that a variable IMF alone has difficulty explaining the observed scatter in the Hα-to-FUV ratio"*. They favour that the variability in the star formation history is responsible for the observed spread. However, Weisz et al. (2012) utilize a fully populated IMF, which is scaled to match the observed luminosities. This means that even in low-star formation rate galaxies, where the classical picture of the IMF requires the size-of-sample effect, their IMF is fully populated. In their model, dwarf galaxies would have fractions of O stars with the corresponding fraction of their hard-photon flux, which is unphysical. Thus again, this approach at testing the IGIMF theory suffers from an unphysical approach.

Taking the analytical IGIMF theory, according to which all distribution functions are optimally sampled density distributions following the empirical correlations stated above, then the predictions of this theory are precise. In comparing with observational data, such as extragalactic ones, it needs to be remembered that the observational data always carry observational uncertainties (random errors) and biases (systematic errors) which can never be removed. Thus, an observational ensemble of star-forming galaxies will always show a significant scatter which may, however, be unphysical.

The IGIMF-model as an analytic formulation can not account for fluctuations in the upper mass regime of the IMFs in star clusters (e.g. through stochastically occurring stellar mergers). An extended formulation of the IGIMF-theory may be capable to predict physical spreads of Hα luminosities of galaxies in the future. But first it has to be clarified how the upper mass regime of the IMF is populated and which physical processes are responsible for the spread in the high-mass regime of the IMF. The only clear fact at the moment is that this regime can not be populated according to the size-of-sample effect, that is, the IMF is not a pure probabilistic distribution function (Sect. 3.3). As such, the analytical IGIMF theory is based on the observed relation of the most-massive star in a CSFE as a function of the CSFE stellar mass, and on the observed relation of the most-massive young CSFE as a function of the current SFR of a galaxy. These relations determine the typical mass of the most-massive star in a CSFE and the typical mass of the most-massive young star cluster in a galaxy. Consequently, the analytical IGIMF-theory specifies the typical galaxy-wide IMF of a galaxy for a given SFR.

Nevertheless, the published papers discussed here (e.g. Fumagalli *et al.* 2011, Sharma *et al.* 2011, Relaño *et al.* 2012) are wrongly misused as evidence against the basic concept of the IGIMF theory. Detailed perusal of this published work shows that so far there is no rigorous evidence against the IGIMF-theory as being the correct description of star formation in galaxies.

4. SFR determinations from SED modelling

During this meeting we have discussed a lot about different methods of deriving star formation rates in galaxies and what the advantages and disadvantages of using different calibrations are (see the comprehensive review talk by Veronique Buat at this meeting; see also review by Calzetti 2012). During the session on SFR determinations from SED modelling, but also throughout the meeting, two SED modelling approaches have been presented and discussed: the energy balance method and the radiative transfer modelling.

The energy balance method relies on the conservation of energy between the stellar light aborbed by dust and that emitted (by the same dust) in the mid-IR/far-IR/submm. This is by far a superior method to only fitting SED templates in a limited spectral range (see talk by Denis Burgarella). Nonetheless, the energy-balance method is not a self-consistent analysis, since the SED of dust emission is not calculated according to the radiation fields originating from the stellar populations in the galaxy under study, but rather according to some templates. The templates can be either empirical (Xu *et al.* 1998, Devriendt *et al.* 1999, Sajina *et al.* 2006, Marshall *et al.* 2007) or theoretical (Dale & Helou 2002, Draine & Li 2007, Natale *et al.* 2010). At this meeting we saw applications of two energy balance methods, MAGPHYS (da Cuhna *et al.* 2008) and CIGALE (Burgarella *et al.* 2005, Noll *et al.* 2009). The energy balance method is a useful tool when dealing with large statistical samples of galaxies for which little information is available regarding morphology/type, orientation and overall size. This advantage comes nevertheless with the disadvantage that the energy balance methods cannot take into account the effect on the dust attenuation and therefore also on the dust emission of the different geometries of stars and dust present in galaxies of different morphological types, neither can it take into account the anisotropies in the predicted stellar light due to disk inclination (when a disk geometry is present). These methods can therefore only be used in a statistical sense, when dealing with overall trends in galaxy populations.

The radiative transfer method is the only one that can self-consistently calculate the dust emission SEDs based on an explicit calculation of the radiation fields heating the dust, consequently derived from the attenuated stellar populations in the galaxy under study. At this meeting we saw applications of the RT model of Popescu *et al.* (2011). This method can take advantage of the constraints provided by available optical information like morphology, disk-to-bulge ratio, disk inclination (when a disk morphology is present) and size. As opposed to the energy balance method, this advantage comes with the disatvantage that such information is not always easily available. Nonetheless these radiation transfer methofs could perhaps be adapted to incorporate this information (when missing) in the form of free parameters of the model, though such attempts have not yet been made. Another drawback of these methods was that radiative transfer calculations are notorious for being computationally very time consuming, and as such, detailed calculations have been mainly used for a small number of galaxies (Popescu *et al.* 2000, Misiriotis *et al.* 2001, Popescu *et al.* 2004, Bianchi 2008, Baes *et al.* 2010, MacLachlan *et al.* 2011, Schechtman-Rook *et al.* 2012, de Looze *et al.* 2012a,b). This situation has been recently changed, with the creation of large libraries of radiative transfer model SEDs, as performed by Siebenmorgen & Krugel (2007) for starburst galaxies,

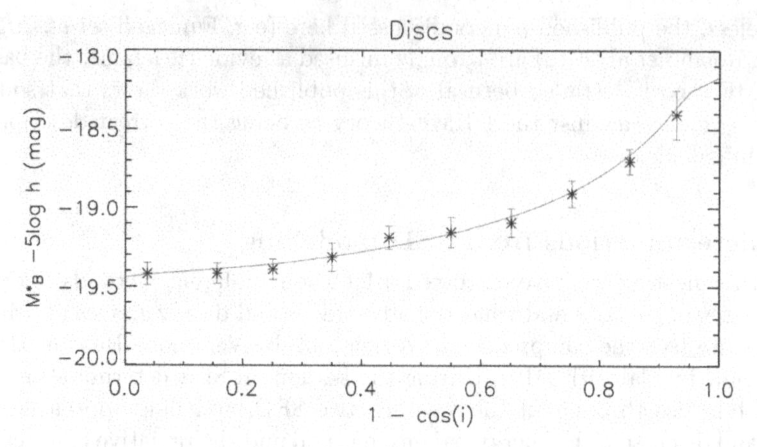

Figure 9. The attenuation-inclination relation from Driver *et al.* (2007). The symbols deliniate the empirical relation derived for disks from the Millenium Galaxy Survey while the solid line is the prediction from the model of Tuffs *et al.* (2004).

Groves *et al.* (2008) for star-forming regions/starburt galaxies and Popescu *et al.* (2011) for spiral galaxies.

Reviews on determination of star-formation in galaxies have so far not included discussions on the use of radiative transfer methods, with the exception of the review of Kylafis & Misiriotis (2006). With the new developments resulting in the creation of libraries of RT models, we can now start to include radiative models as main topics of discussion. Indeed, at this meeting we emphasised that they are in fact the most realible way of deriving star formation rates in galaxies. In this way the SFRs are derived self-consistently using information from the whole range of the electromagnetic spectrum, from the UV to the FIR/submm, incorporating information with morphological constraints (primarily from optical imaging). Here we did not consider radio and Xray emission, though these emissions have been discussed in other sessions of this meeting (e.g. talk by Bret Lehmer). In particular the SED modelling has been discussed in conjunction with the most difficult cases, namely those of translucent galaxies: galaxies with both optically thin and thick components. In one way optically thin galaxies are more easily dealt with, since most of the information on SFR can be derived from the UV. Very optically thick cases are also easy from this point of view, since SFR can be derived from their FIR emission, providing one can isolate the AGN powered emission. But the most difficult cases are the translucent galaxies. These are essentially the spiral galaxies in the Local Universe, and probably a large fraction of the star forming dwarf galaxies - which dominate the population of galaxies in the Local Universe, and which also host most of the star formation activity taking place in the local Universe. We showed in this meeting how important it is to quantify this star formation activity. We also discussed how important it is to quantify the star formation activity in the high redshift Universe; it is just that for the moment we do not have enough detailed information to be able to do the same type of analysis that we can now do for the Local Universe. Here I will summarise the main points we addressed:

1. *Why is it so difficult to calibrate SFR in spiral galaxies and why do we need to follow the fate of photons with radiative transfer calculations?*

• Fundamentally, any fixed observed luminosity in dust emission can be powered either by a small fraction of a large quantity of optical light from older stellar populations, or by a large fraction of a small quantity of UV light from younger stellar populations. Only radiation transfer techniques can unravel this dichotemy.

• Because of the disky nature of these systems, the direct UV and optical light is highly anisotropic, and the attenuation of the stellar photons will depend on the **viewing angle** and **wavelength**. Fig. 9 illustrates the strong dependence of the observed luminosity of stellar disks on the viewing angle.

• Dust in galaxies has a very complex structure, containing both **diffuse** components on kiloparsec scales, as well as **localised** components, at the pc scales, associated with the star forming regions. The escape of radiation from these two components is very different, as it is the heating of dust in the diffuse medium and in the star-forming clouds.

• Disk galaxies have different morphological components, in particular **disks** and **bulges**. The attenuation characteristics of disks is very different from those of bulges (see Fig. 10) and these need to be properly taken into account when dealing with the integrated emission from galaxies.

• Different stellar populations have **different spatial distributions** with respect to the dust distribution, and again their attenuation characteristics will differ, as will their contribution to heating the dust. Fig. 10 illustrates the different behaviour of the variation of attenuation of light coming from different stellar populations with inclination and dust opacity. Fig. 11 also shows how the dust and PAH emission SEDs are changed for various contributions coming from the old and young stellar populations in the disk, as well as from the old stellar populations in the bulge.

2. *Why do SFR calibrators work?*

SED modelling tools based on self-consistent radiative transfer calculations can be used to predict the scatter in the SFR calibration relations, as a function of the main intrinsic parameters that can affect these relations. Several of these relations have been presented. In Fig. 12 we only show predictions for the SFR calibration based on monochromatic FIR luminosities when the dust opacity changes. The predictions are based on the model of Popescu *et al.* (2011). The figure shows a very large scatter in the correlations. A similar large scatter is predicted for correlations corresponding to various contributions coming from the old stellar populations, or for the clumpiness of the ISM. For the UV calibrators large scatters are also predicted when some of the relevant parameters (viewing angle, dust opacity, bulge-to-disk ratio and clumpiness of the ISM) vary. Overall it is apparent from these plots that the predicted scatter in the SFR correlations due to a broad range in parameter values is larger than observed in reality. The question then arises of why are the SFR calibrators working, despite, for example, the very crude dust corrections that have been so far used in the community? A possible answer is the existence of some scaling parameters, which do not allow a continuous variation in parameter space, in particular for dust opacity or stellar luminosity. Recent work from Grootes et al. (2013) proved the existence of a well-defined correlation between dust opacity and stellar mass density. The correlation was derived on data coming from Galaxy and Mass Assembly (GAMA) survey (Driver *et al.* 2011) and the Herschel ATLAS survey (Eales *et al.* 2011), in combination with the model of Popescu *et al.* (2011). These finding give support to the interpretation of the existence of fundamental physical relations that reduce the scatter in the SFR correlations.

3. *A word of caution*

We have identified some points where things should be treated more carefully in the future:

• The energy balance method should not be used on scales smaller than the scalelength

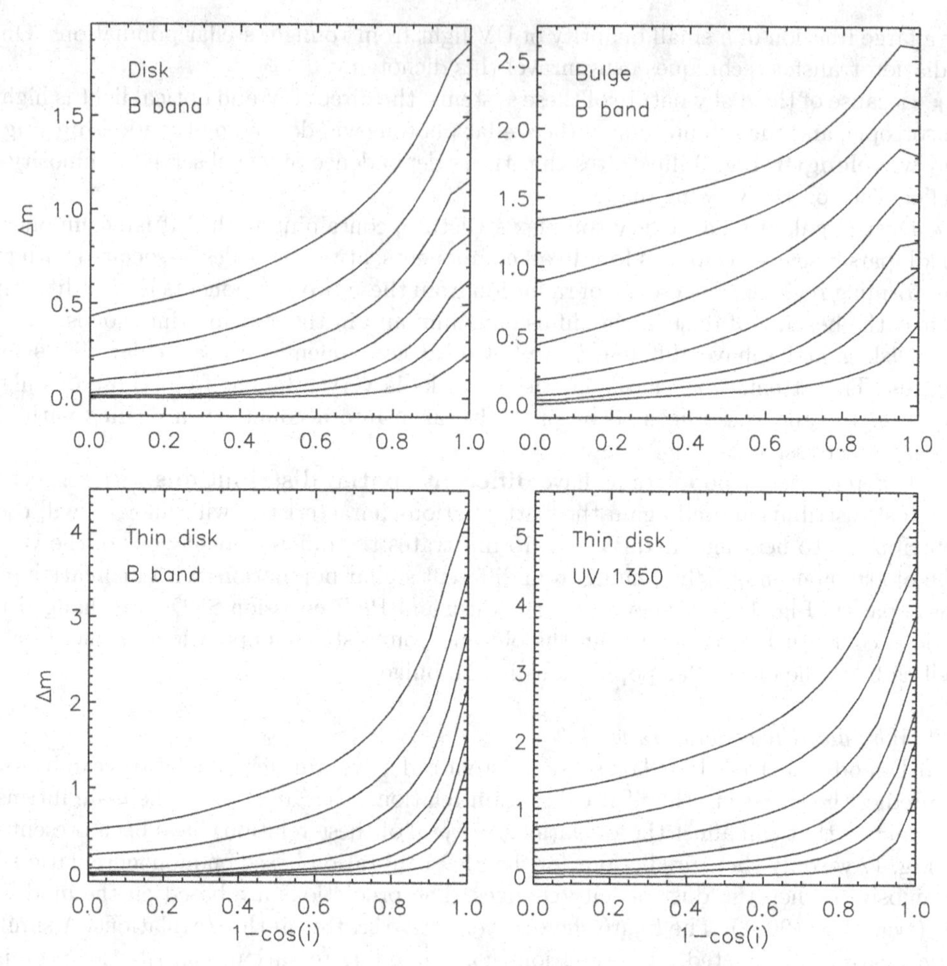

Figure 10. Predictions for the attenuation-inclination relation for different stellar components from Tuffs *et al.* (2004). From bottom to top the curves correspond to central face-on B band optical depth τ_B^f of 0.1,0.3,0.5,1.0,2.0,4.0,8.0

of the disk, as the energy is not conserved below these scales. The role of long range photons in the diffuse ISM should not be underestimated, in particular by considering an average free path of photons in the disk. This is because the free path of photons firstly depends on radial position in the galaxy (disk), where dust opacity is known to decrease monotonically with radius (e.g. Boissier *et al.* 2004, Popescu *et al.* 2005). Secondly, there is also a vertical distribution of dust, and the free path of photons in vertical direction will be different from that in radial direction. One also needs to add the contrast between arm and interarm regions. Finally, the escape of photons from star-forming clouds is strongly anisotropic and fragmented, because of the fragmentation of the clouds themself. Thus, in some directions the stellar light is completely absorbed by dust, while there are lines of sight from which the radiation freely escapes in the surrounding diffuse medium. The multiple facets of the transfer of radiation in galaxies, including the effect of scattered light, means that energy balance method should not be applied on a pixel by pixel basis, as sometimes employed in the literature.

• Mid-IR emission should not only be identified with "small grain" emission, where by small grain we mean stochastically-heated grains. In fact in the range $24 - 60\mu$m most

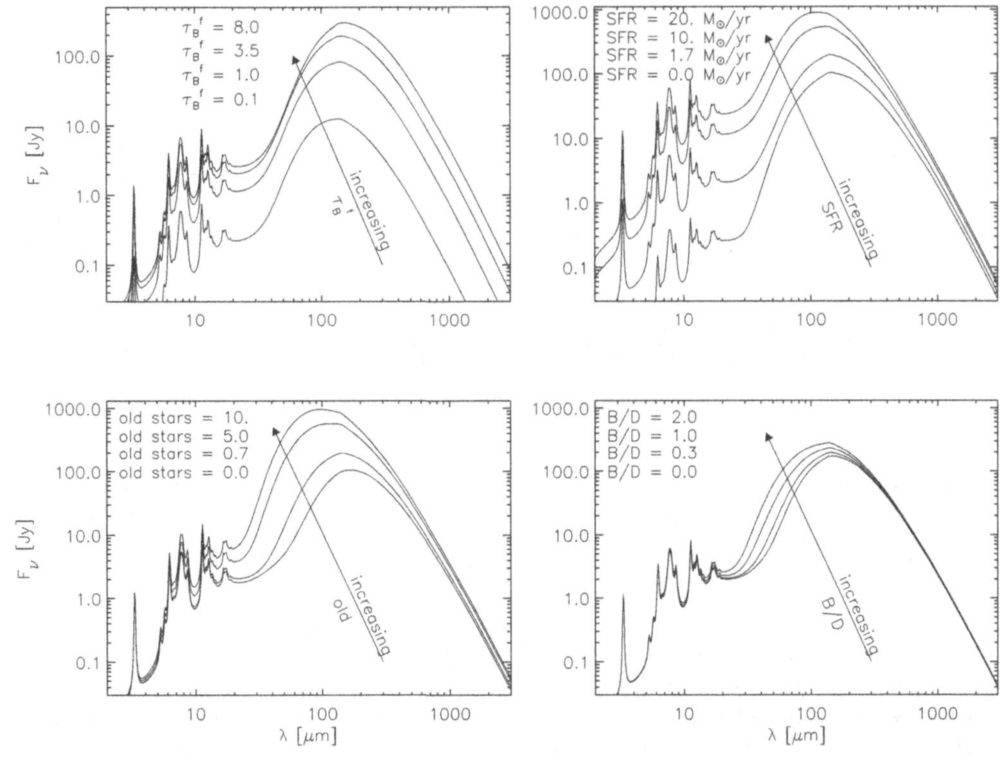

Figure 11. Predictions for dust and PAH emission SEDs based on the model of Popescu *et al.* (2011). In going clock-wise from the top-left, the different panels show the effect of changing the dust opacity, the luminosity of the young stellar populations (SFR), the luminosity of the old stellar populations (old) and the bulge-to-disk (B/D) ratio. In each panel only one parameter at a time is changed, while keeping the remaining ones fixed.

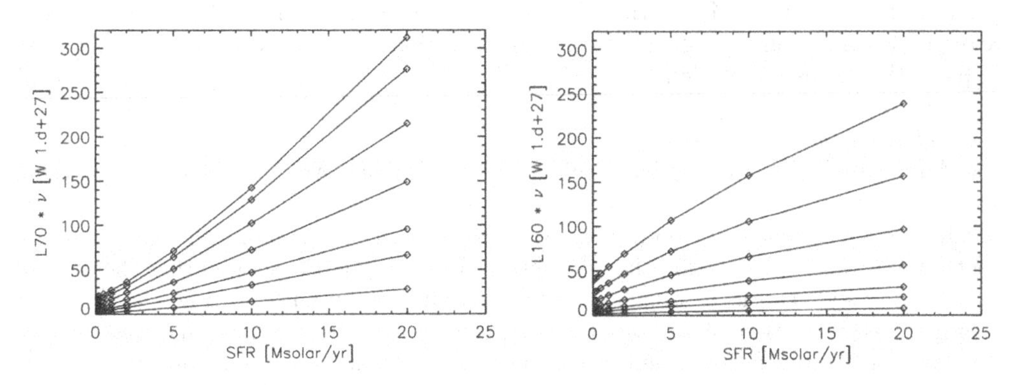

Figure 12. Predictions for the relation betwween the 70 μm (left) and 160 μm luminosity versus SFR based on the model of Popescu *et al.* (2011). From bottom to top the curves correspond to central face-on B band optical depth values of 0.1,0.3,0.5,1.0,2.0,4.0 and 8.0.

of the dust emission is powered by big grains heated at equilibrium temperatures by the strong radiation fields in the star-forming complexes.

• The ratio between mid-IR (PAH range) emission to FIR emission cannot be interpreted only in terms of relative abundances of PAH to big grains. One should also take

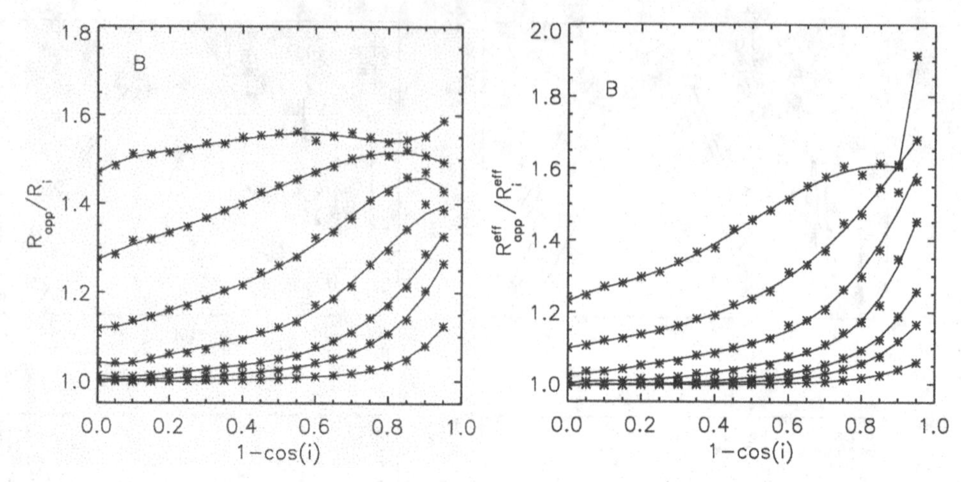

Figure 13. Dust effects on the derived scalelength of disks fitted with exponential functions (left) and on the derived effective radius of disks fitted with variable index Sérsic functions, from Pastrav *et al.* (2013). Both plots are for the B band. From bottom to top curves are for a central face-on dust opacity in the B band of 0.1,0.3,0.5,1.0,2.0,4.0,8.0.

into account the change in the colour and intensity of the radiation fields heating the dust, which also result in strong variations of mid-IR to FIR emission.

• Do we really need to accurately know the absolute star-formation rates in galaxies? Perhaps we can live with some approximations, which would be good enough to allow us to derive trends in galaxy populations over cosmic time. A definitive **no** has been given to this suggestion. An inability to measure absolute SFR would severely limit our ability to constrain physical models of galaxies and the evolving universe. For example, only if we have absolute measurements of SFR will we be able to relate measurements of SFR to measurements of gas content of galaxies in terms of physical models predicting the amount of gas in the ISM and the efficiency of conversion of the ISM into stars. Pavel Kroupa also gave convincing statistical argumentation on the need to measure accurate SFRs in the discussion session.

4. *What else have we learned?*

• Applications of self-consistently calculated model SEDs strongly rely on scaling them according to measurements of the surface area of the stellar disk of the modelled galaxy, which, in turn, depends on an accurate decomposition of the main morphological components of galaxies as observed in the UV/optical. Bogdan Pastrav showed that the derived scale-sizes of stellar disks of galaxies are strongly affected by dust (see Fig. 13), and that a proper determination of the intrinsic distributions of stellar emissivity, and thus of star-formation rates, needs to self-consistently take into account these effects.

• Several panchromatic surveys, with detailed information on bulge-to-disk ratio, inclination and disk size are underway, making these databases ideal for determinations of SFR using radiative transfer models. Andreas Zezas presented "The Star-Formation Reference Survey" (Ashby *et al.* 2011), a unique statistical sample of 369 galaxies selected to cover all types of star-forming galaxies in the nearby universe. The survey overlaps with the SDSS and NVSS areas and has GALEX, SDSS, 2MASS, Spitzer and NVSS multiband photometry and planned bulge-disk decompositions of optical images, which will make it ideal for self-consistent and systematic determinations of SFRs. It will also asses the influence of AGN fraction and environment on SFR.

• Denis Burgarella presented applications of the CIGALE SED fitting method on the Lyman break galaxies at $2.5 < z < 4$ detected in the Far-infrared with Herschel and implication for star formation determinations at high redshift.

• Andrew Hopkins showed results on SFRs derived from applications of the energy balance method on the GAMA survey. Applications of radiative transfer techniques to the de-reddening of GAMA galaxies by Grootes *et al.* (2013) were presented in the review of C. Popescu.

Acknowledgements

PK and JPA thank the organisers for an impressive and memorable conference. BDL acknowledges financial support from the Einstein Fellowship program. AZ acknowledges financial suppor from NASA in the form of an ADAP program, and the EU IRG grant 224878. Space Astrophysics at the University of Crete is supported by EU FP7-REGPOT grant 206469 (ASTROSPACE).

References

Allen, R. J., *et al.* 1997, *ApJ*, 487, 171
Allen, L., Megeath, S. T., Gutermuth, R., *et al.* 2007, *Protostars and Planets V*, 361
Alonso-Herrero, A., *et al.* 2006, *ApJ*, 652, L83
Andersen, M., Meyer, M. R., Robberto, M., Bergeron, L. E., & Reid, N. 2011, *A&A*, 534, A10
André, P., Men'shchikov, A., Bontemps, S., *et al.* 2010, *A&A*, 518, L102
Ashby, M. L. N., Mahajan, S., Smith, H. A., *et al.* 2011, *PASP* 123, 1011
Baes, M., Fritz, J., Gadotti, D. A., *et al.* 2010, *A&A* 518, L39
Banerjee, S., Kroupa, P., & Oh, S. 2012, *MNRAS*, 426, 1416
Basu, S. & Vorobyov, E. I. 2012, *ApJ*, 750, 30
Basu-Zych, A. R., Lehmer, B. D., Hornschemeier, A. E., *et al.* 2013, *ApJ*, 762, 45
Bianchi, S. 2008, *A&A*, 490, 461
Bigiel, F., *et al.* 2008, *AJ*, 136, 2846
Blitz, L. & Rosolowsky, E. 2006, *ApJ*, 650, 933
Bochanski, J. J., Hawley, S. L., Covey, K. R., *et al.* 2010, *AJ*, 139, 2679
Boissier, S., Boselli, A., Buat, V., *et al.* 2004, A&A 424, 465
Boissier, S., *et al.* 2007, *ApJS*, 173, 524
Boissier, S. 2012, "Star Formation in Galaxies", in *Planets, Stars and Stellar Systems*, Springer-Verlag GmbH Berlin Heidelberg, Oswalt, T. D., McLean, I. S., Bond, H. E., French, L., Kalas, P., Barstow, M. A., Gilmore, G. F., Keel, W. C. (Eds.), Volume 6, chapter 111
Boroson, B., Kim, D.-W., & Fabbiano, G. 2011, *ApJ*, 729, 12
Boselli, A., *et al.* 2009, *ApJ* 706, 1527
Braine, J., *et al.* 2010, *A&A*, 518, L69
Brandt, W. N. & Hasinger, G. 2005, *ARAA*, 43, 827
Brandt, W. N. & Alexander, D. M. 2010, *Proceedings of the National Academy of Science*, 107, 7184
Braun, R. 1997, *ApJ*, 484, 637
Bruzual, G. & Charlot, S. 2003, *MNRAS*, 344, 1000
Buat, V., Takeuchi, T. T., Iglesias-Páramo, J., *et al.* 2007, *ApJS*, 173, 404
Burgarella, D., Buat, V., & Iglesias-Páramo, J. 2005, *MNRAS* 360, 1413
Calzetti, D., Kennicutt, R. C., Jr., Bianchi, L., *et al.* 2005, *ApJ*, 633, 871
Calzetti, D., Kennicutt, R. C., Engelbracht, C. W., *et al.* 2007, *ApJ*, 666, 870
Calzetti, D., Chandar, R., Lee, J. C., *et al.* 2010, *ApJ* (Letters), 719, L158
Calzetti, D., Wu, S.-Y., Hong, S., *et al.* 2010, *ApJ*, 714, 1256
Calzetti, D. 2012, *Proceedings of the XXIII Canary Islands Winter School of Astrophysics: 'Secular Evolution of Galaxies'*, edited by J. Falcon-Barroso and J. H. Knapen

Cappellari, M., McDermid, R. M., Alatalo, K., *et al.* 2012, *Nature*, 484, 485
Casoli, F., *et al.* 1998, *A&A*, 331, 451
Cenarro, A. J., Gorgas, J., Vazdekis, A., Cardiel, N., & Peletier, R. F. 2003, *MNRAS*, 339, L12
Chabrier, G. 2003, *PASP*, 115, 763
Colbert, E. J. M., *et al.* 2004, *ApJ*, 602, 231
Combes, F. 2012, Journal of Physics Conference Series, 372, 012041
Corbelli, E., *et al.* 2011, *A&A*, 528, A116
Cowie, L. L., Barger, A. J., & Hasinger, G. 2012, *ApJ*, 748, 50
Crowther, P. A., Schnurr, O., Hirschi, R., *et al.* 2010, *MNRAS*, 408, 731
da Cunha, E., *et al.* 2008, *MNRAS* 388, 1595
da Silva, R. L., Fumagalli, M., & Krumholz, M. 2012, *ApJ*, 745, 145
Dabringhausen, J., Kroupa, P., & Baumgardt, H. 2009, *MNRAS*, 394, 1529
Dabringhausen, J., Kroupa, P., Pflamm-Altenburg, J., & Mieske, S. 2012, *ApJ*, 747, 72
Daddi, E., Dickinson, M., Morrison, G., *et al.* 2007, *ApJ* 670, 156
Dale, D. A. & Helou, G. 2002, *ApJ* 576, 159 *MNRAS* 288, 1595
Dale, D. A., *et al.* 2009, *ApJ* 703, 517
Davé, R. 2008, *MNRAS* 385, 147
Devriendt, J. E. G., Guiderdoni, B., & Sadat, R. 1999, *A&A* 350, 381
de Looze, I., Baes, M., Bendo, G., *et al.* 2012a, *MNRAS* 427, 2797
de Looze, I., Baes, M., Fritz, J., & Verstappen, J. 2012b, *MNRAS* 419, 895
De Marchi, G., Paresce, F., & Pulone, L. 2007, *ApJ* (Letters), 656, L65
de Wit, W. J., Testi, L., Palla, F., Vanzi, L., & Zinnecker, H. 2004, *A&A*, 425, 937
de Wit, W. J., Testi, L., Palla, F., & Zinnecker, H. 2005, *A&A*, 437, 247
Dib, S., *et al.* 2011, *MNRAS*, 415, 3439
Dijkstra, M., Gilfanov, M., Loeb, A., & Sunyaev, R. 2012, *MNRAS*, 421, 213
Domínguez Sánchez, H., Mignoli, M., Pozzi, F., *et al.* 2012, *MNRAS*, 426, 330
Draine, B. T. & Li, A. 2007, *ApJ* 657, 810
Driver, S. P., Hill, D. K., & Kelvin, L. S. *et al.* 2011, *MNRAS* 413, 971
Eales, S., Dunne, L., Clements, D., *et al.* 2010, *PASP* 122, 499
Elbaz, D., Daddi, E., Le Borgne, D., *et al.* 2007,*A&A* 468, 33
Elmegreen, B. G. & Parravano, A. 1994, *ApJ* (Letters), 435, L121
Engargiola, G., *et al.* 2003, *ApJS*, 149, 343
Engelbracht, C. W., Gordon, K. D., Rieke, G. H., *et al.* 2005, *ApJ* (Letters), 628, L29
Fabbiano, G. 1989, *ARAA*, 27, 87
Fabbiano, G. 2006, *ARAA*, 44, 323
Figer, D. F. 2005, *Nature*, 434, 192
Fragos, T., Lehmer, B., Tremmel, M., *et al.* 2013, *ApJ*, 764, 41
Fumagalli, M., *et al.* 2011, *ApJ* 741, L26
Gardan, E., *et al.* 2007, *A&A*, 473, 91
Gilfanov, M. 2004, *MNRAS*, 349, 146
Gratier, P., *et al.* 2012, *A&A*, 542, 108
Gratier, P., *et al.* 2010, *A&A*, 522, 3
Grootes, M., Tuffs, R. J., Popescu, C. C., Pastrav, B. A., Andrae, E. *et al.* 2013, *ApJ*, submitted
Groves, B., Dopita, M. A., Sutherland, R. S., Kewley, L. J., Fischera, J., Leitherer, C., Brandl, B., & van Breugel, W. 2008, *ApJS* 176, 438
Gvaramadze, V. V., Weidner, C., Kroupa, P., & Pflamm-Altenburg, J. 2012, *MNRAS*, 424, 3037
Heavens, A., *et al.* 2004, *Nature*, 428, 625
Hennebelle, P. 2012, *A&A*, 545, A147
Hopkins, A. M. 2004, *ApJ*, 615, 209
Hao, C.-N., *et al.* 2011, *ApJ* 741, 124
Hirashita, H., *et al.* 2003, *A&A* 410, 83
Hunter, D. A., *et al.* 1986, *ApJ*, 303, 171
Hsu, W.-H., Hartmann, L., Allen, L., *et al.* 2012, *ApJ*, 752, 59
Iglesias-Páramo, J., Buat, V., Takeuchi, T. T., *et al.* 2006, *ApJS*, 164, 38

Iwasawa, K., *et al.* 2009, *ApJ* (Letters), 695, L103
Kennicutt, R. C. 1989, *ApJ*, 344, 685
Kennicutt, R. C. 1998, *ARAA* 36, 189
Kennicutt, R. C., *et al.* 2009, *ApJ* 703, 1672
Kennicutt, R. C. & Evans, N. J. 2012, *ARAA* 50, 531
Koen, C. 2006, *MNRAS*, 365, 590
Köppen, J., Weidner, C., & Kroupa, P. 2007, *MNRAS*, 375, 673
Kroupa, P. 2001, *MNRAS*, 322, 231
Kroupa, P. 2002, *Science*, 295, 82
Kroupa, P., Aarseth, S., & Hurley, J. 2001, *MNRAS*, 321, 699
Kroupa, P. & Bouvier, J. 2003a, *MNRAS*, 346, 343
Kroupa, P. & Bouvier, J. 2003b, *MNRAS*, 346, 369
Kroupa, P., Bouvier, J., Duchêne, G., & Moraux, E. 2003, *MNRAS*, 346, 354
Kroupa, P. & Gilmore, G. F. 1994, *MNRAS*, 269, 655
Kroupa, P. & Weidner, C. 2003, *ApJ*, 598, 1076
Kroupa, P., Weidner, C., Pflamm-Altenburg, J., *et al.* 2013, in Planets, Stars and Stellar Systems, Vol 5.: Stellar Systems and Galactic Structure, Gilmore, G. (Ed.), Springer (ArXiv:astro-ph/1112.3340)
Kylafis, N. D. & Misiriotis, A. 2006, in "The many scales in the Universe", JENAM 2004 Astrophysics Reviews, eds. J. C. Del Toro Iniesta, E. J. Alfaro, J. G. Gorgas, E. Salvador-Sole, & H. Butcher, p. 111
Lada, C. J. & Lada, E. A. 2003, *ARA&A*, 41, 57
Laird, E. S., Nandra, K., Hobbs, A., & Steidel, C. C. 2006, *MNRAS*, 373, 217
Lamb, J. B., Oey, M. S., Werk, J. K., & Ingleby, L. D. 2010, *ApJ*, 725, 1886
Lee, J. C., *et al.* 2009, *ApJ* 706, 599
Lehnert, M. D. & Heckman, T. M. 1996, *ApJ*, 472, 546
Lehmer, B. D., *et al.* 2008, *ApJ*, 681, 1163
Lehmer, B. D., *et al.* 2010, *ApJ*, 724, 559
Leroy, A., Bolatto, A., Walter, F., & Blitz, L. 2006, *ApJ*, 643, 825
Leroy, A. K., *et al.* 2008, *AJ* 136, 2782
Li, H.-N., *et al.* 2007, *AJ*, 134, 1315
Li, Y., Crocker, A. F., Calzetti, D., *et al.* 2013, *ApJ*, 768, 180
Linden, T., Kalogera, V., Sepinsky, J. F., *et al.* 2010, *ApJ*, 725, 1984
Liu, G., *et al.* 2011, *ApJ* 735, 63
MacLachlan, J. M., Matthews, L. D., Wood, K., & Gallagher, J. S. 2011, *ApJ* 741, 6
Marks, M. & Kroupa, P. 2012, *A&A*, 543, A8
Marks, M., Kroupa, P., Dabringhausen, J., & Pawlowski, M. S. 2012, *MNRAS*, 422, 2246
Marshall, J. A., Herter, T. L., Armus, L., Charmandaris, V., Spoon, H. W. W., Bernard-Salas, J., & Houck, J. R. 2007, *ApJ* 670, 129
Martin, C. L. & Kennicutt, Jr., R. C. 2001, *ApJ*, 555, 301
Maschberger, T. & Clarke, C. J. 2008, *MNRAS*, 391, 711
Maschberger, T., Clarke, C. J., Bonnell, I. A., & Kroupa, P. 2010, *MNRAS*, 404, 1061
Meurer, G. R., *et al.* 1995, *AJ* 110, 2665
Meurer, G. R., *et al.* 2009, *ApJ* 695, 765
Mineo, S., Gilfanov, M., & Sunyaev, R. 2012a, *MNRAS*, 419, 2095
Mineo, S., Gilfanov, M., & Sunyaev, R. 2012b, *MNRAS*, 426, 1870
Mineo, S., Gilfanov, M., Lehmer, B. D., Morrison, G. E., & Sunyaev, R. 2014, *MNRAS*, 437, 1698
Misiriotis, A., Popescu, C. C., Tuffs, R. J., & Kylafis, N. D. 2001, *A&A*, 372, 775
Molinari, S., *et al.* 2010, *A&A*, 518, L100
Moore, C. A., *et al.* 2010, *AJ* 140, 253
Murgia, M., *et al.* 2002, *A&A*, 385, 412
Natale, G., Tuffs, R. J., Xu, C. K., Popescu, C. C., & Fischera, J. *et al.* 2010, *ApJ* 725, 955
Neininger, N., *et al.* 1998, *Nature*, 395, 871

Nieten, C., *et al.* 2006, *A&A*, 453, 459
Noll, S., *et al.* 2009, *A&A* 507, 1793
Norman, C., Ptak, A., Hornschemeier, A., *et al.* 2004, *ApJ*, 607, 721
Oosterloo, T., *et al.* 2007, *AJ*, 134, 1019
O'Sullivan, E., Forbes, D. A., & Ponman, T. J. 2001, *MNRAS*, 328, 461
Oey, M. S. & Clarke, C. J. 2005, *ApJ* (Letters), 620, L43
Oh, S. & Kroupa, P. 2012, *MNRAS*, 424, 65
Padoan, P. & Nordlund, Å. 2002, *ApJ*, 576, 870
Padoan, P., Nordlund, Å., Kritsuk, A. G., Norman, M. L., & Li, P. S. 2007, *ApJ*, 661, 972
Pastrav, B. A., Popescu, C. C., Tuffs, R. J., & Sansom, A. E. 2013, *A&A*, 557, A137
Pereira-Santaella, M., *et al.* 2011, *A&A*, 535, A93
Persic, M. & Rephaeli, Y. 2007, *A&A*, 463, 481
Pflamm-Altenburg, J. & Kroupa, P. 2009, *ApJ*, 706, 516
Pflamm-Altenburg, J. & Kroupa, P. 2010, *MNRAS*, 404, 1564
Pflamm-Altenburg, J., Weidner, C., & Kroupa, P. 2007, *ApJ*, 671, 1550
Pflamm-Altenburg, J., Weidner, C., & Kroupa, P. 2009, *MNRAS*, 395, 394
Popescu, C. C., Misiriotis, A., Kylafis, N. D., Tuffs, R. J., & Fischera, J., 2000, *A&A* 362, 138
Popescu, C. C., Tuffs, R. J., Kylafis, N. D., & Madore, B. F. 2004, *A&A* 414, 45
Popescu, C. C., Tuffs, R. J., Madore, B. F., *et al.* 2005, *ApJ* 619, L75
Popescu, C., *et al.* 2011, *A&A* 527, 109
Prescott, K. M., *et al.* 2007, *ApJ* 668, 182
Ptak, A., *et al.* 2007, *ApJ*, 667, 826
Ranalli, P., Comastri, A., & Setti, G. 2003, *A&A*, 399, 39
Ranalli, P., Comastri, A., Zamorani, G., *et al.* 2012, *A&A*, 542, A16
Recchi, S., Calura, F., & Kroupa, P. 2009, *A&A*, 499, 711
Reddy, N. A. & Steidel, C. C. 2004, *ApJ* (Letters), 603, L13
Reggiani, M. M. & Meyer, M. R. 2011, *ApJ*, 738, 60
Relaño, M., Kennicutt, Jr., R. C., Eldridge, J. J., Lee, J. C., & Verley, S. 2012, *MNRAS*, 423, 2933
Rosolowsky, E. 2005, *PASP*, 117, 1403
Rosolowsky, E., *et al.* 2008, *A&AR*, 15, 189
Sajina, A., Scott, D., Dennefeld, M., Dole, H., Lacy, M., & Lagache, G. 2006, *MNRAS* 369, 939
Schaerer, D. & de Barros, S. 2013, *A&A* 549, 4
Schaye, J. 2004, *ApJ*, 609, 667
Schechtman-Rook, A., Bershady, M. A., & Wood, K. 2012, *ApJ* 746, 70
Schilbach, E. & Röser, S. 2008, *A&A*, 489, 105
Schuller, F., *et al.* 2009, *A&A*, 504, 415
Sharma, S., Corbelli, E., Giovanardi, C., Hunt, L. K., & Palla, F. 2011, *A&A*, 534, A96
Siebenmorgen, R., Krügel, E., & Spoon, H. W. W. 2004, *A&A*, 414, 123
Siebenmorgen, R. & Krügel, E. 2007 *A&A*, 461, 445
Silva, L., *et al.* 1998, *ApJ* 509, 103
Stamatellos, D., Hubber, D. A., & Whitworth, A. P. 2007, *MNRAS*, 382, L30
Symeonidis, M., Georgakakis, A., Seymour, N., *et al.* 2011, *MNRAS*, 417, 2239
Testi, L., Palla, F., & Natta, A. 1998, *A&A*, 133, 81
Testi, L., Palla, F., Prusti, T., Natta, A., & Maltagliati, S. 1997, *A&A*, 320, 159
Thies, I. & Kroupa, P. 2007, *ApJ*, 671, 767
Thies, I. & Kroupa, P. 2008, *MNRAS*, 390, 1200
Thies, I., Kroupa, P., Goodwin, S. P., Stamatellos, D., & Whitworth, A. P. 2010, *ApJ*, 717, 577
Tremonti, C. A., Heckman, T. M., Kauffmann, G., *et al.* 2004, *ApJ*, 613, 898
Toomre, A. 1964, *ApJ*, 139, 1217
Tuffs, R. J., Popescu, C. C., Völk, H. J., Kylafis, N. D., & Dopita, M. A. 2004, *A&A* 419, 821
Tzanavaris, P. & Georgantopoulos, I. 2008, *A&A*, 480, 663
van Dokkum, P. G. & Conroy, C. 2010, *Nature*, 468, 940
van Dokkum, P. G. & Conroy, C. 2011, *ApJ* (Letters), 735, L13

Vattakunnel, S., Tozzi, P., Matteucci, F., *et al.* 2012, *MNRAS*, 420, 2190
Walcher, J., *et al.* 2010, *Ap&SS* 331, 1
Weidner, C. & Kroupa, P. 2004, *MNRAS*, 348, 187
Weidner, C. & Kroupa, P. 2005, *ApJ*, 625, 754
Weidner, C. & Kroupa, P. 2006, *MNRAS*, 365, 1333
Weidner, C., Kroupa, P., & Bonnell, I. A. D. 2010, *MNRAS*, 401, 275
Weidner, C., Kroupa, P., & Larsen, S. S. 2004, *MNRAS*, 350, 1503
Weidner, C., *et al.* 2011, *MNRAS*, 412, 979
Weisz, D. R., Johnson, B. D., Johnson, L. C., *et al.* 2012, *ApJ*, 744, 44
Whitworth, A., Bate, M. R., Nordlund, Å., Reipurth, B., & Zinnecker, H. 2007, Protostars and Planets V, 459
Whitworth, A., Stamatellos, D., Walch, S., *et al.* 2010, in IAU Symposium, Vol. 266, *IAU Symposium*, ed. R. de Grijs & J. R. D. Lépine, 264–271
Whitworth, A. P., Bhattal, A. S., Chapman, S. J., Disney, M. J., & Turner, J. A. 1994, *MNRAS*, 268, 291
Wilkins, S. M., *et al.* 2008, *MNRAS*, 385, 687
Wilson, C. D. 1997, *ApJ* (Letters), 487, L49
Wu, H., Cao, C., Hao, C.-N., *et al.* 2005, *ApJ* (Letters), 632, L79
Wu, H., Zhu, Y.-N., Cao, C., & Qin, B. 2007, *ApJ*, 668, 87
Wuyts, S., *et al.* 2011, *ApJ* 738, 106
Xu, C., Hacking, P. B., Fang, F., Shupe, D. L., Lonsdale, C. J., Lu, N. Y., Helou, G., Stacey, G. J., & Ashby, M. L. N.. 1998, *ApJ* 508, 576
Young, J. S. & Knezek, P. M. 1989, *ApJ* (Letters), 347, L55
Yuan, F.-T., *et al.* 2011, *PASJ*, 63, 1207
Zhu, Y.-N., *et al.* 2008, *ApJ* 686, 155

Highlights of Astronomy, Volume 16
XXVIIIth IAU General Assembly, August 2012
T. Montmerle, ed.

doi:10.1017/S1743921314011946

SpS9: Future Large Scale Facilities

No contribution was received from this Joint Discussion.

Highlights of Astronomy, Volume 16
XXVIIIth IAU General Assembly, August 2012
T. Montmerle, ed.

doi:10.1017/S1743921314011958

SpS10: Dynamics of the star-planet relations

No contribution was received from this Joint Discussion.

Highlights of Astronomy, Volume 16
XXVIIIth IAU General Assembly, August 2012
T. Montmerle, ed.

doi:10.1017/S1743921315001635

Preface: Special Session SpS11 Strategic Plan and the Global Office of Astronomy for Development

The purpose of this Special Session 11 was to bring together the many individuals from around the world who are active in realising the spirit of the IAU strategic plan. Appropriately entitled IAU Strategic Plan and the Global Office of Astronomy for Development, this special session invited contributions relating to: (i) IAU Strategic Plan, (ii) The IAU Global Office of Astronomy for Development, (iii) Regional nodes for "Astronomy for Development" activities, (iv) Sector Task Forces, and (v) Volunteers and volunteer opportunities.

The details of the IAU Strategic Plan and the Office of Astronomy for Development can be found in the papers by George Miley and Kevin Govender in this session as well as on the OAD website www.astro4dev.org.

The response to the call for contributions was overwhelming with many excellent contributions proposes. It was decided to allow more Oral contributions by reducing the amount of time per speaker. Requests for Oral contributions which could not be squeezed into the tight 1.5 day schedule were asked to bring their contributions in the form of posters. In the end we had 25 Oral contributions and 25 Poster contributions. A few invited speakers set the scene in terms of the bigger picture, followed by discussions on each of the three Task Forces (universities and research; children and schools; and public outreach), and then discussions about regional developments.

Unfortunately due to the many talks and the space constraints in the proceedings, only some presenters were allowed longer papers (up to 2 pages for invited talks) while the others were asked to submit only a revised abstract not exceeding a half page. We also decided to feature one particular effort in the Democratic People's Republic of Korea, which was awarded more space.

Poster presenters unfortunately could not be given space for their abstracts so I list the titles and first authors of the Posters here in order to recognise their contributions:

1. Exploring the unknown: the extremely large telescopes of the future (Annalisa Calamida, Rome Observatory INAF)
2. Astronomy Development in Africa Driven by Collaborative Projects (James Chibueze, Kagoshima University)
3. The Kazan lunar investigation to the Russian Scientific - Education Project Kazan-GeoNa-2015+ (Alexander Gusev, Kazan Federal University)
4. Astro Book Drive sharing materials to improve astronomy education in developing countries (Thilina Heenatigala, Sri Lanka Astronomical Association)
5. The Needs for New Astronomical Observatory in Indonesia (Taufiq Hidayat, Bosscha Observatory)
6. How to increase volunteers activity and impact? (Alisher S. Hojaev, National University of Uzbekistan)

7. Reform Of Astronomical Education In Tajikistan (Ibadinov Kh.I., Rahmonov A.A., The Tajik National University)

8. Improvement of science education through Astronomy education in Iran (Sara Khalafinejad, Leiden University)

9. The Development of International Olympiad on Astronomy and Astrophysics 2007 - 2011 (Chatief Kunjaya, Institut Teknologi Bandung)

10. Centre of space astrometry and perspective technologies on the basis of Engelhardt Astronomical Observatory (Yura Nefedyev, Engelhardt Astronomical Observatory)

11. How to promote astronomy in non-developed countries with the help of amateur astronomers? (Ana Maria Nicuesa Guelbenzu, Thringer Landessternwarte Tautenburg)

12. Universe in a Box: A Low-Cost Educational Kit for Astronomy Educators Across the Globe (Jaya Ramchandani, Leiden Observatory)

13. How to measure the Impact of Astronomy for Development Activities in Developing Countries (Valerio A.R.M. Ribeiro, University of Cape Town)

14. Popularizing Astronomy in Undeveloped Countries (Eduardo Rubio-Herrera, Institute of Astronomy -UNAM-)

15. EU Universe Awareness Inspiring every child with our wonderful cosmos (Pedro Russo, EU Universe Awareness/Leiden University)

16. Creating Awareness in Astronomy Education: Turkey as a Successful Example (A. Talat Saygac, Istanbul University)

17. The Cape Town - Toronto High School Astronomy Partnership (Linda Strubbe, Canadian Institute for Theoretical Astrophysics)

18. The GalileoMobile Project (Linda Strubbe, Canadian Institute for Theoretical Astrophysics)

19. South Asian Knowledge Exchange Programmes in Astronomy (Aniket Sule, Homi Bhabha Centre for Science Education)

20. Lunar observation from a new viewpoint (Bunji SUZUKI, Kasukabe Girls' High School)

21. Kids fun stars and sky activity at nursery school (Akihiko Tomita, Wakayama University)

22. Astronomical outreach activities carried out in Mexico in the last three years (Silvia Torres-Peimbert, Instituto de Astronomia, Universidad Nacional Autonoma de Mexico)

23. Radio Astronomy In Turkey: The Past And The Future (Mustafa Kursad YILDIZ, Erciyes University)

24. Bibliometric study of the astronomy in developing countries (Ivan Zolotukhin, Sternberg Astronomical Institute of the Moscow State University)

25. Vatican Observatory Summer Schools (Jos G. Funes, SJ, Specola Vaticana)

More information about the implementation of the IAU strategic plan and the activiites of the Office of Astronomy for Development can be found at http://www.astro4dev.org.

Kevin Govender and George Miley, co-editors

Highlights of Astronomy, Volume 16
XXVIIIth IAU General Assembly, August 2012
T. Montmerle, ed.

doi:10.1017/S174392131401196X

The IAU Strategic Plan and its Implementation

George Miley

[1]Leiden Observatory, Leiden, University, The Netherlands; email: `miley@strw.leidenuniv.nl`
[2]IAU Vice President, Education and Development

Abstract. I shall review the content of the IAU Strategic Plan (SP) to use astronomy as a tool for stimulating development globally during the decade 2010 - 2020. Considerable progress has been made in its implementation since the Plan was ratified at the last General Assembly.

---

1. Introduction

The IAU Strategic Plan 2010 – 2020: "Astronomy for Development: Building from IYA 2009" was ratified by the IAU General Assembly in August 2009. It is based on the unique power of astronomy as a tool for furthering human and technological capacity building throughout the world. Astronomy is a fundamental science itself and a gateway to physics, chemistry, biology and mathematics. Astronomy has been an important driver for the development of the most sophisticated technology and provides a link to our deepest cultural roots and origin.

2. The IAU Strategy

The Plan *ftp://ftp.saao.ac.za/outgoing/kg/astro4dev/stratplan_2012update.pdf* is a blueprint for using astronomy as a tool for development. The vision is a global one, namely that eventually all countries should participate at some level in astronomical research and that all children throughout the world will be exposed to knowledge about astronomy and the Universe. The strategy of the SP has several components:

1. A strategic phased integrated approach, including primary, secondary and tertiary education, research and public outreach. This will be based on the potential for astronomy research and education in each country, using objective data augmented by advice from experts in the region.

2. Regional involvement. A bottom-up approach, as pioneered during the International Year of Astronomy (IYA2009), involves regional input, including designation of regional institute nodes to coordinate development efforts throughout their region.

3. Special attention to Sub-Saharan Africa. Because of its relative underdevelopment, sub-Saharan Africa is receiving special attention.

4. Using IYA2009 as a springboard. Several IYA "cornerstones" are being continued and supported (e.g. UNAWE, GTTP) and the network of IYA contacts exploited.

5. Enlarging the number of active volunteers. We are recruiting volunteers from amongst members, doctoral students, postdoctoral trainees, non-member experts on education and outreach and amateur astronomers. Expatriates are particularly important.

6. Initiation of new activities. These include semi-popular lectures on inspirational topics and long-term institute twinning between established astronomy institutes and university departments in less developed countries. Stimulating technological expertise is regarded as crucial in all such programmes..

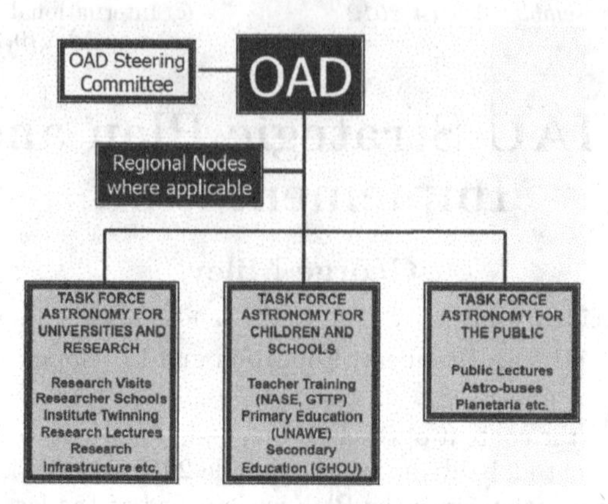

Figure 1. Organisational structure of Astronomy for Development (AfD) activities as envisaged in the SP. Three task forces are being coordinated by the IAU Office of Astronomy for Development (OAD) and regional input ensures a portfolio of demand-driven activities.

7. Exploiting innovative techniques. Innovative approaches to education and development, are being explored, including distance-learning, archives, robotic telescope networks and mobile delivery via astro-buses.

8. Creation of a small global Office of Astronomy for Development (OAD). Mobilising large number of volunteers, implementing new programs and inputting strategic information need professional coordination. Setting up of the IAU OAD has always been regarded as crucial to the success of the SP.

Figure 1 shows the implementation structure as envisaged by the SP. Three Task Forces of experts coordinated by the OAD manage the various AfD activities. The OAD injects a strategic component and safeguards accountability and transparency. Input from the regions ensures that the AfD portfolio matches local needs.

3. Implementation of the IAU Plan

The SP covers the ten-year period 2010 2020 and is being implemented gradually, to match the available funding. The first step was to set up the IAU Office of Astronomy for Development (OAD) to coordinate activities. Following a competitive call for proposals, and the selection of a host organisation by the IAU Executive Committee, the OAD started operation in March 2011. The IAU OAD is a joint venture between the IAU and the South African National Research Foundation (NRF), with Kevin Govender as its first Director. The official launch of the OAD took place on 16th April 2011 by the South African Minister of Science and Technology and the IAU President.

Oversight is provided by a steering committee (Board) with 3 members nominated by both the IAU and the NRF. In considering aspects of the SP most relevant to the IAU, the OAD Steering Committee is enlarged by President and General Secretary of the IAU.

Enormous progress has been made in implementing the IAU Strategic Plan since the last GA. Kevin Govender, the outstanding Director of the OAD, will tell you more about this. Now that the structure envisaged in the Plan is up and running, we are ready to embark on the next phase of its implementation, including fund-raising.

Acknowledgements

I thank Bob Williams and Ian Corbett for their strong support, the many volunteers in Commissions 46 who pioneered AfD activities over many years, all those involved in the success of IYA 2009, including members of Commission 55 and of course Kevin Govender.

Highlights of Astronomy, Volume 16
XXVIIIth IAU General Assembly, August 2012
T. Montmerle, ed.

doi:10.1017/S1743921314011971

The IAU Office of Astronomy for Development

Kevin Govender

IAU Office of Astronomy for Development; E-mail: kg@astro4dev.org

Abstract. The IAU Office of Astronomy for Development (OAD) †, established in March 2011 as part of the implementation of the IAU Strategic Plan, is currently located in South Africa and serves as a global coordinating centre for astronomy-for-development activities. In terms of structure the OAD is required to establish regional nodes (similar offices in different parts of the world which focus on a particular geographic or cultural region) and three task forces: (i) Astronomy for Universities and Research, (ii) Astronomy for Children and Schools, and (iii) Astronomy for the Public. This paper will describe the progress of the OAD towards the realisation of the vision 'Astronomy for a better world'.

1. Introduction

The OAD is hosted at the South African Astronomical Observatory in Cape Town. The activities of the OAD got underway in March 2011 with an official launch by the South African Minister of Science and Technology and the President of the IAU in April 2011.The first year of operation of the OAD was mainly occupied by appointing staff, setting up the office and online infrastructure, building the necessary networks, recruiting volunteers and promoting the existence of the OAD in various forums around the world. During this time the OAD also coordinated the establishment of its three Task Forces and began the process of setting up regional nodes globally.

The role of the OAD is fundamentally that of a strategic coordinating centre. In order to carry out its mission the OAD must fulfil several roles relating to this. Primarily the OAD was set up to implement the IAU Strategic Plan which provides the broad guidelines in terms of realizing developmental benefits from astronomy. The OAD, guided by this plan and a global view of development activities, should provide strategic advice where needed to individuals and organizations involved in similar activities. As such the OAD should be the first port of call for development activities using astronomy. It should also coordinate and facilitate global activities in line with its mission. Such efforts do not imply carrying out activities on the ground but rather sourcing partners or volunteers and providing them with the contacts, assistance and guidance necessary for them to implement a project. In terms of specific projects, the OAD should provide or assist in the acquiring of funding and infrastructure as required by its partners. There may also be specific developmental projects which will be managed and administered by the OAD.

2. Regional Nodes and Task Forces

In accordance with the 'bottom-up' strategy outlined in the strategic plan, an Announcement of Opportunity was issued in January 2012 for the establishment of Regional Nodes of the OAD (ROADs) and Language Expertise Centres for the OAD (LOADs). ROADs would be offices similar to the OAD established within host institutions and

† http://www.astro4dev.org

employing a full time coordinator, with a focus on activities in a specific geographic region. LOADs would have a similar structure but with a focus on a particular language or cultural region, which could sometimes stretch across the entire world. There was an excellent response with 31 Expressions of Interest and 14 full proposals received until July 2012. Evaluation of this first round of proposals was conducted by the EDOC. Currently two regional nodes have been established: one in China for the East Asia region and mainly Chinese language and culture (this would serve as both a ROAD and LOAD); and one (ROAD) in Thailand for the South East Asian region. Numerous negotiations are ongoing regarding the establishment of other nodes across the world.

Following the OAD Stakeholders Workshop in December 2011 (attended by 56 participants from 28 countries), three Task Forces have been established to lead the OAD's activities. These are TF1 (Universities and Research), TF2 (Children and Schools) and TF3 (Public Outreach). Task Forces are groups of international 'experts' in their fields who advise and assist the OAD in the implementation of the strategic plan. These are not necessarily IAU members but individuals who have been nominated and selected from a global science community. The Chairs, Vice Chairs and Management Teams of the TFs have been appointed by the EDOC and chosen to include a combination of veterans from the Commission 46 and 55 Organising Committees and 'new blood' from the many volunteers who have stepped forward thus far, as well as names received from an open call for nominations. In August 2012 the first open call for proposals was released with the IAU allocating funds towards projects relating to each of the Task Forces. The principle of the call was that each Task Force would evaluate proposals received and decide which were to be funded and which were to be put onto a wish list - the OAD would then continue to fundraise for those wish list projects.

3. Collaborations

The OAD has established several collaborations with partners sharing similar objectives. The Royal Astronomical Society in the UK provides a travel grant for UK-based experts to travel abroad on OAD activities. A similar agreement for Netherlands-based experts exists with the Netherlands Organization for Scientific Research. The International Science Programme of Uppsala University in Sweden has supported a pilot project on using astronomy tools to enhance university Physics teaching. The International Centre for Theoretical Physics in Italy launched a very close partnership with the OAD on a range of activities including associate positions, academic schools, building of networks, and individual grants. The Inter-university Centre for Astronomy and Astrophysics in India has agreed to host young astronomers at their facility as well as evaluate new educational materials. The University of Central Lancashire has offered 12 astronomy distance learning scholarships for African countries. The OAD is grateful to its existing partners and welcomes more collaborations such as these.

4. Future Outlook

As the Task Forces, ROADs and LOADs become fully operational, the OAD will embark on a major fundraising campaign. The proposals received through the Task Force calls, together with the needs of the ROADs and LOADs, will inform this campaign. By the end of 2015, the OAD aims to have active regional nodes on all populated continents; active Task Forces with an annual call for proposals; secured funding for the OAD in the 2016 to 2020 period; a sustainable volunteer programme; and funds raised exceeding those put in by the IAU and South African government.

Highlights of Astronomy, Volume 16
XXVIIIth IAU General Assembly, August 2012
T. Montmerle, ed.

doi:10.1017/S1743921314011983

Using Astronomy to shape a country's science and technology landscape

Khotso Mokhele

IAU Office of Astronomy for Development Steering Committee; E-mail: khotso@mokhele.com

Abstract. There is data abundant to show a positive correlation between a nation's investment in science, engineering and technology and the economic prosperity of that nation. Yet, there remain many countries in the world, particularly in developing countries, where little, if any, serious investment in science, engineering and technology is evident. Even in these countries, policy documents speak positively about the positive correlation between investment in science, engineering and technology and national development and prosperity. Unfortunately these positive policy statements rarely get converted into real investment. When the National Research Foundation was founded in Post-Apartheid South Africa it set out to "...contribute to the improvement of the quality of life of all people..." and its inspiring vision was to achieve "A prosperous South Africa and African continent steeped in a knowledge culture, free of widespread diseases and poverty, and proud contributors to the well-being of humanity." This organisation, with its altruistic vision, succeeded in convincing the emerging government to invest in and support the construction of the Southern African Large Telescope as one of its flagship projects. This decision was subsequently followed by a high level national decision to leverage South Africa's geographical advantage to host major global astronomy facilities such as the Square Kilometer Array. This presentation highlighted the reasons for such decisions and how we went about motivating government organs that investing in astronomy would contribute to addressing societal challenges by stimulating the science and technology landscape.

Highlights of Astronomy, Volume 16
XXVIIIth IAU General Assembly, August 2012
T. Montmerle, ed.

doi:10.1017/S1743921314011995

Astronomy for a Better World: IAU OAD Task Force-1 Programs for Advancing Astronomy Education and Research in Universities in Developing Countries

Edward Guinan[1] and Katrien Kolenberg[2]

[1]Villanova University, USA;
[2]Harvard Smithsonian Center for Astrophysics, USA & University of Leuven, Belgium;
email: `edward.guinan@villanova.edu`

Abstract. We discuss the IAU Commission 46 and Office for Astronomy Development (OAD) programs that support advancing Astronomy education and research primarily in universities in developing countries. The bulk of these operational activities will be coordinated through the OAD's newly installed Task Force 1. We outline current (and future) IAU/OAD Task Force-1 programs that promote the development of University-level Astronomy at both undergraduate and graduate levels. Among current programs discussed are the past and future expanded activities of the International School for Young Astronomers (ISYA) and the Teaching Astronomy for Development (TAD) programs. The primary role of the ISYA program is the organization of a three week School for students for typically M.Sc. and Ph.D students. The ISYA is a very successful program that will now be offered more frequently through the generous support of the Kavli Foundation. The IAU/TAD program provides aid and resources for the development of teaching, education and research in Astronomy. The TAD program is dedicated to assist countries that have little or no astronomical activity, but that wish to develop or enhance Astronomy education. Over the last ten years, the ISYA and TAD programs have supported programs in Africa, Asia, Central America and the Caribbean, the Middle East, South East and West Asia, and South America. Several examples are given.

Several new programs being considered by OAD Task Force-1 are also discussed. Other possible programs being considered are the introduction of modular Astronomy courses into the university curricula (or improve present courses) as well as providing access to "remote learning" courses and Virtual Astronomy labs in developing countries. Another possible new program would support visits of astronomers from technically advanced countries to spend their sabbatical leaves teaching and advising University Astronomy programs in developing countries. Suggestions for new Task Force -1 programs are also welcomed. Useful information about the participation of IAU members and volunteers in these programs will be discussed and practical information will be provided.

1. Introduction

An IAU/OAD Task Force (Task Force 1, TF-1) has been established for the development of Astronomy Education and Research to promote Astronomy teaching and research activities at universities in developing countries. This program builds on the long history of related work within the IAU's Commission 46 Program Groups. Astronomy serves to stimulate research and teaching in related STEM (Science, Technology, Engineering and Mathematics) areas and to develop the programs in regions where there is little or no Astronomy. There is also potential for developing research in the historical and cultural aspects of Astronomy which may prove important for igniting interests in

Astronomy (and in sciences in general) in communities where there is no strong interest in the sciences. The broad appeal of Astronomy across the world makes such a program very attractive to people from diverse cultural and education backgrounds.

2. International School of Young Astronomers (ISYA)

The IAU ISYA program was created in 1967 and has organized 34 schools thus far. It is now hosted by OAD TF-1. The ISYA program provides concentrated expert instruction and training in basic and special topics of modern astronomy to a number of selected young astronomers or physicists with or without a graduate degree who otherwise would not have such opportunities. The IAU, in collaboration with the Norwegian Academy of Sciences and Letters, pays for the transport of faculty and participants; the host country pays for stay of faculty and participants. Currently there is one ISYA event held each year. Chair: Dr. Jean-Pierre De Greve (Belgium); Vice Chair: Kam-Ching Leung (USA)

3. Education and Research Programs

Several examples of programs supported by TF-1 are listed below. A few of these programs were previously hosted under the IAU C-46 Teaching Astronomy for Development and World Wide Astronomy Programs. Most of them are new programs.

(1) Visiting astronomer programs or exchange programs. (2) National or regional astronomy schools and workshops. (3) Astronomy education /research equipment and laboratory small grant program (4) Sabbatical leave visit program: grants to spend sabbaticals at a university (or institute) in a developing country to provide educational and technical advice. (5) Technology internships, e.g., instrument specialists to visit universities in developing countries. (6) Grants for the development and implementation of undergraduate astronomy courses and labs in developing countries (7) University twinning programs - develop long term (at least 2 years) links and cooperative programs between institutes and universities with those in developing host countries. (8) Innovative university astronomy education and research programs not listed above (e.g., support or enhance the development of innovative teaching, laboratory and undergraduate research programs). Fresh ideas for high impact programs are most welcome.

4. Proposal Selection Criteria

Proposals can be submitted to the IAU/OAD for IAU financial support for education and research programs. Selection criteria and more informaiton are provided on the IAU/OAD Website http://www.astro4dev.org

Highlights of Astronomy, Volume 16
XXVIIIth IAU General Assembly, August 2012
T. Montmerle, ed.

doi:10.1017/S1743921314012009

TWINNING between Institutions in developed and less developed countries: an ideal way to set-up an astrophysics program

Claude Carignan

Dep. of Astronomy, University of Cape Town, Private Bag X3, Rondebosch 7701, South Africa
email: ccarignan@ast.uct.ac.za

Abstract. It is very difficult to start from scratch a new Astrophysics program in a country with very little or no researchers in the field. In 2007, we began to set-up an Astrophysics program by TWINNING the Université de Ouagadougou with the Université de Montréal in Canada, the Université de Provence in France and the University of Cape Town in South Africa. Already, courses are given at the undergraduate and Master levels and a teaching Observatory has been built. A 1m research telescope was also moved from the La Silla Observatory in Chile to Burkina Faso and the infrastructure is being built at the moment on mount Djaogari in the north-eastern part of the country. In the meantime, 6 students are doing their PhD in Astrophysics overseas (Canada, France and South Africa) and will become the core of the research group at the Université de Ouagadougou. An engineer is also doing his PhD in Astronomical Instrumentation to help with the maintenance of the equipment on the Research Telescope.

1. Astrophysics program in Burkina Faso

The initiative of the project came from the Burkina Faso minister of Higher Education. During the time that Burkina Faso students are doing their PhD overseas, the courses are given by professors and PhD students of the 3 partner Universities. Between 2007 and 2012, the undergraduate Astronomy course was given 4 times, each time to groups of more than 100 students, while the 4 courses given at the Msc levels were given to group of $\sim$20 students. Three DEA diplomas in Astronomy were given, from projects done using the teaching Observatory. Those three students have now started their PhD in one of the partner University. The signing of the two tri-partite agreements helped a lot in setting clear objectives for the project, each partner University developing one particular aspect of the program.

The building of the road to the summit of Mount Djaogari, where the research telescope will be built, has started and the building of the Observatory should start beginning of 2013. The dome is not bought overseas but is being built by the technical University in Bobo. This is a pedagogic project of the technical students for 2 semesters. It should be completed in mid-2013, in time to reconstruct the Marly telescope. In this way, when the first PhD students will return to Ouagadougou, there will be a research instrument in the country. The instrumentation on the telescope will be provided by the instrumentation groups at the Université de Montréal, at the Laboratoire d'Astrophysique de Marseille and at the University of Cape Town. The whole project (with the students back) should be completed around 2015 and will have taken $\sim$ 8 years.

References

Carignan, C. & Koulidiati, J. 2011, *The International Academic Research Journal*, Volume 2, Issue II, pp.2-13

Highlights of Astronomy, Volume 16
XXVIIIth IAU General Assembly, August 2012
T. Montmerle, ed.

doi:10.1017/S1743921314012010

Guideline Principles for Designing Astronomy Activities

Linda Strubbe
Canadian Institute for Theoretical Astrophysics
email: linda@cita.utoronto.ca

1. Introduction

We astronomy outreachers are passionate about sharing astronomy widely—but many astronomy outreach activities are designed more by how we *hope* things will work than in a careful intentional way. We also rarely evaluate how well our activities are achieving our goals. As a result, many educational activities must be *significantly less effective* than they could be. Fortunately, there is a large body of education research on how people learn. If we are serious about sharing astronomy widely and effectively, we must treat our teaching like research: (1) Have clear goals for our outreach, evaluate how well we are achieving our goals, and revise our strategies in light of what we learn; and (2) Use appropriate teaching techniques supported by education research wherever possible.

I have summarized a few top teaching principles from education research to help design effective astronomy activities. A more detailed 2-page version is available at: http://www.cita.utoronto.ca/~linda/eduresearch.pdf

(*a*) **Be strongly goal-oriented in designing and evaluating activities:**
- (*a*) Clearly state your goals for your learners (see below).
- (*b*) Determine what evidence would tell you your learners are reaching your goals.
- (*c*) Design your activity to help your learners achieve these goals.
- (*d*) During / after the activity, evaluate how well learners have achieved your goals.
- (*e*) Use what you learned about your students to revise the activity, for the future.

(*b*) **Choose your learning goals carefully:**
Three different important types of goals concern:
- (*a*) Scientific content — **What do we want to students to know/understand?**
- (*b*) Scientific process skills — **What do we want students to be able to do?**
- (*c*) Scientific attitudes — **How do we want students to feel about science / education in general / life?**

References

Chinn C. A. & Malhotra, B. A. 2002, *Epistemologically Authentic Inquiry in Schools: A Theoretical Framework for Evaluating Inquiry Tasks, Science Education*, 86, 175

National Research Council (NRC) 2000, *How People Learn: Brain, Mind, Experience, and School.* (Washington, DC: The National Academies Press).

National Research Council (NRC) 2007, "Chapter 2: Four Strands of Science Learning." *Ready, Set, SCIENCE!: Putting Research to Work in K-8 Science Classrooms.* (Washington, DC: The National Academies Press).

Wiggins, G. & McTighe, J. 2005, *Understanding by Design.* (Alexandria: Association for Supervision and Curriculum Development).

Highlights of Astronomy, Volume 16
XXVIIIth IAU General Assembly, August 2012
T. Montmerle, ed.

doi:10.1017/S1743921314012022

Status of astronomy in Rwanda and volunteer work at Kigali Institute of Education (KIE)

M. Pović[1], P. Nkundabakura[2] and J. Uwamahoro[2]

[1]Instituto de Astrofísica de Andalucía (IAA-CSIC), Granada, Spain; email: mpovic@iaa.es

[2]Kigali Institute of Education (KIE), Maths-Physics Department, Kigali, Rwanda
emails: nkundapheneas@yahoo.fr, uwamahorojean@yahoo.fr

1. Status of Astronomy in Rwanda

Until 2009, astronomy was undeveloped in Rwanda, without astronomy courses at universities and schools, astronomical facilities, or any outreach programmes. With the international year of astronomy in 2009, Dr. Pheneas Nkundabakura and Dr. Jean Uwamahoro from the KIE Maths-Physics department, both graduates from the South African NASSP Programme (http://www.star.ac.za), started a program of implementing the astronomical knowledge at schools and universities. During the same year 2009, IAU donated 100 galileoscopes for the secondary schools, and several astronomy workshops were organised for the teachers. IAU donated also 5 laptops to help students and lecturers to learn and use astronomy software. With this, KIE students have now a possibility to choose astronomy/space science for their undergraduate final year research projects. Moreover, there is an ongoing effort to look for further collaboration towards establishing the first astronomical facility (observatory) in the country.

2. Advancing astronomy in Rwanda through volunteer work

In 2012 Dr. Mirjana Pović spent one month at KIE as a visiting astronomer. During her stay, her activities consisted in giving public talks to students and lecturers, giving a short astronomy courses to students, supervising three students in their research in photometry and help students with linux and latex. Before going to Rwanda people from the IAA-CSIC in Spain recollected a number of scientific books, 2 projectors, 1 computer, and 1 external disc (containing lots of books of science and in particular of Astronomy/Space science in electronic format) for the KIE Maths-Physics department.

3. Thoughts for way forward

Education is one of the most powerful tools to fight against the poverty on the long time-scales. Therefore sharing the knowledge with the less developed countries can be one way of contributing to boost the worldwide development in education and in particular in the area of astronomy. In these regards, the Office of Astronomy for Development (http://www.astro4dev.org/) is one of the possibilities to promote this type of partnerships where using professional skills and astronomy as an entertaining and inspiring science, people can improve children's critical thinking, their motivation for learning, and increase the level of education and research at schools and universities. Moreover, in this era where much of the astronomical data can be accessed online, astronomers can assist in teaching various techniques and software for data reduction and analysis and hence, ensure that universities in undeveloped countries perform consistent and productive astronomy research locally.

Highlights of Astronomy, Volume 16
XXVIIIth IAU General Assembly, August 2012
T. Montmerle, ed.

doi:10.1017/S1743921314012034

IAU Office of Astronomy for Development: Task Force Children and School Education

Pedro Russo[1] and Edward Gomez[2]

[1]EU Universe Awareness - Leiden University, the Netherland
[2]Las Cumbres Observatory Global Telescope Network
E-mail: russo@strw.leidenuniv.nl & egomez@lcogt.net

Abstract. The main mission of the IAU OAD Task Force on Children and School Education is to support the implementation of the pre-tertiary education part of the IAU Strategic Plan 'Astronomy for Development'. In this presentation we will give an overview of the role and programme of the task force as well as a general discussion about the past, present and future IAU education activities and programmes.

Introduction and Scope

The task force on Astronomy for Children and Schools (TF2) for the IAU Office of Astronomy for Development drives activities related to using astronomy to inspire the very young and stimulate education, especially in Mathematics and Science. This task force looks at introducing astronomy in schools where there is little or no astronomy, and ensuring that the subject is used to positively influence the level of education development. Programmes for very young children, in the early childhood development stage, also fall within this task force. Examples of activities are educator training workshops; developing classroom resources; astronomy clubs in schools; etc.

Work Plan

TF2 is in charge of some organisational and community tasks in the framework of the IAU OAD activities. Organisational Tasks include: (i) supporting OAD Regional Nodes activities, (ii) facilitating interactions between pre-tertiary education stakeholders, (iii) working closely with the relevant IAU Commissions, and (iv) supporting the OAD fundraising. Community management level tasks include

• providing a small (2) number of core projects.

• providing input to the OAD database of astronomy education contacts (teachers, informal educators, etc)

• preparing, planning and supporting the implementation of an Annual Astronomy Education Strategy with on-going global/regional educational activities, opportunities for educational collaborations and other relevant information.

• issuing a call and managing a review committee for project proposals made to the OAD TF2 for endorsement and funding.

• delivering regular information (via OAD website, e-newsletter and social media, etc) about ongoing activities and opportunities, in astronomy education to volunteers, partner organisations and community

Core Projects

Core Project 1 deals with a Peer Review Platform for astronomy educational resources. This platform will provide a method for people to submit their educational resources to review and obtain objective guidance on the resources, have successful resources published in a central repository and receive IAU approval/accreditation. Resources will be open and freely available to everyone and will be able to be submitted in any language. Resources will also be made available in many different formats - PDF (print quality and low-res), .odt, HTML, epub, mobi, etc. and will be syndicated through document sharing sites (OER, Issuu, Slideshare, other social media networks).

According to the 4R Resource Repository Model (Wiley 2000), resources need to follow the four criteria: Reuse (the right to reuse the content in its unaltered / verbatim form); Revise (the right to adapt, adjust, modify, or alter the content itself); Remix (the right to combine the original or revised content with other content to create something new; and Redistribute (the right to share copies of the original content, your revisions, or your remixes with others. The repository, currently under development, will need to follow one more criteria: **Review**. Peer review is essential for the sucess of the repository. The resources' content and quality will be reviewed by the community peers, with the aim of improving each resource and following a similiar approach as the peer-review process in the scientific community.

Core Project 2 is the AstroPack. The distribution of physical materials improves the engagement level with the community. Physical materials can inspire people in general, and young children in particular, in ways which online resources cannot. They also provide a physical reminder of the context they received them (e.g. at an astronomy workshop) and helps to sustain their engagement with astronomy. TF2 will manage a physical pack with physical astronomy resources, which can be delivered in a bundle for wide distribution by local representatives. The main idea is that the pack contents is provided by the community and will be curated by TF2 members. The distribution will be coordinated by the OAD and the first mail-drop is expected for early 2013.

Projects Proposals

A call for projects proposal was issued during 2012. These project proposals aligned with the IAU's 10 year strategic plan, with specific focus on Astronomy for Children and Schools. Some examples of proposals include teacher training and development; resource development and vetting; producing guidelines and support for using robotic telescopes for school projects; exploring distance learning options applicable to school level education; etc.

References

Wiley, D. 2000. *The Instructional Use of Learning Objects.* Bloomington, IN: AECT.

Highlights of Astronomy, Volume 16
XXVIIIth IAU General Assembly, August 2012
T. Montmerle, ed.

doi:10.1017/S1743921314012046

The GTTP Movement: Engaging young minds to the beauty of science and space exploration

Rosa Doran

NUCLIO - Núclio Interactvivo de Astronomia
Largo dos Topázios, 48, 3 Frt, 2785-817 S. D. Rana, Portugal

Abstract. The Galileo Teacher Training Program (GTTP) is a living legacy of IYA2009. As a cornerstone of this important moment in the history of Astronomy, GTTP has managed to name representatives in over 100 nations and reached over 15000 teachers at a global level. The model used so far ensures sustainability and a fast growing support network. The task at hand is to engage educators in the use of modern tools for science teaching. Building the classroom of tomorrow is a promising path to engage young minds to the beauty of science and space exploration.

Keywords. Miscellaneous, sociology of astronomy

The mission assigned to GTTP during IYA2009 was to train teachers in the use of astronomy tools and existing resources that are freely available on the internet and its transfer into classroom science curricula; the range of activities varying from cutting edge digital tools to hand-on experiments with readily available materials. The model adopted to build the network was the cascade effect, where teachers train other teachers and are building blocks of a strong and sustainable structure. Having reached over 100 nations and 15000 educators worldwide the challenge lies now in the continuity of efforts.

The philosophy adopted puts teachers at the heart of what we do. The training events are just the first step on the joint road. Continuous support, new resources, promotion of contests and campaigns are examples of stimulus offered to those entering the network. Collaboration with other existing programs is key to the continuation and strength of GTTP. As examples we can name:

- The use of the radio antennas being constructed by the European Hands-on Universe;
- Participation in the International Astronomical Search Collaboration programs where students learn science while working on real research, a Global Hands-on Universe partner project;
- Participation in European Research projects such as: Discover the Cosmos, Open Discovery Space and Go-lab (European Commission funded projects);
- Promotion of joint projects with Dark Skies Awareness, Astronomers without borders, TWAN, UNAWE, GalileoMobile, to name a few;

A strong social powered communication effort is also an investment taking place in GTTP by the use of channels such as social network platforms, newsletter, youtube channel, etc. Reaching nations equally is a concern and the introduction of eTraining lessons, promotion of eScience cafes, or the preparation of e-lessons are plans being structured for the near future.

GTTP lives as a legacy of IYA2009 and is now part of IAU Strategic Plan for the Developing World. The road ahead is of continuous construction and strengthening of the blocks being built so far.

Highlights of Astronomy, Volume 16
XXVIIIth IAU General Assembly, August 2012
T. Montmerle, ed.

doi:10.1017/S1743921314012058

Education for development under the skies of Chile

Cecilia Scorza[1,2] **and Olaf Fischer**[2]

[1]Heidelberg University [2]House of Astronomy

Abstract. We report on an educational program initiated in Chile in the year 2010 on the frame of an excellence research and graduate exchange program between the University of Heidelberg and the Pontfica Catlica University in Chile, funded by the German International Exchange Office (DAAD).

The idea of accomplishing the excellence research initiative with educational activities for the Chilean schools relies on the large amount of observations carried out in Chile, the coming future observatories (e.g. ELT) and the heritage of the IYA 2009 which all together have produced a great impact on the Chilean society. The interest for astronomy education has enormously increased. As a consequence, the educational ministry in Chile has recognized the educational potential of astronomy in stimulating interest in natural, engineering and technical careers, and has introduced the new school subject "The Earth and the Universe" at all school levels.

The school subject spirals up through all levels connecting the contents of several school subjects on the frame of astronomical contexts. This has produced a huge demand of educational materials and teacher trainings in Chile because the schools have an open curricula, and teachers lack a background knowledge on astronomy.

Via the cooperation between the Heidelberg University and the House of Astronomy a first visit of the authors to Chile took place in November 2010. During that opportunity a meeting with authorities of the universities and observatories located in Chile, interested in supporting the educational activities, took place. Directors, professors and PR-experts from ESO, Cerro Tololo, Universidad Catlica, Universidad Pedagogica and Center for Latin America exchanged their ideas. Two teacher trainings ware offered in Santiago at the Planetario de Santiago de Chile and vists to Cerro Tololo in la Serena and schools in Antofagasta near to Paranal were carried out.

In order to support the better qualification of teachers in astronomy, two Chilean teachers were invited for a working stay of three weeks at the House of Astronomy. With our support the teachers Maria Paz Cornejo and Patricio Castro gained insights into the didactical materials developed at the HoA on the frame of the "Science into School" project (WIS). The aim of the WIS-program is to bring actual contents of modern astronomy research into the classroom. In doing so the gaps of activities in the program "The Earth and the Universe" were also identified and concrete ideas for new materials discussed. Among them arose the idea of developing materials related to the Mars exploration given the similarity between the Atacama desert and Mars landscapes. Once back to Chile both teachers implemented the activities at their schools and sent within four weeks a very positive report.

A second visit has been planned in January 2013 with the aim to make available the educational activities to other interested universities and schools located near the big telescopes. In several of these cities and small towns there is a considerable amount of disadvantaged children. Eight teacher trainings will be offered in Santiago, in San

Pedro de Atacama with educational materials related to ALMA and with support of the ESO and the Associated Universities Inc. (AUI), and in Antofagasta and Valdivia (in collaboration with the program government program "Explora"). During these trainings materials related to ALMA and the VISTA-Camera at the VLT (ESO) will be made accessible to the teachers. A meeting in Santiago will bring together all interested parties with the educational ministry. In doing so we hope that pupils of the high schools living inside and outside Santiago get in touch via inquiry based activities with astronomy research made by Chilean astronomers using the best telescopes of the World.

Highlights of Astronomy, Volume 16
XXVIIIth IAU General Assembly, August 2012
T. Montmerle, ed.

doi:10.1017/S174392131401206X

The 'Astronomy for the Public' Task Force

Carolina J. Ödman-Govender[1] and Ian E. Robson[1,2]

[1]OAD Task Force on Astronomy for the Public;
[2]Science and Technology Facilities Council, UK
email: carolina.odman@gmail.com

Abstract. One of the ways in which astronomy can stimulate development is by raising awareness of our place in the universe among the general public. This contributes to inspiring people and brings the scientific community and scientific thinking closer to everyone. The IAU OAD has set up one task force dedicated to 'Astronomy for the Public'. Proposed activities of the task force range from low-tech astronomy outreach to citizen science. We will present the task force, its objectives and potential developmental impacts for the first few years of operation.

The Earth is populated by over 7 billion people. There are of the order of 20,000 astronomers in the world† The International Year of Astronomy in 2009 reached over 800 million people. This is an achievement that this task force can build on. The motivation for using Astronomy for Development lies in the unique characteristics of the science: it attracts people by its beauty and vertiginous perspectives and opens their minds to critical thinking, environmental awareness, diplomacy and peace, and is a source of inspiration for young and old alike.

One outcome of the 1st Communicating Astronomy with the Public (CAP) Conference in 2003 was the Washington Charter that commits endorsing organisations to recognise the value of public outreach and pursue efforts to carry out such outreach. In 2006, the General Assembly of the IAU saw the CAP working group become Commission 55. Was the IAU prepared for the public's response to the reclassification of Pluto? Probably not entirely but efforts were made and authoritative resources created to answer the questions of the general public.

2009 saw the International Year of Astronomy become the most successful coordinated scientific outreach effort ever with a reach of over 800 million people worldwide and many other achievements. That year, the IAU General Assembly adopted the proposed decadal Strategic Plan 'Astronomy for Development'. Since, the Office of Astronomy for Development (OAD)‡, whose purpose is to realise the vision of the Strategic plan, has started operations (2011) and special focus task forces have been set up (2012). Task Force 3 (TF3) concerns itself with 'Astronomy for the Public' in this context. One of the roles of TF3 is to evaluate projects submitted under an annual call for proposals for the OAD. The first such call was opened at the beginning of August 2012. Additionally, TF3 will work closely with the newly set up IAU Office for Public Outreach. In the future, TF3 will help develop and implement activities for the public and collaborate with Commission 55.

A Task Force Management Team has been set up with members representing a diversity of expertise and geographical backgrounds (see the OAD website). The team will evaluate the proposals coming for funding and fundraising by the OAD.

† The IAU membership is of the order of 10,000. Adding graduate students worldwide and non-practising astronomers probably doubles that figure.
‡ OAD website: `http://www.astro4dev.org/`

A broad variety of projects can be carried out under the umbrella of the IAU OAD. It is important to note that the IAU OAD is actively looking to support innovations in the field. The 1st call for proposals was was open for projects requiring between 1000-5000 euro funding with a deadline set for September 30. The TF3 team will evaluate these and create a list of projects recommended for funding. The proposals that are considered worthwhile but cannot be funded immediately will be added to a wish list that the OAD will use for further fundraising efforts.

The criteria for the proposal evaluation can be found on the OAD website. The most important is probably the alignment with the goals of the IAU Strategic Plan. The good news is that the call is open to anyone. TF3 is committed to transparent feedback to the proposers and the call for funding is annual, so anyone can reapply next year.

Astronomy is a deeply human endeavour. The amazement of seeing the universe from a new perspective is something that can be created with the simplest tools (simple telescopes) and most complex initiatives (large citizen science projects). The potential impact of public outreach is of the scale of the public it is trying to reach. The TF3 management team therefore encourages anyone with a good idea to submit their proposal this year and every following year. 'Astronomy for the Public' is a 7-billion-strong opportunity.

Highlights of Astronomy, Volume 16
XXVIIIth IAU General Assembly, August 2012
T. Montmerle, ed.

doi:10.1017/S1743921314012071

Communicating Astronomy with the Public (Youth) as the Gateway to Development

Dennis R. Crabtree

National Research Council Canada; email: Dennis.Crabtree@nrc.ca

Abstract. Astronomy has a unique ability to excite and stimulate the curiosity of children. Because of this, society can use astronomy as a gateway to lead children on a path towards future learning of science and technology, and potentially to careers in these areas.

Keywords. miscellaneous

Children have a natural curiosity about the world. This extends, to varying degrees, to later in life for most people. This curiosity makes youth open to new ideas and experiences and is an ideal time to expose them to the wonders of the Universe.

At this early stage of their life, children build knowledge, life skills and, perhaps most importantly, attitudes that will be with them for life. Evidence shows that this begins well before a child begins formal education †. It is important to engage young children in exploring and learning about the natural, physical and technological world.

One must remember that not all children will end up in science or technological careers. Think of a world without poets, musicians, artists, and yes, even lawyers. The goal is to ensure that children are exposed to the widest possible gamut of experiences and activities to ensure they will find their *calling* in life.

Astronomy is a powerful force to **excite** youth about science and technology. When combined with informal education's ability to **inform**, astronomy outreach has a unique ability to **exite and inform** youth.

It is important to expose children as often as possible to astronomy-related activities so the level of excitement in the natural world is maintained. Children's families are very important factors in determining a child's path in life. The Astronomical Society of the Pacific's *Family Astro* program ‡ is excellent for helping parents get involved their child's science training and to share in the excitement.

Astronomy outreach for youth - *Excite and Inform*

---

† Deadly Moons - Ireland http://unawe.org/resources/education/deadly_moons_guide_158/
‡ http://www.astrosociety.org/education/family.html

Highlights of Astronomy, Volume 16
XXVIIIth IAU General Assembly, August 2012
T. Montmerle, ed.

doi:10.1017/S1743921314012083

NAOJ's activities on Astronomy for Development: Aiding Astronomy Education in Developing Nations

K. Sekiguchi and F. Yoshida

Office of International Relations National Astronomical Observatory of Japan
E-mail: kaz.sekiguchi@nao.ac.jp

Abstract. We summarize NAOJ's efforts to promote astronomy in developing nations. The Office of International Relations, collaborations with the Office of Public Outreach at NAOJ and with the East Asia Core Observatories Association (EACOA), has engaged children, students and educators about astronomy development in the Asia-Pacific region. In particular, we introduce "You are Galileo!" project, which is a very well received astronomy education program for children. We also report on a continuing effort by the Japanese Government in support of astronomy programs in the developing nations.

1. "You are Galileo!" project

"You are Galileo!" project, which is a continuing program from the IYA2009 to attract children's interest in astronomy by looking through a small telescope they built themselves, has been supported by the Japanese National Committee for UNESCO. This extremely successful project has brought astronomy within the grasp of children in developing regions and helped children to pursue careers in science and engineering. Two recent campaigns were held in Indonesia and in Mongolia in 2011.

In Indonesia, we had five workshops, at Jakarta, Yogyakarta, Mataram, Tomohon and Palembang, for teachers of the elementary and the secondary schools. A total of 230 teachers attended. In Mongolia, six workshops for the teachers and children were conducted. A total of 289 teachers and children attended the workshops and over 600 people showed up for the star parties. For more information, please refer to the following URL: http://kimigali.jp/index-e.html

2. Japans Official Development Assistance (ODA) programme for astronomy

Since 1982, the Japanese Government has donated 27 units of astronomical equipment to 22 nations. 7 of the items donated were professional-grade reflecting telescopes with scientific instruments, such as CCD cameras and/or spectrographs, which can be used for photometric and spectroscopic observations. In addition to these, 20 planetarium systems have been installed at universities and space-education museums. Kitamura M. *et al.* gave detailed information about this program. Follow up assistance programs, supported by Japan International Cooperation Agency (JICA) and some public observatories in Japan, to give technical and observation training, have been conducted.

References

Kitamura, M., *et al.*, 2008, *Third UN/ESA/NASA Workshop on the International Heliophysical Year 2007 and Basic Space Science, Earth Moon and Planets*, p49.

Highlights of Astronomy, Volume 16
XXVIIIth IAU General Assembly, August 2012
T. Montmerle, ed.

doi:10.1017/S1743921314012095

Touch the sky with your hands: a special Planetarium for blind, deaf, and motor disabled

Beatriz García, Javier Maya, Alexis Mancilla, Silvina Pérez Álvarez, Mariela Videla, Diana Yelós and Angel Cancio
Instituto de Tecnologías en Deteccíon y Astropartículas Mendoza
Mendoza, Argentina
email: beatriz.garcia@iteda.cnea.gov.ar

The Planetarium for the blind, deaf, and motor disabled is part of the program on Astronomy and Inclusion of the Argentina Pierre Auger Foundation (FOPAA) and the Institute in Technologies and Detection of Astroparticles-Mendoza (ITeDAM).

The base of the proposal is the multi-sensorial perception of nature, with the approach to the starry sky from a new perspective: the Planetarium, where the stars are represented by LEDs and the hemisphere representing the sky is overhead, the implementation of new resources and tools for scientific outreach in general (like Geiger counters) and Astronomy in particular (Solar System models), designed for people with disabilities. The installation is aimed not only to the blind and visually impaired, but also the deaf and people with motor impediments. This space allows an approach to Astronomy, their concepts and discoveries, developing a sense of wonder before the comprehension of the natural world. The perception of a palpable and sensitive space by generating a particular environment, aims to bring the visitor to the conditions of the external natural environment at night. For these reasons, specific objectives are providing people with special educational needs, visually impaired and / or motor disabilities, a participatory learning and accessible resource, at the same time interesting and training, taking into account the accurately and strictly scientific knowledge, but also ensuring a playful interaction framework. This facility is installed at an appropriate area in *Tecnópolis*, the mega exhibition of Science and Technology, supported by the Ministry of Science, Technology and Productive Innovation of Argentina, with about 5000 visitors a day. (See Figure 1).

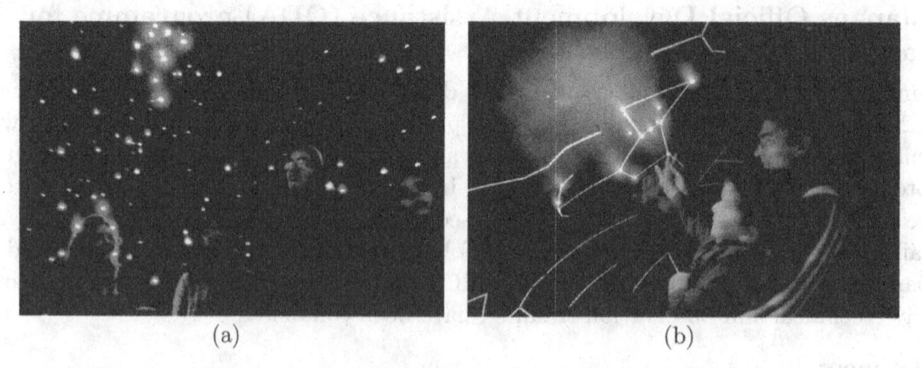

(a) (b)

Figure 1. Touch the sky with your hands: a Planetarium for everyone

Highlights of Astronomy, Volume 16
XXVIIIth IAU General Assembly, August 2012
T. Montmerle, ed.

doi:10.1017/S1743921314012101

Global Astronomy Month - An Annual Celebration of the Universe

Thilina Heenatigala and Mike Simmons
Astronomers Without Borders
E-mail: thilina.heenatigala@yahoo.com

Abstract. One of the most successful global outreach efforts in history was the International Year of Astronomy 2009. With the momentum created by this year long program, it was important to take the efforts to coming years. The Astronomers Without Borders organization captured the energy of the International Year of Astronomy 2009 and refocused it as an ongoing annual celebration of the Universe by organizing Global Astronomy Month, a worldwide celebration of astronomy in all its forms, every April. In 2010, the program saw professionals and amateur astronomers, educators and astronomy enthusiasts from around the globe participating together in the spirit of International Year of Astronomy 2009 and provided a global stage for established programs and a framework for partnerships. The 2011 version of the program saw much bigger participation with several global partner organizations joining in creating more than 40 global level programs throughout the month. Within a short period of two years, Global Astronomy Month has evolved to a much needed global platform after International Year of Astronomy 2009.

Highlights of Astronomy, Volume 16
XXVIIIth IAU General Assembly, August 2012
T. Montmerle, ed.

doi:10.1017/S1743921314012113

Amateur Astronomy Network Development in Indonesia

Avivah Yamani[1] and Hakim L. Malasan[2]

[1]langitselatan, astronomy education and communication media
[2]Astronomy Division and Bosscha Observatory, Institut Teknologi Bandung

Abstract. Indonesia is a very big country with over 238 million people. And we only have one higher learning institution on astronomy, so how do we reach and convey astronomical information effectively to the whole country? The answer lies in Astronomy Clubs who play an increasingly important role to communicate and educate the public. As part of South East Asia, Indonesia is actively involved in the region to develop astronomy.

Keywords. Astronomy Club, astronomy awareness, Indonesia

---

1. The role of Amateur Astronomy to build astronomy awareness

Astronomy in Indonesia has been known since ancient times through maritime and agricultural life. Up until now, we have one Astronomy Study Program, one University Observatory, two Solar Observing Stations, three Small Observatories, four Planetaria, nineteen Astronomy Clubs, four Online Communities, and four Astronomy Media. In Indonesia, we use astronomy as an entry point to stimulate young people to learn science, culture and technology. It is also used to determine and decide the Islamic ritual dates by observing the waxing crescent. We have nineteen astronomy clubs, who introduce astronomy by conducting real astronomical activities among the public such as star parties, teacher training, etc. Astronomy Clubs also realise that collaboration with professional astronomers is very important to teach the basics of astronomy.

2. Result and Discussion

Challenges and problems for Clubs are the lack of instruments, resources and knowledge. Some clubs suggest having an association that conducts regular seminars. It seems that Clubs in Indonesia are not aware that we already have the Indonesia Astronomical Association (HAI) who have been conducting biannual meetings since the 1980s. To solve this problem we suggest that Indonesia needs to create a network to gather key persons from Clubs and Institutions and have suitable communication media to exchange information. Annual astronomy festivals also needed to promote astronomy and train amateurs with proper astronomical knowledge. In terms of the region, Indonesia is part of South East Asia where amateurs also play important roles to educate the public. South East Asian Astronomy Network (SEAAN) includes Public Outreach in their working groups, and South East Asian Young Astronomer Collaboration (SEAYAC) seeks amateurs as potential collaborators.

AY would like to thank LKBF and IAU for the grants to attend the 28th IAU GA.

Highlights of Astronomy, Volume 16
XXVIIIth IAU General Assembly, August 2012
T. Montmerle, ed.

doi:10.1017/S1743921314012125

Astronomy development in Serbia in view of the IAU Strategic Plan

Olga Atanacković

Department of Astronomy, Faculty of Mathematics, University of Belgrade
email: olga@matf.bg.ac.rs

Abstract. An overview of astronomy development in Serbia in view of the goals envisaged by the IAU Strategic Plan is given. Due attention is paid to the recent reform of education at all levels. In the primary schools several extra topics in astronomy are introduced in the physics course. Attempts are made to reintroduce astronomy as a separate subject in the secondary schools. Special emphasis is put to the role and activities of the Petnica Science Center the biggest center for informal education in SE Europe, and to a successful participation of the Serbian team in International astronomy olympiads. Astronomy topics are taught at all five state universities in Serbia. At the University of Belgrade and Novi Sad students can enroll in astronomy from the first study year. The students have the training at the Ondrejov Observatory (Czech Republic) and at the astronomical station on the mountain Vidojevica in southern Serbia. Astronomy research in Serbia is performed at the Astronomical Observatory, Belgrade and the Department of Astronomy, Faculty of Mathematics, University of Belgrade. There are about 70 researchers in astronomy in Serbia (and about as many abroad) who participate in eight projects financed by the Ministry of Education and Science and in several international cooperations and projects: SREAC, VAMDC, Belissima (recruitment of experienced expatriate researchers), Astromundus (a 2-year joint master program with other four European universities), LSST. One of the goals in near future is twinning between universities in the SEE region and worldwide. The ever-increasing activities of 20 amateur astronomical societies are also given.

Highlights of Astronomy, Volume 16
XXVIIIth IAU General Assembly, August 2012
T. Montmerle, ed.

doi:10.1017/S1743921314012137

A project of a two meter telescope in North Africa

Zouhair Benkhaldoun
Oukaimden Observatory, Cadi Ayyad University

Abstract. Site testing undertaken during the last 20 years by Moroccan researchers through international studies have shown that the Atlas mountains in Morocco has potentialities similar to those sites which host the largest telescopes in world. Given the quality of the sites and opportunities to conduct modern research, we believe that the installation of a 2m diameter telescope will open new horizons for Astronomy in Morocco and north Africa allowing our region to enter definitively into the very exclusive club of countries possessing an instrument of that size. A state of the art astrophysical observatory on any good astronomical observation site should be equipped with a modern 2m-class, robotic telescope and some smaller telescopes. Our plan should be to operate one of the most efficient robotic 2m class telescopes worldwide in order to offer optimal scientific opportunities for researchers and maintain highest standards for the education of students. Beside all categories of astronomical research fields, students will have the possibility to be educated intensively on the design, manufacturing and operating of modern state of the art computer controlled instruments. In the frame of such education and observation studies several PhD and dissertational work packages are possible. Many of the observations will be published in articles worldwide and a number of guest observers from other countries will have the possibility to take part in collaborations. This could be a starting point of an international reputation of our region in the field of modern astronomy.

Highlights of Astronomy, Volume 16
XXVIIIth IAU General Assembly, August 2012
T. Montmerle, ed.

doi:10.1017/S1743921314012149

In search of a viable IAU-OAD Regional Node: A case for Africa

B. I. Okere, D. C. Okoh, I. A. Obi, P. N. Okeke and F. E. Opara
Centre for Basic Space Science, University of Nigeria Nsukka. www.cbssonline.com

Abstract. The establishment of the IAU Office of Astronomy for Development (OAD) in Cape Town, South Africa, with the aim of using astronomy to stimulate development at all levels including primary, secondary and tertiary education, science research and the public understanding of science is a welcome development to consolidate the gains of IYA2009. To assist the IAU OAD office in achieving its goal of using astronomy as a tool for development, there is need to have OAD regional nodes. In this paper, we present the astronomy activities/programs required of such a Regional Node in Africa and how the Node can play a significant role to realise the vision of Astronomy for a better world!

Highlights of Astronomy, Volume 16
XXVIIIth IAU General Assembly, August 2012
T. Montmerle, ed.

doi:10.1017/S1743921314012150

Strategic Plan of Development of Astromomy in DPRK

Sok Jong
Pyongyang Astronomical Observatory

Abstract. I would like to outline briefly, at first, an introduction to the Pyongyang Astronomical Observatory(PAO) and the present status of PAO. Next, I will mention about its future strategic plan for the development of astronomy research and education as well as the public outreach in DPRK, and the ways and means for its achievement, mainly emphasizing the international cooperation and support by the IAU such as via ROAD and cooperation programs.

---

1. History of Korean Astronomy and Present State of Astronomy in DPR Korea

Korean Ancient Astronomy: Korea has a long and glorious history of astronomical observation and a large number of astronomical relics still remain in the territory of the DPR Korea.. Before BC 2000 3000, for the first time in human history, our ancestors carved star maps with more than a hundred stars on the lids of dolmen-graves in ancient Korea. More than ten thousand dolmen-graves have been discovered in the DPR Korea, about two hundred of which include star maps. Until now, hundreds of stars in tens of constellations have been identified on dolmen-grave star maps dating from before BC 2000 3000. Astronomical charts carved on stone have also been found that date from AD 400, during the era of the Three Kingdoms. Still later, in 1395, during the Ri-dynasty era, a star map was made called CHON SANG RYOL CHA BUN YA JI DO, that is regarded by our nation with great pride. Some researchers in Pyongyang Astronomical Observatory (PAO) have attempted to clarify inherent evolution in ancient Korean star maps from dolmen-grave, through the astronomical chart carved on stone, up to CHON SANG RYOL CHA BUN YA JI DO. Between BC 3000 and 1000, our ancestors established a professional state apparatus to conduct regular observation of astronomical phenomena. The first recorded Korean observation of an eclipse was in BC 54 and that of a comet was in BC 49. Other ancient records include systemic observation of novae and supernovae.

Pyongyang Astronomical Observatory: The Pyongyang Astronomical Observatory, State Academy of Sciences of the DPR Korea, established in 1957, is the unique national astronomical research institution that plays a leading role in professional research work and public enlightenment in the DPR Korea. It is located 20 km north east of Pyongyang city, on the hill 35 m above sea level. PAO consists of about 40 researchers and engineering staff, including 6 professors and 11 PhD-level astronomers. Its research departments are Theoretical Astrophysics, Solar-terrestrial Physics, Radio Astronomy, Astrometry and Astro-geodynamics, Ephemeris, Astronomical Instruments, Korean Ancient Astronomy and an Astronomical Data Service Centre. PAO also operates its own graduate school, with Masters and PhD courses in solar physics, stellar astrophysics, and astro-geodynamics. Currently, 2 doctoral students and 6 Master students work in the graduate school together with senior researchers from several departments of PAO. PAO operates two observing stations close to Pyongyang: Sunan Geomagnetic Observing

Station and Jaesan Solar Radio Observing Station. Our observatory is also responsible for determining, maintaining and disseminating the national standard time and calculating the almanac. The main observing facilities presently in operation are a 13cm refractor, a photoelectric transit instrument, a horizontal multi-channel spectrograph manufactured by PAO and 2 radio telescope constructed by PAO, with operating frequencies of 723 MHz and 600MHz respectively. The Observatory is in charge of producing of the country-level academic quarterly journal Bulletin of Astronomy, which is the sole astronomical journal of the country and also publishes an Astronomical Calendar and aNautical Almanac annually. PAO also has a planetarium that plays an active role in public outreach for astronomy.

Other Relevant Organizations: The Kim Il Sung University, the largest university in the country, has an astronomy faculty within its physics department. Its astronomy faculty consists of 4 astronomers, who provide lectures on basic astronomy to undergraduate students in the physics department. The Kim Hyong Jik Normal University is equipped with a domed astronomical telescope, devoted to stimulating interest in astronomical education and research. The Pyongyang Childrens Palace has an astronomy circle, including about 30 primary school pupils, equipped with a telescope. This is a center of astronomy education for children. The Cosmic Exhibition in Pyongyang city is a center of public outreach together with PAO Planetarium.

International Collaboration: In recent years, international collaboration with DPRK astronomers has developed substantially. PAO has collaborated with astronomers in foreign countries to perform cooperative research works and train young researchers by means of up-to-date instruments and using observing data of the institutions. During the last decade, dozens of researchers and students have visited astronomical institutions in foreign countries, mainly China. These international contacts include the Cosmology Group, at Huairou Solar Observing Station, Changchun Observatory, Yunnan Astronomical Observatory, of National Astronomical Observatories of Chinese Academy of Sciences (NAOC), National Time Service Center (previous Shanxi Astronomical Observatory), Shanghai Astronomical Observatory, and International Center of Relativistic Astrophysics in Italy. Such visits have helped the development of astronomy in the country substantially. Recently, PAO members have taken active part in several educational activities carried out by the IAU. In 2007 three PAO staff members attended the 29th International School for Yong Astronomers (ISYA). During the school, Prof. Edward Guinan, then chair of the IAU TAD program, suggested that the DPR Korea consider rejoining the IAU. This issue was discussed further between Prof. Edward Guinan and our astronomers through the 8th Pacific Rim Conference on Stellar Astrophysics in Thailand in 2008, and the discussion was followed up by a meeting held between Prof. George Miley, Vice-President of the IAU with me and two other DPRK astronomers at an IAU TAD school in Mongolia during 2008. In 2010, 4 PAO members attended the 31st ISYA, held in China. Recently, the IAU supported a visit of our 2 staffs invited kindly by NAOC, and during the visit an agreement was made between NAOC and PAO for future collaboration and regional cooperation. The IAU also donated about a dozen astronomical books to PAO through Prof Michele Gerbaldi. These books have been of great value in pursuing our work.

2. Strategic Plan for Astronomy Development in DPR Korea

The long-term vision of PAO is to raise the level of astronomy of DPR Korea in all areas of research, education and public outreach to an internationally high level. This includes: (i) Conducting cutting-edge professional research in all of the most important areas of astronomy. (ii) Educating all people in our country in astronomy to a sufficient level to secure the benefit of astronomy as a tool of national capacity building. This objective is based upon appreciation for astronomy as a tool for technical advancement and intellectual capacity building. To achieve this ultimate vision, the first priority of PAO is to enlarge our human resources both quantitatively and qualitatively in all of professional research, school education and public outreach. It is a human-resource-centered policy.

PAOs goals for the next decade (2013-2022) are: (i) To raise the research level of several selected key areas of astronomy to an international standard; (ii) To build an infrastructure that can sustain a full-scale program of school education and public outreach in the DPRK.

The following areas of astronomy have been selected for such special focus: Stellar Astrophysics, Solar Physics, Cosmology and Applied Astronomy. Specific aspects that will be emphasized are: accretion disks on all scales, galactic dynamics, structure formation in the early universe, coronal mass ejections and solar activity and solar and astrophysical dynamos. A prerequisite for achieving these goals is to form a group of at least 10 senior researchers and more than 30 PhDs in the key areas.

In addition we plan to build an infrastructure for education and outreach, with sufficient personnel and material resources involved to achieve the above goals. We envisage continuing and expanding existing efforts and launching new programs.

Elements of the decadal plan are;

1) Increasing the PAO budget by the State Academy of Sciences: Because astronomy is an essential ingredient in the science and technology of the DPR Korea, a continual and increasing support from internal, national sources is indispensable for its stable and further development. This year, the Academy of Sciences finalized a plan for a gradual increase in the PAO budget. Initially these funds will be appropriated for reconstructing the PAO headquarters building, updating existing observational instruments and purchasing computers for our research infrastructure. In order to enlarge human resources to carry forward development and education of astronomy, funds will also be applied to provide as many postgraduate positions as possible.

2) Carrying out Postdoctoral and Doctoral works in foreign advanced institution: The most efficient way for us to enlarge human resources for astronomy quantitatively and improve it qualitatively is to provide senior researchers with postdoctoral courses or extended research visits and to educate PhDs, through doctoral courses in foreign advanced institutions. The IAUs support in negotiating the provision of such postdoctoral and doctoral positions is a main motivation for rejoining the IAU. To achieve our goals we need 2 postdoctoral positions (for less than 2 years periods) every 2 years and 2 doctoral positions (for 3 years periods) every year. An objective of this training will be to produce senior researchers, who will be key members of our staff, lead research groups and be supervisors for domestic postgraduate courses.

Recently, an agreement on collaboration was made between NAOC and PAO, that provides for the initiation of joint PhD programs and advanced training programs. Funds to support this program consist of an in-kind contribution from the NAOC and scholarships from the foreign student program of the Graduate School of Chinese Academy of Sciences (GSCAS). These programs will both provide an important start to the training needed to accomplish the DPRK astronomy strategic plan and are excellent examples of regional collaboration.

We look to the IAU OAD and the IAU East Asia ROAD to help us to find suitable postdoctoral and doctoral positions at up-to-date institutions abroad, including European countries and the East Asia region. We hope that leading astronomers, who are deeply concerned for astronomy development in our country, will exercise their influence in making available non-IAU resources and in-kind contributions from astronomy institutions throughout the world.

3) Taking active part in the IAUs existing and newly launched programs: Previous IAU educational and development activities such as ISYA and TAD have played an important role in awakening and improving astronomy development in our country. We consider the IAUs newly launched programs such as Astronomy Institute Twining and the Endowed Lecturer Program to be very useful in context of our national strategic plan. We hope that twinning of the PAO with appropriate institutions abroad will help provide material resources for research and education, advice on the optimum strategy for DPRK astronomy and help in creating new disciplines and graduate courses.

4) Developing Astronomy Education in Universities More than ten DPRK universities including Kim Il Sung University; University of Natural Sciences and Kim Hyong Jik Normal University have physics departments, and the graduates are a good source of future astronomy researchers. We intend to provide active backing to these universities in intensifying their astronomy education; especially by giving lectures on general astronomy and through supervising graduate students from these universities. During the next decade, PAO will work constructively to introduce astrophysics as a subject in the physics curriculum of DPRK university physics departments.

5) Activating school education and public outreach The 5-member PAO planetarium group is presently involved in school education and public outreach throughout the country. The group is responsible for liaison with relevant organizations including the Ministry of Education and for the development of educational resources. They also devote effort into propagating the knowledge of astronomical phenomena and universe through widespread communication. The group will be expanded to a center for coordinating domestic activities and introducing external advanced resources in school education and public outreach.

Our guiding principle in developing material resources for education and outreach is that of introduction and distribution rather than self-creation. We shall concentrate our efforts on introducing ready-made resources from over the world and distributing them widely to each level of domestic organization rather than making completely new material. Our staff will endeavor to accumulate high quality educational resources in consultation with appropriate organizations throughout the world and to spread these widely to schools and other relevant organizations in the DPRK.

6) Constructing Observational Instruments Nowadays frontier astronomical facilities are too expensive even for rich countries to build alone. For the present, we will endeavor to access international observational data and build a computer infrastructure that can reduce and analyze these data using open-source data reduction packages. We hope to obtain knowledge about the use of such packages and practical techniques of data reduction through the visits to foreign institutes mentioned above In order to gain expertise in the manufacture of instrumentation and telescopes for future work, we also plan to construct some small-scale instruments. In the near future we have a plan to make a telescope for an artificial satellite.

3. Concluding Remark

PAO, the main body for astronomy in the DPRK, has launched an ambitious strategic plan to boost astronomical research, education and public outreach and to use it as a tool for national development. Rejoining the IAU is an important milestone towards achieving this plan. The plan involves training senior researchers and graduate students, who will become our future key astronomers. The plan needs both internal and external support and is based on the assumption that we will obtain maximum support from the IAUs various Astronomy for Development programs including the OAD and the East Asia ROAD/LOAD. We also seek help from the IAU in negotiating long-term visits at foreign institutions needed to realize our goals. We sincerely hope that the IAU and the world astronomy community will support our ambitious efforts.

4. Acknowledgement

On the significant occasion rejoining the IAU, I express my sincere thanks to Prof. George Miley for kind advice and patient effort and Prof. Ed Guinan for his pioneering suggestion and continual care. I thank the Local Organizing Committee for supporting attendance of DPR Korea delegation to IAU General Assembly.

Highlights of Astronomy, Volume 16
XXVIIIth IAU General Assembly, August 2012
T. Montmerle, ed.

doi:10.1017/S1743921314012162

Armenia as a Regional Centre for Astronomy for Development activities

A. Mickaelian

Byurakan Astrophysical Observatory (BAO)
E-mail: aregmick@yahoo.com

Abstract. The Byurakan Astrophysical Observatory (BAO, Armenia, http://www.bao.am) are among the candidate IAU Regional Nodes for Astronomy for Development activities. It is one of the main astronomical centers of the former Soviet Union and the Middle East region. At present there are 48 qualified researchers at BAO, including six Doctors of Science and 30 PhDs. Five important observational instruments are installed at BAO, the larger ones being 2.6m Cassegrain (ZTA-2.6) and 1m Schmidt (the one that provided the famous Markarian survey). BAO is regarded as a national scientific-educational center, where a number of activities are being organized, such as: international conferences (4 IAU symposia and 1 IAU colloquium, JENAM-2007, etc.), small workshops and discussions, international summer schools (1987, 2006, 2008 and 2010), and Olympiads. BAO collaborates with scientists from many countries. The Armenian Astronomical Society (ArAS, http://www.aras.am/) is an NGO founded in 2001; it has 93 members and it is rather active in the organization of educational, amateur, popular, promotional and other matters. The Armenian Virtual Observatory (ArVO, http://www.aras.am/Arvo/arvo.htm) is one of the 17 national VO projects forming the International Virtual Observatories Alliance (IVOA) and is the only VO project in the region serving also for educational purposes. A number of activities are planned, such as management, coordination and evaluation of the IAU programs in the area of development and education, establishment of the new IAU endowed lectureship program and organization of seminars and public lectures, coordination and initiation of fundraising activities for astronomy development, organization of regional scientific symposia, conferences and workshops, support to Galileo Teacher Training Program (GTTP), production/publication of educational and promotional materials, etc.

Highlights of Astronomy, Volume 16
XXVIIIth IAU General Assembly, August 2012
T. Montmerle, ed.

doi:10.1017/S1743921314012174

An exemplary developing astronomy movement in Nepal

Sudeep Neupane

Astronomical Society (NASO) , Astrophysics and Cosmology Research Group, TU

Abstract. Astronomy and space science education had been given least importance by Nepalese government in the past. The modern astronomy movement is believed to have started when an official observation programme of Haley's comet was organized by Royal Nepal Academy of Science and Technology (RONAST) in 1986. Following the huge pressure from the scientific community, the Nepal government (Kingdom of Nepal at that time) established B.P. Koirala Memorial Planetarium, Observatory and Science Museum Development Board in 1992. Initiatives of the project started with observatory set up and the development of astrophysics syllabus for university students. Astrophysics is included as an elective paper in the Physics masters course. The lead astrophysicist of Nepal Dr. Binil Aryal is running a research group in Tribhuvan University since 2005 which has a significant number of international publications. The developing government initiatives and achievements will be discussed.

In 2007, a group of astronomy enthusiastic students along with amateurs working independently in past established Nepal Astronomical Society (NASO), which surprisingly increased the amateur activities and inspired other amateur groups to revive. During IYA 2009, more than 80 outreach and observation events were organized solely by NASO. NASO was able to collaborate with many international programmes and projects like GHOU/GTTP, EurAstro, AWB, UNAWE, SGAC, Star Peace, TWAN etc during and beyond IYA2009. Currently Nepal is recognized as the most eventful country of outreach and astronomy education among the amateur community. The success story of the astronomy movement and the local difficulties while organizing the events will be explained.

Keywords. Outreach, NASO, NAST, Planetarium, Observatory, GHOU, GTTP, UNAWE

Highlights of Astronomy, Volume 16
XXVIIIth IAU General Assembly, August 2012
T. Montmerle, ed.

doi:10.1017/S1743921314012186

Astronomy in Mozambique

Valério A. R. M. Ribeiro[1] and Cláudio M. Paulo[2]
[1]UCT, South Africa [2]UEM, Mozambique

Abstract. We present the state of Astronomy in Mozambique and how it has evolved since 2009 following the International Year of Astronomy. Activities have been lead by staff at University Eduardo Mondlane and several outreach activities have also flourished. In 2010 the University introduced its first astronomy module, Introduction to Astronomy and Astrophysics, for the second year students in the Department of Physics. The course has now produced the first students who will be graduating in late 2012 with some astronomy content. Some of these students will now be looking for further studies and those who have been keen in astronomy have been recommended to pursue this as a career. At the university level we have also discussed on the possibility to introduce a whole astronomy course by 2016 which falls well within the HCD that the university is now investing in. With the announcement that the SKA will be split between South Africa with its partner countries (including Mozambique), and Australia we have been working closely with the Ministry of Science and Technology to make astronomy a priority on its agenda. In this respect, an old telecommunications antenna is being converted by the South Africa SKA Project Office, and donated to Mozambique for educational purposes. It will be situated in Maluana, Mozambique.

Highlights of Astronomy, Volume 16
XXVIIIth IAU General Assembly, August 2012
T. Montmerle, ed.

doi:10.1017/S1743921314012198

Developing Astronomy Research and Education in the Philippines

R. M. D. Sese[1] and M. B. N. (Thijs) Kouwenhoven[2]

[1]University of the Philippines Los Banos, the Philippines
[2]Kavli Institute for Astronomy and Astrophysics, Peking University, P.R. China
email: rmdsese@gmail.com & kouwenhoven@pku.edu.cn

Abstract. In the past few years, the Philippines has been gradually developing its research and educational capabilities in astronomy and astrophysics. In terms of astronomy development, it is still lagging behind several neighboring Southeast Asian countries such as Indonesia, Thailand and Malaysia, while it is advanced with respect to several others. One of the main issues hampering progress is the scarcity of trained professional Filipino astronomers, as well as long-term visions for astronomy development. Here, we will be presenting an overview of astronomy education and research in the country. We will discuss the history and current status of astronomy in the Philippines, including all levels of education, outreach and awareness activities, as well as potential areas for research and collaborations. We also discuss issues that need to be addressed to ensure sustainable astronomy development in the Philippines. Finally, we discuss several ongoing and future programs aimed at promoting astronomy research and education. In essence, the work is a precursor of a possible white paper which we envision to submit to the Department of Science and Technology (DOST) in the near future, with which we aim to further convince the authorities of the importance of astrophysics. With the support of the International Astronomical Union (IAU), this may eventually lead to the creation of a separate astronomy agency in the Philippines.

Highlights of Astronomy, Volume 16
XXVIIIth IAU General Assembly, August 2012
T. Montmerle, ed.

doi:10.1017/S1743921314012204

Some thoughts about the IAU Strategic Plan in Latin America and the Caribbean

Silvia Torres-Peimbert
Instituto de Astronomía, Universidad Nacional Autónoma de México

Abstract. The distribution of overall education and culture among the population in Latin America and the Caribbean has a very wide range; this is also the case for astronomical engagement. The route to follow in the strategic plan needs to address this situation, for the different levels of development. In particular, guidelines should be established for the regional node(s) where achievable goals should be set up and evaluated periodically. I present a set of ideas on this subject.

The long-term vision of the IAU states (a) that all countries will participate at some level in international astronomical research, and (b) that all children throughout the world will be exposed to some knowledge about astronomy and the Universe at school in support of their education. Three aspects that need attention: (i) astronomy for the public, (ii) astronomy education, and (iii) astronomy at universities and research level.

I review some of the activities that have taken place in the continent. At the professional level, 13 Latin American Regional Meetings have been held since 1977; they have been by Argentina (2), Brasil (2), Chile (3) Mexico (3) Uruguay (1) and Venezuela (2); the next one will take place in Brasil. Regarding short courses at university level over the years, several International Schools for Young Astronomers have been held; I can mention: Trinidad Tobago, Paraguay, Brasil, and Mexico. On the public outreach areas, there were widespread activities during the International Year of Astronomy 2009, that have triggered interest in the communities to do follow up actions. About primary or secondary studies, although it is difficult to generalize, since each country has its own peculiarities; at least in Mexico, we have been unable to impact at this level.

To conclude, there are many aspects that can be pursued at all levels so there is plenty of work in this field. Given the large size of the continent, perhaps several OADs should be supported for specific regions, with target goals within the IAU program.

I am grateful to DGAPA-PAPIIT-IN105511 and CONACyT 129753 for their support.

Highlights of Astronomy, Volume 16
XXVIIIth IAU General Assembly, August 2012
T. Montmerle, ed.

doi:10.1017/S1743921315001647

Preface: Special Session SpS12 Modern views of the interstellar medium

The interstellar medium (ISM) plays an important role in the formation and evolution of a galaxy. The ISM provides the material to form stars and stars in turn inject radiation, metals, and mechanical energy into the ISM, altering the physical conditions, abundances, and distribution of the ISM and affecting future generations of star formation. It is thus essential that we understand the physical structure of the ISM, the physical processes that operate in the ISM, and the interplay between stars and ISM.

The physical structure and processes of the ISM are best studied in the Galaxy and nearby galaxies. Many large-scale surveys for different components of the ISM are available: HI – the International Galactic Plane Survey (IGPS) and The HI Nearby Galaxy Survey (THINGS) maps out the HI in nearby galaxies; dust – Spitzer, Herschel and Planck surveys of the Galactic plane and the Magellanic Clouds, Bolocam Galactic Plane Survey (BGPS), and the APEX Telescope Large Area Survey of the Galaxy (ATLASGAL); HII – the Wisconsin H-Alpha Mapper (WHAM) survey of the distribution and velocities of warm ionized gas in the Galaxy; 10^6 K gas – Chandra and XMM-Newton observations; cosmic rays – Fermi Gamma-Ray Observatory's whole sky survey.

In the meantime, considerable progress has been made in the numerical modeling of local and global conditions of the ISM, its morphology, and its time-dependent evolution, owing to rapid advances in the development of hard- and software. It is now possible to follow the full non-linear evolution of a plasma by solving the hydro- or MHD equations in high resolution simulations with adaptive mesh refinement. One of the key results of the past years was to recognize and quantitatively describe the role of compressible turbulence in the ISM and its impact on the distribution of gas into phases, on the mixing of chemically enriched material, on the volume and mass filling factors of the ISM plasma, and on its heating and cooling history.

Special Session 12, "Modern Views of the ISM", is organized to update people on recent advances in the ISM observations and theories. The key topics include (1) physical structure and phase distribution of the ISM in a galaxy, (2) multi-wavelength observations of ISM in the Galaxy and nearby galaxies, (3) recent theory/MHD simulation of ISM in a galaxy, (4) interstellar disk-halo connection in galaxies, and (5) interplay between stars and ISM: star formation and feedback. This special session is dedicated to the late Professor John Dyson (1941-2010), the IAU Div VI president in 2003-2006, who pioneered the dynamical interactions of stellar winds and outflows with the interstellar gas.

Scientific Organizing Committee

D. Breitschwerdt (co-Chair, Germany), Y.-H. Chu (co-Chair, USA), M. de Avillez (Portugal), E. de Blok (South Africa), E. de Gouveia Dal Pino (Brazil), R.-J. Dettmar (Germany), E. Falgarone (France), T. Hartquist (UK), B.-C. Koo (Korea), N. McClure-Griffiths (Australia), E. Ostriker (USA), J. X. Prochaska (USA), L. V. Tóth (Hungary), E. Vazquez-Semadeni (Mexico), K. Wada (Japan), M. Wolfire (USA), J. Yang (China).

You-Hua Chu (ed.)
Astronomy Department, University of Illinois, Urbana, IL 61801, USA; yhchu@illinois.edu

Highlights of Astronomy, Volume 16
XXVIIIth IAU General Assembly, August 2012
T. Montmerle, ed.

doi:10.1017/S1743921314012216

The pre-modern era of the ISM

John M. Dickey

School of Maths and Physics, University of Tasmania, Hobart, TAS 7005, Australia
email: john.dickey@utas.edu.au

Abstract. The history of our understanding of the interstellar medium has a pre-modern era and an early-modern era. Several threads show continuity in observational approach and interpretation, reaching from the first decade of the last century to the present. The first 50 years were a period of pioneering work, done mostly by stellar astronomers, who chanced upon the ISM by a strange variety of pathways.

Keywords. ISM: general, ISM: evolution, ISM: clouds, ISM: structure, Galaxy: evolution

---

The study of the interstellar medium (ISM) began around 1900, and the period from then to about 1950 can be considered the pre-modern era. The early-modern era lasts from about 1950 to the late 1970s. In this written version of the talk I have room only to discuss the former epoch.

Astronomers have recognised bright nebulae, regions of extended emission, and dark nebulae, regions where starlight is obscured by foreground clouds, since before the time of Herschel (1833). Around the turn of the last century, photographic techniques had advanced far enough to allow wide-field plates to record both light and dark nebulae (e.g. Draper 1880, Barnard 1903). Barnard in particular took an interest in the large scale structure of the dark nebulae in the constellation Orion that anticipated the mapping work that came some 40 years later. But telescopes of that period were intrinsically small-field, and surface-brightness sensitivity limitations prevented very large scale photographic imaging of the faint structures that Barnard could see and draw.

In 1904 two papers were published that started the two traditional methods of studying the ISM that continued for much of the 20th Century. In Germany, Johannes Hartmann (1904) accidentally discovered a system of spectral lines, notably Ca, toward a binary star whose orbit he was studying. The Ca lines did not change in velocity as the stellar lines moved back and forth due to the Doppler shift of the binary star δ Orionis. Hartmann correctly interpreted these lines as due to an intervening cloud of gas. Although his conclusion was revolutionary, he spends only one page on it in an 18 page paper on the stellar velocities. He points out that he had seen a similar Ca line in the spectrum of Nova Persei 1901, distinguished from the nova lines both by their different center velocities and their much narrower line-widths.

In the same year, George Comstock (1904) published the results of an extensive survey of stars with distances based on their apparent recoil proper motion due to the Sun's motion relative to the local standard of rest, which had recently been measured. Comstock then computed the expected apparent magnitudes of samples of stars at different distances, and found them to be fainter than his prediction. He ascribes their faintness to an intervening absorbing medium, and he computes a visual obscuration not very different from the modern average of 1.8^m kpc^{-1}. Although Comstock's conclusion drew a bitter refutation from Jacobus Kapteyn (1904), by 1909 Kepteyn seems to have accepted Comstock's absorbing medium. But many other astronomers refused to accept extinction even as the seemingly obvious explanation for the zone of avoidance (e.g. Shapley

1918). After the turbulent years of the 1920s, when our paradigm for the universe completely changed, the role of the ISM as a reddening and obscuring medium was generally accepted (reviewed by Trumpler 1930).

Spectroscopic study of the ISM advanced rapidly in the 1930s, with detection of more absorption systems and more species of atoms and the first interstellar molecules: CN, CH and CH^+ (Dunham 1937, reviewed by Adams 1949). The first ionization equilibrium calculations to explain diffuse hydrogen emission line regions were done by Ambartsumian (1949) and Strömgren (1939), though Strömgren credits Struve and Elvey (1938) with the first computation of the ionization requirements of what we now call H II regions. At this point the spectroscopic study of the ISM divided into studies of emission lines and absorption lines, which usually sample very different solid angles and often detect atoms in different states of excitation, a distinction that continues today, at all wavelengths.

In the late 1930s and 1940s, working with telescopes at the Burakan and Ambastumani Observatories, Ambartsumian and his collaborators made remarkable progress in quantifying the structure of the dark dust clouds that can be traced by their extinction of the light from background stars and galaxies. Both Chandrasekhar and Münch (1950) and Strömgren (1983) credit his work with changing the paradigm for the structure of the cold ISM from a smooth medium to a population of clouds with low filling factor and high internal density, with about six clouds per kpc on a random line of sight. Ambartsumian's results were later picked up by Spitzer in his seminal textbooks, and the results came to be called Spitzer-standard clouds.

By the early 1950s a flood of new results from both theory and observations of the ISM changed the nature of the field from a curiosity indulged in occasionally by stellar astronomers, to a science in its own right. Before the early 1950s astronomers took stars as a celestial given. After their energy source was understood in detail, and their initial mass function was determined (Salpeter 1955), it became clear that stars had to form continuously through the age of the universe, right to the present. So the role of the ISM as the evolutionary partner of stellar nucleosynthesis in the chemical evolution of galaxies was at last appreciated.

References

Adams, W. S. 1949, *ApJ* 109, 354.

Ambartsumian, V. 1932, *MNRAS* 93, 50.

Barnard, E. E. 1903, *ApJ* 17, 77.

Chandrasekhar, S. & Münch, G. 1950, *ApJ* 112, 380.

Comstock, G. C. 1904, *AJ* 24, 139.

Dunham, T. 1937, *PASP* 49, 26.

Draper, H. 1880, *Nature* 22, 583.

Hartmann, J. 1904, *ApJ* 19, 268.

Herschel, J. 1833, *A Treatise on Astronomy*, edited and reproduced by Longmans *et al.* (Cambridge University Press, 2009) http://messier.seds.org/xtra/similar/herschel.html

Kapteyn, J. C. 1904 *AJ*, 24, 115.

Kapteyn, J. C. 1909 *ApJ*, 29, 46.

Salpeter, E. E. 1955, *ApJ* 121, 161.

Shapley, H. 1918, *PASP* 30, 42.

Strömgren, B. 1939, *ApJ* 89, 526.

Strömgren, B. 1983, *ARAA* 21, 1.

Struve, O. and Elvey, C. T. 1938, *ApJ* 88, 364.

Trumpler, R. J. 1930, *PASP* 42, 214.

Highlights of Astronomy, Volume 16
XXVIIIth IAU General Assembly, August 2012
T. Montmerle, ed.

doi:10.1017/S1743921314012228

Modern view of the warm ionized medium

A. Hill[1], R. Reynolds[2], L. Haffner[2], K. Wood[3] and G. Madsen[4]

[1]CSIRO Astronomy & Space Science, Epping, NSW, Australia
[2]Department of Astronomy, University of Wisconsin-Madison, USA
[3]School of Physics and Astronomy, University of St Andrews, Scotland
[4]Sydney Institute for Astronomy, School of Physics, The University of Sydney, NSW, Australia

Abstract. We review the observational evidence that the warm ionized medium (WIM) is a major and physically distinct component of the Galactic interstellar medium. Although up to $\sim$ 20% of the faint, high-latitude Hα emission in the Milky Way may be scattered light emitted in midplane HII regions, recent scattered light models do not effectively challenge the well-established properties of the WIM.

The discovery of free-free absorption of the Galactic synchrotron background led Hoyle & Ellis (1963) to propose the existence of an extended layer of diffuse ionized gas, now known in the Milky Way as the warm ionized medium (WIM). Further evidence for the WIM came from the dispersion of pulsar signals (Guélin 1974). These observations predicted diffuse optical H recombination line emission. A subsequent search using sensitive, high spectral resolution Fabry-Perot spectrometers found Hα emission in agreement with these predictions and a filling factor of $\sim$ 30% (Reynolds *et al.* 1973).

The WIM is now well-established as a major component of the interstellar medium of the Milky Way and other disk galaxies (see review by Haffner *et al.* 2009). It consists of a plasma with nearly fully ionized hydrogen which extends more than 1 kpc from the midplane. The warm and cold neutral media are optically thick to ionizing photons, but only Lyman continuum radiation from hot stars has the energy required to maintain the observed surface recombination rate of the WIM. This challenge has been largely resolved by models in which supernova-driven turbulence and superbubble structures allow Lyman continuum photons to propagate from the midplane to heights $|z| \sim 1$ kpc and ionize the WIM (Wood *et al.* 2010, and references therein).

The discovery of Hα emission from the WIM led to the discovery of other faint nebular emission lines from this warm, widespread plasma. These lines distinguish the WIM from the HII regions surrounding hot stars, as demonstrated in Figure 2 of Madsen *et al.* (2006): along a sightline which contains a classical (locally-ionized) HII region and emission from the WIM at different velocities, the Hα emission is much brighter in the HII region while the [NII]λ6584 and [SII]λ6716 (hereafter [SII]) emission are each brighter in the WIM. This trend is observed in a variety of environments, including at $v_{\rm LSR} = 0$ at high Galactic latitudes, in the Perseus Arm, in M33, and in the edge-on galaxies NGC 891 and NGC 55 at heights $|z| \gtrsim 0.5$ kpc: the [NII]λ6584/Hα and [SII]/Hα line ratios are enhanced in the WIM compared to HII regions (Rand *et al.* 1990; Ferguson *et al.* 1996; Haffner *et al.* 1999; Hoopes & Walterbos 2003; Madsen *et al.* 2006).

In combination with observations of the [NII]λ5755/[NII]λ6584 line ratio, a direct temperature diagnostic, these observations indicate that the temperature in the WIM is $\approx$ 9000 K while the temperature in classical HII regions is $\approx$ 6000 K (Reynolds *et al.* 2001). Simultaneously, weak [OIII]λ5007, strong [OII]λ3726, and weak He I intensities relative to Hα indicate that the ionization state of both He and O are lower in the

WIM than in classical H II regions (Mierkiewicz *et al.* 2006; Madsen *et al.* 2006). Howk & Consiglio (2012) identified similar trends in UV absorption line observations.

Like Reynolds *et al.* (1973), Wood & Reynolds (1999) found that 5 – 20% of the Hα emission not directly associated with classical H II regions could be scattered light from midplane H II regions, not emitted *in situ* from ionized gas. However, the distinct spectral properties of the WIM combined with the evidence from pulsars that free electrons in the solar neighborhood extend to much larger heights than the OB stars are difficult to explain with a significantly larger scattered light contribution. By correlating 100μm and Hα emission and extrapolating from the high-latitude dust cloud LDN 1780, Witt *et al.* (2010) argued that the scattered light contribution could be up to 50% on sightlines with a large dust column and estimate that the most probable scattered light contribution is 0.1 R, about 20% of the most probable high-latitude Hα intensity of 0.52 R. They predict a large scattered light component and therefore low [S II] and [N II] in many sightlines in the southern Milky Way; we will test this hypothesis in new observations with the Wisconsin Hα Mapper, now located at Cerro Tololo in Chile.

Seon & Witt (2012) assert that the emission attributed to the WIM is dominated by scattered light. They have constructed models reproducing the observed [N II]λ6584/Hα and [S II]/Hα line ratios at high latitude through scattered light from late O and B star H II regions. However, their model does not provide any explanation for the similarity of the line ratios in diffuse emission from the solar neighborhood at high latitudes and at altitudes well above the majority of the late OB stars ($|z| \gtrsim 0.5$ kpc) in the Perseus Arm and edge-on galaxies. They note that all O9 and later stars only produce enough ionizing photons to account for half of the diffuse Hα, so this scattered light can at most be one of a number of contributors to the optical emission. Seon & Witt (2012) also argue that the Hα from the WIM may be overestimated due to contamination by Hα from stars. However, this cannot explain the observation that [N II] and [S II] are bright while [O III] is faint (relative to Hα) in the WIM. Moreover, they argue that some of the [S II] emission attributed to the WIM may be emitted in neutral gas. While this is plausible in principle, [O I]λ6300 observations indicate that the H ionization fraction within the optically emitting gas is $\gtrsim 0.9$ (Reynolds *et al.* 1998).

References

Ferguson, A. M. N., Wyse, R. F. G., & Gallagher, J. S. 1996, *AJ* 112, 2567

Guélin, M. 1974, in *IAUS* 60, 51

Haffner, L. M., Reynolds, R. J., & Tufte, S. L. 1999, *ApJ* 523, 223

Haffner, L. M., Dettmar, R.-J., Beckman, J. E., *et al.* 2009, *Rev. Mod. Phys.* 81, 969

Hoopes, C. G. & Walterbos, R. A. M. 2003, *ApJ* 586, 902

Howk, J. C. & Consiglio, S. M. 2012, *ApJ* 759, 97

Hoyle, F. & Ellis, G. R. A. 1963, *Australian Journal of Physics* 16, 1

Madsen, G. J., Reynolds, R. J., & Haffner, L. M. 2006, *ApJ* 652, 401

Mierkiewicz, E. J., Reynolds, R. J., Roesler, F. L., Harlander, J. M., & Jaehnig, K. P. 2006, *ApJ* (Letters) 650, L63

Rand, R. J., Kulkarni, S. R., & Hester, J. J. 1990, *ApJ* (Letters) 352, L1

Reynolds, R. J., Hausen, N. R., Tufte, S. L., & Haffner, L. M. 1998, *ApJ* (Letters) 494, L99

Reynolds, R. J., Roesler, F. L., & Scherb, F. 1973, *ApJ* 179, 651

Reynolds, R. J., Sterling, N. C., Haffner, L. M., & Tufte, S. L. 2001, *ApJ* (Letters) 548, L221

Seon, K.-I. I. & Witt, A. N. 2012, *ApJ* 758, 109

Witt, A. N., Gold, B., Barnes, F. S., *et al.* 2010, *ApJ*, 724, 1551

Wood, K., Hill, A. S., Joung, M. R., *et al.* 2010, *ApJ* 721, 1397

Wood, K. & Reynolds, R. J. 1999, *ApJ* 525, 799

Highlights of Astronomy, Volume 16
XXVIIIth IAU General Assembly, August 2012
T. Montmerle, ed.

doi:10.1017/S174392131401223X

On the origins of the diffuse Hα emission: ionized gas or dust-scattered Hα halos?

Kwang-Il Seon[1] and Adolf N. Witt[2]

[1]Korea Astronomy and Space Science Institute, Daejeon 305-348, Korea
[2]Ritter Astrophysical Research Center, University of Toledo, Toledo, OH 43606, USA

Abstract. We find that the dust-scattering origin of the diffuse Hα emission cannot be ruled out. As opposed to the previous contention, the expected dust-scattered Hα halos surrounding H II regions are, in fact, in good agreement with the observed Hα morphology. We calculate an extensive set of photoionization models by varying elemental abundances, ionizing stellar types, and clumpiness of the interstellar medium (ISM) and find that the observed line ratios of [S II]/Hα, [N II]/Hα, and He I λ5876/Hα in the diffuse ISM accord well with the dust-scattered halos around H II regions, which are photoionized by late O- and/or early B-type stars. We also demonstrate that the Hα absorption feature in the underlying continuum from the dust-scattered starlight ("diffuse galactic light") and unresolved stars is able to substantially increase the [S II]/Hα and [N II]/Hα line ratios in the diffuse ISM.

Keywords. ISM: general, HII regions, scattering

It is known that the diffuse Hα emission outside of bright H II regions not only are very extended, but also can occur in distinct patches or filaments far from H II regions, and the line ratios of [S II]/Hα and [N II]/Hα observed far from bright H II regions are generally higher than those in the H II regions. These observations have been regarded as evidence against the dust-scattering origin of the diffuse Hα emission (including other optical lines), and the effect of dust scattering has been neglected in studies on the diffuse Hα emission. In this talk, we demonstrate that the basic morphological properties of the diffuse H emission can be explained well by the dust-scattered halo surrounding the H II nebula, as opposed to the previous belief that dust scattering does not accord with the H and R-band observations. We show that the optical line ratios of He I/Hα, [N II]/Hα, and [S II]/Hα observed in the diffuse ISM outside of bright H II regions can be reproduced well by the dust-scattered halo scenario, wherein the optical lines originate from H II regions ionized by late O- or early B-type stars in the media with abundances close to WNM and are scattered off by the interstellar dust. The predicted [N II]/Hα and [S II]/Hα line ratios increase with the clumpiness of ISM.

We find that the surface brightness of the individual dust-scattered halos can be approximated well by a formula used by Zurita . *et al.* (2002) and Seon (2009). These results seem to strongly suggest the dust-scattering origin of the diffuse Hα emission in the face-on galaxies. We also show that the Balmer absorption lines in the underlying stellar continuum seem to explain the rise of line ratios in the faint Hα regions. We may need to develop more realistic models of the global Hα morphologies of face-on galaxies in the contexts of not only dust scattering, but also photoionization.

References

Zurita, A., Beckman, J. E., Rozas, M., & Ryder, S. 2002, *A&A* 386, 801
Seon, K.-I. 2009, *ApJ* 703, 1159

Highlights of Astronomy, Volume 16
XXVIIIth IAU General Assembly, August 2012
T. Montmerle, ed.

doi:10.1017/S1743921314012241

Galactic cold cores

M. Juvela, on behalf of the *Planck* and *Herschel* projects on cold cores

Department of physics, University of Helsinki, FI-00014 Helsinki, Finland
email: mika.juvela@helsinki.fi

Abstract. The project *Galactic Cold Cores* is studying the early stages of Galactic star formation using far-infrared and sub-millimetre observations of dust emission. The *Planck* satellite has located many sources of cold dust emission that are likely to be pre-stellar clumps in interstellar clouds. We have mapped a sample of *Planck*-detected clumps with the *Herschel* satellite at wavelengths 100-500 μm. *Herschel* has confirmed the *Planck* detections of cold dust and have revealed a significant amount of sub-structure in the clumps. The cloud cores have colour temperatures in the range of 10–15 K. However, star formation is often already in progress with cold clumps coinciding with mid-infrared point sources. In less than half of the cases, the cloud morphology is clearly dominated by filamentary structures. The sources include both nearby isolated globules and more distant, massive clouds that may be off-the-plane counterparts of infrared dark clouds.

The *Herschel* observations have been completed and the processed maps will be released to the community in 2013.

Keywords. ISM: clouds, stars: formation, infrared: ISM, submillimeter; dust, extinction

---

The main phases of star formation – from molecular clouds, via dense clumps to the formation and collapse of protostellar cores – are already understood (McKee & Ostriker 2007). Our knowledge is based mainly on detailed observations of the nearest star forming regions and on numerical modelling. However, the formation of each star is an individual process. For the full picture, we need to study sources in different environments and in different stages of their evolution.

The conditions within the cold molecular clouds must dictate the characteristics of star formation. These include the star formation efficiency, the mode of star formation (clustered vs. isolated), and the stellar initial mass function (Elmegreen 2011; Padoan & Nordlund 2011). The relationships can be examined in different environments to determine the interplay between turbulence, magnetic fields, kinematics, and gravity. The importance of external triggering has been demonstrated in individual cases but we are still lacking a global picture of its importance for present-day star formation.

The *Planck* all-sky sub-millimetre maps (Tauber *et al.* 2010) provide data for a global census of the coldest component of interstellar medium. The selection of compact cold sources has led to a list of over 10000 objects (Planck collaboration 2011a), mainly of ~1 pc sized clumps and even larger structures. Because temperatures of T <14 K are possible only in very dense and well-shielded cloud regions, *Planck* can pinpoint the regions where star formation is likely to take place. Many *Planck* clumps will correspond pre-stellar or already protostellar cores. This unbiased survey (in terms of sky coverage) provides a good starting point for statistical studies. In the *Herschel* Open Time Key Programme *Galactic Cold Cores*, we have mapped selected *Planck* clumps with the *Herschel* PACS and SPIRE instruments (100–500 μm). The higher spatial resolution of *Herschel* (Poglitsch *et al.* 2010; Griffin *et al.* 2010) makes it possible to study the cloud structure,

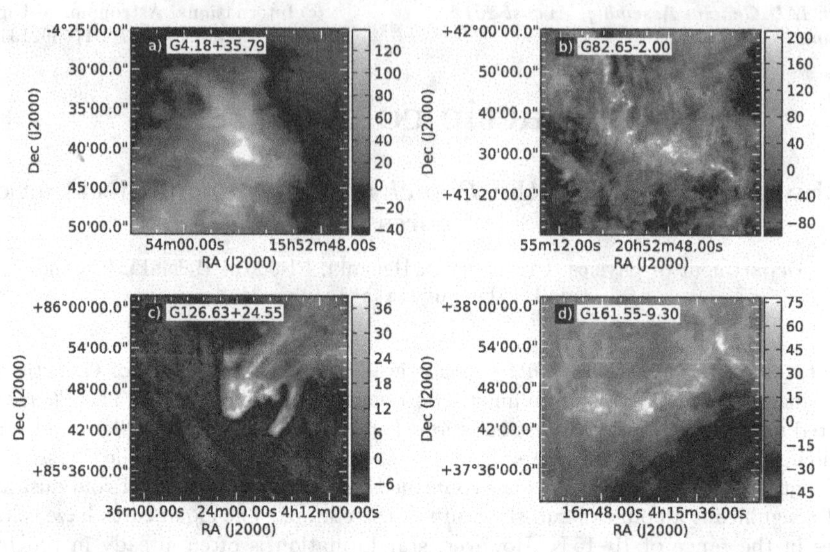

Figure 1. *Herschel* maps of four clouds with cold clumps. There are occasionally signs of dynamical interactions that are directly molding the clouds (see Juvela *et al.* 2012).

often resolving the individual cores. The extension to shorter wavelengths helps the study of dust properties and of the column density and temperature variations.

Our *Herschel* survey includes 117 fields with map sizes between 12 arcmin and one degree. The target selection was done based on *Planck* detections, ensuring the full coverage of clump masses and Galactic locations. The regions covered by the other programmes like Hi-GAL (Molinari *et al.* 2010), the Gould Belt Survey (André *et al.*2010) or HOBYS (Motte *et al.* 2010) were avoided. Our fields cover altogether over 300 individual *Planck* clumps. Their distance distribution extends from 100 pc to over 4 kpc and is similar to the distribution of all *Planck* clumps (Planck collaboration 2011a). Distances have been estimated for ~80% of the fields using various methods (e.g., Marshall *et al.* 2009).

The first Planck results on cold cores were discussed in Planck collaboration (2011a) and Planck collaboration (2011b). Preliminary results of the *Herschel* survey have been presented in papers Juvela *et al.* (2010, 2011, 2012; see Fig.1). The final reduced *Herschel* maps of the 117 fields will be made public in 2013.

References

André, P., Men'shchikov, A., Bontemps, S., *et al.* 2010, *A&A* 518, L102
Elmegreen, B. G. 2011, *ApJ* 731, 61
Griffin, M. J., Abergel, A., Abreu, A.s, *et al.* 2010, *A&A* 518, L3
Juvela, M., Ristorcelli, I., Montier, L., *et al.* 2010, *A&A* 518, L93
Juvela, M., Ristorcelli, I., Pelkonen, V.-M., *et al.* 2011, *A&A* 527, A111
Juvela, M., Ristorcelli, I., Pagani, L., *et al.* 2012 *A&A* 541, A12
Marshall, D. J., Joncas, G., & Jones, A. P 2009, *ApJ* 706, 727
McKee, C. F. & Ostriker, E. C. 2007, *ARAA* 45, 565
Molinari, S., Swinyard, B., Bally, J., *et al.* 2010, *A&A* 518, L100
Motte, F., Zavagno, A., Bontemps, S., *et al.* 2010, *A&A* 518, L77
Padoan, P. & Nordlund, A. A. 2011, *ApJ* 741, L22
Planck collaboration., *et al.* 2011a, *A&A*, 536, A23
Planck collaboration., *et al.* 2011b, *A&A*, 536, A22
Poglitsch, A., Waelkens, C., Geis, N., *et al.* 2010, *A&A* 518, L2
Tauber, J. A., Mandolesi, N., Puget, J.-L., *et al.* 2010, *A&A* 520, A1

Highlights of Astronomy, Volume 16
XXVIIIth IAU General Assembly, August 2012
T. Montmerle, ed.

doi:10.1017/S1743921314012253

A statistical view on the galactic cold ISM distribution

L. Viktor Tóth, Sarolta Zahorecz, Gábor Marton and Erika Verebélyi

Department of Astronomy, Eötvös University, Budapest
email: `l.v.toth@astro.elte.hu`

Abstract. Inhomogenities were found in the distribution of the cold fraction in the interstellar medium (traced by Planck cold clumps). In large scales there is a significant overdensity on some shells, in small scales there are elongated groups. Critical column density for star formation is $N(H_2) = 7\times10^{21}$ cm^{-2}.

Keywords. ISM: dust, ISM: molecules, ISM: bubbles, ISM: clouds, stars: formation

---

Recent Planck (Planck Collaboration 2011a) and Herschel (Pilbratt *et al.* 2010) measurements uncovered the distribution of cold interstellar medium (CISM) in the Galaxy by its FIR and sub-mm radiation. We investigated the all sky distribution of cold clumps (ECC and C3PO, Planck Collaboration 2011b) in all scales using statistical methods, and compared it to the distribution of known galactic structures and that of the young stellar objects (YSOs). Comparing the cold clump distribution to the GMCs as traced by CO we find that besides the overall correlations there are regions with peculiarly low or high CISM content. Large shells and voids traced by HI 21cm radiation (see e.g. Heiles 1984) and 100 μm surface brightness (Könyves *et al.* 2007) are often associated with CISM. About 1/5 of the shells are bearing cold clumps, and some of those are very active sites in the formation of dense CISM. Cold clump clusters were located and investigated using the minimum spanning tree method as described in Gutermuth *et al.* (2009). The distribution of ECC clumps is dominated by elongated groups in the Taurus region. The CISM distribution is well correlated with the distribution of YSOs selected by their WISE (Wright *et al.* 2010) and AKARI FIS (Yamamura *et al.* 2010) colours (Tóth *et al.* 2012). All the ECC clumps that were surveyed by Wu *et al.* (2012) were checked for associated star formation, and a critical column density for star formation of $N(H_2) = 7\times10^{21}$ cm^{-2} was found, i.e. all clumps over that column density have associated young stellar objects. This research was supported by the grants HAS-JSPS118, OTKA K101393 and TAMOP-4.2.1/B-09/1/KMR of the EU and the European Social Fund.

References

Gutermuth, R. A., Megeath, S. T., Muzerolle, J., *et al.* 2009, *ApJS* 184, 18
Heiles, C. 1984, *ApJS* 55, 585
Könyves, V., Kiss, Cs., Moór, A., *et al.* 2007, *A&A* 463, 1227
Pilbratt, G. L., Riedinger, J. R., Passvogel, T., *et al.* 2010, *A&A* 518, L1
Planck Collaboration 2011a, *A&A* 536, 1
Planck Collaboration 2011b, *A&A* 536, 23
Tóth, L. V., Zahorecz, S., Marton, G., *et al.* 2012, *in prep.*
Wright, E. L., Eisenhardt, P. R. M., Mainzer, A. K., *et al.* 2010, *AJ* 140, 1868
Wu, Y., Liu, T., Meng, F., *et al.* 2012, *ApJ* 756, 76
Yamamura., *et al.* 2010, *AKARI/FIS All-Sky Survey Point Source Catalogues (ISAS/JAXA)*

Highlights of Astronomy, Volume 16
XXVIIIth IAU General Assembly, August 2012
T. Montmerle, ed.

doi:10.1017/S1743921314012265

The coupled effects of protostellar outflows, radiation feedback, magnetic fields and turbulence on the formation of massive stars and Orion-like clusters

Richard I. Klein[1,2]

[1]Department of Astronomy, University of California, Berkeley, CA 94720, USA,
email: rklein@astron.berkeley.edu
[2]Lawrence Livermore National Laboratory,
Livermore, CA 95550, USA

Abstract. Feedback processes from massive stars plays a critical role in their formation, destroy the molecular clouds in which they are born and shape the evolution of galaxies. In this talk I will discuss our recent 3D AMR simulations that are the first to include the coupled feedback effects of protostellar outflows combined with protostellar heating and radiation pressure feedback and magnetic fields, in a single computation and their effects on the infalling dusty gas in the surrounding environs of the accreting core envelope. These simulations will address the detailed effects of feedback on the formation of high mass stars and massive clusters with implications for the IMF.

Keywords. star formation, stars: high mass, MHD, radiative transfer, turbulence, outflows

---

1. Introduction

The problem of high mass star formation is complicated by the several physical feedback processes that are at play and may be critically important in determining the final stellar mass of the star which have been observed in mass ranges of $120 - 300 M_{\odot}$. Strong radiative forces communicated to the dusty accretion envelope surrounding the central protostar oppose the force of gravity of the accreting gas as it attempts to make its way onto the accretion disk surrounding the protostar and make its final plunge onto the star as it builds in mass. A delicate balance between these opposing forces results in a tug of war. To further complicate the fate of the final mass of the star, strong protostellar outflows, magnetic fields and ionizing radiation from the central object also contribute to the balance. See Klein (2008) for a detailed discussion of the physical processes. Through the use of highly resolved radiation-magnetohydrodynamic adaptive mesh refinement (AMR) simulations, we can assess the effects of the individual feedback processes and the interaction of the coupled processes on high mass star formation and formation of massive clusters. Our simulations are performed using the state-of-the-art code we have developed at Berkeley, ORION, which is a parallelized 3D AMR Magneto-Rad-Hydro code (Truelove *et al.* 1998; Klein 1999; Krumholz *et al.* 2007c; Cunningham *et al.* 2012).

2. Feedback effects in high mass star formation and clusters

In a series of papers, Krumholz, Klein & McKee (2007a,b,c), have shown the first 3D simulations incorporating the relevant physical processes of radiation feedback to follow the formation of a massive star from the collapse of its embryonic turbulent core

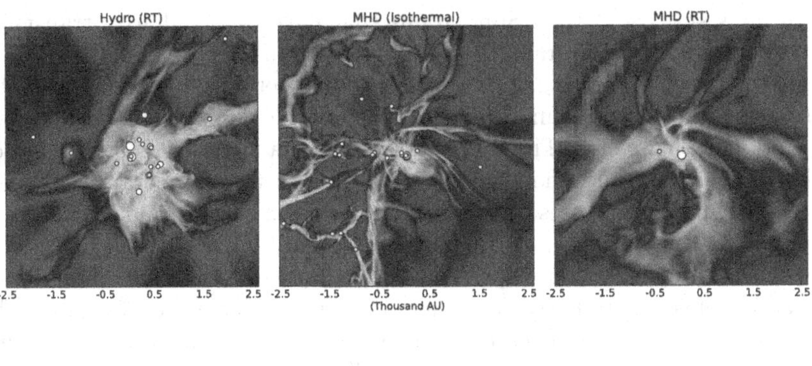

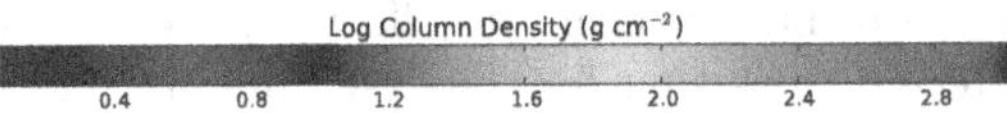

Figure 1. Column density through the simulation volume for simulations with Hydro with radiative transfer (HR); MHD with no radiative transfer in the isothermal approximation (BI) and MHD with radiative transfer (BR). Projections are taken along the x direction and the initial magnetic field is oriented in the positive z direction. Star particles are portrayed as white circles. The smallest circles represent stars with masses between $0.05 M_{\odot}$ and $1.0 M_{\odot}$; the next size up represents masses between $1.0 M_{\odot}$ and $8.0 M_{\odot}$, and the largest size represents stars with $\geqslant 8.0 M_{\odot}$ in each simulation. The figure is zoomed in to show the central 5000 AU around the most massive star in each simulation.

down to the formation of the massive protostar. This work has led to solutions to three longstanding problems in massive star formation theory: fragmentation, angular momentum transport, and radiation pressure. It was shown that the radiation emitted by a accreting high-mass stars can heat up clouds to the point where fragmentation is suppressed (Krumholz, Klein, & McKee 2007a), that angular momentum can be transported very rapidly by large-scale gravitational instabilities in the accretion disks around massive stars (Krumholz, Klein, & McKee 2007b), and that radiation pressure does not halt accretion, because radiation-hydrodynamic instabilities, essentially radiation driven Rayleigh Taylor instabilities, reshape the stellar radiation field, beaming it away from the bulk of the incoming gas (Krumholz *et al.* 2009). These simulations also demonstrated convincingly that the disks around massive stars inevitably fragment to produce massive companions, explaining the observed ubiquity of multiple systems among massive stars. The next important feedback effect on the formation of massive star is the protostellar outflows that interact with the still accreting protostar. In recent work (Cunnngham *et al.* 2011), we made the first simulations to include the feedback effects of protostellar outflows with protostellar radiative heating and radiation pressure. It was shown that feedback from protostellar outflows resulted in the the diminishing of radiative heating and outward radiation force exerted on the protostellar disk and infalling gas thereby reducing the Eddington radiation pressure barrier to high-mass star formation. We also found that radiation focusing in the direction of outflow cavities is sufficient to prevent the formation of radiation pressure-supported circumstellar gas bubbles, in contrast to models which neglect protostellar outflow feedback. In recent work (Myers *et al.* 2012 submitted), we present the first results of 3-dimensional, AMR radiation-magnetohydrodynamic simulations that treat the dust processed radiation from protostars in the flux-limiteod diffusion approximation. The question of how isolated massive cores fragment is a crucial question for any theory of star formation in which the initial mass function (IMF) is set in the gas phase, e.g. the turbulent fragmentation scenario (Padoan & Nordund 2002). In Figure 1 we note that simulations with radiation alone (HR) or magnetic fields alone (BI) result in multiple fragmentation with a low mass cluster of stars and a single high mass star ($\sim 22M_\odot$). Simulations that have both radiative transfer coupled to magnetic fields produce a dramatically different result. As Myers *et al.* (2012) show, considering the thermodynamics of the core, both magnetic fields and radiation feedback are necessary to fully suppress multiple star formation leaving a single high mass star of $\sim 20M_\odot$ and a low mass star. The coupled effects of magnetic fields and radiation (BR) are **critical** in that magnetic fields alone raise the effective temperature in the core in low density gas which suppresses fragmentation in diffuse gas far from central stars in the core. Alternatively, radiation alone raises the effective temperature in high density gas thus suppressing fragmentation in dense gas near the star. However, the interaction of both magnetic fields and radiation together suppress fragmentation far more effectively than either alone since they cover the domain of the core that the other one misses. Radiation feedback can also have a dramatic effect on the evolution and formation of massive star clusters. Recent work that for the first time includes the feedback effects of radiation, protostellar outflows and turbulence on the formation of a massive star cluster (Krumholz, Klein & McKee 2011,2012) are the first simulations published to date that reproduces the observed time invariant IMF in a cluster like the ONC that is large enough to contain massive stars, and where the peak of the mass function is determined by a fully self-consistent calculation of gas thermodynamics.

3. Conclusions

We have briefly discussed recent 3D AMR simulations that for the first time have treated the coupled effects of radiation, protostellar outflows and magnetic fields and their effects on the formation of individual high mass stars and massive clusters.

Acknowledgements

Support for this work was provided by: the US Department of Energy at the Lawrence Livermore National Laboratory under contract DE-AC52-07NA27344, NASA through ATFP grant NNX09AK31G and the National Science Foundation through grant AST-0908553.

References

Cunningham, A. J., Klein, R. I., Krumholz, M. R., & McKee, C. F. 2011, *ApJ* 740, 107

Klein, R. I. 2008, in *Massive Star Formation: Observations Confront Theory*, eds. H. Beauther, H. Linz, & T. Henning, (ASP Conference Series) 38, 306

Klein, R. I. 1999, *J. Comp. App. Math.* 109, 123

Krumholz, M. R., Klein, R. I., & McKee, C. F. 2007a, *ApJ* 656, 959

Krumholz, M. R., Klein, R. I., & McKee, C. F. 2007b, *ApJ* 665, 478

Krumholz, M. R., Klein, R. I., & McKee, C. F. 2007c, *ApJ* 667, 626

Krumholz, M. R., Klein, R. I., McKee, C. F., Offner, S. S. R., & Cunningham, A. J. 2009, *Science* 323, 754

Krumholz, M. R., Klein, R. I., & McKee, C. F. 2011, *ApJ* 740, 74

Krumholz, M. R., Klein, R. I., & McKee, C. F. 2012, *ApJ* 754, 71

Myers, A. T., McKee, C. F., Cunningham, A. J. Klein, R, I., & Krumholz, M. R. 2012, submitted to *ApJ* (arXiv1211.3467M)

Padoan, P. & Nordlund, A. A. 2002, *ApJ* 576,870

Truelove, K. J., Klein, R. I., & McKee, C. F. 1998, *ApJ* 495, 821

Highlights of Astronomy, Volume 16
XXVIIIth IAU General Assembly, August 2012
T. Montmerle, ed.

doi:10.1017/S1743921314012277

Dust and Molecule Formation and Processing in Supernovae and their Remnants

J. Rho[1], M. Andersen[2], A. Tappe[1], H. Gomez[3] and M. Smith[3], J. P. Bernard[4], T. Onaka[5] and J. Cami[6]

[1]SETI Institute, 189 Bernardo Ave, Mountain View, CA 94043;
email: jrho@seti.org
[2]Institut de Planétologie et d'Astrophysique de Grenoble, France
[3]School of Physics and Astronomy, Cardiff University, The Parade, Cardiff, CF24 3AA, UK,
[4]Centre d'Etude Spatiale des Rayonnements, CNRS, 9 av. du Colonel Roche, BP 4346, 31028 Toulouse, France [5]Department of Astronomy, The University of Tokyo, 7-3-1 Hongo, Bunkyo-ku, Tokyo 113-0033, Japan, onaka@astron.s.u-tokyo.ac.jp,
[6]Dept. of Physics & Astronomy, Univ. of Western Ontario, London, ON N6A 3K7, Canada

Abstract. Supernovae (SNe) produce, fragment and destroy dust, molecules and nucleosynthetic elements, and reshape and modify the ISM. I will review recent infrared observations of supernova remnants (SNRs) and SNe which show that SNe are important sites of dust and molecule formation and are major dust creators in the Universe. Detection of carbon monoxide (CO) fundamental band from the young SNR Cas A indicates that astrochemical processes in SNRs interacting with molecular clouds provide astrophysical laboratories to study evolution of the ISM returning material from dense clouds into the more diffuse medium and galactic halo. Two dozen SNRs are known to be interacting with molecular clouds using H_2 and millimeter observations. Recent Spitzer, Herschel and SOFIA observations along with ground-based observations have greatly advanced our understanding shock processing and astrochemistry of dust, H_2, high J CO, and other neutral and ionized molecules and polycyclic aromatic hydrocarbon (PAH). Ionized molecules and warm layer of molecules that are excited by UV radiation, X-rays, or cosmic rays will be described. Finally I will discuss how astrochemical processes of dust and molecules in SNRs impact the large scale structures in the ISM.

Keywords. molecules, supernova remnants, Cas A, ISM, dust, infrared, molecular hydrogen

1. Introduction

Supernovae (SNe) play a key role in the chemical and dust budget of galaxies producing heavy elements and dust in their ejecta and processing dust via strong shocks both in the local and in the early Universe. SN explosions light up regions of stellar birth, trigger further star formation, return processed material to the ISM, providing the elements necessary for life. Recent evidence also suggests core-collapse SNe are dust factories, a dominant source of dust grains in galaxies throughout cosmic history.

Huge quantities of dust observed in high-redshift galaxies raise the fundamental question of the origin of dust in the Universe, as the low-intermediate mass stars that had been believed to be the primary source of dust were not evolved enough in high-redshift galaxies. In contrast, SNe could dominate the dust and molecule production (Nozawa *et al.* 2003; Todini & Ferrara 2001; Cherchneff 2008) significantly altering the the star formation properties of early galaxies as compared to the Population III stars that first cause reionization. Evidence for this was scarce until recently. Our *Spitzer* and *Herschel* observations of young nearby SN remnants (SNRs) demonstrated, for the first time,

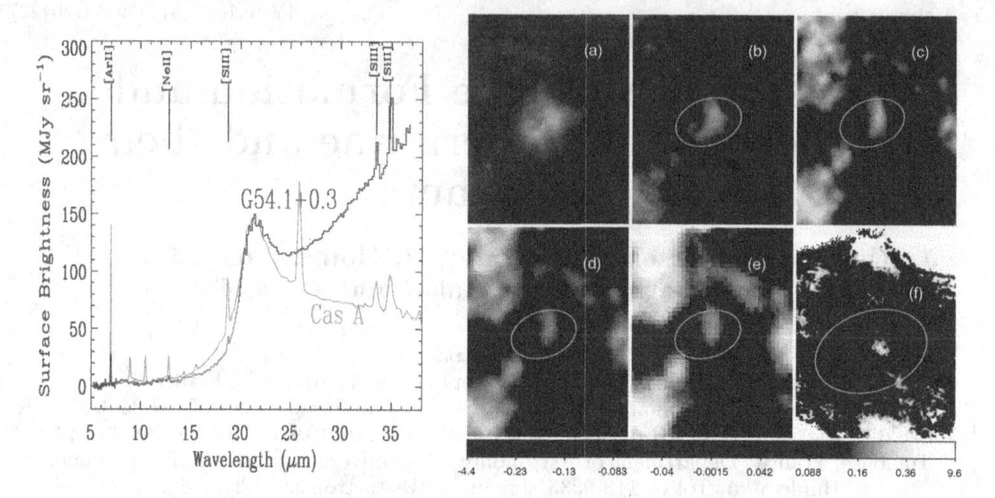

Figure 1. (a: left) *Spitzer* IRS spectrum of G54.1+0.3 shows strong 21 μm dust feature with ionic lines of [Ar II], [Ne II], [S III], [Si II], and [S II]. (b: right) Cold dust of G54.1+0.3; Herschel PACS at 70 (a) and 160 μm (b) and SPIRE at 250 (c), 350 (d) and 500 (e) μm and CSO 350 μm SHARCII (f) images. The size of the entire SNR is marked as an ellipse.

significant amounts of dust in SN ejecta, orders of magnitude more than previously detected (Rho *et al.* 2008; 2009a; Barlow *et al.* 2010; Gomez *et al.* 2012). These results imply that SNe could be responsible for the large dust masses detected in high redshift galaxies and in galaxies today, but only a handful of such observations exist.

We serendipitously discovered a dust feature peaking at 21 μm from G54.1+0.3, and the dust feature as shown in Figure 1a is remarkably similar to that of Cas A from Rho *et al.* (2008). Strong correlation between 21-μm dust and Ar ejecta has been observed in Cas A. Ar ejecta is detected from the IRS spectra of G54.1+0.3 and IRAC 8 μm map shows shell-like morphology indicating the distribution of Ar ejecta. We present detection of cold dust from G54.1+0.3 using CSO SHARCII 350 μm map and Herschel PACS and SPIRE maps as shown in Figure 1b. G54.1 was observed with SHARC II on 2006 April 18 and 2009 May 3 and 4. SHARC II is a 12× 32 bolometer array with field of view of $2.6' \times 0.97'$, observing at 350 μm on the Caltech Submillimetre Observatory in Hawaii. The weather conditions were good during both runs, with τ_{225} ranging from 0.04 to 0.05. Twenty-two scans were observed in total using the boxscan format scanning with on-source integration time of 3.7 hours. We identified cold dust from G54.1+0.3 using HIGAL survey (PI, S. Molinari) on the *Herschel* Space Telescope. In Figure 1b, the PACS 70 and 160 μm and SPIRE 250 and 350 μm images show clear detection of G54.1+0.3 and a shell-like morphology, that is similar to the MIPS 24 μm map (Temim *et al.* 2010). The SPIRE map at 500 μm also detected the SNR; however, that is somewhat confused with the background emission. The SHARCII image with a higher ($8''$) spatial resolution than the Herschel images has detected the brightest part of the SNR in the Herschel 350 μm map, showing consistent detection between CSO and Herschel. The cold dust detection from G54.1+0.3 will be one of a few cases from SNRs.

Our current understanding of molecule and dust chemistry and composition is limited. The current models assume that very little carbon is locked up in CO and almost all of carbon is available for dust formation. As a tracer of gas properties and dust formation, CO is potentially an important diagnostic for any supernova. We detected near-infrared 2.29μm first overtone CO emission from Cas A with the Wide field InfraRed Camera (WIRC) on the Hale 200 inch telescope at Mount Palomar (Rho *et al.* 2009b).

Subsequently we have spectroscopically confirmed the presence of the fundamental band at 4.5μm using the AKARI IRC spectrograph as shown in Figure 2a (Rho *et al.* 2012).

SNRs interacting with molecular clouds (MCs) provide critical laboratories to study shock physics, grain destruction, metal enrichment, and gas heating and cooling in dense interstellar gas. The most significant coolants of the shocked gas are predicted to be fine-structure and metastable lines of atoms and ions, as well as H_2 line emission (Hollenbach & McKee 1989). We used archival *Spitzer* data to systematically search for IR counterparts of all 96 known SNRs covered by the GLIMPSE/IRAC Legacy survey (Reach *et al.* 2006), and many IRAC-detected SNRs reveal an interaction with dense gas through a 4.5μm excess, which indicates H_2 and CO line emission. We also performed follow-up observations with IRS (5-40μm) and detected rotational H_2 (S(0) to S(7)) lines and rich IR ionic lines including O, N, S, Fe, and Ne (Hewitt *et al.* 2009; Andersen *et al.* 2011). The dust model fitting of the SEDs is done using the method described in Bernard *et al.* (2008) and the updated version from Compiégne *et al.* (2008). Three dust components are adopted, PAH molecules, Very Small Grains (VSG) and Big Grains (BG). The abundance of each dust species (Y_{PAH}, Y_{VSG}, Y_{BG}) and the strength of the radiation field (X_{ISRF}) are taken as free parameters in the fit. The fit results are shown in Figure 2b (Andersen *et al.* 2011), showing that abundances of PAHs and VSGs are higher than those in the Milky Way and the LMC (Bernard *et al.* 2008). Typically the radiation field is 10-100 times larger than normal interstellar radiation ISRF, the strength of which is consistent with being created from the shock.

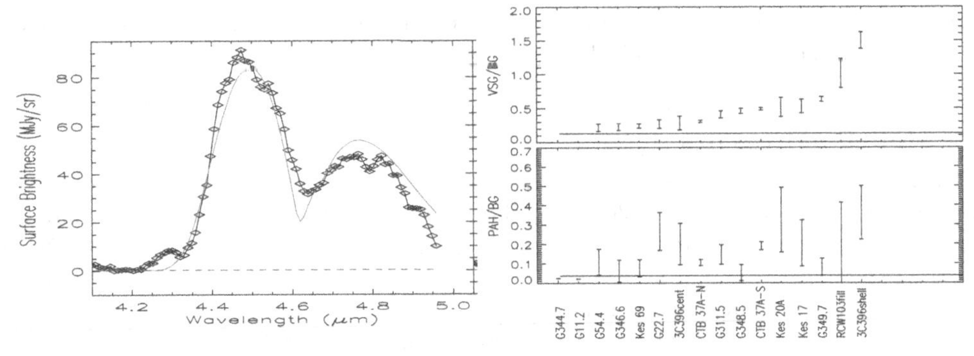

Figure 2. (a: left) High resolution grism spectrum (black) towards the southern part of Cas A. Superposed is the best fit CO model (blue) using a single component (left; velocity of -4800 km s^{-1}), suggesting that the CO may be at a location different from the ejecta in this case.
(b: right) The ratios of PAHs and very small grains to big grains are higher than those of the Milky Way (solid line) except two young SNRs.

References

Andersen, M., Rho, J., Reach, W. T., *et al.* 2011, *ApJ* 742, 7
Barlow, M. J., Krause, O., Swinyard, B. M., *et al.* 2010, *A&A* 518, L138
Bernard, J.-P., Reach, W. T., Paradis, D., *et al.* 2008, *AJ* 136, 919
Compiègne, M., Abergel, A., Verstraete, L., & Habart, E. 2008, *A&A* 491, 797
Hewitt, J. W., Rho, J., Andersen, M., & Reach, W. T. 2009, *ApJ* 694, 1266
Hollenbach, D. & McKee, C. F. 1989, *ApJ* 342, 306
Reach, W. T., Rho, J., Tappe, A., *et al.* 2006, *AJ* 131, 1479
Rho, J., Jarrett, T. H., Reach, W. T., *et al.* 2009, *ApJ* (Letters) 693, L39
Rho, J., Kozasa, T., Reach, W. T., *et al.* 2008, *ApJ* 673, 271
Rho, J., Onaka, T., Cami, J., & Reach, W. T. 2012, *ApJ* (Letters) 747, L6
Rho, J., Reach, W. T., Tappe, A., *et al.* 2009, *ApJ* 700, 579
Temim, T., Slane, P., Reynolds, S. P., *et al.* 2010, *ApJ* 710, 309

Highlights of Astronomy, Volume 16
XXVIIIth IAU General Assembly, August 2012
T. Montmerle, ed.

doi:10.1017/S1743921314012289

Stellar wind and supernova feedback from massive stars

Julian M. Pittard and Hazel Rogers

School of Physics and Astronomy, The University of Leeds, Leeds, LS2 9JT, UK
email: jmp@ast.leeds.ac.uk

Abstract. We have constructed three-dimensional hydrodynamical models to simulate the impact of massive star feedback, via winds and SNe, on inhomogeneous molecular material left over from the formation of a massive stellar cluster. We are studying the timescales for the molecular material to be removed from the environment of a massive stellar cluster and the mass and energy fluxes into the wider environment.

Keywords. hydrodynamics, ISM: bubbles, HII regions, supernova remnants

1. Introduction

Massive stars blow powerful winds and explode as supernovae. However, the degree to which these inputs couple to the clumpy and inhomogenous molecular clouds which initially surround a cluster of massive stars is currently ill-determined. We are using three-dimensional hydrodynamical simulations to elucidate this.

2. Model and results

In our model 3 O-stars are located at the centre of a giant molecular cloud (GMC) clump of radius 4 pc and mass 3000 $M_{\odot}$. The stars collectively drive a cluster wind, and transition through 3 evolutionary phases (main sequence - red supergiant - Wolf-Rayet) before exploding.

We find that hot, high speed gas flows away from the cluster in low-density channels opened up by the stellar winds. Mass is loaded into these flows from the ablation of dense clouds embedded within them and from material stripped from the dense gas which confines and directs the flows. Dense parts of the clump can shield and protect less dense material in their "shadow". The rate of ablation appears to be roughly constant during the first few Myr of the cluster evolution, and the mass flux into the wider environment can be 3 orders of magnitude greater than the combined mass-loss rates of the stars. About one third of the input wind mechanical energy is radiated away. In contrast, almost all of the energy from each SN explosion escapes into the wider medium. This is due to the high porosity of the dense molecular material at this stage which allows the SN blast to rip through the cluster in a largely unimpeded fashion. A detailed analysis of our results is currently underway, and in future work we plan to study a range of clump and cluster parameters.

Highlights of Astronomy, Volume 16
XXVIIIth IAU General Assembly, August 2012
T. Montmerle, ed.

doi:10.1017/S1743921314012290

HII radiative transfer revealed by ionization parameter mapping

M. S. Oey[1], E. W. Pellegrini[1]†, P. F. Winkler[2], S. D. Points[3], R. C. Smith[3], A. E. Jaskot[1] and J. Zastrow[1]

[1]Astronomy Department, University of Michigan, Ann Arbor, MI 48109-1042, USA

[2]Department of Physics, Middlebury College, Middlebury, VT 05753, USA

[3]Cerro Tololo Inter-American Observatory, Casilla 603, La Serena, Chile

Abstract. We develop the technique of ionization parameter mapping (IPM) to probe the optical depth of HII regions, applying our method to the Magellanic Clouds. Our results dramatically clarify the radiative transfer in these galaxies. Based on [SII], [OIII], and Hα imaging from the Magellanic Clouds Emission Line Survey, we find that the frequency of optically thin objects correlates strongly with Hα luminosity and correlates inversely with HI column density. The aggregate escape fraction for the Lyman continuum is sufficient to ionize the diffuse, warm ionized medium, but the galactic escape fraction is dominated by the few largest HII regions. The quantitative trends are similar in both the LMC and SMC in spite of their different star formation and HI properties.

Keywords. radiative transfer, stars: early-type, HII regions, galaxies: ISM

1. Introduction

The diffuse, warm ionized medium (WIM) is the dominant component of ionized gas in the interstellar medium (ISM). While its existence has been known for decades, we still lack a quantitative understanding of its origin. The WIM comprises about 60% of the total Hα luminositiy in star-forming galaxies, generally independent of galaxy morphology, star-formation rate, and star-formation intensity (e.g., Walterbos 1998). A notable exception is the observation of much lower WIM fractions in starburst galaxies (Oey *et al.* 2007). This may be related to the short starburst timescale relative to ISM recombination (Hanish *et al.* 2010), and underscores the need to better understand the relationship between star formation and WIM properties. Ultimately, clarifying the fate of ionizing photons will reveal whether and/or when they escape from their host galaxies altogether, a question of vital importance to cosmic reionization and evolution.

Based on energetic arguments, OB stars are understood to be the ionizing source for the WIM, with an important contribution from field OB stars (e.g., Hoopes & Walterbos 2000; Oey *et al.* 2004) and the rest from stars within optically thin HII regions, but the relative contributions are poorly known. In particular, the ionizing fluxes from OB stars are not known well enough to clarify this problem (e.g., Voges *et al.* 2007; Zastrow *et al.* 2012). It is therefore essential to obtain a comprehensive understanding of the radiation transfer of the HII regions in galaxies. The WIM emission-line spectra provide important constraints, implying long photon path lengths and significant contributions from leaking HII regions (e.g., Castellanos *et al.* 2002; Wood & Mathis 2004; see also A. Hill in these Proceedings). The WIM morphology also points to low opacities and the importance of dust scattering (Seon 2009; and in these Proceedings).

† Present address: Dept of Physics & Astronomy, University of Toledo, Toledo, OH 43606-3390

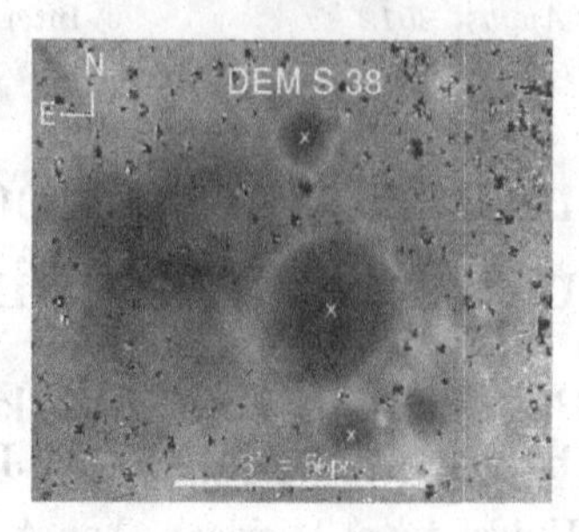

Figure 1. SMC HII regions in [SII]/[OIII], with high values white (Pellegrini *et al.* 2012).

2. Ionization parameter mapping

Here, we present our technique of ionization-parameter mapping, a new application of emission-line imaging to directly reveal the optical depth of ionized gas to the Lyman continuum (Pellegrini *et al.* 2012). Figure 1 shows the [SII]/[OIII] ratio map of some HII regions in the Small Magellanic Cloud (SMC) from the Magellanic Clouds Emission-Line Survey (MCELS; Smith *et al.* 2005). The large, round object, DEM S38, is dominated by the higher-ionization species, [OIII] (black); while the limb is dominated by the lower-ionization species, [SII] (white). The same effect is apparent in other objects marked with 'X' in the image. This ionization structure strongly suggests that these are classical Strömgren spheres, since the ionization must show a transition zone to lower-ionization species between the highly ionized center and the neutral environment. The round morphology is also consistent with that of the simple, Strömgren sphere.

In contrast, the object directly to the east of DEM S38 does not show the [SII]-dominated transition zones, and is instead dominated by [OIII] emission throughout. The object also has highly irregular morphology, which is consistent with radiation-hydrodynamic simulations by Arthur *et al.* (2011) that are characteristic of highly-ionized, optically thin objects subject to champagne flows and other gas instabilities. Thus, ionization-parameter mapping (IPM) offers a vivid, visual technique to directly estimate the nebular optical depth. We also note that IPM can be applied in the line of sight: since the low-ionization envelope must also exist in the line of sight, a minimum threshold emission from that species must be present across an optically thick object. Its absence therefore implies a low optical depth.

Thus, IPM provides a powerful way to evaluate the optical depth of HII regions. Pellegrini *et al.* (2012) discuss the reliability of this technique by carrying out photoionization modeling and comparison with objects having measured estimates of the optical depth. With only two radially varying ions, IPM technically tends to set upper limits to the optical depth, but in general, objects that look like Strömgren spheres with low-ionization envelopes indeed seem to be optically thick. And mapping three radially sensitive ions allows actual measurement of the optical depths.

3. Optical depth of Magellanic Clouds HII regions

We apply IPM to the Magellanic Clouds, using data from MCELS. IPM reveals the ionization structure, providing physical criteria for defining HII region boundaries. We use this to redefine all the HII regions in these two galaxies, compiling new nebular catalogs for both the LMC and SMC (Pellegrini *et al.* 2012).

The MCELS data provide only two radially varying species, [OIII] and [SII], and so optical depth estimates for individual objects are technically upper limits. However, the statistical properties for the nebular populations of the entire galaxies are revealing.

Classifying the objects simply as optically thick or thin based on IPM, we find that the frequency of optically thick objects correlates strongly with HI column density. While this is consistent with expectations, it is interesting to note that optically thin objects dominate in frequency at the very lowest HI columns. We also see a strong correlation in the frequency of optically thin objects with Hα luminosity L, although it is important to note that optically thick objects are found at almost all luminosities. Optically thin objects dominate in both galaxies above 10^{37} erg s^{-1}. Beckman *et al.* 2000 had suggested that high-luminosity HII regions tend to be density bounded, but we find that the characteristic luminosity is much lower. Indeed, 10^{37} erg s^{-1} corresponds to objects ionized by individual O stars, implying that most (but not all) of the bright HII regions that are readily apparent in star-forming galaxies are optically thin. It is especially interesting that trends for optical depth in both galaxies are so similar, given that their HI morphologies differ greatly; the LMC is a disk with strongly shredded HI, while the SMC has amorphous and less porous HI.

Overall, we find that about 40% of the HII regions in the LMC are optically thin, as are 30% of the objects in the SMC. In the aggregate, about 37% of the Lyman continuum radiation from their ionizing stars escapes to ionize the WIM in both galaxies. Given the total WIM luminosities and accounting for field star ionization, our results are consistent with escape fractions from the galaxies of 4% and 11%, respectively, for the LMC and SMC, although these values are quite uncertain (Pellegrini *et al.* 2012). Our results show that the galactic escape fractions are extremely sensitive to the location and HI environment of the few, most luminous objects.

In summary, ionization-parameter mapping is powerful technique for evaluating the optical depth of photoionized gas. Our results for the Magellanic Clouds yield fundamental, quantitative insights on the relation between optical depth, nebular luminosity, and HI properties. We are also applying the technique to starburst galaxies, which revealed the presence of an ionization cone in NGC 5253 (Zastrow *et al.* 2011).

Acknowledgements

We acknowledge an AAS Travel Grant, a U. Michigan Margaret and Herman Sokol Faculty Award, and NSF grants AST-0806476 to MSO and AST-0908566 to PFW.

References

Arthur, S. J., Henney, W. J., & Mellema, G., *et al.* 2011, *MNRAS* 414, 1747
Beckman, J. E., Rozas, M., Zurita, A., Watson, R. A., & Knapen, J. H. 2000, *AJ* 119, 2728
Castellanos, M., Díaz, A. I., & Tenorio-Tagle, G. 2002, *ApJ* (Letters) 565, L79
Hanish, D. J., Oey, M. S., Rigby, J. R., de Mello, D. F., & Lee, J. C. 2010, *ApJ* 725, 2029
Hoopes, C. G. & Walterbos, R. A. M. 2000, *ApJ* 541, 597
Oey, M. S., King, N. L., & Parker, J. Wm. 2004, *AJ* 127, 1632
Oey, M. S., Meurer, G. R., & Yelda, S., *et al.* 2007, *ApJ* 661, 801
Pellegrini, E. W., Oey, M. S., Winkler, P. F., Points, S. D., Smith, R. C., Jaskot, A. E., & Zastrow, J. 2012, *ApJ* 755, 40; Erratum, astro-ph/1202.3334.
Seon, K.-I. 2009, *ApJ* 703, 1159
Smith, R. C., Points, S. D., & Chu, Y.-H., *et al.* 2005, *BAAS* 37, 1200
Voges, E. S., Oey, M. S., Walterbos, R. A. M., & Wilkinson, T. M. 2007, *AJ* 135, 1291
Wood, K. & Mathis, J. S. 2004, *MNRAS* 353, 1126
Walterbos, R. A. M. 1998, *PASA* 15, 99
Zastrow, J., Oey, M. S., Veilleux, S., McDonald, M., & Martin, C. L. 2011, *ApJ* (Letters) 741, L17
Zastrow, J., Oey, M. S., & Pellegrini, E. W. 2012, *ApJ*, submitted

Highlights of Astronomy, Volume 16
XXVIIIth IAU General Assembly, August 2012
T. Montmerle, ed.

doi:10.1017/S1743921314012307

Formation of structures around HII regions: ionization feedback from massive stars

P. Tremblin[1], E. Audit[1,2], V. Minier[1], W. Schmidt[3] and N. Schneider[1,4,5]

[1]Laboratoire AIM Paris-Saclay (CEA/Irfu - Uni. Paris Diderot - CNRS/INSU), Centre d'études de Saclay, 91191 Gif-Sur-Yvette, France
email: pascal.tremblin@cea.fr

[2]Maison de la Simulation, CEA-CNRS-INRIA-UPS-UVSQ, USR 3441, Centre d'étude de Saclay, 91191 Gif-Sur-Yvette, France

[3]Institut für Astrophysik der Universität Göttingen, Friedrich-Hund-Platz 1, D-37077 Göttingen, Germany

[4]Univ. Bordeaux, LAB, UMR 5804, F-33270, Floirac, France

[5]CNRS, LAB, UMR 5804, F-33270, Floirac, France

Abstract. We present a new model for the formation of dense clumps and pillars around HII regions based on shocks curvature at the interface between a HII region and a molecular cloud. UV radiation leads to the formation of an ionization front and of a shock ahead. The gas is compressed between them forming a dense shell at the interface. This shell may be curved due to initial interface or density modulation caused by the turbulence of the molecular cloud. Low curvature leads to instabilities in the shell that form dense clumps while sufficiently curved shells collapse on itself to form pillars. When turbulence is high compared to the ionized-gas pressure, bubbles of cold gas have sufficient kinetic energy to penetrate into the HII region and detach themselves from the parent cloud, forming cometary globules.

Using computational simulations, we show that these new models are extremely efficient to form dense clumps and stable and growing elongated structures, pillars, in which star formation might occur (see Tremblin *et al.* 2012a). The inclusion of turbulence in the model shows its importance in the formation of cometary globules (see Tremblin *et al.* 2012b). Globally, the density enhancement in the simulations is of one or two orders of magnitude higher than the density enhancement of the classical "collect and collapse" scenario. The code used for the simulation is the HERACLES code, that comprises hydrodynamics with various equation of state, radiative transfer, gravity, cooling and heating.

Our recent observations with Herschel (see Schneider *et al.* 2012a) and SOFIA (see Schneider *et al.* 2012b) and additional Spitzer data archives revealed many more of these structures in regions where OB stars have already formed such as the Rosette Nebula, Cygnus X, M16 and Vela, suggesting that the UV radiation from massive stars plays an important role in their formation. We present a first comparison between the simulations described above and recent observations of these regions.

Keywords. methods: numerical, methods: data analysis, ISM: globules, ISM: HII regions, ISM: kinematics and dynamics

---

References

Schneider, N., Csengeri, T., & Hennemann, M., *et al.* 2012a, *A&A* 540, L11
Schneider, N., Güsten, R., Tremblin, P., *et al.* 2012b, *A&A* 542, L18
Tremblin, P., Audit, E., Minier, V., & Schneider, N. 2012a, *A&A* 538, 31
Tremblin, P., Audit, E., Minier, V., Schmidt, W., & Schneider, N. 2012b, *A&A* 546, A33

Highlights of Astronomy, Volume 16
XXVIIIth IAU General Assembly, August 2012
T. Montmerle, ed.

doi:10.1017/S1743921314012319

Different structures formed at HII boundaries

Jingqi Miao†, Paul Cornwall and Tim Kinnear

The School of Physical Sciences, University of Kent, Canterbury, CT2 7NH, UK
email: J.Miao@kent.ac.uk

Abstract. Hydrodynamic simulations on the evolution of molecular clouds (MCs) at HII boundaries are used to show that radiation driven implosion (RDI) model can create almost all of the different morphological structures, such as a single bright-rimmed cloud (BRC), fragment structure and multiple elephant trunk (ET) structures.

Keywords. ISM: molecular cloud, HII regions, ISM: structure

The interaction of UV radiation with MCs creates a diversity of morphological structures at HII boundaries which prompted several theoretical models, such as the RDI model (Lefloch & Lazareff 1994) for the formation of a single BRC (Sugitani & Ogura 1994), collect & collapse (C&C) model (Elmegreen & Lada 1977) for HII bubbles or fragments (Deharveng, Zavagno, & Caplan 2005) and hydrodynamic instability model (HD) for a forest of ETs (White *et al.* 1999). A mixture of the above mentioned structures are often found in one HII region (Chauhan *et al.* 2011), so it is desirable to use one model to consistently interpret the overall structure.

Using an existing numerical (Miao *et al.* 2009) code based on RDI mechanism and smoothed particle hydrodynamic (SPH) method, and considering the clumpiness and non-spherical nature of a MC, we found that a prolate uniform MC could evolve into a bar-like structure with condensed cores embedded along the fragment when its major axis is perpendicular to the UV radiation field. An asymmetrical BRC with one side compressed more strongly than the other side will form, if its major axis is initially inclined to an UV radiation flux by an angle. Triggered single or multi star(s) are found embedded at the centre of the head of the BRCs. These simulation results well interpret the physical features of some of the BRCs observed. Further investigation shows that a clumpy MC with proper initial conditions could form various ET structures.

Our simulation results, combined with that of Walch *et al.* (2011) on the formation of HII bubble structure state that RDI mechanism is a versatile model to deal with almost all of the morphological structure formation at HII boundaries.

References

Chauhan, N., Ogura, K., Pandey, A. K., *et al.* 2011, *PASJ* 63, 795
Deharveng, L., Zavagno, A., & Caplan, J. 2005, *A&A* 433, 565
Elmegreen, B. G. & Lada, C. J. 1977, *ApJ* 214, 725
Lefloch, B. & Lazareff, B. 1994, *A&A* 289, 559
Miao, J., White, G. J., Thompson, M. A., & Nelson, R. P. 2009, *MNRAS* 692, 382
Sugitani, K. & Ogura, K. 1994, *ApJS* 92, 163
Walch, S., Whitworth, A., Bisbas, T., Hubber, D. A., & Wuensch, R. 2011, *2011arXiv* 1109.3478W
White, G. J., Nelson, R. P., Holland, W. S., *et al.* 1999, *A&A* 343, 233

† Present address: The Centre for Astrophysics & Planetary Science, CT2 7NH UK

Highlights of Astronomy, Volume 16
XXVIIIth IAU General Assembly, August 2012
T. Montmerle, ed.

doi:10.1017/S1743921314012320

Molecular cloud structure and star formation in the W43 complex

P. Carlhoff[1], P. Schilke[1], F. Motte[2] and Q. Nguyen Luong[3]

[1]1. Physikalisches Institut, Universität zu Köln, Germany
email: carlhoff@ph1.uni-koeln.de

[2]Laboratoire AIM, CEA/IRFU – CNRS/INSU – Université Paris Diderot, France
[3]Canadian Institute for Theoretical Astrophysics – CITA, University of Toronto, Canada

Abstract. The W43 region is one of the most massive star forming regions in our Galaxy. It is subject to a large IRAM 30m project that observes high spectral resolution maps of the complete complex in the ^{13}CO (2–1) and C^{18}O (2–1) lines. We find a variety of different sources of which we calculate excitation temperature, H_2 column density and mass. We find the total mass of dense clouds in the complex to be $1.2 \times 10^6\ M_\odot$.

Keywords. ISM: molecules, ISM: structure, ISM: kinematics and dynamics, molecular data

---

The W43 region is one of the most massive star forming regions in the Galaxy, situated at the junction point of the Galactic bar and the spiral arms. In ATLASGAL (Schuller *et al.* 2009) dust emission two separate clouds can be distinguished at this location, but Nguyen Luong *et al.* (2011) show that both components are indeed connected and thus form one giant complex.

To study the dynamics of the molecular gas in this region a large IRAM 30m has been initiated. The resulting data cubes of ^{13}CO (2–1) and C^{18}O (2–1) line emission show numerous separated sources and filaments thanks to a spectral resolution of $<0.2\,\mathrm{km\,s^{-1}}$. We find three separated velocity complexes along the line of sight, located in three different spiral arms. An integrated map of the ^{13}CO (2–1) line is shown in Fig. 1.

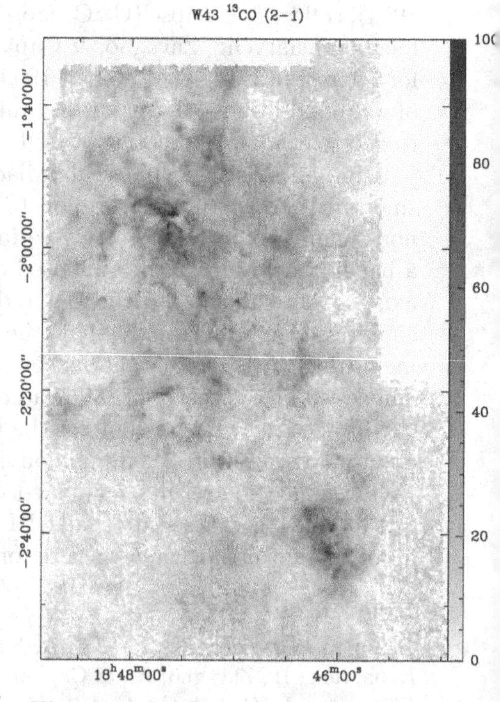

Figure 1. Integrated intensity map of the ^{13}CO (2–1) line.

From these maps we calculate the optical depth of the CO gas, as well as its excitation temperature. Temperatures are in the range from 8 to 20 K. We can also estimate the H_2 column density and the total mass contained in this gas. We find typical masses for single filaments of $10^4\ M_\odot$. We calculate a total mass of the dense molecular gas in the W43 region of $1.2 \times 10^6\ M_\odot$.

References

Nguyen Luong, Q., Motte, F., Schuller, F., *et al.* 2011, *A&A* 529, A41
Schuller, F., Menten, K. M., Contreras, Y., *et al.* 2009, *A&A* 504, 415

Highlights of Astronomy, Volume 16
XXVIIIth IAU General Assembly, August 2012
T. Montmerle, ed.

doi:10.1017/S1743921314012332

Physics and chemistry of UV illuminated gas: the Horsehead case

V. Guzmán[1], J. Pety[1], P. Gratier[1], J. R. Goicoechea[2], M. Gerin[3], E. Roueff[4] and D. Teyssier[5]

[1]IRAM, 300 rue de la Piscine, 38406 Saint Martin dHe'res, France
email: [guzman;pety;gratier]@iram.fr

[3]LERMA-LRA, UMR 8112, Observatoire de Paris and Ecole normale Supérieure, Paris, France

[2]Centro de Astrobiología. CSIC-INTA. 28850 Madrid, Spain

[4]LUTH UMR 8102, CNRS and Observatoire de Paris, Meudon, France

[5]ESA, PO Box 78, 28691 Villanueva de la Cañada, Madrid, Spain

Abstract. Molecular lines are used to trace the physical conditions of the gas in different environments, from high-z galaxies to proto-planetary disks. To fully benefit from the diagnostic power of the molecular lines, the formation and destruction paths of the molecules must be quantitatively understood. This is challenging because the physical conditions are extreme and the dynamic plays an important role. In this context the PDR of the Horsehead mane is a particularly interesting case because the geometry is simple (almost 1D, viewed edge-on; Abergel *et al.* 2003), the density profile is well constrained and we are making several efforts to constrain the thermal profile. The combination of small distance to Earth (at 400 pc, $1''$ corresponds to 0.002 pc), low illumination ($\chi = 60$) and high density ($n_{\rm H} \sim 10^5$ cm^{-3}) implies that all the interesting physical and chemical processes can be probed in a field-of-view of less than $50''$ (with typical spatial scales ranging between $1''$ and $10''$). Hence, the Horsehead PDR is a good source to benchmark the physics and chemistry of UV illuminated neutral gas.

In our recent work on the ISM physics and chemistry in the Horsehead we have shown the importance of the interplay between the solid and gas phase chemistry in the formation of (complex) organic molecules, like H_2CO, CH_3OH and CH_3CN, which reveal that photo-desorption of ices is an efficient mechanism to release molecules into the gas phase (Guzmán *et al.* 2011, Gratier *et al.* in prep, Guzman *et al.* in prep). We have also provided new diagnostics of the UV illuminated matter. For example, we detected CF^+ and resolved its hyperfine structure (Guzman *et al.* 2012b). We propose that CF^+, which is observable from the ground, can be used as a proxy of C^+ (Guzman *et al.* 2012). Finally, we reported the first detection of the small hydrocarbon C_3H^+, which sheds light on the formation pathways of other observed small hydrocarbons, like C_3H and C_3H_2 (Pety *et al.* 2012). Part of these results were possible thanks to a complete an unbiased line survey at 1, 2 and 3 mm performed with the IRAM-30m telescope (Horsehead WHISPER), where approximately 30 species (plus their isotopologues) are detected.

Keywords. astrochemistry, ISM: clouds, ISM: molecules

References

Abergel, A., Teyssier, D., Bernard, J. P., *et al.* 2003, *A&A* 410, 577
Gratier, P., Pety, J., Guzmán, V., *et al.* 2012, in prep
Guzmán, V., Goicoechea, J. R., Pety, J., *et al.* 2012, in prep
Guzmán, V., Pety, J., Goicoechea, J. R., *et al.* 2011, *A&A* 534, 49
Guzmán, V., Pety, J., Gratier, P., *et al.* 2012a, *A&A* 543, L1
Guzmán, V., Roueff, E., Gauss, J., *et al.* 2012b, *A&A* 548, A94
Pety, J., Gratier, P., Guzmán, V., *et al.* 2012, *A&A* 548, A68

Highlights of Astronomy, Volume 16
XXVIIIth IAU General Assembly, August 2012
T. Montmerle, ed.

doi:10.1017/S1743921314012344

Origin of cosmic rays

V. A. Dogiel

P.N.Lebedev Institute of Physics, Leninskii pr, 53, 119991 Moscow, Russia,
email: dogiel@lpi.ru

Abstract. Cosmis rays are an essential component of the interstellar medium because of their high energy density. This paper reviews the origin of cosmic rays.

Keywords. cosmic rays, acceleration of particles, magnetic fields, ISM: bubbles

---

Cosmic rays (CRs) in spite of their small density, $\sim 10^{-10}$ cm^{-3} in comparison with the average gas density in the Galactic disk, ~ 1 cm^{-3}, are an essential component of the interstellar medium because their energy density is high. It is about ~ 1 eV cm^{-3}, that is comparable with that of interstellar magnetic fields and turbulent motions of gas/plasma in the Galactic disk. In this respect, analyses of interstellar plasma dynamics, ionization, heating etc. in the Galaxy is impossible without taking into consideration of CR influence.

Effect of CRs was found in 1785 by de Coulomb who observed a spontaneous discharge of electroscopes. Though the origin of this effect was unclear. The explanation of this phenomenon came only in the beginning of the 20th century and paved the way to one of mankind's revolutionary scientific discoveries. Exactly one hundred years ago Victor Hess concluded from his balloon experiments that the effect of discharge is due to an unknown radiation of extraterrestrial origin: "a radiation of very high penetrating power enters our atmosphere from above and produces .... ionization in closed vessels".

Further experiments showed that this radiation consisted of high energy particles, mainly protons, with the energies higher than above 1 GeV. But it took almost another hundred years in order to understand the origin of radiation.

The first model of CR origin was suggested by Baade and Zwicky in 1934. They assumed that CRs uniformly filled the whole volume of the Universe, and they were produced by supernovae (SNe) which exploded once per 10^3 yr in each (300 pc)^3 volume of the Universe. This idea of extragalactic origin of CRs was criticized by Ginzburg (1972) who suggested a test to prove the galactic CR origin from γ-ray data which ought to show fluctuations of CR density in galaxies. His idea was confirmed from subsequent measurements on the Compton and Fermi γ-ray satellites.

The main problem for the galactic theory of CRs was to explain very specific parameter of their flux, that the spectrum is power-law over six decades, from $\sim 10^9$ to 3×10^{15} eV with a single spectral index, $\propto E^{-2.7}$. The first idea about CR acceleration in the Galaxy was suggested by Fermi (1949, 1954), who assumed that the acceleration took place in the whole volume of the Galaxy by collisions of charged particles with moving fluctuations of magnetic field (magnetic clouds). Advantages of this mechanism was that it produced CRs everywhere in the Galaxy with a power-law spectrum. However, the spectrum is too hard, $\propto E^{-1}$, in comparison with the spectrum of galactic CRs. Secondly, acceleration efficiency of this mechanism is quite low. Hundred million years is required in order to increase the energy of particles only in three times. So, further ideas about acceleration processes in the Galaxy were required.

From the analysis of CR chemical composition it was estimated that the total luminosity of CR in the Galaxy is about $L_{\rm CR} \sim 3 \times 10^{40}$ erg s^{-1}. Then from energetic considerations it was concluded that the most probable CR sources in our Galaxy could be galactic SNe because their energy output is about 10^{42} erg s^{-1}, and if only a few per cents of these energy is transformed into CR, it is enough to produce the observed CR flux (see for details Ginzburg & Syrovatskii 1964 and Berezinskii *et al.* 1990). The question was how SNe being in different parts of the Galaxy generate a universal spectrum with the index close to -2. The answer was obtained in 1977 when Axford *et al.* (1977), Krymskii (1977) and others suggested a model of CR acceleration near shock fronts. Advantages of this model was that the efficiency of acceleration was much higher than that of the classical Fermi acceleration. The mechanism produces a universal power-law spectrum of CRs, $\propto E^{-2}$, independently of medium conditions. And just SN explosions produce shock fronts in the Galactic disk. However, the maximum energy of accelerated particles is determined by the shock lifetime and cannot exceed $E_{\max} \sim 10^{14}$ eV (see e.g. Berezhko & Völk 2000). This means that there are other sources of CR production in the Galaxy. Recent Fermi-LAT observations found indeed regions where particles can be accelerated up to energies higher than 10^{14} eV. This telescope found a cocoon of freshly accelerated cosmic rays in the Cyg OB association (see Ackermann *et al.* 2011) that suggests that OB associations and their superbubbles are likely the source of a substantial fraction of galactic cosmic rays from the collective action of multiple shocks from supernovae and the winds of massive stars where particles can be accelerated up to energies 10^{18} eV (see Bykov & Toptygin 2001). Recent discovery of a mysterious, diffuse giant gamma-ray structures, Fermi bubbles, (see Dobler *et al.* 2010 and Su *et al.* 2010) may indicate another source of CR generation in the Galaxy. These structures may result from star captures by the central black hole. Each capture produces a shock propagating into the Galactic halo where shocks accelerate electrons (up to $E \sim 1$ TeV) and protons with energies up to $E \sim 10^{18}$ eV (see Cheng *et al.* 2011, 2012).

Thus, though many aspects of CR origin are clear now, this theory cannot be considered complete. Future experiments may modify or even change our ideas about CR origin.

Acknowledgements

VAD is partly supported by the RFFI grant 12-02-00005-a and ISSI.

References

Ackermann, M., Ajello, M., Allafort, A., *et al.* 2011, *Science* 334, 1103
Axford,W. I., Leer, E. & Skadron, G. 1977, *Proc. 15th Int. Cosmic Ray Conf.* 11, 132
Baade, W. & Zwicky, F. 1934, *Phys. Rev.* 46, 76
Bell, A. R. 1978, *MNRAS* 182, 147
Berezhko, E. G. & Völk, H. J. 2000, *A&A* 357, 283
Berezinskii, V. S., Bulanov, S. V., Dogiel, V. A., Ginzburg, V. L., & Ptuskin, V. S. 1990, *Astrophysics of Cosmic Rays*, ed. V. L. Ginzburg, Norht-Holland: Amsterdam
Blandford, R. D. & Ostriker, J. P. 1978, *ApJ* (Letters) 221, L29
Bykov, A. M. & Toptygin, I. N. 2001, *Astron. Lett.* 27, 625
Cheng, K.-S., Chernyshov, D. O., Dogiel, V. A., *et al.* 2011, *ApJ* (Letters) 731, L17
Cheng, K.-S., Chernyshov, D. O., Dogiel, V. A., *et al.* 2012, *ApJ* 746, 116
Dobler, G., Finkbeiner, D. P., Cholis, I., *et al.* 2010, *ApJ* 717, 825
Fermi, E. 1949, *Phys. Rev.* 75, 1169
Fermi, E. 1954, *ApJ* 119, 1
Ginzburg, V. L. 1972, *Nature Phys. Sci.* 239, 8
Ginzburg, V. L. & Syrovatskii, S. I. 1964, *The Origin of Cosmic Rays*, New York: Macmillan
Krymskii, G. F. 1977, *Soviet Physics - Doklady* 22, 327
Su, M., Slatyer, T. R., & Finkbeiner, D. P. 2010, *ApJ* 724, 1044

Highlights of Astronomy, Volume 16
XXVIIIth IAU General Assembly, August 2012
T. Montmerle, ed.

doi:10.1017/S1743921314012356

Gas in galactic halos

Ralf-Jürgen Dettmar
Astronomical Institute, Ruhr-University Bochum, 44780 Bochum, Germany
email: dettmar@astro.rub.de

Abstract. The interstellar medium in galactic halos is described as a consequence of feedback mechanisms from processes related to star-formation in the disk. The presence of gas in galactic halos is also expected due to accretion of gas from the circumgalactic environment. The observational evidence for gas in galactic halos - from the hot X-ray emitting coronal phase to cool molecular gas and dust - is reviewed and discussed in the context of current models of the ISM and the "infall vs. outflow" debate.

Keywords. galaxies: ISM, galaxies: evolution, galaxies: halo

1. Introduction

The presence of gas at significant height above the stellar disks in galaxies has been discussed in the past as part of the cycle of interstellar matter between the disk and the halo, e.g. in the framework of galactic fountain flows. This exchange of matter is expected from the feedback of energy and momentum caused by the young stellar populations through supernovae and stellar winds and is thus linked to the star formation rate in the disk. While evolutionary models of galaxies require infalling primordial gas to sustain the observed reservoir of gas and to explain the observed chemical abundances outflows seem to be required in galaxy evolution since galaxy formation models without feedback do not reproduce the properties of galaxies as observed in the local universe. In view of these concepts we are expecting both, gas in halos due to infall as well as due to outflows.

2. Observations of gas in galactic halos

More than half a century ago HI-21 cm observations of High Velocity Clouds (HVCs) in the Milky Way (MW) stimulated the discussion of gaseous galactic halos. The review by Putnam *et al.* (2012) describes the state of the art in this field and discusses consequences with regard to the infall vs. outflow discussion. For extragalactic objects NGC 891 is frequently considered the "typical" disk galaxy (and MW twin). The detection of a large HI halo (Oosterloo *et al.* 2007) led to the expectation that extended HI halos are typical for disk galaxies. However, the preliminary analysis (reported in this SpS) of HI data observed for a sample of disk galaxies in the HALOGAS project (Heald *et al.* 2012) rather leads to the conclusion that such HI halos are rare. This preliminary analysis also suggests that the mass of HI in the halo is correlated with the star formation in the disk. Similar relations have also been found for the presence of Warm or Diffuse Ionized Gas (WIM, DIG) in galactic halos (Rossa & Dettmar 2003; Miller & Veilleux 2003).

Cosmological simulations such as the GIMIC project (Crain *et al.* 2010) predict hot coronae as the consequence of infall from intergalactic filaments. The parameters of the model coronae are in general agreement with X-ray observations for galaxies over a wide range of masses. However, the X-ray observations for disk galaxies correlate equally well with the star formation rate in the disks if global parameters are considered (Tüllmann *et al.* 2006; Li & Wang 2012). The spatial correlation of the observed X-ray halos with star formation regions in the disk is better explained in the outflow scenario.

The general presence of outflows is also suggested by observations of the cold ISM components in galactic halos: dust as well as molecular gas is present in the disk-halo interface of galaxies (Howk & Savage 2000; Veilleux *et al.* 2009). These observations are, however, currently limited to a handful of cases only.

All the above mentioned components of the ISM in galactic halos are typically accompanied by the presence of cosmic rays and magnetic field as shown by radio continuum observations (e.g., Dettmar & Soida 2006). This is again corroborating the case for galactic outflows being responsible for the presence of gas in galactic halos.

3. Conclusions

All constituents of the interstellar medium observed in the disks of spiral galaxies - from the cold molecular gas and the hot X-ray emitting plasma to magnetic fields and cosmic rays - are also observed in halos of galaxies. The presence of all these different gas phases seems to correlate with the starformation rate (per unit area) in the disk and is thus explained by outflows caused in an ISM driven by starformation activity. However, the observed samples of galaxies are not representative but rather biased to those with higher star formation rate. In addition, it has to be mentioned that for some gas phases such as the X-ray emitting HIM the observations are still limited by instrumental sensitivity.

Acknowledgements

Research in this field at RUB is supported by DFG and BMBF/DESY-PT.

References

Crain, R. A., *et al.* 2010, *MNRAS* 407, 1403
Dettmar, R.-J. & Soida, M. 2006, *AN* 327, 495
Heald, G. H., *et al.* 2012, *A&A* 544, 1 1
Howk, J. C. & Savage, B. D. 2000, *AJ* 119, 644
Li, J.-T. & Wang, Q. D. 2012, *arXiv:* 1210.2997
Miller, S. T. & Veilleux, S. 2003, *ApJS* 148, 383
Putman, M. E., Peek, J. E. G., & Joung, M. R. 2012, *ARAA* 50, 491
Oosterloo, T, *et al.* 2007, *AJ* 134, 1019
Rossa, J. & Dettmar, R.-J. 2003, *A&A* 406, 439
Tüllmann, R, *et al.* 2006, *A&A* 457, 779
Veilleux, S., *et al.* 2009, *ApJ* (Letters) 700, L149

Highlights of Astronomy, Volume 16
XXVIIIth IAU General Assembly, August 2012
T. Montmerle, ed.

doi:10.1017/S1743921314012368

HVCs, infall and the Galactic Fountain

Bart P. Wakker

supported by NASA/NSF; affiliated with Department of Astronomy
University of Wisconsin-Madison
475 N Charter St, Madison, WI 53705, USA
email: wakker@astro.wisc.edu

Abstract. High-velocity clouds (HVCs) consist of gas that does not take part in normal Galactic rotation, having velocities deviating by up to 400 $\mathrm{km\,s^{-1}}$ from those expected from rotation. Over the past five decades, studies have shown they these clouds trace a number of different processes, including the Galactic Fountain, tidal streams, and infall. Here, I summarize some recent results concerning measurements of cloud metallicities and distances and how these are used to understand individual clouds and derive the conclusion above.

Keywords. ISM: general, ISM: HVCs, Galaxy: corona

Observations using the ultraviolet spectrographs on the *Hubble Space Telescope* (HST), as well as with the *Far Ultraviolet Spectroscopic Explorer* (FUSE) have now yielded measurements of the S II $\lambda\lambda$1250, 1253, 1259 and many O I lines. Because the ionization potentials of O I and H I are similar and because most of the oxygen is in the gas phase, the O I/H I ratio is usually the same as the O/H ratio. Similarly, SII/H I is usually close to S/H, although sometimes a (small) ionization correction is needed.

Most published results so far concentrate on measurements toward HVC complex C (see Fox *et al.* 2004 for a summary) and the Magellanic Stream (Gibson *et al.* 2000; Fox *et al.* 2005). From our HST Cycle 18 program, as well as an analysis of the FUSE+HST archive we now have 17 different clouds toward which the metallicity of a HVC has been measured, with values ranging from 1/20th solar to solar. There is no exact correlation between velocity and metallicity. Values of 1/20th, 1/10th and 1 times solar are found for intermediate-velocity clouds (HVCs with $|v|$ < 90 $\mathrm{km\,s^{-1}}$); values ranging from 1/10th to 1/4 solar are found in the Magellanic Stream; while for clouds with $|v|$ < 90 $\mathrm{km\,s^{-1}}$ the metallicities range from 1/10th solar to solar. The cloud metallicities allow us to associate individual clouds with one of the main processes summarized above.

To find distances to HVCs requires a set of stellar targets situated both behind and in front of the star. Such targets need to have clean spectra and known distance. Starting with the SDSS and 2MASS surveys, we defined a sample of 3000 candidate probes toward 60 HVCs. We now have good distances for some 15 HVCs, using intermediate-resolution (1 Å) spectra from the SDSS (~800 stars), spectra taken by us using the APO (~1200 stars), photometry from the SDSS (~500 stars), our own photometry program on the WIYN 0.9m (160 nights, ~100 stars), as well as high-resolution (~6 $\mathrm{km^{-1}}$) spectra for 125 stars taken with Keck/HIRES and the VLT/UVES. These distances reveal that the lower-velocity clouds ($|v|$<90 $\mathrm{km\,s^{-1}}$) tend to be located up ~1-5 kpc above the Galactic disk, while higher-velocity clouds range in distance from 3 to 25 kpc (some of these results were shown in Wakker *et al.* (2007, 2008). The three most massive HVCs for which we determined a distance are about 7 kpc above the plane. Fig. 1 shows an image of the Milky Way, with the locations of the HVCs with known distances overlaid.

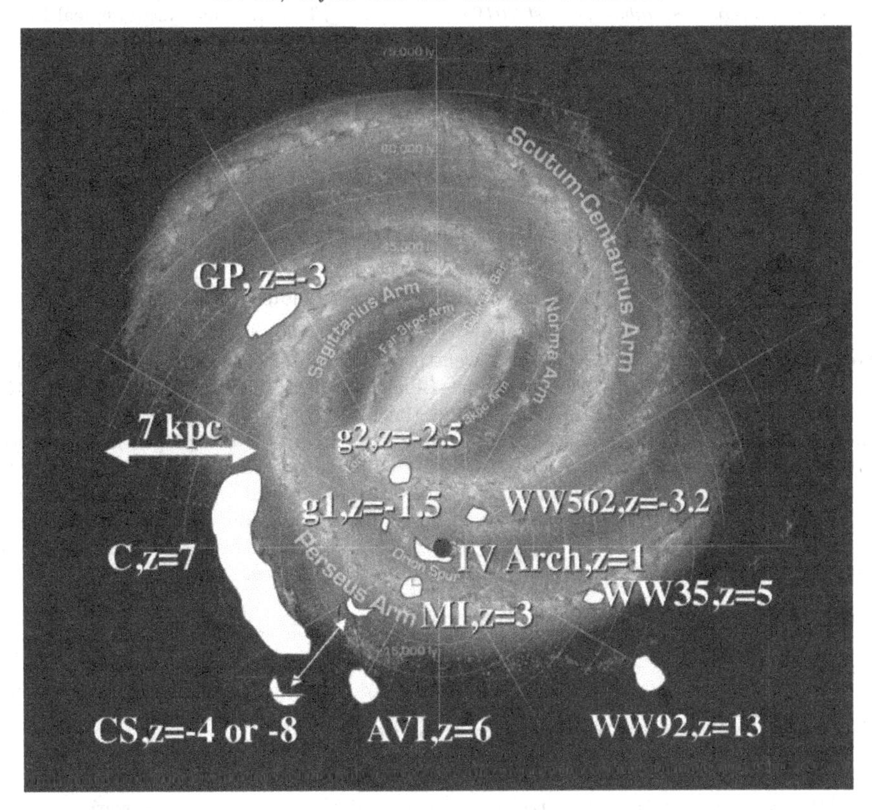

Figure 1. Artist impression image of the Milky Way, with the locations of the HVCs with known distances overlaid

References

Fox, A., Savage, B., Wakker, B., *et al.* 2004, *ApJ* 602, 738
Fox, A., Wakker, B., Smoker, J., *et al.* 2010, *ApJ* 718, 1046
Gibson, B., Giroux, M., Penton, S., *et al.* 2000, *AJ* 120, 1830
Wakker, B., York, D., Howk, J., *et al.* 2007, *ApJ* (Letters) 670, L113
Wakker, B., York, D., Wilhelm, R., *et al.* 2008, *ApJ*, 672, 298

Highlights of Astronomy, Volume 16
XXVIIIth IAU General Assembly, August 2012
T. Montmerle, ed.

doi:10.1017/S174392131401237X

Observational constraints on the multiphase ISM

Mark G. Wolfire

Astronomy Department, University of Maryland, College Park, MD 20742
email: mwolfire@astro.umd.edu

Abstract. In recent years we have seen a wealth of new observations and analysis that sheds light on the distribution and physical properties of various ISM phases. In particular the thermal pressure from C I (Jenkins & Tripp 2011) shows the bulk of the CNM phase with a log normal pressure distribution. It appears that thermal instability is important for phase separation, but with with a thermal pressure variation about the mean driven by turbulence. In additional, there is evidence from C I, H_2, and complex molecules, of both higher and lower pressure environments. An additional "phase" that is of increasing interest for high z, low metallicity galaxies is the C^+/H_2 gas that is not traced by H I or CO. This review presents the observational evidence for the existence and physical properties of these various ISM phases.

Keywords. ISM: general, ISM: clouds, ISM: structure

1. The cold and warm neutral gas phases

1.1. *Thermal pressure*

The term "phase" generally refers to gas within a distinct temperature regime. The cold neutral atomic gas or CNM is at about 100 K, and the warm neutral atomic gas or WNM is about 8000 K. The thermal pressure in the CNM has recently been re-evaluated by Jenkins & Tripp (2011; hereafter JT) using UV absorption spectroscopy of C I. In a mass-weighted PDF they find a median thermal pressure of about $P/k \approx 3800$ K cm^{-3}. The distribution is fit by a log-normal distribution between $\log P/k$ 3.2 and 4. A log-log plot of the PDF shows deviations from the log-normal distribution at low column densities. There is both a high pressure wing (up to $\log P/k \sim 4.6$) and low pressure wing (down to $\log P/k \sim 2.0$). By estimating the UV radiation field along each line of sight, JT argue that the high pressure clouds are close to massive stars and the pressures are affected by mechanical processes such as winds and shocks. The higher pressure clouds are not characteristic of the diffuse ISM. For the low pressure clouds JT suggest that at sizes < 1000 AU the eddy turnover times are shorter that the radiative cooling times and thus the gas is acting closer to adiabatically or $\gamma > 1$. Passot & Vázquez-Semadeni (1998) found that for adiabatic indexes greater than 1, the distribution will be skewed towards lower values. similar to that found by JT.

The width and shape of the pressure distribution is produced by turbulence but what sets the median thermal pressure? Kim *et al.* (2011) have carried out simulations of a multiphase galactic disk. In a thermal pressure versus density (phase) diagram, most of the mass lies along the thermal equilibrium curve. They find a distinct CNM phase and a distinct WNM phase with some mass at thermally unstable temperatures. The median pressure is close to the "two phase" pressure $P_{2p} = \sqrt{P_{min} P_{max}}$ where P_{min} and P_{max} are the minimum and maximum pressures allowed for two phases (Wolfire *et al.* 2003). Thus, the distribution of thermal pressures in the CNM appear to be set by turbulence with a median pressure set by the two phase pressure.

There are also constraints on the CNM gas temperature from UV absorption studies of the ground state H_2 Ortho and Para column densities. For diffuse gas ($A_V < 1$) the mean temperature is about 80 K (Rachford *et al.* 2009).

1.2. *Hot pockets*

In addition to the log-normal distribution in thermal pressures JT required a small (0.05%) mass of gas at high pressure ($P/k \sim 3 \times 10^5$ K cm^{-3}) along each line of sight. This could be a manifestation of small scale turbulence or turbulent dissipation. Additional observations have inferred pockets of small scale heating or density enhancements. For example 1)warm cloud chemistry that might produce HCO^+ and CH^+ (Godard *et al.* 2009; Falgarone *et al.* 2010), 2)Tiny Scale Atomic structure TSAS that appears in H I absorption on 10s of AU scales (Heiles 1997). 3)Warm diffuse H_2 seen in emission (Falgarone *et al.* 2005). 4)High H_2/PAH emission ratios seen in high latitude clouds (Ingalls *et al.* 2011), and 5)warm H_2 seen towards molecular cloud surfaces illuminated by low UV fields (Goldsmith *et al.* 2010; Habart *et al.* 2011).

There are plenty of H_2 observations, however, that can be fit with grain photoelectric heating that do not require mechanical heating. For example, Sheffer *et al.* (2011) fit the observed H_2 rotational line emission from NGC2023 with a PDR model with $n = 10^5$ cm^{-3} and $\chi = 4 \times 10^3$. In this high radiation field intensity, grain photoelectric heating is sufficient to reproduce the line emission.

1.3. *Thermally unstable gas*

There is some disagreement over the fraction of warm neutral medium that is thermally unstable. Based on the results by Heiles & Troland (2003) it is often "quoted" that 50% of the gas mass is in thermally unstable temperatures. First note that they found that locally 60% of the gas is WNM and 40% CNM (by mass). The fraction of thermally unstable gas applies to the WNM (60%) portion, and not the total. Note also that their Figure 2 shows two different distributions: in plane ($|b| < 10$) and out of plane ($|b| > 10$). For the in-plane distribution only $\sim$ 25% of the warm gas is outside the 7000 – 9000 K range expected for thermally stable gas. Thus the thermally unstable gas is only about 15% of the the total in-plane gas mass when the cold phase is included. The out-of-plane distribution is quite odd with very little gas at 7000 – 9000 K. It would appear that the in-plane gas is dominated by thermal instability plus turbulence while the out-of-plane gas is dominated by dynamical processes. Note also that the in-plane uncertainties are large and the statistics are poor. The in-plane results are based on only 8 lines of sight while the out-of-plane results are based on 79 lines of sight. Current (GALFA; Peek *et al.* 2010) and future (GASKAP; Dickey *et al.* 2012) H I surveys will certainly improve the statistics.

1.4. *Phase distribution*

Dickey *et al.*(2009) used an H I emission/absorption technique to measure the mean H I emission per unit length, the mean absorption per unit length, and the ratio of the two. They find that the ratio is nearly constant from the solar circle out to 25 kpc. This means that the ratio of CNM to WNM is nearly constant to 25 kpc. They also find that the volume averaged density and thus the thermal pressure drops by about a factor of $\sim$ 30 from the solar circle to 20 kpc. Wolfire *et al.* (2003) calculated a phase diagram for 18 kpc and found a $P_{\min}$ of about 10 lower than at the solar circle. Thus the inferred thermal pressure from H I observations is much lower than the calculated $P_{\min}$. How can the CNM/WNM fraction stay constant if the thermal pressure is less than $P_{\min}$? One way this might happen is if turbulence bumps up the pressure above $P_{\min}$ to maintain CNM gas. The constant ratio of CNM/WNM to large Galactic radii provides a strict model test.

2. Power spectrum of diffuse and dense gas

There are a number of Herschel PACS and SPIRE key projects that examine the structure and power spectrum of diffuse and dense gas. Miville-Deschênes et al. (2010) presented a SPIRE map at 250 μm of the Polaris Flare region. At $25''$ beam size numerous clumps and filaments are seen in this high latitude translucent cloud. The power spectrum was measured with a slope -2.7 from $\sim$ 0.8 pc down to 0.01 pc. The turbulent cascade extends to quite small sizes at least in these high latitude clouds. Filamentary structure is also seen in the Hi-Gal survey (Molinari *et al.* 2010). Image processing highlights the filamentary structure revealing star formation occurring along the filaments. What forms the filaments and clumps? Probably gravity, turbulence, and magnetic fields all play a role in creating filaments followed by fragment into cores via gravitational instability.

3. OVI column density

Another constraint on phase distributions comes from OVI column densities and mean abundances derived from FUV (FUSE) absorption line spectroscopy (Bowen *et al.* 2008). In collisional ionization equilibrium, the OVI abundance peaks at about 3×10^5 K. This is cooler than X-ray emitting material but warmer than WNM. The OVI comes from conductive interfaces or turbulent mixing layers. As Don Cox has pointed out many times, the line-of-sight averaged OVI is only a few 10^{-8} cm^{-3}, so there cannot be too many interfaces or else the abundance will be higher than observed. The FUSE observations are fit with 1.3×10^{-8} cm^{-3} . The emerging picture of a turbulent ISM is more complicated than having cold clouds embedded in either a hot or warm medium. Recent hydrodynamic modeling by de Avillez & Breitschwerdt (2005) have taken the OVI constrain into account.

4. Dark molecular gas

Dark molecular gas is is gas that has C^+ and H_2 but no or very little CO (Grenier *et al.* 2005; Wolfire *et al.* 2010). Of course it is not really dark but emits in C^+, CI, and IR continuum. The model calculation by Wolfire *et al.* (2010) find about 30% dark gas fraction over a range of giant molecular cloud masses - a value consistent with the gamma-ray observations reported in Grenier *et al.* (2005). More recent gamma-ray observations using FERMI Abdo *et al.* 2010) and IR observations from *Planck* (Planck collaboration 2011) find slightly higher fractions ($\sim$ 50%) but with large cloud-to-cloud variations. The dark gas may in fact not be molecular depending on optical depth effects in H I (Braun *et al.* 2009). Hydrodynamic models may account for the cloud-to-cloud variation and constrain the atomic fraction of dark gas.

Acknowledgements

M.G.W would like to acknowledge partial support for this work provided by NASA through awards issued by JPL/Caltech 1426973 and 1457797.

References

Abdo, A. A., Ackermann, M., Ajello, M., *et al.* 2010, *ApJ* 710, 133
Bowen, D. V., Jenkins, E. B., Tripp, T. M., *et al.* 2008, *ApJS* 176, 59
Braun, R., Thilker, D. A., Walterbos, R. A. M., & Corbelli, E. 2009, *ApJ* 695, 937
de Avillez, M. A. & Breitschwerdt, D. 2005, *ApJ* (Letters) 634, L65
Dickey, J. M., McClure-Griffiths, N., Gibson, S. J., *et al.* 2012, arXiv:1207.0891
Dickey, J. M., Strasser, S., Gaensler, B. M., *et al.* 2009, *ApJ* 693, 1250
Falgarone, E., Godard, B., Cernicharo, J., *et al.* 2010, *A&A* 521, L15
Falgarone, E., Verstraete, L., Pineau Des Forêts, G., & Hily-Blant, P. 2005, *A&A* 433, 997
Godard, B., Falgarone, E., & Pineau Des Forêts, G. 2009, *A&A* 495, 847
Goldsmith, P. F., Velusamy, T., Li, D., & Langer, W. D. 2010, *ApJ* 715, 1370
Grenier, I. A., Casandjian, J.-M., & Terrier, R. 2005, *Science* 307, 1292
Habart, E., Abergel, A., Boulanger, F., *et al.* 2011, *A&A* 527, A122
Heiles, C. 1997, *ApJ* 481, 193
Heiles, C. & Troland, T. H. 2003, *ApJ* 586, 1067
Miville-Deschênes, M.-A., Martin, P. G., Abergel, A., *et al.* 2010, *A&A* 518, L104
Molinari, S., Swinyard, B., Bally, J., *et al.* 2010, *A&A* 518, L100
Ingalls, J. G., Bania, T. M., Boulanger, F., *et al.* 2011, *ApJ* 743, 174
Jenkins, E. B. & Tripp, T. M. 2011, *ApJ* 734, 65
Kim, C.-G., Kim, W.-T., & Ostriker, E. C. 2011, *ApJ* 743, 25
Passot, T. & Vázquez-Semadeni, E. 1998, *Phys. Rev. E* 58, 4501
Planck collaboration 2011, Planck early results 17, *A&A* 536, 17
Peek, J. E. G., Begum, A., Douglas, K. A., *et al.* 2010, *ASPC* 438, 393
Rachford, B. L., Snow, T. P., Destree, J. D., *et al.* 2009, *ApJS* 180, 125
Wolfire, M. G., Hollenbach, D., & McKee, C. F. 2010, *ApJ* 716, 1191
Wolfire, M. G., McKee, C. F., Hollenbach, D., & Tielens, A. G. G. M. 2003, *ApJ* 587, 278

Highlights of Astronomy, Volume 16
XXVIIIth IAU General Assembly, August 2012
T. Montmerle, ed.

doi:10.1017/S1743921314012381

Molecular richness of the diffuse interstellar medium: a signpost of turbulent dissipation

Edith Falgarone[1], **Benjamin Godard**[2], **Guillaume Pineau des Forêts**[3] **and Maryvonne Gerin**[1]

[1]LERMA/LRA, Ecole Normale Supérieure & Observatoire de Paris,
24 rue Lhomond, 75005 Paris, France
email: edith.falgarone@ens.fr

[2]LUTh, Observatoire de Paris, 92195 Meudon, France
email: benjamin.godard@obspm.fr

[3]Institut d'Astrophysique Spatiale, 91405 Orsay, France

Abstract. The *Herschel*/HIFI absorption spectroscopy surveys reveal the unexpected molecular richness of the Galactic diffuse ISM, even in gas of very low average H_2 molecular fraction. In particular, two hydrides, CH^+ and SH^+ with highly endoenergetic formation routes have abundances that challenge models of UV-driven chemistry. The intermittent dissipation of turbulence appears as a plausible additional source of energy for the diffuse ISM chemistry. We present recent results of the so-called models of Turbulent Dissipation Regions (TDR). The abundances of many of the molecules observed in the diffuse ISM, including CO that is used as a tracer of the molecular clouds mass, may be understood in the framework of the TDR models.

Keywords. Astrochemistry, Turbulence, Magnetic Fields, ISM: molecules, ISM: kinematics and dynamics, ISM: general, ISM: evolution

1. The puzzles raised by the cold ISM

The cold diffuse interstellar medium (ISM), as defined in the review of Snow & McCall (2006), makes up the mass of nearby molecular clouds. This is best seen on the probability distribution functions (PDF) of their extinction (Kainulainen *et al.* 2009). The cloud mass is comprised in the log-normal part of the PDFs, *i.e.* the transparent and turbulent part.

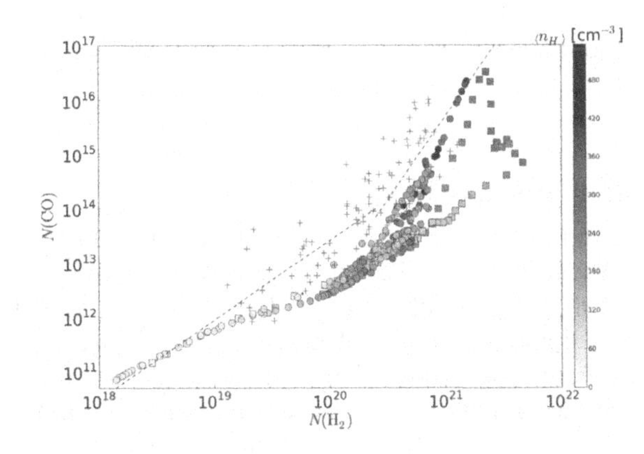

Figure 1. Comparison of observed CO column densities (crosses), derived from absorption lines against nearby stars (see references in Levrier *et al.* (2012)) with state-of-the-art computed values combining the photo-dissociation regions (PDR) model of Le Petit *et al.* (2006) and bi-phasic MHD turbulence simulations of Hennebelle *et al.* (2008).

Molecular abundances of the diffuse gas raise resilient puzzles. For 70 years, the CH^+ abundances have been known to exceed model predictions by two orders of magnitude. This is so because the route to CH^+ is highly endoenergetic and, once formed, CH^+ is rapidly destroyed by collisions with H_2. An additional source of energy is thus required to efficiently form CH^+ in diffuse gas. The observed CO abundances in a broad range of H_2 column densities also exceed model predictions by more than one order of magnitude (Fig. 1 from Levrier *et al.*, 2012).

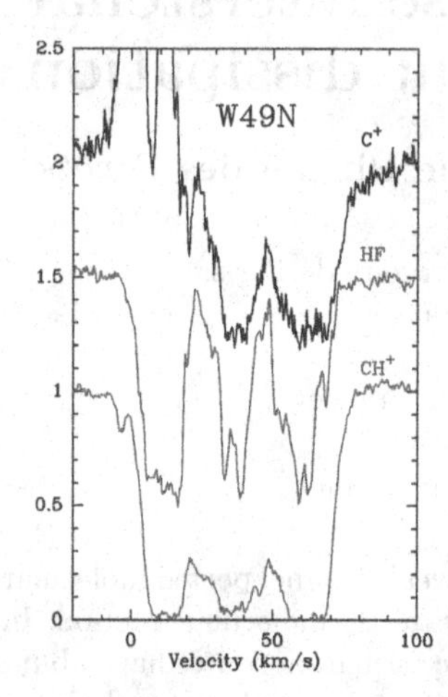

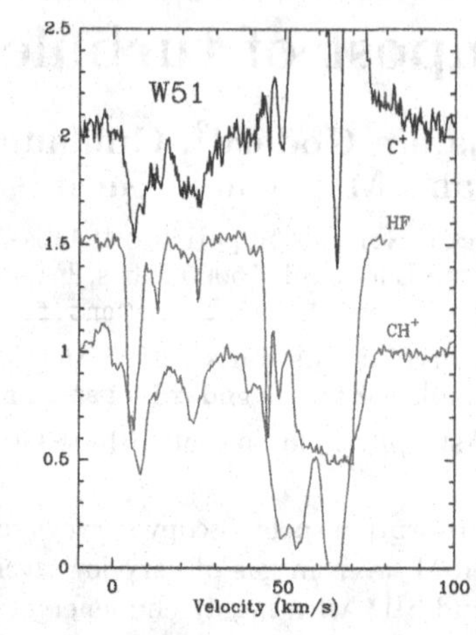

Figure 2. CII, HF and CH^+ spectra observed in the direction of W49 and W51. Note the similarity of the velocity coverage of the CII and CH^+ absorptions away from those of the star forming regions.

Herschel/HIFI has deepened these puzzles. We have conducted an absorption spectroscopy survey against bright star forming regions of the inner Galaxy (PRISMAS key-project, PI Gerin). Each line of sight samples kiloparsecs of gas in the Galactic plane. We detected saturated CH^+(1-0) (and $^{13}CH^+$(1-0)) in absorption on all the sight lines, (Falgarone *et al.* 2010a, 2010b) and SH^+ that has a formation endothermicity twice as large as that of CH^+ (Godard *et al.*, 2012). Last, C^+ is detected in absorption over the same velocity intervals as CH^+ (Fig. 2) and we show that C^+ and CH^+ absorptions occur in the cold neutral medium (CNM) (Gerin *et al.*, in prep.). Using HF as a tracer of molecular hydrogen (Neufeld *et al.* 2010), and e-VLA atomic hydrogen spectra (Menten *et al.* in prep.), we infer the mean H_2 molecular fraction of the absorbing gas: it is low on average and has a large scatter $0.04 < f_{H_2} < 1$ (Godard *et al.*, in prep.). Hence, CH^+ and SH^+ are detected with large abundances even in gas components with very low average H_2 fractional abundance.

UV-driven chemistry is not able either to reproduce these large CH^+ abundances nor the broad range of observed SH^+/CH^+ ratios. The alternative is a warm chemistry that opens the route $C^+ + H_2 \rightarrow CH^+ + H$ and leads to the formation of the pivotal species, CH_3^+. In particular, CH_3^+ reacts with O to form HCO^+, the precursor of CO.

2. Chemistry driven by turbulent dissipation

Turbulence and magnetic fields that support the ISM in the gravitational well of the Galaxy (Cox 2005) are a formidable reservoir of energy. Turbulent dissipation is intermittent (see the review of Anselmet *et al.*, 2001). In the diffuse ISM, the bursts of turbulent dissipation are *locally and temporarily* a dominant source of heating for the gas, large enough to excite the H_2 pure rotational lines by collisions (Falgarone *et al.* 2005; Ingalls *et al.* 2011) and trigger a specific "warm" chemistry. These space-time bursts are modeled as low-velocity MHD shocks (Lesaffre *et al.*, 2012) and/or thin coherent vortices, (*i.e.* the TDR model, for Turbulent Dissipation Regions, Godard *et al.*, 2009) temporarily heating a small fraction of the gas (a few %) to temperatures up to 10^3 K. The heated gas eventually cools down once the dissipation burst is over. The free parameters of the TDR model are constrained by the known large-scale properties of turbulence.

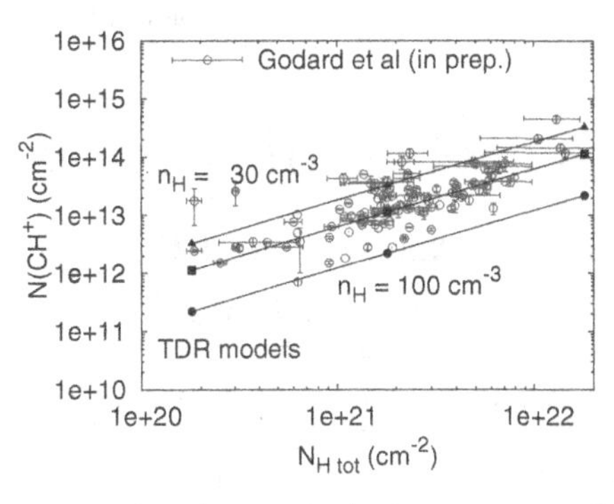

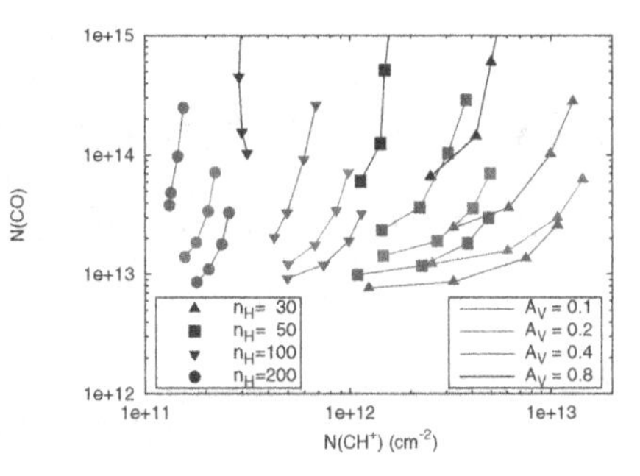

Figure 3. *(Left)* The CH^+ data compared to TDR models. *(Right)* CO and CH^+ column densities computed in TDR models for different densities and UV-shieldings and a total gas column density $N_H = 1.8 \times 10^{21}$ cm^{-2}. The free parameter along each curve, is the rate-of-strain (Godard *et al.* in prep.)

Dissipation is due to both viscosity and ion-neutral friction induced by the decoupling of the neutral fluid from the magnetic fields. The chemical and thermal inertia are large. The chemical relaxation times span a broad range, from 200 yr for CH^+ up to 5×10^4 yr for CO. A random line of sight through the medium therefore samples 3 phases: *(i)* actively dissipating regions, *(ii)* relaxation phases, and *(iii)* the ambient medium.

The main successes of the TDR model are: *(i)* the agreement of CH^+ and SH^+ observations with model predictions (Fig. 3, left). *(ii)* the scaling of CH^+ abundances with the turbulent dissipation rate, *(iii)* the rotational excitation of H_2 in diffuse gas, and *(iv)* the CO abundance in diffuse molecular gas (Fig. 3, right). A fraction as small as a few percent of warm gas is sufficient to reproduce the observations (H_2 excitation diagram, CH^+, SH^+, but also CO abundances). The comparison with data tend to favor models in which dissipation is dominated by ion-neutral friction.

In summary, many of the molecules we observe in the diffuse medium, including CO that is used as a tracer of the molecular mass in galaxies, are too abundant to be explained by state-of-the-art chemistry models driven by the UV-field. A plausible alternative is that they are the outcome of a specific non-equilibrium chemistry triggered by the bursts of turbulent dissipation.

References

Anselmet, F., Antonia, R., & Danaila, L. 2001, *P&SS* 49, 1177
Cox, D. 2005, *ARAA* 43, 337
Falgarone, E., Godard, B., Cernicharo, J., *et al.* 2010a, *A&A* (Letters) 521, L15
Falgarone, E., Ossenkopf V., Gerin M., *et al.* 2010b, *A&A* (Letters) 518, L118
Falgarone, E., Verstraete, L., Pineau des Forêts, G., & Hily-Blant, P. 2005, *A&A* 433, 997
Godard, B., Falgarone, E., Gerin, M., *et al.* 2012, *A&A* 540, 87
Godard, B., Falgarone, E., & Pineau des Forêts, G. 2009, *A&A* 495, 847
Hennebelle, P., Banerjee, R., Vazquez-Semadeni E., *et al.* 2008, *A&A* (Letters) 486, L43
Ingalls, J., Bania, T., Boulanger, F., *et al.* 2011, *ApJ* 743, 174
Kainulainen, J., Beuther, H., Henning, T., & Plume, R. 2009, *A&A* (Letters) 508, L35
Le Petit, F., Nehmé, C., Le Bourlot, J., & Roueff, E. 2006, *ApJS* 164, 506
Lesaffre, P., Pineau des Forêts, G., Godard, B., *et al.* 2012, *A&A* in press
Levrier, F., Le Petit, F., Hennebelle, P., *et al.* 2012, *A&A* 544, 22
Neufeld, D., Sonnentrucker, P., Phillips, T., *et al.* 2010, *A&A* (Letters) 518, L108
Snow, T. & McCall, B. 2006, *ARAA* 44, 367

Highlights of Astronomy, Volume 16
XXVIIIth IAU General Assembly, August 2012
T. Montmerle, ed.

doi:10.1017/S1743921314012393

ISM simulations: an overview of models

M. A. de Avillez[1,2], D. Breitschwerdt[2], A. Asgekar[3] and E. Spitoni[1]

[1]Dept. of Mathematics, University of Évora, 7000 Évora, Portugal
email: `mavillez,spitoni@galaxy.lca.uevora.pt`

[2]Dept. of Astronomy & Astrophysics, Technical University Berlin, D-10623 Berlin, Germany
email: `breitschwerdt@astro.physik.tu-berlin.de`

[3]ASTRON, P.O. Box 2, 7990 AA Dwingeloo, The Netherlands

Abstract. Until recently the dynamical evolution of the interstellar medium (ISM) was simulated using collisional ionization equilibrium (CIE) conditions. However, the ISM is a dynamical system, in which the plasma is naturally driven out of equilibrium due to atomic and dynamic processes operating on different timescales. A step forward in the field comprises a multi-fluid approach taking into account the joint thermal and dynamical evolutions of the ISM gas.

Keywords. ISM: general, ISM: structure, atomic processes, turbulence; MHD

---

1. Introduction

The attempts to model the supernova-driven ISM can be traced to the seminal models of Cox & Smith (1974; CS74) and McKee & Ostriker (1977, MO77). In the former supernovae (SNe) maintain an interconnected tunnel network filled with X-ray emitting gas, while in MO77 the gas is distributed into three phases in global pressure equilibrium. In both models the Galactic volume (50% in CS74 and 70-80% in MO77) is filled with hot ($> 10^5$ K) low-density gas. Further ramifications include the break-out of the hot intercloud medium, cooling and condensing into clouds (galactic fountain; Shapiro & Field 1976) or escaping as a wind (e.g., chimney model; Norman & Ikeuchi 1989).

Although these early works capture some of the essential physics, more complex and sophisticated models were devised by taking advantage of numerical simulations. These comprise the evolution of a patch of the Galactic disk in two dimensions (2D) (hydrodynamical (HD): Bania & Lyon 1980; Chiang & Prendergast 1985; Chiang & Bregman 1988; Rosen *et al.* 1993; Magnetohydrodynamical (MHD): Vazquez-Semadeni *et al.* 1995), and in three-dimensions (3D), e.g.. the MHD evolution of a 200^3 pc^3 region (Balsara *et al.* 2004) and the cosmic-rays driven amplification of the field in a differentially rotated domain ($0.5 \times 1 \times [-0.6, 0.6]$ kpc^3; Hanasz *et al.* 2004). The first disk-halo evolution models (2D HD) were developed by Rosen & Bregman (1995). With increasing of computer power, 3D HD (de Avillez model in 2000 and upgrades - see Avillez & Breitschwerdt 2007; Joung & Mac Low 2006) and MHD (Korpi *et al.* 1999; Avillez & Breitschwerdt 2005; Gressel *et al.* 2008; Hill *et al.* 2012) models have been developed.

In general the disk-halo models consider parameters according to observations (e.g., initial matter distribution with height, SN rates, background UV radiation field). Differences are found in the number of physical processes included (magnetic fields, cosmic rays, heat conduction, etc.), numerical techniques, type of grid (fixed or differentially rotated using the shear box technique), and grid resolutions and sizes. Resolutions are fixed or benefit from the use of the adaptive mesh refinement (AMR) technique (Berger & Oliger 1984). The highest resolutions cover a wide range from 0.5 pc to 10 pc, passing through 2 and 8 pc. The grid sizes in the vertical direction range from 0.1 kpc to 15 kpc on either side of the Galactic midplane. However, grids extending up to 2 kpc imply that the disk-halo-disk cycle can neither be established nor tracked - the simulations are valid for a small period of time before the gas escapes from the top and bottom boundaries.

These simulations showed that: (i) the ISM does not become saturated by SN activity, (ii) the disk expands and relaxes dynamically as SN rate fluctuates in time and space, (iii) the turbulent field builds up exponentially within 20 Myr of disk evolution, (iv) the magnetic field does not strongly correlate with density, except for the densest regions, (v) the magnetic field does not prevent the matter escape into the halo as it only briefly delays the disk-halo cycle, (vi) the volume filling factor of hot gas in the Galactic disk is only $\sim$20 %, (vii) there are large pressure variations in the disk in contrast to MO77 with the thermal pressure dominating at high temperatures ($T > 10^6$ K), magnetic pressure at $T < 200$ K, and ram pressure elsewhere.

2. Thermal and dynamical evolution of the ISM

All models referred previously assumed the ISM plasma to be in CIE, represented by a unique and general cooling function (CF) taken from different sources (e.g., Dalgarno & McCray 1972 (DM72); Sutherland & Dopita 1993; Gnat & Sternberg 2007). CIE assumes that the number of ionizations is balanced by recombinations from higher ionization stages. However, CIE is only valid provided the cooling timescale (τ_{cool}) of the plasma is larger than the recombination times scales of the different ions ($\tau_{rec}^{Z,z}$), something that occurs at $T > 10^6$ K (see references above). For lower temperatures $\tau_{cool} < \tau_{rec}^{Z,z}$, and deviations from CIE are expected (see, e.g., DM72). These departures affect the local cooling, which is a time-dependent process that controls the flow dynamics, feeding back to the thermal evolution by a change in the density and internal energy distribution, which in turn modifies the thermodynamic path of non-equilibrium cooling.

A major improvement in ISM studies is therefore to carry out time-dependent multi-fluid calculations of the joint thermal and dynamical evolution of the plasma, i.e. to follow each fluid element's thermal history by determining its ionization structure and CF at each time-step. Radiative losses are folded into the energy equation with the internal energy including also the potential energies associated to the different ionization stages.

Historically, there have been a number of simulations, which have included part of the ionization history into HD simulations, (Cox & Anderson 1982; Innes *et al.* 1987; Borkowski *et al.* 1994; Smith & Cox 2001; among others), misty tailored for specific astrophysical problems. The effect of delayed recombination has been emphasized by Breitschwerdt & Schmutzler (1994), who have modelled the soft X-ray background. Melioli *et al.* (2009), following the formation and evolution of HI clouds, only considered the time evolution of selected ions (HI, HII, CII-CIV, and OI-OIII) for temperatures below 10^6 K, using a fit to the Sutherland & Dopita (1993) CF for $T \geqslant 10^6$ K. This setup has severe implications in the cooling of the gas as their calculation does not trace the relevant ions recombining to CIV, and OIII.

Recently, owing to the development of the Atomic+Molecular Plasma Emission Code (EA+MPEC) and its coupling to a PPM based AMR code, it has been possible to carry out multi-fluid calculations of the ISM tracing both the thermal and dynamical evolutions of the gas self-consistently. The ionization structure, cooling and emission spectra of H, He, C, N, O, Ne, Mg, Si, S, and Fe ions (with solar abundances; Asplund *et al.* 2009) are traced on the spot at each time step assuming an equal Maxwellian temperature for electrons and ions (see details and references in Avillez & Breitschwerdt 2012).

These simulations showed several interesting effects: (i) in a dynamic ISM, the ionization structure and, therefore, the CF, varies with space and time, depending on the initial conditions and its history, (ii) the cooling paths in general do not follow the one predicted by static plasma emission calculations, (iii) non-equilibrium ionization X-ray emission in the ~ 0.25 keV band of gas with $T < 10^5$K can dominate the corresponding CIE emission at even $T = 10^{6.2}$ K as a result of delayed recombination, (iv) the presence of O VI ions at temperatures $< 10^5$ K corresponding to 70% of the total O VI mass, and (v) a large fraction of electrons are found in the thermally unstable regime and have a log normal distribution with similar properties (mean and dispersion) to those derived from observations against pulsars with known distances.

3. Conclusions

The dynamical and thermal evolution of the ISM are strongly coupled, because the ionization structure determines the CF, which in turn controls the dynamics and thereby the ionization structure, closing a feedback loop. Consequently, strong deviations from CIE occur due to severe mismatches between the different ionization/recombination and dynamical time scales of the plasma. Similar effects due time-dependent cooling are expected in other astrophysical contexts.

Acknowledgements

M.A.A. thanks the IAU and specifically Y.-H. Chu and V. Reuter for financial support.

References

Asplund, M., Grevesse, N., Sauval, A. J., & Scott, P. 2009, *ARAA* 47, 481
Balsara, D. S., Kim, J., Mac Low, M.-M., & Mathews, G. J. 2004, *ApJ* 617, 339
Bania, T. M. & Lyon, J. G. 1980, *ApJ* 239, 173
Berger, M. J. & Oliger, J. 1984, *JCP* 484
Borkowski, K. J., Sarazin, C. L., & Blondin, J. M. 1994, *ApJ* 429, 710
Breitschwerdt, D., Schmutzler T. 1994, *Nature* 371, 774
Chiang, W.-H. & Prendergast, K. H. 1985, *ApJ* 297, 507
Chiang, W.-H. & Bregman, J. N. 1988, *ApJ* 328, 427
Cox, D. P. & Anderson, P. R. 1982, *ApJ* 253, 268
Cox, D. P. & Smith, B. W. 1974, *ApJ* (Letters) 189, L105
de Avillez, M. A. & Breitschwerdt, D. 2005, *A&A* 436, 585
de Avillez, M. A. & Breitschwerdt, D. 2007, *ApJ* (Letters) 665, L35
de Avillez, M. A. & Breitschwerdt, D. 2012, *ApJ* (Letters) 761, L19
Gnat, O. & Sternberg, A. 2007, *ApJS* 168, 213
Gressel, O., Elstner, D., Ziegler, U., & Rödiger, G. 2008, *A&A* 486, 35
Hanasz, M., Kowal, G, Otmianowska-Mazur, K., & Lesch, H. 2004, *A&A* (Letters) 605, L33
Hill, A. S., Joung, M. R., Mac Low, M.-M., *et al.* 2012, *ApJ* 750, 104
Innes, D. E., Giddings, J. R., & Falle, S. A. E. G. 1987, *MNRAS* 227, 1021
Joung, M. K. R. & Mac Low, M.-M. 2006, *ApJ* 653, 1266
Korpi, M. J., Brandenburg, A., Shukurov, A., *et al.* 1999, *ApJ* (Letters) 514, L99
Melioli, C., Brighenti, F., D'Ercole, A., & Gouveia Dal Pino, E. M. 2009, *MNRAS* 399, 1089
McKee, C. F. & Ostriker, J. P. 1977, *ApJ* 218, 148
Norman, C. A. & Ikeuchi, S. 1989, *ApJ* 345, 372
Rosen, A., Bregman, J. N., & Norman, M. L. 1993, *ApJ* 413, 137
Rosen, A. & Bregman, J. N. 1995, *ApJ* 440, 634
Shapiro, P. R. & Field, G. B. 1976, *ApJ* 205, 762
Smith, R. K. & Cox, D. P. 2001, *ApJS* 134, 283
Sutherland, R. S. & Dopita, M. A. 1993, *ApJ* 88, 253
Vazquez-Semadeni, E., Passot, T., & Pouquet, A. 1995, *ApJ* 441, 702

Highlights of Astronomy, Volume 16
XXVIIIth IAU General Assembly, August 2012
T. Montmerle, ed.

doi:10.1017/S174392131401240X

Numerical modeling of multiphase, turbulent galactic disks with star formation feedback

Chang-Goo Kim[1], Eve C. Ostriker[2,3] and Woong-Tae Kim[4]

[1]Department of Physics & Astronomy, Western University, Canada
email: ckim256@uwo.ca

[2]Department of Astronomy, University of Maryland, USA

[3]Astrophysical Sciences, Princeton University, USA

[4]Department of Physics & Astronomy, Seoul National University, Republic of Korea

Abstract. Star formation is self-regulated by its feedback that drives turbulence and heats the gas. In equilibrium, the star formation rate (SFR) should be directly related to the total (thermal *plus* turbulent) midplane pressure and hence the total weight of the diffuse gas if energy balance and vertical dynamical equilibrium hold simultaneously. To investigate this quantitatively, we utilize numerical hydrodynamic simulations focused on outer-disk regions where diffuse atomic gas dominates. By analyzing gas properties at saturation, we obtain relationships between the turbulence driving and dissipation rates, heating and cooling rates, the total midplane pressure and the total weight of gas, and the SFR and the total midplane pressure. We find a nearly linear relationship between the SFR and the midplane pressure consistent with the theoretical prediction.

Keywords. stars: formation, ISM: kinematics and dynamics, methods: numerical, turbulence

---

Star formation feedback is a key ingredient to model the star formation rates (SFR) in galactic disks. The interstellar medium (ISM), the raw material for star formation, is highly dissipative so that within a timescale comparable to the dynamical timescales, such as the gravitational free-fall time and the vertical oscillation period, the thermal energy is radiated away, and the turbulent energy is dissipated by shocks and nonlinear cascades. Since the gas depletion time, the time required to convert all the gas into stars, is much longer than the dynamical timescales (e.g., Krumholz *et al.* 2012), there must be continuous and efficient mechanisms for energy replenishment. Feedback from massive stars is the most probable source that provides a substantial amount of momentum via expanding supernova remnants and photoelectric heating by FUV radiation.

The turbulent and thermal pressures maintained by energy balance support galactic disks against the total weight of the gas under the gravity of gas, stars, and dark matter. In equilibrium, the SFR is thus self-regulated to supply the appropriate amount of thermal and turbulent energy that meets the needs of the ISM. Recently, Ostriker *et al.* (2010) and Ostriker & Shetty (2011) have developed an analytic theory for regulation of SFRs based on the equilibrium model, which successfully explains observed relationships among the SFR surface density, the total and molecular gas surface densities, and the stellar surface density.

To directly test the analytic theory quantitatively, we utilize a series of numerical hydrodynamic simulations that resolve vertical dynamics of the ISM and include self-gravity, cooling and heating, and star formation feedback (see also Kim *et al.* 2011). In our models, star formation feedback is realized by time-dependent heating and by expanding supernova remnants. We focus on outer disk regions where the diffuse ISM

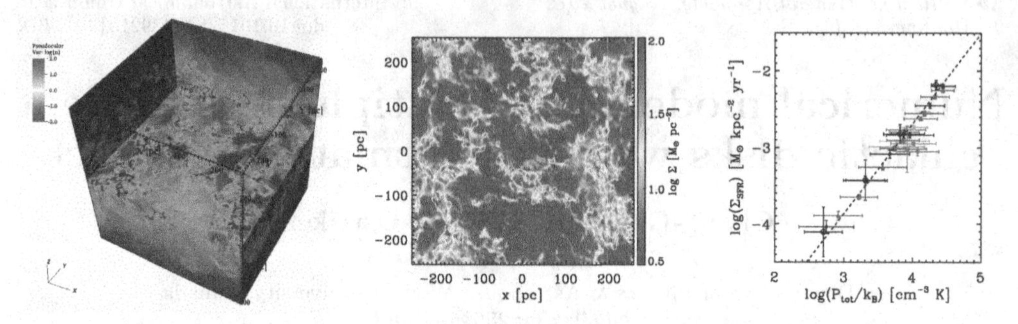

Figure 1. *Left* and *Middle*: Saturated state snapshots of the fiducial model with $\Sigma = 10\ \mathrm{M_\odot\ pc^{-2}}$ and $\rho_{\rm sd} = 0.05\ \mathrm{M_\odot\ pc^{-3}}$. The volume density slices (*left*) and the surface density seen along the vertical direction (*middle*) are drawn in logarithmic color scales. *Right*: SFR surface density as a function of the total midplane pressure at a saturated state for all models. The dashed line is our best fit with a slope of 1.18.

dominates, with gas surface densities $\Sigma = 3 - 20\ \mathrm{M_\odot\ pc^{-2}}$ and star-plus-dark matter volume densities $\rho_{\rm sd} = 0.003 - 0.5\ \mathrm{M_\odot\ pc^{-3}}$.

Our model disks undergo a quasi-periodic cycle of vertical oscillations: the disk expands vertically due to feedback, reducing the SFR, which in turn causes the disk to contract back, increasing the SFR and feedback. After one or two vertical oscillations of the disk, the overall physical properties are fully saturated. Figure 1 shows a morphology of the gas in the fiducial model with $\Sigma = 10\ \mathrm{M_\odot\ pc^{-2}}$ and $\rho_{\rm sd} = 0.05\ \mathrm{M_\odot\ pc^{-3}}$ at saturation in the left and middle panels; the ISM is multiphase, turbulent, and highly-structured, which is qualitatively similar to that seen in recent HI observations (see McClure-Griffith's contribution in this volume). We directly measure for all models the midplane thermal ($P_{\rm th}$) and turbulent ($P_{\rm turb}$) pressures as well as the SFR surface density ($\Sigma_{\rm SFR}$) that are averaged over one orbital period corresponding typically to three or four vertical oscillations. The thermal and turbulent pressures are approximately linearly proportional to the SFR surface density as the equilibrium model predicts; $P_{\rm th} \propto \Sigma_{\rm SFR}^{0.86}$, and $P_{\rm turb} \propto \Sigma_{\rm SFR}^{0.89}$. At the same time, the total (thermal *plus* turbulent) midplane pressure ($P_{\rm tot}$) is in excellent agreement with the dynamical equilibrium pressure, resulting in $P_{\rm tot} \propto \Sigma\sqrt{\rho_{\rm sd}}$. Finally, the fundamental relationship between the SFR surface density and the total pressure naturally reflects the match between supply and demand of the ISM, $\Sigma_{\rm SFR} \propto P_{\rm tot}^{1.18}$ (see the right panel of Figure 1).

Acknowledgements

This work was made possible by the facilities of the Shared Hierarchical Academic Research Computing Network (SHARCNET:www.sharcnet.ca) and Compute/Calcul Canada. C.-G. K. is supported by a CITA National Fellowship.

References

Kim, C.-G., Kim, W.-T., & Ostriker, E. C. 2011, *ApJ* 743, 25

Krumholz, M. R., Dekel, A., & McKee, C. F. 2012, *ApJ* 745, 69

Ostriker, E. C., McKee, C. F., & Leroy, A. K. 2010, *ApJ* 721, 975

Ostriker, E. C. & Shetty, R. 2011, *ApJ* 731, 41

Wolfire, M. G., Hollenbach, D., McKee, C. F., Tielens, A. G. G. M., & Bakes, E. L. O. 1995, *ApJ* 443, 152

Highlights of Astronomy, Volume 16
XXVIIIth IAU General Assembly, August 2012
T. Montmerle, ed.

doi:10.1017/S1743921314012411

Stability properties of phase transition layers in the diffuse ISM revisited

Jennifer M. Stone[1], Shu-ichiro Inutsuka[1], and Ellen G. Zweibel[2]

[1]Department of Physics, Nagoya University, Furo-cho, Chikusa-ku, Aichi 464-8602, Japan
email: stone@nagoya-u.jp

[2]Departments of Phyics and Astronomy and Center for Magnetic Self-Organization, University of Wisconsin-Madison, 475 N Charter Street, Madison, WI 53706, USA

Abstract. In a thermally bistable medium, cold, dense gas is separated from warm, rarified gas by thin phase transition layers, or fronts, in which radiative heating/cooling, thermal conduction, and convection of material are balanced. While these fronts have received only scant attention in the literature, and are not resolved by most current numerical simulations, they have been shown to have important ramifications for transport processes and structure formation in the diffuse interstellar medium. Here, we discuss calculations of their hydrodynamic and magnetohydrodynamic stability properties.

Keywords. hydrodynamics, ISM: structure, instabilities, (magnetohydrodynamics:) MHD, methods: numerical

---

The corrugational instability in phase transtion layers is an intriguing mechanism for sustaining turbulence (e.g. Koyama & Inutsuka 2002) and generating small scale structure (e.g. Heiles 1997) in the diffuse interstellar medium. Previous attempts at calculating the full hydrodynamic stability behavior of thermal fronts have encountered mathematical difficulties in that conventional methods for solving eigenvalue problems do not converge. As a result, all prior investigations have relied upon simplifying approximations to make the calculation more tractable (Inoue *et al.* 2006). However, while such approaches enable conceptual progress, they do not allow for an accurate description of the overall hydrodynamic stability properties of thermal fronts. Furthermore, corresponding approximations are not valid in the magnetohydrodynamic regime. Work is ongoing to develop a numerical method to calculate the full hydrodynamic stability properties of thermal fronts, without the need for approximation. Ultimately, we will extend these efforts to study the effects of magnetic fields and ambipolar diffusion (Stone & Zweibel 2009, 2010) on phase transition layer stability.

Acknowledgements

J.M.S. gratefully acknowledges support from the Global COE Program "Quest for Fundamental Principles in the Universe: from Particles to the Solar System and Cosmos" of Nagoya University.

References

Heiles, C. 1997, *ApJ* 481, 193
Inoue, T., Inutsuka, S.-i., & Koyama, H. 2006, *ApJ* 652, 1331
Koyama, H. & Inutsuka, S.-i. 2002, *ApJ* (Letters) 564, L97
Stone, J. M. & Zweibel, E. G. 2009, *ApJ* 696, 233
Stone, J. M. & Zweibel, E. G. 2010, *ApJ* 724, 131

Highlights of Astronomy, Volume 16
XXVIIIth IAU General Assembly, August 2012
T. Montmerle, ed.

doi:10.1017/S1743921314012423

Planck's view of the intestellar medium

Planck Collaboration, presented by J. A. Tauber

Astrophysics Division, Research and Scientific Support Dpt, European Space Agency,
Keplerlaan 1, 2201AZ Noordwijk, The Netherlands
email: `jtauber@rssd.esa.int`

Abstract. *Planck* is a cosmology experiment, but significant interstellar dust and gas emission exists in the band where the CMB peaks. Therefore, *Planck*'s all-sky surveys provide new views of the ISM and magnetic fields in the Galaxy, as well as the dust and gas in galaxies.

Keywords. ISM: general, Galaxy: general, cosmic microwave background, space vehicles

---

The *Planck*† satellite (Tauber *et al.* (2010)), launched on 14 May 2009, has been surveying the sky continuously since 13 August 2009. It carries a scientific payload consisting of an array of detectors arranged in nine broad-band channels sensitive to a range of frequencies between ~25 and ~1000 GHz, which scan the sky simultaneously and continuously with an angular resolution of $5'$ to $30''$. The *Planck* satellite, its payload, and its performance at the time of launch are described in Volume 520 of *A&A*.

The main objective of *Planck* is to measure the spatial anisotropies of the temperature of the Cosmic Microwave Background (CMB) over the whole sky, with an accuracy set by fundamental astrophysical limits. Its level of performance will enable *Planck* to extract essentially all the information in the CMB temperature anisotropies. *Planck* will also measure to high accuracy the polarisation of the CMB anisotropies, which encodes not only a wealth of cosmological information, but also provides a unique probe of the thermal history of the Universe during the time when the first stars and galaxies formed. The scientific objectives of *Planck* are described in detail in Planck Collaboration (2005).

Planck is by design a cosmology experiment, but very significant Galactic and extragalactic emission exists in the band where the CMB peaks. Thus, its all-sky surveys also produce a wealth of information on the extragalactic sources and on the dust and gas in our own Galaxy. This fact can be clearly appreciated in Fig. 1, which is a composite of data acquired during *Plancks* first complete all-sky survey. Galactic emission dominates a large part of the sky, both at low frequencies (by a mixture of synchrotron, free-free and other non-thermal radiation) and at high frequencies (mainly by thermal dust emission), and has to be measured accurately and removed to gain access to the CMB.

The *Planck* survey's frequency range, all-sky coverage, high sensitivity, ability to measure polarization, and high calibration accuracy have enabled numerous ISM studies:

(*a*) The higher frequency maps allow for the first time to measure accurately the characteristics, amount and distribution of the coldest dust present in the ISM, both in the diffuse ISM (e.g. *Planck* Early Results XXIV) and in dense molecular clouds (e.g. *Planck* Early Results XXV), not only in our own Milky Way, but also in nearby objects such as the Magellanic Clouds (*Planck* Early Results XVII). A spectacular early result by *Planck* is the detailed mapping of so-called "dark" gas that is not spatially correlated with

† *Planck (http://www.esa.int/Planck)* is a project of the European Space Agency – ESA – with instruments provided by two scientific Consortia funded by ESA member states (in particular the lead countries: France and Italy) with contributions from NASA (USA), and telescope reflectors provided in a collaboration between ESA and a scientific Consortium led and funded by Denmark.

Figure 1. The microwave sky as seen by *Planck*. This multi-frequency all-sky image of the microwave sky has been composed using data from *Planck* covering 30 GHz to 857 GHz.

known tracers of neutral and molecular gas (*Planck* Early Results XIX). *Plancks* ability to detect very cold dust has revealed the widespread presence of dense and compact clumps of gas, certainly the sites of future star formation (*Planck* Early Results XXIII).

(*b*) The wide frequency range allows *Planck* to detect and study components of the ISM with uncommon spectral characteristics, for example the anomalous excess emission possibly arising from small spinning grains (*Planck* Early Results XX).

(*c*) The all-sky coverage allows to conduct global studies of the radial distribution of molecular, neutral, and ionised ISM in the Milky Way (*Planck* Early Results XXI).

(*d*) The combination of all-sky coverage and cm-to-submm wavelength range allow *Planck* to survey key parts of the spectral energy distribution of external galaxies (*Planck* Early Results VII), and has revealed many spectrally flat or rising radio galaxies (*Planck* Early Results XIII) and a more-than-expected excess of dusty galaxies, due to the detection of previously unaccounted cold dust (*Planck* Early Results XVI).

(*e*) *Planck*'s ability to measure polarization between 30 and 350 GHz promises an exciting magnetic view of the ISM. The Galactic ISM is threaded by magnetic fields whose morphology is largely unknown; this field in turn induces polarization of both synchrotron and dust emission. *Planck* will uniquely be able to estimate the properties of both the ordered and turbulent components of the Galactic magnetic field.

(*f*) The extremely accurate calibration of *Planck*, for both diffuse emissions and compact sources, will provide highly precise photometric standards in this frequency range (*Planck* Early Results XIV) and Herschel/SPIRE, which is very beneficial to ISM science.

The above examples illustrate the wide range of ISM studies enabled by the *Planck*. When the *Planck* data products are released to the public, starting in March 2013, they will join a wide range of surveys from other observatories, which together provide a broad view of the phenomenology of the Interstellar Medium in our own and other galaxies.

References

Planck Collaboration 2011, *A&A* 536, A1
Planck Collaboration 2005, *ESA Publication*, ESA-SCI(2005)1
Tauber, J. A., Mandolesi, N., Puget, J. L., *et al.* 2010, *A&A* 520, A1

Highlights of Astronomy, Volume 16
XXVIIIth IAU General Assembly, August 2012
T. Montmerle, ed.

doi:10.1017/S1743921314012435

Gravitational fragmentation of the Carina Flare supershell

Richard Wünsch
Astronomical Institue of the ASCR, Boční II 1401, 141 00 Prague, Czech Republic
email: richard@wunsch.cz

Abstract. We study the gravitational fragmentation of a thick shell comparing the analytical theory to 3D hydrodynamic simulations and to observations of the Carina Flare supershell. We use both grid-based (AMR) and particle-based (SPH) codes to follow the idealised model of the fragmenting shell and found an excellent agreement between the two codes. Growth rates of fragments at different wavelength are well described by the pressure assisted gravitational instability (PAGI) - a new theory of the thick shell fragmentation. Using the APEX telescope we observe a part of the surface of the Carina Flare supershell (GSH287+04-17) in the ^{13}CO(2–1) line. We apply a new clump-finding algorithm DENDROFIND to identify 50 clumps. We determine the clump mass function and we construct the minimum spanning tree connecting clumps positions to estimate the typical distance among clumps. We conclude that the observed masses and distances correspond well to the prediction of PAGI.

Keywords. ISM: bubbles, ISM: molecules, stars: formation

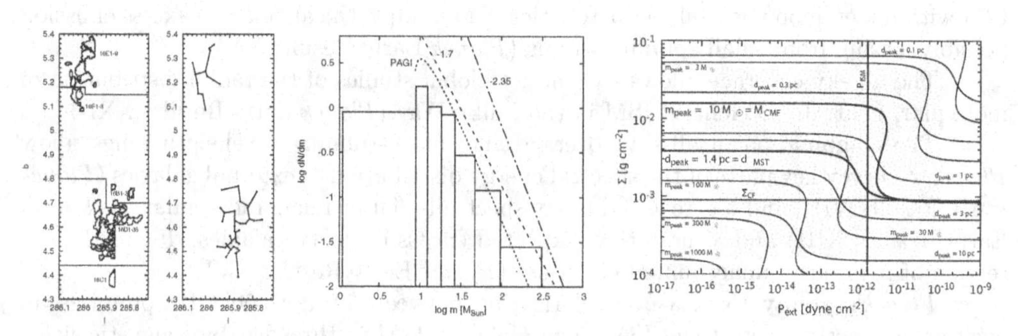

Figure 1. From left to right: (i) contours of clumps identified in the observed region; (ii) minimum spanning tree used to measure the mean distance among clumps; (iii) clumps mass spectrum compared with the prediction of the PAGI theory (Wünsch *et al.* 2010); and (iv) position of the CMF peak (solid contours) and the typical distance among clumps (dashed contours) as functions of the supershell surface density and the external pressure according to PAGI. Contours for the observed values intersect close to values expected for the Carina Flare supershell (dash-dotted lines). See Wünsch *et al.* (2012) for details.

Acknowledgements

This work is supported by the project P209/12/1795 of the Czech Science Foundation and by the project RVO:67985815.

References

Wünsch, R., Dale, J. E., Palouš, J., & Whitworth, A. P. 2010, *MNRAS* 407, 1963
Wünsch, R., Jáchym, P., Sidorin, V., *et al.* 2010, *A&A* 539, A116

Highlights of Astronomy, Volume 16
XXVIIIth IAU General Assembly, August 2012
T. Montmerle, ed.

doi:10.1017/S1743921314012447

The resolved magnetic fields of the quiescent cloud GRSMC 45.60+0.30

Michael D. Pavel, Robert C. Marchwinski and Dan P. Clemens
Institute for Astrophysical Research, Boston University, Boston, MA 02215, USA
email: pavelmi@astro.as.utexas.edu

Marchwinski *et al.* (2012) mapped the magnetic field strength across the quiescent cloud GRSMC 45.60+0.30 (shown in Figure 1 subtending 40x10 pc at a distance of 1.88 kpc) with the Chandrasekhar-Fermi method (CF; Chandrasekhar & Fermi 1953) using near-infrared starlight polarimetry from the Galactic Plane Infrared Polarization Survey (Clemens *et al.* 2012a,b) and gas properties from the Galactic Ring Survey (Jackson *et al.* 2006). The large-scale magnetic field is oriented parallel to the gas-traced 'spine' of the cloud. Seven 'magnetic cores' with high magnetic field strength were identified and are coincident with peaks in the gas column density. Calculation of the mass-to-flux ratio (Crutcher 1999) shows that these cores are exclusively magnetically subcritical and that magnetostatic pressure can support them against gravitational collapse.

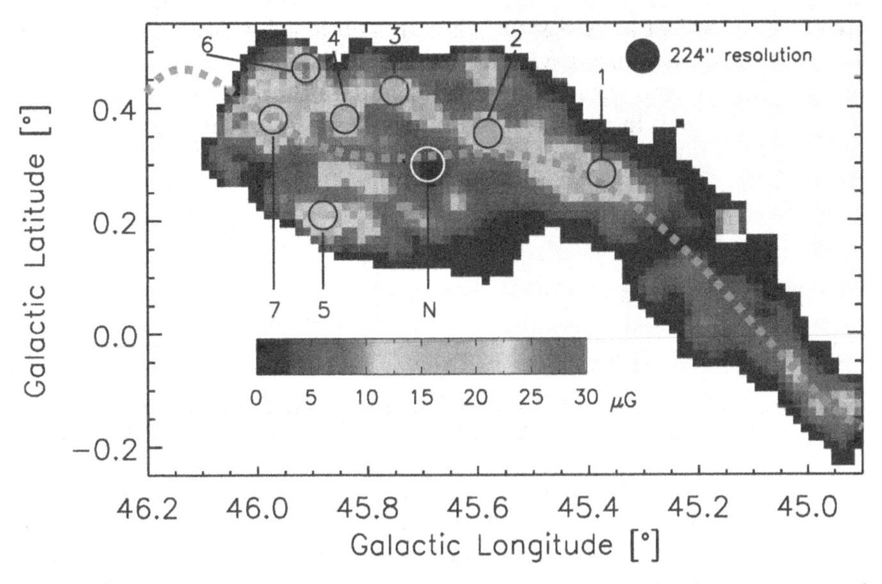

Figure 1. Resolved magnetic field strength across the face of GRSMC 45.60+0.30. The grey dashed line traces the spine of the cloud. Seven magnetic cores are numbered and a region where the CF method breaks down (N) has been identified.

References

Chandrasekhar, S. & Fermi, E. 1953, *ApJ* 118, 113
Clemens, D. P., Pavel, M. D., & Cashman, L. R. 2012a, *ApJS* 200, 21
Clemens, D. P., Pinnick, A. F., Pavel, M. D., & Taylor, B. W. 2012b, *ApJS* 200, 19
Crutcher, R. M. 1999, *ApJ* 520, 706
Jackson, J. M., Rathborne, J. M., Shah, R. Y., *et al.* 2006, *ApJS* 163, 145
Marchwinski, R. C., Pavel, M. D., & Clemens, D. P. 2012, *ApJ* 755, 130

Highlights of Astronomy, Volume 16
XXVIIIth IAU General Assembly, August 2012
T. Montmerle, ed.

doi:10.1017/S1743921314012459

Size distribution of SNRs and the ISM

Abdul I. Asvarov

Institute of Physics, Baku AZ1143, Azerbaijan Republic.
email: asvarov@physics.ab.az

Abstract. Based on the very general assumptions on evolution, we have modeled the statistics of SNRs evolving at various initial and environmental conditions. The method is applied to M33, for which the value of the filling factor of the hot phase of the ISM is estimated to be $\sim 10\%$.

Keywords. SNRs, (ISM:) supernova remnants, galaxies: M33

The large scatter in the observational properties of SNRs can be explained partly by the large differences in the initial parameters of SNe and partly by the large differences in the characteristics of the ISM (density, pressure, etc.). Moreover, it is unclear what fraction of really existing SNRs we observe. To answer these questions we have constructed a simple model describing the statistics of SNRs, occurring with randomly distributed initial and environmental parameters. In our model, the density of the ISM is the most important parameter as its value changes over a wide range - for this parameter we have used three-phase model of McKee & Ostriker. It is obtained, that the filling factor of the hot phase is the main parameter which determines the shape of the size distribution of SNRs. For the comparision with the observational data the catalogue of SNRs in M33 of Long *et al.*(2010) is used. In the list of X-Ray SNRs we select the subset of objects with hardness ratio $H \sim -0.5$ as the subset which has a high degree of completeness.

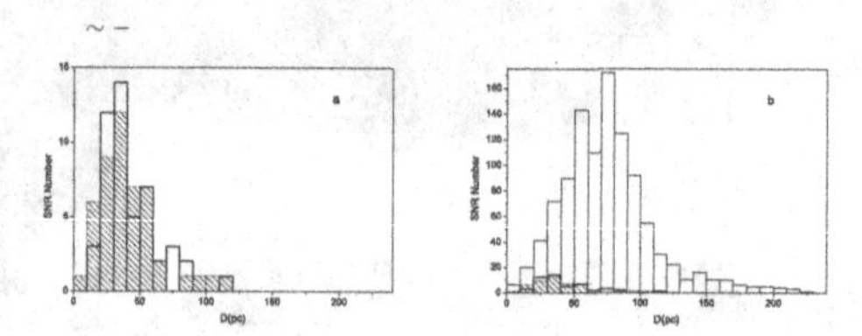

Figure 1. Distributions of modeled (solid lines) and observed (dotted lines) in M33 SNRs with $\mathrm{H} = -0.5 \pm 0.2$ (a) and their position in the distribution of all modeled SNRs in M33 (b).

An excellent coincidence between distributions of this subset of SNRs and the subset of modeled SNRs with the same value of H was reached at the filling-factor of the hot phase of the ISM in M33 of $\sim$ 10% (Fig.1a). The position of this subset of SNRs in the general distribution of generated SNRs (Fig.2a) shows that the observed SNRs represents very small fraction of really existing SNRs. The SNe birthrate for M33 is estimated to be 120-150 yr^{-1}, the full number of unseen but with Mach number $\geqslant 2$ estimated $\geqslant 1200$ (this value depends on the total pressure of ISM).

This work was supported by the Science Development Foundation under the President of the Republic of Azerbaijan Grant No EIF-2010-1(1)-40/05-M-23.

Reference

Long, K. S., Blair, W. P., Winkler, P. F., *et al.* 2010, *ApJS* 187, 495

Highlights of Astronomy, Volume 16
XXVIIIth IAU General Assembly, August 2012
T. Montmerle, ed.

doi:10.1017/S1743921314012460

Dust emission from the atomic and molecular gas in M 33: a changing β

Jonathan Braine[1], Fatemeh Tabatabaei[2] and Manolis Xilouris[3]

[1]Univ. Bordeaux, Laboratoire d'Astrophysique de Bordeaux, CNRS, LAB, UMR 5804, F-33270, Floirac, France. email`braine@obs.u-bordeaux1.fr`

[2]MPIA, Königstuhl 17, 69117, Heidelberg, Germany

[2]National Observatory of Athens, P. Penteli, 15236, Athens, Greece

Abstract. We use the very recently completed high-resolution IRAM CO survey of M33 with the high-resolution HI observations (published by Gratier *et al.* 2010, *A&A*, 522, 3) and Herschel Far-IR and submillimeter mapping observations to study how the dust behaves in the molecular and atomic gas phases of the interstellar medium (ISM). M33 is a "young" object in that it is gas-rich with a young stellar population and low metallicity as compared to large spirals like the Milky Way or Andromeda. Nonetheless, it is very clearly a spiral galaxy with a thin and reasonably axisymmetric disk. As such, it can be viewed as a stepping stone towards less evolved objects like magellanic irregulars (including the LMC and SMC) and perhaps distant objects in the early universe. More specifically, we look for radial variations in the dust emission spectrum (β parameter) as well as comparing regions dominated by either H_2 or HI. *The grey-body emission spectrum flattens (lower β) with galactocentric distance and generally is flatter in the atomic medium as compared to the molecular gas.*

Keywords. Galaxies: Individual: M 33 – Galaxies: Local Group – Galaxies: ISM – ISM: Clouds

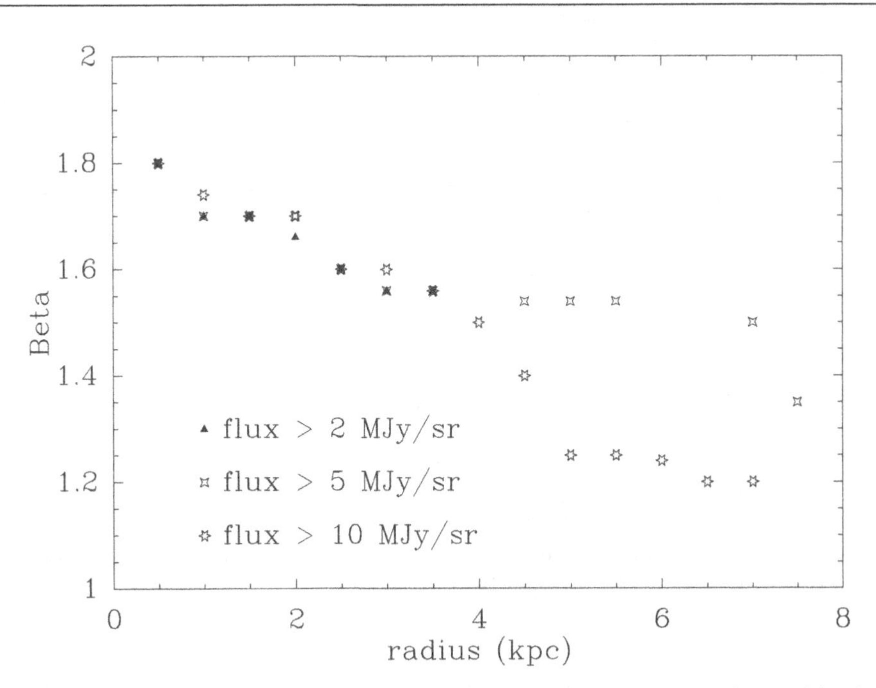

Figure 1. Radial decline in β for varying flux thresholds in M 33. β was obtained by fitting a grey-body ($S_\nu \propto (\nu/\nu_0)^\beta B_{\nu,T}$) to each pixel in M33 with a 250μm flux above the thresholds indicated.

Highlights of Astronomy, Volume 16
XXVIIIth IAU General Assembly, August 2012
T. Montmerle, ed.

doi:10.1017/S1743921314012472

The cool and warm molecular gas in M82 with *Herschel*-SPIRE

J. Kamenetzky[1], J. Glenn[1], N. Rangwala[1], P. Maloney[1], M. Bradford[2], C. D. Wilson[3], G. J. Bendo[4], M. Baes[5] A. Boselli[6], A. Cooray[7], K. G. Isaak[8], V. Lebouteiller[9], S. Madden[9], P. Panuzzo[9], M. R. P. Schirm[3], L. Spinoglio[10] and R. Wu[9]

[1]U. Colorado, Boulder, [2]JPL/NASA, [3]McMaster U., [4]UK ALMA, [5]Universiteit Gent, [6]LAM Marseille, [7]UC, Irvine, [8]ESA Astrophysics Mission Div., [9]Laboratoire AIM, [10]INAF

Abstract. We present *Herschel*-SPIRE imaging spectroscopy (194-671 μm) of the bright starburst galaxy M82. We use RADEX and a Bayesian Likelihood Analysis to simultaneously model the temperature, density, column density, and filling factor of both the cool and warm components of molecular gas traced by the entire CO ladder up to J=13-12. The high-J lines observed by SPIRE trace much warmer gas ($\sim$500 K) than those observable from the ground. The addition of ^{13}CO (and [C I]) is new and indicates that [C I] may be tracing different gas than ^{12}CO. At such a high temperature, cooling is dominated by molecular hydrogen; we conclude with a discussion on the possible excitation processes in this warm component. Photon-dominated region (PDR) models require significantly higher densities than those indicated by our Bayesian likelihood analysis in order to explain the high-J CO line ratios, though cosmic-ray enhanced PDR models can do a better job reproducing the emission at lower densities. Shocks and turbulent heating are likely required to explain the bright high-J emission.

Keywords. galaxies: individual (M82), galaxies: starburst, ISM: molecules

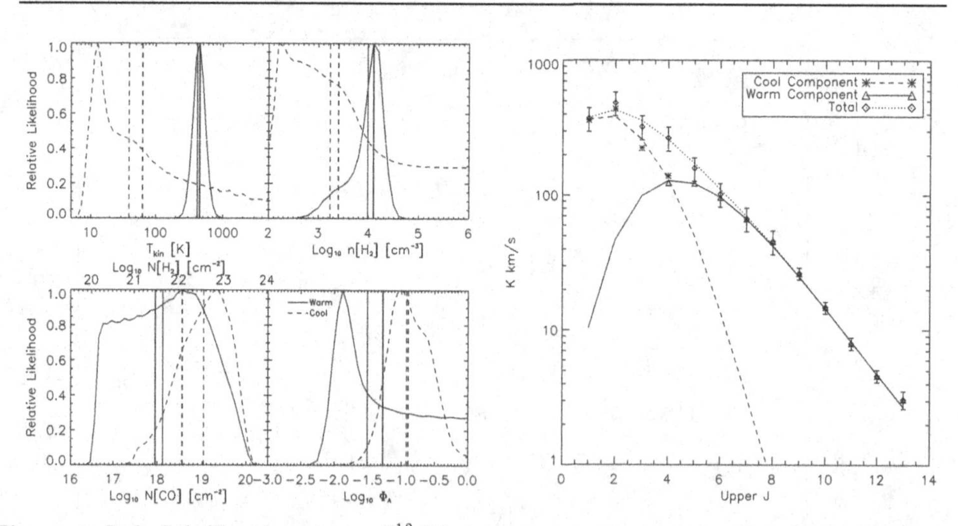

Figure 1. Left: Likelihood analysis of ^{12}CO only, marginalized results. Regular/thick vertical line = median/best fit. Right: Best fit spectral line energy distributions.

Reference

Kamenetzky, J., Glenn, J., Rangwala, N., *et al.* 2012, *ApJ* 753, 70

Highlights of Astronomy, Volume 16
XXVIIIth IAU General Assembly, August 2012
T. Montmerle, ed.

doi:10.1017/S1743921314012484

Gas density histograms of galaxies: the observational density probability function of the interstellar gas density

Handa Toshihiro[1], Yoda Takahiro[2] and Kuno Nario[3]

[1]Depart. Astron. & Phys., Kagoshima Univ., Korimoto 1-21-35, Kagoshima 890-0065, Japan
email: handa@sci.kagoshima-u.ac.jp

[2]Yamanashi Gakuin highschool, [3]National Astronomical Observatory of Japan

Abstract. In the steady state, the probability density function (PDF) of the gaseous interstellar matter (ISM) can be observed as a gas density histogram (GDH) of all cells in the system. We made GDHs of the Milky Way Galaxy (MWG) using Galactic plane surveys in CO lines. We found that the GDH in the MWG is log-normal which suggests that the density structure of the molecular gas is a result of many stochastic processes. Using the Nobeyama CO atlas, we made GDHs of nearby galaxies but in column density. Although some galaxies show log-normal, the others show completely different shapes, suggesting that the density structure of galaxies may be different from galaxy to galaxy.

Keywords. ISM: molecules, ISM: structure, galaxies: ISM

The PDF of the gas density is one of keys to address the origin of density structure of the ISM. In the steady system, PDF can be observed as a GDH of all cells in the system. Using the large survey data we made GDHs of the Milky Way Galaxy (MWG) and nearby galaxies. Using the AMANOGAWA-2SB survey in ^{12}CO (2-1) & ^{13}CO (2-1) (Yoda *et al.* 2010; Handa *et al.* 2013) and the CO (1-0) survey (Dame *et al.* 2001), we made that the GDH of molecular gas in the MWG. All of them are log-normal and the peak of the GDH is about $10^{-1.9} M_{\odot}$ pc^{-3}. The shape of the derived GHD is less affected by models and parameters used. It suggests that the density structure of the molecular gas in the galactic disk of the MWG is a result of many stochastic processes each of which modifies the gas density randomly (Vazquez-Semadeni 1994). Using the Nobeyama CO atlas of nearby galaxies (Kuno *et al.* 2007), we made their GDHs of the column density instead of the volume density. Some galaxies show log- normal like the MWG. However, the others show completely different shapes. These results suggest the major process to make the density structure of ISM is different from galaxy to galaxy. We can address it using the GDH, although we do not know what causes the difference at present.

References

Dame, T. M., Hartmann, D., & Thaddeus, P. 2001, *ApJ* 547, 72
Kuno, N., Sato, N., Nakanishi, H., *et al.* 2007, *PASJ* 59, 117
Handa, T., *et al.* 2013, in: M. Bureau & (eds.) *"Molecular Gas, Dust, and Star Formation in Galaxies"*, Proc. IAU Symp. 292 (CUP) in press
Vazquez-Semadani, E. 1994, *ApJ* 423, 681
Yoda, T., Handa, T., Kohno, K., *et al.* 2010, *PASJ* 62, 1277

Highlights of Astronomy, Volume 16
XXVIIIth IAU General Assembly, August 2012
T. Montmerle, ed.

doi:10.1017/S1743921314012496

Statistical study of the ISM of GRB hosts

A. de Ugarte Postigo[1,2], J. P. U. Fynbo[2], C. C. Thöne[1], L. Christensen[2], J. Gorosabel[1] and R. Sánchez-Ramírez[1]

[1]IAA-CSIC, Glorieta de la Astronomía s/n, E-18008, Granada, Spain

[2]Dark Cosmology Centre, NBI, Juliane Maries Vej 30, Copenhagen Ø, D-2100, Denmark

Abstract. Gamma-ray burst (GRB) afterglows shine, during a brief period of time as the most luminous objects that can be detected in the Universe. They have been observed at almost any redshift, from our nearby environment (the nearest one, at z = 0.08) to the very distant Universe (the current record holder at z = 9.4). Their optical spectra are well reproduced by a clean, simple power law, making them ideal light houses to probe the interstellar medium of their host galaxies at any redshift. We have used the largest sample of GRB afterglow spectra collected to date to perform a statistical study of the interstellar medium in their host galaxies. By analysing the distribution of equivalent widths of the most prominent absorption features we evaluate the different types of environments that host GRBs and study their diversity.

Keywords. gamma rays: bursts, galaxies: ISM

This work is based on the sample presented by Fynbo *et al.* (2009) and pubished by de Ugarte Postigo *et al.* (2012). We analyse the distribution of rest-frame equivalent widths (EWs), which is only possible for the most prominent absorption features in GRB afterglow spectra. Our sample is limited to those absorption features that have a rest-frame EW of at least 0.5 Å in the composite spectrum of Christensen *et al.* (2011), which add to a total of 22 features.

To compare an individual GRB with the sample, we develop *EW diagrams* as a graphical tool. We introduce a *line strength parameter* (*LSP*) that allows us to quantify the strength of the absorption features in a GRB spectrum as compared to the sample by a single number. Using the distributions of EWs of single-species features, we derive the distribution of their column densities through a curve of growth (CoG) fit.

We find correlations between the *LSP* and the extinction of the GRB, the UV brightness of the host galaxies and the neutral hydrogen column density. However, we see no significant evolution of the *LSP* with the redshift. There is a weak correlation between the ionisation of the absorbers and the energy of the GRB, indicating that galaxies with high-ionisation media produce more energetic GRBs. Features in GRB spectra are, on average, 2.5 times stronger than those seen in QSO intervening damped Lyman-α (DLA) systems and slightly more ionised. In particular we find a larger excess in the EW of CIV$\lambda\lambda$1549 relative to QSO DLAs, which could be related to an excess of Wolf-Rayet stars in the environments of GRBs. From the CoG fitting we obtain an average number of components in the absorption features of GRBs of $6.00^{+1.00}_{-1.25}$. The most extreme ionisation ratios in our sample are found for GRBs with low neutral hydrogen column density, which could be related to ionisation by the GRB emission.

References

de Ugarte Postigo, A., Fynbo, J. P. U., Thöne, C. C., *et al.* 2012, *A&A* 548, A11
Christensen, L., Fynbo, J. P. U., Prochaska, J. X., *et al.* 2011, *ApJ* 727, 73
Fynbo, J. P. U., Jakobsson, P., Prochaska, J. X., *et al.* 2009, *ApJS* 185, 526

Highlights of Astronomy, Volume 16
XXVIIIth IAU General Assembly, August 2012
T. Montmerle, ed.

doi:10.1017/S1743921314012502

A kinematical catalogue of HII regions and superbubbles in the LMC

P. Ambrocio-Cruz[1], E. Le Coarer[2], M. Rosado[3], D. Russeil[4], P. Amram[4], A. Laval[4], B. Epinat[4], M. Ramírez[1], M. Odonne[5] and G. Goldes[5]

[1]AA Ciencias de la Tierra y Materiales Universidad Autónoma del Estado de Hidalgo, México
email: patricia@astro.unam.mx

[2]Laboratoire d'Astrophysique de Grenoble, France

[3]Instituto de Astronomía, UNAM, México
email: margarit@astro.unam.mx

[4]Laboratoire d'Astrophysique de Marseille, France

[5]Observatorio Astronómico Laprida, Argentina

Abstract. We report the results of a kinematic Hα survey of the Large Magellanic Cloud (LMC) in the form of a kinematic and photometric catalogue of 210 HII regions, the radial velocity field of the ionized hydrogen in this galaxy, and the LMC Rotation Curve obtained from the velocity field. These data aim at understanding the LMC HII regions, bubbles and superbubbles in a global (galactic) scale so that we could have a 3D view and separate the rotation due to gravitational potential from other motions such as expansions.

Keywords. galaxies: Magellanic Clouds, ISM: kinematics and dynamics, ISM: bubbles, supernova remnants

---

We have carried out a kinematic Hα survey of the LMC. The observations were made with a scanning Fabry-Perot interferometer which produced 3D data cubes for 69 pointings over the LMC, each with a FOV of 38 arcmin. The data cubes were flux calibrated through HII regions with known Hα intensity. Wavelength calibration and astrometry have also been carried out. Each data cube provides radial velocity fields of ionized gas in the LMC and in the Galaxy.

We find a bimodal distribution of the Hα luminosity of LMC HII regions, possibly due to a transition between radiation-bounded and density-bounded HII regions. From the analysis of this large sample of HII regions, we suggest that triggered star formation in the LMC could amount to 48% and that this mechanism seems to be more active in small (young) and density-bounded HII regions. We also derive the local star formation rate of the LMC that agrees quite well with those derived from other tracers.

There is a very good agreement between the rotation curve (RC) we derived from the ionized gas and the RCs derived from other tracers (stars, planetary nebulae, neutral gas), except that for R $>$ 2.5 kpc, HI kinematics starts to deviate towards lesser rotation velocities. Most of the catalogued nebulae follow the RC well, except 30 Dor, N119, LMC 4 and N 44D. These deviations from circular motions may have different causes, such as out of disk location (as it has been proposed for the 30 Dor Nebula), the presence of a prominent bar and inflowing gas, and expansion motions due to supernova explosions. We find no correlation among the velocity dispersion, the size, and the Hα intensity of the HII region. Finally, we note that nebulae with large velocity dispersions correspond to known supernova remnants.

Highlights of Astronomy, Volume 16
XXVIIIth IAU General Assembly, August 2012
T. Montmerle, ed.

doi:10.1017/S1743921314012514

Very deep spectroscopy of NGC 7009

Xuan Fang[1], Xiaowei Liu[1,2] and Peter J. Storey[3]

[1]Department of Astronomy, School of Physics, Peking University, Beijing 100871, P. R. China
email: fangx@pku.edu.cn

[2]Kavli Institute for Astronomy and Astrophysics, Peking University,
Beijing 100871, P. R. China

[3]Department of Physics and Astronomy, University College London,
Gower Street, London WC1E 6BT, UK

Abstract. We report new calculations of the effective recombination coefficients for the nebular N II and O II lines and very deep spectroscopy of the bright planetary nebula (PN) NGC 7009.

Keywords. atomic data, line: formation, ISM: abundances, planetary nebulae: individual

---

New *ab initio* calculations of the effective recombination coefficients for the N II and O II recombination spectra were carried out in intermediate coupling by Fang, Storey & Liu (2011) and P. J. Storey (unpublished), respectively. Both calculations have taken into account the density dependence of effective recombination coefficients arising from the distribution of population among the ground fine-structure levels of the recombining ions. Also new is including the effects of dielectronic recombination via high-n resonances lying between the ground-term thresholds of the recombining ions. The two calculations are valid down to unprecedentedly low temperatures (e.g. $\sim$100 K for N II), and allow plasma diagnostics based on the N II and O II optical recombination lines (ORLs).

Very deep spectroscopy of planetary nebula NGC 7009 (Fang & Liu 2011; Fang & Liu 2012, in preparation) yields very rich ORLs of heavy element ions. Accurate fluxes of the ORLs of C II, N II, O II and Ne II were obtained using multi-Gaussian profile fits. The electron temperatures derived from the N II and O II ORLs, using the new effective recombination coefficients, were close to 1000 K, which is lower than those derived from the collisionally excited lines (CELs) by nearly one order of magnitude, indicating that the heavy element ORLs originate from very cold region. Various plasma diagnostics yield a temperature sequence, $T_{\rm e}$(CELs) $\gtrsim$ $T_{\rm e}$(H I BJ) $\gtrsim$ $T_{\rm e}$(He I) $\gtrsim$ $T_{\rm e}$(N II & O II ORLs), which is consistent with the predictions from the bi-abundance nebular model postulated by Liu *et al.* (2000). The C^{2+}/H^+, N^{2+}/H^+, O^{2+}/H^+ and Ne^{2+}/H^+ ionic abundance ratios derived from ORLs (at $T_{\rm e}$ = 1000 K) are consistently higher, by about a factor of 5, than the corresponding values derived from CELs, an abundance discrepancy first found for this object by Liu *et al.* (1995) from deep spectroscopy. For N II, as well as for O II, the ionic abundances derived from different fine-structure components of a multiplet, or from the transitions of different multiplets, agree with each other, indicating that the current theoretical treatment of the recombination spectrum of N II and O II is reliable.

References

Fang, X. & Liu, X.-W. 2011, *MNRAS* 415, 181
Fang, X., Storey, P. J., & Liu, X.-W. 2011, *A&A* 530, A18
Liu, X.-W., Storey, P. J., Barlow, M. J., & Clegg, R. E. S. 1995, *MNRAS* 272, 369
Liu, X.-W., Storey, P. J., Barlow, M. J., Danziger, I. J., Cohen, M., & Bryce, M. 2000, *MNRAS* 312, 585

Highlights of Astronomy, Volume 16
XXVIIIth IAU General Assembly, August 2012
T. Montmerle, ed.

doi:10.1017/S1743921314012526

Chemical enrichment of the ISM by stellar ejecta

Sun Kwok

Faculty of Science, The University of Hong Kong, Hong Kong, China
email: sunkwok@hku.hk

Abstract. The family of unidentified infrared emission features, consisting of discrete and plateau features in the mid-infrared, are now observed in distant galaxies. A significant fraction of the total energy output of some infrared galaxies is emitted in these features. Comparisons of these features with those observed in the circumstellar and interstellar media suggest that these organic species are synthesized and ejected by evolved stars. Models of possible chemical structures of the carrier of these features are discussed.

Keywords. ISM: dust, planetary nebulae: general, ISM: lines and bands, ISM: molecules

---

Observations from the *Spitzer Space Telescope* have shown that the family of unidentified infrared emission (UIE) features are prominently present in many starburst galaxies (Smith *et al.* 2007). The strengths of the features are so strong that it is estimated over 10% of the total energy of the galaxies are emitted in these features. These results suggest that the carriers responsible for the emission of the UIE features must be abundantly present in galaxies. Some of the UIE features, such as the 3.3, 6.2, 7.7 and 11.3 μm features, have been identified as arising from aromatic compounds. Since UIE features are seen in galaxies with $z > 2$, this suggests that complex organics are already present in the early Universe (Kwok 2011).

The two main questions that remain unanswered are (i) what is the chemical structure of the carrier of the UIE features; and (ii) from where are these carriers synthesized? Although the UIE carriers are popularly assumed to be PAH molecules in the astronomical community, recent work has shown that the observational properties of UIE features are more consistent with they originating from mixed aromatic/aliphatic organic nanoparticles (MAON, Kwok & Zhang 2011). The aliphatic component is manifested in the 3.4 and 6.9 μm features, as well as in strong plateau emissions around 8 and 12 μm.

While the UIE features are widely observed in the diffuse ISM of the Milky Way galaxy, the only definitely known site of formation of the UIE features are evolved stars. UIE features are seen to emerge over $\sim 10^3$ yr time scales in the late stages of stellar evolution. They first appear in the proto-planetary nebulae phase and become prominent in the planetary nebulae phase (Kwok 2004). Since every year one star in the Milky Way enters the planetary nebulae phase, the Galaxy is constantly being enriched by large injections of stellar organics.

References

Kwok, S. 2004, *Nature* 430, 985
Kwok, S. 2011, *Organic Matter in the Universe*, Wiley
Kwok, S. & Zhang, Y. 2011, *Nature* 479, 80
Smith, J. D. T., Draine, B. T., Dale, D. A., *et al.* 2007, *ApJ* 656, 770

Highlights of Astronomy, Volume 16
XXVIIIth IAU General Assembly, August 2012
T. Montmerle, ed.

doi:10.1017/S1743921314012538

Modeling deuterium chemistry of interstellar space with large chemical networks

T. Albertsson[1], D. A. Semenov[1], A. I. Vasyunin[2], Th. Henning[1] and E. Herbst[2]

[1]Max-Planck-Institut für Astronomie, Königstuhl 17, 69117 Heidelberg, Germany
email: albertsson@mpia.de

[2]Department of Chemistry, University of Virginia, Charlottesville, VA 22904, USA

Abstract. Observations of deuterated species are essential to probing the properties and thermal history of various astrophysical environments, and the ALMA observing facilities will reveal a multitude of new deuterated molecules. To analyze these new vast data we have constructed a new up-to-date network with the largest collection of deuterium chemistry reactions to date. We assess the reliability of the network and probe the role of physical parameters and initial abundances on the chemical evolution of deuterated species. Finally, we perform a sensitivity study to assess the uncertainties in the estimated abundances and D/H ratios.

Keywords. astrochemistry, molecular processes, methods: numerical, ISM: clouds, ISM: molecules, stars: circumstellar matter, protostars

---

The life cycle of molecules covers a wide range of environments, starting from the sparse interstellar medium (ISM) that eventually evolves into stars and planets, and interstellar chemistry is the key to probing these environments. As molecular hydrogen cannot be observed in the cold ISM, other molecular tracers are employed to probe the relevant physical conditions and chemical composition. In this regard the study of deuterium chemistry has proven useful in constraining the ionization fraction, density and thermal history of the ISM and protoplanetary disks.

In the light of the high-sensitivity observations we can expect from ALMA, we present a up-to-date, extended, multi-deuterated chemical network consisting of 55 000+ reactions. We assess its reliability by the modeling of the deuterium fractionation in various phases of the ISM and comparing it with observed D/H ratios of a variety of mono–, doubly–, and triply-deuterated species in distinct astrophysical environments. A list of the most promising, potentially detectable deuterated species with ALMA is also provided.

We find a good agreement with previous model studies, such as those by Roberts & Millar (2000a) and Roberts & Millar (2000b). Our model successfully explains the observed D/H ratios for many molecules in the ISM, including water, methanol, ammonia and many hydrocarbons, and offers an improvement in reproducing the observations of DCO^+, HDCO, and D_2CO as compared to previous models.

Lastly, we studied uncertainties in calculated abundances, and hence column densities, as well as D/H ratios for deuterated species. Abundance uncertainties for up to triply-deuterated species with relative abundances $> 10^{-25}$ are shown in the left plot of Figure 1, and the same for D/H ratios is shown in the right plot. We find that, on a 1σ confidence level, the typical uncertainties for the calculated abundances, and hence column densities, are a factor of 1.25–5, 3–10 and approximately 10 for species made of $\lesssim 3$, 4–8 atoms and > 8 atoms, respectively. We find also that uncertainties of deuterated species are in general a factor of $\sim$ 2–3 larger than those of their un-deuterated analogues. A lot of chemically simple species with high abundance uncertainties contain either Mg, Na, or

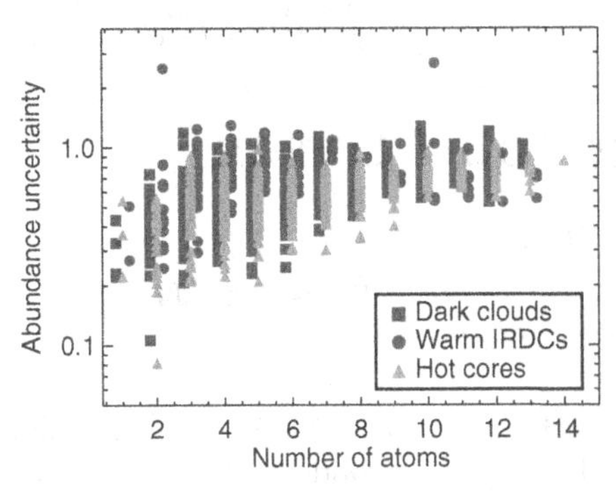

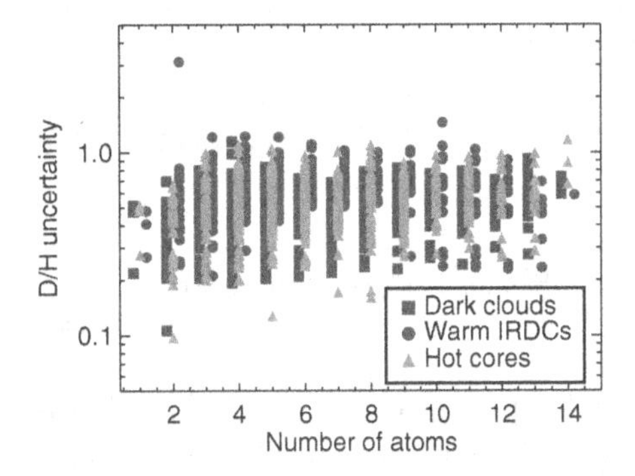

Figure 1. The 1σ uncertainties of abundances (left) and D/H ratios (right) for up to triply-deuterated species with relative abundances $> 10^{-25}$ as a function of molecule size. Dark clouds are shown with (blue) squares, warm infrared clouds with (red) circles and hot cores with (green) triangles.

Si, since the respective chemical pathways remain poorly investigated. We also find that large hydrocarbons (C_nH_m, n$\gtrsim$ 4), despite being abundant (with relative abundances up to $10^{-9} - 10^{-7}$), have large error bars in the computed abundances.

Overall the uncertainties in D/H ratios are lower compared to that of the abundances of the corresponding H- and D-bearing isotopologues. This is because for a majority of deuterated reactions there are no measured rate coefficients, and instead estimates has to be made based upon their un-deuterated analogues. Hence, often abundances of the H- and D-bearing isotopologues both are similarly affected by the rate, which is approximately half an order of magnitude, and vary between a factor of 2 and 10. As in the case of the abundance uncertainties, we find the largest D/H uncertainties for large hydrocarbons (C_nH_m, n$\gtrsim$ 4) and species containing Mg, Na, and Si.

We isolate the most problematic reactions for the isotopologues and isomers of H_3^+, HCO^+, HCN, H_2O, CH_3OH, H_3O^+, CH_3^+, $C_2H_2^+$ and CO, and find that this reaction list is dominated by ion-neutral and dissociative recombination reactions, most of them connected to the chemical evolution of water and methanol. Other essential reactions include the cosmic ray ionization of H_2 and He and a small number of neutral-neutral reactions.

Acknowledgements

T. A. acknowledges the support of the European Community's Seventh Framework Programme [FP7/2007-2013] under grant agreement no. 238258. D.S. acknowledges support by the Deutsche Forschungsgemeinschaft through SPP 1385: "The first ten million years of the solar system – a planetary materials approach" (SE 1962/1-1 and SE 1962/1-2). E. H. acknowledges the support of the National Science Foundation (US) for his astrochemistry program, and the support of NASA for studies in the evolution of pre-planetary matter.

References

Roberts, H. & Millar, T. J. 2000, *A&A* 361, 388
Roberts, H. & Millar, T. J. 2000, *A&A* 364, 780

Highlights of Astronomy, Volume 16
XXVIIIth IAU General Assembly, August 2012
T. Montmerle, ed.

doi:10.1017/S174392131401254X

Commemorating John Dyson

Julian M. Pittard

School of Physics and Astronomy, The University of Leeds, Leeds, LS2 9JT, UK
email: jmp@ast.leeds.ac.uk

John Dyson was born on the 7th January 1941 in Meltham Mills, West Yorkshire, England, and later grew up in Harrogate and Leeds. The proudest moment of John's early life was meeting Freddie Trueman, who became one of the greatest fast bowlers of English cricket. John used a state scholarship to study at Kings College London, after hearing a radio lecture by D. M. McKay. He received a first class BSc Special Honours Degree in Physics in 1962, and began a Ph.D. at the University of Manchester Department of Astronomy after being attracted to astronomy by an article of Zdenek Kopal in the semi-popular journal New Scientist. John soon started work with Franz Kahn, and studied the possibility that the broad emission lines seen from the Orion Nebula were due to flows driven by the photoevaporation of neutral globules embedded in a HII region. John's thesis was entitled "The Age and Dynamics of the Orion Nebula" and he passed his oral examination on 28th February 1966.

John's subsequent career took him to Wisconsin (with a Fulbright Fellowship), Manchester, Munich and Leeds. After studying radio recombination lines at Wisconsin (where John also learnt that the Green Bay Packers were the greatest American football team), in December 1967 John returned to Manchester to take up a lectureship and renewed his interest in dynamics.

One of John's earliest research themes was wind-blown-bubbles (WBBs). Motivated by John Meaburn's discovery of high velocities in HII regions, John led the 1972 Dyson & de Vries paper "The Dynamical Effects of Stellar Mass Loss on Diffuse Nebulae", a key work in the development of the theory of WBBs. It contains the first similarity solutions describing their dynamics and the first comparison of such models with observations. John applied the basic picture of WBBs to a wide range of sources throughout much of his subsequent career. Though John possessed the ability to attack problems purely analytically, he took interest in observational results, and was very successful at developing models and theories which could be used to make predictions and be tested by data.

John's work in the middle 1980s with Judith Perry on the formation of broad emission-line regions in active galactic nuclei (AGN) due to wind-obstacle interactions was among his most influential. He also wrote papers on AGN emission. His 1985 paper with Pedlar and Unger attributed the optical forbidden line emission in radio Seyfert galaxies to regions shocked by the expansion of the radio emitting components. In related research described in a 1992 paper (Taylor, Dyson & Axon), John suggested that jets play a role in the formation of some narrow-line regions in Seyferts. In this and other work, John recognized the similarities between wind-driven and jet-driven dynamics. John's other notable work on jet-driven dynamics included the introduction in 1987 of the working surface model of Herbig-Haro objects. He used it to explain observations described in several papers, which he co-authored with observers, on these objects and the jets associated with them.

From the mid-1980s onwards, John contributed greatly to the development of models of flows in multi-phase media. John's thinking on multi-phase WBBs started off by work he did on RCW 58, where it was discovered that the observed correlation of the velocity of absorption components with the ionization potential of the ions could not be explained by standard models of wind-blown bubbles. The 1986 Hartquist et al. paper which John co-authored was the first in a series devoted to mass-loaded flows. Such flows are the motions of diffuse material with a large volume filling factor that are influenced by the injection of mass stripped or evaporated from embedded clumps. A series of papers on RCW 58 then followed where models of mass-loaded bubbles were developed and refined. John continued working on developing an understanding of the properties of these large-scale flows throughout the 1990s and into the 2000s primarily with Jane Arthur, Robin Williams, and then latterly with myself. I was indeed privileged to have been John's last close collaborator.

John also considered the intermediate-scale structures arising within mass-loaded flows. For instance, in the Dyson *et al.* (1993) paper he used his physical insight together with simple analytical arguments to elucidate how tails form around the clumps embedded in a subsonic diffuse flow. John was also interested in the changes that occur to a flow when clumps are sufficiently close together that they start to affect the flow in a collective fashion. This is work that we are continuing to investigate at Leeds.

John also did much to bring attention to the potential importance of what have become known as turbulent mixing layers, which he and colleagues referred to as turbulent boundary layers. A 1988 paper by Hartquist & Dyson contains a number of considerations concerning the mixing in such layers, including ones that are magnetized.

So over a period of more than two decades, John wrote important papers on the general properties of mass-loaded flows, and the intermediate-scale structures and boundary layers within them. He also applied the models to a number of real objects, including Wolf-Rayet nebulae, PNe, SNe, ultracompact HII regions, starburst regions, and the wind of the galactic centre source Sgr A*. By doing so he provided considerable stimulation to several observers. In the last decade of his research he collaborated with Matt Redman and Robin Williams on the nature of magnetohydrodynamic ionization fronts, a problem with the same classic nature as that concerning WBBs which had attracted John's interest 3 decades earlier.

In 1996, Sam Falle, who collaborated with John for more than three decades, managed to convince John to return to West Yorkshire as the Professor of Astronomy in the University of Leeds. John subsequently built up a group which now contains six permanent academic staff, all of whom share star formation as a common interest. Some of the group members also pursue John's other interests in diffuse media. John became an Emeritus Professor in 2006.

John played many roles in the astronomical community. He served on many funding council committees, was on the RAS Council (1975-77), was editor-in-chief of *Astrophysics and Space Science* for one-and-a-half decades starting in 1993, and was President of IAU Division VI and Commission 34 (2003-6). John also co-authored with David Williams the popular textbook *The Physics of the Interstellar Medium.*

John was very much a "people person". John, his wife Rita and their children, frequently opened their home to astronomers from many nations, whether short-term guests or people who had just arrived and needed somewhere to stay as they found homes. John was always encouraging and patient, with collaborators and students alike. Most of all, John was always fun to be around, and was known for his rapid and incisive wit.

John passed away peacefully in his sleep at home on the morning of 3rd February 2010.

Acknowledgements

This commemoration is based on an obituary in the August 2010 issue of *Astronomy & Geophysics* written by Tom Hartquist and David Williams.

Highlights of Astronomy, Volume 16
XXVIIIth IAU General Assembly, August 2012
T. Montmerle, ed.

doi:10.1017/S1743921314012551

SpS13: High-precision tests of stellar physics from high-precision photometry

No contribution was received from this Joint Discussion.

Highlights of Astronomy, Volume 16
XXVIIIth IAU General Assembly, August 2012
T. Montmerle, ed.

doi:10.1017/S1743921315001659

Preface: Special Session SpS14 Communicating Astronomy with the Public for Scientists

Astronomers "have a responsibility to communicate their results and efforts with the public for the benefit of all", especially since most are publically funded. The importance of science communication has never been greater nor have the opportunities for communication of science to the public been larger. While astronomers are highly trained in carrying out scientific research, in general they are poorly equipped to communicate the excitement and results of their research to a broader audience. Generally, they do not have information about the various opportunities for communicating astronomy nor on the resources to help them and support them in communicating astronomy.

The primary goal of this Special Session 14 was to provide the interested astronomers with helpful hints and tips into effective communication with the public. The IAU General Assembly provided an ideal environment to attract astronomers interested in the public communication of astronomy.

Journalists and scientists often operate at opposite ends of the communication spectrum. Often scientists are disappointed about media coverage of their research, and journalists report frustration with the difficulties of describing and understanding important scientific findings. There is plenty of room for mistrust to build and problematic issues to arise. Some scientists are uncomfortable about participating in science communication but there are many good reasons why scientists should participate in public science communication:

- To expose the work of his/her specific community.
- To highlight a specific result.
- To highlight the work of an institution.
- To highlight the work of a group.
- To highlight individual efforts (which is perfectly all right!).
- To acknowledge a sponsor.
- To do a favour to the scientific community as a whole (a sense of duty).

SpS 14 featured key speakers knowledgeable in various areas of science communication. These invited speakers shared their best practices with the attendees. SpS 14 also featured many contributed talks from an extremely wide geographic area. These talks exposed the audience to the multitide of communication opportunities that are available to anyone interested in communicating astronomy with the public.

A very powerful example of astronomy communication was presented early on the first day. Members of the Tenpla project, visited Iwate prefecture, one of the areas most severely damaged by the Great Tohoku earthquake. They held star parties, workshops, plays and talks at several evacuation centers. Even in such difficult circumstances, people responded very positively to their efforts.

The same group also provided samples of astronomical toilet, in Japanese of course!, which provided much comic relief.

Dennis Crabtree and Lars Lindberg Christensen , co-chairs SOC

Highlights of Astronomy, Volume 16
XXVIIIth IAU General Assembly, August 2012
T. Montmerle, ed.

doi:10.1017/S1743921314012563

The IAU Office of Astronomy for Development

Kevin Govender

IAU Office of Astronomy for Development, P.O. Box 9, Observatory, 7935, South Africa
email: kg@astro4dev.org

Abstract. On 16 April 2011 the IAU's Office of Astronomy for Development (OAD) was launched jointly by the President of the IAU and the South African Minister of Science and Technology, at the South African Astronomical Observatory in Cape Town. This OAD was set up to realise the IAU's strategic plan which aims to use astronomy as a tool for development. Communicating astronomy with the public is one of the OAD's focus areas.

1. Background

The IAU adopted the strategic plan 2010-2020 at its 2009 General Assembly in Rio de Janeiro, Brazil. The basis of this strategic plan was to use astronomy to stimulate global development. With the slogan 'Exploring our Universe for the benefit of humankind' this plan aimed not only to stimulate the field of astronomy but also importantly to realize its benefits to humanity. The elements at the heart of the plan are primary, secondary and tertiary education, scientific research and public outreach. In order to realize this strategic plan the IAU established the Office of Astronomy for Development (OAD), which, after a lengthy international bidding process, was located in Cape Town, South Africa, at the South African Astronomical Observatory, and began operations in 2011.

2. The Office of Astronomy for Development (OAD)

The OAD is overseen by a steering committee made up of 3 nominations from the IAU and 3 nominations from the host organization in South Africa (the South African Astronomical Observatory, a facility of the National Research Foundation). Currently the steering committee consists entirely of IAU members and is led by George Miley. Although the OAD itself has only 3 full time staff, global expertise is sought in the form of its Task Forces. Task Forces are groups of international 'experts' in their fields who advise and assist the OAD in the implementation of the strategic plan. These are not necessarily IAU members but nominated and selected from a global science community.

Following an international stakeholder workshop in December 2011 and an open call for nominations, the three task forces were established. They are: (i) Astronomy for Universities and Research, led by a veteran of the IAU Commission 46, Edward Guinan; (ii) Astronomy for Children and Schools, led by the ex-IYA-coordinator, Pedro Russo; and (iii) Astronomy for the Public, led by one of the former presidents of IAU Commission 55 and current organising committee member, Ian Robson.

In addition to the Task Forces the OAD structure includes the establishment of regional nodes and language expertise centers. Regions covered by these nodes could be geographic (e.g. neighbouring countries) or cultural (e.g. globally distributed language or culture). This would allow for greater impact in that region and a more informed conversation with the IAU about particular local considerations when implementing the strategic plan in that region. Currently two regional nodes have been established: one in China for the East Asia region and mainly Chinese language and culture; and one in Thailand for the South East Asian region. Numerous negotiations are ongoing regarding the establishment of other nodes across the world.

The OADs vision 'Astronomy for a better world' captures the essence of the IAU strategic plan and allows the freedom to innovate in terms of the impact that astronomy can have on society. The OAD as such aims to further the use of astronomy as a tool for development by mobilizing the human and financial resources necessary in order to realize its scientific, technological and cultural benefits to society. The OAD has embarked on a recruitment drive to find volunteers willing to contribute to the implementation of the strategic plan. To date over 400 volunteers have signed up via the OAD website listing their skills and potential contributions. These volunteers are called upon as specific opportunities arise. They are also encouraged to participate in the OAD calls for project proposals so that their ideas can be supported.

In August 2012 the first open call for proposals was released with the IAU allocating funds towards projects relating to each of the Task Forces. The principle of the call was that each Task Force would evaluate proposals received and decide which were to be funded and which were to be put onto a wish list - the OAD would then continue to fundraise for those wish list projects. Under Task Force 3 (Astronomy for the Public) proposals were sought for activities such as journalist/blogger/communicator and amateur astronomer training; citizen science projects; models for outreach activities; traditional/cultural astronomy; databases of existing celebrations; creation of images and multimedia resources; stargazing events; astro-tourism activities; outreach awards; evaluation guidelines; etc.

In order to support the anticipated demand for projects the OAD has already established relationships with several organizations including the Royal Astronomical Society, the Netherlands Organisation for Scientific Research, the International Science Programme (Uppsala University), the International Centre for Theoretical Physics, the Inter-University Centre for Astronomy and Astrophysics and the University of Central Lancashire.

3. Public Outreach within the IAU

The OADs Task Force 3 (Astronomy for the Public), is certainly not the only activity of the IAU in terms of Public Outreach, nor should it be, given the wide reaching nature of the field and the many activities from IYA that need to be sustained. The IAU has also established the Office for Public Outreach which is based at the National Astronomical Observatory of Japan. This office is led by the Public Outreach Coordinator (Sarah Reed) and its purpose is to 'Organize and conduct public outreach activities of the IAU by coordinating and sharing responsibilities with the IAU Office of Astronomy for Development.'

In addition to this office the IAU still has within its structures Commission 55 (Communicating Astronomy with the Public). This Commission has been responsible for the very successful series of Communicating Astronomy with the Public (CAP) conferences as well as the publication of the CAP Journal, amongst its other activities such as the Washington Charter and the Virtual Astronomy Multimedia Project. The work of this Commission will thus certainly still be needed in the mix of IAU public outreach. In fact it will be essential that all three structures (the OAD Task Force 3, the Office of Public Outreach, and IAU Commission 55) work closely together and synergise for effective communication of astronomy with the public.

Highlights of Astronomy, Volume 16
XXVIIIth IAU General Assembly, August 2012
T. Montmerle, ed.

doi:10.1017/S1743921314012575

Communicating Astronomy in a Metropolis and Disaster Area – Activities of the Tenpla Project

K. Kamegai[1], N. Takanashi[2], M. Hiramatsu[3] and S. Naito[3]

[1]Dept. of Industrial Administration, Faculty of Science and Technology, Tokyo University of Science, Yamazaki 2641, Noda, Chiba 278-8510, Japan email: kamegai@rs.tus.ac.jp

[2]Inst. of Industrial Science, The Univ. of Tokyo, 4-6-1 Komaba Meguro, Tokyo 153-8505, Japan

[3]National Astronomical Observatory of Japan, 2-21-1 Osawa, Mitaka, Tokyo 181-8588, Japan

Abstract. We present recent activities delivering astronomy to the public by the Tenpla project in Japan. One is voluntary activities in the disaster area of the Great East Japan Earthquake. The other is holding tens of star parties and public lectures in the central area of Tokyo.

Keywords. communicating astronomy, star party, lectures

The Tenpla project (tenpla.net) is a cooperative network among researchers, students, museum staffs, educators, designers, and those who engage in popularization of astronomy in Japan. The aims are providing a chance to enjoy the latest astronomy for the public and making relationship among astronomy, society, and us (individual persons) by communicating each other. Since it was founded in 2003, various activities have been carried out for example developing science goods such as Astronomical Toilet Paper, holding science cafes, lectures and workshops, some of which were reported by Hiramatsu *et al.* (2008) and Kamegai *et al.* (2008) in SpS2 of 26th IAU General Assembly in Prague.

We have performed voluntary activities in the disaster area of the Great East Japan Earthquake since May 2011. Some Tenpla members visited Iwate prefecture where is one of the most heavily suffered area, and held star parties, workshops, plays and talks at several evacuation centers in cooperation with a few science volunteer groups. Even in such tough situation, attendees enjoyed our events. When we focused on distant astronomical phenomena, we all were able to communicate each other based on equal point of view. Through these experiences, we confirm that astronomy is an effective communication tool even in such a situation.

On the other hand, we have arranged various events at the central area of Tokyo. The purpose is to offer opportunities to enjoy astronomy for people who work in the center of city and/or who have no chance to communicate with astronomy in their daily life. For example, more than 60 star parties have been held at rooftop deck of one of the Tokyo's highest skyscraper. On some night of the star party we have a public astronomy lecture in the same building. In addition, a series of talks by young astronomers has been held at small shop in the center of city. Participants of the lectures are mainly office workers in their 30's – 50's , which means the events are successfully accepted by targeted people.

References

Hiramatsu, M., *et al.* 2008, *Innovation in Astronomy Education*, 198
Kamegai, K., *et al.* 2008, *Innovation in Astronomy Education*, 199

Highlights of Astronomy, Volume 16
XXVIIIth IAU General Assembly, August 2012
T. Montmerle, ed.

doi:10.1017/S1743921314012587

School Workshops on Astronomy

J. Molenda-Żakowicz[1] and G. Żakowicz[2]

[1]Uniwersytet Wrocławski, Instytut Astronomiczny, ul. Kopernika 11, 51-622 Wrocław, Poland
email: molenda@astro.uni.wroc.pl

[2]XIII Liceum Ogólnokształcące im. Aleksandra Fredry, ul. gen. Józefa Haukego-Bosaka 33-37, 50–447 Wrocław, Poland, email: gzakowicz@o2.pl

Abstract. Do you want to know how to make students volunteer to stay all night long watching the stars with their telescopes freezing? Or how to inspire decent adults to prepare a 'queue-list to Jupiter', wait for their turn for hours, and control that no one approaches the telescope bypassing the line? Or how to attract people of all age to forget their laziness and duties, and to get up at 3 a.m. to watch the transit of Venus? If your answer is 'yes', then come and see what can be done at the School Workshops on Astronomy.

Keywords. Public Outreach, Education

The outreach programme School Workshops on Astronomy† (SWA) is held at the Orle Mountain Station in the Izera Dark-Sky Park‡ in Poland every spring and autumn since 2009. These Workshops started as a few-days-long event addressed to the secondary and high-school pupils but soon enriched their offer by giving the pupils an opportunity to participate in various current astronomical events.

SWA have been already attended by around 300 pupils many of whom took part in those Workshops more than once. So far, each edition of SWA has been different. Our programme includes events like exhibitions, observations of the Moon, the stars, the Sun, and the atmospheric phenomena, talks given by the scientists and by pupils, courses of astrophotography and orientation on the sky, practical exercises, scientific measurements, discussions, and documentaries. The pupils learn how to measure the darkness, how to prepare the site to building a copy of an ancient observatory there, how to prepare themselves for a trip to Mars, how to work in a group and to present the results of their work in a concise and interesting way.

As a part of their activities, SWA participated in the opening of the Planetary Path in the Izera Mountains in 2010 and their representatives were invited to the Astronomers' Ball at the Royal Observatory Greenwich in 2011. In the frame of SWA, we organized watching the Transit of Venus in 2012 and several other public astronomical activities.

As a result, SWA have attracted enough interest of the media to be mentioned in the regional press and TV, and at different Polish and international web sites. This initiative without doubt belongs to the public astronomical highlights in the region of the Lower Silesia in Poland and the enthusiastic reception of each edition of the Workshops make us believe that this programme is a good case study for showing how to attract the general public to astronomy.

Acknowledgements

J. M-Ż acknowledges the Polish Ministry grant N N203 405139 and the IAU grant GA 1041.

† http://www.swa.edu.pl
‡ http://www.izera-darksky.eu

Highlights of Astronomy, Volume 16
XXVIIIth IAU General Assembly, August 2012
T. Montmerle, ed.

doi:10.1017/S1743921314012599

The Inflativerse - The University of Nottingham's inflatable planetarium

J. R. Ownsworth†, B. Haeussler[1]‡, E. Johnston¶ and N. Hatch||
University of Nottingham, School of Physics and Astronomy, Nottingham, NG7 2RD, U.K.

The Inflativerse is a mobile planetarium run by volunteers lead by the Physics and Astronomy departments at the University of Nottingham. This is a new project set up in 2011 to offer free of charge planetarium shows and activities for schools and public events.

Within this first year of running we have assembled a wide range of Astronomy related talks and activities targeted at an age range of 7 – 16. These include talks to younger children about the myths and legends behind the constellations from a variety of different cultures to more advanced talks on why the stars are different colours. It is also possible to talk about subjects outside of astronomy such as ocean currents and plate tectonics. Along side the talks that take place inside the planetarium we also run activities outside of the planetarium. Such activities include planet recognition exercises in the form of a solar system memory game for younger children and using balls and marbles to visualise the scales of sizes and distances in the solar system.

The school visit program is targeted at under performing schools in the local area. These are mostly schools that cannot afford the commercial travelling planetariums and other such education enrichment schemes. This program has been very successful within the first year of operation. We have visited 12 different schools and provided talks and activities to over 1200 children in that time. In this program we offer half and full day visits providing the chance for many children to participate in the activities that we run. We offer this service free of charge to the school.

We also run a public talk program that caters for public events in the local area. This program is for people of all ages to attend. This normally entails a series of talks throughout the event and a team of people outside of the dome to take any questions that the public may want to ask. In our first year we have attending 3 public events including the BBC Stargazing Live. In these events we were able to host over 700 people attending talks within the planetarium.

These programs have been very well received by schools and the public alike and we hope to continue this success into the future. We are now starting to assemble a wide range of activities both in and out of the planetarium and we hope to continually expand this range as time goes on. We are happy to share these activities and receive any ideas for new activities from the wider outreach community.

† E-mail: ppxjo1@nottingham.ac.uk
‡ E-mail: boris.haeussler@nottingham.ac.uk
¶ E-mail: ppxej@nottingham.ac.uk
|| E-mail: nina.hatch@nottingham.ac.uk

Highlights of Astronomy, Volume 16
XXVIIIth IAU General Assembly, August 2012
T. Montmerle, ed.

doi:10.1017/S1743921314012605

Strategies for the public communication of eclipses

P. S. Bretones

Departamento de Metodologia de Ensino/UFSCar, Brazil email: bretones@ufscar.br

Eclipses are among the celestial events that draw the attention of the public. This paper discusses strategies for using eclipses as public communication opportunities in the media. It discusses the impact of articles written by the author and analysis of published material for 25 observed eclipses over the last 30 years by mass media in the state of São Paulo, Brazil. On each occasion, a standard article was posted on the Internet and sent to newspapers, radio and TV with information, such as: date, time and local circumstances; type of the eclipse; area of visibility; explanation; diagram of the phenomenon, and the Moon's path through Earth's shadow; eclipses in history; techniques of observation; getting photographs; place and event for public observation. Over the years, direct contact was maintained with the media and jounralists by the press offices of the institutions.

The importance and methodology of the evaluation of newspaper articles about astronomy are mentioned by Fonseca *et al.* (*CAPjournal*, **8**, 27–29). The Budd score (*Journalism Quarterly*, **41**, 259–262) gives a news play measure, within the newspaper context. The most important information is presented in the title and first paragraph. The following information is considered less important in the inverted pyramid model. Evaluating the articles and studying the focus given by journalists indicates aspects considered to be more important for communication with the public. It was analyzed the approach taken by journalists using the sent material in the published articles about titles and images. Were published and analyzed 19 articles signed by the author (credited) and 71 articles not signed by the author (uncredited). The most used titles refer to the date and time of the phenomenon and place of event for public observations and the most common images of the eclipsed Moon and photo of the author. It is also discussed the interviews for TVs (14) and radios (15) about: the need for availability for interviews, inside or outside the studios, before, during, and after the events given the extensive agenda of the reporters. Frequently asked questions refer to the time, the place of observation, use of instruments, appearance of Sun and Moon during the event and eye safety.

In conclusion, sending articles on eclipses to the media have a high yield in publications being a "hot" subject. The publication or dissemination depended on the time of the phenomenon, the day of the week and events for public observation. The titles, illustrations and approaches indicate a trend to inform directly the date and time of the event and the use of images of the eclipsed object. Sending material to news agencies of great newspapers allow the spread of news to more distant places nationwide using of the Internet in order to make the best use of new media. Finally, it is important to be available for interviews.

Acknowledgements

I would like to express my deepest thanks to Jorge E. Horvath and Jay Pasachoff for the suggestions. I also thank the IAU, CAPES and FAPESP for the travel grants and financial support.

Highlights of Astronomy, Volume 16
XXVIIIth IAU General Assembly, August 2012
T. Montmerle, ed.

doi:10.1017/S1743921314012617

Communicating the science of the 11-year sunspot cycle to the general public

A. R. Choudhuri

Department of Physics, Indian Institute of Science, Bangalore - 560012, India
email: arnab@physics.iisc.ernet.in

Astrophysics is one branch of science which excites the imagination of the general public. Pioneer science popularizers like George Gamow and Fred Hoyle wrote on different aspects of astrophysics. However, of late, we see a trend which I find disturbing. While it has become extremely fashionable to write popular science books on cosmology, other areas of astrophysics are grossly neglected.

The 11-year sunspot cycle, which is essentially the magnetic cycle of the Sun, is one of the most intriguing cyclic phenomena known to us. Magnetic fields of large sunspots can cause such explosive phenomena like solar flares and coronal mass ejections, which are capable of affecting our lives—by disrupting radio communications, by making polar airline routes and space flights dangerous, by even producing power blackouts in extreme situations. While talking to my non-physicist friends, I find that many of them are quite curious about the sunspot cycle and would be interested in reading a popular science book explaining the science behind this cycle. To the best of my knowledge, no such book exists in the English language. Recently I have signed a contract with Oxford University Press for writing such a book.

I had the good fortune of doing my PhD thesis under the supervision of Professor E. N. Parker, the most influential solar physicist of our time, and have personally known many of the major players in this field. I have also been involved in the development of the flux transport dynamo model (Choudhuri, Schüssler & Dikpati 1995; Nandy & Choudhuri 2002), the currently most promising model of the sunspot cycle. This model was used by us to predict the strength of the cycle 24 (Choudhuri, Chatterjee & Jiang 2007). Our prediction now appears to be coming true. A recent survey of our field has been given by Choudhuri (2011).

After an introduction to the phenomenology of the sunspot cycle, my book will explain the basic ideas of standard solar model and magnetohydrodynamics. Then the central part of the book will be devoted to the core topics of sunspot formation by magnetic buoyancy and dynamo generation of the solar magnetic field, giving an introduction to the flux transport dynamo. The Sun-Earth connection will be discussed at the end. Throughout the book, there will be accounts of the historical development of this field, sometimes giving my personal reminiscences.

References

Choudhuri, A. R. 2011, *Pramana*, 77, 77
Choudhuri, A. R., Chatterjee, P., & Jiang, J. 2007, *Phys. Rev. Lett.*, 98, 131103
Choudhuri, A. R., Schüssler, M., & Dikpati, M. 1995, *A&A*, 303, L29
Nandy, D. & Choudhuri, A. R. 2002, *Science*, 296, 1671

Highlights of Astronomy, Volume 16
XXVIIIth IAU General Assembly, August 2012
T. Montmerle, ed.

doi:10.1017/S1743921314012629

Communicating ALMA with the Public in Japan

M. Hiramatsu

National Astronomical Observatory of Japan, 2-21-1, Osawa, Mitaka, Tokyo, 181-8588 Japan
email: hiramatsu.masaaki@nao.ac.jp

Abstract. I present the strategy and activities of the public outreach and communication of ALMA in Japan. Since most of the public is not familiar with the radio astronomy, we present the human side of ALMA to attract the public interest, as well as by showing the science results. To evoke the public interest on the radio astronomy, it is also effective to show the radio astronomy research topics on the planets, the Sun, and bright stars so that they can connect the daily night sky with the radio universe explored by ALMA.

---

1. ALMA Outreach Strategy in Japan

ALMA is the powerful radio telescope constructed in the Chilean Andes with the international partnership of Europe, North America and East Asia in cooperation with the Republic of Chile. Public outreach activities become more and more important with the progress of the construction and start of science operation in September 2011. Here I describe the outreach and public communication of ALMA in Japan.

The outreach of ALMA in Japan is difficult in the following points; Chile is distant from Japan, the public is not familiar with radio astronomy, and even less with radio interferometer. To overcome these difficulties, our strategy is not only to show the science results but also to "tell the stories" of the cool universe, ALMA staff members, and high technology employed by ALMA. Showing the human side of ALMA, for example "What do astronomers want to know with ALMA?", "How is the life in the ALMA observatory?", "How did the engineers build the great telescope with cutting-edge technologies?" initiates the interest of ALMA among those who are not interested in astronomy or science.

2. ALMA Outreach Activity in Japan

Along with the public lectures and science cafes, we make use of the online tools; web site, twitter, facebook, and mail magazine. To showing the "human face", we have the photo reports of the construction progress, the serial online articles written by the NAOJ ALMA staff members to introduce their daily life and work in Chile, and short movies of the antenna construction factories with the interviews of the craftsmen.

With twitter, we tweet the news updates of ALMA construction and scientific results, announcements of public talk, as well as notices of celestial events such as eclipses and planet conjunctions. The number of retweets and favorates is the largest for the last topic, which shows many people are interested in the visible sky and want to share those information. They look unrelated to ALMA, but the planets and the Sun are also ALMA's important targets. Together with the guidance of celestial events, we tweet the research topics on the Sun, planets, exoplanets, and so on, which would be explored by ALMA. Those tweets obtain larger number of "retweets" and "favorates", and this helps the public connect the daily night sky with the radio universe.

Highlights of Astronomy, Volume 16
XXVIIIth IAU General Assembly, August 2012
T. Montmerle, ed.

doi:10.1017/S1743921314012630

Knowing the people who come to public astronomical observatories: The case of Akita prefecture, Japan

N. Kawamura
Akita University, Tegata-gakuen, Akita, 010-8502 Japan
email: norihito@ed.akita-u.ac.jp

The purpose of this research is to know and gain a better understanding of people who come to astronomical observatories and to find out more about their experiences and thoughts on astronomy. To find some of the issues about science communication in astronomy, the author carried out questionnaire research studies involving high school students and junior high school and elementary school teachers.

One of the questions to high school students dealt with what heavenly bodies they wanted to observe when they had a chance to use a telescope. Generally, the students tend to want to see the heavenly bodies in our solar system.

Some questions to teachers dealt with actual situations in teaching astronomy, dealing with any problems, and having telescopes as teaching instruments in school. Almost all elementary school teachers were assigning homework to observe the stars, but almost all the teachers who were assigning these homework projects found some difficulties. E.g. one of the major problems was that teachers could not instruct students directly. About 70

The author's co-worker carried out questionnaire research at every astronomical event with the visitors who came to our observatory center. One of his questions dealt with what heavenly bodies they wanted to observe. The visitors tend to want to see planets, nebulae, and clusters using a telescope. In addition, they also tend to want to watch temporary events such as eclipses. In short, people hope that a telescope will give them a better view of the moon, planets and stars. Understanding astronomy may not be the main reason why they visit an observatory.

Highlights of Astronomy, Volume 16
XXVIIIth IAU General Assembly, August 2012
T. Montmerle, ed.

doi:10.1017/S1743921314012642

Communicating astronomy with the public for scientists

R. Girola

Universidad Nacional de Tres de Febrero and Planetario Galileo Galilei. Buenos Aires, Argentina email: rafaelgirola@yahoo.com.ar

This article intends to convey the improvement regarding the knowledge exchange in the astronomical field through an improvement in the quality of professional communication between researchers, teachers and the like whose job is to broadcast astronomical concepts. It has been a couple of years since the difficulty of communicating astronomical concepts decreased due to institutional projects, schools and education systems. Inside the education system, the need to include astronomy as an innovative element in curricula has become obvious. Outside, an informal public interested in astronomy became greater in number and began to be fostered by different organizations which spread their astronomical knowledge via workshops and demonstrations. An example of a place where courses, learning projects and informal activities relating to astronomy are held is the Planetario Galileo Galilei. The planetarium is the empirical proof that it is possible to bring together researchers, teachers and others who are focused on the popularization of Astronomy around a project, in which they combine their skills to transmit their knowledge to the public. The Planetario offers various means (multimedia/rooms) which enable visitors to have access to scientific material. The course called the universe, stars and galaxies: Is there more matter than the visible one? although inspired by scientific texts, it was changed into a stimulating medium which facilitated the interaction with the public and permitted them having a better comprehension of some subjects. The success of the course was possible because of the collaboration of various professionals in the field: Dr Josefa Perez, Dr Hector Vucetich as well as the Planetario staff in charge of the course. The combination of knowledge and professionalism of all the actors (researchers, teachers and others) led to very positive results for it successfully captured the attention of a large public. During the process of data collection, I employed material from the internet and libraries in order so as to encourage analysis and study in a public which were as diverse as their fields of expertise from lawyers to science, biology and architecture professors. My intention was to convert the scientific terminology in the texts into a more educational register in order to change the recipients of the texts from a group of researchers to a wider public. Due to the lack of experience among the public, the techniques applied in the workshop were different from the ones utilized by the experts. Notwithstanding, the public did not attend classes to become astronomers. Participants were merely willing to acquire a general knowledge about astronomy, in this particular case, to comprehend astronomical phenomena and theories as well as to incorporate information. The impression that the public had formed about scientists gained importance in this workshop. As it was mentioned before, the purpose of this course was to arise the dynamic of curiosity in the public by the use of simple techniques, readings, problem setting and solving. The idea is that the public would not be satisfied with immediate and passive solutions. On the contrary, exercises were aimed at putting the public into action and inviting participants to assume their roles as researchers. Each class was centered on a theme which in turn corresponded with the one utilized in the research project. Even though the background of each participant varied, the groups managed to maintain their unity when working together. An example of this behavior was when they did not necessarily know the meanings of scientific terms. Despite this, they implicitly understood them by means of the context. Other topics covered in the course were the stars, galaxies and the models of the universe which are currently accepted. The scientists also had to deal with misconceptions, myths and errors which were caused either because of the lack of information or because of the influence of media in which an expert opinion is rarely consulted. It is important to note that scientists do not spread the information only taking into account the popular aspect of science. Information has to be founded on scientific grounds rather than resembling a piece of news. The tactics when reaching the public may vary, but they always have to convey the essence of the message. Is the public aware of the heterogeneous nature of its participants? Developing the conscience of a unified group in the public requires that the science broadcaster as well as the scientist show their existence different from a passive role, in which preconceptions about scientific notions or the ideas that are created after treating a topic are generated and rejected. As the public lacks the ability to control the means in which the communication is established, it is at risk of being manipulated by the science broadcaster. The latter has to be transparent about the sources of the ideas conveyed. It is also important that the public understands that information of any kind of subject always comes from a source and that comprehension is achieved through different means such as the narration of a story, the display of images and the use of outreach techniques. As regards the scientists who had participated in the informational classes, they sought examples that were easy to understand so as to encourage the publics interest in the topic. In relation to the scientific jargon, it is complicated to understand from an outsiders view. Semantics are communicated effectively between scientists. But this is not the case of the public. The successful delivery of the contents will chiefly depend on the experience the scientist has at addressing the public and the role he assumes in the communication process. By eliminating the language barriers and mingling with the public, the scientists will be able to perceive reality from the receivers point of view. What matters at the moment of undertaking the receivers role is to display the obscurities and challenges that may appear at the time of facing problems that arise from a scientific model. Needless to say, the scientist has to be in good spirits and has to take his time to dexterously accomplish his role as a communicator. In this case, both scientists who collaborated with the workshop, Dr. Hctor Vucetich and Dr. Josefa Prez intended to achieve a good communication with the public. The purpose was to deliver assorted knowledge to the public through experience. They both showed respect towards the public and were able to create an integrated environment, which meant to destroy class differences for the sake of cognitive skills. One of the keys for capturing the public was being modest. It was meant to discard any verticality of knowledge acquisition that may widen the gap between the public and the astronomical science. At the end, scientists and public create an active and serene language. The difficulty to understand a concept still existed but the friction between the public and the comprehension was reduced since the public had a better disposition. Although the length of the course was short and only covered the general contents without acknowledging the public in advance, scientists were successful at achieving communication. In reply, the public displayed an active attitude. These proved that prejudices and myths about scientists were wrong. By the use of a simple though accurate language, a rigorous method and implementing the apposite technology in the research, the public was able to interact and to improve their intellectual capacity. Science popularization offers the public the possibility of discovery, method and knowledge acquisition. If the scientist is capable of delivering these skills, the public would be able to distinguish when the source and the intermediator apply their tools to create an image so as to maintain the publics confidence. In this particular case, the purpose was achieved as the course dropout was rather low at the end.

Highlights of Astronomy, Volume 16
XXVIIIth IAU General Assembly, August 2012
T. Montmerle, ed.

doi:10.1017/S1743921314012654

Working with Journalists: Media Access and Why You May Need It

R. T. Fienberg and S. P. Maran

American Astronomical Society, email: rick.feinberg@aas.org

The Washington Charter for Communicating Astronomy with the Public says that "individuals and organizations that conduct astronomical research - especially those receiving public funding for this research - have a responsibility to communicate their results and efforts with the public for the benefit of all." Aside from a sense of obligation, though, there are other reasons why astronomers ought to communicate with the broader citizenry. Among them: it is encouraged by the leaders of government funding agencies, it helps further public recognition and support for science in general and for astronomy in particular, and it can enhance one's career prospects.

One of the most effective ways to reach the public is through the news media. But scientists and reporters don't necessarily agree on what constitutes "news". In practice, news is what reporters want to cover, not necessarily what scientists and institutions want to publicize.

In *Science and Journalists - Reporting Science as News* (Free Press, 1986), Sharon Friedman writes, "Editors and reporters tend to value stories that contain drama, human interest, relevance, or application to the reader, criteria that don't always map easily onto scientific importance."

If scientific importance doesn't guarantee newsworthiness, what other criteria might apply? In *The Hands-On Guide for Science Communicators* (Springer, 2007), Lars Lindberg Christensen offers a list that includes such things as *relevance* (the issue has influence on people's lives or on the way they think about the world), *conflict* (the discovery involves a hotly debated topic or resolves a hotly contested issue), a *record* (the discovery is the first, last, oldest, youngest, biggest, smallest), and aesthetics (the finding is accompanied by an exceptionally beautiful image). But in *Making the News* (Westview, 1998), Jason Salzman offers a list of things that will send reporters running in the other direction. Of particular relevance to science news, that list includes *complexity.*

Here's a useful way to tell if you have a newsworthy story: In a single paragraph of no more than 75 words, answer the questions Who?, What?, When?, Where?, Why?, How?, and, most importantly, So what?, then show it to a couple of people who aren't astronomers. If they are intrigued, you've got news. If not, you probably don't.

When is the right time to announce new findings? Upon presentation at a conference? Upon submission of a paper? Upon acceptance of a paper? Upon publication of a paper? Each of these options has pros and cons. Worse, scientists don't necessarily have control of the timing, because reporters have ready access to "unreleased" results, such as presentations at meetings, posts to online preprint servers, advance electronic publication of journal papers, talks at university colloquia, and even such informal sources as astronomers' websites, blogs, Facebook pages, and Twitter feeds.

A key question when thinking about announcing something to the news media is whether or not your results are embargoed. Embargoes are used by journals such as *Nature* and *Science* to make results available to journalists in advance with the understanding that their news reports will not be published until a specified date and time, usually a few days hence. On the plus side, embargoes give reporters time to do their homework before writing their stories, potentially resulting in higher-quality coverage of the science. On the minus side, scientists may refuse to discuss *un*embargoed results (often because they misunderstand embargo rules), and journalists have a harder time getting a"scoop". Some of the main embargo policies affecting astronomers are:

- http://www.nature.com/nature/authors/policy/embargo.html (Nature)
- http://www.sciencemag.org/site/help/authors/embargo.xhtml (Science)
- http://aas.org/press/embargo_policy (American Astronomical Society)

Who do you call with your news? *The New York Times*? No, you call your institutional PIO (public-information officer). PIOs are (usually) experienced science writers; they're experts on their "beat" (subject area), they have established relationships with local/regional/national media, and they know how to get institutional approval for press releases and other types of announcements.

Depending on the nature of your project and the makeup of your team, you may need to work with PIOs at your institution, observatories, space missions, funding agencies, and so on. How do you find these people? Media contacts are listed on institutions' websites, under headings such as "media relations", "press", or "news".

Your PIO may choose to get the word out via a press release, via your participation in a press conference (if the result is being presented at a scientific meeting and/or if the result is extraordinary), and/or via personal pitches to individual journalists. In any case, your PIO will probably offer to prepare you to talk with reporters, perhaps even providing media training to enable you to find the right words, relax on camera, and hone your message into appropriate "sound bites".

It's very important not to wait till the last minute to contact your PIO about a forthcoming newsworthy science result, because it typically takes several weeks to write, edit, and illustrate a press release and to get it approved for distribution by all the institutions involved in the announcement.

Press releases used to go only to bona fide journalists (by mail, then by fax, then by email) and were used as the jumping-off point for researching and writing original news articles. Now, with journalists being laid off in alarming numbers and press releases posted online for easy aggregation, often the only "news article" that appears - all over the Internet - is the press release itself. Thus it is vital that the release be well written, error free, and approved by all who have a stake in it.

There are numerous services that distribute press releases to qualified journalists, among them EurekAlert!, AlphaGalileo, and PR Newswire. In astronomy, the most effective way to get your announcement into the hands of reporters, editors, broadcasters, and others worldwide who regularly cover astronomy and space science is to have your PIO send it to the AAS at press-release@aas.org.

For more information, ask your PIO if he or she has any guidelines for preparing press releases and/or participating in press conferences. Ask, too, if your institution offers media training. If it doesn't, you can take advantage of media-training programs offered by scientific societies, funding agencies, and private consultants. Here's a good place to start: http://communicatingscience.aaas.org/Pages/newmain.aspx

Finally, there are many good books available for scientists who want to do a better job of communicating with the public. Two of the best are *Am I Making Myself Clear? A Scientist's Guide to Talking to the Public* by Cornelia Dean (Harvard, 2009) and *The Hands-On Guide for Science Communicators: A Step-by-Step Guide to Public Outreach* by Lars Lindberg Christensen (Springer, 2007).

Highlights of Astronomy, Volume 16
XXVIIIth IAU General Assembly, August 2012
T. Montmerle, ed.

doi:10.1017/S1743921314012666

The challengers of an astronomer being a journalist

N. Podorvanyuk[1,2]

[1]Sternberg Astronomical Institute, Moscow State University email: nicola@sai.msu.ru
[2]Iinternet newspaper "Gazeta.Ru", "Science" department (http://www.gazeta.ru/science)

Abstract. As the weakness of russian astronomers in observational astronomy became chronic Russia should enter European Southern Observatory. But the Russian government is still not providing any financing of the entrance of Russia to ESO. The author states this situation as an example of his experience of work as an astronomer and as a journalist at the same time.

Keywords. Telescopes, notices, miscellaneous

The russian science is underfunded that's why a lot of young researchers have the general challenge: **to be a scientist in Russia and to survive**. Russian astronomy has yet had a good reputation but the majority of its successes were made in USSR. 6-meter russian telescope BTA in Special Astrophysical Observatory of the Russian Academy of Sciences (SAO RAS) was the largest telescope in the world in 1975-1993. It is the opinion of the head of SAO RAS Yuriy Balega:"We won't be able to conduct astrophysical research on high level with modern instruments soon. Russian astronomers should enter ESO because our weakness in observational astronomy became chronic" (*http://www.gazeta.ru/science/2012/02/23_a_4010781.shtml*).

The author works at the same time as an astronomer and as a journalist. Being involved in media he has got **the possibility to speak with famous scientists**. In 2010 he made the controversial interview with the president of Russian Academy of Sciences Yuriy Osipov about the very low citation rate for articles published in Russian-language science journals (*http://www.nature.com/nature/journal/v464/n7286/full/464141b.html*).

In 2011 Mr. Osipov talked about the entry Russia to ESO with prime-minister Vladimir Putin who supported it and ordered to create a working group for this. But at the beginning of 2012 there were still no news about Russia and ESO. The author's next challenge was **to find out news about it**. So he interviewed Mikhail Kovalchuk, the secretary of scientific board under the president of Russia. Kovalchuk told: "You, russian astronomers, are just asking for money. The time of pinches is finished. Nobody will give you money just for fun. You should present the special program and vision which would be understood by people which take decisions and which are interested in the future of country" (*http://www.gazeta.ru/science/2012/02/07_a_3991449.shtml*).

The fee for Russia to enter ESO is about 150 million Euros. Russia bears about 250 million Euros of the construction cost of X-ray Free-Electron Laser (*http://www.xfel.eu/overview/in_brief/*). Future stadium of the football club Zenit St. Petersburg (*http://www.kommersant.ru/doc-y/2029104*) costs about one billion Euros, similar to the 39-meter E-ELT which will be built by ESO's members. USA, Chile, Korea and Australia are building the 24.5-meter GMT. USA, Canada, Japan, China and India are going to build the 30-meter TMT. But Russia now (at the middle of 2012) is not going to participate in building of any extremely large telescope. **Trying to change this situation** is the challenge for me as an astronomer and a journalist.

The author thanks "Gazeta.Ru", Rashid Sunyaev, ESO and his wife Olga Alekseeva.

Highlights of Astronomy, Volume 16
XXVIIIth IAU General Assembly, August 2012
T. Montmerle, ed.

doi:10.1017/S1743921314012678

Las Cumbres Observatory: Building a global telescope network from the ground up

E. L. Gomez

Las Cumbres Observatory Global Telescope Network, Suite 102, 6740 Cortona Dr, Goleta, CA 93117, USA
email: egomez@lcogt.net

Abstract. Las Cumbres Observatory Global Telescope Network (LCOGT) are building a global network of telescopes which will be available to both professional scientists and the science curious public. This telescope network will be global and so will the community, therefore all aspects of the endeavour must be online and self-sustaining - from the observing software to the analysis tools. During 2012 LCOGT have deployed the first 1-meter telescopes, and launched a citizen science project using LCOGT data, Agent Exoplanet, as well as many other online resources for anyone to use as they explore astronomy.

---

1. Building a global telescope network

One of the primary missions of LCOGT is to *"inspire critical thinking by pursuing scientific investigations with robotic telescopes"*. During 2012-2015 LCOGT will create a homogeneous network of over 30 telescopes which will be used by professional scientists and the general public. To ensure this network can be accessed as widely as possible all observing tools will be available through an online web portal.

Whilst the majority of time available on the network will be for scientific use, a large fraction will also be available for informal or citizen science projects. At full deployment LCOGT hopes to be able to offer up to 100 hours of observing time each day on the 0.4-meter network of telescopes for citizen science. LCOGT will run and support a small number of projects leaving the rest of the time available for the community to propose projects. Key to the success of a these proposals will be how well supported the user-base will be while they follow the project.

As the telescope network is being deployed a diverse range of tools and resources have been developed for astronomy education and engagement. All are available online at the LCOGT website (`http://lcogt.net`). Some examples are:

- Virtual Sky - A customisable planetarium which can be embedded into any webpage. VirtualSky shows the positions of planets, meteor showers, the Moon and Sun, constellations, as well as many different sky projections. `http://lcogt.net/virtualsky`
- Observation Archive - The full archive of data from 2004 to the present, is available to browse and query. All images and data may be used under a Creative Commons licence. `http://lcogt.net/observations`
- SpaceBook - An online textbook of astronomy, written by LCOGT containing hundreds of pages on a wide variety of astronomical topics. `http://lcogt.net/spacebook`
- Agent Exoplanet - Explore extra-solar planet data using the transit detection method. All the data analysis and graphing tools are built into this online citizen science investigation. `http://portal.lcogt.net/agentexoplanet`

Highlights of Astronomy, Volume 16
XXVIIIth IAU General Assembly, August 2012
T. Montmerle, ed.

doi:10.1017/S174392131401268X

Exploring science and technology through the *Herschel* space observatory

V. Minier[1] and M. Rouzé[2]

[1]CEA/Irfu, Service d'astrophysique, 91191 Gif-sur-Yvette, France
email: vincent.minier@cea.fr

[2]CNES, CDT/ME/EU, 18 av. Edouard Belin, 31401 Toulouse, France

Abstract. Because modern astronomy associates the quest of our origins and high-tech instruments, communicating and teaching astronomy explore both science and technology. We report here on our work in communicating astronomy to the public through Web sites (www.herschel.fr), movies on Dailymotion (www.dailymotion.com/AstrophysiqueTV) and new ITC tools that describe interactively the technological dimension of a space mission for astrophysics.

Keywords. communication, history and philosophy of astronomy, techniques: photometric

---

We report on a project aiming to communicate *Herschel* space observatory results to the French-speaking public. This public outreach effort is complementary to the European Space Agency (ESA) communication on *Herschel*. The project started in 2008 at CEA in France and then became national (CNES – CEA – CNRS) in 2009 following the successful launch of *Herschel* on May, 14th 2009. The project is mainly based on Web resources with the creation of a dedicated Web site: www.herschel.fr. The Web site includes 3 levels of reading: a home page with videos and news board; sections dedicated to science topics and instrumentation; applications, videos and texts to go further in detail in the *Herschel* mission. An important effort has been made to produce videos of high quality with professional motion designers. Many videos include interviews of scientists presenting their goals, and historical background given by decision-makers and mission principal investigators. Other videos propose 3D animation of the *Herschel* observations (Picturing Star Motion) using colour-coding and effects in precise agreement with scientific data (intensity, distance, depth, coordinates). These videos are available in the Multimedia area as well as in streaming on the Astrophysique TV Dailymotion channel. These resources provide background for science journalists, educators and general public, and have been used outside the Internet, in exhibitions, teaching, public talks, movie theatres... Based on statistics, we estimate that more than 200,000 people have been reached via Internet. The secondary impact is difficult to estimate. All the archives used for this project feed an historical study of *Herschel* invention led by V. Minier.

Acknowledgements

We would like to acknowledge the work in this project of Thierry Morin (CEA Irfu) and Lucille Colombel (CEA DCom) on the Web design and development, Pierre-Francois Didek (Karamoja Productions) on the Web documentaries, Paco Abelleira and Laurent Fouilloux (Novae Factory) on 3D animations and interactive application, the support of Philippe Chauvin (CNRS/INSU) and all the astronomers who contributed directly to the production of contents or indirectly through their scientific results. This project has been funded mainly by CNES, the French Space Agency, and CEA. Web applications summarising the studies in history of techniques conducted at CEA Larsim were partly supported by the Campus Spatial – Université Paris Diderot (project Astro Innov).

Highlights of Astronomy, Volume 16
XXVIIIth IAU General Assembly, August 2012
T. Montmerle, ed.

doi:10.1017/S1743921314012691

The Venus Transit, the Mayan Calendar and Astronomy Education in Guanajuato, Mexico

H. Bravo-Alfaro, C. A. Caretta, E. M. S. Brito, P. Campos and F. Macias

Universidad de Guanajuato, Mexico,
email: `hector@astro.ugto.mx`

In this work we present two aspects of the Astronomy education activities carried out in 2012 by a multidisciplinary group at Universidad de Guanajuato, including specialists in Astronomy, Social Sciences and Environmental Engineering. The first program linked the Venus Transit, occurred in June 2012, with a national campaign of vulgarization of both modern and ancient (Mayan) Astronomy. Professional astronomers all around the country took advantage of the recent myth linked to the end of a large Mayan calendar cycle (13 *baktuns*, or some 5125 years) happening, after certain authors, in December 2012. In Guanajuato, the Astronomy Department organized live observations of the Venus Transit at two different locations, and complemented with conferences about astronomical events and the fake predictions of disasters linked to the "end" of the Mayan calendar. This program was very successful not only in Guanajuato but throughout the country, with several thousands of people attending live observations, conferences, expositions, etc.

The second program described in this work started back in 2009, in collaboration with researchers of the Culture and Society Studies Dept., who are specialized in rural development. After the great success of the first national star gazing (*Noche de las Estrellas 2009*), being part of the International Year of Astronomy in Mexico, several communities across the central region of Mexico (in the state of Guanajuato) requested our support in order to organize their own star gazing. Nowadays, we have visited more than a dozen towns, with several hundred visitors at each run (mainly children), most of them watching the sky through a telescope for the very first time. Now we have the collaboration of people from the Environmental Engineering Group, who are actively working to improve the conditions of water, soil and the forests in general, of those spots that have been included in our program.

More information about both these programs may be seen at the web site, http://www.astro.ugto.mx and at: *El sujeto cultural y los studios multidisciplinarios*, Macias & Campos eds. Universidad de Guanajuato. 2012 in press.

Highlights of Astronomy, Volume 16
XXVIIIth IAU General Assembly, August 2012
T. Montmerle, ed.

doi:10.1017/S1743921314012708

Hinode, the Sun, and public outreach

K. Yaji[1], H. Tonooka[2], M. Shimojo[2], N. Tokimasa[3], D. Suzuki[4], A. Nakamichi[5] and I. Shimoikura[6]

[1]Rikkyo University, 3-34-1, Nishi-Ikebukuro, Toshima-ku, Tokyo, Japan
email: yaji@rikkyo.ac.jp

[2]National Astronomical Observatory in Japan, [3]Nishi-Harima Astronomical Observatory, [4]Kawaguchi Science Musium, [5]Kyoto Sangyo University, [6]Tokyo Gakugei University

Extended Abstract

Hinode is a solar observation satellite in Japan and its launch was in September 2006. Its name means "SUNRISE" in Japanese. It has three instruments onboard in visible light, X-ray, EUV to solve mystery of coronal heating and origins of magnetic fields.

Hinode has been providing us with impressive solar data, which are very important for not only investigating solar phenomena but also giving new knowledge about the sun to the public. In order to efficiently communicate Hinode data to the public, we organized working group for public use of Hinode data. which are composed of both researchers and educators in collaboration. As follow, we introduce our activities in brief.

For the public use of Hinode data, at first, we produced two DVDs introducing Hinode observation results. In particular, second DVD contains a movie for kids , which are devloped to picturebook. Now, it is under producing an illustrated book and a planetarium program. It turn out that the DVDs help the public understand the sun from questionnaire surveys. Second, we developed teaching materials from Hinode data and had a science classroom about the sun, solar observations, practice with PC such as imaging software at junior high school. As the results, they had much interests in Hinode data. Third, we have joint observations with high school students and so on in a few years. The students compare their own data with Hinode data and have a presentation at science contests. The joint observations make their motivation higher in their activities.

It is important to record and report our activities in some ways. So, we positively publish papers and have presentions in domestic/international meetings. Though we are supported in budget, resources and so on by NAOJ Hinode Team, we apply research funds for promoting our EPO activities and acquire some funds such as NAOJ Joint Research Expenses and Grands-Aid for Scientific Research Funds since the launch.

This way, since its launch, we have continued various and constant EPO activities for the public use of Hinode data and have been giving intense impacts and high interest to the public. As the result, our activities contribute in further extension of Hinode Mission. Those are quite unique and would be reference of other similar ones. Hinode is now operating and solar activities might get more higher.

As long as SUN RISE, we would GO FORWARD!!

Keywords. Hinode, Sun, DVD, teaching material, joint observation

Acknowledgements

We thank all the members of Hinode Science Center/NAOJ and solar scientists for encouraging us in our EPO activities. K.Y. is supported by the Grant-in-Aid for Scientific Research (No. 23501027) from the Ministry of Education, Culture, Sports, Science and Technology of Japan.

Highlights of Astronomy, Volume 16
XXVIIIth IAU General Assembly, August 2012
T. Montmerle, ed.

doi:10.1017/S174392131401271X

Mitaka "Taiyokei" (solar system) walk; a collaborative science outreach program by institues, local government, and shopping stores

T. Handa[1], H. Agata[2], S. Ooasa[3], K. Karasaki[3], N. Kitahori[3], M. Arai[4], A. Ohta[4], K. Nishino[4], T. Ishii[4], C. Yoshida[4], T. Taguchi[5], E. Totsuka[5], S. Watanabe[5], H. Fukaya[6], Y. Kakihana[7], A Inoue[7], K. Itabashi[8], E. Yoshida[8], K. Ikeda[9], K. Saito[10] and T. Kamoshita[11]

[1]Depart. Astron. & Phys., Kagoshima Univ., Korimoto 1-21-35, Kagoshima 890-0065, Japan
email: `handa@sci.kagoshima-u.ac.jp`

[2]National Astronomical Observatory of Japan, [3]Mitaka Network University Organization
[4]Mitaka society of commerce and industry, [5]Mitaka-city tourism association
[6]Ghibli museum Mitaka, [7]Mitaka Municipal Office, [8]Mitaka Town Management Organization
[9]Mitaka international society for hospitality, [10]Mitaka city arts foundation
[11]NPO Flowers and Green City Mitaka Creative Association, [12]MT-planning Co. Ltd.

It is difficult to get a real scale image of the solar system through lecture. A scale model is a classical and one of good solutions (e.g. Handa *et al.* 2003, Handa *et al.* 2008). Through this model, people living in or visiting to the city can physically understand the scale of the solar system. This scale gives 1 cm for Earth's diameter and 115 m for 1 AU. However, some gadget is required to make it attractive for public citizens.

We have, therefore, started a program "Mitaka TAIYOKEI walk" since 2009. In this program we made a scale model of the solar system with 1:1,300,000,000 over Mitaka city, Tokyo and link it to many spots in the city. More than 100 shopping stores and facilities in and around Mitaka city join this program every year. We set the Sun at the railway station and put a pictorial flag of the Sun in the same scale. The city area is divided into 11 annual zones centered on the "Sun" and each zone is assigned to a planet, the Sun, Ceres, or Pluto. Each shop to entries has a pictorial flag and a stamp of the *planet* of the zone. At several spots in the city, a visitor can get a booklet. It shows a figure of sizes of all *planets* in the same scale, and astronomical data of the planets, the concept of the program, and a map of the shops. A person who collects the stamps with some criteria can get some novelties with prizes after famous astronomers. During the program several science-café talks on astronomy were also held. Through this program shop keepers, public servants, residents, and visitors are stimulated the interest to astronomy. The program can become a tourist attraction and is a good outreach with the widest entrance. We will continue the program for many years.

The official web page is `http://www.taiyokei-walk.jp/`. Please visit there.

References

Handa T., Matsuura K., & Koike K. 2003, in: A. -L. Melchior & R. Ferlet (eds.) *Global Hands-on Universe 2002*, (Frontier Group, ESA), ISBN 2914601107, p. 221

Handa T., Matsuura K., & Koike K. 2008, in: T. Handa & M. Okyudo (eds.) *Global Hands-on Universe 2007*, Frontiers Science Serise 54, (Tokyo: Univsersal Academy Press), ISBN978-4-904164-05-1, p. 209

Highlights of Astronomy, Volume 16
XXVIIIth IAU General Assembly, August 2012
T. Montmerle, ed.

doi:10.1017/S1743921314012721

A Global view of the Eclipse over the Earth (GEE) in 2009 and 2012

T. Handa[1], K. Hata[2], T. Hara[3], T. Horaguchi[4], M. Hiramatsu[5], T. Arai[6], Y. Sato[7] and K. Ohnishi[8]

[1]Depart. Astron. & Phys., Kagoshima Univ., Korimoto 1-21-35, Kagoshima 890-0065, Japan
email: handa@sci.kagoshima-u.ac.jp

[2]Okayama Shoka University Highschool , [3]Toyooka Highschool, Saitama

[4]National Museum of Nature and Science, Tokyo, [5]National Astron. Obs. of Japan

[6]Katsushika City Museum, [7]Hokkaido University, [8]Nagano National College of Technology

A solar eclipse is one of the most popular events in astronomy. Although it is the single astronomical event, it gives different images from place to place and changes in time. It is the most important message from astronomy to the public that an apparent face depends on the viewpoint and we should get the comprehensive view.

A map made of many movies taken from many locations can be helpful to understand what happens actually and to get the comprehensive view of the eclipse. Therefore, we promoted a campaign to take synchronized sequential images of the eclipsing sun and to draw a "big picture" made from them. The campaign was done both in 1997, 2009, and 2012. In 2012 we successfully got many shots in every 5-min interval or more frequently from more than 20 locations over Japan and Taiwan.

From our archive we have made a sample movie of the annular eclipse through Taiwan and Japan, which can be watched via our web site located at `http://milkyway.sci.kagoshima-u.ac.jp/ handa/eclipse/`. Combined with a CG movie watched from the space for instruction, we can make a better talk on the solar eclipse with the comprehensive view.

From our archive we will choose images enough accurate to estimate a geometrical parallax and/or moving velocity of the Moon shadow on Earth (e.g. Handa *et al.* 2000). Using these data, we can make a lesson for students to derive the mass of Earth.

Using the same observation skill and instruments we observed the Venus transit on this June. We coordinated synchronous observations at different locations. A pair of images from two distant sites gives geometrical parallax to estimate 1 AU. At present we have succeeded to get at least one pair from USA and Japan baseline with longer than 8000 km baseline. As a preliminary result we got the distance from Earth to the Sun to be about 1.6×10^8 km. We should offer many available pairs, because redundancy is essential to check the consistency and it is very important for science.

The astronomical unit is the most basic length in astronomy and a good introduction to approach any modern astrophysics (see proceedings of IAU Symp. 289 held in Beijin General Assembly). We will make a textbook for high school students to estimate 1 AU from these images as the next step with the help of GHOU and GTTP members.

Reference

Handa T., Agata H., & Fukushima T. 2000, *Astronomical Herald*, 93, 432

Highlights of Astronomy, Volume 16
XXVIIIth IAU General Assembly, August 2012
T. Montmerle, ed.

doi:10.1017/S1743921314012733

Communicating through Vernacular Media: Scope and Challenges

A. Sule

Homi Bhabha Centre for Science Education, Tata Institute of Fundamental Research, Mumbai, India, email: aniket.sule@gmail.com

India is a country with a large number of languages which not only differ in scripts but are essentially part of different language families. "Marathi" is one such Indian regional language spoken by nearly 70 million people and is the native language of the author. Like all major regional languages, there is a strong and vibrant media in Marathi with 45 odd newspapers and 6 television news channels.

India has a large student population with about half a million graduating with bachelors degree. However, India has only 220 IAU members. Also, in the demographics of professional astronomers, some languages are under-represented either due to cultural differences in communities or due to brain-drain or due to difference in opportunities and geographic distribution of astronomy institutions. In short, we have astronomers, we have press, but no communication between the two groups. Many astronomers don't know how to explain astronomy to non-experts. Further, the media bosses exhibit apathy towards science due to superstitious beliefs prevalent in society as well as fear of lack of viewers for serious content. Most journalists perceive astronomers as snobbish and they lack expertise to make sense of what the astronomer is saying.

The author has been residing for last six years in Mumbai, capital city of Marathi speaking Maharashtra state and home to studios of most major news channels (English, Hindi and Marathi). Astronomy expertise is available for media at TIFR ($\sim$20 astronomers) and a planetarium. However, none of these experts are native Marathi speakers. Thus, media mostly depend on college physics teachers and amateur astronomers. After relocating to Mumbai the author has tried popularising astronomy at every possible opportunity like eclipses, the transit of Venus, India's moon mission and other satellite launches. The author spent about 20 hours on Marathi television channels, wrote more than 25 popular articles in Marathi for newspapers and magazines and gave about 30 popular talks (in Marathi and English) at schools and colleges.

A few key lessons can be learnt from these interactions with media. First, we should remain accessible and approachable and develop personal rapports with journalists. Second, the expert should understand the intended audience. Each media outlet has a target audience in mind and city readers/viewers have different sensibilities than hinterland readers/viewers. Similarly, viewers of vernacular language news channels expect content in a different format than English language news. Further, the choice of format of interaction and choice of media outlet is also important. Live newsbites require more skill, but give you more direct control over content. What you say cannot be re-edited by others. One should take into account past track record of a media outlet on science reporting and if it distorts the truth for sensationalisation. It helps if the expert also has some experience of working with camera/editing. Experts also must remember audiences for local media prefer local cultural context.

To summarize, it is important to reach audience glued to local languages and dialects. A stark pointer of this reality is the fact that most students prefer conversing in their native language rather than language of instructions at the school. Strategies for teaching can be universal but localisation is essential and science communication resources developed in local languages have a possible application in respective migrant communities in developed countries.

Highlights of Astronomy, Volume 16
XXVIIIth IAU General Assembly, August 2012
T. Montmerle, ed.

doi:10.1017/S1743921314012745

Astro Talk in Social Media - Indonesia

A. Yamani and W. Soegijoko

langitselatan, astronomy education and communication media

Abstract. Social media is a new trend in communicating and connecting to people. It is also a good choice to build awareness of astronomy as issues spread easily and quickly, creating hot topics. This paper will analyze the trend of astro talk in Indonesia and hope to inspire astronomers to use social media in raising awareness.

Keywords. astronomy, social media

---

1. Astronomy and Social Media

Indonesia right now is witnessing a high adoption rate of social media where news and issues spread easily among public. As of August 2012, Indonesia has become the 4th largest Facebook user (Socialbakers, 2012) in the world and 5th largest on Twitter (Infographic Labs, 2012). Social media is a platform to communicate but also a tool to introduce astronomy. An astronomer would gain advantages from participating in social media, such as 1) simplifying ideas and give better understanding to the people; 2) giving the correct information and/or deal with misconception directly; 3) to show that science is doable; 4) share scientific result and build trust with the public. Indonesia has 19 clubs and 3 online media (astronomy related) and most of them active. Additionally, there are 5 astronomy Pages and 11 Groups on Facebook from Indonesia. On Twitter, there are 16 astro-tweeps whom actively tweet about their activities and provide basic information. Astronomy has become famous when there is an event such as a meteor shower, eclipses, Venus transit and near the beginning of Ramadhan to determine and decide the Islamic dates by observing the waxing crescent of the Moon. Statistics gathered by PeopleBrowsr in 2012 (PeopleBrowsr, 2012) show that for indonesia 1) Public curiosity is high, 2) People need facts from trusted source, 3) Social media provides a platform to bridge the ivory towers, 4) Social media can be used as collaboration tool between astronomers and amateurs, and 5) (public) Reach becomes unlimited.

Aside from peer reviewed journals and press releases, an astronomer can deliver their findings through social media to build public awareness and begin a conversation with the public. When astronomers talk, 1) Laymen know the forefront of science instantly, 2) feedback is unfiltered and in real time (compared to a newspaper article), 3) affording less miscommunication, 4) increased trust between the scientist and public, as well as closer scrutiny for tax funded projects. 5) public feels closer to the source. Having the astronomers or scientist talk directly with them will inspire the public to learn more about science.

AY would like to thank LKBF and IAU for the grants to attend the 28th IAU GA.

References

PeopleBrowsr 2012. retrieved 28.08.2012 http://www.peoplebrowsr.com/
Socialbakers 2012. retrieved 29.08.2012. http://www.socialbakers.com/facebook-statistics/
Infographic Lab 2012. retrieved 29.08.2012. http://infographiclabs.com/news/twitter-2012/

Highlights of Astronomy, Volume 16
XXVIIIth IAU General Assembly, August 2012
T. Montmerle, ed.

doi:10.1017/S1743921314012757

Communicating astronomy by the Unizul Science Centre

A. Beesham[1] and N. Beesham[2]

[1]Department of Mathematical Sciences, University of Zululand, Kwa-Dlangezwa 3886, South Africa, email: abeesham@pan.uzulu.ac.za

[2]Paul Sykes Primary School, Earlsfield, Durban 4037, South Africa

The University of Zululand, situated along the east coast of KwaZulu-Natal, has a thriving Science Centre (USC) situated in the developing port city of Richards Bay. Over 30 000 learners visit the centre annually, and it consists of an exhibition area, an auditorium, lecture areas and offices. The shows consist of interactive games, science shows, competitions, quizzes and matriculation workshops. Outreach activities take place through a mobile science centre for schools and communities that cannot visit the centre.

The USC offers a programme in astronomy both at the centre as well as outside upon request to nearby areas. This 4 hour show has to be booked in advance. It is usually booked by schools, but also by old age homes and for birthday parties, etc. Roughly 3 shows per month are held. The programme involves looking at the southern skies in an inflatable Starlab planetarium (Zululand's only one) where about 30 people can be accommodated, a tour of the USC, analemmatic sundial, telescopes, optics, history of South African space science and an astronomy quiz game. In addition, one of the following award-winning shows can be selected: the solar system, powers of 10 (distances in astronomy), science of astronomy, from Sputnik to Sumbandila science show, robotics in space or the earth and the sun.

The USC hosts a regional Astro Quiz, which is an astronomy competition run nationally by the South African Agency for Science & Technology Advancement. It is aimed at grade 7 primary school learners, who are sent material and then asked questions. The school with the most points is the winner and then goes on to participate at national level.

A unique feature of our USC is the incorporation of indigeneous knowledge (IK). Mr Mdumiseni Nxumalo, who is in charge of the astronomy shows, uses IK to help develop a better understanding of astronomy. He endeavours to communicate astronomy through an IK approach. Many constellations are named from a northern "western" perspective and in the southern hemisphere make little sense to many people. The USC shows now describe traditional star lore, e.g., the appearance of Pleiades marks the beginning of the ploughing season. Such an IK approach has elicited a positive response from visitors, who now want more traditional astronomy. The USC plans to build a planetarium in the shape of a traditional Zulu beehive hut, which will be unique in the world.

The USC works on the idea that they use only aspects of IK that form part of science, and not myths, etc, extracting relevant aspects that encourage science literacy and scientific thinking. The view is adopted that both science and IK can work together in communicating astronomy (and science in general) more effectively to the public.

Highlights of Astronomy, Volume 16
XXVIIIth IAU General Assembly, August 2012
T. Montmerle, ed.

doi:10.1017/S1743921314012769

One World, One Sky: Outreach in a Multicultural, Multilingual Metropolis

M. Reid

Dunlap Institute for Astronomy and Astrophysics, University of Toronto
email: mike.reid@utoronto.ca

As cities around the world grow more and more diverse, we must take this diversity into account in developing outreach activities and materials. The International Year of Astronomy in 2009 brought a lot of attention to the needs of underserved communities and developing countries, emphasizing the ideal of widespread access to astronomy outreach. Increasingly, however, we find that some of the same challenges facing underserved communities and developing countries are also present in modern metropolises. Conveniently, the linguistic and cultural diversity of our cities is more and more accurately reflected among the astronomy community. The diversity of the astronomical community itself creates opportunities for effective multicultural, multilingual outreach. The Dunlap Institute is located in Toronto, the largest city in Canada and one of the most diverse in the world. The Canadian emphasis on multiculturalism has produced in Toronto a unique blending of the world's cultures. According to *Statistics Canada*, nearly half of Torontonians are foreign-born and furthermore, half of Torontonians speak a language other than English as their mother tongue . Many of those whose first language is not English do not speak English at all, or not well enough to comfortably discuss astronomy in English. We have found that traditional forms of outreach are ineffective for reaching this diverse audience. Lectures and observing nights in English alone, hosted at English-speaking schools, universities, and science centers do not tend to attract a diverse audience.

For our 2012 Transit of Venus celebration, we hosted 6,000 people in a sports stadium at the University of Toronto to watch the transit. We provided outreach materials in ten languages. The original materials were written in English and the translations were sourced from among our own research group. Translations were contributed by graduate students, postdocs, faculty, and staff. We have also begun directing our outreach efforts genuinely outward, taking our events into the diverse communities of Toronto rather than hoping that members of those communities will come to us. We have found that setting up a solar telescope on a busy street corner at lunch hour is a fantastic way to reach a diverse audience and easily attracts as many people in an hour as an on-campus lecture would. When we do sidewalk astronomy of this form, we try to have materials to hand out in multiple languages.

We hope to turn this newfound capacity for outreach in multiple languages and contexts into an ongoing commitment to developing and distributing multilingual, multicultural outreach materials, starting locally and expanding globally.

Highlights of Astronomy, Volume 16
XXVIIIth IAU General Assembly, August 2012
T. Montmerle, ed.

doi:10.1017/S1743921314012770

Astronomy Outreach Adventures in Rural Guatemala

L. Strubbe
Canadian Institute for Theoretical Astrophysics
email: linda@cita.utoronto.ca

Astronomy can be an inspirational gateway to learning science and analytical reasoning, and to careers in STEM fields—particularly important in developing countries where educational opportunities can be scarce. Following this idea and my interest in learning about other cultures, I decided to spend 6 weeks in late 2011 (between Ph.D. and postdoc) doing astronomy public outreach in Guatemala. I volunteered through a Spanish language school embedded in a poor rural community (typical earning $\sim$ \$3/day), working mostly with children. My teaching goals were primarily attitudinal: to encourage people to observe and ask questions about the world around them, and to show them that phenomena have explanations that we can understand. My tools were a Galileoscope, Earth ball, diffraction grating glasses, and pictures of planets and galaxies. People were really excited and curious about astronomy, and it was rewarding for me to get to know people from a different cultural background. I share my experiences to show how a short gap between Ph.D. and postdoc can be taken for doing public outreach, to offer ideas for sharing astronomy in rural underdeveloped areas, and to encourage others in bringing astronomy to educationally disadvantaged parts of the world as well.

Figure 1. 8-year-old Luis (and Linda) with the Earth ball, outside Luis's family's house (in a rural community an hour outside Quetzaltenango, Guatemala). Luis couldn't read at all and had only a very basic understanding of arithmetic, but he loved learning—especially the locations of Guatemala, California and England on the Earth ball, which he proudly showed anyone who walked by down the street.

I have written an article about my experiences in Guatemala published in the American Physical Society's Forum on Education Newsletter, Spring 2012, available at `http://www.aps.org/units/fed/newsletters/spring2012/strubbe.cfm`.

Highlights of Astronomy, Volume 16
XXVIIIth IAU General Assembly, August 2012
T. Montmerle, ed.

doi:10.1017/S1743921314012782

Astronomical Education for public and its future development in Mongolia

R. Tsolmon, V. Oyudari and A. Dulmaa

NUM-ITC-UNESCO Space Science and Remote Sensing Laboratory ,National University of Mongolia email: tzr112@psu.edu, tsolmon91@yahoo.com

International activities for astronomy began when Mongolia joined the IAU at the General Assembly held in Prague in August 2006, because space scientists, astronomers and researchers in Mongolia are coming to understand that astronomy can help Mongolian socioeconomic development. For instance, astronomy can increase general interest and encourage public engagement in the sciences.

Mongolia is the only one of the ancient nomad states to retain the tenets of its original nomadic civilization, including the classic migration of livestock and closeness to nature. Like other Asian nations, the astronomy has been developing since ancient times as a science for Mongols with deep sacral worship of the blue sky and eternal heavens.

Today, Mongolia does not have a planetarium for public viewing. Since the collapse of the communist system due to economic difficulties, little attention has been paid to science and astronomy education. There is a lack of training opportunities for Astronomy amateurs and the public. The educational system of Mongolia has no astronomy textbook in Mongolian language. Astronomy education falls even further behind other developing countries. If we will have planetarium the younger generation as well as the general public will find planetariums fascinating. We hope that IAU will help us to raise awareness by giving us the means to provide a public planetarium. We desperately need more research opportunities and education in astronomy in Mongolia. Mongolia has a climate most conducive for observational astronomy, has a good advantageous to observe moon and stars the air is mostly dry and industrial and light pollution are mainly almost absent. In order to develop astronomy and space science in Mongolia we need many years of cooperation with international educational organizations in Asian region.

For developing graduate education and training programs we should involve international school and activities with help international astronomy community. Taking advantage of Mongolia's dry climate, comparatively less light pollution that promise precise chances of astronomical research and study we aim to bring science of astronomy to new qualitative level of development. Thus it creates a demand for new astronomers in Mongolia.

Highlights of Astronomy, Volume 16
XXVIIIth IAU General Assembly, August 2012
T. Montmerle, ed.

doi:10.1017/S1743921314012794

Australian sites of astronomical heritage

T. Stevenson[1,2] and N. Lomb[1]

[1]Sydney Observatory, Powerhouse Museum, Sydney
email: toner.stevenson@phm.gov.au

[2]The University of Sydney, Faculty of Arts, Museum Studies, NSW, Australia

Abstract. The heritage of astronomy in Australia has proven an effective communication medium. By interpreting science as a social and cultural phenomenon new light is thrown on challenges, such as the dispersal of instruments and problems identifying contemporary astronomy heritage. Astronomers are asked to take note and to consider the communication of astronomy now and in the future through a tangible heritage legacy.

"I was thrilled to be close to the historic 1874 large refracting lens telescope! The modern computer controlled reflecting telescope was interesting although it did not bring me an equal level of excitement as the 1874 telescope" commented a recent visitor to Sydney Observatory. Heritage can be a relevant, humanising and highly effective form of astronomy communication. In Australia astronomical heritage has encountered many obstacles including the inability of research organisations to identify and preserve important instruments. This includes the Yale Columbia Telescope remains from Mt Stromlo used for an artistic work.

Heritage collections are a medium through which scientists can communicate astronomy for the benefit of all. The legacy of astronomical results and methods is not confined to published papers. According to French sociologist, Bruno Latour, it is the instruments, the sites and even the workplaces of scientific endeavour that have agency in the production of the results. Furthermore historic objects have great potential to engage the public by presenting science within humanity through narratives of successes and failures. There is clear evidence for this at Sydney Observatory where bringing back the historic collection has helped to escalate visitor attendances in the past thirty years by a factor of eight.

Nineteenth century astronomers kept notebooks, correspondence, and well-used brass, polished glass and timber instruments. Unfortunately contemporary technology is not easily identifiable as heritage. Artefacts, such as the first Apple Computer and the Image Photon Counting System, from the Anglo Australian Telescope, are both innovative technology artefacts almost discarded in the past. These are two of the 100 527 artefacts on the popular Powerhouse Museum collections database attracting over 900 000 visits per year. In Australia, museums adopt the Significance 2.0 model for assessing collections. Its four primary criteria are: 'the historic; artistic; scientific and social or spiritual values that items and collections have for past, present and future generations.' There is a further four comparative criteria: 'provenance (who used it, where and when), rarity or representativeness, completeness and interpretive capacity'.

Astronomers need to carefully consider how and where their heritage remains will be kept, interpreted and displayed. Although museums cannot keep everything, using a sound methodology they can ensure that astronomy is represented in society. The future communication of astronomy through material culture depends on what astronomers do today.

References: Powerhouse Museum Collections database: http://from.ph/157946; Significance 2.0: http://www.environment.gov.au/heritage/publications/significance2-0/

Highlights of Astronomy, Volume 16
XXVIIIth IAU General Assembly, August 2012
T. Montmerle, ed.

doi:10.1017/S1743921315001660

Summary: Special Session SpS15: Data Intensive Astronomy

A new paradigm in astronomical research has been emerging – "Data Intensive Astronomy" that utilizes large amounts of data combined with statistical data analyses.

The first research method in astronomy was observations by our eyes. It is well known that the invention of telescope impacted the human view on our Universe (although it was almost limited to the solar system), and lead to Keplerfs law that was later used by Newton to derive his mechanics. Newtonian mechanics then enabled astronomers to provide the theoretical explanation to the motion of the planets. Thus astronomers obtained the second paradigm, theoretical astronomy. Astronomers succeeded to apply various laws of physics to reconcile phenomena in the Universe; e.g., nuclear fusion was found to be the energy source of a star. Theoretical astronomy has been paired with observational astronomy to better understand the background physics in observed phenomena in the Universe. Although theoretical astronomy succeeded to provide good physical explanations qualitatively, it was not easy to have quantitative agreements with observations in the Universe. Since the invention of high-performance computers, however, astronomers succeeded to have the third research method, simulations, to get better agreements with observations. Simulation astronomy developed so rapidly along with the development of computer hardware (CPUs, GPUs, memories, storage systems, networks, and others) and simulation codes.

It has been well known that we need to conduct "statistical" analysis among and/or comparisons with various celestial objects to better understand astrophysical processes. However the limited sensitivity and amount of data in the past prohibited us to do so.

The rapid development of computer hardware depends strongly on semiconductor technologies, which, in turn, leads to large sensitive detectors that enabled astronomers to easily survey large sky areas. There are several challenging projects in the world to get large amount of data: ALMA with a few Petabytes of data product per year, LSST with expected product of 200 Petabytes for over ten years, Pan-STARRS that will produce several Terabytes per "night", together with the VISTA, ELT, TMT, SKA, and others. These projects cover a wide range of scientific themes: cosmology, the large-scale structure of the Universe, formation of galaxies, star formation, variable stars, transient phenomena such as the Gamma-ray bursts, small bodies in the solar system, extrasolar planets, life in the Universe, dark matter and dark energy, and others.

Thus a new era of astronomical research utilizing large amounts of data will soon come, and astronomers need to be well-prepared for this new era. Since the data production rate will be 100 to 1,000 times larger than the past, it will be crucial to have a combination of advanced machine learning technologies with immediate access to extant, distributed, multi-wavelength databases. Such an approach is necessary to make these assessments and to construct event notices that will be autonomously distributed to robotic observatories for near-real-time follow-up. Advanced data analyses combined with statistics and data mining will be essential to derive general "rules" and/or "knowledge" on various phenomena in the Universe, as the data volumes will make human inspection and analysis of the data impossible. The most important and exciting astronomical discoveries of the coming decade will rely on research and development in data science disciplines (including data management, access, integration, mining, visualization and analysis algorithms) that enable rapid information extraction, knowledge discovery, and scientific decision support for real-time astronomical research facility operations.

Significant scientific results are expected to be obtained from data-intensive astronomical research in the very near future and beyond under the fourth paradigm in astronomy – Data Intensive Astronomy.

— M. Ohishi, Past President of Comm. 5, October 2012

Summary of the Special Session

Special Session 15: Data Intensive Astronomy was held during the IAU General Assembly in Beijing in eight sessions from August 28 to August 31, 2012. The program began with three excellent review presentations that placed the age of data-intensive astronomy in context, focusing on the optical, radio, and X-ray spectral ranges. Subsequent talks made it abundantly clear that the data deluge is upon us now, from instruments and surveys such as SDSS, LAMOST, ALMA, VISTA, APERTIF, and LOFAR, and that the magnitude of the challenge will only get worse in the future, with, e.g., LSST, the SKA precursors, and SKA itself.

T. Tyson made a prescient observation in his presentation on LSST, in which he described LSST as an integrated survey system. That is to say, the future of large surveys is not about a telescope, detector, or data processing system, but about the complete integration of these subsystems. The data flows and complexities are such that all components must be designed from the outset as a balanced whole (with, we might add, the software and data management components developed in concert with the hardware, not as an under-funded afterthought). We are, he said, entering the age of data abundance and the era of the data-driven astronomer.

The data volumes and the complexities and interdependencies in the data being produced now and in the coming decade necessitate innovation in our software tools and data management infrastructure:

- Algorithmic research and cross-fertilization with other fields such as computer science, statistics, knowledge discovery, and data mining, with clever implementations that defeat the standard scaling constraints.
- Automated and semi-automated classification.
- Common workflow development environments and shared workflows.
- Common data representations; where FITS may no longer be sufficient in terms of data volume or structural complexity, consider the CASA measurement set or the HDF5 formats.
- Access to atomic and molecular databases for spectral line analysis.
- Access to numerical simulations and comparison of simulations to observations, through "virtual observations" that take into account telescope, detector, and possible atmospheric effects.
- Parallel computing, including parallel implementations of large (peta-scale) databases, and adaptations of algorithms to exploit these architectures.

- Server-side and cloud-based computation; moving the algorithm to the data rather than the data to the algorithm.
- GPUs and other innovative computer architectures.
- Efficient visualization of high-dimension data (with server-side and client-side implementations).
- Versatile, global data discovery, access, and interoperability through the Virtual Observatory.

A closing session concerning public outreach and education emphasized how astronomical discoveries and the data underlying them are having a huge societal impact. Millions of students, educators, and members of the general public explore the universe with tools such as World Wide Telescope and Google Sky. Thousands of people contribute to real-world research problems through citizen science and crowd-sourcing initiatives such as the Zooniverse. Such efforts help to promote logical thinking and create an informed society, and an informed society is one that will be more supportive of scientific research in general.

Special Session 15 drew a diverse audience. Members of the SOC recognized a number of attendees, but more important was the number that we did not recognize. The community recognizes the challenges we face and wishes to help in finding solutions.

Special thanks are due to M. Ohishi for his extraordinary efforts as chair of the SOC, to C. Cui (China) for special support with computer networking, and to all of the contributors to the program.

Special Session 15 was sponsored by Commission 5: Astronomical Data and Documentation, which as part of the reorganization of the IAU will move to the new Division B: Facilities, Technologies, and Data Science (D. Silva, president). We believe this is a good move for Comm. 5, and will provide excellent opportunities for dialog amongst facility and instrument developers, data processing experts, data managers, and the virtual observatory.

We must embrace our data-intensive present–and even more-data-intensive future–with energy and enthusiasm, as this is the way new astronomical discoveries will emerge, from planetary science and stellar astrophysics to cosmology and all fields in between.

— R. Hanisch, President of Comm. 5, October 2012

Scientific Organizing Committee

Masatoshi Ohishi (Japan, Chair)
Kirk Borne (United States)
Janet Drew (United Kingdom)
Robert Hanisch (United States)
Melaine Johnston-Hollitt (New Zealand)
Nick Kaiser (United States)
Ajit Kembhavi (India)
Oleg Malkov (Russian Federation)
Bob Mann (United Kingdom)
Raffaella Morganti (Netherlands)
Paolo Padovani (Germany)
Hu Zhan (China Nanjing)

Robert J. Hanisch
Space Telescope Science Institute, 3700 San Martin Drive,
Baltimore, MD 21218, USA
email: hanisch@stsci.edu

Masatoshi Ohishi
National Astronomical Observatory of Japan, 2-21-1,
Osawa, Mitaka, Tokyo, 181-8588, Japan
email: masatoshi.ohishi@nao.ac.jp

Highlights of Astronomy, Volume 16
XXVIIIth IAU General Assembly, August 2012
T. Montmerle, ed.

doi:10.1017/S1743921314012800

Optical Surveys of Galaxies: Past, Present, and Future

Sadanori Okamura

Department of Advanced Sciences, Faculty of Science and Engineering, Hosei University, 3-7-2 Kajino-cho, Koganei city, Tokyo, 184-8584 Japan
email: sadanori.okamura@hosei.ac.jp

Abstract. A brief history is given of wide area optical surveys of galaxies and resulting catalogs, starting from the Shapley-Ames Catalog through POSS and CfA surveys to modern surveys. Scientific impacts of large surveys are described in terms of the complete sample, large homogeneous samples, and new discoveries. Upcoming and future ambitious surveys are also mentioned. A recent review of surveys in various wavelength regions is given by Djorgovski *et al.* (2012).

Keywords. surveys, catalogs, galaxies

1. Early Photometric Surveys and Resulting Catalogs

Shapley-Ames catalog (SA) (Shapley & Ames 1932) is the first magnitude-limited catalog of galaxies based on more or less uniform photographic plates instead of the naked-eye observation. Three essential factors of the survey, i.e., completeness, homogeneity, and wide sky coverage, were explicitly mentioned for the first time in the SA. Another important early survey was the number count of about 210,000 faint galaxies carried out at Lick observatory by Shane & Wirtanen (1954),which led to the first secure measurement of two-point correlation function of galaxy distribution in the form of $\xi(r) \propto r^{-1.8}$ (Totsuji & Kihara 1969; see also Peebles 2012).

Eye-inspection of wide-field photographic plates produced from sky surveys with large Schmidt telescopes led to large galaxy catalogs. Among others, Catalogue of Galaxies and Clusters of Galaxies (CGCG) by Zwicky *et al.* (1961-68) and Uppsala General Catalogue of Galaxies (UGC) by Nilson (1973) are noted. CGCG and UGC had been the major data sources of observational cosmology until recently because of their reasonable completeness. Entries of all the manually produced catalogs in these days are less than $\sim 30,000$, which may reflect the man-power limit.

Three issues of the Reference Catalogues (RC) by de Vaucouleurs and colleagues were the state-of-the-art compilations of galaxy data including 2599 (RC1, 1964), 4364 (RC2, 1976), and 23024 (RC3, 1991) galaxies. Digitization of Schmidt plates by dedicated scanning machines were intensively carried out in 1980s-90s, resulting in far larger catalogs such as APM Catalog which contains 2 million galaxies (Maddox *et al.* 1990).

2. Redshift Surveys and Modern Surveys

Inspired by the pioneering work by Gregory & Thompson (1978), many redshift surveys were carried out in 1980s-90s. CfA-I ($m \leqslant 14.5$) and CfA-II ($m \leqslant 15.5$), representative early wide surveys which used CGCG and UGC as input catalogs, led to the discovery of the large scale structure of the universe (e.g., de Lapparent *et al.* 1986). The CfA survey motivated many wider and/or deeper redshift surveys (e.g., Kontizas 1997).

Modern wide spectroscopic surveys went down to $m \sim 19$ mag, more than three magnitudes deeper than those before 2000. New photometric catalogs deeper than CGCG were necessary as the input catalog for these surveys. APM catalog was used in 2dfGRS which measured redshifts of 220,000 galaxies (Colless *et al.* 2001). In the Sloan Digital Sky Survey (SDSS) which measured redshifts of more than a million galaxies, photometric survey to produce the input catalog and spectroscopic survey based on the input catalog were carried out in parallel using the same 2.5-m telescope (York *et al.* 2000).

3. Impacts of Large Surveys

Large surveys which collected an unprecedentedly large amount of data of unprecedentedly high quality (in terms of accuracy and homogeneity) always made a large impact on astronomy, by allowing us to construct a complete samle free from biases and a statistically large homogheneous sample, and often leading us to new discoveries.

SA and its successor, A Revised Shapley-Ames Catalog by Sandage and Tammann (1981) became the basis of the studies of the structure of the Local Supercluster (Yahil *et al.* 1980) and luminosity function of field galaxies (Binggeli *et al.* 1988). SDSS turned data-poor studies of galaxy properties into precision science. Large high-quality samples from SDSS enabled the detection of extremely weak signals which had been buried in the noise in previous data (*e.g.*, Eisenstein *et al.* 2005). The Next Generation Virgo Survey (Ferrarese 2012) will surpass the Las Campanas Survey 30 years ago (Binggeli *et al.* 1985) by 8 mag. in the integrated magnitude and 3 mag. in the surface brightness.

4. Upcoming and Future Surveys

Astronomy is changing from a data-starving science to a data-rich science. There are a wealth of existing and upcoming digital surveys in optical and near-infrared (see Sec. 4.1 of Djorgovski *et al.* 2012). The Large Synoptic Sky Survey (LSST) is a very ambitious future survey using a dedicated 8-m telescope in Chile, which will produce data of 100 Petabyte scale. Another survey with 8-m Subaru Telescope (HSC Survey; Takada 2010) is about to start in 2013.

References

Binggeli, B., Sandage, A., & Tammann, G. A. 1985 *AJ*, 90, 1681
Binggeli, B., Sandage, A., & Tammann, G. A. 1988, *ARA&A*, 26, 509
Colless, M. M. *et al.* 2001, *MNRAS*, 328, 1039
de Lapparent, V., Geller, M. J., & Huchra, J. P. 1986, *ApJ*, 302, L1
Djorgovski, S. G. *et al.* 2012, in *Astronomical Techniques, Software, and Data*, H. Bond (ed.)
Eisenstein, D. *et al.* 2005, *ApJ*, 633, 560
Ferrarese, L. *et al.* 2012, *ApJS*, 200, 4
Gregory, S. A. & Thompson, L. A. 1978, *ApJ*, 222, 784
Kontizas, E., *et al.* (eds.) 1997, *Wide-Field Spectroscopy, Astrophys. Space Sci. Library* (Kluwer)
Maddox, S. J. *et al.* 1990, *MNRAS*, 246, 433
Peebles, P. J. E. 2012, *ARA&A*, 50, 1
Shane,C. D. & Wirtanen, C. A. 1954, *AJ* 59, 285
Shapley, H. & Ames, A. 1932, *HCO Bulletin No.887, p.1; Annals of HCO, vol. 88, p.41*
Takada, M. 2010, *AIP Conf. Ser.*, 1279, 120
Totsuji, H. & Kihara, T. 1969, *PASJ* 21, 221
Yahil, A., Sandage, A., & Tammann, G. A. 1980, *ApJ*, 242, 448
York, D. G. *et al.* 2000, *AJ*, 120, 1579

Highlights of Astronomy, Volume 16
XXVIIIth IAU General Assembly, August 2012
T. Montmerle, ed.

doi:10.1017/S1743921314012812

Radio Surveys: an Overview

Raffaella Morganti[1,2]

[1] Netherlands Institute for Radio Astronomy, Postbus 2, 7990 AA, Dwingeloo, The Netherlands
email: morganti@astron.nl

[2] Kapteyn Astronomical Institute, University of Groningen, Postbus 800, 9700 AV Groningen, The Netherlands

Abstract. Radio astronomy has provided important surveys that have made possible key (and sometimes serendipitous) discoveries. I will briefly mention some of the past continuum and line (H I) radio surveys as well as new, on-going surveys and surveys planned for the near future. This new generation of large radio surveys is bringing extra challenges in terms of data handling but also great new possibilities thanks to the wider range of data products that they will provide.

Keywords. surveys, radio continuum: galaxies, radio lines: galaxies

---

Surveys (and deep fields) have played an extremely important role in radio astronomy. Now we are at a turning point, entering a new era for radio surveys. This is due to the new technologies developed for the upgrade of existing facilities (e.g. VLA, WSRT) and for the new facilities (e.g. LOFAR, ASKAP, Meerkat) on the way to SKA. Systems with wide field-of-view and/or broad bandwidth will soon allow larger and deeper surveys, in some cases providing simultaneously radio continuum and line (H I) data. In summary, the new generation of radio surveys will deliver a broader variety of products and give the chance for serendipitous discovery (see e.g. Kellerman *et al.* 2009). These surveys are starting to revolutionize the field and will provide exciting databases for the community.

Some of the characteristics of the available radio continuum surveys (and deep fields) are illustrated in Fig. 1 and discussed in more detailers in Norris *et al.* (2012 and ref. therein). For the largest continuum surveys, the number of sources detected ranges from 1.7×10^6 for NVSS at 1.4 GHz to $\sim 10^5$ for low-frequency surveys like SUMSS, WENSS and VLSS.

The "blind" H I surveys have been carried out so far with single-dishe instruments, thus providing limited spatial and morphological information. Also the sensitivity of these surveys is limited, thus being able to trace H I only in the local Universe (see also Fig. 2). One of the main "blind" H I surveys is the H I Parkes All Sky Survey (HIPASS) that covers $\sim 2/3$ of the sky, out to 12700 km/s ($z \sim 0.04$) with a spatial resolution of ~ 15 arcmin and rms noise of 13 mJy/b/ch. This surveys had about 5300 extragalactic detections of H I emission. The other large H I survey is the Arecibo Legacy Fast ALFA (ALFALFA, Giovanelli *et al.* 2005) covering 7000 deg^2 up to $z \sim 0.06$, with a spatial resolution of ~ 3 arcmin and rms noise of 1.6 mJy/ch. About 30000 extragalactic detections H I emission are expected when the survey will be completed. A number of smaller samples - a few hundred objects - have been studies in more details (WHISP, THING, ATLAS3D etc.), providing important information about the morphology and kinematics of the H I and allowing major step forward in the study of ISM in nearby galaxies. All together, only ~ 100 objects have been studied in H I emission above $z = 0.1$.

In the near future, a number of new (continuum and line) surveys will be carried out and they will partly overcome some of the limitations related to spatial resolution, sensitivity and frequency coverage.

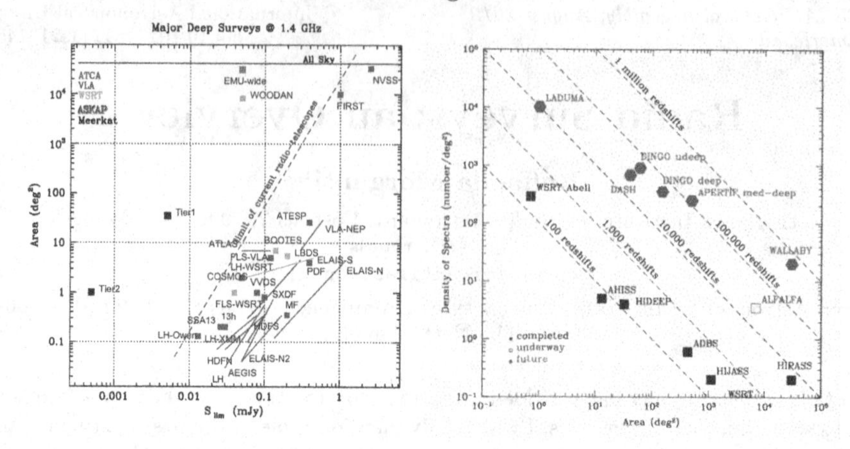

Figure 1. *Left* - Summary of the sensitivity and area covered by radio continuum surveys (and deep fields) done so far (from Norris *et al.* 2012, kindly provided by I. Prandoni). *Right* - Summary of the number of objects versus the area for H I surveys (Staveley-Smith *et al.* 2012, http://www.astron.nl/ewass2012/Programme.html/Staveley-Smith.pdf).

At low frequencies, LOFAR and GMRT are expecting to deliver very interesting databases. The TIFR GMRT Sky Survey (TGSS) survey at 150 MHz covers about 32,000 sq. deg north of declination $-30°$ and is reaching an rms noise of 7-9 mJy/beam at an angular resolution of about 20 arcsec. The survey is expected to detect more than 2 million sources (http://tgss.ncra.tifr.res.in). For LOFAR, in addition to the planned large and deep surveys (Rottgering *et al.* 2011), the LOFAR Multifrequency Snapshot Sky Survey (MSSS, Heald *et al.* 2011) is expected to provide fluxes between 30 and 180 MHz for sources in the cover the northern sky and reach an rms noise of ~ 10 mJy/b, albeit with poor spatial resolution. Continuum all-sky surveys at 1.4 GHz from the SKA pathfinders/precursors at 20-cm (ASKAP/Apertif) are estimated to detected (Norris *et al.* 2011) of the order of 100 million sources (for rms 10 μJy over the entire sky, see Fig. 1).

ASKAP and Apertif together will survey H I over the entire sky in an uniform way (Fig. 2). The expected number of detections of H I emission is of the order of 10^6 galaxies (out to redshift $z \sim 1$ for H I absorption). In addition, most of these H I detections will be spatially resolved. This will provide new possibilities for the science that will be done. The other major improvement will be the synergy between continuum and line surveys in the 20-cm band. It will be possible, by exploiting the broad band of the new receivers, to extract continuum and line information and study simultaneously e.g. the properties of the non-thermal activity and of the gas. One survey will fit all! This is bringing major challenges in the handling of the many products that these surveys will provide but it also promises to offer a amazing database for the community to exploit.

References

Barnes D. G., *et al.* 2001, *MNRAS*, 322, 486
Giovanelli R., *et al.* 2005 AJ 130, 2598
Heald G., *et al.* 2011, *JApA*, 32, 589
Johnston S., *et al.* 2008, ExA, 22, 151
Kellermann K., *et al.* 2009, PoS 99, 5
Norris R. P., *et al.* 2012, arXiv, arXiv:1210.7521
Oosterloo T., *et al.* 2009, "SKA 2009", PoS 132, 70
Röttgering H., *et al.* 2011, *JApA*, 32, 557

Highlights of Astronomy, Volume 16
XXVIIIth IAU General Assembly, August 2012
T. Montmerle, ed.

doi:10.1017/S1743921314012824

The Role of Wide Field X-ray Surveys in Astronomy

Richard D. Saxton

Telespazio VEGA / XMM SOC, ESAC, Apartado 78, 28691 Villanueva de la Cañada, Madrid, Spain
email: richard.saxton@sciops.esa.int

Abstract. We review the history of X-ray sky surveys from the early experiments to the catalogues of 10^5 sources produced by ROSAT, Chandra and XMM-Newton. At bright fluxes the X-ray sky is shared between stars, accreting binaries and extragalactic sources while deeper surveys are dominated by AGN and clusters of galaxies. The X-ray background, found by the earliest missions, has been largely resolved into discrete sources at soft (0.3-2 keV) energies but at higher energies an important fraction still escapes detection. The possible identification of the missing flux with Compton-thick AGN has been probed in recent years by Swift and Integral.

Variability seen in objects observed at different epochs has proved to be an excellent discriminator for rare classes of objects. The comparison of ROSAT All Sky Survey (RASS) and ROSAT pointed observations identified several Novae and high variability AGN as well as initiating the observational study of Tidal Disruption events. More recently the XMM-Newton slew survey, in conjunction with archival RASS data, has detected further examples of flaring objects which have been followed-up in near-real time at other wavelengths.

Keywords. X-rays: general, surveys

---

1. Introduction

In common with other wavelengths, X-ray surveys have two broad goals (i) to construct large samples to understand the global properties of a class of objects (ii) to find rare objects which help to fill in the details. X-ray astronomy started in 1962 with the launch of an experiment aboard an Aerobee 150 sounding rocket (Giacconi *et al.* (1962)). The discovery of strong emission from a point source (SCO X-1) and the cosmic X-ray background (CXB), by this mission, were highly influential in shaping the goals of future missions. Subsequent high-energy surveys of the galactic plane (e.g. by Integral (Winkler *et al.* (2003)) or MAXI (Matsuoka *et al.* (2009))) have now found ~100 Low-Mass X-ray binaries (LMXB), of which SCO X-1 is the brightest member, which can vary their output by two orders of magnitude in a matter of weeks. The CXB was later shown to be a summation of extragalactic point sources (e.g. Setti & Woltjer (1973)). Many missions have investigated their properties; ASCA, Beppo-Sax and Einstein resolved 25-30%, ROSAT resolved 75% in soft X-rays while deep surveys by XMM-Newton and Chandra increased this to 80-90% (Brandt & Hasinger (2005) and references therein). At soft X-ray energies the sky is dominated by unobscured AGN while at higher energies the contribution from the more numerous self-absorbed AGN becomes increasingly important. Synthesis models fitting the CXB can be used to constrain the evolution of AGN and predict a significant fraction of highly absorbed (Compton-thick) objects (e.g. Gilli, Comastri & Hasinger (2007)). Deep surveys revealed a luminosity-dependent evolution of number density, with high-luminosity QSO peaking at z=2-3 while lower luminosity Seyfert galaxies are more prevalent at z< 1 (La Franca *et al.* (2005), Hasinger, Miyaji &

Schmidt (2005)). The elusive Compton thick population has begun to be revealed in the Swift BAT (Burlon *et al.* (2011)).

2. Rare objects

Large surveys allow rare objects to be found from their X-ray colours, time signatures or celestial location. Seven isolated neutron stars were identified from their soft X-ray spectrum and faint optical companion in the RASS (Voges *et al.* 1999). All belong to the Gould belt at a distance of 100–500 pc (Haberl (2006)). Transient events in Galactic nuclei led to the discovery by ROSAT of Tidal Disruption events (see Komossa 2002 for a review) which have also been found in small numbers in the XMM-Newton slew survey (Esquej *et al.* (2007), Saxton *et al.* (2012)). Ultra Luminous X-ray sources (ULX) with $L_X > 10^{39}$ ergs s^{-1} have been found in the outskirts of a few hundred galaxies in the XMM-Newton serendipitous survey (2XMM; Watson *et al.* (2009)). A selection of these which may harbour intermediate mass black holes ($10^2 < M_{BH} < 10^5 M_\odot$) have been proposed by Walton *et al.* (2011). The most extreme of these, HLX-1, with $L_X = 10^{42}$ ergs s^{-1}, appears to have $M_{BH} \sim 10^3 M_\odot$ (Farrell *et al.* (2009)). Coming from the other end, a small number of AGN with $10^4 < M_{BH} < 10^5 M_\odot$ have also been found with XMM-Newton from their very soft X-ray spectra (e.g. Terashima *et al.* (2012)).

3. Future prospects

eRosita (Predehl *et al.* (2010)), has the potential to revolutionise wide-field X-ray surveys in the same way that ROSAT did in the 1990s. It is expected to see more than 10^6 AGN and 10^5 clusters of galaxies. Tight constraints on cosmological and dark energy parameters may be available from cluster counts. In addition, transient detection will likely open up new research areas. For example, the mission should detect ~ 1 tidal disruption event each week. At higher energies, NuSTAR (Harrison *et al.* (2010)) and HXMT will provide new constraints on highly absorbed AGN.

References

Brandt, W. & Hasinger, G. 2005, *Annu. Rev. of Astron. and Astroph.* 43, 827
Burlon, D., Ajello, M., Greiner, J., Comastri, A., Merloni, A., & Gehrels, N. 2011, *ApJ* 728, 58
Esquej, P., Saxton, R., Freyberg, M., *et al.* 2007, *A&A* 462, 49
Farrell, S., *et al.* 2009, *Nat.* 460, 73
Giacconi, R., & Gursky, H., Paolini, F., & Rossi, B. 1962, *Phys. Rev. Lett.* 9, 439
Gilli, R., Comastri, A., & Hasinger, G. 2007, *A&A* 463, 79
Haberl, F. 2006 *Space Sci.* 308, 181
Harrison, F., *et al.* 2010, *SPIE* 7732, 27
Hasinger, G., Miyaji, T., & Schmidt, M. 2005, *A&A* 441, 417
Komossa, S. 2002, *RvMA* 15, 27
La Franca, F. 2005, *ApJ* 635, 864
Matsuoka, M., *et al.* 2009 *PASJ* 61, 999
Predehl, P., Boehringer, H., Brunner, H., *et al.* 2010 *SPIE* 7732, 23
Saxton, R., Read, A., Esquej, P., *et al.* 2012, *A&A* 541, 106
Setti, G. & Woltjer, L. 1973, *IAU Symposium No. 55* 208
Terashima, Y., Kamizasa, N., Awaki, H., Kubota, A., & Ueda, Y 2012, *ApJ* 752, 154
Voges, W., Aschenbach, B., Boller, T., *et al.* 1999, *A&A* 349, 389
Walton, D., Roberts, T., Mateos, S., & Heard, V. 2011, *MNRAS* 416, 1844
Watson, M., *et al.* 2009, *A&A* 493, 339
Winkler, C., Courvoisier, T., Di Cocco, G., *et al.* 2003 *A&A* 411, L1

Highlights of Astronomy, Volume 16
XXVIIIth IAU General Assembly, August 2012
T. Montmerle, ed.

doi:10.1017/S1743921314012836

LAMOST and China-VO

Yongheng Zhao

National Astronomical Observatories, Chinese Academy of Sciences, Beijing 1000012, China
email: yzhao@bao.ac.cn

Abstract. LAMOST is a special reflecting Schmidt telescope used to observe ten million spectra of celestial objects. There are about half million spectra released in the pilot survey of LAMOST. In the "Big Data" era of astronomy, Virtual Observatory will play an important role to make use of those massive spectral data of LAMOST.

Keywords. telescopes: LAMOST, techniques: spectroscopic, astronomical data bases: virtual observatory

1. LAMOST

The Large Sky Area Multi-Object Fiber Spectroscopic Telescope (LAMOST) is a special reflecting Schmidt telescope (Cui, Zhao, Chu *et al.* 2012). LAMOST breaks through the bottleneck of the large scale spectroscopic survey observation with both large aperture (effective aperture of 3.6m to 4.9m) and wide field of view (5 degrees). It is an innovative active reflecting Schmidt configuration achieved by the active optics technique. The focal surface has precisely positioned 4000 fibers connected to 16 spectrographs with the distributive parallel-controllable fiber positioning system, and thus the LAMOST can observe up to 4000 objects simultaneously. LAMOST is the telescope of the highest spectrum acquiring rate.

LAMOST project was approved as a national major science project by Chinese government in April 1997. LAMOST started to be constructed in September 2001. The installation of all LAMOST subsystems was completed in August 2008. The telescope is located in the Xinglong station of National Astronomical Observatories, Chinese Academy of Sciences. After the two year commission period from 2009, it did a pilot spectroscopic survey with LAMOST from October 2011 to June 2012 (Zhao, Zhao, Chu *et al.* 2012).

More than 480,000 spectra of objects are being released to Chinese astronomers and international collaborators (Newberg, Carlin, Chen *et al.* 2012). About 320,000 high quality spectra of the pilot survey data was released to public in August 2012 (Luo, Zhang, Zhao *et al.* 2012).

The key scientific goals of LAMOST including: (1) the extragalactic spectroscopic survey for the large scale structure of the Universe and the physics of galaxies; (2) the stellar spectroscopic survey on the structure of the Galaxy; (3) the cross identification of multi-waveband surveys (Zhao 1999). The spectroscopic survey carried out by the LAMOST of near ten millions of galaxies and quasars will make substantial contribution to the study of extragalactic astrophysics and cosmology, such as the large scale structure of the Universe, the baryon acoustics oscillations, the dark energy and dark matter, the formation and evolution of galaxies, the accretion process of the massive black holes in the active galactic nuclei, and so on. Its spectroscopic survey of near ten millions of stars will make substantial contribution to the study of the stellar astrophysics and the structure of the Galaxy, such as the spheroid substructure of the Galaxy, the galactic gravitational potential and the distribution of the dark matter in the Galaxy, the extremely metal poor stars and hypervelocity stars, the 3D extinction in the Galaxy, the structure of thin and thick disks of the Galaxy, and so on. Its spectroscopic survey combining with the surveys in other wavebands, such as radio, infrared, ultraviolet, X-ray and gamma-ray, will make substantial contribution to the cross-identification of multi-waveband of celestial objects.

2. China-VO

Virtual Observatory (VO) is a data-intensively on-line astronomical research and education environment, taking advantages of advanced information technologies to achieve seamless, global access to astronomical information. VO aims to make multi-wavelength science and large database science as seamless as possible, and it will act as a facility class data infrastructure (Lawrence 2009).

The organization to coordinate global development on the VO and cultivate its awareness and usage is the International Virtual Observatory Alliance (IVOA), which has 17 members currently. It emphasizes the role of VO for future astronomers and requires all new projects data to be VO-compliant.

Chinese Virtual Observatory (China-VO) is the national VO project in China initiated in 2002 and as a national member of IVOA. The China-VO is focusing on developing a platform which could be used for unified access to on-line astronomical resources and services, VO-ready projects and facilities, VO-based astronomical research activities, and VO-based public education (Cui & Zhao 2004) .

Now China-VO is the data access platform of LAMOST data archive.

3. Summery

The spectroscopic survey of LAMOST will obtain ten million spectra. To rise the scientific productivity of LAMOST, it needs to develop advanced methods such as spectral fitting, photometric search, catalogs cross-matching and data mining technology to identify unknown relations and patterns in large-scale data and to discover new classes of extremely rare astrophysical objects (Zhang & Zhao 2006; Peng, Zhang, Zhao, Wu 2012). And the data archive of LAMOST could be used as a good scientific resources for data archiving, data fusion, data mining, machine learning with the VO technology and methods of astroinformatics.

References

Cui, C. & Zhao, Y. 2004, *Public. of NAOC* 1(3), 203

Cui, X., Zhao, Y., Chu, Y. *et al.* 2012, *Research of Astron. & Astrophys.* 12,1197

Lawrence, A. 2009, in: Simpson R.J., Ward-Thompson. Length D. (eds.), *Astronomy: Networked Astronomy and the New Media*

Luo, A., Zhang, H., Zhao, Y. *et al.* 2012, *Research of Astron. & Astrophys.* 12,1243

Newberg, H. J., Carlin, J. L., Chen, L., *et al.* 2012, *ASPC* 458, 405

Peng, N., Zhang, Y. , Zhao, Y., & Wu, X. 2012, *MNRAS* 425, 2599

Zhang, Y. & Zhao, Y. 2006, *ASPC* 351, 173

Zhao, Y. 1999, *Public. of Yunnan Obs.* Suppl., p. 1

Zhao, G., Zhao, Y., Chu Y., *et al.* 2012, *Research of Astron. & Astrophys.* 12, 723

Highlights of Astronomy, Volume 16
XXVIIIth IAU General Assembly, August 2012
T. Montmerle, ed.

doi:10.1017/S1743921314012848

Taming the ALMA Data Avalanche

Felix Stoehr

ESO, Karl-Schwarzschild-Str 2, 85748 Garching, Germany
email: fstoehr@eso.org

Abstract. The Atacama Large Millimeter/submillimeter Array (ALMA) is nearing its phase of full operations. ALMA will collect about 200TB/year of astronomical data which will be reduced by an automatic pipeline and turned into fully calibrated science-ready data products. We present design choices, challenges and solutions from data capturing over data reduction and data distribution to archival research, that allow to deal with the large amounts of data and, hopefully, achieve the maximum amount of science return.

Keywords. Data intensive astronomy, ALMA, data management, archive, pipeline

---

1. Introduction

With its 66 antennas in the Atacama Desert in Chile and with baselines up to 16 km, ALMA will allow an unprecedented view of the the mm/sub-mm sky. Currently Early Science observations are underway, while the array continues to be completed. The start of the phase of full operations is expected to be in the second half of 2013. Out of the many challenges that had to be faced, we will present here those related to the large amounts of data that ALMA will produce.

2. Data Challenges

2.1. *Data rate*

Based on the predicted data-rate requirements of the expected type of science observations, the official ALMA data rate has been set to 200TB/yr (6MB/s) with peak values reaching 10 times these values for extended periods. These values have been recently revised using experience from the first Early Science observations (Lacy & Halstead, 2012) concluding that the expected average will be larger, up to 700TB/yr (23MB/s). The ALMA data flow system was designed with scalability in mind and was build without hard bottlenecks out of commodity hardware, so that this new data rate and future data rate increases can be coped with.

2.2. *Data storage*

Once written out by the ALMA correlator, the data is stored following an "All data taken have to be stored forever" policy. The storage solution is required to be cost-effective, scalable, future-proof, PB-scale, allow for high read/write performance despite large variation in file size, contain a built-in file-management layer, do automatic consistency checks and support globally distributed archives. NGAS, the Next Generation Archive System (Wicenec & Knudstrup, 2007) developed at ESO was chosen. It is is a combined hardware-software solution, is portable, provides a powerful plugin architecture, uses commodity hardware and allows for online processing capabilities.

2.3. *Archive copies*

The main ALMA archive is located in Santiago in Chile. Three full copies of that archive exist, one at each of the three ALMA Regional Centres (ARC): in Charlottesville at NRAO in USA, in Mitaka at NAOJ in Japan and in Garching at ESO in Germany. The copies serve as backup of the main archive and allow the ARCs to provide user-support, reprocessing, quality control, as well as phase 3 work. The data transfer happens over the network through VPN tunnels.

2.4. *Operation*

ALMA is a general purpose telescope with standard calls for proposals once a year. Also, ALMA will support a large variety of different types of observations (continuum, line, spectral sweep, mosaic, solar) with different observing settings (12m array/ACA/TP) in different configurations). The choice was to create a fully-integrated data-flow system with the Archive in the centre. A sophisticated observing tool and the ALMA Science Data Model (ASDM) have been developed in order to cope with this complexity.

2.5. *Pipeline*

The amount of data of a single project can easily exceed the data-reduction resources available to a typical user today. A fully automatic pipeline is therefore required for ALMA delivering science-grade products (a first in radio astronomy). The policy is that all ALMA science data will be reduced by the project. Indeed, as telescopes will become more complex and deliver more data, the pressure on data providers will rise to deliver science-grade products. An analysis software package CASA (Common Astronomy Software Applications) was written (nearly) from scratch. A pipeline (heuristics + CASA) was developed. Commissioning of the pipeline is currently ongoing.

2.6. *Support*

If users want to re-analyze the data e.g. an optimized reduction, for their particular science case, they need help. ALMA is meant to be a telescope that can be used by all astronomers, not only those that have a radio background. The choice was therefore, to provide extensive support to users including face-to-face support. This support is provided by the three ARCs and is already now in the early stages of the project highly appreciated.

3. Summary

High data rates and the long-term nature of ALMA are a real challenge and a huge effort was deployed at all ends to build a system that can handle the data intake. Hard and software solutions have been built to be scalable and flexible to allow to adapt to change. The goal of ALMA is to help the scientists wherever possible from proposal preparation over the science-grade data products and data-reduction to archival research and to work towards a great end-to-end user-experience.

Acknowledgements

The ALMA construction team deserves all the credit for the work presented here.

References

Lacy M. & Halstead D. 2012, *NAASC Memo* 110
Wicenec A. & Knudstrup J. 2007, *The Messenger* 129, 27

Highlights of Astronomy, Volume 16
XXVIIIth IAU General Assembly, August 2012
T. Montmerle, ed.

doi:10.1017/S174392131401285X

LSST Data Management: Entering the Era of Petascale Optical Astronomy

Mario Juric[1] and Tony Tyson[2]

[1]Large Synoptic Survey Telescope,
950 North Cherry Ave, Tucson, AZ 85719,
email: mjuric@lsst.org

[2]University of California, Davis,
Physics Department, University of California, One Shields Avenue, Davis, CA 95616 USA
email: tyson@physics.ucdavis.edu

Abstract. The Large Synoptic Survey Telescope (LSST; Ivezic *et al.* 2008, http://lsst.org) is a planned, large-aperture, wide-field, ground-based telescope that will survey half the sky every few nights in six optical bands from 320 to 1050 nm. It will explore a wide range of astrophysical questions, ranging from discovering killer asteroids, to examining the nature of dark energy. LSST will produce on average 15 terabytes of data per night, yielding an (uncompressed) data set of 200 petabytes at the end of its 10-year mission. Dedicated HPC facilities (with a total of 320 TFLOPS at start, scaling up to 1.7 PFLOPS by the end) will process the image data in near real time, with full-dataset reprocessing on annual scale. The nature, quality, and volume of LSST data will be unprecedented, so the data system design requires petascale storage, terascale computing, and gigascale communications.

1. Introduction

By visiting each patch of 18,000 square degrees of sky over 800 times in pairs of 15 sec exposures, LSST will explore a wide range of astrophysical questions, from studies of the Solar System, to examining the nature of dark energy, and will revolutionize time domain astrophysics. LSST is an integrated survey system. The observatory, telescope, camera and data management systems will be built to conduct the LSST survey and will not support a 'PI mode' in the classical sense. Instead, the ultimate, science-enabling, deliverable of LSST will be the fully reduced data – catalogs and images. The machine learning algorithms developed for automated discovery will find wide application.

2. LSST Data Products

The data management challenge for the LSST Observatory is to provide fully-calibrated public databases to the user community to support the frontier science expected of LSST (LSST Science Book, 2009), while simultaneously enabling new lines of research not anticipated today. The nature, quality, and volume of LSST data will be unprecedented, so the data management system (DMS) design features petascale storage, terascale computing, and gigascale communications. The computational facility and data archives of the LSST DMS will rapidly make it one of the largest and most important facilities of its kind in the world (see Table 1). New algorithms will have to be developed and existing approaches refined in order to take full advantage of this resource, so "plug-in" features in the DMS design and an open data/open source software approach enable both science and technology evolution over the decade-long LSST survey.

Table 1. LSST Data Management Computing System Size

Quantity	Size	Comment
Cumulative Image Archive	345 PB	Total over all Data Releases, including virtual data
Cumulative Catalog Archive	46 PB	Total over all Data Releases, incl. database indices
Final Image Collection	75 PB	In final data release (DR 11), including virtual data
Final Database	32 trillion rows (9 PB)	In final data release (DR 11)
Final Disk Storage	228 PB (3700 drives)	Archive Site only (at NCSA)
Final Tape Storage	83 PB (3800 tapes)	Single copy only
Number of Nodes	1800	Archive Site, includes both compute and database nodes
Alerts Generated	6 billion	Alerts generated over the life of the survey

LSST Data and Computing at a Glance: The sizes of various components of LSST data management systems and data products. "Virtual data" is data that is dynamically recreated on-demand from provenance information.

LSST will deliver or enable three different classes of data products (LSST Science Requirements Document, 2011):

- *Level 1* ("nightly") data products will be generated continuously every observing night and include measurements such as alerts to objects that have changed brightness or position. They will be broadcast world-wide using VO protocols.
- *Level 2* ("yearly") data products will be made available as annual Data Releases and will include images and measurements of quantities such as positions, fluxes, and shapes, as well as variability information such as orbital parameters for moving objects and an appropriate compact description of light curves. The exact contents of Level 2 products will be set by the desire to minimize the necessity to independently reprocess the image data.
- The LSST will enable the creation of *Level 3* ("user-created") data products, by making available approximately 10% of its computing capability to the community. These "community cycles" will be used to perform custom analyses not fully enabled by Level 1/2, taking advantage of co-location of computation with the entire LSST data set.

3. Open Data, Open Source – Community Resource

All LSST research will be done by the community using LSST data products, and not by the LSST Project. In particular, LSST data products, including images and catalogs, will be made available with no proprietary period to the astronomical communities of the United States, Chile, and to LSST's international partners. Transient alerts will be made available for world-wide distribution within 60 seconds, using standard VO protocols.

LSST data processing stack, including the image processing stack, pipeline middleware, and a highly distributed database, will be free software. Highly advanced prototypes are already available at http://dev.lsstcorp.org/cgit. A novel database architecture is being developed, enabling fast efficient search in space and time.

Acknowledgements

The authors would like to acknowledge the support of the National Science Foundation under Grant No. AST-1227061, *"LSST - Concept & Development"*."

References

Ivezic, Z., Tyson, J. A., Acosta, E., *et al.* 2008, arXiv:0805.2366
LSST Science Collaboration, Abell, P. A., Allison, J., *et al.* 2009, arXiv:0912.0201
LSST Science Requirements Document, http://www.lsst.org/files/docs/SRD.pdf

Highlights of Astronomy, Volume 16
XXVIIIth IAU General Assembly, August 2012
T. Montmerle, ed.

doi:10.1017/S1743921314012861

Data Intensive Radio Astronomy en route to the SKA: The Rise of Big Radio Data

A. R. Taylor

Department of Physics and Astronomy, University of Calgary, 2500 University Dr. N.W., Calgary, Alberta, Canada T2N 1N4
email: russ@ras.ucalgary.ca

Abstract. Advances in both digital processing devices and in technologies to sample the focal and aperture planes of radio antennas is enabling observations of the radio sky with high spectral and spatial resolution combined with large bandwidth and field of view. As a consequence, survey mode radio astronomy generating vast amounts of data and involving globally distributed collaborations is fast becoming a primary tool for scientific advance. The Square Kilometre Array (SKA) will open up a new frontier in data intensive astronomy. Within the next few years SKA precursor telescopes will demonstrate new technologies and take the first major steps toward the SKA. Projects that path find the scientific journey to the SKA with these and other telescopes are currently underway and being planned. The associated exponential growth in data require us to explore new methodologies for collaborative end-to-end execution of data intensive observing programs.

Keywords. methods: analytical, instrumentation: miscellaneous, surveys

---

1. Introduction

The first decade of this century has seen a tremendous advance in information and digital technologies which have driven a commensurate advance in the data capacities of radio telescopes. The instantaneous observing bandwidths of both single-dish and array radio telescopes have increased by two orders of magnitude. High dynamic range imaging over wide fields with radio array antennas, and the need to mitigate narrow-band radio frequency interference has driven observing programs to high resolution spectral channelization over broad continuum bandwidths. In combination with the ultra-wide instantaneous fields of view now provided by focal plane horn arrays and phased-array feeds, and by many-element aperture plane arrays, the data rates to the observer sustained by current observing programs are $10^3 - 10^4$ times larger than typical only a few years ago.

Many of our most pressing astrophysical questions require synthesis of information covering large areas of the sky and over a significant range of cosmic history. Survey mode observations, in which large amounts of observing time are devoted to major programs that create vast data sets is becoming an essential approach to observational astronomy at both radio and optical wavelengths. Survey mode observing combined with the new instrumental data capacities is driving an exponential growth in both the rate and the volume of data in radio astronomy. Survey projects dominate the science program for the SKA precursor telescopes in Australia (ASKAP) and South Africa (MeerKAT). Figure 1 illustrates this data trend by showing data output rates and volumes for major observing programs over the course of the past decade and projected into the future en route to the Square Kilometre Array.

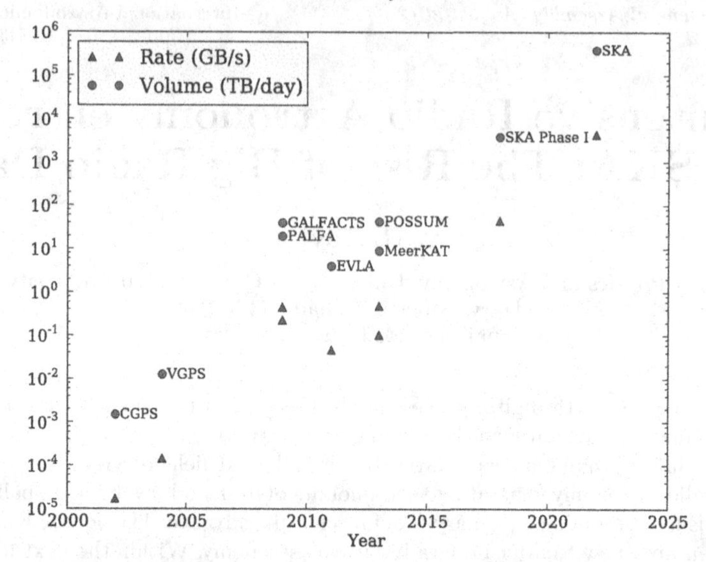

Figure 1. The rise of big radio data. The output data rates to the observer are shown for several survey mode observing projects starting with the Canadian Galactic Plane Survey (Taylor *et al.* 2003) to the SKA post 2020. The data rate is increasing on average by 150% per year.

2. Cyber Solutions to Data Intensive Radio Astronomy

Survey mode observing is changing the way that science programs are executed and creating new challenges associated with management of the very large data volumes produced, the complexity of processing and analysis of data that may support multiple scientific goals, and collaborative execution of science by large, globally distributed teams.

As we move into the era of the SKA, these challenges may be met with cyber infrastructure platforms that wed emerging web technologies with distributed global resources of cloud-enabled high performance computing and smart data infrastructure. The cyberSKA initiative (Kiddle *et al.* 2011, Grimstrup *et al.* 2012) is a research project to develop a scalable and distributed cyber infrastructure platform to meet the evolving needs of data intensive radio astronomy en route to the SKA. A collaboration between government agencies, universities and industries, cyberSKA is being co-developed with SKA pathfinding survey programs in time-domain and imaging astronomy with Arecibo, the JVLA and GMRT. The platform is accessed via a web portal (www.cyberska.org) and is building e-science infrastructure for collaboration, smart distributed data management, interface to distributed computing for data processing, data visualization and analytics for very large data sets, and an API for community contributed applications that interface to the collaboration metadata and the distributed data system.

References

Taylor, A. R. *et al.* 2003, *AJ*, 125, 3145.

Grimstrup, A., Mahadevan,V., Eymere, O., Anderson, K., Kiddle, C., Simmonds, R., Rosolowsky, R. & Taylor, A. R. 2012, in *Software and Cyberinfrastructure for Astronomy II*, N. M. Radziwill and G. Chiozzi (eds.), Proc. SPIE 8541, 15.

Kiddle, C., Taylor, A. R., Cordes, J., Eymere, O., Kaspi, V., Pigat, D., Rosolowsky, E., Stairs, S. & Willis, A. G. 2011, in *GCE'11: Proceedings of the 2011 ACM Workshop on Gateway Computing Environments*, (ACM: New York, NY, USA), p 65.

Highlights of Astronomy, Volume 16
XXVIIIth IAU General Assembly, August 2012
T. Montmerle, ed.

doi:10.1017/S1743921314012873

Real-time Visualisation and Analysis of Tera-scale Datasets

Christopher J. Fluke†

Centre for Astrophysics & Supercomputing, Swinburne University of Technology, Hawthorn, Victoria, 3122, Australia ,
email: cfluke@swin.edu.au

Abstract. As we move ever closer to the Square Kilometre Array era, support for real-time, interactive visualisation and analysis of tera-scale (and beyond) data cubes will be crucial for on-going knowledge discovery. However, the data-on-the-desktop approach to analysis and visualisation that most astronomers are comfortable with will no longer be feasible: tera-scale data volumes exceed the memory and processing capabilities of standard desktop computing environments. Instead, there will be an increasing need for astronomers to utilise remote high performance computing (HPC) resources. In recent years, the graphics processing unit (GPU) has emerged as a credible, low cost option for HPC. A growing number of supercomputing centres are now investing heavily in GPU technologies to provide O(100) Teraflop/s processing. I describe how a GPU-powered computing cluster allows us to overcome the analysis and visualisation challenges of tera-scale data. With a GPU-based architecture, we have moved the bottleneck from processing-limited to bandwidth-limited, achieving exceptional real-time performance for common visualisation and data analysis tasks.

Keywords. methods:data analysis, techniques:image processing

1. Introduction

Next generation facilties such as the Large Synoptic Survey Telescope and the Square Kilometre Array, and its pathfinders, will survey more of the sky, more often, with more pixels covering more wavelengths – leading to more data that needs to be processed, analysed and visualised before more science will result. This petascale astronomy data era brings with it many new computational challenges, as existing desktop-based solutions are unlikely to scale. Instead, a move away from analysis of data on desktop computers will need to occur, with greater reliance on remote data processing. A critical ingredient for the success of remote processing solutions will be their ability to support real-time, interactive data analysis and visualisation tasks for individual data products measured in terabytes. High performance computing facilities that leverage the processing power of graphics processing units (GPUs) are showing great promise as a solution to the interactive data analysis and visualisation of terabyte-scale data cubes.

2. Graphics processing units

GPUs have emerged as low-cost computational accelerators, capable of providing 10×–100× speed-ups for algorithms with "massive data parallelism" (e.g. single instruction multiple data) and high arithemic intensity (ratio of floating point operations to memory access). While existing codes cannot be used directly on a GPU, there are a growing number of programming options including: the CUDA and OpenCL application programming

† Present address: Mail H30, PO Box 218, Hawthorn, Victoria, 3122, Australia.

interfaces; code libraries including cuFFT, cuBLAS, and Thrust (C++ template library with a focus on algorithms); PyCUDA for Python; support within IDL and Matlab; and the OpenACC compiler-based directives. Prior to choosing a specific implementation, it is essential to determine whether an algorithm is suitable for a GPU – see Barsdell *et al.* (2010) and Fluke *et al.* (2011) for practical GPU-adoption strategies.

3. GPU-accelerated data analysis and visualisation

The WALLABY all-sky, extragalactic HI survey, to be performed with the Australian Square Kilometre Array Pathfinder (ASKAP), will produce over 1000 spectral data cubes (right ascension, declination and frequency or line-of-sight velocity) with a likely resolution of $4096 \times 4096 \times 16384$ voxels, requiring ~ 1 TB of storage each. Compare this to the 12 GB data volume of the entire HI Parkes All-Sky Survey [HIPASS; Barnes *et al.* (2001)] where the 387 individual cubes contribute to a single $1721 \times 1721 \times 1024$ voxel super-cube.

Using Australia's gSTAR national facility, Hassan *et al.* (submitted) have benchmarked several common data analysis and visualisation tasks for a 542 GB spectral data cube (replication of the HIPASS supercube 48 times = $6884 \times 6884 \times 3072$ voxels). Once in GPU memory of the cluster (576 GB total memory available using 90× Tesla C2070 and 6× Tesla C2090 GPUs, each with 6 GB/GPU) it takes approximately:

- 0.14 seconds to provide a three-dimensional volume rendering;
- 2 seconds to calculate the global mean and standard deviation;
- 4 seconds to calculate a histogram;
- 45 seconds to iteratively calculate the global median; and
- 20 milliseconds to extract a one-dimensional spectrum from an arbitrary orientation of the cube, a step towards interactive quantitative investigation.

4. Conclusions

Graphics processing units are a practical solution to the real-time data analysis and visualisation needs of the petascale astronomy era. They provide a low-cost option to O(100) Tflop/s high performance computing, compared to the equivalent number of CPUs. The majority of the benchmarked data analysis and visualisation tasks were bandwidth limited – the GPUs have more than enough processing capability. Future advances in GPU cluster computing will benefit from more memory per GPU, more GPUs per compute node (currently limited by the PCI Express bus), and faster inter-node communication. Astronomy is headed towards an exciting, massively-parallel future.

Acknowledgements

CJF thanks his collaborators Amr Hassan (Swinburne), David Barnes (Monash), and Virginia Kilborn (Swinburne) for their contributions. This work was performed on the gSTAR national facility at Swinburne University of Technology. gSTAR is funded by Swinburne and the Australian Governments Education Investment Fund.

References

Barsdell, B. R., Barnes, D. G., & Fluke, C. J., 2010, *MNRAS*, 408, 1936
Barnes, D. G., *et al.* 2001, *MNRAS*, 322, 486
Fluke, C. J., Barnes, D. G., Barsdell, B. R., & Hassan, A. H., 2011, *PASA*, 28, 15
Hassan, A. H., Fluke, C. J., Barnes, D. G., & Kilborn, V. A., *MNRAS*, under review

Highlights of Astronomy, Volume 16
XXVIIIth IAU General Assembly, August 2012
T. Montmerle, ed.

doi:10.1017/S1743921314012885

Knowledge Discovery Workflows in the Exploration of Complex Astronomical Datasets

Raffaele D'Abrusco[1], Giuseppina Fabbiano[1], Omar Laurino[1] and Francesco Massaro[2]

[1]Harvard-Smithsonian Center for Astrophysics - Cambridge (MA), 02138 - Garden Street 60
[2]SLAC National Laboratory and Kavli Institute for Particle Astrophysics and Cosmology, 2575 Sand Hill Road, Menlo Park, CA 94025, USA

Abstract. The massive amount of data produced by the recent multi-wavelength large-area surveys has spurred the growth of unprecedentedly massive and complex astronomical datasets that are proving the traditional data analysis techniques more and more inadequate. Knowledge discovery techniques, while relatively new to astronomy, have been successfully applied in several other quantitative disciplines for the determination of patterns in extremely complex datasets. The concerted use of different unsupervised and supervised machine learning techniques, in particular, can be a powerful approach to answer specific questions involving high-dimensional datasets and degenerate observables. In this paper I will present CLaSPS, a data-driven methodology for the discovery of patterns in high-dimensional astronomical datasets based on the combination of clustering techniques and pattern recognition algorithms. I shall also describe the result of the application of CLaSPS to a sample of a peculiar class of AGNs, the blazars.

Keywords. surveys, data-mining, AGNs, blazars, WISE, γ-ray

1. CLaSPS

The Clustering-Labels-*Score* Pattern Spotter (CLaSPS) method (D'Abrusco *et al.* (2012)) for the discovery of patterns in complex astronomical *feature* spaces is based on unsupervised clustering techniques (Hastie *et al.* (2009)), complemented by additional data, the *labels*, employed to characterize the content of different clusters. The *labels* are used to characterize the content of the set of the clusters determined in the *feature* space. Previously, some of the authors ((D'Abrusco *et al.* (2009)) and (Laurino *et al.* (2011))) have used the same approach for the selection of optical candidate quasars from photometric datasets and the determination of photometric redshifts employing. CLaSPS has generalized this method extended to multiple labels, both numerical and categorial. The originality of CLaSPS lies in the criterion used to select the interesting aggregation of sources in the *feature* space that show correlations with the *labels* distribution. The quantitative diagnostics used to select such clusterings is called the *score*. For a generic clustering of the dataset in the *feature* space, the *score* of the i-th cluster is defined as:

$$S_i = \sum_{j=1}^{M^{(L)}-1} \|f_{ij} - f_{i(j-1)}\| \tag{1.1}$$

where f_{ij} is the fraction of the i-th cluster members with values of the *label* falling in the j-th bin (or class) and the sum is over all *labels* bins. The total *score* s of the clustering follows:

$$S_{\mathrm{tot}} = \frac{1}{N_{\mathrm{clust}}} \cdot \sum_{i=1}^{N_{\mathrm{clust}}} S_i = \frac{1}{N_{\mathrm{clust}}} \sum_{i=1}^{N_{\mathrm{clust}}} \Big(\sum_{j=1}^{M^{(j)}-1} \|f_{ij} - f_{i(j+1)}\| \Big) \tag{1.2}$$

The values of the *scores* are then used to select the clustering(s) showing the largest degree of correlation between the *label* classes and clustering membership distributions (left plot in Fig. 1). The effectiveness of the *score* has been assessed on simulated clusterings before the application to real astronomical datasets.

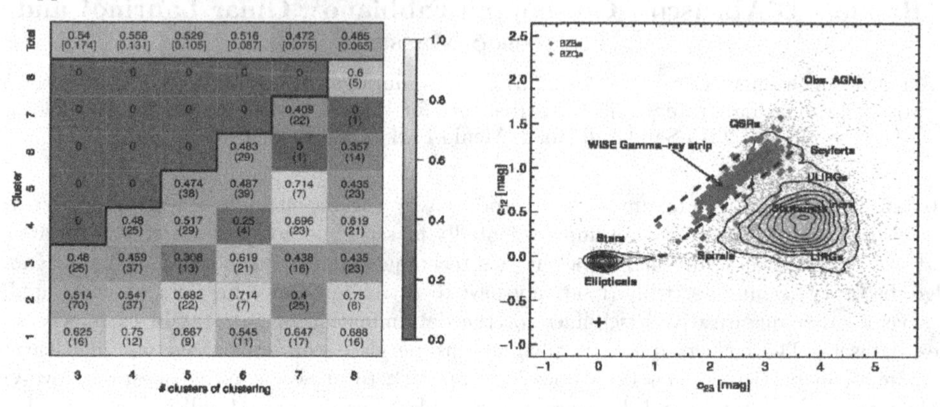

Figure 1. Left: map of the *scores* for the clusterings of the blazars experiment described in Sec. 2 as a function of the total number of clusters in each clustering. Right: projection of the blazars WISE *locus*, discovered by CLaSPS, onto the WISE [4.6]−[12] vs [3.4]−[4.6] color-color plane.

2. Application to Blazars

One of the first applications of CLaSPS, involving a sample of *bona fide* blazars in a *feature* space generated by broad-band color from mid-infrared to far ultraviolet, has led to the discovery of a previously unknown pattern followed by the blazars in the mid-infrared color space generated by the WISE magnitudes (D'Abrusco *et al.* (2012)). CLaSPS has been applied to the distribution of blazars in the 9-dimensional colors *feature* space using as *labels*, among other observables, the blazars spectral classification in BL Lacs and Flat Spectrum Radio Quasars, and the detection of γ-ray emission. CLaSPS determined a significant pattern for γ-ray emitting blazars of both spectral types in the three dimensional WISE colors space, revealing that this class of extragalactic sources occupy a peculiar and narrow *locus* in this *feature* space. The projection of the 3D WISE blazars *locus* color plane is shown in the right plot in Fig. 1. This discovery has also been used to devise a method for the selection of WISE candidate blazars that has been already applied to different sample of high-energy unidentified sources (the application to the unidentified γ-ray sources from Fermi in (Massaro *et al.* (2012)).

References

D'Abrusco, R., Longo, G., & Walton, N. A. 2009, *MNRAS*, 396, 223
D'Abrusco, R., Massaro, F., Ajello, M., *et al.* 2012, *ApJ*, 748, 68
D'Abrusco, R., Fabbiano, G., Djorgovski, G., *et al.* 2012, *ApJ*, 755, 92
Hastie, T., Tibshirani, R., & Friedman, J. 2009, *The Elements of statistical learning*, Springer.
Laurino, O., D'Abrusco, R., Longo, G., & Riccio, G. 2011, *MNRAS*, 418, 2165
Massaro, F., D'Abrusco, R., Tosti, G., *et al.* 2012, *ApJ*, 750, 138

Highlights of Astronomy, Volume 16
XXVIIIth IAU General Assembly, August 2012
T. Montmerle, ed.

doi:10.1017/S1743921314012897

Kernel PCA for Supernovae Photometric Classification

Emille E. O. Ishida

IAG, Universidade de São Paulo, Rua do Matão 1226, Cidade Universitária, CEP 05508-900, São Paulo, SP, Brazil
email: emilleishida@usp.br

Abstract. In this work, we propose the use of Kernel Principal Component Analysis (KPCA) combined with k = 1 nearest neighbor algorithm (1NN) as a framework for supernovae (SNe) photometric classification. It is specially recommended for analysis where the user is interested in high purity in the final SNe Ia sample. Our method provide good purity results in all data sample analyzed, when SNR$\geqslant$5. As a consequence, we can state that if a sample as the *Supernova Photometric Classification Challenge* were available today, we would be able to classify $\approx$ 15% of the initial data set with purity higher than 90%. This makes our algorithm ideal for a first approach to an unlabeled data set or to be used as a complement in increasing the training sample for other algorithms. Results are sensitive to the information contained in each light curve, as a consequence, higher quality data (low noise) leads to higher successful classification rates.

Keywords. (stars:) supernovae: general, methods: statistical, techniques: photometric

1. Introduction

Type Ia supernovae (SNe Ia) observations provided the first evidence towards an accelerated cosmic expansion in late 1990's. Following this first result the scientific community is deeply engaged in improving these cosmological results in order to have a better idea about the properties and nature of dark energy. Such line of thought has lead to implementation of a large number of on going and future SNe surveys, aimed at observing SNe Ia for cosmological use. The result of such effort can already be experienced. According to Sako *et al.* (2011), within 3 observational seasons, the *Sloan Digital Sky Survey* (SDSS) supernova survey was able to obtain $\approx$ 3500 SNe light curves. However, given the limited spectroscopic follow-up available, for only $\approx$500 of them the spectroscopic confirmation was obtained. At this point, only this small sub-set is used in cosmological analysis. This situation will become more drastic with the advent of surveys now under construction.

In order to put different SNe classifiers in the same grounds, a *Supernova Photometric Classification Challenge* (SNPCC) was proposed. It consisted of sample of synthetic light curves, build according to what is expected from the 5 years of operations of the *Dark Energy Survey* (DES). The data was divided into a spectroscopically confirmed sample ($\approx$1000), for which the labels were provided, and a photometric only sample ($\approx$20000). Results were reported in Kessler *et al.* (2010).

2. Kernel PCA application

In this work, we approach the SNe photometric classification problem using a 2 steps machine learning algorithm. In a first moment, we applied *Kernel Principal Component Analysis* (KPCA) to the light curves in the spectroscopic sample. KPCA is a statistical

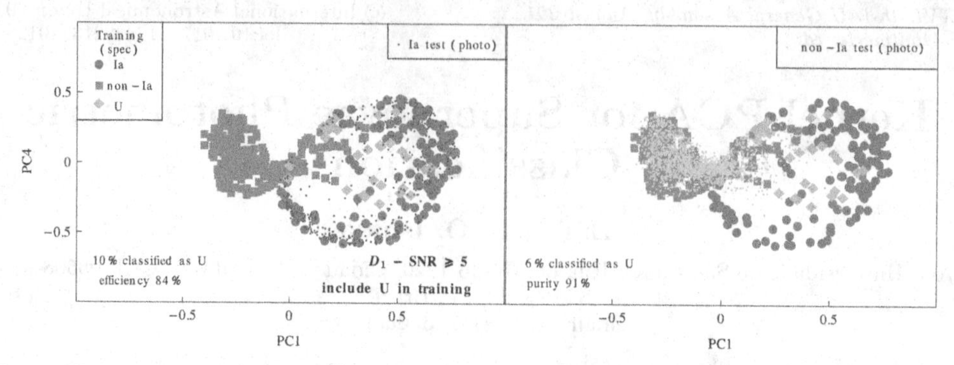

Figure 1. Example of low dimensional representation. D_1+SNR⩾5 means we used data between -3 and +24 days since maximum brightness demanding SNR⩾5 for at least epochs in each filter. The pink triangles correspond to SNe in the boundaries of Ia/nonIa regions.

tool aimed at reducing the dimensionality of a given parameter space. In our suggested framework, all available information for a given SNe (flux measurements in different epochs and filters) are mapped into a 2-dimensional parameter space where the *locus* occupied by Ia are as separate as possible to that occupied by nonIa. Once this lower dimensional representation is constructed, we take one photometric light curve at a time and project it in this new parameter space. A *k=1 Nearest Neighbor* algorithm is then used to address a classification to this photometric point based on the geometrical distribution of the spectroscopic sample (see example of point projections in figure 1).

The method was applied to the SNPCC data set. Although our results are sensitive to low SNR observations, leading to a non-ideal efficiency at high redshifts, it achieved impressive purity values when compared to results from other algorithms reported within the *Challenge*, specially in redshifts within $0.2 \leqslant z < 0.5$. The good purity values are important given that, for a cosmological analysis, nonIa contamination in a Ia sample can lead to wrong cosmological models. A detailed analysis of these results and comparisons to others already available in the literature can be found in Ishida & de Souza (2012).

3. Conclusions

We proposed the use of KPCA+1NN algorithm as a framework for SNe photometric classification. From applying our method to the SNPCC data set, we are able to state that, if a sample as the one expected from the DES were available today, we would be able to classified 15% of the original set with purity levels ⩾90%. This makes our method ideal as a first approach to the already available light curves samples without spectroscopic confirmation, as the one within the SDSS data.

Acknowledgements

E.E.O.I. is partially supported by the Brazilian financial agency FAPESP through grant number 2011/09525-3.

References

Ishida, E. E. O.. & de Souza, R. S. 2012, arXiv:astro-ph/1201.6676
Kessler, R., *et al.* 2010, *PASP* 122, 898
Sako, M., *et al.* 2011, *ApJ* 738, 162

Highlights of Astronomy, Volume 16
XXVIIIth IAU General Assembly, August 2012
T. Montmerle, ed.

doi:10.1017/S1743921314012903

Virtual Atomic and Molecular Data Centre: Level 3 Service and Future Prospects

M. L. Dubernet[1], G. Rixon[2], M. Doronin[1] and VAMDC Collaboration†

[1] LPMAA, UMR7092, Université Pierre et Marie Curie, and Observatory of Paris, France
email: `marie-lise.dubernet@obspm.fr`

[2]Institute of Astronomy, University of Cambridge, Madingley Road, Cambridge, CB30HA, UK

Abstract. The Virtual Atomic and Molecular Data Centre (VAMDC, http://www.vamdc.eu) is an international Consortium that has created an interoperable e-science infrastructure for the exchange of atomic and molecular data. The VAMDC defines standards for the exchange of atomic and molecular data, develop reference implementation of those standards, deploys registries of internet resources (yellow pages), designs user applications in order to meet the user needs, builds data access layers above databases to provide unified outputs from these databases, cares about asynchronous queries with workflows and connects its infrastructure to the grid. The paper describes the current service deployment of the VAMDC data infrastructure across our registered databases and the key features of the current infrastructure.

Keywords. atomic data, molecular data, standards, data bases

1. Level 3 Release of VAMDC, September 2012

A level-n release is a combined release of: standards for data access, the VAMDC nodes, each containing a database and web service following the standards, a registry of the services, a web portal as UI for the system, softwares to prepare the databases (Node Software) and implement the web services, web services to process VAMDC results into other formats and presentations (Consumer Service), update of VAMDC tools. Rixon *et al.* (2011) and Doronin *et al.* (2012) describe Level 1 and Level 2 services respectively, thus providing the definition of the above concepts, while Dubernet *et al.* (2010) is the funding paper describing the VAMDC consortium.

The level-3 release contains the resources made available to users in VAMDC on September 2012, based on the 2011.12 release of standards. The standards for the 2011.12 release (http://vamdc.eu/standards) specify: the data model and format XSAMS, in the variant currently supported by VAMDC, called VAMDC-XSAMS version 0.3, the web-service protocol VAMDC-TAP, the query language VSS2, the VAMDC dictionary of standard terms, the method of registering nodes, and the protocol for XSAMS-consuming services. The level 3 databases are registered in the level 3 registry (http://registry.vamdc.eu/registry-11.12/) which is separate from the registries for other releases and from the registry used for development of the system. The VAMDC portal (http://portal.vamdc.eu) always reads the latest released registry.

2. What can you do with VAMDC services

• Use our graphical User Interface, called portal (http://portal.vamdc.eu) to query all currently registered VAMDC databases, and download or visualize the data. The portal

† See References Dubernet *et al.* (2010), Rixon *et al.* (2011), Doronin *et al.* (2012)

usage is supported by different e-tutorials that can be found at http://vamdc.eu/usersupport. It should be noted that the "consumer services" which transform the VAMDC-XSAMS files into other formats are independent of the portal and users may offer to connect their own consumer service to the portal.

• Implement your own database within VAMDC using the Node Software in Python or Java (Doronin *et al*, 2012)

• Implement our librairies and/or standards in your own application software : you may use the on-line documentation, download the different libraries and software.

• Use our currently available tools (http://vamdc.eu/software): TAPValidator to check the output of VAMDC nodes, SPECTCOL to combine collisional data with spectrocopic data for the study of non-LTE interstellar media.

• Get support for all these features and propose new features or tools

3. Support for Users

The section User Support (http://vamdc.eu/usersupport) provides the access to the support mailing list, and to training materials currently including the use of Portal, the use of Taverna workflow, the use of the GRID. The documentation for standards are found at http://vamdc.eu/standards, the documentation related to software are included with the software (http://vamdc.eu/software). For any kind of request, bug report, question on the operation, support to user, or general enquiries, please e-mail support@vamdc.eu. Your email will be placed as a ticket to be looked at by the VAMDC support-community and an answer will come back as soon as possible.

4. Future Prospect

At the time of writing this paper the standards version 2012.07 have been released, the node software in Java version 12.07r1, the TAPValidator Tool v12.07, the client SPECTCOL version 12.07 have been released, and Level-4 release, based on standards v12.07, is planned for mid-december 2012. From December 2012 the VAMDC consortium will provide new n-level releases every year at most, so that stability of software is garantied.The VAMDC is further suppported for 2 years by a new european grant "SUP@VAMDC" (http://sup.vamdc.eu) that will allow to support the already existing infrastructure and to extend its usage by new producers of data in a wider range of countries, as well as by new users including high education, schools, citizens and SMEs.

Acknowledgements

VAMDC and SUP@VAMDC are funded under the Combination of Collaborative Projects and Coordination and Support Actions Funding Scheme of The Seventh Framework Program. Call topic: INFRA-2008-1.2.2 and INFRA-2012 Scientific Data Infrastructure. Grant Agreement numbers: 239108 and 313284.

References

Dubernet, M. L., Boudon, V., Culhane, J. L., *et al.* 2010, *JQSRT* 111, 2151

Doronin, M. & Dubernet, M. L., al 2012, *in ASP Conf. Ser., ADASS XXI, ed. P. Ballester, D. Egret, and N. P. F. Lorente (San Francisco: ASP)* 461, p.331

Rixon, G., Dubernet, M. L., Piskunov, N., *et al.* 2011, *in American Institute of Physics Conference Series, American Institute of Physics Conference Series, ed. A. Bernotas, R. Karazija, & Z. Rudzikas*, 1344, p.107–115

Highlights of Astronomy, Volume 16
XXVIIIth IAU General Assembly, August 2012
T. Montmerle, ed.

doi:10.1017/S1743921314012915

Variable Stars and Data-Intensive Astronomy

Nikolay N. Samus[1,2] and Sergey V. Antipin[2,1]

[1]Institute of Astronomy (Russian Academy of Sciences), 48, Pyatnitskaya Str., Moscow 119017, Russia; email: samus@sai.msu.ru

[2]Sternberg Astronomical Institute, M.V. Lomonosov Moscow State University, 13, University Ave., Moscow 119991, Russia; email: antipin@sai.msu.ru

Abstract. Extensive discoveries of variable stars in ground-based photometric surveys and in observations from space missions that provide unprecedented accuracy demonstrate that, till the most recent time, we knew only a tiny fraction of all detectable variable stars of our Galaxy. As a result, our knowledge on stellar variability types, related physical processes, variable-star statistics turn out to be based on an unrepresentative sample and are expected to be radically revised in the near future. The flow of new discoveries results in quite difficult problems for catalogs of variable stars, making it impossible to proceed in their compilation in the traditional way. Regretfully, automatic solutions are still not completely satisfying. Though not able to suggest a perfect way out, we present our vision of the future of variable-star research.

Keywords. catalogs, stars: statistics

1. Introduction

Since 1940s, the Moscow team of variable-star researchers is working, on behalf of the IAU, on catalogs of variable stars. This work encounters unprecedented difficulties in the era of data-intensive astronomy.

Probably the most fundamental problem is that there is no generally accepted brightness-variation amplitude limit the star must exceed to be considered variable. The General Catalogue of Variable Stars (GCVS; Samus *et al.* 2012) currently contains stars with peak-to-peak amplitudes from $> 19^m$ (V1500 Cyg) down to 0.004^m (α Aql). The number of known variables drastically increases with development of observing techniques.

2. Current Statistics and Future Prospects

The currently largest stellar catalog, USNO-B1.0 (Monet *et al.* 2003), contains about 1 billion stars down to $20{-}21^m$. Experience shows that one of 80–100 stars varies at the $0.03{-}0.05^m$ level, and thus some 10 million stars are potentially detectable as variables with a ground-based 1-m telescope, ordinary CCD detector, and standard software for automatic search for variable stars. The largest modern lists of known galactic variable stars (like VSX, Watson *et al.* 2007, supported by the AAVSO) contain about 200000 objects; it means that we know no more than 2% of all detectable variables of the Galaxy, a percentage lower than that of barion matter in the Universe! New discoveries can seriously change our ideas concerning frequency of different variability types. Even information contained in astronomical plate archives, if extracted and analyzed in digital form, changes statistics of variability types and period distributions (cf. Kolesnikova *et al.* 2010). Surveys performed with very small or moderate-sized CCD telescopes, like ASAS-3 (Pojmanski 1997), ROTSE-I/NSVS (Woźniak *et al.* 2004) , SuperWASP (Pollacco *et al.*

2006), Catalina (Drake *et al.* 2009), and others result in thousands of new variable-star discoveries, made by the survey teams and, mainly, by the astronomical community via data mining. Some of these surveys present their own lists of variable-star discoveries. It is difficult even to simply merge them with the GCVS because of different standards of positional accuracy and different level of detail of variable-star classification schemes.

If we turn to observations from outer space, amplitudes as low as 0.0001^m become detectable. According to results of just one month from the *Kepler* mission, about 60,000 periodic variables and 34,000 marginally periodic variables could be detected among some 150,000 targets (Basri *et al.* 2011). Such statistics is clearly incompatible with a concept of variable-star catalogs different from star catalogs in a more general sense; it makes impossible the traditional star-by star basis (with human expertise) our team uses, for decades, in our GCVS work.

Catalogs of variable stars contain classification of variables according to the existing scheme. The GCVS system clearly needs improvement; however, it seems too early to introduce a new scheme since dramatic changes are expected in the near future. Until recently, there were no automatic classification systems permitting to find the star's type from photometric data (see, however, Blomme *et al.* 2011 for a rather promising new development). Unfortunately, photometry alone is often insufficient for variable-star classification, additional (spectroscopic, X-ray, kinematical) information can be needed. The problem of really effective automatic classification needs further efforts to solve it.

3. Brief Conclusions

With full-scale results from Kepler (and then from similar missions) approaching, the number of known variable stars will drastically increase and everything we know about variability statistics will have to be revised. We are forced to admit that the era of traditional variable-star catalogs is probably near its end, despite traditional Argelander-style variable-star names being still popular.

Perhaps the only possible solution for the predictable future could be compilation of universal star catalogs, with variability information contained in them as a minor but important part.

Acknowledgements

We thank all participants of the GCVS team for their many-year effort aimed at high-quality catalogs of variable stars. Our research is supported by the Program "Non-Stationary Processes of Objects in the Universe" of the Presidium of Russian Academy of Sciences and by the Russian Foundation for Basic Research (grant 11-02-00495).

References

Basri, G., Walkowicz, L. M., Batalha, N., *et al.* 2011, *AJ*, 141, 20
Blomme, J., Sarro, L. M., O'Donovan, F. T., *et al.* 2011, *MNRAS*, 418, 96
Drake, A. J., Djorgovski, S. J., Mahabal, A., *et al.* 2009, *ApJ*, 696, 870
Kolesnikova, D. M., Sat, L. A., Sokolovsky, K. V., *et al.* 2010, *ARep*, 54, 1000
Monet, D. G., Levine, S. E., Canzian, B., *et al.* 2003, *AJ* 125, 984
Pojmanski, G. 1997, *AcA*, 47, 467
Pollacco, D., Skillen, I., Cameron, A. C., *et al.* 2006, *PASP*, 118, 1407
Samus, N. N., Durlevich, O. V., Kazarovets, E. V., *et al.* 2012, General Catalogue of Variable Stars (GCVS Database, Version 2012 Jan.), CDS B/gcvs
Watson, C. L., Henden, A. A., & Price, A. 2007, *JAAVSO* 35, 414
Woźniak, P. R., Vestrand, W. T., Akerlof, C. W., *et al.* 2004, *AJ*, 127, 2436

Highlights of Astronomy, Volume 16
XXVIIIth IAU General Assembly, August 2012
T. Montmerle, ed.

doi:10.1017/S1743921314012927

Galaxy Zoo: Outreach and Science Hand in Hand

Karen L. Masters[1,2] and the Galaxy Zoo Team

[1]Institute for Cosmology and Gravitation, University of Portsmouth, Dennis Sciama Building, Burnaby Road, Portsmouth, PO1 3FX, UK
email: karen.masters@port.ac.uk

[2]South East Physics Network, www.sepnet.ac.uk

Abstract. Galaxy Zoo (www.galaxyzoo.org) is familiar to many as a hugely successful public engagement project. Hundreds of thousands of members of the public have contributed to Galaxy Zoo which collects visual classifications of galaxies in Sloan Digital Sky Survey and Hubble Space Telescope images. Galaxy Zoo has inspired a suite of similar Citizen Science projects known as "The Zooniverse" (www.zooniverse.org) which now has well over half a million participants. Galaxy Zoo has also shown itself, in a series of peer reviewed papers, to be a fantastic database for the study of galaxy evolution. In this invited talk I described how that public engagement via citizen science is not only an effective means of outreach from data intensive surveys, but if done right can and must also increase the scientific output of the survey.

---

1. Introduction

Galaxy Zoo was launched in July 2007 to solve the problem of how to visually classify the ~1 million galaxies in the Sloan Digital Sky Survey Main Galaxy Sample (Strauss *et al.* 2002). The original site asked volunteers to classify galaxies as either spiral or elliptical - the most basic morphological split. Something in Galaxy Zoo resonated extraordinarily with the general public. By April 2008, when the Galaxy Zoo team submitted their first paper (Lintott *et al.*2008), over 100,000 volunteers had classified each of the images an average of 38 times.† The popularity of Galaxy Zoo and the number of classifications received enabled science which simply would not have been possible otherwise. Not only does Galaxy Zoo give a classification for each galaxy, but also by collecting ~ 40 independent classifications of each galaxy out can produce an estimate of how reliable that classification is, and calibrate individual clicks with weightings to remove volunteers who disagree with most other classifiers. Galaxy Zoo was so popular that it is now in its fourth version, asking for detailed classifications of both SDSS and HST images of galaxies.

The scientific results coming out from Galaxy Zoo have been essential to its success. A survey which studied the motivations of the citizen scientists participating in Galaxy Zoo, (Raddick *et al.* 2010) showed that the majority of people identified a desire to help with scientific research as their main motivation for spending time on the site. Galaxy Zoo was designed with specific and immediate questions to answer. The first publications from Galaxy Zoo came out within a year of launch (Lintott *et al.* 2008, Land *et al.* 2008), and the total number of peer reviewed papers based on Galaxy Zoo data is now more than 30 (e.g see Table 1 of Masters 2012). They include among them work separating colour and morphology in the galaxy population (e.g. Bamford *et al.* 2009, Skibba *et al.* 2009,

† All of the data from this first phase of Galaxy Zoo were published in Lintott *et al.* (2011) and are available to download via the SDSS servers at http://skyserver.sdss3.org/CasJobs/. For more information on the history of the Galaxy Zoo project see Masters (2012) in this volume, or Fortson *et al.* (2012).

Schawinski et a. 2009, Masters *et al.* 2010a,b), on the interplay between supermassive black hole activity and host galaxies (e.g. Lintott *et al.* 2009, Schawinski *et al.* 2010a,b, Teng *et al.* 2012, Simmons *et al.* 2012) and on the role bars play on galaxy evolution (e.g. Masters *et al.* 2011, Hoyle *et al.* 2011, Skibba *et al.* 2012, Masters *et al.* 2012). Galaxy Zoo classifications have also been used to calibrate machine learning and other automated methods of galaxy classification, illustrating a complementarity with these methods which will become even more important as galaxy catalogues grow even larger (e.g. Banerji *et al.* 2010).

Galaxy Zoo was a pioneer in what has become a new methodology of involving "citizen scientists" in research. The Zooniverse† was launched in December 2009 to provide a framework to collect a variety of similar projects. This now includes several other astronomically themed projects. A review of citizen science (including Zooniverse project) contributions to astronomy research can be found at Christian *et al.* (2012).

Astronomers (and other scientists) involved in big data projects can now propose to develop their own Zooniverse project‡. The philosophy of citizen science in the Zooniverse has been built based on the experience with Galaxy Zoo. Central to this is that all projects in the Zooniverse exist to analyse data and produce science (or research) which would not be possible without the help of citizen scientists.

Involving the public in the scientific method in this way provides a fantastic opportunity for science education across a wide variety of audiences and at many levels, but scientists embarking on a Zooniverse project must commit resources to produce scientific results out of the contributions. We can all benefit when we work together.

Acknowledgements

This publication has been made possible by the participation of more than 200,000 volunteers in the Galaxy Zoo project. Their contributions are individually acknowledged at http://www.galaxyzoo.org/volunteers.

KLM acknowledges funding from the Peter and Patricia Gruber Foundation as the 2008 Peter and Patricia Gruber Foundation IAU Fellow, and from a 2010 Leverhulme Trust Early Career Fellowship, as well as support from the Royal Astronomical Society to attend the 28th GA of the IAU.

The Citizen Science Alliance acknowledges funding from the Alfred P. Sloan Foundation to enable open calls for Zooniverse projects.

References

Bamford, S. P., *et al.* 2009, *MNRAS*, 393, 1324

Banerji, M., Lahav, O., Lintott, C. J., *et al.* 2010, *MNRAS*, 406, 342

Christian, C., Lintott, C., Smith, A., Fortson, L., & Bamford, S. 2012, *Organizations, People and Strategies in Astronomy Vol. 1*, Edited by Andre Heck, Venngeist, Duttlenheim (2012) pp. 183-197, 183 (arXiv:1202.2577)

Fortson, L., Masters, K., Nichol, R., *et al.* 2012, *Advances in Machine Learning and Data Mining for Astronomy*, CRC Press, Eds.: Michael J. Way *et al.*, p. 213-236, 213 (arXiv:1104.5513)

Hoyle, B., Masters, K. L., Nichol, R. C., *et al.* 2011, *MNRAS*, 415, 3627

Land, K., Slosar, A., Lintott, C. J., *et al.* 2008, *MNRAS*, 388, 1686

Lintott, C. J., *et al.* 2008, *MNRAS*, 389,1179

Lintott, C. J., Schawinski, K., Keel, W., *et al.* 2009, *MNRAS* 399, 129

Lintott, C. J., *et al.* 2011, *MNRAS*, 410, 166

Masters, K. L., *et al.* 2010a, *MNRAS* 404, 792.

† `www.zooniverse.org`

‡ See `http://www.citizensciencealliance.org/proposals.html`

Masters, K. L., *et al.* 2010b, *MNRAS* 405, 783.
Masters, K. L., *et al.* 2011, *MNRAS*, 411, 2026
Masters, K. L., Nichol, R. C., Haynes, M. P., *et al.* 2012, *MNRAS*, 424, 2180
Masters, K. L. 2012, Highlights of Astronomy, Volume 16, Ed Thierry Montmerle.
Raddick, M. J., Bracey, G., Gay, P. L., *et al.* 2010, Astronomy Education Review, 9, 010103
Skibba, R. A., *et al.* 2009, *MNRAS*, 399, 966
Skibba, R. A., Masters, K. L., Nichol, R. C., *et al.* 2012, *MNRAS*, 423, 1485
Schawinski, K., Evans, D. A., Virani, S., *et al.* 2010a, *ApJL*, 724, L30
Schawinski, K., Urry, C. M., Virani, S., *et al.* 2010b, *ApJ*, 711, 284
Simmons, B. D., Lintott, C., Schawinski, K., *et al.* 2012, *MNRAS* (submitted; arXiv:1207.4190)
Strauss, M. A., *et al.* 2002, *AJ*, 124, 1810
Teng, S. H., Schawinski, K., Urry, C. M., *et al.* 2012, *ApJ*, 753, 165

Highlights of Astronomy, Volume 16
XXVIIIth IAU General Assembly, August 2012
T. Montmerle, ed.

doi:10.1017/S1743921314012939

Discover the Cosmos - Bringing Cutting Edge Science to Schools across Europe

Rosa Doran

NUCLIO - Núclio Interactivo de Astronomia
Largo dos Topázios, 48, 3 Frt, 2785-817 S. D. Rana, Portugal

Abstract. The fast growing number of science data repositories is opening enormous possibilities to scientists all over the world. The emergence of citizen science projects is engaging in science discovery a large number of citizens globally. Astronomical research is now a possibility to anyone having a computer and some form of data access. This opens a very interesting and strategic possibility to engage large audiences in the making and understanding of science. On another perspective it would be only natural to imagine that soon enough data mining will be an active part of the academic path of university or even secondary schools students. The possibility is very exciting but the road not very promising. Even in the most developed nations, where all schools are equipped with modern ICT facilities the use of such possibilities is still a very rare episode. The Galileo Teacher Training Program GTTP, a legacy of IYA2009, is participating in some of the most emblematic projects funded by the European Commission and targeting modern tools, resources and methodologies for science teaching. One of this projects is Discover the Cosmos which is aiming to target this issue by empowering educators with the necessary skills to embark on this innovative path: teaching science while doing science.

Keywords. Miscellaneous, sociology of astronomy

The Galileo Teacher Training Program (GTTP), a cornerstone of IYA2009, and a strong legacy, aims to build a strong support network for educators around the world. Partnering with skilled trainers and scientists we are establishing a sustainable structure to act as a 24 hour helpdesk to teachers all over the world. The scheme is to carefully build the basis of this structure in every nation and scaffold the effort via an effective cascade mechanism where teachers train other teachers. GTTP has such building blocks in 100 nations having reached over 15 000 teachers so far.

The possibility to participate in projects such as:

- Discover the Cosmos (DtC) (http://www.discoverthecosmos.eu)
- European Hands-on Universe (EUHOU) (http://www.euhou.net)
- Open Discovery Space (ODS) (http://www.opendiscoveryspace.eu)

allows a more quickly and effective growth of the basic structure and training opportunities as well as gathering precious information regarding the usability of modern tools and resources for science teaching. By understanding the limitations faced by educators: curriculum wise, lack of ICT competences or equipment, etc., as well as existing opportunities in terms of school infrastructure and freedom to introduce innovation in classroom, DtC, EUHOU and ODS are constructing and documenting the roadmap for the classroom of tomorrow.

The above mentioned projects open the possibility to introduce students and educators to data repositories and science facilities and invite them to engage in real research exercises while learning curriculum content. The suggested resources and projects are built using a modern approach to science teaching, based on replication the scientific method, the inquiry method.

Resources are being created making use of data archives and structuring lessons that have as an end result a scientific outcome. One such example is the Asteroid Search, a program promoted by the International Astronomical Search Collaboration that involves students in the discovery or follow-up of small bodies of the Solar System (asteroids, comets, etc.). Data mining programs such as: CosmoQuest (http://cosmoquest.org/) , ZooUniverse (http://zooniverse.org/), Aladin (http://aladin.u-strasbg.fr/) or World Wide Telescope (http://www.worldwidetelescope.org/) are also integrated in our suggestions to educators. The use of infrastructures for science teaching is another possibility being presented to educators such as radio antennas (http://euhou.obspm.fr/public/) or robotic telescopes (http://www.faulkes-telescope.com/).

Despite the fact that the resources are freely available and some of them very successful among the public in general, statistics show that educators are not using them as a teaching tool. Several are the reasons for this: lack of training, inadaptability to curriculum, lack of classroom time, etc. GTTP aim is precisely to empower educators to use such tools and resources, to help them localize and adapt materials to their educational panorama. Our training events are the entry door to a joint path where further training, new tools and resources, sharing spaces and peer support is always available.

GTTP takes the mission further by engaging the science community to support this enterprise and commit to the production of useful material, and eventually receiving useful contributions to their own research. This collaborative effort (schools x science community) is a win-win situation where students awake for the beauty of science and learn science while doing science. This fast growing community of stakeholders in each of the participating nations: trainers, teachers, students, scientists and education authorities is a promising vision for construction of a community of educators with innovative competences, capable of taking part of this new trend for education: the construction of the classroom of tomorrow.

Highlights of Astronomy, Volume 16
XXVIIIth IAU General Assembly, August 2012
T. Montmerle, ed.

doi:10.1017/S1743921315001672

Preface: Special Session SpS16 Unexplained Spectral Phenomena in the Interstellar Medium

There are several outstanding mysteries in interstellar medium spectroscopy which have remained unsolved after decades of effort. The diffuse interstellar bands (DIBs) have been known for almost a century (Heger 1922). Although more than 400 bands from the near UV to near infrared have been detected, none of them has been identified. In the Milky Way Galaxy, DIBs have been seen towards over one hundred stars. In the Magellanic Clouds, DIBs have been seen in the spectrum of SN 1987A as well as in the spectra of reddened stars (Ehrenfreund *et al.* 2002). DIB carriers in the interstellar medium of external galaxies can be probed by supernovae (Sollerman *et al.* 2005), and DIBs have been detected in external galaxies with redshifts up to 0.5 (Sarre 2006).

The 217.5 nm extinction feature has been known for about 45 years (Stecher 1965). It was extensively observed by the *International Ultraviolet Explorer (IUE)* satellite and is found to have remarkable constancy in its peak wavelength of 217.5 nm, corresponding to 5.7 eV. This is not just a local phenomenon as the feature has been detected in galaxies as distant as redshift $z>2$ (Elíasdóttir *et al.* 2009).

A family of unidentified infrared emission (UIE) features was discovered over 30 years ago (Russell *et al.* 1977) and the number of features has been expanding as the result of infrared spectroscopic observations from *ISO* and *Spitzer*. The UIE phenomenon include aromatic bands at 3.3, 6.2, 7.7, 8.6, and 11.3 μm, aliphatic features at 3.4 and 6.9 μm, broad emission plateaus at 8, 12, and 17 μm, as well as a host of weaker features that are too broad to be atomic or molecular lines. The UIE features are seen in very different radiation environments. The energy source responsible for the excitation of the features ranges from tens of thousands of degrees in planetary nebulae, $\sim$30000 K in H II regions, to thousands of degrees in reflection nebulae and proto-planetary nebulae. Although the UIE phenomena have been widely suggested to be due to polycyclic aromatic hydrocarbon (PAH) molecules, other forms of carbonaceous materials have also been under discussion.

The observation of the Extended Red Emission (ERE) also goes back 30 years. ERE is commonly seen in reflection nebulae (Witt & Schild 1988, Witt & Boroson 1990). It has also been detected in dark nebulae, cirrus clouds, planetary nebulae, H II regions, the diffuse interstellar medium, and in haloes of galaxies. The central wavelength of the emission shifts from object to object, or even between locations within the same object. Other than the fact that it may be due to photoluminescence, exact nature of its carrier is still unknown. Other unidentified optical emissions include a set of bright visible bands seen alongside ERE in the Red Rectangle nebula (Schmidt & Witt 1991) and Blue Luminescence which is seen both in this object and a number of other sources (Vijh, Witt & Gordon 2004).

The 21 and 30 μm unidentified infrared features are generally associated with objects in the late stages of stellar evolution (Kwok *et al.* 1989, Forrest *et al.* 1981), and can be responsible for a significant fraction of the energy output of the stellar sources (Hrivnak *et al.* 2000). High resolution *ISO* observations have found that the 21 μm features have the same intrinsic profile and peak wavelength (20.1 μm) (Volk *et al.* 1999). There is no evidence for any discrete sub-structure due to molecular bands in the observed spectra, suggesting that the 21-μm feature is either due to a solid substance or a mixture of many similarly structured large molecules.

It is interesting to note that these phenomena have been observed not only in the diffuse ISM, but also in circumstellar environments, in the galactic halo, and in external galaxies. In some cases they have been observed in galaxies with high redshifts, suggesting that the carriers responsible for these features were already present in the early Universe.

We now recognize that all these UV, optical and infrared spectral features are ubiquitous in the Universe implying that their carriers must be a substance made of common elements, most likely carbon. The detection of over 170 different molecular species (many of them organic) in the millimeter-wave and infrared spectral regions through their rotational and vibrational transitions. Complex organics are also found in meteorites, interplanetary dust particles, and planetary surfaces. These new developments have shown that the Universe is filled with organics (Kwok 2011). There is a possibility that the carriers of are of organic composition but the relevant materials are not naturally present in the terrestrial environment. The recent detections of C_{60} and C_{70} in planetary nebulae (Cami *et al.* 2010, Garćia-Hernández *et al.* 2010, Zhang & Kwok 2011) and reflection nebulae (Sellgren *et al.* 2010) have also raised interest in other carbon allotropes in the ISM.

The Special Session 16 was held on August 27-28 at the IAU GA in Beijing. We were able to bring together observers who reported on the latest measurements of these features, modelers who use molecular data to interpret the observations, and laboratory spectroscopists who provide expertise in helping the identification. Some significant new results were reported. We are hopeful that this Special Session will be remembered as a major step towards the solution to these long standing mysteries.

References

Cami, J., Bernard-Salas, J., Peeters, E., & Malek, S. E. 2010, *Science* 329, 1180
Ehrenfreund, P. *et al.* 2002, *ApJ* 576, L117
Elíasdóttir, Á. *et al.* 2009, *ApJ* 697, 1725
Forrest, W. J., Houck, J. R., & McCarthy, J. F. 1981, *ApJ* 248, 195
Garćia-Hernández, D. A. *et al.* 2010, *ApJ* 724, L39
Heger, M. L. 1922, *Lick Observatory Bulletin* 337, 141
Hrivnak, B. J., Volk, K., & Kwok, S. 2000, *ApJ* 535, 275
Kwok, S. 2011, *Organic Matter in the Universe*, Wiley
Kwok, S., Volk, K. M., & Hrivnak, B. J. 1989, *ApJ* 345, L51
Russell, R. W., Soifer, B. T., & Willner, S. P. 1977, *ApJ* 217, L149
Sarre, P. J. 2006, *J. Mol. Spectrosc.* 238, 1
Schmidt, G. D., & Witt, A. N., 1991, *ApJ* 383, 698
Sellgren, K. *et al.* 2010, *ApJ* 722, L54
Sollerman, J. *et al.* 2005, *A&A* 429, 559
Stecher, T. P. 1965, *ApJ* 142, 1683
Vijh, U. P. *et al.* 2004, *ApJ* 606, L65
Volk, K., Kwok, S., & Hrivnak, B. J. 1999, *ApJ* 516, L99
Witt, A. N. & Boroson, T. A. 1990, *ApJ* 355, 182
Witt, A. N. & Schild, R. E. 1988, *ApJ* 325, 837
Zhang, Y. & Kwok, S., 2011, *ApJ* 730, 126.

I want to thank Anisia Tang for her assistance in organizing this Special Session and the Local Organizing Committee on their help in the logistical arrangements.

Sun Kwok, Chair SOC,
Hong Kong, China, November 30, 2012

Highlights of Astronomy, Volume 16
XXVIIIth IAU General Assembly, August 2012
T. Montmerle, ed.

doi:10.1017/S1743921314012940

Unexplained Spectral Phenomena in the Interstellar Medium

Sun Kwok
Faculty of Science, The University of Hong Kong, Hong Kong, China
email: sunkwok@hku.hk

Abstract. We present an overview of the present observational status of unexplained spectral phenomena in the ISM. The possibility of organic molecules and solids as the carrier of the DIB, 217 nm feature, ERE, UIR,and the 21 and 30 μm features is discussed.

Keywords. (ISM:) dust, extinction, (ISM:) planetary nebulae: general, ISM: lines and bands, ISM: molecules

---

The origin of diffuse interstellar bands (DIB), 217.5 nm feature, extended red emission (ERE), unidentified infrared emission (UIE) features, and the the 21 and 30 μm features have remained unsolved after decades of efforts. Although the nature of the carrier is unknown, the strengths of the features suggest that their respective carriers are abundantly present in the ISM. These are also not isolated phenomena as they are observed in a wide range of celestial objects throughout the Universe. Could these phenomena be related to each other in any way? Is there a possibility that there exists a common carrier?

1. Carriers of the spectral features

Because of the strengths and ubiquitous nature of the features, the carrier must be made up of common, abundant elements. The element carbon is naturally assumed to play a major role. While the DIBs are widely accepted to be due to electronic transitions of gas-phase C-based molecules, the carrier of the 217 nm feature is more likely to be a carbonaceous solid such as amorphous carbon (Mennella *et al.* 1998), carbon onions (Iglesias-Groth 2004), hydrogenated fullerences (Cataldo & Iglesias-Groth 2009), or polycrystalline graphite (Papoular & Papoular 2009). Many proposals for the origin of the 21 μm feature have been proposed, including hydrogenated fullerenes (Webster 1995), SiC (Speck & Hofmeister 2004), and thiourea groups attached to aromatic/aliphatic structures (Papoular 2011). Since ERE is the result of photoluminescience, the carrier is likely a semiconductor with a nonzero band gap and silicon nanoparticles have been suggested as a candidate (Ledoux *et al.* 1998, Witt *et al.* 1998). Other possibilities include QCC (Sakata *et al.* 1992), C_{60} (Webster 1993), and nanodiamonds (Chang *et al.* 2006).

In astronomical literature, PAH molecules are widely assumed to be the carrier of the UIE features. The PAH hypothesis states that the UIE features are the result of infrared fluorescence from small (~50 atoms) gas-phase PAH molecules being pumped by far-UV photons (Tielens 2008). The central argument for the PAH hypothesis is that single-photon excitation of PAH molecules can account for the 12 μm excess emission observed in cirrus clouds in the diffuse ISM. However, the PAH hypothesis suffers from a number of problems: (i) PAH molecules have well-defined sharp features but the UIE features are broad; (ii) PAHs are primarily excited by UV, with little absorption in the visible but UIE features seen in PPN and reflection nebulae with no UV radiation; (iii) the predicted strong and narrow gas phase features in the UV are not seen in interstellar extinction curves; (iv) no PAH molecules have been detected in spite of the fact that the vibrational and rotational frequencies are well known; (v) no PAH emission spectrum has been able to reproduce the UIE spectrum w.r.t. either band positions or relative intensities (Cook *et al.* 1998); (vi) the shapes and peak wavelengths of the UIE features are independent of temperature of the exciting star, which raises doubts about radiative excitation mechanism; and (vii) in order to fit the astronomical observations, the PAH model has to appeal of a mixture of PAHs of different sizes, structures, and ionization states, as well as artificial broad intrinsic line profiles.

2. Complex organics with disorganized structures

In addition to the sp^2 (graphite and PAH) and sp^3 (diamonds) structures, carbon compounds can also be in amorphous forms. Some carbonaceous compounds (HAC, coal, petroleum) are known to have a mixed sp^2/sp^3 structures with no well-defined geometric structural patterns. It has been suggested that mixed aromatic/aliphatic organic nanoparticles (MAON) may be responsible for the UIE features (Papoular 2001, Kwok & Zhang 2011). Discussions on how such structures can contribute to our understanding of the ERE, 217 nm, and the UIE features be found in the papers of Cataldo, Duley, Jones, Pino in these proceedings.

3. Organics in the Solar System

Although the traditional picture is that the Solar System is primarily made up of minerals, metals, and ices, organic compounds are now increasingly recognized as a common component in the Solar System. Complex organics have been found in meteorites, asteroids, comets, interplanetary dust particles, and planetary satellites. Although space scientists use different terms to refer to these organics (IOM in meteorites, tholins in Titan), any similarity in the chemical structures of these organics with the organics in the interstellar medium may suggest an interstellar origin of the Solar System organics.

Acknowledgements

This work is supported in part by a grant from the HKRGC (HKU 7027/11P).

References

Cataldo, F., & Iglesias-Groth, S. 2009, *MNRAS* 400, 291
Chang, H.-C., Chen, K., & Kwok, S. 2006, *ApJ* 639, L63
Cook, D. J., *et al.* 1998, *JPCA* 102, 1465
Iglesias-Groth, S. 2004, *ApJ* 608, L37
Kwok, S., Zhang, Y. 2011, *Nature* 479, 80
Ledoux, G., *et al.* 1998, *A&A* 333, L39
Mennella, V., Colangeli, L., Bussoletti, E., Palumbo, P., & Rotundi, A. 1998, *ApJ* 507, L177
Papoular, R. 2001, *A&A* 378, 597
Papoular, R. 2011, *MNRAS* 415, 494
Papoular, R. J., & Papoular, R. 2009, *MNRAS* 394, 2175
Sakata, A., *et al.* 1992, *ApJ* 393, L83
Speck, A. K. & Hofmeister, A. M. 2004, *ApJ* 600, 986
Tielens, A. G. G.. M. 2008, *ARA&A* 46, 289
Webster, A. 1993, *MNRAS* 264, L1
Webster, A. 1995, *MNRAS* 277, 1555
Witt, A. N., Gordon, K. D., & Furton, D. G. 1998, *ApJ* 501, L111

Highlights of Astronomy, Volume 16
XXVIIIth IAU General Assembly, August 2012
T. Montmerle, ed.

doi:10.1017/S1743921314012952

Unidentified Infrared Emission Features

Christine Joblin[1,2]

[1]Université de Toulouse; UPS-OMP; IRAP; Toulouse, France
[2]CNRS; IRAP; 9 Av. Colonel Roche, BP 44346, F-31028 Toulouse cedex 4, France
email: christine.joblin@irap.omp.eu

Abstract. When referring to unidentified infrared emission features, one has in mind the series of aromatic IR bands (AIBs) between 3.3 and 15 μm that are observed in emission in many environments where UV photons irradiate interstellar matter. These bands are now used by astronomers to classify objects and characterize local physical conditions. However, a deep analysis cannot proceed without understanding the properties of the band carriers. Large polycyclic aromatic hydrocarbon molecules are attractive candidates but interstellar species are still poorly characterized. Various studies emphasize the need for tackling the link between molecular aromatic species, aliphatic material and very small carbonaceous grains. Other unidentified emission features such as the 6.9, 21 and 30 μm bands could be involved in the evolutionary scenario.

Keywords. Infrared: ISM, ISM: lines and bands, ISM: molecules, stars: carbon

1. AIBs as a tool for astronomers

The AIBs at 3.3, 6.2, 7.7, 8.6, 11.3 and 12.7 μm are major features in the cycle of dust in the Milky Way, from evolved stars (planetary nebulae), to the diffuse interstellar medium (ISM) and bright photodissociation region (PDRs) associated to molecular clouds and to the circumstellar environment of young stars including their disks. They are also well observed in external galaxies including luminous IR galaxies (e.g. Smith *et al.* 2007; Fadda *et al.* 2010). The observed spectra exhibit variations in band positions, profiles and relative intensities and also differ in their satellite features. Peeters *et al.* (2002) and van Diedenhoven *et al.* (2004) have performed spectral classification as a function of object type and Smith *et al.* (2007) have published the spectral decomposition tool PAHFIT that provides a band by band analysis. Galliano *et al.* (2008) have derived an empirical law between the 6.2 over 11.3 μm band intensity ratio and the ionisation parameter $\gamma = G_0\sqrt(T)/n_e$ where G_0 is the UV field intensity, T the gas temperature and n_e the electron density. Further spectral analysis requires a better understanding of the nature of the AIB carriers and their physical and chemical evolution due to environmental conditions. In a series of papers (Rapacioli *et al.* 2005; Berné *et al.* 2007; Joblin *et al.* 2008; Berné *et al.* 2009), we have shown that the AIB spectrum can be decomposed into four template spectra, the various mixtures of which create the observed spectral variations in most UV-irradiated environments. Three of the spectra carry only band emission and are assigned to PAH cations, PAH neutrals and large ionized PAHs. The other spectrum has band and continuum emission and it is assigned to evaporating very small grains (eVSGs) by the authors (see discussion in Pilleri *et al.* 2012). A drop of the eVSG emission is observed at the UV irradiated surface of clouds and is correlated with an increase in the PAH emission, strongly suggesting that free PAHs are produced by destruction of eVSGs under UV irradiation. The fitting tool PAHTAT includes the four template spectra described above. It is publicaly available and can be used to derive local physical parameters such as G_0 or the extinction along the line of sight (Pilleri *et al.* 2012; http://userpages.irap.omp.eu/~cjoblin/PAHTAT).

2. Which carriers for the AIBs

The best carriers for the AIBs are polycyclic aromatic hydrocarbon (PAH) molecules. The emission mechanism following excitation by single UV photons can account for the observational constrains such as the broadness and profile of the bands, without requiring a (too) large diversity of species (Pech *et al.* 2002). The evolution of relative band intensities with the ionisation parameter is consistent with the molecular properties of PAHs (e.g., Ricca *et al.* 2012 and references therein). The recent identification of C_{60} (see conference by J. Cami) supports the presence of large carbon molecules in space. Yet, there has been no identification of a single PAH in space. The mid-IR range reflects mainly chemical bonds and, therefore, provides information on the composition and some trends about the size and structure, but is not well suited to identify individual species. Still, some bands are challenging to be accounted for by pure PAHs (e.g the presence of N heteroatoms has been proposed by Hudgins *et al.* 2005 to account for the 6.2 μm band). To better characterise interstellar PAHs, one has therefore to progress in the understanding on how these species are formed in the envelopes of evolved stars. At the protoplanetary nebula stage, gas-phase hydrocarbon photochemistry (Cernicharo *et al.* 2001) is observed as well as signatures of hydrogenated amorphous carbon grains (e.g., Chiar *et al.* 1998), aromatics and aliphatics in molecular form and/or in the form of pristine/hot small grains (Kwok *et al.* 1999; Peeters *et al.* 2002). Irradiation by UV photons in more advanced evolutionary stage of the central star leads to destruction of the fragile aliphatic bonds and further aromatisation of the material, although the involved species and the mechanisms leading to the formation of PAHs have still to be clarified. For instance it has been proposed that PAHs are generated by photodestruction of carbonaceous grains by UV photons and possibly shocks from the central star (Kwok *et al.* 2001; Joblin *et al.* 2008; García-Hernández *et al.* 2011), but this has still to be proven. In solving this evolutionary scenario it is important to consider other emission features than AIBs, such as the 3.4 and 6.9 μm bands due to aliphatics, the broad 8 and 12 μm bands, as well as the 21 and 30 μm features (see conference by K. Volk).

References

Berné, O., Joblin, C., Deville, Y., Smith, J. D., *et al.* 2007, *A&A* 469, 575
Berné, O., Joblin, C., Fuente, A., & Menard, F. 2009, *A&A* 495, 827
&Cernicharo, J. and Heras, A. M., Pardo, J. R., *et al.* 2001, *ApJ* 546, L127
Chiar, J. E., Pendleton, Y. J., Geballe, T. R., & Tielens, A. G. G. M. 1998, *ApJ* 507, 281
Fadda, D., Yan, L., Lagache, G., *et al.* 2010, *ApJ* 719, 425
Galliano, F., Madden, S. C., Tielens, A. G. G. M., Peeters, E., & Jones, A. P. 2008, *ApJ* 679, 310
&García-Hernández, D. A. and Iglesias-Groth, S., Acosta-Pulido *et al.* 2011, *ApJ* 737, L30
Hudgins, D. M., Bauschlicher, Jr., C. W., & Allamandola, L. J. 2005, *ApJ* 632, 316
Joblin, C., Szczerba, R., Berné, O., & Szyszka, C. 2008, *A&A* 490, 189
Kwok, S., Volk, K., & Hrivnak, B. J. 1999, *A&A* 350, L35
Kwok, S. and Volk, K., & Bernath, P. 2001, *ApJ* 554, L87
Pech, C., Joblin, C., & Boissel, P. 2002, *A&A* 388, 639
Peeters, E., Hony, S., Van Kerckhoven, C., *et al.* 2002, *A&A* 390, 1089
Pilleri, P., Montillaud, J., Berné, O., & Joblin, C. 2012, *A&A* 542, A69
Rapacioli, M., Joblin, C., & Boissel, P. 2005, *A&A* 429, 193
Ricca, A., Bauschlicher, Jr., C. W., Boersma, C., Tielens, A. G. G. M., & Allamandola, L. J. 2012, *ApJ* 754, 75
Smith, J. D. T., Draine, B. T., Dale, D. A., *et al.* 2007, *ApJ* 656, 770
van Diedenhoven, B., Peeters, E., Van Kerckhoven, C., *et al.* 2004, *ApJ* 611, 928

Highlights of Astronomy, Volume 16
XXVIIIth IAU General Assembly, August 2012
T. Montmerle, ed.

doi:10.1017/S1743921314012964

Carbon Star Dust Features: the 21 and 30 μm Features

Kevin Volk

Space Telescope Science Institute, 3700 San Martin Drive, Baltimore, Maryland, U.S.A.
email: volk@stsci.edu

Abstract. Some characteristics of the 21 and 30 μm features observed in carbon star dust shells are briefly discussed.

---

1. The 30 μm Feature

This feature was discovered by Forrest, Houck & McCarthy (1981) from *Kuiper Airborne Observatory* observations of carbon stars in the 16 to 30 μm wavelength range. The feature is generally observed in higher optical depth dust shells, or in objects that have evolved off the asymptotic giant branch (AGB). The feature is particularly strong in the G-type and F-type post-AGB object spectra, for unknown reasons.

The feature cannot be observed from the ground due to atmospheric absorption, so most of the data we have on the feature comes from *Infrared Space Observatory (ISO)* or *Spitzer Space Telescope (Spitzer)* spectroscopy. We do not have a survey of objects with the feature, only pointed observations of individual stars that have been observed by these satellites. *ISO* observed the feature in the spectra of 63 carbon-rich objects. *Spitzer* provided the spectra of $\sim$140 more objects with the feature, including some S-type star spectra with weak 30 μm features. We observe variations in the feature shape and peak wavelength, particularly in the post-AGB phase, which are not understood.

The carrier was proposed to be a mixture of simple metal sulphides, primarily MgS, in Nuth *et al.* (1985). This has been the favoured explanation since then, but there are unresolved concerns about whether the S abundance is high enough to produce this strong feature. We are severely hampered in assessing these sulphide carriers because the optical constants (n, k) have not been measured for them at wavelengths <1 μm, and so realistic radiative transfer models cannot be made to compare with the available observations. It has been suggested that MgS does not form grains by itself, but rather that it coats other grains (SiC, amorphous carbon) which already are present in carbon star dust shells. There is no trace of such coatings on pre-solar carbon grains isolated from meteorites, so it not clear to me that this is a viable idea.

Whatever the carrier is, observations suggest that the feature becomes weaker or is absent in carbon stars in lower metallicity systems such as the Small Magellanic Cloud. The carrier also does not survive in the interstellar medium, as there is no trace of the feature in the extinction curve. We also know of some highly obscured carbon stars which lack the 30 μm feature, and it is not clear why the carrier is absent.

My opinion is that MgS is probably not the carrier of the feature. The few alternatives which have been proposed in the literature have not been given much attention. I believe that we need to get proper optical constants for MgS and related grains, and make realistic models to see whether these can match the observations. We will not make fundamental progress while we lack basic information about the proposed carriers.

2. The 21 μm Feature

This is a rare feature that is only present in carbon-rich post-AGB stars of G-type to late B-type. It was discovered from the *Infrared Astronomical Satellite* Low Resolution Spectrograph data by Kwok, Volk & Hrivnak (1989) in four objects. Although the feature can be observed from the ground, the best observations we have of the feature are from *ISO* and *Spitzer*. There are a total of only 26 objects known to definitely have the feature: 19 in the Galaxy, 6 in the Large Magellanic Cloud, and 1 in the Small Magellanic Cloud. There are approximately 15 more possible detections of the feature, but these are very weak or otherwise questionable. The feature does not appear to show the kind of variations in peak wavelength or shape that the 30 μm feature does.

The feature is unusual in that it appears to be transient. There is no convincing detection of the feature in any carbon star or planetary nebula spectrum. It appears early in the post-AGB evolution and disappears by the time the star evolves to early B-type. It always seems to be accompanied by the 30 μm feature and the unidentified infrared (UIR) bands, with wide variations in the relative strengths. No clear correlations between the features have been demonstrated. In the discovery paper it was suggested that the feature is due to a transient molecule, but no good candidate for this has been found. There have also been suggestions that the feature is due to dust grains, such as TiC or FeO dust, but in such a case it is not clear why the feature disappears as the star evolves. There are around a dozen carriers proposed in the literature, none of which has gained general support in the community. Many of the proposed carriers are related to polycyclic aromatic hydrocarbons, since these are generally thought to be the carriers of the UIR bands which accompany the 21 μm feature.

Abundance analysis has been carried out for many of the stars whose spectra show the 21 μm feature. They are found to generally be of lower metallicity than average in their host system, and have strong s-process element enhancements. They are also generally carbon-rich, although in a few cases the measured C/O ratios in the stars are $\sim$1 with moderate uncertainties. I assume that the stars are all carbon-rich because the circumstellar dust shells have carbon-based dust and no trace of silicate dust. While it has sometimes been stated in the literature that all carbon-rich post-AGB stars of G- and F-type with strong s-process element enhancement have the 21 μm feature, there are at least two exceptions to this that I know of. It is not clear whether the strong dredge-up experienced by these stars is needed to produce the feature.

At this time the identity of the 21 μm feature carrier is an open question. I tend to believe that the carrier is a molecule rather than a dust grain. If we could take spectral polarimetry observations across the feature we should be able to definitively determine whether the carrier is a dust grain. I think that finding more early post-AGB objects in the hope of observing the feature as it appears would be extremely important in determining the carrier. If we can observe changes in the feature over time (decades? centuries?) that may also give us the key to finding the true nature of the carrier.

References

Forrest, W. J., Houck, J. R., & McCarthy, J. F. 1981, *ApJ* 248, 195

Kwok, S., Volk, K., & Hrivnak, B. J. 1989, *ApJ* 345, L51

Nuth, J. A., Moseley, S. H., Silverberg, R. F., Goebel, J. H., & Moore, W. J. 1985, *ApJ* 290, L41

Highlights of Astronomy, Volume 16
XXVIIIth IAU General Assembly, August 2012
T. Montmerle, ed.

doi:10.1017/S1743921314012976

Near-Infrared Spectroscopy of the Diffuse Galactic Emission

T. Onaka[1], I. Sakon[1], R. Ohsawa[1], T. I. Mori[1], H. Kaneda[2], M. Tanaka[3], Y. Okada[4], F. Boulanger[5], C. Joblin[6] and P. Pilleri[7]

[1]Department of Astronomy, The University of Tokyo, Tokyo 113-0033, Japan
email: onaka@astron.s.u-tokyo.ac.jp

[2]Graduate School of Science, Nagoya University, Aichi 464-8602, Japan

[3]Center for Computational Sciences, University of Tsukuba, Ibaraki 305-8577, Japan

[4]I. Physikalisches Institut, Universität zu Köln, 50937 Köln, Germany

[5]Institut d'Astrophysique Spatiale (IAS), UMR 8617, CNRS & Université Paris-Sud 11, Bâtiment 121, 91405 Orsay Cedex, France

[6]Université de Toulouse, UPS-OMP, IRAP & CNRS, IRAP, 31028 Toulouse Cedex 4, France

[7]Centre de Astrobiología (INTA-CSIC), 28850 Torrejón de Ardoz, Spain & Observatorio Astronómico Nacional, Apdo. 112, 28803 Alcalá de Henares, Spain

Abstract. The near-infrared (NIR) spectral range (2–5 μm) contains a number of interesting features for the study of the interstellar medium. In particular, the aromatic and aliphatic components in carbonaceous dust can be investigated most efficiently with the NIR spectroscopy. We analyze NIR spectra of the diffuse Galactic emission taken with the Infrared Camera onboard *AKARI* and find that the aliphatic to aromatic emission band ratio decreases toward the ionized gas, which suggests processing of the band carriers in the ionized region.

Keywords. ISM: general, (ISM:) dust, extinction, infrared: ISM, infrared: general

---

The near-infrared (NIR) spectral range from 2 to 5 μm contains a number of important features of gaseous and solid species in the interstellar medium (ISM). Absorption features of several major ices are observed at 3 μm (H_2O), 4.3 μm (CO_2), and 4.6 μm (CO) (e.g., Shimonishi *et al.* 2010). In this spectral range, there are also the fundamental vibration mode of CO gas (e.g., Rho *et al.* 2012) and several molecular hydrogen lines, from which we can estimate the physical conditions of the warm gas (e.g., Lee *et al.* 2011). In addition, it contains emission features of hydrocarbon dust around 3 μm, which allow us to investigate aromatic and aliphatic components (Boulanger *et al.* 2011; Kwok & Zhang 2011; Jones 2012). The 3.3 μm emission band, which is commonly attributed to polycyclic aromatic hydrocarbons (PAHs), is useful to study the size distribution and the processing of the band carriers in the ISM (e.g., Mori *et al.* 2012). Deuterated hydrocarbon features are also expected to be present in 4.4–4.7 μm (e.g., Peeters *et al.* 2004, Onaka *et al.* 2011). Despite these interests, however, this spectral range has barely been explored by instruments with high sensitivity.

The Infrared Camera (IRC: Onaka *et al.* 2007) onboard *AKARI* (Murakami *et al.* 2007) enabled for the first time high-sensitivity spectroscopy in the NIR, with a spectral-resolution of $\lambda/\Delta\lambda \sim 100$ (Ohyama *et al.* 2007). Even after the exhaustion of the cryogen, the IRC continued to carry out NIR observations and obtained a large number of NIR spectra in various celestial objects (Onaka *et al.* 2010).

As part of the Interstellar Medium in our Galaxy and Nearby Galaxies program (ISMGN: Kaneda *et al.* 2009), we carried out NIR spectroscopy towards more than 100 positions in the Galactic plane. We describe here the 3 μm data. We fit the observed

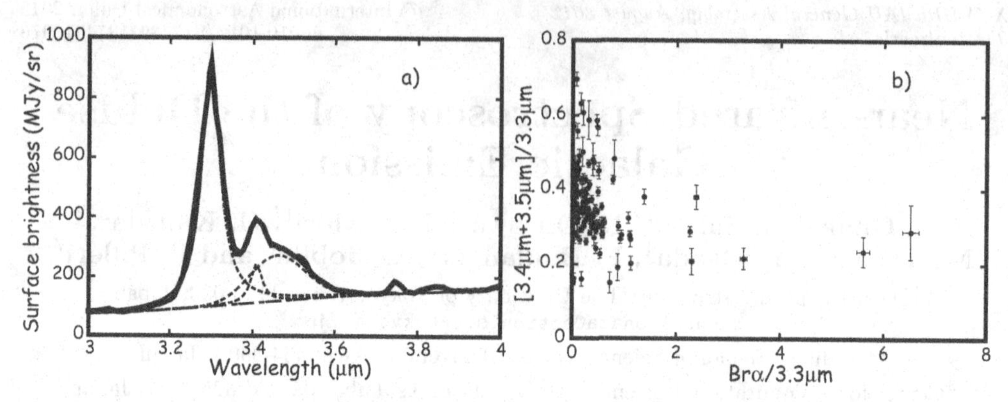

Figure 1. a) Example of *AKARI*/IRC NIR spectrum. The dotted lines indicate the decomposition of the 3.3, 3.4, and 3.5 μm bands (see text). The thick and thin solid lines represent the observed and fitted spectra, respectively. The dot-dashed line shows the continuum. b) The aliphatic to aromatic band ratio against the Brα to the 3.3 μm band intensity ratio (see text).

features with two Lorentzians (3.3 and 3.4 μm) and one Gaussian (3.5 μm). This fit is merely technical due to the low spectral-resolution of the instrument, but correctly estimates the aromatic (the 3.3 μm band) and the aliphatic (the summation of the 3.4 and 3.5 μm bands) band intensities. Figure 1a shows an example of the IRC spectrum together with the fit. Figure 1b plots the ratio of the aliphatic to aromatic band intensities against the ionized gas indicator, the ratio of Brα to the 3.3 μm band intensities. A trend is seen that the aliphatic to aromatic ratio decreases towards the ionized region. This implies a preferential destruction of aliphatic bonds relative to aromatic bonds in the ionized gas as described in Joblin *et al.* (1996) or that the aliphatic component resides mostly in small band carriers, which are quickly destroyed in the ionized gas (Mori *et al.* 2012).

Acknowledgements

This work is based on observations with *AKARI*, a JAXA project with the participation of ESA. The author thanks all the members of the *AKARI*/ISMGN team for their continuous help and encouragements. This work is supported in part by a Grant-in-Aid for Scientific Research from the Japan Society for the Promotion of Science.

References

Boulanger, F., Onaka, T., Pilleri, P., & Joblin, C. 2011, *EAS Publ. Ser.* 46, 399
Joblin, C., Tielens, A. G. G.. M., Allamandola, L. J., & Geballe, T. R. 1996, *ApJ* 458, 610
Jones, A. P. 2012, *A&A* 540, A2
Kaneda, H., Koo, B.-C., Onaka, T., & Takahashi, H. 2009, *Adv. Sp. Res.* 44, 1038
Kwok, S. & Zhang, Y. 2011, *Nature* 479, 80
Lee, H.-G., Moon, D.-S., Koo, B.-C., *et al.* 2011, *ApJ* 740, 31
Mori, T. I., Sakon, I., Onaka, T., Kaneda, H., Umehata, H., & Ohsawa, R., 2012, *ApJ* 744, 68
Murakami, H., Baba, H., Barthel, P., *et al.* 2007, *PASJ* 59, S369
Ohyama, Y., Onaka, T., Matsuhara, H., *et al.* 2007, *PASJ* 59, S411
Onaka, T., Matsuhara, H., Wada, T., *et al.* 2007, *PASJ* 59, S401
Onaka, T., Matsuhara, H., Wada, T., *et al.* 2010, *Proc. of SPIE* 7731, 77310M
Onaka, T., Sakon, I., Ohsawa, R., *et al.* 2011, *EAS Publ. Ser.* 46, 55
Peeters, E., Allamandola, L. J., Bauschlicher, C. W., Jr., *et al.* 2004, *ApJ* 604, 252
Rho, J., Onaka, T., Cami, J., & Reach, W. T. 2012, *ApJ* 747, L6
Shimonishi, T., Onaka, T., Kato, D., *et al.* 2010, *A&A* 514, A12

Highlights of Astronomy, Volume 16
XXVIIIth IAU General Assembly, August 2012
T. Montmerle, ed.

doi:10.1017/S174392131401299X

Fullerenes in Circumstellar and Interstellar Environments

Jan Cami

Department of Physics & Astronomy, The University of Western Ontario,
London ON N6A 3K7, Canada
email: jcami@uwo.ca

Abstract. In recent years, the fullerene species C_{60} (and to a lesser extent also C_{70}) has been reported in the mid-IR spectra of various astronomical objects. Cosmic fullerenes form in the circumstellar material of evolved stars, and survive in the interstellar medium (ISM). It is not entirely clear how they form or what their excitation mechanism is.

Keywords. Astrochemistry, (stars:) circumstellar matter, ISM: molecules, infrared: ISM

1. Fullerenes in astrophysical environments

Fullerenes (such as the buckminsterfullerene C_{60}) are large carbonaceous molecules in the shape of a hollow sphere of ellipsoid. They are very stable, and thus it was suggested early on that they could also form in space and be abundant and widespread in the Universe (Kroto *et al.* 1985). Astronomical searches for the electronic bands of neutral C_{60} were unsuccessful though (for an overview, see Herbig 2000); and the detection of two diffuse interstellar bands near the predicted wavelengths of C_{60}^{+} awaits confirmation from a gas-phase laboratory spectrum (Foing & Ehrenfreund 1994). C_{60} also has 4 IR active vibrational modes at 7.0, 8.5, 17.4 and 18.9 μm. Dedicated searches for these bands did not result in a detection either (Clayton *et al.* 1995; Moutou *et al.* 1999).

Recently, we reported the detection of the IR active modes of C_{60} and C_{70} in the *Spitzer*-IRS spectrum of the young, low-excitation planetary nebula (PN) Tc 1 (Cami *et al.* 2010). Since then, fullerenes have been found in many more PNe (García-Hernández *et al.* 2010, 2011a), a proto-PN (Zhang & Kwok 2011), a few R Cor Bor stars (García-Hernández *et al.* 2011b; Clayton *et al.* 2011) and even in O-rich binary post-AGB stars (Gielen *et al.* 2011). In addition, fullerenes have turned up in interstellar environments (Sellgren *et al.* 2010; Rubin *et al.* 2011; Boersma *et al.* 2012) and in young stellar objects (Roberts *et al.* 2012) as well. From these detections it is clear that fullerenes are formed in the circumstellar environments of evolved stars. They then either survive in the ISM (possibly incorporated into dust grains), or form there when conditions are right.

2. The fullerene excitation mechanism

To explain the IR emission of cosmic fullerenes, two mechanisms have been considered that offer quite different predictions about the relative band strengths (for a detailed comparison, see Bernard-Salas *et al.* 2012). Thermal C_{60} emission models show large variations in the relative strength of all bands as a function of temperature; for $T \leqslant 300$ K, the 7.0 and 8.5 μm bands are very weak compared to the 17.4 and 18.9 μm bands. For fluorescence on the other hand, the band strengths only depend on the average absorbed photon energy; in that case, the 17.4/18.9 μm band ratio is roughly constant (for reasonable photon energies) while the 7.0 and 8.5 μm bands should be fairly strong.

Observationally, the 7.0 and 8.5 μm bands are often very weak or even undetectable, while there are considerable variations in the 17.4/18.9 μm band ratio; this is more easily explained by thermal models than by fluorescence models. However, these variations could also be the consequence of contamination by PAH emission. In the three known uncontaminated fullerene-rich PNe on the other hand, the 7.0 μm band is far *too strong* to be explained by even fluorescence from C_{60} alone. As pointed out to the careful reader by Bernard-Salas *et al.* (2012), the 7.0 μm emission in those sources includes a significant contribution from C_{70}, provided at least that the emission is due to fluorescence. For one object (Tc 1), fluorescence is further supported by the observation that the C_{60} emission peaks at large distances ($\sim$ 8000 AU) from the central star. If fluorescence is also the excitation mechanism for the other astronomical sources where fullerenes have been detected, then the weak 7.0 and 8.5 μm bands indicate that the fullerene emission is not due to isolated, free C_{60} molecules in the gas-phase; there might be contributions from other species as well and/or the emission may be due to fullerene clusters or nanocrystals.

3. The formation of cosmic fullerenes

Several routes have been proposed to explain the formation of fullerenes in astrophysical environments. Densities in circumstellar and interstellar environments are too low for bottom-up fullerene formation on reasonable timescales (Micelotta *et al.* 2012). Fullerenes could form from the processing of PAHs (Berné & Tielens 2012), but this requires fine-tuned initial conditions. A promising route starts from *arophatics* – large clusters of aromatic rings with aliphatic and olefinic bridges that originate from a:C-H grains (Micelotta *et al.* 2012). UV irradiation first dehydrogenates and aromatizes such structures; subsequent C_2 ejection then shrinks down the resulting cages to C_{60}. Further shrinking is inhibited by a high energy barrier. The spectral imprint of the parent a:C-H grains in the IR spectra of fullerene-rich PNe offers some observational support for this mechanism (Bernard-Salas *et al.* 2012).

References

Bernard-Salas, J., Cami, J., Peeters, E., *et al.* 2012, *ApJ* 757, 41
Berné, O. & Tielens, A. G. G. M., 2012, *PNAS* 109, 401
Boersma, C., Rubin, R. H., & Allamandola, L. J. 2012, *ApJ* 753, 168
Cami, J., Bernard-Salas, J., Peeters, E., & Malek, S. E. 2010, *Science* 329, 1180
Clayton, G. C., De Marco, O., Whitney, B. A., *et al.* 2011, *AJ* 142, 54
Clayton, G. C., Kelly, D. M., Lacy, J. H., *et al.* 1995, *AJ* 109, 2096
Foing, B. H. & Ehrenfreund, P. 1994, *Nature* 369, 296
García-Hernández, D. A., Iglesias-Groth, S., Acosta-Pulido, J. A., *et al.* 2011a, *ApJ* 737, L30
García-Hernández, D. A., Kameswara Rao, N., & Lambert, D. L. 2011b, *ApJ* 729, 126
García-Hernández, D. A., Manchado, A., García-Lario, P., *et al.* 2010, *ApJ* 724, L39
Gielen, C., Cami, J., Bouwman, J., Peeters, E., & Min, M. 2011, *A&A* 536, A54
Herbig, G. H. 2000, *ApJ* 542, 334
Kroto, H. W., Heath, J. R., Obrien, S. C., Curl, R. F., & Smalley, R. E. 1985, *Nature* 318, 162
Micelotta, E., Jones, A. P., Cami, J., *et al.* 2012, *ApJ* 761, 35
Moutou, C., Sellgren, K., Verstraete, L., & Léger, A. 1999, *A&A* 347, 949
Roberts, K. R. G., Smith, K. T., & Sarre, P. J. 2012, *MNRAS* 421, 3277
Rubin, R. H., Simpson, J. P., O'Dell, C. R., *et al.* 2011, *MNRAS* 410, 1320
Sellgren, K., Werner, M. W., Ingalls, J. G., *et al.* 2010, *ApJ* 722, L54
Zhang, Y. & Kwok, S. 2011, *ApJ* 730, 126

Highlights of Astronomy, Volume 16
XXVIIIth IAU General Assembly, August 2012
T. Montmerle, ed.

doi:10.1017/S1743921314013003

Amorphous Hydrocarbon Optical Properties

Anthony Jones

Université Paris Sud & CNRS, Institut d'Astrophysique Spatiale, UMR8617, Orsay, F-91405
email: Anthony.Jones@ias.u-psud.fr

Abstract. Hydrogenated amorphous carbon materials, a-C(:H), whose optical properties evolve in response to UV irradiation processing are promising candidate materials for cosmic carbonaceous dust. The optical properties of a:C(:H) particles have been derived as a function of size, band gap and hydrogen content over a wide wavelength range (EUV-cm) and can be used to investigate the size-dependent evolution of a-C(:H) material properties in the ISM.

Keywords. (ISM:) dust, extinction, ISM: molecules, ISM: general

---

1. Modelling amorphous hydrocarbon optical properties

The *exact* nature of interstellar carbonaceous dust is still something of a mystery. The evolution a-C:H, a-C or HAC solids is a complex subject that presents a particular challenge because these materials appear to be rather vulnerable to interstellar processing (*e.g.*, Serra Díaz-Cano & Jones 2008; Jones 2009; Jones & Nuth 2011) and to undergo complex, size-dependent evolution arising, principally, from UV photon-driven processing. Here we introduce the optEC$_{(s)}$(a) model for the optical properties of amorphous hydrocarbons, a-C(:H), from hydrogen-poor, a-C, to hydrogen-rich, a-C:H, carbonaceous solids. These data provide a tool that can be used to explore carbonaceous dust and its observable characteristics. The optEC$_{(s)}$(a) model for a-C(:H) materials is presented in a series of papers (Jones 2012b,c,a), which derive their size-dependent structure and their complex refractive indices, $m(n,k)$). These data are publicly-available through the CDS (see the links in papers). We note that the derived data are strongly-constrained by the available laboratory data and have *not* been adjusted to fit astronomical observations. The upper panels in Fig. 1 show the parent or bulk material optEC$_{(s)}$(a) values of n and k, from EUV to mm wavelengths, as a function of the bulk material Tauc band gap, E_g, and the lower panels in Fig. 1 show the equivalent n and k data for 1 nm radius particles. The major changes that occur as particle size decreases (see Jones 2012a) are: 1) increased surface hydrogenation and CH$_n$ IR band intensities, and 2) a "collapse" of the continua for $\lambda \gtrsim 0.5\,\mu$m and $E_g \lesssim 1.5$ eV. As shown by Jones (2012a), the latter effect is due to a reduction in the maximum-allowable, particle-radius-determined aromatic domain sizes as the particle radius decreases, which is clearly seen in the lower panels in Fig. 1. For particles with radii < 1 nm the optical properties begin to look rather similar.

As has been shown the derived optEC$_{(s)}$(a) data are qualitatively consistent with: the FUV-UV bump-visible extinction (non-)correlations, the IR absorption and emission bands in the $3.3 - 3.6\,\mu$m region, variations in the FIR-mm emissivity index (Jones 2012b,c,a). Further, the model predicts that: the $3.28\,\mu$m aromatic CH band will always be accompanied by aliphatic CH$_n$ bands and/or a plateau in the $3.35 - 3.55\,\mu$m region, the end of the road evolution for small a-C(:H) particles is probably aromatic/aliphatic cage-like structuresthat could provide a route to fullerene formation (Bernard-Salas *et al.* 2012; Micelotta *et al.* 2012) the UV-photolytic fragmentation of small a-C:H grains will lead to the formation of small hydrocarbon molecules (CCH, c-C$_3$ H$_2$, C$_4$H, etc.) in PDR

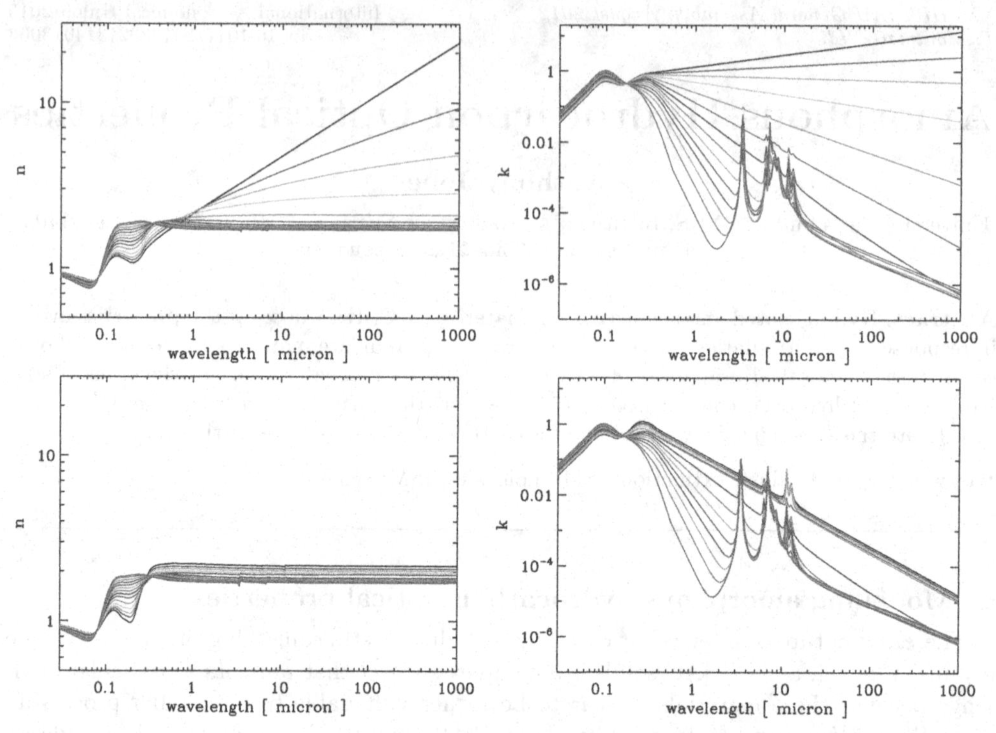

Figure 1. The optEC$_{(s)}$(a) model real and imaginary parts of the complex refractive index, n and k, for bulk a-C:H materials ($\equiv a > 100$ nm, upper) and for $a = 1$ nm particles (lower). The bulk material band gap, E_g, increases from -0.1 eV (top) to 2.67 eV (bottom) at $\lambda \sim 2\,\mu$m.

regions, and "pure" graphite grains and "perfect" PAHs are probably not important components of dust in the ISM (Jones 2012b,c,a).

2. Concluding remarks

A new data-set for the size-dependent optical properties of amorphous hydrocarbon particles, for the first time, provides a means to a detailed exploration of the evolution of these complex materials in the ISM. The principal drivers of their evolution would appear to be photon-induced processing, which results in de-hydrogenation and band gap closure that are coupled to significant changes in their IR spectra and their long wavelength emission.

References

Bernard-Salas, J., Cami, J., Peeters, E., *et al.* 2012, *ApJ* 757, 41

Jones, A. P. 2009, in *Astronomical Society of the Pacific Conference Series*, Vol. 414, Cosmic Dust - Near and Far, ed. T. Henning, E. Grün, & J. Steinacker, 473–+

Jones, A. P. 2012a, *A&A* 542, A98

Jones, A. P. 2012b, *A&A* 540, A1

Jones, A. P. 2012c, *A&A* 540, A2

Jones, A. P. & Nuth, J. A. 2011, *A&A* 530, A44

Micelotta, E. R., Jones, A. P., Cami, J., *et al.* 2012, *ApJ* 761, 35

Serra Díaz-Cano, L. & Jones, A. P. 2008, *A&A* 492, 127

Highlights of Astronomy, Volume 16
XXVIIIth IAU General Assembly, August 2012
T. Montmerle, ed.

doi:10.1017/S1743921314013015

Prebiotic Matter in Space

Pascale Ehrenfreund, Andreas Elsaesser and J. Groen

Leiden Institute of Chemistry, P O Box 9502, 2300RA Leiden, The Netherlands
email: p.ehrenfreund@chem.leidenuniv.nl

Abstract. A significant number of molecules that are used in contemporary biochemistry on Earth are found in interstellar and circumstellar regions as well as solar system environments. In particular small solar system bodies hold clues to processes that formed our solar system. Comets, asteroids, and meteorite delivered extraterrestrial material during the heavy bombardment phase ~3.9 billion years ago to the young planets, a process that made carbonaceous material available to the early Earth. In-depth understanding of the organic reservoir in different space environments as well as data on the stability of organic and prebiotic material in solar system environments are vital to assess and quantify the extraterrestrial contribution of prebiotic sources available to the young Earth.

Keywords. Prebiotic matter, comets, asteroids, meteorites, carbonaceous material, early Earth

---

The cycle of organic molecules in space

Recent observations obtained from space missions and Earth-based telescopes have led to important discoveries in the field of molecular detection. The carbon chemistry seems to follow common pathways throughout the Universe. Interstellar dense molecular clouds and circumstellar envelopes are astrophysical environments providing the conditions for complex molecular synthesis (Ehrenfreund & Charnley 2000, Kwok 2009, Herbst & van Dishoeck 2009). But also our solar nebula was chemically active and dynamic as evidenced by results from the *Stardust* mission and recent theoretical calculations (Brownlee *et al.* 2006, Ciesla & Sandford 2012). Currently 175 molecules are detected in the interstellar and circumstellar gas; 53 molecules are found in extragalactic sources (http://www.astro.uni-koeln.de). A lot of progress has been made to reveal the nature of the polycyclic aromatic hydrocarbon (PAHs) distribution, dust and carbonaceous grains in space. New results on the abundance and composition of organic molecules detected in the cometary coma and through laboratory analyses of meteorites have improved our knowledge on the fraction of material that reached the early Earth during the extended Late Heavy Bombardment LHB triggered by rapid migration of giant planets (Gomes *et al.* 2005, Bottke *et al.* 2012). This material may have provided raw material useful for protocell assembly on Earth (Chyba *et al.* 1992, Ehrenfreund *et al.* 2002).

Carbonaceous extract of meteorites have shown 66 extraterrestrial amino acids in addition to many other numerous organic compounds in the soluble phase although the major carbon component in meteorite samples is composed of a macromolecular organic fraction. Ultra high resolution molecular analysis of the solvent accessible organic fraction of Murchison shows high molecular diversity and many molecules still remain to be detected. The highest abundance of amino acids is currently measured in primitive SR meteorites found in Antarctica (Martins *et al.* 2007). Extraterrestrial nucleobases have been confirmed by isotopic measurements (Martins *et al.* 2008). The Murchison and Lonewolf Nunataks 94102 meteorites contained also a diverse suite of nucleobases, which included three unusual and terrestrially rare nucleobase analogs. Parallel experiments on the formation mechanism showed the formation of nucleobases in reactions of ammonium cyanide (Callahan *et al.* 2011). There is a consensus that prebiotic compounds such as amino acids and nucleobases are formed by aqueous alteration within their parent bodies and in the solar nebula. The abundance, distribution and ratio of amino acids (e.g. the β-Alanine/AIB ratio) shows a consistent dependence on aqueous alteration ruling out radiation processing (Glavin & Dworkin 2009). In order to understand how this material could have been useful for the origin of life on Earth it is crucial to determine the stability of organic compounds under elevated radiation and temperature conditions. Amino acid destruction has been monitored by many different techniques and shows a degradation mechanism leading to decarboxylation. A slower destruction is observed when amino acids are embedded in ice (Gerakines *et al.* 2012). The consensus is that amino acids and nucleobases must be shielded by ice or minerals in order to survive harsh conditions (in space) and on the young Earth. Two decades of successful experiments on the *International Space Station (ISS)* and free-flying satellites have provided new information about the evolution of organic and biological material in space and planetary environments; some of them could not be reconciled with ground truth experiments (Guan *et al.* 2010, Mattioda *et al.* 2012).

Another important question is how prebiotic molecules can assemble on the early Earth? The prebiotic availability of fatty acids or their precursors and their properties of encapsulation, high permeability, and membrane growth, make them ideal model systems for investigating primitive forms of life. We have recently shown that PAH derivatives can be incorporated into fatty acids membranes and measured the first indication of a cholesterol-like stabilizing effect of oxidized PAH derivatives in a simulated prebiotic membrane (Groen *et al.* 2012).

In order to understand the link between extraterrestrial chemistry, the delivery of organic material to the young planets via small bodies and the origin of life on Earth it is crucial to investigate the inventory and the stability of organic material in different astronomical environments and on the early Earth. Contemporary biomolecules are not very stable against radiation or temperature cycles which invoke the presence of possible more stable precursor molecules like aromatic molecules (Ehrenfreund *et al.* 2006).

References

Bottke, W. F., *et al.* 2012, *Nature* 485/7396, 78–81
Brownlee, D., *et al.* 2006, *Science* 314, 1711-1716
Callahan, M. P., *et al.* 2011, *PNAS* 108(34), 13995-13998
Chyba, C. & Sagan, C. 1992, *Nature* 355, 125-132
Chiesla, F. J. & Sandford, S. A. 2012, *Science* 336/6080, 452
Ehrenfreund, P. & Charnley, S. B. 2000, *ARAA* 38, 427-483
Ehrenfreund, P., *et al.* 2002, *Reports on the Progress in Physics* 65, 1427-1487
Ehrenfreund, P., Rasmussen, S., Cleaves, J. H., & Chen, L., 2006, *Astrobiology* 6/3, 490-520
Gerakins, P. A., Hudson, R. L., Moore, M. H., & Bell, L. 2012 *Icarus* 220, 647-659
Glavin, D. P. & Dworkin, J. P. 2009, *PNAS* 106, 5487-5492
Gomes, R., Levison, H. F., Tsiganis, K., & Morbidelli, A. 2005, *Nature* 435, 466-469
Groen, J., Deamer, A., Kros, A., & Ehrenfreund, P. 2012, *OLEB* 42, 295-306
Guan, Y., *et al.* 2010, *Planetary and Space Science* 58, 1327-1346
Herbst, E. & van Dishoeck, E. F. 2009, *ARAA* 47, 427-480
Kwok, S. 2009, *Int. Journal of Astrobiology* 8/3, 161-167
Martins, Z., *et al.* 2007, *Meteoritics & Planetary Science* 42/12, 2125-2136
Martins, Z., *et al.* 2008, *Earth and Planetary Science Letters* 270, 130-136
Mattioda, A., *et al.* 2012, *Astrobiology* 12, 841-853

Highlights of Astronomy, Volume 16
XXVIIIth IAU General Assembly, August 2012
T. Montmerle, ed.

doi:10.1017/S1743921314013027

Carbon Nanoparticles and Carbonaceous Solids

W. W. Duley

Department of Physics and Astronomy, University of Waterloo,
200 University Ave. W. Ontario, Canada
email: wwduley@uwaterloo.ca

Abstract. This paper reports on the preparation of hydrogenated amorphous carbon nanoparticles whose spectral characteristics include an absorption band at 217.5 nm with the profile and characteristics of the interstellar 217.5 nm feature. Vibrational spectra of these particles also contain the features commonly observed in IR absorption and emission from dust in the diffuse interstellar medium. These materials are produced under "slow" deposition conditions by minimizing the flux of incident carbon atoms and by reducing surface mobility.

Keywords. ISM: molecules, ISM: lines and bands, ISM: clouds

Samples of hydrogenated amorphous carbon (HAC) prepared by single atom deposition of carbon at 77 K results in the formation of nano-particles whose spectroscopic properties include the appearance of a 217.5 nm absorption band. The characteristics of this band reproduce those of the interstellar feature as catalogued by Fitzpatrick & Massa (1986). Chemical analysis of this material obtained from XPS, Raman and SERS spectra show that it consists of a network of chains and rings. The ring component is limited to small sp^2 bonded molecules, including a variety of naphthalene derivatives, which we suggest are the source of the 217.5 nm absorption band. The spectra contain evidence for acetylenic, aliphatic and olefinic chains containing up to six carbon atoms, as well as hybrid structures involving chains bonded to rings. One of these is diphenylacetylene. Compositions that result in the enhancement of the 217.5 nm band correspond to an sp^2 bonded content of ~ 0.3 and an sp^3 bonded content of ~ 0.7, implying that deposition under these conditions tends to form carbon chains, rather than rings. These experiments suggest that HAC, in the form of nanoparticles, forms in situ on the surface of cool silicate grains in the DISM. This material has a rich vibrational spectrum that can

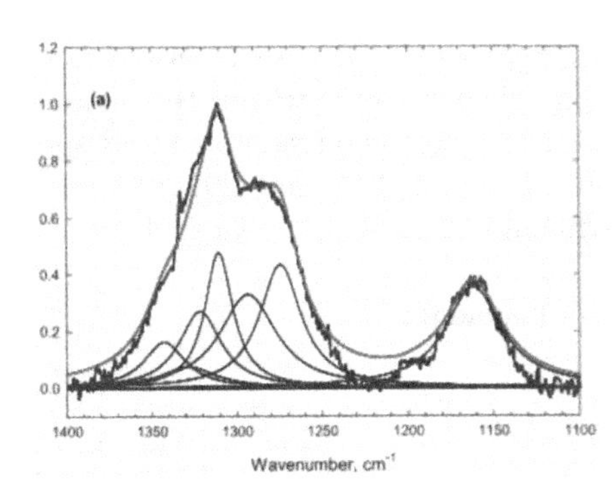

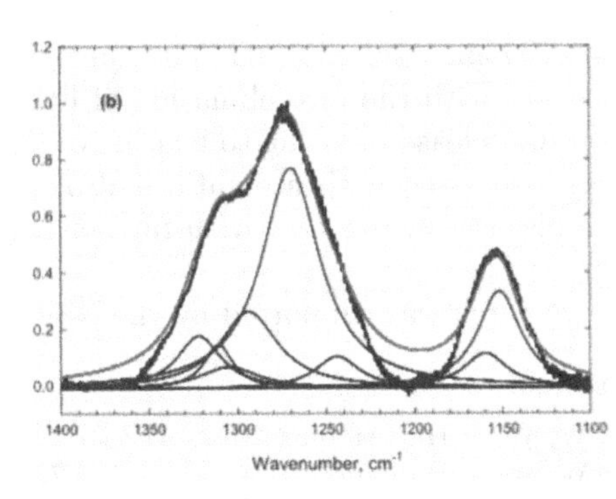

Figure 1. Type A and Type B emission in the middle IR (van Diedenhoven *et al.* 2004) with fits derived from laboratory spectra. Fit parameters are listed in Table 1. There is a clear distinction between these fits, indicating that the material responsible for this emission is different in the two types of source.

Table 1. Spectral components appearing in the fits to Type A and Type B emission spectra. These components are seen in spectra of laboratory samples.

Type A			Type B		
Wavelength μ m	Energy cm^{-1}	FWHM cm^{-1}	Wavelength μ m	Energy cm^{-1}	FWHM cm^{-1}
7.446	1343	15	7.446	1343	15
7.524	1329	20			
7.564	1322	15	7.564	1322	15
7.628	1311	10	7.628	1311	10
			7.662	1305	20
7.728	1294	20	7.728	1294	20
7.843	1275	15			
			7.874	1270	20
			8.039	1244	15
8.382	1193	10			
8.606	1162	20	8.621	1160	15
			8.681	1152	15

be resolved into a large number of components corresponding to specific modes of the chemical structures contained in these nanoparticles. This spectroscopic data has been used to produce detailed fits to emission in type A and B sources as determined by van Diedenhoven *et al.* (2004) (Fig. 1) as well as that of the PPN IRAS 01005+7910 (Zhang & Kwok 2011). We believe that these represent the first accurate fits to astronomical spectra of this kind using purely experimental measurements on carbonaceous materials produced in the laboratory. Further investigation of the properties of these materials is warranted, as they would seem to represent an accurate simulation of interstellar carbon dust.

The model that is indicated from this analysis is that of the in situ formation of hydrogenated amorphous carbon nanoparticles by accretion of individual carbon atoms from the gas phase under DISM conditions. The overall morphology is shown in Duley & Hu (2012). This accretion occurs on the surface of irregular porous silicate particles and initially leads to the formation of small chains. Above a certain length, these chains become unstable with respect to cyclic structures, particularly to the formation of phenyl. Reactions between radicals also promote ring formation, and as both of these effects occur, the composition evolves into a mixture of ring and chain compounds. Phenyl and naphthyl radicals act as key templates for the formation of a variety of substituted rings and larger structures, but the chemistry continues to be driven by addition reactions involving ethynyl groups being created on the surface by the continued accretion of incident carbon atoms from the ambient gas. This results in HAC particles whose internal composition is enriched in small aromatic rings, and whose outer surface contains a higher concentration of chains and radicals. The chemical energy associated with reactions between radicals and with the recombination of hydrogen atoms trapped within this material is periodically released leading to heating of the particle and to emission of the IR spectral features associated with aromatic hydrocarbons (Duley & Williams 2011). A comprehensive discussion, together with further experimental data will appear in Duley & Hu (2012).

This research was supported by the NSERC of Canada.

References

Duley, W. W. & Hu, Anming 2012, *ApJ* 761, 115
Duley, W. W. & Williams, D. A. 2011, *ApJ* 737, L44
Fitzpatrick, E. L. & Massa, D. 1986, *ApJ* 307, 286
van Diedenhoven, B., Peeters, E., Van Kerckhoven, C., *et al.* 2004, *ApJ* 611, 928
Zhang, Y. & Kwok, S. 2011, *ApJ* 730, 126

Highlights of Astronomy, Volume 16
XXVIIIth IAU General Assembly, August 2012
T. Montmerle, ed.

doi:10.1017/S1743921314013039

Laboratory Simulations of Physico-chemical Processes under Interstellar Conditions

Guillermo M. Muñoz Caro

Center of Astrobiology (INTA-CSIC), Ctra. de Ajalvir km 5, Torrejón de Ardoz, Madrid, Spain
email: munozcg@cab.inta-csic.es

Abstract. The accretion and desorption of gas molecules on cold dust grains play an important role in the evolution of dense clouds and circumstellar regions around YSOs. Some of the gas molecules detected in interstellar clouds were likely synthesized in icy dust grains and ejected to the gas. But in dark cloud interiors, with temperatures as low as 10–20 K, thermal desorption is negligible and a non-thermal mechanism like ice photodesorption is required. Reactions in the ice matrix are driven by energetic processing such as photon and ion irradiation. In circumstellar regions the photon flux (UV and X-rays) is expected to be significantly higher than in dense cloud interiors, icy grain mantles present in the outer parts will experience significant irradiation. The produced radicals lead to the formation of new species in the ice, some of them of prebiotic interest. Laboratory simulations of these processes are required for their understanding. The new ultra-high vacuum set-ups introduce some important improvements.

Keywords. Astrochemistry, methods: laboratory, molecular processes, techniques: spectroscopic, ISM: molecules, infrared: ISM

1. Introduction

In the last five decades, numerous works were dedicated to the study of thermal desorption of ice and bulk ice processes including energetic (UV and ion) processing. Only in the last decade, the use of ultra-high vacuum (UHV) techniques, with base pressures in the 10^{-10} to 10^{-11} mbar range, has become increasingly important. This allows to keep the sample surface clean from background water contamination during the experiment, and enables the study of surface phenomena like photodesorption and reactions occurring at the solid-gas interface. The absence of water contamination in the chamber also serves to better constrain the role played by water in the chemistry of ice mantles and helps to detect products with very low abundances. Some recent results are presented below.

2. Recent experimental results on the photoproduction and photodesorption of species in ice mantles

The case of ice photodesorption underlines the importance of studying ices at the monolayer (ML) level. Most molecular species present in inter- and circumstellar ice mantles (H_2O, CO_2, CH_3OH, NH_3, etc.) dissociate efficiently for photon energies between 7 and 10.5 eV, corresponding to the emission spectrum of the hydrogen lamp commonly used in experimental astrochemistry. But some species like CO are not dissociated directly by Lyman-alpha photons (10.2 eV), it appears that the main dissociation pathway in CO ice is via the photoexcitation of a molecule reacting with another molecule leading to $CO^\star + CO \rightarrow CO_2 + C$. This reaction is not efficient, only about 5% of the absorbed photons lead to formation of CO_2 and other species like C_3O and C_3O_2. The remaining 95% of the photons leads to photodesorption. The photodesorption of CO ice was the most studied so far, and it was found to be very efficient (Öberg *et al.* 2007,2009; Muñoz Caro *et al.* 2010; Fayolle *et al.* 2011; Bertin *et al.* 2012). It is now established that for these photon energies only the top 5–6 CO monolayers in the ice contribute to the photodesorption, photons absorbed at deeper layers do not lead to a photodesorption event (Muñoz Caro *et al.* 2010; Fayolle *et al.* 2011). Indeed, for CO ice thicknesses above 5–6 ML the photodesorption rate is constant, but for thinner ices it decreases exponentially with decreasing ice thickness. The use of monochromatic light showed that the photodesorption rate measured at different wavelengths is directly proportional to the UV absorption at that wavelength (Fayolle *et al.* 2011). Since recently, our group at the Center of Astrobiology can measure the UV lamp emission spectrum in situ during the experiment and obtain the UV absorption cross section of ices. Finally, the deposition temperature of the CO ice affects the photodesorption very significantly, which was associated to the different ice morphology, i.e. more or less amorphous ice (Öberg *et al.* 2007,2009; Muñoz Caro *et al.* 2010).

Ice processing due to more energetic photons, X-rays, is poorly known. Larger icy particles up to one micron, formed by coagulation of small grains, can be processed efficiently by X-rays in circumstellar disks. The formation of X-ray photoproducts in the laboratory is commonly very scarce because the X-ray absorption cross section of the ice is low and soft X-ray sources have a low flux. Nevertheless, under UHV conditions X-ray irradiation of ice was studied. In the case of pure CO ice irradiated with soft X-rays, some of the products formed are not common to UV irradiation experiments, these are C_2O, C_3, C_4O, and CO_3/C_5 (Ciaravella *et al.* 2012). In the case of CH_3OH and H_2S ices, X-ray irradiation at low doses leads to formaldehyde, H_2CO, and H_2S_2 formation, respectively (Ciaravella *et al.* 2010, Jiménez-Escobar *et al.* 2012).

In conclusion, physico-chemical processes that occur in interstellar ice mantles are more complex than previously thought. Nowadays, we have the technology required to study this phenomena in detail.

Acknowledgements

The author wishes to thank the research team that contributed to the results reported in this summary, in particular A. Jiménez-Escobar, G. A. Cruz-Diaz, E. Dartois, A. Ciaravella, C. Cecchi-Pestellini, and Y.-J. Chen.

References

Bertin, M., Fayolle, E. C., Romanzin, C., Öberg, K. I., Michaut, X., Moudens, A., Philippe, L., Jeseck, P., Linnartz, H., & Fillion, J.-H. 2012, *PCCP* 14, 9929

Ciaravella, A., Jiménez-Escobar, A., Muñoz Caro, G. M., Cecchi-Pestellini, C., Candia, R., Giarrusso, S., Barbera, M., & Collura, A. 2012, *ApJ*(Letters) 746, L1

Ciaravella, A., Muñoz Caro, G. M., Jiménez-Escobar, A., Cecchi-Pestellini, C., Giarrusso, S., Barbera, M., & Collura, A. 2010, *ApJ*(Letters) 722, L45

Fayolle, E. C., Bertin, M., Romanzin, C., Michaut, X., Öberg, K. I., Linnartz, H., & Fillion, J.-H. 2011, *ApJ*(Letters) 739, L36

Jiménez-Escobar, A., Muñoz Caro, G. M., Ciaravella, A., Cecchi-Pestellini, C., Candia, R., & Micela, G. 2012, *ApJ*(Letters) 751, L40

Muñoz Caro, G. M., Jiménez-Escobar, A., Martín-Gago, J.Á., Rogero, C., Atienza, C., Puertas, S., Sobrado, J. M., & Torres-Redondo, J. 2010, *A&A* 522, 108

Öberg, K. I., Fuchs, G. W., Awad, Z., Fraser, H. J., Schlemmer, S., van Dishoeck, E. F., & Linnartz, H. 2007, *ApJ*(Letters) 662, L23

Öberg, K. I., van Dishoeck, E. F., & Linnartz, H. 2009, *A&A* 496, 281

Highlights of Astronomy, Volume 16
XXVIIIth IAU General Assembly, August 2012
T. Montmerle, ed.

doi:10.1017/S1743921314013040

Synthesis and Transformation of Carbonaceous Nanoparticles

Vito Mennella

INAF - Osservatorio Astronomico di Capodimonte, via Moiariello, 16 80131 Napoli, Italy
email: mennella@na.astro.it

Abstract. The physical properties of carbonaceous nanoparticles depend on the production conditions. In addition, these properties are modified by heat, UV and ion irradiation and gas interaction. We will discuss the synthesis and transformation of carbon nanoparticles that have been proposed as carriers of aromatic and aliphatic spectroscopic features observed in the interstellar medium.

---

1. Carbon particles in space

The presence in the diffuse interstellar medium of aromatic and aliphatic carbon components is inferred from the UV interstellar extinction bump and the 3.4 μm absorption band, respectively. Interstellar and circumstellar objects in our Galaxy and many extragalactic objects emit a spectrum characterized by features generally attributed to aromatic carbon molecules such as polycyclic aromatic hydrocarbon molecules (PAHs). Fullerene have recently been confirmed in many sources of different nature. The variety of carbon species present in space is due to the nature of the carbon atom. It can have three bonding configurations sp^3, sp^2 and sp that give rise to allotropic forms (diamond, graphite, fullerene and carbine or carbonoid material) with different optical and physical properties. Besides these forms there is great variety of disordered materials that are characterized by mixtures of carbon bonding configurations. Their properties depend on the prevailing bonding configuration. In carbon materials the atomic organization is closely related to the electronic structure and, consequently, to the optical properties.

2. Carbon particles in the laboratory

To get insight into the nature of the interstellar carbon components, analogue materials have been studied. Different methods, such as arc discharge, laser ablation, chemical vapor deposition, flame synthesis and energetic processing of carbon containing molecules at low temperature, have been applied for the production of carbon nanoparticles. The produced samples have been studied with different analytic techniques to obtain information on structure and optical properties. Laboratory simulation of grain processing in space is very important to interpret the changes of grains in different interstellar environments through an evolutionary physical-chemical scheme of materials. In fact, dust grain composition evolves through exposure to UV photons, heat, gas and cosmic rays.

On the basis of the experimental results of UV irradiation of hydrogenated carbon grains a model for the UV interstellar extinction bump was proposed (Mennella *et al.* 1998). The model relies on the extrapolation of the laboratory trend of the dielectric function as function of the UV dose to represent the UV processing pathway for hydrogenated carbon grains in the diffuse medium. Recent experimental work confirms that the UV extinction bump is consistent with UV irradiated hydrogenated carbon grains (Gadallah *et al.* 2011).

Energetic processing by UV photons and ions of aliphatic materials destroys their C-H bonds. UV irradiation of hydrogenated carbon grains and hydrocarbon molecules under simulated dense and diffuse medium conditions is indeed characterized by a strong decrease of the 3.4 μm band (Mennella *et al.* 2001, Muñoz Caro *et al.* 2001, Gadallah *et al.* 2012). On the basis of the experimental results one can conclude that the interstellar radiation field should destroy the C-H bonds aliphatic component of diffuse regions in a very short time scale (10^4 years). A similar destruction of the aliphatic C-H bonds is observed during ion irradiation of hydrogenated carbon grains under simulated dense and diffuse interstellar medium conditions (Mennella *et al.* 2003, Godard *et al.* 2011).

Simulation of carbon dust processing under diffuse and dense medium conditions indicates that the key process for the evolution of the interstellar aliphatic carbon component is its interaction with H atoms. This interaction is able to counteract the destruction of C-H bonds by UV photons and cosmic rays and it activates the 3.4 μm band in diffuse interstellar regions and the 3.47 μm feature in dense clouds (Mennella *et al.* 2002, Mennella 2008). The interesting aspect is that the same carbon grain population can absorb at the two wavelengths as a consequence of evolutionary transformations caused by processing. The transformations are compatible with the time-scale required by fast cycling of materials between dense and diffuse regions of the interstellar medium.

Moreover, this carbon component, after the inclusion in a comet during the formation process in the cold outer edge of the solar nebula, may evolve to develop the CH_2 and CH_3 groups (as suggested by laboratory simulations) and contribute to the aliphatic band at 3.4 μm of Interplanetary Dust Particles and particles of comet Wild 2 collected by the Stardust mission (Mennella 2010).

Acknowledgements

This work has been supported by ASI research contracts.

References

Gadallah, K. A. K., Mutschke, H., & Jager, C. 2011, *A&A* 528, A56
Gadallah, K. A. K., Mutschke, H., & Jager, C. 2012, *A&A* 544, A107
Godard, M., *et al.* 2011, *A&A* 529, 146
Mennella, V. 2008, *ApJ* 682, L101
Mennella, V. 2010, *ApJ* 718, 867
Mennella, V., Colangeli, L., Bussoletti, E., Palumbo, P., & Rotundi, A. 1998, *ApJ* 507, L177
Mennella, V., Muñoz Caro, G., Ruiterkamp, R., Schutte, W. A., Greenberg, J. M., Brucato, J. R., & Colangeli, L. 2001, *A&A* 367, 355
Mennella, V., Brucato, J. R., Colangeli, L., & Palumbo, P. 2002, *ApJ* 569, 531
Mennella, V., Baratta, G. A., Esposito, A., Ferini, G., & Pendleton, Y. J. 2003, *ApJ* 587, 727
Muñoz Caro, G., Ruiterkamp, R., Schutte, W. A., Greenberg, J. M., & Mennella, V. 2001, *A&A* 367, 347

Highlights of Astronomy, Volume 16
XXVIIIth IAU General Assembly, August 2012
T. Montmerle, ed.

doi:10.1017/S1743921314013052

Laboratory Analogues of the Carbonaceous Dust: Synthesis of Soot-like Materials and their Properties

T. Pino[1], Y. Carpentier[1]†, G. Féraud[1], Ph. Bréchignac[1], R. Brunetto[2], L. d'Hendecourt[2], E. Dartois[2] and J.-N. Rouzaud[3]

[1]Institut des Sciences Moléculaires d'Orsay, CNRS - Univ Paris-Sud, F91405 Orsay Cedex, France
email: `thomas.pino@u-psud.fr`

[2]Institut d'Astrophysique Spatiale, CNRS - Univ Paris-Sud, F91405 Orsay Cedex, France
[3] Laboratoire de Géologie, Ecole Normale Supérieure, F75231 Paris Cedex 5, France

Abstract. Carbonaceous cosmic dust is observed through infrared spectroscopy either in absorption or in emission and the details of the spectral features are believed to shed some light on its structure and finally enable the study of its life cycle. Other spectral domains also contain some information, as does the UV bump at 217 nm. In order to progress on the understanding of these spectral features, many laboratory works are devoted to the production and characterization of laboratory analogues. Generally several analytical tools are used in combination to better analyse the intimate structure of the analogues and the influence of the nanostructuration on the spectral properties. In this proceeding We will focus on the elaboration of new spectral parameters that enables the nanostructuration of the carriers of the AIBs to be traced.

Keywords. ISM: general, astrochemistry, (ISM:) dust, extinction, infrared: ISM

---

1. Introduction

Carbonaceous materials are observed in many objects in space (Henning & Salama 1998, Ehrenfreund & Charnley 2000) through their specific spectral features. Some are found in the visible to UV range (Désert *et al.* 1990, Draine 2003, Zubko *et al.* 2004), but most of the spectral information is obtained in the infrared range of wavelength. In particular, the aromatic infrared bands (AIBs) (Gillett *et al.* 1973) trace carbonaceous particles whose inferred size vary from large molecules, the polycyclic aromatic hydrocarbons (PAHs), to nanoparticles of a few nanometers (Léger & Puget 1984, Crawford *et al.* 1985, Tielens 2008). Three main classes A, B and C of astrophysical spectra have been proposed based on the observed spectral characteristics in the 6 to 9 μm wavelength region (Peeters *et al.* 2002). If the band positions, in most sources, are clearly attributed to aromatic materials (Goto *et al.* 2007), however, the exact nature of the emitters still escape. In the case of the class A emitters, the carbonaceous species emit through their vibrational bands thanks to the stochastic heating mechanism (Puget & Léger 1989, Allamandola *et al.* 1989, Draine & Li 2001). In the other cases, thermal emission may occur, particularly for the class C objects. In order to study these materials, many laboratory astrophysics works are perfomed on various analogues. At Orsay (France), we focus on the polyaromatic carbons with a special emphasize on the soot-like nanoparticles. Such materials provide interesting laboratory analogues of cosmic dust. The soot samples are

† Present address: Laboratoire de Physique des Laser, Atomes et Molécules, Bât P5 - USTL, F-59655 Villeneuve d'Ascq cedex, France

produced using a low-pressure flame, flat and premixed. The nanostructural characterisation and their spectral signatures through the positions of infrared bands are described and shown to provide a spectral index of the shape of the polyaromatic units. The astrophysical implications of such index are thus discussed. The detailed analysis is reported elsewhere (Carpentier *et al.* 2012).

2. A spectral index for the shape of the polyaromatic units

Only spectroscopic parameters extracted from the infrared bands can be used to decipher the intimate structure of the carriers of the AIBs. Using our soot, we explore the details of the variations of the infrared bands combining Raman spectroscopy and HRTEM images (not shown here) in order to link the spectral features to specific structural properties of the polyaromatic units within the soot. We have recently shown that the 6.2 μm band position traces the intimate cross-linkage of the polyaromatic units composing the soot (Pino *et al.* 2008). Such evolution is accompanied by an evolution of the band at about 8 μm. The correlation between the 6.2 and 8 μm band is shown in Fig. 10 in Carpentier *et al.* 2012. We tentatively attribute this band to defect-like modes of the polyaromatic units. The defects are located either at the edge and cross-linking the units together, or within the unit and shaping the unit away from planarity. Such nanostructural characteristics are, in combustion processes, related to the growth mechanism of the soot (D'Anna 2009). Similarly, different interstellar emitters could have experienced different formation pathways.

In most analysis, the spectral parameters are used to build an aromatization index when dealing with polyaromatic materials. It is necessary because most of the properties are determined by the size of the polyaromatic units and the sp^2 to sp^3 carbon ratio, and the aromatic index increases with both. However, in the present soot samples, size

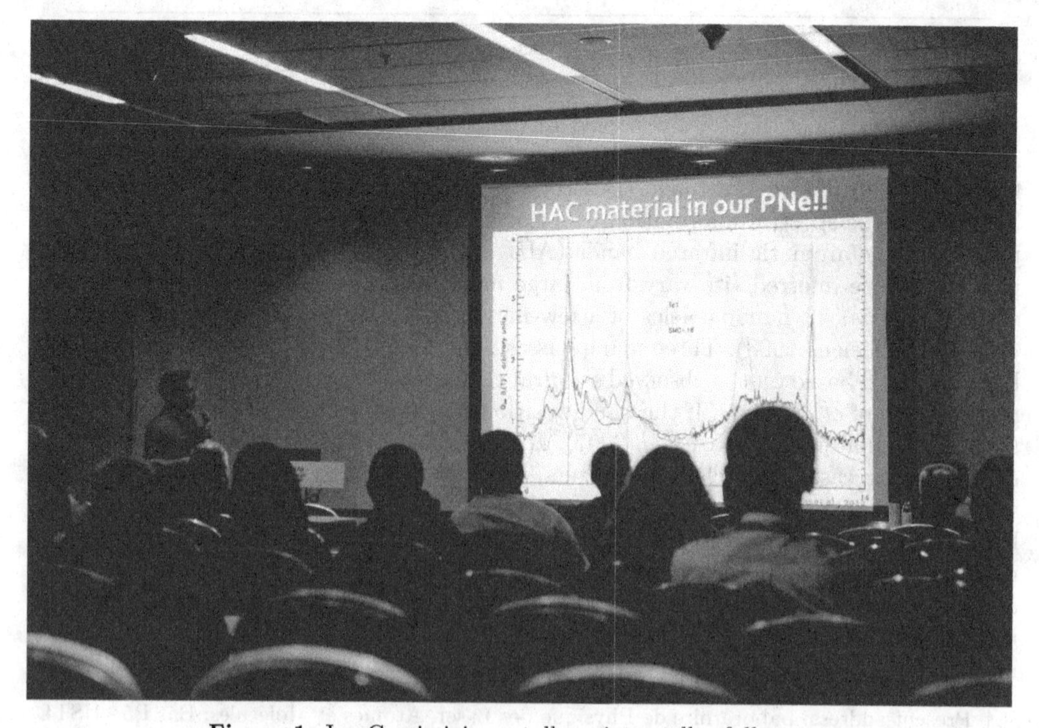

Figure 1. Jan Cami giving a talk on interstellar fullerenes.

variation is not important. In addition, the mean size is always below 2 nm and precludes the use of any reliable index, whatever the experimental methods. For such polyaromatic samples, it is more important to consider the shapes of the polyaromatic units which are found from the study of their nanostructuration. It appears that the 8 μm band position is a good tracer of the defects that build curvature and tortuosity for the polyaromatic units. The correlation of its position with that of the 6.2 μm band shows that the latter band position is also strongly influenced by these defects. Thus the positions of the two bands can be seen as a (nano)structural index.

3. Conclusions

Thanks to soot analogues of the cosmic dust, the meaning of the spectral characteristics of these AIBs, classified from A to C, can be deeply studied in the laboratory. The role of the defects in the carbonaceous particles appears to be at the core of the variations of the spectral features, in particular those related to the aromatic carbon skeleton: the band at about 6.2 μm and the band at about 8 μm. Their positions are sensitive to the (nano)structuration. Progress on the understanding of the nanostructuration of the emitters will be essential to decipher the growth and lifecycle of this component of the interstellar medium, as proposed in Carpentier *et al.* 2012.

Acknowledgements

This work was supported by the French national program "Physique et Chimie du Milieu Interstellaire" and the French "Agence Nationale de la Recherche" (contract ANR-10 BLAN-0502).

References

Allamandola, L. J., Tielens, A. G. G. M., & Barker, J. R. 1989, *ApJS* 71, 733

Carpentier, Y., Féraud, G., Dartois, E., Brunetto, R., Charon, E., Cao, A.-T., d'Hendecourt, L., Bréchignac, P., Rouzaud, J. N., & Pino, T. 2012, *A&A* 548, 40

Crawford, M., Tielens, A., & Allamandola, L. 1985, *ApJ* 293, L45

D'Anna, A. 2009, in: H. Bockhorn, A. D'Anna, A. F. Sarofim & H. Wang (eds.), *Combustion Generated Fine Carbonaceous Particles*, Proc. of an International Workshop (Villa Orlandi, Anacapri: KIT Scientific Publishing), p. 289

Désert, F.-X., Boulanger, F., & Puget, J. L. 1990, *A&A* 237, 215

Draine, B. T. 2003, *ARAA* 41, 241

Draine, B. T. & Li, A. 2001, *ApJ* 551, 807

Ehrenfreund, P. & Charnley, S. B. 2000, *ARAA* 38, 427

Gillett, F. C., Forrest, W. J., & Merrill, K. M. 1973, *ApJ* 183, 87

Goto, M., Kwok, S., Takami, H., Hayashi, M., Gaessler, W., Hayano, Y., Iye, M., Kamata, Y., Kanzawa, T., Kobayashi, N., Minowa, Y., Nedachi, K., Oya, S., Pyo, T.-S., Saint-Jacques, D., Takato, N., Terada, H., & Henning, T. 2007, *ApJ* 662, 389

Henning, T. & Salama, F. 1998, *Science* 282, 2204

Léger, A. & Puget, J. 1984, *A&A* 137, L5

Peeters, E., Hony, S., van Kerckhoven, C., Tielens, A. G. G. M., Allamandola, L. J., Hudgins, D. M., & Bauschlicher, C. W. 2002, *A&A* 390, 1089

Pino, T., Dartois, E., Cao, A.-T., Carpentier, Y., Chamaillé, T., Vasquez, R., Jones, A. P., d'Hendecourt, L., & Bréchignac, P. 2008, *A&A* 490, 665

Puget, J. L. & Léger, A. 1989, *ARAA* 27, 161

Tielens, A. G. G. M. 2008, *ARAA* 46, 289

Zubko, V., Dwek, E., & Arendt, R. G. 2004, *ApJS* 152, 211

Highlights of Astronomy, Volume 16
XXVIIIth IAU General Assembly, August 2012
T. Montmerle, ed.

doi:10.1017/S1743921314013064

A Review on Carbon-rich Molecules in Space

Franco Cataldo[1,2], D. Aníbal García-Hernández[3,4] and Arturo Manchado[3,4,5]

[1]Istituto Nazionale di Astrofisica - Osservatorio Astrofisica di Catania, Via S. Sofia 78, 95123, Italy

[2]Actinium Chemical Research, Via Casilina 1626/A, 00133 Rome, Italy

[3]Instituto de Astrofisica de Canarias, Via Lactea s/n, E-38200, La Laguna, Tenerife, Spain

[4]Universidad de La Laguna (ULL), Dept. de Astrofísica, E-38205, La Laguna, Tenerife, Spain

[5]CSIC, Madrid, Spain

Abstract. We present and discuss carbon-rich compounds of astrochemical interest such as polyynes, acetylenic carbon chains and the related derivative known as monocyanopolyynes and dicyanopolyynes. Fullerenes are now known to be abundant in space, while fulleranes - the hydrogenated fullerenes - and other carbon-rich compounds such as very large polycyclic aromatic hydrocarbons (PAHs) and heavy petroleum fractions are suspected to be present in space. We review the synthesis, the infrared spectra as well as the electronic absorption spectra of these four classes of carbon-rich molecules. The existence or possible existence in space of the latter molecules is reported and discussed.

Keywords. Astrochemistry, molecular data, infrared: ISM

The review regards a series of carbon-rich molecules which were recently studied by the authors. First we will discuss the first class of carbon-rich molecules: the polyynes and the related molecules known as monocyanopolyynes and dicyanopolyynes. These molecules were detected by radioastronomy in carbon-rich Asymptotic Giant Branch (AGB) stars, in dense and dark interstellar clouds like TMC-1 and in hot molecular cores. Although polyynes are highly unstable in terrestrial conditions, Cataldo (2004, 2006a,b) has developed a synthesis of these molecules by using a carbon arc and trapping the molecules in a solvent. When the arc is struck in liquid hydrocarbons, all the polyynes homologue series from $H\text{-}(C{\equiv}C)_3\text{-}H$ to $H\text{-}(C{\equiv}C)_9\text{-}H$ is obtained and the electronic absorption spectrum of each individual polyyne molecular specie has been recorded. The formation of polyynes is always accompanied by the formation of PAHs and carbon soot. Indeed, polyynes are considered the precursors of PAHs and carbon soot. When the carbon arc is struck in liquid ammonia or in acetonitrile, then a mixture of polyynes series and monocyanopolyynes series $H\text{-}(C{\equiv}C)_n\text{-}CN$ is obtained. Dicyanopolyynes $NC\text{-}(C{\equiv}C)_n\text{-}CN$ can be synthesized by striking the carbon arc in liquid nitrogen. It was proposed that the laboratory conditions of polyynes synthesis through the carbon arc are comparable to those existing in the circumstellar medium of certain AGB stars since both in laboratory conditions and in the circumstellar environment the mechanism of carbon chain formation follows a free radical path (Cataldo 2006a,b). More recent work has shown that polyynes are also formed in the gas phase by arcing graphite in a flow of argon, in which the addition of methane to Ar strongly enhances the formation of polyynes, which are then accompanied by PAHs such as naphthalene (Cataldo 2007). Polyynes are an endothermal compound with a positive free energy of formation from the elements at 298 K. However, their Gibbs free energy of formation becomes negative just above 4400 K, which is the temperature of the carbon arc. The polyynes are thermodynamically stable and form quite easily at such high temperatures, but are then quenched to a lower temperature by the hydrocarbon solvent surrounding the arc and become trapped there, thus permitting their manipulation, separation and analysis (Cataldo 2007).

Another interesting carbon-rich class of molecules are fullerenes, which were recently detected in a series of astrophysical objects (Cami *et al.* 2010; García-Hernández *et al.* 2010, 2011a,b; Sellgren *et al.* 2010; Zhang & Kwok 2011). Fullerenes show considerable stability toward the action of high energy radiation so that they could survive for billions of years in the ISM under certain conditions (Cataldo *et al.* 2009). This fact may explain why fullerenes have been found in completely different astrophysical objects. To facilitate the search of C_{60} and C_{70} fullerenes in space, we have studied the dependence of infrared band pattern and band intensity of these molecules with temperature and measured the molar absorptivity for the quantitative determination of their abundance (Iglesias-Groth *et al.* 2011). Furthermore, the electronic absorption spectra of the radical cation of C_{60} and C_{70} in a very high dielectric constant medium has been determined (Cataldo *et al.* 2012). Fullerenes are very reactive with atomic hydrogen, and form fulleranes, which are hydrogenated fullerenes (Cataldo & Iglesias-Groth 2010). When heated or photolyzed, the fulleranes release molecular hydrogen yielding back the parent fullerene. Therefore, fullerenes may play a key role in the molecular hydrogen formation starting from atomic hydrogen. The photolysis rate constant of fulleranes appears of the same order of magnitude as that of polyynes (Cataldo & Iglesias-Groth 2009). Curiously, the fullerane $C_{60}H_{36}$ shows an electronic absorption spectrum with a unique peak at 217 nm, exactly matching the UV "bump" of the interstellar light extinction curve (Cataldo & Iglesias-Groth 2009). The infrared spectra of a series of reference fulleranes have been recorded in the laboratory by Iglesias-Groth *et al.* (2012) together with the relative molar absorptivity. Consequently, since the reference spectra are now available, a search for fulleranes can now made in space, and their possible detection in astrophysical environments is now only a matter of time and luck.

Coal was proposed by Papoular and collagues (1989) as a possible model for matching the infrared band pattern of the unidentified infrared bands (UIBs) of certain planetary nebula (PNe) and proto-PNe (PPNe). However, anthracite, "mature" coal, is eminently aromatic with minor aliphatic components. More recently, Cataldo *et al.* (2002, 2004) and Cataldo & Keheyan (2003) have shown that certain heavy petroleum fractions are also able to match the band pattern of coal and of PPNe and PNe. Thus, petroleum fractions should be used as realistic model compounds since they are composed by a "core" of 3 to 4 condensed aromatic rings surrounded by cycloaliphatic (naphthenic rings) and chains of aliphatic sp^3 hybridized carbon. Thus, instead of searching for pure PAHs, the heavy petroleum fraction model offers molecules where the aromatic, naphthenic and aliphatic moieties co-exist, and match certain PPNe spectra where a mixture of aliphatic/cycloaliphatic and aromatic structures are evident.

Very large PAHs (VLPAHs) are not easily accessible but of high interest as reference molecules for the explanation of the diffuse interstellar bands (DIBs) of the interstellar medium (ISM). Using the Scholl reaction, we have synthesized a series of VLPAHs ranging from dicoronylene to quaterrylene to hexabenzocoronene. Dicoronylene was also obtained by the thermal dimerization of coronene. If the thermal treatment of coronene is prolonged, the oligomerization of coronene proceeds further, yielding a black oligomer, which is probably a trimer or a higher homologue. It is shown that from coronene oligomerization it is possible to build a sheet of graphene.

References

Cami, J., Bernard-Salas, J., Peeters, E., & Malek, S. E. 2010, *Science* 329, 1180

Cataldo, F. 2004, *Int. J. Astrobiol.* 3, 237
Cataldo, F. 2006a, *Int. J. Astrobiol.* 5, 37
Cataldo, F. 2006b, *OLEB* 36, 467
Cataldo, F. 2007, *Fullerenes Nanotubes Carbon Nanostructures* 15, 291
Cataldo, F. & Iglesias-Groth, S. 2009, *MNRAS* 400, 291
Cataldo, F. & Iglesias-Groth, S. 2010, *Fulleranes: The Hydrogenated Fullerenes*, (Dordrecht: Springer)
Cataldo, F. & Keheyan, Y. 2003, *Int. J. Astrobiol.* 2, 41
Cataldo, F., Keheyan, Y., & Heymann, D. 2002, *Int. J. Astrobiol.* 1, 79
Cataldo, F., Keheyan, Y., & Heymann, D. 2004, *OLEB* 34, 13
Cataldo, F., Strazzulla, G., & Iglesias-Groth, S. 2009, *MNRAS* 394, 615
Cataldo, F., Iglesias-Groth, S., & Manchado, A. 2012, *Fullerenes Nanotubes Carbon Nanostructures* 20, 656
García-Hernández, D. A., Manchado, A., García-Lario, P., Stanghellini, L., Villaver, E., Shaw, R. A., Szczerba, R., & Perea-Calderón, J. V. 2010, *ApJ* 724, L39
García-Hernández, D. A., Iglesias-Groth, S., Acosta-Pulido, J. A., Manchado, A., García-Lario, P., Stanghellini, L., Villaver, E., Shaw, R. A., & Cataldo, F. 2011a, *ApJ* 737, L30
García-Hernández, D. A., Kameswara Rao, N., & Lambert, D. L. 2011b, *ApJ* 729, 126
Iglesias-Groth, S., Cataldo, F., & Manchado, A. 2011, *MNRAS* 413, 213
Iglesias-Groth, S., Garcia-Hernandez, A., & Cataldo, F., Manchado A. 2012, *MNRAS* 423, 2868
Papoular, R., Conrad, J., Giuliano, M., Kister, J., & Mille, G. 1989, *A&A* 217, 204
Sellgren, K., Werner, M. W., Ingalls, J. G., Smith, J. D. T., Carleton, T. M., & Joblin, C. 2010, *ApJ* 722, L54
Zhang, Y. & Kwok, S. 2011, *ApJ* 730, 126

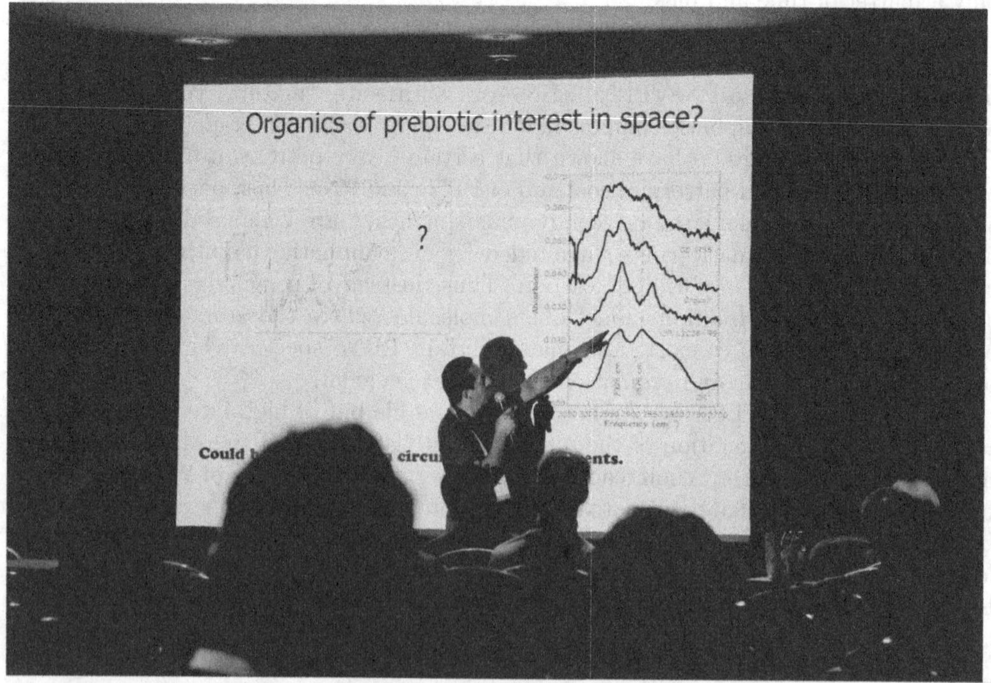

Guillermo Muñoz Caro addressing the problem of formation of prebiotic organics on interstellar grains.

Highlights of Astronomy, Volume 16
XXVIIIth IAU General Assembly, August 2012
T. Montmerle, ed.

doi:10.1017/S1743921315001684

Preface: Special Session SpS17 Light Pollution: Protecting Astronomical Sites and Increasing Global Awareness Through Education

The issue of Light Pollution was a major concern of the International Astronomical Union; during the IAU General Assembly in the city of Rio de Janeiro in 2009. A resolution was unanimously adopted (Resolution B5) to support the need to preserve the night sky and the right to see stars. With the increasing use of artificial light at night posing a growing threat to the visibility of the night sky, this Special Session at the XXVIIIth General Assembly highlighted technical aspects of astronomical site protection and the educational aspects of increasing global awareness on issues concerning light pollution.

The primary responsibility for the disappearance of stars is peoples lack of knowledge. Recognizing the problem, studying its causes, and assessing options is part of the solution. Several topics for the session examined the impact of various approaches of education and public outreach on issues concerning light pollution and its impact on our world. To accomplish this, we present invited presentations by ecologists, amateur and professional astronomers, professional educators at schools, universities, planetaria, science centers and nature centers, as well as people in the media, landscape astrophotography, the medical field and lighting design. Contributions on how to improve our handling of language and cultural barriers, review progress being made by outreach and citizen-science programs such as GLOBE at Night, Global Astronomy Months Dark Skies Awareness, One Star at a Time, the Starlight Initiative and the educational resources offered through the International Dark-Sky Association are also present in these pages.

Protecting sites (including observatories) and slowing the encroachment of light pollution requires engaging the public on several levels. These actions include producing long-term sky brightness data intercomparable with broader public monitoring programs; taking opportunities to educate the public about the value of dark sky preservation; interacting with policy makers and public agencies ranging from localities to the UN to provide legal protection and enforcement for dark sky zones; and interacting with lighting engineers to define dark-sky preserving products and to encourage their deployment. Rapidly advancing technology and associated promotion is leading to deployment of blue-rich artificial light sources that threaten to impact a spectral region previously left relatively untouched. Exploring the nature, possible impact, and potential mitigation of this trend has been a timely aspect of this session.

The knowledge of starry sky represented for our ancestors the basis for survival on Earth, contributed to the development of science in all areas and was a source of inspiration for humanity. This heritage is being lost; to recover it for us and future generations is everyone's responsibility, because we know that individuals approach to the Cosmos today is still of the same importance and magnitude as in antiquity.

Richard Green, and Beatriz García, Constance Walker and Xue Sui Jian, co-chairs SOC,
Rosa M. Ros, Elizabeth Álvarez, Magda Stavischi and Scott Kardel, SOC-Editors

December 31, 2012

Highlights of Astronomy, Volume 16
XXVIIIth IAU General Assembly, August 2012
T. Montmerle, ed.

doi:10.1017/S1743921314013076

China in Action – Starry Sky Project of China

Xiaohua Wang, Yifeng Wang and Hui Ren

Room 501 Rongsheng Tower, No.5 Financial Street, Xicheng District, Beijing, China, 100032
email: mr_wxh@hotmail.com

Abstract. Two pilot Starlight Parks and Protection Areas are under planning.

Keywords. Light Pollution, Starlight Park

Starry Sky Protection is an inevitable subject in the human pursuit of high-quality environment. China also needs a clear night sky. Light pollution first threatens astronomical observation, but to control light pollution needs the action of the whole society. Starry Sky Protection is a long-term task and needs a sustainable model. To combine it with tourism is an effective approach. Based on the above basic understanding, we started to initiate and promote the Project of Starry Sky in China.

We introduced the topic of light pollution in the Chinese National Geography (CNG) magazine of March 2012. There is a total of 4 articles and 50 charts covering light pollution, star observation, changes of the observatory, and dark sky park in altogether 50 pages. The CNG is one of China's most influential magazines, with a monthly circulation of one million copies. We then provided articles of starlight tourism to the Tourism World magazine of April 2012, which took up 1/3 of the magazine.

Taking reference from the programs and standards of Dark Sky Parks and Starlight Reserves, we are planning two pilot Starlight Parks & Protection Areas at the Starlight Lake in Mt. Yimeng and the Seven Star Mesa in Mt. Tai in Shandong Province. We plan to inject ecological, scientific and cultural elements into the tourism areas and create the basis for appreciation, education and protection of the starry sky. Our work has been supported by the operators of these tourism areas and local governments. We worked out the program and prepared to establish standards and draw experience from the scenic area planning, astronomical facilities, lighting control, staff training and tourism operation. Our idea is this: the Starlight Park & Protection Area will combine with the tourism area, natural and cultural heritage, and astronomical observatory, to create a replicable model, so that the starry sky protection can move on rapidly and gain sustainable development.

The Starry Sky Protection is a cross-industry project and requires the whole society to participate. We are pleased to have a group of people to work for it in China, including astronomy professionals, tourism and investment entrepreneurs, media and law practitioners, and young volunteers. They are from different sectors and contribute their efforts to the common cause. This cause should be recognized and protected by law. In this regard, we have also carried out research and preparation. China is a big country with unique social and cultural patterns. We are actively contacting relevant agencies, to form a joint force at a higher level, and when conditions permit, we plan to establish the Starry Sky Protection Agency of China.

In Beijing and other big cities, young people born in the 80s and 90s are already away from the splendid starry sky. How shall we face our children, and children of our children?!

Highlights of Astronomy, Volume 16
XXVIIIth IAU General Assembly, August 2012
T. Montmerle, ed.

doi:10.1017/S1743921314013088

Losing the Dark: A Planetarium PSA about Light Pollution

Carolyn Collins Petersen[1] and Constance Walker[2]

[1]Loch Ness Productions, P.O. Box 924, Nederland, CO 80466, USA, email: carolyn@lochnessproductions.com

[2]National Optical Astronomy Observatories, 950 N. Cherry Ave., Tucson, AZ 85719, USA, email: cwalker@noao.edu

Abstract. *Losing the Dark* is a six-minute PSA video created for fulldome theaters by Loch Ness Productions, the International Dark Sky Association Education Committee headed by Dr. Constance Walker of the National Optical Astronomy Observatories, Dome3, Adler Planetarium, and Babak Tafreshi (The World at Night). It explains light pollution, its effects, and ways to implement "wise lighting" practices to mitigate light pollution. The show is also made in flat-screen HD format for classical planetariums, non-dome theaters, and for presentatons by IDA speakers.

Keywords. light pollution, planetarium show, wise lighting practices

---

The Project

Light pollution interferes with our ability to study the sky in visual wavelengths. Residents of many communities live under light-polluted skies, affecting their abilities to see the stars. Light pollution is also known to have impacts on health, safety, and energy costs. Organizations such as the International Dark-Sky Association (IDA) work to educate the public about ways to reduce wasteful lighting. The IDA Education Committee determined that planetariums and science centers are useful venues to teach members of the public about light pollution. Working with fulldome production company Loch Ness Productions and using visual materials from Adler Planetarium, NASA, and others, they have crafted a short "public service announcement" type presentation called *Losing the Dark* for planetarium theaters and other venues. Early in the production process, the group decided to make the show available not just for fulldome, but also in "flat screen" movie file format for "classic" planetarium theaters (those without fulldome video technology), educators and othersto use in public lectures and presentations. The show is produced in English and several other languages. The official premiere was at the International Astronomical Union meeting in Beijing, China in August, 2012.

Thematic Points of the Show

The points about the light pollution's threats to astronomy and its costs and health impacts were used to inform the script for *Losing the Dark*. In addition, we offer three simple "wise use" steps to help mitigate light pollution: turn off unwanted lights, use shielded lights, place lights only where they are needed.

Losing the Dark utilizes striking views of light pollution on Earth and from space to illustrate the main points. It provides planetarium professionals, educators and other outreach professionals a multimedia approach to inform the public about this important topic. The show will be available in early 2013 for download by any facility that wishes to use it. For more information on *Losing the Dark*, please visit www.lochnessproductions.com or www.darksky.org.

Highlights of Astronomy, Volume 16
XXVIIIth IAU General Assembly, August 2012
T. Montmerle, ed.

doi:10.1017/S174392131401309X

Espinho Planetarium's Public Outreach on Light Pollution

Lina Canas, Pedro B. Silva and Antonio Pedrosa

Navegar Foundation, Centro Multimeios Espinho, Av. 24, n800, 4500-202 Espinho, Portugal
email: lina.canas@multimeios.pt, pedro_silva@multimeios.pt, apedrosa@multimeios.pt

Abstract. Navegar Foundation has dedicated the past twelve years to astronomy public outreach by engaging the community in activities provided by the planetarium and observatory. The activities developed range from exhibitions to telescope observing sessions and planetarium shows. Whether partnering with local entities or with a variety of joint national and international activities, Navegar always had a proactive policy on raising awareness of light pollution issues. The outcomes of these activities are discussed.

Keywords. planetarium, observatory, light pollution

Planetarium: an oasis in the urban centers

Through our years of experience we have effectively used the planetarium to show the potential of a true night sky to schools and general population from urban areas in the northern region of Portugal. Located near sites of great population density centers and surrounded by an increasingly brighter night sky, the planetarium reveals itself as a unique vehicle to show how a true sky can be.

On direct feedback received from our visitors we can infer that although there is a true sense of awe from the audience when the star field is on, the sense of recognition of what the real sky is like is regarded more and more as science fiction! Many actually think that there are no such places on Earth where we actually see the sky as shown in the Planetarium. Younger audiences have no recollection of ever seeing the night sky like that. Older audiences regard it like a faraway memory and not as a present experience.

International campaigns and dedicated light pollution awareness events [2][3][4] prove to be effective contributors to raise and increase the interest of the community to light pollution issues. The sense of common goal and belonging to a worldwide campaigns, effectively draw attention. Another solution encountered was to relate light pollution problems to other sciences such as biology [1] and chemistry. Interlinking astronomy to other subjects proves to be a major asset in increasing the attendance number and raising global awareness, as this is a problem that the community can relate to beyond astronomy, and be more sensitive to wildlife impact problems or even health issues. Finally, there is a need of a constant and steady policy in introducing this subject on a regular basis to lead to habit change. The planetarium here proved and still proves to be one of the most powerful of tools to help us all remember the true sky and to strengthen the movement to reclaim the right to a real and a true dark sky for all.

References

Kwok, S. 2001, Cosmic Butterflies: The Colorful Mysteries of Planetary Nebulae, Cambridge University Press

GLOBE at Night. http://www.globeatnight.org/ (accessed July 27, 2012).

Dark Skies Awareness. http://www.darkskiesawareness.org/ (accessed July 27, 2012).

GAM Dark Skies Awareness Programs. Astronomers Without Borders. http://astronomerswithoutborders.org/gam2012-programs/dark-skies-awareness.html (accessed July 28, 2012).

Highlights of Astronomy, Volume 16
XXVIIIth IAU General Assembly, August 2012
T. Montmerle, ed.

doi:10.1017/S1743921314013106

Media and Light Pollution Education for the Public

Julia Romanowska

Polaris OPP Association
Sopotnia Wielka 174, 34-340 Jeleśnia, Poland
email: `j.e.romanowska@gmail.coml`

Abstract. As the problem of light pollution becomes more and more threatening, there is a need to educate the society about it. Methods of doing it differ in various parts of the world. There are shown examples of methods from developing European countries, in particular Poland.

Keywords. media, light pollution, education

1. Overview

Twenty years ago, when International Dark-Sky Associaion was established, nobody in Poland had heard about light pollution. There were other problems and the government and municipalities didn't have the money to spend on street lighting. Nowadays, as the country is rapidly developing and new highways are being built, people expect more outdoor lighting. Furthermore, there are funds to provide it. That means that raising awareness about light pollution is crucial. However, in the era of very robust flow of information, educating people about light pollution is becoming harder and harder, as they filter the news.

The main goal of my presentation was to depict various ways of raising awareness, including educating common people during rural fairs, showing them that "astrotourism" is a new kind of tourism and that light pollution is not helping; educating people in cities through various events (even happenings in the supermarkets), banners, conferences; informing municipalities that installing street lamps of proper types and limiting the number of them may cut the expenses; talking with the lighting and energy companies about improving their reputation by reducing light pollution, as well as informing people through the media – Internet, radio and newspapers.

In the presentation I have also tried to show what effects the methods mentioned above produced and I will mention what other interesting ways of educating people were successful in other countries. More information can be viewed at Ciemne Niebo and Polaris.

References

Ciemne Niebo `www.ciemneniebo.pl`
Polaris `www.polaris.org.pl`

Highlights of Astronomy, Volume 16
XXVIIIth IAU General Assembly, August 2012
T. Montmerle, ed.

doi:10.1017/S1743921314013118

TV shows on Light Pollution Education for the Public

Valentin Grigore[1,2]

[1]Societatea Astronomic Romn de Meteori (Romanian Society for Meteors and Astronomy), Str. Tineretului nr. 1, Trgovite 130020, Dmbovia, Romania sarm.ro@gmail.com

[2]Noi i Cerul (Us ans the Sky) TV show, Columna TV, Romania

Abstract. TV shows have the biggest impact for the public, so we can use them to inform and educate the public about light pollution and the importance of the dark sky for humanity and for the contemporary society. Some examples used in the TV show Us and the Sky at Columna TV, Romania, are presented.

Keywords. Light Pollution, Citizen Science, Public Outreach

As you saw on TV! Could we use this phrase when talking about light pollution? Visual impact is most effective for education. The subject of light pollution should be a part of regular TV programming, not only on specialized TV channels. The role of TV:

A. TO INFORM Provide fair and complete information! 1. About light pollution. It affects directly the quality of our life (health, budget, comfort, and security), the world climate, wildlife and the natural night sky and our right to see the stars, the astronomy and research. 2. Ways to combat light pollution. It is not necessary to turn off the public illumination. We just need to streamline it and light responsibly! Suggestion: Invite onto TV shows specialists from different fields: doctors, biologists, ecologists, astronomers, city managers, tourism workers, politicians (for LP laws), etc.

B. TO EDUCATE How does light pollution appear? Who is responsible for that? What do we need to do to combat light pollution? Suggestions: 1. Use common objects to explain the effects of the LP. See: http://www.youtube.com/watch?v=GfzXLzeuo2A 2. Use animations and cartoons. We need more good multimedia resources, easy to understand by the public and even by children. IAU, IDA, NASA, ESO and others should create more short multimedia products (video, animations) to be used in TV shows. See: http://planetgreen.discovery.com/videos/my-place-starry-night.html

C. TO ADVERTISE the global efforts to combat light pollution. The global campaigns such as GLOBE at Night, Earth Hours, Dark-Sky Week, World Night in Defence of the Starlight, GAM, etc. Dedicated meetings and conferences: IAU SpS17, European Symposium for the Protection of the Dark Sky, Dark-sky Parks, etc. Suggestions: 1. Make these events as important TV News subjects. 2. Invite journalists and workers of public and state institutions at all important dark-sky events.

D. TO STRESS night sky and astronomy importance for humanity, for culture and progress. Suggestions: 1. Show practical aspects of astronomy (calendar, navigation, communications even mobile phones!). 2. Delight people with spectacular views of the night sky! 3. Use space music, astrofolk music, cosmopoetry.

In the future we should have partnerships with media institutions, including TV. A TV Network with affiliates in every IAU country is useful. Dark-sky and light pollution campaigns should create more video spots and contacts with local TVs. Noi si Cerul (Us and the Sky) is a weekly TV show produced by Valentin Grigore,President of the Romanian Society for Meteors and Astronomy (SARM) at Columna TV, a Romanian TV channel, streaming live online via Internet at http://www.columnatv.ro/tv.

Highlights of Astronomy, Volume 16
XXVIIIth IAU General Assembly, August 2012
T. Montmerle, ed.

doi:10.1017/S174392131401312X

Knowing What is Best

R. E. M. Griffin

Dominion Astrophysical Observatory, Victoria,V9E 2E7, Canada
email: elizabeth.griffin@nrc-cnrc.gc.ca

Abstract. Already, and with good reason, light pollution is recognized as one of the most damaging legacies that current "civilization" is bequeathing to its children. Denying them the opportunity, even the right, to experience visually the majesty and awe of the universe has obvious repercussions for our science, and other contributors to this meeting are addressing those eloquently. But it is also critically important to place light pollution in the cadre of the general environmental degradation which unbridled technology is causing. The amount of power consumed by one outdoor light is only a minuscule drop in the ocean, but enough of those drops make an ocean. Using low-wattage bulbs, and getting more power out of them via halogen or LED technology, can ease the drain on the supply of power, but when several can be run, and run brighter, than a single tungsten lamp and cost is the only goal, the consumer simply installs more of them.

We all hope for a restored and sustainable environment, but the challenge is first to learn and practice the essential difference between "want" and "need". A more specific challenge is to educate the affluent countries about the deleterious effects of nighttime lights on human health and on other bio-systems and species, and to explain the truth about "security" issues. If astronomers place the needs of their own science too foremost they risk the criticism of selfishness: why are our own scientific requirements more important than the pleasure and health benefits of a whole town in pursuing sports activities outdoors after sunset? How can the need to illuminate streets, intersections, parking lots and to deter intruders be less necessary for personal and community safety and wellbeing than some rather esoteric scientific opportunities for a small population of astronomers?

Dark Sky Preserves are a lovely concept but they are not the full answer; reducing light pollution is the responsibility of every citizen, and ensuring good astronomical conditions in a few remote parks must does not excuse the prevailing cavalier attitudes that are causing light pollution to be blithely accelerated everywhere else.

The solution is Education, and for astronomers the key is to tone down the astronomically important and to emphasize the many general benefits of reducing light pollution. It is only when the non-astronomical community starts offering the astronomical advantages too of its own volition that we will see true progress in fighting this most pervasive of modern environmental disasters.

Highlights of Astronomy, Volume 16
XXVIIIth IAU General Assembly, August 2012
T. Montmerle, ed.

doi:10.1017/S1743921314013131

More Observations in Schools for Promoting Astronomy and Sky Protection

Rosa M. Ros

Applied Mathematics Dept, Technical University of Catalonia, Jordi Girona 1-3, modul C3, ES-08034 Barcelona, Spain
email: ros@ma4.upc.edu

Abstract. In astronomy it is important to promote observation and the quality of the sky is essential for a good observation impact. It is important that children have a nice memory of their observations in a non-polluted sky. Using students as agents of change it is possible to promote good practice for sky protection in society.

Keywords. Education, Motivation, Observation, Light Pollution.

1. Contents

It is imperative to transfer the enthusiasm of the teachers to their students or pupils. It is not always possible, but we have to present our lessons in an attractive format. It is essential to grab and keep the attention of the students. If it is feasible, we can use surprise as a tool to get and maintain interest: observation is often a useful resource.

All the schools have a sky above their buildings, it must be used to observe and take measurements. If the school does not have instruments to carry out observations, we can encourage students to produce their own devices. We probably lose precision, but we gain motivation, and this is much more relevant. They learn by doing, observing and reasoning. It is important to promote rationality, curiosity and creativity. In our society which has a great presence of pseudo-sciences, it is necessary to endorse the scientific spirit.

What are the aspects that people best remember? Without doubt, memories linked to positive emotions. If we feel a passion related to some astronomical observation, we will keep it in mind and we will add a positive connotation to astronomy. This relationship will stay with the person for some time and it will be a tie that will not be broken easily. It is good that the experiences, and in special the first experiences with astronomy, are exciting. We must get the students to feel the thrill!

People are beginning to enjoy astronomical tourism, because they cannot see the deep sky in their cities. Of course light pollution is a serious problem for astronomy, for ecology, but it is also important to us because we have lost a part of our heritage: the sky that humanity saw for centuries. We should demand a clear sky for our enjoyment and our emotions.

We should use all the above mentioned strategies such as causing surprise, enthusiasm, encouraging the participation of everybody and to get passionate about the sky above us and to fight against the light pollution.

References

Berthomieu, F., da Costa, A., Deustua, S., Fierro, J., Garca, B., Hemenway, M. K., Moreno, R., Pasachoff, J. M., Percy, J., Ros, R. M., & Stavinschi, M. 2012, *14 steps to the Universe*

Highlights of Astronomy, Volume 16
XXVIIIth IAU General Assembly, August 2012
T. Montmerle, ed.

doi:10.1017/S1743921314013143

Dark Skies Rangers - Fighting light pollution and simulating dark skies

Rosa Doran[1], **Nelson Correia**[1,2], **Rita Guerra**[1] **and Ana Costa**[1]

[1]NUCLIO - Núcleo Interactivo de Astronomia
Largo dos Topázios, 48, 3 Frt, 2785-817 S. D. Rana, Portugal
email: geral@nuclio.pt

[2]Escola Secundária Maria Lamas
Rua 25 de Abril, 2350-786 Torres Novas, Portugal

Abstract. Dark Skies Rangers is an awareness program aimed at students of all ages to stimulate them to make an audit of light pollution in their school/district. The young light pollution fighters evaluate the level of light pollution, how much energy is being wasted, and produce a report to be delivered to the local authorities. They are also advised to promote a light pollution awareness campaign to the local community targeting not only the dark skies but also other implications such as effects in our health, to the flora and fauna, etc.

Keywords. miscellaneous, sociology of astronomy, data analysis, surveys.

A common debate in astronomy and space science gatherings is the fact that light pollution is growing too fast and our efforts to fight it are far from being enough. In order to efficiently fight this we need to recruit an army of literate soldiers, with a good understanding of our disappearing skies and the consequences it may bring. Recruiting schools to help astronomers in this enterprise might be a quick and effective solution. It is a win-win situation where teachers can address topics of the curricula using modern trends for science education and students may gain a deeper knowledge about our symbiosis with nature.

Dark Skies Rangers (DSR) is a joint collaboration between the National Optical Astronomy Observatory of the USA (NOAO) and the Galileo Teacher Training Program (GTTP). A very good repository of materials was created and is available in www.darkskiesawareness.org/DarkSkiesRangers. In Portugal a pilot effort was implemented in the municipalities of Cascais and Torres Novas to test the adaptability of the tools and resources in classroom, the receptivity of the communities and impact in the school. All the information of the efforts in Portugal can be found in dsr.nuclio.pt.

In Cascais the program started with 50 students analysing the light pollution in the streets surrounding their schools. During the school year of 2012/2013 ten more schools will integrate the program and the objective is to have a good map of the light pollution in the major areas of town. In Torres Novas the program was adopted by 93 students and 10 teachers in a cooperative effort. Students from 7th to 12th grade were involved in a wide range of activities. This project received an "Science in School" award from the Portuguese Foundation Ildio Pinho.

In the school year of 2012/2013 the program will be included in the Eco-Schools program and will be publicised in 1500 Portuguese eco-schools. The objective is to build a sustainable community of light pollution rangers. This program is using the IBSE model to share the good practices with the participants as part of the pilot efforts of the European Commission's funded project "Discover the Cosmos", under the scope of the FP7.

Highlights of Astronomy, Volume 16
XXVIIIth IAU General Assembly, August 2012
T. Montmerle, ed.

doi:10.1017/S1743921314013155

The Impact of Light Pollution Education through a Global Star-Hunting Campaign & Classroom Curricula

Constance E. Walker[1] **and Sanlyn Buxner**[2]

[1]National Optical Astronomy Observatory, Tucson, AZ, USA, `cwalker@noao.edu`

[2]Planetary Science Institute, Tucson, AZ, USA, `buxner@psi.edu`

Abstract. Results of a survey assessing the impact on respondents and their students from the worldwide, citizen-science light pollution campaign called GLOBE at Night and the accompanying environmental-astronomy-based curricula called Dark Skies Rangers are presented.

Keywords. Light Pollution, Citizen-Science, Public Outreach

Hosted by the U.S. National Optical Astronomy Observatory over the last 7 years, GLOBE at Night (`www.globeatnight.org`) is an international campaign to raise public awareness of the impact of light pollution by inviting citizen-scientists to measure their night sky brightness. As a follow-on to GLOBE at Night, Dark Skies Rangers (`www.globeatnight.org/dsr/`) is a set of environmental-astronomy-based lesson plans for grades 3 through 12. The activities teach the importance of proper lighting technology, light pollutions effects on wildlife and energy, and how to monitor the darkness of your sky. An externally administered and analyzed survey was conducted among educators and students to determine the effectiveness of the campaign and lessons and identify where improvements could be made. 17%(or 102 of 585) educators completed the online survey in Spring 2012. Some survey results are described here.

Dark Sky Rangers activities are most often used for in-class astronomy units or in club or afterschool programs, but also in planetaria, nature-science centers, state/national parks and star parties. Survey respondents stressed the versatility of the activities for different audiences and venues. The top reported benefit from the Dark Skies Education workshops was increased awareness of light pollution issues (98%), followed by teaching resources (81%) and resources for dark sky advocacy (68%). The Demonstrating Light Pollution and Shielding activity was the top-rated activity (62%). When asked about the impact of doing the Dark Skies Rangers activities, 96% of the respondents and 76% of their students were likely to take action to reduce light pollution in their community.

67% of the survey respondents and 81% of their students who participated in the GLOBE at Night citizen-science campaign made changes in their use of lighting in their homes. Changes included upgrading fixtures, turning off outside lights, reducing usage, using more energy efficient lights, utilizing motion sensors, being an advocate and educating others about light usage, becoming involved in policy making, and becoming more aware of community lighting. All but 1 respondent indicated they believed that engaging students in GLOBE at Night activities increased their understanding of light pollution.

Following the framework (`http://insci.org/resources/Eval_Framework.pdf`) used by the U.S. National Science Foundation for evaluating impacts of informal science education projects, there is evidence that Dark Sky Rangers and GLOBE at Night programs have long-term impact on several categories including 1) awareness, knowledge or understanding, 2) engagement or interest, and 3) behavior. For more details, see the authors' article by the same title in the 2013 ASP Conference Series on "Communicating Science".

Highlights of Astronomy, Volume 16
XXVIIIth IAU General Assembly, August 2012
T. Montmerle, ed.

doi:10.1017/S1743921314013167

GLOBE at Night in China

Hongfeng Guo
National Astronomical Observatories, Chinese Academy of Sciences
email: ghf123@yahoo.cn

Abstract. The GLOBE at Night citizen-science campaign was introduced in China in 2010. Observations and works made by students are presented. The students were guided to participate in this meaningful international activity by 1) taking light pollution observations of the night sky at different locations, 2) becoming aware of the severity of the effects of light pollution, and 3) making the whole society aware of the importance to save energy by reducing light pollution.

Keywords. light pollution

China Hands On Universe introduced GLOBE at Night into China in 2010. We organized students to join the program during astronomy class and after class. We also set up a web page on the China Hands On Universe web site so that other students, parents and the public could learn about the GLOBE at Night program and send in the reports of their observations. We received 160 reports the first year (statistics in Table 1).

We found that some students in the first year who took part in the activity had high interest, but in the second year of GLOBE at Night in China there was reduced interest. Therefore, we improved the method and increased the activity content. For example, we advised the students to prepare the following items: a city map from Google Earth or GPS,) a path from city center to suburb, 4-5 stops (every 10 km) when out of town, a sky map (Orion), paper, pen etc. We instructed them to gather information for each observation at each location: seek and mark stars in Orion (e.g., make or take pictures); check latitude and longitude; and record the weather.

Once the observations were completed, we helped the students with their logs and data reduction and writing small scientific papers to contribute to popular science magazines. Students had more and more interest when they did more hands-on activities. More students wanted to join the program every year. The enthusiasm of the students affected their teachers and guardians enthusiasm. As a result, we received more reports in 2012.

In addition to the reporting observations for the GLOBE at Night program, we also received a lot of students' works, such as photographs, hand-painted pictures of the night sky, logs, maps, etc. The following is one of our students papers. He said that he went to a hill on the outskirts of Beijing. On the top of the hill he could see very clearly many stars toward Orion. But when he arrived at the foot of the hill by the highway, he could see only a few stars. When he came to the center of Beijing, he could see almost no stars. He personally experienced how severe the effects of light pollution can be and called on the city to minimize unneeded lighting and thereby reduce light pollution.

Table 1. Statistics of GLOBE at Night Reports from China 2010-2012

Year	Schools	Students	Reports	Public	Reports	Total Reports
2010	10	200	150	100	10	160
2011	10	100	90	100	10	100
2012	20	500	400	200	100	500
Total	40	800	640	400	120	760

Highlights of Astronomy, Volume 16
XXVIIIth IAU General Assembly, August 2012
T. Montmerle, ed.

doi:10.1017/S1743921314013179

Citizen Science Programs on Light Pollution Awareness: Where Do We Go with the Data?

Constance E. Walker[1] **and Christopher C. M. Kyba**[2]

[1]National Optical Astronomy Observatory, Tucson, AZ, USA, cwalker@noao.edu
[2]Institute for Space Sciences, Freie Universität Berlin,
Carl-Heinrich-Becker-Weg 6-10, 12165 Berlin, Germany, christopher.kyba@wew.fu-berlin.de

Abstract. Once data from a citizen-science program on light pollution is verified, what research projects, on-line analytical tools and tutorials should be developed, and what ways can results and acknowledgements be provided to the public? These and other questions are explored.

Keywords. Light Pollution, Citizen Science, Public Outreach

The past 7 years have seen several citizen-science light pollution campaigns: GLOBE at Night, UK's Orion Star Count, the Great World Wide Star Count, How Many Stars, the Great Indian Star Count, Brazil's Milky Way Marathon, & the Big Aussie Star Hunt. Most data are taken visually by citizen-scientists of all ages without Moon or cloud cover.

Data has also been taken with various skyglow monitors. The LE, LU and DL models of Unihedron's Sky Quality Meter have >50 stations globally and 1000s of handheld models. The IYA Lightmeter has >30 stations worldwide. The International Dark-Sky Association has 20 Night Sky Brightness Monitor stations, most at major observatories and U.S. national parks. There are also Henk Spoelstras DigiLum devices and STEM Laboratory Inc.s sky brightness meters.

Until recently, the lack of a common measurement standard has hampered efforts to compare measurements from different locations and to develop long-term databases. Efforts through the Cabauw Lightmeter InterComparison workshop and subsequent public comment period produced a new standard officially adopted at the Protection of the Night Sky Symposium in 2012. See www.darksky.org/night-sky-conservation/248.

In the GLOBE at Night data we find a strong relationship between average estimated naked-eye limiting magnitude (NELM) reported by citizen scientists and approximate skyglow predictors, such as the upward-directed radiance measured by the Defense Meteorological Satellite Program (DMSP), 2010 and the skyglow simulation in "The World Atlas of the Artificial Night Sky Brightness", 2001. Analysis of the GLOBE at Night data over the last 7 years indicates a growing trend for brighter skies worldwide.

Research projects for the public could include comparison of sky brightness over time with population density or with data from satellites. Analyses could be compared with data on health, wildlife and energy. People could take lighting inventories or search for dark sky oases. Understanding the behavior of sky brightness over time is essential for making cases for lighting ordinances to city officials or changing peoples mindsets. As examples, student projects with the GLOBE at Night data have been done on the 'lesser long nosed' bat and on monitoring the sky brightness overnight for several months.

On-line tutorials and analytical tools are needed to help the public render valid analysis. Some tutorials have taken place at in-situ workshops or in on-line forums, videoconferences, teleconferences with Powerpoints and Skype sessions. More effort is needed in getting results and acknowledgements to the public. Efforts to date have seen results published in popular science magazines, social networks, Facebook, Twitter, blogs and through national partners and press releases to local and international media.

Highlights of Astronomy, Volume 16
XXVIIIth IAU General Assembly, August 2012
T. Montmerle, ed.

doi:10.1017/S1743921314013180

Night of Darkness Campaign: Make Light Pollution Something Everybody Cares About

Friedel Pas
European Liaison Officer
International Dark-Sky Association, Aarschotsebaan 29, B-3191, Hever, Belgium
email: europe@darksky.org

Abstract. To battle light pollution issues, influence the market so that customers request only good lighting. An important factor in doing this is to convince decision makers that light pollution is something the whole society cares about. This article gives an overview of an existing campaign that does this and emphasizes some important factors to consider in your campaign.

Keywords. education and public outreach, light pollution, citizen-science campaign

---

Awareness campaigns have a common goal. Increase awareness with the public, but also convince politicians to take action – especially when campaigning against light pollution.

To maximize the impact of your campaign amongst politicians, it is important to increase your critical mass. Campaigns against light pollution are mostly done from the point of view of astronomy. Light pollution is an important issue to amateur and professional astronomers, but astronomers are small in number. Their reasons for preserving a dark night sky give the public the impression the attention is to a selective hobby or a profession done on far-away mountaintops – not affecting them.

But the impact of light pollution affects much more than astronomy. It impacts the environment, human health, energy consumption, wildlife and our culture. To increase your critical mass for your campaign, it is important that you involve organizations representing these areas. It is important to involve a broad range of interests to demonstrate that light pollution is something that affects everybody.

The more people participate in an event, the more you increase the critical mass of your campaign. The night of Darkness is an example of such a campaign that offers the public the opportunity to participate. The annual event take place on October 20, 2012 for the 17th time in Flanders, Belgium and has expanded to other regions in Europe. On that night, municipalities are requested to switch off their unnecessary lighting. About 2/3 of the municipalities are participating in some way – including switching off monument lighting, reducing the number of illuminated light poles and turning off lighting in a region or sometimes a whole municipality or city. Local activities are being organized: star gazing parties, night nature promenades, storytelling and concerts in the dark. Local municipalities, astronomy clubs, environmental organizations and cultural organizations are collaborating to organize these events. In 2011, the campaign events attracted more than 26,000 participants.

At the same time, the events are unique in providing astronomy outreach to people who do not typically experience astronomy. With astronomy events, you attract participants that are already interested in astronomy. With an event that is a combination of different interests, you can present the beauty of the sky to people who have a variety of reasons for being at the event. Apart from raising awareness to general public and politicians, providing further education in light pollution prevention is also an important aspect of a successful campaign.

Highlights of Astronomy, Volume 16
XXVIIIth IAU General Assembly, August 2012
T. Montmerle, ed.

doi:10.1017/S1743921314013192

A new Starlight Reserve for the central South Island of New Zealand

John Hearnshaw
Dept. of Physics and Astronomy
University of Canterbury, Christchurch, New Zealand
email: john.hearnshaw@canterbury.ac.nz

Abstract. The Aoraki Mackenzie International Dark Sky Reserve is a new reserve created in 2012 by the International Dark-Sky Association in the central South Island of New Zealand, and covers over 4300 square kilometres around Mt John University Observatory. It is the first such reserve to be recognized at gold tier level and is the largest dark sky reserve in the world. Astro-tourism in the new reserve will be a prominent activity in the coming years.

Keywords. Dark sky reserve, combatting light pollution, astro-tourism

1. Overview

We recently made a proposal for a new Dark Sky Reserve in the centre of the South Island of New Zealand, to be known as the Aoraki Mackenzie International Dark Sky Reserve. The area includes Mt John University Observatory and all of Aoraki/Mt Cook National Park and covers 4367 square kilometres. There are three small centres of population in the Reserve, with a total population of about 1500 people. Most of this large area has had light pollution controlled by a lighting ordinance since 1981 (possibly the first in the southern hemisphere), so as to limit light pollution through filtering, shielding and the time of operation of outdoor illumination. The National Park also has lighting controls in place. The sky is exceptionally dark (averaging about 21.7 V magnitudes per square arcsec in the core regions of the reserve) and the air is very clear and unpolluted, making it an ideal place for stargazing. A number of astro-tourism companies are now thriving in this region, and over 20,000 visitors annually, coming from all over the world, are benefitting from guided night sky tours.

The reserve was recognized in April 2012 by the International Dark-Sky Association in Tucson AZ and was granted gold tier status, the first in the world at gold tier level and just the third dark sky reserve to be recognized by IDA. It is also the first reserve to be declared in the southern hemisphere and the largest reserve area. Full details of the application document to IDA can be viewed on-line (Austin *et al.* 2012). A management board is now being established for the reserve, so as to monitor compliance with the lighting ordinance, to recommend improvements to outdoor lighting and to organize activities in the reserve to promote astro-tourism. Our plans are to stay at the forefront of the fast-growing international dark sky reserve movement, and to set a standard for dark skies and light pollution controls for others to follow.

Reference

Austin, M. E., Hearnshaw, J. B., Butler, S., & Loveridge, A. 2012, *An Application to the International Dark-Sky Association for a Starlight Reserve in the Aoraki/Mt Cook National Park and the Mackenzie Basin of the central South Island of New Zealand*, pp xii + 167, http://www.saps.canterbury.ac.nz/starlight/

Highlights of Astronomy, Volume 16
XXVIIIth IAU General Assembly, August 2012
T. Montmerle, ed.

doi:10.1017/S1743921314013209

Astro tourism: Astro Izery project

Tomasz Mrozek[1,2], Sylwester Kołomański[1], Grzegorz Żakowicz[3], Stanisław Kornafel[4], Tomasz L. Czarnecki[5], Pavel Suchan[6] and Zbigniew Kamiński[7]

[1]Astronomical Institute, University of Wrocław, ul. Kopernika 11, 51-622 Wrocław, Poland
email: mrozek@astro.uni.wroc.pl
[2]Solar Physics Division, Space Research Centre, Polish Academy of Sciences
[3]XIIIth High School in Wrocław
[4]Tourist Station "Orle"
[5]teleskopy.net
[6]Astronomical Institute, Academy of Sciences of the Czech Republic
[7]Forestry Commision Świeradów

Abstract. The Astro Izery project is carried by several institutions from Poland and Czech Republic. Its aim is to educate and inform tourists, who visit the Izery Mountains, about astronomy and light pollution. The project consists of two activities: permanent (sundials, planetary path etc.) and periodic (meetings, workshops). After five years the project is in good health and will gain more elements in next years.

Keywords. Keyword1, keyword2, keyword3, etc.

1. Overview

An idea of the Astro Izery project was born during the first edition of School Workshop on Astronomy (SWA) organized in the Tourist Station Orle. We intend to connect the tourist and natural value of the Izera Mountains with an astronomical education under the dark starry sky. The project extends the present tourist and education offer of the Izera Mountains. It consists of several components that may be divided into two groups: permanent (pt) and periodic (pc):

- The scaled model of the Solar System (pt) - 4.5 km long (scale 1:1 bln)
- Izera Dark Sky Park (IDSP)(pt) - opened on November 4th, 2009 as the first dark sky park in Poland, Czech Republic and in Europe. At present it is still the one and only international DSP in the world.
- Sun dial and a gnomon (pt) - located in the Orle settlement.
- School Workshop of Astronomy (pc) - for high school students. Ten events.
- All-Poland Astronomical Meeting (pc) - for amateur astronomers. Seven events.
- Astronomical Day in Izera Mountains (pc) - meeting for everyone. Five events.

The IDSP gives an opportunity to educate society in the field of light pollution and to include nocturnal darkness conservation in the existing nature conservation of this area. The educational part of the Astro Izery project is active in the form of meetings with "living astronomers". In the near future we plan the next events of meetings and we are going to realize several new permanent components. The planetary path will be supplemented by dwarf planets. Thanks to that, the path will run across the entire IDSP. The next element will be the cafeteria on the Stóg Izerski summit. We plan to create a ceiling with 2000 points of light made of optical fibers that will constitute a map of the sky that is visible in the Izera Mountains in winter months. Additionally, permanent exhibitions of meteorites will be opened in the cafeteria and in the "Orle".

Highlights of Astronomy, Volume 16
XXVIIIth IAU General Assembly, August 2012
T. Montmerle, ed.

doi:10.1017/S1743921314013210

Angular distribution of uplight at 10,000 ft over Berlin

Christopher C. M. Kyba[1,2], Thomas Ruhtz[1], Carsten Lindemann[1], Jürgen Fischer[1], Franz Hölker[2] and Christian B. Luginbuhl[3]

[1]Institute for Space Sciences, Freie Universität Berlin,
Carl-Heinrich-Becker-Weg 6-10, 12165 Berlin, Germany
email: christopher.kyba@wew.fu-berlin.de

[2]Leibniz Institute of Freshwater Ecology and Inland Fisheries
Müggelseedamm 310, 12587, Berlin, German
email: hoelker@igb-berlin.de

[3]US Naval Observatory Flagstaff Station
10391 West Naval Observatory Road, Flagstaff, Arizona 86001-8521 USA
email: cbl@nofs.navy.mil

Abstract. The upward emission direction of artificial light from cities is unknown, and is the most important systematic uncertainty in simulations of skyglow. We present a technique for measuring the emission for zenith angles up to 70°.

Keywords. instrumentation: detectors, radiative transfer

1. Overview

The airmass from Earth's surface to the top of atmosphere is greatest for horizontally propagating light, so the scattering probability is far higher for horizontally directed light. Understanding the angular distribution of artificial light emitted by cities is thus of crucial importance for simulating skyglow accurately. Because the distribution has not been measured, until now it has only be inferred from comparing skyglow observations to simulations (Luginbuhl *et al.* 2009).

By mounting two cameras on an aerial measurement platform, we performed measurements of the angular distribution of uplight at an elevation of 10,000 ft over Berlin, Germany. This system is able to measure the upwards emission from zenith to an angle of 70°, for any given azimuthal direction. Testing whether the upward emitted light is azimuthally symmetric is of interest, because it could explain the polarization of skyglow observed by Kyba *et al.* (2011). Azimuthally symmetric emission is generally assumed by skyglow models (e.g. Aubé & Kocifaj (2012), Falchi & Cinzano (2012)).

This technique is described more completely in "Two Camera System for Measurement of Urban Uplight Angular Distribution", a forthcoming proceedings paper from the 2012 International Radiation Symposium.

References

Aubé, M. & Kocifaj, M. 2012, *MNRAS*, 422, 819
Kyba, C. C. M., Ruhtz, T., Fischer, J., & Hölker, F. 2011, *J Geophys Res*, 116, D24106
Falchi, F. & Cinzano, P. 2012, *MNRAS*, in press
Luginbuhl, C. B., Duriscoe, D. M., Moore, A., Richard, G. W., & Lockwood, G. W. 2009, *PASP*, 121, 204

Highlights of Astronomy, Volume 16
XXVIIIth IAU General Assembly, August 2012
T. Montmerle, ed.

doi:10.1017/S1743921314013222

NIXNOX project: Sites in Spain where citizens can enjoy dark starry skies

J. Zamorano[1], A. Sánchez de Miguel[1], E. Alfaro[2], D. Martínez-Delgado[3], F. Ocaña[1], J. Gómez Castaño[1] and M. Nievas[1]

[1]Depto. Astrofísica y CC. de la Atmósfera, Universidad Complutense de Madrid, Facultad de Físicas, Ciudad Universitaria 28040 Madrid, Spain [2]Instituto de Astrofísica de Andalucía (IAA-CSIC), Granada, Spain [3]Max Plank Intitut für Astronomie, Heidelberg, Germany
email: `jzamorano@fis.ucm.es`

Abstract. The NIXNOX project, sponsored by the Spanish Astronomical Society, is a Pro-Am collaboration with the aim of finding sites with dark skies. All sky data of the night sky brightness is being obtained by amateur astronomers with Sky Quality Meter (SQM) photometers. We are not looking for remote locations because the places should be easily accessible by people with children. Our goal is to motivate citizens to observe the night sky. NIXNOX will provide information to answer the question: where can I go to observe the stars with my family?

Keywords. Light Pollution, Night Sky Brightness, Photometry

The NIXNOX project.
This is a Pro-Am collaborative effort promoted and sponsored by the Spanish Astronomical Society (SEA) to find and characterize open air observatories. The observations and additional information are being provided by amateur astronomers. It contributes to outreach in Astronomy and it is a help to dark skies fights but it is also a scientific project. Our objectives are to locate sites with dark skies with easy access, to encourage local authorities to preserve them and finally to help citizens to enjoy the starry skies.

Night sky brightness maps. All-sky maps of the night sky brightness are being obtained by amateur astronomers associations in places selected by them. They are using 12 SQM-L photometers on loan from SEA that were cross-calibrated by us. The observations should be made on clear and moonless nights. The spatial sampling is a trade-off between resolution and the time needed to complete a map. Since SQM-L photometers have a field of view of 20 degrees (FWHM), we select to observe, besides zenith, 12 positions in azimuth at 20, 40, 60 and 80 degrees of altitude. The resulting all-sky maps (in units of magnitudes per square arcsecond) are similar to calibrated fish eye pictures of the sky and they inform us of the sources of light pollution. Evolution of the light pollution will be measured with repeated observations over the next years.

Additional information. More than 50 amateur astronomer associations of Spain are collaborating with SEA. They are also gathering additional information. For each site the SEA webpage will also provide a brief description with panoramic pictures, lodging and meteo information, how to reach the site, etc. The Spanish Astronomical Society will publish these information to help the citizens to choose where to go to observe the stars with their family or friends. More info about NIXNOX can be obtained at SEA webpage `http://www.sea-astronomia.es/`.

Highlights of Astronomy, Volume 16
XXVIIIth IAU General Assembly, August 2012
T. Montmerle, ed.

doi:10.1017/S1743921314013234

The Night Sky Monitoring Network in Hong Kong

Chun S. J. Pun[1], Chu W. So[2] and Chung F. T. Wong[3]

[1]Department of Physics, The University of Hong Kong,
Pokfulam, Hong Kong
email: `jcspun@hku.hk`

[2]email: `socw@hku.hk`

[3]email: `terryfai@hku.hk`

Abstract. The Night Sky Monitoring Network is a project that aims to study the extent, distribution, and properties of the light pollution condition in the populous metropolis of Hong Kong. Continuous measurements of the Night Sky Brightness (NSB) at strategically chosen locations that cover a wide range of population density and land usage were made, with over 2.5 million NSB readings collected in 18 months up to June 2012. Results from the project are presented, with focus on the contrast between the urban and rural night sky profiles, and light pollution contributions from artificial lightings. This project is supported by the Environment and Conservation Fund of the Hong Kong SAR government (ECF 10/2009, ECF 1/2007).

Keywords. Measurement: night sky brightness, light pollution, atmospheric effects.

The positions of the 18 measuring stations (equiped with Sky Quality Meters - Lens Ethernet) are overlayed on the night sky image taken by a camera mounted on the *International Space Station* taken in 2003 in Figure 1 (*left*), clearly indicating the contrast between the urban (*red*) and rural (*blue*) night sky profiles. The long-term trend of the median NSB at selected stations are shown in Figure 1 (*right*). The observed monthly fluctuations of NSB at any particular site was found to be due mostly to seasonal variations in meteorological factors such as cloud and rain (Please refer to contribution by So *et al.* in this volume). On the other hand, the relative ranking of different sites remained steady, with the NSB of rural sites consistently darker than those in urban locations by 3–4 mag, under similar meteorological conditions. This clearly indicated the polluting effects of artificial lightings.

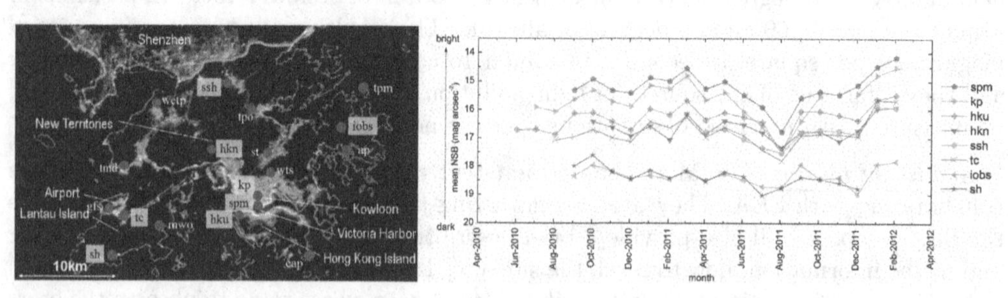

Figure 1. (*Left*) Locations of urban (*red*) and rural (*blue*) monitoring sites were overlaid on the night time picture of Hong Kong taken from the International Space Station; (*Right*) Monthly variations of the median NSB in selected sites.

For details, please refer to the project website: `http://nightsky.physics.hku.hk`

Highlights of Astronomy, Volume 16
XXVIIIth IAU General Assembly, August 2012
T. Montmerle, ed.

doi:10.1017/S1743921314013246

SKYMONITOR: A Global Network for Night Sky Brightness Measurements

Dan McKenna,[1] Donald Davis,[2] and Paul Boley[2]

[1] 1Palomar Observatory Po Box 200, Palomar Mountain CA 92060
email: dmck@astro.caltech.edu

[2] International Dark-Sky Association, 3223 N. First Ave., Tucson, Arizona 85719
email: drd@psi.edu and pboley@gmail.com

Abstract. We have deployed a network of autonomous photometers that continuously measures the night sky brightness in the visual region at two sky positions simultaneously, typically near the zenith and the second at an elevation angle of 20 degrees. The Photometers are calibrated as a network to better than 5.

Keywords. night sky brightness, networked photometers, SKYMONITOR

1. Overview

The SKYMONITOR network uses two, 4 degree field of view, silicon photo detectors filtered by Hoya CM500 glass to cover a wavelength range of 320nm to 750 nm. The brightness data are temperature corrected and calibrated before leaving the manufacturing facility for an estimated variation in calibration of less than 3%. Comparing several photometers co-located and pointed, we find that the variation between photometers of the observed sky brightness is primarily due to differences in the field of view. We conclude that our photometers need to be pointed better than 1/10 the field of view especially when dealing with areas of the sky with high gradients such as near the galactic plane or near the horizon.

2. Current Results

Photometers as tested in the field by on sky inter comparison vary by less than 5%. Variations in field of view are mainly due to detector packaging and alignment between optical and mechanical axis and are the largest components of uncertainty. The current preliminary photometric zero point is about +0.52 Magnitudes different from a calibration based on Vega. Our estimated sky brightness for the darkest skies is currently 21.84 corrected to V magnitude. Long term stability will be monitored by periodically retesting photometer sensors in the lab, the observation of bright stars and the galactic plane. The current generation of Photometers transmits raw brightness measurements and sensor temperature data every 30 minutes to the IDA collection center in Tucson where the calibration is applied. The data may be viewed on line at nsbm.darksky.org.

Highlights of Astronomy, Volume 16
XXVIIIth IAU General Assembly, August 2012
T. Montmerle, ed.

doi:10.1017/S1743921314013258

A standard format for measurements of skyglow

Christopher C. M. Kyba[1], Dorien E. Lolkema[2] and Constance E. Walker[3]

[1]Institute for Space Sciences, Freie Universität Berlin,
Carl-Heinrich-Becker-Weg 6-10, 12165 Berlin, Germany
email: christopher.kyba@wew.fu-berlin.de

[2]National Institute for Public Health and the Environment
A. van Leeuwenhoeklaan 9, 3720 BA Bilthoven, The Netherlands
email: dorien.lolkema@rivm.nl

[3]National Optical Astronomy Observatory
950 North Cherry Ave. Tucson, AZ 85719 USA
email: cwalker@noao.edu

Abstract. A standard format for recording skyglow measurements is needed to allow for effective data exchange. A proposal for such a format was discussed at the IAU Symposium.

Keywords. standards, instrumentation: detectors, atmospheric effects

---

1. Overview

The development of the International Year of Astronomy Lightmeter and Sky Quality Meter have resulted in a great increase in the number of locations worldwide for which skyglow is periodically or constantly observed (Biggs *et al.* 2012, Kyba *et al.* 2011, Kyba *et al.* 2012, Lolkema *et al.* 2011, Müller *et al.* 2011). Exchanging such data would benefit all researchers, particularly for validating skyglow models such as those of Aubé & Kocifaj (2012) and Falchi & Cinzano (2012). Unfortunately, sharing such data has been hampered by the different data formats used.

At the IAU Symposium, a proposal for a standard format was discussed. Following the meeting, the format was officially adopted by the light at night research community during the 12th European Symposium for the Protection of the Night Sky, and was announced in Kyba & Lolkema (2012). The final definition of the format is archived by the International Dark Sky Association at http://www.darksky.org/measurements.

References

Aubé, M. & Kocifaj, M. 2012, *MNRAS*, 422, 819

Biggs, J. D., Fouch, T., Bilki F., & Zadnik, M. G. 2012, *MNRAS*, 421, 1450

Kyba, C. C. M.. & Lolkema, D. E. 2012, *ASTRON GEOPHYS*, 53, 6.17

Kyba, C. C. M., Ruhtz, T., Fischer, J., & Hölker, F. 2011, *PLoS ONE*, 6, e17307

Kyba, C. C. M., Ruhtz, T., Fischer, J., & Hölker, F. 2012, *MNRAS*, 425, 701

Falchi, F. & Cinzano, P. 2012, *MNRAS*, in press

Lolkema, D., Haaima, M., den Outer, P., & Spoelstra H. 2011, *Technical Report RIVM #680151002, Effects of meteorological and atmospheric parameters on night sky brightness. Netherlands National Institute for Public Health and the Environment, Bilthoven, Netherlands*

Müller, A., Wuchterl, G., & Sarazin, M. 2011 *in Serie de Conferencias Vol. 41 of RevMexAA, Measuring the night sky brightness with the lightmeter* 4649

Highlights of Astronomy, Volume 16
XXVIIIth IAU General Assembly, August 2012
T. Montmerle, ed.

doi:10.1017/S174392131401326X

Assessing the contribution from different parts of Canary islands to the hemispheric spectral sky radiance levels over European Northern Observatories

Martin Aubé

Département de physique, Cégep de Sherbrooke , Shebrooke, Canada
email: martin.aube@cegepsherbrooke.qc.ca

Abstract. In this paper, we suggest to use a sky radiance model which accounts for heterogeneous distribution of light fixtures, their photometry, the ground reflectance and topography, to infer the point to point contribution of Canary Islands to the artificial sky radiance at Observatorio del Teide (Tenerife) and Observatorio Roque de los Muchachos (La Palma). In-situ hyperspectral sky radiance measurements, acquired on site in 2010, have been used to calibrate the model and to evaluate its inherent error. We aim to identify and characterize zones at which any lighting level increase or decrease may have a larger impact on light pollution at both European Northern Observatory sites, and then help to control and/or reduce their light pollution levels. This innovative methodology, can then be seen as a high level decision tool to help local authorities to restrict or reduce light pollution with the objective of protecting research class astronomical sites.

Keywords. sky radiance, modelling, measurements, Canary Islands

---

An overview of artificial sky brightness modeling and measuring experiment in both European Northern Observatories is presented. The numerical radiative transfer model ILLUMINA (Aubé *et al.* 2005; Aubé 2007) has been used to calculate artificial sky radiance. During this experiment, a large amount of sky spectral radiance measurements have been acquired on both observatories to calibrate modeled radiances. As a standard output, the model delivers the sky radiance value, the relative contribution map, and the relative sensivity map for each set of input parameters (5 key wavelengths, 4 aerosol optical depths, many viewing angles, 2 observatories, before/after midnight). Sky radiance relative contribution map gives the contribution of each square km to the modeled sky radiance in percentage. Basically it says from where the sky radiance is coming at the given site, viewing angle and atmospheric conditions. The relative sensitivity map is the contribution map divided by the installed lumen in each square km and then renormalized. The map indicates how each pixel will impact the sky radiance when one removes or adds a standard light fixture. So basically the highest levels on that map indicates the first place to make changes to the light fixture inventory in order to have the highest reduction. Various maps can be seen in our prototype portal at galileo.graphycs.cegepsherbrooke.qc.ca/atlas

References

Aubé, M., Franchomme-Foss, L., Robert-Staehler, P., & Houle, V, 2005, *Proc. SPIE*, 5890, 248.
Aubé, M. 2007, *Marin C., Jafari J., eds, Proc. Starlight a common heritage, La Palma, Spain*, 351.

Highlights of Astronomy, Volume 16
XXVIIIth IAU General Assembly, August 2012
T. Montmerle, ed.

doi:10.1017/S1743921314013271

CTA site characterization: a contribution on Sky Background Brightness

Gonzalo de la Vega[1], **Beatriz García**[1], **Javier Maya**[1], **Alexis Mancilla**[1] **and Ezequiel Rosemblat**[2]

[1]ITeDA Mendoza, CNEA,CONICET, UNSAM
Azopardo 313 5501 Godoy Cruz, Mendoza, Argentina
email: beatriz.garcia@iteda.cnea.gov.ar
[2]Universidad Mendoza, Mendoza, Argentina

Abstract. The Cherenkov Telescope Array (CTA) is an international project for which more than 1000 scientists from 27 countries work together to install two observatories, one in the southern hemisphere and the other in the north, to study gamma-ray radiation. Argentina is a candidate site for one of the observatories, and to characterize proposed sites, one must study the sky background brightness. This critical point must be studied with care. Nearest cities can be a decisive factor in determining where to locate an observatory in our times.

Keywords. Sky Background Brightness, Sky Background Brightness: Observatories site selection

---

Background sky emission is studied for site characterization for the installation of astronomical observatories. Atmospheric emission has two main sources, one from Earth (air glow, activity by auroras and thermal conditions) and one from outer space (contributions from the moon, stars or galaxies, zodiacal light, and diffuse background from our galaxy). The first depends on different variables (seasons, solar activity, geographical location), but the second is connected to the telescope pointing position. The extraterrestrial sources are due to unresolved stars and galaxies and, the most important, sunlight scattered by interplanetary dust. (Patat (2006), Patat (2008)).

In recent years, many observatories and groups devoted to site selection to install telescopes or astronomical instrumentation have used commercial devices such as the Sky Quality Meter (SQM) manufactured by Unihedron (*http : //www.uniheadron.com*). This equipment is a very simple photometer, easy to install, inexpensive and can be used in different sites at the same time.

In 2012, we installed two SQM-LR devices in Argentina at the CTA candidate sites (El Leoncito in San Juan, and San Antonio de los Cobres in Salta). The instruments were calibrated at the laboratory and cross-calibrated with adapted SQM-L instruments installed at the same sites since March 2011. The results from the instruments are comparable, but cross calibration shows that a correction (of about 0.4 mag/arcsec^2) must be applied to the data. This implies that if different devices are used (i.e. SQM-L and SQ-LR) and if precise results are needed, cross calibration between them is always necessary.

We determine that the effect of a single star of low magnitude is negligible in SQM measurements, but 15 stars of magnitude 4 would increase the background by 0.23 mag/arcsec^2. This situation is worst during the transit of the Milky Way across the instrument's field of view; the bulge of the Galaxy contributes about one magnitude.

References

Patat, F., *ESO Internal Report*, 2006. (http://www.eso.org/~fpatat/science/skybright/)
Patat, F., *A & A*, 481, 575-591, 2008

Highlights of Astronomy, Volume 16
XXVIIIth IAU General Assembly, August 2012
T. Montmerle, ed.

doi:10.1017/S1743921314013283

Protection of Northern Chile as an ICOMOS/IAU "Window to the Universe"

Malcolm G. Smith

Cerro Tololo Interamerican Observatory,
Casilla 603, La Serena, Chile
email: msmith@ctio.noao.edu

Abstract. Over the last two decades, La Serena's population has increased by about 70 percent. A site description of the AURA Observatory in Chile as a "Window to the Universe" is now available on the recently-launched UNESCO-IAU Astronomical Heritage Web Portal, www.astronomicalheritage.net This can serve as an example of possible material for the Chilean authorities, should they wish to propose the dark skies over much of northern Chile for protection as a World Scientific Heritage site. Some of the steps involved are discussed briefly here.

Keywords. Site protection, Light pollution, astrotourism, national parks

---

1. Overview

On 24th August 2012, during this General Assembly, a press release was issued by the IAU announcing the launch of the UNESCO-IAU Astronomical Heritage Web Portal. This effort - led on the IAU side by Clive Ruggles and the Astronomy and World Heritage Working Group and by the IYA2009 Astronomy and World Heritage Cornerstone project - is a response to an accord set up between UNESCO and the IAU in 2008. This launch follows the publication in 2010 of the Thematic Study on Astronomical Heritage produced by the IAU working together with ICOMOS. ICOMOS is UNESCO's advisory body for cultural sites. The press release indicates "Endorsed by UNESCO's World Heritage Committee in 2010, the Thematic Study provides guidelines for UNESCO member states on the inscription of astronomical properties. Much of its content has been incorporated onto the portal".

Astronomy is being seen once again as an ally in putting the "S"' back into UNESCO - orginally put there after much work by astronomer Bart Bok. Work is currently in progress on two-way links between the World Heritage Centre's own website (http://whc.unesco.org) and the new portal. Ruggles is quoted "The portal is not only aimed at the public, but also at UNESCO National Commissions seeking to nominate astronomical heritage sites for inscription onto the World Heritage List. It will provide guidance and comparative material. Scientific heritage is not just important in itself, but often relates to sustainablity, landscape conservation, tourism and even biodiversity". The press release goes on to mention that the portal will not only feature sites and monuments, but will also include dark-sky places.

These studies cover not only ancient heritage sites but also the sites of modern, large, international observatories. "...For example, members of the AWHWG have been working over the past year to develop nine much more detailed Extended Case Studies which will be brought on line one they have been discussed and approved at the IAU General Assembly. Several of them, it is hoped, will be a direct help in stimulating new World Heritage List nominations."

Included among the Extended Case Studies are three modern observatory sites, designated as "Windows to the Universe" - Mauna Kea, Hawaii, the Canary Islands and Northern Chile. These are all places with large numbers of international telescopes with at least one having a primary mirror of 6.5m or larger equivalent diameter. The international community has selected these sites with their pocketbooks. The international community will want the IAU to continue to help protect them in the future

Windows to the Universe.

Of the exisiting major international observatory sites in Northern Chile, that occupied by the AURA Observatory in Chile is the most threatened by light pollution. AURA was the first of the large optical astronomy organizations to build an observatory in Chile - it was a bold venture, siting an observatory at the southern edge of the Atacama desert in the early 1960s. The AURA property was already beyond the limit of paved roads at that time. The town of La Serena, was quite small and, at a line-of-sight distance of about 50km, it was not a significant source of light pollution over Cerro Tololo or any other part of the AURA property. The recent boom years of growth in northern Chile - particularly that derived from the copper-mining industry - has driven the current population of the La Serena/Coquimbo conurbation to well over 400,000; 70 percent of that growth has occured in the last two decades. Thanks to an intervening coastal range of mountains, most of the source of this light is not directly visible from Cerro Tololo or Cerro Pachon. This may change as a proposed new tunnel under the Andes helps open up trade still further between China and Brazil and the development boom moves up into the Elqui valley towards the observatory.

The useful sky accessible above normal limit-switch settings on telescopes on Cerro Tololo and Cerro Pachon (around 75-degree zenith distance) is still unpolluted (Kriscuinas *et al.* 2008). The observatory is thus on an excellent, but threatened site which serves as an appropriate Extended Case Study. We will learnmuch over the next three or four decades about attempting to control light pollution in Chile. At the very least, these lessons will be valuable for observatories futher north in Chile such as the Carnegie Southern observatory and the European Southern Observatory sites on Cerro La Silla, Cerro Paranal and Cerro Armazones. Models of the development of the light pollution as population increases have correctly predicted that light pollution of the sky over the AURA observatory site will not be a problem for some time, even without taking into account the beneficial effect of screening by the coastal moutain range. However, if bluerich lighting such as delivered by LEDs is permitted by Chilean authorities, more careful modelling will be advisable.

Reference

Kriscuina, K., Bogglio, H., Sanhueza, P., & Smith, M. G. 2010, *PASP*, 122, 373

Highlights of Astronomy, Volume 16
XXVIIIth IAU General Assembly, August 2012
T. Montmerle, ed.

doi:10.1017/S1743921314013295

Dark Sky Collaborators: Arizona (AZ) Observatories, Communities, and Businesses

Elizabeth Alvarez del Castillo[1], Christopher Corbally[2], Emilio E. Falco[3], Richard F. Green[4], Jeffrey C. Hall[5] and G. Grant Williams[6]

[1]AdC Consulting, Tucson, AZ, USA email: ema26@cornell.edu
[2]Vatican Obs., Vatican City State email: corbally@as.arizona.edu
[3]Smithsonian Astrophysical Obs, Cambridge, MA, USA email: falco@cfa.harvard.edu
[4]Large Binocular Telescope Obs., Tucson, AZ, USA email: rgreen@lbto.org
[5]Lowell Obs., Flagstaff, AZ, USA email: jch@lowell.edu
[6]MMT Obs., Univ. of Arizona, Tucson, AZ, USA email: ggwilli@mmto.org

Abstract. With outdoor lighting ordinances in Arizona first in place around observatories in 1958 and 1972, then throughout the state since 1986, Arizonans have extensive experience working with communities and businesses to preserve our dark skies. Though communities are committed to the astronomy sector in our state, astronomers must collaborate with other stakeholders to implement solutions. Ongoing education and public outreach is necessary to enable ordinance updates as technology changes. Despite significant population increases, sky brightness measurements over the last 20 years show that ordinance updates are worth our efforts as we seek to maintain high quality skies around our observatories. Collaborations are being forged and actions taken to promote astronomy for the longer term in Arizona.

Keywords. Site Protection, Light Pollution, Lighting Ordinances, Education, Public Outreach

---

Given astronomys significant economic impact in the state, AZ supports protecting our resource of dark skies, but communities, astronomy, and businesses must collaborate and revisit the adequacy and enforcement of existing ordinances in a new effort to reduce light pollution, especially that which is associated with changes due to growth and new technologies. Ongoing education, relationships and collaboration are essential to success.

To implement solutions, astronomers collaborate with stakeholders. Statewide, relationships with stakeholders were critical when seeking to control a recently proposed proliferation of LED billboards. Northern AZ around Flagstaff has cultivated community pride in their dark skies, and astronomers continue to network and maintain relationships. A recent ordinance update in southern AZ around Tucson was almost torpedoed by a national interest group who arrived late in the process, but prior discussions, collaboration and compromise with local stakeholders saved the day.

The biggest challenge is anticipating future developments. Catching an issue while its young offers early opportunities to affect decisions; coming to the table late is complicated and may result in ill will. Astronomers must monitor lighting technologies and develop research-based recommendations to be prepared to address the effects of developing technology on our dark skies. Strong collaborations with stakeholders like the International Dark-Sky Association and Commission Internationale de l'Eclairage support these efforts.

To preserve our window to the universe: Designate a leader to cultivate community relationships. Remain vigilant, watching for developments. Communicate early and continue, as often as possible. Express concerns to the public and government entities. Maintain scientific ethics and standards; be credible. Collaborate and leave room for compromise.

Highlights of Astronomy, Volume 16
XXVIIIth IAU General Assembly, August 2012
T. Montmerle, ed.

doi:10.1017/S1743921314013301

The Selection and Protection of Optical Astronomical Observing Sites in China

Jin Wenjing[1], **Jinming Bai**[2] **and Yongqiang Yao**[3]

[1]Shanghai Astronomical Observatory, CAS, China
email: jwj@shao.ac.cn

[2]Yunnan Astronomical Observatory, CAS, China

[3]National Astronomical Observatory, CAS, China

Before 1950 there are two observatories, Shanghai and Purple Mountain Astronomical Observatories (SHAO and PMO), and two observing stations, Qingdao and Kunming stations in China. With the requirements of astronomical research, two observatories, Beijing and Shaanxi Astronomical Observatories (BAO and SXAO) and two artificial satellite stations, Urumqi and Changchun, were established about 1960. Based on the current management, now there are 4 observatories, SHAO, PMO, NAOC(National Astronomical Observatories), which was grouped from BAO, YNAO and 2 others, as well as XAO (Xinjiang Astronomical Observatory). The optical 1-2 m class telescopes are being operated at former four observatories. SXAO is changed as National Time Service Center. Because of city expansion as well as the traveling and economic developments, these observatories are suffered severe light pollution. For example, Zo Ce is located at the suburb of Shanghai city. A 40 cm double astrograph was installed in 1900 and a 1.56 m optical reflector have been operated since November 1987. In 1994 the seeing is better than 1 and the night sky brightness in V is about 19 $mag/arcsec^2$, stars fainter than 20 mag with CCD are visibles. In 2007 a large playground was built in Zô Cè area. The light pollution is severe gradually. The night sky brightness has been increased to 15.8 $mag/arcsec^2$. The other observatories have similar situation. New site surveys and found new stations to solve the problem. Except the solar and radio stations of each Astronomical Observatory, now there are 3 optical observing sites at PMO (Hong-He, Xu-Yi and Yaoan), 2 at SHAO (Zô Cè and Tian Huang Ping) and 2 at YNAO (Kunming and Gao-Mei-Gu) as well as 1 optical observing site at BAO (Xing-Long). The best observing site is Gao-Mei-Gu, which is selected as the optical observing site of YNAO and where atmospheric turbulence distribution is 0.11 near ground with heights from 6.5m to 2.7m during night. Sky brightness in B and V band are 22.34 and 21.54. The extinction coefficient K'bv and K'v are 0.298 and 0.135. The seeing measurement is 0.72',. In the recent years a new 2.4m telescope the second largest telescope in China, was installed there.

The protection of Optical Astronomical Observing Sites is carrying out in three ways: (a) conscious education through popular astronomy given by Planetarium, Astronomical museums and network; (b) making agreement between observatory and local government. For example, at Gao-Mei-Gu the first level protected area is radius of 5 km at the center of optical telescope. 75 watt lamps should be had suitable shade. Light source can not be seen out of 4 m. The second level protected area is between 5-20 km. There have no factories; (c) Special design for lamps along the road to mountain summit, such as inductive lamps are adopted as the illumination on the mountain load. Finally, each observatory has their own optical observing site in accordance with the optical observing condition. The current site survey program in China is being carried out in the provinces of Tibet, Xinjiang, Qinhai, Yunnan, especially in Ali area of Tibet about 5000m above sea level.

Highlights of Astronomy, Volume 16
XXVIIIth IAU General Assembly, August 2012
T. Montmerle, ed.

doi:10.1017/S1743921314013313

Light pollution in Beijing and effects on Xinglong Station of National Astronomical Observatory

Ligen Lu[1], Baozhou Zhang[1,2,*], Jian Liu[3] and Shanshan Zeng[1]

[1]Dept. of Astronomy, Beijing Normal University,
19 XinJieKouWai St., Beijing, China

[2]Beijing Key Laboratory of Applied Optics,
19 XinJieKouWai St., Beijing, China

[3]Dept. of Physics, Beijing Normal University,
19 XinJieKouWai St., Beijing, China

[*]email: zhangbzh@bnu.edu.cn

Abstract. A night-sky luminance survey was carried on in Beijing to assess the level of light pollution. The luminance of the zenith night sky and skies in four directions at six sites with different distances from the city center was measured by using a photometric luminance meter. The Xinglong Station of National Astronomical Observatory was included to study the impacts of city lights on an astronomical observatory. The survey shows that the night-sky luminance decreases with increasing distance from the city center. Measurement results indicate that outdoor lighting in the Xinglong county town which is close to the observatory has non-negligible influence on the night sky at Xinglong Station.

Keywords. night-sky luminance, astronomical observatory, outdoor lighting, light pollution

---

Measurements were made to estimate the quality of the Xinglong Station, a major optical/IR site for astronomical observations in China. A small and easy-to-use device called the Night Sky Luminance Meter was used in the survey, which gives the luminance in linear (cd/m^2) units. The zenith night sky and skies at 45 degrees altitude in four directions of north, south, east and west are all brighter before midnight though the luminance decreases with time. After midnight, the night-sky luminance reduces to a low value and remains generally stable until twilight appears. Measurement results show that night skies of urban, suburban and even rural sites in Xinglong have been brightened by artificial outdoor lighting of Beijing. The zenith luminance of the night sky decreases with increasing distance from the city center. At the Xinglong station the quality of the zenith night sky is about 21.4 mag/arcsec^2. Compared with luminance data taken at Lingshan, Xinglong Station proves to be suffering from light pollution from the Xinglong county town which is close to the observatory. Diffusely distributed suburban towns in the northeast of Beijing also contribute to the problem.

Measurements provide evidence to support for the development of regulations on the usage of outdoor lighting. Moreover, much more education outreach is needed as lack of awareness and apathy are the main problems.

Future plans to continue our work include extending the survey by constructing a network of observing stations in which the night-sky luminance will be simultaneously and automatically monitored at multiple astronomical sites in China at a high frequency for a long period of time. Data from this project is expected to provide a more comprehensive historical record of the light pollution situation at observatories.

Highlights of Astronomy, Volume 16
XXVIIIth IAU General Assembly, August 2012
T. Montmerle, ed.

doi:10.1017/S1743921314013325

Legal protection of the night sky in Andalusia (Western Europe)

David Galadí Enríquez[1] **and Ángela Ranea-Palma**[2]

[1]Centro Astronómico Hispano-Alemán (CAHA)
Apartado 2010, ES-04080-Almería, Spain
email: dgaladi@caha.es

[2]Consejería de Agricultura, Pesca y Medio Ambiente, Junta de Andalucía
Avenida Eritaña 1, ES-41071-Sevilla, Spain
email: angela.ranea@juntadeandalucia.es

Abstract. Andalusia (Spain) houses several astronomical observatories, among them the main observational facility in continental Europe: Calar Alto Observatory. In recent years, the regional government of Andalusia has been setting up a regulation to protect the natural conditions of darkness at night all over the region. This regulation includes several outstanding features and poses specific rules to protect the influence area of Calar Alto Observatory.

Spain is a federal state composed of 19 entitites called *communities*, three in Africa and 16 in Europe. Andalusia (87268 km^2) marks Europe's southernmost extreme and Spain's most populous community (8.2 million people). Its official language is Spanish.

Andalusian legislation on light pollution has been evolving since 2007 in several phases and is close to being fully operational. It protects all the territory of the community (eg. max. ULORinst 1% everywhere), with special attention on the huge area of environmentally protected land and with a specific focus on major astronomical sites: Sierra Nevada Observatory (OSN) and Calar Alto Observatory (CAHA). The influence areas of these observatories have been approved and their most outstanding feature is the restriction on white and blue light inside these zones (3877 km^2 for Calar Alto Observatory).

Those in charge of the development of this law are well aware of the need to interact to enforce the regulation successfully. To this end, the release of the law is accompanied by an *unprecedented* effort of communication, education, public outreach and direct action: *Dissemination*: leaflets, digital documents, electronic and phone hotlines, specific actions and materials on blue- and white-light restriction. *Local councils*: training programs for local politicians and technicians, legal and technical support and advice, help to search for funding and financing. *Universities*: training sessions for university students (architecture, environmental sciences, lighting engineers). *Educational community*: short film and art (posters, poetry, graffiti) contests, teaching unit for secondary schools (useable worldwide). *General public*: awareness campaigns in many different contexts. *Amateur astronomers*: annual course on multi-band photometry, participation in their meetings and events, support of their awareness campaigns. *Events*: fairs, participation in congresses on dark-sky protection, organization of the 1st International Meeting of Administrations with Legislation on Light Pollution (2011). *Measurement campaigns*: multi-band CCD-based and one-band SQM-based measurement campaigns to produce the Andalusian map of sky glow for diagnosis and to assess the future effects of the law.

The Andalusian law is not only one of the most advanced, but is a very outstanding example of what can be done when administrations take light pollution seriously. All this has been done by a small team of four specialists with external help from many Andalusian and non-Andalusian individuals and entities, both public and private.

Highlights of Astronomy, Volume 16
XXVIIIth IAU General Assembly, August 2012
T. Montmerle, ed.

doi:10.1017/S1743921314013337

SAAO small telescopes, capabilities and Challenges

Ramotholo Sefako

South African Astronomical Observatory, PO Box 9, Observatory, 7935, South Africa
email: rrs@saao.ac.za

Abstract. The SAAO is at a geographically crucial site in the southern hemisphere between South America and Australasia. SAAO has a long history of involvement in infrared and optical astronomy that dates back almost two hundred years. The observatory expects to continue contributing to astronomical research for many years to come, using its small (0.5m, 0.75m, 1.0m and 1.9m) telescopes and their various instruments (ranging from spectroscopy to polarimetry and high-speed photometry), together with the Southern African Large Telescope (SALT) and other hosted international telescopes. In this paper, I discuss the capabilities and uses of the SAAO small telescopes, and the challenges that threaten astronomical research at the observatory, including light pollution and other emerging threats to the usually dust-free and dark-night-sky site at Sutherland. This is mitigated by the legislation called the Astronomy Geographic Advantage (AGA) Act of 2007 that protects the observatory from these threats.

Keywords. Telescopes, Instruments, Protection

Overview of astronomy and related challenges at SAAO

Background. The South African Astronomical Observatory (SAAO) with its headquarters in Cape Town and the observing station in Sutherland, about 300 km from Cape Town, is the main custodian of optical astronomy in South Africa. It has operated four (1.9m, 1.0m, 0.75m and 0.5m) telescopes at Sutherland for 40 years, as well as a number of other international facilities on behalf of, or in partnership with, international partner(s). Currently, it is also home to the SALT, the largest single telescope in the Southern Hemisphere, and over half a dozen other small and mostly robotic telescopes belonging to international partners/collaborators.

SAAO telescopes' main objectives are in (i) support of SALT observations; (ii) time domain astronomy, spectroscopy and polarimetry; (iii) long-term monitoring, and (iv) postgraduate training. The main instruments include the spectrograph (SpCCD), High-speed Photo POlarimeter (HIPPO), Sutherland High-speed Optical Cameras (SHOC) and Fibre-Fed Echelle spectrograph (GIRAFFE). The Observatory is protected under the AGA Act of 2007 against mainly light and dust pollution.

Challenges. The main challenge relates to the emergence of wind energy developments that are required to have lighting on their turbines to address other legislation, the Civil Aviation Act of 1962, which requires that any structure taller than 45m be marked in order to warn incoming aircraft of such structures. A number of these wind energy farms are planned around Sutherland. Hydraulic fracturing in search of shale gas is another possible threat to the clear and air pollution free skies around Sutherland, given that the South African Cabinet has lifted the moratorium on hydraulic fracturing (or fracking), which is seen as a potential source that can address the country's energy demands.

Highlights of Astronomy, Volume 16
XXVIIIth IAU General Assembly, August 2012
T. Montmerle, ed.

doi:10.1017/S1743921314013349

Night Sky Protection Initiatives in Argentina

Beatriz García[1], Silvina Pérez Álvarez[1], Victor Bibé[2], Andrés Risi[3] and Lisandro Gino[4]

[1]ITeDAM, CNEA,CONICET, UNSAM, Mendoza, Argentina
email: beatriz.garcia@iteda.cnea.gov.ar

[2]Asociación Sigma Octantis, Ushuaia, Tierra del Fuego, Argentina

[3]Planetario Malargüe, Mendoza, Argentina

[4]UCSE, Rafaela, Santa Fe, Argentina

Abstract. Light Pollution is a global problem. Some local actions carried out by a network of professional and amateur astronomers in Argentina are changing the way to attack this problem, taking into account measurements, education, public activities, planetarium shows and legislation proposals.

Keywords. Light Pollution control, Light Pollution: Education and Outreach, Light Pollution: Normatives

---

1. Introdution

Light pollution is one of the subjects of our time. The growing population of our planet has little awareness of this kind of problem, which in addition with other forms of contamination, will impact the quality of life. Malargüe, in Mendoza; Rafaela, in Santa Fe; and Ushuaia, in Tierra del Fuego; can be the corners of a work to modify habits about light use, showing that is possible avoid light pollution, and proving that a few people can be a factor of change to transform the behavior of the community where they live.

2. Objectives and Main Goals

The initiatives have as main objectives:

- *Demonstrate that the light pollution modify our life and the life of the whole ecosystems.*
- *Show that we can control light pollution to recover the night landscape.*
- *Propose and implement local legislation which can be transformed to a National Law on night sky protection and rational use of energy.*

A team of professional and amateur astronomers, engineers, designers, and lawyers are working to implement different actions to create a network of citizens devoted to protect the sky, and also, the environment and life. As part of a long term program of study of light pollution in Malargüe, where the Cosmic Rays Pierre Auger Observatory is installed, several actions, like global campaigns, planetarium shows, videos and contests were performed. These programs have spread to other cities in the country. As a consequence, a National Law of Sky Protection is actually under study. The community of Malargüe is planning to offer this destination as part of a program of astronomical tourism. All the recent initiatives in Argentina on light pollution control and sky protection have two main lines of action: education and communication with the public.

Highlights of Astronomy, Volume 16
XXVIIIth IAU General Assembly, August 2012
T. Montmerle, ed.

doi:10.1017/S1743921314013350

An Introduction to IAU 2009 Resolution B5

Malcolm G. Smith
Cerro Tololo Interamerican Observatory,
Casilla 603, La Serena, Chile
email: msmith@ctio.noao.edu

Abstract. For the purposes of the discussion of IAU follow up on Resolution B5, adopted at the GA in Rio de Janiero in 2009, reference is made here to the resolution itself and the background which led to its proposal and adoption by the IAU.

Keywords. Light pollution, IAU resolution

IAU 2009 Resolution B5 was distributed to participants at the beginning of the discussion session here in Beijing. The orginal resolution was initially drafted by Pedro Russo and Connie Walker. It can be found at

http://www.iau.org/static/resolutions/IAU2009_English.pdf

It refers to the Starlight Initiative http://www.starlight2007.net/ and recognises a series of principles raised in a declaration approved by that Initiative in 2007. It recognises the competent bodies within the IAU to work on such matters - namely Commission 50 and its WG for Controlling Light Pollution, as well as the IYA Cornerstone Project "Dark Skies Awareness". It then lists a series of considerations concerning the adverse effects of light pollution and urges IAU members to raise public awareness of the aims of the Starlight Initiative, "in particular the educational, scientific, cultural, health and recreational importance of preserving access to an unpolluted night sky for all humankind".

Returning focus to the scientific needs of professional astronomy, Resolution B5 resolves that "Protection of the astronomical quality of areas suitable for scientific observation of the Universe should be taken into account when developing and evaluating national and international scientific policies, with due regard to local cultural and natural values."

Following on from the first meeting of the Starlight Initiative, many aspects of Resolution B5 have been taken up in collaboration with IAU Commission 41, History of Astronomy, in particular with its Astronomy and World Heritage Working Group, under the leadership of Clive Ruggles. On 24th August 2012, (following planning carried out a couple of months earlier at a meeting in New Zealand) a new UNESCO-IAU online Portal to the Heritage of Astronomy was launched at the General Assembly. As stated in the press release, "The site, which resides at www.astronomicalheritage.net, is a dynamic, publically accessible database, discusssion forum and document repository on astronomical heritage sites throughout the world, even if they are not on UNESCO's World Heritage List."

In addition to considering ancient heritage sites such as Stonehenge, the concept of starlit skies is advanced as a Scientific and Cultural heritage for future generations - which will be lost if not protected now. Included among such sites are "Windows to the Universe", one of which (Northern Chile) is discussed here in the subsession on "Light Pollution Legislation and Protecting Observatory Sites". These "Windows" are ground-based sites in the northern and southern hemisphere considered particularly suitable for professional, international, optical observatories.

Highlights of Astronomy, Volume 16
XXVIIIth IAU General Assembly, August 2012
T. Montmerle, ed.

doi:10.1017/S1743921314013362

IAU Resolution 2009 B5 - Commission 50 Draft Action Plan - Presentation and Discussion

R. F. Green
Large Binocular Telescope Observatory,
933 N. Cherry, Tucson, AZ, USA
email: rgreen@lbto.org

Abstract. IAU Resolution 2009 B5 calls on IAU members to protect the public's right to an unpolluted night sky as well as the astronomical quality of the sky around major research observatories. The multi-pronged approach of Commission 50 includes working with the lighting industry for appropriate products from the solid state revolution, arming astronomers with training and materials for presentation, selective endorsement of key protection issues, cooperation with several other IAU commissions for education and outreach, and provision of clear quantitative priorities for outdoor lighting standards.

Keywords. Site protection, Light pollution, Outdoor lighting

1. Key Excerpts from the Resolution

An unpolluted night sky that allows the enjoyment and contemplation of the firmament should be considered a fundamental socio-cultural and environmental right, and that the progressive degradation of the night sky should be regarded as a fundamental loss; IAU members [should] be encouraged to take all necessary measures to involve the parties related to skyscape protection in raising public awareness of the educational, scientific, cultural, health and recreational importance of preserving access to an unpolluted night sky for all humankind. Protection of the astronomical quality of areas suitable for scientific observation of the Universe should be taken into account when developing and evaluating national and international scientific and environmental policies, with due regard to local cultural and natural values.

2. General Means to Accomplish the Objectives and Practical Implementation

The Commission can promote engagement of more of the astronomy community to include the message in their public outreach activities. There is still momentum from the IYA. Providing one or two effective slides in every public talk could be very powerful. At more depth, a package could include vision and background statements with supporting material to those astronomers who are prepared to be more engaged in educating the public. This is where the International Dark-Sky Association can be a prime source, because of their strong approach to implementing sky protection goals. During discussion, the IDA executive officer proposed receiving direct support to develop the information to comprise an IAU dark skies standard outreach package available to astronomers. It would cover areas of value of night sky protection, outdoor lighting design issues, and model ordinance, among others. Promotion of availability would be through the IAU Newsletter and through direct contact with national astronomical societies.

As a key technical activity of Commission 50, it should support continuing work with the lighting industry, so that there are products available to support astronomer goals of full cut-off and limited spectral pollution. This area is where Commission 50s highly beneficial and productive engagement with the International Commission on Illumination, CIE (http://cie.co.at/), is worth amplifying to keep the positive interaction at a high level. Identify and prioritize the key upcoming meeting opportunities with the lighting industry, recruit participants, and work with their institutional leaders to generate support for their participation. We also note the CIE is influential, particularly in Europe, with their published outdoor lighting standards. They are not yet fully consistent with the goals of the resolution, but the CIE has encouraged astronomer participation in their Working Groups with studies that impact revisions.

Protection of astronomical sites is often a lonely battle in an isolated area. Commission 50 can help those astronomers who are engaged with local, national, and international authorities to protect astronomical sites. It is widely noted that past presidents of Commission 50 have been doing yeoman's service in their work to get sites into the World Heritage category for protection. The broader themes of the Resolution include supporting astronomers who are engaging local and national authorities for lighting regulation to protect natural, historic, and ordinary local areas from light pollution encroachment. Commission 50 can provide a forum for information exchange among those working on dark sky protection. Production of IAU letters of endorsement to relevant entities attesting to the astronomical value of a site can have real value in such activities. A stronger result could be multi-agency (or IAU) funding requests for support of specific protection measures.

It is clear that education and outreach are key to making progress on the objectives of the Resolution, both to astronomers and by astronomers to the general public. Close collaboration with Commission 46 (co-sponsor of this Special Session) is obviously critical, as well as with Commission 41 on communicating astronomy with the public. Engaging with Commission 55 and their working group on Astronomy and World Heritage will be vital for protection of natural and historic areas.

3. Commission 50 priorities for outdoor lighting

1. Full cut-off shielding. Light emitted just above the horizontal has the most deleterious impact. As a rule without exception, this full shielding approach is not yet accepted in CIE standards, particularly for roadway lighting.

2. Spectral management to minimize blue-light threat. Preservation of spectral access to the night sky in the near zones around research observatories is best achieved with narrow-band amber LEDs, the next generation replacement for low-pressure sodium. The near-term issue is that they do not produce the same level of energy savings as broader-band sources. Urban lighting requirements are different for distant observatories and for the general public in the context of the Resolution. As shown in these proceedings by Luginbuhl, blue light from urban centers has largely been scattered away before reaching distant research facilities; however, the publics right to a dark night sky requires minimal blue light for outdoor lighting. A conservative figure is limiting the output of a luminaire to no more than 15% of its radiated energy at wavelengths shorter than 500 nm.

3. Zone-appropriate lighting levels, including curfews. These principles are in place in both CIE and International Energy Conservation Code guidelines. Astronomers will need to exert downward pressure on the absolute lighting levels prescribed based on zone descriptions.

Highlights of Astronomy, Volume 16
XXVIIIth IAU General Assembly, August 2012
T. Montmerle, ed.

doi:10.1017/S1743921314013374

The Effects of Lamp Spectral Distribution on Sky Glow over Observatories

C. B. Luginbuhl[1], P. A. Boley[2], D. R. Davis[3] and D. M. Duriscoe[4]

[1]U.S. Naval Observatory Flagstaff Station, Flagstaff, Arizona, U.S.A. email: `cbl@nofs.navy.mil` [2]Max-Planck-Institut für Astronomie, Heidelberg, Germany [3]Planetary Science Institute, Tucson, Arizona, U.S.A. [4]National Park Service, Bishop, California, U.S.A.

Abstract. Using a wavelength-generalized version of the Garstang (1991) model, we evaluate overhead sky glow as a function of distance up to 300 km, from a variety of lamp types, including common gas discharge lamps and several types of LED lamps. We conclude for both professional, and especially cultural (visual), astronomy, that low-pressure sodium and narrow-spectrum amber LED lamps cause much less sky glow than all broad-spectrum sources.

Keywords. atmospheric effects, scattering, site testing

We have modified the Garstang (1991) model to include wavelength-dependent scattering and absorption. We evaluate overhead ($z \leqslant 60°$) sky glow at distances from 0.1-300km from low-pressure sodium (LPS), amber LED (ALED; peak 590nm, FWHM 15nm), high-pressure sodium (HPS), white LED with CCT of 2400K (wLED) and 5100K (cLED), metal halide with CCT of 4100K (MH), and a white LED with a 500nm filter (FLED). All lamp types are set to emit equal luminous flux. Results are summarized in Fig. 1.

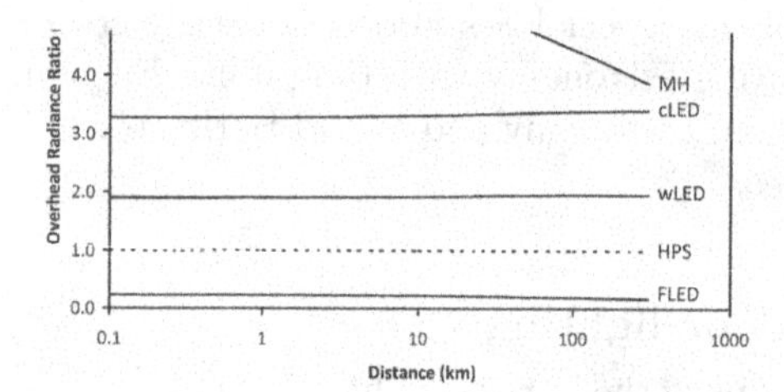

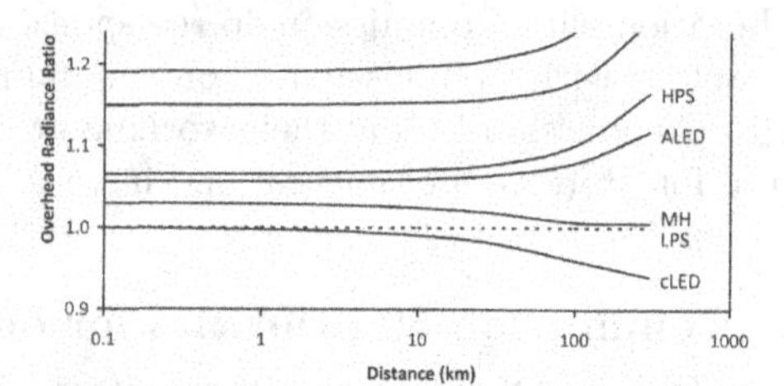

Figure 1. Ratio of overhead radiant sky glow as a function of distance. The left panel, relative to HPS, is for λ350-500nm (LPS/ALED have no emission here); the right panel, relative to LPS, is for λ500-650nm. (Does not include natural sky glow)

In the range λ500-650nm, wLED and FLED cause 15-35% more overhead radiant sky glow (RSG) than LPS or ALED. Increased scattering at short wavelengths is balanced by increased extinction when observed from < 10km. At greater distances, RSG from MH and cLED decreases relative to LPS, while that from wLED, FLED and HPS increases. In the range λ350-500nm, wLED, cLED and MH contribute ~2-5x more to the RSG than HPS. FLED has reduced blue RSG compared to broad-spectrum sources, but substantially greater blue and red RSG than LPS/ALED, especially at large distances.

Due to the Purkinje shift, narrow-spectrum yellow sources like LPS cause dramatically less visible sky glow (1/2 – 1/9) than all broad-spectrum sources, including FLED.

References

Garstang, R. H. 1991, *PASP*, 103, 1109

Highlights of Astronomy, Volume 16
XXVIIIth IAU General Assembly, August 2012
T. Montmerle, ed.

doi:10.1017/S1743921314013386

Light Pollution and Protecting Astronomical Sites in China

Richard F. Green
933 N. Cherry, Tucson, AZ, USA
email: rgreen@lbto.org

Abstract. Prof. Baozhou Zhang hosted a meeting of IAU Commission 50 members with Chinese academics and lighting professionals at Beijing Normal University.

Keywords. Site protection, Light pollution, Outdoor lighting

Participants were

Xiaju Chen	National Institute of Metrology
Yongqiang Yao	National Astronomical Observatories, Chinese Academy of Sciences
Tao Luo	China Academy of Building Research
Shuxiao Wang	China Academy of Building Research
Yuan Li	China Academy of Building Research
Jia Liu	Tsinghua University
Xumei Zhang	Tsinghua University
Gang Liu	Tianjin University
Ligen Lu	Beijing Normal University
Baozhou Zhang	Beijing Normal University
Chris Luginbuhl	US Naval Observatory
Ramotholo Sefako	South African Astronomical Observatory
Richard Green	Large Binocular Telescope Observatory

City lighting planners are charged with reducing light pollution, but it isnt easy. Luminaires dont always have photometrics for directions above horizontal. Sometimes manufacturers will respond. When the planners have some control, they can get contractors to comply, at the level of line managers and installers, but they cant control others projects. There is concern about energy efficiency of full shielding. In China, they observe CIE guides of 1%, 4% uplight, etc. and find it difficult to illuminate roadways uniformly with full cutoff. They have 2 candela/m^2 protocol for major arteries; 0.75 for feeder streets.

Results are measured through night sky brightness meters and calibration of satellite data. Lu calibrates their measurements of sky brightness around Beijing out to Xinglong Station with a standard lamp every 3 months. Luginbuhl calibrates his CCD images with well sampled stars in a panoramic image with 45 images stitched together. Wang has a technique to measure roadway luminance with a 35-mm lens and CCD.

Narrow-band amber LED is best for spectral protection around observatories. Mixing yellow and 10% white LEDs in a common luminaire can provide color rendition. The white light can be shut off at some point during the night. Color rendition is needed outdoors primarily for pedestrian areas with commerce; it is most important for interior lighting. The preference is for warmer color (lower CCT) in white light rather than cooler.

To protect observatory sites in China, the first step is for astronomers to work with local governments to convince them of the importance of night sky protection. The nighttime brightness of a city is perceived to reflect its economic importance.

Highlights of Astronomy, Volume 16
XXVIIIth IAU General Assembly, August 2012
T. Montmerle, ed.

doi:10.1017/S1743921314013398

Summaries of SpS17 Discussions IAU GA 2012 Special Session on Light Pollution

Constance E. Walker[1], Bob Parks[2], Dan McKenna[3], Ramotholo Sefako[4], Malcolm Smith[5] and David Galadí-Enríquez[6]

[1]National Optical Astronomy Observatory, Tucson, AZ, USA, cwalker@noao.edu
[2]International Dark-Sky Association, USA, bparks@darksky.org
[3]Palomar Observatory, PO Box 200, Palomar Mountain, CA 92060, USA, dmck@astro.caltech.edu
[4]South African Astronomical Observatory, PO Box 9, Observatory, 7935, South Africa, rrs@saao.ac.za
[5]Cerro Tololo Interamerican Observatory, Casilla 603, La Serena, Chile, msmith@ctio.noao.edu
[6]Centro Astronómico Hispano-Alemán (CAHA), Apartado 2010, ES-04080-Almería, Spain, email: dgaladi@caha.es

1. Introduction

To address light pollution issues, IAU Commissions 41, 46, 50, and 55 are involved in getting the word out to the public and IAU members via cultural, educational, technical; however, efforts can always improve and evolve. To carry out a successful light pollution abatement program supported by the IAU, it takes a diversity of groups, professions, and disciplines with their collective knowledge and experience. In manifesting dark skies awareness effectively, we are stronger together than we are alone; therefore, combining efforts of Commissions 41, 46, 50 and 55 with organizations like the International Dark-Sky Association, Astronomers Without Borders, The World at Night and partnering with events like Earth Hour or GLOBE at Night is a good step forward.

2. Light Pollution and the Public

- **The Role of Media, Planetaria and Amateur Astronomers in Light Pollution Education**
- **Light Pollution Education in Schools and in Cultures**
- **Global Star-Hunting and "Nights of Darkness" Campaigns**

The presentations and discussion on September 29 for the IAU GA 2012s Special Session on Light Pollution, SpS17, focused on the role of planetaria, amateur astronomy associations and media to change perspective, global initiatives in light pollution education, and global campaigns on dark skies awareness. In the discussion at the end of the day, a few key points and actions were identified, hopefully as items IAU members (collectively and individually) can support:

(*a*) Teach through the students with well-trained educators:
It is our responsibility to influence and inspire the next generation, since it is on their shoulders to make a difference in society. If they feel convinced something is important, they will act to change society. This process is accomplished in part by means of teachers we train. More light pollution education can be done through the IAUs GTTP, NASE

and UNAWE programs through the professional development for teachers and learning experiences for students.

(*b*) Find a way to reach the public through the media and global events:
If you have never seen the Milky Way Galaxy, you are less inclined to invest in a problem that you are not convinced exists. Until you can get most people aware there is a problem, then the problem cannot be changed. We (IAU) have to have a better way to break into the news media. The media generates a consensus among the people to push the government to do something. We need to package light pollution (as sound bites) in terms of its relation to many different types of issues (the environment, health, air pollution, climate change, etc), so that the news media is interested and can easily write about it. One suggestion was to start a campaign similar to past successful campaigns demoting cigarette smoking and littering or to collaborate with the organizers of a successful campaign like Earth Hour, for example. For an easy recommendation, more people could include light pollution issues in their oral presentations and in planetarium shows. Losing the Dark is a 5-minute planetarium public service announcement that is free to all planetaria that will help accomplish this task. Also, astronomy is easier to visualize and thereby appreciate than other sciences. By offering more night sky images to the news media, people can see for themselves what they are missing (e.g., The World at Night and its annual Earth and Sky Photo Contest).
With the public in various countries, astro-tourism is on the rise and is another excellent venue for raising dark-skies awareness. This topic is addressed in a later session of the IAU GA 2012 SpS17 on Light Pollution.

(*c*) Acquire more data to state factually what the numbers are:
Numbers support why steps need to be taken to redress light pollution issues. Raising awareness is good but the bottom line is that we need to measure whether we are protecting the night sky. Documenting the degradation of the night sky or its improvement is important, in addition to improving legislation. Numbers for instance would support why we should conserve energy and how that affects cost. Other session participants felt it is the sense of wonder not wallets.
In some areas there is not yet a critical awareness that excites people to do enough research to back up the effects of light pollution on things like crime levels or the increased use of blue lights. We better be ready to address proactively these issues the day after we have promoted our stance in the media.

(*d*) Convince authorities through IAU:
If, for instance, the IAU can address the European Parliament, and as a community, we use this as momentum to organize action around the world, this may prove an effective means to an end. Officers of Commissions 46 and 50 (on Education and Observatory Site Protection) were present at the IAU GA 2012 Special Session on Light Pollution, SpS17. In the next three years they hope to carry these ideas forward by working together with Commissions 41 and 55 (on World Heritage and Communicating with the Public) and the participants of SpS17 and other identified stakeholders in a variety of areas (health, environment, engineering, astronomy, etc) and from a variety of organizations (IDA, Astronomers Without Borders, CIE, etc). Not only does the topic of light pollution touch upon different areas, but people involved in these different areas are stronger working together than separately.

3. Social Event at the Beijing Planetarium (Premier of Losing the Dark)

The IAU SpS17 hosted an evening program focused on light pollution and night sky protection at the Beijing Planetarium. There were about 100 people in attendance, on Wednesday August 29th. The event was informative, well attended and gave attendees an opportunity to interact in a casual environment.

The session began with an award ceremony for Dr. Malcolm Smith, from the National Optical Astronomy Observatory (NOAO), He was honored for his substantial contributions to light pollution abatement on behalf of astronomical observatories and the community at large. Dr. David Silva (NOAO Director) and Bob Parks (International Dark-Sky Association Executive Director) presented Dr. Smith with the Dr. David Crawford Lifetime Achievement Award. This award, in honor of the International Dark-Sky Associations (IDA) co-founder and first executive director, recognizes those who have made a sustained effort to raise awareness of light pollution and to help implement solutions to light pollution.

After the award ceremony, the International Dark-Sky Association previewed a full-dome planetarium show entitled Losing the Night. This short educational program explains in simple terms what light pollution is, its impact on the environment and how it can be reduced. It was produced by Loch Ness Productions for IDA and is intended to be shown as a preview to regular planetarium shows. It is hoped that this program will help raise international awareness of issues related to light pollution and reach new audiences. It will be available for free to planetariums and a flat screen version will also be available for educational purposes.

After the showing of Losing the Dark, the attendees were treated to a showing of the documentary film The City Dark. The City Dark is an award-winning film documentary, which was directed by independent filmmaker Ian Cheney. It documents his experience of moving to New York City from rural Maine. The film covers a wide range of issues related to light pollution in a very accessible fashion. Mr. Cheney has interviewed individuals with unique perspectives on how the loss of the night sky impacts society and the environment.

4. Dark Sky Places, Starlight Reserves and Astro-Tourism

The effort to promote the preservation of Dark Skies has the greatest appeal in numbers based on emotional, rather than scientific needs. Astro-tourism spans all cultures that have different views of the night sky as well as varied context defined by history, geography and climate.

The interests in promoting tourism in general can be at odds with astro-tourism due to the lighting of architectural features and public areas. Solutions to optimize the partnership with astro-tourism will not be universal, but on individual bases. Two strategies seem to emerge: create the largest area to protect, such as the U.S National Park Service and identifying small, close-by viewing areas. One such solution on the island of Malta is to provide night sky tours by boat.

Promotion of dark-sky awareness has a day-time component that should not be overlooked. Parks that have adopted and developed astronomy-based themes such as solar system scaled trails with sponsored markers, sundials, or gnomon provide activities that will naturally lead into night-time sky awareness.

5. Dark Skies Measurements and Site Monitoring

The IDA has been promoting dark-sky awareness and advocating the protection of the night sky for over twenty years. An estimation that 75% of the current population of the Earth has never seen the Milky Way leads to the thought that if people do not know what they have lost, it is hard to convince them to get it back. The IDA and the IAU could develop a stronger partnership that would strengthen both organizations. The IDA is active in park certification, promoting research, developing technical standards, informing on solid-state lighting technology, supporting legislation, monitoring of night sky brightness, and has a focus on youth. The IDA emphasized that this developed expertise could be of great value to the IAU commission 50.

Measurements of light sources from air and spacecraft contribute valuable data for monitoring and modeling the characteristics of ground-based light sources. Measurements by aircraft can provide highly calibrated measurements of the low angle scattering, the elevation, azimuth, and spatial as well as spectral distribution. Measurements from the International Space Station using digital cameras that record color images in the RAW mode have been found to be very linear and provide high-quality data when combined with aircraft ground truth. The ISS data combines calibration from ground truth and stellar background fields.

The need for accurate long-term night sky brightness measurements is critical and urgent. Even with well-calibrated instrumentation, a change in sky brightness due to the atmosphere or light pollution is difficult to untangle. As time progresses the technology used to measure night sky brightness will evolve. It is therefore necessary to adopt standards in reporting measurements that will provide the necessary information to unify datasets. Presently there are two types of night sky photometers used to collect data. the Sky Quality Meter (SQM) family manufactured by Unihedron and the Night Sky Brightness Monitor (NSBM) built by Matrix Development under contract from the SKYMONITOR project. Both devices use the same optical filter and detector and thus have the same spectral response. The NSBM has a narrower field of view, about 4 degrees with a baffle tube for nearby light rejection. SQMs are producing the majority of the new, sky brightness, data and it is clear that the calibration must be understood both in terms of short and long-term stability.

6. Light Pollution Legislation and Protecting Observatory Sites

There is a need to protect the night sky against light pollution as both a heritage of citizens and a scientific resource. Low-pressure sodium (LPS) is recommended as an efficient form of lighting that is less detrimental to astronomy, and has potential for energy savings. It is important to consider impacts that light pollution would cause to the environment and astronomy, but also consider the potential benefits of using efficient lighting or using lighting only when necessary.

A question was asked about how do we help change the culture of the lighting industry to be more sensitive to light pollution issues?

The following were noted:

- to recognize the need for light pollution model
- to recognize the need for one model of legislation regarding lighting
- to recognize that light pollution increases every day and
- to recognize the need for concrete results that everyone is likely to believe

Different groups of people are concerned with light pollution; however, we as a group are not necessarily (as with astronomy) the main role player. We should therefore not compromise without consulting other role players and we should be aware of the importance of the concerns of the general populous regarding light pollution. We should consider using the IAUs status as a large organization to influence governments decisions on lighting legislation. We should also support our statements with factual statistics and numbers that make sense and that are acceptable when discussing the effects of light pollution on human beings and other animals.

Given that scientists don't write laws, but engineers do have a hand in standardizing lighting regulations, we need to make sure that lighting industries and lighting engineers are closer to reality and that contradictions with what astronomers want are minimized.

There is a need for understanding the technical aspects of lighting as defined by the CIE. Focus should shift towards use of lighting, not dark sky. We should only use the night sky as a reference point. It should be pointed out that in the strictest sense, all light is bad, and LEDs are worse and do not beat High-Pressure Sodium and Low-Pressure Sodium lights when it comes to efficiency.

7. Progress and Action Plan for Implementing IAU 2009 Resolution B5

This subsession began with a distribution of copies of B5. The resolution can be located at: *http : //www.iau.org/static/resolutions/IAU2009_English.pdf*

Consideration 4 urges that "IAU members be encouraged to take all necessary measures to involve the parties related to skyscape protection in raising public awareness - be it at local, regional, national, or international levels - about the contents and objectives of the International Conference in Defence of the Quality of the Night Sky and the Right to Observe Stars (http://www.starlight2007.net/), in particular the educational, scientific, cultural, health and recreational importance of preserving access to an unpolluted night sky for all humankind.

Led by the IAC, the "Starlight" Initiative and the IDA, astronomers, biologists, medical researchers, lawyers, school teachers and tourism organizations around the world have started a significant, joint, international effort to work with each other. This effort gained significant momentum during 2009 - the International Year of Astronomy - momentum that has largely been maintained since then.

Key components include:

• the education of the public, with special emphasis on working with school teachers

• the education of authorities at local, regional, national and international levels with a view to improving and extending appropriate legislation and enforcement

• alerting lighting engineers of the need to direct light where it is needed, when it is needed as well as the especially damaging impact of blue-rich light sources. A further key effort is work with UNESCO, ICOMOS, UCN and others on protection of the heritage of the natural night sky. Here the IAU's institutional support is proving particularly influential.

8. Spectra of Artificial Blue-Rich Sources

A dialogue was raised on which lamp should be recommended from an astronomical point of view. Amber LED, spectrally preferable, is less efficient, but there is the risk of judging tomorrow's technology according to today's standards. Both from the energetic and the spectral sides, Low-Pressure Sodium light seems the obvious choice. A balance among environmental impact and efficiency seems necessary. Taking into account other advantages of LED lights (they can be regulated and switched on/off rapidly), amber LEDs are not to be discarded.

Yellow light is better: it is less Rayleigh-scattered and human scotopic vision is much less sensitive to a residual yellowish skyglow.

The question is made whether not having requested full-cutoff at Hawaii from the very beginning can be considered a mistake in retrospect. The answer underlines the difficulty of negotiating under pressure and the lack of realistic choices at the time.

9. Astronomical Input to Lighting Industry Development: Prospects for Success

It is said that industry states cool LEDs are good for the environment, but this does not tell the entire story. The industry often does not even seem to be aware of the environmental impact of blue light (e.g., Rayleigh scattering and chronodisruption are less familiar to them).

The question was raised: Are we ready, as the IAU, to propose any specific recommendation or resolution addressed to industry?

The opinions given may be summarized: 'The narrower the band, and the redder the spectrum, the better'. Bluish light affects all of the nighttime environment, but only humans determine whether or not to put this light into the night; however, it is not clear if the IAU should dare to enter into biological or medical argumentations. Some doubts on the biological universality of blue light as a melatonin inhibitor were posed but it was stated that there are studies showing that this universality holds.

The case was made that general, but solid, physical and astronomical arguments defend a recommendation to say no to broad spectra and to blue-rich light. Then society will make a decision based on a compromise among all aspects: energy, health, biology, astronomy, but our task as astronomers is to make the case for everyone's sky.

Finally the moderator (Green) assumed the task to produce a draft based on these elementary principles: yellower, narrower.

Author Index